GWYBOD MEDR IACHAU

D1429190

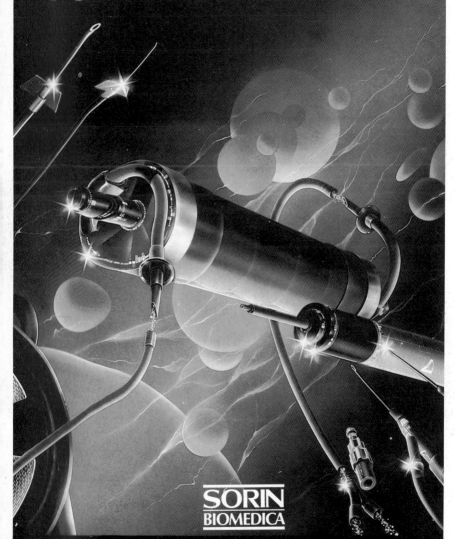

Our latest generation
is no small
achievem

Remove Fluids Effectively with Amicon's DIAFILTER® Hollow Fiber Hemofilters

In Acute Renal Failure . . .

Gentle, Continuous Control of Body Water

The DIAFILTER hemofilter transports plasma water and microsolutes through the membranes by convection (ultrafiltration), not by diffusion (dialysis). It is directly connected to the patient's artery and vein. The resultant pressure gradient is sufficient for gentle, continuous, slow fluid removal, due to the hemofilter's low transmembrane resistance. No blood pump, vacuum, nor special machinery needed. Ideal for intensive care. With appropriate I.V. fluids, drugs and hyperalimentation, acid-base, electrolyte balance, uremic toxin concentration and body water can be controlled.

- 0.4–1.2 l/hr fluid removal. Uninterrupted operation is possible and extensively reported.
- Effective at upward of 60 mm Hg systolic pressure, lower than normally used for dialysis. Blood flow of over 20 ml/min sufficient.
- Small priming volume: 20 or 40 ml.

Reported connections to artery and vein:

1. Cannulation of femoral blood vessels with the Kramer-Seldinger technique.
2. Arteriovenous Scribner shunt.
3. Double-lumen subclavian cannula (Uldall, pending); requires blood pump.

Partial Reference List:

P. Kramer, G. Kaufhold, H. J. Grone, W. Wigger, J. Rieger, D. Matthaei, T. Stokke, H. Burchardi, and F. Scheler. "Management of Anuric Intensive-Care Patients with Arteriovenous Hemofiltration," *Int J of Artif Organs* 3:225–230(1980).

E. P. Paganini and S. Nakamoto, "Continuous Slow Ultrafiltration in Oliguric Acute Renal Failure," *Trans Am Soc Artif Int Organs*, 26:201–204(1980).

In Chronic Renal Failure . . .

Fast Reduction of Severe Fluid Overload

The use of DIAFILTER hemofilters with a blood pump permits fast, effective removal of excess body fluid.

- 2–5 kg weight loss in only 3 hours.
- Low residual blood volume: 2–4 ml.
- Filtration rate easily regulated.
- Fewer complications and side effects reported than with dialysis.
- Central venous and pulmonary artery pressures reduced, even after a brief period of fluid removal.
- Simple extracorporeal circuit. Single outlet port. No vacuum. No sealing of extra outlets.
- Existing A-V fistulas or equivalent vascular access can be used.

Partial Reference List:

A. M. Rainone, E. Litman, "The Use of the Amicon Filter in Dialysis Related Complications," *The American Association of Nephrology Nurses and Technicians* 8:332–37(1981).

M. E. Silverstein, C. A. Ford, M.J. Lysaght, L. W. Henderson, "Treatment of Severe Fluid Overload by Ultrafiltration," *New England Journal of Medicine* 291:747–51(1974).

J. Lamar, W.A. Briggs, F.D. McDonald, "Effective Fluid Removal with the Amicon Diafilter", *Proceedings Dialysis Transplant Forum*, 127-129(1978).

Technical Data

	Diafilter-20	Diafilter-30
Size:	12.5 × 4.5 cm	20 × 4.5 cm
Membrane Area:	0.25 m²	0.55 m²
Priming Volume:	20 ml	40 ml
Residual Blood Volume:	2 ml	4 ml
Maximum Transmembrane Pressure:	500 mm Hg	500 mm Hg
Filtration Rate*:	10–40 ml/min.	10–80 ml/min.
Outlet Dimensions:	.64 cm 1/4″	.64 cm 1/4″

*Depends on blood flow, transmembrane pressure and viscosity.

Upper Mill
Stonehouse
Glos., U.K.
Tel: 045 382 5181

amicon®

NOTE: For more detailed information describing intended use, warnings, precautions and contraindications, refer to the instructions provided with each device.

Publication No. 414

August 1982

2568

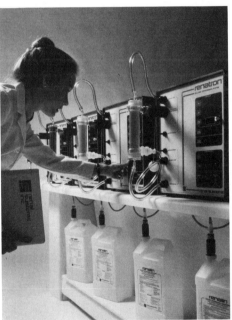

FB-T SERIES

NIPRO HOLLOW FIBER DIALYZER

Highest performances have been obtained through use of the new cellulose acetate membrane as thin as 15 micron of improved quality.

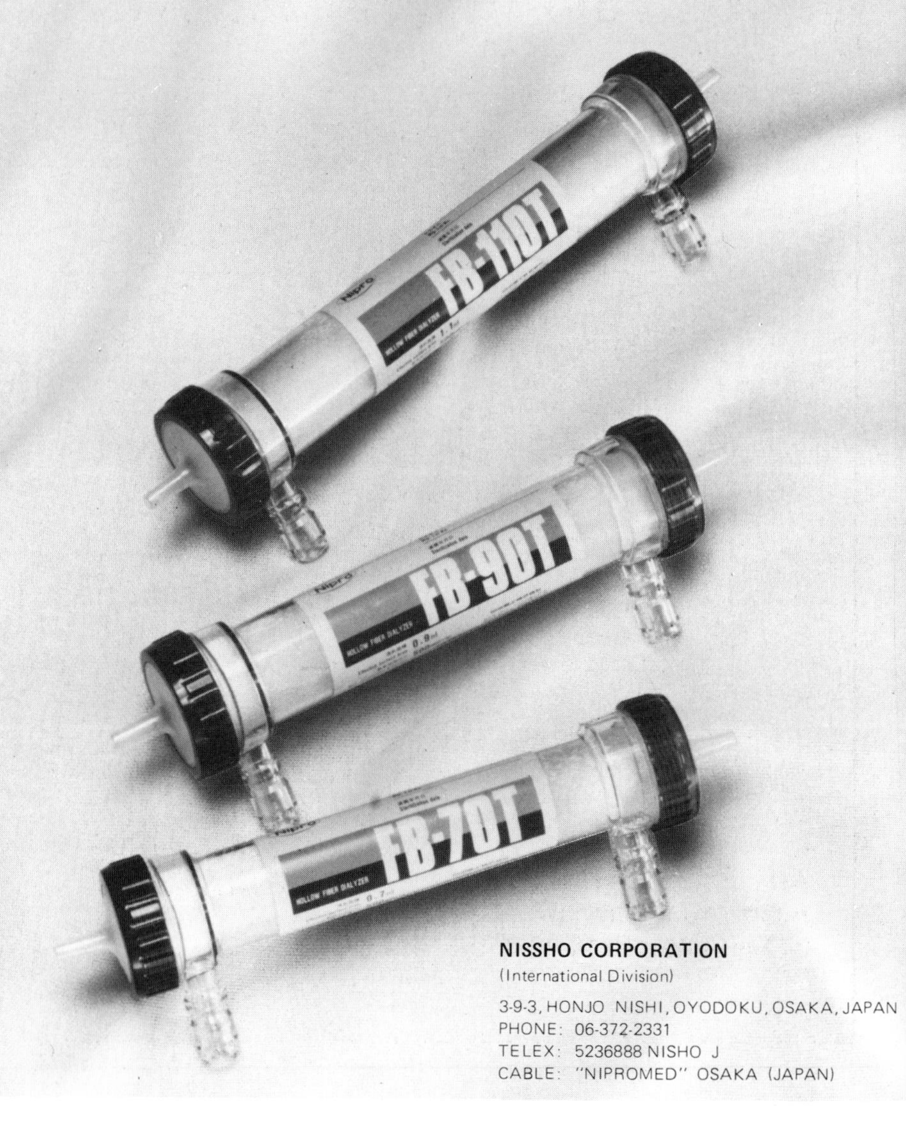

gambro®

Lundia 10
– the modern
plate dialyzer.

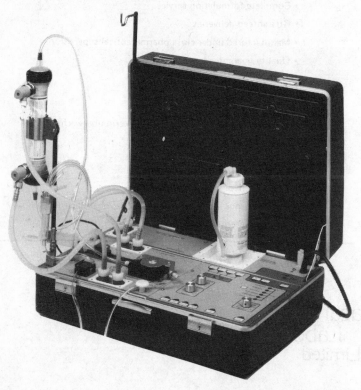

PROCEEDINGS OF
THE EUROPEAN DIALYSIS AND TRANSPLANT ASSOCIATION
– EUROPEAN RENAL ASSOCIATION

PROCEEDINGS OF THE EUROPEAN DIALYSIS AND TRANSPLANT ASSOCIATION — EUROPEAN RENAL ASSOCIATION

Twenty-first Congress held in Florence, Italy, 1984

Editor
Alex M Davison

Associate Editor
Pierre J Guillou

PITMAN

PITMAN PUBLISHING LIMITED
128 Long Acre, London WC2E 9AN

Associated Companies

Pitman Publishing Pty Ltd, Melbourne
Pitman Publishing New Zealand Ltd, Wellington

Co-published by Urban & Schwarzenberg, Baltimore

First published 1985

ISBN 0-272-79804-5
ISSN 0308-9401

Printed and bound in Great Britain at The Pitman Press, Bath

EUROPEAN DIALYSIS AND TRANSPLANT ASSOCIATION
EUROPEAN RENAL ASSOCIATION

OFFICERS AND MEMBERS OF THE COUNCIL
1984/1985

OFFICERS

President of the Association	J S CAMERON, United Kingdom
President of the Congress	S M G RINGOIR, Belgium
President Elect of Congress	I TARABA, Hungary
Secretary/Treasurer	S T BOEN, The Netherlands*

COUNCIL MEMBERS

J BOTELLA, Spain
H BRYNGER, Sweden
V CAMBI, Italy
A DAL CANTON, Italy
C JACOBS, France
H KLINKMANN, German Democratic Republic
A J WING, United Kingdom
C VAN YPERSELE, Belgium

EDITOR

A M DAVISON, United Kingdom

ASSOCIATE EDITOR

P J GUILLOU, United Kingdom

* Department of Nephrology and Dialysis
Sint Lucas Hospital, Jan Tooropstraat 164
1061 AE Amsterdam, The Netherlands
Telephone: 020-510 83 90

1985 CONGRESS

22nd CONGRESS

OF

THE EUROPEAN DIALYSIS AND TRANSPLANT ASSOCIATION
EUROPEAN RENAL ASSOCATION

BRUSSELS
BELGIUM

25–29 June 1985

President of Congress

S M G Ringoir

Vice Presidents of Congress

C Toussaint
C van Ypersele di Strihou

Congress Secretariat

Mr M Wouters
Brussels International Conference Centre
Place de Belgique, Belgieplëin
B-1020 Brussels, Belgium

Telephone: (32-2) 478 48 60

FOREWORD

The 21st annual Congress of the European Dialysis and Transplant Association (EDTA)–European Renal Association (ERA) was held in Florence from 23rd to 26th September 1984. The permanent record of the Congress is given in this volume.

With the Congress in Florence I have reached the end of my Presidency of the Association. It has been for me a great pleasure to be back in Florence, the city where the 9th Congress of the Society was held in 1972; I was Congress Secretary at that time, and in Florence I was elected a Council member. Since then I have also been Secretary-Treasurer and finally President of the Society, having devoted 13 years of my life to the success of EDTA, now European Renal Association. In the last year of my Presidency a successful cooperation between our Association and the National Societies of Nephrology in Europe has been clearly established to really make our Society *the* European Society of Nephrology. On 28th October 1983, in Milan I chaired the first meeting of the Presidents of National Societies of Nephrology in Europe which was sponsored by our Society. On that occasion the following important decisions were taken:

1. Every year a meeting will be organised between the Council of our Association and the Presidents of National Societies.
2. The President of the National Society of the host country will be invited to Council meetings one year before the Congress to attend for matters related to the Congress.
3. Two years before a Congress the President of the Congress will write to the Presidents of the National Societies asking for topics, lectures, workshops, speakers, chairmen and (very important indeed) referees for the different nephrological topics.

Our Association has now an outstanding advisory panel composed of the Presidents of National Societies of Nephrology. I must express my deep appreciation to all Presidents, and in particular to Professors Suc, Crosnier and Richet, for their great contribution in such a difficult achievement.

Another important decision has been taken this year; it concerns the relationship between our Association and the EDTNA. All of us agree that no other field in medicine exhibits such a close cooperation between physicians, nurses and technicians as in nephrology. This is why EDTNA was created in a very tight relation with our Society.

I am always looking with great sympathy at the EDTNA, which was born in Florence, at the 9th Congress of our Society in 1972. Since then the EDTA Congress and the EDTNA Conference have always been held simultaneously.

Both Societies, however, have grown considerably in the past few years. The attendance at our annual Congress has greatly increased. In 1972, in Florence, the 9th Congress of the Association had 1,200 registered members including nurses. For the 21st Congress this year about 3,000 were registered, an overall attendance greater than that in Los Angeles for the Congress of the International Society of Nephrology. This development of both Societies is making it more and more difficult to find adequate space for future Congresses, particularly if we consider that more and more nephrologists are now participating in our scientific activity.

The Council of our Society and the EDTNA Council have faced this problem together, and have planned, for the future, that the EDTNA Conference will follow immediately the Congress of EDTA-ERA. This will maintain a link between our Society and EDTNA, while allowing more possibilities for simultaneous sessions during our Congress.

My experience as an official and particularly as President of the Society has been unique. I am greatly honoured and deeply pleased to have devoted my best efforts for the success of our Society. I am grateful to all Council members and to the Secretary-Treasurer, Dr S T Boen, for their continuous help and support during my term of office. I am also pleased to acknowledge the excellent activity of the Registration Committee and of its Chairman, Dr A J Wing. The Registry of our Society remains unique in the world.

I am finally happy to express the gratitude of the whole Association and myself, in particular, to my friend Professor Vincenzo Cambi, President of the 21st Congress, for organising this most successful Congress in Florence. In Florence Professor Luigi Migone and Professor D N S Kerr were elected honorary members of the Society for their outstanding activity in the field of nephrology.

I acknowledge the excellent and unbelievably rapid publication of these Proceedings, due to continuous great enthusiasm of Dr A M Davison, our Editor, and Pitman Publishing Ltd, our Publisher.

Vittorio E Andreucci
President of EDTA–European Renal Association

PREFACE

This year Florence provided a memorable setting for the 21st Congress of the European Dialysis and Transplant Association–European Renal Association. The weather could have been better and many of us were unaware that the monsoons spread as far as Tuscany. This did not dampen our spirits and the Congress was a lively exchange of clinical and scientific information. This year the colour of the Proceedings, Chianti red, has been chosen to represent the social programme, so well arranged by the organising committee.

The Congress was bigger this year with the Registry reports, nine guest lectures, one symposium, 66 free communications and 73 poster presentations. The Proceedings has, therefore, increased in size and this has provided a great increase in workload to achieve publication within such a tight schedule. We are grateful to the majority of authors who provided their manuscripts by the deadline, 115 of the 139 accepted papers were received by the deadline date. It is interesting to note that those who were late in submitting manuscripts last year were the same this year, an observation which has been made previously. Some communications have not been included in the Proceedings, one because the manuscript did not arrive in time, one because it had been accepted for publication elsewhere and one because, in spite of the manuscript being received in good time, it was not presented at the Congress. This year only a few authors have included more than 10 references but in such cases these have been deleted in order to be fair to those authors who adhered to the editorial instructions.

In a number of free communications it will be noted that the text does not fully record the discussion comments. The reason for this was considerable technical problems with recording and some of our tapes have proved impossible to transcribe. In some instances, for completeness, we have used the question sheets and asked the author for his replies. This is not entirely satisfactory but it is the best compromise in the situation. Again this year we have attempted to reference the discussion comments. We hope we have found the appropriate reference and again have been surprised to be unable to find some, which we were sure would be easily available, in spite of rigorous searching. The general view seems to be that the discussion adds little to the value of the Proceedings and, therefore, thought must be given to dropping this, an action which would be very cost saving and which would reduce significantly the editorial workload.

It is a pleasure to record our thanks to Professor Vincenzo Cambi, and Dr Fabbio Bono and their organising committee for all the help given to the editorial staff before and during the Congress. The facilities provided were excellent and the local organising committee worked hard throughout the meeting. The editorial office has again been organised by Mrs Gillian Howard and the speedy publication of this volume is due in considerable part to her efforts. Mrs Betty Dickens and Miss Claire Vinycomb provided secretarial help in typing the discussion sessions. This year the major part of the proof-reading has been undertaken by Iain Davison and we are grateful to his eagle-eyes in correcting the proofs. As previously the rapid publication of this volume is due to the co-ordinated efforts of many people but we are especially indebted to Mrs Betty Dickens of the Medical Conference Unit of Pitman Publishing for publishing this work with such rapidity.

Alex M Davison
Editor,
Department of Renal Medicine,
St James's University Hospital,
Beckett Street, Leeds LS9 7TF

Pierre J Guillou
Associate Editor,
University Department of Surgery,
St James's University Hospital,
Beckett Street, Leeds LS9 7TF

CONTENTS

Part XXVIII – TRANSPLANTATION POSTERS

ADVISORY COUNCIL OF THE PRESIDENTS OF NATIONAL SOCIETIES

The Council of the European Dialysis and Transplant Association – European Renal Association is grateful to the Advisory Council of the Presidents of the National Societies for their help in formulating the programme of this Congress.

Bulgarian Society of Nephrology	Professor Angel Astrug
Czechoslovakian Society of Nephrology	Professor Albert Valek
Danish Society of Nephrology	Professor Steen Olsen
Dutch Society of Nephrology	Professor Lembert Van Es
Finnish Society of Nephrology	Dr Erkki Happanen
French Society of Nephrology	Professor J Crosnier
GDR Society of Nephrology	Professor K Precht
German Society of Nephrology	Professor Dr med W Thoenes
Hellenic Society for Nephrology	Professor Theodore Mountoulakis
Hungarian Society of Nephrology	Professor I Taraba
Irish Society of Nephrology	Dr John Donohoe
Italian Society of Nephrology	Professor Carmelo Giordano
Scottish Renal Association	Dr John Anderton
Spanish Society of Nephrology	Dr Emilio Rotellar
Swedish Society of Nephrology	Professor Ulla Bengtsson
Swiss Society of Nephrology	Dr J Diezl
Turkish Society of Nephrology	Professor Kemal Onen
UK Renal Association	Dr Mary G McGeown
Yugoslavia Society of Nephrology	Professor S Luzar

REFEREES SUGGESTED BY PRESIDENTS OF NATIONAL SOCIETIES AND BY COUNCIL FOR THE FLORENCE CONGRESS 1984

The Kidney in Systemic Disease

Professor Gabriel Richet, France
Dr John Donohoe, Ireland
Dr Eberhard Ritz, Federal Republic of Germany
Dr Guiseppe D'Amico, Italy
Dr Netar Mallick, United Kingdom

Diagnostic Techniques

Dr Amos Pasternak, Finland
Professor Mattias Aurell, Sweden
Dr A Tourkantonis, Greece
Dr C van Ypsersele de Strihou, Belgium
Dr G Lubec, Austria

Transplant Immunology

Dr Jean Francois Bach, France
Dr Graham Catto, United Kingdom
Dr Jan van Rood, Belgium
Dr H Pichlmayr, Federal Republic of Germany
Dr E Curtoni, Italy

PART I

STATISTICAL REPORTS

Chairmen: V E Andreucci
V Cambi

Professor Peter Kramer, who presented the Report of the Registration Committee at the Congress in Florence, died suddenly on 7th October, 1984. The members of the Registration Committee acknowledge his very great contribution to their work during the last six years and dedicate this year's Report to his memory and to his wife, Heidi, and their three sons.

COMBINED REPORT ON REGULAR DIALYSIS AND TRANSPLANTATION IN EUROPE, XIV, 1983

Members

M BROYER	Hôpital Necker Enfants Malades, Paris, France
F P BRUNNER	Departement für Innere Medizin, Universität Basel, Switzerland
H BRYNGER	Department of Surgery I, Sahlgrenska Sjukhuset, Göteborg, Sweden
P KRAMER	Medizinische Universitätsklinik, Göttingen, Federal Republic of Germany
R OULÈS	Centre Hospitalier Regional et Universitaire de Nimes, France
G RIZZONI	Clinica Pediatrica dell'Universitá, Ospedale Civile di Padova, Italy
N H SELWOOD	UK Transplant, Bristol, United Kingdom
A J WING*	St Thomas' Hospital, London, United Kingdom

* Chairman

Director

S CHALLAH	St Thomas' Hospital, London, United Kingdom

Research Fellow

É A BALÁS	EDTA Registry, St Thomas' Hospital, London, United Kingdom and Semmelweis Medical School, Budapest, Hungary

CONTENTS

Acknowledgments

This work was supported by grants from the Governments or National Societies of Nephrology of Austria, Belgium, Bulgaria, Cyprus, Denmark, Egypt, the Federal Republic of Germany, France, the German Democratic Republic, Greece, Iceland, Ireland, Israel, Italy, Luxembourg, the Netherlands, Norway, Sweden, Switzerland and the United Kingdom.

Generous grants were made by Asahi Medical GmbH, B Braun, Melsungen AG, Bellco S.p.A., Cobe Laboratories Inc, Cordis Dow B.V., Enka AG, Fresenius AG, Gambro AB, Hospal Ltd, Sorin Biomedica S.p.A., Travenol Laboratories Ltd.

The post of Research Fellow to the EDTA Registry, held during the past year by Dr Norbert Gretz, Mannheim, Federal Republic of Germany and Dr András Balás, Budapest, Hungary, was funded by the National Kidney Research Fund, United Kingdom. A grant was also received from the Wellcome Foundation Ltd.

We acknowledge the co-operation of UK Transplant, Bristol, United Kingdom.

We particularly thank those doctors and their staff who have completed questionnaires. Without their collaboration, this Report could not have been prepared.

Proc EDTA–ERA (1984) Vol 21

COMBINED REPORT ON REGULAR DIALYSIS AND TRANSPLANTATION IN EUROPE, XIV, 1983

P Kramer, M Broyer, F P Brunner, **H Brynger**, S Challah, R Oulès, G Rizzoni, N H Selwood, A J Wing, E A Balás

Introduction

The fourteenth Combined Report for the European Dialysis and Transplant Association covers the year ending 31st December, 1983 and includes information from both centre and individual patient questionnaires. Table I shows the number of known centres in each of 32 European countries, and the proportion of them completing returns in 1983. Of the 1,798 known centres in Europe, 85.5 per cent returned patient questionnaires.

Centre questionnaires were returned by 83.7 per cent of units, and information from these is presented in Tables II, III, IV and V. National Keymen supplemented the information provided by the centre questionnaires and their contribution varied from country to country depending on the response rate of the respective centres. Table II shows the acceptance rates for new patients in 1983 per million population (pmp). The acceptance rates reached 72.5 pmp in Luxembourg and exceeded 50 in eight other countries, compared to 34 pmp for the total Registry. The numbers of patients on the various forms of haemodialysis and haemofiltration are shown in Table III. At the end of 1983 there were 70,845 patients on haemodialysis in Europe, of whom 1,740 were on haemofiltration; 5,563 patients were on CAPD (Table IV) and 16,489 were alive with a functioning graft (Table V). Centre questionnaires recorded 43,853 grafts performed over all years with over 6,000 done in 1983.

Tables VI to IX contain results obtained from the individual patient questionnaires. Discrepancy with results based on the centre questionnaire is partly due to patients who had not been updated for 1983 at the time of the analysis. The number of these records not yet updated is shown by country in Table VI, which also contains information on patients lost to follow-up. The treatment modes of patients known to be alive on 31st December, 1983 are shown in the same table; it is likely that the proportion of patients on the various types of treatment in each country will not change very much once updating is complete.

Tables VII, VIII and IX show respectively the number of patients per million population who have come onto treatment, died or been transplanted in each of

TABLE I. Summary of centres known to the EDTA Registry, the number per million population (pmp) and the proportion (%) returning centre and patient questionnaires

Country	Population in millions	Known centres	Known centres pmp	% replied Centre questionnaire	% replied Patient questionnaire
Algeria	18.3	4	0.2	50.0	50.0
Austria	7.5	26	3.5	100.0*	100.0
Belgium	9.8	57	5.8	89.5*	93.0
Bulgaria	9.0	32	3.6	90.6*	93.8
Cyprus	0.6	2	3.3	100.0	100.0
Czechoslovakia	15.2	24	1.6	100.0	100.0
Denmark	5.1	12	2.4	100.0*	91.7
Egypt	38.9	27	0.7	81.5*	85.2
Fed Rep Germany	61.2	301	4.9	87.7*	88.4
Finland	4.8	25	5.2	96.0*	96.0
France	53.4	206	3.9	91.8	94.2
German Dem Rep	16.8	52	3.1	100.0	100.0
Greece	9.3	51	5.5	64.7	64.7
Hungary	10.7	10	0.9	100.0	100.0
Iceland	0.2	1	5.0	100.0	100.0
Ireland	3.3	5	1.5	60.0	60.0
Israel	3.8	26	6.8	100.0*	100.0
Italy	56.8	362	6.4	63.4*	66.6
Lebanon	2.7	7	2.6	0	0
Libya	2.9	3	1.0	33.3	33.3
Luxembourg	0.4	5	12.5	100.0	100.0
Netherlands	14.0	48	3.4	89.6*	89.6
Norway	4.1	18	4.4	100.0	100.0
Poland	35.4	50	1.4	92.0	94.0
Portugal	9.8	34	3.5	88.2*	88.2
Spain	37.0	183	5.0	92.9*	95.6
Sweden	8.3	32	3.9	100.0	96.9
Switzerland	6.5	34	5.2	94.1*	91.2
Tunisia	6.2	7	1.1	100.0*	100.0
Turkey	44.2	15	0.3	93.3	93.3
United Kingdom	55.9	63	1.1	85.7*	90.5
Yugoslavia	22.1	76	3.4	69.7	76.3
Total Registry	574.2	1,798	3.1	83.7	85.5

* Data supplemented by Keymen

TABLE II. Summary of new patients accepted onto renal replacement therapy during 1983 in Europe, based on data from centre questionnaire

Country	1983 N	1983 Per million population
Algeria	25	1.4
Austria	402	53.6
Belgium	598	61.0
Bulgaria	254	28.2
Cyprus	19	31.7
Czechoslovakia	315	20.7
Denmark	206	40.4
Egypt	301	7.7
Fed Rep Germany	3,416	55.8
Finland	220	45.8
France	2,366	44.3
German Dem Rep	466	27.7
Greece	379	40.8
Hungary	128	12.0
Iceland	1	5.0
Ireland	80	24.2
Israel	254	66.8
Italy	2,586	45.5
Libya	0	0
Luxembourg	29	72.5
Netherlands	640	45.7
Norway	222	54.1
Poland	285	8.1
Portugal	399	40.7
Spain	2,267	61.3
Sweden	510	61.4
Switzerland	358	55.1
Tunisia	50	8.1
Turkey	319	7.2
United Kingdom	1,865	33.4
Yugoslavia	708	32.0
Total Registry	19,668	34.3

TABLE III. Stock of patients alive on haemodialysis/haemofiltration in Europe on 31st December, 1983, based on data from centre questionnaire

Country	Hospital HD	Home HD	Total	Bicarb. HD	Haemo-filtration	Haemodia-filtration
Algeria	24	2	26	0	0	0
Austria	867	49	916	85	20	22
Belgium	1,974	110	2,084	576	67	33
Bulgaria	702	23	725	20	35	3
Cyprus	95	0	95	0	0	0
Czechoslovakia	696	1	697	0	46	27
Denmark	434	99	533	11	25	69
Egypt	452	0	452	0	21	0
Fed Rep Germany	12,853	1,822	14,675	947	728	408
Finland	229	3	232	10	1	0
France	8,458	2,334	10,792	1,336	186	39
German Dem Rep	1,275	0	1,275	81	2	1
Greece	853	0	853	22	35	14
Hungary	225	0	225	2	0	0
Iceland	12	0	12	0	0	0
Ireland	140	33	173	0	0	0
Israel	879	64	943	10	10	74
Italy	14,572	959	15,531	1,499	364	187
Libya	0	0	0	0	0	0
Luxembourg	87	3	90	3	5	0
Netherlands	1,850	151	2,001	97	34	6
Norway	220	6	226	1	9	4
Poland	668	16	684	72	25	9
Portugal	1,428	90	1,518	0	0	0
Spain	7,449	375	7,824	382	44	57
Sweden	647	114	761	17	21	4
Switzerland	953	193	1,146	97	22	15
Tunisia	205	0	205	0	0	41
Turkey	201	0	201	0	0	10
United Kingdom	1,543	2,190	3,733	31	17	41
Yugoslavia	2,147	70	2,217	35	23	63
Total Registry	62,138	8,707	70,845	5,334	1,740	1,127

HD=haemodialysis

TABLE IV. Stock of patients alive on peritoneal dialysis in Europe on 31st December, 1983, based on data from centre questionnaire

Country	Intermittent peritoneal dialysis	Continuous ambulatory peritoneal dialysis	Total	Continuous cycling peritoneal dialysis	Peritoneal dialysis combined with HD
Algeria	0	8	8	1	0
Austria	4	15	19	0	0
Belgium	3	161	164	1	0
Bulgaria	0	0	0	0	0
Cyprus	0	0	0	0	0
Czechoslovakia	3	12	15	1	0
Denmark	39	140	179	1	0
Egypt	44	7	51	0	9
Fed Rep Germany	109	289	398	15	0
Finland	13	151	164	0	0
France	218	693	911	52	5
German Dem Rep	3	3	6	1	0
Greece	2	89	91	0	7
Hungary	22	4	26	0	7
Iceland	0	0	0	0	0
Ireland	0	48	48	0	0
Israel	36	70	106	0	1
Italy	134	1,093	1,227	17	14
Libya	0	0	0	0	0
Luxembourg	0	0	0	0	0
Netherlands	0	234	234	1	0
Norway	3	20	23	0	0
Poland	41	16	57	2	10
Portugal	42	11	53	0	1
Spain	98	545	643	21	3
Sweden	41	194	235	2	1
Switzerland	3	211	214	2	0
Tunisia	4	15	19	0	0
Turkey	17	13	30	0	6
United Kingdom	59	1,496	1,555	31	8
Yugoslavia	46	25	71	0	11
Total Registry	984	5,563	6,547	148	83

HD=haemodialysis

TABLE V. Summary of grafts performed during 1983, based on data from centre questionnaire. The total number of grafts performed in all years, and the stock of functioning grafts on 31st December, 1983 are also shown. The table includes supplementary information collected by National Keymen

Country	CADAVER			LIVING DONOR			TOTAL		Total grafts performed all years	Functioning on 31st December 1983
	1st graft	Patients <15	Total CAD grafts	1st graft	Patients <15	Total LD grafts	All grafts	Per million population		
Algeria	0	0	0	0	0	0	0	0	0	0
Austria	128	2	154	0	0	0	154	20.5	816	420
Belgium	160	16	198	25	3	31	229	23.4	2,623	1,245
Bulgaria	0	0	0	0	0	0	0	0	0	0
Cyprus	0	0	0	0	0	0	0	0	0	0
Czechoslovakia	126	3	133	1	0	1	134	8.8	845	227
Denmark	66	4	111	3	0	5	116	22.8	1,997	381
Egypt	0	0	0	11	0	14	14	0.4	54	29
Fed Rep Germany	706	49	983	30	5	35	1,018	16.6	5,843	2,039
Finland	81	0	89	6	2	6	95	19.8	1,381	651
France	479	47	958	13	5	33	991	18.6	3,825	1,954
German Dem Rep	96	5	102	0	0	1	103	6.1	939	371
Greece	4	0	5	24	0	25	30	3.2	280	154
Hungary	49	0	49	3	1	3	52	4.9	265	118
Iceland	0	0	0	0	0	0	0	0	0	0
Ireland	33	0	38	8	1	11	49	14.9	510	239
Israel	40	9	58	19	1	20	78	20.5	634	131
Italy	222	13	410	22	4	25	435	7.7	1,535	865

TABLE V. *Continued*

Country	CADAVER			LIVING DONOR			TOTAL			
	1st graft	Patients <15	Total CAD grafts	1st graft	Patients <15	Total LD grafts	All grafts	Per million population	Total grafts performed all years	Functioning on 31st December 1983
Libya	0	0	0	0	0	0	0	0	0	0
Luxembourg	3	0	3	0	0	0	3	7.5	6	5
Netherlands	136	15	350	9	1	13	363	25.9	1,161	625
Norway	73	1	97	41	5	44	141	34.4	1,241	660
Poland	93	0	94	5	0	5	99	2.8	461	19
Portugal	65	2	73	1	1	3	76	7.7	134	81
Spain	541	27	665	54	11	61	726	19.6	2,615	1,203
Sweden	97	1	202	40	2	76	278	33.5	2,788	489
Switzerland	83	3	165	0	0	2	167	25.7	1,332	445
Tunisia	0	0	0	0	0	0	0	0	0	0
Turkey	102	0	107	40	0	40	147	3.3	352	176
United Kingdom	723	47	1,212	87	6	117	1,329	23.8	11,850	3,810
Yugoslavia	16	0	16	25	0	26	42	1.9	366	152
TOTAL REGISTRY	4,122	244	6,272	467	48	597	6,869	12.0	43,853	16,489

CAD=Cadaver
LD=Living donor

TABLE VI. Summary of patients alive on 31st December, 1983 according to treatment modality, based on data from patient questionnaire. Patients not updated or lost to follow-up are shown separately

Country	PATIENTS ON TREATMENT AT 31st DECEMBER, 1983							OTHER REGISTERED PATIENTS		
	on Haemodialysis/ Haemofiltration									
	Hospital	Home	IPD	CAPD	With functioning transplant	Total	Per million population	Recovered	*	†
Algeria	21	1	0	10	0	32	1.7	3	28	22
Austria	1,082	58	3	15	279	1,437	191.6	12	19	66
Belgium	1,825	111	4	147	900	2,987	304.8	46	32	173
Bulgaria	579	19	0	0	6	604	67.1	1	13	125
Cyprus	99	0	0	0	24	123	205.0	4	7	3
Czechoslovakia	774	2	2	12	204	994	65.4	6	8	33
Denmark	384	61	39	131	354	969	190.0	9	12	324
Egypt	537	2	9	4	37	589	15.1	5	110	66
Fed Rep Germany	11,730	1,784	128	262	1,904	15,808	258.3	176	263	2,248
Finland	216	4	14	128	502	864	180.0	6	9	70
France	8.776	2,027	218	692	1,910	13,623	255.1	142	248	2,256
German Dem Rep	1,233	0	6	3	410	1,652	98.3	6	11	97
Greece	1,003	3	23	83	155	1,267	136.2	17	89	486
Hungary	270	0	23	4	102	399	37.3	4	2	31
Iceland	13	0	0	0	7	20	100.0	0	0	0
Ireland	140	30	1	36	184	391	118.5	1	4	85
Israel	772	63	38	55	215	1,143	300.8	5	18	41
Italy	9,737	908	115	992	1,269	13,021	229.2	145	224	2,983

TABLE VI. *Continued*

Country	PATIENTS ON TREATMENT AT 31st DECEMBER, 1983							OTHER REGISTERED PATIENTS		
	on Haemodialysis/ Haemofiltration									
	Hospital	Home	IPD	CAPD	With functioning transplant	Total	Per million population	Recovered	*	†
Lebanon	–	–	–	–	–	–	–	0	8	94
Libya	48	0	0	0	4	52	17.9	0	9	35
Luxembourg	81	3	1	0	5	90	225.0	3	0	1
Netherlands	1,720	158	1	176	602	2,657	189.8	61	36	814
Norway	211	5	3	18	560	797	194.4	17	8	42
Poland	718	2	46	12	183	961	27.1	7	11	78
Portugal	1,227	0	1	4	68	1,300	132.7	12	11	214
Spain	6,712	333	88	483	1,123	8,739	236.2	60	143	754
Sweden	559	93	25	125	774	1,576	189.9	42	13	454
Switzerland	880	201	3	177	465	1,726	265.5	25	25	353
Tunisia	173	0	1	8	1	183	29.5	1	5	16
Turkey	201	1	2	16	131	351	7.9	7	59	41
United Kingdom	1,488	1,992	54	1,228	3,797	8,559	153.1	92	119	1,580
Yugoslavia	2,085	23	11	15	140	2,274	102.9	30	59	308
TOTAL REGISTRY	55,294	7,884	859	4,836	16,315	85,188	146.3	945	1,623	13,903

* Patients reported as lost to observation or who have moved to a non-reporting country.
† Patients known to be alive when last reported to the Registry but whose record has not been subsequently updated. Treatment status on 31st December, 1983 was therefore uncertain.

TABLE VII. Acceptance rates for new patients onto renal replacement therapy 1974–1983 in 32 European countries shown per million population. Data based on patient questionnaire

Country	New patients per million population									
	1974	1975	1976	1977	1978	1979	1980	1981	1982	1983*
Algeria	–	–	–	–	0.5	1.3	2.7	0.6	0.7	1.1
Austria	19.2	26.5	23.6	32.3	29.2	37.5	36.8	40.5	44.7	46.4
Belgium	31.5	31.4	34.1	41.0	37.3	42.6	45.4	50.7	53.3	53.1
Bulgaria	4.9	6.7	5.8	6.2	9.3	11.3	9.1	19.8	28.0	20.9
Cyprus	15.0	16.7	30.0	30.0	30.0	18.3	51.7	58.3	45.0	38.3
Czechoslovakia	5.8	8.0	8.9	9.6	11.6	13.4	15.1	17.0	19.5	18.8
Denmark	39.0	28.4	36.5	39.0	32.7	30.4	31.2	41.8	36.5	31.8
Egypt	–	0.1	0.4	1.0	1.5	1.6	1.7	6.1	9.2	7.4
Fed Rep Germany	25.7	32.1	34.4	36.8	38.8	42.5	46.7	50.7	52.4	47.8
Finland	19.4	26.0	22.9	27.1	32.5	28.8	35.0	30.2	35.2	42.1
France	24.6	31.4	34.2	36.1	35.5	35.8	41.6	43.6	40.9	37.2
German Dem Rep	9.8	10.6	13.1	15.5	16.1	18.3	19.6	21.7	25.7	24.3
Greece	17.4	19.6	22.4	27.0	26.2	30.2	27.8	32.3	31.7	31.3
Hungary	4.5	4.6	7.3	6.2	5.9	6.3	7.7	8.4	12.7	10.2
Iceland	15.0	15.0	15.0	20.0	10.0	20.0	10.0	25.0	35.0	5.0
Ireland	14.8	14.2	17.0	15.5	18.8	20.3	21.5	22.1	20.6	20.9
Israel	28.4	35.0	45.3	38.9	39.2	51.3	58.2	51.8	62.4	49.2
Italy	23.9	28.6	30.8	31.9	33.8	34.3	37.7	43.2	42.7	37.7
Lebanon	0.7	0.7	2.6	4.4	5.6	4.8	13.0	2.6	5.6	0
Libya	–	–	0.7	1.0	0	4.5	12.4	11.4	7.9	3.4
Luxembourg	15.0	35.0	25.0	30.0	40.0	57.5	47.5	27.5	45.0	65.0
Netherlands	18.1	20.4	22.5	26.6	26.3	26.9	35.4	35.8	34.0	34.3
Norway	22.7	27.3	27.6	31.5	29.5	29.3	39.5	42.9	40.7	32.2
Poland	1.9	2.5	3.0	3.7	5.0	5.8	6.1	5.6	6.5	7.7
Portugal	2.9	4.4	3.3	5.9	6.7	15.6	22.6	32.3	37.6	30.8
Spain	10.5	15.7	21.2	26.6	32.8	32.0	37.6	40.3	42.3	39.3
Sweden	24.5	28.0	29.3	34.8	33.4	38.8	46.7	47.0	53.0	36.6
Switzerland	32.0	36.5	35.5	40.9	43.2	40.9	48.8	44.0	51.8	44.9
Tunisia	0.2	0.8	1.0	1.9	1.5	1.3	1.8	6.9	9.0	7.9
Turkey	0.3	0.7	1.7	1.3	1.2	1.4	1.9	3.0	4.9	3.1
United Kingdom	15.4	16.7	16.6	17.8	20.2	22.4	24.9	27.8	31.2	25.2
Yugoslavia	9.8	11.3	15.8	16.1	19.1	20.0	22.0	23.5	22.7	22.9
TOTAL REGISTRY	14.0	16.8	18.5	20.3	21.6	23.0	26.0	28.7	30.1	26.7

* data for 1983 provisional

TABLE VIII. Deaths of patients on renal replacement therapy 1974–1983 in 32 European countries shown per million population. Data based on patient questionnaire

Country	Deaths per million population									
	1974	1975	1976	1977	1978	1979	1980	1981	1982	1983*
Algeria	–	–	–	–	0.1	0.6	0.8	0.7	0.3	0.3
Austria	8.8	13.9	10.3	15.2	15.1	17.6	20.0	20.9	24.5	25.5
Belgium	11.8	11.5	14.1	17.0	15.8	21.0	24.9	27.1	29.7	28.2
Bulgaria	1.1	3.0	4.0	2.8	3.7	4.1	5.4	7.8	8.1	8.4
Cyprus	11.7	8.3	6.7	8.3	16.7	18.3	15.0	20.0	25.0	26.7
Czechoslovakia	3.9	5.3	3.5	6.0	8.0	6.4	9.4	9.8	11.6	11.3
Denmark	19.8	17.1	19.0	16.5	23.3	24.9	23.3	22.5	22.7	23.5
Egypt	–	0	0	0.2	0.8	0.9	0.8	2.4	2.9	2.5
Fed Rep Germany	8.7	10.7	12.8	12.5	15.8	19.2	21.2	24.8	26.0	24.8
Finland	11.9	13.5	14.2	18.1	16.0	19.6	20.6	21.3	16.5	18.5
France	6.6	8.3	10.2	12.1	13.6	16.0	17.8	22.4	21.7	22.5
German Dem Rep	5.8	7.3	8.5	10.1	10.2	9.9	10.7	12.1	13.0	12.9
Greece	4.2	6.9	9.6	9.7	13.7	13.7	16.0	17.5	16.7	14.4
Hungary	2.6	3.9	2.6	4.8	4.6	4.0	4.2	4.8	4.8	5.8
Iceland	0	15.0	0	0	5.0	30.0	5.0	25.0	5.0	15.0
Ireland	8.5	7.0	5.5	7.0	10.0	10.9	7.9	9.4	7.6	4.5
Israel	7.6	8.9	14.5	21.3	21.3	27.6	28.4	34.2	35.0	30.0
Italy	5.2	7.8	9.0	10.4	11.2	13.0	15.7	17.1	19.9	19.2
Lebanon	0	0	0	0	0	0	0.7	2.6	3.7	0
Libya	–	–	0	0	0	0.7	3.1	3.4	2.4	0.7
Luxembourg	5.0	10.0	17.5	27.5	20.0	22.5	10.0	40.0	20.0	25.0
Netherlands	5.0	6.5	7.0	8.6	9.4	12.6	13.4	13.6	15.7	14.8
Norway	10.0	13.9	17.3	16.3	19.5	22.4	22.7	18.8	22.0	21.7
Poland	0.8	1.5	1.9	1.5	1.9	2.7	3.0	2.8	3.3	3.1
Portugal	0.3	0.8	0.3	0.5	0.8	3.1	4.0	4.8	5.4	7.0
Spain	1.8	2.6	3.4	4.6	6.5	9.1	10.6	12.1	12.9	14.7
Sweden	11.6	17.2	15.8	19.9	20.1	22.8	21.9	26.3	28.7	23.5
Switzerland	9.5	12.6	16.6	19.5	19.2	16.9	23.8	26.5	23.2	22.9
Tunisia	0	0	0	0.3	1.3	1.5	1.1	1.3	1.9	3.1
Turkey	0.1	0.2	0.8	0.7	0.6	1.0	1.0	1.4	2.2	2.5
United Kingdom	6.9	7.2	7.9	7.7	8.4	8.7	9.3	10.4	12.1	11.1
Yugoslavia	2.9	4.8	5.1	6.7	7.1	8.8	11.1	13.4	13.5	12.7
TOTAL REGISTRY	4.6	5.9	6.7	7.6	8.6	10.1	11.3	13.0	13.9	13.4

* data for 1983 provisional

TABLE IX. New grafts per million population 1974–1983 in 32 European countries. Data based on patient questionnaire

Country	Grafts per million population									
	1974	1975	1976	1977	1978	1979	1980	1981	1982	1983*
Algeria	–	–	–	–	0	0.1	0	0	0	0
Austria	12.7	11.5	8.1	10.4	8.3	11.1	8.8	9.5	11.5	11.1
Belgium	10.2	9.6	9.4	9.8	11.4	18.4	18.6	16.4	15.7	11.9
Bulgaria	0.1	1.0	0.9	0.8	0.2	0.1	1.0	0.6	0.1	0
Cyprus	1.7	1.7	8.3	6.7	3.3	15.0	3.3	1.7	1.7	0
Czechoslovakia	1.8	2.6	3.1	4.1	5.6	4.5	4.9	5.0	6.1	6.9
Denmark	29.0	24.3	22.7	25.9	25.9	22.9	24.3	22.9	16.7	12.9
Egypt	–	0	0.1	0.2	0.4	0.4	0.3	0.4	0.2	0.3
Fed Rep Germany	2.2	3.5	4.2	5.9	6.2	7.1	7.9	8.1	9.5	6.4
Finland	19.4	16.3	19.2	23.8	24.2	27.9	29.2	28.1	19.2	12.5
France	4.9	5.4	6.8	6.8	8.9	9.0	10.4	9.6	11.5	6.8
German Dem Rep	2.1	2.9	4.5	6.5	6.0	6.5	7.1	7.4	9.3	3.8
Greece	4.1	1.7	2.9	3.1	3.9	5.2	5.8	4.8	4.7	2.6
Hungary	1.6	0.7	1.1	1.4	1.8	3.3	3.3	2.6	3.1	2.7
Iceland	10.0	0	10.0	25.0	0	10.0	0	10.0	5.0	0
Ireland	8.2	6.4	6.1	9.1	10.6	17.9	13.0	15.2	19.1	6.1
Israel	5.0	6.1	9.7	11.6	11.6	13.9	25.0	23.7	23.9	9.7
Italy	2.6	2.6	3.8	4.7	3.1	3.3	4.6	5.2	5.9	6.6
Lebanon	0	0	0	0	0	0	0	0	0	0
Libya	–	–	0	0.3	0.7	0.7	0.3	0	0	0
Luxembourg	0	0	7.5	0	0	0	12.5	0	2.5	7.5
Netherlands	7.9	9.4	11.4	14.2	14.9	13.1	12.2	15.0	12.5	9.9
Norway	21.0	25.6	19.8	23.9	22.4	21.2	21.0	29.3	27.1	41.0
Poland	0.1	0.4	0.3	1.4	1.1	0.9	1.5	1.1	1.3	1.8
Portugal	0.2	0.2	0.2	0.2	0	0	0.8	1.7	2.6	3.7
Spain	0.7	1.0	1.3	1.9	2.9	5.2	7.1	9.8	9.9	12.1
Sweden	19.9	21.8	19.0	24.8	25.9	24.3	27.0	24.7	24.6	18.6
Switzerland	16.9	16.2	18.3	20.2	20.2	17.1	22.9	22.5	18.3	13.8
Tunisia	0	0	0	0	0.3	0	0.3	0.2	0.3	0
Turkey	0.1	0.1	1.0	0.5	0.4	0.5	0.3	0.7	1.0	1.5
United Kingdom	10.7	12.2	12.2	14.1	17.2	15.3	18.4	16.4	19.8	16.4
Yugoslavia	1.4	2.3	1.6	1.6	1.3	1.8	0.8	1.4	1.6	0.7
TOTAL REGISTRY	4.0	4.4	4.8	5.8	6.3	6.5	7.5	7.5	8.2	6.6

* data for 1983 provisional

the years 1974 to 1983. Transplants are shown according to the country of residence of the patients, rather than by country of operation (Table IX).

The age distribution of patients on the Registry has continued to change (Figure 1). In 1983 almost half (44%) of the new patients were aged over 55

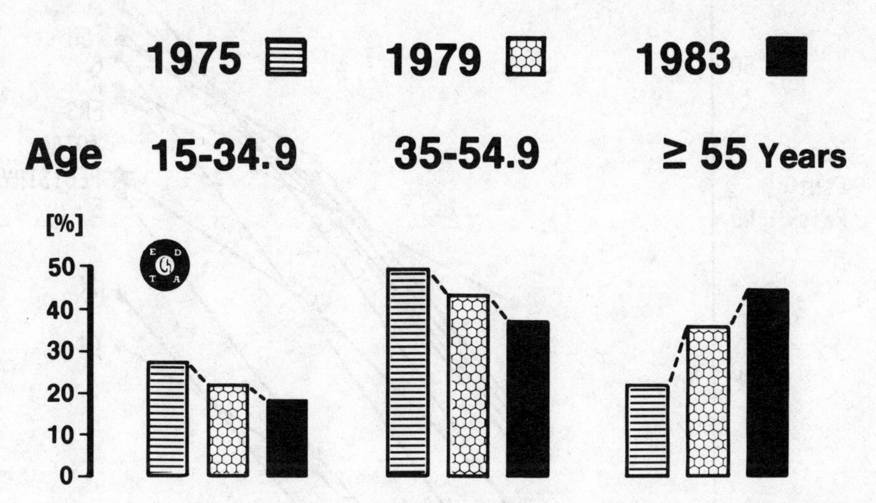

Figure 1. Age distribution of new patients commencing renal replacement therapy in 1975, 1979 and 1983. Proportion of patients in each of three age groups shown as a percentage

years at start of therapy, although the actual numbers of patients accepted in the younger age groups remained constant. Over 7,000 patients on treatment at the end of 1983 were then over 65 years of age. The proportion of new patients aged over 55 years accepted onto treatment for the years 1975, 1979 and 1983 is compared in 11 countries in Figure 2. With the increasing age of patients on treatment, the balance between deaths and new patients has altered. In 1979 the number of deaths was equivalent to 42 per cent of the total of new patients; in 1980 this figure reached 44 per cent and in 1983, 52 per cent.

Data consistency

The final version of the data files containing 1982 patient information held over 150,000 patient records. A study was undertaken to investigate the integrity and completeness of our data base. It was found that the treatment status of just over 6,000 patients, who were alive when last reported to the Registry, was uncertain at 31st December, 1982 because of failure of centres to update. Individual questions were well answered. Sex was reported in 99.7 per cent of 1982 questionnaires, date of birth in 97.1 per cent, primary renal disease in 98.7 per cent, blood group in 94.6 per cent and body weight in 88.1 per cent (Figure 3). Body height which was first requested in the 1983 questionnaire was supplied in 75.3 per cent of returns. Analyses which depend

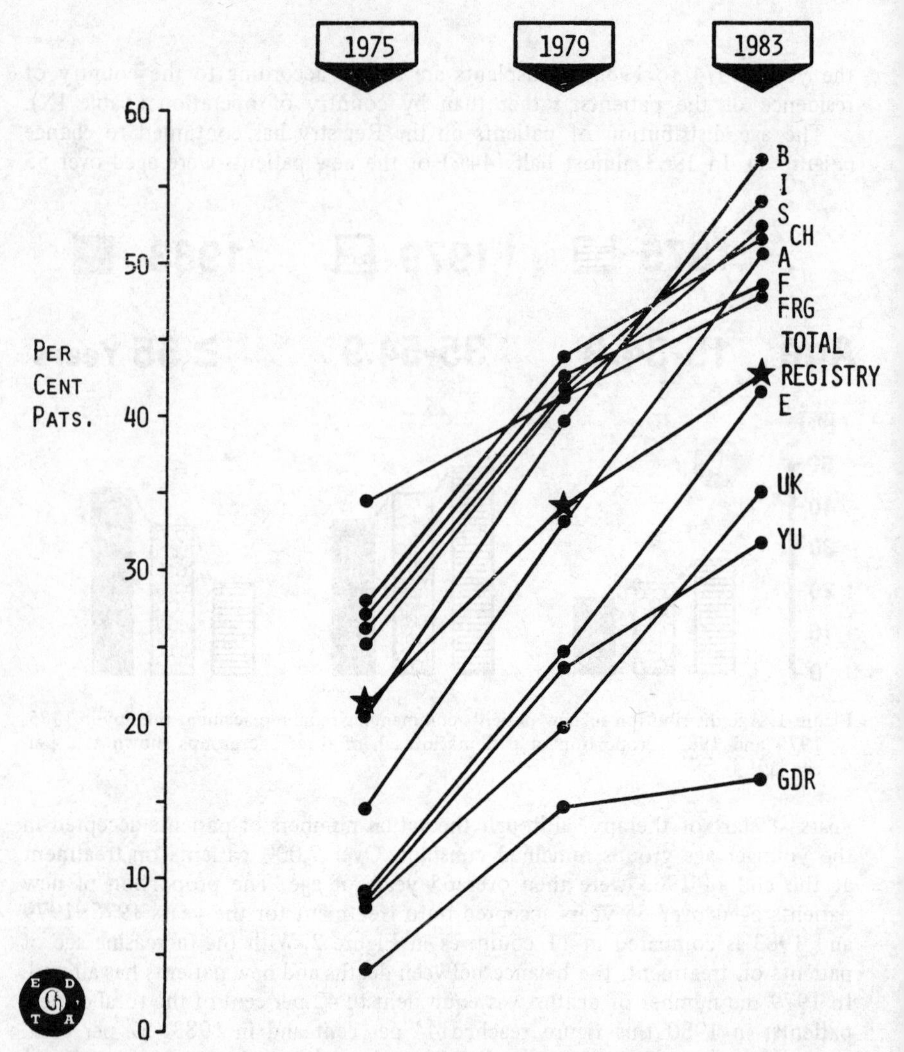

Figure 2. Proportion of new patients aged over 55 years accepted onto renal replacement therapy in 1975, 1979 and 1983. Data for eleven countries and total Registry shown. B=Belgium; I=Italy; S=Sweden; CH=Switzerland; A=Austria; F=France; FRG=Federal German Republic; E=Spain; UK=United Kingdom; YU=Yugoslavia; GDR=German Democratic Republic

on these items of data must discard patients with incomplete answers; fortunately, this is only a small proportion.

Amongst patients for whom haemodialysis or haemofiltration was recorded in the treatment sequence for 1982 (Figure 4), technical information was missing from nine per cent of the forms. On the other hand, in three per cent the technical section was filled in but no corresponding entry was made in the treatment sequence. The treatment sequence is used to identify events and is therefore essential data for all stock and flow analyses. Incomplete reporting

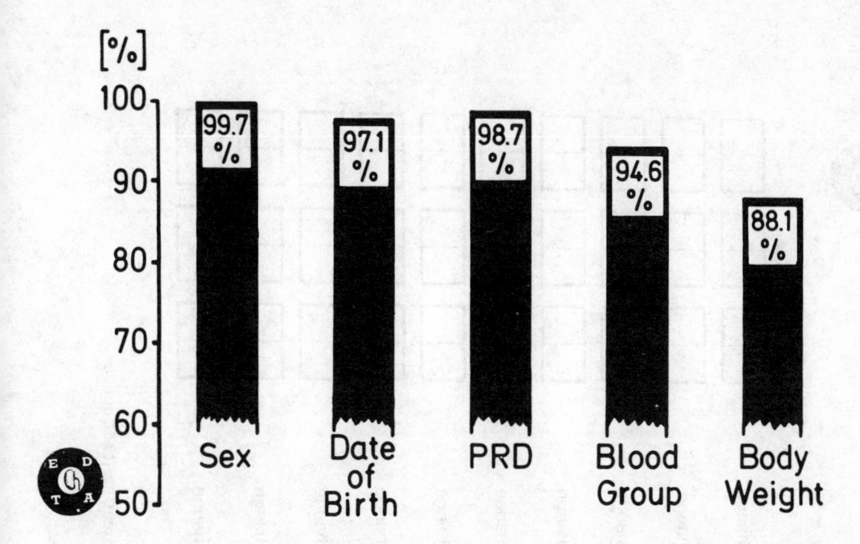

Figure 3. Proportion of 1982 patient questionnaires returned to the Registry which contained answers to selected questions

of technical information means that analyses of these data are carried out on a subset of all cases.

Patient survival

Survival on integrated therapies

In Figure 5, haemodialysis, pictured as a bridge, is used to illustrate how cumulative survival is calculated. Ten patients commence the follow-up as if they started haemodialysis on the same date; at zero time, survival is 100 per cent. The bridge is divided into four equal time intervals. During the first interval, one patient died (illustrated by the coffin) and one patient was transplanted (indicated by the patient climbing down into the motor boat). At the beginning of the second interval only eight patients were left: one patient died and one patient was not updated or was untraced by his doctors – this is illustrated by the one disappearing through the hole. Other patients become 'lost to follow-up' because they reach the limit of their period on treatment so far. During the third interval, out of six patients, two died and one started CAPD (illustrated by the one climbing down into a raft). During the fourth interval one patient died; thus only two patients reached the other end of the bridge, that is, remained alive on haemodialysis. A simple statment of patient survival would be 20 per cent. The percentage, shown in the line above, however, is much higher (33%). This is because the actuarial method (Figure 6) is based on the numbers at risk in each interval and has to adjust for the patients lost to follow-up, transplanted, untraced and changing to CAPD. This it does by assuming cases to have been kept alive by haemodialysis for 50 per cent of the interval

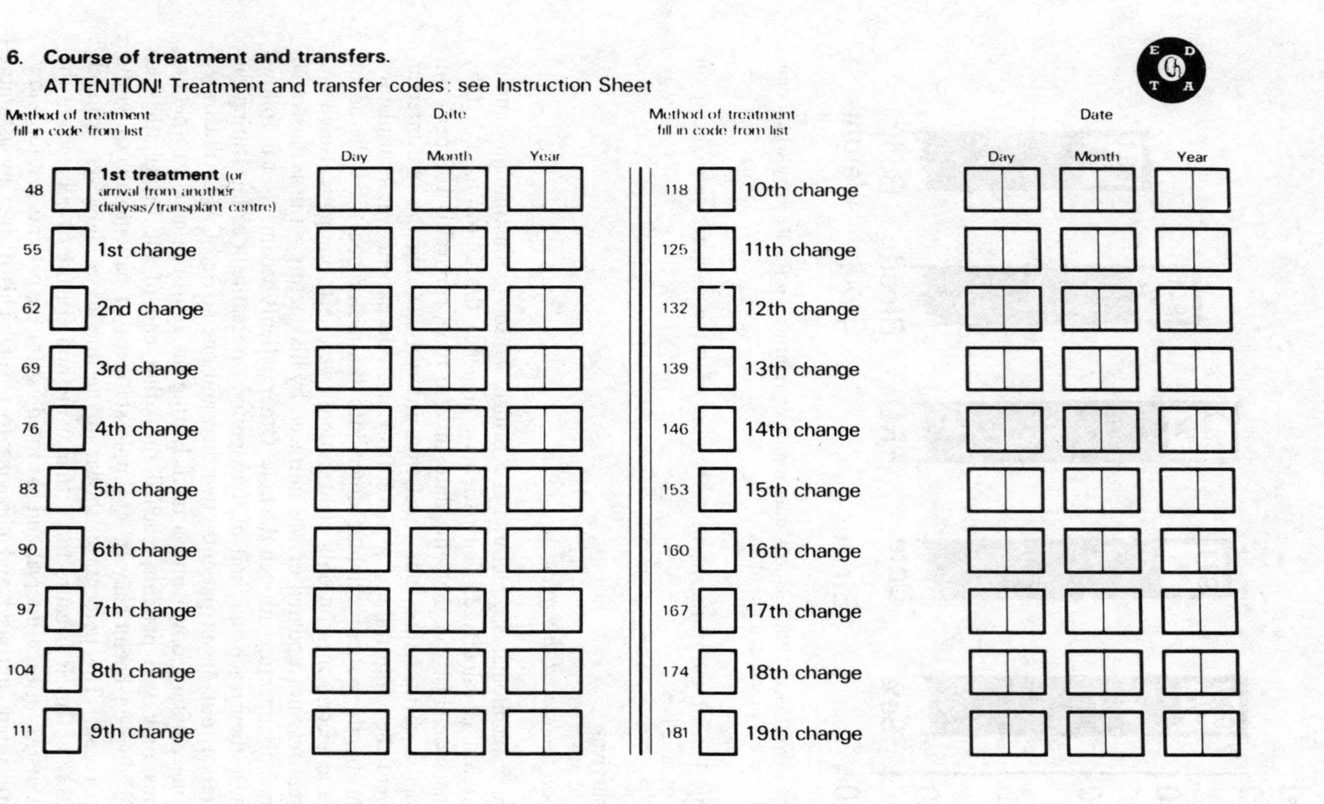

6. **Course of treatment and transfers.**

ATTENTION! Treatment and transfer codes: see Instruction Sheet

E D T A

Method of treatment fill in code from list		Date Day	Month	Year	Method of treatment fill in code from list		Date Day	Month	Year
48	**1st treatment** (or arrival from another dialysis/transplant centre)				118	10th change			
55	1st change				125	11th change			
62	2nd change				132	12th change			
69	3rd change				139	13th change			
76	4th change				146	14th change			
83	5th change				153	15th change			
90	6th change				160	16th change			
97	7th change				167	17th change			
104	8th change				174	18th change			
111	9th change				181	19th change			

Figure 4. Section of patient questionnaire dealing with treatment sequence

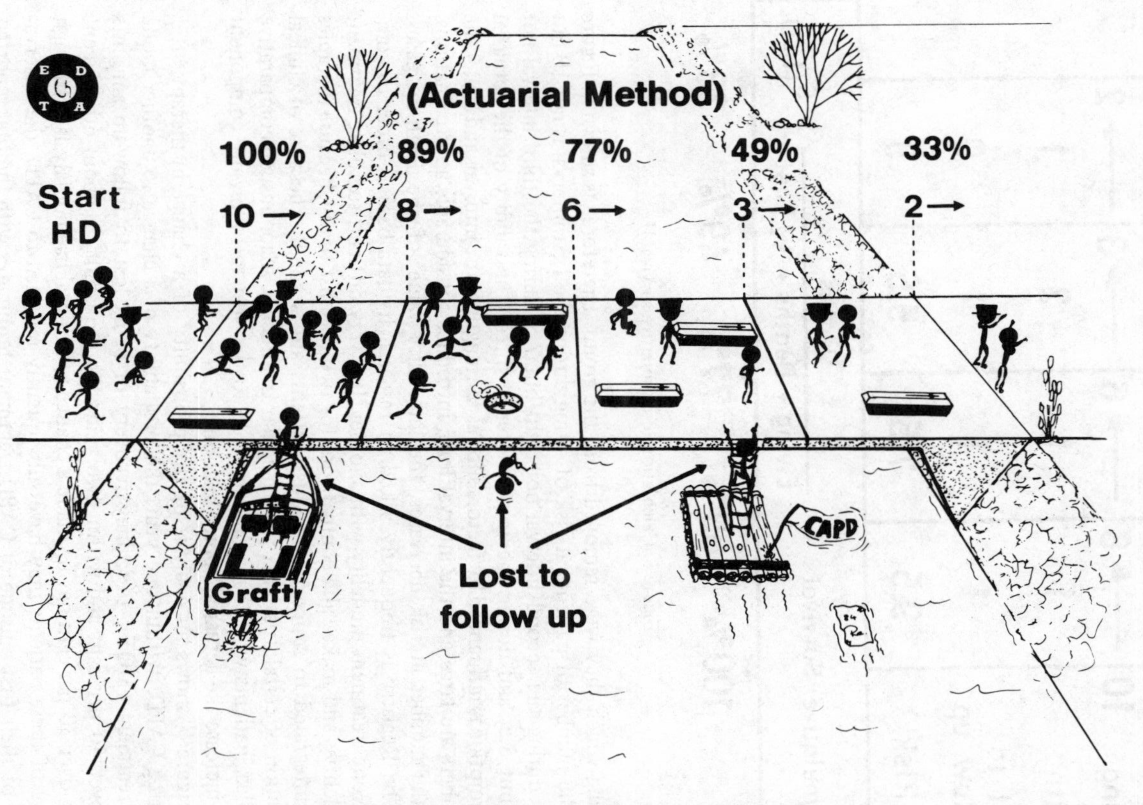

Figure 5. Calculation of cumulative survival I

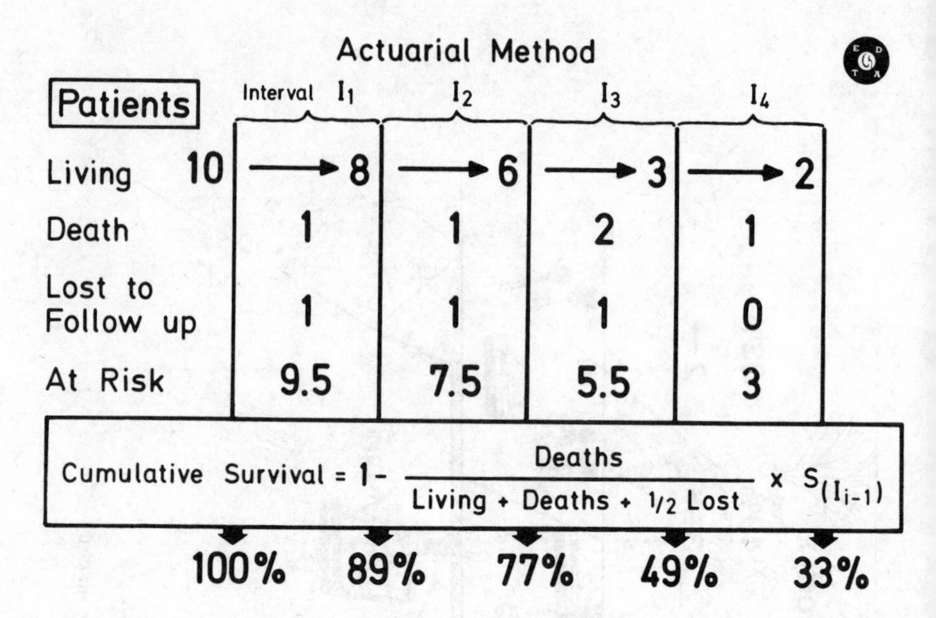

Figure 6. Calculation of cumulative survival II

during which they were removed from the group. Therefore, as shown in Figure 6, the number of patients at risk of dying during the first interval is not 10 but 9.5, during the second interval not eight but 7.5, during the third interval not six but 5.5, and during the fourth interval three. The validity of the actuarial assumption is influenced by factors such as the length of the interval, and can, if the numbers are large and the interval long, introduce considerable approximations. As the number at risk decreases, the 'mortality value' of one death increases.

The Registry is frequently asked to research its data base in order to compare one treatment modality with another. It is very difficult to isolate treatment modalities and make valid comparisons for two reasons: firstly, because therapies are integrated to support individual patients and secondly, because of clinical selection of certain patients for particular treatments. Also, in such comparative studies, patients who change therapy add to the 'lost to follow-up' component and increase the actuarial approximation.

Figure 7 shows an analysis of the probability of a change in therapy after starting CAPD. Within one year, the probability of a change to another mode of treatment (MOT), to kidney recovery or being lost to follow-up amounts to over 40 per cent and within two years the probability is about 60 per cent with over 30 per cent having made a change to hospital haemodialysis. Although many of these patients may have subsequently returned to CAPD, the observation of their first episode of CAPD treatment terminates with the first reported change in therapy. It is therefore very difficult to consider CAPD in isolation from haemodialysis. Similar considerations are relevant to transplantation because it is integrated with various methods of dialysis which contribute to patient survival before grafting and when the graft fails.

EDTA

	1 year	2 years
➡ Hosp. HD	22.7%	32.0%
➡ Home HD	0.9%	1.1%
➡ IPD	3.8%	4.9%
➡ Transplantation	7.5%	11.1%
➡ Kidney recovered	1.7%	2.2%
➡ Not updated (lost)	5.9%	9.2%
All changes	42.5%	60.4%

Figure 7. Probability of change in therapy within one and two years after starting CAPD

The Selwood analysis

We are indebted chiefly to Dr Neville Selwood for the development of a new approach to survival on integrated therapies. This is addressed to the problem of 'MOT crossover' which is illustrated by the three patients whose sequence of treatments is shown in Figure 8. It is the convention of actuarial methodology that when a patient changes his method of treatment, the number at risk in the interval during which he changes is altered by 0.5. In the Selwood analysis the number at risk on any treatment is diminished by those leaving the therapy and augmented by those entering it. The number at risk (as shown in the small circles) may therefore increase. Thus, this new approach does not follow individual patients on single methods of therapy but looks at the mode of therapy and calculates, according to actuarial conventions, the contribution of that therapy to overall survival within the total time on substitution treatment.

In conventional actuarial analysis the number of patients at risk always decreases from the first day of treatment. With this approach, however, the number of patients at risk on a single mode of therapy may go up as well as down (Figure 9). The number of patients at risk on IPD was high during the first half year and decreased considerably; at the same time the number of patients on home haemodialysis and CAPD increased and also the number of transplanted patients. The big number of patients on conventional haemodialysis increased during the first two months and thereafter decreased slowly because a large proportion of patients stayed on hospital haemodialysis for their whole treatment history.

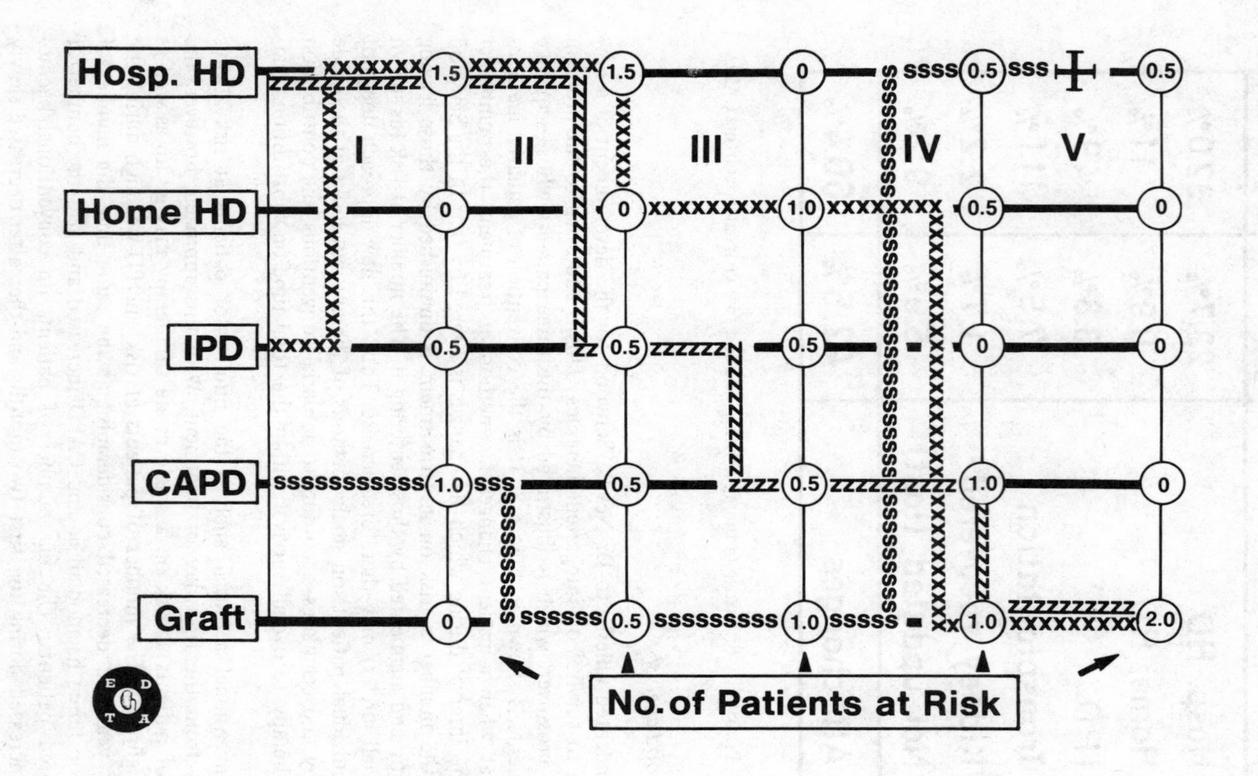

Figure 8. Diagram to illustrate the problem that arises with changes in method of treatment. Three different treatment sequences are shown. Roman numerals refer to intervals

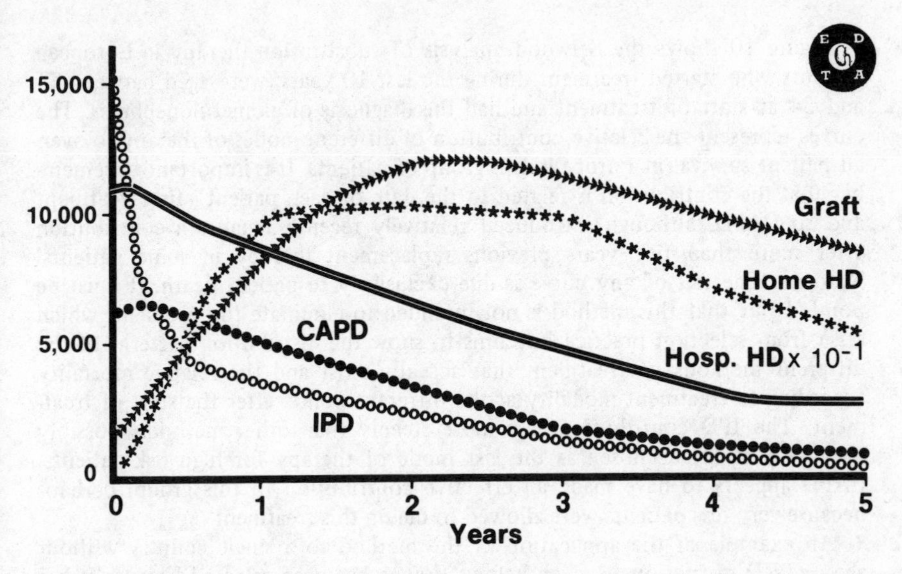

Figure 9. Graph to show change in number of patients at risk, for calculation of cumulative survival on different modes of treatment in Europe, using the Selwood method

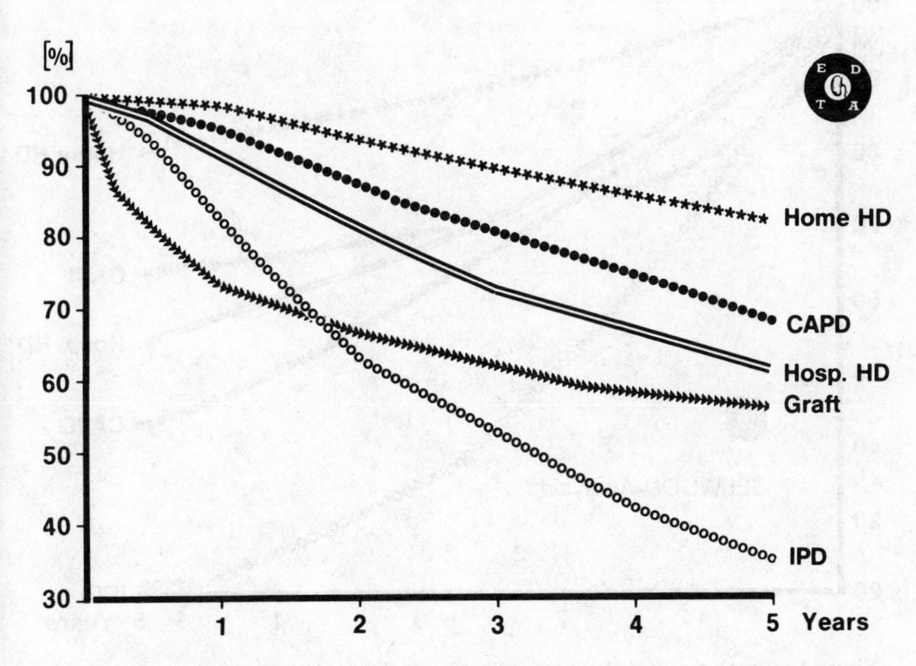

Figure 10. Graph to show relative contribution of different modes of therapy to overall survival of patients with glomerulonephritis aged 35–54 years (Selwood method)

Figure 10 shows the Selwood analysis of substitution therapy in European patients who started treatment during the last 10 years, were aged between 35 and 54 at start of treatment and had the diagnosis of glomerulonephritis. The curves represent the relative contribution of different modes of therapy to overall patient survival in Europe in this group of patients. It is important to remember that the contribution is related to the date of each patient's first treatment and so CAPD, although introduced relatively recently, makes a contribution after more than five years previous replacement therapy in some patients' histories. The fall of any curve is due exclusively to patient death. It must be pointed out that this method is not intended to eliminate the problems which arise from selection practice but aims to show the distribution patterns of the different methods of treatment that actually exist and the relative mortality according to treatment modality at the different points after the start of treatment. The IPD contribution sags more steeply than other methods, possibly because it is often chosen as the last mode of therapy for high risk patients. CAPD appears to have made an effective contribution to this group, perhaps because very few patients were allowed to die on this treatment.

An example of the application of the method to a single country without age or PRD restrictions is given below: France has been selected because it has good facilities for haemodialysis, an active transplantation programme and many patients on CAPD. Figure 11 shows that transplantation supersedes hospital

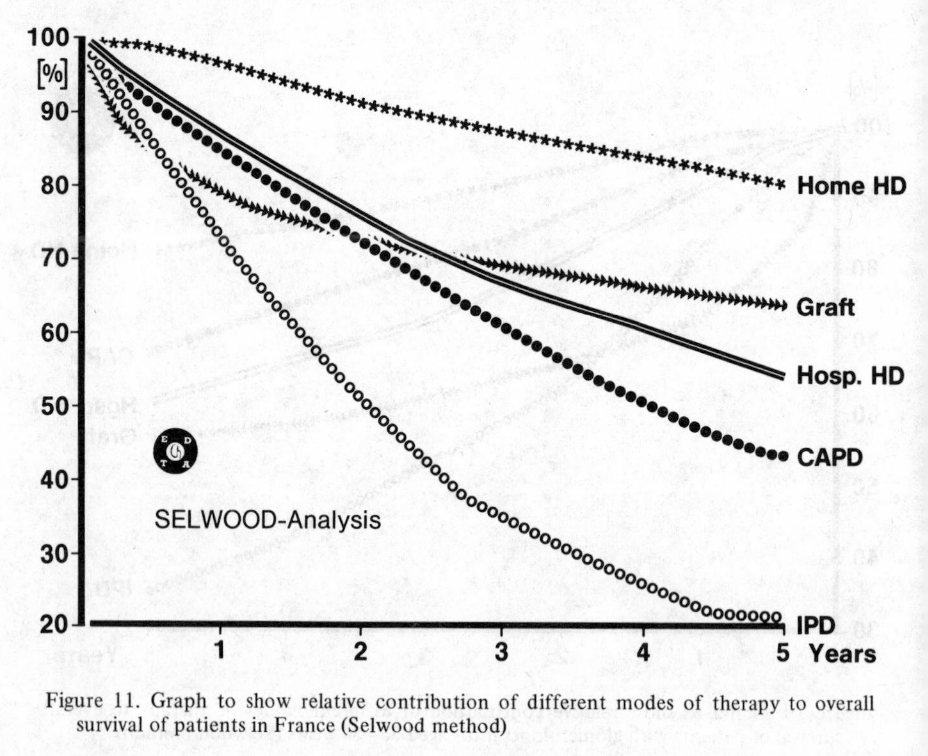

Figure 11. Graph to show relative contribution of different modes of therapy to overall survival of patients in France (Selwood method)

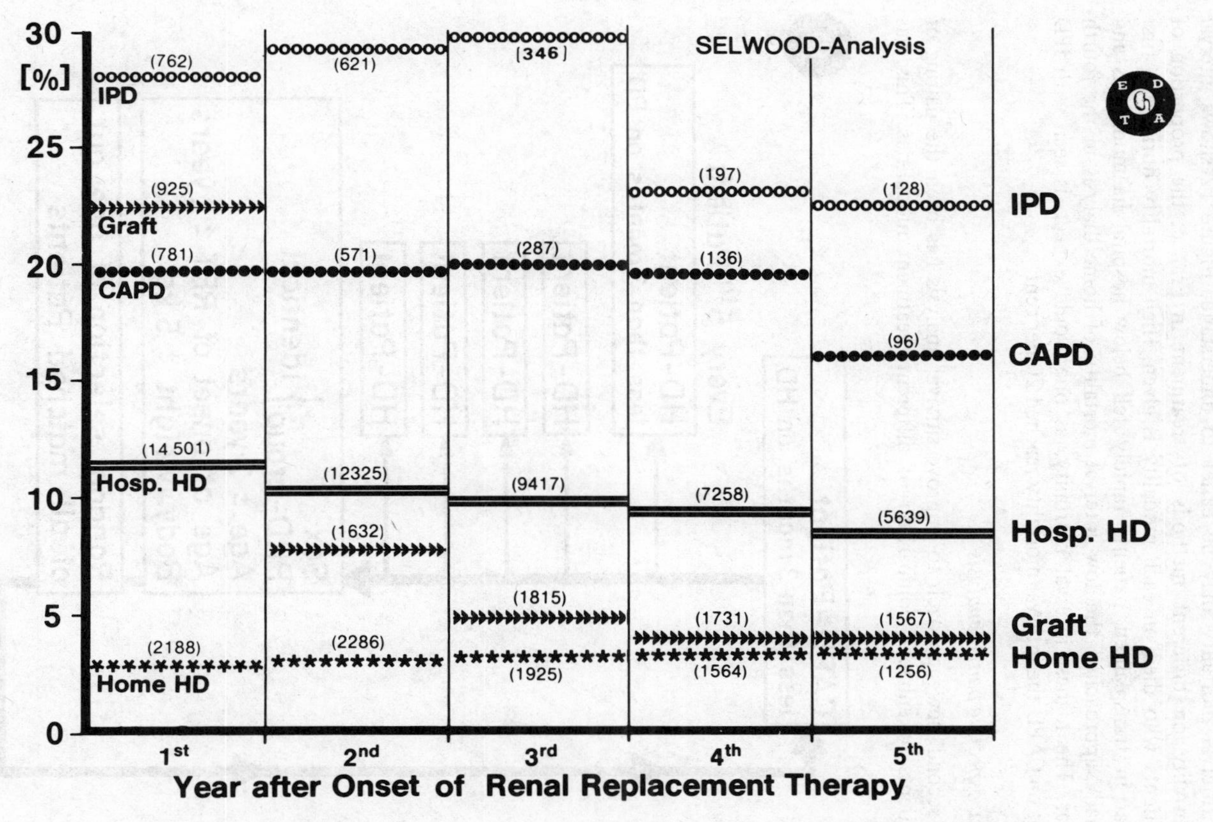

Figure 12. Interval mortality on different methods of treatment in France (Selwood method)

haemodialysis from the third year after start of renal replacement therapy onwards and that mortality on CAPD is greater than that on hospital haemodialysis.

Interval mortality may be easier to understand. Figure 12 shows interval mortality on different methods of treatment in France; the proportion of patients who died on each modality is given. High mortality during the first year in transplanted patients rapidly fell below hospital haemodialysis and finally approached the low interval mortality of home dialysis in the fourth year. The highest interval mortality, as one would expect, is seen with IPD but on CAPD the interval mortality reached 20 per cent.

Retrospective paired analysis

A second new approach to improve survival analysis has been the pairing of patients to enhance comparison of different treatment modalities. This was

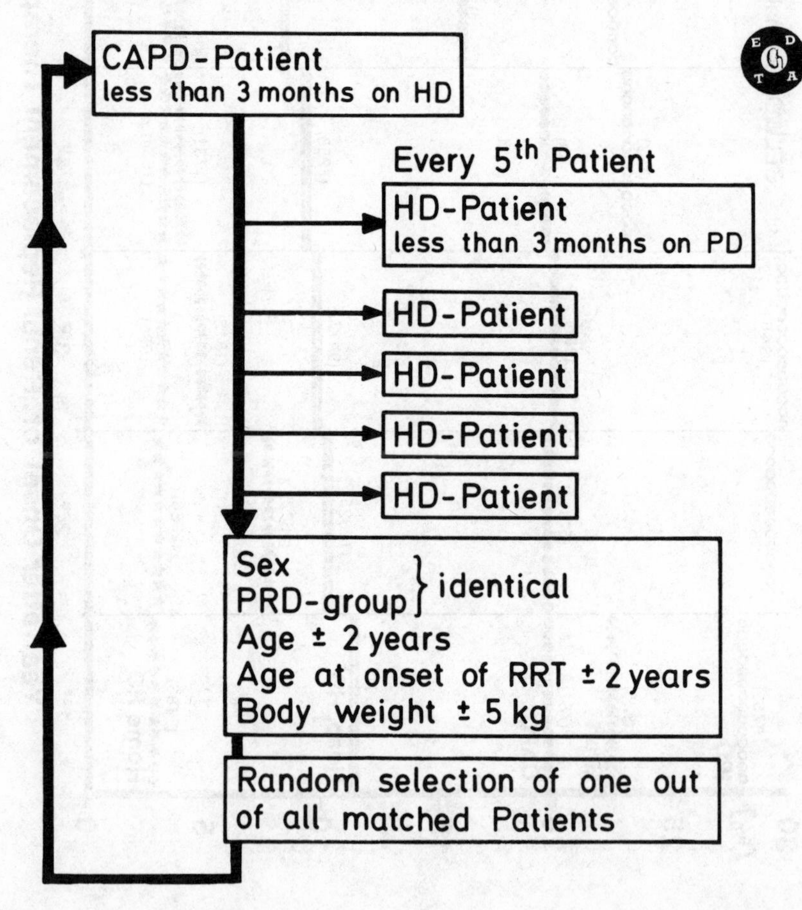

Figure 13. Diagram to illustrate pairing of haemodialysis and CAPD patients from the same country for retrospective comparison of survival. Patients with a graft at any time or history of malignancy are excluded

made possible with a computer program developed for the Registry by Dr Andras Balás. The program first identified CAPD patients never grafted and without a history of malignant disease (Figure 13); it then searched the files and checked every fifth haemodialysis patient who had less than three months on PD and who came from the same country. To qualify as a possible paired case, the haemodialysis patient had to have the same sex, the same primary renal disease group, to have started treatment within ± two years, age at start of treatment had to be within ± two years and the body weight not more than ± five kilograms that of the CAPD patient. Having collected all patients who satisfied these criteria the program then picked out one of them randomly for pairing. The process was then repeated for the next suitable CAPD patient. Construction of the file consumed a large amount of computing resources, but the number of CAPD patients who could be satisfactorily paired was not large.

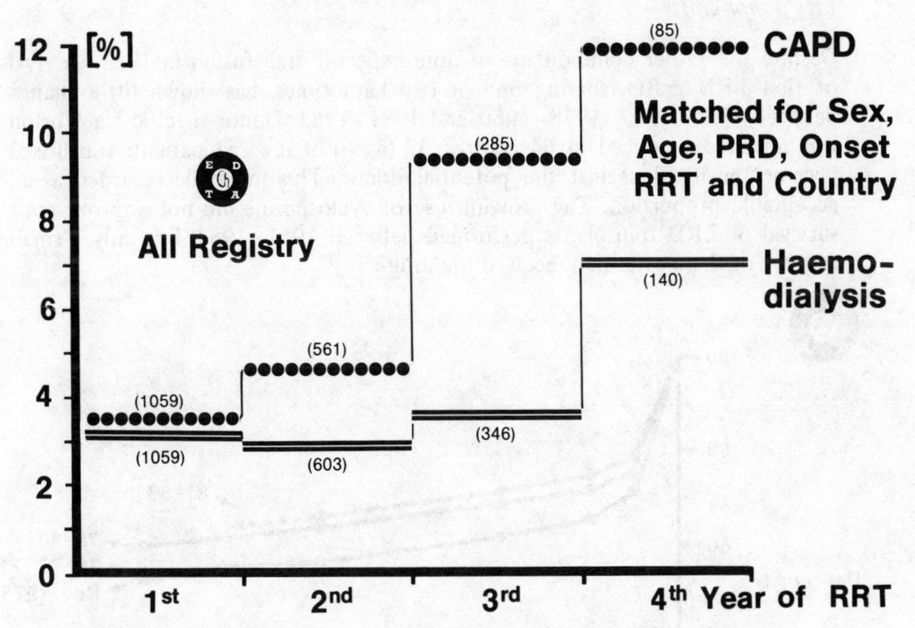

Figure 14. Interval mortality in paired patients on CAPD and haemodialysis

Figure 14 compares interval mortality in the paired groups on CAPD and haemodialysis. For the first year results were similar, but thereafter haemodialysis had a lower mortality. This means either that CAPD achieved comparable results only in the short term or that patients who began CAPD two or three years ago were not so well treated as those who began more recently.

This study is a preliminary example of a new technique devised to refine comparison of survival on different therapies. It should not be seen as an approximation to a controlled trial. Dr Balás' program pairs cases retrospectively in respect of certain limited features. It does not take into account the characteristics which may have governed the choice of CAPD or haemodialysis for a particular patient in the first place and which are a source of bias in comparing survival on the two therapies.

Transplantation

The development of transplantation in the decade 1974 to 1983 is shown in Table IX (page 16). The total number of transplants performed in 1983 was more than three times higher than 10 years earlier. The main contribution was from cadaveric transplantation, with live donor transplantation contributing only about 10 per cent of all grafts. Although there were still large variations in transplant rates, the overall picture was one of increasing activity in most countries since 1981.

Live donor grafts

Despite the rather common use of donor specific transfusion in 1983, survival of first LRD grafts, sharing one or two haplotypes, has shown little change between 1975–1977, 1978–1980 and 1981–1983. Donor specific transfusion was reported practiced in 64 centres; 37 (15%) of the 251 patients transfused became sensitised against the potential donor. This might be regarded as an acceptable proportion. The growing use of cyclosporine did not improve graft survival of LRD transplants performed between 1981–1983, but only a small number of these cases had received the drug.

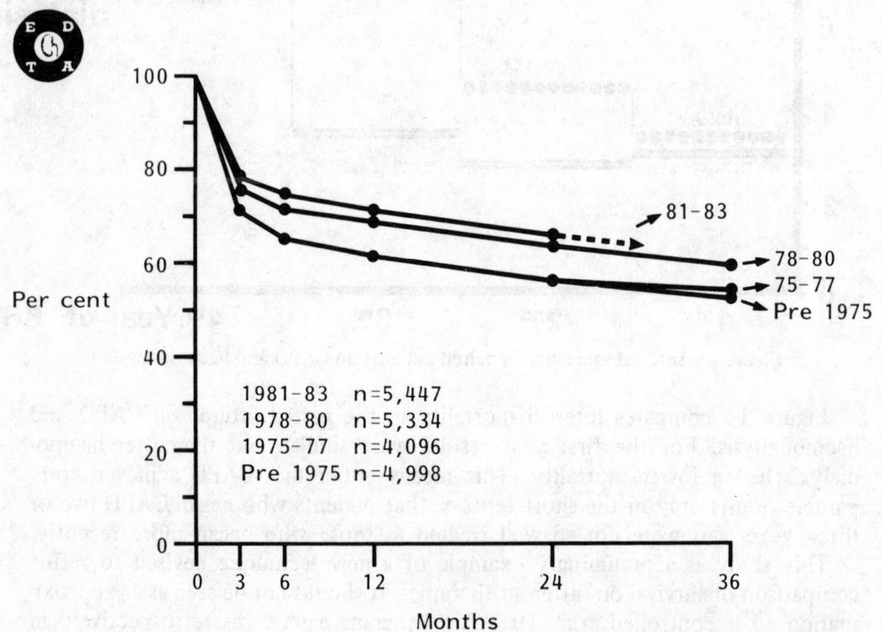

Figure 15. Comparison of survival of 1st cadaver grafts performed prior to 1975, 1975–1977, 1978–1980, 1981–1983 in patients aged 15 to 45 years. Diabetic patients excluded

Cadaver grafts

Period comparison of first cadaver graft survival in a standard group of patients, defined as those aged 15 to 45 years excluding diabetic patients, showed an improvement between 1975–1977 and 1978–1980 with a further slight improvement in 1981–1983 (Figure 15). Only a small proportion of patients received cyclosporine between 1981 and 1983, and therefore its influence on survival cannot be evaluated. There has been a substantial improvement in patient survival of the same standard group (Figure 16), except in the last period, 1981–1983 where it was almost identical to that in 1978–1980.

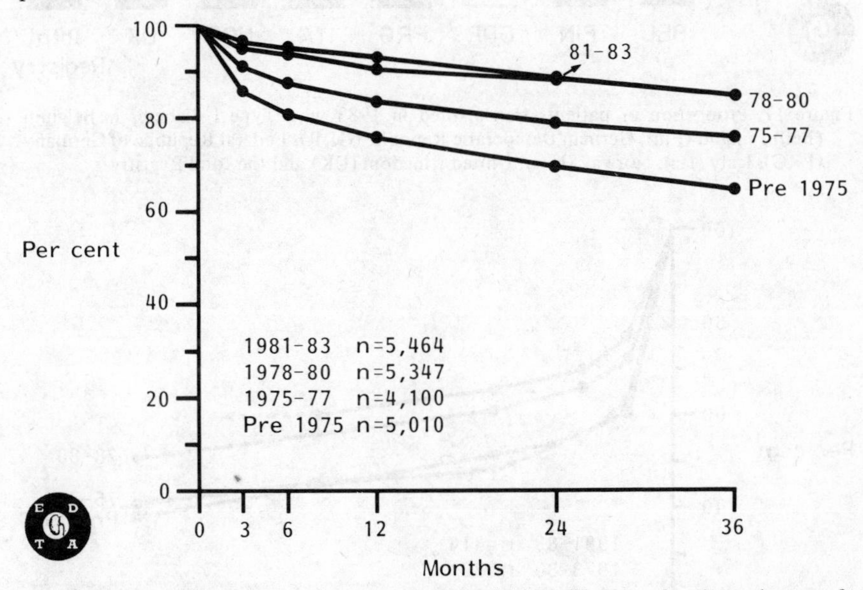

Figure 16. Comparison of survival in patients aged 15 to 45 years after 1st cadaver graft performed prior to 1975, 1975–1977, 1978–1980, 1981–1983. Diabetic patients excluded

Diabetic patients

Transplantation in diabetic patients has been analysed separately. Figure 17 gives the proportion of insulin-dependent or Type I diabetic patients in the transplant programmes of seven selected countries in 1983. There were large differences between countries, with lowest transplant activity in diabetic patients in the German Democratic Republic (0%) and the highest in Finland (27%). The variation may in part be due to a difference in the incidence of nephropathy in Type I diabetes across Europe, a hypothesis supported in previous research done by the Registry [1], but there is undoubtedly considerable reluctance to transplant diabetic patients in several countries.

Survival of first CAD grafts in all types of diabetic patients showed improvement in 1978–1980 and 1981–1983 compared with 1975–1977 and pre-1975

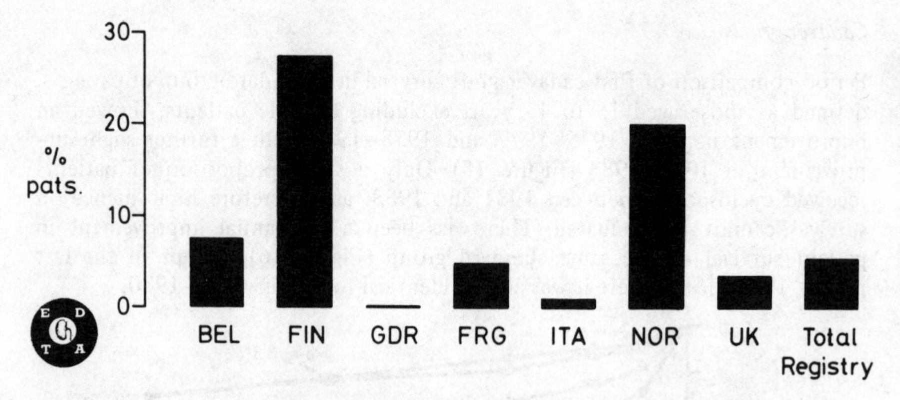

Figure 17. Proportion of patients transplanted in 1983 with Type I diabetes in Belgium (Bel), Finland (Fin), German Democratic Republic (GDR), Federal Republic of Germany (FRG), Italy (Ita), Norway (Nor), United Kingdom (UK) and the total Registry

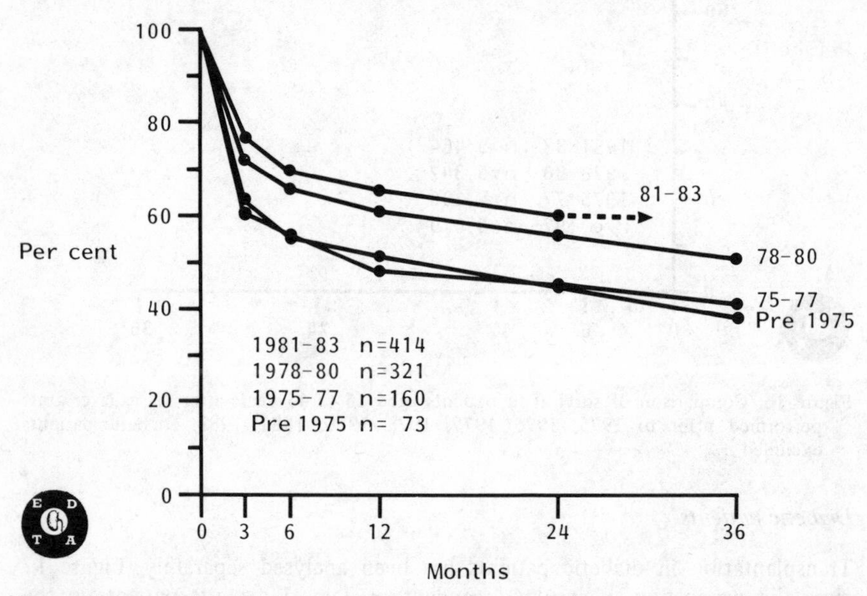

Figure 18. Comparison of survival of 1st cadaver grafts in diabetic patients performed prior to 1975, 1975–1977, 1978–1980, 1981–1983

(Figure 18). The two year graft survival of 60 per cent in the period 1981–1983 must be considered very good indeed, and hopefully will stimulate more nephrologists and transplant surgeons to take on patients with diabetes for their transplant programmes. A substantial improvement has been recorded in patient survival after first CAD graft since 1977 but there has been no further improvement in the latest time period studied (Figure 19).

Information on pancreatic transplantation was obtained in the 1983 centre

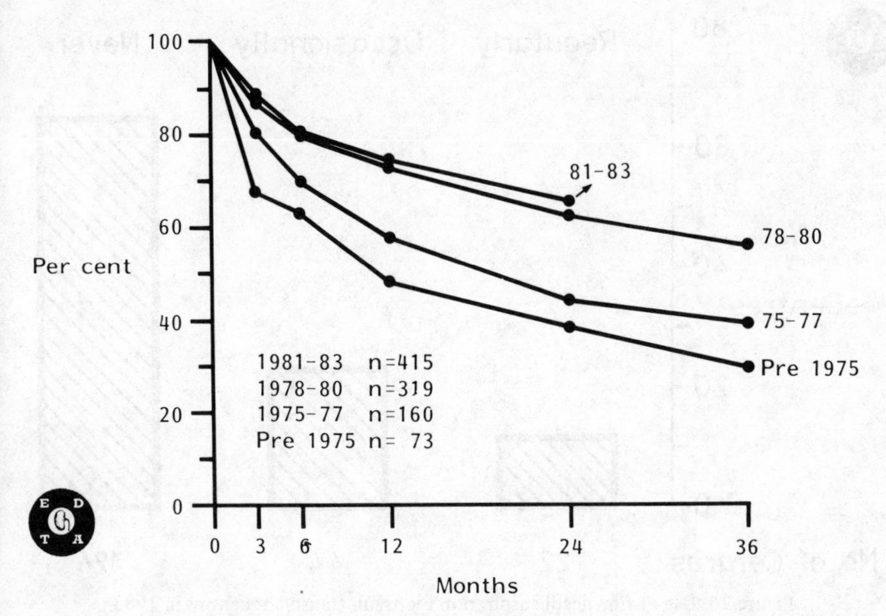

Figure 19. Comparison of survival of diabetic patients after 1st cadaver grafts performed prior to 1975, 1975–1977, 1978–1980, 1981–1983

questionnaire. Of the 1,477 centres who submitted data, 14 reported 58 grafts in total. Four of the centres were in the United Kingdom, two each in Belgium and Spain, and one in Austria, Czechoslovakia, Federal Republic of Germany, Italy, Norway and Sweden.

Fine needle aspiration biopsy

The 1983 centre questionnaire contained a question on the use of fine needle aspiration biopsy to monitor immunosuppression. Of reporting centres, 60 per cent never used it while 40 per cent used it either regularly (22 centres) or occasionally (44 centres) (Figure 20). These answers were linked to the main patient file in order to evaluate first cadaver graft survival in centres using fine needle aspiration biopsy in all patients or in no patients, and to compare cyclosporine versus conventional therapy (Figure 21). Irrespective of immunosuppression, graft survival tended to be better in centres using fine needle aspiration biopsy. However, the numbers were small and should be interpreted with caution. The use of fine needle aspiration biopsy may only be a marker for a well equipped centre, with the experience or other diagnostic resources to achieve better results.

Cyclosporine

In 1983 approximately half of all transplant centres used cyclosporine (Figure 22). Eighteen centres used cyclosporine alone whilst the majority, 79 centres,

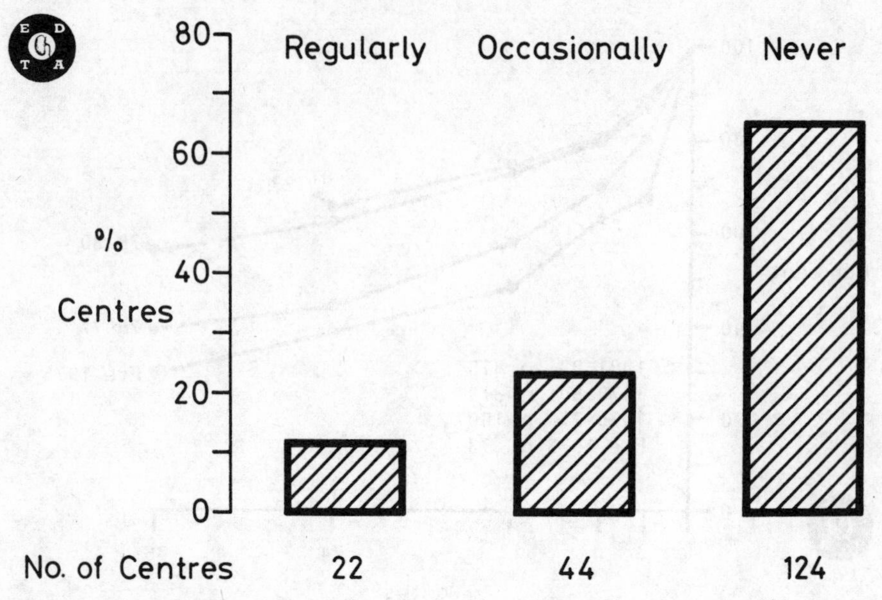

Figure 20. Use of fine needle aspiration biopsy in transplant centres in 1983

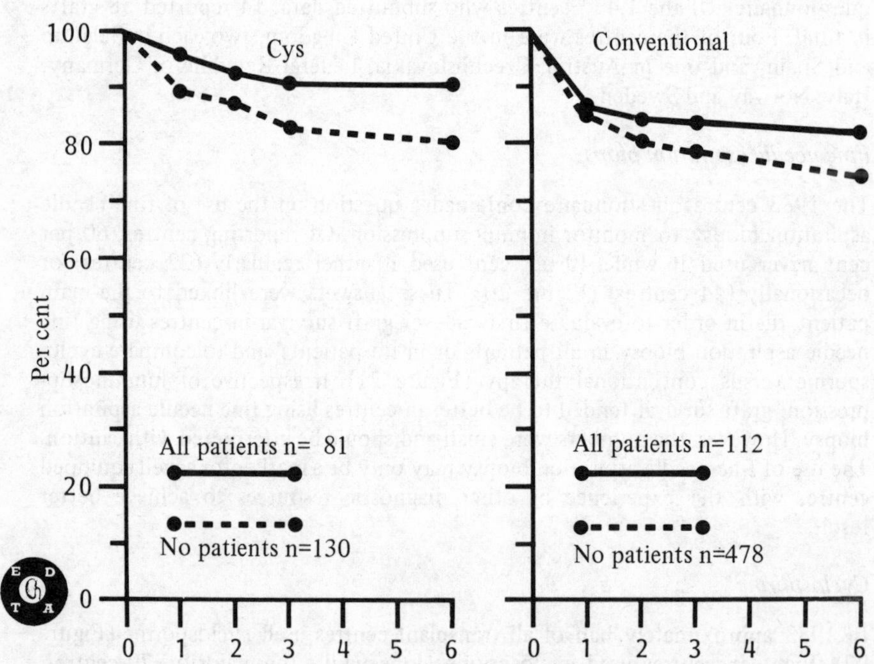

Figure 21. Comparison of 1st cadaver graft survival according to use of fine needle aspiration biopsy in 1983. Results shown separately for cyclosporine (Cys) and conventional immunosuppressive therapy

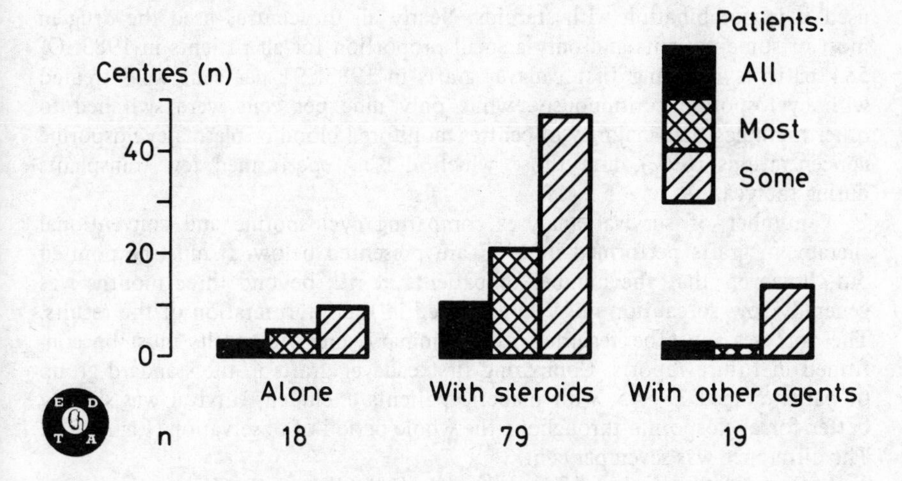

Figure 22. Use of cyclosporine in transplant centres in 1983. Cys 'Yes'=100 centres. Cys 'No'=103 centres

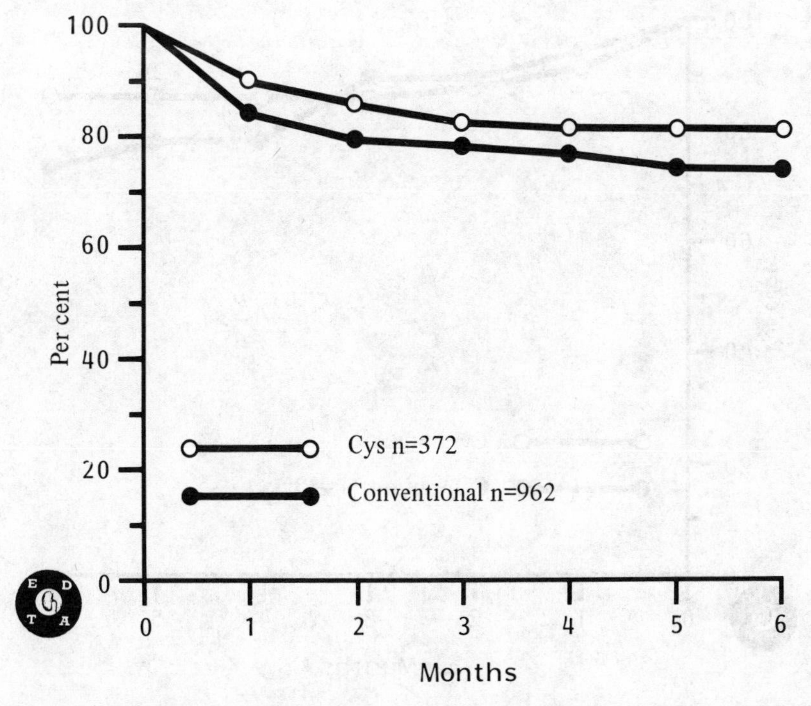

Figure 23. Comparison of survival of 1st cadaver grafts performed in 1983 according to use of cyclosporine. Diabetic patients excluded from analysis

used it in combination with steroids. Nearly all the centres used the drug in most or some patients and only a small proportion for all patients in 1983. Of 584 patients receiving first cadaver grafts in 1983, 91 per cent were treated with cyclosporine continuously, while only nine per cent were switched to other therapies. The majority of centres monitored blood or plasma cyclosporine concentrations (89%), and those which did not performed few transplants during the year.

A number of survival analyses comparing cyclosporine and conventional therapy in grafts performed in 1983 are presented below. It must be pointed out, however, that the number of patients at risk beyond three months was generally low so caution must be exercised in the interpretation of the results. The analyses must be regarded as preliminary and the results must be confirmed in future reports. Comparing first cadaver grafts in the standard group of patients aged 15–45, with diabetic patients excluded, survival was slightly better for cyclosporine throughout the whole period of observation (Figure 23). The difference was seven per cent.

Analysis of the results of first cadaver grafts in diabetic recipients (Figure 24) shows that the cyclosporine curve becomes a straight line from three months

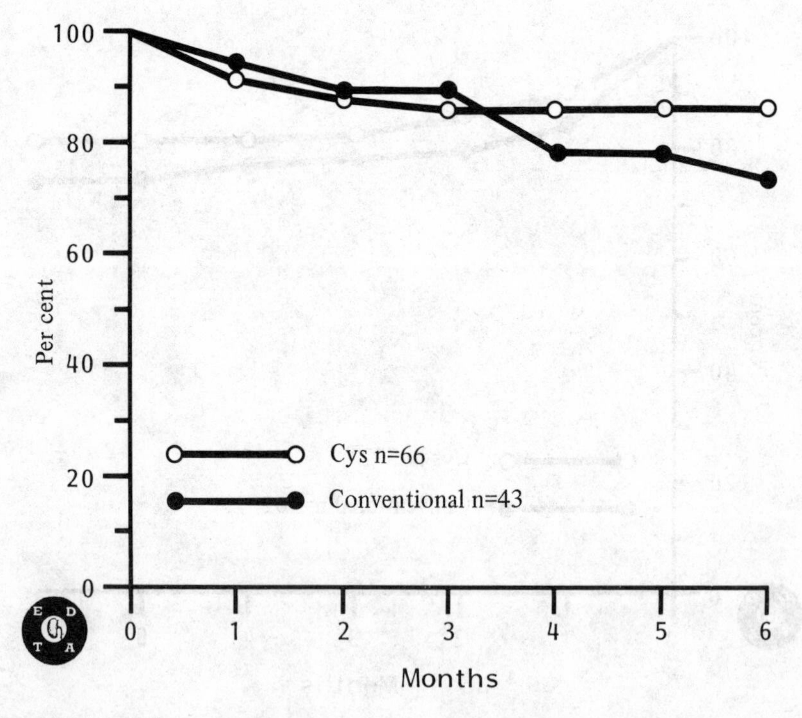

Figure 24. Comparison of survival of 1st cadaver grafts performed in diabetic patients during 1983 according to use of cyclosporine

emphasising that numbers were small and caution in interpretation is needed.

Survival of first cadaver grafts in the small number of elderly patients seemed slightly better in the cyclosporine group (Figure 25).

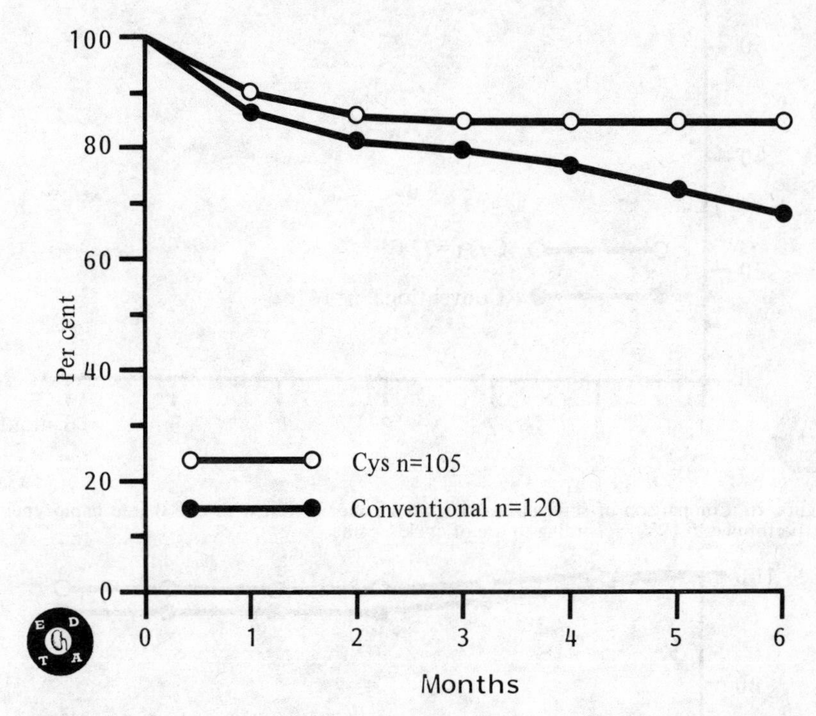

Figure 25. Comparison of survival of 1st cadaver grafts performed in patients aged 55 and over during 1983 according to use of cyclosporine

In transplantation of live related donor grafts with one shared haplotype (Figure 26), the outcome for the cyclosporine group was excellent with 95 per cent graft survival at six months compared to 79 per cent for the conventional group.

Study of patient survival showed the differences to be small, with a slight advantage in favour of those given cyclosporine in the standard group of patients (Figure 27).

Patient survival for diabetic patients showed a similar slight difference in favour of cyclosporine (Figure 28), while the beneficial tendency for elderly patients seems to be slightly stronger (Figure 29).

Graft function

The cause of graft failure in first cadaver grafts performed in 1983 was compared in those on conventional therapy and those receiving cyclosporine (Figure 30).

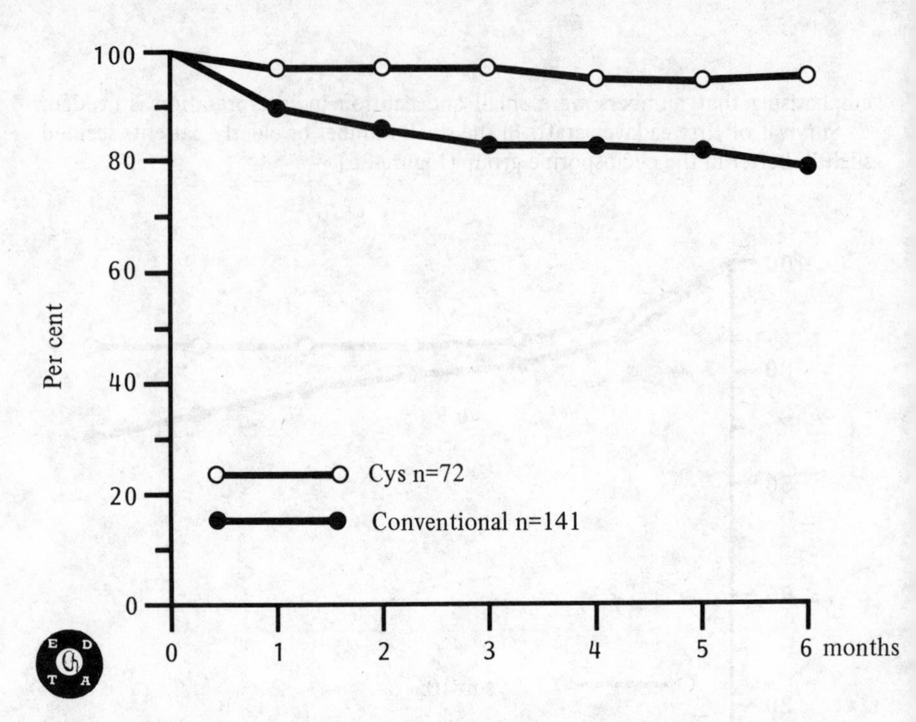

Figure 26. Comparison of survival of 1st related donor grafts (with one shared haplotype) performed in 1983, according to use of cyclosporine

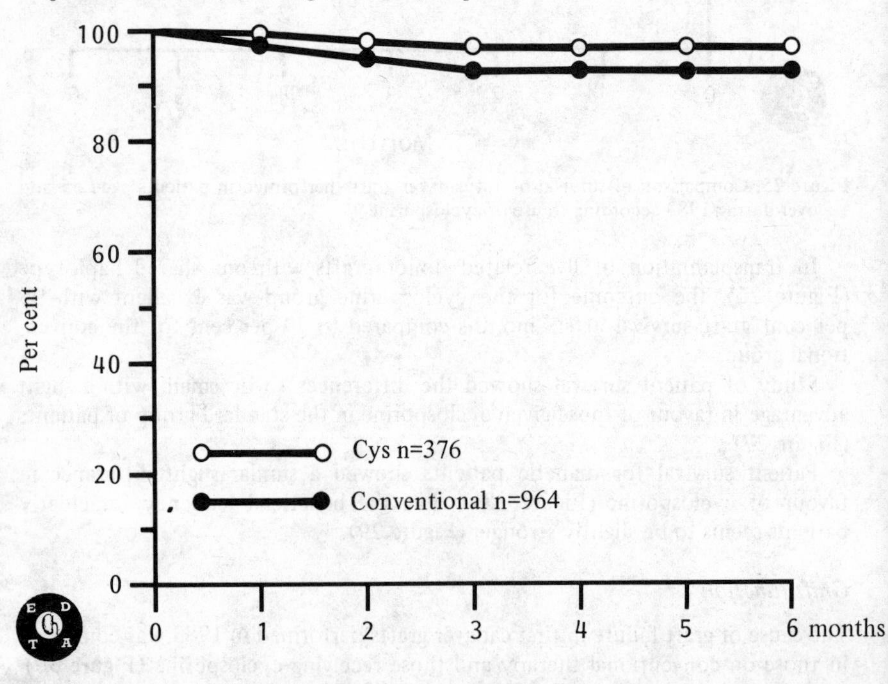

Figure 27. Comparison of survival of patients aged 15 to 45 years after 1st cadaver graft performed in 1983, according to use of cyclosporine. Diabetic patients excluded

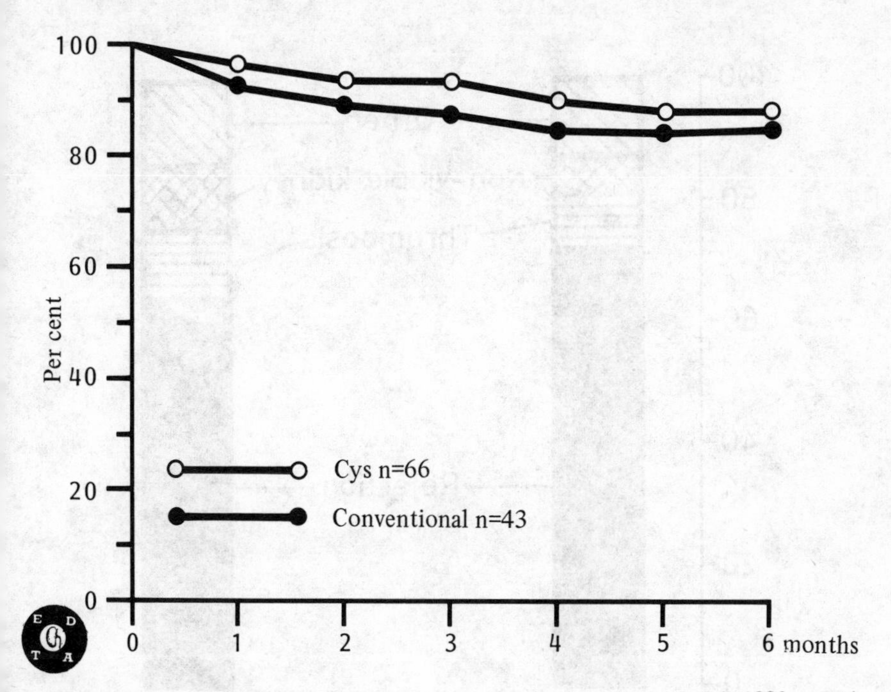

Figure 28. Comparison of survival of diabetic patients after 1st cadaver graft in 1983, according to use of cyclosporine

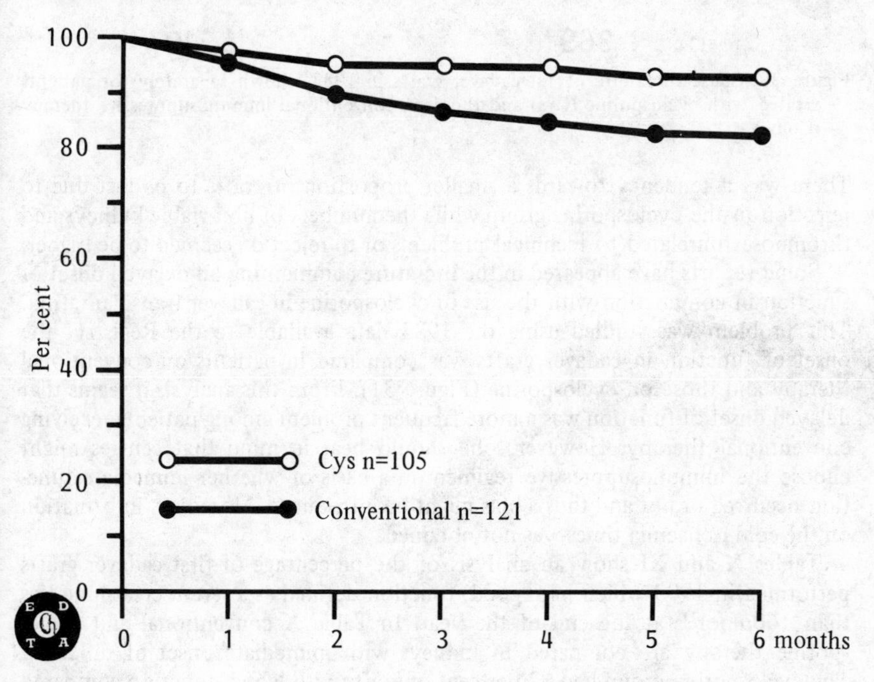

Figure 29. Comparison of survival of patients aged 55 years and over after 1st cadaver graft performed in 1983, according to use of cyclosporine. Diabetic patients excluded

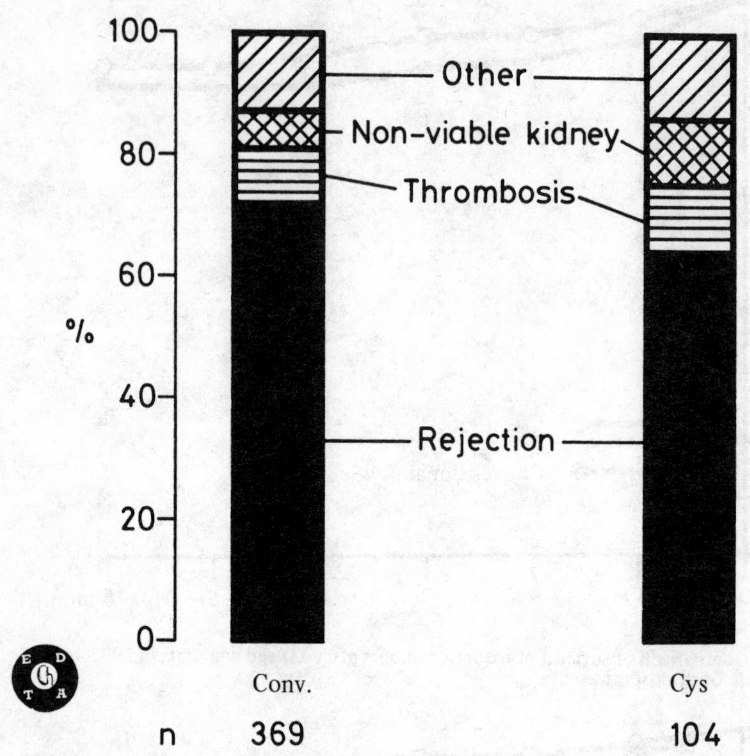

Figure 30. Cause of failure of 1st cadaver grafts in 1983, shown separately for patients treated with cyclosporine (Cys) and those on conventional immunosuppressive therapy (Conv.)

There was a tendency towards a smaller proportion of grafts to be lost due to rejection in the cyclosporine group while the numbers of non-viable kidneys and thromboses unrelated to technical problems or to rejection seemed to be higher.

Some reports have appeared in the literature commenting on delayed onset of function in conjunction with the use of cyclosporine in cadaver transplantation. This problem was studied using the 1983 data available to the Registry: the onset of function in cadaver grafts was compared in patients on conventional therapy and those on cyclosporine (Figure 31). From this analysis it seems that delayed onset of function was a more frequent problem among patients receiving conventional therapy. However, one should bear in mind that centres might choose the immunosuppressive regimen on a basis of whether immediate function occurred or not and thus a bias might be introduced. Moreover, information on the cold ischaemia times was not obtained.

Tables X and XI show an analysis of the percentage of first cadaver grafts performed in 1983 which had 'good' function defined as a serum creatinine less than 160μmol/L at the end of the year. In Table X conventional and cyclosporine therapy are compared in kidneys with immediate onset of function. The cyclosporine group had 62 per cent of grafts with 'good' function compared

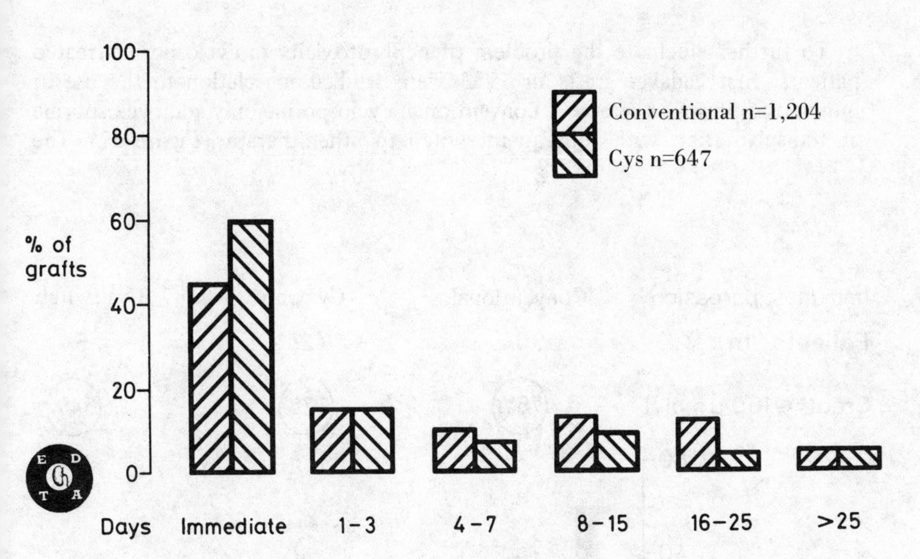

Figure 31. Onset of function of 1st cadaver grafts performed in 1983 in patients given cyclosporine (Cys) or conventional immunosuppressive therapy

TABLE X. Proportion of 1st cadaver grafts with 'good' function shown according to treatment regime. Only grafts with immediate onset of function included

	Immediate onset of function		
Treatment regime	'Good' function	Impaired function	N
Conventional	82%	18%	407
Cyclosporine	62%	38%	245

TABLE XI. Proportion of 1st cadaver grafts with 'good' function shown according to treatment regime. Only grafts with delayed onset of function included

	Delayed onset of function		
Treatment regime	'Good' function	Impaired function	N
Conventional	72%	28%	369
Cyclosporine	47%	53%	117

to 82 per cent in the conventionally treated patients. The difference was even more pronounced in grafts with delayed onset; only 47 per cent with 'good' function in the cyclosporine group versus 72 per cent in patients who received conventional immunosuppression (Table XI). These differences may reflect the well-known nephrotoxicity of cyclosporine, but the rather clear detrimental impact of delayed onset of function should be noted.

To further elucidate the problem of nephrotoxicity in cyclosporine-treated patients, first cadaver grafts in 1983 were studied in relation to the use of immunosuppressive therapy – conventional, cyclosporine only and cyclosporine at transplantation with subsequent switch to other therapy (Figure 32). The

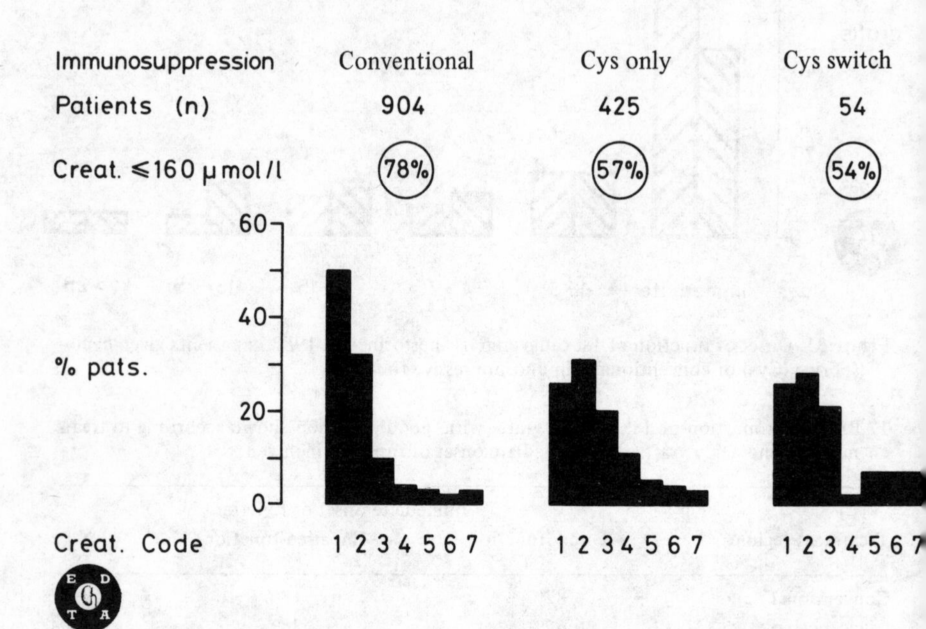

Figure 32. Creatinine values at end of 1983 in patients with 1st cadaver grafts performed in that year, according to immunosuppressive regime. Creatinine codes (μmol/L): 1=≤120; 2=121–160; 3=161–200; 4=201–240; 5=241–300; 6=301–400; 7=>400

codes for the creatinine values are given in the caption and it is clear that patients treated with conventional therapy had a substantially higher proportion of grafts with 'good' or excellent function. The rather large proportion of grafts with poor function in the switched group is not too surprising, since indication for change of therapy in many cases might be poor graft function.

Figure 33 shows an analysis of the function of first cadaver grafts at the end of 1983 according to length of function. The proportion of grafts with 'good' function increased with time after transplantation, but on the other hand the number of grafts went down with years. This is partly because fewer grafts were done in earlier years but also suggests that grafts with impaired function were lost with time, leaving those with 'good' function to continue to do well.

In Figure 34, the distribution of serum creatinine in recipients of a first cadaver graft performed between 1974 and 1983 is shown in relation to body

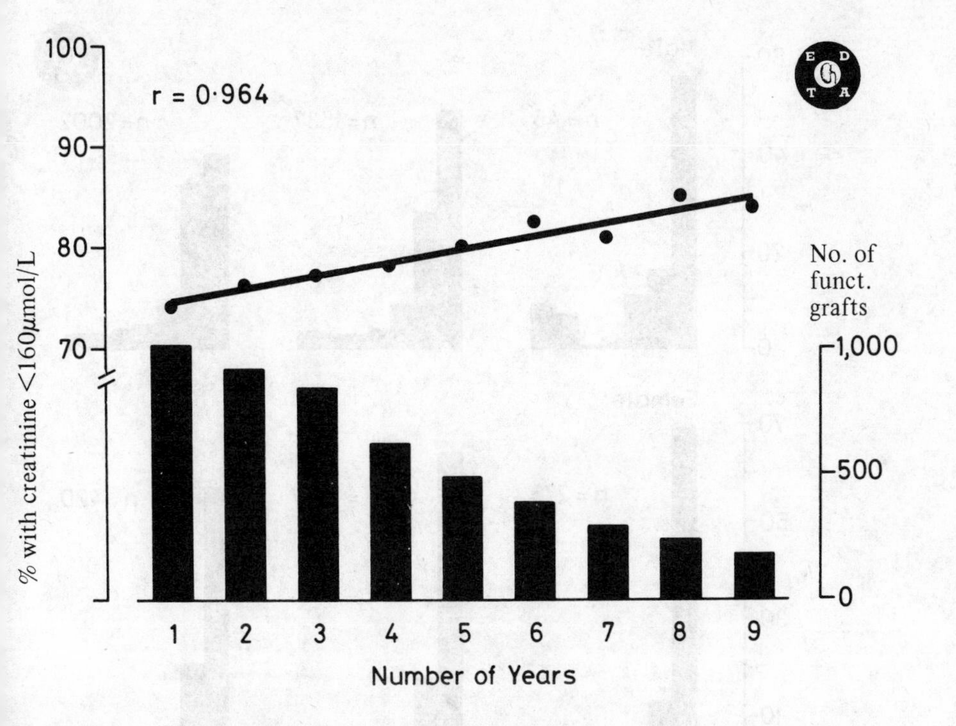

Figure 33. Proportion of patients with a 1st cadaver graft having a serum creatinine less than 160μmol/L at the end of 1983 according to length of function

weight at the end of 1983. Patients on cyclosporine were excluded from the analysis which was done separately for male and female patients. The same convention for creatinine codes as Figure 32 is used. The proportion of patients with a serum creatinine less than 120μmol/L falls with increasing weight, while the percentage with serum creatinine between 120 and 160μmol/L rises. This is true for both sexes.

Body mass index (BMI)

Figure 35 illustrates the importance of relating body weight to body height. Two patients may have the same weight of 70kg but the difference in body height determines whether they are considered to be slim or overweight. The body mass index (BMI) is the body weight in kilograms divided by the body weight in metres^2. For the same weight the slim tall patient on the left would have a BMI of 18.4 and the overweight patient on the right would have a BMI of 27.3

Figure 36 shows the BMI distribution of over 41,000 patients reported to the Registry. The mean BMI was 22.7 ± 3.9. The figure also shows the range of acceptable BMI as defined by the Royal College of Physicians [2]. BMI does not change much with age and the sex difference is only minimal. Table XII

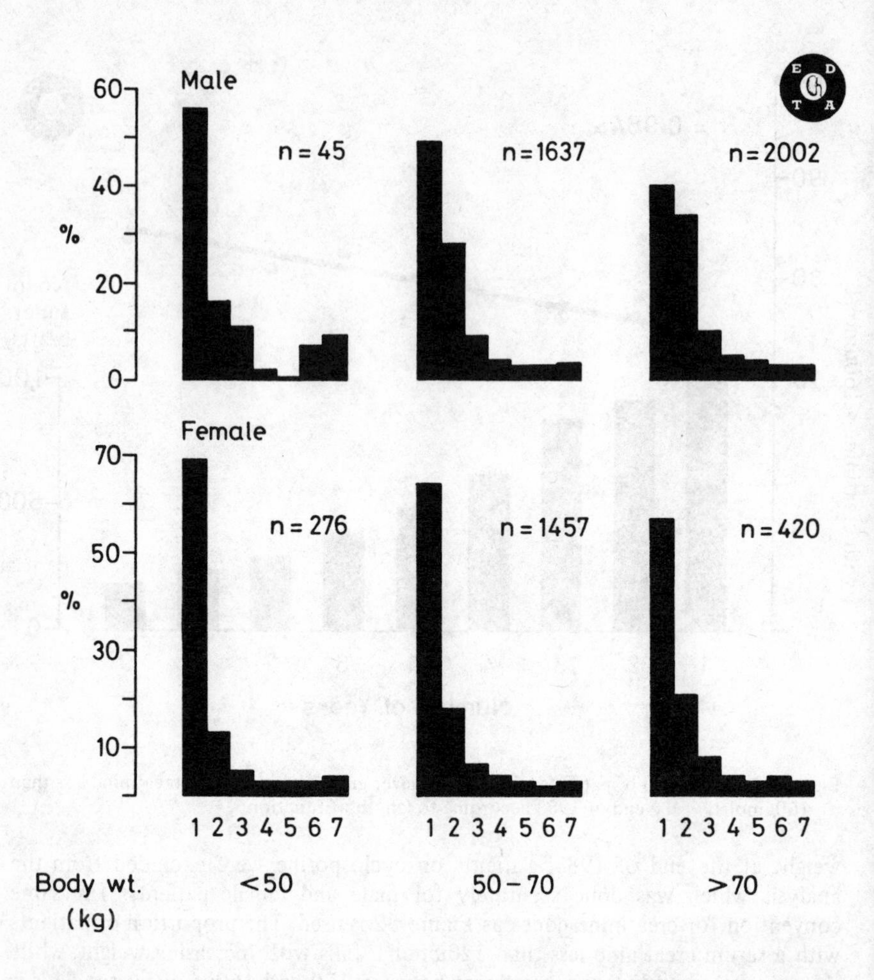

Figure 34. The relation of serum creatinine to body weight in patients with a 1st cadaver graft performed between 1974 and 1983. Creatinine codes as Figure 32

TABLE XII. Mean body mass index (BMI) of patients with glomerulonephritis shown according to age

Age group	Number of patients	Mean BMI
15–34	8,707	21.3 ± 3.8
35–54	18,239	23.0 ± 4.0
55–99	14,108	22.9 ± 3.8

shows the BMI of patients with glomerulonephritis according to age.

Age at onset of renal replacement therapy was studied in relation to BMI determined from the 1983 questionnaire, i.e. from weight and height at the end

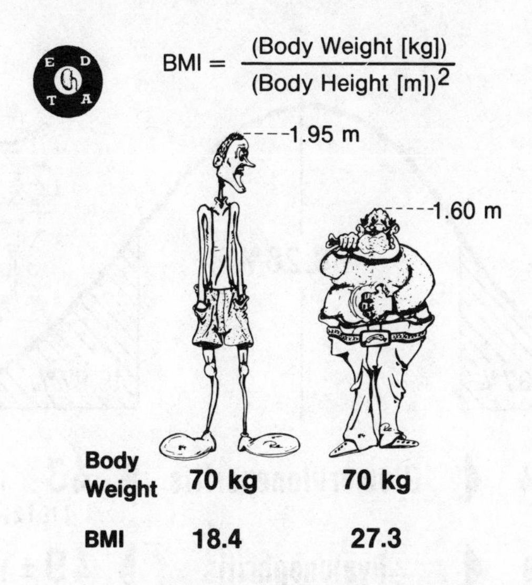

Figure 35. The calculation of body mass index (BMI)

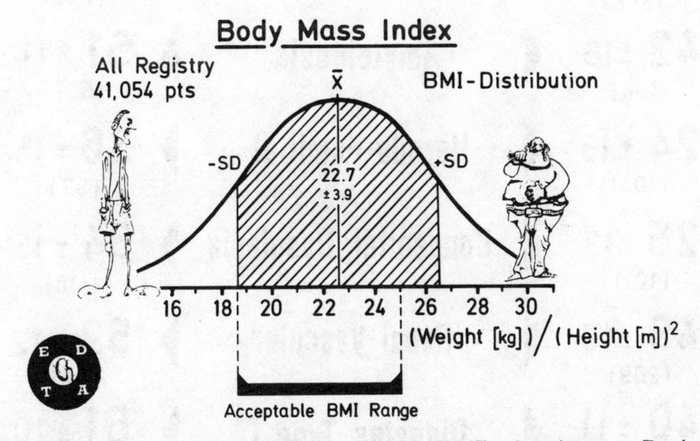

Figure 36. Distribution of body mass index (BMI) of patients on Registry

of 1983. Patients with a BMI lying more than one standard deviation away from the mean for the Registry were defined as thin or obese. This choice was arbitrary, and means that 16 per cent of patients were designated slim and 16 per cent overweight. Age at onset of renal replacement therapy was analysed for eight primary renal diseases and with the exception of drug nephropathy and renal vascular disease, patients currently underweight were found to have commenced treatment 10 years earlier than obese patients (Figure 37). This finding was contrary to what we had anticipated; it was thought that obese patients might consume a protein rich diet and thus, according to Brenner's hypothesis [3] suffer more rapidly progressive renal failure. However, the

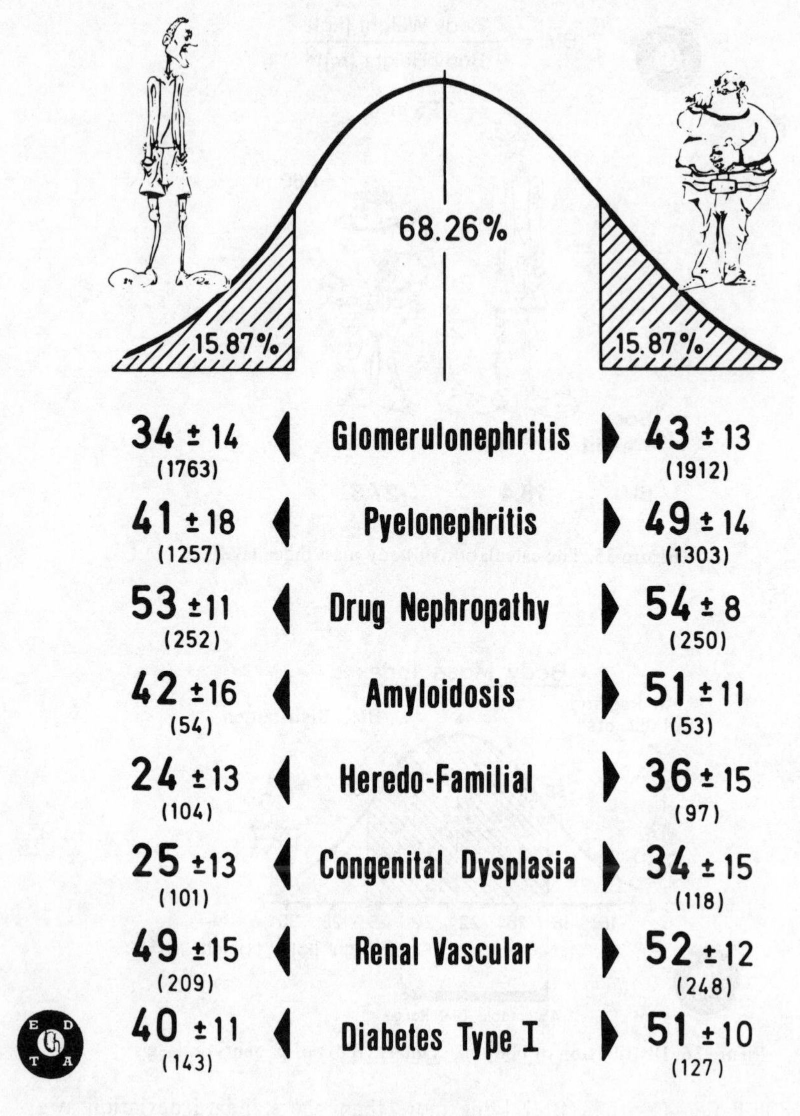

Figure 37. Relation of age at onset of renal replacement therapy (RRT) to body mass index in 1983 for selected primary renal diseases

evidence that obese individuals eat a protein rich diet is sparse and there is no evidence at all about the diets spontaneously selected by patients with chronic renal failure. The data presented above suggest that slim patients commence renal replacement therapy earlier than overweight subjects, and although flawed by the assumption that current BMI reflects regular weight prior to the development of terminal renal failure, do not support Brenner's hypothesis [3].

Rare causes of death

Some of the rarer causes of death have been studied this year. Table XIII shows the proportion of deaths from liver failure between 1980 and 1983 in three groups of patients: those with a successful kidney graft, those with a failed graft and finally those on haemodialysis never transplanted. In the first two groups the deaths are divided according to time after kidney transplantation; in the third, the interval is related to start of haemodialysis. Prolonged immunosuppression appears to be associated with a greater proportion of deaths due to liver failure.

TABLE XIII. The number and proportion (shown as % in brackets) of deaths due to liver failure between 1980 and 1983, compared with deaths from all causes. Deaths are shown separately for patients with a functioning graft, those with a failed graft and those never grafted on haemodialysis

Graft status	Cause of death	Time after transplantation* <1 year	≥1 year
Functioning	Liver failure	22 (3.7%)	81 (8.6%)
	All causes	600	940
Failed	Liver failure	4 (0.9%)	37 (2.9%)
	All causes	433	1,261
Never grafted on haemodialysis	Liver failure	18 (1.2%)	161 (2.5%)
	All causes	1,515	6,367

* For those never grafted time refers to interval after start of haemodialysis

Death from sclerosing peritonitis, mesenteric infarction or perforation of peptic ulcer was investigated in relation to peritoneal dialysis (Table XIV). The proportions of deaths due to these causes were particularly high in patients who had been on peritoneal dialysis at some time although many changed back to haemodialysis because of abdominal complications. Sclerosing peritonitis caused 14 deaths in 1983 and was more frequent in patients who had peritoneal dialysis as sole therapy compared to haemofiltration or haemodialysis as sole therapy. It is possible that sclerosing peritonitis may be a late complication in patients who had peritoneal dialysis at some time, causing death only after return to haemodialysis. Similarly perforation of peptic ulcer was more common in patients who had been on peritoneal dialysis at some time compared to haemodialysis as sole therapy.

Malignancies

Table XV shows the prevalence of malignancies among patients alive on renal replacement therapy on 31st December, 1982 according to mode of treatment,

TABLE XIV Proportion (%) of deaths due to sclerosing peritonitis, mesenteric infarction, perforated peptic ulcer, and perforated colon in each of three therapeutic groups

Cause of death	% deaths in each therapeutic group		
	PD sole therapy	PD at some time	HD/HF sole therapy
Sclerosing peritonitis	10.8	6.1	0.2
Mesenteric infarction	9.4	9.8	5.4
Perforation peptic ulcer	2.7	4.9	0.8
Perforation colon	4.0	3.7	3.5
Total number of deaths	741	818	4,818

PD=peritoneal dialysis; HD/HF=haemodialysis/haemofiltration

TABLE XV. Prevalence of malignancies among patients alive on renal replacement therapy on 31st December, 1982 according to mode of therapy. Tumours diagnosed before and after start of treatment shown separately

Mode of therapy	Malignancies/1,000 patients	
	Diagnosis before start RRT	Diagnosis after start RRT
Haemodialysis	13	10
IPD and CAPD	18	3
Functioning 1st transplant	4	10
Functioning transplant >1st	2	14

RRT=renal replacement therapy; IPD=intraperitoneal dialysis; CAPD=continuous ambulatory peritoneal dialysis

the distinction is made between neoplasms diagnosed before and after the start of renal replacement therapy. The site of malignancies diagnosed prior to start of haemodialysis in patients treated exclusively in this way is given in Table XVI for males and females separately. Myeloma accounts for the highest proportion of malignancies in both sexes followed by hypernephroma in males and breast cancer in females. In contrast the most common tumours among those diagnosed after start of treatment in the same narrowly defined group are bronchus and lung in males (15.4%) and breast in females (20.9%) (Table XVII). A similar breakdown is shown for malignancies diagnosed before and after transplantation in Tables XVIII and XIX.

TABLE XVI. The site of malignancies diagnosed prior to start of treatment in patients treated only by haemodialysis. Males and females shown separately

Malignancies in male patients (n=698)		Malignancies in female patients (n=528)	
Sites	%	Sites	%
Myeloma	20.2	Myeloma	17.8
Hypernephroma	17.6	Breast	16.7
Bladder	13.0	Cervix	13.6
Prostate	10.0	Hypernephroma	9.8
Colon and rectum	7.6	Renal pelvis	6.1
All other sites	31.6	All other sites	36.0

TABLE XVII. The site of malignancies diagnosed after start of treatment in patients treated only by haemodialysis

Malignancies in male patients (n=667)		Malignancies in female patients (n=452)	
Sites	%	Sites	%
Bronchus and lung	15.4	Breast	20.9
Colon and rectum	10.2	Cervix	8.1
Bladder	8.7	Colon and rectum	8.1
Hypernephroma	6.5	Bladder	7.3
Prostate	6.3	Renal pelvis	5.3
All other sites	52.9	All other sites	50.3

TABLE XVIII. The site of malignancies diagnosed prior to transplantation

Malignancies in male patients (n=59)		Malignancies in female patients (n=54)	
Sites	%	Sites	%
Hypernephroma	30.5	Cervix	16.7
Wilm's tumour	8.5	Breast	11.1
Testis	6.8	Wilm's tumour	11.1
Urinary bladder	6.8	Hypernephroma	9.2
Colon and rectum	6.8	Renal pelvis	5.5
All other sites	40.6	All other sites	46.4

TABLE XIX. The site of malignancies diagnosed after transplantation

Malignancies in male patients (n=214)		Malignancies in female patients (n=167)	
Sites	%	Sites	%
Skin	29.0	Skin	16.0
Kaposi's sarcoma	5.0	Cervix	16.0
Bronchus and lung	4.0	Breast	15.0
Stomach	4.0	Colon and rectum	5.0
Lip	4.0	Ovary	5.0
All other sites	53.0	All other sites	43.0

Parathyroidectomy

Incidence of a first parathyroidectomy in 1983 and prevalence of parathyroidectomy in patients alive on renal replacement therapy at the end of December, 1982 were analysed in relation to time on renal replacement therapy and to primary renal disease. Information on whether a parathyroidectomy had been performed was available for 70 per cent of almost 100,000 patients. The prevalence of a first parathyroidectomy (i.e. the number of patients at any one time with the operation) among patients alive at the end of 1982 on renal replacement therapy was 45 per 1,000. This compares with 16 per 1,000 for a first parathyroidectomy performed after grafting in patients alive with a functioning first cadaver graft at the end of 1982. Table XX shows the prevalence of

TABLE XX. Prevalence of parathyroidectomy per thousand patients according to primary renal disease and time on dialysis. Grafted patients excluded from the analysis

	Prevalence parathyroidectomy/1,000					
Years on dialysis	Glomerulo-nephritis	Pyelo-nephritis	Analgesic nephropathy	Polycystic disease	Diabetes mellitus	All primary renal diseases
<2	7	11	8	7	2	8
2–5	20	22	29	22	6	21
5–10	74	73	98	64	31	72
>10	192	171	0	141	0	159

parathyroidectomy among patients on any form of dialysis according to time on treatment; patients with a graft at any time are excluded. Prevalence of parathyroidectomy rose from eight per 1,000 in patients with less than two years' treatment, to 160 per 1,000 in the group on renal replacement therapy for 10 years or more. Patients with diabetic nephropathy showed a strikingly lower prevalence of parathyroidectomy than those with other primary renal diseases.

TABLE XXI. Incidence of first parathyroidectomy in 1983 per thousand patients alive on treatment at the end of 1982 according to graft status and time on renal replacement therapy

Years on RRT	Incidence parathyroidectomy/1,000 (Number of patients shown in brackets)		
	Never grafted	At any time	1st CAD graft functioning 31st December, 1982
<2	4 (18,992)	10 (3,050)	2 (874)
2–5	12 (14,554)	12 (5,283)	2 (2,645)
5–10	24 (8,675)	13 (5,613)	3 (2,921)
>10	32 (3,734)	17 (3,950)	7 (2,172)

Incidence of a first parathyroidectomy in 1983 (i.e. the number of patients with the operation done that year) increased with time on renal replacement therapy (Table XXI). Successful transplantation may reduce the need for parathyroidectomy.

Hepatitis

The number of new cases of hepatitis in patients and staff in 1983 is shown in Table XXII. There were 2,286 new cases in patients and 328 in staff. Figure 38

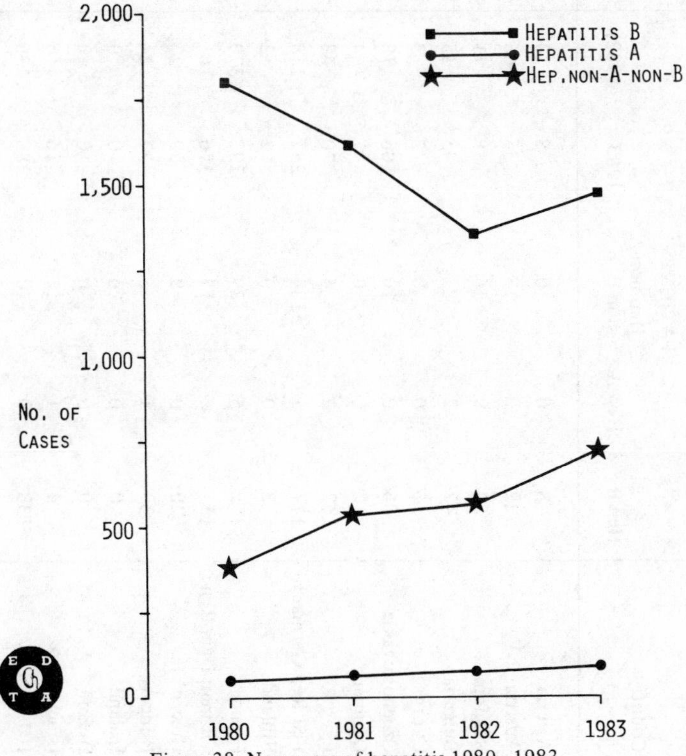

Figure 38. New cases of hepatitis 1980–1983

TABLE XXII. New cases of hepatitis in patients and staff in 1983 by country. A summary of 1982 data is shown for comparison

Country	PATIENTS					STAFF					Deaths (staff)
	Hep B	Hep A	Hep non-A -non-B	Total 1983	Total (1982)	Hep B	Hep A	Hep non-A -non-B	Total 1983	Total (1982)	
Algeria	8	0	0	8	20	2	0	0	2	13	0
Austria	22	0	13	35	66	1	0	0	1	8	0
Belgium	52	0	16	68	71	1	1	3	5	4	0
Bulgaria	42	2	1	45	15	11	3	0	14	13	0
Cyprus	3	0	0	3	3	0	0	0	0	0	0
Czechoslovakia	86	10	10	106	78	7	1	0	8	13	0
Denmark	2	0	0	2	4	1	0	0	1	0	0
Egypt	62	2	6	70	48	8	0	0	8	3	0
Fed Rep Germany	117	1	134	252	218	19	2	2	23	54	0
Finland	2	0	0	2	2	0	0	0	0	0	0
France	159	5	129	293	370	15	0	4	19	57	0
German Dem Rep	151	6	19	176	161	23	0	1	24	35	0
Greece	20	0	9	29	27	5	0	0	5	5	0
Hungary	30	1	6	37	24	4	0	2	6	8	0
Iceland	0	0	0	0	0	0	0	0	0	0	0
Ireland	0	0	0	0	0	0	0	0	0	0	0
Israel	4	1	11	16	16	2	0	0	2	3	0
Italy	138	40	110	288	236	43	19	6	68	53	0

TABLE XXII. *Continued*

Country	PATIENTS					STAFF					Deaths (staff)
	Hep B	Hep A	Hep non-A -non-B	Total 1983	Total (1982)	Hep B	Hep A	Hep non-A -non-B	Total 1983	Total (1982)	
Lebanon	0	0	0	0	1	0	0	0	0	0	0
Libya	0	0	0	0	0	1	0	0	1	0	0
Luxembourg	2	0	0	2	4	1	0	0	1	1	0
Netherlands	3	2	3	8	17	1	0	1	2	0	0
Norway	4	0	0	4	4	0	0	0	0	0	0
Poland	81	0	5	86	57	45	0	0	45	24	0
Portugal	61	3	47	111	38	10	2	0	12	7	0
Spain	162	6	159	327	265	27	1	3	31	34	0
Sweden	2	0	4	6	10	0	0	0	0	0	0
Switzerland	11	1	14	26	24	0	0	2	2	5	0
Tunisia	28	0	0	28	11	1	0	0	1	5	0
Turkey	34	3	21	58	41	5	0	0	5	9	0
United Kingdom	6	1	7	14	29	1	0	0	1	1	0
Yugoslavia	181	1	4	186	137	41	0	0	41	24	0
TOTAL REGISTRY	1,473	85	728	2,286	1,997	275	29	24	328	379	0

TABLE XXIII. Summary of successful pregnancies reported to the Registry up to the end of 1982

	Cumulative total to 31 December 1978	1979	1980	1981	1982	Cumulative total to 31 December 1982
Total number of successful pregnancies*	125	33	48	52	47	305
Transplanted patients	109	29	46	44	43	271
Dialysed patients	16	4	2	8	4	34
Total number of babies	131	33	48	52	48	312
Male babies	54	16	31	17	23	141
Female babies	63	17	17	33	25	155
Sex not recorded	14	0	0	2	0	16
Average duration of pregnancy (wks)						
Transplanted patients (N)	35.0 (105)	36.4 (28)	36.0 (46)	36.0 (44)	36.3 (42)	35.7 (265)
Dialysed patients (N)	33.3 (14)	34.3 (4)	34.0 (2)	31.9 (8)	34.3 (4)	33.2 (32)
Average birth weight (kg)						
Transplanted patients (N)	2.5 (103)	2.6 (28)	2.6 (46)	2.6 (43)	2.6 (43)	2.6 (263)
Dialysed patients (N)	1.9 (12)	1.8 (4)	1.6 (2)	1.7 (8)	1.8 (4)	1.8 (30)
Number of abnormalities reported						
Transplanted patients	7	2	3	3	1	16
Dialysed patients	0	0	0	1	1	2
Four weeks postpartum						
Alive	†	31	46	47	45	167
Dead	†	2	2	1	0	5

TABLE XXIII. *Continued*

	Cumulative total to 31 December 1978	1979	1980	1981	1982	Cumulative total to 31 December 1982
Number breast fed	†	5	8	12	4	29
Number of 2nd or subsequent babies	6	1	3	2	6	12
Number of twin pregnancies	4	0	0	1**	1	6
Number of triplet pregnancies	1	0	0	0	0	1
Body length (cm)	††	††	††	††	47.9 (27)	47.9 (27)
Head circumference (cm)	††	††	††	††	32.5 (28)	32.5 (28)
Chromosome studies (yes)	††	††	††	††	2	2

† These questions were not asked until 1980 (1979 data)
†† These questions were not asked until 1983 (1982 data)

* Includes mothers who have had 2nd or subsequent babies
** 1 twin pregnancy = live birth x 1, still birth x 1

shows the total number of cases of hepatitis B, hepatitis A and non-A-non-B for the years 1980–1983.

Successful pregnancies

Table XXIII gives the numbers of successful pregnancies reported to the Registry on a mini-questionnaire. Results for 1979, 1980, 1981 and 1982 are shown separately, together with a cumulative total for the years up to the end of 1978. Between 1979 and 1982, 180 successful pregnancies were reported.

Dialysis equipment

The results of a question on dialysis equipment, which appeared in the centre questionnaire, are given in Table XXIV. Overall 81 per cent of centres provided information on dialysis machines. There were over 1,100 haemofiltration machines in use in Europe during 1983 of which 6.4 per cent were in homes. Figure 39

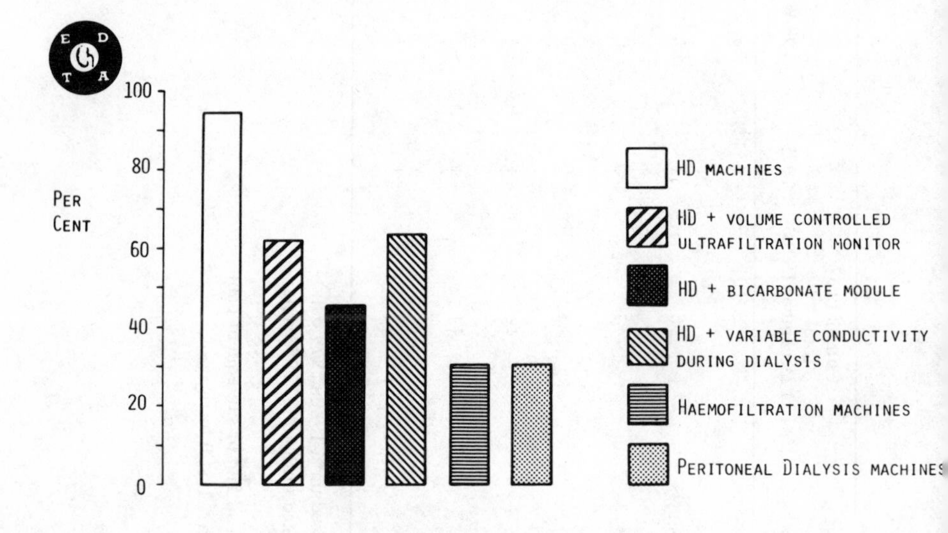

Figure 39. Proportion of 1,452 centres using different types of dialysis equipment in 1983

shows the proportion of centres using the various types of dialysis equipment. Haemodialysis machines with a volume controlled ultrafiltration monitor were used in 60 per cent of 1,452 centres. Haemofiltration machines were in use in about 30 per cent of units.

The average number of hospital patients per hospital haemodialysis machine varied from one in the United Kingdom to 4.5 in Czechoslovakia. Not surprisingly, there is a strong correlation between the numbers of machines and the numbers of patients on haemodialysis in different European countries

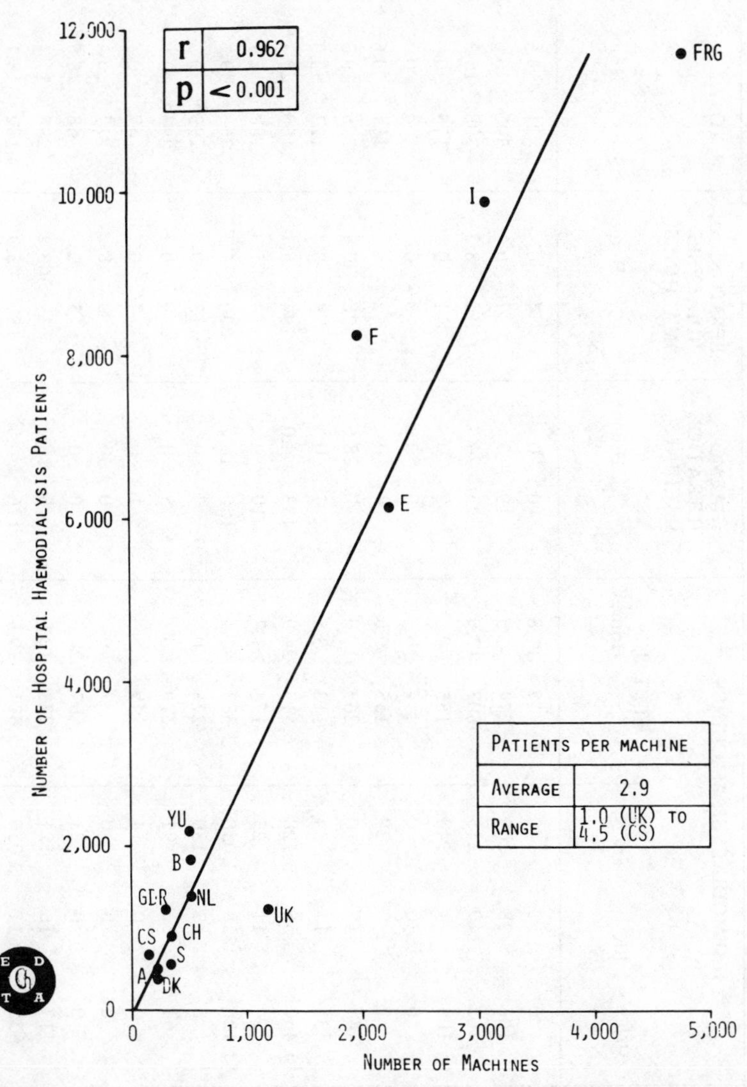

Figure 40. Relation between numbers of haemodialysis machines in hospital centres and numbers of hospital haemodialysis patients in countries with at least 30 machines. Only 14 countries shown, but regression performed with data from 27, using centre questionnaire. A=Austria; B=Belgium; CS=Czechoslovakia; DK=Denmark; FRG=Federal Republic of Germany; F=France; GDR=German Democratic Republic; E=Spain; I=Italy; NL=Netherlands; S=Sweden; CH=Switzerland; UK=United Kingdom; YU=Yugoslavia

(Figure 40). The proportion of centres using haemofiltration and haemodialysis with bicarbonate dialysate module machines is shown for selected countries in Figure 41. The percentage of haemofiltration machines is highest in the Federal Republic of Germany and lowest in Poland.

TABLE XXIV. Summary of dialysis equipment in use in Europe at the end of 1983. The proportion of machines in homes is shown

Country	Centres replied %	HAEMODIALYSIS MACHINES					HAEMO-FILTRATION MACHINES		PERITONEAL DIALYSIS MACHINES		TOTAL	
		VCU N	BIC N	Vary Na N	TOTAL N	TOTAL Home %	N	Home %	N	Home %	N	Home %
Algeria	50	0	0	0	17	11.8	0	0	4	0	21	9.5
Austria	92	106	28	209	274	17.5	19	0	7	0	300	16.0
Belgium	89	262	149	215	603	13.9	93	24.7	12	8.3	708	15.3
Bulgaria	91	105	1	143	165	0	9	0	1	0	175	0
Cyprus	50	0	0	0	4	0	0	0	0	0	4	0
Czechoslovakia	92	7	0	29	155	0.6	4	0	2	0	161	0.6
Denmark	100	41	10	10	263	17.9	7	28.6	26	30.8	296	19.3
Egypt	63	38	11	34	113	3.5	7	0	3	0	123	3.3
Fed Rep Germany	85	1,467	455	3,728	6,513	24.9	453	4.0	194	5.7	7,160	23.1
Finland	96	24	11	23	174	1.7	10	0	4	25.0	188	2.1
France	90	1,128	558	1,050	3,383	41.1	150	16.7	230	24.3	3,763	39.1
German Dem Rep	85	8	7	130	273	0	7	0	13	0	293	0
Greece	63	61	32	91	279	0	5	0	9	0	293	0
Hungary	100	4	0	14	49	0	1	0	7	0	57	0
Iceland	100	0	0	0	7	0	0	0	0	0	7	0
Ireland	60	9	0	21	63	28.6	0	0	5	0	68	26.5
Israel	96	117	16	121	273	13.9	6	16.7	19	63.2	298	17.1
Italy	63	635	611	2,056	3,821	19.5	176	0.6	185	5.9	4,182	18.1

TABLE XXIV. *Continued*

Country	Centres replied %	HAEMODIALYSIS MACHINES					HAEMO-FILTRATION MACHINES		PERITONEAL DIALYSIS MACHINES		TOTAL	
		VCU N	BIC N	Vary Na N	TOTAL N	TOTAL Home %	N	Home %	N	Home %	N	Home %
Libya	33	0	0	0	10	0	0	0	0	0	10	0
Luxembourg	100	3	1	0	40	7.5	5	0	0	0	45	6.7
Netherlands	85	136	42	273	701	23.1	32	0	16	6.3	749	21.6
Norway	100	37	1	51	155	3.9	7	0	3	0	165	3.6
Poland	88	49	25	95	241	0	4	0	9	0	254	0
Portugal	85	177	13	168	395	0	0	0	11	0	406	0
Spain	91	527	152	878	2,571	11.9	31	0	96	2.1	2,698	11.4
Sweden	94	49	6	84	378	6.6	9	0	16	6.3	403	6.5
Switzerland	91	79	44	87	556	35.1	14	0	10	40.0	580	34.3
Tunisia	100	14	0	0	61	0	0	0	4	0	65	0
Turkey	67	40	0	32	78	0	30	0	0	0	108	0
United Kingdom	81	353	75	1,283	3,330	64.3	14	7.1	158	10.1	3,502	61.7
Yugoslavia	67	234	24	346	548	7.1	16	0	8	0	572	6.8
TOTAL REGISTRY	81	5,710	2,272	11,171	25,493	27.0	1,109	6.4	1,052	11.8	27,654	25.6

VCU = volume controlled ultrafiltration monitor
BIC = bicarbonate dialysate module
Vary Na = variable dialysate sodium concentration

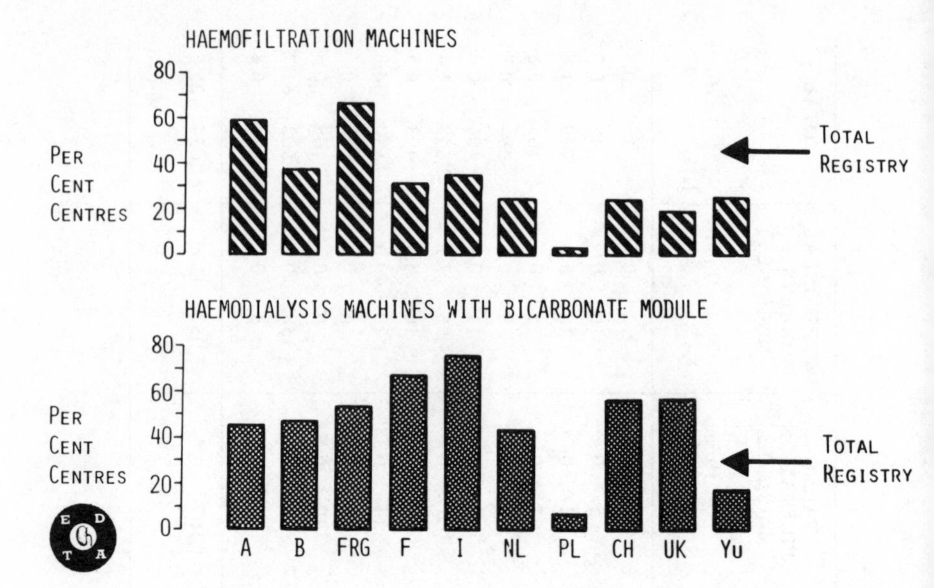

Figure 41. Proportion of centres using haemofiltration and haemodialysis with bicarbonate dialysate module machines in selected countries. A=Austria; B=Belgium; FRG=Federal Republic of Germany; F=France; I=Italy; NL=Netherlands; PL=Poland; CH=Switzerland; UK=United Kingdom; YU=Yugoslavia

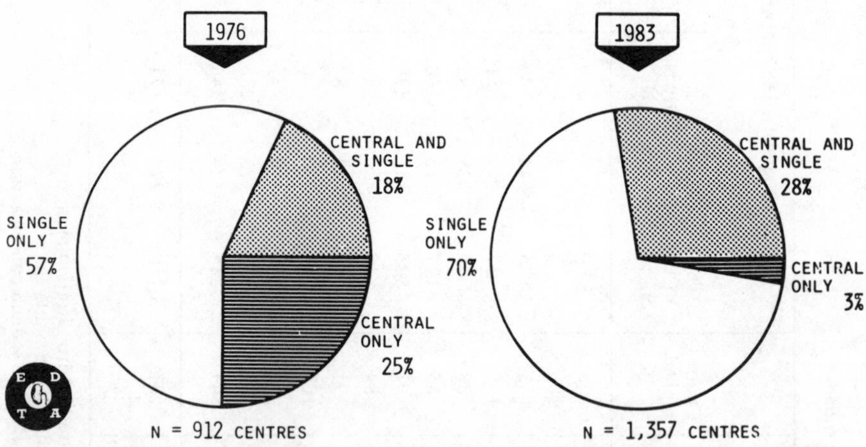

Figure 42. Proportion of centres using central and single dialysis fluid delivery systems in 1976 and 1983. Data from centre questionnaire

Figure 42 compares the percentage of centres using central and single dialysis fluid delivery systems in 1976 and 1983. The proportion using a central system alone has fallen from 25 to three per cent, while use of a single or mixed system has grown. The situation in individual countries is shown in Table XXV.

TABLE XXV. Proportion of centres (%) using central dialysis fluid delivery systems in 1983 by country

Country	Total replies	Central system used	% of all replies
Algeria	2	0	0
Austria	24	4	16.7
Belgium	51	31	60.8
Bulgaria	29	8	27.6
Cyprus	2	0	0
Czechoslovakia	23	10	43.5
Denmark	9	7	77.8
Egypt	17	3	17.7
Fed Rep Germany	241	47	19.5
Finland	20	5	25.0
France	175	59	33.7
German Dem Rep	47	27	57.5
Greece	27	10	37.0
Hungary	9	5	55.6
Iceland	1	0	0
Ireland	3	0	0
Israel	22	8	36.4
Italy	218	82	37.6
Libya	1	0	0
Luxembourg	5	1	20.0
Netherlands	41	8	19.5
Norway	18	0	0
Poland	45	12	26.7
Portugal	27	8	29.6
Spain	157	35	22.3
Sweden	28	3	10.7
Switzerland	30	5	16.7
Tunisia	6	2	33.3
Turkey	12	3	25.0
United Kingdom	44	7	15.9
Yugoslavia	48	20	41.7
TOTAL REGISTRY	1,382	410	29.7

Computers

The use of computers in renal units is growing throughout Europe. About 10 per cent of units reporting to the Registry have their own computer; another 15 per cent have access to one. The type of computing facilities used by centres

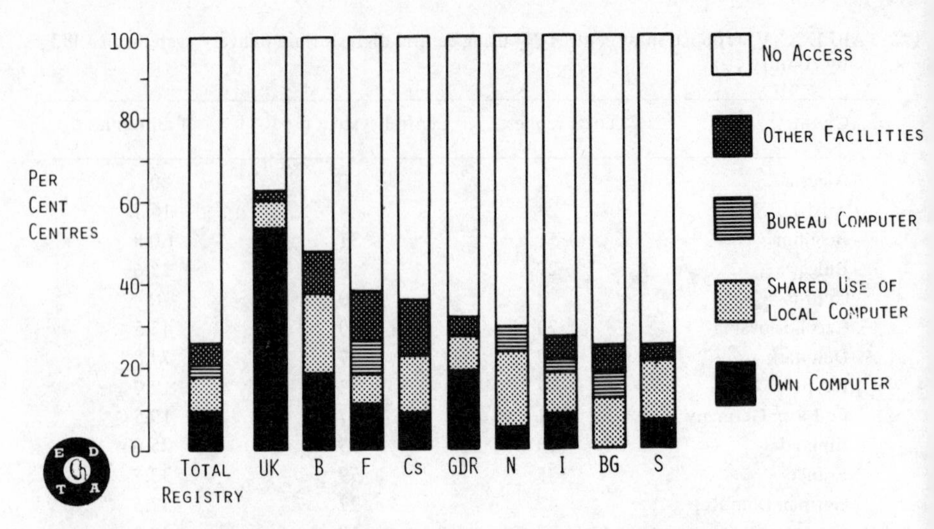

Figure 43. Type of computing facilities available to renal units in selected countries. Those shown returned more than 15 replies, and contained at least 25 per cent of centres with facilities. Total Registry, UK=United Kingdom; B=Belgium; F=France; Cs=Czechoslovakia; GDR=German Democratic Republic; N=Norway; I=Italy; BG=Bulgaria; S=Sweden

in nine selected countries for which adequate data was provided is given in Figure 43.

Conclusions

1 The EDTA Registry recorded nearly 100,000 live patients in Europe at the close of 1983. The increment of live patients 1982–1983 was eight per cent and the overall mortality eight per cent. Nearly half the patients were older than 55. Analysis of data quality showed that basic information was remarkably complete.

2 The two year probability of CAPD patients changing to some other mode of therapy was 60 per cent.

3 Retrospective comparison of survival on different methods of therapy is difficult because of selection. A new method has been described for comparing the contribution of different methods to overall survival on RRT in defined groups.

4 An attempt has been made to pair patients in order to enhance comparison of treatment modes.

5 Patients with a low BMI appear to reach end-stage renal failure 10 years earlier than those with a high BMI.

6 Organ transplantation is opening up new prospects for sufferers from cardiac, hepatic and diabetic diseases and is now the treatment goal for a large proportion of patients with end-stage renal failure.

7 Availability of cadaveric organs has improved so that over 6,000 kidney grafts were performed in 1983 in Europe.

8 The new immunosuppressive drug, cyclosporine, has gained wide acceptance and the retrospective analysis of pooled results collected by the Registry suggests that rejection of grafts has been brought under better control.

9 There were raised serum creatinine values in a higher proportion of patients on cyclosporine than in those on conventional therapy and this difference was particularly marked in the group with late onset of function.

10 Pooled results of live related donor transplantation were excellent and those in diabetic patients extremely encouraging.

11 Sclerosing peritonitis is an important cause of death in patients who have had peritoneal dialysis, and death due to liver diseases was frequent in patients with successful transplants.

12 The prevalence of a first parathyroidectomy was 45 per thousand in patients on any form of renal replacement therapy and its incidence was lower in grafted patients. Successful transplantation appears to reduce the need for parathyroidectomy.

References

1 Jacobs C, Brunner FP, Brynger H et al. *Diabetic Nephropathy 1983; 2:* 12
2 Royal College of Physicians. *J Roy Coll Physicians Lond 1983; 17:* 4
3 Brenner BM, Meyer TW, Hostetter TM. *N Engl J Med 1982; 307:* 625

Appendix

1. Survival rates and sample size

The mathematical formulae used for the calculation of survival rates have been stated in appendices to our previous Reports (see Appendix, Report V, 1974).

2. Computer programs

The programs for analysing the questionnaires have been written in the FORTRAN programming language and are used on a VAX 11/750 under the VMS operating system.

Open Discussion

SHALDON (Montpellier) I think it is high time that we had less non-statistics and I must congratulate you on your lucidity. My question relates to the Selwood method and the arbitrary time of change of treatment, for example, a failing transplant and the patient lives for one month afterwards on haemodialysis. You did not indicate what the time intervals would be and is this not a weakness of the method?

SELWOOD The effects of the timing of change in therapy is dealt with in exactly the same way as in all actuarial methods, so the actuarial approximation is used. However, we are planning to implement the product limit methods to make sure that we get a more accurate picture by having the exact time of change implemented in the analysis so that we no longer will have to make the actuarial assumption.

KERR (London) You presented the data on overweight patients as if the only possible explanation was that being fatter makes you go into renal failure later. It is equally possible that developing renal failure later makes you fatter. The population gets fatter as they get older unless they take up jogging and if you have renal failure you cannot jog. Do your patients with renal failure get fatter as they get older any faster than the general population?

KRAMER The difference we found is considerable, it is not a slight difference. It is the difference between the two standard deviations: these are extreme patients and the difference which we expect by age is rather small. There is an increase in body mass index with age but it is minimal: it increases from 20 to 21.5. There was little difference in sex. I expect that we have a representative group of slim and overweight patients, not influenced by age.

DOOLAN (Connecticut, USA) Having been interested in body composition for years it appears you really studied ectomorphs and endomorphs.

KRAMER Sure, but it is not an obese group, it is an overweight group because beyond the standard deviation are still people who would be considered only to be overweight.

DOOLAN Is the body mass index substantiated by body specific gravity or total body water measurements?

KRAMER No, it is well accepted by most as the representative measurement for body mass.

DOOLAN I do not want to labour the point but the Washington Red Skin football team was not eligible for draft in World War II because of high weight compared to standard tables. We weighed them under water to give the lean body mass.

KRAMER But would you agree that it is better to check the body mass index than the weight?

DOOLAN Well the weight and height. I am asking how your body mass index correlates with more substantial measures such as total body water and body specific gravity? What you really have are two groups of people who you could just guess endomorphy or ectomorphy.

KRAMER I think that is a wide discussion, but the body mass index is well accepted by most scientists and it is not related to the body surface, it is related to the height in metres squared, that is the difference.

ANDREUCCI (Chairman) Sometimes we postpone dialysis in fat patients because we have difficulty in blood access, but I do not think for 10 years. You cannot establish a CAPD programme without having haemodialysis first.

KRAMER Sure, we included patients who were three months on haemodialysis.

ANDREUCCI Concerning the nephrotoxicity of cyclosporine it is something we should check, probably by looking at the renal biopsy. As far as I know this appears to be a functional renal failure and when you decrease the dose you have improvement in renal function.

BRYNGER I am not quite sure it is as simple as that. If the cyclosporine dose in the long term is too high I agree that serum creatinine will come down with the reduction of cyclosporine. In the very early period of transplantation with an ischaemically damaged kidney and too high a cyclosporine dose there may be a different kind of pathology. It is not yet clear but some data which we have with not too well functioning grafts and those with delayed function suggests that there is such a problem.

PORT (Michigan, USA) Dr Brynger, you have shown the improvement of transplantation survival for graft and patients over the year of transplantation. Has the Registry done any covariate analysis of multiple factors, such as transfusion, to sort out whether it was the year of transplantation as our technique improves or do known factors have the major impact?

BRYNGER Unfortunately we do not have the data for all of the years. We have to restrict ourselves to limited data to get all the questionnaires back which means we have had to drop questions on transfusion so we cannot run such an analysis.

ANDREUCCI Thank you very much.

COMBINED REPORT ON REGULAR DIALYSIS AND TRANSPLANTATION OF CHILDREN IN EUROPE, XIII, 1983

Members

M BROYER Hôpital Necker Enfants Malades, Paris, France

F P BRUNNER Departement für Innere Medizin, Universität Basel, Switzerland

H BRYNGER Department of Surgery I, Sahlgrenska Sjukhuset, Göteborg, Sweden

P KRAMER Medizinische Universitätsklinik, Göttingen, Federal Republic of Germany

R OULÈS Centre Hospitalier Regional et Universitaire de Nimes, France

G RIZZONI Clinica Pediatrica dell'Universitá, Ospedale Civile di Padova, Italy

N H SELWOOD UK Transplant, Bristol, United Kingdom

A J WING* St Thomas' Hospital, London, United Kingdom

* Chairman

Director

S CHALLAH St Thomas' Hospital, London, United Kingdom

Research Fellow

É A BALÁS EDTA Registry, St Thomas' Hospital, London, United Kingdom and Semmelweis Medical School, Budapest, Hungary

CONTENTS

For Appendix please see page 63 of the Adult Report

Acknowledgments

This work was supported by grants from the Governments or National Societies of Nephrology of Austria, Belgium, Bulgaria, Cyprus, Denmark, Egypt, the Federal Republic of Germany, France, the German Democratic Republic, Greece, Iceland, Ireland, Israel, Italy, Luxembourg, the Netherlands, Norway, Sweden, Switzerland and the United Kingdom.

Generous grants were made by Asahi Medical GmbH, B Braun, Melsungen AG, Bellco S.p.A., Cobe Laboratories Inc, Cordis Dow B.V., Enka AG, Fresenius AG, Gambro AB, Hospal Ltd, Sorin Biomedica S.p.A., Travenol Laboratories Ltd.

The post of Research Fellow to the EDTA Registry, held during the past year by Dr Norbert Gretz, Mannheim, Federal Republic of Germany and Dr András Balás, Budapest, Hungary, was funded by the National Kidney Research Fund, United Kingdom. A grant was also received from the Wellcome Foundation Ltd.

We acknowledge the co-operation of UK Transplant, Bristol, United Kingdom.

We particularly thank those doctors and their staff who have completed questionnaires. Without their collaboration, this Report could not have been prepared.

Proc EDTA–ERA (1984) Vol 21

COMBINED REPORT ON REGULAR DIALYSIS AND TRANSPLANTATION OF CHILDREN IN EUROPE, XIII, 1983

G Rizzoni, **M Broyer**, F P Brunner, H Brynger, S Challah, P Kramer, R Oulès, N H Selwood, A J Wing, É A Balás

Introduction

The thirteenth combined report of children on dialysis and after kidney transplantation is based on information available from the 1983 patient and centre questionnaires, and the 1982 special paediatric mini-questionnaire. The special paediatric enquiry was principally concerned with pubertal growth after transplantation.

Numbers of patients and methods of treatment

At the time of writing, the number of paediatric patients on the register was 4,153 of whom 2,703 were alive on treatment on 31st December, 1983. In addition, 33 had recovered function of their own kidneys and 51 were reported lost to observation. There were 551 records of patients, alive when last reported to the Registry, which had not been updated at time of going to press. The above numbers include all patients under the age of 15 years at the time of first treatment. Table I shows the stock of paediatric patients reported alive on treatment on 31st December in each of the years 1974 to 1983. The figures for 1982 have been updated since the last Report [1], while those for 1983 are provisional. Of those commencing treatment at age less than 15 years and were alive at the end of 1983 on a known therapy, 53 per cent were over 15 at the end of the year of report.

The distribution of paediatric patients according to treatment modality is shown in Table II. The table includes all patients alive on 31st December, 1983 who commenced treatment under 15 years of age, and shows the numbers not updated for 1983 at the time of reporting.

The 1983 acceptance rates for paediatric patients are shown in Table III, together with the numbers of deaths in the same year. Figure 1 shows the number of new patients accepted per million child population (pMCP) in the five largest European countries for each of the years 1974 to 1983 [2]: the broken lines emphasize that the data for 1983 is incomplete. Acceptance rates

TABLE I. Stock of paediatric patients registered alive on treatment at the end of each year, 1974–1983. Both absolute numbers and figures per million child population (pMCP) shown for 31 countries. Data from individual patient questionnaire

	PATIENTS ALIVE ON TREATMENT									
	1974		1975		1976		1977		1978	
Country	N	pMCP	N	pMCP	N	pMCP	N	pMCP	N	pMCP
Algeria	0	–	0	–	0	–	0	–	0	–
Austria	12	6.8	13	7.4	16	9.1	22	12.6	29	16.6
Belgium	26	11.8	25	11.3	34	15.4	39	17.6	51	23.1
Bulgaria	5	2.6	3	1.6	5	2.6	6	3.1	8	3.1
Cyprus	2	11.3	2	11.3	2	11.3	4	22.6	4	22.6
Czechoslovakia	0	–	1	0.3	3	0.9	5	1.5	3	0.9
Denmark	20	17.4	19	16.5	22	19.1	28	24.4	30	26.1
Egypt	0	–	0	–	0	–	0	–	0	–
Fed Rep Germany	101	7.6	138	10.4	187	14.1	220	16.6	250	18.8
Finland	6	5.7	8	7.6	8	7.6	11	10.5	16	15.2
France	149	12.1	194	15.8	236	19.2	277	22.6	321	26.2
German Dem Republic	10	2.8	13	3.7	13	3.7	18	5.1	24	6.8
Greece	6	2.8	7	3.2	8	3.7	10	4.6	14	6.5
Hungary	1	0.5	1	0.5	1	0.5	2	0.9	5	2.3
Iceland	–	–	–	–	–	–	–	–	–	–
Ireland	2	2.1	3	3.1	4	4.1	5	5.1	10	10.3
Israel	10	8.8	11	9.7	17	15.0	21	18.5	23	20.3
Italy	81	6.0	107	8.0	122	9.1	142	10.6	177	13.2
Libya	0	–	0	–	0	–	0	–	0	–
Luxembourg	0	–	0	–	0	–	0	–	0	–
Netherlands	38	11.0	45	13.0	58	16.7	77	22.2	91	26.3
Norway	8	8.4	9	9.4	11	11.5	14	14.7	15	15.7
Poland	2	0.2	4	0.5	7	0.9	7	0.9	11	1.3
Portugal	0	–	0	–	0	–	1	0.4	1	0.4
Spain	26	2.7	35	3.6	52	5.4	78	8.1	117	12.2
Sweden	20	11.8	24	14.2	25	14.7	27	15.9	33	19.5
Switzerland	16	11.4	23	16.0	25	17.8	30	21.4	39	27.8
Tunisia	0	–	0	–	0	–	0	–	0	–
Turkey	1	–	2	–	5	–	7	–	5	–
United Kingdom	132	10.1	160	12.2	178	13.6	203	15.5	242	18.5
Yugoslavia	7	1.3	12	2.2	19	3.5	18	3.3	18	3.3
TOTAL	681		859		1058		1272		1535	

pMCP=per million child population

TABLE I. *Continued*

	PATIENTS ALIVE ON TREATMENT									
	1979		1980		1981		1982		1983	
Country	N	pMCP	N	pMCP	N	pMCP	N	pMCP	N	pMCP
Algeria	0	–	0	–	0	–	1	–	1	0.2
Austria	32	18.3	34	19.4	29	16.6	32	18.3	36	20.9
Belgium	69	31.2	73	33.0	93	42.0	105	47.4	108	50.5
Bulgaria	8	4.1	9	4.7	13	6.7	12	6.2	14	7.2
Cyprus	4	22.6	4	22.6	4	22.6	3	17.0	4	26.7
Czechoslovakia	6	1.8	9	2.7	16	4.8	17	5.1	16	4.6
Denmark	32	27.9	34	29.6	41	35.6	36	31.3	43	37.7
Egypt	0	–	0	–	0	–	4	0.2	11	1.0
Fed Rep Germany	294	22.1	329	24.8	413	33.6	370	30.1	376	30.2
Finland	18	17.1	20	19.0	23	21.9	25	23.8	26	25.5
France	364	29.7	423	34.5	524	41.7	569	45.3	591	48.2
German Dem Republic	26	7.3	34	9.6	39	11.0	52	14.7	57	16.5
Greece	19	8.8	22	10.2	30	13.8	25	11.5	27	12.4
Hungary	4	1.9	4	1.9	3	1.4	1	0.5	1	0.4
Iceland	–	–	–	–	–	–	–	–	1	16.7
Ireland	16	16.4	19	19.5	21	21.6	21	21.6	22	22.0
Israel	30	26.4	32	28.2	41	36.1	53	46.7	57	47.5
Italy	213	15.9	231	17.2	301	22.4	303	22.5	280	21.0
Libya	0	–	0	–	0	–	6	–	3	2.8
Luxembourg	3	41.7	6	83.3	4	57.1	3	42.8	2	28.6
Netherlands	102	29.5	117	33.8	148	42.7	139	40.1	140	41.8
Norway	17	17.8	21	22.0	29	30.4	34	35.6	34	36.2
Poland	17	2.0	18	2.2	30	3.6	24	2.9	33	4.1
Portugal	1	0.4	2	0.8	7	2.6	16	5.9	33	12.2
Spain	139	14.5	179	18.6	218	22.7	238	24.8	236	24.5
Sweden	32	18.9	34	20.0	42	24.8	48	28.3	43	25.3
Switzerland	42	30.0	44	31.4	52	37.1	54	38.5	45	34.1
Tunisia	0	–	0	–	0	–	3	–	3	1.2
Turkey	6	–	4	–	6	–	4	–	10	9.4
United Kingdom	275	21.0	325	24.9	399	30.5	425	32.5	409	32.0
Yugoslavia	17	3.1	28	5.1	43	7.8	44	8.0	41	7.5
TOTAL	1786		2055		2569		2667		2703	

pMCP=per million child population

TABLE II. Stock of patients, commencing renal replacement therapy under 15 years of age, alive at 31st December, 1983, according to treatment modality. Patients with recovery of kidney function, those lost to observation and those not updated for 1983 at time of writing are shown separately

Country	PATIENTS ON TREATMENT AT 31st DECEMBER, 1983							OTHER REGISTERED PATIENTS		
	on haemodialysis/ haemofiltration Hospital	Home	IPD	CAPD	With functioning transplant	Total	pMCP	Recovered	*	†
Algeria	0	0	0	0	0	1	0.2	1	1	2
Austria	15	1	2	1	17	36	20.9	0	0	2
Belgium	28	1	1	1	77	108	50.5	1	0	11
Bulgaria	13	0	0	0	1	14	7.2	0	1	0
Cyprus	0	0	0	0	4	4	26.7	0	0	0
Czechoslovakia	11	0	0	0	5	16	4.6	0	0	1
Denmark	15	0	0	10	18	43	37.7	0	0	14
Egypt	9	0	0	1	1	11	1.0	0	1	1
Fed Rep Germany	145	20	4	13	194	376	30.2	5	4	77
Finland	7	0	0	3	16	26	25.5	0	0	2
France	278	38	0	21	254	591	48.2	2	10	97
German Dem Rep	36	0	0	0	21	57	16.5	1	0	6
Greece	17	1	0	1	8	27	12.4	0	6	9
Hungary	1	0	0	0	0	1	0.4	0	0	2
Iceland	1	0	0	0	0	1	16.7	0	0	0
Ireland	10	0	0	0	12	22	22.0	0	0	2
Israel	25	0	0	3	29	57	47.5	2	2	5
Italy	149	16	2	25	88	280	21.0	6	9	86

TABLE II. *Continued*

Country	PATIENTS ON TREATMENT AT 31st DECEMBER, 1983							OTHER REGISTERED PATIENTS		
	On haemodialysis/ haemofiltration Hospital	Home	IPD	CAPD	With functioning transplant	Total	pMCP	Recovered	*	†
Lebanon	–	–	–	–	–	–	–	–	–	3
Libya	3	0	0	0	0	3	2.8	0	0	3
Luxembourg	1	0	0	0	1	2	28.6	0	0	0
Netherlands	35	2	0	10	93	140	41.8	3	1	24
Norway	4	0	0	0	30	34	36.2	1	0	1
Poland	18	0	3	5	7	33	4.1	3	2	9
Portugal	28	0	1	0	4	33	12.2	1	0	1
Spain	141	8	1	10	76	236	24.5	3	7	53
Sweden	12	2	1	2	26	43	25.3	0	0	13
Switzerland	11	3	0	7	24	45	34.1	1	0	14
Tunisia	3	0	0	0	0	3	1.2	0	0	0
Turkey	9	0	0	0	1	10	9.4	0	1	1
United Kingdom	69	39	0	57	244	409	32.0	3	4	103
Yugoslavia	37	0	0	0	4	41	7.5	0	2	9
TOTAL REGISTRY	1,131	131	15	171	1,255	2,703	3.8	33	51	551

* Patients reported as lost to observation or who have moved to a non-reporting country.

† Patients known to be alive when last reported to the Registry but whose record has not been subsequently updated. Treatment status on 31st December 1983 was therefore uncertain.

TABLE III. Acceptance rates for new paediatric patients onto renal replacement therapy, per million child population (pMCP) in 1983 by country. The number of deaths in 1983 of patients commencing therapy under the age of 15 years is also shown, according to age at time of death

Country	Deaths in 1983		New patients in 1983	
	<15	⩾15		
	N	N	N	pMCP
Algeria	0	0	2	0.4
Austria	1	0	4	2.3
Belgium	0	1	7	3.3
Bulgaria	0	0	3	1.5
Cyprus	0	0	0	0
Czechoslovakia	6	1	7	2.0
Denmark	1	1	6	5.3
Egypt	2	0	8	0.7
Fed Rep Germany	6	4	44	3.5
Finland	1	0	4	3.9
France	10	1	70	5.7
German Dem Rep	3	1	9	2.6
Greece	1	0	7	3.2
Hungary	0	0	0	0
Iceland	0	0	1	16.7
Ireland	0	0	3	3.0
Israel	3	0	10	8.3
Italy	6	4	37	2.8
Lebanon	0	0	0	0
Libya	0	0	0	0
Luxembourg	1	0	0	0
Netherlands	2	0	14	4.2
Norway	1	0	2	2.1
Poland	3	0	11	1.4
Portugal	0	0	10	3.7
Spain	3	4	33	3.4
Sweden	0	0	6	3.5
Switzerland	1	1	6	4.5
Tunisia	1	0	1	0.4
Turkey	1	0	7	6.6
United Kingdom	3	7	67	5.2
Yugoslavia	6	2	11	2.0
TOTAL REGISTRY	62	27	390	0.5

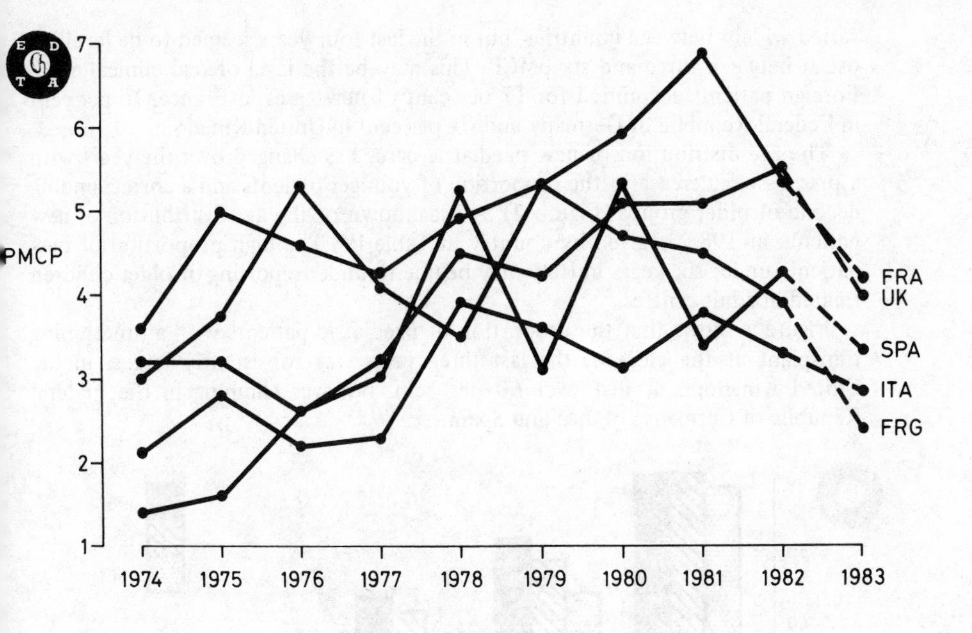

Figure 1. Rate of acceptance (per million child population) of paediatric patients onto renal replacement programmes in five European countries (France, United Kingdom, Spain, Italy, Federal Republic of Germany), 1974–1983

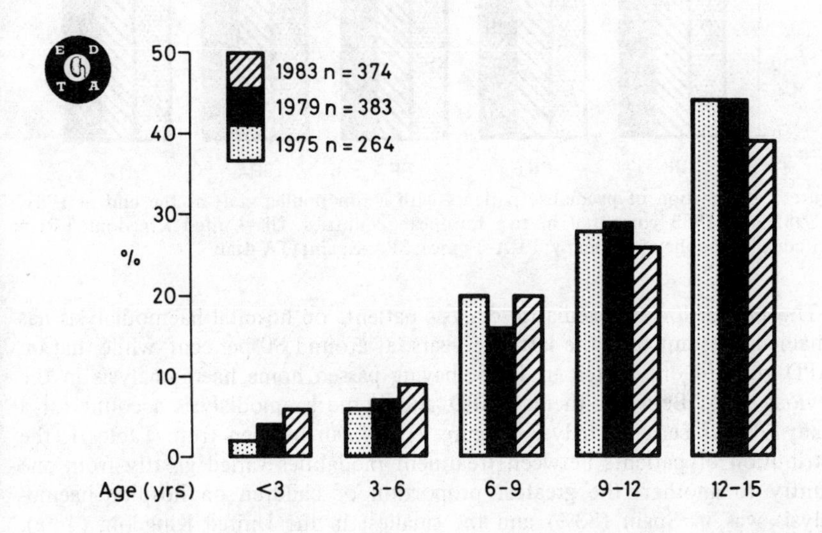

Figure 2. Comparison of age distribution of paediatric patients accepted onto renal replacement programmes in Europe in the years 1975, 1979 and 1983. Proportion (%) of children in each of five age groups is shown

varied widely between countries, but in the last four years seemed to be levelling off at between three and six pMCP. This may be the level of real clinical need. Foreign patients accounted for 17 per cent of new cases in France, 16 per cent in Federal Republic of Germany and six per cent in United Kingdom.

The age distribution of new paediatric cases has changed over the years with a progressive increase in the proportion of younger patients and a corresponding decline of older groups (Figure 2). A breakdown of the age distribution of new patients in 1983 is given by country in Table IV. The high proportion of new patients under six years in Italy may be due to under-reporting of older children treated in adult centres.

Figure 3 shows that the proportion of paediatric patients with a functioning transplant at the close of the last three years was consistently largest in the United Kingdom, at just over 60 per cent, but was climbing in the Federal Republic of Germany, France and Spain.

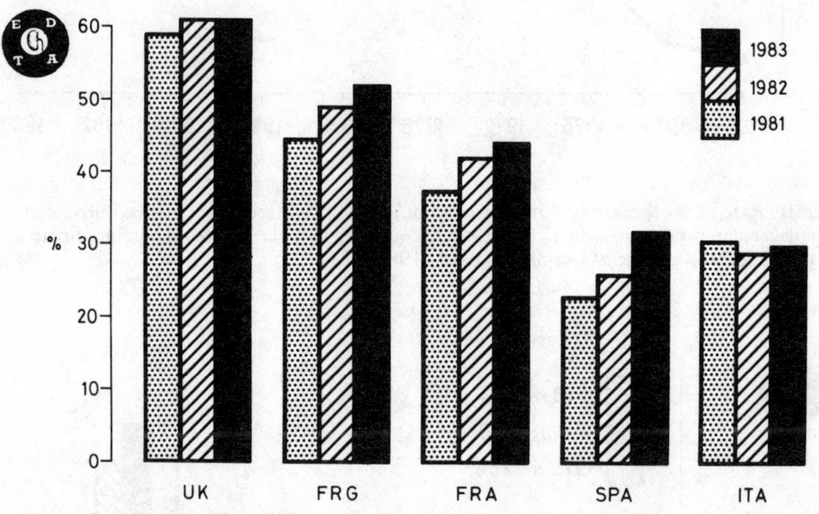

Figure 3. Proportion of paediatric patients with a functioning graft at the end of 1981, 1982 and 1983 compared in five European countries. UK=United Kingdom; FRG= Federal Republic of Germany; FRA=France; SPA=Spain; ITA=Italy

The proportion of paediatric dialysis patients on hospital haemodialysis has remained constant over the last five years, at around 80 per cent, while that on CAPD grew to one in ten in 1983, having passed home haemodialysis in the previous year. Between them, CAPD and home haemodialysis account for a steady 20 per cent of dialysis (Figure 4). As can be seen from Table II, the distribution of patients between treatment modalities varied greatly from one country to another; the greatest proportion of children on hospital haemodialysis was in Spain (83%) and the smallest in the United Kingdom (48%), which had the largest contribution from CAPD and home haemodialysis (Figure 5).

TABLE IV. The age distribution (%) of paediatric patients commencing renal replacement therapy in 1983 by country. Data from individual patient questionnaire

Country	<3.0 %	3–5.9 %	6–8.9 %	9–11.9 %	12–14.9 %	Total N
Algeria	0	0	0	50.0	50.0	2
Austria	0	0	0	0	0	0
Belgium	16.7	16.7	0	33.3	33.3	6
Bulgaria	33.3	0	0	33.3	33.3	3
Cyprus	0	0	0	0	0	0
Czechoslovakia	0	0	14.3	28.6	57.1	7
Denmark	0	0	50.0	16.7	33.3	6
Egypt	0	12.5	12.5	50.0	25.0	8
Fed Rep Germany	5.1	7.7	28.2	23.1	35.9	39
Finland	0	25.0	25.0	50.0	0	4
France	4.8	9.5	19.0	30.1	36.5	63
German Dem Rep	0	0	33.3	33.3	33.3	9
Greece	0	0	0	57.1	42.9	7
Hungary	0	0	0	0	0	0
Iceland	0	0	100.0	0	0	1
Ireland	0	0	0	33.3	66.7	3
Israel	0	15.4	15.4	15.4	53.8	13
Italy	16.7	9.5	19.0	16.7	38.1	42
Lebanon	0	0	0	0	0	0
Libya	0	0	0	0	0	0
Luxembourg	0	0	0	0	0	0
Netherlands	14.3	21.4	35.7	7.1	21.4	14
Norway	0	0	0	100.0	0	2
Poland	8.3	16.7	8.3	25.0	41.7	12
Portugal	10.0	0	40.0	20.0	30.0	10
Spain	5.3	10.5	13.2	34.2	36.8	38
Sweden	0	0	40.0	40.0	20.0	5
Switzerland	0	16.7	33.3	16.7	33.3	6
Tunisia	0	0	0	0	100.0	1
Turkey	0	0	0	0	100.0	7
United Kingdom	3.6	9.1	14.5	25.5	47.3	55
Yugoslavia	9.1	9.1	54.5	9.1	18.2	11
TOTAL REGISTRY	6.2	9.1	20.3	25.9	38.5	374

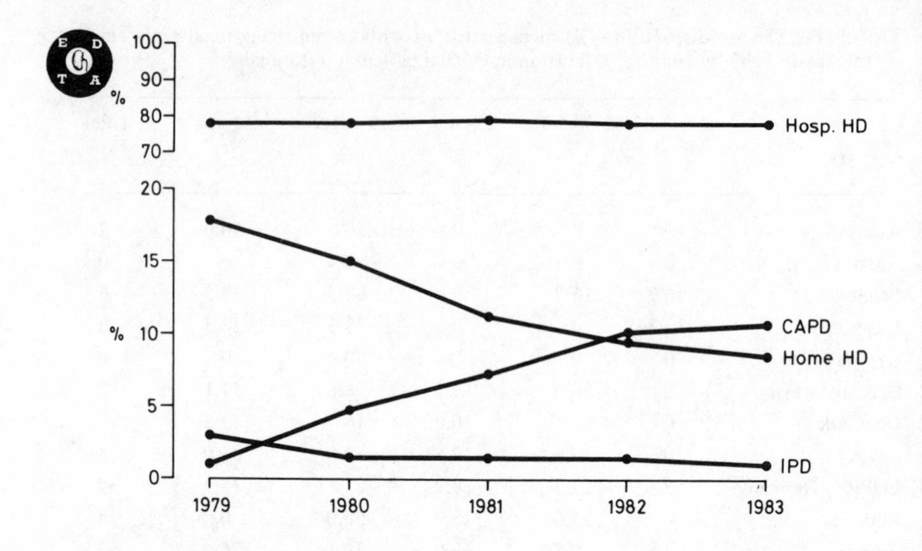

Figure 4. Proportion (%) of paediatric dialysis patients on different forms of dialysis on 31st December of the years 1979–1983

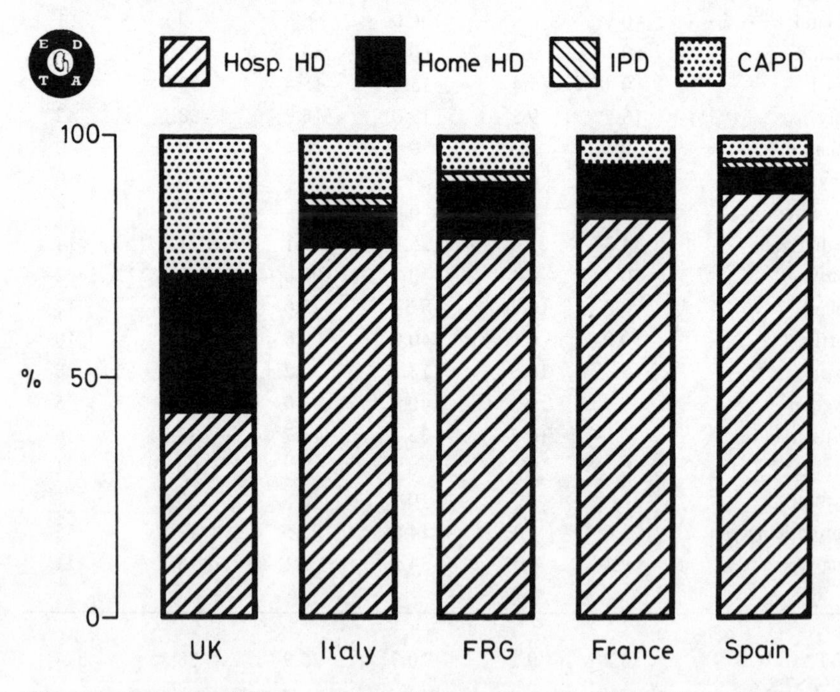

Figure 5. Proportion (%) of paediatric dialysis patients on different forms of dialysis on 31st December, 1983 in five European countries

Specialised paediatric centres

Since 1980, units treating children have been designated as specialised paediatric centres or not by their directors. The number of specialised units so defined is compared for the years 1980 and 1983 in Table V. Discrepancies with previously published data are due to redesignation of centres. The number of specialised centres increased slightly between 1980 and 1983, and reached 0.8 to 1.0 pMCP in most countries.

TABLE V. Comparison of the number of specialised paediatric centres identified to the Registry in 1980 and 1983 by country. The number of specialised paediatric centres per million child population (pMCP) in 1983 is also shown

Country	1980 N	1983 N	1983 pMCP
Austria	1	2	1.14
Belgium	3	4	1.80
Czechoslovakia	1	2	0.60
Denmark	1	1	0.87
Fed Rep Germany	11	12	0.82
France	11	14	1.14
German Dem Rep	3	3	0.84
Greece	1	1	–
Ireland	1	1	–
Israel	2	2	1.76
Italy	5	8	0.59
Netherlands	4	4	1.15
Poland	3	6	0.73
Portugal	–	2	–
Spain	7	8	0.83
Switzerland	2	3	2.14
Turkey	–	3	–
United Kingdom	8	9	0.68
Yugoslavia	1	2	0.36
TOTAL REGISTRY	65	87	–

The specialised paediatric centres appear to have made an important contribution. The proportion of children grafted in 1983 was greater in the specialised than non-specialised units; transplantation – as measured by the ratio of transplants performed in 1983 to the total number of patients alive on dialysis at the end of the year – was more active in specialised centres in the United Kingdom, Italy and Spain than in the non-specialised units (Table VI). Furthermore, the annual death rate among paediatric patients is lowest in those countries with the highest provision of specialised centres in 1983 (Figure 6).

TABLE VI. Comparison of the activity of specialised (spec) and non-specialised (non-spec) paediatric centres in 1983

Country	No. of children on dialysis on 31st December, 1983			No. of children transplanted in 1983			Ratio of grafts in 1983 to patients on dialysis in December, 1983	
	spec	non-spec	in spec %	spec	non-spec	in spec %	spec	non-spec
Fed Rep Germany	77	8	91	44	5	90	0.57	0.63
France	177	94	65	37	24	61	0.22	0.25
Italy	72	78	48	31	16	66	0.43	0.21
Netherlands	38	–	100	19	–	100	0.50	–
Spain	62	45	58	28	11	72	0.45	0.24
United Kingdom	59	54	52	43	11	80	0.73	0.20

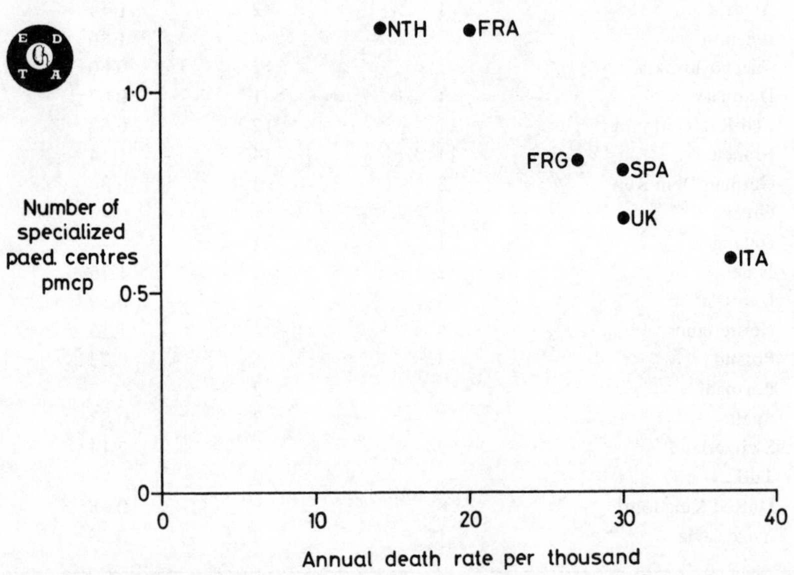

Figure 6. Mean annual death rate of paediatric patients, 1979–1983, plotted against provision of specialised paediatric centres (per million child population) in selected European countries in 1983. NTH=Netherlands; FRA=France; FRG=Federal Republic of Germany; SPA=Spain; UK=United Kingdom; ITA=Italy

Haemodialysis

Analysis of haemodialysis practice (Tables VII and VIII) in children showed that 80 per cent of cases were dialysed three times a week, and that the most popular regimen was 11–13 hours weekly, suggesting that most patients are treated for four hours three times a week. There was no substantial difference in the frequency or duration of dialysis for different age groups.

TABLE VII. Proportion of children (%) treated by haemodialysis 1, 2, 3, 4 or 5–7 times per week in 1983 according to age

Age group (years)	Number of patients	Times treated per week (%) 1	2	3	4	5–7
<5	55	1.8	14.5	74.5	7.3	1.8
5 – 9	151	0.7	11.2	86.7	0.7	0.7
10 – 14	406	3.0	15.5	79.6	1.7	0.2
>15	593	0	9.4	88.9	1.5	0.2
All ages	1,205	1.2	12.0	84.8	1.7	0.3

TABLE VIII. Hours per week on haemodialysis in 1983 according to age group. Number of children on each dialysis regime shown as percentage

Age group (years)	Number of patients	Hours per week on dialysis (%) <5	5–7	8–10	11–13	14–16	17–19	>20
<5	55	1.8	10.9	23.6	40.0	21.8	0	1.8
5 – 9	151	1.3	2.6	25.8	52.3	16.6	0.7	0.7
10 – 14	406	3.0	6.4	18.5	51.0	19.0	1.7	0.5
>15	593	2.2	2.4	9.6	55.4	24.1	5.1	1.2
All ages	1,205	2.3	4.1	15.3	52.9	21.3	3.2	0.9

TABLE IX. Re-use of haemodialysers/filters in 1983 reported for paediatric patients

Number of re-uses	N	%
0	1,093	93.1
1 – 2	38	3.2
3 – 5	32	2.7
>5	11	0.9

Re-use of haemodialysers/filters (Table IX) has become unusual in children: 93 per cent of patients were treated with single use in 1983.

Excess weight gain, defined as a gain exceeding eight per cent of body weight at least once a week, occurred in one in 10 children, and was a particular problem in those aged between five and 10 years (Table X). The proportion of children who died during 1983 was higher among children with excess weight gain (7.5%) than in those without (2.5%) (Figure 7). This difference was greatest in those aged 10–15 years. The observation underlines the importance of controlling fluid weight gain between haemodialysis sessions.

TABLE X. Comparison of the number of paediatric patients with and without excess weight gain between haemodialysis sessions in 1983 by age. Proportion (%) with excess weight gain in each age group also shown

Age group (years)	Excess weight gain N	%	No excess N
<5	5	9.4	48
5 – 9	19	13.1	125
10 – 14	35	8.9	356
>15	63	10.9	514
All ages	122	10.4	1,043

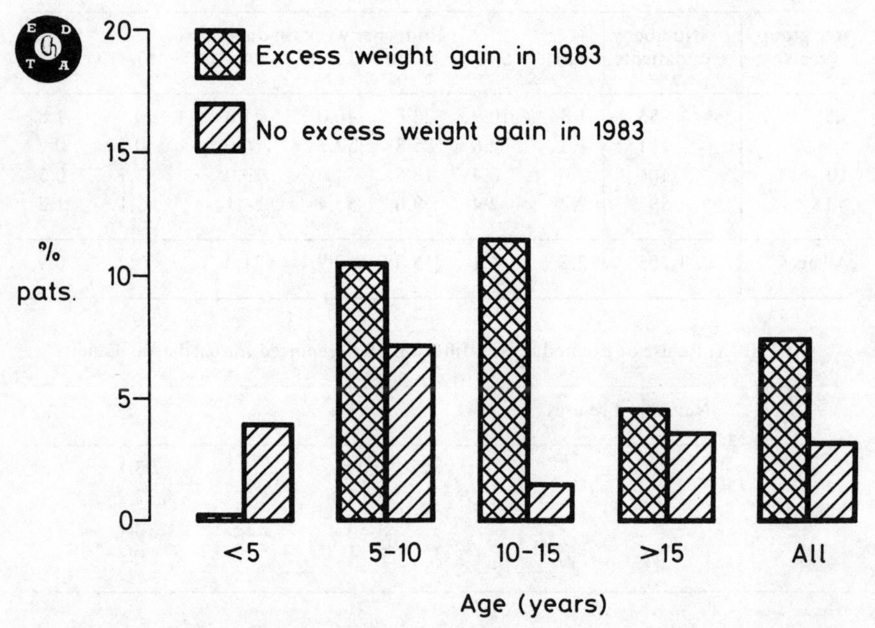

Figure 7. Percentage of children with excess weight gain who died in 1983 compared to the percentage without excess weight gain who died in the same year, shown for all children and broken down into four different age groups

Continuous ambulatory peritoneal dialysis (CAPD)

Seventeen per cent of new paediatric patients in Europe in 1983 started with CAPD as first treatment (Table XI). The proportion was highest in those aged 3–6 years (27%). The use of CAPD for children, aged less than 15 at the end

TABLE XI. The total number of new paediatric patients in Europe in 1983 by age group and the proportion (%) of them started on CAPD as first treatment

Age group (years)	New patients	% new patients treated by CAPD
<3	23	22
3 – 5	34	27
6 – 8	76	17
9 – 11	97	22
12 – 15	144	12
All ages	374	17

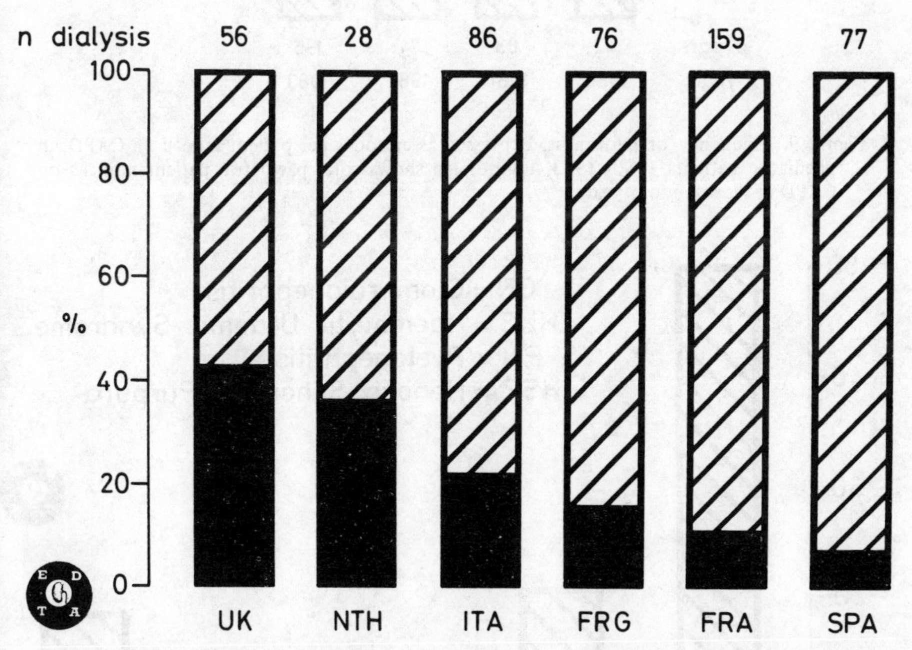

Figure 8. Proportion (%) of paediatric dialysis patients on CAPD on 31st December, 1983 in six selected countries. Absolute number (n) also shown. UK=United Kingdom; NTH= Netherlands; ITA=Italy; FRG=Federal Republic of Germany; FRA=France; SPA=Spain

of 1983, differed between countries, ranging from eight per cent of dialysis patients in Spain to 43 per cent in the United Kingdom (Figure 8).

The frequency of peritonitis did not fall below 0.2 per patient month on treatment, despite greater experience in 1983 (Figure 9). Preliminary results suggest that the incidence of peritonitis does not differ markedly between younger and older children.

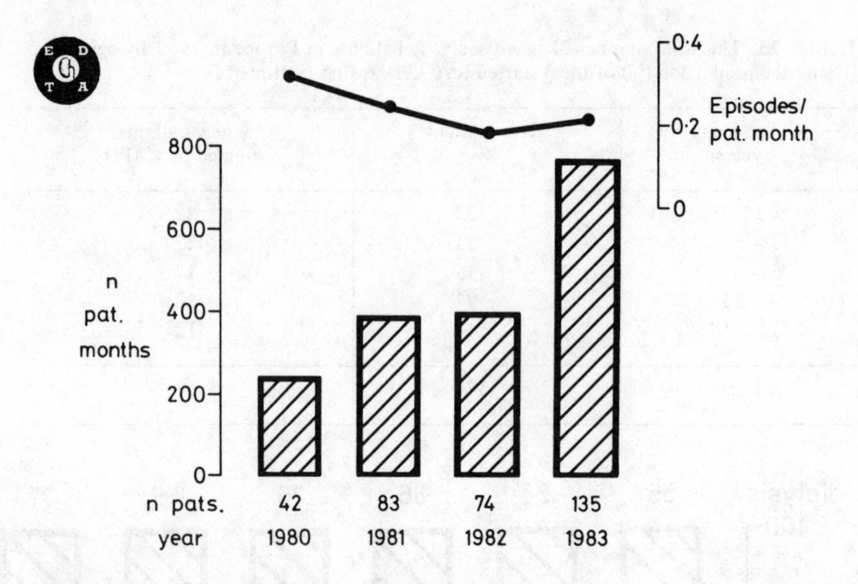

Figure 9. Frequency of peritonitis, expressed as episodes per patient month on CAPD, in paediatric patients 1980–1983. Figure also shows total paediatric patient months on CAPD in the same four years

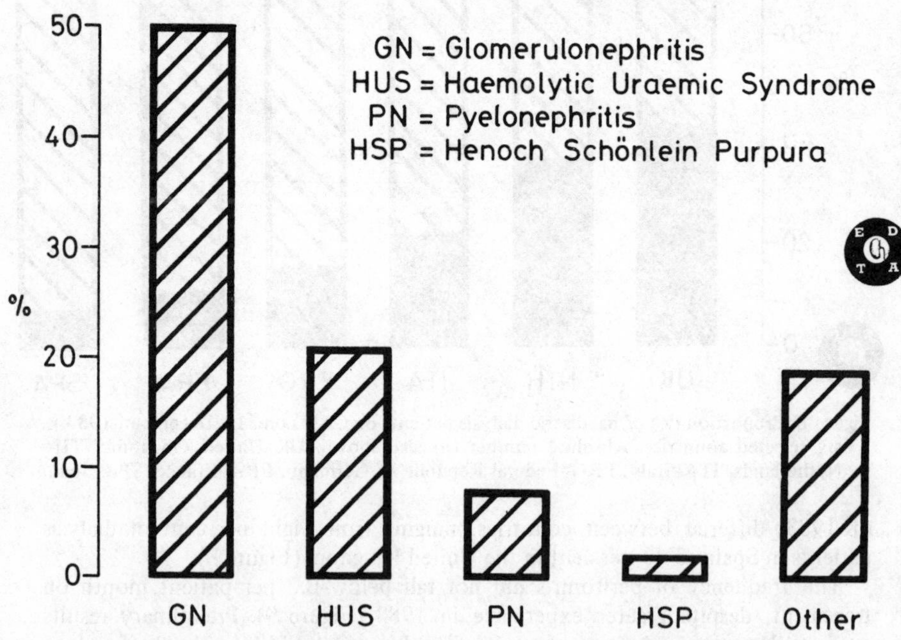

Figure 10. Proportion (%) of children with recovery of renal function with selected primary renal diseases

Recovery of function

Half of the 48 children whose kidneys recovered function after at least six weeks' renal replacement therapy had glomerulonephritis (Figure 10). Recovery can sometimes be diagnosed after a long time on renal replacement therapy, and about 20 per cent of cases were reported after a year's treatment (Table XII).

TABLE XII. Time on renal replacement therapy (RRT) of the 48 children reported with recovery of kidney function. Both the number and proportion (%) of children in each RRT interval shown

	Time on RRT (months)			
	1.5–6	6–12	12–24	24–36
N	28	10	9	1
%	58.3	20.8	18.8	2.1

The probability of maintained recovery (as measured by actuarial methods) is not high, and by one year about half would have returned to dialysis (Figure 11).

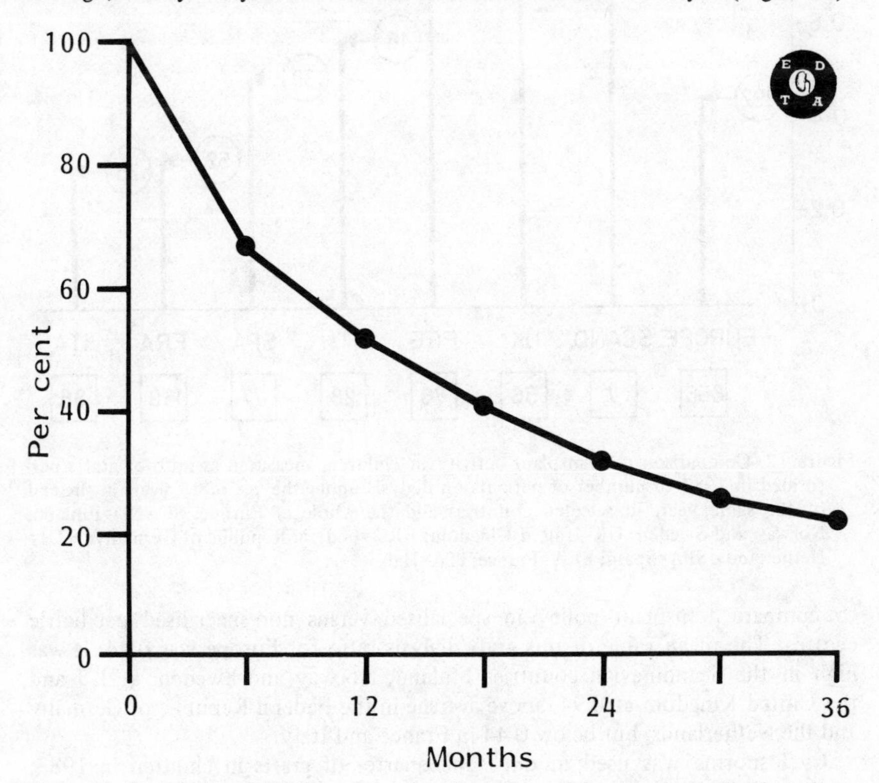

Figure 11. Cumulative probability of continued recovery of function in paediatric patients

Transplantation

In 1983, 292 transplants were recorded in children aged less than 15 years at time of transplant. Of these, 46 (16%) were living related donor transplants; in the United Kingdom, France and the Federal Republic of Germany, the countries with the most active transplant programmes, the proportion was only 10 per cent.

Transplant activity for 1983 was compared in different countries using the ratio of grafts performed in 1983 to the number of children on dialysis on 31st December of that year (Figure 12). The same ratio was used earlier

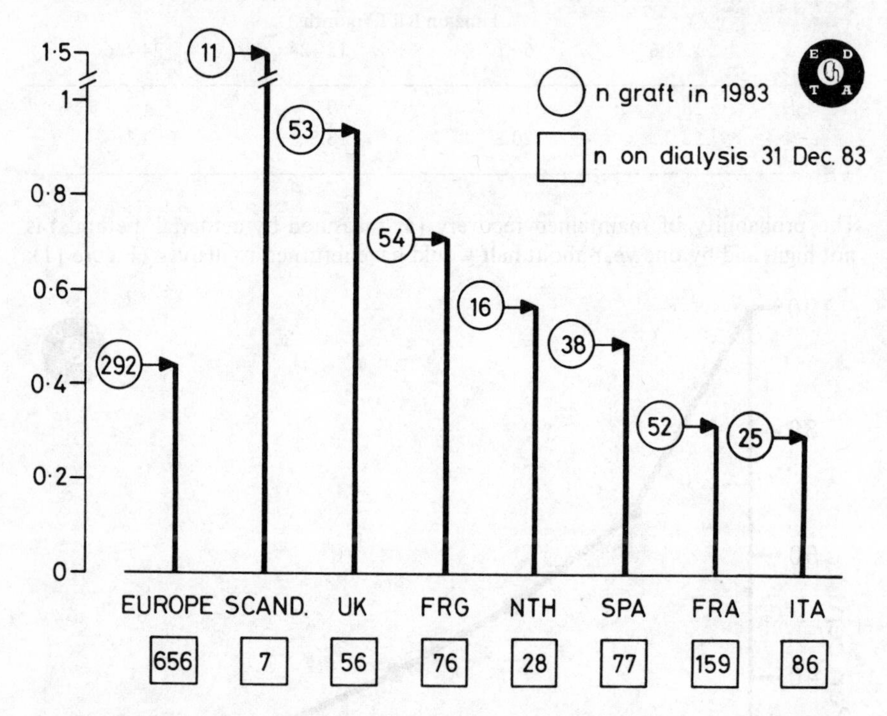

Figure 12. Comparison of transplant activity in children, measured as ratio of grafts performed in 1983 to number of patients on dialysis under the age of 15 years at the end of the same year, in selected countries and the whole of Europe. SCAND=Finland, Norway and Sweden; UK=United Kingdom; FRG=Federal Republic of Germany; NTH=Netherlands; SPA=Spain; FRA=France; ITA=Italy

to compare transplant policy in specialised versus non-specialised paediatric centres. The mean value of this graft/dialysis ratio for Europe was 0.44; it was high in the Scandinavian countries Finland, Norway and Sweden, at 1.5 and the United Kingdom at 0.94, above average in the Federal Republic of Germany and the Netherlands, but below 0.44 in France and Italy.

Cyclosporine was used in only one quarter of grafts in children in 1983;

42 (18%) received it continuously and in one per cent it was switched to other therapies. Six per cent of children started on other therapy after transplantation were switched to cyclosporine and were still on it at the end of 1983. Creatinine levels at the close of 1983 in grafts performed in that year were surprisingly not higher in those cases treated with cyclosporine; Figure 13

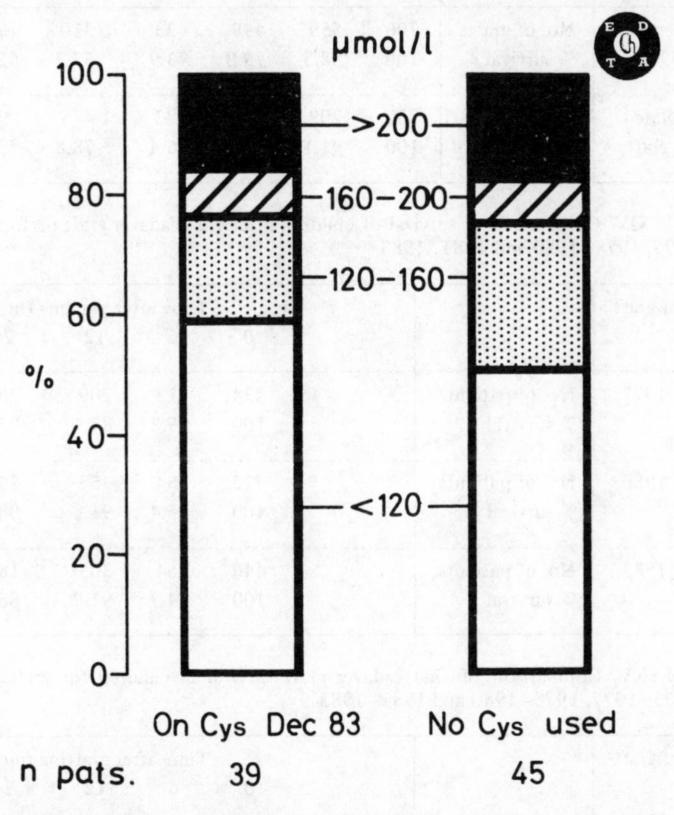

Figure 13. Proportion (%) of children who received a first cadaver graft in 1983, with different creatinine levels (µmol/litre). Children on cyclosporine (Cys) at the end of 1983 and those who did not receive the drug shown separate

shows that the proportion of children with low creatinine levels was greater in the group which had received cyclosporine than in the group which had not. Graft survival was better in the cyclosporine group over the first six months, but the number of cases collected so far is very small and further studies will be required.

Graft survival after first cadaver and first living related donor grafts performed between 1979 and 1983 are given in Table XIII. Period comparison of patient survival after first cadaver grafting (Table XIV) showed some improvement between 1975–1977 and 1978–1980, but no change in 1981–1983. Graft survival likewise had not improved very much in the last three years

TABLE XIII. Survival of first cadaver and first living related donor grafts performed in children between 1979 and 1983

Type of graft		Time after grafting (months) 0	3	6	12	24	36	48
Cadaver graft	No. of grafts	706	559	499	423	310	192	98
	% survival	100	88.3	79.0	73.9	67.7	63.9	57.3
Live related donor graft	No. of grafts	232	209	201	182	1,475	98	53
	% survival	100	91.1	88.9	86.4	78.8	73.2	71.8

TABLE XIV. Comparison of survival of children after first cadaver graft performed 1975–1977, 1978–1980 and 1981–1983

Date of graft		Time after grafting (months) 0	6	12	24	36
1975–1977	No. of patients	234	213	209	204	195
	% survival	100	89.3	87.1	83.3	79.4
1978–1980	No. of patients	372	354	352	337	313
	% survival	100	95.4	91.8	88.3	87.2
1981–1983	No. of patients	446	364	305	187	55
	% survival	100	94.7	91.9	88.5	88.5

TABLE XV. Comparison of first cadaver graft survival in children for grafts performed 1975–1977, 1978–1980 and 1981–1983

Date of graft		Time after grafting (months) 0	6	12	24	36
1975–1977	No. of grafts	233	162	154	141	121
	% survival	100	73.7	68.4	59.6	55.2
1978–1980	No. of grafts	373	303	285	264	235
	% survival	100	82.1	77.5	70.7	66.2
1981–1983	No. of grafts	443	293	232	135	40
	% survival	100	78.4	70.9	66.7	63.3

(Table XV). First cadaver graft survival according to selected primary renal diseases is reported in Table XVI: it exceeded 80 per cent at one year for cystinosis but was lower for oxalosis.

TABLE XVI. Comparison of first cadaver graft survival in children with selected primary renal disease

Primary renal disease		Time after grafting (months) 0	3	6	12
Cystinosis	No. of grafts	39	34	32	29
	% survival		86.9	84.2	81.2
Nephronophthisis	No. of grafts	63	51	50	42
	% survival		85.6	82.1	78.2
Hypoplasia	No. of grafts	71	52	50	44
	% survival		79.6	76.4	72.9
Haemolytic uraemic syndrome	No. of grafts	39	25	22	17
	% survival		72.8	66.0	66.0
Glomerulonephritis	No. of grafts	205	160	148	122
	% survival		78.8	75.0	68.9
Focal sclerosis	No. of grafts	56	39	37	29
	% survival		81.7	75.0	67.3
Pyelonephritis ± urinary tract malformation	No. of grafts	214	146	138	120
	% survival		71.3	66.1	59.5
Oxalosis	No. of grafts	10	5	4	3
	% survival		50.0	50.0	35.7

TABLE XVII. Causes of failure of first cadaver grafts performed in 1983: children compared with all Registry

Cause graft failure	Children N	Children %	All Registry %
Hyperacute rejection	2	5.4	5.5
Rejection while taking immunosuppressive drugs	16	43.2	58.0
Rejection after stopping all immunosuppressive drugs	3	8.1	2.8
Recurrent primary renal disease	2	5.4	1.2
Vascular or ureteric operative problems	3	8.1	7.1
Vascular thrombosis	10	27.0	9.2
Infection of graft	0	0	4.1
Removal of functioning kidney	0	0	0.2
Non-viable kidney	1	2.7	6.9
Other	0	0	5.1
Total	37	100.0	100.0

The causes of first cadaver graft failure in children are compared with total Registry data in Table XVII for transplants done in 1983. The most important differences were the greater proportion of failure in children due to rejection after stopping immunosuppressive drugs, recurrence of primary renal disease and vascular thrombosis.

Growth

Data on growth in transplanted pubertal patients is reported from the 1982 patient and special paediatric questionnaires. Patients with cystinosis and oxalosis were excluded from the analysis. The information relates to children grafted before 1st January, 1982 who kept the transplant through the year until 31st December, 1982. In Figures 14 and 15, data is plotted according to the patient's age on 31st December, 1982. Figure 14 shows that the height of girls was below that of the healthy population, although some late growth may have occurred between 16–19 years of age. A similar observation was made in boys who were relatively smaller than girls (Figure 15). Pubertal status on Tanner's scale was generally lower than in the normal population. In Figure 16, the mean of Tanner's four criteria, taken as a rough index of pubertal status, is plotted against age for adolescents with a functioning graft (as defined above). The majority of 14-year old girls have not yet attained stage 3, while 97 per cent of the normal population would have done so. The same was true for 15-year old boys.

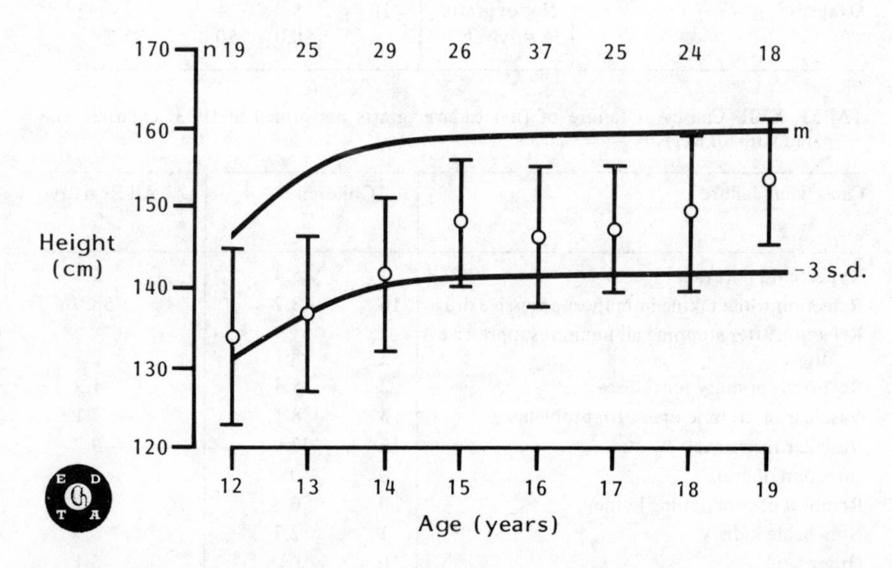

Figure 14. Mean height of pubertal girls (± 1SD) with a functioning graft for more than a year, according to age in December, 1982. Mean (m) and three standard deviations below the mean (-3SD) of healthy population shown as dark lines for comparison

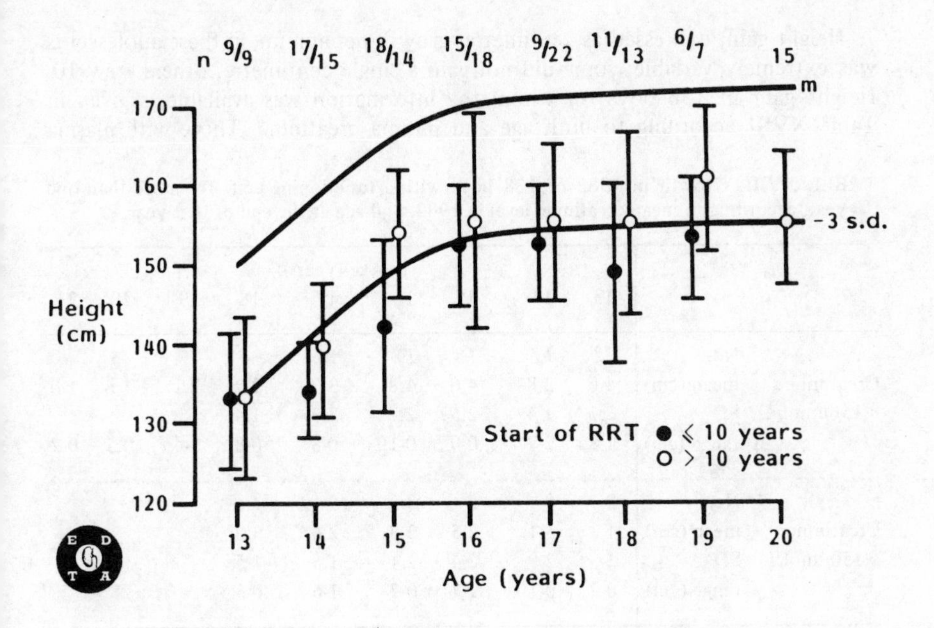

Figure 15. Mean height of pubertal boys (± 1SD), with a functioning graft for more than a year, according to age in December, 1982, and time on renal replacement therapy (RRT). Mean (m) and three standard deviations below the mean (−3SD) of healthy population shown as dark lines for comparison

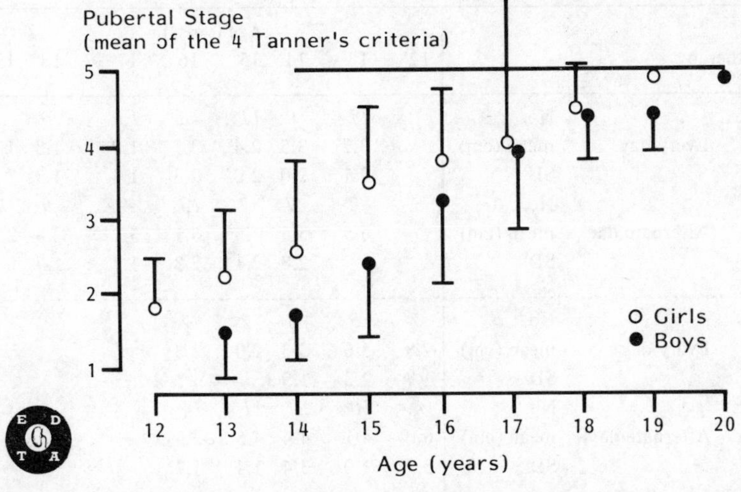

Figure 16. Mean pubertal stage in adolescents (± 1SD), with a functioning graft for more than one year, according to age in December, 1982 and sex

Height gain, expressed as centimetres growth per annum, in these adolescents was extremely variable; some did not gain a single centimetre, others grew 10. Height gain in 138 boys for whom the information was available is given in Table XVIII according to both age and plasma creatinine. Those with plasma

TABLE XVIII. Growth in 1982 of 138 boys with a functioning graft for more than one year, according to mean creatinine level in 1982, and age at the end of that year

		Age (years)								
		13	14	15	16	17	18	19	20	21
	No.	*12*	*17*	*18*	*17*	*15*	*7*	*7*	*5*	*4*
Creatinine	mean (cm)	4.0	3.8	4.0	4.8	3.7	2.7	2.1	1.6	1.0
$<150\mu$mol/L	SD	2.1	1.4	2.5	2.7	2.3	1.4	1.2		
	range (cm)	1-8	2-7	0-9	0-10	0-8	0-8	0-4	0.5	0-2
	No.	*2*	*9*	*7*	*8*	*6*	*4*			
Creatinine	mean (cm)	1	3.1	2.5	3.3	2.6	3.0			
$>150\mu$mol/L	SD		2.9	2.4	2.1	1.6	1.5			
	range (cm)	0-2	0-10	1-7	0-7	1-6	0-5			

creatinines above 150μmol/L showed a lower mean growth than patients with a level below 150μmol/L, but the standard deviations were large. The data for girls were not analysed. In contrast, one year growth according to administration

TABLE XIX. Growth in 1982 of 196 boys and girls with a functioning graft for more than one year, according to age at the end of 1982 and steroid regimen

			Age (years)							
Steroid therapy			12	13	14	15	16	17	18	19
		No.		*11*	*9*	*11*	*8*	*13*	*6*	*3*
	Every day	mean (cm)		3.2	3.5	2.4*	3.1	1.9**	1.8	1.0
		SD		2.1	2.0	2.0	2.3	1.4	1.5	
MALE		No.		*2*	*7*	*7*	*13*	*9*	*9*	*7*
	Alternate day	mean (cm)		6.5	3.5	4.8*	5.1	5.2**	3.4	2.0
		SD			2.8	2.1	2.3	1.7	1.7	
		No.	*7*	*8*	*8*	*2*	*2*			
	Every day	mean (cm)	2.8	3.6	3.3	2.0	2.3			
		SD	1.7	2.3	1.9					
FEMALE		No.	*4*	*10*	*9*	*11*	*14*			
	Alternate day	mean (cm)	6.0	4.1	4.4	4.5	2.5			
		SD	3.9	1.0	3.4	1.4	1.2			

* $p<0.005$; ** $p<0.001$ (t test)

of corticosteroids was studied in both boys and girls (Table XIX). Those receiving alternate day prednisone grew better than children on a daily regime, and this difference was statistically significant for boys aged 15 and 17.

The impact of impaired pubertal growth on ultimate height was also investigated in 196 cases for which information was available. The patients studied were those over 20 years of age who had two identical height measurements a year apart, and therefore considered to have reached ultimate stature (Table XX). Adult height varied according to age at start of renal replacement therapy – for example, males who started at 11 years of age attained a mean adult height of 152 centimetres compared to 163 centimetres in those starting aged 14.

TABLE XX. Ultimate height of patients aged over 20 in 1982, with a functioning graft for more than one year, according to age at start of renal replacement therapy (RRT)

		Age at start of RRT (years)				
		8-10	11	12	13	14
	No.	*6*	*8*	*9*	*15*	*21*
MALE	mean (cm)	155.6	151.8	154.2	158.8	162.6
	range (cm)	133-180	132-179	138-165	150-178	148-185
	No.	*5*	*10*	*9*	*11*	*31*
FEMALE	mean (cm)	146.6	147.7	152.0	155.9	157.4
	range (cm)	140-152	135-162	142-167	147-170	134-172

Parathyroidectomy

Information on parathyroidectomy was available for 1,999 of the paediatric patients alive on treatment on 31st December, 1982. A total of 82 were found to have had parathyroid tissue removed (41 per thousand). The prevalence of parathyroidectomy according to time on renal replacement therapy is shown in Table XXI for patients never grafted, and for those grafted at any time. The

TABLE XXI. Prevalence of parathyroidectomy in 1982 among paediatric patients, alive at the end of that year, never grafted (ungrafted), and those grafted at any time (grafted), shown according to time on renal replacement therapy (RRT)

		Years on RRT			
		<2	2-5	5-10	>10
Ungrafted	Number of patients	291	192	92	39
	Parathyroidectomy/1000	3	42	87	359
Grafted	Number of patients	251	457	427	250
	Parathyroidectomy/1000	4	6	66	76

prevalence of the operation was lower among patients transplanted at some time compared with those never grafted except in the group who received less than two years renal replacement therapy. Prevalence of parathyroidectomy was particularly high in patients never grafted on renal replacement therapy for more than 10 years.

Conclusions

1. On 31st December, 1983, there were 2,703 patients alive on dialysis or with a functioning transplant in Europe who had commenced their treatment before the age of 15.

2. Acceptance rates for new paediatric patients seem to be levelling off in recent years to between three and six per million child population, with 15 per cent of the children aged less than six years.

3. Specialised paediatric centres have encouraged transplantation and may have had an influence on patient mortality.

4. The most common haemodialysis schedule in 1983 for children was 12 hours per week over three sessions.

5. Excess weight gain between dialysis sessions was a problem in 10 per cent of paediatric dialysis patients. The percentage of children who died in 1983 was higher among those with excess weight gain than those without.

6. Continuous ambulatory peritoneal dialysis was the initial treatment offered to 17 per cent of new paediatric patients in Europe in 1983, and 0.2 episodes of peritonitis per patient month were recorded.

7. Forty-eight children on renal replacement therapy for at least six weeks were found with recovery of renal function in the paediatric register. Amongst them, glomerulonephritis was the most common primary renal disease.

8. The contribution of transplantation to the treatment of children continued to increase in 1983, but transplant activity varied from country to country.

9. Cyclosporine was used in 25 per cent of paediatric transplants in 1983.

10. The causes of graft failure in 1983 found in a greater proportion of children than in the total Registry were: rejection after stopping all immunosuppressive therapy, recurrence of primary renal disease and vascular thrombosis.

11. Preliminary data on pubertal growth after transplantation suggests that

although variable it is most often insufficient. Growth velocity may be slowed by reduction of renal function and daily versus alternate day corticosteroid therapy.

12 Ultimate height of patients aged more than 20 years was generally below normal and seems to be related to age at start of renal replacement therapy.

13 The prevalence of parathyroidectomy in children never grafted increases with time on renal replacement therapy from 42 per thousand at five years to more than 300 per thousand after 10 years.

References

1 Broyer M, Donckerwolcke RA, Brunner FP et al. *Proc EDTA–ERA 1983; 20:* 76
2 *UN Demographic Year Book 1979*

Open Discussion

ANDREUCCI (Chairman) Thank you Dr Broyer for this interesting presentation concerning the behaviour of children on dialysis and transplantation. Can I ask you how you can explain the different behaviour of cyclosporine in children and in adults? This looks like a functional renal failure. In children you have a better serum creatinine than in adults despite the fact that the kidney source is the same because usually it is from an adult to a child and probably you use a lower dosage in children than in adults.

BROYER I cannot give you a simple explanation, it is a fact. Maybe children are more carefully followed for the blood concentration of cyclosporine but this is only one hypothesis.

ANDREUCCI Thank you very much and congratulations to all the Registration Committee for their help in the excellent work.

PART II

GUEST LECTURES ON HAEMODIALYSIS

Chairmen: F Locatelli
B Redaelli

I Taraba
N De Santo

Proc EDTA–ERA (1984) Vol 21

DIALYSIS MEMBRANES

E F Leonard

Columbia University, New York, USA

The essential function of an artificial kidney is the extraction of selected metabolites from blood across a blood-wetted surface. Occasionally the surface has been a sorbent particle in which case it is necessary for the medium to exhibit both selectivity and capacity to store metabolites. In almost all clinical situations the surface is that of a membrane, in which case capacity is irrelevant. Three aspects of a membrane determine its suitability:

1. Its spectrum of transmission coefficients. In haemodialysis these are the diffusive permeabilities of the solutes and the hydraulic permeability of the solvent, water [1]. In haemofiltration the important coefficients are, for the solvent, again the hydraulic permeability and for the solutes the product of this permeability by each solute's 'sieving coefficient' [1,2]. Some devices and treatment strategies are mixed in that solute transport is determined by both kinds of transmission coefficients.
2. The practical burdens it places on the system and thence the patient and physician.
3. The absence of negative effects on the patient's well-being: its biocompatibilities.

In this paper an assessment according to these aspects of membranes in current artificial kidney devices is presented. The principal purpose is to identify the further developments that are likely and to determine whether they are those that are most desirable. It is concluded that important changes in the qualitative and quantitative aspects of transport properties are not mandated by defined medical needs and will not occur; that cost will continue to dominate membrane development and patterns of membrane use, including re-use; that efforts should and will continue to remove bioincompatibilities, but only within rather strict limits imposed by costs, and that variegated therapeutic approaches will be more commonly specified to limit special membranes and devices along with the costs they carry to patients and situations that require them.

Transmission coefficients (diffusive and hydraulic permeabilities, sieving coefficients)

A finite transmission coefficient causes solute movement only if a driving force is present. The driving force of dialysis is a difference between the chemical activity of the solute in plasma and in dialysate. For most, but not all, solutes this difference is well approximated by a difference between plasma and dialysate concentration. The principal exceptions are substances that exist in multiple forms, e.g. free and bound calcium, multivalent anions such as the phosphates, and small molecules that bind significantly to larger ones such as proteins. The driving force of haemofiltration is a difference in pressure across the membrane. It is very important to remember that for clinical situations one must consider a *spectrum* of transmission coefficients. The luxury of designing a membrane to deal optimally with one substance without regard for how it behaves with other substances is not permissible. No artificial kidney membrane – nor that of any dialyser, haemofilter, or even the peritoneum – has an appropriate spectrum of transmission coefficients, and it is thus necessary to compensate for the inappropriate spectrum. The procedure is universal: find a membrane and use enough of it to remove a sufficient amount of the substance most difficultly removed, using a maximum driving force. For dialysis this driving force will be the plasma concentration minus the minimum concentration of substance attainable in dialysate, essentially zero. Then compensate for the excessive amounts that would be removed of other substances by setting finite, non-zero concentrations of these substances in the dialysate, thus decreasing their driving forces. Following this universal approach leaves one with a more-or-less expensive solution that moderates the permeabilities of dialysis or compensates for the excessive values of the product of hydraulic permeability by sieving coefficient in haemofiltration.

It is interesting to divide all metabolites into five categories in order to examine the spectrum of transmission coefficients. These are:

1. The solvent water
2. The major ('stoichiometric') bearer of nitrogen, urea
3. The small, ionised species
4. The 'middle molecules', both those one wishes to remove and those one wishes to retain
5. Large molecules.

Given the few parameters adjustable in any given membrane system the approach to achieving a spectrum that is workable – albeit only with the driving force compensation just described – has been to stipulate urea as the substance whose quantity and transmission coefficient considered together define it to be the substance most difficult to remove and then to maximise the coefficient for urea subject to maintaining the large molecule transmission at essentially zero. In dialysis, maximisation of urea permeability has almost always set the hydraulic permeability to water in a range where water movement could be controlled by adjusting the transmembrane pressure. (However, in a few cases an upper limit on hydraulic permeability has been the factor limiting how high urea permeability could be raised because of technical difficulty in maintaining transmembrane pressure near zero.) With urea permeability sufficiently high, permeability to salts has been high enough that their movement could be controlled by decreasing ionic driving forces through the addition of salts to dialysate. In some instances phosphate becomes an exception to this generalisation, and aluminium has been described as an exception although, in fact, its transport is controlled not by its permeability but by concentrations in dialysate that are small but significantly different from zero. After many years of research, agreement still remains to be reached on the importance of removing any middle molecules larger than creatinine and uric acid [3].

It is clear that the use of pressure as a driving force in haemofiltration gives rise to a spectrum of effective permeabilities different from that achieved in dialysis and, for some patients, a consequent clinical improvement. The theory of diffusion in membranes shows that for a given membrane the decrease in product of hydraulic permeability by sieving coefficient that occurs with increasing molecular weight will occur rather more sharply at higher molecular weights in comparison to the steady changes in diffusive permeability that take place over a wider range of molecular weights. However, both for dialysis and for haemofiltration, the concept of a 'cut-off molecular weight' that separates solutes that will pass a membrane from those that will not is a vast over-simplification. Figure 1 is a sketch showing qualitatively how one might expect the diffusive permeability and the product of hydraulic permeability by sieving coefficient to vary with solute molecular weight for a series of chemically similar, penetrating solutes. The figure is based upon approximate calculations using the data of reference 4.

Many commentators have wished for haemodialysis membranes in which the entire spectrum of molecular weights was raised so that less membrane area would be required to cause a given removal rate. Developments in this direction have been principally physical, that is the membranes have been made more permeable not by changing their chemical composition and thus their intrinsic diffusive properties, but rather by decreasing their thickness. Thus far these overall improvements in membrane permeability have been beneficial to designers who understood them and adjusted appropriately the configuration of the devices using the more permeable membranes. However, substantial further progress may not be possible. Here is the problem: higher permeabilities are useful only if employed to increase the rate of solute removal per unit of dialyser area, the so-called solute flux. If membrane permeabilities are increased to allow higher solute fluxes one must either arrange also to improve transport coefficients in the adjacent blood and dialysate (especially the former) or one must expect that conditions in these fluids will increasingly take control of the permeability spectrum away from the membrane. The permeability spectrum of contemporary membranes, even though imperfect, is far more desirable than that which transport controlled by blood, alone, would dictate. Transport coefficients in blood *can* be increased to avoid this problem, but only by increasing, through design changes, the shear rate of the blood [5]

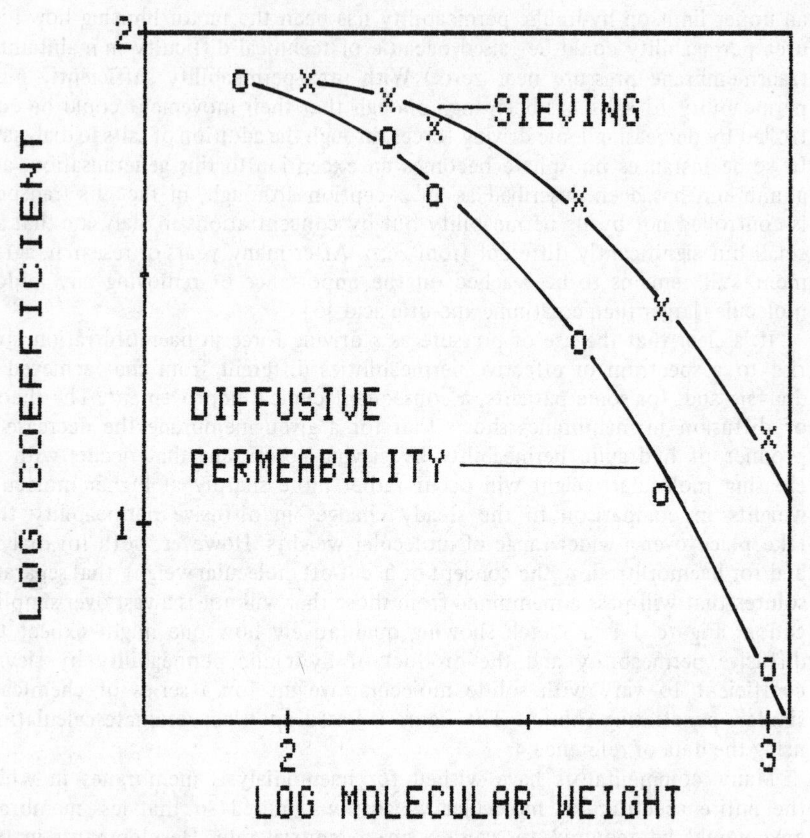

Figure 1. The logarithm of transmission coefficient is plotted versus the logarithm of molecular weight for a series of polyethylene glycol molecules acting as solutes passing through a membrane of polyvinyl alcohol [4]. The upper data set is *calculated* from the lower set by dividing the respective points by a factor proportional to the free-liquid diffusivity of each solute molecule; they and the line fitting them represent the sieving coefficient one would expect to observe with this membrane. The lower set of points are obtained directly from measured data and, with the line drawn through them, give the diffusive permeability as a function of solute molecular weight. If the 'cut-off' is defined as the transmission coefficient that is half that for a small molecule, the value for the sieving coefficient is about 460 while that for the diffusive permeability is about 250

and thus, as a necessary consequence, the blood-side pressure drop. Increasing the pressure drop may lead to excessive ultrafiltration unless the hydraulic permeability is lowered. Thus the desiderata of membrane design, even considered only with respect to transport, are several and probably conflicting at the level of membrane design. To repeat, substantial further improvement in the overall permeability of dialysis membranes is unlikely. It is also unlikely that the *shape* of the membrane permeability spectrum (viewed, say, as a plot of permeability versus solute molecular weight) will be significantly changed in the foreseeable future. There is not a clear enough goal to be addressed by doing so, especially since other aspects of membranes, considered below, deserve more attention.

The practical burdens a membrane imposes

The practical burdens imposed by a membrane are not widely recognised, but they are very real.

Associated with every membrane is a direct cost: the price of the membrane itself. Also associated with every membrane are hidden costs: these include the design features necessary to hold it in proper contact with the blood stream. For example, some membranes cannot be well formed as hollow fibres; the requirement of using them in a more expensive flat-plate configuration is a hidden cost. As already mentioned, dialysing or replacement solution is used to compensate for the imperfect aspects of a membrane's permeability spectrum; the price of this solution is also a hidden cost. Some membranes may require the use of greater quantities of drugs or the use of more time and materials at the clinic to prepare them (e.g. by rinsing); these materials and procedures generate other hidden costs. The true cost of a membrane is the sum of its direct and hidden costs. This cost will not be the same everywhere because it depends not only on the economy of the factory, but also on the value assigned to the patient's time loss during treatment and the value ascribed to labour in the clinic. Calculating this cost must include taking account of the pressure that may or may not exist to give other responsibilities than dialyser preparation to clinical staff. These observations are particularly true and their consequences particularly difficult to assess when dialyser re-use is considered. The direct membrane cost per dialysis is clearly reduced by re-use. The effect of re-use on true cost depends enormously on complex calculations involving the perceived cost and availability of labour and the availability and cost of capital to automate re-use when labour cost is an issue. These factors vary greatly from area to area, among kinds of institutions within a given area, and sometimes even among different clinics in the same institution.

It is perfectly clear that the true cost of a membrane will be the major factor that determines the degree to which it is accepted – when the situation is observed over a long period of time. Over a shorter period of time, other considerations such as technical novelty, the ability of a membrane to address a particular medical problem that is of concern at one time, the role of nationalism and other, political, non-economic forces at work in the market place will all exert a force. However, only one factor will remain as important as true cost for any membrane that has a basically acceptable permeability spectrum. This factor has acquired a very fancy name over the past few years: biocompatibility. Except for a few summarising remarks, the balance of this paper will be devoted to a consideration of this aspect of membranes. Consideration will be given to the quality of biocompatibility as it applies not only to membranes but also, necessarily for the purposes of this article, to whole artificial kidney systems.

Biocompatibility and Bioincompatibility

It is impossible to consider 'biocompatibility' in any reasonable technical detail as it would be so to consider 'health'. Biocompatibility can only be reached by eliminating recognised bio-IN-compatibilities just as health can only be reached by focusing attacks on particular diseases. In artificial kidney systems, bioincompatibilities can arise from any of the following components: in the dialyser – membrane, case material, potting compound, and dialysate; in the rest of the system – tubing, needles, and injectables. The great emphasis placed upon the bioincompatibility of membranes in contrast to the rest of the dialysis system arises because many deleterious phenomena depend upon area, and the area of contact between blood and membrane far exceeds that of any other component of an artificial kidney. Each source of bioincompatibility must be examined for the principal mechanisms by which it may do damage. These sources are: leaching, protein transformation at the interface, cell adhesion and aggregation, and mechanical (shear) effects [6].

Leaching is the slow dissolution of some entity from a solid phase into surrounding liquid. It is a major potential mechanism of bioincompatibility. Membranes have been suspected of leaching oligomer [7] (low-molecular-weight fractions of the membrane-forming polymer), sterilants such as ethylene oxide [8] and formaldehyde [9,10], and trace contaminants both organic and inorganic. In addition to acting as a possible source of contamination itself, the membrane determines what potential contaminants from dialysate may reach blood. Studies have repeatedly shown that the membranes commonly used block the passage of organisms and neither serve as the source of, nor allow the passage of, endotoxins [11,12]. Looking beyond membranes, leaching of phthalate esters from tubing [13] has also been reported, to a degree that may have long-term significance for patient health. By some stretching of the definition, one may include under leaching the release of *particulate matter* into blood. Such matter may be particles inadvertently introduced into a device at the time of manufacture or may be products of mechanical degradation occurring during processing, storage, or use of any part of the system. The particulates that have attracted greatest attention are fragments detached from the tubing segment of a roller pump [14,15]. Particulates are also potentially (and actually) present in all parenteral solutions and all injected drugs.

By *protein transformation* is meant the heterogeneous chemical reaction of a plasma protein at a surface so that the product has changed biological activity. The major transformations that have been recognised so far are activation of clotting factors [16,17] and complement [18]. Activation may lead to clinically significant events that occur either at the activating surface (when the activated substance remains adsorbed), at a distal point in the system, or in the bulk, systemic blood. Studies of activation are difficult because clinically recognisable effects vary widely, even when activation may be occurring at the same rate; this variability arises because clinically observable phenomena probably involve overload of a compensating mechanism in the patient. When activation rate exceeds capacity for deactivation, which may vary substantially from patient to patient and even from day to day in the same patient, clinical symptoms build rapidly. Below this threshold they may be almost totally absent.

One of the most insidious bioincompatibilities to have been encountered in artificial kidneys is rare but life-threatening anaphylactic reaction consequent to initiation of dialysis. The phenomenon is still not well understood but may involve both leaching and protein transformation. It is suspected that ethylene oxide leaches from the mass of polyurethane potting compound used to immobilise membrane in many hollow-fibre dialysers. The ethylene oxide then transforms one or many plasma proteins or possibly some of the soluble oligomer in the membrane in such a way that a potent allergic reaction is induced in a small sub-population of all patients. It is clear that the dominant reservoir of ethylene oxide in dialysers following gas sterilisation is the potting compound [20]. A very similar physical problem exists with respect to the use of formaldehyde in dialyser re-use although the biological consequence is different. Studies – which are underway – are urgently needed to find the principal reservoirs for formaldehyde, perhaps so that dialysers that rinse more effectively can be designed specifically for re-use.

Cells are frequently involved in bioincompatibility phenomena. The clinically dominant effect may be either depletion of cells – e.g. leukopenia – or their activation followed by such non-physiological behaviour as adhesion to artificial surfaces, bulk aggregation, and sequestration in particular capillary beds [21–24]. The mechanism of cellular interaction may be direct – contact with a surface which affects the cell inherently or through some molecule (most likely a protein) absorbed upon it – or indirect. Indirect mechanisms include reaction of cells in bulk blood with proteins transformed at a surface and subsequently returned to the blood. Activated complement probably affects cells in this manner. However, the indirect mechanisms involving cells may be even more complex; some cells that interact directly with a surface may subsequently release substances into bulk blood so as to affect cells that never directly contact the surface. Both leucocytes and thrombocytes have this capability [25].

Separately and as a modifier of the phenomena described above, *mechanical effects* are also a cause of bioincompatibility. Two basic flow phenomena account for the great preponderance of mechanical effects. These are flow separation [26] and shear [27]. Flow separation occurs in rigid artificial systems that have abrupt shape changes. Its effect is to place a small part of a device's blood volume into relative isolation and sustained contact with a surface. This volume then becomes a small reactor whose contents interact without dilution and in intimate contact with an artificial surface for long periods of time. The damage arising from separated flows may be local, e.g. the formation of a thrombus in the separated region or it may be systemic: the separated flow can serve as a nidus from which systemic toxins are generated. The elimination of separated flows is a matter of engineering although it may not always be simple engineering.

Shear is an ignored and misappreciated phenomenon that occurs everywhere in flowing blood [28]. Shear is the sliding of fluid layers over one another. It has many effects. The simplest is the distorting force it exerts on a cell trapped between two layers. These forces can result in lysis but more often they result in sublethal cell damage that causes release of metabolites from a cell, initiation of intracellular processes (like the combination of actin and myosin and the subsequent shape changes that occur in platelets), and increases in the tendency of cells to adhere and aggregate. Sometimes the potentiation of aggregability is temporarily concealed by the very shear that caused it. Sometimes it is only when the stresses that prevent two cells from staying next to each other long enough to adhere are removed, that the actual aggregation occurs [29,30]. Shear is greatest in fluid that is immediately adjacent to a surface. A cell that has adhered to such a surface will be more damaged by shear flow than it would be if it were free. The cell may be torn away from the surface, leaving a part of itself behind. Alternately, it may be provoked very quickly to empty itself of metabolites.

By far the largest shear levels encountered during haemodialysis occur in the arterial and venous cannulae [29]. In general the situation is worse in single-needle, single-lumen systems and worst of all in single-needle, double-lumen systems [30]. While the time of exposure is short in this part of the system the rates of shear may exceed $20{,}000 sec^{-1}$, perhaps 50 times that encountered in the juxta-membrane volumes within a dialyser. As already noted one does not see the effects of this shear in the needle but rather in distal, low-shear regions. If the dialyser operated at a low shear rate and with narrow spaces, the result will be extracorporeal deposition of cells; otherwise the cells may only manifest their damage intracorporeally.

Shear may also cause cells to tumble and thence to migrate, usually so as to create a cell-free zone near the wall. However, these processes may result in increased (rather than decreased) exposure of susceptible cells because it may be precisely they that are marginated on the outer surface of the cellular core formed by migration.

The largest effect of shear, however, is upon the exchange of molecular species between a fluid and its bounding wall. The effect of shear on molecular exchange pertains primarily to the artificial kidney as an efficient transport device for metabolites (see above), and is not our present subject. However, the effect of shear on transport may also enhance or ameliorate an underlying bioincompatibility. Moderately high shear will cause more cells and molecules of blood to contact a reactive surface and may thus result in a more widespread effect of the bioincompatibility. This fact notwithstanding, it is more likely that the dominant effect of moderately high shear will be the dilution of reaction products and the control of local events such as thrombogenesis so that they do not eventually force shutdown of the system.

The pursuit of biocompatibility, like the pursuit of health is a complex matter, and probably an unending scientific Odyssey. The conquest of one problem only sets the stage for an attack upon the next. For example, in all the foregoing discussion, no mention has been made of the clinically mandatory use of heparin to overcome one of the most fundamental bioincompatibilities encountered in extracorporeal therapy: the activation of clotting. So long as systems require heparinisation – and it is systems, not membranes, that must be addressed – they can hardly be considered to be biocompatible. However this issue has as many economic as therapeutic overtones; elimination of the heparin requirement will save money and when it occurs will be strongly affected not by a purely technical solution, which may already exist, but rather by whether the cost exceeds the saving. In like mode, elimination of complement activation, transient leukopenia, and pulmonary distress at the onset of dialysis (whether or not these phenomena are connected), will depend upon identification of a real and ultimately cost-related problem and upon discovery of a cost-effective modification of membranes, devices, or procedures to eliminate the real problem.

What, then, does the future hold for dialysis membranes? In the best of all possible worlds: *variety*. In the hands of membrane manufacturers such variety already exists. Economics, variations in a patient's requirements, different perspectives in the world regarding the value of capital and labour, and the very process of continuing discovery in the research laboratory and the clinic demand that the users and choosers of dialysers *not* strive for the unique best, the monolithic solution. For example, the needs of developing countries will be addressed wholly differently from those where artificial kidney therapy is already established as a right. To give another example, in any area, there will be found patients and situations that require special devices and membranes, but there is no need to impose these on a whole patient population in order to provide what is necessary or highly desirable for only a few. Some members of society will embrace wholly portable artificial kidneys and one may expect to see effective devices of this type in use before the present decade has passed, but not for the majority of patients. Re-use has not yet become a settled, ensconced activity, and it is likely that manufacturers will offer disposable dialysers cheap enough that users in areas where the cost of capital and labour is explicity considered will think twice before espousing re-use, at least if their principal motivation is economic. All these compelling reasons for a varied approach will place great responsibilities – as free choice always does – upon the clinical staff. Both the chances to mismatch, even tragically, and the chances to provide truly superior rehabilitation will be much enhanced.

Whether regenerated cellulose will continue to be the dominant membrane for haemodialysis will be decided, firstly, on the basis of its total cost per dialysis relative to other possible materials and, secondly, on its perceived biocompatibility relative to other materials, with the necessary proviso that as current research will change these perceptions and possible chemical modifications to both cellulose and competing materials may modify their reactions with biological molecules.

Perhaps the most exciting thing to be said about dialysis membranes in 1984 is that a simple prediction about them is impossible.

Acknowledgment

The author has benefited over many years from the gentle and humble teaching of Dr Werne Henne of Enka AG. He wishes to dedicate these thoughts, some of whose imperfections were removed by Dr Henne, to him upon the occasion of his retirement from full-time work in the field.

References

1 Klein E, Villarroel F, eds. *Evaluation of Hemodialyzers and Dialysis Membranes.* Washington: US Government Printing Office. 1977: 7–27
2 Lysaght M, Ford C, Colton C et al. In Frost TH, ed. *Technical Aspects of Renal Dialysis.* Tunbridge Wells: Pitman Medical. 1978: 81
3 Gotch FA. In Frost TH, ed. *Technical Aspects of Renal Dialysis.* Tunbridge Wells: Pitman Medical. 1978: 1
4 Odian M, Leonard EF. *Trans ASAIO 1968; 14:* 19
5 Cooney DO, Kim S-S, Davis EJ. *Chem Eng Sci 1974; 29:* 1731
6 Klinkmann H, Wolf H, Schmitt E. *Contr Nephrol 1984; 37:* 70
7 Pearson F, Bohon J, Lee W et al. *Artif Organs 1984; 8:* 291
8 Dolovich J, Bell BJ. *Allergy Clin Immunol 1978; 62:* 30
9 Shaldon S. *Contro Nephrol 1983; 36:* 46
10 Fassbinder W, Koch K-M. *Contr Nephrol 1983; 36:* 51
11 Port FK, Berrnick JJ. *Contr Nephrol 1983; 36:* 100
12 Dinarello CA. *Contr Nephrol 1983; 36:* 90
13 Kevy S, Jacobson M. *Contr Nephrol 1983; 36:* 82
14 Leong ASA, Path MRC, Disney APS, Gove DW. *N Engl J Med 1982; 306:* 135
15 Bommer J, Waldherr R, Ritz E. *Contr Nephrol 1983; 36:* 115
16 Forbes CD. *Clin Haematol 1981; 10:* 653
17 Gasparotto ML, Bertoli M, Vertolli U et al. *Contr Nephrol 1984; 37:* 96
18 Chenoweth DE. *ASAIO J 1984; 7:* 44
19 Dolovich J, Marshall CP, Smith EKM et al. *Artif Organs 1984; 8:* 334
20 Henne W, Schulze H, Pelger M et al. *Artif Organs 1984; 8:* 299
21 Hoenich NA, Woffindin C, Qureshi M, Kerr DNS. *Contr Nephrol 1983; 36:* 1
22 Bauer H, Brunner H, Franz HE, Bütmann B. *Contr Nephrol 1983; 36:* 9
23 Böler J, Kramer P, Götze O et al. *Contr Nephrol 1983; 36:* 15
24 Craddock PR, Hammerschmidt DE. *ASAIO J 1984; 7:* 50
25 Tetta C, Jeantet A, Camussi G et al. *Proc EDTA–ERA 1984:* this volume
26 Goldsmith HL, Karino T. *Ann NY Acad Sci 1977; 283:* 241
27 Grabowski EF, Friedman LI, Leonard EF. *IEC Fundamentals 1972; 11:* 255
28 Leonard EF. In Colman, Hirsch, Marder, Salzman, eds. *Textbook of Hemostasis and Thrombosis.* Philadelphia: Lippincott. 1982: 755–765
29 Leonard EF, VanVooren C, Hauglustaine D, Haumont S. *Contr Nephrol 1983; 36:* 34
30 Leonard EF. *Artif Organs 1984:* in press

Open Discussion

GOTCH (San Francisco) You indicated that there is now adequate technical knowledge to fabricate a non thrombogenic artificial kidney circuit. Would you please elaborate on this comment.

LEONARD I think the developments in Sweden by Drs Larssen, Olsen and their colleagues could be brought to the point where a totally heparin free dialysis could be effective. I think that most people here do not want a heparin free dialysis under the present conditions because of the burden it places on staff. You would end up taking care of a few patients without heparin rather than the large number of patients you could take care of with 'automatic' heparin.

DI GIULIO (Rome) What are the relationships between biocompatibility and the electrical charge of the dialysis membranes and what are the effects of the membrane electrical charge on ion transport?

LEONARD I think that the subject of biocompatibility has a certain lunatic fringe around it in the sense that people state that one aspect of the membrane determines its biocompatibility. The fact is that we have been unable to separate out the specific aspects of the membrane which determines its biocompatibility. If one chooses a particular system and then proceeds to alter the charge, paying attention to which ionic groups are placed on the membrane as the charges alter, one can get some kind of a biocompatibility correlation. We are very ignorant on this subject. If the biocompatibility is based on platelet adhesion we will get some kind of a ranking but if it is based on activation of factor XII we will get a different ranking and if it is based on complement activation or leucocyte adhesion we will get yet an even different ranking. In systemic studies we will find large variations from patient to patient because some patients will be near their leucocyte activation thresholds while others will be near their thrombocytopenia thresholds. I think we are simply not at a place where we can say, except for those with very simplistic view of life, the charge is separable as a factor.

As far as ion transport is concerned a probable rule of thumb is to make the membrane as neutral as possible because the moment one puts a charge on a membrane it does more to drive away the counter-ion than to induce the movement of the ion whose charge is on the membrane. In general if one wants to optimise the membrane for ion transfer at a given overall porosity the membrane should be as charge free as possible. The subject of specific ion transfer across membranes will be revisited but my guess is that what we will have will not be membranes that have better ion transfer but artificial kidneys with built in capacity to remove specific difficult ions such as phosphate by sorption. However I do not know of any developments in this field as yet.

PORT (Ann Arbor, Michigan) You have beautifully outlined issues of biocompatibility due to shear. Yesterday a paper presented proposed the use of blood flow rates of 600ml/min and as clinicians we do not see obvious problems when using flows of say 400 versus 150ml/min. Could you comment on this apparent tolerance?

LEONARD The issue can be sorted in three parts. Firstly is to take very great care of the blood access and a short 14g needle with a blood flow of 600ml/min is going to really rip up the cells and the major biocompatibility problem will be the strong activation of susceptible cells. It is unlikely to show as haemolysis but more likely as platelet and leucocyte damage. Secondly there are very few substances whose removal is flow limited and there are much better ways of removing more of what is presented to the dialyser than by increasing flow. Finally the other part of the system where we might see trouble under these conditions is at the inlet and outlet to the fibres of the dialysers. When dialysers are made the knife cut must be made across the surface of the potted fibres and it is difficult for manufacturers to control the quality of that cut.

HENNE (Wuppertal) I congratulate you on your excellent survey. There is but one statement which I cannot follow. Many of the facts you mentioned in respect of biocompatibility were surface related. Why did you exclude then the decrease of surface area as a means to improve biocompatibility? If you would reduce the surface of the whole system by a very small amount, holding constantly the performance, in our opinion that would improve the system. In the last decade hollow fibre devices have been miniaturised from about $1.6m^2$ to about $0.8m^2$. During this period of time survival rates and morbidity have been improved as was shown by Kjellstrand from the US statistics. This encourages us to follow that way furthermore.

LEONARD In the last ten years you have straightened me out on many occasions and without you I would have made many more mistakes in my presentation. I agree that if you can reduce the membrane surface area of the artificial kidney holding performance the same there will be an improvement in biocompatibility as many of the problems are area related. However, I feel we have reached the mass transfer limits of the membranes and if we try to have higher fluxes on the membrane we are going to have more and more rejected molecules and cells on the surface. As a matter of engineering we are going to be unable to handle this accumulated material and the result will be protein denaturation, increased cell membrane interaction, and an inability to permeate this barrier. I think it will be a long time before we can significantly reduce surface areas. I think we stand to gain a lot more by increasing areas slightly and making the conditions on the surface a little more gentle.

Proc EDTA–ERA (1984) Vol 21

THE IMPORTANCE OF WATER TREATMENT IN HAEMODIALYSIS AND HAEMOFILTRATION

P Keshaviah, D Luehmann

Hennepin County Medical Center, South Minneapolis, Minnesota, USA

"Till taught by pain, men really knew not what good water's worth"
Lord Byron in "Don Juan"

The need for water treatment

Water, depending on the source, will contain varying amounts of dissolved inorganic and organic substances, particulates (clay, sand, iron, etc.) and micro-organisms and pyrogens. Agricultural runoff, mining operations, and industrial and municipal pollution all play a role in the contamination of the water supply. Not all municipal supplies are regulated and while they may be safe for drinking purposes, the majority will require some form of treatment to make them safe for haemodialysis or haemofiltration.

Drinking water regulations are based on a weekly exposure of 14 litres. The patient on haemodialysis is exposed each week to between 300 and 400 litres and the patient on haemofiltration to between 70 and 100 litres. Further, these patients have diminished renal function and therefore, compromised urinary excretion of toxic substances. To compound matters further, the dialysis membrane is less selective than gastrointestinal absorption so that diffusion of toxic substances directly into the patient's blood stream occurs. Some of these toxic substances, such as the heavy metals, bind to plasma proteins thereby sustaining the gradient for diffusion from dialysate to blood so that even small gradients result in a significant toxic load. In the case of haemofiltration, there is direct infusion into the patient's blood without the restriction of a separating membrane.

The requirements for water used in haemodialysis and haemofiltration are, therefore, more stringent than for drinking water, necessitating additional water supply treatment. Some substances documented as toxic in haemodialysis are not regulated in the drinking water, a prime example being aluminium. There are reports from dialysis centres all over the world [1–10] correlating the incidence of 'dementia dialytica' to high concentrations of aluminium in the water used to prepare dialysate. Table I lists those substances which have been

TABLE I. Contaminants with documented toxicity in haemodialysis

Contaminants	Toxic effects
Aluminium	Dialysis encephalopathy, and renal bone disease
Calcium/Magnesium	Hard water syndrome: nausea, vomiting, muscular weakness, flushed feeling, and hyper/hypotension
Chloramines	Haemolysis, anaemia, and methaemoglobinaemia
Copper	Nausea, chills, headache, liver damage, and fatal, haemolysis
Fluoride	Osteomalacia, osteoporosis, and other bone diseases
Nitrate	Methaemoglobinaemia with cyanosis, hypotension and nausea
Sodium	Hypertension, pulmonary oedema, confusion, vomiting, headache, tachycardia, shortness of breath, seizures, coma and death
Sulfate	Nausea, vomiting, and metabolic acidosis
Zinc	Anaemia, nausea, vomiting, and fever
Microbial	Pyrexial reactions – chills, fever, nausea, hypotension, and cyanosis

documented as toxic in haemodialysis and their toxic effects. Exposures to some of these substances in haemodialysis have resulted in fatalities.

As stated earlier, most municipal supplies require additional treatment to make them safe for haemodialysis and haemofiltration. There are variations in water quality that may be seasonal, related to variations in municipal treatment or to the use of multiple water sources. Certain chemicals such as fluoride and aluminium that are added by municipal authorities to make the water safe for drinking make the water unsafe for haemodialysis and haemofiltration. With the addition of such toxic chemicals, there always exists the possibility of the accidental addition of much larger quantities than desired, an accidental overfluoridation of a municipal supply in Maryland affected eight dialysis patients with one death [11]. In addition the distribution system may itself contribute to contamination, toxic materials such as copper, brass, zinc and lead may leach into the water supply during distribution. All of these factors make additional treatment of the municipal supply essential for the safety of the haemodialysis/haemofiltration procedure. Further, a continuing dialogue between the dialysis centre and the municipal supplier needs to be established in order to anticipate and adequately compensate for variations in the municipal supply.

In this paper we deal with the various water treatment options available to the dialysis practitioner along with details concerning the nature of the treatment, the principles involved, the monitoring necessary and other features of the treatment options. We will also discuss the quality of product water required to ensure safety of the patient and the monitoring requirements to ensure such water quality. The responsibilities of the manufacturer/supplier as well as those of an informed consumer will also be discussed. It is our hope that this paper will provide the practitioner with some of the basic technical information necessary to make informed choices concerning the type and configuration of water treatment necessary for a given water supply to safeguard the well-being of their dialysis patients.

Water treatment options

The water treatment options available for haemodialysis and haemofiltration include sediment filtration, softening, activated carbon filtration, deionisation, reverse osmosis, ultrafiltration and distillation.

Sediment filtration

Sediment filtration is used to remove large particulates ($>1\mu$) from the water supply. If these particulates are not removed, there exists the possibility of plugging of equipment downstream with patient consequences depending on the mode of equipment failure. Figure 1 is a schematic representation of a sediment filter. Filtration is accomplished by size exclusion as the water percolates through screens, closely packed fibres or a porous matrix. Sediment filters are rated by 'nominal' pore size for particulate exclusion and the maximum flow

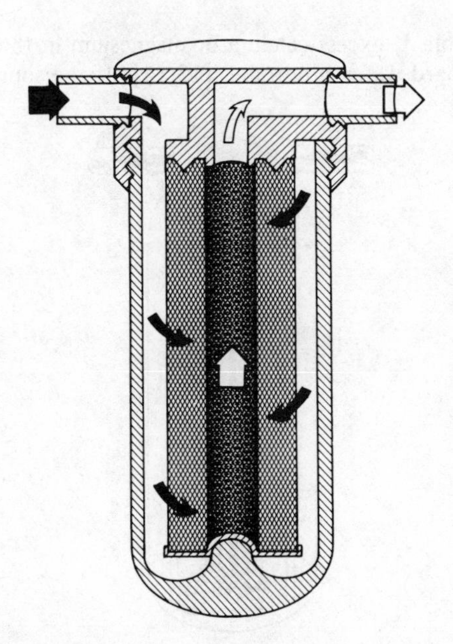

Figure 1. Schematic representation of a sediment filter

rate at which they may be operated. Once the exclusion capacity of the filter is exceeded, there may be 'break-through' of particulates into the effluent of the filter, as well as decreased flow in the system. It is, therefore, important to monitor the pressure drop across the filter by suitable pressure gauges, filters being discarded once the pressure drop exceeds the value specified by the manufacturer as a criterion for filter replacement.

One problem associated with sediment filters is the growth of microbial organisms on the filter media with consequent contamination of equipment downstream and the possibility of bacteraemias and pyrogen reactions in the patient. Disinfection of the water treatment system regularly and replacement of filters at appropriate intervals will minimise the problem.

In some filters, there may be a tendency for sloughing of the filter medium, thereby defeating the purpose of the filter. Further, it should be ensured that toxic residues are not released into the water supply from passage through the sediment filter. In one case [12], the filter was composed of cotton fibres held together by thermosetting polymers prepared from a resin of melamine and formaldehyde. Formaldehyde leached from the filter causing an outbreak of haemolytic anaemia involving 12 patients.

Sediment filters may be used singly or in series or parallel configurations and may be used at several points in the water treatment system. These include prefiltration of the municipal supply, following carbon filters to protect equipment downstream from particulates released from the carbon filters, at the inlet of the dialysis machine, etc.

Softening

As indicated in Table I, excess calcium or magnesium in the municipal supply may result in the 'hard water syndrome' [13,14]. If the source water is high in

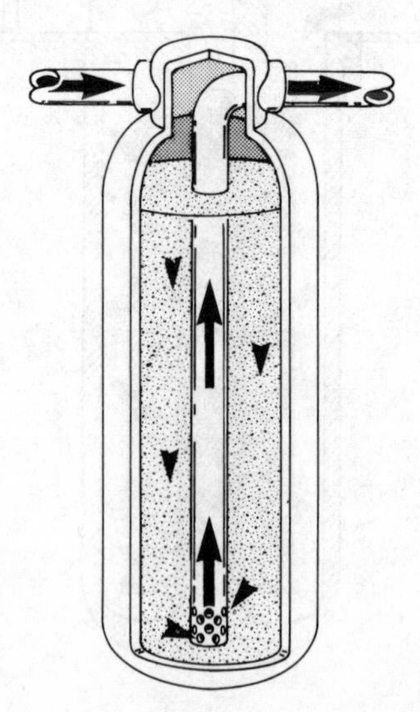

Figure 2. Schematic representation of a water softener

calcium or magnesium, water softening is essential. Water softeners are also used as pre-treatment for reverse osmosis devices to protect the membranes of these devices from scale build up and consequent failure.

Water softeners (Figure 2) are ion-exchangers, the resin of the water softener being a cationic resin that exchanges Na^+ ions for Ca^{++} and Mg^{++} ions as well as other polyvalent cations such as iron and manganese. When all of the available Na^+ ions in the resin have been exchanged, the resin is said to be exhausted and has to be regenerated. A brine of sodium chloride is used for regeneration, the ion-exchange process being thus reversed and Na^+ ions being reinstated on the resin. Regeneration may be accomplished on-site or in a central commercial facility. Water softeners have to be sized appropriately, factors including the degree of hardness of the water, the rate of water consumption, and the ion-exchange capacity of the resin as well as the frequency of regeneration cycles. The effectiveness of the softener may be monitored by measuring the hardness of the effluent using suitable titration kits that have a resolution to 1mg/L.

In water softeners that are regenerated on-site and are not equipped with bypass valves, care must be taken to see that regeneration does not occur during the dialysis event. In a case described in the literature [15], accidental interruption of power to the unit caused a mistiming of the regeneration cycle with the development of hypernatremic symptoms (flushing, thirst, vomiting, back pain, headache, disorientation, etc.). If the water softener is provided with an automatic bypass valve, the supply water will not percolate through the resin during regeneration, thereby preventing the risk of hypernatremia from the brine used for regeneration.

As with sediment filters, the possibility of microbial growth on the softener resin exists. Back washing of the resin during regeneration will minimise this problem. Such back washing will also prevent the accumulation of particulates in the resin bed and consequent loss of ion-exchange capacity.

If the resin is regenerated at a central commercial facility, the facility must ensure no admixing of resin from the dialysis centre with resins of other non-medical users to prevent microbial and toxic contamination of the resin. In one case [16], cross-contamination of the resin between urban and rural users occurred with heavy microbial contamination of the urban user's resin.

The risk of microbial growth in a water softener is to some extent inherent to the design of the resin bed. The problem may, however, be exacerbated if the effluent flow path incorporates dead ends and areas of flow stagnation. Rigorous disinfection procedures and periodic culturing can alleviate this problem. Ideally, a softener is followed by a membrane process, such as reverse osmosis (vide infra), which effectively removes bacteria from the water.

Activated carbon filtration

Activated carbon filters remove chlorine, chloramines, and dissolved organics (60–300 daltons) from the water supply and are used not only to meet the product water standards for these substances, but also as pre-treatment for deionisers and reverse osmosis devices. As shown in Figure 3, activated carbon

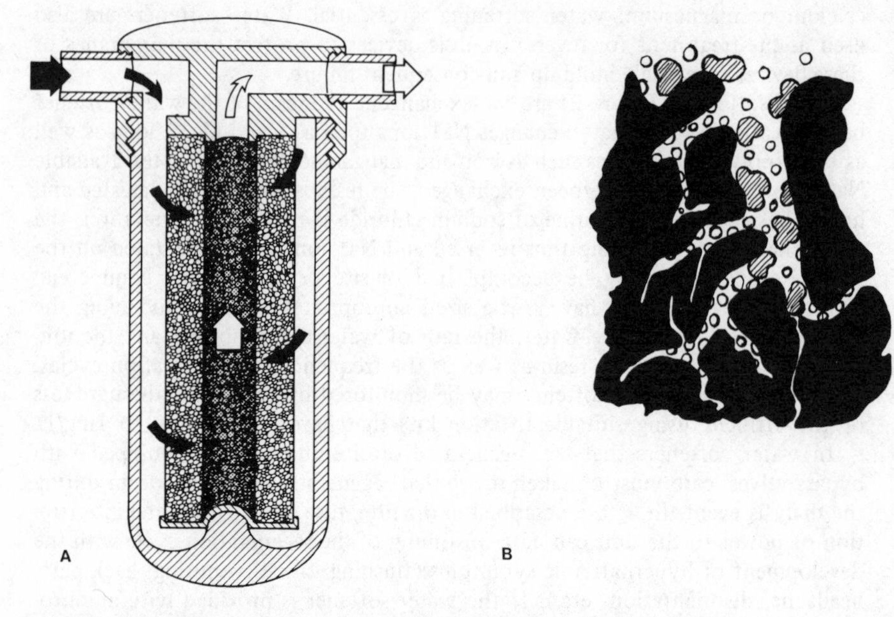

Figure 3. A. Schematic of an activated carbon filter. B. Microporous structure of activated carbon

has a microporous structure and a very large surface area to weight ratio that facilitates adsorption of chlorine, chloramines, and smaller organic contaminants. The adsorption capacity and the rate of removal varies with the source of the carbon and the nature of the activation process. It is important to recognise this in choosing a carbon filter and in sizing it for a particular water supply. If the adsorption capacity is exceeded, there will be 'spill-over' of chlorine, chloramine, and dissolved organics into the effluent. Monitoring the concentration of chlorine or chloramine in the effluent is, therefore, a convenient method for monitoring the efficacy of a carbon filter. Carbon filters cannot be regenerated effectively and must be used as disposables or be repacked with fresh activated carbon. Activated carbon, because of its microporous structure, has a tendency to release particles of carbon called 'fines'. Sediment filters are, therefore, used downstream of carbon filters to trap these fines and prevent them from plugging equipment downstream.

Chlorine removal is necessary with some types of reverse osmosis membranes. However, removal of chlorine promotes microbial growth. Further, the porosity of activated carbon and its affinity for organics make carbon filters susceptible to microbial contamination. Carbon filters do contribute to increased levels of gram-negative bacteria and endotoxins. Appropriate disinfection of the downstream water treatment system is, therefore, important when carbon filters are used.

Chloramines are increasingly being used as bactericidal substitutes for chlorine in municipal treatment of the water supply because chlorine reacts with human acids and other naturally occurring organic compounds in the water supply to form trithalomethones which have been shown to be carcinogenic in animals [17]. The US EPA has promulgated a maximum level of 0.1mg/L for total trihalomethone in drinking water necessitating the use of chloramines in municipalities. Chloramine in concentrations as low as 0.25mg/L can cause haemolysis, anaemia, and methaemoglobinaemia especially in patients with hexose monophosphate shunt deficiency [18–23]. The improper sizing of a carbon filter or absence of monitoring can therefore be extremely hazardous in situations where chloramines are used as bactericidal agents. As activated carbons have a varying ability to remove chloramines, the choice of the right type of carbon filter is also important in such situations.

Deionisation

Deionisers (Figure 4), as the name implies, are used for the removal of dissolved inorganic ions in order to meet product water standards. As with softeners, ion-exchange is the basis for deionisation. There are two types of resin used, cationic and anionic. Cationic resins exchange hydrogen ions for cations while anionic resins exchange hydroxyl ions for anions. In a mixed bed deioniser, cationic and anionic resins are in a mixture so the exchanged hydrogen and

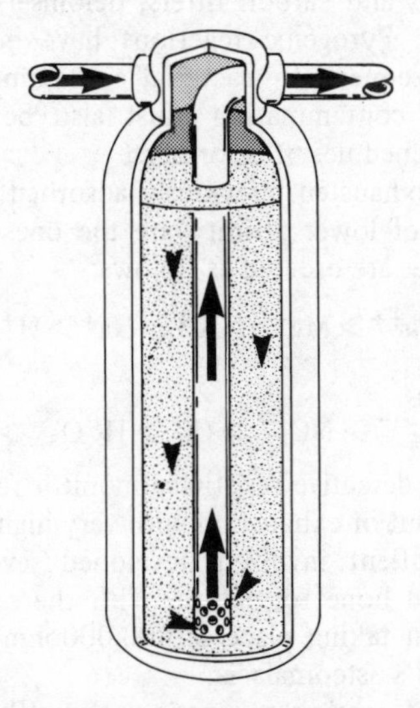

Figure 4. Schematic representation of a deioniser

hydroxyl ions combine to form water. Mixed bed deionisers produce water of very high quality, the resistivity of the water (a measure of deionisation efficacy) being in excess of 1megohm-cm. Ion exchange resins are usually made by polymerising a mixture of styrene and divinylbenze to form a polystyrene matrix. Cationic resins result from sulfonation of this matrix while anionic resins are made by chloromethylating and aminating the matrix. A dual-bed deioniser, unlike a mixed-bed deioniser, consists of two separate resin beds, one cationic, and the other anionic, in series. Such deionisers are less effective than the mixed-bed design. As with softeners, when the exchange capacity of the cationic and anionic resins is exhausted, regeneration is possible. Cationic resins are regenerated with strong acids, while anionic resins are regenerated with strong alkalis. The sizing of deionisers is important and is based on the concentration of dissolved ions in the feed and the volume of flow through the system between regeneration cycles. If the feed water has a high content of total dissolved solids (inorganic), deionisers are expensive to operate. Deionisers are sometimes used to 'polish' the product water of reverse osmosis devices when it alone is not adequate for removing certain ions such as fluoride and nitrates. Used thus, the costs of deioniser operation are reduced by an order of magnitude as reverse osmosis typically removes more than 90 per cent of the dissolved solids.

Deionisers are monitored by measuring the resistivity of the effluent. Such resistivity monitors must be temperature compensated for appropriate monitoring of deioniser function. Regeneration of the deioniser is based on the resistivity of the effluent, a value of 1megohm-cm being usually used as the criterion for regeneration.

Deioniser resins are subject to fouling if the organic content of the feed water is high because of the large surface area and porous structure of the resins. As with softeners and carbon filters, deionisers are also susceptible to bacterial contamination. Pyrogenic reactions have been documented in the literature [24–29] consequent to bacterial contamination of deionisers. The potential for microbial contamination must also be considered in deciding sizing and regeneration schedules of deionisers.

When a deioniser is exhausted, previously adsorbed ions may be eluted into the effluent. The ions of lower affinity are the ones most readily displaced. With cations, the affinities are ordered as follows:

$$Ca^{++} > Mg^{++} > K^{+} > Na^{+} > H^{+}$$

With anions, the ordering is:

$$NO_3^{-} > SO_4^{--} > NO_2^{-} > Cl^{-} > HCO_3^{-} > OH^{-} > F^{-}$$

In one incident [30], a defective resistivity monitor resulted in the use of the deioniser beyond the point of exhaustion with very high fluoride concentrations in the effluent. The patient involved developed severe osteomalacia, bone resorption, and decreased bone formation. With the replacment of the faulty monitor and regeneration taking place at 800,000ohm-cm, there was improvement noted in the patient's osteomalacia.

When a deioniser is nearing exhaustion, the effluent water may become acidic because the imbalance between cation and anion exchange capacities usually favours cations. Two types of hazards have been documented as a result of acidic effluents from deionisers. One is that of leaching of copper [31–33] from the piping and components of the dialysis machine with copper-induced toxicity (haemolytic anaemia, nausea, vomiting, chills, anorexia, etc.). The other is inactivation of heparin by acidic dialysate with the consequent clotting of dialysers [34,35].

As with softeners, it must be ensured during commercial regeneration, that resins from dialysis centres are not intermixed with resins of non-medical users. In industry, deionisers are used for recovery of plating metals such as chromium and silver. Traces of these toxic metals may remain bound to the resin and be eluted into the water used for dialysis, with toxic consequences, if intermixing of resins occurs. Further, the chemicals used for regenerating the resins of a dialysis centre must not contain excessive levels of toxic impurities such as the heavy metals. Chemicals used for disinfection must be adequately rinsed from the deioniser before use.

With certain municipal supplies, the effluent from deionisers have been found to contain nitrosodimethylamine [36], a suspected carcinogen. In such situations, the feed water must be appropriately pre-treated or a carbon filter used in combination with the deioniser. Anion resins may also be susceptible to hydrolytic cleavage of amines which are dialysable and may constitute a health risk. As with carbon filters, deioniser resins may slough particles and a sediment filter downstream of the deioniser may be necessary.

Reverse osmosis

This form of treatment is effective for dissolved inorganics, dissolved organics, bacteria, pyrogens and particulates. Reverse osmosis is a membrane process that is based on molecular sieving (>200 daltons) and ionic exclusion with 90–98 per cent rejection of monovalent ions and 95–99 per cent rejection of divalent ions. When two solutions of different ionic concentrations are separated by a semipermeable membrane, solvent flows from the less concentrated to the more concentrated side. This phenomenon is defined as reverse osmosis and osmotic pressure is defined as the pressure that must be applied to the concentrated solution side to prevent such a flow. If the applied pressure is greater than the osmotic pressure, there is reversal of solvent flow with solvent moving from the more concentrated to the less concentrated side. This process is called reverse osmosis.

There are several types of membranes suitable for reverse osmosis – cellulosic, aromatic polyamides, polymimides, and polyfuranes. Thin-film composite membranes have also been recently developed for reverse osmosis. The geometrical configurations of the membrane in the reverse osmosis module include plate and frame, tubular, helical tube, spiral wound, and hollow fibre configurations. The spiral wound (Figure 5) and hollow fibre (Figure 6) configurations are usually used in water treatment for haemodialysis and haemofiltration. However, hollow fibre modules are susceptible to plugging and fibre leaks.

Reverse osmosis membranes are subject to premature failure if the feed water has not been appropriately pre-treated. If calcium, magnesium, iron, and manganese are present in high concentrations, there is scale formation on the membrane with loss of efficacy. Some membranes are sensitive to chlorine and chloramines. Cellulose acetate membranes are sensitive to pH>8. Cellulosic membranes are also susceptible to bacterial degradation. Pre-treatment is, therefore, essential to appropriate performance of the reverse osmosis device. The nature of the

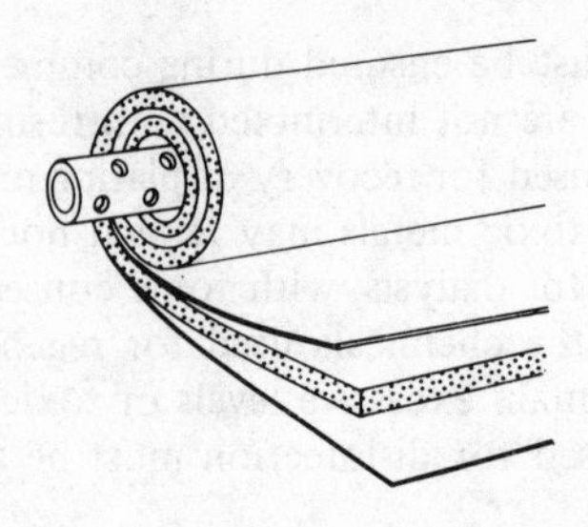

Figure 5. A spiral wound reverse osmosis module in schematic

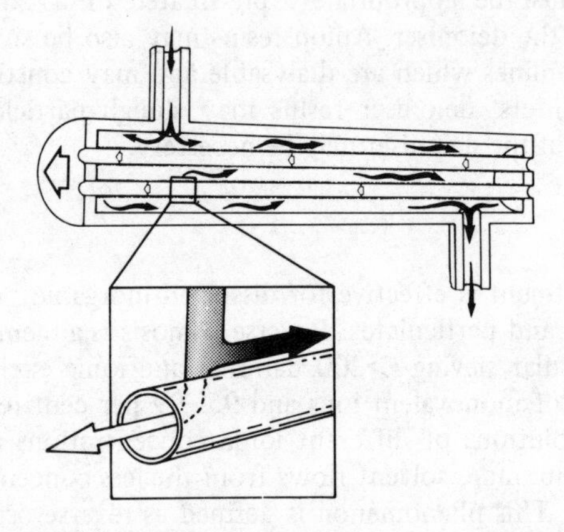

Figure 6. Representation of a hollow fibre reverse osmosis module in schematic showing water permeation through a single fibre

pre-treatment will depend on the quality of the feed water and the type of membrane being used. Reverse osmosis membrane modules are expensive, and once damaged, ability to regenerate the membrane is limited.

While reverse osmosis is effective for removal of bacteria, viruses, and pyrogens, microbial infestation of the reverse osmosis module can occur with micro-organisms and pyrogens penetrating small defects in the membrane or leaky seals. Bacterial contamination of the product water is, therefore, possible and appropriate disinfection schedules must be planned to control the problem.

Beyond the initial capital cost, water treatment by reverse osmosis is relatively inexpensive (barring premature membrane failure) and is a mode of treatment that effectively removes all four types of contaminants – organic, inorganic, particulate, and microbial. In situations where the ionic concentrations in the feed are very high (e.g. fluoride, and nitrate), a deioniser may be used to 'polish' the product water of reverse osmosis. Such a scheme can create a problem with microbial contamination and a downstream ultrafilter may be required. Reverse osmosis devices are monitored by resistivity measurements of the feed water and product water streams and by calculating the percent rejection from these measurements. In some instances, the effluent may also be analysed for specific contaminants.

Ultrafiltration

Ultrafiltration is a membrane process like reverse osmosis, but the molecular cutoff is much higher, being between that of reverse osmosis and sediment filtration. Ultrafiltration is effective for removing micro-organisms, pyrogens, colloids, and particulates (sub-micron). Thin, polymeric membranes are used in ultrafilters, in a sheet or tubular configuration. Ultrafilters are available in different size ranges, relative to nominal pore size and flow capacity. Their effectiveness may be monitored either by measuring the pressure drop across the filter or by measuring bacterial and pyrogen levels in the effluent filtrate.

Ultrafilters may be used to protect the more expensive reverse osmosis membranes from bacterial contamination and fouling by colloids and particulates. Regeneration by chemical cleaning and back-flushing may be possible with certain ultrafilter designs, others being disposable. Ultrafilters are particularly useful as the last stage of treatment to control bacteria and pyrogens in the product water when upstream devices introduce such contamination.

Distillation

Distillation removes non-volatile organic and inorganic substances, particulates, colloids, micro-organisms, and pyrogens. It is an effective but very expensive

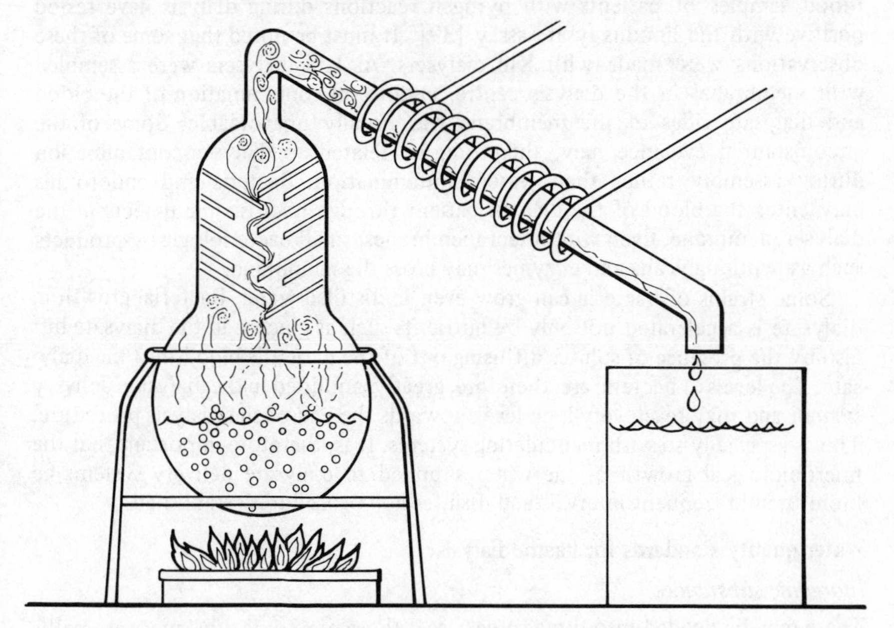

Figure 7. Schematic depiction of the distillation process

mode of treatment that is not commonly encountered in water treatment for haemodialysis. Liquid water is converted, with expenditure of energy, to the vapour phase with subsequent condensation of the vapour back to the liquid phase by cooling (Figure 7). This form of treatment necessitates the use of storage tanks and distribution pumps. The overall system may, therefore, be prone to bacterial proliferation even though distillation, itself, eliminates bacteria and pyrogens. Monitoring of efficacy is accomplished by measuring the concentration of relevant species in the distillate for organics and by resistivity measurements for ionic contaminants. Distillation is not effective with volatile contaminants because of carry-over. Distillation systems are continuously reusable, but maintenance requirements are stringent.

Bacterial contamination of water

As indicated above, sediment filters, softeners, carbon filters and deionisers may all support microbial growth. The following types of bacteria have been identified in dialysis systems: pseudomonas, acinetobacter, flavobacterium, achromobacter, seratia, moraxella, aeromonas, mycobacterium chelonei, fortuitum, erwinia and alcaligenes. While bacteria and endotoxins ($10^5 - 10^6$ daltons) cannot cross intact dialysis membranes, there is circumstantial evidence linking dialysate contamination and pyrogen reactions. Sudden outbreaks of febrile reactions have been associated with high levels of gram negative bacteria in dialysate [24,28]. Antibodies to bacterial endotoxin have been detected in the blood of dialysis patients [37] and electron micrographs have revealed bacteria on both blood and dialysate sides of the dialyser membrane [38]. Blood samples of patients with pyrogen reactions during dialysis have tested positive with the limulus lysate assay [39]. It must be noted that some of these observations were made with Kiil dialysers. As Kiil dialysers were assembled with membranes in the dialysis centre, endotoxin contamination of the blood and dialysate sides of the membrane was usually unavoidable. Some of the circumstantial evidence may, therefore, be related to dialyser contamination during assembly rather than water contamination. Bacteria and endotoxins may enter the blood of the dialysis patient through microscopic defects in the dialyser membrane. Even with intact membranes, small bacteriologic by-products such as peptidoglycans and enzymes may cross the membrane.

Some strains of bacteria can grow even in distilled water. Bacterial growth in dialysate is accelerated not only by nutrients such as glucose in the dialysate but also by the presence of solutes diffusing out of the patient's blood into the dialysate. The levels of bacteria are, therefore, greatly amplified in the dialysate delivery system and may reach very high levels towards the end of the dialysis procedure. This is especially so with recirculating systems. It is, therefore important that the microbiological growth of the water supplied to dialysate delivery systems be monitored at frequent intervals and disinfection procedures established.

Water quality standards for haemodialysis

Inorganic substances

These may be divided into three categories – those non-toxic substances normally included in dialysate, toxic substances described in the dialysis literature and toxic substances regulated by the US EPA in drinking water [40] which have not yet been implicated as toxic in dialysis. The maximum recommended concentrations for these substances are listed in Table II. These recommendations were formulated by us in a study on the risks and hazards associated with haemodialysis systems under contract to the FDA [41]. They have since been adopted by the Association for the Advancement of Medical Instrumentation (AAMI) and the AAMI standard has been approved by the American National Standards Institute, Inc. [42].

TABLE II. Water quality standard for haemodialysis

Substance	Maximum Concentration (Mg/L)	Rationale
Toxic Substances Described in the Dialysis Literature		
Aluminium	0.01	Based on lowest toxic levels reported in the dialysis literature allowing a margin of safety
Chloramines	0.10	
Copper	0.10	
Fluoride	0.20	
Nitrate	2.0	
Sulfate	100	
Zinc	0.10	
Non-toxic Substances Normally Included in Dialysate		
Calcium	2 (0.1mEq/L)	Based on allowable margin for error in final dialysate
Magnesium	4 (0.3mEq/L)	
Potassium	8 (0.2mEq/L)	
Sodium	70 (3.0mEq/L)	
Toxic Substances Regulated by EPA in Drinking Water		
Arsenic	0.005	10% of EPA level or 'no-transfer' level, whichever is higher
Barium	0.01	
Cadmium	0.001	
Chromium	0.014	
Lead	0.005	
Mercury	0.0002	
Selenium	0.09	
Silver	0.005	

The maximum recommended concentrations for substances normally included in dialysate are based on the allowable margin for error in the final dialysate. The recommended levels for substances identified as toxic in dialysis are based on the lowest toxic levels reported in the dialysis literature with an appropriate margin of safety. For substances regulated in drinking water, one should ideally set the level at the 'no transfer' level, the level at which no transfer occurs from dialysate to blood. 'No-transfer' levels have been established for some of these substances based on experiments with the radio-isotopes of these substances [43]. Some of the 'no-transfer' levels are below the detection limit of analytical methods established by the EPA for water [44]. Another approach to setting these limits would be based on the 25-fold exposure of the dialysis patient to water relative to drinking water exposure. A 25-fold reduction in the levels of these substances may be too restrictive in light of the successful application in haemodialysis of reverse osmosis treatment of water. As reverse osmosis devices typically remove more than 90 per cent of the total dissolved solids (inorganic) in the feed water, we have proposed maximum contaminant concentrations that are either one-tenth of the EPA drinking water concentrations or the 'no-transfer' concentrations, whichever is higher. In addition to the above recommendations, AAMI has suggested a maximum level of 0.5mg/L for chlorine because of its potential toxicity.

Organic and radioactive substances

The EPA drinking water standards regulate the concentrations of chlorinated hydrocarbons (pesticides), chlorophenoxys (herbicides) and man-made nuclides emitting α,β, and photon radioactivity. As adequate information is not available at this time to set maximum contaminant levels for these substances in the water used for haemodialysis, no recommendations are possible except to note that activated carbon filters will remove organic material in a range of 60–300 daltons and reverse osmosis and ultrafiltration membranes will remove substances with a molecular weight greater than 200 daltons. A combination of activated carbon and membrane filtration is therefore recommended for removal of organic compounds. Radioactive substances are expected to be removed by mechanisms similar to those for their non-radioactive counterparts.

Microbiology

Based on studies conducted by the Centres for Disease Control, AAMI has recommended a maximum total viable count of 200 CFU/ml for the water used to prepare dialysate. For dialysate, a maximum of 2,000 CFU/ml at the end of the dialysis procedure is recommended.

Special considerations concerning water quality used for haemofiltration

Haemofiltration involves the ultrafiltration of large quantities of plasma water (convective removal of uraemic toxins) and replacement with intravenous quality substitution fluid volumes of 25–35 litres/treatment are required. If substitution fluid is commercially produced, the cost may be prohibitive. On-site preparation of substitution fluid has, therefore, been undertaken by several investigators. Henderson et al [45] first established that substitution fluid could be prepared by membrane filtration of dialysis-quality fluid to ensure removal of micro-organisms and pyrogens. Shaldon et al [46] reduced such a scheme to practice with the design of appropriate equipment. Our group has established a successful large-scale long-term experience with batch on-site preparation of substitution fluid [47].

While the volume of exposure is less in haemofiltration, the fluid is directly infused into the patient's blood without the restrictive diffusion barrier of the dialysis membrane. As most of the chemical contaminants of concern are small molecular weight solutes and as small solute clearances of haemodialysis are of the same magnitude or slightly higher than in haemofiltration, we are of the opinion that haemodialysis product water standards for chemical contaminants may be safely applied to the haemofiltration setting.

Haemofiltration involves direct infusion of substitution fluid into the patient's blood, the fluid must be of intravenous quality, sterile and non-pyrogenic. Water treatment processes that include reverse osmosis and ultrafiltration along with membrane filtration of the final substitution fluid can achieve this quality of fluid. Figures 8, 9 and 10 depict schematically the pre-treatment of water, the

SCHEMATIC OF WATER TREATMENT SYSTEM

Water from Well → Chlorination → Softener → Heater → Carbon Filter → Sediment Filter → Reverse Osmosis → To Proportioning Pump

Figure 8. The water pre-treatment scheme used at the Regional Kidney Disease Program (RKDP), Minneapolis, USA for on-site preparation of substitution fluid for haemofiltration

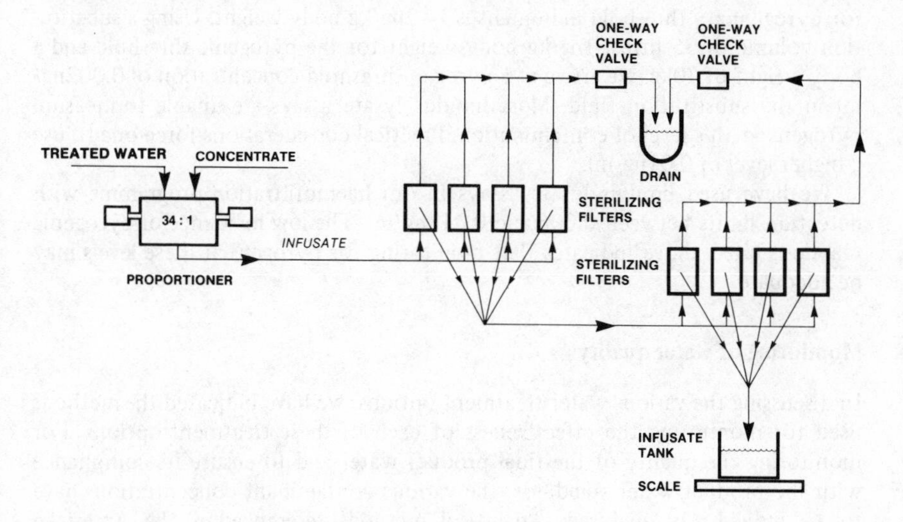

Figure 9. Schematic of central system (RKDP, Minneapolis, USA) for preparation of substitution fluid for haemofiltration

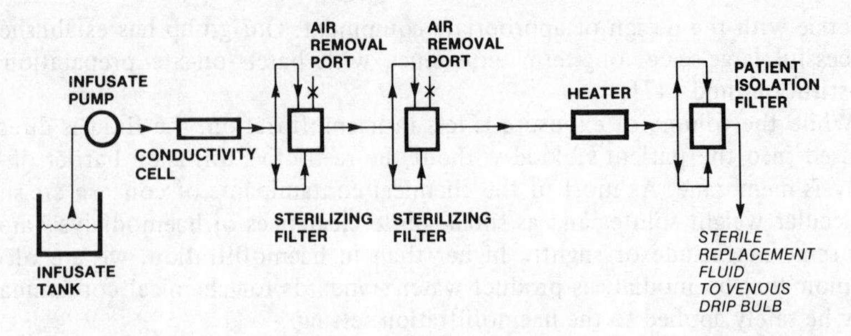

Figure 10. Schematic depiction of the system used for delivery of haemofiltration therapy (Gambro AB), the patient isolation filter being used to prevent machine contamination with blood

central production of substitution fluid, and the delivery of this fluid into the patient's blood through membrane filters as practised in our haemofiltration programme.

We have reported [47] an incidence of 0.2 per cent for pyrogenic reactions with this scheme of treatment and our total experience with on-site preparation is of the order of 6,000 treatments. The pyrogenic reactions observed were associated with low morbidity and easily resolved with hydrocortisone and Benadryl. The fluid invariably tested negative with the limulus lysate assay, even when a pyrogenic episode occurred, indicating very low levels of pyrogen contamination.

In arriving at a standard for pyrogen levels in substitution fluid, one must consider the large infused volume, the patient's pyrogenic threshold and the practical aspects of limulus lysate test sensitivity. The commonly cited threshold for pyrogenicity threshold in humans is 1–2ng/kg body weight. Using a substitution volume of 35 litres, 1ng/kg body weight for the pyrogenic threshold and a body weight of 70kg, we arrive at a pyrogen threshold concentration of 0.002ng/ml in the substitution fluid. Most limulus lysate assays are unable to measure pyrogens at this level of contamination. Practical considerations force one to use a higher level of 0.01ng/ml.

We have used limulus lysate assays in our haemofiltration programme with detection limits between 0.006 and 0.013ng/ml. The low incidence of pyrogenic reactions cited above indicates that monitoring for pyrogens at these levels may be adequate.

Monitoring of water quality

In discussing the various water treatment options, we have indicated the methods used for monitoring the effectiveness of each of these treatment options. For monitoring the quality of the final product water and to ensure its compliance with the product water standards, the various contaminant concentrations have to be individually analysed. Analytical methods referenced in the American Public Health Association's "Standard Methods for the Examination of Water and Waste Water" [48] or the US EPA's "Methods for Chemical Analysis of

Water and Waste" [44] are recommended. For assessing the microbiological quality of the water, conventional microbiological techniques, such as pour or spread plates or the membrane filter technique may be used. The 1ml calibrated loop technique should not be used. Samples must be assayed within 30 minutes of collection or within 24 hours if stored at 5°C. Standard methods agar, blood agar, or tryptic soy agar may be used for culture media and the colonies should be counted after 48 hours of incubation at 37°C. These methods may have to be modified for mycobacteria.

The frequency of monitoring is an important consideration. Unfortunately, no simple guidelines can be established that are universally applicable. The quality of the municipal supply is variable and the performance of water treatment equipment may change with time. The ideal of very frequent monitoring has to be tempered by the realities of analytical costs. A complete water analysis can cost as much as 300 US dollars. The dialysis unit must consult with state and local agencies to determine seasonal variations and the nature of variability of municipal treatment methods before a monitoring schedule can be determined. Further, the nature and configuration of the water treatment system must be considered in determining the schedule. The manufacturer/supplier of the system should advise the dialysis centre regarding schedules of regeneration and replacement of various components. However, monitoring of water quality is the responsibility of the dialysis practitioner.

AAMI has established some guidelines concerning frequency of monitoring. Such guidelines should be viewed with caution in light of the remarks above. AAMI recommends analysis of chemical contaminants every 12 months if reverse osmosis and/or deionisers are used. A frequency of three months and additional monitoring at times of expected high concentrations is recommended for other types of treatment. AAMI also recommends checks of microbiological quality on a monthly basis and when warranted by pyrogenic reactions or bacteraemia. They add that such frequent monitoring may not be suited to the home dialysis setting. AAMI also recommends that, initially, monthly testing should be done and a log maintained of such testing. Once a historical record has been developed, testing can be done less frequently and some discrimination used in deciding which contaminants need to be monitored and how often.

Responsibilities of the manufacturer/supplier of water treatment equipment

In the United States, the FDA has now classified water treatment equipment as medical devices. The Good Manufacturing Practices Act does, therefore, apply to the manufacturer of water treatment equipment. There are additional responsibilities that we feel are incumbent upon the manufacturers.

The manufacturer should obtain a certified laboratory analysis of the feed water from the dialysis centre, local water authorities, or EPA certified laboratories. Based on this analysis, the manufacturer should recommend the type and capacity of the system needed, including any pre-treatment, in order to meet either the product water quality standard or other standards specified by the dialysis centre. The recommendations should consider known or anticipated seasonal variations. A disclosure document should be prepared indicating expected contaminant concentrations at various points of the treatment system.

In recommending certain components for the water treatment system, the manufacturer should provide details such as inlet water temperature, pressures and flow rates, generic nature of materials contacting water and whether they leach, oxidise, or otherwise alter water composition, chemicals that are not compatible with materials of construction, proof of non-toxicity of chemicals such as flocculants required for the water treatment system, etc.

Upon installation of the system, the manufacturer should validate the system for a duration of time adequate to determine the capacity of components such as filters and deionisers. For compounds that require replacement or regeneration, typical life and schedules for replacement/regeneration should be specified by the manufacturer. Methods of system disinfection should be provided along with maintenance guidelines and schedules, trouble-shooting procedures, and lists of spare parts.

Responsibility of the dialysis centre

Beyond installation and initial validation of the system, everything else is the responsibility of the dialysis centre. This includes monitoring of water quality, periodic disinfection and system maintenance. Untoward consequences resulting from inadequate water quality are ultimately the responsibility of the dialysis practitioner. If the aims of haemodialysis and haemofiltration are to detoxify and purify the blood of the patient, the water used to accomplish these aims cannot be of poor quality without jeopardising the patient's well-being. The dialysis practitioner should bear in mind the Italian proverb: "Acqua torbida non lava" (Dirty water does not wash clean).

References

1 Alfrey AC, Mischell M, Burke SR et al. *Trans ASAIO 1972; 18:* 257
2 Platts MM, Goode GC, Hislop JS. *Br Med J 1977; 2:* 657
3 Ward MK, Ellis HA, Feest TG et al. *Lancet 1978; i:* 841
4 Rozas VV, Port FK, Rutt WM. *Arch Intern Med 1978; 138:* 1357
5 Alfrey AC. *Ann Rev Med 1978; 29:* 93
6 Dunea G, Mahurkar SD, Mamdani B, Smith EC. *Ann Intern Med 1978; 88:* 502
7 Berkseth RO, Anderson DC, Mahowald MW, Shapiro FL. *Kidney Int 1978; 14* 670
8 Elliot HL, Dryburgh F, Fell GS et al. *Br Med J 1978; 1:* 1101
9 Alfrey A. *Proc 10th Annual Contr Conf 1977:* 37
10 Flendrig JA, Kruis H, Das HA. *Lancet 1976; i:* 1235
11 *Mortality and Morbidity Weekly Report 1980:* 134
12 Orringer EP, Matter WD. *N Engl J Med 1976; 294:* 1416
13 Freeman RM, Lawton RL, Chamberlain MA. *N Engl J Med 1967; 276:* 1113
14 Evans DB, Slapak M. *Br Med J 1975; 3:* 748
15 Nickey WA, Chinitz VL, Kim KE et al. *JAMA 1970; 214:* 915
16 Stamm JM, Engelhard WE, Parsons JE. *Appl Microbiol 1969; 18:* 376
17 Wilkins JR, Reiches NA, Kruse CW. *Am J Epidemiol 1979; 110:* 420
18 Botella J, Traver JA, Sanz-Guajardo D et al. *Proc EDTA 1977; 14:* 192
19 Eaton JW, Kolpin CF, Swofford HS et al. *Science 1973; 18:* 463

20 Kjellstrand CM, Eaton JW, Yawata Y et al. *Nephron 1974; 13:* 427
21 Neilan BA, Ehlers S, Kolpin CF, Eaton JW. *Clin Nephrol 1978; 10:* 105
22 Yawata Y, Kjellstrand C, Buselmeir T et al. *Trans ASAIO 1972; 18:* 301
23 Yawata Y, Howe R, Jacob HS. *Ann Intern Med 1973; 79:* 362
24 Favero MS, Peterson NJ, Boyer KM et al. *Trans ASAIO 1974; 20:* 175
25 Favero MS, Petersen NJ, Carson LA et al. *Health Lab Sci 1975; 12:* 321
26 Spatz DD. *Methods of Water Purification. Presentation. AANNT Joint Conference Seattle, Washington. April 1972*
27 Otten G, Brown GD. *Am Laboratory 1973:* 49
28 Favero MS, Carson LA, Bond WW, Petersen NJ. *Appl Microbiol 1974; 28:* 822
29 Petersen NJ, Boyer KM, Carson LA, Favero MS. *Dial and Transplant 1978; 7:* 52
30 Johnson WJ, Taves DR. *Kidney Int 1974; 5:* 451
31 Ivanovich P, Manzler A, Drake R. *Trans ASAIO 1969; 15:* 316
32 Manzler AD, Schreiner AW. *Ann Intern Med 1970; 73:* 409
33 Matter BJ, Pederson J, Psimenos G, Lindeman RD. *Trans ASAIO 1969; 15:* 309
34 Schwarzbeck A, Wagner L, Squarr HU, Strauch M. *Clin Nephrol 1977; 7:* 125
35 Schwarzbeck A, Wagner L, Squarr HU, Strauch M. *Dial and Transplant 1978; 7:* 740
36 Kirkwood RG, Dunn S, Thomasson L, Simenhoff ML. *Trans ASAIO 1981; 27:* 168
37 Gazenfield GE, Eliahou HE. *Israel J Med Sci 1969; 5:* 1032
38 Jans H, Bretlaie P, Nielsen B. *Nephron 1978; 20:* 10
39 Raij L, Shapiro FL, Michael AF. *Kidney Int 1973; 4:* 57
40 Environmental Protection Agency, Office of Water Supply. *National Interim Primary Drinking Water Regulations.* Washington, DC: US Government Printing Office. 1978
41 Keshaviah P, Luehmann D, Shapiro F, Comty C. *Investigation of the Risks and Hazards Associated with Hemodialysis Systems. Technical Report, Contract No 223-78-5046.* Silver Spring, Md: US Department of Health Service, Food and Drug Administration Bureau of Medical Devices. 1980
42 Renal Disease and Detoxification Committee. *American National Standard for Hemodialysis Systems.* Arlington, VA: Association for the Advancement of Medical Instrumentation. 1982
43 NIH Report. *Trace Metal Protein Binding.* Gulf South Research Institute. 1979
44 Trace Metal Protein Binding. *Methods for Chemical Analysis of Water and Wastes.* US Environmental Protection Agency, Research and Development. 1979: 4–24
45 Henderson L, Beans E. *Kidney Int 1978; 14:* 522
46 Ramperez D, Beau MG, Deschodt G et al. *Proc EDTA 1981; 18:* 293
47 Luehmann D, Hirsch D, Ebben J et al. *Trans ASAIO 1984:* in press
48 *Standard Methods for the Examination of Water and Waste Water, 15th Edition.* Washington, DC: APHA, AWWA, WPCF. 1980

Open Discussion

TARABA (Chairman) Thank you Dr Keshaviah for this excellent presentation. I think this showed that this subject is very large and it was impossible to go into all the details which would be very important. In your lecture you did not mention, I think, about iron, do you think it is not important to measure the iron content of the water?

KESHAVIAH I think it is very important because it can foul equipment downstream, such as deionisers, resins and the membranes. I showed a diagram of our own water treatment system and that showed iron filters. Very often if you do not remove iron you will see rust stains in other parts of the downstream equipment.

BONSDORFF (Helsinki) What kind of pipe material should be used to transport the treated water to the dialysis point, glass, steel or PVC?

KESHAVIAH Till recently, especially in the USA, PVC was not used for purified water. It was only used for sewage systems, but the codes have now been changed so you can use PVC for transporting pure water. Very often lined glass is used when you use distallation because the high purity of distilled water can attack other substances such as copper and lead. PVC is quite adequate and copper alright as long as you are careful that you do not have acidic water, but brass should be avoided.

SALVADORI (Florence) When you speak about nitrates having a maximum concentration of 2mg/L do you mean as N or as NO_3^-?

KESAVIAH Yes, as nitrogen.

PECCHINI (Cremona, Italy) Is reverse osmosis able to remove nitrosodimethylamine?

KESAVIAH Yes, you can use reverse osmosis. The DMSO is associated only with deionisers and certain municipal supplies. If you have a system with build up of DMSO in the effluent of the deionisers it is advisable to use reverse osmosis.

FOURNIER (Amiens) Do you advocate 'finishing' the treatment of the water after reverse osmosis by deionisation? If so, do you think it is valid with regard to the increased risk of bacterial infection after the reverse osmosis?

KESAVIAH That is a good point. I do not recommend this. I think it may be necessary in some instances such as when there is very high levels of fluoride. If the water supply fluoride is very high reverse osmosis may not be adequate. Similarly if aluminium is very high reverse osmosis may not be adequate. If you monitor the effluent and find high levels of fluoride and aluminium then you are forced to use a deioniser in that situation.

For economy you use it after the RO membrane which introduces the problem of bacteria so you have to follow that up with some form of membrane filtration.

FOURNIER (Amiens) As regards the limit for water aluminium do you think that 10μg/L is a safe limit? According to the protein binding of aluminium you may expect a positive balance until your patients have plasma values of 50μg/L.

KESAVIAH There is no such thing as a safe level really for many of these toxic heavy metals. The reason why 10μg/ml has been specified is that you have to be practical because the detection limit of even the most sophisticated methods is between one and three parts per billion. You can say that it should be less than three, but if you cannot measure it you might as well dream up the numbers. For practical reasons it is best at 10, and if you look at the literature there has been no toxicity reported when water levels have been below

50 parts per billion. There is no such thing as a safe level and the only safe level is the no transfer level.

BRANCACCIO (Milan) If you had to manage a new dialysis centre today what water treatment would you suggest?

KESAVIAH As I stated water supplies are very variable so I wish I could give you a simple answer but I will give you a simple guideline at least. If you were to combine reverse osmosis with activated carbon, the reverse osmosis takes care of most of the contaminants and also the organics greater than 200–300 daltons. The activated carbon takes care of the organics smaller than 200 daltons so that sort of design is very good and usually a very safe system. If you have got high fluoride or aluminium you are forced to add other systems and if you have got high calcium and magnesium you have to add pre-treatment softening. I cannot give you a simple answer but reverse osmosis is a very effective treatment.

Proc EDTA–ERA (1984) Vol 21

ACUTE PULMONARY HYPERTENSION, LEUCOPENIA AND HYPOXIA IN EARLY HAEMODIALYSIS

J F Walker, R M Lindsay, W J Sibbald, A L Linton

Victoria Hospital Corporation, London, Ontario and University of Western Ontario, London, Ontario, Canada

Summary

A sheep model is described which produces acute pulmonary hypertension, leucopenia and hypoxia after blood, previously placed in contact with a Cuprophan hollow fibre artificial kidney, re-enters the circulation. Relationships between these manifestations (acute pulmonary hypertension, leucopenia and hypoxia) were examined in normal leucopenic and Indomethacin pre-treated sheep. The degree of pulmonary vascular response, and severity of leucopenia and hypoxia were all directly interrelated and were dependent upon the volume of blood injected. The induction of leucopenia did not affect the pulmonary hypertension or hypoxia. Pre-treating the animals with the cyclo-oxygenase inhibitor, Indomethacin, abolished both the pulmonary hypertension and the hypoxia without any effect on the development of neutropenia. These results suggest that leucocytes do not play a role in the haemodynamic response nor in the hypoxia; activation of the cyclo-oxygenase system is necessary for the development of acute pulmonary hypertension which causes hypoxia subsequent to alterations in ventilation perfusion relationships.

Introduction

Blood-foreign surface interactions, as occur with the use of extracorporeal devices such as the artificial kidney, may result in clinically significant problems. Activation of the alternate pathway of complement, with the generation of C3a and C5a anaphylotoxins results in pulmonary sequestration of leucocytes which has been held responsible for the peripheral leucopenia and arterial hypoxia that frequently occur during the first 30 minutes of dialysis [1,2]. Cuprophan (or dialyser) hypersensitivity is an infrequent but serious consequence that arises following blood-dialyser contact [3,4]. Characteristically, acute chest and back pain, dyspnoea, diaphoresis and occasionally flushing and oedema of the skin occur within the first 10 minutes of commencing haemodialysis. This has resulted in cardiopulmonary arrest and death [3]. The majority of reports describe the phenomenon in patients dialysed for the first time by a new (not reused) artificial kidney containing a Cuprophan membrane (Enka, West Germany) in a hollow fibre configuration.

We have recently reported the haemodynamic manifestations of blood-dialyser interactions in an animal model [5], some of which showed striking similarities to 'Cuprophan hypersensitivity'. Acute pulmonary hypertension occurred within the first few minutes of haemodialysis and was associated with a fall in cardiac output, together with electrocardiographic evidence of myocardial ischaemia and arrhythmias [5]. At the same time, peripheral leucopenia and hypoxia were noted; the changes in pulmonary artery pressure and peripheral leucocyte count occurred simultaneously but preceded the maximal decrease in partial pressure of oxygen. In this model, the pulmonary vascular response depended both on the surface area of the artificial kidney as well as its chemical structure and configuration. Membranes containing regenerated Cellulose or Cuprophan had a significantly greater effect than with Cellulose Acetate, or Polyacrylonitrile [5]. In clinical studies dialysis with membrane containing Cellulose Acetate or Polyacrylonitrile results in less complement activation, leucopenia and arterial hypoxia than with the use of Cuprophan surfaces [6,7]. This raises the possibility that the cardiopulmonary manifestations of 'Cuprophan hypersensitivity' may reflect a more severe expression of the phenomena of leucopenia and hypoxia and that changes in pulmonary vascular tone are common. Preliminary results from clinical studies support this concept and have shown that increases in pulmonary vascular resistance index occur [8].

This paper examines the relationship between these three manifestations of blood foreign surface contact.

Methods

Specific details of the sheep model have been described elsewhere [5]. In summary, Suffolk sheep of either sex, six to 12 months old, were prepared for haemodynamic monitoring by inserting, via a jugular vein, a triple lumen thermodilution Swan-Ganz catheter (Edwards Lab, Santa Ana, Calif) for subsequent measurement of the pulmonary artery pressure. A carotid artery was cannulated with a polyvinyl catheter (O D 3mm) for blood sampling. A Quinton-Scribner SAF-T shunt (Extracorporeal Medical Specialties, King of Prussia, Pa) was inserted into the femoral artery and vein. The Scribner shunt was connected to a dialyser by standard haemodialysis blood lines. The dialyser could be bypassed by directing flow through a bypass line, or by changing arterial clamps introduced into the circuit. Blood was pumped through a circuit using a Sarns semi-occlusive pump (model 550, Sarns Inc, Ann Arbour, Mich). The circuit was primed using 0.9% saline and Heparin 3,000 IU (Hepalean, Organon, Toronto, Canada) was given into the arterial line at commencement of blood flow. Blood was allowed to fill the dialysis circuit containing a Cuprophan hollow fibre artificial kidney (Erika 200 HPF, Erika Corp, Rockleigh, NJ) which was then clamped off. After 10 minutes static contact with this dialyser, 50ml of blood was withdrawn. The mean pulmonary artery pressure (PAP) was monitored continuously and the neutrophil count, and partial pressure of oxygen were measured prior to and every 20 seconds following reinjection of increasing volume (1,2,5,10 and 15ml) of this static contacted blood. The experiments were performed in normal, leucopenic or Indomethacin pretreated sheep (n=5 in each group). Leucopenia was induced with intravenous Mustine Hydrochloride (Merck, Sharpe and Dohme, West Point, Pa) 0.4mg/kg on days one and four. The experiments were performed on days six and seven after confirmation that the total white blood cell count was less than 1 x 10^9/L. Indomethacin sodium trihydrate (2.0mg/kg) was reconstituted in sterile water and given intravenously over 20 minutes.

The basic data are expressed as mean ± SEM using the change from baseline for each parameter (Δ PAP, ΔPaO_2 and per cent fall in neutrophil count). The analysis of variance was used to examine the effects of volume and treatment (leucopenia, Indomethacin) on ΔMean PAP, ΔPaO_2 and per cent fall in neutrophil count.

Results

Figure 1 shows a typical response that occurs following reinjection of a volume (10ml) of blood after static contact with Cuprophan. The pulmonary artery pressure increased from 13 to 43mmHg and coincided with a fall in neutrophil count (5.9 x 10^9/L) within 20 seconds of reinjection of static contacted blood. Maximal hypoxia occurred at 60 seconds (100 to 66mmHg PaO_2). All these events returned to baseline by 300 seconds.

Figure 2 shows the change in mean pulmonary artery pressure (ΔMean PAP), per cent fall in neutrophil count, and the change in partial pressure of oxygen (ΔPaO_2) following reinjection of increasing volumes (1,2,5,10 and 15ml) of blood after 10 minutes static contact with a Cuprophan hollow fibre kidney in normal, leucopenia, or Indomethacin pretreated sheep.

In normal sheep, 1 and 2ml of blood reinjected after contact with Cuprophan had little effect on the pulmonary artery pressure; however, reinjection of 5ml after contact caused a 17mmHg increase in mean PAP. The injection of 10 and 15ml after contact caused an even greater elevation in mean PAP (28 and 32mmHg respectively). The changes in arterial oxygenation (ΔPaO_2) and percentage fall in neutrophil count mirrored the degree of pulmonary hypertension with each volume reinjected in normal animals after contact with Cuprophan. The analysis of variance showed a significant effect of the volumes reinjected on the Δ mean PAP ($p<0.001$), degree of hypoxia ($p<0.0005$), and per cent fall in neutrophils ($p<0.05$).

In leucopenic animals, a similar degree of pulmonary hypertension and arterial hypoxia occurred as demonstrated in normals. One ml reinjected caused little effect on pulmonary artery pressure or on arterial oxygenation. Five ml reinjected caused an increment of 34mmHg in mean PAP and a fall in PaO_2 of 32mmHg. Analysis of variance showed a significant effect of the volumes reinjected on the Δmean PAP ($p<0.001$) and on the degree of hypoxia ($p<0.0005$).

The use of Indomethacin blocked both the pulmonary hypertensive and

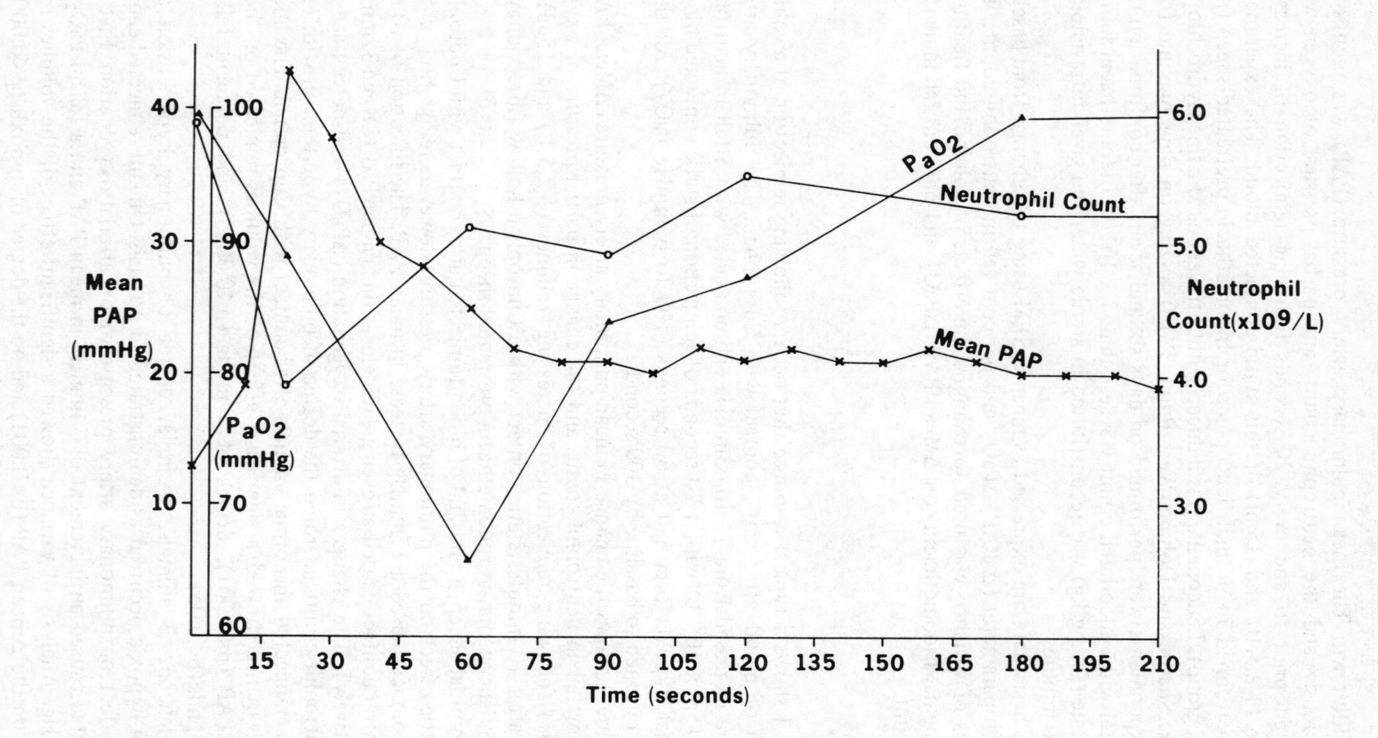

Figure 1. Mean PAP (mmHg), PaO_2 (mmHg), and neutrophil count (x 10^9/L) following reinjection of 10ml of blood after 10 minutes static contact with a Cuprophan hollow fibre dialyser

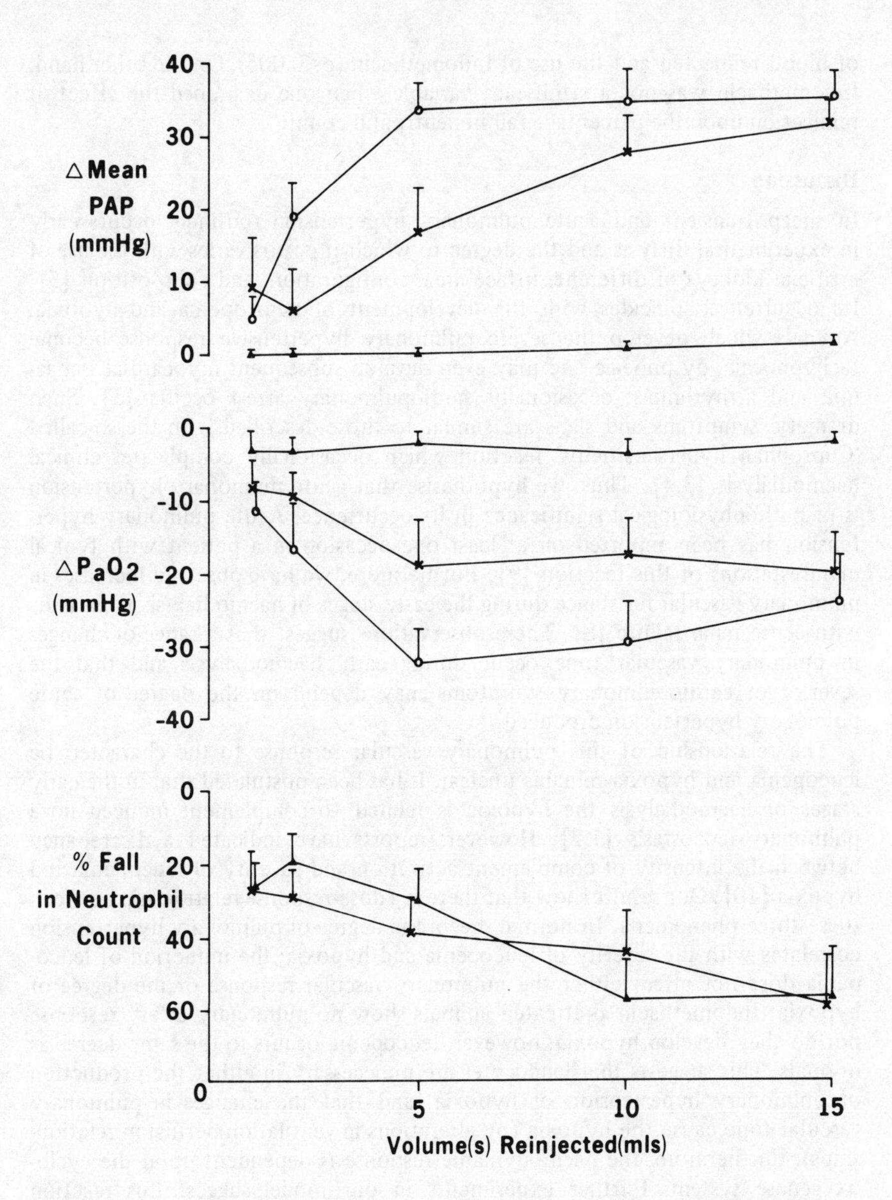

Figure 2. Changes in mean PAP (mmHg), PaO_2 (mmHg), and per cent fall in neutrophil count in normal (x–x), leucopenic (○–○) and Indomethacin pretreated sheet (△–△). Results are expressed as mean ± SEM

hypoxic responses with all blood volumes reinjected while the percentage fall in neutrophil count occurred to a similar degree as demonstrated in non-Indomethacin pretreated sheep. Examination of the pulmonary and hypoxic responses using analysis of variance indicated that interactions existed between the volumes of blood reinjected and the use of Indomethacin ($p<0.005$). On the other hand, Indomethacin was not a significant variable when one examined the effect of reinjection upon the percentage fall in neutrophil count.

Discussion

In sheep transient and acute pulmonary hypertension routinely occurs early in experimental dialysis and the degree to which it occurs varies with the use of artificial kidneys of different surface area, configuration, and composition [5]. Its occurrence coincides with the development of neutropenia and hypoxia. Animals which develop the severe pulmonary hypertensive response become tachypnoeaic, dyspnoeaic and may even develop subsequent myocardial ischaemia and arrhythmias; occasionally cardiopulmonary arrest occurs [5]. Such dramatic symptoms and signs are similar to those described with the so-called 'Cuprophan hypersensitivity' reaction which occasionally complicates clinical haemodialysis [3,4]. Thus, we hypothesise that acute pulmonary hypertension is of pathophysiological significance in its occurrence. Acute pulmonary hypertension has been reported on at least one occasion in a patient with typical manifestations of this reaction [9]. Furthermore, we have observed increases in pulmonary vascular resistance during the early stages of haemodialysis in patients with acute renal failure [8]. These observations suggest that a range of changes in pulmonary vascular tone occur during early haemodialysis and that the severity of cardiopulmonary symptoms may depend on the degree of acute pulmonary hypertension produced.

The relationship of this pulmonary vascular response to the characteristic leucopenia and hypoxia remains unclear. It has been postulated that in the early stages of haemodialysis the hypoxia is related to complement induced intra pulmonary leucostasis [1,2]. However, reports have indicated a discrepancy between the intensity of complement activation and severity of leucopenia and hypoxia [10]. Our results show that there is a dose-response relationship between these three phenomena. In normal sheep, the degree of pulmonary hypertension correlates with the severity of leucopenia and hypoxia; the induction of leucopenia does not affect either the pulmonary vascular response or the degree of hypoxia; Indomethacin pretreated animals show no pulmonary artery response nor do they develop hypoxia; however, leucopenia occurs to the same degree as normals. This suggests that leucocytes are unnecessary in either the production of pulmonary hypertension or hypoxia, and that the changes in pulmonary vascular tone cause the hypoxia (by alterations in ventilation perfusion relationships). Furthermore, the haemodynamic response is dependent upon the cyclo-oxygenase system. Further experiments in our model suggest this reaction requires the presence of a plasma factor which is heat labile, calcium and complement dependent. Further work is necessary to confirm these preliminary observations.

Acknowledgments

This study was supported by a grant from Erika Inc, Rockleigh, New Jersey. The authors appreciate the help of Mrs E Wood for secretarial assistance, Ms R

Wainwright and Mrs A Neal for invaluable technical help, and the Department of Visual Medical Records for the preparation of figures.

References

1 Craddock PR, Fehr J, Brigham KL et al. *N Engl J Med 1977; 296:* 769
2 Craddock PR, Fehr J, Dalmasso AP et al. *J Clin Invest 1977; 59:* 879
3 Popli S, Ing TS, Kheirbek AO et al. *Int J Artif Organs 1982; 6:* 312
4 Nicholls AJ, Platts MM. *Br Med J 1982; 285:* 1607
5 Walker JF, Lindsay RM, Peters SD et al. *ASAIO J 1983; 6:* 123
6 Jacob AI, Gavellas G, Zarco R et al. *Kidney Int 1980; 18:* 505
7 Ivanovich P, Chenoweth DE, Schmidt R et al. *Kidney Int 1983; 24:* 758
8 Walker JF, Lindsay RM, Sibbald WJ et al. *Trans ASAIO 1984.* In press
9 Agar JW, Hull JD, Kaplan M et al. *Ann Intern Med 1979; 90:* 792
10 Aljama P, Bird PAE, Ward MK et al. *Proc EDTA 1978; 15:* 144

Open Discussion

RITZ (Heidelberg) Could you tell us how this corresponds to the situation in the human? Can you give us some information on blood pressure and left ventricular function. The reason I m asking this is because there was a presentation at the International Congress of Nephrology* where blood pressures with this model of up to 200mmHg or more have been observed and this has certainly not been seen in dialysed patients.

WALKER If you look in the literature at the few described cases of hypersensitivity reaction you will find that both hypotension, normotension and hypertension have been described. The reasons for this are not known. In the animals we see the same thing but most commonly we see profound hypertension. I am not sure of the reason for this but it is quite possible that there may be some effect of the cyclo-oxygenase system or of the alternate pathway of complement to the anaphylactotoxin C_3a or C_5a on systemic vasculature. The other alternative possibility is that there is a compensatory mechanism with release of catecholamines which may cause the systemic hypertension. I have no data on left ventricular function in sheep or in humans. We believe that it is not a primary left ventricular problem but a primary right ventricular problem, in that you have physical obstruction to blood flow through the lungs which causes acute right ventricular failure and therefore a decreased pre-load on the left ventricle. There is other animal models in the same situation which correlate with this particular problem. In the human I can tell you that we have done invasive haemodynamic monitoring in acute renal failure patients. We have found approximately 30 per cent of them develop acute pulmonary hypertension in the first 15 minutes. In our chronic stable patients we did find a significant number developed a dramatic fall in right ventricular ejection fraction using the radionucleotide equilibration technique without any significant change in left ventricular ejection fraction. This was in asymptomatic patients. The

*Cheung A, Le Winter M, Chenoweth D et al. *International Congress of Nephrology.* Los Angeles. 1984

primary effect is on the right ventricle rather than the left ventricle, and the left ventricle suffers only because the right ventricle has failed.

UNKNOWN I ask why you see hypoxaemia occurring with slight complement activation which occurs with polysulphone membranes?

WALKER The problem is that the complement activation or I should say hypoxaemia during haemodialysis, is multifactorial. We believe that the initial hypoxaemia that occurs during haemodialysis is related to complement activation and is subsequent to changes in pulmonary vascular tone which causes acute ventilation perfusion mismatches which results in the hypoxia. There are many causes of hypoxaemia during dialysis such as microemboli or whether the dialysate buffer is acetate or bicarbonate. They tend to lose CO_2 across the dialyser as well and so there are many different mechanisms. I am not sure if that answers your question or not.

Proc EDTA–ERA (1984) Vol 21

EFFECT OF EOSINOPHILIA ON THE HETEROGENEITY OF THE ANTICOAGULANT RESPONSE TO HEPARIN IN HAEMODIALYSIS PATIENTS

A Santoro, **P Zucchelli**, E Trombetti, E Degli Esposti, A Sturani, A Zuccalà, C Chiarini, M Casadei-Maldini, A Mazzuca

Ospedale M Malpighi, Bologna, Italy

Summary

The anticoagulant response to heparin was determined, during haemodialysis, in a group of seven patients with eosinophilia and in a control group. The heparin half-life was similar in the two groups, but the heparin effect index was lower in patients with eosinophilia. The dose-response curve showed a reduced sensitivity to heparin in patients with eosinophilia. In patients with eosinophilia a significant reduction in eosinophil count was observed during cuprophan dialysis, but not during polyacrylonitrile dialysis. The hyposensitivity to heparin might be related to eosinophil degranulation, during cuprophan dialysis, with release of a major basic protein that neutralises heparin.

Introduction

It has been shown that the heparin anticoagulant effect, in both normal subjects and in uraemic patients, does not only depend on the heparin concentration but also on other factors: concentration of the antithrombin III, platelet factor 4 and α_1-acid glycoprotein [1,2]. During haemodialysis an additional factor could be the aggregation and degranulation of granulocytes in the early phase of dialysis [3]. Eosinophils, which are frequently increased in dialysis patients [4], contain granule products that can neutralise the heparin anticoagulant activity [5]. The present study was undertaken to determine whether the presence of eosinophilia in dialysis patients was accompanied by modifications in the anticoagulant response to heparin during conventional dialysis.

Patients and methods

Two groups of haemodialysis patients entered the study. Group I consisted of seven patients (3 females, 4 males; mean age 53.1±12.9 years) with an eosinophil count greater than 500/mm^3 and group II of seven patients (1 female, 6 males; mean age 53.8±8.8 years) with a normal peripheral eosinophil count.

The time on dialysis was 77.7±46 months in group I and 72±21 in group II (p=NS).

The study was carried out during a regular dialysis session. Cuprophan plate dialysers were used and except for heparin no drugs were administered during the study. Blood samples were drawn immediately prior to initiation of dialysis to determine haematocrit, albumin, globulin fractions, differential leucocytes count, platelet count, fibrinogen, activated partial thromboplastin time (APTT) and antithrombin III (AT-III). A single loading dose (70U/kg) of acqueous sodium heparin (Liquemin, Roche) was injected into the arterial tubing before connecting the patient to the dialyser. Blood samples were then drawn from the dialyser arterial line at 5,10,15,30,60 and 120 minutes after the start of dialysis for APTT, heparin concentration and differential leucocytes count assay. In five patients in group I the same protocol study was repeated using a polyacrylonitrile flat plate dialyser. Heparin and antithrombin III were assayed by COA-TEST (Ortho Diagnostics), white cell counts by Coulter Counter (H 6000, Technicon). The heparin half-life was determined by conventional methods of interpolation on curve, expressing the relationship between the logarithm of heparin concentration and the time after drug injection. The relationship between the heparin concentration and the corresponding APTT was calculated by subtracting the pre-treatment APTT from the APTT obtained on samples drawn after heparin injection and dividing the result (ΔAPTT) by the heparin concentration in the samples. This was called heparin-effect index (HEI).

Results

The main clinical and biochemical pre-dialysis parameters in the two groups of patients are summarised in Table I. The mean leucocyte and eosinophil count of

TABLE I. Biochemical and clinical parameters

Parameter		Group I mean ± SD	Group II mean ± SD	p
Haematocrit	(%)	29.01±5.7	31.12±8.2	0.58
Albumin	(gm/dl)	3.8±0.33	4.18±0.32	0.113
α_1-globulin	(gm/dl)	0.18±0.06	0.17±0.01	0.72
α_2-globulin	(gm/dl)	0.63±0.13	0.54±0.08	0.16
β-globulin	(gm/dl)	0.66±0.05	0.66±0.07	0.96
γ-globulin	(gm/dl)	1.50±0.56	1.20±0.21	0.08
Leucocytes	(cell/mm^3)	9214±2764	6028±1372	*0.017*
Eosinophils	(cell/mm^3)	2205±1441	100±55	*0.0021*
Platelets	(cell/mm^3)	185±54x10^3	176±32x10^3	0.705
Baseline APTT	(sec)	31.7±3.8	28.5±3.9	0.252
Fibrinogen	(mg/dl)	236±55	267±50	0.293
Antithrombin III	(%)	87±23	88±18	0.90
Body weight	(kg)	58.7±10.5	64.6±14.4	0.10

patients in group I was significantly greater (p<0.05) than the corresponding value of group II; no other statistically significant differences were found in the other parameters.

The APTT response to the intravenous bolus dose of heparin, during haemodialysis, is shown in Figure 1. The pre-dialysis APTT was 31.7±3.8 seconds in group I and 28.5±3.9 seconds in group II (p=NS). During haemodialysis the APTT values of group I were significantly lower than the corresponding value of group II. The differences were statistically significant (p<0.05) at 5,15,30 and 60 minutes from the start of dialysis. The heparin half-life was 83.8±46.3 minutes in group I and 70.5±35.6 minutes in group II (p=NS).

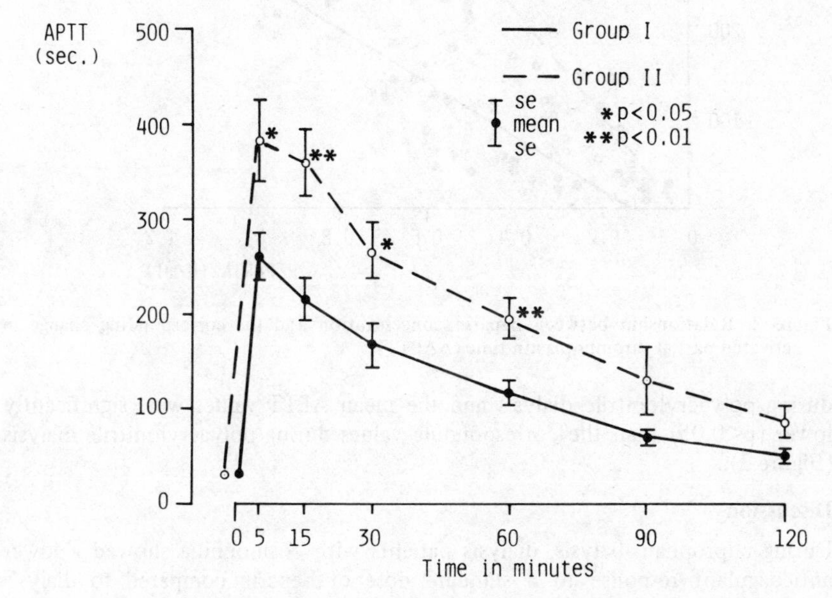

Figure 1. APTT changes after heparin injection during haemodialysis

The relationship between APTT changes and plasma heparin concentration are shown in Figure 2. A linear dose response curve, within a heparin concentration range of 0.1 to 1.25U/ml, was found in the two groups of patients, but the slope of regression line was significantly lower in patients in group I than those in group II. The mean HEI, during the two hours of the study, was significantly lower in group I (206±117sec/U/ml) compared to group II (523±244 sec/U/ml, p<0.001).

In the patients in group I, during cuprophan dialysis but not during polyacrylonitrile dialysis, a significant reduction of circulating neutrophils and eosinophils occurred during the first 15 minutes of haemodialysis. Then, leucocytes and neutrophils returned to pre-dialysis values. Circulating eosinophils also increased but remained at a lower level compared to pre-dialysis values. During cuprophan dialysis, the anticoagulant heparin activity was lesser than

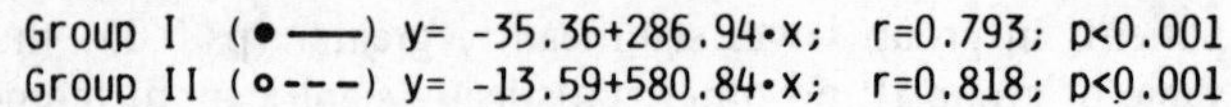

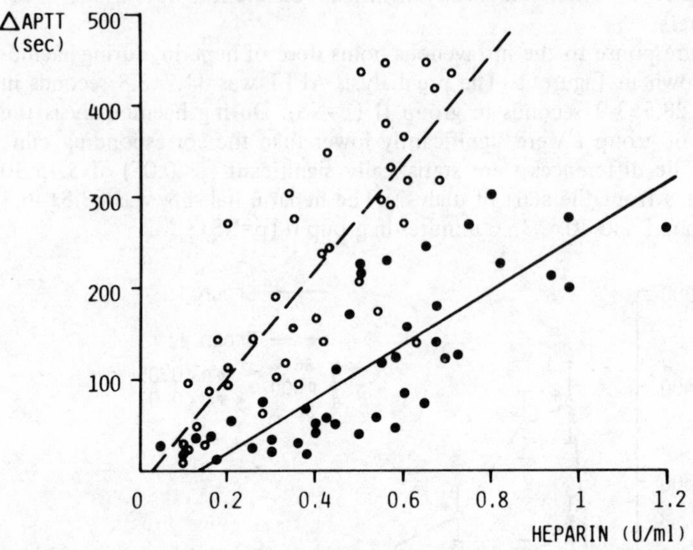

Figure 2. Relationship between heparin concentration and the corresponding change in activated partial thromboplastin time (ΔAPTT)

during polyacrylonitrile dialysis and the mean APTT values were significantly lower ($p<0.05$) than the corresponding values during polyacrylonitrile dialysis (Figure 3).

Discussion

During cuprophan dialysis, dialysis patients with eosinophilia showed a lower anticoagulant response to a standard dose of heparin compared to dialysis patients without eosinophilia.

Prior to dialysis, haematocrit, platelets and AT-III levels were similar in patients with and without eosinophilia. This result indicates that eosinophilia per se, rather than other biochemical factors, is responsible for the different anticoagulant action of heparin. The hyposensitivity to heparin appears to be a membrane-related phenomenon: a normal anticoagulant response to heparin was obtained after changing the cuprophan-containing dialysers to polyacrylonitrile dialysers. Both neutrophils and eosinophils were also unaffected by polyacrylonitrile membrane. Hallgren et al [6] reported that dialysers containing cuprophan may induce both aggregation and degranulation of neutrophils and eosinophils. The result of eosinophil degranulation is the release of specific granule constituents such as eosinophil cationic protein (ECP) and major basic protein (MBP) which possesses heparin neutralising activity [5,7].

Therefore it seems likely that haemodialysis using cuprophan membrane results in eosinophil degranulation with a release of MBP neutralising heparin in patients with eosinophilia.

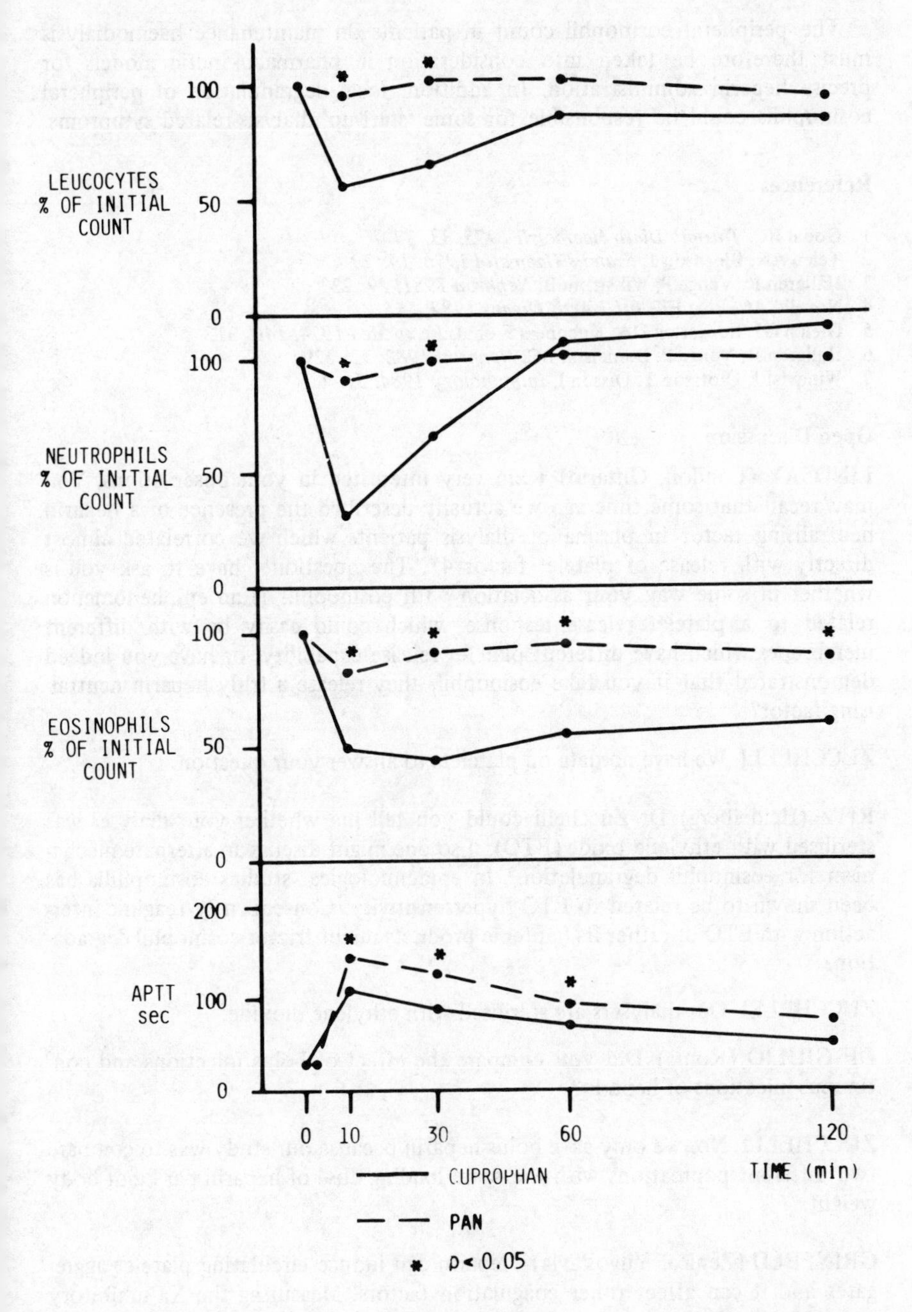

Figure 3. Changes in leucocytes, neutrophils, eosinophils count and activated thromboplastin time (APTT) during cuprophan and polyacrylonitrile dialysis in patients in group I

The peripheral eosinophil count in patients on maintenance haemodialysis must therefore be taken into consideration in pharmacokinetic models for precise heparin administration. In addition, local degranulation of peripheral eosinophils could be responsible for some 'start-up' dialysis-related symptoms.

References

1 Godal HC. *Thromb Diath Haemorrh 1975; 33:* 77
2 Teien AN, Biornson J. *Scand J Haematol 1976; 17:* 29
3 Hallgren R, Venge P, Wikstrom B. *Nephron 1981; 29:* 233
4 Novello AC, Port FK. *Int J Artif Organs 1982; 5:* 5
5 Gleich GJ, Loegering DA, Kueppers F et al. *J Exp Med 1974; 140:* 313
6 Hallgren R, Venge P, Danielson BG. *Nephron 1982; 32:* 329
7 Winqvist I, Olofsson T, Olsson I. *Immunology 1984; 51:* 1

Open Discussion

LINDSAY (London, Ontario) I am very interested in your observations. You may recall that some time ago we actually described the presence of a heparin neutralising factor in plasma of dialysis patients which we correlated almost directly with release of platelet factor 4*. The question I have to ask you is whether in some way your association with eosinophils is an epiphenomenon related to a platelet release response, which could easily be with different membranes which have different platelet release capability, or have you indeed demonstrated that if you take eosinophils they release a truly heparin neutralising factor?

ZUCCHELLI We have no data on platelets to answer your question.

RITZ (Heidelberg) Dr Zucchelli could you tell me whether your dialyser was sterilised with ethylene oxide (ETO), if so one might discuss an alternate mechanism for eosinophil degranulation? In epidemiological studies eosinophilia has been shown to be related to ETO hypersensitivity. Consequently reaginic interaction with ETO or rather its haptenic products might trigger eosinophil degradation.

ZUCCHELLI Our dialysers are sterilised with ethylene dioxide.

DE GIULIO (Rome) Did you compare the effect of bolus injections and continuous injections of heparin?

ZUCCHELLI No, we only gave bolus heparin because our study was to compare two different populations with the same loading dose of heparin per kg of body weight.

GRINFELD (Zenica, Yugoslavia) Heparin can induce circulating platelet aggregates and it can affect other coagulation factors. Measuring the Xa inhibitory

*Lindsay RM, Rourke J, Reid B et al. *Trans ASAIO 1976; 22:* 292

factor gives information on plasma free heparin and we can conclude that eosinophilia diminishes plasma heparin.

ZUCCHELLI I agree with you, the aim of our study was not to study the effect of heparin on platelets but to compare two different groups of patients, one with eosinophilia and one without.

Proc EDTA–ERA (1984) Vol 21

DIRECT INTERACTION BETWEEN POLYMORPHONUCLEAR NEUTROPHILS AND CUPROPHAN MEMBRANES IN A PLASMA-FREE MODEL OF DIALYSIS

C Tetta, A Jeantet, G Camussi, A Thea, L Gremo, P F Martini, R Ragni, A Vercellone

Ospedale Maggiore S.G. Battista-Molinette, Torino, Italy

Summary

Plasma-free, purified, normal, human polymorphonuclear neutrophils (PMN) were recirculated for 60 minutes in an experimental model of dialysis using cuprophan membranes and acetate or bicarbonate dialysate. At different time intervals, the intracellular contents of PMN-derived cationic proteins (NCP), the release of lysosyme, β-glucuronidase and PAF as well as the occurrence of PMN and platelet aggregating activities in the supernatants were evaluated. The formation of PMN aggregates, the depletion of intracellular contents of NCP together with the release of lysosomal constituents occurred early (5–10 min) in the course of recirculation. These events were concomitant with the occurrence of PMN aggregating activity in the supernatants due to the release of NCP, as it was antagonised (30–40%) by a rabbit anti-human NCP, and to the release of PAF which also accounted for the platelet aggregating activity that was independent from both adenosine diphosphate and cyclo-oxygenase inhibitors. These data suggest that direct interaction occurs between human PMN and cuprophan in in vitro conditions in the absence of plasma factors and point to a role for cellular mediators in the pathogenesis of the intravascular alterations occurring early in haemodialysis.

Introduction

The early transient neutropenia that commonly occurs during haemodialysis using cuprophan membranes has been related to activated complement components generated by plasma-membrane interaction [1]. However, in vitro and in vivo studies have questioned complement activation as being the only factor involved in the pathogenesis of haemodialysis-associated neutropenia [2,3]. The latter does not occur during haemodialysis using different membranes such as polyacrylonitrile [2] and polysulphone [3], despite signs of complement activation. In vitro, short-term incubation of purified, normal human polymorphonuclear neutrophils (PMN) with cuprophan membranes elicits the release of lysosomal constituents such as neutrophil-derived cationic proteins (NCP), and β-glucuronidase) and platelet-activating factor (PAF). Both NCP and PAF have a potent PMN aggregating activity in vitro and cause dose-dependent neutropenia when intravenously injected in the rabbit [4].

In the present study, we investigated the possibility of a direct interaction between cuprophan membranes and human PMN. To this purpose, plasma-free, purified, normal human PMN were recirculated in an experimental model of dialysis using cuprophan membranes with acetate or bicarbonate dialysate, and PMN aggregation as well as the release of NCP, lysosyme, β-glucuronidase and PAF were evaluated.

Materials and methods

Preparation of purified normal human PMN

Human PMN were purified as described previously in detail [4], by differential centrifugations, gelatine sedimentation and washing in Tris-Tyrode's (TT) supplemented with 0.25% bovine serum albumin (BSA), (TT-BSA, pH 7.4), without Ca^{++} and Mg^{++}. Finally, the PMN preparations, from 90 to 95% pure, were resuspended in TT-BSA with Ca^{++} and Mg^{++} and adjusted to a concentration of 7,000 cells/mm^3. Contaminating cells were 1–10% lymphomononuclear cells. Cell viability was assessed by lactic dehydrogenase (LDH) determination and by eosin exclusion.

Experimental dialysis model

Plasma-free, purified normal human PMN were recirculated in hollow fibre cuprophan membrane dialysers (Travenol CF1511) using a Cobe Centry 2 machine for acetate dialysis, and, in other experiments, using the same machine modified for bicarbonate dialysis (Sorin 1B 8011, Infusor). The composition and temperature (37°C) of the acetate and bicarbonate dialysate were kept constant as follows: for acetate, Na^+ 142mEq/L, K^+ 1.5mEq/L, Ca^{++} 4mEq/L, Mg^{++} 1.5mEq/L, acetate 40mM/L, pH 7.14; for bicarbonate, Na^+ 142mEq/L, K^+ 2mEq/L, Ca^{++} 3.5mEq/L, Mg^{++} 1mEq/L, bicarbonate 39mEq/L, pH 7.26.

A sample of the PMN preparation was taken immediately before the infusion to the dialysis model and served as basal control. The sampling time was 1, 5, 10, 20, 30 and 60 minutes after pump start using a T device connected to a syringe to avoid interruption of recirculation.

Recirculating PMN aggregometry

Two hundred and fifty microlitres of the basal control and of the samples obtained at the different time intervals were added to an aggregometer (Elvi 840) cuvette in continuous stirring (800rpm) and the changes in light transmission were recorded for at least two minutes.

Morphological studies

At the end of the experiment (60min) the circuit was opened, and the dialyser extensively washed with isotonic saline. A solution of toluidine blue, prepared as described previously [4] was used to stain cells entrapped in the capillaries. A number (5–10) of capillaries obtained from near the inlet and from the outlet of the dialyser were examined microscopically for the presence of cell aggregates.

Assay of intracellular NCP

Intracellular NCP were detected by direct immunofluorescence using fluorescein rabbit anti-human NCP antiserum which has been characterised previously [5], on acetone-fixed human PMN from the basal control and from the samples obtained at different time intervals. The extent and positivity of the staining were graded from 0 to 3+ on at least 100 cells counted by two independent examiners.

Assay of PMN-derived constituents released in the supernatants

The samples obtained at different time intervals were immediately centrifuged (700g, 20 min, 4°C). The basal control was first sonicated and served as the 100 per cent of lysosomal constituent release. All supernatants were kept frozen at −70°C until analysis. Lysosyme was quantitated by the method of Prockop and Davidson [6] and β-glucuronidase according to Talay et al [7] (Sigma Chemical Co, St Louis, Mo). The release of both lysosyme and β-glucuronidase were expressed as the percentages of the values in respect to the basal control. LDH was determined and was used as lysis indicator. LDH release in the basal control was always below two per cent. The release of NCP was assayed on the basis of the antagonising effect of the rabbit anti-human NCP antiserum on the in vitro PMN aggregation performed as described previously [4], after addition of 10–50μl of the supernatants from the basal control and from the samples obtained at different time intervals. Platelet aggregating activity in the supernatants (5–20μl) was assayed on washed rabbit platelets as described previously [8]. The purification and characterisation of PAF were performed as described previously [9].

Results

PMN aggregation occurred early (5 min) in the course of recirculation in a dialyser using cuprophan membranes as indicated by the increased light transmission in the aggregometer and confirmed by the morphological evidence of PMN aggregates being firmly adherent to the capillary surfaces. Depletion of intracellular NCP was manifest 5–10 minutes after the start of recirculation and was temporally related to the release of lysosyme (25%) and β-glucuronidase (55%) in the supernatants. LDH values did not rise above two per cent within the first 10 minutes of recirculation. However, probably due to the continuous traumatic impact on recirculating cells, LDH increased above two per cent after 20 minutes, reflecting a lytic release (Table I).

TABLE I. Quantitation of intracellular NCP, release of lysosyme, β-glucuronidase and PAF from plasma-free, purified normal human PMN recirculated in a dialyser with cuprophan membranes using acetate or bicarbonate dialysate

Time of sampling	IF[1] Anti-NCP	Lysosyme %[2]	β-glucuronidase %[2]	LDH %[2]	PAF ng/ml
Basal control	3+	0	0	2	0
1 min	2+	20 – 25[3]	45 – 51[3]	2[3]	0[3]
	2+	19 – 23[4]	43 – 50[4]	2[4]	0[3]
5 min	1+	22 – 25	45 – 54	2	5.1 ± 2.3[5]
	1+	23 – 25	44 – 55	2	4.3 ± 2.6[5]
10 min	0/1+	24 – 25	43 – 54	2	7.2 ± 3.1
	0/1+	23 – 24	41 – 55	2	6.8 ± 4.0
20 min	0/1+	20 – 23	42 – 52	2	4.1 ± 1.2
	0/1+	19 – 21	40 – 53	2	4.3 ± 2.0
30 min	0	18 – 22	39 – 50	2	0.4 ± 1.3
	0	20 – 23	43 – 53	2	0.2 ± 0.8
60 min	0	21 – 25	45 – 54	2	ND
	0	20 – 23	43 – 51	2	ND

[1] Direct immunofluorescence (IF) using a fluorescein rabbit anti-human NCP antiserum was carried out on acetone-fixed human PMN.

[2] Percentages in respect of the sonicated basal control assumed as 100 per cent.

[3] These are ranges of values obtained during recirculation in a dialyser using acetate dialysate.

[4] These are ranges of values obtained during recirculation in a dialyser using bicarbonate dialysate.

[5] Mean ± SD on five experiments performed on different occasions. ND=not done.

A PMN aggregating activity was detected in the supernatants obtained at 5, 10, 20 and 30 minutes (Figure 1A). This activity could be inhibited (20–40%) by the rabbit anti-human NCP antiserum (Figure 1A). Furthermore, these supernatants had also a potent aggregating activity on platelets (Figure 1B). Both the PMN and platelet aggregating activities were unaffected by indomethacin and CP/CPK and could be in part related to the release of PAF (Table I). No PMN or platelet aggregation could be observed in the supernatants of the basal control. Furthermore, no significant differences in terms of PMN aggregometry, release of NCP, and lysosomal constituents as well as of PAF occurred using bicarbonate instead of acetate dialysate.

Discussion

The present study indicates that recirculation of plasma-free, purified normal human PMN in an experimental acetate or bicarbonate dialysis model using a

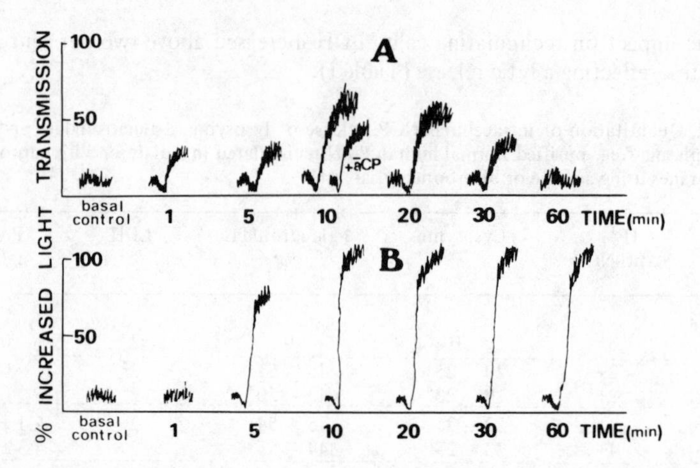

Figure 1. A. 5 x 10^6 human PMN, stirred at 900rpm in Tris-Tyrode's (250μl) supplemented with 0.25% gelatine (Difco Laboratories, Detroit, Mich) were tested by aggregometry to 10–50μl of supernatants from the basal control and from the samples obtained at different time intervals. In some experiments, human PMN were pre-incubated with a rabbit anti-human NCP antiserum (aCP). The aggregometric profiles are representative of five experiments performed on different occasions.
B. Two 5 x 10^7 washed human platelets, stirred at 900rpm in Tris-Tyrode's (250μl) supplemented with 0.25% gelatine in the presence of indomethacin (1 x 10^{-5} M) (Sigma) and of the creatine phosphate (CrP)–creatine phosphokinase enzymatic system (312.5μg of CrP, and 152.5μg of CPK, Sigma) were tested to 5–20μl of the supernatants from the basal control and obtained at different time intervals. The aggregometric profiles are representative of five experiments performed on different occasions

dialyser with cuprophan membranes leads to: 1) the aggregation and the sequestration of PMN in the capillaries; 2) the simultaneous occurrence of a PMN and platelet aggregating activity in the supernatants. The PMN aggregating activity could be related either to the release of NCP as shown by: a) the depletion of the intracellular contents of NCP, and b) the partial inhibitory effect of the anti-NCP serum or to the generation and release of PAF.

Using this experimental dialysis model in plasma-free conditions, the role of the direct interaction between PMN and cuprophan membranes could be studied without interference from the effects on PMN due to activated complement components. In fact, in in vivo conditions, in which both cells and plasma factors may come into contact and interact with cuprophan membranes, the relative roles of cellular (NCP, PAF) or humoral (complement) mediators appear closely interwoven, and, therefore, difficult to be discriminated. Nonetheless, during the early course of haemodialysis, the release of NCP has been shown to occur simultaneously with the fall of circulating PMN. Our studies provide further indications for a role of cellular mediators such as NCP and PAF, with a high PMN aggregating activity, in the pathogenesis of the intravascular alterations occurring early in haemodialysis using cuprophan membranes.

Acknowledgment

This work was supported by C.N.R. Rome No 82.002214.04 and M.P.I.

References

1 Craddock PR, Fehr J, Brigham KI et al. *N Engl J Med 1977; 296:* 769
2 Henderson IW, Miller NE, Hamilton RW et al. *J Lab Clin Med 1975; 85:* 191
3 Aljama P, Bird PAE, Ward MK et al. *Proc EDTA 1978; 15:* 144
4 Camussi G, Tetta C, Bussolino F et al. *Int Arch Allergy Appl Immun 1980; 62:* 1
5 Camussi G, Tetta C, Segoloni G et al. *Clin Immunol Immunopathol 1982; 24:* 299
6 Prockop DJ, Davidson ND. *N Engl J Med 1964; 270:* 269
7 Talay P, Fishman WH, Huggins C. *J Biol Chem 1964; 166:* 757
8 Camussi G, Mencia-Huerta JM, Benveniste J. *Immunology 1977; 33:* 523
9 Camussi G, Aglietta M, Malavasi F et al. *J Immunol 1983; 131:* 2397

Open Discussion

KOCH (Chairman) Did you perform similar studies, as a control, with more biocompatible membranes or were all the studies done on cuprophan?

TETTA We did studies in vitro with polysulphone and polymethacrylate membranes, and these were reported in the Internation Journal of Artificial Organs in 1978*. We showed that both membranes were not able to induce PMN activation. We did not study the effects on PMN after recirculation in this experimental model of dialysis.

*Canussi G, Segoloni G, Rotunno M, Vercellone A. *Int J Artif Organs 1978; 1:* 123

Proc EDTA–ERA (1984) Vol 21

EFFECTS OF BLOOD–DIALYSER INTERACTION (SHAM-DIALYSIS) ON HAEMODYNAMICS AND OXYGEN TENSION IN HEALTHY MAN

A Danielsson, U Freyschuss, J Bergström

Karolinska Institute, Huddinge University Hospital, Huddinge, Sweden

Summary

To evaluate the normal haemodynamic and respiratory responses to blood-membrane contact sham-dialysis, i.e. with blood flowing through a dialyser but without dialysate and ultrafiltration, was performed on healthy young men during 150 minutes. Heart rate, cardiac output, arterial blood pressure and pulmonary arterial blood pressure (continuously recorded during the initial 30 minutes) did not change significantly. The white blood cell count fell markedly to a minimum after 20 minutes of blood-membrane contact, then returning to above the baseline values, but PaO_2 did not change significantly.

Introduction

Increased interest has been focused on dialyser biocompatibility and its contribution to respiratory and haemodynamic responses during haemodialysis. Research so far has been performed during standard haemodialysis which makes it difficult to separate the effects of blood-membrane contact from those of diffusion and ultrafiltration [1–4]. Some protocols have exclusively tried to study the blood-membrane interaction but procedures have differed a great deal from the haemodialysis procedures [5,6]. In order to evaluate the respiratory and haemodynamic responses to blood contact with a cuprophan dialyser membrane we have performed sham-dialysis (SHD), i.e. with blood flowing through the dialyser but without diffusion or ultrafiltration taking place. Healthy young men have been studied because one aim of the study was to assess the normal physiological responses to membrane interaction, i.e. not changed by the uraemic state.

Material and methods

Eleven healthy men (mean age 27 ± 1 years) gave their informed consent and volunteered for the study.

Blood access for SHD was obtained by introducing a catheter to a femoral vein and a needle into a brachial vein. SHD was performed with a dialysis monitor (Gambro AK 10 UDM) and a cuprophan hollow fibre dialyser (1.2m^2, Gambro 120 M). The blood pump speed was 200ml/min. The blood compartment was primed with isotonic saline. The dialysate compartment was first primed with dialysate followed by ultrafiltration of 2L saline from the blood compartment, thereby rinsing the dialysate compartment free from dialysate; the inlet and outlet dialysate ports were then closed. The closed dialysate compartment contained about 100ml isotonic sodium chloride. At start of SHD a loading dose of 5000IU heparin was given which occasionally was followed by an additional dose of 2500IU. Blood temperature was followed with a thermodilution catheter and was held constant by clothing the dialyser with a metal foil and placing the inlet blood tubing in a blood warmer bath.

Right heart catheterisation was performed with a Swan-Ganz thermodilution catheter which was introduced percutaneously into an antecubital vein and positioned in the pulmonary artery to measure cardiac index (CI), stroke index (SI), pulmonary arterial blood pressure (PAP) and pulmonary capillary wedge pressure (PCW). A short Teflon catheter was introduced percutaneously into a brachial artery for blood pressures (AP) and blood sampling. Pressures were measured with strain gauge transducers (EMT 34, Siemens, Elema, Sweden) and recorded together with ECG on a UV-recorder (SE 3006, DI, SE lab, EMI, Feltham, England). Mean arterial blood pressure was calculated by electronic damping. Systemic vascular resistance index (SVR) was calculated and calf vascular resistance (CVR) was assessed by venous occlusion plethysmography. SHD was performed during 150 minutes on eight subjects with blood pressure measurements every 30 minutes. On two of these eight and on three additional subjects blood pressures were recorded continuously during the first 30 minutes of SHD.

Arterial blood samples were analysed for PaO_2 and $PaCO_2$ every 10 minutes the first 30 minutes, and then every 30 minutes of SHD (conventional electrode technique, ABL2, Radiometer). White blood cell count (WBC) from the arterial side of the dialyser was sampled every 10 minutes the first 30 minutes and at the end of SHD (Coulter counter; if less than 3 x 10^9/L verified manually.

Statistical methods: two-way analysis of variance was performed to test overall changes over time. Only if the analysis showed a significant result paired 't' test was performed. Values are presented as mean ± SEM.

The investigation was approved by the Ethical Committee of the Karolinska Institute and Huddinge University Hospital.

Results

No severe side effects or complications were observed in any of the subjects.

Haemodynamics

There were no significant changes of heart rate, CI, and SVR. CVR seemed to decrease between 30 and 150 minutes, but analysis of variance did not show a

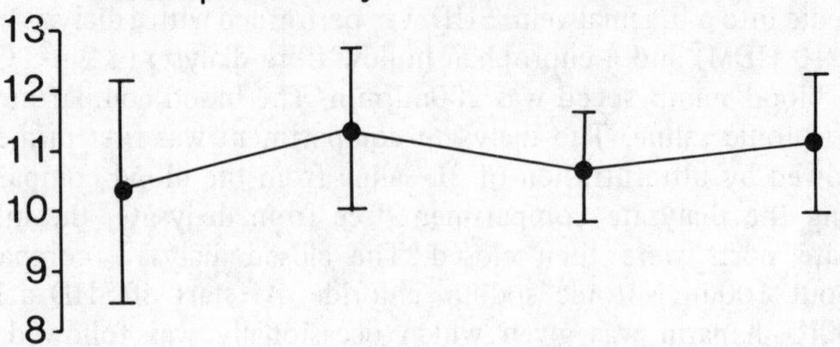

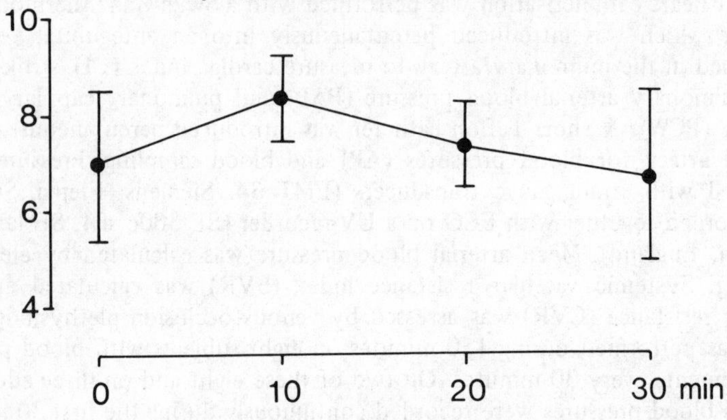

Figure 1. Mean pulmonary arterial blood pressure and pulmonary capillary wedge pressure at 10, 20 and 30 minutes of blood-membrane contact in three healthy men. Mean ± SEM. The pulmonary arterial pressure was continuously recorded

significant change over time. Systolic, diastolic and mean AP were constant as were systolic, diastolic and mean PAP. AP and PAP measured continuously in five subjects and PCW recorded intermittently in three subjects between 0–30 minutes of blood-membrane contact did not change. Values in three subjects are shown in Figure 1.

Blood gases and white blood cell count (Figure 2)

WBC decreased by 59 per cent at 20 minutes, thereafter increasing to 35 per cent above baseline at 150 minutes. $PaCO_2$ increased slightly during the first 20 minutes but PaO_2 did not change significantly.

Discussion

It is known that blood-membrane contact activates the complement system [7,8] which may be the reason for granulocyte sequestration in the pulmonary

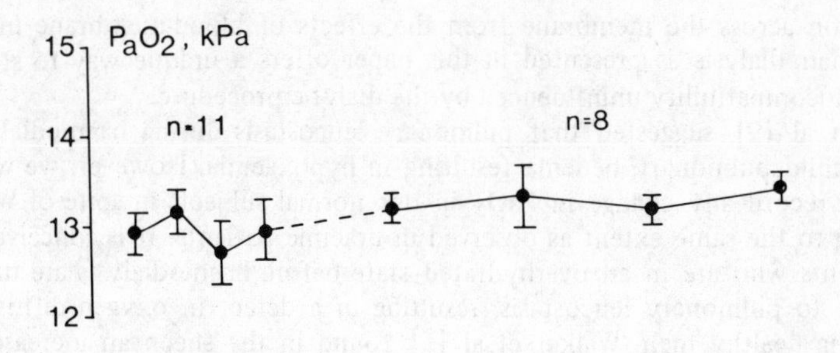

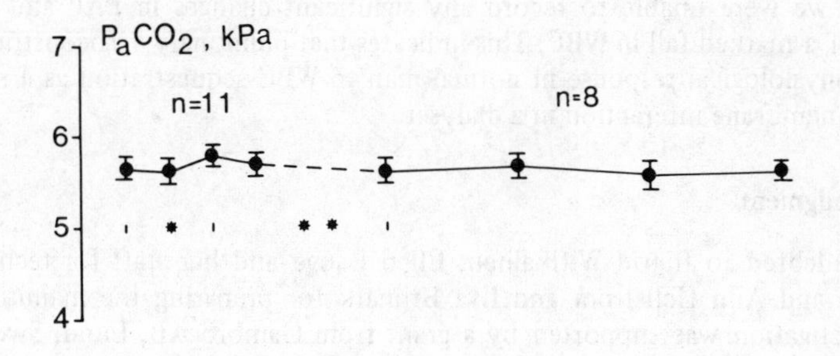

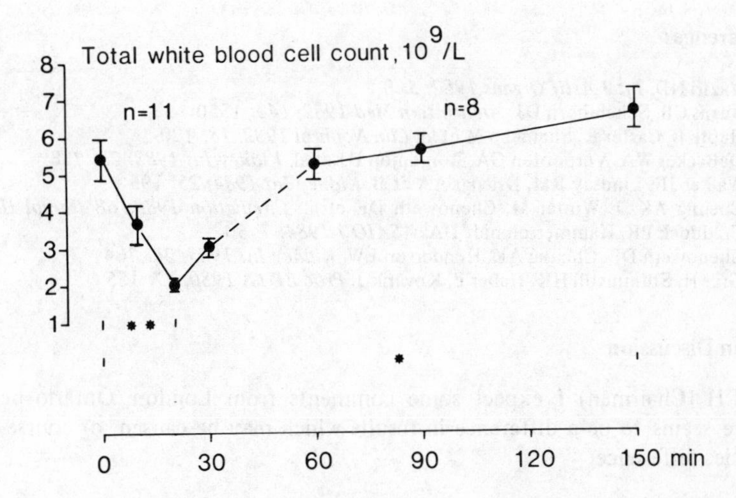

Figure 2. PaO_2, $PaCO_2$ and total white blood cell count in 11 healthy men (11 during 30 minutes, 8 during 150 minutes). Mean ± SEM. * and ** indicate significant values at $p<0.05$ and $p<0.01$ levels respectively

vascular system resulting in neutropenia early during dialysis. There is a great controversy concerning the possible relationship between the above mentioned phenomenon and the fall in PaO_2 during haemodialysis [1–4]. One reason for this is that in standard haemodialysis it is not possible to separate the influence of diffusion across the membrane from the effects of blood-membrane interaction. Sham-dialysis as presented in this paper offers a unique way to study dialyser biocompatibility uninfluenced by the dialytic procedure.

Graf et al [9] suggested that pulmonary leucostasis during haemodialysis evokes a mild pulmonary oedema resulting in hypoxaemia. However, we were unable to record any change in PaO_2 in our normal subjects in spite of WBC decreasing to the same extent as observed in uraemic patients. It is conceivable that patients who are in an overhydrated state before haemodialysis are more vulnerable to pulmonary leucostasis, resulting in a defect in oxygen diffusion than young healthy men. Walker et al [5] found in the sheep an increase in pulmonary arterial pressure after injection of plasma incubated with cuprophan. However, we were unable to record any significant changes in PAP and PCW in spite of a marked fall in WBC. This indicates that pulmonary vasoconstriction is not a physiological response in normal man to WBC sequestration as a result of blood-membrane interaction in a dialyser.

Acknowledgment

We are indebted to Ingrid Witikainen, Ellen Bauge and her staff for technical assistance and Ann Hellström and Eva Brimark for preparing the manuscript. This investigation was supported by a grant from Gambro AB, Lund, Sweden.

References

1 Vaziri ND. *Int J Artif Organs 1982; 5:* 8
2 Burns CB, Scheinhorn DJ. *Arch Intern Med 1982; 142:* 1350
3 Habte B, Carter R, Shamebo M et al. *Clin Nephrol 1982; 18:* 120
4 DeBacker WA, Verpooten GA, Borgonjon DJ et al. *Kidney Int 1983; 23:* 738
5 Walker JF, Lindsay RM, Driedgr AA et al. *Kidney Int 1984; 25:* 195
6 Cheung AK, LeWinter M, Chenoweth DE et al. *Circulation 1983; 68 (Suppl III):* 44
7 Craddock PR, Hammerschmidt DA. *ASAIO J 1984; 7:* 50
8 Chenoweth DE, Cheung AK, Henderson LW. *Kidney Int 1983; 24:* 764
9 Graf H, Stummvoll HK, Haber P, Kovarik J. *Proc EDTA 1980; 17:* 155

Open Discussion

KOCH (Chairman) I expect some comments from London, Ontario because there seems to be a difference in results which may be caused, of course, by a species difference.

LINDSAY (London, Ontario) I am fascinated by your work and may I first of all congratulate you because I do not think that we would be able to do those experiments in North America. It seems to me that you have to do the same study in patients with end-stage renal failure requiring dialysis. I really cannot understand why the difference is not there because as Dr Walker said earlier this morning we have seen changes in pulmonary vascular tone, occurring in patients with acute renal failure during monitoring, from a slight increase in pulmonary vascular resistance index*. It is either something that is different in the healthy person from the uraemic patients, not just the species difference as Dr Koch was suggesting, or I suppose that Gambro have a very good dialyser and you should be looking at a different manufacturer's cuprophan.

GOTCH (Chairman) May I ask a question of both of you? Is there any possibility that there is a different amount of membrane contact, your sheep are 25–50kg and there is a stagnant time of 10 minutes in the system. Have you ever done your experiments with single pass flow through the dialyser and do you obtain the same effects? The other question is, do people have to become sensitised to the membrane? If you study a dialysis patient on his very first single dialysis would you expect to have the same results or do you become sensitised?

LINDSAY Frankly I don't know the answer to the latter part of the question. Certainly with sheep you will see the same phenomenon with single pass, but obviously not to the same extent. Whatever is turned on causes an end-organ response which obviously is dependent upon the amount of factor or factors, let's assume it is C_5a or something else, that gets into the circulation on a unit time basis. I was talking just last week to Lee Henderson discussing this particular issue and he brought up some very interesting observations. He told me that with some material he has been looking at, the sheep is more sensitive than the pig, which is more sensitive than the dog, which is probably more sensitive than the humans, so there is a clear species difference there. Thus if one is looking at this phenomena it may well be that the sheep is the best animal model to use, but I can't really say any more than that.

DANIELSON We only investigated the surface once.

DI GIULIO (Rome) Blood vascular access was different in Dr Walker's study since he used arteriovenous while you used a veno-venous access. Do you think that this may account for the discrepancy of results between yours and Dr Walker's study?

DANIELSON I think this is possible.

KOCH (Chairman) In your abstract you mentioned the measurement of catecholamines.

DANIELSON Yes, we unfortunately have not analysed this yet.

*Walker JF. *Proc EDTA–ERA (1984) 21:* 141

Proc EDTA–ERA (1984) Vol 21

BLOOD LINES SHOULD BE CONSIDERED IN STUDIES OF BIOCOMPATIBILITY DURING ACETATE HAEMODIALYSIS

B Branger, R Oulès, D Treissède, J P Balducchi, F Michel, A Crastes de Paulet, M Treille

Centre Hospitalier Régional Universitaire de Nimes et Montpellier, France

Summary

In this one-year prospective study, the biocompatibility of blood lines used for acetate haemodialysis treatment was evaluated in 12 patients. These blood lines, polyvinyl chloride (PVC) and polyurethane extruded PVC (PE) respectively were compared in a schedule of four alternating three-month periods of treatment: PVC/PE/PVC/PE. White blood cell count, complement, IgE and thromboxane values were recorded monthly. A reduction in white blood cell and polymorphonuclear counts after 30 minutes of dialysis was significantly less during period 2 than period 1. Pre-dialysis eosinophil counts varied in a seasonally dependent pattern. We conclude that in spite of their small area, blood lines have some effect on biocompatibility and that the season of the year has to be considered in biocompatibility studies involving eosinophils.

Introduction

Interactions between blood and dialyser membranes have been extensively studied [1–3] in short-term protocols, but so far no one has studied the biocompatibility of blood lines. One reason could be the small surface area of these blood lines in contact with the blood relative to that of the dialyser membrane. Nevertheless the first contact of blood with the extracorporeal haemodialysis circuit is with the blood lines and so we have evaluated the possibility of differences in biocompatibility between two types of blood lines.

Materials and methods

This one-year prospective study was performed on 12 patients – six males and six females – whose mean age was 51 years (range 32–75 years) and who had been on chronic haemodialysis for 47.5 months (6–96 months). They were randomly selected from among those non-smokers (40 of 46 patients) in our dialysis centre population who were not taking any drug that interacts with prostaglandins (35 of the 40).

Acetate haemodialysis was performed three times weekly (dialysate acetate 37mmol/L and Na^+ 140mmol/L) with non-reused cuprophan membrane dialysers of $1.2m^2$ area rinsed with 21 of saline before each dialysis. Continuous heparinisation was carried out with the standard amount of 800IU/hr. Blood lines of polyvinyl chloride (PVC) and polyurethane extruded PVC (PE) with similar internal areas of $0.036m^2$, were used alternately for three-month periods: PVC/PE/PVC/PE. The mean length of time between the production and clinical use of the PE blood lines was 5.5 months (4.5 to 6.5) and 11 months (9.5 to 11.5) for the first and second periods, respectively.

The following counts were made monthly: white blood cells (WBC), polymorphonuclear leucocytes, eosinophils, lymphocytes and platelets using a Coulter counter. When the relative count of eosinophils was below two per cent, it was determined from 500WBC instead of 100WBC; these counts, the measurement of haemoglobin and the assay of complement fractions (CH_{50}, C_3, C_4) Institut Pasteur, Beckman) were performed at dialysis time 0, 30 minutes and 180 minutes. Total immunoglobulin E (IgE) and thromboxane (TxB_2) were measured by radioimmunoassay before and after haemodialysis.

For nine of 12 patients, a one-year retrospective study of pre-dialysis eosinophil count was added to this protocol.

Calculations: all values were corrected for any haemoconcentration or haemodilution determined by haemoglobin variation occurring during haemodialysis, and were expressed as the percent change from the initial value (mean ± 1SD). The Wilcoxon rank sum test for paired observation was used for statistical analysis.

Results

Variations in WBC and platelet counts, complement, IgE, and TxB_2 are shown in Table I. The mean correction for haemodilution at 30 minutes was 2.0% ± 2.0 and for haemoconcentration at 180 minutes 12.3% ± 6.7. No significant change of pre-haemodialysis values except for the eosinophil count was observed: the WBC 5685 ± $1410/mm^3$, polymorphonuclear 2583 ± $1159/mm^3$ (63%), lymphocytes 1616 ± $420/mm^3$ (28%), monocytes 183 ± $63/mm^3$ (3%). The absolute eosinophil count varied significantly according to season with a peak during spring (period 4 in Table II). Additional retrospective data available for nine of 12 patients showed a similar pattern of variation during the year before the protocol (Table II). The mean complement values before haemodialysis were 35.8 ± $4.8UH_{50}$ (n=50 ± 9) for CH_{50}; 89.4 ± 11.0mg/100ml (n=80 to 150) for C_3; 26.6 ± 6.3mg/100ml (n=20 to 50) for C_4.

IgE and TxB_2 values were respectively 102.7 ± 269.0IU/L (n<20) and 15.4 ± 3.1pg/L (n=12–18).

During haemodialysis, there was a significant drop of WBC, polymorphonuclear, eosinophil and lymphocyte counts at 30 minutes, and a significant rise of polymorphonuclear count at 180 minutes. The drop of WBC and polymorphonuclear counts at 30 minutes was significantly less during period 2 (PE) than during period 1 (PVC). No such difference was found in period 4 relative to period 3. No correlation was found between absolute eosinophil count and

TABLE I. Changes in blood during haemodialysis, expressed as mean ± SD

Period	PVC 1	PE 2	PVC 3	PE 4
% change between time 0 and 30 minutes				
WBC	−62.5 ± 10.9†,*	−55.1 ± 11.9†	−60.1 ± 14.9†	−58.8 ± 11.0†
pn	−77.5 ± 10.7†,**	−68.1 ± 13.7†	−70.4 ± 16.0†	−67.9 ± 12.9†
eo	−57.5 ± 11.0†	−61.2 ± 13.8†	−62.9 ± 20.2†	−47.3 ± 24.9†
l	−24.8 ± 3.8†	−26.1 ± 8.6†	−10.4 ± 37.4†	−31.2 ± 25.2†
pl	− 6.6 ± 5.9	− 5.1 ± 5.6	− 7.2 ± 3.4	− 8.1 ± 4.1
CH_{50}	2.9 ± 10.3	1.8 ± 5.7	4.7 ± 11.5	0.9 ± 10.4
C_3	− 0.7 ± 4.7	1.4 ± 7.3	1.8 ± 6.1	3.2 ± 16.7
C_4	− 0.6 ± 9.3	− 0.1 ± 13.4	7.5 ± 14.8	6.3 ± 27.2
% change between 0 and 180 minutes				
WBC	7.4 ± 17.7	8.9 ± 15.1	9.4 ± 19.1	11.2 ± 15.1
pn	21.7 ± 22.7†	26.9 ± 21.7†	30.0 ± 27.0†	29.0 ± 28.4†
eo	5.4 ± 28.3	3.4 ± 35.0	−12.9 ± 36.7	22.5 ± 74.2
l	− 8.4 ± 14.0	− 7.4 ± 11.8	−15.2 ± 17.3	−15.1 ± 16.1
pl	0.6 ± 13.6	2.7 ± 13.1	0.1 ± 9.8	2.3 ± 16.2
CH_{50}	0.2 ± 18.1	2.8 ± 7.1	0.5 ± 9.1	− 4.6 ± 13.3
C_3	− 0.2 ± 16.1	− 5.7 ± 9.9	− 1.1 ± 5.7	− 4.6 ± 17.3
C_4	− 4.7 ± 10.9	3.6 ± 12.3	10.7 ± 9.3	12.0 ± 22.6
TxB_2	18.3 ± 27.8	10.7 ± 52.2	31.1 ± 72.6	29.5 ± 27.4
IgE	− 6.8 ± 13.9	− 6.9 ± 17.5	− 2.8 ± 35.0	19.9 ± 53.5

WBC=white blood cells; pn=polymorphonuclear leucocytes; eo=eosinophils; l=lymphocytes; pl=platelets; CH_{50}, C_3 and C_4 are complement and its fraction 3 and 4.
† Significantly different from pre-dialysis values $2\alpha<0.01$.
* $2\alpha<0.02$ period 2 compared to 1.
** $2\alpha<0.01$ period 2 compared to 1.

TABLE II. Pre-dialysis eosinophil count according to periods. Period 4 was during spring

	PVC 1	PE 2	PVC 3	PE 4
Prospective study n=12	238 ± 199	307 ± 254	271 ± 181	365 ± 331**
	PVC	PVC	PVC	PVC
Retrospective study n=9	252 ± 279	291 ± 226	229 ± 309	381 ± 367*

* $2\alpha<0.02$ 4 versus 1, 4 versus 3
** $2\alpha<0.01$ 4 versus 1, 4 versus 3

its change during haemodialysis. Changes in IgE were not significant and were not correlated with either eosinophil count or its changes. The rise of TxB_2 was less with PE than with PVC but the difference was not significant.

Discussion

The use of polyurethane coating inside the PVC blood line should theoretically enhance biocompatibility, because polyurethane is considered one of the more biocompatible materials as evaluated by cell-culture technique [4]. In our study, the comparison of period 2 (PE) to period 1 (PVC) shows a significantly less pronounced drop of WBC and polymorphonuclear counts both of which are considered reliable markers of biocompatibility during HD. The absence of this difference between periods 3 and 4 could be due to progressive loss of the polyurethane barrier related to the migration of plasticiser from the PVC as found by Bommer et al [5] when PE blood lines are used more than six months after manufacture. Another cause could be a variable season-dependent biological response as suggested by the significant rise of the pre-haemodialysis eosinophil count during the spring. This hypereosinophilia is considered by some authors [6,7] to be a treatment-related phenomenon but no long-term follow-up studies have been published so far and our results suggest a season-dependent variation irrespective of the composition of the blood lines. TxB_2 variations can be considered a stimulated macrophage index [5] and therefore a good marker of biocompatibility. In a previous 'acute' study [8] we found a significant rise of TxB_2 during acetate haemodialysis with PVC blood lines and the date of the present study show a non significant rise of TxB_2 in this larger population.

In conclusion, analysis of the biological response of uraemic patients to haemodialysis material should be done with a long-term protocol in order to include time-dependent changes. A large population is also necessary because scattered inter-individual responses to haemodialysis. The use of PE blood lines during acetate haemodialysis seems to induce less leucopenia than PVC blood lines and therefore blood line effects should be considered in studies of biocompatibility during haemodialysis.

References

1 Kaplow LS, Goffinet JA. *JAMA 1968; 203:* 1135
2 Jacob A, Gavellas G, Zarco R et al. *Kidney Int 1980; 18:* 505
3 Amadori A, Candi P, Sasdelli M et al. *Kidney Int 1983; 24:* 775
4 Imai Y, Watanabe A, Masuhara E. *Trans ASAIO 1979; 25:* 299
5 Bommer J, Ritz E, Andrassy K. In Crosnier J, Funk-Brentano JL, Bach JF, Grunfeld JP, eds. *Actualités néphrologiques de l'Hôpital Necker.* Flammarion: Paris. 1984: 357–379
6 Spinowitz BS, Simpson M, Manu P et al. *Trans ASAIO 1981; 27:* 161
7 Bodner G, Peer G, Zabuth V et al. *Nephron 1982; 32:* 63
8 Branger B, Deschodt, Grandolleras C et al. *Néphrologie 1984; 4:* 190

Open Discussion

GOTCH (Chairman) Thank you very much and I congratulate you on being able to statistically demonstrate these relatively small changes. The differences you show are really quite small, do you think that they would significantly change the outcome of membrane studies? How important do you think they truly are when you are comparing two different membrane materials?

BRANGER It is really quite difficult to be certain because very few studies have been performed with long-term protocols. I think there is some variation in individuals with regard to the dialyser membranes. I think it is difficult to observe such minimal correlations and I don't know if they have any real importance with regard to biocompatibility. It exists, but I don't know if a five or 10 per cent change of leucopenia is important.

Proc EDTA–ERA (1984) Vol 21

THE EFFECT OF DIALYSATE TEMPERATURE ON HAEMODIALYSIS LEUCOPENIA

G Enia, C Catalano, F Pizzarelli, G Creazzo, F Zaccuri, A Mundo, D Iellamo, Q Maggiore

Centro di Fisiologia Clinica del C.N.R. e Divisione Nefrologica, Reggio Calabria, Italy

Summary

We assessed the influence of dialysate temperature on intra-dialytic leucopenia. Lowering dialysate temperature from 38°C to 20.5°C caused a decrease in the dialysis associated white blood cell reduction from 82±6 per cent to 32±19 per cent. The degree of leucopenia bore a highly significant relationship with dialyser blood temperature suggesting that a further lowering of blood temperature (to about 20°C) would almost entirely prevent intra-dialytic leucopenia.

Introduction

There is increasing evidence that leucopenia occurring early during Cuprophan haemodialysis results from blood-membrane contact leading to alternative pathway complement activation, anaphylatoxin release, leuco agglutination, and pulmonary sequestration of granulocytes [1–7]. The intra-dialytic activation of complement is similar to that observed in vitro on incubating serum with zymosan [1,7]. In their original investigation Pillemar et al found that zymosan activation of the complement can be prevented if the temperature of reaction is kept at 16°C, instead of 37°C [8]. If complement activation of the alternative pathway is inhibited at low temperatures, so should the attendant leucopenia. We have, therefore, investigated whether intra-dialytic leucopenia can be prevented by lowering the temperature of blood-membrane interaction.

Materials and methods

We studied a total of 13 male patients on regular dialysis treatment during acetate haemodialysis with cuprophan hollow fibre dialysers. Patients were free of clinically manifest infections and none of them were receiving drugs known to affect white blood cell count. The mean age was 46 years (range 30–59), duration of dialysis treatment 7.8 years (range 2–10). Renal disease was chronic glomerulonephritis in five, polycystic kidneys in one, nephropathy associated with Lawrence Moon-Biedl syndrome in one, undetermined in six.

Study protocol 1. Six patients underwent three haemodialyses each, with dialysate temperature kept at 33.8±(SD) 0.3°C, 36.9±0.2°C and 38.1±0.2°C respectively.

Study protocol 2. Nine patients underwent two haemodialyses each, one with dialysate at room temperature (20.5±0.7°C), and the other at 36.8±0.5°C. In the cool procedure, before returning to the patient the blood was warmed through a serpentine inserted in the venous line and immersed in a thermostatic bath. Blood lines of similar length and conformation were employed in both procedures.

Leucocyte and differential counts were determined in duplicate by manual methods on blood drawn from the arterial line at 0, 15, 30, 60, 120 and 240 minutes of haemodialysis. Temperature was monitored by means of thermocouple needles (Ellab) placed in the blood and dialysate circuits. Dialyser urea and creatinine clearance at a mean dialysate flow of 500ml/min and blood flow of 260ml/min, were determined in duplicate.

Statistical analysis was performed by paired Student's 't' test and Spearman's rank correlation test.

Results

Study protocol 1. After 15 minutes of haemodialysis white blood cell count fell by 59±15 per cent at a dialysate temperature of 33.8°C, 65±15 per cent

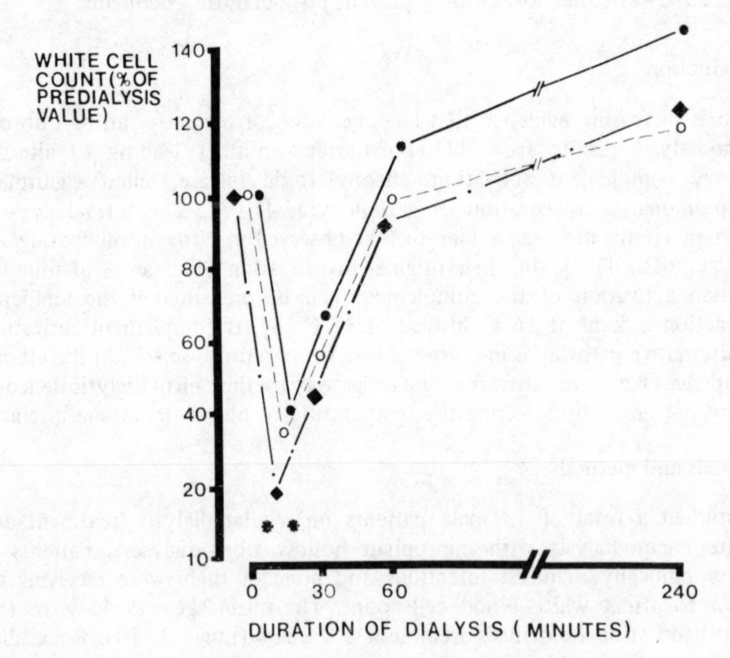

Figure 1. White blood cell count at: •33.8±0.3°C; ○ 36.9±0.2°C; ♦38.1±0.2°C; * 33.8 vs 38.1°C p<0.02; *36.9 vs 38.1 p<0.02

(SD) at the dialysate temperature of 36.9°C and 82±6 per cent at the dialysate temperature of 38°C (Figure 1). The maximal fall in white blood cell count bore a highly significant relationship with dialysate temperature (Spearman's test; r= -0.69; p<0.01). Pre- and post-dialyser blood temperature corresponding to each dialysate temperature and white blood cell count are also reported in Table I.

TABLE I. Dialysate temperature, blood temperature and white blood cell count in the two studies

	n	Dialysate temperature °C	Pre-dialyser blood temperature °C	Post-dialyser blood temperature °C	White blood cell x 10^3 mm^3 pre-HD	at max.fall
1st study	6	38.1±0.2	36.0±0.4	36.4±0.5	6.9±1.7	1.28±0.48
	6	36.9±0.2	36.2±0.5	35.5±0.7	6.2±1.2	1.87±0.58
	6	33.8±0.3	35.6±0.7	33.0±0.3	6.2±1.9	2.34±0.45
2nd study	9	36.8±0.5	35.7±0.4	35.1±0.4	8.2±2.6	1.97±0.57
	9	20.5±0.7	34.9±0.6	20.6±0.9	8.2±4.2	5.15±2.7

HD=haemodialysis

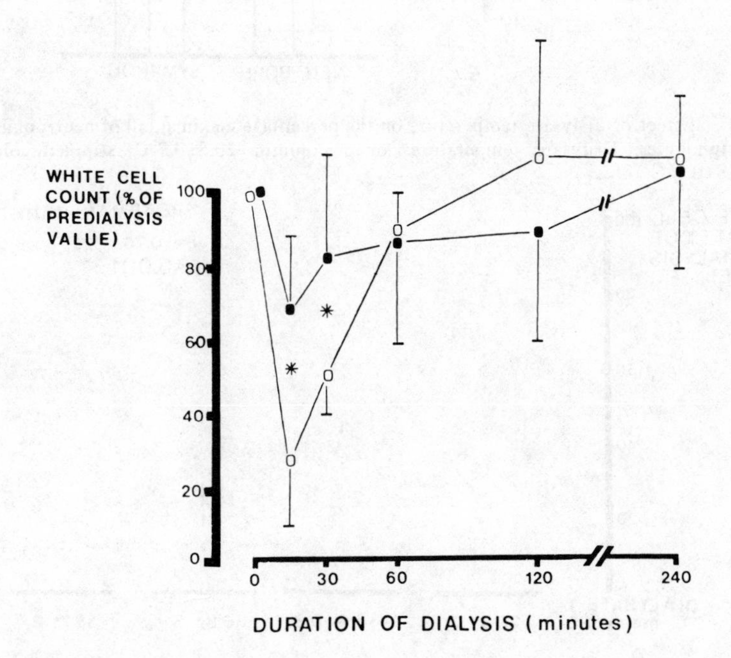

Figure 2. White blood cell fall at dialysate temperature of 36.8±0.5°C (○) and 20.5±0.7°C (●). *p<0.01

Study protocol 2. Use of dialysate kept at room temperature (20°C) caused a 15°C fall in blood temperature along the dialyser (Table I). At this temperature the fall in white blood cell count was less than half that occurring in the control procedure, i.e. 32±19 per cent versus 73±17 per cent ($p<0.01$) (Figure 2). Leucopenia was sustained primarily by the fall in total neutrophils, which averaged 94 per cent in the control procedure, and was reduced to 43 per cent in the dialysis at room temperature (Figure 3) ($p<0.001$).

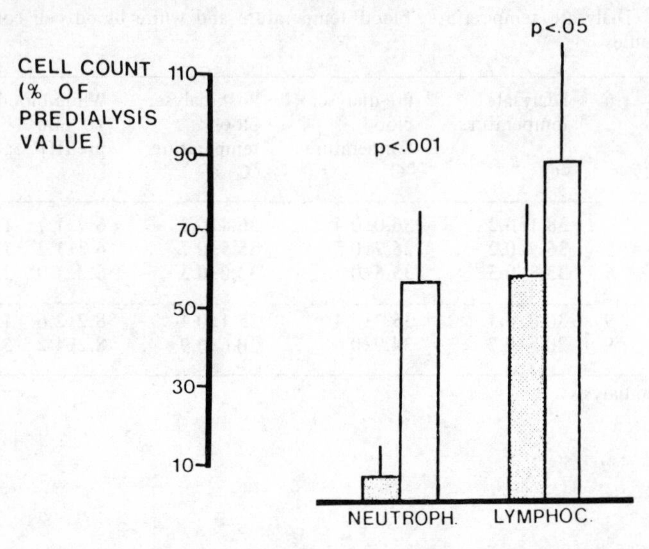

Figure 3. Effect of dialysate temperature on the percentage maximal fall of neutrophils and lymphocytes. Dialysate temperature: open column: 20.5±0.7°C; stippled column: 36.8±0.5°C

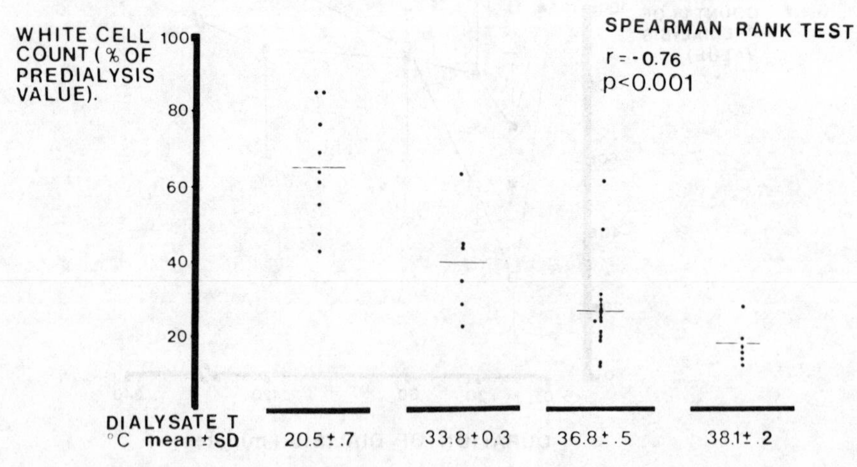

Figure 4. Relationship between Δ temperature and maximal percentual fall in white blood cell count

Analysing together all the studies performed in the two protocols, the relationship between dialysate temperature and maximal white blood cell fall remained highly significant (Figure 4) (Spearman's test; $r = -0.76$; $p < 0.001$). A highly significant relationship was also observed between blood temperature (average of pre- and post-dialyser blood temperature) and maximal white blood cell count fall ($r = -0.77$; $p < 0.001$).

Lowering the dialysate temperature caused only minor decrease in dialysis efficiency. In fact in the cooler procedure (20.5°C) the dialyser clearance for small solutes diminished by less than 10 per cent, the urea clearance from 186±27ml/min to 172±23ml/min, the creatinine clearance from 146±22 to 135±22ml/min.

Discussion

Our studies show that lowering the dialysate temperature to 20°C markedly lessens the degree of intra-dialytic leucopenia. This effect can be obtained at the expense of only a slight decrease in small molecule clearance.

A likely explanation of our findings is that cooling of blood in contact with the dialyser membrane causes a lesser degree of complement activation than in standard dialysis. The lesser degree of complement activation would in turn produce a lesser degree of leucopenia.

We conclude that intra-dialytic leucopenia is a temperature dependent phenomenon. Basing ourselves on the regression line between blood temperature and maximal fall in white blood cells, we predict that further decreasing blood temperature to 20°C or less would almost entirely prevent complement activation with its attendant leucopenia.

Acknowledgments

Partially supported by C.N.R. specialised subproject on 'Plasma Treatment'

References

1 Craddock PR, Fehr J, Dalmasso AP et al. *J Clin Invest 1977; 59:* 878
2 Craddock PR, Fehr J, Brigham KL et al. *N Engl J Med 1977; 296:* 769
3 Chenoweth DE, Cheung AK, Henderson LW. *Kidney Int 1983; 24:* 764
4 Chenoweth DE, Cheung AK, Ward DM et al. *Kidney Int 1983; 24:* 770
5 Ivanovich P, Chanoweth DE, Schmidt R et al. *Kidney Int 1983; 24:* 758
6 Amadori A, Candi P, Sasdelli M et al. *Kidney Int 1983; 24:* 775
7 Jacob HS. *Q J Med 1983; 207:* 289
8 Pillemer L, Blum L, Lepow IH et al. *Science 1954; 120:* 279

Open Discussion

UNKNOWN (Antwerp) Do you think this temperature phenomenon could be the explanation of the different results in the human study? Could it be that in the human study there was no hypoxaemia or increase in vascular resistance because there was no or lesser complement activation?

ENIA I don't know if we can explain the differences on the basis of complement activation and we found that at this level of blood cooling complement activation decreased, so it is a possibility.

RITZ (Heidelberg) Are you suggesting that your present findings are an explanation for the improved vascular stability you saw during low temperature dialysis?

ENIA There are some benefits of lowering dialysate temperature down to 34°C, but I don't know if at 34°C there is a lesser degree of complement activation. We studied activation only at extreme conditions and we are now going to study complement activation at 34°C.

Proc EDTA–ERA (1984) Vol 21

ANAPHYLACTOID REACTIONS DURING HAEMODIALYSIS ARE DUE TO ETHYLENE OXIDE HYPERSENSITIVITY

A J Nicholls, M M Platts

Royal Hallamshire Hospital, Sheffield, United Kingdom

Summary

Anaphylactoid reactions during haemodialysis are unusual, but increasingly recognised. It has recently been reported that there is a significant association between such reactions and the presence of IgE to albumin exposed to ethylene oxide (ETO). Review of the clinical features and epidemiology of dialysis anaphylaxis in the light of this new data suggests that these reactions are due to ETO hypersensitivity.

Introduction

In 1982 we reported a series of patients who had experienced severe reactions resembling anaphylaxis during haemodialysis, or more rarely, during haemofiltration and membrane plasma separation [1]. The same adverse reactions have also been widely reported from various centres in the United States [2], and in abstract form from several other European dialysis units (Cledes J. *Abstracts EDTA–ERA 1983:* 58; Perez-Garcia R. *Abstracts EDTA–ERA 1983:* 134; Zirovannis P. *Abstracts EDTNA 1983:* 21).

Recently published laboratory data on our patients suggested an association between these reactions and IgE specific to albumin exposed to ETO (ETO-HSA) [3]. The purpose of this communication is to review the clinical features of these anaphylactoid reactions to dialysis (the 'first-use syndrome' [2]) in the light of this new information, and to propose that the reactions are due to ETO hypersensitivity.

Clinical features of anaphylactoid reactions to dialysis

These attacks occurred in 16 long-term haemodialysis patients. Most frequently a single brand of flat-plate dialyser or another manufacturer's hollow-fibre dialyser was involved, but identical and extremely serious anaphylaxis occurred on one occasion with a haemofilter and a membrane plasma separator. Seven patients had an isolated reaction, six had two or three reactions, often with different dialysers; while three patients had repeated stereotyped attacks on every use of a new dialyser, which could be avoided only by treating a new dialyser with formalin before its first use. These last three patients had persistent blood eosinophilia. It is normal practice for our patients to rinse dialysers and sterilise them with formalin, but we have never had a reaction reported with a formalin sterilised dialyser.

The attacks all began within a minute or two of blood returning from a new ETO sterilised extracorporeal device to the patient, and were witnessed by a doctor or dialysis nurse in the majority of cases. The commonest symptom (in over two thirds of cases) was wheezing, while 50 per cent had urticaria and nearly 40 per cent chest pain. Less commonly cardiovascular collapse, sneezing, a runny nose, flushing, diarrhoea, vomiting and watery eyes were present. Most patients suffered several simultaneous anaphylactoid symptoms, and all remarked that the onset coincided with the initial flow of blood from the dialyser. There were no deaths; recovery from severe attacks occurred rapidly and spontaneously when haemodialysis was stopped, while milder attacks resolved spontaneously within two hours if dialysis was continued.

Possible causes

We have argued elsewhere [1] that these reactions are clearly distinguishable from other adverse reactions to haemodialysis, including dialysis-triggered asthma [4], pulmonary leucostasis with hypoxaemia [5], and febrile rigors related to endotoxin [6]. The illness bore all the hallmarks of generalised histamine release [7] and could thus have been due either to direct mast cell degranulation, or to an IgE mediated response to a foreign antigen. Ethylene oxide, used in the sterilisation of these devices, seemed a likely culprit as antigen (or hapten) in view of the fact that it had been incriminated in anaphylaxis before [8], can be readily eluted by blood from ETO-sterilised products [9], and was the only agent common to all the reactions observed.

In vitro studies

Laboratory investigations to investigate a possible association between these reactions and ETO hypersensitivity have been published elsewhere [3]. In brief, we obtained serum from seven patients who had suffered reactions (cases 4, 5, 9, 10, 11, 12 and 15 from the series reported previously [1]), and from age and sex matched controls who had been on regular haemodialysis with identical devices, for a similar length of time. Serum was also obtained from three non-atopic non-dialysed controls with no known exposure to ETO. The following measurements were made: total IgE, total antibody binding to ETO-HSA, and IgE against ETO-HSA.

It was found that six of seven reactors had detectable amounts of IgE to ETO-HSA compared with only one of six non reactors ($p<0.05$) and that the geometric mean value of IgE to ETO-HSA was 2.0ng ETO-HSA bound by IgE per ml of serum in reactors compared with 0.2ng/ml in non reactors ($p<0.05$).

No significant differences in total IgE or total antibody binding to ETO-HSA were found.

Discussion

We felt that these reactions were unlikely to be due to the dialysis membrane or type of extracorporeal device as they occurred with four different devices utilising membranes made of cuprammonium cellulose ('cuprophan', Enka AG), cellulose acetate, and anisotropic polysulphone. Furthermore, as they only occurred during first use, blood lines, heparin or saline could not be incriminated. Our suspicions therefore, fell upon ethylene oxide as the likely partial antigen triggering immediate-type hypersensitivity.

The above results are consistent with this hypothesis, the very close association between the presence of IgE to ETO-HSA and a history of dialysis anaphylaxis being in favour of a causal link. We suggest that repeated exposure to trace amounts of ETO in new dialysers can sensitise susceptible individuals, who may then exhibit an anaphylactoid reaction on subsequent exposure. It is not clear however, why these reactions affected patients in such an apparently sporadic and haphazard fashion, with only three patients suffering repeated reactions.

The one reactor with undetectable IgE to ETO-HSA had only a very mild reaction to dialysis (sneezing and runny nose) which might not have been a true anaphylactoid reaction; and she only had reactions with a hollow-fibre dialyser whereas all the other six patients reported here had reactions to a flat-plate dialyser of different manufacturer. It is also possible that the antigen in this patient was ETO linked to a serum protein other than HSA.

It is interesting to note that one control patient had a moderate amount of IgE to ETO-HSA but denied reactions. This could be due to various causes; dialysers containing little ETO; poor releasability of patient's mast cells; or perhaps protection by the rather elevated amount of total antibody binding to ETO found in that patient.

Although we believe that dialysis anaphylaxis is mediated through IgE to ETO-HSA, measurement of total antibody binding indicates exposure to hapen-conjugated proteins. As yet there are too few patients to be certain if there is a significant link between total antibody binding and first use reactions.

It has been suggested that the incidence of anaphylactoid reactions to the first use of a dialyser is 0.0035 per cent [2]. However, lack of awareness of the syndrome has probably led to many reactions being unreported, unrecognised, or attributed to other causes such as disequilibrium, hypotension and cardiac ischaemia.

These first use reactions can certainly result in significant morbidity and even mortality [2], so understanding the mechanism and predicting patients at risk is desirable. If it can be confirmed that these are due to ETO hypersensitivity, additional measures may be necessary in rinsing new dialysers before use to ensure the absence of even trace amounts of ETO.

Finally, if exposure to ETO can evoke specific IgE antibodies, this might account for the commonly observed and hitherto unexplained phenomenon of eosinophilia in haemodialysis patients [10] which was present in some of our patients.

Acknowledgments

We should like to thank Professor R Patterson, Dr L Grammer and Ms Roberts RN, (Northwestern University Medical School, Chicago, USA) for collaboration in laboratory studies.

References

1 Nicholls AJ, Platts MM. *Br Med J 1982; 285:* 1607
2 Ing TS, Daugirdas JT, Popli S, Gandhi VC. *Int J Artif Organs 1983; 6:* 235
3 Grammer LC, Roberts M, Nicholls AJ et al. *J Allergy Clin Immunol.* In press
4 Aljama P, Brown P, Turner P et al. *Br Med J 1978; 2:* 251
5 Craddock PR, Fehr J, Dalmasso AP et al. *J Clin Invest 1979; 59:* 879
6 Robinson PJA, Rosen SM. *Br Med J 1971; 1:* 528
7 Neugebauer E, Lorenz W. *Behringer Institute Mitteilungen 1981; 68:* 102
8 Poothullil J, Shimizu A, Day RP, Dolovich J. *Ann Intern Med 1975; 82:* 58
9 Lindop CR, Willcox TW, McKegg PM, Harris EA. *J Thorac Cardiovasc Surg 1980; 79:* 845
10 Scheurmann EH, Fassbinder W, Frei U et al. *Contr Nephrol 1983; 36:* 133

Open Discussion

BONADONNA (Padua) Have you any information on the pH of the dialysate? In our experience of three cases of anaphylactoid reaction we found that the pH of the dialysate was very low.

NICHOLLS No, I'm afraid I haven't.

WALKER (London, Ontario) There has been reports in the literature of people who have had apparent ethylene oxide sensitivity and when they were changed from cuprophan hollow fibre dialysers to cellulose acetate hollow fibre dialysers, sterilised with ETO, they did not have any further allergic symptoms. In addition Hakim* in Boston has looked at patients who have had the first use syndrome and/or the hypersensitivity phenomenon and found that the degree of complement activation is significantly higher in people who have hypersensitivity reactions than in people who do not have hypersensitivity reactions. I just wonder whether what we are seeing here is in fact just a person who has a very responsive immunological system and this may just be a marker rather than an aetiological agent.

NICHOLLS To deal with your first point, I think this is extremely important, if you recall what Dr Leonard was saying about leachability of potential noxious agents from dialysers† I think we can incriminate ethylene oxide to account for differences between dialysers on different occasions. It may be that either the

*Hakim RM, Breillatt J, Lazarus JM, Port FK. *N Engl J Med 1984; 331:* 878
†Leonard EF. *Proc EDTA-ERA 1984; 21:* 99

membrane or the potting compounds release ethylene oxide at different rates. I'm suggesting that ethylene oxide is an agent and it may require certain potting or certain membranes to produce this reaction. So far as your other point is concerned all I can say is that it is an interesting hypothesis but I have no particular data.

GOTCH (Chairman) Can I ask what technical steps such as changes in dialyser processing were required to terminate your incredible epidemic?

NICHOLLS We changed to priming dialysers with two litres of saline and ensuring all patients ran the dialysate for half an hour before starting dialysis. It was very interesting that the reactions occurred at the time when our unit was being rebuilt and we had to go over to single use. On many of these occasions, owing to pressure on staff, dialysate had only been run for as little as 10 minutes through the dialysate side and the priming had been done with possibly slightly less than one litre. In addition the throughput of disposable dialysers in our unit was extremely fast and so the dialysers were not lying on the shelves as they previously had been.

KOCH (Chairman) Did you ever use non ETO sterilised disposables in your reactors?

NICHOLLS No. The difficulty we had in studying patients who had repeated reactions was that one of them could not tolerate the reactions any longer and opted to switch to CAPD and refused to be studied on dialysis, the other two patients were home dialysis patients who spontaneously started formalising their dialysers, even when they were new, and they also refused to be studied in hospital. I also feel that it would have been unethical to study these patients in hospital in view of the deaths that have been recorded in the States.

Proc EDTA–ERA (1984) Vol 21

IMPAIRED REGULATION OF β-ADRENOCEPTORS IN PATIENTS ON MAINTENANCE HAEMODIALYSIS

A E Daul, A M Khalifa, N Graven, O-E Brodde

Medizinische Klinik and Poliklinik, University of Essen, FRG

Summary

In patients on maintenance haemodialysis the number of lymphocyte β_2-adrenoceptors (determined by (±)-^{125}iodocyanopindolol binding) was not different from that in healthy controls; lymphocyte cyclic AMP responses to (−)-isoprenaline (10^{-8} – 10^{-4}M) or NaF (10 and 50mM), however, were significantly reduced. Dynamic exercise on a bicycle (80% of maximum heart rate) for 15 minutes caused in 10 healthy volunteers a fourfold increase in plasma catecholamines; concomitantly lymphocyte β_2-adrenoceptor number increased by about 55 per cent. In contrast, in patients on maintenance haemodialysis exercise induced only a twofold increase in plasma catecholamines and did *not* affect β_2-adrenoceptor number. It is concluded that in chronic uraemia regulation and responsiveness of β-adrenoceptors is impaired.

Introduction

Several signs of reduced sympathetic activity have been observed in patients on maintenance haemodialysis treatment. A defective function of sweat glands, reduced elevation of blood pressure in response to sustained hand-grip exercise, non-volume responsive chronic hypotension [1], elevated plasma norepinephrine (NE) and reduced responsiveness to NE-infusion [2] have been reported. In order to find out whether these disturbances may be due to an impairment of the β-adrenoceptor (β-R)/adenylate cyclase system we determined in lymphocytes of chronic haemodialysis patients β_2-receptor number (by (±)-^{125}iodocyanopindolol (ICYP) binding) and responsiveness (by cyclic AMP responses to isoprenaline stimulation). In addition, acute regulation of lymphocyte β_2-receptor was assessed by the effects of dynamic exercise (15 minutes on a bicycle at 80 per cent of maximal heart rate) on β_2-receptor number and responsiveness.

Subjects and methods

Twenty-eight patients on maintenance haemodialysis treatment (9 females, 19 males, mean age 48 ± 2 (24–68) years) and 41 healthy volunteers (12 females, 29 males, mean age 36 ± 2.8 (20–81) years) participated in the study after having given informed written consent. All subjects were not on antihypertensive therapy. Blood samples were always taken between 9.00 and 10.00 am, i.e. 10–20 hours after the previous haemodialysis treatment. After 30 minutes of rest 20ml venous blood was drawn with the subjects in a sitting position and anticoagulated with 500 IU heparin/10ml blood. Lymphocytes were isolated and β_2-receptor density was determined by ICYP binding as recently described [3]. Changes in intracellular cyclic AMP in response to isoprenaline or NaF stimulation were measured as described by Dillon et al [4].

Exercise protocol

Seven male patients on maintenance haemodialysis (mean age 30.1 ± 3.2 (20–40) years) and 10 male healthy controls (mean age 28.5 ± 1.7 (23–36) years) participated in this study. Exercise was carried out in a quiet air-conditioned room always between 10 and 12 am. Subjects assumed a supine position and a cannula was inserted into an anticubital vein. After one hour of rest exercise was performed on a bicycle ergometer (Bosch, Berlin, FRG) in a supine position starting with an initial work load of 25W (in haemodialysis patients) or 50W (in controls). Work load was increased by 25W every two minutes until 80 per cent of the maximal heart rate (200-age) was reached. This final work load (75–125W in the haemodialysis patients, 100–150W in the controls, respectively) was kept constant until a total exercising time of 15 minutes was reached.

Blood samples were obtained immediately prior to exercise, at the end of exercise and one hour after exercise for determination of lymphocyte β_2-receptor number and responsiveness and for radioenzymatic measurement of plasma catecholamines [5].

Blood pressure and heart rate were recorded automatically by a Tonomed® (Speidel & Keller, Jungingen, FRG) and by an electrocardiogram.

Statistical evaluations

The experimental data given in the text and in the figures are means ± SEM on N experiments. The maximal number of ICYP binding sites and the equilibrium dissociation constant (K_D) were calculated from plots according to Scatchard [6]. The significance of differences was estimated by Student's 't' test. A p-value less than 0.05 was considered to be significant.

Results

The mean number of β_2-receptors in lymphocytes of patients on maintenance haemodialysis (956 ± 81 specific ICYP binding sites/cell, n=28) was not significantly different from that in an age-matched control group (774 ± 49 ICYP binding sites/cell, n=41). The β-receptor agonist isoprenaline (10^{-8}-10^{-4}M) produced in both groups a concentration-dependent increase in cyclic AMP content. In haemodialysis patients, however, this response was significantly attenuated at each concentration (Figure 1). In addition, NaF (10 and 50mM,

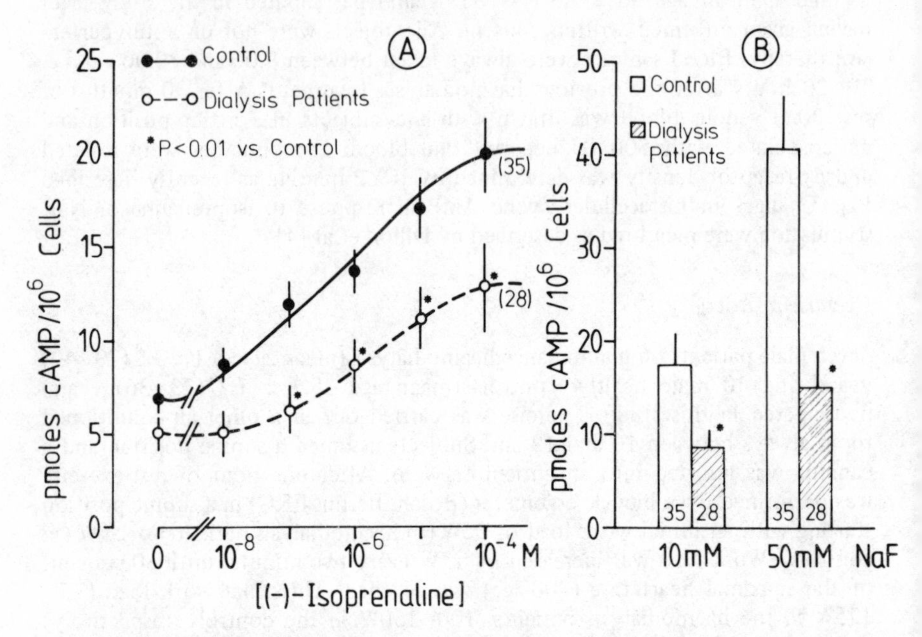

Figure 1. Effects of (–)-isoprenaline (A) and NaF (B) on the intracellular cyclic AMP in healthy volunteers and in patients on maintenance haemodialysis

respectively) caused at both concentrations in lymphocytes of healthy controls an increase in cyclic AMP content, which was nearly three times as high as in lymphocytes of haemodialysis patients (Figure 1). Exercise on a bicycle in a supine position (at 80% of the maximal heart rate for 15 minutes) led to an increase in systolic blood pressure from 120mmHg to 190mmHg in both groups (Figures 2 and 3). Immediately after exercise in controls plasma catecholamine levels were increased about fourfold (0.47 ± 0.11ng/ml to 1.83 ± 0.23ng/ml, Figure 2), while in haemodialysis patients this increase was significantly less (0.79 ± 0.21 to 1.65 ± 0.31ng/ml, Figure 3). In controls, lymphocyte β_2-receptor number increased significantly from 463 ± 36 prior to exercise to 720 ± 64 following exercise. This increase in β_2-receptor was accompanied by an exaggerated response of the lymphocyte cyclic AMP system to isoprenaline (10μM) stimulation (Figure 2).

In haemodialysis patients, on the contrary, dynamic exercise failed to produce any significant changes in lymphocyte β_2-receptor number. Accordingly, isoprenaline evoked increases in intracellular cyclic AMP content were not changed (Figure 3).

One hour after exercise systolic blood pressure, β_2-receptor number and plasma catecholamine had reached pre-exercise values in both groups.

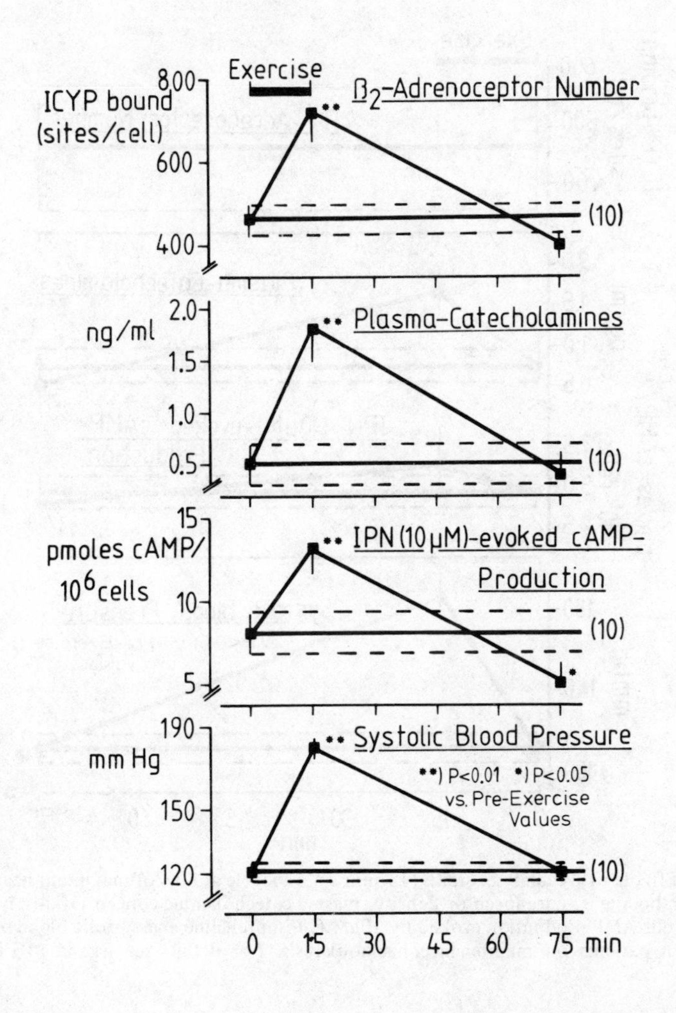

Figure 2. Effects of dynamic exercise (15 min on a bicycle at 80% of maximum heart rate) on lymphocyte β_2-adrenoceptor density, plasma catecholamine concentrations, lymphocyte cyclic AMP production evoked by 10μM (–)-isoprenaline and systolic blood pressure in 10 healthy volunteers. *Ordinates (from top to bottom)*: β_2-adrenoceptor density in ICYP binding sites/cell; plasma catecholamines in ng/ml; increase in cyclic AMP content induced by 10μM (–)-isoprenaline in pmoles cyclic AMP/10^6 cells and systolic blood pressure in mmHg. *Abscissa*: time in minutes. Given are means ± SEM. Solid horizontal lines and broken lines: means of pre-exercise values ± SEM assessed after one hour of rest in a supine position

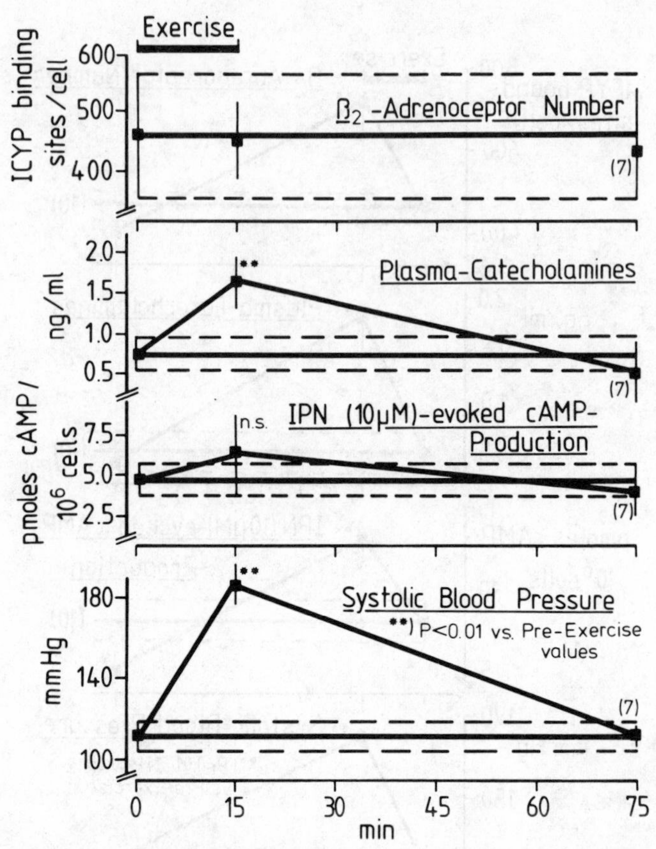

Figure 3. Effects of dynamic exercise (15 min on a bicycle at 80% of maximum heart rate) on lymphocyte β_2-adrenoceptor density, plasma catecholamine concentrations, lymphocyte cyclic AMP production evoked by 10μM (–)-isoprenaline and systolic blood pressure in seven patients on maintenance haemodialysis. For details see legend to Figure 2

Discussion

In the present study, isoprenaline (via β_2-receptor stimulation) produced in lymphocytes of patients on maintenance haemodialysis significantly less increases in intracellular cyclic AMP content than in controls. These results favour the idea that in patients undergoing chronic haemodialysis treatment the responsiveness of β_2-receptor is reduced. Since the number of β_2-receptors in haemodialysis patients was not different from controls, the decreased responsiveness of the β-adrenergic system seems to be due to a post-receptor defect. In haemodialysis patients the stimulatory effect of NaF was markedly attenuated. Since NaF is known to activate the adenylate cyclase directly, i.e. non-receptor mediated [7], it may be concluded that in chronic haemodialysis patients reduced responsiveness of β_2-receptor is due to an impairment of the adenylate cyclase activity.

In healthy subjects dynamic exercise (80% of maximal heart rate of 15 min) led to an increase (about fourfold) in plasma catecholamine and lymphocyte β_2-receptor number (approximately 55%). The increase in β_2-receptor number was accompanied by an elevation of isoprenaline-evoked cyclic AMP production indicating an enhanced β_2-receptor responsiveness. The mechanism of this concomitant rise in plasma catecholamines and β_2-receptor number and responsiveness is not yet clarified. In rat reticulocyte ghosts, however, it has been shown that stimulation of β-receptor by catecholamines increases enzymatic methylation of phospholipids and that enhanced phospholipid methylation may unmask cryptic β-receptor resulting in an increasing receptor number. It may be possible, therefore, that in healthy subjects exaggerated release of endogenous catecholamines evoked by dynamic exercise may unmask cryptic β_2-receptor by a similar mechanism. Taking this into consideration, the lack of effect of dynamic exercise on β_2-receptor number in haemodialysis patients may further support the view that the responsiveness of the β-adrenergic system is reduced.

Conclusions

In patients undergoing maintenance haemodialysis treatment the responsiveness of lymphocyte β_2-receptor is diminished. This reduced responsiveness might be due to a post-receptor defect, most likely to an impairment of the adenylate cyclase activity. Since the properties of lymphocyte β_2-receptor are very similar to β_2-receptor in other tissues [9], including human heart [10], it may be concluded that the responsiveness of peripheral β-receptor is also reduced in chronic uraemia. Such a reduced β-receptor responsiveness may be the cause of end-organ resistance to adrenergic stimulation in haemodialysis patients.

Acknowledgments

This work was supported by the Landesamt für Forschung NRW and the SANDOZ-Stiftung für Therapeutische Forschung.

References

1 Campese VM, Romoff MS, Levitan D et al. *Kidney Int 1981; 20:* 246
2 Botey A, Gaya J, Montoliu J et al. *Proc EDTA 1981; 18:* 586
3 Brodde O-E, Engel G, Hoyer D et al. *Life Sci 1981; 29:* 2189
4 Dillon N, Chung S, Kelly JG, O'Malley K. *Clin Pharmacol Ther 1980; 27:* 769
5 Nagel M, Schümann HJ. *Clin Chem Clin Biochem 1980; 18:* 431
6 Scatchard G. *Proc NY Acad Sci 1949; 51:* 660
7 Perkins JP. In Greengard P, Robison GA, eds. *Advances in Cyclic Nucleotide Research; 3.* New York: Raven Press. 1973: 1–64
8 Strittmatter WJ, Hirata F, Axelrod J. In Dumont JE, Greengard P, Robison GA, eds. *Advances in Cyclic Nucleotide Research; 14.* New York: Raven Press. 1981: 83–91
9 Brodde O-E, Kuhlhoff F, Arroyo J, Prywarra A. *Naunyn-Schmiedeberg's Arch Pharmacol 1983; 322:* 20
10 Brodde O-E, Karad K, Zerkowski H-R et al. *Circ Res 1983; 53:* 752

Open Discussion

WIZEMANN (Giessen) Firstly, did you incubate lymphocytes from healthy persons with uraemic serum and secondly is there evidence that adrenoceptors on platelets are representative for the heart and the vasculature?

DAUL At the moment we can't say what changes in lymphocyte β-adrenoceptors mean in patients. In normotensive subjects the properties of lymphocyte β_2-receptors are very similar to the properties of receptors in other tissue including heart and lung.

RITZ (Chairman) May I ask two questions. The first relates to the mechanism of the putative post-receptor defect? Some years ago there was a paper from the Hopital Necker measuring the isoproterenal heart rate response showing it was impaired in uraemic patients*. This was reversible with parathyroidectomy. Do you have any information on this?

DAUL No, but let me say that we have two patients with chronic hypertension which was reversible after dialysis.

RITZ The second question relates to your findings with sodium chloride which are suggestive of a coupling protein defect. Did you test other hormones such as PTH or glucogen to see whether they had a similar attenuated cyclic AMP response?

DAUL No.

ZOCCALI (Reggio, Calabria) Your results are at variance with a previous paper by Galeazzi† who found that the fall in heart rate caused by the β-adrenoreceptor blocking drug Pindolol is higher in dialysis patients than in normal subjects. Is there any way to reconcile these results with yours?

DAUL I think it is very difficult to compare these two studies.

*Ulmann A, Drueke T, Zingraff J, Crosnier J. *Clin Nephrol 1977; 7:* 38
†Galeazzi RL, Gugger M, Weidmann P. *Kidney Int 1979; 15:* 661

Proc EDTA–ERA (1984) Vol 21

CHANGES IN LEFT VENTRICULAR ANATOMY DURING HAEMODIALYSIS, CONTINUOUS AMBULATORY PERITONEAL DIALYSIS AND AFTER RENAL TRANSPLANTATION

A Deligiannis, E Paschalidou, G Sakellariou, V Vargemezis, P Geleris, A Kontopoulos, M Papadimitriou

University of Thessaloniki, 'Aghia Sophia' Hospital, Thessaloniki, Greece

Summary

The changes in left ventricular anatomy in 30 patients with end-stage renal disease and stable cardiac function, undergoing regular haemodialysis (10 patients), continuous ambulatory peritoneal dialysis (10 patients) and after successful renal transplantation (10 patients) were evaluated by M-mode echocardiography. Initially all had evidence of left ventricular hypertrophy and dilatation. Re-evaluation after a mean follow-up of 22 months on each mode of treatment showed that in the haemodialysis group the left ventricular mass and volume were increased, while in continuous ambulatory peritoneal dialysis (CAPD) and, especially renal transplantation, the hypertrophy and dilatation were reversed. This improvement was probably due to a reduction of cardiac workload.

Introduction

It is well known that chronic renal failure and its maintenance treatment procedures predispose to cardiovascular disease [1]. Although chronic haemodialysis is successful in making productive life possible, 25 to 40 per cent of patients have evidence of left ventricular hypertrophy and dysfunction [1,2]. Several factors such as long-standing hypertension, accelerated atherosclerosis, hyperdynamic circulation and uraemic 'cardiomyopathy' have been incriminated [1,2]. However in the recent literature there is some evidence that successful transplantation and continuous ambulatory peritoneal dialysis (CAPD) have a beneficial effect on cardiac size and function in patients with chronic renal failure [3,4]. This study further defines the changes in left ventricular anatomy and function by M-mode echocardiography in patients on long-term haemodialysis or CAPD and after successful renal transplantation.

Materials and methods

Thirty patients with end-stage chronic renal failure were studied, group A: 10 patients on regular haemodialysis, group B: 10 patients on CAPD and group C:

10 patients who were on regular haemodialysis and underwent successful renal transplantation. None of them had evidence of other systemic disease, severe hypertension, cardiac failure or pericarditis during the study. In group C the arterio-venous fistula was patent. Measurements were made within one month after the initiation of regular haemodialysis or CAPD and before renal transplantation and 22 months (range 20–24 months) following the application of each procedure. M-mode echocardiograms were obtained with an Ekoline 20A (Smith, Kline & French) machine attached to a VR-6 Electronics for Medicine osciloscopic recorder. The patients were studied in the left decubitus position by the standard technique.

The left ventricular internal diastolic dimension (LVIDd), systolic dimension (LVISd), left ventricular posterior wall thickness (PW), interventricular septal thickness (IVS), aortic root dimension (ARD) and left atrial dimension (LAD) were calculated according to the recommendations of the American Society of Echocardiography [5]. From these data the following parameters were calculated: a) Per cent shortening fraction of the left ventricle (%ΔD=LVIDd-LVISd/LVIDd x 100); b) LAD/ARD ratio; c) left ventricular mass ($LVM=0.77 [(LVIDd+PW+ISV)^3-LVIDd^3]+2.4$); d) the end-diastolic volume ($EDV=LVIDd^3$), stroke volume (SV=EDV−ESV) and cardiac index ($CI=SV \times heart\ rate/m^2$); e) the relative left ventricular wall thickness (R/Th ratio=LVIDd/2PW). All measurements represent the average value of five cardiac cycles.

Data were analysed using the Student's 't' test for paired data.

Results

Clinical data from the three groups are shown in Table I. There were no significant differences in the mean age, the mean concentration of serum creatinine, haematocrit, mean blood pressure and the mean heart rate at the time of the initial evaluation. The changes of the above clinical features during the follow-up period are given in the same table. As during the study the calculated body surface area of the patients did not change the absolute values for chamber size and cardiac performance were compared.

Table II shows that the mean LVIDd increased by 4.8 per cent ($p<0.01$) in haemodialysis patients but decreased by 2.5 per cent and 2.9 per cent ($p<0.01$) in CAPD and renal transplantation patients respectively. There was also a significant reduction of the mean values of LVISd and IVS by 3.7 per cent and 17 per cent ($p<0.01$) in group C. Although the initial mean values of LVIDd, LVISd, PW and IVS were significantly greater than those of the control group, a tendency towards normal in groups B and C was observed.

In all groups the initial left ventricular mass was greater than normal. Following haemodialysis the left ventricular mass increased by 24.3 per cent ($p<0.01$), but decreased during CAPD and, especially, after renal transplantation by 16.3 per cent and 20.7 per cent ($p<0.01$) respectively. The mean left ventricular end-diastolic volume increased by 15.3 per cent ($p<0.01$) during haemodialysis, but decreased after renal transplantation by 6.6 per cent ($p<0.05$). Although during CAPD there was a reduction in end-diastolic volume, this was not statistically significant. Finally, R/Th ratio showed a significant increase during CAPD and after renal transplantation.

TABLE I. Clinical features of 30 chronic renal failure patients

Data	Haemodialysis	CAPD	Renal transplantation
Male/female	6/4	7/3	5/5
Age (years)	42±11	46±13	40±8
Duration of chronic renal failure (years)	8	6	9
Serum creatinine (mg/100ml)	13.1±1.9 → 12.1±2.5*	12.4±3.8 → 7.5±1.6	11.8±3.4 → 1.7±0.8
Haematocrit (%)	23.4±5.8 → 28.2±5.2*	21.3±2.5 → 31.3±4.7	26.8±6.2 → 38.5±7.6
Mean blood pressure (mmHg)	131±16 → 122±12*	123±12 → 109±10	125±8 → 106±11
Heart rate (beats per minute)	77±11 → 73±13*	74±6 → 71±9	75±8 → 72±10

* Comparison of initial and follow-up mean values (±SD)

TABLE II. Echocardiographic data of left ventricular anatomy and performance

Data	Control group	Haemodialysis		CAPD		Renal transplantation	
		initial	follow-up	initial	follow-up	before	follow-up
Left ventricular internal diastolic dimension (mm)	48±4	52±3	54±3*	53±4*	51±4*	52±4	50±4*
Left ventricular internal systolic dimension (mm)	32±4	34±4	36±3†	34±6	34±5†	35±4	33±4*
Left ventricular posterior wall thickness (mm)	9.5±1	10±2	11±2†	11±2	10±1†	10±2	9.0±1†
Interventricular septal thickness (mm)	10±1	12±2	13±2†	11±2	10±2†	12±2	10±2*
Left atrial dimension/aortic root dimension	1.1±0.2	1.3±0.3	1.2±0.3†	1.1±0.2	1.1±0.3†	1.2±0.3	1.1±0.2†
%ΔD	34±5	34±6	35±6†	37±8	36±7†	33±5	34±5†
Relative left ventricular wall thickness	2.5±0.3	2.5±0.5	2.4±0.4†	2.5±0.3	2.7±0.4**	2.7±0.5	2.9±0.5**
Left ventricular mass (g)	158±25	209±38	260±38*	214±44	179±25*	203±57	161±38*
End diastolic volume (ml)	111±20	140±26	162±30*	149±39	139±35†	141±33	131±32**
Stroke volume (ml)	79±16	100±19	115±20*	112±20	101±17†	99±21	94±17†
Cardiac index ($L/min \cdot m^2$)	3.7±0.4	5.1±0.5	5.4±0.6**	5.3±0.7	4.6±0.5*	4.8±0.5	4.4±0.5*

Mean values ±SD; * $p<0.001$; ** $p<0.005$; † Not significant

In all groups the initial values of stroke volume and cardiac index were greater than normal. During haemodialysis the stroke volume and cardiac index increased by 15.2 per cent and 6.5 per cent ($p<0.01$); CAPD and renal transplantation reduced stroke volume by 9.6 per cent and 4.9 per cent (NS), and cardiac index by 14.3 per cent and 7.8 per cent ($p<0.01$) respectively.

Discussion

The results of this study show that patients with end-stage chronic renal failure have evidence of left ventricular hypertrophy and dilatation. This is most probably due to the increased cardiac work imposed in chronic renal failure. While in patients undergoing regular haemodialysis the left ventricular mass, the end-diastolic dimension and volume and the cardiac index increase with time, CAPD and, especially, successful renal transplantation lead to an opposite effect by significantly reducing the aforementioned indices. This should be regarded as a beneficial effect of the last two procedures.

The increased workload faced by the heart in chronic renal failure is primarily caused by anaemia, hypertension and the presence of the arterio-venous fistula [1]. This may lead to left ventricular hypertrophy and cardiac failure [1,2]. It has also been suggested that uraemia itself can cause myocardial dysfunction (uraemic 'cardiomyopathy') possibly through a direct effect of some uraemic toxins [1]. These changes are reversible, since it has been documented that left ventricular hypertrophy could regress after correction of left ventricular overload as has been shown after aortic valve replacement and in experimental hypertension [6]. Regression of left ventricular hypertrophy was also observed after effective parathyroidectomy in uraemic patients with secondary hyperparathyroidism [7].

The observed deterioration of left ventricular hypertrophy and dilatation following long-term haemodialysis has been previously reported [2,8]. The main factors contributing to left ventricular hypertrophy and dilatation are the persistence of chronic anaemia, the presence of the arterio-venous fistula and the hypertension which is common in patients with end-stage chronic renal failure even on long-term haemodialysis [1]. Under the above circumstances which lead to volume and pressure overload the left ventricle reacts with hypertrophy and dilatation [2,8]. It was found that the behaviour of the heart during haemodialysis depends on the pre-dialysis status of left ventricular function. Patients with depressed left ventricular function show an improvement of left ventricular ejection fraction while patients without heart failure remain stable or even deteriorate [9]. There is still controversy regarding uraemic toxins and uraemic cardiomyopathy which may have a negative inotropic effect on left ventricular contractility contributing independently to left ventricular dilatation [1,2].

The improved left ventricular hypertrophy and function in patients on long-term CAPD deserves special comments. It is supported that the beneficial effect of CAPD is mainly due to a reduction in volume and pressure overload [3,10]. This is reflected by a significant reduction of LVIDd, left ventricular mass and cardiac index. It is most probably effected by improvement of anaemia, better control of uraemia and hypertension, absence of the arterio-venous fistula and avoidance of the rapid haemodynamic fluctuations common during haemodialysis [3,10]. However further investigations are needed as far as the role of CAPD in removing uraemic toxins is concerned. Reversibility of left ventricular hypertrophy and dysfunction after successful renal transplantation has been recently reported [4,8]. Our findings are similar to those described in the literature and furthermore show that renal transplantation is superior to CAPD regarding the improvement of left ventricular hypertrophy and dilatation. This should be expected since after successful renal transplantation most of the previously described contributing factors to left ventricular hypertrophy and dysfunction are effectively corrected.

It is concluded that left ventricular hypertrophy and dilatation in patients with end-stage chronic renal failure and stable function deteriorate with regular haemodialysis, while they regress with CAPD and especially after successful renal transplantation.

References

1 Eiser AR, Swartz CE. In Lowenthal DT, Pennock RS, Likoff W, Onesti G, eds. *Management of the Cardiac Patient with Renal Failure.* Philadelphia: F A Davis Co. 1981: 69
2 Lewis BS, Milne FJ, Goldberg B. *Br Heart J 1976; 38:* 1229
3 Leenen F, Smith D, Khanna R et al. *Peritoneal Dialysis Bulletin 1983; 3:* S26
4 Cueto-Garcia L, Herrera J, Arriaga J et al. *Chest 1983; 83:* 56
5 Sahn DS, de Maria A, Kisslo J et al. *Circulation 1978; 58:* 1072
6 Schuler G, Peterson KL, Johnson A et al. *Am J Cardiol 1979; 44:* 585
7 Deligiannis A, Kontopoulos A, Memmos D et al. *Hellenik Cardiol Rev.* In press
8 Montague TJ, MacDonald RP, Boutilier FE et al. *Chest 1982; 82:* 441
9 Hung J, Harris PJ, Uren RF et al. *N Engl J Med 1980; 302:* 547
10 Deligiannis A, Vargemezis V, Tsobanelis T et al. *Life Support Systems 1983; 1:* S179

No discussion owing to technical difficulties with recording.

Proc EDTA–ERA (1984) Vol 21

THE ROLE OF ENDOGENOUS OPIOIDS IN THE BAROREFLEX DYSFUNCTION OF DIALYSIS PATIENTS

C Zoccali, M Ciccarelli, F Mallamaci, Q Maggiore, *M Stornello, *E Valvo, *L Scapellato

*Centro Fisiologia Clinica CNR, Reggio Calabria, *Ospedale Umberto I, Siracusa, Italy*

Summary

We studied the effect of the opiate antagonist naloxone on the response to Valsalva manoeuvre in nine dialysis patients, in six diabetics with normal renal function whose response to Valsalva manoeuvre was similar to that of dialysis patients and in eight healthy subjects.

Naloxone caused a progressive increase in the subnormal Valsalva ratio in dialysis patients but it did not cause any change in diabetics nor in healthy subjects. The increase in Valsalva ratio observed in dialysis patients was due to restoration of the parasympathetically mediated reflex bradycardia of the release phase of the manoeuvre.

Endogenous opioids may be responsible for the baroreflex dysfunction of dialysis patients.

Introduction

It is well established that baroreflex control of heart rate is often impaired in patients with chronic renal failure [1,2]. However, the factors involved in the pathogenesis of this dysfunction remain unclear.

It has recently been proposed that accumulation of unknown middle molecular weight substances interfering with baroreflex control might account for the baroreflex dysfunction in patients with chronic renal failure [3]. We thought that one of such substances might be the endogenous opioid (Met)-enkephalin, indeed this compound depresses baroreflex sensitivity, has a MW of 576 daltons and is retained in chronic renal failure [4].

The present study was designed to find out whether the retention of (Met)-enkephalin is involved in the baroreflex dysfunction in uraemic patients on chronic haemodialysis. For this purpose we have tested the effect of the opiate antagonist naloxone on the response to Valsalva manoeuvre in a group of dialysis patients, in a group of diabetic patients whose response to Valsalva manoeuvre was similar to that of dialysis patients and in a group of normal subjects.

Methods

Subjects

We studied nine uraemic non diabetic males (age 36–58 years; mean 44), six diabetic males (19–64 years; mean 43) and eight healthy males (31–53 years; mean 41). The uraemic patients had been dialysed three times a week for periods ranging from one to 11 years. They were well at the time of the study with no evidence of heart failure or pericardial effusion. The six diabetic patients had normal serum creatinine and had been treated with insulin (4 cases) or with oral hypoglycaemic drugs (2 cases) for periods ranging from two to 13 years. None of the dialysis or diabetic patients had symptoms of peripheral or autonomic neuropathy. No patient was taking antihypertensive drugs or other medications known to affect the autonomic nervous system. The nature of the study was explained to all participants who gave their informed consent.

Protocol

Each subject was studied on two occasions 48 hours apart, uraemic patients being investigated the day between dialyses. In the first study, at 9a.m. subjects rested quietly for 15–20 minutes and their supine mean arterial pressure (MAP= pulse pressure/3+diastolic) and heart rate were recorded at five minute intervals with a Dinamap 845 monitor. The response to Valsalva manoeuvre was then tested and the Valsalva ratio calculated as the ratio of the longest RR interval of the release phase to the shortest RR interval of the strain phase [5]. After these baseline measurements (nil time) subjects received intravenously either naloxone (.07mg/kg) or placebo (10ml of saline) in randomised, balanced and single blind fashion. Supine mean arterial pressure and heart rate and the response to Valsalva manoeuvre were revaluated one, two and three hours after the injection. On the second study the treatments were crossed over and the same experimental sequence repeated.

All results are expressed as mean ± SEM and their changes with time analysed with repeated measures. To compare the effects within the groups of naloxone with placebo the paired 't' test was employed.

Results

Before injection of either naloxone or placebo (Figure 1, nil time) there were no significant differences between the three groups as for supine mean arterial pressure and heart rate. By contrast, Valsalva manoeuvre ratio was significantly less in dialysis and diabetic patients than in normal subjects ($p<0.05$ or less). In both dialysis and diabetic patients the impaired Valsalva ratio was due to a reduced response during the release phase ($p<0.01$ or less), the response during the strain phase being no different from that of healthy subjects (Figure 2, nil time).

Naloxone had no effect on supine mean arterial pressure and heart rate in any of the groups. However, in dialysis patients the drug caused a progressive

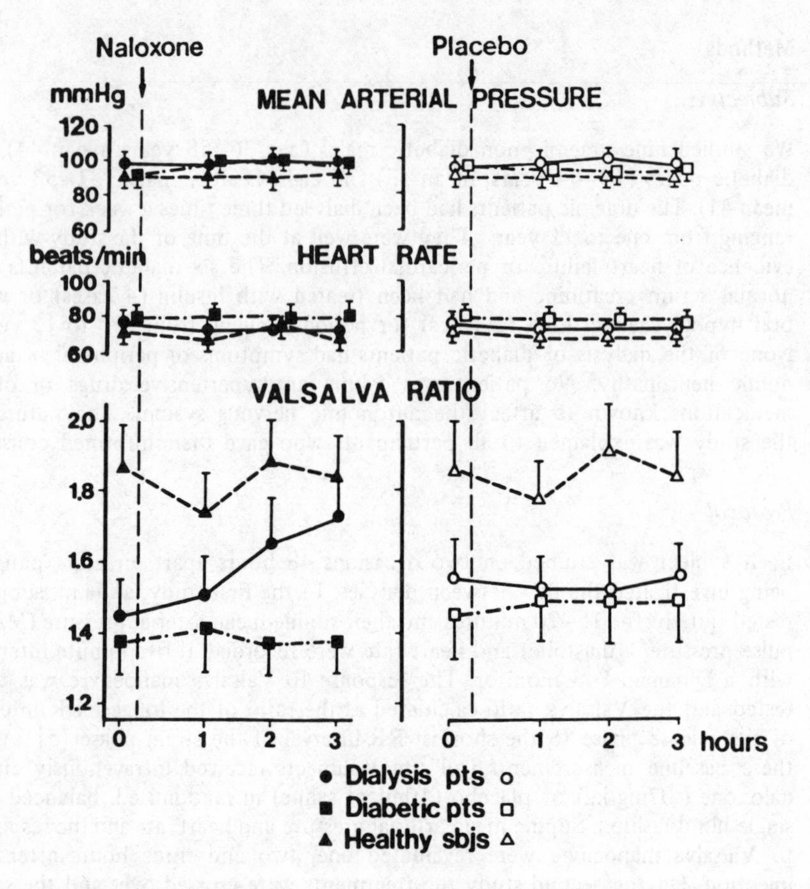

Figure 1. Effects of naloxone and placebo on supine mean arterial pressure and heart rate and the response to Valsalva manoeuvre

increase in Valsalva ratio from a baseline value of 1.46 ± 0.09 to a final value of 1.73 ± 0.011 ($p<0.0125$). Placebo administration did not alter Valsalva ratio in any of the groups. As shown in Figure 2, the improvement in Valsalva ratio observed in dialysis patients after naloxone was almost entirely attributable to a significant lengthening of the RR interval during the release phase ($p<0.01$), the RR interval being unaffected during the strain phase.

Discussion

Abnormal baroreflex control of heart rate to stimuli such as Valsalva manoeuvre and phenylephrine injection were first reported in chronic uraemics by Hennessy [1] and Pickering [2]. Subsequently several authors [7,8], including ourselves [5], confirmed these observations. This autonomic defect has been attributed to a parasympathetic dysfunction probably resulting from an afferent or central lesion in the baroreflex arc [5]. In agreement with these studies we also found

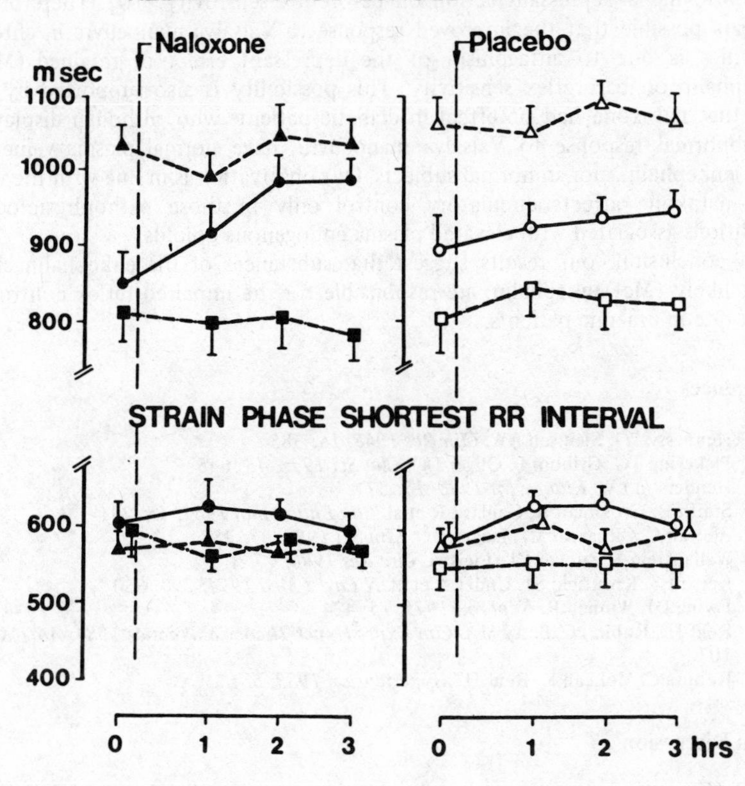

Figure 2. Effects of naloxone and placebo on the release phase and the strain phase of Valsalva manoeuvre. Symbols are as identified in Figure 1

that the reduced Valsalva ratio in dialysis patients was due to a parasympathetic dysfunction. Indeed this abnormality was sustained mainly by an impaired heart rate response during the release phase which is known to depend on parasympathetic activation.

By restoring the response during the release phase naloxone almost entirely reversed the abnormal Valsalva ratio in dialysis patients. The drug had no similar effect in diabetic patients nor in normal subjects. The favourable effect of naloxone in chronic uraemics most likely results from antagonism of endogenous opioids. Endogenous opioids have been recently measured in uraemic patients by Smith et al [4] who found a slight increase in plasma beta-endorphin and markedly elevated levels of (Met)enkephalin. Enkephalins are important modulators of baroreflex control: in rabbits, intracisternal injection of an enkephalin analogue attenuates the reflex heart rate response to hypotensive and hypertensive stimuli such as sodium nitroprusside infusion and phenylephrine injection while these effects are prevented by intravenous naloxone [9]. That enkephalins participate in the regulation of baroreflex activity in man is suggested by the observation that intravenous infusion of a (Met)enkephalin analogue,

DAMME, has a depressant action on baroreflex sensitivity [10]. Therefore, it appears possible that the improved response to Valsalva manoeuvre in chronic uraemics is due to antagonism of the depressant effect of retained (Met)-enkephalin on baroreflex sensitivity. This possibility is also supported by the fact that naloxone had no effect in diabetic patients who, although displaying an abnormal response to Valsalva manoeuvre, have normal plasma values of (Met)enkephalin, nor in normal subjects. Our observation is in line with the view that naloxone affects circulatory control only in those pathophysiological conditions associated with elevated plasma endogenous opioids.

In conclusion, our results suggest that substances of the enkephalin class, most likely (Met)enkephalin, are responsible for the impaired reflex control of heart rate in uraemic patients.

References

1 Hennessy WJ, Siemsen AW. *Clin Res 1968; 16:* 385
2 Pickering TG, Gribbin B, Oliver DO. *Clin Sci 1972; 43:* 645
3 Henderson LW. *Kidney Int 1980; 17:* 571
4 Smith R, Grossman A, Gaillard R et al. *Clin Endocrinol 1981; 15:* 291
5 Zoccali C, Ciccarelli M, Maggiore Q. *Clin Sci 1982; 63:* 285
6 Wallanstein S, Zucker CL, Fleiss J. *Circ Res 1980; 47:* 1
7 Kersh ES, Kronfield SJ, Under A et al. *N Engl J Med 1974; 290:* 650
8 Ewing DJ, Winney R. *Nephron 1975; 15:* 424
9 Reid JL, Rubin PC, Petty MA. *Clin Exper Hyper Theory & Practice 1984; A6(1 & 2):* 107
10 Rubin PC, McLean K, Reid JL. *Hypertension 1983; 5:* 535

Open Discussion

RITZ (Chairman) Could you briefly comment on whether there was nausea in your patients on the administration of naloxone? This of course could heighten parasympathetic tone.

ZOCCALI No, there were no side effects.

RITZ Your observations would suggest that there is a reversible functional component to baroreflex dysfunction. There is also some evidence I would like to mention: Dr Röckel's* paper in the European Journal of Clinical Investigations five years ago stated that even after transplantation there are still subtle abnormalities pointing to autonomic nerve dysfunction. Have you had the opportunity of studying patients after transplantation in order to see whether the abnormality has gone or is still responsive to naloxone?

ZOCCALI Yes, we read the paper about the effect of renal transplantation on autonomic insufficiency and we found that the autonomic dysfunction is at least in part reversed by renal transplantation, but a subtle defect remains although it is a minor one.

*Röckel A, Hennemann H, Sternagel-Haase A, Heidland A. *Eur J Clin Invest 1979; 9:* 23

Proc EDTA–ERA (1984) Vol 21

ABSENCE OF A BENEFICIAL HAEMODYNAMIC EFFECT OF BICARBONATE VERSUS ACETATE HAEMODIALYSIS

R Vanholder, M Piron, S Ringoir

University Hospital, Gent, Belgium

Summary

The present study compares data on blood pressure and clinical tolerance, obtained consecutively in the same patients during acetate and bicarbonate haemodialysis. Twenty-one patients were followed over an equal period of acetate and bicarbonate dialysis, averaging more than 30 months per patient. Absolute and relative blood pressure changes were noted. Contrary to what often has been claimed previously, it is concluded from the present long-term study that, bicarbonate haemodialysis has no specific beneficial effect on blood pressure in stabilised chronic patients. As far as vomiting and nausea are concerned, clinical tolerance is, however, significantly better than for acetate haemodialysis.

Introduction

The influence of bicarbonate haemodialysis on blood pressure and haemodynamically mediated symptoms remains a controversial issue. During the last eight years an impressive series of studies with contradictory results has been published (Table I). Of 17 studies, only six revealed no different blood pressure behaviour in association with acetate and bicarbonate dialysis. The remaining 11 studies showed less hypotensive episodes when a bicarbonate dialysate was used.

The reasons for these divergent results are multiple. Studies were often performed in pre-selected patients either with a low acetate tolerance [1,2] and/or on high efficiency dialysis [1,3,4]. In some studies there were significant differences in intra-dialytic ultrafiltration and inter-dialytic weight gain, despite a similar behaviour of the blood pressures [4,5]. Most studies were performed in small patient groups, covering only one or a few haemodialyses [1,3,5–7]; hypotension was often defined as an absolute value, rather than a relative change towards the pre-dialysis figure [1,2,4,8].

Finally, blood pressure changes during haemodialysis were not always taken

TABLE I. Comparison of bicarbonate and acetate haemodialysis* (literature review)

	Hypotension	Cramps	Headache	Nausea	Vomiting
1. Aizawa et al (1977)	–	NS	NS	NS	NS
2. Graefe et al (1978)	+	NS	+	+	+
3. Wehle et al (1978)†					
Na 145	–	NS	NS	NS	NS
Na 133	+	NS	NS	NS	NS
4. Fournier et al (1979)	+	+	+	NS	+
5. Nagai et al (1979)	NS	NS	+	+	–
6. Van Stone and Cook (1979)	–	NS	NS	NS	NS
7. Cambi et al (1980)	+	NS	NS	NS	NS
8. Iseki et al (1980)	+	NS	NS	NS	NS
9. Borges et al (1981)	–	–	–	–	
10. Deroussent et al (1981)	+	NS	NS	NS	NS
11. Hampl et al (1982)	+	NS	NS	NS	NS
12. Landwehr and Okusa (1982)	+	+	–	+	+
13. Leenen et al (1982)	+	NS	NS	NS	NS
14. Lefebvre et al (1983)	+	+	+	+	+
15. Vincent et al (1983)	–	NS	NS	NS	NS
16. Wolf et al (1983)	+	–	+	NS	–
17. Thieler et al (1984)	–	–	+	+	+

+: results better with bicarbonate dialysis than with acetate.

–: results not better with bicarbonate than with acetate.

NS: not studied.

†: these authors used two types of dialysate with a different sodium concentration.

*: complete list of references can be obtained from the authors.

into account [3,5,6,9] and practically no attention was paid to eventual hypertensive episodes [4,10].

Subsequently, in spite of the available literature on the subject, long-term comparative data between acetate and bicarbonate dialysis in unselected chronic renal failure patients are scarce. The present study was undertaken to compare the long-term influence of acetate and bicarbonate haemodialysis on intradialytic blood pressure and symptomatology of chronic renal failure patients. Except for nausea and vomiting, no significant differences between both regimens were observed.

Patients and methods

The study was performed in 21 clinically stable chronic dialysis patients, 14 males and 7 females, with an average age of 50.1±12.5 years (range 23 to 72 years). They had been on single needle dialysis with double head pump for a period of 50.9±25.1 months (range 4 to 105 months) at the start of the study.

In all patients an approximately identical period of acetate and bicarbonate dialysis was compared. Only patients with a follow-up of six months or more with each type of dialysate were included. The average follow-up was 15.8±5.5 months on acetate and 16.1±5.6 months on bicarbonate dialysis, corresponding to a total number of 4,233 and 4,279 dialysis sessions respectively.

Five patients were on single needle haemodiafiltration, with 10 litres of substitution fluid, the remaining patients being on conventional single needle haemodialysis. All patients were treated four hours three times weekly. The dialysis strategy was not changed during the study period. Cuprophan or polyacrylonitrile dialysers with a surface area of 1.0 to 1.2m² were used. The blood flow and the dialysate flow averaged 292.7±32.7 and 507.2±44.6ml/minute respectively.

The dialysate composition (mmol/L) was as follows:

	Acetate	Bicarbonate
Na^+:	138.0	138.0
K^+:	2.0	2.0
Ca^{++}:	1.8	1.8
Mg^{++}:	0.5	0.5
Cl^-:	108.6	108.6
Bic.:	none	34.0
Acet.:	35.0	1.0

Bicarbonate solution, concentrate and water were mixed in a central delivery system (4009-1, multi-patient, Drake Willock, USA). None of the patients took antihypertensive medication. The pre- and post-dialysis body weight, systolic and diastolic blood pressures, and the lowest and highest blood pressure of each dialysis session were registered.

The relative differences with the pre-dialysis values were calculated. Apart from these relative changes, hypotensive and hypertensive episodes were also registered as absolute values. They were defined either as a decrease and an increase of systolic blood pressure by 30 per cent, or as a systolic blood pressure lower than 100mmHg and a diastolic pressure higher than 95mmHg. The frequency of symptomatic hypotension, nausea, vomiting, headache and muscle cramps was registered and calculated per 100 dialysis sessions.

Results were calculated as means ± standard deviation (M±SD). Statistical comparison was performed with Wilcoxon's test for paired values. Significance was accepted for $p<0.05$.

Results

There were no significant differences in post-dialysis body weight during the entire study period, the average values ranging from 61.9±9.0kg at the start to 62.2±8.1kg at the end of the study. The inter-dialytic weight gain was also similar in both acetate and bicarbonate groups (Δ%: +2.50±1.22 and +2.58±1.16% respectively, $p>0.05$).

Figure 1 illustrates the evolution of the blood pressures in both groups. The absolute values, as well as the percentual changes towards pre-dialysis were not significantly different.

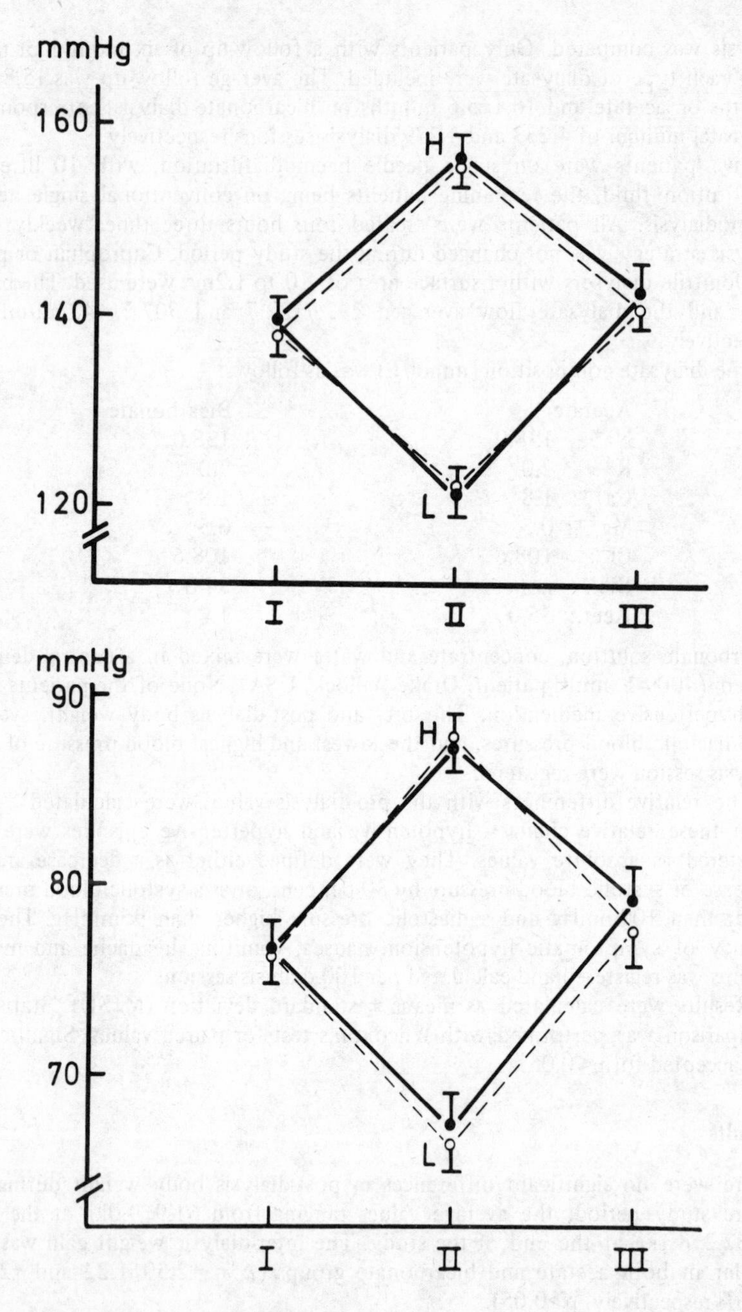

Figure 1. Evolution of systolic (upper panel) and diastolic blood pressures (lower panel); I: pre-dialysis; II: intra-dialytic; III: post-dialysis blood pressure; H: highest and L: lowest intra-dialytic blood pressure; •–•: acetate; ○- -○: bicarbonate. No significant differences were observed

TABLE II. Complications (per 100 dialyses)

	Acetate	Bicarbonate	P
Hypotension			
Systolic blood pressure ↘≥ 30%	5.94	5.84	NS
Systolic blood pressure <100mmHg	6.31	6.57	NS
Symptomatic hypotension	1.17	0.57	NS
Hypertension			
Systolic blood pressure ↗≥ 30%	6.38	5.41	NS
Diastolic blood pressure >95mmHg	12.06	9.61	NS
Headache	0.77	0.47	NS
Cramps	3.34	3.19	NS
Nausea	1.03	0.24	p<0.01
Vomiting	3.15	0.26	p<0.01

Table II summarises the complication profiles. There were no significant differences in the frequency of hypotensive and hypertensive episodes, either defined as an absolute value or as a relative change towards pre-dialysis. Symptomatic hypotensive episodes (i.e. associated with nausea, vomiting, headache or cramps) were twice as frequent during acetate than during bicarbonate dialysis. The difference was, however, not significantly different. The frequency of headache and cramps was similar in both groups. Nausea and vomiting were significantly less frequent during bicarbonate dialysis.

Discussion

The present study evaluates the influence of acetate and bicarbonate containing dialysate on blood pressure and dialysis tolerance in a large patient population, studied over a prolonged period.

A first conclusion is that intra-dialytic hypotensive episodes, either defined as an absolute value or as a relative change towards pre-dialysis, do not occur more frequently during acetate than during bicarbonate dialysis. This result is in agreement with a number of earlier studies [4–7,9,10]. Two of these studies had, however, been conducted in acute renal failure patients [5,10]. Moreover, the remaining studies had been performed in small patient groups and/or during short study periods [4,6,7,9]. The present study comes to a similar conclusion, but is based on a large group of chronic renal failure patients with a prolonged follow-up, averaging more than 15 months per patient and per type of dialysate.

The frequency of hypertensive episodes during acetate and bicarbonate dialysis is similar with both techniques. As far as we know this problem has rarely been studied, except for a study by Borges et al [10] who evaluated this eventuality in acute renal failure patients, and a study by Thieler et al [4] performed in chronic renal failure patients. In both studies no significant differences between both groups were found.

Finally, the present study compares the occurrence of a number of symptoms that traditionally have been associated with haemodynamic instability during haemodialysis. Although headache and muscle cramps occurred at the same rate, the frequency of nausea and vomiting was significantly lower during bicarbonate than during acetate haemodialysis. It must be stressed, however, that even in acetate dialysis a very low rate of complications was observed in these single needle dialysis patients. The fact that dialysis tolerance is improved during bicarbonate haemodialysis is in agreement with a substantial number of earlier studies (Table I), but most of these studies had been carried out in preselected patients, either with low acetate tolerance and/or on high efficiency dialysis [1–4,8]. The present long-term study seems to confirm earlier suggestions of better clinical tolerance for nausea and vomiting with bicarbonate dialysis, but in a controlled group of unselected patients.

In conclusion, this study reveals no differences in the evolution of the pre- and post-dialysis and intra-dialytic blood pressures during acetate and bicarbonate dialysis, despite a definition of hypo- and hypertension in both absolute and relative terms. On the other hand, there is a better clinical tolerance of dialysis with bicarbonate, at least as far as nausea and vomiting are concerned.

References

1 Graefe U, Milutinovich J, Follette WC et al. *Ann Int Med 1978; 88:* 332
2 Wolf C, Rebaiz K, Mauperin P et al. *Néphrologie 1983; 4:* 204
3 Nagai K, Pagel M, Rattazzi T et al. *Proc EDTA 1979; 16:* 122
4 Thieler H, Muller K, Schmidt U et al. *Nieren- und Hochdruckkrankheiten 1984; 13:* 106
5 Vincent JL, Vanherweghem JL, Degaute JP et al. *Kidney Int 1983; 22:* 653
6 Wehle B, Asaba H, Castenfors J et al. *Clin Nephrol 1978; 10:* 62
7 Van Stone JC, Cook J. *Dialysis and Transplantation 1979; 8:* 703
8 Lefebvre A. *Néphrologie 1983; 4:* 200
9 Aizawa Y, Ohmori T, Imai K et al. *Clin Nephrol 1977; 8:* 477
10 Borges HF, Fryd DS, Rosa AA et al. *Am J Nephrol 1981; 1:* 24

Open Discussion

VAN GEESON (FRG) Dr Vanholder, I know that you have a large experience of haemodiafiltration so did you also follow symptoms during haemodialysis?

VANHOLDER In this group there were some haemodiafiltration patients included. In any case when we made a similar comparison for haemodiafiltration and conventional haemodialysis on a large scale we have not found major differences in blood pressure although this study was not as elaborate as the present one.

KESHAVIAH (Minneapolis) I noticed from your numbers that these patients were relatively stable patients with a low incidence of symptomatic hypotension and also no instance of cramps. I think you cannot show beneficial effects of bicarbonate in such stable patients. If you were to make the patients unstable and then do the study then possibly you may get changes.

VANHOLDER Yes.

KESHAVIAH We have recently done this by shortening the treatment every time, and report each patient that has been shortened to the level of his instability and then by using bicarbonate you see great changes.

VANHOLDER Yes I agree with you. I think the point here is that we are not trying to say that bicarbonate in general for every patient has no advantages. I think that there are some patients who do gain advantage from bicarbonate or acetate but on the other hand there are some patients who certainly do not. The question is whether you should in a normal dialysis situation look after the patient and bring them to instability?

GOTCH (San Francisco) The comment I would like to make is these were all done with single needle systems and relatively inefficient dialysers, that is in the sense that the clearances were probably 140–150, and I think it might be very different if you had clearances of 200–230 which are easily obtained: There I think you might have seen a really high acetate.

VANHOLDER I agree with that, but the first publication on that subject, also done on high efficiency dialysers, showed the same thing.

Proc EDTA–ERA (1984) Vol 21

CARDIOVASCULAR RISK FACTORS AND CARDIOVASCULAR DEATH IN HAEMODIALYSED DIABETIC PATIENTS

C Strumpf, F Katz, *A J Wing, E Ritz

Department Internal Medicine, University of Heidelberg, FRG,
**EDTA Registry, St Thomas' Hospital, London, United Kingdom*

Summary

In a retrospective study, causes of death and the cardiovascular risk conferred by established risk factors were analysed in 200 diabetic and 200 matched non-diabetic uraemic patients admitted for haemodialysis. Total and cardiovascular mortality was considerably higher in diabetics, both type I (4.8-fold) and type II (3.0-fold). Fifty-eight per cent of deaths in diabetics, but only 38 per cent of deaths in non-diabetics were due to cardiovascular causes; myocardial infarction and stroke accounted for <15 per cent of deaths, the majority being due to sudden death. Cardiovascular death in diabetes was predicted by both history of hypertension and by cardiomegaly, but to a much lesser extent by clinical evidence of macroangiopathy. The results are compatible with an important role of non-coronary cardiomyopathic mechanisms of cardiovascular death in dialysed diabetics.

Introduction

High cardiovascular mortality is the single most important factor for poor survival of diabetics on haemodialysis. Little information is available on the risk factors associated with such elevated cardiovascular mortality. Quantitative information on the importance of hypertension or other risk factors and risk estimates in relation to non-diabetics are currently not available.

The present retrospective multicentre study was designed specifically to answer these questions. The records of diabetics having started dialysis between 1972 and 1983 were analysed; potential cardiovascular risk factors at entry were noted and related to the subsequent course of the patients on dialysis.

Patients and methods

Patient population

From the records of the European Dialysis and Transplant Association (EDTA) all diabetics who entered uraemia treatment programmes in 17 German dialysis centres between 1.1.1972 and 31.5.1983 were studied. A total of 228 diabetic patients were reported to the EDTA registry. The documents of 24 patients could not be traced, four patients were erroneously reported as having diabetes; this leaves a total of 200 patients for further analysis. Patients who subsequently were either transplanted or transferred to peritoneal dialysis were excluded from further analysis. Of the remaining 182 haemodialysed patients, 2.7 per cent were on limited care dialysis, 2.7 per cent on home dialysis and 94.6 per cent on centre haemodialysis. Included in the final analysis were 58 patients with type I diabetes, 111 patients with type II diabetes and 13 patients with non-classified diabetes.

Case controls were obtained by a manual hierarchical search of the EDTA files. Patients were listed for each centre according to their dates of first treatment. This list was searched alternately on either side of the diabetic case until two cases were found with the same sex and age (±5 years). The search was extended to only one year on either side. The first of the two controls whose notes were retrieved was used as the control case. Of the 200 control patients entry treatment programmes, 14 control patients were transplanted and three transferred to peritoneal dialysis, leaving a total of 183 patients for analysis. In unbalanced pairs where either the diabetic or matched control patient was lost for analysis because of transplantation or transition to peritoneal dialysis, the remaining partner was used for further analysis. Of the 183 control patients included in the final analysis, 16 (8%) were on limited care dialysis, 10 (6%) on home dialysis and 157 (86%) on centre haemodialysis.

Analysis of cardiovascular risk factors

Taking the data of admission to the dialysis programme as the reference point, information on the following risk factors was obtained from available written patient records. Documented history of long-standing (>5 years) hypertension (>160/95mmHg on three separate occasions or antihypertensive therapy); hypertension at time of admission into dialysis; ECG evidence of left ventricular hypertrophy – LVH); cardiomegaly at time of first dialysis (cardiothoracic ratio >0.5); absence of one or more peripheral pulses on physical examination; recorded amputation; history of myocardial infarction (ECG or enzyme changes); stroke or transient ischaemic attacks (TIA).

Endpoints

The time of death and causes of death were taken from records, consultation of family members or private physicians.

Autopsy records were available for 23 of 93 deceased diabetics (25%) and 13 of 45 deceased controls (29%).

Statistical evaluation

Actuarial survival was calculated using standard techniques. The relative risk was calculated as $R = \frac{A/N1}{C/N2}$ from a contingency table. To avoid a Bonferoni type of error from multiple testing, the findings are considered significant only when $p<0.01$.

Results

Survival

Figure 1 shows actuarial rate of loss of patients due to cardiovascular death for type I and type II diabetics and their matched controls. It is obvious that the rate of cardiovascular death in diabetics of all groups vastly exceeds that of their respective controls. Cardiovascular mortality progressively increased with age (data not given). In all type I diabetics, the overall proportion of patients dying from cardiovascular causes was 4.8-fold higher than in their matched controls ($p<0.001$); in type II diabetics, the respective proportion was 3.0-fold higher ($p<0.01$).

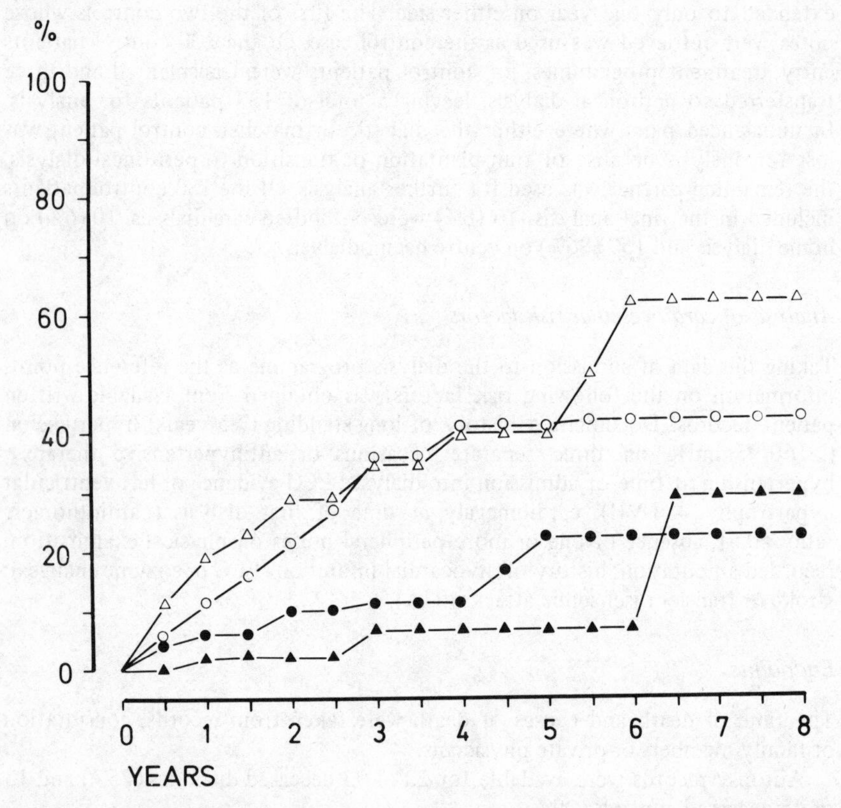

Figure 1. Cardiovascular mortality. △–△=type I; ○–○=type II; ▲–▲=control to type I; ●–●=control to type II

The causes of death in diabetics and controls are given in Table I. Cardiovascular causes accounted for a higher proportion of total death in diabetics (58% of all deaths) than in matched controls (38%), but myocardial infarction

TABLE I. Causes of death in diabetics and controls

	type I	type II	non-classified	total diabetics
Diabetics				
Sudden death	17/34	21/50	5/9	(43)
Myocardial infarction	0/34	5/50	2/9	(7)
Stroke	2/34	2/50	0/9	(4)
Total cardiovascular	19/34	28/50	7/9	54 (58%)
Infection	2/34	7/50	1/9	10 (11%)
Other	7/34	2/50	0/9	9 (10%)
Unknown cause	6/34	13/50	1/9	20 (21%)
Total dead	34	50	9	93 (100%)
(Total diabetics observed	58	111	13	182)
Controls				
Sudden death	2/11	7/27	3/7	(12)
Myocardial infarction	1/11	2/27	0/7	(3)
Stroke	1/11	1/27	0/7	(2)
Total cardiovascular	4/11	10/27	3/7	17 (38%)
Infection	0/11	1/27	0/7	1 (2%)
Other	4/11	5/27	2/7	11 (24%)
Unknown cause	3/11	11/27	2/7	16 (36%)
Total dead	11	27	7	45 (100%)
(Total non-diabetics observed)	58	111	14	183)

and stroke were responsible for only a minority of cardiovascular deaths, the majority being due to sudden death.

Relation between cardiovascular risk factors at admission for dialysis and subsequent cardiovascular death

Table II gives the relative risk for subsequent cardiovascular death in relation to risk factors present at entry into haemodialysis. The prevalence of risk factors (given in brackets) was generally higher in diabetics than in the control population.

A documented history of long-standing hypertension increased the relative risk in diabetics and to a lesser extent in non-diabetics (difference statistically not significant). Normotension at start of dialysis diminished and hypertension increased the relative risk. Electrocardiographic evidence of left ventricular hypertrophy, cardiomegaly at the time of first dialysis and particularly a combination of cardiomegaly and ECG abnormalities (other than left ventricular hypertrophy) considerably increased the relative risk.

In contrast, evidence of peripheral macroangiopathy, i.e. absence of lower

TABLE II. Relative risk for cardiovascular death (number in brackets: prevalence of risk factors (or their absence) in diabetic and control population)

	Total diabetics (n=151)	Total controls (n=152)
1. History of hypertension >5 years	2.3 (89.7%)	1.3 (80.4%)
2. Blood pressure at start of dialysis <160mmHg	0.4* (21.9%)	0.4 (35.1%)
3. Blood pressure at start of dialysis >160mmHg	2.5* (78.1%)	2.5 (64.9%)
4. Left ventricular hypertrophy in ECG	1.5 (52.1%)	1.9 (30.7%)
5. Cardiomegaly	1.7* (63.5%)	0.8 (51.2%)
6. Combination of cardiomegaly and ECG changes other than left ventricular hypertrophy	2.3** (23.5%)	***
7. Absence of lower extremity pulses	1.4 (37.1%)	4.4** (25.4%)
8. Amputation	1.7 (18.4%)	

Significant increase (or decrease) of relative risk (in diabetics or controls respectively)

* nominal $p<0.05$

** nominal $p<0.01$

*** insufficient number of patients

extremity pulses and amputation as well as (data not given) a history of myocardial infarction, stroke or TIA significantly increased the cardiovascular risk in non-diabetics but much less so in diabetics.

Discussion

Although the authors are painfully aware of the limitations of retrospective studies, several important conclusions can be drawn from the present results.

Hypertension is a risk factor of overriding importance in diabetic patients and a history of hypertension appears to be even more deleterious in diabetic as compared to non-diabetic uraemic patients.

Surprisingly, and in contrast to results of prospective studies in non-uraemic diabetics, evidence of peripheral macroangiopathy had little predictive value for cardiovascular death. This finding, in association with the rarity of myocardial infarction as a documented cause of death, may point to the importance of non-coronary cardiac mechanism of death (although admittedly the high proportion of 'sudden death' may include individuals with suicide, hyperkalaemia or severe autonomous polyneuropathy).

The notion of important non-coronary mechanisms of cardiac death would be consistent with the finding that cardiomegaly, particularly in association with ECG abnormalities, predicted cardiovascular death in diabetic and not in non-diabetic patients. The extent to which such presumed cardiomyopathy is the result of hypertension or evidence of specific diabetic cardiomyopathy remains to be defined. Of particular concern to the authors is the possibility of aggravation of cardiomyopathic damage by hypotensive episodes during haemodialysis.

Acknowledgment

We thank Professor Hennemann (Coburg), Professor Höffler (Darmstadt), Professor Schöppe (Zentrum Innere Medizin, Frankfurt), Dr Vlachojanis (St Markus-Krankenhaus, Frankfurt), Dr Piazolo (Friedrichshafen), Professor Wizemann (Giessen), Dr Huber (Rehabilitationsklinik, Heidelberg-Wieblingen), Dr Kluge (Heilbronn), Professor Strauch (Mannheim), Professor Kopp (Krankenhaus rechts der Isar, München), Professor Gessler (Nürnberg), Professor Heinze (Offenburg), Dr Streicher (Katharinen-hospital, Stuttgart), Dr Cremer (Trier), Dr Walb (Wiesbaden) and Professor Heidland (Würzburg) for permission to examine their patients.

The help of Dr Richard Moore in selecting patients from the EDTA registry is gratefully acknowledged.

We thank Ms Stelz for secretarial help.

Open Discussion

SHALDON (Montpellier) Did you have in any of your unexplained sudden deaths post mortem evidence to suggest that the clinical diagnosis was invalid? What concerns me is your assumption that sudden death is necessarily a cardiovascular death, obviously it is if you have a ventricular arrhythmia and the heart stops. If you have hyperkalaemia in an acidotic patient you would anticipate this might be a more frequent occurrence than in a patient with normal potassium. Do you have any evidence that the diabetic population was less or more acidotic?

RITZ No, we posed this question ourselves to avoid falling into the trap of artifact due to hyperkalaemia or severe autonomic polyneuropathy. I think the best evidence against this would be that the cardiovascular death can be predicted by certain features present at the start of dialysis. I cannot rule out the possibility that some patients died from hyperkalaemia or some died from autonomic problems but the very fact of the predictability argues against this for the major proportion of subjects.

I think one intriguing possibility which emerges from here is the finding of Quellhorst* that in diabetics on haemofiltration survival is much better than in

*Quellhorst E. In Brann J, Pilgrim R, Gessler U, Seybred D, eds. *Die Behandlung von Herzrhythmusstörungen bie Nierenkranken.* Basel: Karger. 1984: 23–35

non-randomised diabetic patients on haemodialysis. The finding he obtained was that the diabetic as well as the non diabetic dialysis patients had on close monitoring a greater incidence of arrhythmia and this was related to intradialytic hypotension. The patients with hypotension during dialysis had arrhythmias after dialysis and this would fit nicely with the predicted value of cardiomegaly. The concept would be that patients with cardiomegaly are more prone to hypotension and as a consequence more prone to a sudden cardiovascular death. This, of course, needs to be tested by a prospective study.

ZOCCALI (Italy) You found that cardiomegaly is not a predictor of sudden death in the normal population but only in diabetic patients. In the literature you will find many references to cardiomegaly as a bad prognostic factor. Is there any reason why you do not find cardiomegaly as a predictor of sudden death in non diabetic patients?

RITZ I think there is a very good reason, what was analysed was the cardiothoracic ratio at the time of the first dialysis. What you are referring to are reports of the patient on dialysis who is at his dry weight and has cardiomegaly. We are talking about two different groups.

Proc EDTA–ERA (1984) Vol 21

EFFECT OF DIALYSATE BUFFER ON POTASSIUM REMOVAL DURING HAEMODIALYSIS

A J Williams, J N Barnes, J Cunningham, F J Goodwin, F P Marsh

The London Hospital, London, United Kingdom

Summary

There is a linear relationship between potassium removal during haemodialysis and plasma potassium (Kp). Kp falls rapidly during the first hour of dialysis but very little during the last two hours of a five hour dialysis. There is a fairly constant movement of potassium from the intracellular to extracellular space throughout dialysis. Total potassium removal is best predicted by pre-dialysis Kp, but change in Kp is related to the impact of dialysis on acid-base status. The choice of acetate or bicarbonate buffered dialysate does not effect potassium removal during dialysis.

Introduction

Control of plasma potassium (Kp) is a major objective in treating patients with chronic renal failure. In haemodialysed patients the balance between intracellular and extracellular potassium is often disturbed. Potassium homeostasis is maintained by increased faecal excretion, dietary restriction and removal during haemodialysis.

Small molecules and ions, such as urea and potassium, pass across haemodialysis membranes readily: the net transfer depends on, amongst other things, the concentration gradient between plasma and dialysate. As plasma concentration falls, a change in the gradient between intracellular and extracellular solute concentration develops and solute moves out of cells. For substances such as urea, which pass freely across cell membranes, this movement maintains equal intracellular and extracellular concentrations [1]. However, in the case of potassium, additional factors influence distribution across cell membranes and prediction of the amount removed by haemodialysis is more difficult [2,3].

In animal studies there is a greater fall in Kp during intravenous acetate infusion than during infusion of an equimolar quantity of bicarbonate, despite similar changes in arterial blood pH [4]. If a similar difference were to occur in haemodialysed patients when using acetate as opposed to bicarbonate as a buffer, increased entry of potassium into cells would reduce its gradient across the haemodialysis membrane and, presumably, its removal during dialysis.

The aims of this study were: a) to evaluate the kinetics of potassium removal during haemodialysis when using bicarbonate or acetate buffered dialysate, b) to compare the effect of bicarbonate buffered dialysate with that of acetate buffered dialysate on potassium removal during haemodialysis and c) to evaluate the relationships between potassium removal during haemodialysis, Kp, red blood cell potassium (Kc) and acid-base status.

Patients and methods

Nine patients (8 males, 1 female; age 43 ± 15 (SD) years; mean weight 65.5 ± 13.3kg) treated by chronic haemodialysis were each studied on two occasions. On the first occasion patients were randomised to either bicarbonate or acetate haemodialysis and on the second occasion dialysis was performed under identical conditions using the alternative buffer. After a breakfast of toast and coffee patients were dialysed for five hours using Gambro AK10 (bicarbonate or acetate) haemodialysis monitors and extracorporeal hollow fibre dialysers (Triex 1) with two needle vascular access in radial arteriovenous fistulae. Blood and dialysate flows were constant at 200ml/min and 500ml/min respectively. The dialysate composition was: sodium 140mmol/L, potassium 1.5mmol/L, calcium 1.5mmol/L, magnesium 0.5mmol/L and dextrose 2g/L. In addition bicarbonate dialysate contained chloride 107mmol/L, acetate 4mmol/L and bicarbonate 40mmol/L while acetate dialysate contained chloride 104mmol/L and acetate 40mmol/L.

Blood was taken anaerobically from the arterial limb of the dialysis circuit before dialysis and then at half, one, two, three, four and five hours after starting dialysis. pH, pCO_2 and PO_2 were measured immediately using an ABL 1 blood gas analyser (Radiometer, Copenhagen) and bicarbonate concentration calculated from the Henderson Hasselbach equation. Plasma potassium and sodium were measured by flame photometry. Plasma urea, creatinine, phosphate, albumin and uric acid were measured using standard autoanalyser techniques (Technicon). Haematocrit was measured using a Coulter counter. Red blood cell potassium (Kc) was determined using the formula:

$$Kc = \frac{Kt - Kp\,(1 - PCV)}{PCV}$$

where Kt is whole blood potassium concentration (mmol/L) and PCV is the haematocrit expressed as a fraction of 1. Kt was determined by flame photometry after the cells had been lysed by addition of deionised water (4ml) to whole blood (2ml).

Samples of dialysate were collected from the input (Di) and output (Do) channels of the dialyser immediately after starting dialysis, half and one hour later and then at hourly intervals throughout dialysis. The concentration of sodium and potassium in the dialysate was estimated by flame photometry. The hourly potassium removal (HKR) was calculated on the basis of a dialysate flow of 30 litres per hour.

$$HKR = \frac{((K_{Do}-K_{Di_t}) + (K_{Do}-K_{Di_{t-1}})) \times 30\text{mmol}}{2}$$

where t is the number of hours dialysis. The dialysate effluent was collected in two containers and total potassium removal (TKR) calculated.

$TKR = ((K_{Do}-K_{Di}) \times 150) + (K_{Do} \times \text{wt loss})$ mmol where 150 represents total dialysate volume in a five hour dialysis and weight (wt) loss of the patient is measured in kg. The results are expressed as means ± SD. Student's 't' test was used to assess the significance of correlation coefficients.

Results

There was no significant difference in the acid-base (Table I) and biochemical status of the patients before the two forms of dialysis: acetate (bicarbonate): plasma urea 25.5 ± 5.2 (23.8 ± 4.3) mmol/L; plasma creatinine 1045 ± 200 (992 ± 171) μmol/L; plasma phosphate 1.69 ± 0.73 (1.78 ± 0.69) mmol/L; plasma uric acid 485 ± 73 (463 ± 77)μmol/L.

Plasma potassium fell rapidly during the first hour of dialysis and then gradually during the subsequent four hours (Figure 1a). The changes were similar with bicarbonate and acetate buffered dialysis (Table I). The rate of change in Kp (ΔKp/hr) during dialysis and removal of potassium by dialysis (HKR) appeared to be linearly related to Kp (Figures 1b and 1c). Total potassium removal (TKR) during a five hour dialysis was the same when using acetate (64.8 ± 15.0mmol/L) or bicarbonate (62.1 ± 14.2mmol/L) buffered dialysate. The fall in Kp (ΔKp) during a bicarbonate dialysis did not differ significantly from that during acetate dialysis (Table I). There was no change in Kc during dialysis.

To assess the factors that influence potassium removal, the studies using acetate and bicarbonate dialysis were considered together. Potassium removal by dialysis (TKR) correlated with pre-dialysis Kp ($r=0.60$; $p<0.01$) particularly when adjusted for body weight (TKR/kg: $r=0.79$; $p<0.001$). It did not correlate significantly with pre-dialysis arterial pH or bicarbonate nor with the change in pH, bicarbonate or base excess. ΔKp during dialysis correlated with pre-dialysis Kp ($r=0.86$; $p<0.001$) and TKR/kg ($r=0.63$; $p<0.01$) as well as with the change in base excess ($r=0.53$; $p<0.05$), Δbicarbonate ($r=0.5$; $p<0.05$) and ΔpH ($r = 0.54$; $p<0.05$). ΔKp did not correlate with pre-dialysis bicarbonate or pH nor with TKR. During the course of dialysis, Kp became progressively less dependent on pre-dialysis Kp (1 hr, $r=0.93$; 2 hr, $r=0.74$; 3 hr, $r=0.46$; 4 hr $r=0.13$; 5 hr, $r=0.10$). The decrease in correlation between pre- and post-dialysis (5 hr) Kp was less marked with creatinine ($r=0.56$; $p<0.02$), phosphate ($r=0.66$; $p<0.01$), uric acid ($r=0.36$; $p<0.1$) and urea ($r=0.34$; $p<0.1$).

Discussion

The ability to predict the kinetics of solute removal by haemodialysis is of value when planning treatment. Previous studies have shown that the volume of distribution of urea consists of a 'single pool' and that after a short dialysis

TABLE I. Plasma and erythrocyte potassium and acid-base status during a 5-hour dialysis using either acetate (Ac) or bicarbonate (HCO_3) buffered dialysate. Results are expressed as mean ± SD. No statistical significance was observed with respect to the different dialysate buffers

Hours of dialysis	Plasma potassium (mmol/L)		Red cell potassium (mmol/L)		Arterial pH		Arterial pCO_2 (kPa)		Bicarbonate (mmol/L)		Base excess	
	Ac	HCO_3	Ac	HCO_3	Ac	HCO_3	Ac	HCO_3	Ac	HCO_3	Ac	HCO_3
0	4.6 ± 0.6	4.4 ± 0.6	105.5 ± 4.4	105.7 ± 8.2	7.39 ± 0.05	7.39 ± 0.02	4.57 ± 0.37	4.87 ± 0.38	20.5 ± 2.2	21.7 ± 1.5	−3.6 ± 2.6	−2.6 ± 1.4
½	4.0 ± 0.6	3.9 ± 0.5	104.7 ± 3.2	105.4 ± 6.3	7.41 ± 0.04	7.41 ± 0.03	4.52 ± 0.37	4.98 ± 0.43	21.1 ± 1.4	23.4 ± 1.3	−2.8 ± 1.7	−0.7 ± 1.2
1	3.6 ± 0.5	3.6 ± 0.5	104.4 ± 3.2	105.8 ± 6.4	7.43 ± 0.04	7.43 ± 0.02	4.55 ± 0.26	5.11 ± 0.40	22.4 ± 1.4	24.9 ± 1.4	−1.4 ± 1.6	0.7 ± 1.3
3	3.0 ± 0.2	3.2 ± 0.2	105.2 ± 4.6	105.1 ± 6.7	7.48 ± 0.03	7.49 ± 0.02	4.54 ± 0.10	4.96 ± 0.34	25.3 ± 1.8	28.3 ± 2.0	1.9 ± 2.0	4.7 ± 1.8
5	2.7 ± 0.2	3.0 ± 0.2	104.5 ± 3.8	104.6 ± 7.1	7.50 ± 0.03	7.50 ± 0.02	4.79 ± 0.33	5.08 ± 0.37	27.7 ± 1.4	29.6 ± 1.6	4.3 ± 1.5	6.0 ± 1.5

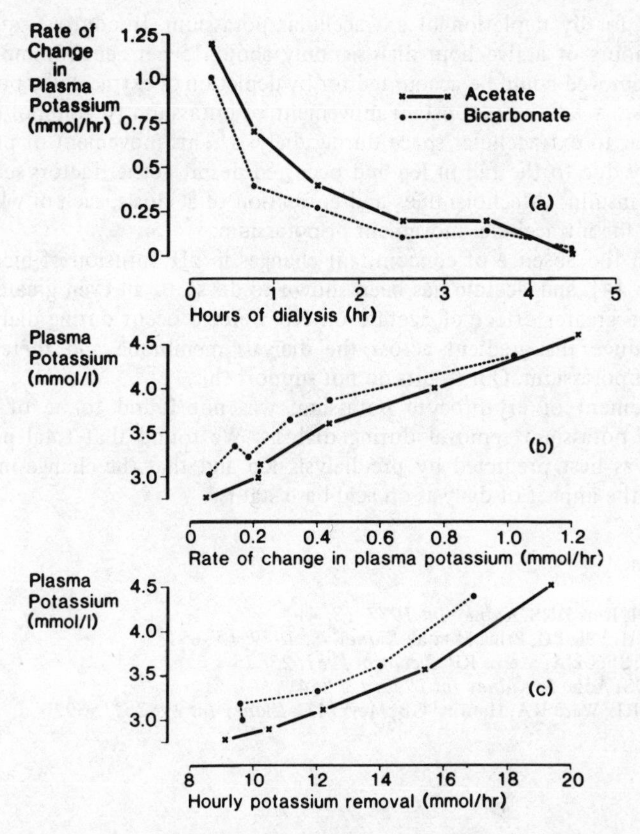

Figure 1. Kinetics of potassium removal during dialysis when using acetate and bicarbonate buffered dialysate. The rate of change of plasma potassium in Figure 1(a) was derived at 15 minutes, 45 minutes, 90 minutes and then hourly. No statistical significance was observed with respect to the different dialysate buffers

(3 hr) there is a direct relationship between pre- and post-dialysis plasma urea [3]. In contrast, the distribution of potassium cannot be represented by a 'single pool' model and pre- and post-dialysis Kp do not correlate.

Potassium removal occurs at a rate proportional to the gradient of potassium between plasma and dialysate throughout a 5 hour dialysis (Figure 1c). Plasma potassium fell rapidly during the first hour of dialysis but only slowly after the second hour. During the fifth hour of the five hour dialysis against a dialysate potassium of 1.5mmol/L, Kp fell by less than 0.1mmol/L when either acetate or bicarbonate was used as buffer (Figure 1a). With constant extracellular volume, Kp during dialysis depends both on the loss of potassium across the dialysis membranes and on the flux across cell membranes. Assuming that the extracellular volume of our patients was 13 litres (20% of body weight) and that potassium was distributed equally throughout this volume, about 12mmol of the 18mmol of potassium removed during the first hour of dialysis may be accounted for by depletion of extracellular potassium. In contrast, during the last two hours of a five hour dialysis, only about 15 per cent (3mmol) of the 20mmol removed could be accounted for by depletion of extracellular potassium. This implies a relatively constant movement of potassium (6–8mmol/hr) from intracellular to extracellular space during dialysis. This movement of potassium was mainly due to the fall in Kp and occurred despite other factors such as the release of insulin, catecholamines and correction of acidosis, each of which may encourage the intracellular movement of potassium.

Even in the absence of concomitant changes in pH, infusion of bicarbonate lowers Kp [4] and acetate has been shown to do so to an even greater degree [5]. If the greater effect of acetate on Kp were to occur during dialysis, this should reduce the gradient across the dialysis membrane and therefore the removal of potassium. Our results do not support this.

Measurement of erythrocyte potassium was not found to be of value in predicting potassium removal during dialysis. We found that total potassium removal was best predicted by pre-dialysis Kp and that the change in Kp was related to the impact of dialysis on acid-base status.

References

1 Frost TH, Kerr DNS. *Kidney Int 1977; 12:* 41
2 Sterns RH, Feig PU, Pring M et al. *Kidney Int 1979; 15:* 651
3 Feig PU, Shook A, Sterns RH. *Nephron 1981; 27:* 25
4 Fraley DS, Adler S. *Kidney Int 1977; 12:* 354
5 Wathen RL, Ward RA, Harding GB, Meyer LC. *Kidney Int 1982; 21:* 592

Proc EDTA–ERA (1984) Vol 21

FAVOURABLE EFFECTS OF BICARBONATE DIALYSIS ON THE BODY POOL OF PHOSPHATE

F Mastrangelo, S Rizzelli, V De Blasi, C Corlianò, L Alfonso, M Napoli, A M Montinaro, M Aprile, P Patruno

Ospedale "V Fazzi", Lecce, Italy

Summary

An intravenous infusion of 3,430mg of PO_4^- has been given to 11 patients on acetate and to 11 patients on bicarbonate haemodialysis. The 'phosphate spaces' and dialytic removal were determined. The bicarbonate dialysis causes lower values of phosphate pool, total phosphate space, cellular space and phosphate cellular clearance. There is also a greater phosphate removal during bicarbonate dialysis. The better correction of metabolic acidosis and the absence of acetate metabolism are two factors which may be responsible for these phenomena.

Introduction

In a trial with bicarbonate dialysis (BiHD) it has been demonstrated [1,2]: 1) lower values of serum phosphate (P) when we compared acetate haemodialysis (AcHD) and BiHD in the same patients; 2) the possibility of reducing or suspending the use of phosphate binders in 70 per cent of patients. The explanation of the phenomena is not clear. In animals Wathen [3] showed that the infusion of acetate produced a reduction in plasma phosphate. In previous studies in man [4], we suggested that AcHD leads to a shift of PO_4^- from the extracellular compartment (ExC) and consequently produces a lower phosphate removal.

Patients and methods

Analysis of the body pool of phosphate

The study was performed on 22 patients (11 on AcHD and 11 on BiHD) (Table I) and eight patients with normal glomerular filtration rate (GFR) and eight uraemic patients (GFR <4ml/min) as control groups. No patients were receiving vitamin D or calcium supplements.

The patients were fasted overnight. An intravenous infusion of a solution

TABLE I. Body storage capacity for phosphate on uraemic and haemodialysis patients in basal conditions (t_0) and after intravenous phosphate load (t_{180}). The same reduction of serum calcium and Ca^{++} was observed in AcHD patients and BiHD patients

		Normal (n=8)	Uraemic (n=8)	p	BiHD (n=11)	p	AcHD (n=11)
P_{pl} (mg%)	t_0	3.3 ± 1.4	3.7 ± 1.7	–	3.6 ± 0.8	–	4.5 ± 1.1
	t_{180}	14.8 ± 1.2	19.3 ± 4.3	–	22.5 ± 4.8	–	20.1 ± 3.7
Ca_{pl} (mg%)	t_0	–	9.1 ± 1	–	10.1 ± 0.4	–	10 ± 0.7
	t_{180}	–	7.2 ± 0.7	–	8 ± 0.7	–	8.3 ± 0.6
Ca^{++}_{pl} (mg%)	t_0	–	–	–	4.8 ± 0.8	–	4.7 ± 0.4
	t_{180}	–	–	–	3.2 ± 0.4	–	3.5 ± 0.6
Blood pH	t_0	–	7.36 ± 0.01	–	7.38 ± 0.04	<0.05	7.32 ± 0.06
	t_{180}	–	7.36 ± 0.02	–	7.38 ± 0.05	<0.05	7.32 ± 0.08
Basal PO_4 Pool: P_{pl}xS (mg)		473 ± 102	823 ± 101	–	692 ± 137	<0.05	1006 ± 291
Total PO_4 space: $S=(½t\ K_i)/K_p$ (L)		16.8 ± 2.1	24.1 ± 4	<0.01	19.2 ± 2.7	<0.01	22.7 ± 2.1
PO_4 cell space: S_c=S–0.2 body weight (L)		2.8 ± 1.9	10.1 ± 4	<0.01	5.8 ± 3	<0.01	8.6 ± 2.1
'Extra-renal clearance': S_c/½t (ml/min)		32.8 ± 20	112 ± 51	<0.01	62 ± 30	<0.01	96 ± 23
Fractional turnover time: (S_c/S)x(1/½t)x100		0.18 ± 0.10	0.44 ± 0.12	<0.05	0.29 ± 0.11	<0.05	0.41 ± 0.07
Fractional cell uptake: (KpSct/343)x100		10.37 ± 7	44.5 ± 16	= 0.05	29.9 ± 14	<0.05	41.8 ± 9.7
PTH		–	3.5 ± 3.2	<0.05	9 ± 3	–	9.3 ± 5.1
Age (years)		46.2 ± 8	57.1 ± 16	–	54.4 ± 14	–	50.7 ± 8.7
Body weight (kg)		67.7 ± 6	65.5 ± 11	–	66.6 ± 7	–	64.4 ± 9

containing 3,430mg of PO_4^- in 250ml of water (as $Na_2HPO_4 \cdot 2H_2O$ 65.7g/L and $NaH_2PO_4 \cdot 2H_2O$ 11.5g/L buffered at pH 7.35) was started [5,6]. The infusion rate was progressively increased every 30 minutes from a basal value of 10.1mg/min to 25mg/min by an electric pump. Venous samples were performed at regular 30 minute intervals.

As the total amount of phosphate infused increases, at a constant ratio, the phosphate concentration in the 'phosphate space' (S), we derived our indices on the basis of the equation:

$$\tfrac{1}{2}Ki \times t^2 = [Kp \times S \times t] + [\tfrac{1}{2}Ku \times t^2]$$

where Ki=acceleration coefficient of infusion (mg/min^2); Kp=rate of increase in plasma PO_4^- (mg/ml x min); Ku=acceleration coefficient of urinary PO_4^- excretion (mg/min^2); t=infusion time.

Dialytic kinetics of phosphate

At the end of the phosphate load the patients were submitted to a three hour haemodialysis session with a large surface area dialyser, keeping constant Q_B, Q_D and the dialysate concentration (only buffers changed: acetate 38mEq/L or bicarbonate 35mEq/L). The phosphate removed with the dialysate (P_D) was tested on paired samples.

Because P_D is function of the amount of phosphate available for dialytic removal, i.e. pre-dialysis values of extracellular phosphate pool ($Pool_{ex}$) and/or of whole body pool ($Pool_{PO_4}$), we calculated the amount of extracellular pool cleared ($P_D/Pool_{ex}$), the amount of body pool cleared ($P_D/Pool_{PO_4}$), the dialytic change of $Pool_{ex}$ ($\Delta Pool_{ex}$), and the amount of cellular pool cleared ($P_D-\Delta Pool_{ex}$).

As there was the probability that the phosphate load had produced an anomalous situation, another study (protocol B) was carried out on 15 patients (7 in AcHD and 8 in BiHD). These patients were subjected to AcHD and BiHD at a 48-hour interval with unchanged Q_B, Q_D, t and dialyser (Table II).

TABLE II. PO_4 kinetics in AcHD and BiHD: the same 15 patients were subjected at a 48-hour interval to both sessions of AcHD and BiHD (*p significant)

	AcHD		BiHD	
	T_0	T_{180}	T_0	T_{180}
Body weight	61.5 ± 10.2	59.5 ± 10.2	61.5 ± 10.1	59.1 ± 10.1
P_{pl} mg%	4.16 ± 0.89	1.7 ± 0.59	3.86 ± 1.1	1.8 ± 0.48
PO_4 $Pool_{ex}$ mg	516 ± 120*	210 ± 66	476 ± 150*	217 ± 63
ΔPO_4 $Pool_{ex}$ mg	307 ± 98*		260 ± 115*	
$PO_{4\,D}$	438 ± 153		447 ± 160	
$PO_{4\,D} - \Delta Pool_{ex}$	130 ± 123		181 ± 127	
$PO_{4\,D}/Pool_{ex}\ t_0$	0.84 ± 0.27		0.95 ± 0.22	
$PO_{4\,D}/\ Pool_{ex}$	1.6 ± 0.5*		2.2 ± 0.9*	

Results

Analysis of body pool PO_4

The increments in plasma phosphate are different in AcHD (y=3.5+0.8x) and in BiHD (y=26.9±1.02x). The phenomenon is an expression of the different space of distribution (Table I).

The $Pool_{PO_4}$ is increased in uraemic patients. The plasma phosphate is not an expression of the pool state in that it is in the normal range in all patients. $Pool_{PO_4}$ depends on the plasma phosphate and on the phosphate space. The phosphate space is really increased: both uraemic patients and the haemodialysis patients present pathological values. But the BiHD patients present significantly lower values. The expansion of the phosphate spaces depends mostly on a remarkable increase of the cellular space of phosphate (Sc). This index is greatest in the uraemic patients and in the AcHD patients.

The behaviour of other indices is directly linked with the increase in the cellular phosphate space and consequently results changed less in the BiHD patients.

Correlation between phosphate spaces and PTH or acid-base balance

No significant correlation has been shown between serum iPTH (C_{65-84}) and the different indices of PO_4^- body spaces (S, Sc, etc). But a direct correlation is shown between blood pH and $Pool_{PO_4}$ (y=−2.9x+22.2, r=0.607), blood pH and cellular clearance (y=−20.8x+1625.6, r=0.444), and blood pH and cell uptake (y=−83x+643, r=0.448).

Dialytic kinetics of phosphate

The comparison between P_D and $Pool_{ex}$ during the two haemodialysis treatments permits the following remarks (Table II).

The pre-dialysis $Pool_{ex}$ was greater in AcHD, while the $Pool_{ex}$ post-dialysis is not different, so $\Delta Pool_{ex}$ is greater in AcHD ($p<0.05$). $\Delta Pool_{ex}$ expresses the amount of PO_4^- leaving the extracellular space during dialysis: so it should be found in dialysate since the phosphate clearance is not different by the two methods. But P_D is equal in AcHD and BiHD. This indicates that in AcHD there is an amount of phosphate which leaves the extracellular space but which is not eliminated into dialysate.

Other data confirm a relatively higher phosphate removal during BiHD. We found a direct correlation between:
$Pool_{ex}$ and P_D (y=140+0.68x, r=0.66) and $\Delta Pool$ and P_D (y=245.5+0.696x, r=0.504).

These positive correlations support that $Pool_{ex}$ may be a valid expression of the amount of phosphate subjected to dialytic removal and its change ($\Delta Pool_{ex}$) during dialysis providing that there are no changes in extracellular phosphate. Consequently the ratio $P_D/Pool_{ex}$ could be considered an adequate index of dialysis phosphate removal. After the phosphate load, we also found a direct correlation between: $P_D/Pool_{ex}$ and S (y=9624+158.5x, r=0.75) and $P_D/Pool_{ex}$ and Sc (y=−4366+158.5x, r=0.78).

In protocol B, during BiHD, higher values of ($P_D/Pool_{ex}$), (P_D- $\Delta Pool_{ex}$) and ($P_D/\Delta Pool_{ex}$) were found (Table II). Moreover, we found the correlations between: $Pool_{ex}$ and P_D (y=114+0.75x, r=0.78) and $\Delta Pool_{ex}$ and P_D (y= 256.4+0.86x, r=0.68) significant only during BiHD, but not during AcHD (r=0.46 and r=0.23).

Discussion

As shown in Table I, BiHD exerts a favourable effect on phosphate kinetics. We found that: 1) in uraemic patients both body and cellular spaces of phosphate are increased, in agreement with Mioni et al [6]; 2) AcHD does not change this state; 3) BiHD modifies such a pathological state.

The interpretation of these findings is not easy. The difference between AcHD and BiHD could be attributed to several factors: a) different serum PTH values; b) different acid-base status; c) different dialytic clearance of phosphate.

The excessive circulating PTH will enhance the cellular uptake of inorganic phosphate [6–8]. It increases the passive distribution of phosphate across the cell membranes, and causes abnormal phosphate spaces. In the two groups of our haemodialysis patients the values of PTH were not different and they were not significantly related to S, Sc, cellular clearance and cell uptake.

The BiHD induced a better correction of acidosis and this could explain the differences found in these patients. We documented a direct correlation between blood pH and body pool of phosphate, blood pH and Sc, blood pH and cell uptake.

In AcHD there is a relatively lower clearance of phosphate (Table II) and this phenomenon could be due to an intracellular shift of phosphate or to a smaller depletion of cellular space, owing to metabolism of acetate [3,4]. In further support of this hypothesis, there are both the highest values of ($P_D/Pool_{ex}$), (P_D- $\Delta Pool_{ex}$) and ($P_D/\Delta Pool_{ex}$) found in BiHD and the existence of a correlation between P_D and $Pool_{ex}$ and P_D and $\Delta Pool_{ex}$ only during BiHD.

Because of the absence of these correlations with acetate dialysis we can suppose that during AcHD some events interfere with the cellular efflux of phosphate from the extracellular space to dialysate, producing a smaller clearance. Therefore we believe that in AcHD, phosphate intake being equal, a greater body load of phosphate will occur, contributing to the greater expansion of phosphate spaces.

In conclusion there is evidence that BiHD permits a better correction of the abnormal phosphate spaces in dialysis patients. This occurs by various mechanisms which include the better acid-base status and the greater removal of phosphate.

References

1 Mastrangelo F, Rizzelli S, De Blasi V et al. *8th International Congress ISN Athens 1981.* Poster DI-130)

2 Mastrangelo F, Rizzelli S, Corlianò C et al. In Petrella E, ed. *Uremic Acidosis Extracorporeal Treatment.* Wichtig. 1983: 185

3 Wethen RL, Ward RA, Neuhanser M et al. *Kidney Int 1979; 16:* 995
4 Mastrangelo F, Rizzelli S, De Blasi V et al. In La Greca G, Ghezzi PM, Fabris A, eds. *Asymmetrical Cellulose Hydrate.* Wichtig. 1982: 129
5 Anderson J, Parsons V. *Clin Sci 1963; 25:* 431
6 Mioni G, Ossi E, D'Angelo A et al. *Biomedicine 1973; 18:* 491
7 Geary GP, Cousins FB. *Aust J Exp Biol Med Sci 1971; 49:* 463
8 Rasmussen H, Arnaud C. In *L'Osteomalacie.* Paris: Masson. 1967: 230

Proc EDTA–ERA (1984) Vol 21

SODIUM ACETATE, AN ARTERIAL VASODILATOR: HAEMODYNAMIC CHARACTERISATION IN NORMAL DOGS

M A Saragoca, Angela M A Bessa, R A Mulinari, S A Draibe, A B Ribeiro, O L Ramos

Escola Paulista de Medicina, Sao Paulo, Brazil

Summary

Sodium acetate (SA) has been implicated in hypotensive episodes of haemodialysis because of its vasodilatory effects. The haemodynamic correlates of the changes in blood pressure, cardiac output (CO) and total peripheral resistance (TPR) are well known but the site of action of SA (i.e. arteriolar, venular or both) is not yet clarified. We thus studied the changes in CO, TPR and mean arterial pressure (MAP) induced by four graded doses of SA (0.034 to 0.300 mEq/kg/min) in seven normal dogs. To evaluate the site of vasodilation we also measured the changes in cardiopulmonary volume (CPV), mean pulmonary artery pressure (MPAP) and mean transit time (MTT). From control to the highest infusion rate, CO increased from 1.63±0.20 to 3.59±0.38L/min ($p<0.001$), TPR decreased from 78.2±11.3 to 36.4±4.8A.U. ($p<0.001$). MAP rose significantly from 107.2±4.0 to 116.5±8.5mmHg ($p<0.05$) and stroke volume was maintained (17.2±2.3 to 19.6±2.1ml, NS) in spite of the marked tachycardia observed (heart rate from 106.1±7.6 to 194.8±9.1bpm, $p<0.001$). This was associated with increases in MPAP (from 13.3±0.7 to 19.6±2.1mmHg, $p<0.01$) and CPV (from 195.0±21.3 to 224.4±24.3ml, $p<0.01$) and marked decrease in MTT (from 7.74±0.73 to 3.78±0.22sec, $p<0.01$). Our data show that marked increases in CO occur associated with vasodilation induced by SA and that MAP increases significantly; the decrease in MTT shows that hyperkinetic circulation is induced by the infusion. This may be due to a sympathetic reflex activity originated by the increase in CPV and MPAP. This haemodynamic pattern strongly suggests an arterial site of action of acetate as a vasodilator. Because MAP increased during infusion, a stimulatory effect of acetate itself must also be postulated to explain the hyperkinetic circulation observed.

Introduction

Hypotensive episodes during haemodialysis have been frequently attributed to to the mass of sodium acetate buffer transferred to the patients [1,2]. Although the vasodilatory effects of this substance have been documented [3,4] the comparative mode of action in the arteriolar and venular circulations, as well as its cardiac effects still remain to be fully described. Thus, in order to provide the characterisation of its haemodynamic effects we studied the pattern of vasodilation to graded doses of sodium acetate in normal dogs, using standard haemodynamic techniques.

Material and methods

We studied the haemodynamic responses of 10 normal mongrel dogs, anaesthetised with morphine (2mg/kg) and sodium pentobarbital (15mg/kg), to graded infusions of sodium acetate at the rates of 0.037, 0.075, 0.150 and 0.300mEq/kg/min, during 20 minutes at each dose. At the end of each period of infusion, haemodynamic measurements were performed. We used dye-dilution techniques in a Gilford cuvette densitometer for haemodynamic measurements and a Gould 2400S writing recorder with Statham pressure transducers for direct blood pressure recordings. To study the actions of sodium acetate in the systemic circulation, the following indices were obtained: heart rate (HR), mean arterial pressure (MAP), cardiac output (CO), total peripheral resistance (TPR) and mean transit time (MTT). Cardiopulmonary blood volume (CPV), obtained from the dye-dilution curve, and mean pulmonary artery pressure (MPAP), were determined to evaluate the actions of the drug in the pulmonary circulation. The ratio between cardiac output and the cardiopulmonary volume (CO/CPV) was determined and used as an index of cardiac performance.

TABLE I. Changes in haemodynamic parameters induced by four graded doses of sodium acetate

		Sodium acetate infusion rate			
	Control	0.037	0.075	0.150	0.300*
MAP	107.2 ±4.0	109.3 ±3.7	111.2 ±5.6	117.6 ±6.2	116.5 ±8.5
HR	106.1 ±7.6	135.2 ±8.9	150.4 ±11.7	178.3 ±13.3	194.8 ±9.1
CO	1.63±0.20	1.82±0.25	2.36±0.30	2.90±0.31	3.59±0.38
TPR	78.2 ±11.3	69.4 ±7.8	54.1 ±6.1	43.1 ±4.0	36.4 ±4.8
SV	17.2 ±2.3	15.1 ±2.3	18.1 ±2.6	17.7 ±2.3	19.6 ±2.1
MPAP	13.3 ±0.7	15.8 ±0.8	17.6 ±1.5	19.1 ±1.5	19.7 ±1.7
CPV	195.0 ±21.3	188.8 ±22.6	209.0 ±22.2	223.5 ±23.6	224.4 ±24.3
MTT	7.74±0.65	6.63±0.51	5.57±0.40	4.96±0.45	3.78±0.22
CO/CPV	8.33±0.73	9.66±0.82	11.37±0.86	13.33±0.95	16.22±1.02

MAP=mean arterial pressure; HR=heart rate; CO=cardiac output; TPR=total peripheral resistance; SV=stroke volume; MPAP=mean pulmonary artery pressure; CPV=cardiopulmonary volume; MTT=mean transit time.

* mEq/kg/min

Analysis of variance was used to assess significance of the changes from control values. Results are expressed as mean ± standard error of the mean and are summarised in Table I.

Results

The infusions of sodium acetate induced dose-dependent changes in most of the haemodynamic variables studied. From control to the fourth dose, TPR decreased (from 78.2±11.3 to 36.4±4.8AU, $p<0.001$), CO increased (from 1.63±0.20 to 3.59±0.38L/min, $p<0.001$), and HR also increased (from 106.1±7.6 to 194.8±9.1bpm, $p<0.001$). Mean arterial pressure increased during the infusion (from 107.2±4.0 to 116.5±8.5mmHg, $p<0.05$), while SV did not change significantly (from 17.2±2.3 to 19.6±2.1ml, NS). We also noted a marked, dose-dependent decrease in MTT (from 7.74±0.65 to 3.78±0.22sec, $p<0.001$).

The cardiopulmonary volume increased significantly during the infusion (from 185.0±21.3 to 224.4±24.3ml, $p<0.01$) and so did MPAP (from 13.3±0.7 to 19.7±1.7mmHg, $p<0.01$). Also, CO/CPV increased significantly during the infusion (from 8.33±0.73 to 16.22±1.02, $p<0.001$). Haematocrit, however, did not change significantly during the experiments (see Table I).

Discussion

Our results clearly show that dose-dependent vasodilation was induced by the infusion of the graded doses of sodium acetate. This effect was accompanied by marked increases in CO and in heart rate. As a result, stroke volume did not decrease, but rather, it tended to increase with the drug. This is in accordance with previous reports suggesting an arteriolar site of action of sodium acetate in normal dogs [3]. In the present work we could document significant increases in MPAP and in CPV, associated with hyperkinetic circulation, as shown by the marked decreases observed in MTT. This pattern of action is indeed consistent with a predominantly arteriolar site of action of sodium acetate as a vasodilator with little effects, if any, on the venous circulation [5,6]. Because haematocrit did not significantly change during the infusion, it is reasonable to assume that the total blood volume was not affected [7] and thus the increased cardiopulmonary volume and pulmonary pressure must be attributed to shifts of blood from the periphery to the central circulation. In this case, the increases in CO observed during the experiments may be explained by reflex sympathetic activity elicited by distension of pressure and volume receptors of the pulmonary circulation. On the other hand, a venoconstrictor action of sodium acetate as suggested by Molnar et al [8] cannot be ruled out, since this effect also helps to explain the marked increases observed in cardiac output through increases in venous return.

We also noted marked shortening of MTT associated with increases, rather than decreases in MAP. Also, stroke volume was not decreased during the infusion as it might be expected because of the marked tachycardia induced by the drug. These facts strongly suggest that increased sympathetic support to the heart elicited by reflex activity may be a major determinant of the increased cardiac pump performance observed during the infusion of sodium acetate. This cardiostimulatory action was further documented by the significant increase observed in the CO/CPV ratio. However, the possibility still exists that this action may be related to a direct effect of acetate on the heart by providing energy substrate [9]. This question can be resolved with appropriate sympathetic blockade studies to obviate reflex activity. Nevertheless, the overcompensation observed in MAP during sodium acetate infusion suggest that this latter mechanism can be at least in part involved in the overall response. Further investigation is also still necessary to clearly differentiate the haemodynamic effects of acetate from those of infused hyperosmolar sodium solutions.

References

1 Iseki K, Onoyama K, Maeda T et al. *Clin Nephrol 1980; 14:* 294
2 Graefe U, Milutinovich J, Follete WC et al. *Ann Intern Med 1978; 88:* 332
3 Kirkendol PL, Robie NW, Gonzalez FM et al. *Trans ASAIO 1978; 24:* 714
4 Olinger GN, Werner PH, Bonchek LI. *Ann Surg 1979; 190:* 305
5 Chatterjee K, Drew D, Parmley WW et al. *Ann Intern Med 1976; 85:* 467
6 Koch-Weser J. *N Engl J Med 1976; 295:* 320
7 Migdal S, Alexander EA, Bruns FJ et al. *Circ Res 1975; 36:* 71
8 Molnar J, Scott JB, Frohlich ED et al. *Am J Physiol 1962; 203(1):* 125
9 Ling CS, Lowenstein JM. *J Clin Invest 1978; 62:* 1029

Proc EDTA–ERA (1984) Vol 21

EFFECT OF DIALYSATE SODIUM ON THE DOPAMINE, ADRENALINE AND NORADRENALINE CONCENTRATION IN HAEMODIALYSIS PATIENTS

A Ksiażek, J Solski

School of Medicine, Lublin, Poland

Summary

In eight stable patients, plasma dopamine (DA), adrenaline (A) and noradrenaline (NA) was measured during haemodialysis with high sodium (148mEq/L) and low sodium (131mEq/L) dialysate. During haemodialysis and ultrafiltration with high dialysate sodium an increase of DA, A and NA was not observed in response to the cold pressor test (CD). However, with the low sodium dialysate an increase in DA and NA was observed during CD. We suggest that high sodium dialysate is connected with the sympathetic dysfunction of haemodialysis patients.

Introduction

Haemodialysis patients are said to have a widespread dysfunction of the autonomic system including abnormal plasma catecholamines (CA) [1]. One of the important factors determining catecholamine uptake and plasma values is sodium concentration [2]. There is a possibility that in haemodialysis patients the dialysate sodium concentration interferes with sympathetic function. This study evaluates the effect of high and low dialysate sodium on plasma catecholamines during conventional haemodialysis and sequential ultrafiltration.

Patients and methods

The study was performed on eight haemodialysis patients. All patients were anuric or had negligible excretory function and had been established on haemodialysis for a minimum of 20 months. None had clinical evidence of peripheral neuropathy as assessed by peroneal nerve conduction velocity. They had a normal diet with a free protein and salt intake. All patients were dialysed for 18 hours weekly with Travenol apparatus using Vita 2, $1m^2$ dialyser. The dialysate composition was: Na^+ 131 or 148mEq/L, K^+ 1.5mEq/L, Ca^{++} 3.5 mEq/L, Mg^{++} 1.5mEq/L, Cl^- 103 or 119mEq/L, acetate 35mEq/L and osmolality 255 or 280mOsm/kg.

The patients were dialysed for two weeks by either of the two different dialysates before the studies were started. The experimental procedure was carried out in the morning between 7a.m. and 8a.m. at a constant temperature. During the study patients were dialysed, first with the dialysate containing high sodium for six hours. The next dialysis was started with one hour ultrafiltration at a pressure of 200mmHg. The same haemodialysis and ultrafiltration was undertaken using low sodium dialysate. Plasma DA, A and NA determined by sensitive radioenzymatic method [3] were measured in response to the cold pressor test [4]. Mean values were tested for significant differences by the paired 't' test.

Results

Plasma NA and DA before and after dialysis were significantly higher with high sodium dialysate compared to during low sodium dialysate. After the cold pressor test, NA and DA increased significantly when low sodium dialysate was used, and remained constant with high sodium dialysate (Table I). Plasma DA, A and NA before and after ultrafiltration were significantly higher with high sodium dialysate compared to low sodium dialysate (Table II). Increased plasma NA and DA was observed before dialysis and ultrafiltration in patients on low sodium dialysate in response to the cold pressor test (Tables I and II). When high sodium dialysate concentration was used increases in CA were not observed after the cold pressor test (Tables I and II). Increase in plasma NA, DA and A was observed in response to the cold pressor test after dialysis and ultrafiltration in patients on low sodium dialysate (Tables I and II), but when high sodium dialysate was used no increase in catecholamines was observed in response to the cold pressor test (Tables I and II).

Discussion

The mechanism of the sodium effect on sympathetic function is not clear. Some authors suggest from in vitro studies that sodium is essential for the processes that store, accumulate and release CA in synaptosomes [2,5]. We confirm their suggestion, in vivo, and have observed that stable haemodialysis patients show different sympathetic reflex function to the cold pressor test depending on the dialysate sodium concentration. In patients treated with high sodium dialysate, increase in plasma catecholamines after the cold pressor test was not observed, despite a high basal NA and DA concentration. These abnormalities may be explained by the influence of the dialysate sodium on sympathetic function. Brunvels [6] noted that in vitro a high sodium concentration in the synaptic cleft diminished the release of monoamine neurotransmitters. It is possible that decreased release of catecholamines from the sympathetic endings as a consequence of the higher sodium dialysate causes an inhibition of the increase of plasma catecholamines during the cold pressor test. Higher dialysate sodium disturbs not only the release of catecholamines from synaptosomes but also the re-uptake [7] and in this way the basal plasma noradrenaline concentration may be increased.

TABLE I. Effect of the cold pressor test on plasma dopamine (DA), adrenaline (A) and noradrenaline (NA) concentration in haemodialysis patients treated with high (148mEq/L) and low (131mEq/L) sodium dialysate

Dialysate	CA pg/ml	Before haemodialysis Cold pressor pre $\overline{x} \pm SD$	Before haemodialysis Cold pressor post $\overline{x} \pm SD$	p_1	After haemodialysis Cold pressor pre $\overline{x} \pm SD$	After haemodialysis Cold pressor post $\overline{x} \pm SD$	p_1
	NA	224.82 ± 24.98	230.53 ± 28.28	<0.001	236.27 ± 32.53	330.59 ± 24.50	<0.001
Na^+ 131	A	95.89 ± 50.04	112.21 ± 25.59	NS	90.04 ± 19.18	209.84 ± 154.62	<0.05
	DA	70.54 ± 17.81	94.91 ± 14.02	<0.02	67.67 ± 19.21	114.39 ± 28.32	<0.001
	NA	347.22 ± 43.55 p_2 <0.001	334.89 ± 40.58	NS	374.92 ± 56.26 p_2 <0.001	366.21 ± 31.24	NS
Na^+ 148	A	132.00 ± 23.33 NS	126.28 ± 24.51	NS	136.68 ± 36.29 p_2 <0.01	137.71 ± 32.21	NS
	DA	106.44 ± 20.82 p_2 <0.001	108.53 ± 18.42	NS	109.91 ± 21.32 p_2 <0.001	116.65 ± 28.42	NS

p_1 in reference to the value pre cold pressor test
p_2 in reference between dialysis with Na^+ 131mEq/L and Na^+ 148mEq/L

TABLE II. Effect of cold pressor test (CD) on plasma dopamine (DA), adrenaline (A) and noradrenaline (NA) concentration in ultrafiltration (UF) patients treated with high (148mEq/L) and low (131mEq/L) sodium dialysate

Dialysate	CA pg/ml	Before ultrafiltration Cold pressor pre $\overline{x} \pm SD$	post $\overline{x} \pm SD$	p_1	After ultrafiltration Cold pressor pre $\overline{x} \pm SD$	post $\overline{x} \pm SD$	p_1
	NA	221.35 ± 34.52	300.53 ± 42.53	<0.001	362.21 ± 68.53	836.21 ± 156.59	<0.001
Na^+ 131	A	96.15 ± 52.34	129.31 ± 15.30	NS	94.01 ± 37.27	138.00 ± 42.31	<0.001
	DA	68.96 ± 22.31	96.31 ± 32.41	<0.01	69.00 ± 11.51	83.80 ± 31.21	<0.001
	NA	363.53 ± 72.53 $p_2 < 0.001$	349.93 ± 64.56	NS	474.24 ± 84.95 $p_2 < 0.001$	479.33 ± 94.51	NS
Na^+ 148	A	135.06 ± 43.28 $p_2 < 0.05$	134.22 ± 39.01	NS	163.95 ± 29.43 $p_2 < 0.001$	151.90 ± 39.28	NS
	DA	115.80 ± 21.36 $p_2 < 0.001$	123.90 ± 36.21	NS	117.00 ± 18.21 $p_2 < 0.001$	112.30 ± 23.31	NS

p_1 in reference to the value pre cold pressor test
p_2 in reference between dialysis with Na^+ 131mEq/L and Na^+ 148mEq/L

In conclusion, our work suggests that high sodium dialysate causes dysfunction of the sympathetic system.

References

1 Ksiazek A. *Nephron 1979; 24:* 170
2 Haddy FJ. *Ann Intern Med 1983; 93:* 781
3 Peuler JD, Johnson GA. *Life Sci 1977; 21:* 625
4 Naik RB, Mathias CJ, Wilson CA et al. *Clin Sci 1981; 60:* 165
5 Bogdański DF, Tissari A, Brodie BB. *Life Sci 1968; 7:* 419
6 Bruinvels J. *Nature 1975; 257:* 606
7 Haddy FJ, Pamnani MB, Clough DL. *Life Sci 1979; 24:* 105

Proc EDTA–ERA (1984) Vol 21

THE CARDIOVASCULAR AND METABOLIC EFFECTS OF MIXTURES OF ACETATE AND SUCCINATE: A POTENTIAL IMPROVEMENT IN DIALYSATE SOLUTIONS

F M Gonzalez, P L Kirkendol, J E Pearson, E Reisin

Louisiana State University Medical Center, New Orleans, Louisiana, USA

Summary

Various mixtures of acetate (AC) and succinate (SUCC) were studied for their metabolic and cardiovascular (CV) effects in 10 dogs. The CV effects seen with all mixtures were similar to those reported for SUCC alone with the only changes being increased cardiac output and decreased total peripheral resistance. The 50:50 per cent AC/SUCC offered the advantage of a rapid but less marked and more sustained HCO_3 production. The pH changes in all mixtures followed HCO_3 values. Since the addition of SUCC seems to reduce the untoward CV changes seen with AC alone, the ratio of AC/SUCC should be based on metabolic considerations. The present data suggest that a 50:50 per cent AC/SUCC mixture is optimal for metabolic and CV effects.

Introduction

Sodium acetate has been used as a substitute for sodium bicarbonate in dialysis solutions since 1964 [1]. Since bicarbonate has a low solubility and tends to precipitate in the presence of calcium and magnesium, the preparation of concentrated solutions presented major difficulties. Thus, the use of bicarbonate required proportioning pumps and/or delicate pH adjustment. The shift to acetate as a source of fixed base circumvented the physical problems associated with bicarbonate. Since acetate is rapidly metabolised and generates bicarbonate in the process [2], it can restore bicarbonate in patients undergoing haemodialysis.

Sodium acetate, however, is not a pharmacologically inert substance. It has been shown to have significant cardiovascular effects like hypotension and myocardial depression when given intravenously either as a bolus [3] or by continuous infusion [4]. There is, therefore, a need for a compound that displays fewer cardiovascular effects than acetate but will generate bicarbonate.

Kirkendol et al [5,6] previously reported that bolus injections and infused succinate produced less pronounced cardiovascular changes than acetate. This study relates to the cardiovascular and metabolic effects of sodium acetate: sodium succinate mixtures as a source of fixed base in dialysate solutions.

Materials and methods

In three mongrel dogs of either sex weighing from 18 to 24kg each were anaesthetised intravenously with 30mg/kg of pentobarbital sodium. A polyethylene cannula was inserted in the right femoral vein for infusion of the sodium salts. The mean arterial blood pressure was measured using a Statham P23DC transducer via a polyethylene cannula inserted into the right femoral artery. Heart rate was continuously recorded using a Grass cardiac tachograph. Cardiac output was determined by the right heart thermal washout technique. A Swan-Ganz thermistor-tipped catheter was placed in the pulmonary artery via the jugular vein, with the saline indicator injected at the level of the right atrium through a port in the same catheter. The cardiac output was calculated by an Instrumentation Laboratory Cardiac Output Computer Model 601. The cardiac output values used in this study were the average of three determinations per observation period. A catheter was placed in the left ventricle via the left carotid artery for the measurement of left ventricular pressure. The maximum rate of rise of the left ventricular pressure (dp/dt), an index of myocardial contractility, was obtained using a Grass Polygraph differentiator (7P20). Total peripheral resistance was calculated as mean blood pressure (mmHg) divided by cardiac output (L/min), and femoral vascular resistance was calculated as mean blood pressure divided by femoral blood flow (ml/min). After preparation of the animal and recording of the control parameters, mixtures of sodium acetate, sodium succinate, 25:75 acetate/succinate, 50:50 acetate/succinate or 75:25 acetate/succinate were infused at doses of 0.125, 0.25, 0.5, and 1.0mEq/kg/min for 10 minutes at each dose level. At the end of each 10 minute infusion period, cardiac output was determined. All other parameters were continuously recorded, but the values reported are those obtained at the end of each infusion period.

In another series of experiments, the mixtures of sodium acetate:succinate were infused at the rate of 0.25mEq/L/min in a volume of 0.25ml/kg for one hour in seven acutely nephrectomised dogs. Blood samples were taken every 20 minutes for the first two hours and every hour for the next two hours. Blood pH on each sample was determined using a Beckman Micro Sensor assembly and a Corning digital 110 pH meter. Plasma bicarbonate levels were determined by the titration method using standard acid and base.

Results

The cardiovascular effects of the solutions were determined. Figure 1 shows the effects of these solutions on total peripheral resistance. All three mixtures produced similar decreases in TPR with the maximum being about a 50 per cent reduction. Along with the decreases in TPR there were dose-related increases in cardiac output with the 25:75 AC/SUCC mixture increasing the CO from 4 to 8.5L/min. The infusion of these solutions resulted in no changes either in blood pressure, heart rate or dp/dt.

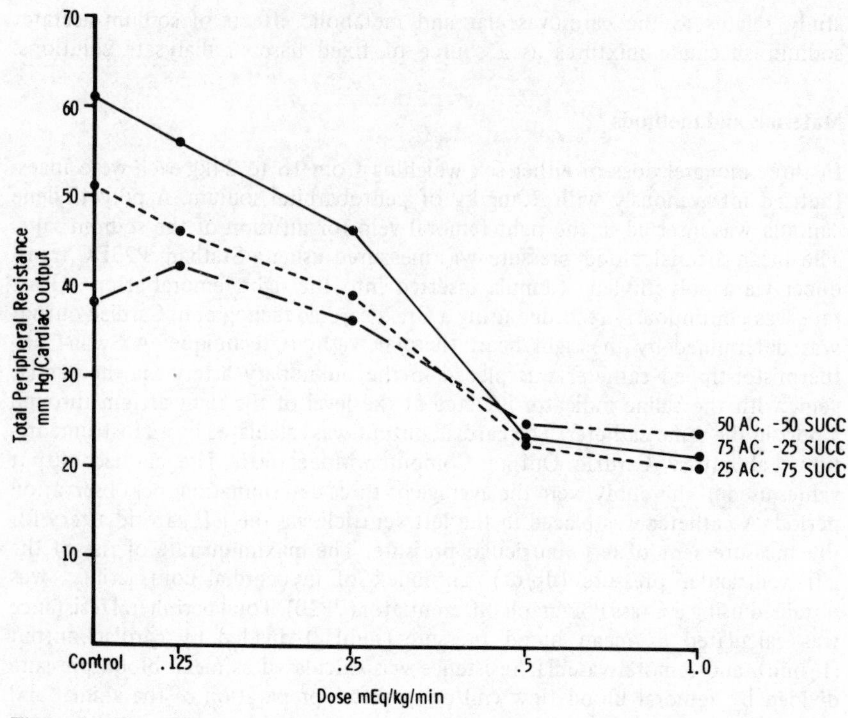

Figure 1. The effects of mixtures of acetate and succinate on total peripheral resistance

Figure 2 shows the effects of the various mixtures on plasma bicarbonate. Succinate, alone, produced bicarbonate at the slowest rate and lowest amounts. As the amounts of acetate were increased, the rate of generation of bicarbonate increased with the highest rate seen with acetate alone. The changes in blood pH in these animals, in general, followed the bicarbonate. All the solutions produced similar increases in plasma sodium with the concentrations increasing by about 40mEq/L. The solutions of acetate:succinate at various concentrations did not produce changes in plasma potassium or in the haematocrit.

Discussion

Metabolic acidosis is a feature of end-stage renal disease which must be corrected during dialysis. Sodium bicarbonate was first used as a replacement buffer system in dialysis but later was replaced in most of the cases by sodium acetate to avoid its low solubility and relative low stability [1]. Sodium acetate has been shown in animal and human studies [3,7] to be associated with a reduction in peripheral resistance due to vasodilation and myocardial depression. As a consequence hypotension frequently develops as a complication of the use of acetate in the uraemic patient undergoing haemodialysis. A number of sodium salts of organic substances which should generate bicarbonate were evaluated for their effects on the cardiovascular system using doses equivalent to those used

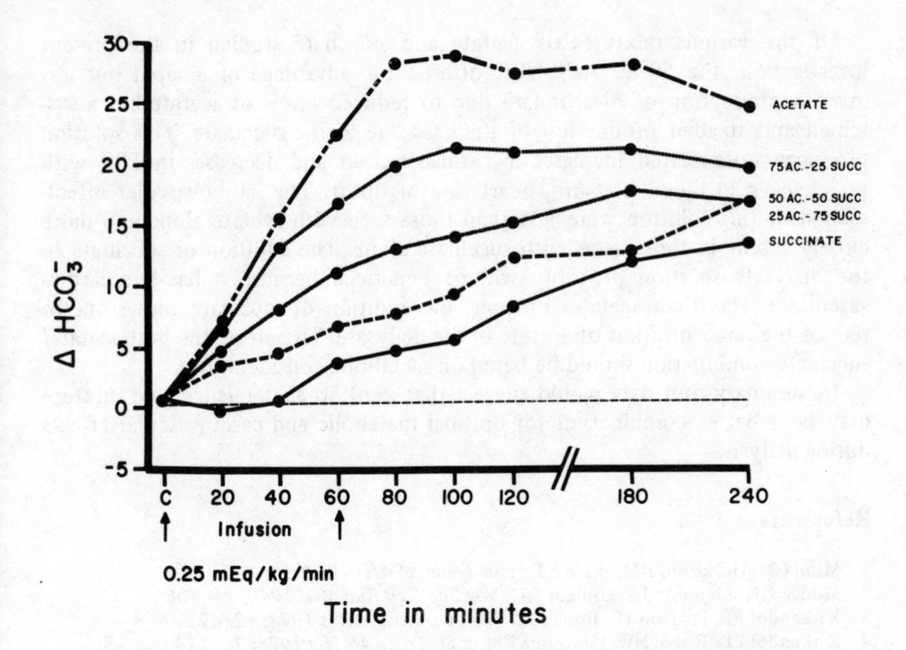

Figure 2. The effects of mixtures of acetate and succinate on changes in plasma bicarbonate

with acetate. Several compounds previously tested in animal studies have included lactate, pyruvate, alpha-keto-butyrate, succinate, glutamate, gluconate, fumarate, malate, oxaloacetate and glyoxylate. All these compounds were compared with equivalent doses of bicarbonate [5,8,9]. The cardiovascular effects produced by fumarate, malate, oxaloacetate and glyoxylate were much more marked than those seen with acetate and some even proved to be lethal to the experimental animals [5]. Lactate and pyruvate produced myocardial depression and decreases in blood pressure similar to those seen with acetate and appeared to offer no advantage over the currently used acetate [5]. Glutamate, aspartate, gluconate and succinate all produced fewer cardiovascular complications than acetate and therefore merited additional evaluations as possible sources of fixed bases.

The effect of succinate on the various cardiovascular parameters show that it more closely resembles bicarbonate than acetate. Succinate has no effect on cardiac output, dp/dt, heart rate or stroke volume and the blood pressure was not significantly affected with dose-related decreases in total peripheral resistance.

From the metabolic point of view, succinate generates bicarbonate at a slower rate than acetate, producing less marked changes. However, this slower rate of bicarbonate production with succinate results in an initially increased anion gap acidosis. This observation led to the evaluation of mixtures of acetate and succinate in an attempt to find a solution with less cardiovascular effects but taking advantage of the bicarbonate generating characteristics of both acetate and succinate.

Of the various mixtures of acetate and succinate studied in the present investigation, the 50:50 AC/SUCC offered the advantage of a rapid but less marked production of bicarbonate due to reduced levels of acetate but a sustained and steadier production of the base due to the succinate. This solution produced dose-related increases in cardiac output and decreases in TPR with no changes in blood pressure, heart rate or dp/dt. The cardiovascular effects seen with this solution were less than those seen with acetate alone and more closely resemble those seen with succinate alone. The addition of succinate to the dialysate solution probably will be beneficial because it has less cardiovascular effects than acetate. Further, the addition of succinate allows one to reduce the concentration of acetate in the dialysate. Therefore, the best acetate/succinate combination should be based on metabolic considerations.

In summary, our data would suggest that a 50:50 acetate/succinate mixture may be a better combination for optimal metabolic and cardiovascular effects during dialysis.

References

1 Mion CM, Hegstrom RM, Boen ST et al. *Trans ASAIO 1964; 10:* 110
2 Mudge GH, Manning JA, Gilman JA. *Proc Soc Exp Biol Med 1949; 71:* 136
3 Kirkendol PL, Pearson JE, Bower JD et al. *Cardiovasc Res 1978; 12:* 127
4 Kirkendol PL, Robie NW, Gonzalez FM et al. *Trans ASAIO 1978; 24:* 714
5 Kirkendol PL, Devia CJ, Bower JO et al. *Trans ASAIO 1977; 23:* 399
6 Kirkendol PL, Pearson JE, Gonzalez FM. *ASAIO Journal 1980; 3:* 57
7 Graefe U, Milutinovich J, Follette WC et al. *Ann Intern Med 1978; 88:* 332
8 Kirkendol PL, Pearson JE, Robie NW. *Clin Exp Pharm Physiol 1980; 7:* 617
9 Kirkendol PL, Starrs J, Gonzalez FM. *Trans ASAIO 1980; 26:* 323

Proc EDTA–ERA (1984) Vol 21

CHANGES IN BODY COMPARTMENTS ON DIFFERENT TYPES OF HAEMODIALYSIS

M García, M Carrera, *C Piera, R Deulofeu, X Company, J Ma Pons, J Montoliu, *J Setoain, L Revert

*Servicio de Nefrología, *Servicio de Medicina Nuclear, Hospital Clinic, Barcelona, Spain*

Summary

Changes in plasma volume (PV), extracellular volume (ECV) and intracellular volume (ICV) were studied in seven patients on conventional haemodialysis (HD) and in six patients on stable hypertonic HD. Weight loss and ultrafiltration were similar in both groups.

Before HD the spaces of ^{125}RISA (PV), $^{35}SO_4Na_2$ (ECV) and $^3H_2$) (total body water, TBW) were simultaneously determined ICV = TBW – ECV. At the end of HD the space of $^{35}SO_4Na_2$ was again tested.

PV and ECV diminished more on conventional HD than on hypertonic HD, whereas ICV increased on conventional HD and decreased on hypertonic HD.

The handling of plasma osmolality during HD is an effective method for modifying transcompartmental body fluid shifts in HD by distributing weight loss between intracellular and extracellular spaces allowing for a better maintenance of plasma volume.

Introduction

Intradialytic morbidity is a common phenomenon attributed to plasma volume removal and haemodynamic changes during haemodialysis (HD). Rapid decrease in plasma osmolality is in part responsible for this phenomenon [1] and a slower decline in plasma osmolality ameliorates clinic symptomatology [2] and preserves plasma volume (PV) [3].

In acute dialysis studies the use of high sodium dialysate showed that volume removal in HD was due to intracellular (ICV) and extracellular (ECV) loss [4]. Chronic hypertonic HD with alternate and sudden changes in sodium dialysate (cell-wash dialysis) achieves intracellular dehydration but produces thirst and greater weight gain between dialyses which limits its use on short-time routine programmes [5]. Haemodiafiltration employing hypotonic dialysate and hypertonic haemofiltration solution allows a reduction in dialysis time by altering the rate of decline of plasma osmolality which favours vascular stability without changes or weight gain [6].

In order to avoid high end-dialysis plasma sodium concentrations associated with greater thirst and weight gain, we have performed short-time chronic HD using a low sodium dialysate and perfusing into blood a small volume of hypertonic sodium chloride. The aim of this study has been to evaluate the changes in PV, ECV and ICV on conventional HD and hypertonic HD.

Methods

Thirteen male patients were studied. Seven patients aged 43.7±10.2 years and weighing 65.4±9.5kg (dry weight), had been on conventional HD for 74.8±40 months, 4x3hr/week, using dialysate containing: Na 138mEq/L, K 2mEq/L, Ca 4mEq/L, Mg 1.5mEq/L, acetate 40mEq/L, glucose 3g/L and osmolality 307mOsm/L.

Six patients aged 45.6±14.2 years, weighing 58.5±5.3kg and having been on HD for 55.8±32.7 months were placed for one month on a stable regimen of hypertonic HD, 3x3hr/week, using dialysate containing: Na 130mEq/L, K 2mEq/L, Ca 3mEq/L, Mg 1.5mEq/L, acetate 32mEq/L, glucose 4g/L and osmolality 295mOsm/L. During the first and last 20 minutes of the first hour of HD, 171mEq of NaCl (50ml of 20% NaCl) and for the last two hours of HD, 60mEq of $NaHCO_3$ (60ml of 1M $NaHCO_3$), were infused into the venous return-line.

All patients were dialysed with an UF control device using an AN-69 membrane dialyser (Hospal, H.12–10) with blood flow of 250ml/min and dialysate flow rate of 500ml/min.

The day on which body compartments were evaluated, no food or fluid intake were permitted prior to starting the study (8 and 3 hours respectively) until the completion of any given volume determination. In order to avoid fluid replacement during HD, patients were kept ½kg above their usual dry weight.

One hour before HD, patients were injected with 90μCi of ^3H_2O, 75μCi of $Na_2\,^{35}SO_4$ and 5μCi of ^{125}RISA for TBW, ECV and PV determination respectively [7]. Equilibration time was considered for ^{125}RISA at 10 minutes and for ^3H_2O and $Na_2\,^{35}SO_4$ at 45 minutes. Initial ICV was calculated: TBW-Initial ECV. Changes in hourly and post-HD PV were determined by changes in serum albumin concentration (Alb): PV post = PV_O x Alb_O/Alb post (o=initial). Sodium and plasma osmolality were measured hourly.

At the end of HD, 75μCi $Na_2\,^{35}SO_4$ were again given for repeat ECV determination. Final TBW was assumed to be equal to: Initial TBW − weight loss in HD. Final ICV was calculated: Final TBW − final ECV.

Data were statistically analysed by the Student's 't' test for unpaired data.

Results

Changes in serum sodium concentration, plasma osmolality and PV during and after both types of HD are shown in Figure 1. It is important to note that PV decrease was less on hypertonic HD than on conventional HD.

The group of patients on conventional HD had a weight loss of 1900±381g with an ultrafiltration volume of 2484±314ml. Initial TBW and PV were 38426±3173ml and 3795±823ml respectively. Pre and post HD ECV and ICV were

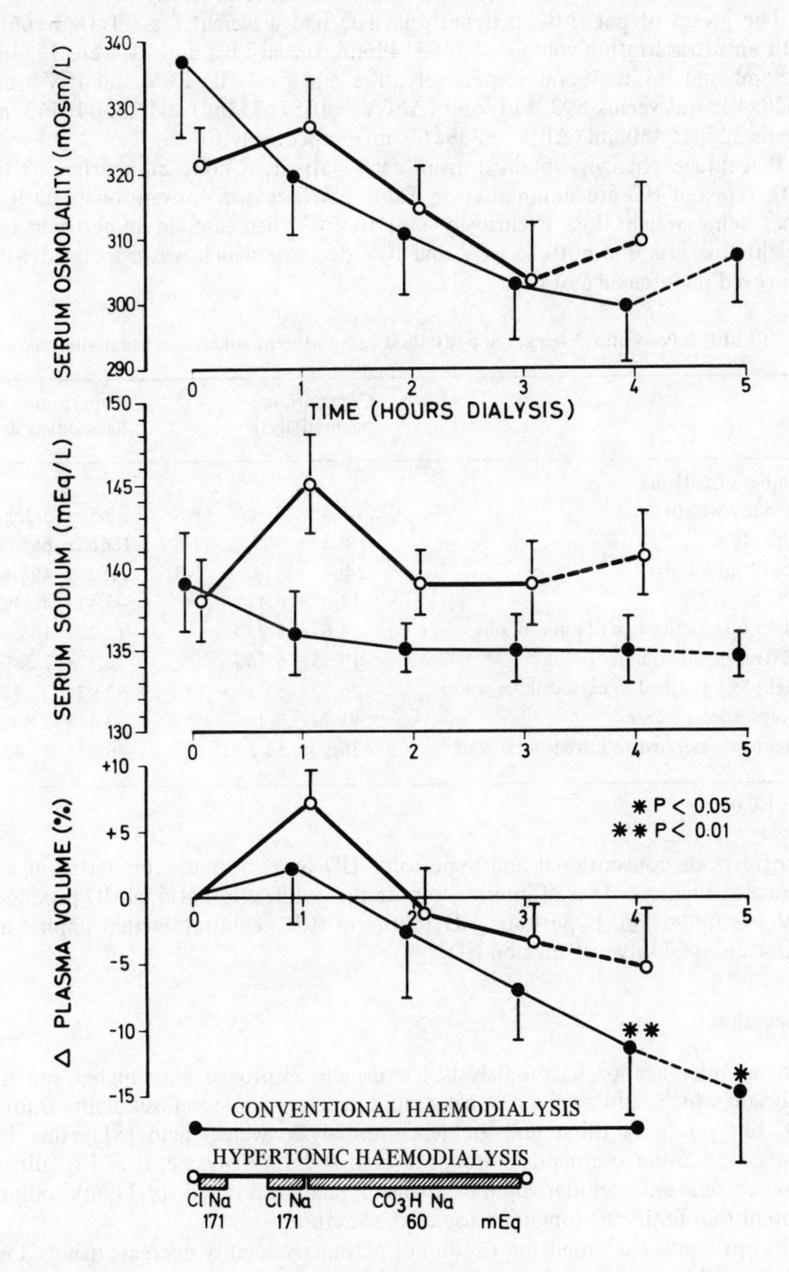

Figure 1. Changes in serum concentration, osmolality and plasma volume during and after haemodialysis

12233±1922ml versus 9920±1960ml (ΔECV: −2312±652ml) and 26193±3789ml versus 26604±3529ml (ΔICV: +411±794ml) respectively.

The group of patients on hypertonic HD had a weight loss of 1866±668g with an ultrafiltration volume of 2453±496ml. Initial TBW and PV were 35546±3388ml and 3426±396ml respectively. Pre and post HD ECV and ICV were 9829±1334ml versus 8939±1146ml (ΔECV: −1057±734ml) and 23504±4632ml versus 22701±4603ml (ΔICV: −839±685ml) respectively.

Percentage changes obtained from each patient in body compartments on both types of HD are summarised in Table I. Whereas in conventional HD ICV rose, being weight loss exclusively due to ECV decrease, in hypertonic HD weight loss was due both to ECV and ICV decrease which was associated with improved maintenance of PV.

TABLE I. Percentage changes in body fluid compartment volume on haemodialysis

	Conventional haemodialysis		Hypertonic haemodialysis
Number of patients	7		6
Dry body weight	65.4 ± 9.5kg	NS	58.5 ± 5.3kg
Weight loss	1900 ± 387g	NS	1866 ± 668g
Ultrafiltration	2484 ± 314ml	NS	2453 ± 496ml
Δ Plasma volume	−14.48± 6.41%	*	−5.41± 4.53%
Weight loss ascribed to plasma volume	−24.61± 8.93%	*	−12.22± 10.65%
Δ Extracellular volume	−19.35± 6.26%	*	−10.37± 7.36%
Weight loss ascribed to extracellular water	−126.53± 51.23%	*	−53.93± 37.43%
Δ Intracellular water	+1.74± 3.28%	*	−3.44± 3.80%
Weight loss ascribed to intracellular water	+26.53± 51.23%	*	−46.07± 37.43%

* $p<0.05$

Effects of conventional and hypertonic HD on each body compartment are shown in Figure 2. It is of interest to note that while conventional HD produces ICV overhydration, hypertonic HD results in ICV dehydration that implies an important qualitative change on HD.

Discussion

Clinical tolerance to haemodialysis has usually improved with higher sodium dialysate which reduces the relatively rapid decrease in plasma osmolality during HD, but produces thirst and greater interdialytic weight gain [8]. Thus, by increasing plasma osmolality through a high sodium dialysate, it is difficult to avoid chronic extracellular volume expansion related to raised total body sodium content that limits the long-term use of this method.

In our study we varied the rhythm of plasma osmolality decrease using a low sodium dialysate and perfusing into blood hypertonic NaCl at the beginning of HD to increase plasma osmolality above basal values during the first hour of

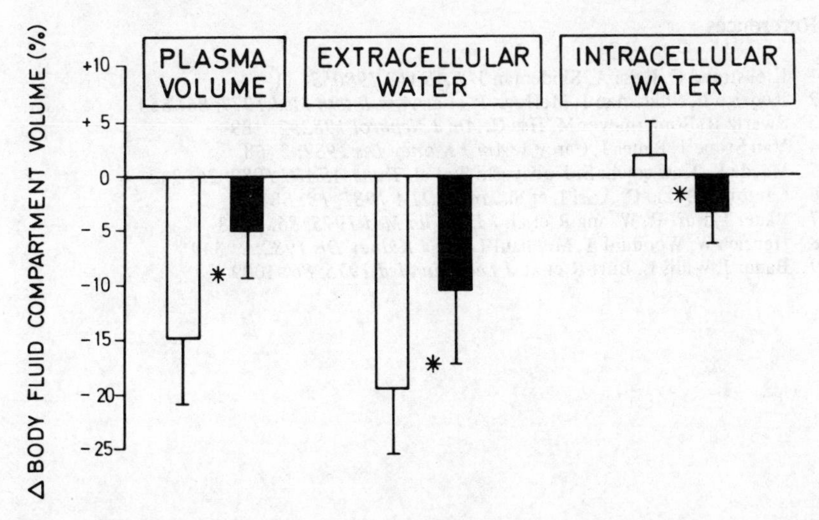

Figure 2. Distribution of body compartment volume removal in haemodialysis. □=conventional haemodialysis (n=7). ■=hypertonic haemodialysis (n=6); * $p<0.05$

dialysis and achieving a negative sodium balance at the end of HD. By this method, plasma osmolality at the end of HD is similar to that of conventional HD, but whereas in hypertonic HD volume removal comes from both intracellular and extracellular compartments with improved maintenance of PV, in conventional HD volume removal is exclusively due to ECV with subsequent ICV hydration and decrease in PV stability.

Intradialytic weight gain did not significantly change over this study; this fact allowed good clinical tolerance to ultrafiltration on short-time HD. In hypertonic HD and specially during its first hour and as in cell-wash dialysis [5], an intense ultrafiltration rate between intra and extracellular compartments must account for eliminating cellular overhydration, thereby improving microcirculation and facilitating solute transfer from cells to extracellular compartment.

The use of isotopic techniques pre and post HD allows simultaneous determination of changes in body compartments. Radioactive tracers, having a combined radiation exposure of a routine chest X-ray [9], appear to be a good method for evaluating the source of fluid removed and fluid shifts during HD, and therefore in this way should contribute to the improvement of dialysis efficiency and comfort.

Acknowledgments

Supported in part by Grant No 1065-81 from the CAICYT (Comisión Asesora de Investigación Cientifica y Técnica).

References

1 Kjellstrand C, Rosa A, Shideman J. *J ASAIO 1980; 3:* 11
2 Rodrigo F, Shideman J, McHugh R et al. *Ann Intern Med 1977; 86:* 554
3 Swartz R, Somermeyer M, Hsu C. *Am J Nephrol 1982; 2:* 189
4 Van Stone J, Bauer J, Carey J. *Am J Kidney Dis 1982; 2:* 58
5 Maeda K, Kawaguchi S, Kobayashi S et al. *Trans ASAIO 1980; 26:* 213
6 Cambi V, Buzio C, Arisi L et al. *Proc EDTA 1981; 18:* 681
7 Bauer J, Burt R, Whang R et al. *J Lab Clin Med 1975; 86:* 1003
8 Henrich W, Woodard T, McPhaul J. *Am J Kidney Dis 1982; 2:* 349
9 Bauer J, Willis L, Burt R et al. *J Lab Clin Med 1975; 86:* 1009

Proc EDTA–ERA (1984) Vol 21

RAISED SERUM NICKEL CONCENTRATIONS IN CHRONIC RENAL FAILURE

M Drazniowsky, I S Parkinson, M K Ward, Susan M Channon, *D N S Kerr

*The Medical School, Newcastle upon Tyne, *Royal Postgraduate Medical School, Hammersmith Hospital, London, United Kingdom*

Summary

We have measured serum nickel concentrations using flameless atomic absorption spectrophotometry. In 71 normals the median concentration was 1.0μg/L, range <0.6–3.0μg/L. Increased concentrations ($p<0.05$) were found in patients with chronic renal failure (CRF) treated conservatively (median 1.6μg/L, range <0.6–3.6μg/L).

Significantly increased concentrations ($p<0.001$) were found in patients treated by continuous ambulatory peritoneal dialysis (CAPD) (median 8.6μg/L, range 5.4–11.4μg/L) and haemodialysis. In patients on haemodialysis, post-dialysis concentrations (median 8.8μg/L, range 3.0–21.4μg/L) were significantly higher ($p<0.001$) than pre-dialysis values (median 8.6μg/L, range 0.6–16.6μg/L).

Introduction

In 1971 McNeely et al [1] reported diminished concentrations of nickel in patients with chronic renal failure (CRF) compared with healthy adults and they speculated that this was associated with reduced concentrations of serum proteins. Indeed, they found a positive correlation between the serum concentrations of nickel and albumin. Nickel was one of seven trace metals which Salvadeo et al [2] found to be reduced in dialysis fluid leaving the dialyser, suggesting transfer of nickel from dialysis fluid to blood. However, they were unable to demonstrate raised post-dialysis blood nickel concentrations, probably because of the low sensitivity of the analytical methods employed.

Interest in nickel as a cause of symptoms in renal failure was stimulated by Webster et al [3] who reported that symptoms of nausea, vomiting, weakness, headache and palpitations during haemodialysis were associated with nickel intoxication. This was attributed to the elution of nickel from a nickel-plated stainless steel heater in the water supply used to prepare dialysis fluid.

This study of serum nickel in patients with CRF, particularly those treated by CAPD or regular haemodialysis, was undertaken to determine if long-term pertubation of serum nickel concentration is of clinical significance in renal failure.

Materials and methods

Nickel concentrations in normal subjects and patients with CRF were determined on a Perkin-Elmer 603 atomic absorption spectrophotometer fitted with an HGA-76B furnace and AS-1 autosampler. This was superseded by a Perkin-Elmer 3030 spectrophotometer fitted with an HGA-500 furnace and AS-40 autosampler for the determination of nickel concentrations in patients undergoing CAPD or haemodialysis.

Glass distilled water (nickel content $<0.7\mu g/L$) was used for washing and standard preparation. Aristar grade nitric acid was used to acid-leach containers. A dilute solution of Triton X-100 in distilled water (20ml/L) was used to prepare serum-based standards and to pre-treat serum samples. The diluted wetting agent was found to have a mean nickel content of $0.6\mu g/L$.

Acid-leached 10ml polypropylene tubes for blood collection and storage, acid-leached 60ml polyethylene bottles for reagent storage, acid-leached volumetric flasks, syringes, venepuncture needles, autosampler cups and micropipette tips were all free of nickel contamination as judged by exposure to water, serum or whole blood for a time in excess of that expected in clinical or laboratory practice.

Samples of blood (5ml) from normal subjects, patients with CRF and patients on CAPD were obtained by venepuncture using stainless steel needles. Blood from haemodialysed patients was obtained from the arterial line before the patient was connected to the dialyser and at the end of dialysis. As a further precaution against contamination, 5 x 10ml aliquots of blood were withdrawn and reinjected through the line to flush the needle and syringe before blood collection.

The blood was allowed to clot in polypropylene tubes for 24 hours at room temperature. The tubes were centrifuged at 850g for 15 minutes, and serum decanted into polypropylene tubes and stored at $-40°C$. After vortex mixing for 30 seconds, $200\mu l$ aliquots of serum were mixed with an equal volume of Triton X-100 solution prior to analysis to prevent carbon accumulation in the furnace. Nickel concentrations were determined in triplicate by reference to a serum standard curve.

The instrumental conditions for the 603 spectrophotometer were as follows:

Wavelength: 232.0nm	Slit: 0.2nm
Lamp current: 15mA	Mode: Peak Area
Recorder: Integrate	Expansion: 0.4
Background correction: Yes	Sample volume: $50\mu l$

The conditions for the 3030 spectrophotometer were similar except a lamp current of 25mA and an expansion of 2.5 were used.

The temperature programme used for the 603 spectrophotometer was as follows (3030 programme in parenthesis):

	Temperature °C	Hold time Sec	Ramp time Sec
Dry	120 (90)	60	67 (60)
Char	700 (1300)	20	76 (60)
Atomise	2200	9 (5)	–
Clean	2500	5	–

Pyrocoated graphite tubes were used with both instruments. During the atomisation stage 'maximum power' and 'gas stop' were used. The detection limit was 0.6μg/L.

There was no significant difference in the nickel concentrations in 31 samples determined on both the 603 and 3030 spectrophotometers (paired 't' test, $p>0.05$).

Results

The results are displayed in Figures 1 and 2. Statistical calculations were performed using the Mann-Whitney U test unless otherwise stated.

In 71 normal subjects (39 males, 32 females, age range 19–64 years) the median serum nickel concentration was 1.0μg/L, lower quartile 0.6μg/L, upper quartile 1.4μg/L, range <0.6–3.0μg/L. There was no significant difference ($p>0.05$) between the concentration of nickel in males (median 1.0μg/L) and females (median 1.1μg/L).

Increased concentrations ($p<0.05$) were found in 31 patients with CRF treated conservatively. The median value was 1.6μg/L, lower quartile 1.0μg/L, upper quartile 2.0μg/L, range <0.6–3.6μg/L. The serum creatinine range in these patients was 135–1116μmol/L.

No significant correlation was found between serum nickel and the extent of renal failure as assessed by serum creatinine concentrations (Spearman's rank correlation coefficient $r_S=0.10$, $p>0.05$). There was no significant correlation between serum nickel and serum albumin concentrations ($r_S=0.01$, $p>0.05$).

In 13 patients on CAPD the median serum nickel concentration was 8.6μg/L, lower quartile 7.5μg/L, upper quartile 9.7μg/L, range 5.4–11.4μg/L; a significant increase ($p<0.001$) over patients with CRF.

Increased concentrations ($p<0.001$) of serum nickel compared to patients with CRF were found in 25 haemodialysed patients. Post-dialysis serum nickel concentrations (median 8.8μg/L, lower quartile 7.3μg/L, upper quartile 13.8μg/L, range 3.0–21.4μg/L) were significantly higher (paired 't' test, $p<0.001$) than pre-dialysis values (median 8.6μg/L, lower quartile 5.9μg/L, upper quartile 12.1μg/L, range 0.6–16.6μg/L).

Initial analysis of nickel in tap water, water treated by deionisation and reverse osmosis and dialysis fluid indicated that the highest concentrations were present in the dialysis fluid (2–3μg/L).

Discussion

We conclude that serum nickel is elevated in patients with CRF, especially those treated by haemodialysis or CAPD. Preliminary observations suggest

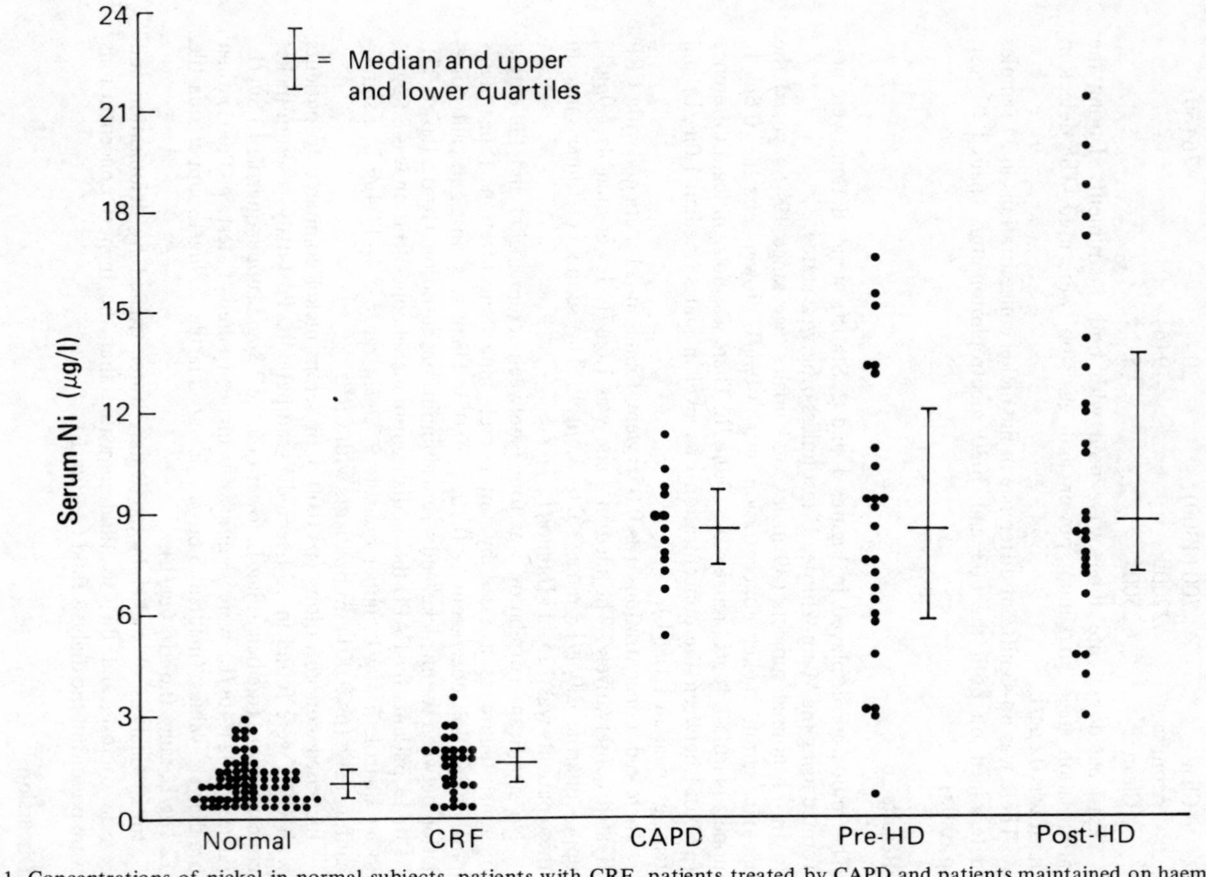

Figure 1. Concentrations of nickel in normal subjects, patients with CRF, patients treated by CAPD and patients maintained on haemodialysis, before (Pre-HD) and after (Post-HD) one dialysis

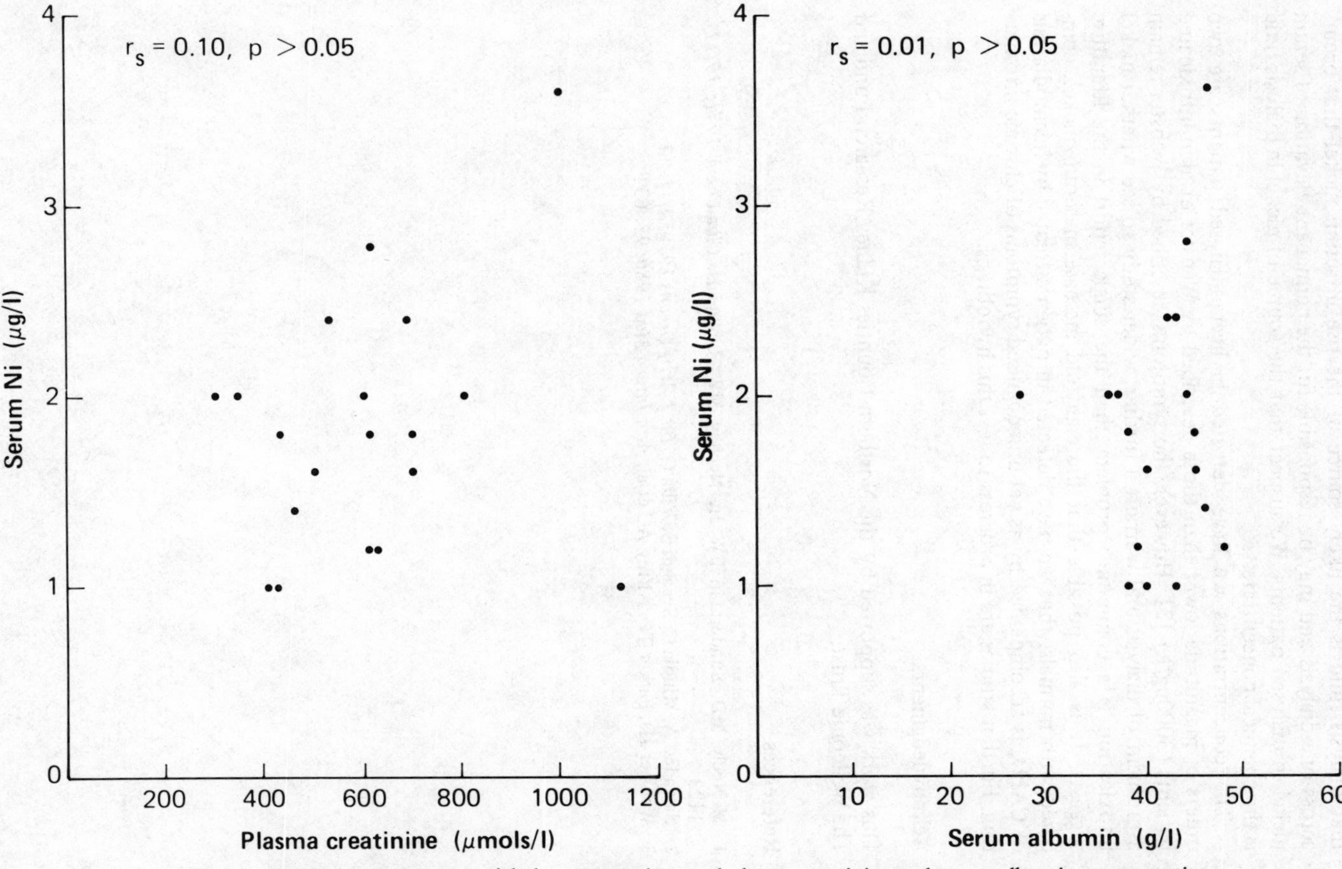

Figure 2. Correlation between serum nickel concentrations and plasma creatinine and serum albumin concentrations

that dialysis fluid is the likely source of the rise in serum nickel that occurs across one dialysis and may be responsible for the permanent elevation of serum nickel in dialysed patients. We suspect that the source of nickel in dialysis fluid is the chemical concentrate.

The concentrations we have detected in haemodialysed patients are two orders of magnitude lower than those described by Webster et al in intoxicated patients (3000μg/L) [3]. However, the symptoms described by Webster remain a problem of dialysis and warrant a further, wider study to see whether nickel intoxication is a commoner problem than the single report in the literature suggests. It is also possible that the eightfold increase in serum nickel, compared with normals, that we have detected in patients treated by haemodialysis or CAPD, is accompanied by as yet unrecognised symptoms of chronic intoxication. Further studies are in progress to test this hypothesis.

Acknowledgments

This work was supported by the Northern Counties Kidney Research Fund and The Wellcome Trust.

References

1 McNeely MD, Sunderman FW Jr, Nechay MW, Levine H. *Clin Chem 1971; 17(11):* 1123
2 Salvadeo A, Minoia C, Segagni S, Villa G. *Int J Artif Organs 1979; 2(1):* 17
3 Webster JD, Parker TF, Alfrey AC et al. *Ann Intern Med 1980; 92:* 631

Proc EDTA–ERA (1984) Vol 21

CHANGES IN COPPER AND ZINC IN CHRONIC HAEMODIALYSIS PATIENTS

S Hosokawa, H Nishitani, *T Imai, *T Nishio, *T Tomoyoshi, †K Sawanishi

*Utano National Hospital, *Shiga University, †Kyoto University, Kyoto, Japan*

Summary

We studied the behaviour of copper (Cu) and zinc (Zn) during haemodialysis (HD) in 65 chronic renal failure patients. Serum Cu, Zn and total protein were measured before and after dialysis. The dialyser membrane contained Cu and Zn. However, when the dialyser was washed with normal saline, Zn was removed, while Cu was liberated from the membrane during HD. We conclude that during HD serum Cu and Zn concentrations increased significantly ($p<0.01$), Cu from 98 to 120μg/dl by haemoconcentration and liberation from the membrane, and Zn from 78 to 91μg/dl by haemoconcentration.

Introduction

Abnormalities in copper (Cu) and zinc (Zn) metabolism have been described in chronic haemodialysis patients [1–5]. Copper intoxication causes copper fever, haemolytic anaemia and liver diseases; clinical symptoms of chronic Zn deficiency in chronic haemodialysis patients are hypogonadism in males, anorexia, and impair wound healing. The mechanisms of the abnormality of these trace metal concentrations in the serum of chronic dialysis patients are not well understood. Therefore, we studied the behaviour of Cu and Zn during haemodialysis.

Materials and methods

We examined 65 chronic renal failure patients (40 males, 25 females, average age 39±16 years) who were undergoing 5-hour haemodialysis (HD), three times a week, by various kinds of dialysers. Serum Cu, Zn total serum protein (TP) were measured before and after dialysis. Ultrafiltrate fluid (UF) was obtained by extracorporeal ultrafiltrate method (ECUM) at five minutes after the start of HD and concentrations of Cu and Zn in the ultrafiltrate fluid (Cu(uf), Zn(uf)) were measured. At 30 minutes after the beginning of HD, we measured serum Cu, Zn and TP (Cu(in), Zn(in), TP(in)) in the blood, Cu and Zn (Cu(Din), Zn(Din)) in the dialysate at the inflow site of the dialyser, serum Cu(out), Zn(out) and TP(out) in the blood, and Cu(Dout) and Zn(Dout) in the dialysate at the outflow site of the dialyser. The dialyser membrane contains Cu and Zn. However, Zn was removed by washing with 500ml normal saline. We examined the amount of Cu liberated from the membrane during HD. Copper values in normal saline (1,500ml) were measured before and after the saline was used to wash dialysers of various kinds. After 5-hour dialysis, the dialyser membrane was washed out with 100ml normal saline, and the copper in the saline measured. We also studied the copper concentration in saline after fresh dialysers were washed out with 1,000ml normal saline and then filled with normal saline for five hours to determine how much copper was liberated from the dialyser membranes. On examination days, patients ate only after HD was completed. The average BUN, serum creatinine, and haematocrit values of the 65 patients were 85±15μg/dl, 9.2±1.6μg/dl and 23.6±1.8% before HD, respectively. No patients were receiving zinc or vitamin supplements, or blood transfusions in the month prior to examination. Serum, ultrafiltrate fluid and dialysate were all handled in the same manner. Cu and Zn concentrations were determined by a standard dilution method using a flameless atomic absorption spectrophotometer (Hitachi, Japan). We used Student's 't' test, regarding a 95 per cent level of confidence ($p<0.05$) as significant.

Results

In our patients, we obtained the following results for copper:

	Cu(b)	98 ± 10μg/dl
	Cu(a)	120 ± 13μg/dl
	Cu(uf)	6.3 ± 0.8μg/dl
	Cu(in)	99.6 ± 8.4μg/dl
	Cu(out)	118 ± 11μg/dl
	Cu(Din)	2.7 ± 0.4μg/dl
	Cu(Dout)	3.3 ± 0.5μg/dl
for zinc:	Zn(b)	78 ± 5.0μg/dl
	Zn(a)	88.9 ± 6.4μg/dl
	Zn(uf)	8.2 ± 0.9μg/dl
	Zn(in)	79.4 ± 4.4μg/dl
	Zn(out)	86 ± 3.6μg/dl
	Zn(Din)	10.8 ± 1.1μg/dl
	Zn(Dout)	9.8 ± 1.0μg/dl
for total serum protein:	TP(b)	6.2 ± 0.6g/dl
	TP(a)	7.0 ± 0.5g/dl
	TP(in)	6.3 ± 0.5g/dl
	TP(out)	6.9 ± 0.4g/dl

Copper liberation from dialyser membranes: after dialyser was washed with 1,000ml normal saline, 700±300μg/1m^2; after dialysis, 500±200μg/1m^2; during 5-hour examination of fresh dialyser, 1,500±400μg/1m^2.

Discussion

We examined copper and zinc transfer during dialysis. Approximately 96 per cent of the Cu in serum binds to protein; non-protein bound copper in serum is free diffusible copper. The ultrafiltrate copper concentration is equivalent to free diffusible non-protein bound copper in serum. When the dialysate Cu concentration is lower than the free diffusible Cu in serum, Cu moves from blood to dialysate. In our patients non-protein bound Cu in serum was 6.3±0.8μg/dl and the concentrations of Cu in the dialysate (Cu(Din)) 2.7±0.4μg/dl. According to the principle of diffusion, Cu moved from blood to dialysate through the dialyser membrane. Serum Cu at the inflow site of the dialyser (Cu(in)) was 99.6±8.4μg/dl and Cu at the outflow site (Cu(out)) increased significantly to 118±11μg/dl ($p<0.01$). The total protein value significantly increased from 6.3±0.5g/dl to 6.9±0.4g/dl ($p<0.01$). The values of TP(out)/TP(in) were 1.09±0.04 and those for Cu(out/Cu(in) were 1.19±0.06. We found a significant difference ($p<0.05$) between the ratios of TP(out)/TP(in) and Cu(out)/Cu(in). TP(a)/TP(b) ratios were 1.13±0.03 and the Cu(a)/Cu(b) ratios were 1.22±0.05. There was a significant difference ($p<0.01$) between TP(a)/TP(b) and Cu(a)/Cu(b). At that time the Cu of the dialysate was 2.7±0.4μg/dl at the inflow site and 3.3±0.5μg/dl at the outflow site. There was no significant difference between Cu(Din) and Cu(Dout). The Cu moved from blood to dialysate; however, Cu contained in the dialyser membrane was liberated from the membrane to blood and dialysate. Therefore Cu(out) values were higher than Cu(in) due to haemoconcentration and liberation from the membrane, despite the decrease in serum copper (Cu(out)) at the outflow site due to diffusion. These results indicate that serum copper post-dialyser exceeded the serum copper pre-dialyser because haemoconcentration and liberation had a greater influence than did diffusion. It was for this reason that the serum copper significantly increased ($p<0.01$) from 98±10μg/dl before HD to 120±13μg/dl after HD (Figure 1).

Ultrafiltrable zinc (non-protein binding zinc in serum) was 8.2±0.9μg/dl and dialysate zinc concentrations were 10.8±1.1μg/dl. The dialysate zinc values were significantly ($p<0.05$) higher than ultrafiltrable zinc. Therefore, zinc moved from dialysate to blood by diffusion.

There was no significant difference between dialysate zinc concentrations before the dialyser (Zn(Din)=10.8±1.1μg/dl) and after (Zn(Dout)=9.8±1.0μg/dl). Serum zinc pre-dialyser (Zn(in)) was 79.4±4.4μg/dl and post-dialyser 86.0±3.6μg/dl ($p<0.05$). There was no significant difference between TP(out)/TP(in) (1.09±0.04) and Zn(out)/Zn(in) (1.08±0.06). These results show that the changes in serum zinc concentration mainly depended on haemoconcentration during HD.

There was no difference between the ratio of total serum protein before HD to that after HD (TP(a)/TP(b)=1.13±0.03) and the ratio of Zn(a)/Zn(b) (1.14±0.06).

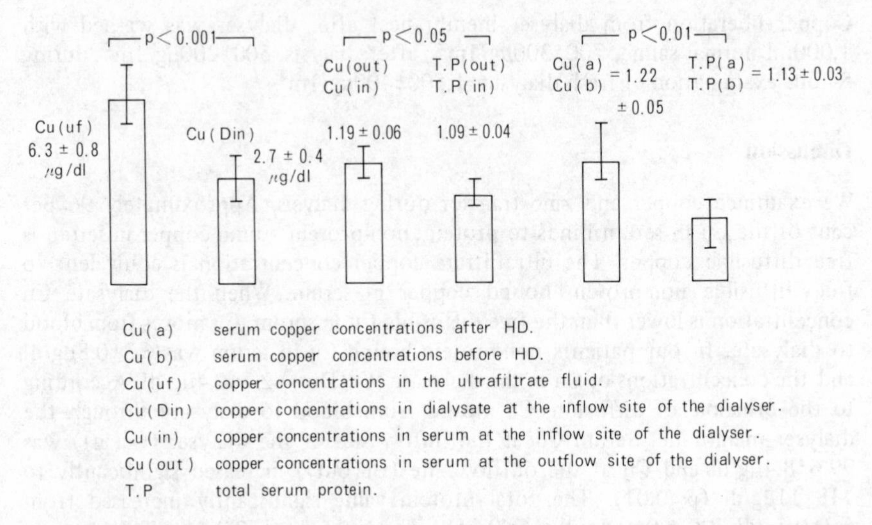

Figure 1. Copper concentrations in ultrafiltrate fluid, dialysate, and serum (at the inflow site and outflow site of the dialyser); ratios of serum copper concentrations before and after HD; and ratios of total serum protein before and after HD

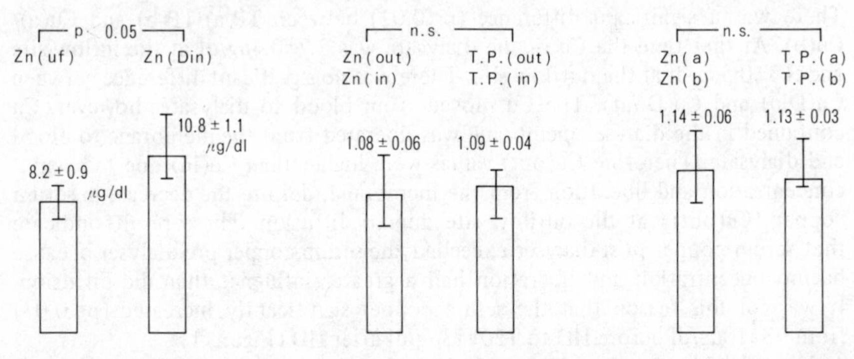

Figure 2. The zinc concentrations in ultrafiltrate fluid (Zn(uf)), dialysate and serum (at the inflow site and outflow site of the dialyser; Zn(Din), Zn(Dout), Zn(in), Zn(out)); ratios of serum zinc concentration before and after HD (Zn(a)/Zn(b)); and ratios of total serum protein before and after HD (TP(a)/TP(b))

Serum zinc significantly increased from 78±5.0μg/dl before HD to 88.9±6.4μg/dl after HD, due to haemoconcentration and to a lesser extent to zinc diffusion from dialysate to blood (Figure 2).

References

1 Mahajan SK, Gardiner WH, Abbasi AA et al. *Trans ASAIO 1978; 24:* 50
2 Mahler DJ, Walsh JR, Haynie GD. *Am J Clin Path 1971; 56:* 17
3 Mansouri K, Halsted JA, Gombos AE. *Arch Intern Med 1970; 125:* 88
4 Lyle WH, Payton JE, Hui M. *Lancet 1976; i:* 1324
5 Bustamante J, Martin Mateo MC, Paula de Pedro A et al. *Nephron 1978; 22:* 312

Proc EDTA–ERA (1984) Vol 21

MYOCARDIAL STUDIES IN HAEMODIALYSIS PATIENTS

R Dudczak, L Fridrich, K Derfler, K Kletter, H Frischauf, L Marosi, P Schmidt, J Zazgornik

I Medizinische Universitätsklinik, Vienna, Austria

Summary

Myocardial perfusion was studied using Thallium-201 (Tl-201) after dipyridamol in 33 patients on maintenance haemodialysis. It could be shown that coronary artery disease was underestimated by clinical symptoms, 55 per cent of patients had abnormal Tl-201 scintigrams, whereas typical or atypical chest pain was present in only 33 per cent of the patients. Eleven patients died within a year of the scintigraphic study, which resulted in an average mortality rate of 7.7 per cent/year. The risk of developing fatal cardiovascular complications was higher in patients with an abnormal Tl-201 perfusion (7 of 18) than in those with a normal scintigram (1 of 15). Thus nuclear medicine procedures appear to be of diagnostic value in haemodialysis patients, which in addition may have prognostic implications.

Introduction

Chronic haemodialysis therapy has altered the course of the disease in patients with end-stage renal failure, improving overall life span. Nevertheless cardiac disorders pose major problems in many patients. Various factors are implicated in the pathogenesis of heart failure during chronic renal insufficiency. The commonly associated conditions of volume overload, anaemia, hypertension, 'uraemic toxins', and coronary artery disease may individually or in combination affect left ventricular function [1–4]. Commonly coronary artery disease is diagnosed when angina or myocardial infarction occurs. However, the advent of non-invasive techniques has permitted a more comprehensible evaluation of heart disease in patients on maintenance haemodialysis treatment.

Our study examines the incidence of coronary artery disease by the use of Tl-201 dipyridamole stress scintigraphy [5]. We also assessed, in a subset of patients, left ventricular ejection fraction at rest and after isometric exercise [6].

Patients

Thirty-three patients on haemodialysis for 10 to 117 months (37.8±19.6 months) were studied, 21 males and 12 females, aged 27 to 77 years (51.8±13.3 years).

None of the patients exhibited signs of congestive heart failure at the time of the study. However, eight patients had signs of circulatory congestion during the previous one to two years. Five patients had a history of pericarditis.

Ejection murmurs were present in 11 patients, in three of them due to valvular heart disease. Eight patients had normal ECG findings, one patient had pacemaker rhythm. Two patients had ECG features of myocardial infarction. Left ventricular hypertrophy was present in 11 and left atrial enlargement in two. Non-specific ST-segment and/or T-wave abnormalities were seen in 21 patients. Cardiomegaly was found on chest x-ray in 16 patients. All patients were anaemic and 28 patients were hypertensive. Seven patients were diabetic (type I: n=3; type II: n=4). The mean serum triglyceride was 3.03±1.3nmol/L, elevated in 84 per cent of patients. The mean serum cholesterol was normal (5.4±1.9nmol/L), but elevated in 24 per cent of patients. The serum uric acid was elevated in 70 per cent (0.51±0.02nmol/L).

In all patients myocardial perfusion was evaluated by Tl-201 scintigraphy.

In addition, following haemodialysis, left ventricular ejection fraction was measured at rest by radionuclide angiography in 31 patients. In 13 of these left ventricular ejection fraction was also assessed after isometric exercise.

For isometric exercise a hand dynamometer was squeezed to one-third of a predetermined maximum for four minutes. Left ventricular ejection fraction was determined during the last two minutes.

Radionuclide studies

All radionuclide studies were undertaken with a LFOV gamma camera (Siemens) interfaced to a computer (DEC PDP 11/34) using an all purpose low energy parallel whole collimator in Tl-201 studies, and a high sensitivity low energy parallel whole collimator for determination of left ventricular ejection fraction.

Thallium-201 scintigraphy Patients were studied supine after an overnight fast. The patients were given 0.50mg dipyridamole/kg body weight intravenously within five minutes under continuous ECG monitoring. Immediately thereafter 2mCi Tl-201 was injected intravenously. No patients had to be given aminophylline because of severe anginal pain. Imaging was started five minutes after Tl-201 injection. Three views were taken (LAO 45°, anterior, left lateral), each with a preset count rate of 500kc. Four hours later redistribution scintigrams were obtained.

Visual interpretation of analogue and digitised images from multiple views were used to locate and determine the size of the defects, thereby attempting to identify coronary artery disease.

Radionuclide angiography Left ventricular function studies were performed using ECG synchronised blood pool imaging of ^{99m}Tc-labelled albumin (20mCi).

The camera was positioned in the LAO projection, obliquity and the caudal tilt were varied (35°–50° and 5°–15°, respectively) depending on optimal visualisation of the interventricular septum. For gated ventriculograms either 1000 cycles were collected, or data accumulated for a preset time (two minutes). Each heart beat was divided into 24 frames. The data were recorded in zoom mode and recorded in a 64 x 64 matrix. Left ventricular ejection fraction and wall motion were analysed, as described previously.

Results

The clinical and scintigraphic findings are summarised in Table I. Twelve of 33 patients had typical or atypical chest pain. Two of these had a history of myocardial infarction before onset of haemodialysis therapy. In the remaining

TABLE I. Clinical and scintigraphic findings in 33 patients on maintenance haemodialysis

	Haemodialysis months	Age years	Sex	Symptoms of angina	Tl-201 scan	LVEF %	Cause of death
No. 1	10	61	M	0–1 yr HD	+	68	stroke
No. 2	14	66	M	–	+	50	cardiac failure
No. 3	15	55	F	0–1 yr HD	+	50	–
No. 4	16	43	M	–	–	76	–
No. 5	18	59	M	–	–	66	–
No. 6	19	69	M	–	–	57	pneumonia
No. 7	21	39	M	–	–	60	–
No. 8	22	52	F	–	–	70	–
No. 9	23	38	M	–	–	–	–
No. 10	25	47	F	–	+	–	–
No. 11	28	68	M	–	+	52	myocardial infarct
No. 12	30	52	M	–	–	75	coma hepaticum
No. 13	33	73	M	1–2 yr HD	+	65	cardiac failure
No. 14	36	50	M	–	–	63	–
No. 15	37	40	M	0–1 yr HD	+	63	myocardial infarct
No. 16	39	65	M	before HD	+	60	myocardial infarct
No. 17	39	77	M	1–2 yr HD	+	68	–
No. 18	40	59	M	–	+	60	–
No. 19	40	51	F	0–1 yr HD	+	58	–
No. 20	40	43	M	–	–	58	–
No. 21	41	37	F	–	–	62	–
No. 22	41	40	M	1–2 yr HD	–	57	myocardial infarct
No. 23	43	65	M	1–2 yr HD	+	60	–
No. 24	44	49	M	–	–	63	–
No. 25	46	50	F	2–3 yr HD	+	76	–
No. 26	47	28	M	–	–	62	–
No. 27	49	40	F	–	–	70	coma hepaticum
No. 28	49	64	M	before HD	+	54	–
No. 29	54	30	M	–	+	63	–
No. 30	55	68	F	1–2 yr HD	+	66	–
No. 31	55	52	F	–	+	60	cardiac failure
No. 32	60	53	F	–	+	75	–
No. 33	117	27	F	–	–	64	–

LVEF=Left ventricular ejection fraction; HD=haemodialysis

patients symptoms appeared in the first (n=4), second (n=5), or third (n=1) year of haemodialysis therapy.

In myocardial studies with Tl-201, perfusion defects (n=22) after dipyridamole were found in 12 symptomatic and six asymptomatic patients 30 to 77 years of age.

In 16 patients there was complete redistribution of initially seen defects after four hours. In the two patients with previous myocardial infarction no redistribution was seen in the scintigraphic defects. There was a trend for the appearance of an abnormal Tl-201 perfusion in older patients, but also with duration of haemodialysis treatment. The most common risk factors in these patients were hypertension (n=15) and hypertriglyceridaemia (n=16).

Thus 55 per cent of the patients had abnormal Tl-201 scans, whereas coronary artery disease was underestimated from clinical symptoms alone (36%). Eleven patients died within a year of the scintigraphic examination. From the 15 patients with normal Tl-201 scan 27 per cent (4 of 15) died. Myocardial infarction was the cause of death in one patient; there were serious myocardial complications in the remaining three. Thirty-nine per cent of patients with abnormal Tl-201 scans died (4 symptomatic, 3 asymptomatic). The causes of death were cardiovascular complications (stroke; myocardial infarction, n=3; cardiac failure, n=3). In all patients dying of cardiovascular complications at autopsy coronary narrowing of 50 per cent or more could be found. Thus in one symptomatic patient (at autopsy three vessel disease) Tl-201 scintigraphy was false negative.

The cumulative survival was 67 per cent at five years, and the annual mortality rate averaged 7.7 per cent. These figures are higher compared to asymptomatic men who develop unstable angina or myocardial infarction, showing a cumulative survival of 80 per cent at five years [7]. However, it resembles mortality in patients with three vessel disease and depressed left ventricular ejection fraction [8]. Since left ventricular ejection fraction was normal in our patients (62.4± 6.1%) one might not expect an increased rate of cardiovascular failure as cause of death.

We therefore evaluated the influence of isometric exercise on left ventricular ejection fraction in patients with normal as well as with abnormal Tl-201 findings (Table II). The studies were done after haemodialysis, as haemodialysis is known to improve left ventricular function, mainly by a decrease in end-diastolic volume and by an increase in the contractile state [9].

Isometric exercise was accompanied by an increase in heart rate ($p<0.01$) and blood pressure ($p<0.01$). In six of seven patients with abnormal Tl-201 scans left ventricular ejection fraction decreased by a mean of 12.9±5.5 per cent (9.4–24%). Wall motion abnormalities were seen in five patients during isometric exercise.

However, in four of six patients with normal Tl-201 findings left ventricular ejection fraction also decreased by 9.8±1.4 per cent (8–11%).

This finding possibly represents initial alterations in ventricular performance unmasked by isometric exercise. Thus besides coronary artery disease uraemic compounds may induce alterations in left ventricular function, slightly more pronounced in patients with an abnormal Tl-201 finding.

TABLE II. Left ventricular ejection fraction (LVEF) at rest and during isometric exercise

	LVEF % at rest	LVEF % exercise
Abnormal Tl-201 scintigram – symptoms		
No. 1	50	38*
No. 2	63	57*
No. 3	64	56*
No. 4	68	68*
No. 5	75	68*
No. 6	63	56
No. 7	68	60
Normal Tl-201 scintigram		
No. 8	63	56
No. 9	63	57
No. 10	75	67
No. 11	76	70
No. 12	66	68
No. 13	70	70

*Wall motion abnormality

Consequently, it is quite likely that the detrimental influences of coronary artery stenosis on the myocardium may be augmented by 'uraemic toxins', particularly between periods of haemodialysis therapy.

Discussion

Cardiac abnormalities detectable by non-invasive methods are common in patients on maintenance haemodialysis therapy. Our findings show that coronary artery disease is underestimated from clinical symptoms alone. This is in line with recently reported findings which have shown that myocardial ischaemia was considerably underestimated by symptoms. In addition, the risk for fatal cardiovascular complications appears to be greater in patients with an abnormal Tl-201 perfusion finding, apparently not different for symptomatic or asymptomatic haemodialysis patients.

The incidence for ischaemic heart disease appears to be higher in our patients compared to previous reported studies. Yet commonly coronary artery disease was diagnosed when clinical symptoms of angina or myocardial infarction occurred. However, in the general population ischaemic heart disease mortality trend shows geographic variability, while declining in USA it has increased in Eastern Europe.

Difficulty in predicting coronary artery disease is accentuated in this population due to muscular weakness, hypertension, and neuropathy, which also prevent adequate exercise testing. Consequently, Tl-201 scintigraphy after pharmacological stress should also prove of diagnostic value in patients with atypical chest pain as well as in asymptomatic patients who present with several risk factors implicated in the pathogenesis of coronary artery disease.

Patients with abnormal Tl-201 perfusion findings should be under more rigorous control with medical treatment initiated, thus eventually reducing the number of ischaemic events.

References

1 Friedmann HS, Shah BN, Kim HG et al. *Clin Nephrol 1981; 16:* 75
2 Lindner A, Charra B, Sherrard DJ, Scribner BH. *N Engl J Med 1974; 200:* 697
3 Rostand SG, Gretes JC, Kirk AK et al. *Kidney Int 1979; 16:* 600
4 Scheuer J, Stezoki SW. *J Mol Cell Cardiol 1973; 5:* 287
5 Schmoliner R, Dudczak R, Kronik G et al. *Z Kardiol 1981; 70:* 111
6 Bodenheimer MM, Banka VS, Foshee CM et al. *J Nucl Med 1979; 20:* 724
7 Kent KK, Rosing DR, Ewels CJ et al. *Am J Cardiol 1982; 49:* 1823
8 Proudfit WJ, Bruschke AVG, MacMillan JP et al. *Circulation 1983; 68:* 986
9 Nixon JV, Mitchell JH, McPhaul JJ, Heurich WL. *J Clin Invest 1983; 71:* 377

Proc EDTA–ERA (1984) Vol 21

ALTERNATIVE VASCULAR ACCESS IN PATIENTS LACKING VEINS FOR STANDARD ARTERIOVENOUS FISTULAE

W P Reed, P D Light, J H Sadler, E Ramos

University of Maryland School of Medicine, Baltimore, Maryland, USA

Summary

We have evaluated the Hickman catheter and Hemasite access port as means to re-establish vascular access in patients lacking veins for conventional arterio-venous fistulae. The Hemasite is more convenient, but is also costlier, requires more surgical skill to implant, and is more frequently associated with major infections. One-half of the Hemasites have failed because of infection. As a result, the long-term survival rate is lower for Hemasite graft, although the differences noted have not yet reached statistical levels of significance.

Introduction

The vascular access needs of most patients requiring long-term haemodialysis will be adequately met by the creation of a standard arteriovenous fistula of the type described by Brescia and Cimino [1]. Unfortunately, not every patient needing chronic haemodialysis will have veins adequate to develop or maintain a functional fistula. When obesity, prior scarring or small vessel size precludes suitable venous maturation for useful access, alternative forms of access will be necessary if haemodialysis is to continue as a therapeutic option.

Over the past three years, we have evaluated two new forms of vascular access as means of maintaining haemodialysis in problem patients. This report details our experience with each device at the University of Maryland Hospital.

Materials and methods

The Hemasite access port (Renal Systems, Minneapolis, MN, USA) consists of a T-shaped titaneum tube which is covered externally with Dacron velour to promote tissue ingrowth for implant stabilisation and inhibition of bacterial entry [2]. Each end of the tube is connected to polytetrafluoroethylene (PTFE) tubing to facilitate arteriovenous incorporation of the implant. The side arm, or button, contains a silastic septum with pre-formed slits which are designed

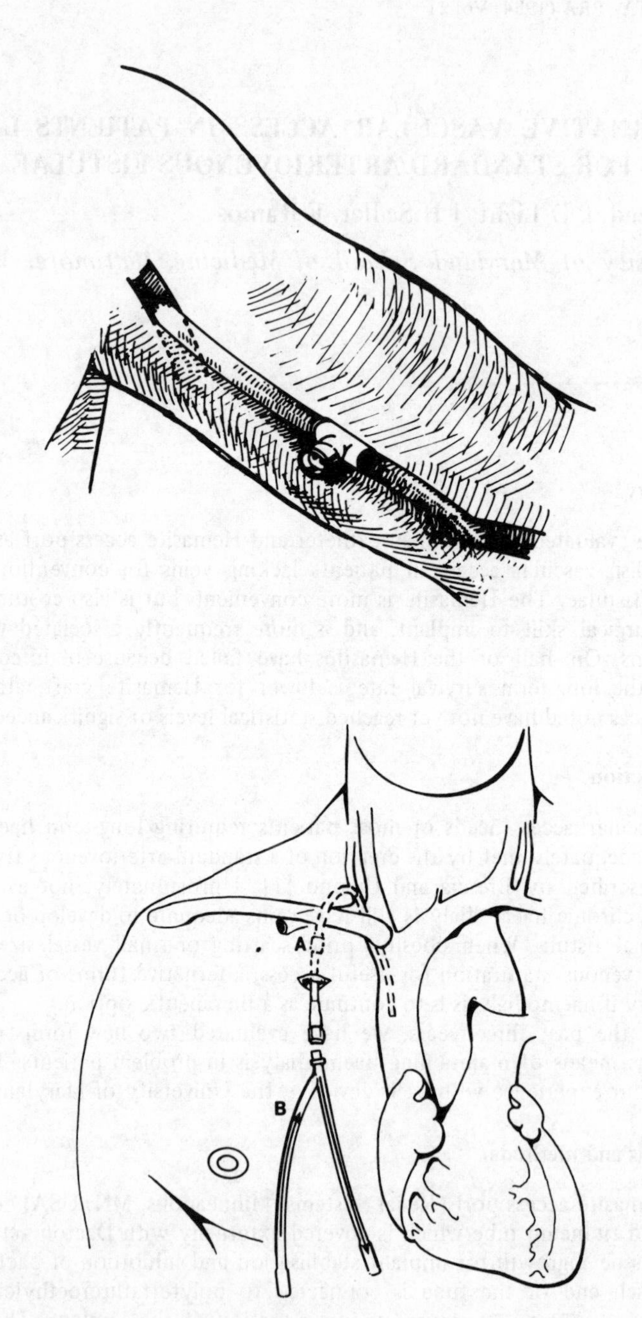

Figure 1. Top: Hemasite in preferred location. Bottom: (A) Hickman in preferred location. (B) Y-connector used for single needle dialysis

to accept introduction of a blunt-tipped double needle access set. Automatic closure of these slits upon removal of the needle set assures haemostasis between periods of haemodialysis. The Hemasite is usually implanted within a subcutaneous tunnel along the medial aspect of the arm (Figure 1, top), allowing the button to exit midway between a brachial artery anastomosis at the elbow and a basilic vein anastomosis just distal to the axilla. Povidone iodine solution is retained in contact with the septum by a silastic cap when the device is not in use. Operative time for implantation is 60–120 minutes. Device cost is 3000.00 US dollars.

The Hickman catheter (Evermed, Medina, WA, USA) used for dialysis is a shorter (26cm), larger bore (26mm internal diameter) version of the right atrial catheter used for chemotherapy [3]. An enveloping Dacron cuff attached to its outer surface secures the catheter in place and limits bacterial ingrowth. Implantation is usually via the right external or internal jugular vein (Figure 1, bottom). Cannulation of the inferior vena cava through the greater saphenous vein can be used as an alternate route. Single needle dialysis is carried out by attaching a Y-connector (Figure 1, bottom) to the luer-lok adapter on the external catheter end as previously described [3]. When not in use the catheter is filled with 2000 units sodium heparin in 2ml normal saline and sealed with a threaded cap. Operative time for insertion is 15–30 minutes. Device cost is 40.00 US dollars.

Survival curves were calculated by the method of Kaplan and Meier [4]. Data were analysed by the log rank test as outlined by Savage [5].

Results

Since July 1981, 42 Hemasite access ports have been implanted in 32 haemodialysis patients at the University of Maryland (Table I). Twenty-three patients received one implant, eight received two and one received three. Twelve Hemasites continue to function 20–1050 days (mean 364 days) after implantation.

TABLE I

	Hemasite	Hickman
Number	42	38
Patients	32	35
Age-range (mean)	23–78 (54)	26–77 (55)
Sex (M/F)	17/15	10/25
Years on dialysis (mean)	0–11 (4.4)	0–11 (3.8)
Number prior access (mean)	0–12 (4.8)	1–15 (5)

Three functioned for 163–349 days (mean 232 days) prior to patient death from causes unrelated to renal function or access. The remaining 27 Hemasites have failed from 1–799 days (mean 370 days) after insertion. In 21 instances, this failure was due to infection developing around the graft and in six instances to thrombosis which could not be reversed by graft revision. Most graft infections were obvious, but three patients died of sepsis that developed insidiously without external signs of graft infection. Accumulation of pus about the graft, with corresponding micro-organisms isolated from other tissues, confirmed the Hemasite to be the source of sepsis in the only patient of the three to be examined by autopsy.

Over the same time period, 38 Hickman catheters were implanted in 35 haemodialysis patients (Table I). Thirty-three have received one catheter and three have received two. Twenty-four catheters continue to function 12–595 days (mean 242 days) after insertion. Eleven catheters functioned for 6–126 days (mean 36 days) until patient death from unrelated causes (six cases), recovery of renal function (two cases) or the maturation of more conventional arteriovenous access (three cases), prompted discontinuation of the Hickman dialysis. Only three catheters have failed to date. One failed at eight days because of poor surgical positioning through the left jugular vein. Two failed at 15 and 81 days when the stabilising cuff migrated along the subcutaneous tunnel toward the exit site, leading to poor flows due to catheter tip withdrawal from its vena caval position. No episode of catheter related sepsis was documented. Several exit site infections were easily controlled with oral antibiotics. The main difficulty encountered with these catheters was the daily replacement of heparin that was required to maintain patency [6]. Some patients were able to accomplish this at home without difficulty, but most required visiting nurse assistance to assure compliance. When thrombosis did occur, it was always partial, allowing infusion to continue but limiting withdrawal. Patency could be restored with low dose streptokinase infusion (1000 units per hour in 10cc D_5W for 12–48 hours).

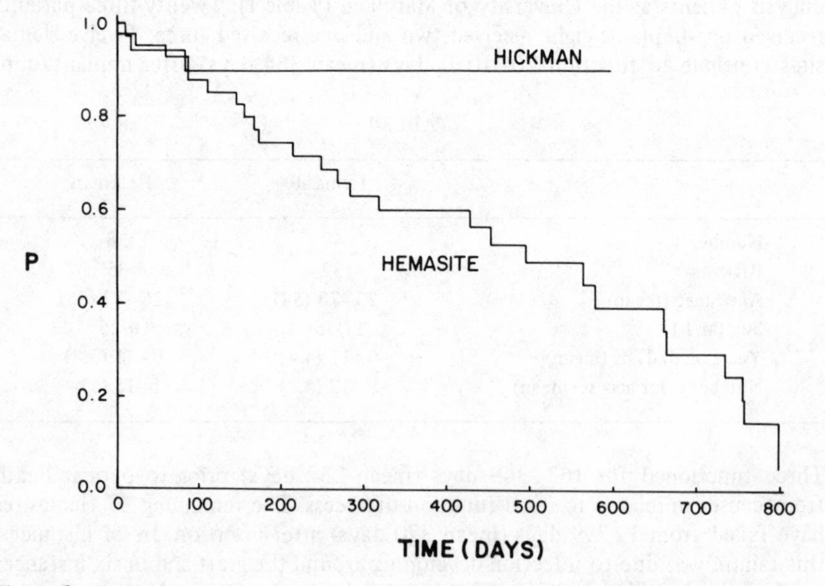

Figure 2. Kaplan-Meier plot of Hickman dialysis catheter (upper curve) and Hemasite access port survivals

The Kaplan-Meier survival curves for the two devices are illustrated in Figure 2. There is a clear trend in favour of the Hickman catheter, but the difference is not yet significant (p=0.08). It should be noted that 13 patients were placed on venous dialysis after one to three Hemasites failed. Only one patient was implanted with a Hemasite because of patient preference after first receiving a Hickman.

Discussion

The Hemasite and Hickman are both attractive devices for trial in the patient who has unsuitable veins for conventional access. Both devices provide access that is immediately available by exploiting veins that are usually still untouched after multiple standard fistula procedures. Both devices can also be inserted under local anaesthesia and demonstrate reasonably long-term patency for a difficult group of patients. Access survival was 60 per cent at one year and 30 per cent at two years for the Hemasite, and even higher for the Hickman, in patients whose average of five prior access channels had each failed after less than one year of function.

Hickman catheters are technically simpler to insert, but require daily instillation of heparin and a single needle dialysis technique. For these reasons, we initially limited our use of the Hickman to a less favourable group of candidates. In fact, 13 Hickman patients received this type of access only after the failure of one or more Hemasites. It came as a pleasant surprise that their catheter survival rates have been even higher than those of the Hemasites. That these differences have not yet reached the levels of statistical significance is due to the small numbers in place beyond 18 months, which reflects our early reluctance to employ the Hickman except in desperately ill patients with abdominal sepsis who needed some quick means of haemodialysis that could be reused as long as needed.

Infection has presented a more serious problem than anticipated with the Hemasite device. One-half of these ports have failed because of infection. Most disturbing have been the three fatal cases of sepsis where the onset of symptoms was insidious and no clinically evident graft infection existed. Although intravenous antibiotics were useful in temporarily controlling these graft infections, often prolonging Hemasite survival for many weeks or even months, our experience with insidious sepsis has cautioned us not to persist in the antibiotic treatment of repeated Hemasite infections. This will undoubtedly increase further the survival difference being noted between the two devices.

References

1 Brescia MJ, Cimino JE, Appel K, Hurwich BJ. *N Engl J Med 1966; 275:* 1089
2 Collins AJ, Shapiro FL, Keshaviah KI et al. *Trans ASAIO 1981; 27:* 308
3 Reed WP, Light PD, Sadler JH. *Kidney Int 1984; 25:* 838
4 Kaplan EL, Meier P. *J Am Stat Assoc 1958; 53:* 457
5 Savage IR. *Ann Math Stat 1956; 27:* 590
6 Reed WP, Newman KA, de Jongh CA et al. *Cancer 1983; 52:* 185

Proc EDTA–ERA (1984) Vol 21

INITIAL EXPERIENCE WITH HEMASITE VASCULAR ACCESS DEVICE FOR MAINTENANCE HAEMODIALYSIS

U Bonalumi, G A Simoni, D Friedman, E Borzone, F Griffanti-Bartoli

Patologia Chirurgica 1, University of Genoa, Italy

Summary

Thirteen uraemic patients having undergone chronic haemodialysis from a minimum of 16 months to a maximum of 15 years (mean 6.5 years) with unsuitable peripheral vessels for standard arteriovenous fistulae, received Hemasite®, a new vascular access device which provides vascular access without needle puncture.

Eight devices are still being used routinely with enthusiastic acceptance by the patients. Three subjects died because of unrelated causes, two of whom had a functioning device. Nine thromboses occurred in five patients. Thrombectomy was successful in three subjects. There were two cases of infection with loss of one device. In conclusion, the main advantage of Hemasite is the possibility of performing haemodialysis without needles, thus potentially maintaining the longevity of graft fistula. The only disadvantage of the device is its cost.

Introduction

The arteriovenous fistula (AVF) [1] is the method of choice for gaining vascular access in uraemic patients undergoing chronic haemodialysis. There is general agreement that the operation should be performed in the distal forearm in order to spare vessels which could be available for successive operations in case of failure of the first procedure. Provided the operation is performed with adequate vessels and the arterialised vein correctly punctured, the AVF could represent reliable and occasionally long-lasting angioaccess [2]. The vessels, nevertheless, are sometimes unsuitable for constructing an AVF because of different reasons: repeated operations with progressive exhaustion of the vascular surface; inadequate calibre and/or blood flow; or presence of abundant subcutaneous tissue. In these cases the surgeon can utilise prosthetic material of biological and synthetic origin in order to create a new vascular surface utilisable for venepuncture. However, the use of such grafts did not meet with complete success due to complications such as thrombosis, infection, bleeding after needle extraction and true aneurysm or pseudo-aneurysm formation [3].

Recently a new device, namely Hemasite® (Renal System Inc, Minneapolis, MN, USA) has been introduced into clinical practice [4–6], its main advantage being the possibility of performing haemodialysis without needle puncture. The preliminary results utilising Hemasite are challenging. The aim of this paper is to report our experience with this type of vascular access.

Materials and methods

The Hemasite (Figure 1) consists of a carbon coated titanium T-shaped body with a silicone self-sealing septum inserted in the round side arm or 'well' which extends above skin level. The T configuration provides a permanent transcutaneous access through the well. Beneath skin level, the body is surrounded by a Dacron velour which is incorporated into the subcutaneous tissue by fibrous ingrowth thus providing a barrier to bacterial invasion. The body is supplied with or without 6mm venous PTFE and tapered 4mm arterial PTFE grafts. At dialysis, the blunt double cannula access set is inserted in the septum. The cannulae have three holes on opposite sides to allow the blood to flow in and out.

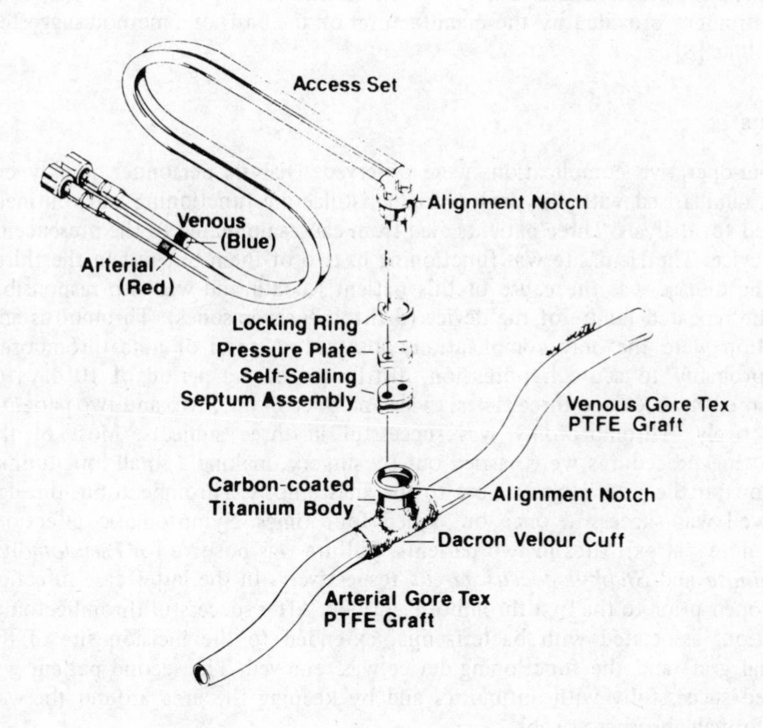

Figure 1. Schematic representation of Hemasite with the septum assembly and the access set

Between January 1983 and April 1984 13 Hemasites were implanted in 13 uraemic patients having undergone chronic haemodialysis for a minimum of 16 months to a maximum of 15 years (mean 6.5 years). The patients numbered 10 females and three males, ranging from 44 to 73 years of age (mean 58 years). Mean follow-up was nine months. Causes of renal failure were chronic glomerulonephritis (6 patients), hypertensive nephropathy (4 patients), polycystic disease (2 patients) and SLE (1 patient). Before undergoing insertion of Hemasites the patients had been submitted to numerous vascular access procedures, ranging from a minimum of two to a maximum of 11 (mean 8.0±3.0). All the devices were implanted in a straight fashion in the upper arm interposed between the brachial artery and the brachial vein. Our operative technique is very similar to the one described very accurately by McIntyre and Putnam [7]. Under local anaesthesia two separate skin incisions are made in order to isolate the vessels. Hemasite is introduced after tunnelling the subcutaneous tissue between the two incisions and a small round segment of the skin is removed at a predetermined exit point and the device is passed through. End-to-side anastamoses are performed in a continuous fashion with 6/0 monofilament suture between graft and vein first and then between graft and artery. Prophylactic antibiotics and antiplatelet drugs are not administered routinely. Fistulae are not used for a period of 10–15 days after operation. Blood flow through the device is measured at operation and four and seven months post-operatively by means of an instrument provided by the manufacturer on the basis of a method suggested by Collins [8].

Results

No per-operative complications were observed. Dialysis personnel quickly became familiarised with the device. Eight fistulae are functioning and routinely utilised for dialysis. Three patients died from causes unrelated to the presence of the device. The Hemasite was functioning in two of them but not in the third. Cardiac disease was the cause of this patient's death and was also responsible for the repeated failure of the device (3 thrombosis episodes). Thrombosis and infection were the only complications observed. A total of nine thromboses, due probably to acute hypotension, distributed over a period of 10 days to eight months, occurred three times, twice and once in one, two and two patients, respectively. Thrombectomy was successful in three subjects. Most of the declotting procedures were carried out by surgery, making a small longitudinal incision on the arterial graft close to the anastomosis. Thrombectomy through the well was successful once but failed four times. Symptomatic infections were noted at exit sites in two patients. Culture was positive for *Pseudomonas aeruginosa* and *Staphylococcus aureus* respectively. In the latter case infection developed prior to the first thrombotic episode. After successful thrombectomy, infection, associated with bacteriaemia, extended to the incision site of the arterial graft and the functioning device was removed. The second patient was treated successfully with antibiotics and by keeping the area around the well dry through absorbent swabs.

Mean blood flow through the Hemasite was: 650±240ml/min (11 subjects), 1157±364ml/min (7 subjects) and 1103±235ml/min (7 subjects) at operation, and four and seven months later, respectively. The difference between per-operative and post-operative values was statistically significant ($p<0.001$). Distal ischaemic complications and/or congestive heart failure never occurred.

Discussion

These preliminary results in a small number of cases, are encouraging and suggest that Hemasite might be a reliable device for providing durable vascular access for haemodialysis. Eight devices have been and are being used routinely. A retrospective comparison between cumulative survival of Hemasite group and cumulative survival of PTFE and bovine grafts showed no significant difference [9]. Patient acceptance has been enthusiastic because of the opportunity of undergoing haemodialysis by means of a needleless device thus eliminating pain when starting the treatment. In our experience, indeed, the reason for utilising the Hemasite in all the subjects was the lack of peripheral vessels and not patient fear of receiving venepuncture at every dialysis. According to Nissenson's classification all the subjects were classified as 'high risk patients' except one ranking at 'moderate risk' [10]. Thrombosis and infection were the only complications noted. Thrombosis was a major concern. Five devices clotted nine times because of hypotensive episodes. The incidence of thrombosis was similar to that reported by Kaplan but much higher if compared with other larger series [4,6]. The cause of the difference cannot be easily discerned. The presence of almost all of our patients in the high risk group might be a plausible explanation. Declotting by the use of a small Fogarty catheter through the well was unsatisfactory because of a high chance of leaving a clot remnant in the conduit close to the arterial anastomotic line.

Contrary to our expectations only one device was lost because of persistent infection which started at the exit site and spread through the arterial graft. The infection rate (15%) was very close to those in Collins's and Dienst's observations [4,6]. As stated by other authors [5] it is very important to maintain the integrity of the skin around the well and take all precautions such as accurate cleaning and preventing any trauma to this area in order to avoid infectious complications.

Blood flow measurements revealed a statistically significant increase in flow four and seven months after operation, but provided no predictability of imminent thrombosis of the device by early detection of a decrease as observed in a few cases by Kaplan.

In conclusion, the main advantages of the Hemasite vascular access device are:

1. Elimination of needles used for cannulation, involving no pain for patients, no trauma of the skin and graft, no bleeding after dialysis, no true aneurysm or pseudo-aneurysm formation.
2. High blood flow, useful for haemofiltration and ultrafiltration procedures.
3. The theoretical possibility of performing a declotting procedure through the well without operation.

We would finally state that the high cost of the device, the only drawback found so far, may be largely offset by its many proven advantages.

References

1 Brescia MJ, Cimino JE, Appel K et al. *N Engl J Med 1966; 275:* 1089
2 Bonalumi U, Civalleri D, Rovida S et al. *Br J Surg 1982; 69:* 486
3 Kootstra G. *Proc EDTA 1982; 19:* 99
4 Collins AJ, Shapiro FL, Keshaviah P et al. In Kootstra G, Jörning PJG, eds. *Access Surgery.* Lancaster: MTP Press. 1983: 297
5 Kaplan AA, Grant J, Galler M et al. *Trans ASAIO 1983; 29:* 369
6 Dienst SG, Oh HK, Levin NW et al. *Trans ASAIO 1983; 29:* 353
7 McIntyre KE, Putnan CW. *Surg Gynecol Obstet 1983; 156:* 805
8 Collins AJ, Shapiro FL, Keshaviah P et al. *Trans ASAIO 1981; 27:* 308
9 Collins AJ, Ilstrup K, Keshaviah P et al. *Trans ASAIO 1983; 29:* 789
10 Nissenson AR. *Trans ASAIO 1983; 29:* 784

Proc EDTA–ERA (1984) Vol 21

THE BIOCARBON® VASCULAR ACCESS DEVICE (DiaTAB®) FOR HAEMODIALYSIS

P J H Smits, M J H Slooff, D H E Lichtendahl, G K v d Hem

University Hospital, Groningen, The Netherlands

Summary

The Biocarbon® vascular access device (DiaTAB®) is a relatively new method in secondary access surgery. Punctures, often the cause of complications can be avoided because it is a no-needle method of dialysis. However thrombosis due to stenosis of the venous anastomosis or of the efferent vein is a continuing problem. A new experience is the fibrin flap formation under the plug of the device, which can be removed easily.

Introduction

Secondary access surgery is often hampered by complications. Several of these complications like bleeding, infection, thrombosis and formation of pseudo-aneurysms are related to the puncturing necessary for access to the bloodstream [1–4]. DiaTAB is a no-needle method of dialysis. The main advantage is that puncture lesions can be avoided. This inherent quality of the DiaTAB should make it less complication-prone compared to other devices in which punctures are necessary for access and dialysis.

This report gives our early experience with this access device in patients eligible for secondary access surgery.

Methods and materials

From July 1982 until July 1984 16 DiaTAB devices were implanted in 11 patients (Figure 1). The indications for placing such a device were multiple failed previous access procedures, unsuitable skin conditions for repeated puncture, needle phobia and home dialysis patients.

Five of these patients were diabetics. Their ages ranged from 37–68 years. The mean dialysis period was 48 months (range 12–74 months). The mean number of preceding access operations was eight (range 4–16). Fourteen straight

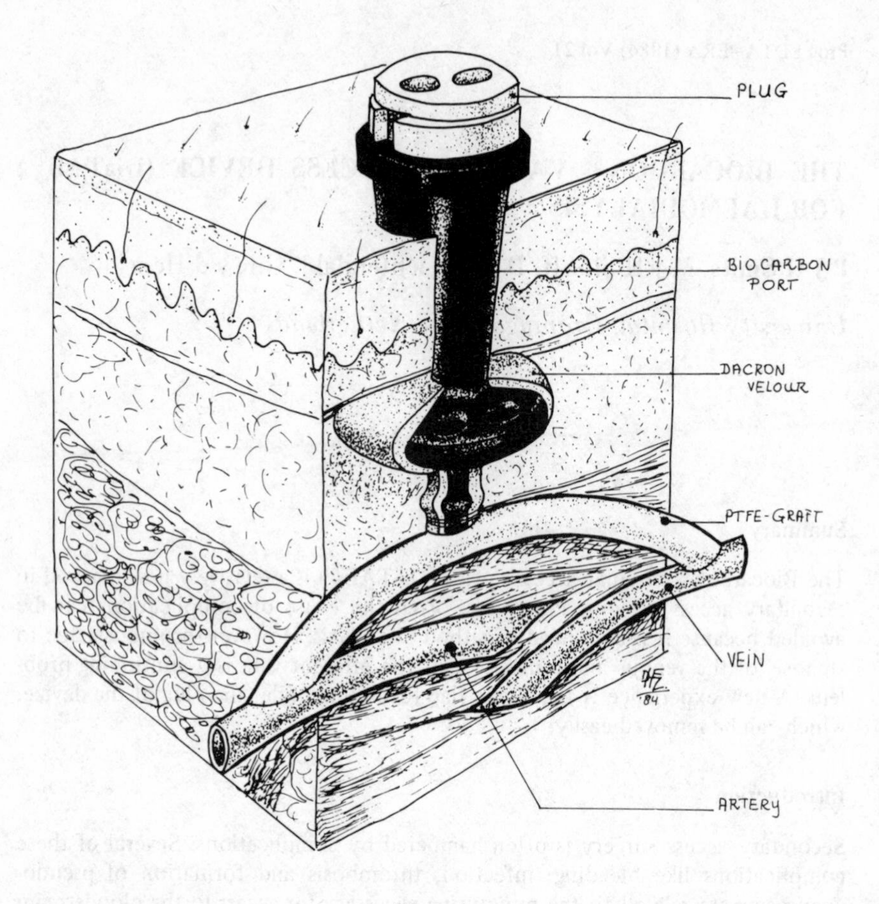

Figure 1

grafts were placed in the upper arm and two loop grafts were positioned in the groin. The mean follow-up period of the grafts was 10 months with a range of two to 24 months. Prophylactic broad spectrum antibiotics and anticoagulants (coumarin derivatives and/or acetylsalicylic acid) were used routinely.

Results

In six DiaTAB devices thrombotic episodes were observed over a total of 4,896 observation days. Thrombectomy was successfully performed in five grafts. In three of these five grafts the thrombosis, due to stenosis of the venous anastomosis, demanded replacement of the device. In the remaining DiaTAB a lesion of the wall opposite the plug was observed. This lesion was caused by inexperience with this method of dialysis. This DiaTAB needed replacement because of irreversible thrombosis.

A new problem in access surgery related to this no-needle method was fibrin flap formation under the plug. In three DiaTAB devices several episodes of fibrin flap formation were observed. Although the graft was patent effective dialysis was not possible. The flap can be removed easily using a Fogarty® catheter through the port of the device. Interestingly enough the fibrin flap formation was noted in combination with venous outflow obstruction in all three cases.

Infection was observed in only one graft. Because this infection occurred three months after the operation and an abcess was observed on the extremity of the graft, the infection was considered to be secondary to the abcess. After removal of the graft and treatment of the primary focus the patient received a new DiaTAB.

A steal syndrome was observed in one graft. Banding of the arterial anastomosis was successfully performed. However the patient died 36 hours after the operation due to a massive myocardial infarction.

Handling of the DiaTAB during operation and replacement was never difficult. The patient and nurse compliance was excellent although extensive instruction to patients and dialysis staff is necessary.

Conclusion

Thrombosis still remains the main problem but is the consequence of using prosthetic vessel material. Prevention is difficult because most often it is a sequela of the arterialisation of the arterialisation of the efferent vein. Arterialisation leads to phlebosclerosis and subsequent stenosis of the efferent vein.

Perioperative mortality was observed in one patient and could not be related to the method described in this preliminary series.

Other types of complications like steal syndromes, fibrin flap formation and even infection were observed relatively infrequently compared to the incidence of puncture-related complications with other methods [5]. The DiaTAB no-needle method of dialysis seems to be a promising tool in secondary access surgery. It is not complication-free but less complication-prone.

References

1 Nissenson AR. *Contemporary Dialysis. October 1981*
2 Golding AL. *Trans ASAIO 1980; 26:* 105
3 Golding AL. *Proc Dial Transplant Forum 1979; 9:* 242
4 Nissenson AR. *Clin Nephrol 1981; 15:* 302
5 Sloof MJH, Smits PJH, Lichtendahl DHE, Van Der Hem GK. *Proc EDTA 1982; 19:* 234

Proc EDTA–ERA (1984) Vol 21

HAEMODIALYSIS IN DOGS WITH A HEPARIN COATED HOLLOW FIBRE DIALYSER

L E Lins, P Olsson, *M-B Hjelte, *R Larsson, †O Larm

*Karolinska Hospital, Stockholm, *IRD Biomaterial, Stockholm, †Swedish University of Agricultural Sciences, Uppsala, Sweden*

Summary

Haemodialysis was performed in non-uraemic dogs with equipment coated with a stable heparin. During a three hour dialysis a constant blood flow of 205ml/min was easily maintained. There was no increase in whole blood coagulation time and no heparin release from the surface. The platelet count was initially reduced by 15 per cent, but remained constant at this value throughout the dialysis. No increase in FPA concentration was detected. Heparin coating on inherently thrombogenic materials enables haemodialysis in the absence of systemic anticoagulation and without measurable activation of the haemostatic mechanism.

Introduction

Conventional haemodialysis requires heparin anticoagulation to prevent clotting in the extracorporeal system, but systemic heparinisation, however, is associated with enhanced bleeding tendency and long-term side-effects of repeated administration of heparin such as osteoporosis [1].

In an attempt to exclude systemic heparin we prepared for haemodialysis a hollow fibre dialyser and PVC tubings with a covalently bonded heparin coating which exhibits thromboresistant properties with regard to platelets and plasma coagulation [2]. In this paper results from dialysis in non-uraemic dogs are presented.

Materials and methods

Low density polyethylene tubings (A/S Surgimed, Oelstykke, Denmark) with an inner diameter of 3mm were used for aortic blood pressure recordings, arterial blood sampling and vascular connection to the extracorporeal circuit.

Cellulose acetate hollow fibre dialysers (CDAK 3500, Cordis Dow, Concord, USA) with a surface area of approximately $0.9m^2$ were used in combination with PVC blood lines (Gambro, Sweden).

Haematocrit was determined by high speed centrifugation (Wifug, Stockholm, Sweden) and platelet count by a platelet counter (Linson 431A, LIC, Stockholm, Sweden). For the calculations only haematocrit corrected values were taken into account. Coagulation time in whole blood was determined by gently tilting 2ml of blood in 10ml glass test tubes every 30 seconds. Solid clot formation was regarded as the end point.

Heparin concentration in plasma was determined by using a factor X_a inhibition test with the aid of a synthetic chromogenic substrate (Coatest, Kabi, Stockholm, Sweden). Standard curves on pooled dog plasma with heparin concentrations from 0.025 to 0.1IU or from 0.1 to 1.0IU heparin/ml were constructed.

Fibrinopeptide A (FPA) in plasma was radioimmunologically assayed essentially according to Nossel et al [3] with the aid of human rabbit antiserum cross-reacting with canine FPA [4,5].

Surface heparinisation of tubings and hollow fibre dialysers

After careful rinsing with water the material was coated with a covalently bonded layer of heparin. The procedure consists of two steps. In the first step polyethyleneimine is attached to the substrate surface. The heparin to be used is partially degraded by nitrous acid whereby terminal residues of 2,5-anhydro-D-mannose with a reactive aldehyde function are formed. In the second step of the coating procedure, these reactive residues are allowed to react with the surface bound primary amine groups to form a Schiff's base, which is reduced by sodium cyanoborohydride to form a stable secondary amine bond. The details of the binding method are described elsewhere [2].

The dialysers were replastisised using glycerine and sterilised, procedures which were kindly performed by Cordis Dow Co, Concord, USA. The dialyser performance with regard to ultrafiltration and dialysance was essentially unchanged after heparinisation procedure: ultrafiltration coefficient 5.9ml/hr/mmHg. In vitro dialysance at Q_B = 200ml/min for urea 128, sodium chloride 115 and uric acid 84ml/min respectively.

Experimental procedure

Five mongrel dogs, 16 to 22kg, were used. The animals were anaesthetised with intravenous sodium pentothal, tracheally intubated and ventilated with a mixture of oxygen and dinitrous oxide in a volume controlled respirator (Mivab, Stockholm, Sweden). A carotid artery was cannulated for aortic blood pressure monitoring and blood sampling. A jugular vein was cannulated for fluid infusion.

The femoral artery and vein were dissected unilaterally and cannulated with surface heparinised tubings. The animals were dialysed for three hours using surface heparinised blood lines and a AK5 monitor (Gambro, Sweden) with an adjusted, not totally occlusive, roller pump. The dialysate compartment of the dialyser was filled with standard dialysis fluid and sealed. The arterial blood flow was measured electromagnetically (Nycotron, Drammen, Norway) with the probe placed on the femoral artery. The arterial blood flow and the aortic blood pressure in the animals were continuously recorded (Grass Polygraph, Quincy, Mass, USA).

Blood samples for determination of haematocrit, platelet count, whole blood coagulation time, plasma heparin concentration and FPA were drawn from the carotid artery immediately after the insertion of the catheter, after 15 minutes and 30 minutes and then every 30 minutes.

Four additional dogs were merely anaesthetised for three hours. In these animals arterial blood samples for FPA determinations were taken from a carotid artery cannula at time intervals as described above.

For statistical analysis Student's 't' test or the Wilcoxon test was used. All results are given as mean ± SD, range or median values, $p<0.05$*, $p<0.01$**, $p<0.001$***.

Results

A three hour haemodialysis session was performed in all animals without any technical complications and with no heparin given systemically.

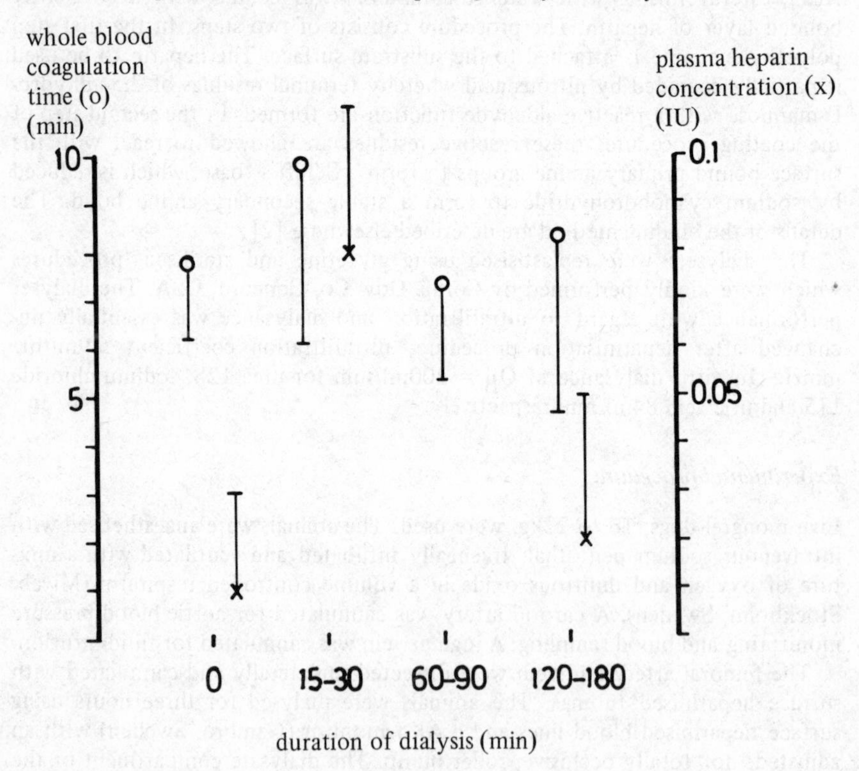

Figure 1. Whole blood coagulation time and plasma heparin concentration during haemodialysis with a heparin coated hollow fibre dialyser (mean, SD)

The roller pump was adjusted to give a blood flow of 160 to 215ml/min (mean 205±24ml/min) which was kept constant in each dog. The systemic aortic blood pressure (164±13mmHg) and the haematocrit (45±3%) were not significantly affected by dialysis.

The whole blood coagulation time did not change during the dialysis (Figure 1). There was a minor but still statistically significant elevation of the plasma heparin concentration ($p<0.001$) during the first 15 minutes of dialysis (Figure 1).

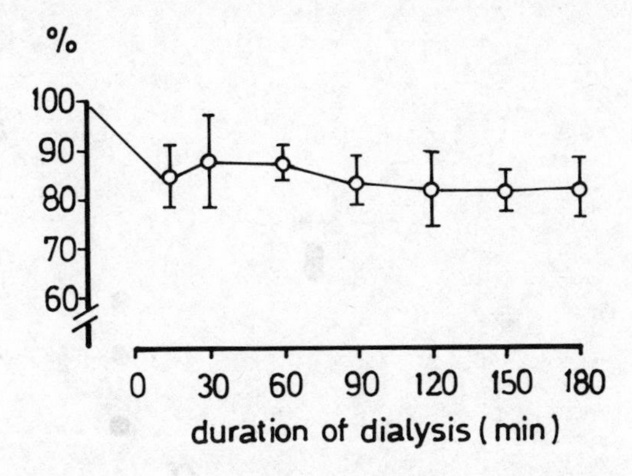

Figure 2. Platelet count during haemodialysis with a heparin coated hollow fibre dialyser (per cent of initial value; mean, SD)

The platelet count, ranging from 222 to 300×10^9/L, decreased on the average by about 15 per cent ($p<0.01$) during the first 15 minutes of dialysis, but was then constant (Figure 2). The FPA concentration in the arterially sampled blood increased during dialysis, but not more than in merely anaesthetised animals (Figure 3).

The total extracorporeal system was after disconnection rinsed with 1000–1500ml of saline. The dialysers and tubings were almost completely cleaned by this procedure. At connections between tubings of different diameters minute clots were occasionally seen. In the air trap a small wall attached clot at the blood-air interface level was always present.

Discussion

It is evident that a stable heparin coating on inherently thrombogenic materials enables extracorporeal circulation in the absence of systemic anticoagulation treatment and without measurable activation of the haemostatic mechanism. The present covalent binding of partially degraded heparin gives a surface heparin concentration in the order of 0.5 to 1.0IU per cm^2 and should from the theoretical point of view result in a completely stable heparin binding [2]. At a first contact with plasma, however, minute amounts of most probably unspecifically bound heparin is released [2]. Although the coagulation time in the present investigation remained unchanged during dialysis a minor elevation of

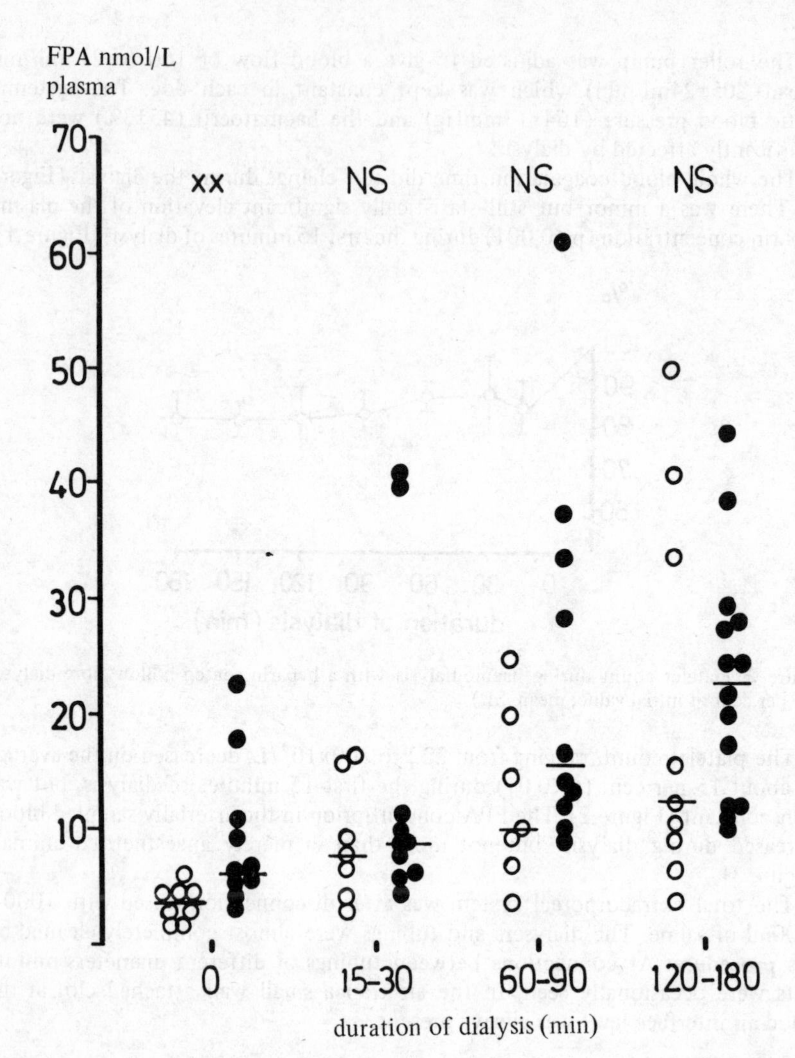

Figure 3. Arterial FPA concentrations in dogs undergoing dialysis with heparin coated dialysers (•) and in merely anaesthetised dogs (○). Individual and median values are presented

the plasma heparin concentration was accordingly seen in the early period of the dialysis session.

Heparin surfaces are, like a number of similarly prepared glycosaminoglycan surfaces, platelet compatible [4]. The reason for this seems to be that fibrinogen is not included in the firmly attached and about 50Å thick protein adsorbate which is formed at contact with plasma [6]. The reduction in the platelet count which was found in the present study is far less than that observed during previous dialysis experiments in dogs given heparin systemically [7]. Determination of the FPA concentration reveals that thrombin digests or recently has digested fibrinogen in blood and it must be considered as the most sensitive quantitative assay available on intravascular coagulation. This implies that FPA generated from minute coagulation in sampling catheters or on catheter induced endothelial lesions may contaminate the blood sample. The extent to which the uneven distribution and the wide range of the FPA determination were owing to such factors can not be settled. We find it rather striking that dialysis with the aid of surface heparinised equipment did not cause fibrinogen-fibrin conversion beyond that in merely anaesthetised animals.

The surface heparinisation procedure did not essentially alter the in vitro performance of the cellulose acetate membrane in the dialyser. Toxicological tests of the heparin surface have been favourable. Trials with surface heparinised equipment seems therefore to be justified in evaluating the possible clinical advantages.

References

1 Griffith GC, Nichols G, Asher JD, Flanagan B. *JAMA 1965; 193:* 91
2 Larm O, Larsson R, Olsson P. *Biomater Med Devices Artif Organs 1983; 11:* 161
3 Nossel HL, Yudelman J, Canfield RE et al. *J Clin Invest 1974; 54:* 43
4 Larsson R, Lindahl U. *Artif Organs1979; 3 (Suppl).* Proceedings Second Meeting International Society Artificial Organs
5 Larsson R, Olsson P, Lindahl U. *Thromb Res 1980; 19:* 43
6 Olsson P, Blombäck B, Lagergren H et al. *Bibliotheca Haematol 1978; 44:* 134
7 Arnander C, Hjelte MB, Lins LE et al. *Proc IX Ann Meeting ESAO 1982:* 312

Proc EDTA–ERA (1984) Vol 21

LOW MOLECULAR WEIGHT HEPARIN IN HAEMODIALYSIS AND HAEMOFILTRATION – COMPARISON WITH UNFRACTIONED HEPARIN

H Renaud, P Moriniere, *J Dieval, Z Abdull-Massih, H Dkhissi, †F Toutlemonde, *J Delobel, A Fournier

*Service de Nephrologie, *Laboratoire d'Hematologie, CHU, Amiens, †Laboratoire Choay, Paris, France*

Summary

The low molecular weight heparin CY 222 (CHOAY) has been compared to unfractioned heparin (UFH) in patients on chronic haemodialysis and haemofiltration at various doses as regards it biological activity (measured by Activated Partial Thromboplastin Time and by anti-Xa activity) and its clinical effect on clot formation in the blood lines and bleeding at the puncture sites or recent wounds. Compared to UFH, CY 222 has a greater anti-Xa activity for a shorter APTT. This biological difference is of clinical advantage since clotting in lines is comparable or less than with UFH whereas compression time at puncture sites is shorter and recently bleeding wounds in 28 patients did not bleed again. The long half life of CY 222 allows its use as a single priming dose of 300 anti-Xa U/kg in haemodialysis and 450 anti-Xa U/kg in haemofiltration.

Introduction

Various methods have been proposed for anticoagulation during haemodialysis (HD) or haemofiltration (HF) in patients at risk of haemorrhage: high flow dialysis without heparin or with low dose heparin, neutralisation of heparin by protamine into the venous line, anti-aggregative drugs, and prostacyclin. Inconsistent results or severe side effects limit their practical use [1].

Low molecular weight heparin (LMWH) such as CY 222 of CHOAY laboratory with a mean molecular of 2,500 has, on a weight basis compared to standard unfractioned heparin (UFH), greater activity against activated factor X (Xa) (250 versus 160 USP respectively) and less activity against activated factor II (IIa or thrombin) (25 versus 160 USP respectively). Thus with CY 222 the transformation of fibrinogen to fibrin is not completely blocked and the haemorrhagic risk is therefore expected to be lower [2].

In a first study we compared in patients on chronic haemodialysis and haemofiltration the clinical and biological efficiency of the usual heparinisation schedule with UFH and heparinisation with various doses of CY 222 given as a priming dose with or without subsequent continuous infusion. In a second study, we compared in patients with haemorrhagic risk the clinical and biological efficiency of low heparinisation schedule with UFH and heparinisation with CY 222 given at a lower priming dose than in the first study but followed by continuous infusion.

Comparison of routine heparinisation with CY 222 heparinisation

In haemodialysed patients

This study was performed in 11 patients (mean haematocrit 26 ± 4%) during 60 dialyses of four hours with hollow-fibre dialysers Travenol 15.11 with 15G needles. Standard heparinisation consisted in a priming dose of 60 ± 13 IU/kg of UFH followed after 120 minutes by the injection of a bolus of 2,000 IU. CY 222 was used either as a priming dose of 75, 150 or 300 anti-Xa U/kg followed by a continuous infusion of 1,000 U per hour or as a single priming dose of 150 – 300 U/kg.

Table I shows that Activated Partial Thromboplastin Time (APTT) (measured by CK-PREST STA60) [3] was significantly shorter with both regimen of CY 222 at 15 and 120 minutes but no difference was found at 240 minutes. In contrast anti-Xa activity (HEPACLOT STA60) [4] was greater with both regimen of CY 222 than with standard heparinisation. The clot generation expressed as -, ±, + was expressed as a percentage of the dialyses, was lower for CY 222 at the highest dosage of each regimen or comparable for the other doses. The mean compression time to stop haemorrhage at needle sites after removal is comparable for CY 222 given with continuous infusion but less for CY 222 given as a single priming dose. Plasma concentration of fibrino-peptide A (determined by STA60 ELISA) [5] at the end of dialysis as a biological marker of fibrin-formation (since it is released into the blood when fibrin is formed) was significantly lower with CY 222 used as a single priming dose of 300 U/kg than with the standard heparinisation (6.2 ± 2.6mg/ml versus 15 ± 2.5mg/ml; normal <3.5mg/ml), suggesting less coagulation in the extracorporeal circuit with CY 222.

In patients on haemofiltration

Nine patients (mean haematocrit 28 ± 2%) were studied during 30 haemofiltrations (3 x 15 litres per week) of 180 minute duration performed with ASAHI PAN 200 and two 15 G needles. Standard heparinisation consisted in a priming dose of 66 ± 15 IU/kg followed by a continuous infusion of 1,000 IU per hour. CY 222 was given at various priming doses followed by continuous infusion of 1,000 anti-Xa U/hour. Table II shows that APTT is significantly shorter for each dose of CY 222 than for standard heparinisation whereas anti-Xa activity is higher with CY 222 at doses of 150 or higher. Clot generation is lower with CY 222 only for the higher dose of 450 anti-Xa U/kg. Compression time to stop bleeding at puncture sites is slightly shorter with each dose of CY 222.

TABLE I. Comparison of CY 222 with standard heparinisation with unfractioned heparin (UFH) in haemodialysis

Heparin (units)	Priming dose /kg	(1) APTT sec. 15 min	120 min	240 min	(2) Anti-Xa Activity U/ml 15 min	120 min	240 min	Clots generation % of run –	±	+	Compression time min at inlet and outlet (min)
UFH (IU) + 2,000 at 120 min	60 ± 13	119	65	39	0.28	0.24	0.16	60	40	–	7
CY 222 (anti-Xa U) + 1,000 U/hr	75	42*	39*	34	0.45	0.81	0.51*	60	40	–	5.6
	150	51*	38*	38	0.6*	0.44*▲▲	0.55*	66	17	17	5
	300	58*	39*	39	0.97**	0.84**	0.98**▲	100	–	–	7.5
CY 222 (anti-Xa U) single dose	150	–	39*	34	–	–	–	55	13	12	2.3
	300	–	40*	38	–	0.98	0.44	100	–	–	2.3

Significance of the comparisons

UFH versus CY 222 : *$p<0.02$; **$p<0.01$

CY 222 single dose versus CY 222 with infusion : ▲ $p<0.02$; ▲▲ $p<0.01$

(1) APTT Activated Partial Thromboplastin Time (normal control = 30 sec; efficient heparinisation with UFH = 60 sec)

(2) Efficient heparinisation with UFH leads to anti-Xa activity of 0.2 – 0.5 U/ml

TABLE II. Comparison of CY 222 with standard heparinisation with unfractioned heparin in haemofiltration

Heparin	Priming dose /kg	APTT sec. 15 min	APTT sec. middle run	APTT sec. end	Anti-Xa activity 15 min	Anti-Xa activity middle run	Anti-Xa activity end	Clots generation % of runs –	Clots generation % of runs ±	Clots generation % of runs +	Compression time at inlet and outlet mean (min)
UFH (IU) + 1,000/hr	66 ± 15	138	90	107	0.7	0.7	0.5	80	20	–	7
CY 222 (anti-Xa U) + 1,000/hr	75	38**	36**	34**	0.4	0.5	0.6	50	50	–	5.5
	150	44**	42**	39**	1.0	0.9	1.0**	60	40	–	4.5
	300	46**	44**	41**	1.1*	1.0*	0.8*	75	25	–	5.5
	450	55**	47**	47**	1.2**	1.2**	1.3**	100	–	–	5.5

Significance of the difference UFH versus CY 222: *$p<0.05$; **$p<0.01$

Comparison of low dose of heparin with low dose of CY 222 in patients with haemorrhagic risk

In patients with haemorrhagic risk (3 recently bleeding gastric ulcers; 2 retro peritoneal haematoma; 1 bleeding colitis; 17 recent surgical operations and 5 bone biopsies), CY 222 was used at a priming dose of 33, 75 or 100 anti-Xa U/kg followed by a continuous infusion of 1,000 anti-Xa U/hour in 107 haemodialyses and 17 haemofiltrations. Massive coagulation in the blood lines was observed only once. Bleeding at the surgical wound or at the ulcer seen by endoscopy just after treatment or in the drainage tubes did not occur. APTT and anti-Xa activity was measured in the middle and at the end of 14 treatments performed with various doses of CY 222 and of three treatments performed with low heparinisation of 33 IU/kg of UFH followed by a continuous infusion of 500 IU/hour. At the dose of 33 anti-Xa U/kg, CY 222 induces the same anti-Xa as UFH but without increasing the APTT.

Discussion and conclusions

Our clinical results confirm that in haemodialysis and haemofiltration CY 222 has a greater anti-Xa activity and lower anti-IIa activity (reflected by shorter APTT) than UFH. This biological difference seems to have clinical advantage since compression time at puncture sites to stop bleeding after needle removal is shorter with CY 222, whereas clot generation in the blood line is either comparable or less for the highest dose of CY 222. Furthermore the use of low dose of CY 222 in patients with haemorrhagic risks is safe since it does not trigger new bleeding and prevents clotting in the blood lines in 98 per cent of cases. The long half life of anti-Xa activity of CY 222 allows its safe administration as a single priming dose of 300 – 450 anti-Xa U/kg which simplifies its use.

References

1 Lindsay RM. In Drukker N, Parsons FM, Maher JF, eds. *Replacement of Usual Function by Dialysis.* Boston: Nijhoff. 1983: 201–222
2 Duclos JP. *L'héparine – Fabrication, structure, propriétés, analyse.* Paris: Masson. 1984
3 Larrieu MJ, Weilland C. *Nouv Press Franc Hemato 1957; 12:* 2
4 Denson KWE, Bonnar J. *Thromb Diath Hoem (Stuttg) 1973; 30:* 471
5 Soria J, Soria N, Rickwaert JJ. *Thromb Res 1980; 20:* 425

Proc EDTA–ERA (1984) Vol 21

HAEMODIALYSIS WITH PROSTACYCLIN (EPOPROSTENOL) ALONE

P B Rylance, M P Gordge, *H Ireland, *D A Lane, M J Weston

*Dulwich Hospital, *Charing Cross Hospital Medical School, London, United Kingdom*

Summary

Dialysis with prostacyclin (Epoprostenol, PGI_2) alone prevents platelet activation and endothelial cell stimulation but not the elevation of fibrinopeptide A (FPA), a sensitive marker of fibrin generation. The generation of FPA may explain why some patients develop clot in the dialysis circuit during PGI_2-only dialysis. In combination with heparin, PGI_2 augments the anticoagulant effect of the heparin as well as providing platelet protection.

Introduction

Patients with chronic renal failure may display a divergence of normal haemostasis with not only a high prevalence of cardiovascular mortality but also a bleeding tendency. Platelet dysfunction may contribute to this bleeding tendency and may be made worse by both functional exhaustion of platelets (due to their activation on the dialysis membrane), and by heparinisation. Reversible protection of platelets may be obtained during haemodialysis by PGI_2 infusion [1], and may minimise risk of bleeding. Patients may be dialysed with PGI_2 alone [2–4] though with increased experience with PGI_2, it has become clear that in some patients fibrin may form resulting in clotting in the dialysis circuit. This study was undertaken to elucidate further the behaviour of platelets and clotting parameters during dialysis with PGI_2.

Patients and methods

In eight stable chronic haemodialysis patients a single dialysis with PGI_2 alone (Epoprostenol, Wellcome) was compared with a dialysis with unfractionated heparin. PGI_2 (0.5mg diluted in 50ml glycine buffer, pH 10.5) was infused into a peripheral vein via an electrically-driven syringe pump at a rate of 5ng/kg/min. Following the commencement of dialysis, the PGI_2 infusion was transferred to the arterial inlet of the dialyser and subsequently the dose was increased provided the blood pressure did not significantly fall (2.5–10ng/kg/min, mean 6.9 ± 0.6ng/kg/min). In the dialysis with heparin, a bolus of 50iu/kg (Leo) was given into the arterial line at the start of dialysis and after two hours an infusion of 30iu/kg/hr was continued to the end of dialysis. All dialyses lasted for four hours and were carried out with flat-plate cuprophan dialysers (Gambro). Blood samples, withdrawn slowly from the arterial line, were taken pre-PGI_2 infusion, pre-dialysis (0 min), 15 min, 60 min, 120 min and 240 min, for platelet count, activated whole blood clotting time (AWBCT, measured by the Hemochron technique), activated partial thromboplastin time (APPT), fibrinopeptide A (FPA, the first peptide cleaved by thrombin from the N-terminus of fibrinogen, and cleavage of FPA signals the formation of fibrin-1), β-thromboglobulin (βTG) and platelet factor 4 (PF4) (both factors released from platelet α granules following activation), and factor VIII related antigen (VIII RAg, a marker of endothelial cell stimulation). Platelet aggregation (maximum) was measured in response to ADP (5μmol) and collagen (1μg/ml) within two minutes of blood sampling, using an electronic impedance whole blood aggregometer. The remaining samples were kept on melting ice, centrifuged at 4°C and stored at −20°C for subsequent analysis. FPA, βTG and PF4 were measured by radioimmunoassay techniques and VIII RAg by radial immunodiffusion. Where values are expressed as per cent changes, pre-dialysis = 100 per cent. Results are expressed as mean ± standard error of mean. Statistical comparisons were made within and between regimes by paired Student's 't' tests.

In a separate study a further eight haemodialysis patients were dialysed with bolus doses of heparin (Leo) 50iu/kg, 30 iu/kg and 20iu/kg given at the commencement of successive dialyses and studied for two hours. The same patients were subsequently dialysed on three further occasions with the same heparin boluses but with the addition of PGI_2 infusion (5ng/kg/min). In this second study the prolongation of thrombin clotting time (TCT) at 15 minutes was plotted against the bolus dose of heparin with and without PGI_2 infusion. FPA was measured during both dialyses with heparin alone and with PGI_2 + heparin, and the correlation between FPA and AWBCT was plotted for PGI_2 + heparin dialysis.

Results

PGI_2 was well tolerated; initial mild facial flushing was transient and at 5ng/kg/min patients experienced no other side effects or significant hypotension compared with heparin dialysis. The results of the first study are shown in Figure 1. During dialysis with PGI_2 there was a small but not significant fall in AWBCT; mean APTT significantly fell at 15 minutes and remained less than pre-dialysis values. Platelet count fell by nine per cent though this was not significantly different from heparin dialysis. βTG rose during heparin dialysis, while with PGI_2 βTG was significantly lower but was significantly elevated above pre-dialysis values at four hours. PF4 was not significantly changed during PGI_2 dialysis, while PF4 with heparin dialysis was markedly elevated at 15 minutes, probably as a result of heparin displacing PF4 from the vascular endothelium. Whereas VIII RAg rose with heparin dialysis, with PGI_2 dialysis there was no

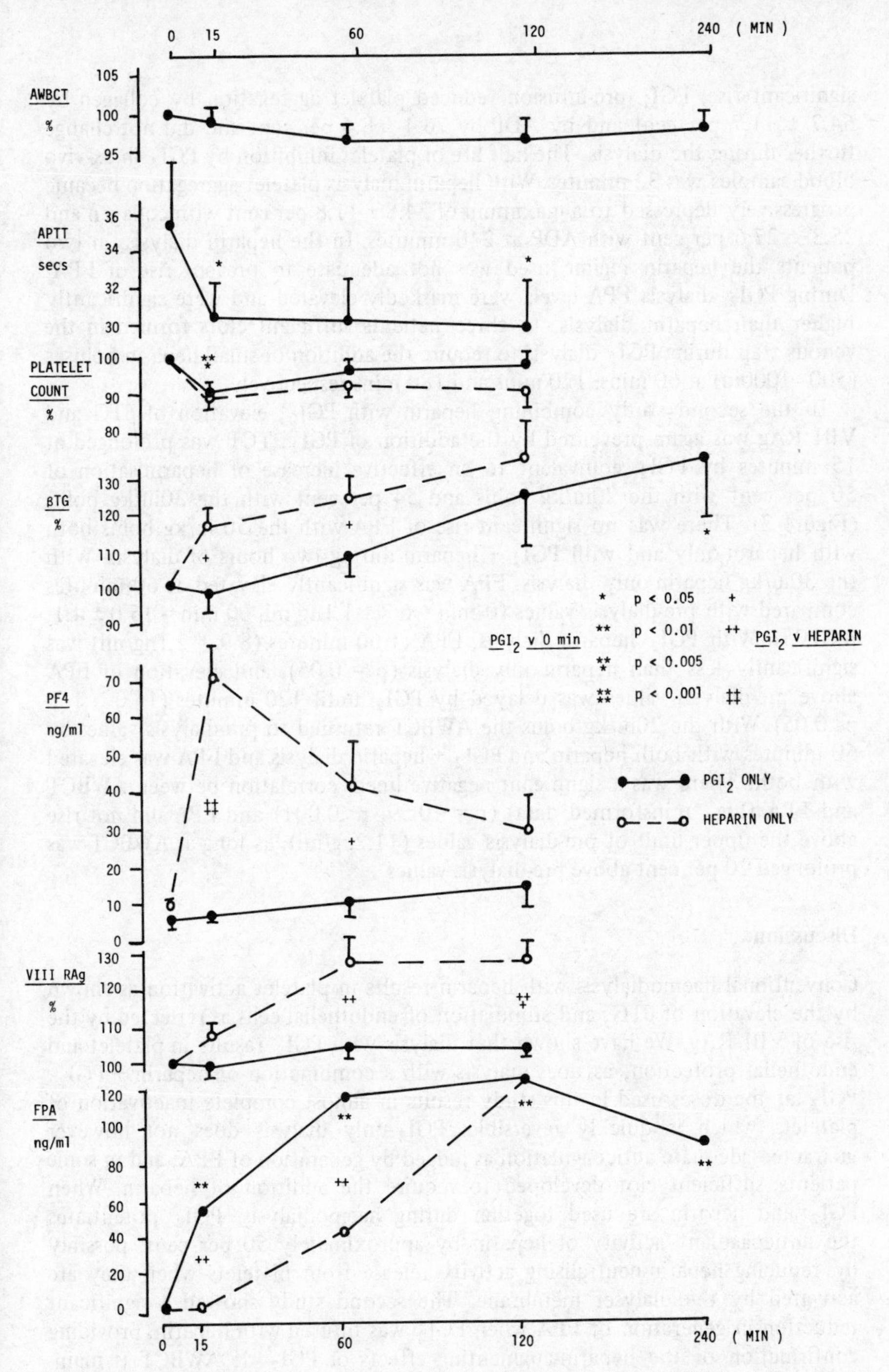

Figure 1. Platelet and clotting parameters during haemodialysis with PGI_2 alone (●——●) compared with dialysis with heparin alone (○– – –○)

significant rise. PGI_2 pre-infusion reduced platelet aggregation by collagen by 64.7 ± 11.7 per cent and by ADP by 76.4 ± 8.4 per cent and did not change further during the dialysis. The half life of platelet inhibition by PGI_2 in ex-vivo blood samples was 32 minutes. With heparin dialysis platelet aggregation became progressively depressed to a maximum of 54.8 ± 11.8 per cent with collagen and 25.3 ± 27.6 per cent with ADP at 240 minutes. In the heparin dialysis, in two patients the heparin regime used was not adequate to prevent rise of FPA. During PGI_2 dialysis FPA levels were markedly elevated and were significantly higher than heparin dialysis. In three patients sufficient clots formed in the venous trap during PGI_2 dialysis to require the addition of small heparin boluses (500–1000iu) at 60 mins, 120 mins and 180 mins respectively.

In the second study combining heparin with PGI_2, elevation of βTG and VIII RAg was again prevented by the addition of PGI_2. TCT was prolonged at 15 minutes by PGI_2 equivalent to an effective increase of heparinisation of 50 per cent with the 20iu/kg bolus and 54 per cent with the 30iu/kg bolus (Figure 2). There was no significant rise of FPA with the 50 iu/kg bolus both with heparin-only and with PGI_2 + heparin during two hours of dialysis. With the 30iu/kg heparin-only dialysis, FPA was significantly elevated at 60 minutes compared with pre-dialysis values (0 min = 6.7 ± 1.1ng/ml, 60 min = 15.0 ± 4.1, $p<0.05$). With PGI_2 heparin dialysis, FPA at 60 minutes (8.9 ± 2.1ng/ml) was significantly less than heparin-only dialysis ($p = 0.05$), and elevation of FPA above pre-dialysis values was delayed by PGI_2 until 120 minutes (15.0 ± 1.2, $p<0.05$). With the 20iu/kg bolus the AWBCT returned to pre-dialysis values by 60 minutes with both heparin and PGI_2 + heparin dialysis and FPA was elevated with both. There was a significant negative linear correlation between AWBCT and FPA ($\log_n$ transformed data) ($r = -0.59$, $p<0.001$) and FPA did not rise above the upper limit of pre-dialysis values (11.2ng/ml), as long as AWBCT was prolonged 20 per cent above pre-dialysis values.

Discussion

Conventional haemodialysis with heparin results in platelet activation as shown by the elevation of βTG, and stimulation of endothelial cells as reflected by the rise of VIII RAg. We have shown that dialysis with PGI_2 results in platelet and endothelial protection, as does dialysis with a combination of heparin + PGI_2. PGI_2 at the doses used in this study results in almost complete inactivation of platelets which is quickly reversible. PGI_2-only dialysis does not however guarantee adequate anticoagulation as judged by generation of FPA, and in some patients sufficient clot developed to require the addition of heparin. When PGI_2 and heparin are used together during haemodialysis, PGI_2 potentiates the anticoagulant activity of heparin by approximately 50 per cent, possibly by reducing heparin-neutralising activity release from platelets when they are activated by the dialyser membrane. The second study showed a significant reduction in generation of FPA when PGI_2 was infused with heparin, providing confirmation of the heparin-augmenting effects of PGI_2. If AWBCT is maintained at 20 per cent above pre-dialysis values by small heparin boluses, FPA levels are not elevated above the upper limit of pre-dialysis values. Thus for a

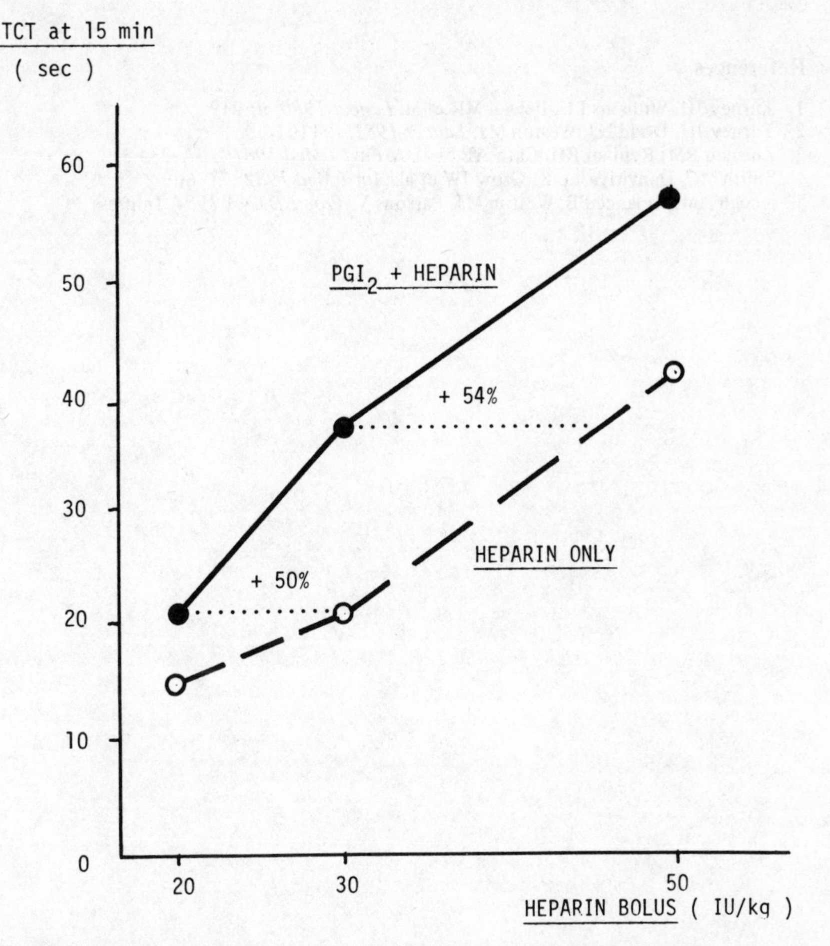

Figure 2. Relationship between prolongation of thrombin clotting time (TCT) at 15 minutes during dialysis with a combination of PGI_2 and heparin (•——•) compared with dialysis with heparin alone (○– – –○)

patient at risk of haemorrhage, PGI_2 can be used to protect platelets from activation and to reduce heparin requirements as judged by TCT and FPA.

Elsewhere we report our experience of 101 dialyses with PGI_2 in patients who were actively bleeding or were at risk of doing so [5]. In none of these did bleeding appear to be aggravated during or after the dialysis but in 30 per cent of dialyses with PGI_2-alone there was significant clotting in the dialyser requiring changing of bubble-trap or circuit and heparinisation. The results of these studies suggest that for dialysis in patients at risk of bleeding it is preferable to combine prostacyclin infusion with small doses of heparin the activity of which is enhanced by PGI_2, to prevent fibrin generation and dialyser clotting.

References

1 Turney JH, Williams LC, Fewell MR et al. *Lancet 1980; ii:* 219
2 Turney JH, Dodd NJ, Weston MJ. *Lancet 1981; i:* 1101
3 Zusman RM, Reuben RH, Cato AE et al. *N Engl J Med 1981; 304:* 934
4 Smith MC, Danviriyasup K, Crow JW et al. *Am J Med 1982; 73:* 669
5 Keogh AM, Rylance PB, Weston MJ, Parsons V. *Proc EDTNA 1984.* In press

Proc EDTA–ERA (1984) Vol 21

REDUCTION OF SILICONE PARTICLE RELEASE DURING HAEMODIALYSIS

J Bommer, *E Pernicka, J Kessler, E Ritz

*University of Heidelberg, *Max-Planck-Institut für Kernphysik, Heidelberg, FRG*

Summary

Spallation of silicone was evaluated in an in vitro system, using a commercial blood pump and dialysis tubing. Silicone particle release was assessed at various occlusion forces (5.5–22 kp). When the occlusion force was reduced from 22 to 5.5 kp, the number of released silicone particles decreased by ~ 80 per cent; in parallel, the amount of silicone retrieved from the recirculation fluid decreased from 1.6mg to less than 0.23mg. It is concluded that reduction of occlusion pressure within the blood pump effectively reduces spallation of silicone tubing.

Introduction

Spallation of silicone dialysis tubing and deposition of silicone particles in the viscera of haemodialysis patients has recently been described by several authors [1–5]. Several clinical abnormalities have been related to particle loading of visceral macrophages, e.g. elevated transaminases and granulomatous hepatitis, hepato-splenomegaly and hypersplenism with pancytopenia [6]. In addition, in experimental studies, alteration of macrophage function with increased prostaglandin synthesis could be demonstrated when rats were loaded with silicone, PVC or polyurethane particles by intravenous or intraperitoneal injection [7].

Studies from this laboratory [8] and from other investigators [3], using scanning electronmicroscopy, demonstrated extensive damage on the luminal surface of dialysis tubing after exposure to a roller pump for five hours. Previous investigators using chloroform extraction procedures gave low figures for silicone release, i.e. 50μg per five hours of haemodialysis; however, particle load may have been underestimated since the procedure used allows only measurement of soluble silicone oligomers [2].

The present in vitro study was designed to further quantitate the amount of silicone release under conditions closely imitating the haemodialysis procedure.

Particle release was quantitated by counting and by direct silicone measurements without prior extraction steps.

Material and methods

A recirculation system with a priming volume of 300ml saline was set up. The system consisted of a commercial roller pump (Fresenius Co), silicone tubing for the roller pump insert (internal diameter 8mm, wall thickness 1.8mm) and PVC tubing for the non-roller pump segments. The pump operated at 25rpm=280ml/min. The occlusion force was varied over a range from 5.5–22 kp by adjusting the distance between the roll and the abuttment using spring coils with defined forces. The pressure was directly measured with commercial calibrated gauges.

After five hours the priming fluid was assessed for silicone particles both by measuring silicone concentration with atomic absorption spectrophotometry and by counting silicone particles microscopically.

Silicone was measured using a Perkin Elmer atomic absorption spectrometer 430 with a pyrrolytically coated graphite tube and programmed temperature control. A 25μl sample was dried (100°C; 60 sec), decomposed (350°C; 60 sec) and atomised (1950°C; 8 sec). Sodium silicate diluted in NaOH was used as standard solution. Appropriate blanks and internal standards were used throughout. All measurements were done in triplicate. Detection threshold was 0.2ppm and the coefficient of variation of replicate measurements was ~ eight per cent at 0.5ppm.

For counting silicone particles, the fluid was filtered through a Millipore filter (exclusion limit 0.025μ). The filter was rendered transparent by silicone oil. Particles were counted using a light microscope (x 500) with a counting grid. The method was compared with counting in a Neubauer chamber which generally yielded much lower values. This was ascribed to particle loss during transfer into the counting chamber. The values given in Table I refer to counting on a Millipore filter.

Results

As shown in Table I, at an occlusion force of 22 kp 20.3 x 10^6 silicone particles >0.3μ per ml recirculation fluid were released; this would be equivalent to a silicone particle load of 6 x 10^9 particles per dialysis session. This figure will

TABLE I. Number of particles and amount of silicone released at different occlusion pressures (n=6 experiments)

	occlusion force		
	22 kp	5.5 kp	p (Wilcoxon)
number of particles (10^6/ml)	20.3 ± 5.1	3.4 ± 2.6	<0.01
total amount of silicone (mg/5hr)	1.6 ± 0.6	<0.23*	<0.01

* detection threshold

even underestimate the total number of particles since fragments $<0.3\mu$ could not reliably be detected microscopically. Reduction of the occlusion force significantly ($p<0.01$) reduced generation of silicone particles.

This finding was validated with an independent technique, i.e. measuring silicone release into the recirculation fluid with flameless atomic absorption spectrometry. Silicone tubing (polydimethylsiloxane) has a silicone content of 26 per cent (information obtained from the manufacturer); this figure was used to calculate silicone release on the basis of silicone measurements.

Reduced fragmentation of dialysis tubing at lower occlusion pressure would be compatible with qualitative electronmicroscopic observations of the luminal surface of the tubing, even when the large sampling error is considered (Figure 1).

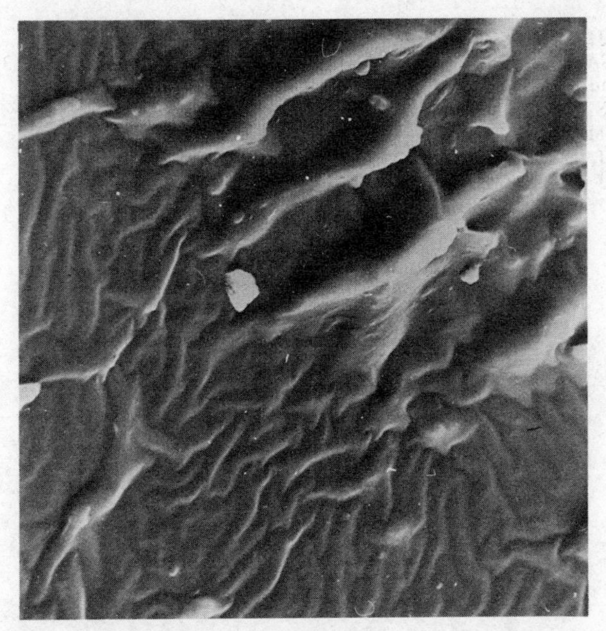

Figure 1. Scanning electronmicrograph of luminal surface of silicone tubing after five hours exposure to a roller pump (x 1000). Note 'hillocks' projecting over and several particles adhering to luminal surface

Discussion

The present experiments clearly demonstrate that appropriate adjustment of the occlusion pressure in the roller pump permits a reduction in the calculated annual silicone load in haemodialysis patients from 250mg/year to 36mg/year (i.e. per 150 dialysis sessions). This observation is easily explicable by reduction of non-elastic deformation of silicone tubing with reduced pressure. Such deformation causes fragmentation of the luminal surface (as demonstrated with SEM) and release of elastomer particles. The effect of occlusion pressure is also illustrated by measurements of the distance between abuttment and roll at varying occlusion forces. For tubings used in the above experiment, the distance was 2.9mm at 22 kp and 3.5mm at 5.5 kp. Since total wall thickness was 2 x 1.7–1.8mm, these figures document considerable deformation of tubing wall at high occlusion pressure.

From the above studies over-occlusion emerges as a major factor responsible for particle release. Only future studies will determine whether it will be sufficient to optimise occlusion pressure in order to avoid particle related clinical problems or whether changes of dialysis equipment will be required, e.g. use of more resilient silicone tubing or introduction of membrane pumps.

References

1 Bommer J, Waldherr R, Ritz E. *Proc EDTA 1981; 18:* 731
2 Leong ASA, Path MRC, Disney APS, Gove DW. *N Engl J Med 1982; 306:* 135
3 Laohapand T, Osman EM, Morley AR et al. *Proc EDTA 1982; 19:* 143
4 Parfrey PS, Paradinas FJ, O'Driscoll JB et al. *Proc EDTA 1982; 19:* 153
5 Morales JM, Colina F, Arteaga J et al. *Proc EDTA 1982; 19:* 265
6 Bommer J, Ritz E, Waldherr R. *N Engl J Med 1981; 305:* 1077
7 Bommer J, Gemsa D, Waldherr R et al. *Kidney Int 1984; 26:* 331
8 Bommer J, Waldherr R, Ritz E. *Contr Nephrol 1983; 36:* 115

Proc EDTA–ERA (1984) Vol 21

CARPAL TUNNEL SYNDROME, SHOULDER PAIN AND AMYLOID DEPOSITS IN LONG-TERM HAEMODIALYSIS PATIENTS

B Charra, E Calemard, M Uzan, J C Terrat, T Vanel, G Laurent

Centre de rein artificiel de Tassin, Tassin, France

Summary

Carpal Tunnel Syndrome (CTS), and Shoulder Pain and Stiffness (SPS) are common in very long-term haemodialysis patients. To know whether this is a fortuitous association or if there is a link between these two manifestations a retrospective analysis of clinical charts, CTS surgical reports, tendon and synovia biopsies (Congo red, Crystal violet and Thioflavin T) was undertaken for 110 patients treated by haemodialysis (HD) for eight years or more.

SPS was less frequent (24%) in 58 patients not operated on for CTS than in 52 operated patients (SPS incidence: 77%). Furthermore the 38 patients with amyloid deposits at carpal biopsy had a very significantly ($p<0.001$) higher incidence of SPS (95%) than the 14 operated patients free of amyloid deposits (SPS incidence: 28%).

Hence, amyloid deposits represent a pathological link between these two correlated manifestations in very long-term haemodialysis patients.

Introduction

When the Carpal Tunnel Syndrome (CTS) was reported first by Warren [1] in haemodialysis (HD) patients in 1975 the HD technique was already 15 years old. CTS was then considered a curiosity and a rarity. Today CTS is reported commonly in this setting, perhaps because of increased diagnosis but also because the HD population has come to age and this complication is connected with duration of HD treatment. When recognised and operated early CTS heals without sequelae.

Similar to Baillod [2] we were impressed by the incidence of Shoulder Pain and Stiffness (SPS) in patients maintained for a long time on dialysis. For this painful and invalidating complication we are short of treatment.

We have [3] like others [4] observed frequent amyloid deposits in the carpal tunnel synovia and tendons. Amyloid substance was also found at autopsy in biceps tendon and in shoulder synovia of two of our patients operated on for

CTS and complaining of SPS. This puzzling observation prompted this study on the correlation between CTS and SPS.

Patients and methods

Fifty-two patients were operated on for median nerve entrapment. Usually signs and symptoms were adequate for diagnosis, only in 14 cases nerve conduction velocity measurement and electromyogram were required to ensure diagnosis. In all cases surgery confirmed the compression; we have considered CTS only if it was surgically proven. In all cases paresthesia and pain disappeared within a few hours while sensory and motor deficit took a few months to improve.

Among 52 such surgically confirmed CTS the delay between first dialysis session and surgery was (mean ± SEM) 11.2 ± 0.4 years. We have used an eight-year dialysis duration to select our 'at risk' population. They numbered 110, all treated with the same unchanged technique ($24m^2$ hours per week on standard Kiil dialysers, acetate-buffered glucose-free dialysate). The clinical features of these patients are presented in Table I. Only one (non-operated) of them had a secondary amyloidosis, none was transplanted or on peritoneal dialysis for more than six months.

TABLE I. Clinical features of 110 patients treated by haemodialysis for eight years or more

Clinical features		CTS operated patients (n=52)	Non-operated patients (n=58)	Total population (n=110)
Sex	Female	23	13	36
	Male	29	45	74
Age* (Mean ± SEM)		56.5 ± 1.4	52.6 ± 1.6	54.4 ± 1.5
Aetiology of CRF	CGN	21	24	45
	CIN	7	11	18
	PKD	8	10	18
	Others	7	4	11
	Unknown	9	9	18

*Age at time of surgery or at eight years of dialysis treatment.

CTS=carpal tunnel syndrome; CRF=Chronic renal failure; CGN=chronic glomerulonephritis; CIN=chronic interstitial nephritis; PKD=polycystic kidney disease

The number of dialysis sessions and the overall treatment time was computed for each patient at risk. Pertinent data were also collected on vascular access history, tobacco consumption, pre-dialysis mean blood pressure, haematocrit, white blood cell count, and immunoglobulin G and M values.

Criteria for SPS included pain, use of pain alleviating drugs, and/or scapular infiltration, limitation of movements, and absence of shoulder soft tissue calcification on X-ray.

Carpal tunnel syndrome was always surgically confirmed. All operations were performed by the same hand-surgeon and with the same technique without tourniquet. It included a great carpal incision, followed by a systematic inspection of the median nerve, local vascular changes, deep tendon and synovial whitish granulations. Biopsies were taken of tendon and synovia, and granulation if any. Then transverse carpal ligament section, neurolysis and epineurotomy were completed.

All biopsies were studied by the same pathologist according to the same protocol. Standard stains (Haemalun, phloxine, saffran) were completed with three special amyloid stains: Congo red, Crystal violet and Thioflavin T read under fluorescent light. The biopsy was considered positive for amyloid deposits whenever the three stains were positive.

All means were expressed with the standard error of the mean (SEM). Statistical analysis included Student's test for unpaired data and Chi-square test. A $p<0.05$ was considered as the limit of significance.

Results

Fifty-eight of 110 patients treated for eight years or more (53%) were not operated on for a CTS (group I) whereas 52 (47%) were operated on (group II). Sixty-four per cent of females were operated vs only 37 per cent of males ($p<0.05$).

TABLE II. Technical, clinical and laboratory data concerning 110 patients treated by haemodialysis for eight years or more

	CTS operated patients (n=52)	Non-operated patients (n=58)	Statistical* significance
Number of dialysis sessions	1436 ± 45	1378 ± 51	NS
Total dialysis duration (hrs) (mean±SEM)	14180 ± 575	13664 ± 503	NS
Mean BP (mmHg)±SEM	93.4 ± 2.5	100.5 ± 2.4	$p<0.05$
Tobacco addiction Number (per cent)	26 (50%)	25 (43%)	NS
Pre-dialysis haematocrit Mean±SEM	0.31 ± 0.01	0.31 ± 0.01	NS
Parathyroidectomy Number (per cent)	13 (25%)	12 (21%)	NS
White blood cells Nbx1000±SEM	6.2 ± 1.9	6.0 ± 2.0	NS
Immunoglobulin G (g/L) Mean±SEM	16.5 ± 0.7	15.8 ± 0.7	NS
Immunoglobulin M (g/L) Mean±SEM	2.1 ± 1.0	1.7 ± 0.6	NS

*Statistical significance according to Student's 't' test for unpaired data or Chi-square test
CTS=carpal tunnel syndrome; BP=blood pressure

There was no difference in mean age or aetiology of chronic renal failure between both groups. Neither was there any difference in HD duration (Table II) expressed in terms of number of sessions or of total dialysis time. Even though it often started on the vascular access side, the CTS was more often operated on both sides (28 patients) than on one side only (24 patients). In Table II none of the contemplated factors show any significant difference between groups apart from mean pre-dialysis blood pressure being slightly lower in group II. Blood pressure was lower ($p<0.001$) in female patients (82.8 ± 2.7mmHg) than in male (100.7 ± 3.6mmHg). SPS affected roughly three out of four CTS operated patients versus only one of four patients in group I (Table III).

TABLE III. Shoulder pain and stiffness (SPS) in relation to carpal tunnel syndrome (CTS) surgically proven with or without amyloid deposits in 110 patients treated for eight years or more (Tassin, 1984)

Patient groups	Patients operated on for CTS (n=52)		Non-operated patients (n=58)
Patients with SPS Number (per cent)	40 (77)		14 (24)
Statistical value*	$p<0.001$		
	Positive amyloid (n=38)	Negative amyloid (n=14)	
Patients with SPS Number (per cent)	36 (95)	4 (28)	
Statistical value*	$p<0.001$		

*Using Chi-square test

CTS=carpal tunnel syndrome; SPS=shoulder pain and stiffness

Amyloid deposits were found in synovia and tendon of 38/52 operated patients (subgroup IIa) whereas 14/52 had negative amyloid stains (subgroup IIb). SPS was observed with a very significantly ($p<0.001$) higher incidence (95%) in subgroup IIa than in subgroup IIb (28%).

Trigger fingers were surgically corrected in 10 patients. Nine out of 10 belonged to subgroup IIa. In four cases removed tissue was sent to pathology: amyloid stains were positive in all four.

Discussion

This retrospective analysis of clinical, surgical, and pathological data of 110 patients treated by HD for eight years or more confirms the already reported [3] frequent coexistence of CTS and SPS.

We did not find any difference in frequency of SPS between the group of patients not operated on (group I) and the group of patients operated on and free of amyloid deposits (subgroup IIb). However there was a very high incidence of SPS in patients with CTS and amyloid deposits (subgroup IIa). This has been observed by Kachel et al [5] who reported that 11/13 patients dialysed for eight years or more and operated on for CTS with amyloid deposits had a scapulo-humeral periarthritis.

Amyloid deposits in synovia and tendon have been reported by several groups [4–6], but are never or rarely found by some others [7,8]. It may be that surgery with a tourniquet does not allow for an easy recognition of the granulations to biopsy but it might as well be that some HD carpal tunnel syndromes are not accompanied by amyloid deposition. It would be of interest to know if there is any significant difference of population or technique between these amyloid-free CTS groups and the amyloid CTS groups.

Other localisations of amyloid deposits have been reported in long-term HD patients, especially in trigger fingers [6] and in skin [5] but not in rectum.

The pathogenic mechanisms are unknown. Ischaemia has probably a facilitating role as advocated by the higher incidence of CTS on the fistule side at the early stage, in hypotensive patients, and in females, but is not absolute.

Calcium and phosphorus metabolism does not seem involved: no shoulder soft tissue calcifications were found, parathyroidectomy (Table II) is not more frequent in group II than in group I.

White blood cell count, immunoglobulin G and M are not different either (Table II). No argument can be found so far for an immunological origin of this syndrome.

The almost universally accepted increasing incidence of CTS observed by groups with numbers of long-term patients surviving is confirmed by our experience. We have a particularly high incidence of CTS (and SPS), but we have also a particularly high proportion of dialysis 'old-timers' (32% of our present population have been on HD for more than eight years).

The very long incubation time is in favour of some slow toxic or incompatibility [5] mechanism.

This experience with several synovial and tendinous localisations, together with that of the German group [5] on skin pleads for a generalised pathological entity increasing with HD treatment duration and characterised histologically by frequent amyloid deposits.

References

1 Warren DJ, Otieno LS. *Postgrad Med J 1975; 597:* 450
2 Baillod RA, Varghese Z, Fernando ON, Moorhead JF. In *Uremia; 8.* Wichtig. 1981: 35
3 Assenat H, Calemard E, Charra B et al. *La Nouv Presse Med 1980; 24:* 1715
4 Durroux R, Benouaich L, Bouissou H et al. *La Nouv Presse Med 1981; 10:* 45
5 Kachel HG, Altmeyer P, Baldamus CA, Koch KM. *Contrib Nephrol 1983; 36:* 127
6 Suzuki M, Hirasawa Y, Muraoka M, Saitoh H. *Abstracts IXth International Congress of Nephrology 1984:* 192A
7 Scardapane D, Halters S, Delisa JA, Sherrard DJ. *Dial Transpl Forum 1979:* 15
8 Spertini F, Wauters JP, Poulenas I. *Clin Nephrol 1984; 21:* 98

Proc EDTA-ERA (1984) Vol 21

REDUCTION OF OSMOTIC HAEMOLYSIS AND ANAEMIA BY HIGH DOSE VITAMIN E SUPPLEMENTATION IN REGULAR HAEMODIALYSIS PATIENTS

K Ono

Ono Geka Clinic, Fukuoka, Japan

Summary

In 15 haemodialysis patients receiving oral vitamin E supplementation, 600mg daily for 30 days, both plasma and RBC vitamin E concentrations were significantly increased, while in unsupplemented patients the values remained unchanged. Mean osmolarity at which in vitro haemolysis occurs at the start and end of haemodialysis decreased from 102.8±0.9 to 98.9±0.7 and 72.1±1.1 to 67.4±0.8mOsm/L, respectively in supplemented patients. In addition the Hct increased from 26.1±1.0 to 28.1±1.2 per cent ($p<0.05$). In conclusion, oral supplementation of vitamin E could be of clinical benefit in correcting anaemia in regular dialysis patients by reducing the fragility of red blood cells.

Introduction

Anaemia remains one of the major problems in patients with end-stage renal failure treated by regular haemodialysis. Vitamin E is known to block the peroxidation of the polyunsaturated fatty acid constituents of cell membranes including red blood cells [1] and the increased sensitivity of red cells to oxidative stress is an index of vitamin E deficiency [2]. In addition, administration of vitamin E can significantly improve the anaemia of sickle cell disease [3]. In spite of these observations, very little information is available on the importance of vitamin E in patients on regular haemodialysis, particularly in relation to anaemia. For these reasons, we have investigated the effect of oral supplementation on plasma and red blood cell vitamin E concentrations and the osmotic fragility of red blood cells.

Materials and methods

Thirty patients aged from 29 to 68 years were included in the study. They had all been dialysed for 5 to 126 months and were on 5 to 5.5 hours thrice weekly haemodialysis using a hollow Cuproammonium rayon fibre kidney. None of the patients were receiving preparations containing vitamin A or E. The patients were randomly divided into two groups: 15 patients received oral supplementation of 600mg vitamin E daily for 30 days in addition to routine medication. The remaining 15 patients were not given vitamin E and served as controls. For the purpose of comparison, 10 healthy subjects were also studied. Blood samples were taken from the arterial line immediately before and after dialysis in all 30 patients on two occasions, just before vitamin E supplementation was started (day 0) and 30 days later (day 30). Plasma vitamin E concentrations were measured by high performance liquid chromatography after heparinised plasma was separated from red cells by centrifugation and storage at -20°C. The measurement of triple washed red cell vitamin E was carried out using a slight modification of Abe's method [4]. The RBC vitamin E results were expressed as micrograms of alpha-tocopherol per ml of packed cells. Since only alpha-tocopherol is biologically active and accounts for about 85 per cent of total tocopherol in both plasma and RBC, the term 'vitamin E' used in this paper represents alpha-tocopherol. The osmotic fragility of RBCs was examined by the coil planet centrifuge method [5]. In this technique a stable linear gradient is created between two different concentrations of sodium chloride solution. Red cells pipetted into the higher concentration (150mOsm) side of the tubes are forced to travel through the gradient to the lower osmotic points where haemolysis occurs. The distribution of haemoglobin indicates the start and end of osmotic fragility of the samples. Red and white cell counts, Hct, reticulocytes, plasma iron, ferritin, total protein, GOT, GPT, LDH, total lipids, BUN, creatinine, uric acid, Na, K, Ca, Cl, and P were determined using routine methods. Each subject served as his or her own control. The statistical analysis of results between the two groups was made using the paired 't' test. Different groups of individuals were assessed by the unpaired 't' test.

Results

The pre-supplemented mean plasma concentrations of vitamin E were 10.67 ± 0.85 and 9.73 ± 0.77μg/ml in the two groups. These values were not lower than the normal range previously reported in haemodialysis patients [6]. Plasma vitamin E was significantly increased by oral vitamin E supplementation in 14 of 15 patients and the mean value of vitamin E almost doubled after 30 days treatment (20.37 ± 1.61μg/ml, $p<0.01$). In contrast RBC vitamin E in regular dialysis patients ranged from 0.23 to 1.10μg of vitamin E per ml of packed red blood with a mean of 0.57 ± 0.05 and 0.45 ± 0.07μg/ml in two groups. These results are significantly lower than the mean for 10 normal control subjects (1.34 ± 0.12μg/ ml packed RBC, $p<0.001$). However, vitamin E oral supplementation increased the mean vitamin E concentration in RBC to 1.56 ± 0.11μg/ml packed cells, which is almost triple the base-line value. RBC vitamin E in unsupplemented patients remained low. Twelve of 15 supplemented patients had a rise in Hct (mean from 26.1 ± 1.0 to 28.1 ± 1.2 per cent) and no subject showed a decrease in Hct after vitamin E administration, while an increase in Hct was not observed in unsupplemented patients. The mean change in Hct of two per cent was significant using the paired 't' test. Mean osmolarities at which in vitro haemolysis occurs at the start and end of haemodialysis decreased from 102.8±0.9 to 98.0±0.7 and 72.1±1.1 to 67.4±0.8mOsm/L, respectively after 30 days of treatment with vitamin E. These differences are statistically significant ($p<0.05$). Unsupplemented patients showed no improvement in osmotic fragility of RBC. There were no other biological or haematological changes. None of the patients received a blood transfusion during the 30 days treatment period.

Discussion

Since the observation by Rose and György that vitamin E can act as antioxidant which can reduce haemolysis of RBC caused by oxidant stress [7], additional evidence of a wide variety of roles in membrane-related activities and pathological processes secondary to a deficiency have been reported [8]. Polyunsaturated fatty acids represent 40 per cent of the total fatty acids in the human erythrocyte membrane. The fact that these polyunsaturated fatty acids do not normally auto-oxidise in cells implies the presence of a highly efficient protective mechanism. However, impairment of this mechanism, or conversely, susceptibility to peroxidation has been found in a variety of haemolytic states. Because of these observations we became interested in evaluating the effect of vitamin E on anaemia in regular haemodialysis patients. There is a paucity of information about the importance of vitamin E in haemodialysis patients, and investigations on the metabolism of vitamin E in such patients is lacking. Because vitamin E is thought not to be lost during dialysis most authors have stated that there does not appear to be a need for supplementation [9]. Vitamin E status has traditionally been determined by measuring plasma vitamin E but there have been relatively few measurements of alpha-tocopherol in red cells owing to technical difficulties of the assay. In contrast to the original report on plasma vitamin E concentration in dialysis patients by Ito who found low values [10], succeeding papers have concluded that the plasma concentration of vitamin E was normal or rather high in dialysis or CAPD patients [9]. On the basis of these findings it has been suggested that the vitamin E requirement of haemodialysis or CAPD patients can be met by the usual dietary intake without the use of supplementation [6]. Our results agree with previous publications indicating that plasma vitamin E is normal in such patients. Although normal, because of the importance of vitamin E on cell membranes, the red cell concentrations need to be directly measured rather than continuing to rely on plasma values. This is particularly important as haemolysis may be more directly related to the tocopherol in RBC than the plasma. In a recent study of vitamin E in regular dialysis patients, it has been stated that since the serum vitamin E was normal and since serum values reflected the concentration of the vitamin E within the erythrocyte membrane, it is reasonable to assume that the concentration of vitamin E in the RBC membrane is normal and that the antioxidant effect would likewise be normal. Our results do not confirm this assumption as the content of vitamin E in RBC is markedly below normal despite the normal plasma values: the reason for this is not clear. Increased utilisation of vitamin E by RBC membranes or a low transfer rate of vitamin E from plasma to RBCs in particular pathological situations, such as uraemia, might be an explanation for the discrepancy in vitamin E in the two compartments. The results of this present study indicate that the administration of oral vitamin E appears to increase both plasma and RBC values. Twelve of 15 supplemented patients showed an increase in Hct and all patients showed a reduced susceptibility to osmotic haemolysis as measured by coil planet centrifuge. It could be argued that the erythrocyte haemolysis test may not be a valid assessment of vitamin E status in haemodialysis patients. Uraemic toxins could interfere with the test particularly as slightly improved osmotic haemolysis can be seen in post-dialysis blood samples. Our results, however, show a very significant improvement after vitamin E administration, suggesting an important role for this substance in RBC fragility. This finding is similar to the observation that the impaired osmotic fragility of thalassaemic erythrocytes returned to near normal following administration of vitamin E. Increasing plasma vitamin E by supplementation could provide an excess of alpha-tocopherol to RBC membranes which may improve the ability of the cells to withstand osmotic stress. A similar conclusion was recently reported after intramuscular administration of vitamin E which decreased the content of malonyldialdehyde in RBCs. As this is an intermediate product of polyunsaturated fatty acid oxidation a decrease could mean an increase in the alpha-tocopherol in RBCs leading to a higher Hct in regular dialysis patients. In conclusion, our study suggests that oral supplementation of vitamin E in regular dialysis patients could be of clinical importance by reducing anaemia and the efficacy of long-term vitamin administration in such patients should be seriously considered.

Acknowledgment

The author is most grateful to Mr Yasumasa Yuguchi, Eisai Company for assistance with measuring vitamin E, and to Ms Yohko Hisasue and other nursing staff of the Ono Geka Clinic for their valuable assistance in this study.

References

1 Tudhope GR, Hopkins J. *Acta Haemat 1975; 53:* 98
2 Horwitt MK, Harvey CC, Harmon EM. *Vitam Horm 1968; 26:* 487
3 Natta CL, Machlin LJ, Brin M. *Am J Clin Nutr 1980; 33:* 968
4 Abe K, Yuguchi Y, Katsui G. *J Nutr Sci Vitaminol (Tokyo) 1975; 33:* 183
5 Ito Y, Weistein M, Aoki I, Harada R. *Nature 1966; 5066:* 985
6 Johnson KS, Hendricks DC, Wyse BW. *Dial Transpl 1983; 12:* 477
7 Rose CS, György R. *Blood 1950; 5:* 1062
8 Nelson JS. In Machlin LJ, ed. *Vitamin E.* New York: Marcel Dekker. 1980: 397–428
9 Blumberg A, Hanck A, Sander G. *Clin Nephrol 1983; 20:* 244
10 Ito T, Niwa T, Matsui E. *JAMA 1971; 217:* 699

Proc EDTA–ERA (1984) Vol 21

VACCINATION AGAINST HEPATITIS B IN PATIENTS WITH RENAL INSUFFICIENCY

J Bommer, M Grussendorf, *W Jilg, *F Deinhardt, H-G Koch, G Darai, G Bommer, M Rambausek, E Ritz

*University of Heidelberg, Heidelberg, *Max von Pettemkofer-Institut, University of Munich, Munich, FRG*

Summary

To define the degree of renal insufficiency at which the immune response to vaccination against hepatitis B is impaired, anti-HB concentrations after vaccination with 20μg HB-Vax® at 0, 1 and 6 months were examined in 76 dialysis patients, 24 patients with incipient renal failure (S-creatinine 1.4–3.5mg/dl) and in 43 controls. Compared with controls, seroconversion rate and anti-HB concentrations were significantly ($p<0.02$) lower in patients with incipient renal failure. The time course of anti-HBs in dialysis patients with successful vaccination, either with three doses (0, 1, 6 months) or with five doses (0, 1, 2, 4, 6 months) of HB-Vax was compared with healthy controls. The proportion of patients losing anti-HBs and the decrease of antibody concentration was significantly greater in dialysis patients immunised with three doses of the vaccine. In dialysis patients vaccinated with five doses, the percentage losing HB antibodies was slightly higher than in controls, but final titres after 24 months were comparable.

Introduction

Patients with end-stage renal failure have an impaired response to vaccines containing hepatitis B surface antigen (HBsAg) [1–6]. Seroconversion rates as well as anti-HB concentrations in these patients have consistently been found to be reduced. Two problems have not been solved. Firstly, at what stage of renal insufficiency is the immune response to the vaccine impaired? Resolution of this problem will have important consequences for vaccination policies in prospective dialysis patients. Secondly, what is the time course of anti-HBs in dialysis patients who had shown seroconversion? Resolution of this problem is important to determine the need for booster vaccinations.

Patients and methods

Hepatitis B vaccine (HB-Vax®, Merck, Sharp & Dohme) was given to 43 staff members (19 male, 24 female) and to 76 dialysis patients (50 male, 26 female).

Staff members received a dose of 20μg HB-Vax per vaccination; dialysis patients received 40μg per vaccination. In one subgroup (passive/active vaccination) 3ml of hepatitis B immunoglobulin (Merck, Sharp & Dohme, Hep B Gamma GE®) was given intramuscularly. The time intervals chosen for vaccination are shown in Table I.

TABLE I. Vaccination Schedules

Group	Vaccination
Staff	month 0, 1, 6
Dialysis patients:	
group I (regular)	month 0, 1, 6
group II (frequent)	month 0, 1, 2, 4, 6
group III (active/passive)	month 0, 1, 2, 4, 6
3ml HBIG intramuscularly	month 0, 1

Patients who produced no anti-HBs received a booster injection after one year. In addition, 24 patients with incipient renal failure (S-creatinine 1.4–3.5mg/dl; 16 male, 8 female) were vaccinated using the same schedule as in staff members. The underlying disease was: glomerulonephritis (n=11), type II diabetes mellitus (n=2), polycystic kidney disease (n=2), reflux nephropathy (n=3), analgesic abuse (n=2), renovascular disease or malignant hypertension (n=4).

Anti-HBs were measured with a radioimmunoassay (Ausab®, Abbott Laboratories, North Chicago, Ill, USA). Anti-HB values are given as mIU/ml; they were calculated according to Hollinger [7].

Results

Anti-HBs in incipient renal failure

As shown in Table II, after eight months 90 per cent of controls, but only 65 per cent of patients with incipient renal failure had seroconverted as compared to 54 per cent of patients on dialysis, although the latter had received the double dose. Median anti-HB concentrations in responders were comparable in controls (333 mIU/ml) and patients with incipient renal failure (394 mIU/ml). The proportion of patients with seroconversion was comparable in glomerulonephritis (8 of 11) and other, i.e. non-immunologically mediated, renal diseases (7 of 13).

TABLE II. Immune response to three doses of hepatitis B vaccine in patients with incipient renal failure compared with healthy controls and dialysis patients

	Vaccine dose	Age (years)	male/female	1 month a) seroconversion b) anti-HB concentrations (mIU/ml) median (range)	6 months a) b)	8 months* a) b)
Staff (n=43)	20μg	31.6±8.2	19/24	a) 13/43 (30%) b) 4.4 (2–1199)	a) 36/43 (84%) b) 62 (2.7–4006)	a) 38/42 (90%) b) 333 (4.2–14,112)
Dialysis patients (n=29)	40μg	47.2±12.1	14/15	a) 0/29 b) –	a) 10/29 (34%) b) 7.9 (2.4–58)	a) 15/28 (54%) b) 25 (2.6–509)
Patients with incipient renal failure (n=24)	20μg	46.1±9.1	16/8	a) 0/24 b) –	a) 10/24 (42%) b) 97 (13–437)	a) 15/23 (65.1%) b) 394 (7.7–13,824)

* one patient in each group lost for follow-up (eight months and long-term observations in Table III).

At eight months seroconversion rate significantly ($p<0.02$) different between staff and incipient renal failure (chi-square); anti-HB concentrations significantly ($p<0.02$) different between dialysis patients and staff or incipient renal failure respectively (Wilcoxon test).

TABLE III. Time course of anti-HB concentrations in dialysis patients

	male/female	age (years)	duration of dialysis (months) median (range)	8 months a) seroconversion b) anti-HB concentration (mIU/ml) median (range)	12 months a) b)	18 months a) b)	24 months a) b)
Staff (n=42)	19/23	31.5±8.6		a) 38/42 (90%) b) 333 (4.2–14,112)	37/42 (88%) 100 (2.7–4,000)	data incomplete	34/42 (81%) 52 (2.4–3,940)
group I (regular) (n=28)	14/14	47.3±12.0	41 (9–85)	a) 15/28 (54%) b) 25 (2.6–509)	14/28 (50%) 30.5 (2.0–323)	data incomplete	5/28 (18%)* 4.2 (3.1–34)
group II (frequent) (n=23)	17/6	46.7±9.5	33 (3–122)	a) 15/23 (65) b) 338 (3–8,516)	14/23 (61%) 67.5 (2.4–192)	12/23 (52%) 555.5 (2.3–159)	11/23 (48%)* 125 (6.3–929)
group III (passive/active) (n=25)	19/6	44.5±9.9	38 (2–112)	a) 15/25 (60%) b) 100 (2–18,157)	14/25 (56%) 29.5 (3.6–158)	13/25 (52%) 12 (5.4–180)	12/25 (48%)* 29.5 (3.4–1,383)

* at 24 months significant ($p<0.01$) difference between staff and respective patient group.

Time course of anti-HBs in dialysis patients

As shown in Table III, at eight months, i.e. two months after the last vaccination, seroconversion was found in 54 per cent of dialysis patients who had obtained three doses of vaccine, 65 per cent of dialysis patients who had obtained five doses and 60 per cent of dialysis patients after active/passive vaccination. In dialysis patients receiving five vaccinations, anti-HB concentrations in responders were higher and corresponded to those observed in controls. After two years, the proportion of patients with anti-HBs had diminished by 36 per cent in patients in group I, by 17 per cent in patients in group II and by 12 per cent in group III as compared with nine per cent in controls. Consequently, after two years only 33 per cent of individuals who had seroconverted remained positive in group I, 73 per cent in group II and 80 per cent in group III. The respective proportion in controls was 89 per cent.

Discussion

The present study clearly documents that patients with incipient renal failure have abnormal anti-HB responses to hepatitis B vaccination. It might be argued that this is due to disturbed immune regulation implicated in the genesis of some renal diseases. However, even in the group with glomerulonephritis, the prototype of immunologically mediated renal diseases, no less than 8 of 11 patients seroconverted. Therefore, it is more likely that abnormal anti-HB response is caused by renal failure. Defective T-cell function has been described in patients with renal failure. Although in the past early vaccination in incipient renal failure has been recommended by several authors [2,4] the present data cast some doubt on whether regular vaccination schedules will be sufficient in such patients.

Our results in dialysis patients are in line with the preliminary findings of Batuman [8] indicating that dialysis patients given regular vaccination schedules (0, 1, 6 months) will rapidly lose anti-HBs despite a double dose of the vaccine (40μg instead of 20μg per vaccination). In healthy people, Jilg et al [9] found a relation between maximal anti-HB concentration and maintenance of anti-HBs in the post-vaccination period. It is therefore of note that in dialysis patients more frequent vaccination not only caused a higher rate of seroconversion but also resulted in higher median anti-HB concentrations. As in healthy controls [9] in dialysis patients vaccinated five times, maximal HB concentrations were higher and anti-HBs were maintained for longer periods of time compared to dialysis patients vaccinated three times.

Our experience with additional booster vaccinations indicates that only about 10 to 20 per cent of dialysis patients who did not seroconvert after up to five vaccinations will seroconvert after an additional vaccination. Anti-HB concentrations achieved by such further (40μg) vaccination tends to be low ($\leqslant$ 10 mIU/ml): out of 14 patients who had never responded or had become negative after seroconversion in response to three vaccinations (group I) only two developed low anti-HB values; of 10 dialysis patients who were negative only three responded to revaccination after one year and of nine patients with passive/active vaccination only two transiently developed low anti-HBs. It is still to be determined whether more frequent vaccinations and/or still higher doses will give better results in such poor responders.

References

1 Crosnier J, Jungers P, Courucé A-M et al. *Lancet 1981; i:* 797
2 Bommer J, Ritz E, Andrassy K et al. *Proc EDTA 1983; 20:* 161
3 Bommer J, Deinhardt F, Jilg W et al. *Dtsch Med Wschr 1983; 48:* 1823
4 Benhamou E, Courouce A-M, Jungers P. *Clin Nephrol 1984; 21:* 143
5 Koehler H, Arnold W, Renschien G et al. *Kidney Int 1984; 25:* 124
6 Stevens CE, Szmuness W, Goldman AI et al. *Lancet 1980; ii:* 1211
7 Hollinger FB, Adam E, Heiberg D, Melnick JL. In Szmuness W, Alter HJ, Maynard JE, eds. *Viral Hepatitis*. Philadelphia: Franklin Inst Press. 1982: S 451–466
8 Batuman V, Curry E, Tecson-Tumang F et al. *Int Congress Nephrology, Los Angeles, 1984:* A 148
9 Jilg W, Schmidt M, Deinhardt F, Zachoval R. *Lancet 1984; ii:* 458

Proc EDTA–ERA (1984) Vol 21

PLATELET LIFE SPAN IN URAEMIA

H Tanaka, K Umimoto, N Izumi, *K Nishimoto, T Maekawa, *T Kishimoto, *M Maekawa

*Tadaoka Municipal Hospital, Osaka, *Osaka City University, Osaka, Japan*

Summary

Platelet life span estimated by the regeneration time of platelet cyclo-oxygenase activity after acetyl salicylic acid (ASA) intake, as measured by the malonyldialdehyde (MDA) production rate was 10.3±1.0 days for healthy volunteers (n=7), 6.7±1.0 days for haemodialysis patients (n=13), 8.0±1.5 days for continuous ambulatory peritoneal dialysis (CAPD) patients (n=6) and 5.0 and 4.9 days for non-dialysed uraemic patients (n=2). In uraemic patients, the platelet cyclo-oxygenase activity was significantly impaired and it correlated with the decline in platelet life span. The restoration of the platelet life span and cyclo-oxygenase activity was achieved better by CAPD.

Introduction

Uraemic patients are known to have platelet abnormalities which are usually improved by haemodialysis [1,2], but these abnormalities are thought to affect the platelet life span. Continuous ambulatory peritoneal dialysis (CAPD) is being used increasingly for treating uraemic patients. Since it has not yet been reported whether CAPD or haemodialysis is better in improving the platelet abnormalities of uraemic patients, we compared the platelet life span in CAPD patients with that in haemodialysis patients. Usually platelet life span is measured by the ^{51}Cr-labelled technique but radioactive substances are harmful for patients, staff and examiners. In this study, we calculated the platelet life span by measuring the regeneration time of platelet cyclo-oxygenase activity after acetyl salicylic acid (ASA) intake.

Materials and methods

The study was undertaken in seven healthy volunteers, 13 haemodialysis patients, six CAPD patients and two non-dialysed uraemic patients just before starting regular dialysis treatment. All patients had normal liver function, none had splenomegaly, and none were diabetic. None had received any blood transfusion for more than six months and none had taken any medication known to affect the platelet function for at least a week before the start of the study. The biochemistry and blood cell counts in the uraemic patients are shown in Table I.

TABLE I. Biochemical findings and blood cell counts

	Haemodialysis patients (n=13)	CAPD patients (n=6)	Non-dialysed patients (n=2)
BUN (mg/dl)	72.0 ± 18.0	67.6 ± 18.4	90.3, 122.9
creatinine (mg/dl)	14.6 ± 2.9	9.9 ± 2.1	11.7, 15.9
uric acid (mg/dl)	8.4 ± 0.9	8.0 ± 1.5	8.8, 12.3
Ca (mg/dl)	9.9 ± 1.4	8.3 ± 0.6	6.8, 7.8
iP (mg/dl)	7.2 ± 1.4	6.5 ± 1.6	7.1, 11.8
WBC (/mm³)	6600 ± 2000	5600 ± 1600	4500, 8300
RBC (x10⁴/mm³)	257 ± 57	312 ± 74	199, 188
Hb (g/dl)	8.2 ± 1.2	9.6 ± 1.9	6.4, 6.6
Platelet (x10⁴/mm³)	15.9 ± 4.6	17.1 ± 3.1	16.3, 27.1

Data are expressed as mean ± SD

The platelet malonyldialdehyde (MDA) production rate was measured in all of the patients before and after ASA intake by the modified Okuma's method [3]. After venepuncture, each patient was given 660mg of ASA to inhibit platelet cyclo-oxygenase completely and irreversibly. Then the MDA was serially measured until it recovered the pre-ASA value. The time required to restore the platelet MDA production rate to the pre-ASA value is the platelet regeneration time, because only platelets formed after the effect of ASA has worn off can produce MDA. Therefore, the platelet life span can be estimated indirectly from the platelet regeneration time [4,5]. The MDA production rate was also measured in an additional 25 healthy volunteers, 17 haemodialysis patients and six CAPD patients.

The statistical analysis was done by the unpaired Student's 't' test.

Results

Table I indicates the biochemical findings and blood cell counts in the three patient groups. The platelet counts were comparable between the haemodialysis and CAPD patients. The platelet MDA production rate (pre-ASA value) was 13.3±2.0nmol/10^9 platelets for the 32 healthy volunteers, 8.3±2.0nmol/10^9 platelets for the 30 haemodialysis patients and 9.6±1.2nmol/10^9 platelets for the 12 CAPD patients (Figure 1). It was significantly lower in uraemic patients

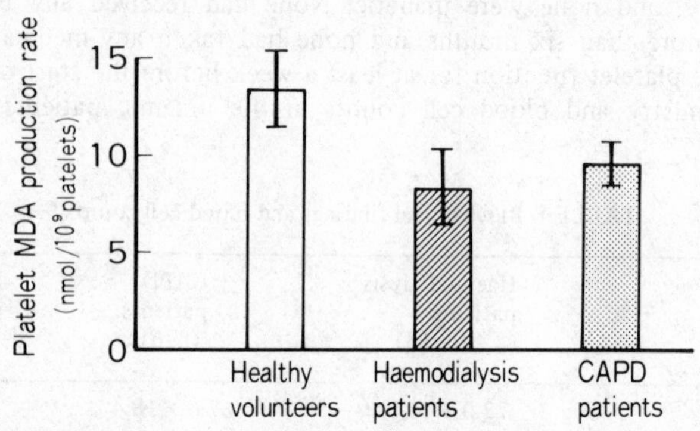

Figure 1. Platelet MDA production rate for the healthy volunteers (open column), haemodialysis (cross-hatched column) and CAPD patients (shaded column). The values represent mean ± SD. There were significant differences between the healthy volunteers and the haemodialysis ($p<0.001$) and CAPD patients ($p<0.001$) and also between the haemodialysis and CAPD patients ($p<0.05$)

than the healthy volunteers and was also lower in haemodialysis patients than in CAPD patients.

The recovery time of the platelet MDA production rate after ASA intake or the platelet life span in the seven healthy volunteers, 13 haemodialysis patients, six CAPD patients and two non-dialysed patients was 10.3±1.0 days, 6.7±1.0 days, 8.0±1.5 days and 5.0 and 4.9 days, respectively (Figure 2). The platelet life span in uraemic patients was significantly shorter than that in healthy volunteers. Furthermore, it was significantly shorter in haemodialysis patients than in CAPD patients.

In all of the healthy volunteers and uraemic patients, there was a significant correlation between the platelet MDA production rate and its life span.

Discussion

The platelet life span was found to be significantly shorter in uraemic patients than in healthy volunteers. There may be many reasons for the shortening of the platelet life span in uraemic patients, such as platelet dysfunction, uraemic toxins and excess consumption of the platelet, but it had not been clarified.

Among uraemic patients, the haemodialysis patients who received extracorporeal circulation had significantly shorter platelet life spans than the CAPD patients. Serum beta-thromboglobulin and platelet factor 4 increases after haemodialysis [6], which may be due to platelet agglutination or aggregation. Platelet counts decrease at the start of haemodialysis, but later recover to their former value [7]. CAPD patients do not need extracorporeal circulation and the platelets are therefore not at risk. In other words, there is a great difference in platelet stress between haemodialysis and CAPD. However, Levin et al reported that many platelets are not destroyed during haemodialysis [8]. We have also shown that platelet MDA production rate is the same before and after haemo-

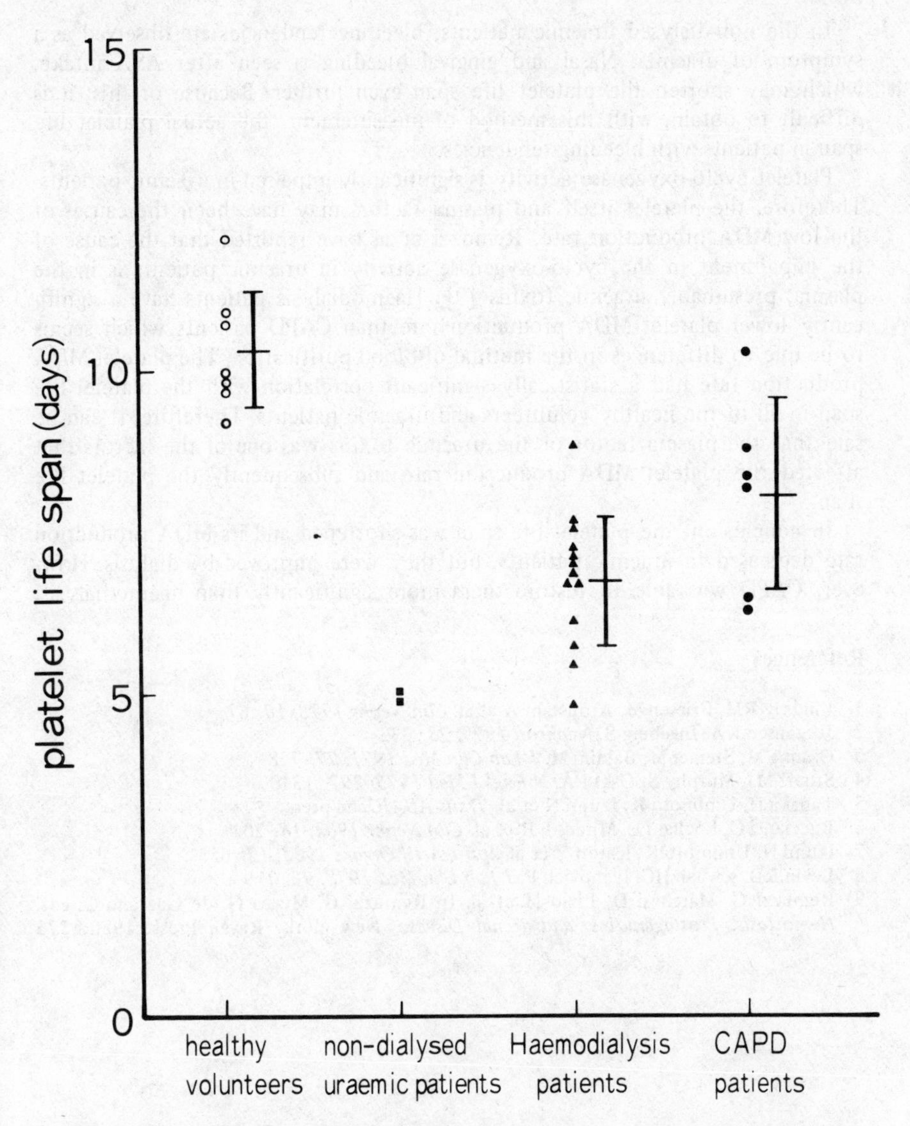

Figure 2. Regeneration time of the platelet cyclo-oxygenase activity after ASA intake (platelet life span) for healthy volunteers (open circles), non-dialysed (squares), haemodialysis (triangles) and CAPD patients (closed circles). Values represent mean ± SD. There were significant differences between the healthy volunteers and the haemodialysis ($p<0.001$) and CAPD patients ($p<0.01$) and between the haemodialysis and CAPD patients ($p<0.05$)

dialysis regardless of ASA intake. Therefore, platelets may not have been consumed by the extracorporeal circulation and the platelet life span may have been shortened by other causes.

In the non-dialysed uraemic patients, bleeding tendencies are observed as a symptom of uraemia. Nasal and gingival bleeding is seen after ASA intake, which may shorten the platelet life span even further. Because of this, it is difficult to obtain, with this method of measurement, the actual platelet life span in patients with bleeding tendencies.

Platelet cyclo-oxygenase activity is significantly impaired in uraemic patients. Therefore, the platelet itself and plasma factors may have been the causes of the low MDA production rate. Remuzzi et al have reported that the cause of the impairment in the cyclo-oxygenase activity in uraemic patients is in the plasma, presumably uraemic toxins [9]. Haemodialysis patients have a significantly lower platelet MDA production rate than CAPD patients which seems to be due to differences in the method of blood purification. The platelet MDA production rate had a statistically significant correlation with the platelet life span in all of the healthy volunteers and uraemic patients. Therefore, it can be said that the plasma factor or the uraemic toxins was one of the factors that affected the platelet MDA production rate and subsequently the platelet life span.

In conclusion, the platelet life span was shortened and its MDA production rate decreased in uraemic patients, but they were improved by dialysis. However, CAPD was able to restore them more significantly than haemodialysis.

References

1 Lindsay RM, Friesen M, Aronstam A et al. *Clin Nephr 1978; 10:* 67
2 Jorgensen KA, Ingeberg S. *Nephron 1979; 23:* 233
3 Okuma M, Steiner M, Baldini M. *J Lab Clin Med 1971; 77:* 728
4 Stuart MJ, Murphy S, Oski FA. *N Engl J Med 1975; 292:* 1310
5 Tanaka H, Umimoto K, Izumi N et al. *Trans ASAIO:* in press
6 Buccianti G, Pogliani E, Miradoli R et al. *Clin Nephr 1982; 18:* 204
7 Izumi N, Umimoto K, Katoh Y et al. *Jpn J Artif Organs 1983; 12:* 653
8 Levin RD, Kwaan HC, Ivanovich P. *J Lab Clin Med 1978; 92:* 779
9 Remuzzi G, Marchesi D, Livio M et al. In Remuzzi G, Mecca G, de Gaetano G, eds. *Hemostasis, Prostaglandins, and Renal Disease.* New York: Raven Press. 1980: 273

Proc EDTA–ERA (1984) Vol 21

MORBIDITY OF PATIENTS WITH ANALGESIC-ASSOCIATED NEPHROPATHY AND END-STAGE RENAL FAILURE

A Schwarz, W Pommer, F Keller, G Kuehn-Freitag, G Offermann, M Molzahn

Steglitz Medical Complex, West Berlin, FRG

Summary

In our haemodialysis centre patients (n=144), we compared 48 aspects of morbidity in patients with analgesic-associated nephropathy (AAN) and patients with other kidney diseases to determine the presence of characteristic diagnostic features of AAN in addition to a history of habitual analgesic intake. The comparison between 48 AAN patients and the control patients revealed statistically significant differences ($p<0.05$) with regard to myocardial infarction (25% vs 7%), angina pectoris (63% vs 32%), atrial fibrillation (21% vs 4%), arteriosclerosis obliterans of the lower extremity (52% vs 33%), anaemia (mean haemoglobin, 8.38 vs 9.16g/dl), renal osteodystrophy (67% vs 41%), carpal tunnel syndrome (23% vs 7%), peptic ulcers and erosive gastritis (54% vs 23%), colonic diverticula (15% vs 4%), and haemorrhoids (67% vs 28%). AAN patients therefore have significantly higher morbidity with a characteristic pattern than do patients with other renal diseases.

Introduction

At onset, analgesic-associated nephropathy (AAN) can easily escape diagnosis because of the patients' tendency to deny or to minimise their regular analgesic intake. These diagnostic difficulties may well be one reason why analgesic abuse is still not generally accepted as an important causal factor in the development of kidney disease [1]. Although the reported incidences of AAN vary from author to author and from country to country, most investigators do agree that AAN is a significant cause of terminal renal failure [2,3,4].

We compared haemodialysis patients with AAN and those with other kidney diseases to determine whether a characteristic morbidity pattern exists for AAN patients which, together with the special personality features of these patients and their habitual analgesic consumption, would facilitate the diagnosis of AAN.

Methods

All 144 patients on maintenance haemodialysis at our ambulatory haemodialysis centre were investigated. Patients with diagnostic signs of AAN were assigned to one group (n=48) and patients with other kidney diseases, to the control group (n=76). After cross-sectional evaluation of patient records and data, the findings of 48 morbidity criteria in the AAN patients were compared with those of patients with other kidney diseases. Patients with some diagnostic signs of AAN and the presence of another kidney disease (n=20) were excluded from the study. The differences were analysed for statistical significance with the χ^2 distribution test and the Student's 't' test. A probability of $p<0.05$ was taken as the limit of statistical significance. Each AAN patient (n=48) was age-matched (± 5 years) with a patient from the control group.

Results

AAN is the most frequently diagnosed kidney disease at our haemodialysis centre. In addition to a history of analgesic abuse, the diagnostic signs indicating AAN were papillary necrosis (60%), intra-renal calcifications (13%), irregular kidney shrinkage with papillary cavities (23%), chronic interstitial nephritis on renal biopsy (8%), and urinary tract carcinoma (4%).

Comparison of 48 morbidity criteria revealed statistically significant differences between AAN patients and the controls for 15 of the criteria (Table I).

TABLE I. Comparison of morbidity in haemodialysis patients with analgesic-associated nephropathy and haemodialysis patients with other kidney diseases. Statistical significance was established with the χ^2 distribution test and the Student's 't' test*

	Analgesic-associated nephropathy (n=48)	Other kidney diseases (n=76)	Statistical significance	Statistical significance (matched for age)
ANAEMIA				
mean haemoglobin (g%)	8.4 ± 1.9	9.2 ± 2.2	p<0.05*	p<0.02*
blood transfusions (preceding year)	38%	18%	p=0.018	NS
RENAL OSTEODYSTROPHY				
mean serum calcium (mmol/L)	2.5 ± 0.2	2.5 ± 0.2	NS*	NS*
mean serum alkaline phosphatase (U/L)	167 ± 99	161 ± 253	NS*	NS*
radiological signs	67%	41%	p=0.005	p=0.004
bone pain and/or myopathic signs	28%	8%	p=0.0038	p=0.016
parathyroidectomy, necessary or performed	32%	22%	NS	NS
GASTROINTESTINAL DISORDERS				
peptic ulcers and/or erosive gastritis	58%	25%	p<0.001	p=0.011
angiodysplasia	2%	1%	NS	NS

(continued)

TABLE I (continued)	Analgesic-associated nephropathy (n=48)	Other kidney diseases (n=76)	Statistical significance	Statistical significance (matched for age)
colonic diverticula	15%	4%	p=0.034	p=0.027
haemorrhoids	67%	28%	p<0.001	p=0.014
haemorrhoidal surgery or obliteration	10%	0%	p=0.004	p=0.022
CARDIAC DYSRHYTHMIA				
sinus rhythm	94%	100%	p=0.027	NS
ectopic beats	4%	5%	NS	NS
paroxysmal atrial tachycardia	4%	1%	NS	NS
paroxysmal or permanent atrial fibrillation	21%	4%	p=0.0028	p=0.014
ectopic ventricular beats requiring treatment	0%	1%	NS	NS
pacemaker	2%	1%	NS	NS
ELECTROCARDIOGRAPHIC CHANGES				
normal ECG	19%	11%	NS	NS
left ventricular hypertrophy	42%	53%	NS	NS
myocardial infarction	21%	11%	NS	NS
first-degree atrioventricular block	10%	13%	NS	NS
right bundle branch block	2%	1%	NS	NS
left bundle branch block	0%	1%	NS	NS
left anterior hemiblock	4%	14%	NS	NS
left posterior hemiblock	0%	0%	NS	NS
multiple block	8%	3%	NS	NS
repolarisation disturbance	75%	72%	NS	NS
HYPERTENSION				
before maintenance haemodialysis	77%	76%	NS	NS
during maintenance haemodialysis	33%	37%	NS	NS
before and during maintenance haemodialysis	33%	33%	NS	NS
hypertensive emergency	10%	7%	NS	NS
ARTERIOSCLEROSIS				
myocardial infarction	25%	7%	p=0.0036	p=0.028
angina pectoris	63%	32%	p=0.001	p=0.014
apoplectic insult	15%	12%	NS	NS
jet phenomenon of carotid artery	17%	8%	NS	NS
arteriosclerosis obliterans of lower extremities	52%	33%	p=0.034	NS

(continued)

TABLE I (continued)	Analgesic associated nephropathy (n=48)	Other kidney disease (n=76)	Statistical significance	Statistical significance (matched for age)
HEART FAILURE (signs on chest film)				
normal	50%	64%	NS	NS
heart enlargement	40%	29%	NS	NS
pulmonary congestion	15%	11%	NS	NS
heart enlargement and pulmonary congestion	8%	7%	NS	NS
pulmonary congestion without heart enlargement	6%	3%	NS	NS
PERIPHERAL POLYNEUROPATHY (electromyographic signs)				
normal EMG	34%	27%	NS	NS
severe EMG changes	8%	20%	NS	NS
CARPAL TUNNEL SYNDROME	23%	7%	p=0.008	p=0.049
URINARY TRACT CARCINOMA / DYSPLASTIC CELLS IN URINE	10%	5%	NS	NS
TUBERCULOSIS DURING MAINTENANCE HAEMODIALYSIS	6%	8%	NS	NS
RECURRENT AND/OR COMPLICATED URINARY TRACT INFECTIONS	40%	9%	p<0.001	p=0.0025

Patients with AAN were significantly older than patients of the control group (60±10 vs 52±15 years). After age-matching, the differences in blood transfusion requirements, lower extremity arteriosclerosis, and sinus rhythm were no longer statistically significant, but the other differences remained significant. No statistically significant difference could be established for the duration of haemodialysis treatment (4.57±3.96 vs 3.94±3.55 years).

Discussion

The higher morbidity of the AAN patients when compaied to the age-matched patients with other kidney diseases was statistically significant. Several well-known aspects of morbidity in AAN described in the literature are anaemia [5], ulcers and erosive gastritis [6], recurrent or complicated urinary tract infections [7], and coronary heart disease [8]. Pronounced hyperlipoproteinaemia has been mentioned as a risk factor of general premature atherosclerosis [3,8,9]. Our AAN patient group also included significantly more smokers than the control group (63% vs 35%). The incidence of atrial fibrillation in the AAN patients was higher than in the control group. The more severe osteodystrophy in the AAN patients could be due to the longer course of renal failure prior to starting haemodialysis or to a direct effect of phenacetin on bone metabolism such as that occurring with anticonvulsant drugs [10]. A carpal tunnel syndrome, which usually complicates the late phase of haemodialysis treatment, appeared earlier in the AAN patients than in the controls. The higher incidence of colonic diverticulosis and haemorrhoids in the AAN patients could be related to laxative abuse.

Our study strongly suggests that patients with AAN, even in cases of undiagnosed kidney disease and denial of regular analgesic intake, can be identified by the characteristic *diagnostic signs* and the additional *morbidity features* of AAN (Table II). Together these two aspects yield a characteristic morbidity pattern that differs significantly from that of patients with other kidney diseases.

TABLE II. Diagnostic signs and morbidity features indicating analgesic-associated nephropathy

DIAGNOSTIC SIGNS

papillary necrosis
- (exclusion of diabetes mellitus, reflux and obstructive nephropathy, sickle-cell anaemia)

renal histology
- (capillary sclerosis of urinary tract, chronic interstitial nephritis)

irregular kidney shrinkage with papillary necrosis

intrarenal calcifications (radiology, ultrasound)

transitional cell carcinoma of urinary tract

MORBIDITY FEATURES

premature atherosclerosis
- coronary heart disease
- arteriosclerosis of lower extremity

atrial fibrillation

gastrointestinal disorders
- peptic ulcers and/or erosive gastritis
- colonic diverticulosis
- haemorrhoids

recurrent and/or complicated urinary tract infections

inadequate anaemia

early and severe renal osteodystrophy

carpal tunnel syndrome

References

1 Murray TG, Stolley PD, Anthony JC et al. *Arch Intern Med 1983; 143:* 1687
2 Dubach UC, Rosner B, Pfister E. *N Engl J Med 1983; 308:* 357
3 Nanra RS. *Br J Clin Pharmacol 1980; 10:* 359S
4 Schreiner GE, McAnally JF, Winchester JF. *Arch Intern Med 1981; 141:* 349
5 Jasinski B. *Schweiz Med Wschr 1948; 78:* 681
6 Kincaid-Smith P. *Second International Congress of Nephrology. Excerpta Medica Intern Congress Series 1963; 67:* 91
7 Laberke HG, Mall A. *Dtsch Med Wschr 1983; 108:* 1671
8 Kaladelfos G, Edwards KDG. *Nephron 1976; 16:* 388
9 Helber A, Wambach G, Böttcher W et al. *Nephron 1980; 26:* 111
10 Jaeger P, Burckhardt P, Wauters JP. In Normal AW, Schäfer K, Herrath v D, Grigoleit H-G, eds. *Vitamin D, Chemical, Biochemical and Clinical Endocrinology of Calcium Metabolism.* Berlin: Walter de Gruyter. 1982: 1012–1015

Proc EDTA–ERA (1984) Vol 21

HERPES VIRUS INFECTION PREVALENCE IN REGULAR HAEMODIALYSIS PATIENTS – A COMPARATIVE EVALUATION OF COMPLEMENT FIXATION, INDIRECT IMMUNOFLUORESCENCE AND ELISA TESTS

F Scolari, N Manca, S Sandrini, L Cristinelli, E Prati, G Setti, S Savoldi, A Turano, R Maiorca

Spedali Civili and University, Brescia, Italy

Summary

The presence and titres of specific serum IgG and IgM antibodies to cytomegalovirus, herpes simplex virus and varicella-zoster virus were evaluated in 50 haemodialysis patients by complement fixation, immunofluorescence and Elisa tests. A second serum sample was tested in 24 patients after four weeks. Specific serum IgG antibodies to Epstein-Barr virus were also measured by immunofluorescence in 26 patients.

By immunofluorescence and Elisa tests, the prevalence of cytomegalovirus, herpes simplex virus and Epstein-Barr virus infection is approximately 100 per cent, and varicella-zoster virus 60 per cent. High titres of IgG specific antibodies found by Elisa tests, detection of IgM antibodies in 16–18 per cent of patients and sero-conversion in 25 per cent of patients, suggests continuous antigenic stimulation.

Introduction

Herpes virus family includes a group of agents (cytomegalovirus, herpes simplex virus, varicella-zoster virus, Epstein-Barr virus) known for their widespread distribution in humans and in numerous other mammals. Herpes viruses are able to persist in infected cells in a latent form and can be reactivated, commonly in immunosuppressed patients. The high rate of herpes virus infection in recipients of renal allografts is well known, occurring in up to 90 per cent of patients [1,2].

In contrast, the frequency and importance of herpes virus infection in haemodialysis patients is incompletely understood, though immunological surveys of chronic renal failure patients and transplant candidates reveal a prevalence of antibodies to cytomegalovirus, herpes simplex virus, varicella-zoster virus and Epstein-Barr virus ranging from 54 to 100 per cent [3–6]. Moreover, unexplained febrile illnesses, pericarditis, pneumonitis and HBs-Ag negative hepatitis have been described in association with a significant rise in antibody titres to cytomegalovirus and Epstein-Barr virus [7–9].

To investigate the prevalence and role of herpes virus infection we conducted a serological survey of haemodialysis patients in our centre.

Patients and methods

The study group included 50 haemodialysis patients (37 males, 13 females) with an average age of 44 years (range 22–78). The average period of haemodialysis treatment was four years (range 1–12).

Serum specimens were drawn from all patients and tested for cytomegalovirus, herpes simplex virus, varicella-zoster virus, IgG antibodies by complement fixation, immunofluorescence and Elisa tests. Sera were absorbed with Rheumatoid Factor Latex suspension, before testing for IgM antibodies to cytomegalovirus, herpes simplex virus, varicella-zoster virus by immunofluorescence and Elisa tests. A second serum sample was examined for cytomegalovirus, herpes simplex virus, varicella-zoster virus, IgG antibodies in 24 patients four weeks later. Specific serum IgG antibodies to Epstein-Barr virus were also measured by immunofluorescence in 26 patients.

Titres equal to or greater than 1:4 by complement fixation, 1.8 by immunofluoresence, 1:640 by Elisa were considered positive (inactive infection). Seroconversion (active infection) was diagnosed on the basis of a four-fold or greater increase in antibody titres by complement fixation, immunofluorescence and Elisa tests.

Results

The percentages of patients with detectable IgG antibodies, the range of antibody titres and geometric mean titres for cytomegalovirus, herpes simplex

TABLE I. Number (%) of seropositive patients, range of Ab titres and GMT for CMV, HSV, VZV, EBV, by CF, IF and EL

	% Seropositive patients	Range of Ab titres	GMT
1. CMV			
CF	30%	4–64	11
IF	96%	8–1024	161
EL	88%	640–20480	2238
2. HSV			
CF	76%	8–64	16
IF	100%	64–4096	446
EL	94%	1280–40960	4160
3. VZV			
CF	6%	4–16	11
IF	60%	8–256	90
EL	56%	640–5120	1795
4. EBV			
IF	92%	16–512	90

Ab = antibodies; GMT = geometric mean titres; CMV = cytomegalovirus; HSV = herpes simplex virus; VZV = varicella-zoster virus; EBV = Epstein-Barr virus; CF = complement fixation; IF = immunofluorescence; EL = Elisa virus, varicella-zoster virus and Epstein-Barr virus, obtained by each test are shown in Table I.

Quantifiable IgM antibodies to cytomegalovirus were detected in five (10%) patients by immunofluorescence and Elisa; IgM antibodies to herpes simplex virus in two (4%) by immunofluorescence and in three (6%) patients by Elisa. One patient (2%) showed both anti-cytomegalovirus and anti-herpes simplex virus IgM antibodies. IgM and anti-varicella-zoster virus antibodies were not found.

Seroconversion was ascertained in three (12%) patients for cytomegalovirus, two (8%) for herpes simplex virus, one (4%) for varicella-zoster virus. Antibody changes were simultaneous and of the same order of magnitude by complement fixation, immunofluorescence and Elisa tests. All patients with active infection were asymptomatic.

Discussion

This study confirms, as observed in other surveys, the high prevalence of herpes virus infections in haemodialysis patients [3–6]. By more sensitive tests, Elisa and immunofluorescence, inactive infection was diagnosed in 88–96 per cent of patients for cytomegalovirus, 94–100 per cent for herpes simplex virus, 56–60 per cent for varicella-zoster virus and 92 per cent for Epstein-Barr virus. Complement fixation test showed lower sensitivity and failed to detect an appreciable number of sera with antibody detectable by Elisa and immunofluorescence.

High anti-cytomegalovirus and anti-herpes simplex virus titres were obtained by Elisa. In animal models, it has been found that persistent antigenic stimulation is necessary for maintenance of high antibody titres to herpes simplex virus [10]. The high antibody titres detected in our patients may actually be a reflection of antigenic load at the time of the study.

IgM anti-cytomegalovirus and/or anti-herpes simplex virus antibodies were detected in 16–18 per cent of patients. Generally, measurable titres of IgM antibodies indicate active or recent immunological stimulus by a specific agent and offer an indication of recent infection. Active infection, defined as significant increase in antibody titres, was detected in 12 per cent of patients for cytomegalovirus, eight per cent for herpes simplex virus and four per cent for varicella-zoster virus. No overt infectious disease was found in these patients.

Our data suggest a continuous antigenic stimulation by herpes virus in our haemodialysis patients. Active infection from herpes virus after renal transplantation almost always seems to be a secondary infection. Prospective serological and epidemiological investigations should be undertaken to determine the extent to which herpes viruses contribute to pneumonitis, pericarditis, unexplained febrile illnesses and Hbs Ag negative hepatitis among haemodialysis patients.

References

1 Pass RF, Long WK, Whitley RJ et al. *J Infect Dis 1978; 137:* 556
2 Simmons RL, Lopez C, Balfour H et al. *Ann Surg 1974; 180:* 623

3 Betts RF, Cestero EVM, Freeman RB et al. *J Med Virol 1979; 4:* 89
4 Ikram H. *Clin Nephrol 1981; 15:* 1
5 Cappel R, Hestermans O, Toussaint Ch et al. *Archs Virol 1978; 56:* 149
6 Naraqi S, Jackson GG, Jonasson O et al. *J Infect Dis 1977; 136:* 531
7 Pabico RC, Hanshaw JB, Yakub YN et al. *Proc Dial Transplant Forum 1971; 1:* 117
8 Pavlica P, Viglietta G. *Radiol Med 1980; 66:* 813
9 Corey L, Stamm WE, Feorino PM et al. *N Engl J Med 1975; 293:* 1273
10 Nahmias AJ, Dowdle WR, Kramer JH. *J Immunol 1969; 102:* 956

Proc EDTA–ERA (1984) Vol 21

NITRATE INDUCED ANAEMIA IN HOME DIALYSIS PATIENTS

M Salvadori, F Martinelli, L Comparini, S Bandini, A Sodi

Department of Nephrology, USL, Florence, Italy

Summary

Many home dialysis patients in Florence and the surrounding area suddenly showed an unusual anaemia. All used a softener for water treatment. They demonstrated methaemoglobinaemia, Heinz bodies and reduction in plasma haptoglobin indicating Hb oxidation. Tap water analysis showed excessive nitrates. The substitution of the softeners with deionisers solved this important and unusual clinical problem.

Introduction

A deterioration in the anaemia of home dialysis patients is a well known problem often due to the pollution of dialysis water. Chloramines seem to be the toxic compound most frequently found [1,2]; however, cases of severe anaemia due to nitrites or nitrates are well known [3,4]. In particular these last compounds are frequently encountered in rural areas and in home dialysis patients using well water for the dialysis treatment.

We report here an outbreak of anaemia in home dialysis patients in a large city and its surrounding area, probably due to an excessive nitrate concentration in the tap water.

Materials and methods

The present study concerns 40 home dialysis patients: 14 using deionisers and carbon filters for water treatment, and 26 using softeners and carbon filters.

All patients had been on home dialysis treatment for between 8 and 48 months. All the patients were previously well with varying degrees of anaemia in relation to their renal failure.

In all patients blood counts, serum iron, serum haptoglobin, methaemoglobin (by spectroscopic method), and erythrocyte Heinz bodies (in blood smears)

were regularly measured. In the dialysis water we repeatedly checked: chlorine, chloramines (by DPD colorimetric method), nitrites, nitrates, and trace elements such as copper, lead, and zinc (by atomic absorption). Nitrates were estimated by U.V. spectrometric examination at 210nm.

Results

In the summer of 1983 during a serious water shortage we observed an important and significant deterioration in the anaemia of 26 patients on regular home dialysis in Florence and surroundings. All patients used softeners for water treatment. We did not observe any change in the anaemia of 14 patients who used deionisers (Figure 1). Together with the worsening of the anaemia we observed

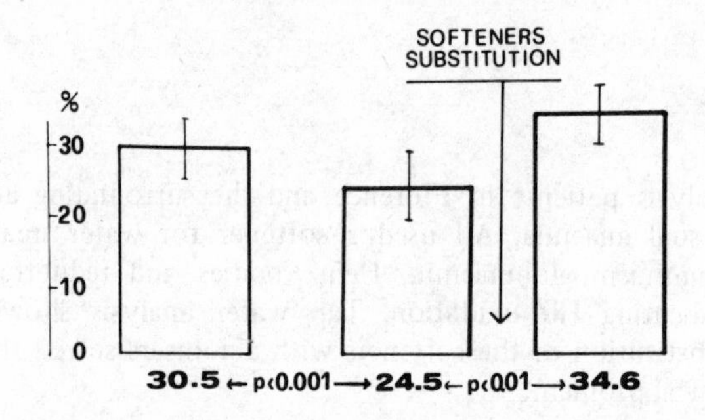

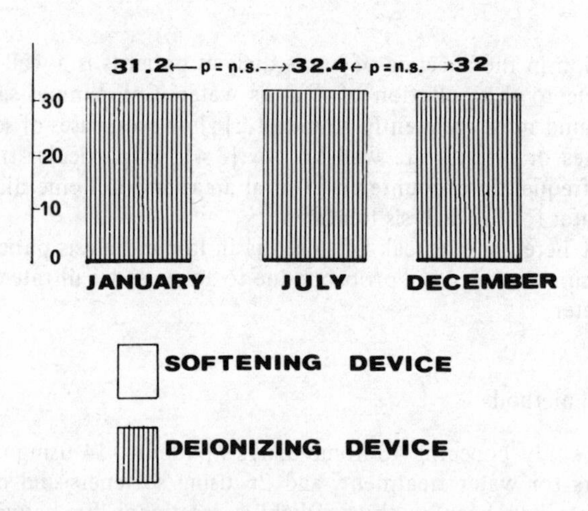

Figure 1. Mean haematocrit values during 1983

the presence of methaemoglobin in the blood, of Heinz bodies in the erythrocytes and a significant decrease of haptoglobin (Figure 2).

Water analysis repeatedly showed no chlorine, chloramines or toxic trace

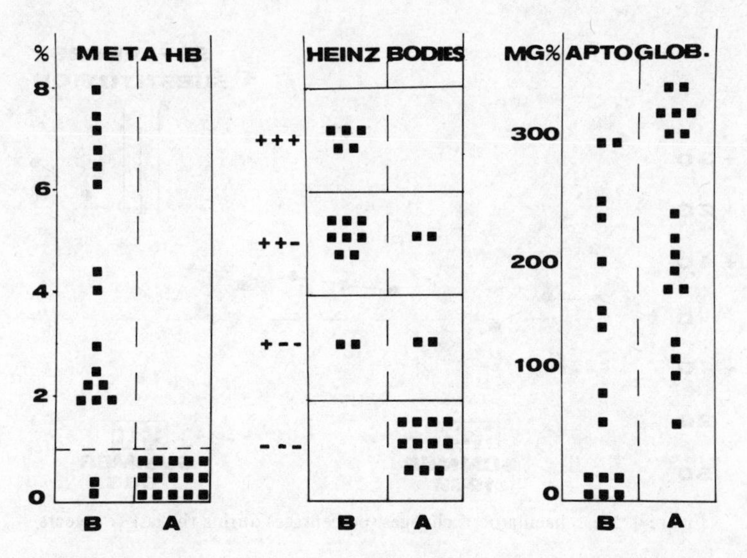

Figure 2. Methaemoglobin, Heinz bodies and haptoglobin before and after substitution of the softeners

elements but in many cases we observed a nitrate concentration exceeding the maximum safe standards proposed by AAMI [5,6].

Nitrate concentration in the water was most severe in rural areas and in those parts of the city where well water was added to tap water, usually during dry spells (Figure 3).

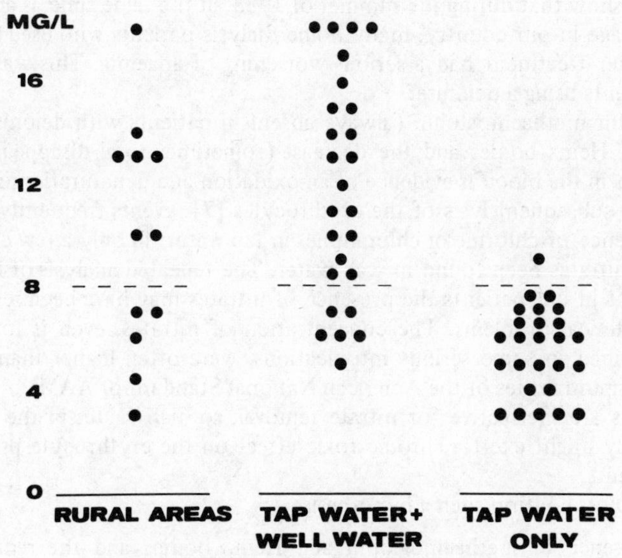

Figure 3. Concentration of nitrates (NO_3) in water

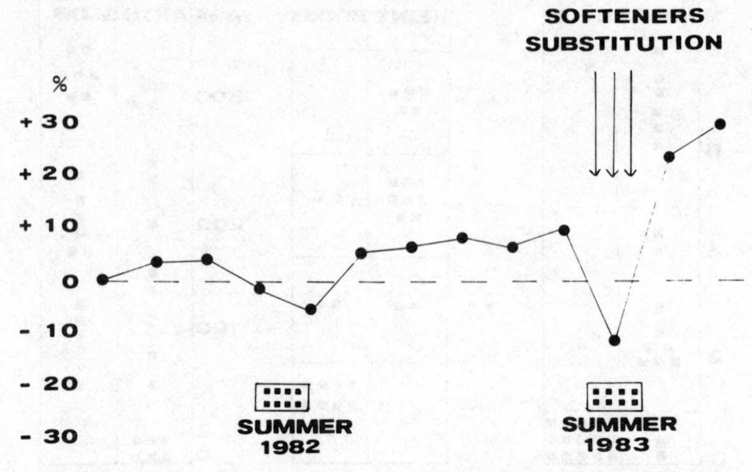

Figure 4. Mean haematocrit changes (percentage) during the last two years

The substitution of the softeners with deionisers solved the clinical problem with a sudden increase of haematocrit, the disappearance of methaemoglobin and Heinz bodies and with a return to normal of the haptoglobin values (Figures 1 and 2). Further, we observed haematocrit values more than exceeding the mean values previously observed in the same patients (Figure 4).

Discussion

Our results show that during the summer of 1983, at the same time as an unusual water shortage in our country, many home dialysis patients who used softeners for the water treatment had a serious worsening of anaemia. This was not the case in patients using a deioniser.

The serum methaemoglobin (always absent in patients with deionisers), the presence of Heinz bodies and the decrease (sometimes total disappearance) of haptoglobin in the blood is evidence of an oxidation and denaturation of haemoglobin with subsequent lysis of the erythrocytes [7], events frequently ascribed to the presence of chlorine or chloramines in tap water: in only a few cases have nitrites or nitrates been found in well water. The repeated analysis of tap water suggests that in our patients the presence of nitrates may have been responsible for these clinical problems. The concentration of nitrates, even if lower than those described in some serious intoxications, were often higher than the suggested maximum values of the American National Standard of AAMI.

Softeners are ineffective for nitrate removal, so high values in the dialysate water supply might exert a chronic toxic effect on the erythrocyte production of our patients.

Three points confirm such a hypothesis:

1. The presence of methaemoglobin and Heinz bodies and the reduction in plasma haptoglobin were found only in patients with softeners. No abnormal

finding was present in patients with deionisers. The abnormal amounts of methaemoglobin, even if not very high, are probably due to two main causes. First our blood samples always refer to the interdialytic period when most of the haemoglobin will have been reduced by enzymatic systems [8] ; second, we think that in most of our patients a chronic rather than an acute illness, happened due to the abnormal, even if not very high, nitrate values.

2. The substitution of the softeners suddenly solved the clinical problems. The mean haematocrit of the patients reached values 30 per cent higher than the mean preceding value. In our opinion this means that some toxic substance not extracted by the softeners steadily produced a worsening of anaemia in our patients. Such compounds are in higher concentration during the summer but probably are always present in other months.

3. The water analyses show that nitrates, more than other substances, may have caused these problems.

We consistently found no chlorine, no chloramines, no toxic trace elements except for nitrates and, in two patients also nitrites. We found higher nitrate concentrations in rural areas, where they are probably due to the use of fertilisers. Dangerous nitrate values were also found in some areas of the city where, usually in dry times, some well water is added to tap water. In such instances we think that nitrates may have come from waste seepage into well water.

Conclusion

In our opinion three main points arise from these observations:

1. In dialysis great attention must be paid to water quality. Probably concentrations of solutes even slightly higher than those proposed by AAMI may be dangerous over long periods.
2. Water treatment is better with deionisers or reverse osmosis than with softeners.
3. The reduced quality of the environment causes water pollution so that the tap water of a large city may be dangerous for dialysis treatment.

References

1 Kjellstrand EA, Eaton JW, Yawata Y et al. *Nephron 1974; 13:* 427
2 Eaton JW, Kolpin CF, Swofford HS et al. *Science 1973; 181:* 463
3 Carlson DJ, Shapiro FL. *Ann Int Med 1970; 73:* 757
4 Compty CM, Shapiro FL. In Drukker W, Parsons FM, Maher JF, eds. *Replacement of Renal Function by Dialysis.* The Hague: Nijhoff. 1983: 142
5 Bok DV, Grondon GD. *Contemp Dial 1980; October:* 28
6 Compty C, Luehmann D, Wathen R et al. *Trans ASAIO 1974; 20:* 189
7 Wintrobe MM et al, eds. In *Clinical Haematology.* Philadelphia: Lea & Febiger. 1981: 1012
8 Wintrobe MM et al, eds. In *Clinical Haematology.* Philadelphia: Lea & Febiger. 1981: 97

NITROGEN BALANCE: AN ADEQUATE TREATMENT PRESCRIPTION INDEX FOR CHILDREN ON HAEMOFILTRATION

M Giani, M Picca, A Saccaggi, *R Galato, A Bettinelli, C Funari, †R Rusconi, L Ghio, A Edefonti

*Clinica Pediatrica II, Università di Milano, *Ospedale C'Granda, Milano, †Clinica Pediatrica I, Università di Milano, Milan, Italy*

Summary

We studied 12 children (10.6±2.5 years; 21±4.8kg body weight) with end-stage renal disease on chronic haemofiltration prescribed using a urea kinetic model. After 12 months, urea generation rate, protein catabolic rate and nitrogen balance were assessed in each patient. Pre-treatment BUN was adequate in all 12 patients but in three nitrogen balance and growth rate were unsatisfactory. These three children were studied for a second year in which plasma water cleared/week was increased; during this year a significant increase in nitrogen balance and growth rate was obtained.

Introduction

One of the goals in the management of patients with end-stage renal disease (ESRD) undergoing haemodialysis is to tailor their dialysis requirements [1]. Recently, low pre-treatment BUN values, obtained using a urea kinetic model, have been reported as an adequate index of dialysis prescription [1,2]. However, low BUN may be due to malnutrition and has been associated with increased morbidity and mortality [1,3]. Therefore, an index is needed which assesses not only treatment effectiveness but also adequacy of dietary protein intake so that the risks associated with malnutrition are avoided. This could be useful in children with ESRD with growth retardation [4] since malnutrition is recognised as a factor in the pathogenesis of growth failure [5].

This study investigates whether nitrogen balance, recently reported as a good nutritional index [5,6], may be also considered as an index of adequate dialysis prescription in children with ESRD.

Materials and methods

Twelve children (10.6±2.5 years; 21.4±4.8kg body weight) treated by haemofiltration were studied. All the children were pre-pubertal without obvious signs of osteodystrophy at the time of the study. The residual kidney function was 0.8±0.7ml/min/1.73m^2. Haemofiltration was prescribed according to a urea kinetic model to obtain pre-treatment BUN <100±20mg/dl.

Caloric and protein intake were assessed from dietary records obtained, during 10-day periods, every two months. All the children were on unrestricted diet. Protein catabolic rate was computed from the urea kinetic analysis. Nitrogen balance was obtained from the difference between dietary protein intake and protein catabolic rate.

Three of 12 children (12±1 years; 24.7±2.4kg body weight) showed adequate pre-treatment BUN but unsatisfactory nitrogen balance and growth rate during the first year of the study. These children were followed for another year during which the treatment prescription was increased.

The plasma water cleared/week (the sum of residual kidney function and ultrafiltrate amount) was considered as an index of dialysis efficacy.

Mid-arm muscle circumference was considered as an index of muscle mass. Growth rate was evaluated every three months in all three children.

We also recorded the number of hospitalisations per patient during the two study periods.

Results

Table I shows the mean values of dietary protein intake, BUN, urea generation rate, protein catabolic rate and nitrogen balance obtained in our patients during the first year. A highly satisfactory nitrogen balance was obtained in nine patients, whereas in the other three it was unsatisfactory.

TABLE I. Dietary protein intake, BUN, urea generation rate, protein catabolic rate and nitrogen balance in 12 patients during one year on haemofiltration (values expressed as mean ± SD)

	Dietary protein intake g/kg/day	BUN mg/dl	Urea generation rate mg/min	Protein catabolic rate g/kg/day	Nitrogen balance mg/kg/day
9 patients	2.2 ±0.57	100±32	3.1±0.87	1.6 ±0.24	127±40
3 patients	1.63±0.49	100±25	3.1±0.75	1.36±0.75	34±25

In Table II the values of BUN, urea generation rate and protein catabolic rate obtained in three patients after increasing plasma water cleared/week from 8.6±0.7 to 10±0.5ml/min/1.73m^2 are shown. During the second year of the study a further decrease of all parameters considered was obtained.

Nitrogen balance increased (Table III) in all three patients from a mean of 34mg/kg/day to a mean of 174mg/kg/day. Mid-arm muscle circumference also increased from 14.04±0.5 to 15.13±0.37cm as well as growth rate (from 3.1± 0.7 to 4.1±0.2cm/year).

TABLE II. Plasma water cleared/week, BUN, urea generation rate and protein catabolic rate in three patients during two years on haemofiltration (values expressed as mean ± SD)

	Plasma water cleared/week ml/min/1.73²	BUN mg/dl	Urea generation rate mg/min	Protein catabolic rate g/kg/day
Haemofiltration Year 1	8.6±0.7	100±25	3.1±0.75	1.36±0.32
Haemofiltration Year 2	10±0.5	90±19	2.6±0.66	1.1 ±0.17

TABLE III. Nitrogen balance, mid-arm muscle circumference and growth rate in three patients during two years on haemofiltration (values expressed as mean ± SD)

	Nitrogen balance mg/kg/day	Mid-arm muscle circumference cm	Growth rate cm/year
Haemofiltration Year 1	34±25	14.04±0.5	3.1±0.7
Haemofiltration Year 2	174±54	15.13±0.37	4.1±0.2

TABLE IV. Dietary protein and caloric intake in three patients during two years on haemofiltration (values expressed as mean ± SD)

	Dietary protein g/kg/day	Recommended daily intake %	Calorie intake Kcal/kg/day	Recommended daily intake %
Haemofiltration Year 1	1.36±0.49	143±35	64±8	84±5
Haemofiltration Year 2	1.96±0.64	170±43	70±8	92±7

Table IV reports the mean values of dietary protein intake and caloric intake for the three patients during the two periods of the study and the percentage recommended daily intake [7]. In particular, protein intake was always over 100 per cent of the recommended values; during the second year of the study a further increase of protein and caloric intake was observed. The number of hospitalisations due to infections or congestive heart failure decreased from 1.33 episodes/patient during year one to 0.33 episodes/patient in year two.

Conclusions

Three patients with an apparently adequate haemofiltration prescription had an unsatisfactory nitrogen balance, an index of nitrogen available for protein synthesis. In fact, young children with no renal disease and with adequate dietary intake have been reported to have a nitrogen balance of 70mg/kg/day [8]. The improved dialysis efficacy, obtained by increasing the amount of plasma water cleared/week, decreased the protein catabolic rate and increased the nitrogen balance.

At the same time a better nutritional state was confirmed by the increase in mid-arm muscle circumference and growth rate.

It should be noted that the protein intake in our children was always over 100 per cent of the recommended daily intake values. Such intakes appeared necessary to obtain a satisfactory nitrogen balance in children with ESRD who had a higher catabolism than normal subjects [9,10]. These high protein intakes did not cause increased uraemic toxicity if BUN were controlled by the treatment prescription.

The decreased number of hospitalisations during the second study period may be considered the result of a more adequate treatment prescription [1].

In conclusion, our data show that haemofiltration prescriptions tailored only to obtain a BUN target of under 100±20mg/dl may be insufficient in paediatric patients. A satisfactory nitrogen balance, that is an indication of both adequate protein intake and treatment effectiveness, appears to be a more reliable index.

References

1 Lowrie EG, Laird NM, Parker TF et al. *N Engl J Med 1981; 305:* 1176
2 Parnell S, Sawyer RF, Meister MJ et al. *Dial Transplant 1981; 10:* 288
3 Degoulet P, Legrain M, Réach I et al. *Nephron 1982; 31:* 103
4 Harmon WE, Spinozzi N, Meyer A et al. *Dial Transplant 1981; 10:* 324
5 Grupe W, Harmon WE, Spinozzi N. *Kidney Int 1983; 24:* 15, S-6
6 Giani M, Picca M, Saccaggi A et al. *Proc EDTA 1983; 20:* 201
7 Wassner SJ. *Ped Clin N Am 1982; 29:* 973
8 Iyengar AK, Narasinga Rao BS. *Am J Clin Nutr 1982; 35:* 733
9 El Bishti M, Burke J, Gill D et al. *Clin Nephrol 1981; 15:* 53
10 Schoenfeld PY, Henry RR, Laird NM et al. *Kidney Int 1983; 23:* 13, S-80

Proc EDTA–ERA (1984) Vol 21

VISUAL FUNCTION, BLOOD PRESSURE AND BLOOD GLUCOSE IN DIABETIC PATIENTS UNDERGOING CONTINUOUS AMBULATORY PERITONEAL DIALYSIS

J Rottembourg, *P Bellio, K Maiga, M Remaoun, *F Rousselie, M Legrain

*Hôpital de la Pitie, Paris, *Hôpital Bichat, Paris, France*

Summary

Over the last five years, 46 insulin dependent diabetic patients (mean age 52 ± 13 years) have been treated by continuous ambulatory peritoneal dialysis (CAPD). Fourteen patients have been on treatment for more than two years. Visual acuity assessed every six months showed that improvement has been observed in 14 eyes (19%), stabilisation in 36 eyes (46%), worsening in 17 eyes (21%), five eyes had a minimal function during the entire follow-up. Systolic blood pressure decreased from 173 ± 42mmHg at start of dialysis to 149 ± 30 and 146 ± 32 after one and two years. Mean fasting and post-prandial blood glucose assessed monthly in 36 patients treated with four daily intraperitoneal injections of insulin (660 determinations) were respectively 7.5 ± 3.5 and 8.5 ± 3.5mmol/L.

Introduction

Continuous ambulatory peritoneal dialysis (CAPD) is now considered as a satisfactory dialysis method for insulin dependent diabetic patients with end-stage renal disease. Encouraging results drawn from a rather large series were reported by Amair [1] and our group [2]. This paper is restricted to the analysis of the evolution of the following three important clinical parameters in insulin-dependent diabetic patients on CAPD treatment: visual acuity, blood pressure and blood glucose control.

Patients and methods

Patients

Between August 1978 and December 1983, 51 insulin-dependent diabetic patients were selected for treatment by CAPD. The CAPD technique was the first dialysis choice in 46 patients and five patients were transferred from haemodialysis, one because of ocular lesions, three because of cardiovascular problems and one because of access difficulties. Six patients, mainly because of visual problems, were changed early to the so-called Continuous Cyclic Peritoneal Dialysis (CCPD). The data of the six patients on CCPD were included in the CAPD series. As of December 1983, the cumulative duration of treatment was 65.6 patient years with an average time per patient of 17.1±11 months (range 1 to 38 months). Fourteen patients had been treated for at least two years. At the beginning of treatment there were 27 males and 24 females whose mean age was 52.3 ± 13.5 years. The mean age when diabetes was discovered was 29.2 ± 15.2 years. The mean delay between discovery of diabetes and start of dialysis was 23 ± 17.6 years. All patients were treated with insulin when dialysis was started.

Extra-renal complications at the start of CAPD included hypertension in 49 cases (94%), proliferative retinopathy in 50 cases (5 patients were totally blind), a previous history of myocardial infarction in 26 cases (41%), severe peripheral vascular disease in 30 (50%) and clinical peripheral neuropathy in 38 cases (75%).

Methods

Technical details have been described elsewhere [3]. CAPD was carried out in most cases with four exchanges daily. Patients and their relatives were trained to measure the blood sugar with the finger prick technique on the glucometer Ames®. At home twice daily measurements were performed routinely. Once every two weeks, patients were asked to perform six serial measurements. In 36 patients on CAPD, insulin was administered four times daily exclusively through the peritoneal route by a special injection port in the line of the bag [4]. Four patients on CAPD treated at the early phase of this series were receiving insulin subcutaneously. The six patients on CCPD were treated with three daily subcutaneous injections of insulin. Frusemide was given to 32 patients during the first year to maintain residual renal function [2].

Results

Five patients were unable to handle the CAPD technique and were transferred early to haemodialysis. On peritoneal dialysis the overall patient survival was 85 per cent, 65 per cent and 57 per cent at one, two and three years respectively. The CAPD technique success rate was 88 per cent after one year and 78 per cent after two years in the younger age group (19 patients under 50 years old, mean age 39.1 ± 8 years) while it was only 70 per cent and 50 per cent in the older age group (27 patients over 50 years old, mean age 61.5 ± 8 years).

Visual status

Visual status was assessed at the start of treatment and every six months during treatment in 46 patients. At each ophthalmic visit the best corrected visual acuity for each eye was determined, a thorough ophthalmic examination completed and retinopathy documented by a multiple field stereophotograph (one every year). Visual acuity was evaluated in five functional categories: 1) reading visual acuity 20/10 to 20/50; 2) impaired visual acuity 20/70 to 20/100; 3) ambulatory visual acuity 10/160 to counting fingers; 4) minimal visual function, hand motions to light perceptions; 5) no visual function, no light perception. All patients, except one, presented a proliferative retinopathy, 64 per cent in stages 1 and 2 and 36 per cent in stage 3. The evolution of visual acuity is summarised in Table I. The results are as follows: in the younger age group

TABLE I. Visual acuity in 46 insulin dependent diabetic patients treated by CAPD

Group	Visual acuity	Baseline		Final	
		N	%	N	%
1	20/10–20/50	30	34	38	42
2	20/70–20/100	26	29	21	23
3	20/160–counting fingers	22	24	16	17
4	light perception	7	8	11	13
5	totally blind	5	5	5	5
	Total	92	100	92	100

(19 patients), 16 eyes were in group 1 and 22 eyes in group 2. During treatment and at the final examination, visual acuity improved in 10 eyes (26%), stabilised in 21 (55%) and worsened in seven (19%). In the older age group (over 50 years old) 40 eyes could be studied over a year: 26 eyes were in groups 2 to 5 and only 14 in group 1. Visual acuity improved in only four eyes (10%), stabilised in 26 (65%) and worsened in 10 (25%).

Visual improvement was due to spontaneous resorption of vitreous haemorrhage and the beneficial effect of photocoagulation and vitrectomy. Deterioration of visual function occurred in 17 eyes due to multiple events, sometimes interrelated: recurrent vitreous haemorrhage (5 eyes), cataract (6 eyes), macular degeneration (3 eyes) and anterior segment neovascularisation (3 eyes).

Blood pressure control

At the start 43 patients (94%) were hypertensive and receiving drugs including diuretics (36 patients), beta blockers (25 patients), central sympathicoplegics (29 patients), vasodilators (18 patients) and converting enzyme inhibitors (5 patients). Mean supine blood pressure was 173 ± 42/90 ± 35mmHg. After one year of dialysis 17 of 31 patients (55%) were normotensive without any treatment (if one excludes frusemide), 10 were normotensive on treatment and four remained hypertensive (systolic 169 ± 36mmHg and diastolic 109 ± 21mmHg). After two years 11 of 14 patients were normotensive without antihypertensive drugs: mean supine blood pressure was 146 ± 32/144 ± 28mmHg and erect 144 ± 28/86 ± 18mmHg. In eight patients with major neurological disorders, severe postural hypotension was observed induced mainly by rapid ultrafiltration following the use of the high glucose concentration solutions (4.2 to 4.5g/L).

Blood glucose control

In the 36 patients treated by CAPD and using intraperitoneal administration of regular insulin the mean total daily dose of insulin was 86 ± 32IU. In this series the 660 monthly fasting and post-prandial blood glucose determinations performed in the out-patient clinic gave respectively the following mean results: 7.5 ± 3.5mmol/L and 8.5 ± 3.5mmol/L. The mean results of 132 serial determinations of blood glucose levels obtained in 21 patients were 7.7 ± 3.2mmol/L at 8 a.m., 8.5 ± 3.5mmol/L at 10 a.m., 7.1 ± 3.0mmol/L at noon, 8.5 ± 3.5mmol/L at 2 p.m., 7.3 ± 2.8mmol/L at 4 p.m., 7.3 ± 3.2mmol/L at 6 p.m., 8.0 ± 3.4mmol/L at 8 p.m. The mean serum glycosylated haemoglobin HBA1c measured during the last year in 22 patients treated by CAPD for at least one year was 8.7 ± 1.3%.

Discussion

Over the past five years it has been shown that CAPD can offer excellent control of both diabetes and uraemia to insulin and non-insulin diabetics with end-stage renal disease [1–4]. For these reasons, since 1978, many institutions have selected CAPD as the first choice method of home dialysis for diabetic patients [5]. Despite some serious complications [6], the technique success rate obtained with CAPD can compete very favourably with those obtained in a diabetic population of the same mean age treated by haemodialysis [7], intermittent peritoneal dialysis [8] and even transplantation if one accepts the excellent results obtained with a living related donor [9]. The drop-out rate (including deaths and transfers) and morbidity are largely influenced by age, the major risk factor.

Good control of blood pressure is commonly observed and in our series the proportion of normotensive patients taking no anti-hypertensive drugs reaches 79 per cent after two years. However the incidence of symptomatic hypotension in few patients should be emphasised.

Rapid deterioration of visual function is one of the major threats to diabetic uraemic patients who undergo maintenance haemodialysis therapy [10]. Until now little data on the progression of visual morbidity among diabetic patients treated by CAPD has been available [1,2,10]. Our results are encouraging and either improvement or stabilisation are observed in about 75 per cent of the patients less than 50 years old. In the older age group improvement is seldom observed but stabilisation can be obtained. Because of the absence of follow-up studies on a large and comparable group, over a sufficient period of time, valid evaluation of the effect of CAPD on visual function, compared to the results obtained by haemodialysis or after transplantation, is too early. We agree with others [10] that visual loss relates to the stage of retinopathy rather than the duration of dialysis even on CAPD. Specialised ophthalmic care, including scatter photocoagulation and vitrectomy, are part of efficient treatment.

The excellent control of blood glucose achieved by the intraperitoneal administration of insulin first claimed by Flynn [4] is confirmed by our own results. The injection of insulin either into the dialysate bag or into the line [3] is favoured by most patients. However the average total daily dose of insulin is commonly two to three times the pre-CAPD dose partly because of adsorption of insulin by the material and losses in the dialysate effluent. In our experience, as in other series [1–3], the use of the intraperitoneal route does not significantly increase the rate of peritonitis.

CAPD should be one choice in an integrated programme including all types of dialysis and transplantation for insulin dependent diabetic patients. It offers many patients, including the older age group, an adequate control of diabetes, uraemia and hypertension.

References

1 Amair P, Khanna R, Lebel B et al. *N Engl J Med 1982; 306:* 625
2 Legrain M, Rottembourg J, Bentchikou A et al. *Clin Nephrol 1984; 21:* 72
3 Rottembourg J, El Shahat Y, Agrafiotis A et al. *Kidney Int 1983; 23:* 40
4 Flynn CT. In Friedman EA, l'Esperance FA, eds. *Diabetic Renal Retinal Syndrome 2.* New York: Grune & Stratton. 1982: 321–330
5 Jacobs C, Brunner F, Brynger M et al. *Diabetic Nephropathy 1983; 2:* 11
6 Rottembourg J, Gahl GM, Poignet JL et al. *Proc EDTA 1983; 20:* 236
7 Kjellstrand C, Whitley K, Comty C et al. *Diabetic Nephropathy 1983; 2:* 5
8 Slingeneyer A, Mion C, Selam JL. In Gahl GM, Kessel M, Nolph KD, eds. *Advances in Peritoneal Dialysis.* Amsterdam: Excerpta Medica. 1981: 378–388
9 Sutherland EDR, Fryd DS, Peters C et al. In Friedman EA, l'Esperance FA, eds. *Diabetic Renal Retinal Syndrome 2.* New York: Grune & Stratton. 1982: 373–383
10 Ramsay RC, Cantril L, Knolbloch WH et al. *Diabetic Nephropathy 1983; 2:* 30

Proc EDTA–ERA (1984) Vol 21

COMPUTERISED PROGRAM TO COMPARE TOLERANCE OF DIALYSIS PATIENTS TO DIFFERENT DIALYSIS SCHEDULES

F Lonati, B Pea, **A Castellani**

Umberto I Hospital, Brescia, Italy

Summary

In the attempt to compare rapidly and easily the tolerance of dialysis patients to different dialysis schedules, we developed a computerised program for use with a personal computer.

As a first application of this program we analysed the effects of long-term substitution of bicarbonate for acetate in reducing dialysis hypotension.

Introduction

When the results of different therapeutic strategies vary slightly and are subject to variation with time or to multifactorial interference, it is necessary to perform long-term studies and to examine all aspects of the question. It is necessary to collect exact multiple data and to elaborate them frequently, rapidly, and without mistake. In such cases the use of the computer is no doubt necessary.

Personal computers have an adequate storage and computational power; however, the software for medical application is frequently inadequate for the task and/or expensive or too difficult to use. It is also important to reduce or to avoid the need to interact with the computer through an intermediary [1].

For these reasons we have developed a program applicable to a personal computer to study the value of different dialysis schedules in reducing dialysis intolerance. Initially we have applied it to the analysis of a still controversial question: the value of substituting bicarbonate for acetate to reduce dialysis-induced hypotension [2–6].

Material and methods

We used an Olivetti Personal Computer M20 ST 256 K Bytes RAM memory. The program is prepared both in English and Italian.

This program may be used also by people not expert in basic language, to compare the tolerance to two or more kinds of treatment in the same population or in different populations.

Once the prospective study is planned for each type of treatment or group of patients, the following data are collected and recorded for each dialysis:

patient name or marker
date
marker of type or phase of treatment
pre-dialysis and lowest observed systolic and diastolic blood pressure (BP)
pre-dialysis and post-dialysis body weight (BWT)
number of symptomatic hypotension episodes (SHY)
time of appearance of first SHY
other clinical incidents (cramps, headache, vomiting, extrasystoles, tachycardia)
saline and hypertonic sodium chloride bolus injected
run duration.

The data are subjected to automatic quality control and then compacted and recorded.

These data, and the ones mathematically deducted by the computer (mean BP; mean Δ BP; Δ systolic BP; Δ diastolic BP; Δ BWT; Δ BWT % dry BWT per run; Δ BWT % dry BWT/hour) can be quickly picked out for a single patient or for a chosen group, for a selected period or for the total period of each schedule.

In a few seconds the computer elaborates, according to the request, the following stages, not necessarily in sequence.

A) Graphically the trend in symptomatic hypotension together with hourly ultrafiltration % BWT; or delay of appearance of first symptomatic hypotension; or systolic and diastolic mean blood pressure absolute or as Δ % of pre-dialysis value; or other incidents; or saline and hypertonic NaCl bolus injected.

B) The numerical elaboration (mean, SD, % frequency) of each parameter and/or statistical comparison ('t' test, 't' test for paired data, Chi square test, F test) of the studied parameters for the different periods, populations or schedules.

C) The linear correlation among the different parameters with visualisation of the scattering of the points, the number of the examined pairs, the regression equation and r.

D) The changes of frequency or of intensity of a phenomenon by the use of CUSUM analysis: the swings from the overall mean in the chosen period and their importance and length are visualised [7,8].

Examples of the above applications are given in Figure 1, from A to D.

The patients

As a first application of our computerised program, we studied 16 patients aged 54 ± 16 (male 6; female 10), with frequent symptomatic hypotension episodes in a standard dialysis schedule (QB 220–300ml/min; QD 500ml/min; flat plate

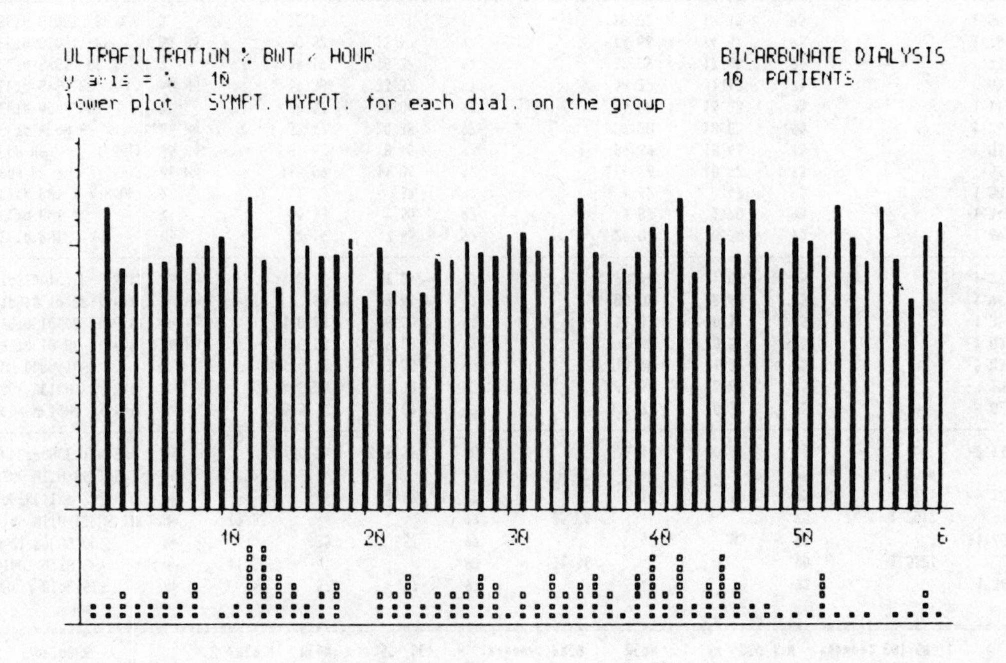

Figure 1A. Example of elaboration stage: for further explanation see section on 'Material and methods'

VARIABLE		C.G. ACETATE DIALYSIS PHASE .1. runs from 270782 to 280283 FREQ.	MEAN	ST. DEV.	N.	######	C.G. BICARBONATE DIAL. PHASE 1. runs from 020383 to 031083 FREQ.	MEAN	ST. DEV.	N.	######	CHI SQ.	t
SHY / DIALYSIS	Nr		.57	.77	97			.42	.70	95			1.36
DIAL. WITH SHY	Nr	41.24			97		31.58			95		1.5381	
NaCl BOLUS/DL.	Nr		.25	.57	97			.42	.85	95			-1.65
DIAL.WITH BOLUS INF.	Nr	17.53			97		25.26			95		1.2812	
SALINE INF./DIAL.	Nr		.14	.52	97			.16	.57	95			-.17
DIAL.WITH SALINE INF.	Nr	7.22			97		8.42			95		.0018	
APPEARANCE 1th SHY	min		150.25	49.35	40			176.17	50.22	30			-2.13
BODY WEIGHT (pre)	Hg		515.34	11.89	97			524.75	10.32	95			-5.82
BODY WEIGHT (post)	Hg		488.20	7.70	97			495.67	6.67	95			-7.15
RUN DURATION	min		194.43	33.50	97			215.79	21.27	95			-5.23
SYSTOLIC BP (pre)	mm Hg		170.00	17.28	97			172.68	18.51	95			-1.03
SYSTOLIC BP (post)	mm Hg		110.62	28.50	97			116.37	30.11	95			-1.35
DIASTOLIC BP (pre)	mm Hg		87.63	9.61	97			85.05	8.63	95			1.94
DIASTOLIC BP (post)	mm Hg		72.22	11.26	97			70.16	11.04	95			1.27
DELTA BWT	Hg		27.14	7.45	97			29.07	8.30	95			-1.69
DELTA BWT %	%		5.56	1.50	97			5.87	1.70	95			-1.34
DELTA BWT % / HOUR	%		1.72	.39	97			1.62	.43	95			1.58
MEAN BP (pre)	mm Hg		115.09	11.06	97			114.26	10.42	95			.53
MEAN BP (post)	mm Hg		81.62	18.45	97			84.39	18.46	95			-1.03
DELTA MEAN BP	mm Hg		33.47	20.38	97			29.88	18.85	95			1.26
DELTA MEAN BP	%		28.57	17.04	97			25.88	16.35	95			1.11
DELTA SYST.BP	mm Hg		59.38	33.37	97			56.32	31.62	95			.65
DELTA SYST. BP %	%		34.19	18.37	97			32.29	17.64	95			.73
DELTA DIAST. BP	mm Hg		20.52	15.57	97			16.66	14.67	95			1.76
DELTA DIAST. BP %	%		22.87	17.11	97			19.02	17.32	95			1.54

Figure 1B. Example of elaboration stage: for further explanation see section on 'Material and methods'

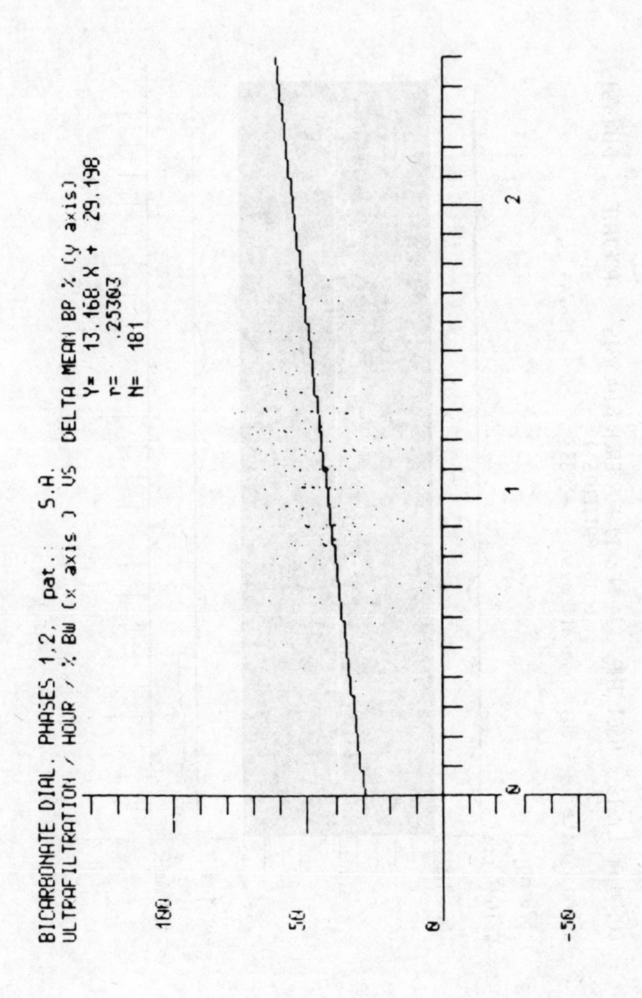

Figure 1C. Example of elaboration stage: for further explanation see section on 'Material and methods'

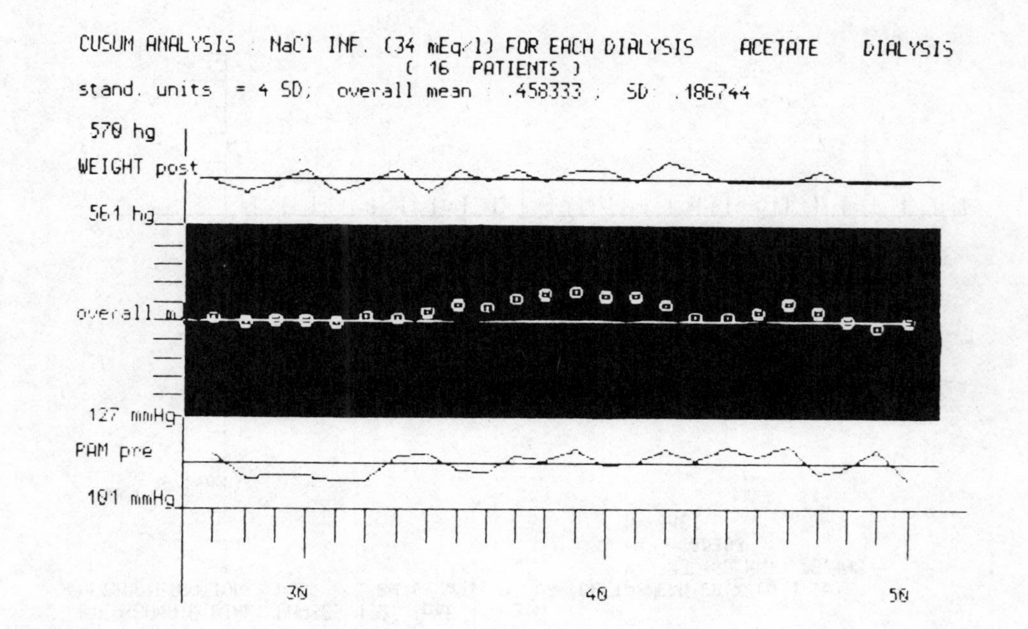

Figure 1D. Example of elaboration stage: for further explanation see section on 'Material and methods'

or hollow fibre dialysers with 0.9–1m^2 effective surface area, 140mmol/L dialysate sodium).

They were analysed in acetate dialysis (38–40mmol/L) for at least six months and then they were switched to bicarbonate dialysis for a mean of 13 months.

QB, QD, run duration, dialysers, drugs, dialysate sodium concentration were the same as for acetate dialysis.

Results

The patients were switched to bicarbonate dialysis only after observing the long-term stabilisation of the following data on acetate dialysis: percentage hypotensive dialyses; number of hypotensive episodes for each dialysis; number of NaCl bolus injected; time of appearance of first symptomatic hypotension; Δ systolic, Δ diastolic, Δ mean blood pressure.

In Figure 2 CUSUM analysis of the number of hypotensive episodes for each dialysis (A) and of Δ systolic blood pressure (C) are presented. The stabilisation in the last 50 acetate dialyses is evident.

When switched to bicarbonate dialysis, CUSUM analysis reveals a biphasic trend for all the above parameters: improvement in the first 15–20 dialyses, followed by a deterioration (Figure 2B and Figure 2D).

If we compare the total acetate dialyses (1,576) with the total bicarbonate dialyses (3,087) no significant differences are detectable for all the analysed parameters.

However, if we examine selected periods or single patients some differences appear.

Table I reports the statistical comparison for the whole population of the last 50 acetate dialyses with the first 20 bicarbonate dialyses, with the bicarbonate dialyses 21–50 and with the last 20 bicarbonate dialyses.

During the four examined periods the pre-dialysis blood pressure, the hourly ultrafiltration % body weight, the pre-dialysis body weight and the dry body weight do not significantly change.

It is noteworthy that the hypotension indices improve in the first 20 bicarbonate dialyses and then worsen. There is no difference among the stabilisation periods in the whole population.

When we studied patient by patient, however, we observed that in two out of 16 patients the improvement remained constant.

Discussion and conclusions

The use of a personal computer with appropriate software prepared by ourselves appears very useful to compare with accuracy the tolerance of a single patient or of a group of patients to different schedules of dialysis.

It is easy to evaluate the trends, and to have in any moment and in a few minutes for instance up-to-date results with statistical analysis.

It is possible to make objective evaluation of the 'stabilisation' and to demonstrate the value of the treatment variation in a patient or in a group of patients,

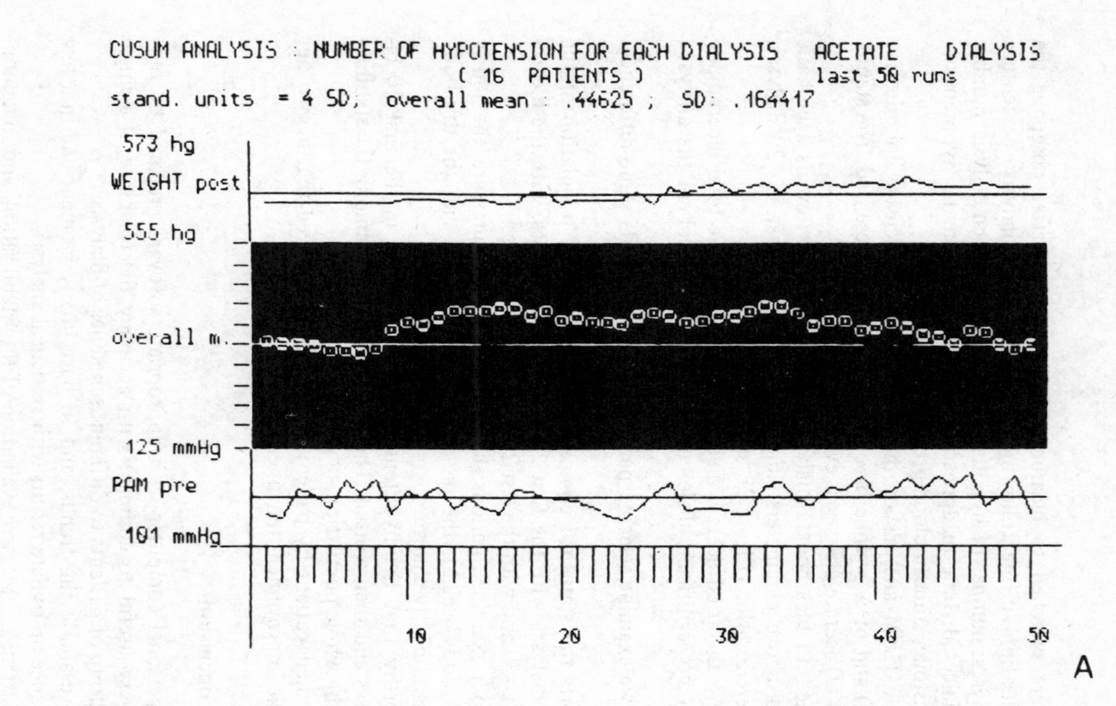

Figure 2. CUSUM analysis of the number of hypotensive episodes for each dialysis (A and B) and of Δ systolic blood pressure (C and D). The last 50 acetate dialyses are shown in A and C, the first 50 bicarbonate dialyses in B and D

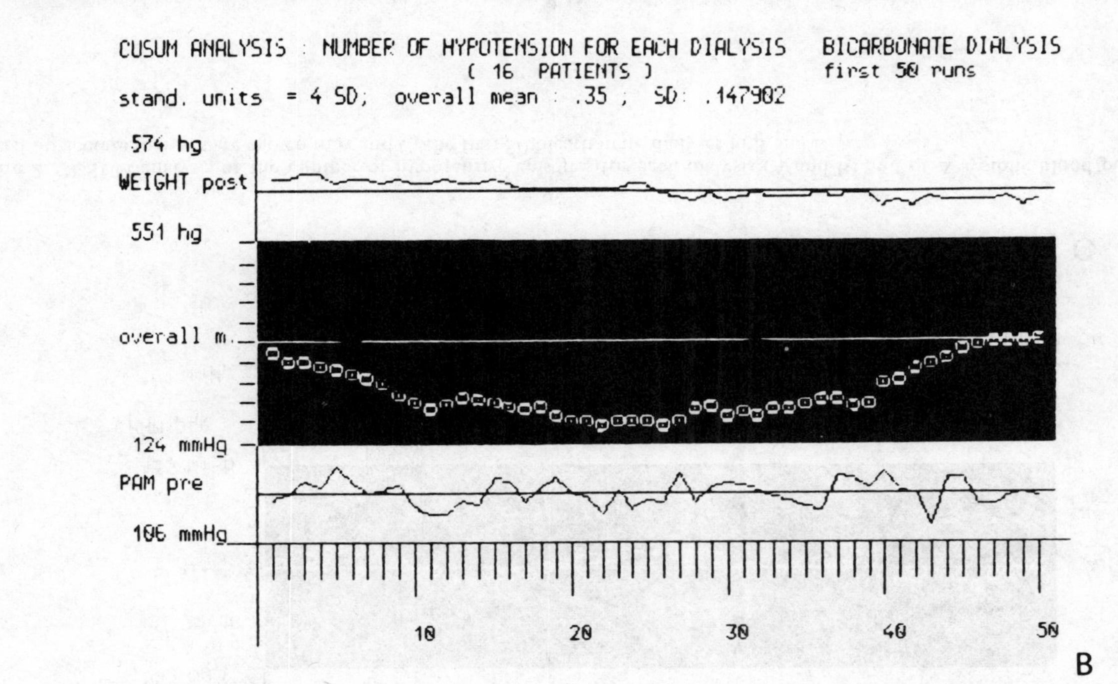

Figure 2. CUSUM analysis of the number of hypotensive episodes for each dialysis (A and B) and of Δ systolic blood pressure (C and D). The last 50 acetate dialyses are shown in A and C, the first 50 bicarbonate dialyses in B and D

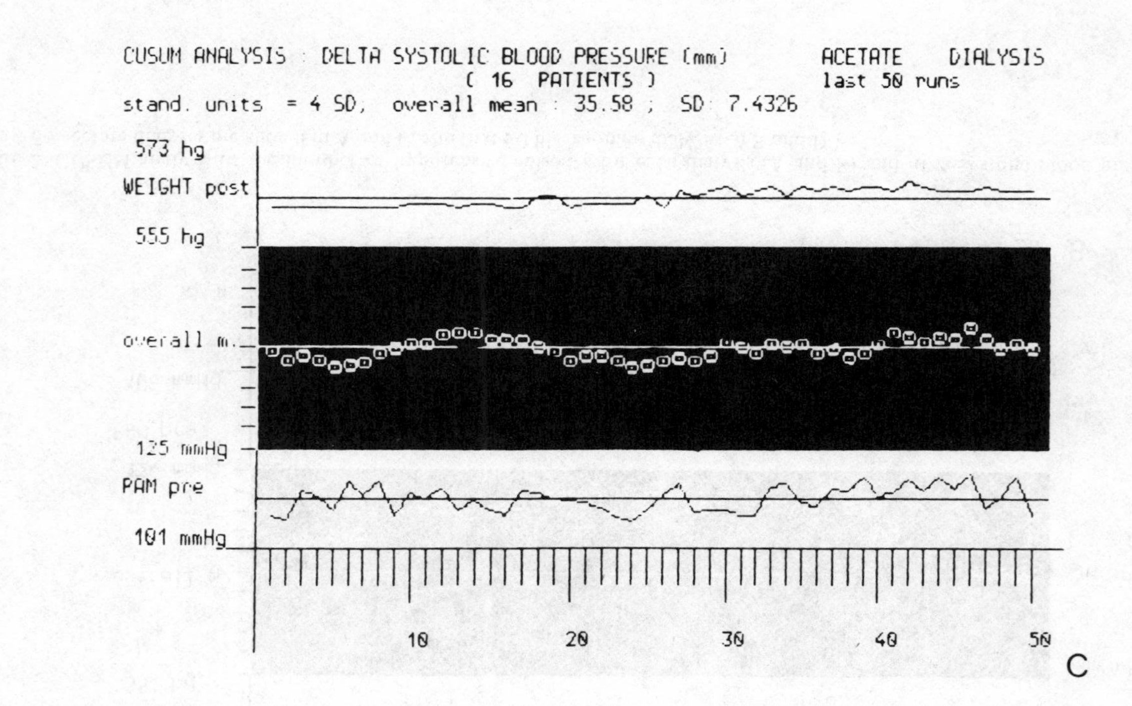

Figure 2. CUSUM analysis of the number of hypotensive episodes for each dialysis (A and B) and of Δ systolic blood pressure (C and D). The last 50 acetate dialyses are shown in A and C, the first 50 bicarbonate dialyses in B and D

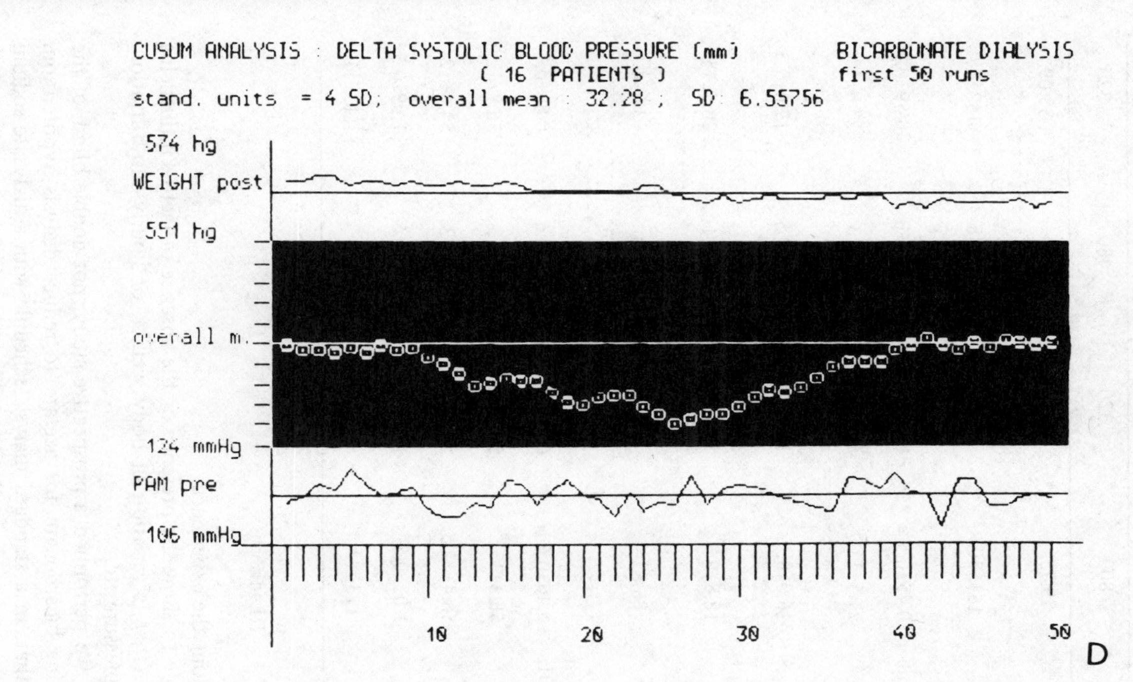

Figure 2. CUSUM analysis of the number of hypotensive episodes for each dialysis (A and B) and of Δ systolic blood pressure (C and D). The last 50 acetate dialyses are shown in A and C, the first 50 bicarbonate dialyses in B and D

TABLE I. Comparison of the last 50 acetate dialyses with the first 20 bicarbonate dialyses, with the bicarbonate dialyses 21–50 and with the last 20 bicarbonate dialyses for the whole population

	ACETATE		BICARBONATE				
	last 50 dialyses		first 20 dialyses		21–50 dialyses		last 20 dialyses
DIALYSIS NR	800		320		480		320
Hypotensive DL % frequency	29.75	***	18.44	*	24.79	*	33.09
			+-------		***-------		-------+
Time of appearance 1st SHY min	161±30	*	177±31	**	155±38		162±37
Hypotension/DL number	0.45±0.5	***	0.26±0.3		0.41±0.6		0.59±0.7
Δ Systolic BP %	21.5±11	*	17.6±10	*	20.6±13		23.8±13
	+-------		**-------		-------+		
Δ Diastolic BP %	16.0±8.1	*	12.6±7.7		15.3±9.5		17.0±9.3
			+-------		**-------		-------+
Δ Mean BP %	18.8±9.3	*	15.2±8.8	*	18.0±11		20.5±11
			+-------		**-------		-------+
DL with NaCl bolus % frequency	26.62	**	18.12	***	30.00		34.34
			+-------		***-------		-------+
	+-------		-------	*-------	-------		-------+
NaCl bolus number/DL	0.46±0.6		0.27±0.4		0.48±0.6		0.60±0.7
Pre-dialysis BWT kg	58.3±13		58.4±13		58.2±13		56.8±12
Post-dialysis BWT kg	56.4±12.5		56.4±13		56.1±13		54.6±12
Δ BWT/hr %	1.03±0.5		1.04±0.5		1.07±0.5		1.09±0.5
Pre-dialysis mean BP	113±13		115±13		115±13		113±11

*** $p<0.001$
** $p<0.01$
* $p<0.05$ DL=dialysis

or the need to prolong the experience.

It is also possible to show the presence or the absence of factors influencing correct evaluation (that is variation of body weight, of hourly ultrafiltration, pre-dialysis blood pressure etc.

As an example we performed a prospective study, not double blind, of the value of substituting bicarbonate for acetate to reduce dialysis hypotension.

We concluded that, in a standard dialysis schedule with a dialysate sodium of 140mmol/L, a short-term substitution of bicarbonate in place of acetate significantly reduces the dialysis hypotension, but that when long-term periods are examined the improvement remains constant in a very few patients (2 out of 16). Therefore a 'placebo' effect cannot be excluded.

Acknowledgments

The authors gratefully acknowledge the secretarial assistance of Miss Laura Pesci.

References

1 Stead WW. *Kidney Int 1983; 24:* 436
2 Friedman EA. *Am J Kidney Dis 1982; 2:* 289
3 Kjellstrand C. *Controv Nephrol 1980; 2:* 12
4 Keshaviah P, Shapiro FL. *Am J Kidney Dis 1982; 2:* 290
5 Castellani A, Lonati F, Bigi L et al. *Proc EDTA 1982; 19:* 340
6 Pagel MD, Ahmad S, Vizzo JE et al. *Kidney Int 1982; 21:* 513
7 Rosa AA, Fryd DS, Kjellstrand CM. *Arch Int Med 1980; 140:* 804
8 Woodward RH, Goldsmith PL. *Cumulative Sum Techniques.* Edinburgh: Oliver and Boyd. 1964

Proc EDTA–ERA (1984) Vol 21

SYSTOLIC TIME INTERVALS BEFORE AND AFTER HAEMODIALYSIS

J Braun, H Rupprecht, U Gessler

University Institute of Nephrology, Nuernberg, FRG

Summary

We investigated changes of haemodynamics during haemodialysis in 107 patients by noninvasive measurement of systolic time intervals. Heart rate and pre-ejection period index increased and systolic and diastolic blood pressure, left ventricular ejection time and isovolumic contraction time decreased. We found, that if the pre-dialysis ejection fraction (EF) was more than 55 per cent, post-dialysis EF decreased by about 10 per cent. However if pre-dialysis EF was less than 55 per cent, post-dialysis EF increased by five per cent. Regarding cardio-thoracic-ratio, Sokolow-Lyon-Index (SLI) and repolarisation disorders we found the most remarkable decrease in EF in patients with normal sized hearts with enlarged SLI and repolarisation disorders.

Introduction

There are various factors implicated in the pathogenesis of heart failure in uraemic patients, for example, coronary heart disease, hypertension, hypervolaemia, disorders of potassium and calcium metabolism, uraemic pericarditis and anaemia.

We investigated the influence of haemodialysis on cardiac function. As patients are already stressed by multiple punctures and the haemodialysis procedure we used a noninvasive measurement of the systolic time intervals.

Patients and methods

We studied 107 patients, 50 women and 57 men, with a mean age of 54±12 years (range: 23–79) immediately before and after haemodialysis. Dialysis time was normally 3 x 4.5hr and all patients had an acetate dialysate and parallel flow (Gambro Lundia major 11.5μ) or hollow fibre artificial kidneys (Gambro HF 120). The lead II ECG, apexcardiogram, phonocardiogram and carotid pulse tracing were recorded on a Hellige Multiscriptor EK 36 with the patient lying on his left side.

We determined the following parameters: the left ventricular ejection time (LVET), from the upstroke to the incisural notch of the carotid pulse tracing, the pre-ejection period ($PEP = QS_2 - LVET$), the A-wave amplitude (in per cent of the entire systolic contraction wave of the apexcardiogram) and the heart rate (HR). PEP-index was calculated by the PEP-HR regression equation $PEP_I = PEP + 0.4 \times HR$, LVET-index by the LVET-HR regression equation $LVET_I = LVET + 1.7 \times HR$ for men and $LVET_I = LVET + 1.6 \times HR$ for women. The ejection fraction was calculated by using Weissler's index in Garrard's formula: $EF = 1.125 - 1.25 \times PEP/LVET$ [1].

Values obtained in five consecutive cardiac cycles were averaged for each determination. In addition patients body weight and blood pressure were measured before and after haemodialysis.

Furthermore we took into account some clinical parameters, the cardiothoracic-ratio (CTR), the Sokolow-Lyon-Index ($SLI = S$ in V_2 + R in V_5) and the presence of repolarisation disorders.

Results

The changes in haemodynamic parameters during haemodialysis are shown in Table I. We found a significant increase in heart rate and pre-ejection period index (PEP_I) and a significant decrease in systolic and diastolic blood pressure, left ventricular ejection time, left ventricular ejection time index, electroacustical systole (QA_2), electroacustical systole corrected for heart rate ($QA_{2\,I}$) and the A-wave amplitude. The pre-ejection period, that was not corrected for heart rate did not change significantly. We divided pre-ejection period into QI and the isovolumic contraction time (IVC = time of left ventricular pressure rise) and found a significant prolongation of QI and a significant shortening of the IVC.

TABLE I. Pre- to post-haemodialysis changes

	before mean	SD	after mean	SD	mean change	significance
HR (b/min)	76.0	12.7	89.7	15.3	+13.7	$p<0.001$
BP_{sys} (mmHg)	138.8	21.5	130.6	25.5	−8.2	$p<0.001$
BP_{dias} (mmHg)	79.3	12.1	75.5	13.8	−3.8	$p<0.005$
LVET (sec)	0.288	0.032	0.239	0.031	−0.049	$p<0.001$
$LVET_I$ (sec)	0.413	0.023	0.387	0.022	−0.026	$p<0.001$
PEP (sec)	0.102	0.017	0.102	0.020	0.000	NS
PEP_I (sec)	0.133	0.016	0.138	0.019	+0.004	$p<0.001$
PEP/LVET	0.361	0.076	0.435	0.116	+0.074	$p<0.001$
QI (sec)	0.065	0.009	0.069	0.011	+0.003	$p<0.001$
IVC (sec)	0.037	0.014	0.032	0.016	−0.005	$p<0.001$
QA_2 (sec)	0.391	0.037	0.341	0.034	−0.049	$p<0.001$
$QA_{2\,I}$ (sec)	0.547	0.024	0.525	0.021	−0.022	$p<0.001$
A-wave (%)	14.3	5.6	11.2	6.2	−3.1	$p<0.001$

Haemodialysis was associated with a significant decrease ($p<0.001$) of the mean value of ejection fraction (EF) from 67.4 per cent to 58.1 per cent (normal range 60 per cent to 75 per cent). Changes in EF depend on the pre-dialysis EF. In patients with a normal pre-dialysis EF we found a decrease of about 10 per cent whereas patients with a pre-dialysis EF of less than 55 per cent showed an increase of about five per cent (Figure 1). The pre-dialysis EF was also independent of patients age (range: 22.8 to 79.2 years), and duration on haemodialysis (range: 1 to 186 months).

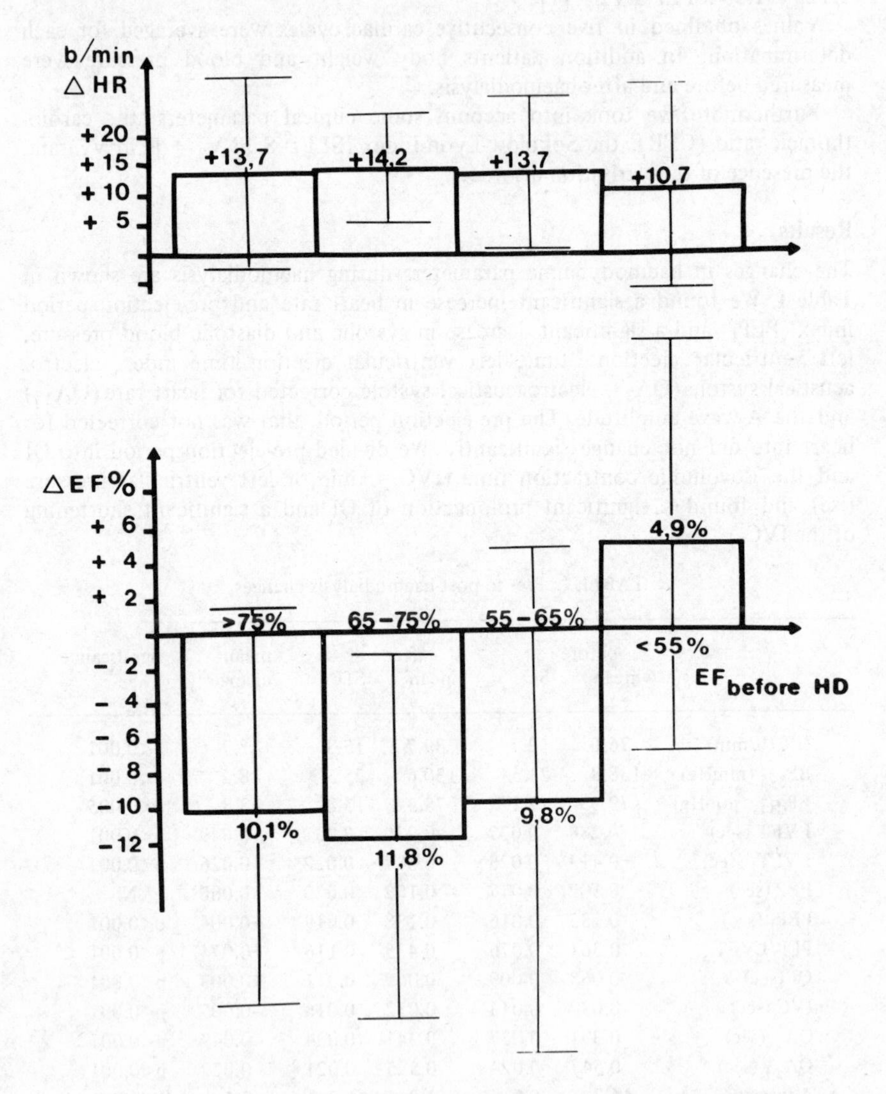

Figure 1. Relationship between pre-dialysis EF and change in EF

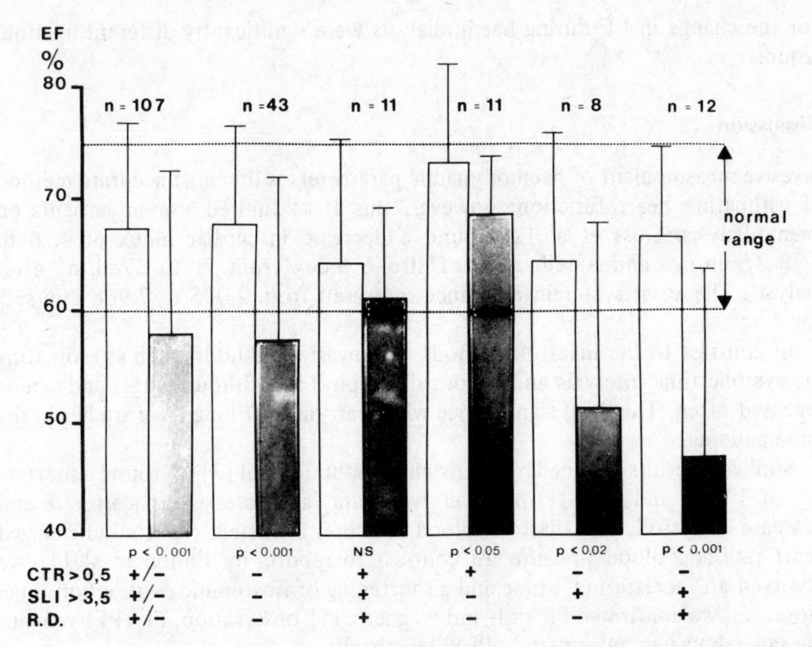

Figure 2. Changes in EF during haemodialysis regarding cardio-thoracic-ratio (CTR), Sokolow-Lyon-Index (SLI) and repolarisation disorders

As shown in Figure 2, we divided all patients into different groups considering the presence of repolarisation disorders, enlarged cardio-thoracic-ratio (CTR >0.5) or Sokolow-Lyon-Index (SLI >3.5mV). Patients that had neither repolarisation disorders, nor an enlarged CTR of SLI (column 2) had a pre-dialysis EF of 67.9 per cent and a post-dialysis EF of 57.4 per cent. These values resemble those of the whole group (column 1). Patients with increased CTR, that had no repolarisation disorders and a normal SLI (column 4) had an EF in the upper normal range before haemodialysis and showed only a slight decrease after haemodialysis (from 73.4 per cent to 68.8 per cent, $p<0.05$). Patients with normal sized hearts on chest X-ray, but increased SLI with or without repolarisation disorders (columns 5 and 6) showed the most considerable decrease in EF (from 64.6 per cent to 51.3 per cent, $p<0.02$ and from 66.8 per cent to 47.0 per cent, $p<0.001$), although the pre-dialysis EF was in the lower normal range.

The CTR seems to be a very important factor influencing the change in EF. We found a positive correlation between the change in EF and the CTR ($p<0.01$), whereas the pre-dialysis EF did not depend on CTR. The decrease in EF is significantly diminished as the CTR increases.

Mean weight loss in our patients during haemodialysis was 1.9kg (range: 0.0 to 5.4kg). The change in the heart rate corrected pre-ejection period was found to correlate to the body weight loss ($PEP_I = 3.692 \times BW - 2.057$, $p<0.01$).

Furthermore, we tried to find differences between digitalised and not digitalised patients (42 were on digitoxin, 65 were not). Neither the pre-dialysis EF, nor the change in EF during haemodialysis were significantly different in either groups.

Discussion

Invasive measurement of haemodynamic parameters is the most accurate method of estimating heart function; however, this is of limited use in patients on haemodialysis. Goss et al [2] found a decrease in cardiac index of 4.26 to 2.98 $1/min/m^2$ and a reduction of stroke index from 51 to $37 ml/m^2$ after dialysis. The total systemic resistance increased from 2,025 to 2,968 dyn/sec/ cm^{-5}/m^2.

In contrast to the invasive methods, noninvasive methods (such as estimating the systolic time intervals and echocardiography) are without danger and can be repeated often. This is of significance when carrying out long-term studies in the same patients.

Similar to results obtained by Bornstein, Zambrano et al [3] we found a shortening of LVET and $LVET_I$; however we found an increase of heart rate and decrease of systolic and diastolic blood pressure. Bornstein reported unchanged heart rate and blood pressure. In contrast to reports by Endou et al [4] we observed an increase in CI-time and a shortening of isovolumic contraction time. Moreover, we confirmed Prakash and Wegner's [5] observation, that PEP remains the same, however, in our study PEP_I lengthens.

A lengthening of PEP_I demonstrates that the time needed to achieve diastolic aortic pressure is also lengthened. This may be due either to a slowed fibre-shortening, or to an increased diastolic aortic pressure, or to both. Because diastolic blood pressure decreases during dialysis, we would expect a shortening of PEP_I. The second factor is a slowed fibre-shortening which may be due to two reasons: a decreased end-diastolic stretch or a reduction of intrinsic contractility. However, shortening of IVC seems to be an argument for an improved intrinsic contractility. When there is a decrease in preload, an increase in IVC is to be expected.

The intensification of contractility could arise from an increase in ionised calcium during haemodialysis and from increased sympathetic activity as a reaction to volume loss and fall in blood pressure.

The lengthening of PEP_I seems to be predominantly determined by a decrease in preload. This is confirmed by the correlation between changes in PEP_I and weight loss.

LVET reduction also seems to be determined by a fall in preload with consequent decrease in stroke volume [2].

An improvement of EF in patients with poor pre-dialysis EF was in contrast to a fall in EF found in patients with normal pre-dialysis EF. We attributed this phenomenon to different starting points on the Starling-curve. Patients with poor pre-dialysis EF are on the declining part of the curve (exceeded preload reserve) and due to preload decline experience a left-shift towards a higher section of the curve. Patients with normal pre-dialysis EF (on the rising part of the curve) experience a definitive deterioration of EF due to left-shift.

Patients with enlarged CTR, without repolarisation disorders and with normal

SLI, show an above average high pre-dialysis EF. This is due to an improved end-diastolic stretch. These patients also show a considerably smaller decrease in EF compared to other patients; this is the result of a left-shift on the top of the Starling-curve, which is determined by the falling preload.

In contrast to other groups, patients with hypertrophic (but not enlarged) hearts, who have an especially poor adaptability to volume changes, seem to have a very marked decrease in EF.

In patients with hypertrophic hearts and repolarisation disorders, a strict volume control is especially important. This is necessary because a rapid volume increase can exceed preload reserve, and on the other hand, a rapid volume decrease during haemodialysis can lead to a sharp fall in EF.

References

1 Garrard CL, Weissler AM, Dodge HT. *Circulation 1970; 42:* 455
2 Goss JE, Alfrey AC, Vogel JHK et al. *Trans ASAIO 1967; 13:* 68
3 Bornstein A, Zambrano S, Morrison RS et al. *Am J Med Sci 1975; 269:* 189
4 Endou K, Kamijima J, Kakubari Y et al. *Cardiology 1978; 63:* 175
5 Prakash R, Wegner S. *Am J Med Sci 1972; 264:* 127

Proc EDTA–ERA (1984) Vol 21

GASTROINTESTINAL ALUMINIUM ABSORPTION: IS IT MODULATED BY THE IRON-ABSORPTIVE MECHANISM?

J B Cannata, C Suarez Suarez, V Cuesta, R Rodriguez Roza, M T Allende, J Herrera, J Perez Llanderal

Hospital General de Asturias, Universidad de Oviedo, Oviedo, Spain

Summary

Gastrointestinal aluminium (Al) absorption has been proved but its mechanism is still unknown. This study investigates the pattern of Al absorption in patients with different degrees of iron stores.

We studied 29 haemodialysis patients forming three groups according to their serum ferritin values. Over seven days all patients received the same dose of aluminium hydroxide after which patients with 'low-normal' and normal serum ferritin increased their serum Al proportionally with the increased aluminium hydroxide intake. By contrast patients with high serum ferritin did not show any change in their serum Al values.

Our results therefore suggest that a 'common pathway' of metal absorption could be implicated in Al absorption. Serum ferritin might be a valuable predictor of different behaviour.

Introduction

The risk of aluminium (Al) absorption from Al-containing phosphate binders is almost universally accepted [1–3]. Moreover, some authors have recently reported that patients not undergoing haemodialysis who were only exposed to oral Al can develop bone toxicity [2,3] as has been previously demonstrated in patients intoxicated with Al through high-Al dialysate exposure.

An Al-free phosphate binder has been synthesised [4] but requires further evaluation, and meanwhile Al-containing phosphate binders remain the standard drug for reducing serum phosphorous in chronic renal failure. In recent years great interest has centred on studying the likely factors modulating gastrointestinal Al absorption which may help us to recognise the Al 'hyperabsorbers'.

This study investigates whether haemodialysis patients with low-normal, normal and high iron stores have similar or different patterns of Al absorption.

Patients and methods

We studied 29 patients dialysed thrice weekly with the same schedule, aged 29 to 65 years (mean 40.6), time on haemodialysis 37.6 ± 25 months, using hollow fibre and plate dialysers 1.0–1.3m^2 and employing deionised water with an Al content of less than 0.3μmol/L throughout the study. All patients were receiving oral water soluble vitamin supplements and a single fasting dose of 100–300mg of ferrous sulphate as has been previously described [5].

We measured serum Al by inductively coupled emission spectrometry, serum ferritin by enzyme-immunoanalysis (ELISA) and parathyroid hormone (PTH) by a human C terminal radioimmunoassay (INC) which is reactive to the 65–84 sequence of human PTH (normal values less than 1.5ng/ml).

The patients were separated into three groups according to their serum ferritin concentrations:

- Group I: Serum ferritin below 100ng/L (low-normal ferritin group)
- Group II: Serum ferritin 100–250ng/L (normal ferritin group)
- Group III: Serum ferritin above 250ng/L (high ferritin group)

During the first seven days (Period I) all patients received ferrous sulphate and aluminium hydroxide ($Al(OH)_3$) in the dose they were currently having to maintain the serum phosphorus (P) between 1.5–1.8mmol/L. At the end of Period I ferrous sulphate was stopped and during a further seven days (Period II) all received the same dose of $Al(OH)_3$ (2.8g/day).

Samples for serum Al were taken at the end of Periods I and II.

Statistical analysis was performed by student's 't' test and all values are expressed as a mean ± SD.

Results

As shown in Table I Group III had a mean serum ferritin almost 10 times higher than Group I and significantly lower ($p<0.05$) haemoglobin concentration.

TABLE I. Details of patients divided into three groups based on the same serum ferritin

	Group I	Group II	Group III
Serum ferritin (ng/L)	100	100–250	250
Mean serum ferritin	48 ± 23	149 ± 40	468 ± 206
Number of patients	13	10	6
(mean age)	(51.6±9.9)	(50.7±9.9)	(45.3±10.1)
Haemoglobin (g/dl)	9.1±2.6	8.47±1.6	7.13±1.6 *

*$p<0.05$ between Group I and Group III

TABLE II. Period I. Baseline results before giving 2.8g/day of $Al(OH)_3$ during Period II to all groups

	Group I	Group II	Group III
Serum ferritin (ng/L)	<100	100–250	>250
$Al(OH)_3$ dose (g/day)	2.04±1.5	1.44±1.9	0.77±1.08**
Serum P (mmol/L)	1.68±0.33	1.69±0.32	1.31±0.31**
Serum Al (μmol/L)	2.94±2.1	3.09±1.9	1.62±0.6*
Serum PTH (ng/ml)	5.01±3.4	3.57±1.9	3.71±1.8
Patients on $1,25(OH)_2D_3$	8 of 13 (61%)	5 of 10 (50%)	1 of 6 (83%)

**$p<0.01$; *$p<0.05$ between Group I and Group III

In Period I (Table II) the significant differences ($p<0.05$) in serum Al results between Groups I and III can not be explained either on the basis of plasma PTH or $1,25(OH)_2D_3$ intake, but by significant differences ($p<0.05$) in $Al(OH)_3$ ingestion. Despite a significantly lower use of binder, Group III achieved a significantly better serum P control ($p<0.01$).

Figure 1 shows that in all groups the change of $Al(OH)_3$ dose from Period I to Period II represented a significant increase, being the highest ($p<0.001$) in the high ferritin group. Nevertheless, this group was the only one who did not show any great increase in serum Al. When we compared the ratio of Serum Al/ $Al(OH)_3$ dose, we found no significant changes in Groups I and II, so, the serum Al increase was proportional to the $Al(OH)_3$ change. By contrast, Group III had a highly significant change from Period I to Period II ($p<0.001$).

Discussion

It took many years to fully convince nephrologists about the risks of toxicity from well known Al-phosphate binders which have been used as antacids for almost half a century.

Berlyne and colleagues drew attention to Al-resin toxicity in chronic renal failure in 1970 [6] yet we are still using similar phosphate binders whose mechanisms of absorption are not certain. Al balance studies are very difficult to perform and have produced controversial and sometimes irreproducible results. Furthermore, the fact we cannot use a non-toxic labelled Al makes the absorption studies more imprecise. These problems are magnified when evaluating likely factors influencing Al absorption.

In clinical studies the best guide we have is the serum Al changes. The results of animal experiments cannot always be extrapolated to the human situation.

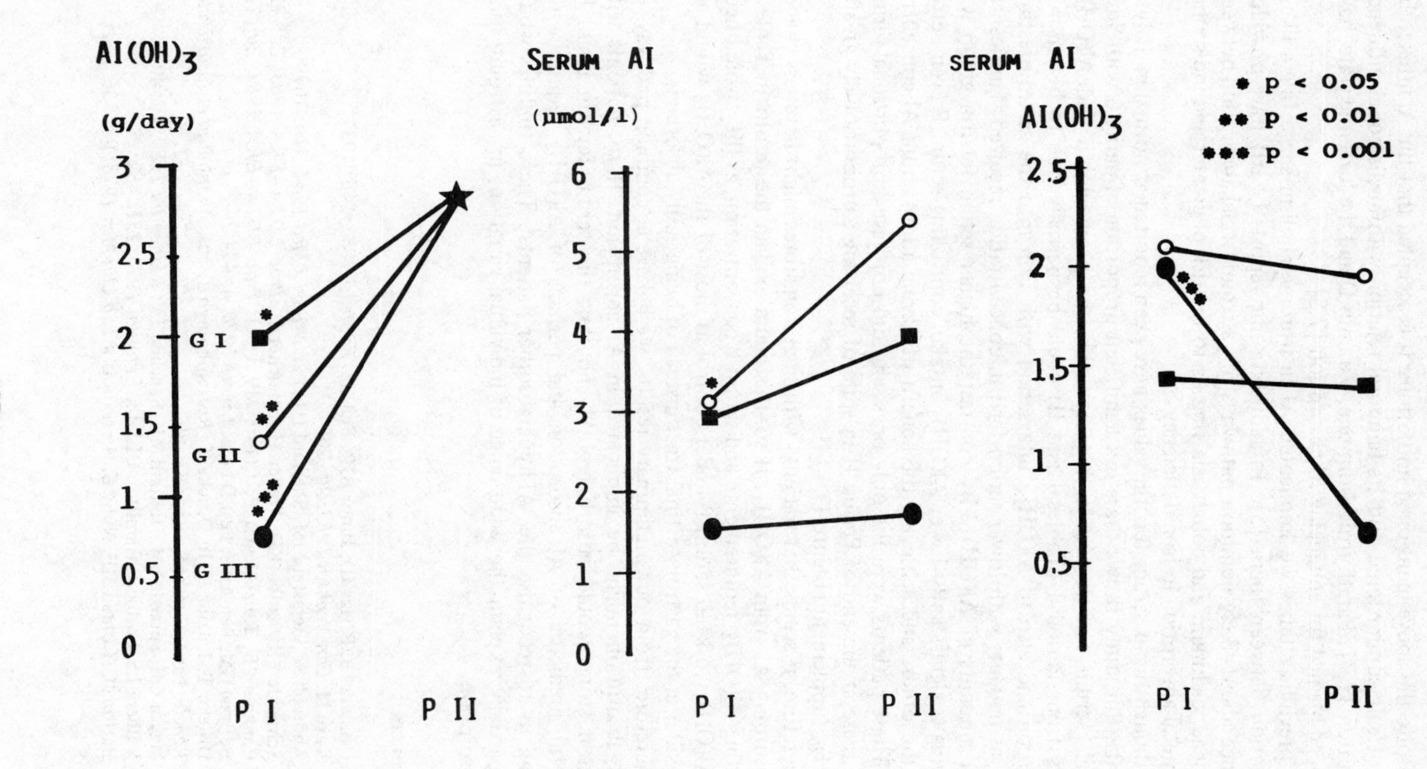

Figure 1. $Al(OH)_3$ dose, Serum Al and Serum Al/$Al(OH)_3$ ratio. Changes from Period I (P I) to Period II (P II)

Despite the above mentioned limitations, it is accepted that the Al intake, the kind of salt administered and its temporal relation with meals could influence absorption [1,7]. Parathyroid hormone, Vitamin D and the 'uraemic state' have also been blamed for enhancing Al absorption [7,8].

In previous studies we implicated Al in iron metabolism suggesting a likely interaction between them [5]. In addition some reports [9,10] have provided evidence about likely common pathways of gastrointestinal absorption between iron, lead, cadmium and cobalt, alerting us to the risk of greater and indiscriminate metal absorption in iron deficiency.

Although mean serum ferritin values have been reported as normal in dialysis patients, this study shows there is a significant proportion of patients with 'low-normal' serum ferritin which, judging from the haemoglobin and $Al(OH)_3$ results from Period I (Tables I and II) could be partially due to better iron utilisation and also to $Al(OH)_3$ interaction with iron into the gastrointestinal tract, as patients with lower serum ferritin concentration received significantly higher amounts of $Al(OH)_3$. By contrast the higher serum ferritin group was receiving a significantly lower $Al(OH)_3$ intake, controlling serum P even better than the others and achieving the benefit of having a low serum Al concentration. These patients were the only ones who did not have a proportional serum Al increase at the end of Period II in spite of receiving a proportionally greater $Al(OH)_3$ supplement (Figure 1).

Therefore, it seems that patients with high iron stores might have less chance of absorbing Al from $Al(OH)_3$. If so, this could explain the results in Table II regarding $Al(OH)_3$ requirements and serum P concentration, as they would need less $Al(OH)_3$ to keep serum P under control as most of the $Al(OH)_3$ would act as a binder in the gastrointestinal tract instead of being easily absorbed.

Therefore, from our preliminary results, we believe a 'common pathway' of metal absorption might be implicated in Al absorption. If so, patients with high-iron stores would absorb less Al and thus the serum ferritin might be a useful predictor of Al absorption and perhaps of $Al(OH)_3$ requirements. helping us to recognise the 'Al-hyperabsorber patients'. These findings would also partially explain the wide range of individual variations in gastrointestinal Al absorption.

References

1 Cannata JB, Briggs JD, Junor BJR, Fell GS. *Br Med J 1983; 286:* 1937
2 Kaye M. *Clin Nephrol 1983; 20:* 208
3 Andreoli SP, Bergstein JM, Sherrard DJ et al. *N Engl J Med 1984; 310:* 1079
4 Schneider H, Kulbe KD, Weber H, Streicher E. *Proc EDTA–ERA 1983; 20:* 725
5 Cannata JB, Ruiz-Alegria P, Cuesta MV et al. *Proc EDTA–ERA 1983; 20:* 719
6 Berlyne GM, Ben-Ari J, Pest D et al. *Lancet 1970; ii:* 494
7 Drüeke T, Lacour B. In Drüeke T, Rottembourgh J, eds. *Aluminium et Insuffisance Rènale.* Paris: Gambro. 1984: 63–69
8 Mayor GH, Sprague SM, Hourani MR, Sanchez TV. *Kidney Int 1980; 17:* 40
9 Valberg LS, Sorbie J, Hamilton DL. *Am J Physiol 1976; 231:* 462
10 Barton JC, Conrad ME, Nuby S, Harrison L. *J Lab Clin Med 1978; 92:* 536

Open Discussion

ROODVOETS (Haarlem, Netherlands) A high ferritin value may be due to repeated blood transfusions and therefore may not reflect the capacity of intestinal iron absorption.

CANNATA Yes, that is correct, patients from group III had received more blood transfusions. I can say from our experience that if your iron stores are repleted there is not a good stimulus to absorb iron and therefore you will not have a strong stimulus to absorb aluminium. I can't say anything about iron absorption because some of these patients have become iron overloaded from transfusions.

KERR (Chairman) Dr Cannata if your thesis is correct it is a little surprising that you have an effect upon phosphate control when you consider how little aluminium is absorbed. You are giving your patients 2.8 grams of aluminium hydroxide daily and so you are talking about gram quantities of aluminium. The changes are, in fact, measured by micro grams per litre. I don't think we know in absolute terms the amount absorbed but it is probably milligrams at the most. It is hard to see how that could alter the effectiveness of aluminium hydroxide as a phosphate binder. Would you like to comment on that?

CANNATA I too was in trouble trying to think of an explanation for that. I don't have a clear explanation and my interpretation is really a hypothesis but I don't have any hard data to support it.

KERR It is possible, surely, that your over-transfused patients differ in some way from other groups. Have you looked at their phosphate intakes to see whether there is any dietary difference?

CANNATA I have not.

Proc EDTA–ERA (1984) Vol 21

PROTEIN BINDING OF ALUMINIUM IN NORMAL SUBJECTS AND IN PATIENTS WITH CHRONIC RENAL FAILURE

H Rahman, S M Channon, A W Skillen, M K Ward, D N S Kerr
Royal Victoria Infirmary, Newcastle upon Tyne, United Kingdom

Summary

Ultrafiltration of serum through YM10 membranes showed that 46 per cent of the aluminium in normal subjects and 33 per cent of the aluminium in patients with chronic renal failure is ultrafiltrable, suggesting that the majority of the aluminium is bound to some serum component(s) having molecular weight greater than 10,000 daltons. After desferrioxamine infusion, both the ultrafiltrable and protein-bound aluminium increases significantly, probably due to mobilisation of aluminium from body tissues. Gel filtration on Sephacryl S-300 and affinity chromatography have shown that transferrin is the major aluminium binding protein.

Introduction

Aluminium accumulation in chronic renal failure has been implicated in the causation of dialysis dementia, dialysis osteodystrophy and microcytic hypochromic anaemia but very little is known about the mechanism(s) by which aluminium causes its toxic effects.

During haemodialysis aluminium is transferred from the dialysis fluid to the patient against a concentration gradient and is difficult to remove subsequently [1]. This is probably due to the tight binding of aluminium to some serum component(s). Although the possibility of removal of aluminium by dialysis with low aluminium-containing dialysis fluid is controversial, a significant amount of aluminium can be removed by haemodialysis after desferrioxamine infusion [2], although the mechanism by which this occurs is not well defined.

Studies are carried out to determine the protein binding of aluminium in normal subjects and in patients with chronic renal failure and to determine the effect of desferrioxamine on the protein binding of aluminium.

Materials and methods

Protein binding of aluminium in serum was studied by ultrafiltration, gel filtration and affinity chromatography.

Ultrafiltration was carried out in a stirred Amicon ultrafiltration cell which was pressurised by 5% CO_2 in air, using YM10 membranes (molecular weight cut off 10,000 daltons). Thirty normal subjects and 30 patients with chronic renal failure on haemodialysis were studied to determine the ultrafiltrable aluminium in the serum. Ten patients with chronic renal failure were studied before and 48 hours after an infusion of 2gm of desferrioxamine to determine the effect of desferrioxamine on the protein binding.

Specific aluminium binding protein(s) were identified by gel filtration and affinity chromatography. Gel filtration was performed in a Pharmacia column, packed with Sephacryl S-300 superfine. Earle's Medium at pH 7.4 was the eluting buffer and the column was calibrated with IgM, IgG, transferrin, albumin and aluminium-desferrioxamine complex markers. The serum of five normal subjects and 15 patients with chronic renal failure on haemodialysis was studied by gel filtration to identify the aluminium binding protein(s) and another five patients with chronic renal failure were studied after a desferrioxamine infusion. Affinity chromatography was carried out with cyanogen bromide activated Sepharose coupled with anti-transferrin. Phosphate buffered saline at pH 7.2 was used in the immobilisation and washing cycle and phosphate citrate buffer at pH 2.8 was used in the dissociation cycle.

Aluminium estimations of the serum, ultrafiltrate and column fractions were performed by flameless atomic absorption spectrophotometry. For the fractions, matrix matched standards were essential and the background correction was needed for the fractions from the affinity column.

Protein estimations of the fractions were carried out by recording the absorbance at 280nm using an ultraviolet spectrophotometer, immunoturbidimetric assay on the centrifugal analyser (Cobas-Bio) and electrophoresis.

Precautions were taken to avoid aluminium contamination during the studies and analysis.

As some of the data were not normally distributed, non-parametric statistics were used for the analysis.

Results

In the normal subjects, the mean serum aluminium was 8μg/L (range 6.4–11.6μg/L). The mean ultrafiltrable aluminium was 3.6μg/L (range 2.8–4.5μg/L). The mean non-ultrafiltrable aluminium was 4.4μg/L (range 2.8–7.1μg/L). Forty-six per cent of the serum aluminium was ultrafiltrable. The remaining 54 per cent could not pass through the membrane and was probably bound to some serum component(s) (Figure 1).

In the patients with chronic renal failure, the serum aluminium (mean ± SD) was 99±54μg/L. The ultrafiltrable aluminium was 30±16μg/L and the protein-bound aluminium was 68±41μg/L. Thirty-three per cent of the serum aluminium was ultrafiltrable (Figure 1). The percentage of ultrafiltrable aluminium showed

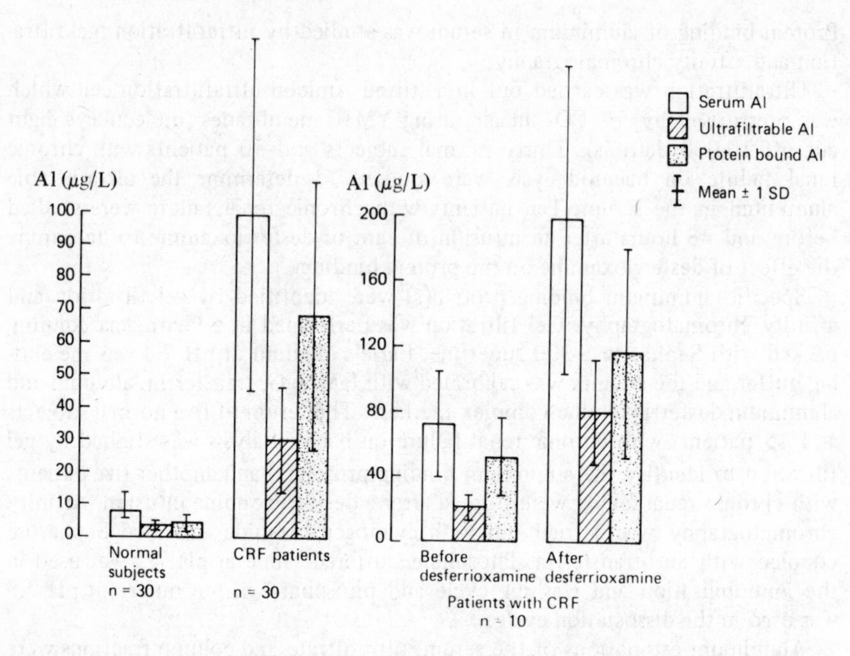

Figure 1. Total serum, ultrafiltrable and 'protein-bound' aluminium in normal subjects, chronic renal failure patients and the effect of desferrioxamine on the protein binding of aluminium in patients with chronic renal failure

a significant inverse correlation with the total serum concentration, both in the normal subjects and in patients with chronic renal failure (Spearman's rank correlation coefficient: $rs = -0.47$, $p<0.01$ normals; $rs = -0.52$, $p<0.01$ chronic renal failure patients).

After desferrioxamine infusion, the serum aluminium (mean ± SD) showed a remarkable increase from 72±31μg/L to 198±94μg/L. Both the ultrafiltrable (21±7–81±33μg/L) and protein bound aluminium (51±25–117±63μg/L) showed a significant rise (Figure 1). The mean percentage ultrafiltrable aluminium showed a significant increase from 32 per cent to 43 per cent after desferrioxamine infusion (paired 't' test: $t=6.81$, $p<0.001$).

Gel filtration of serum from the normal subjects and patients with chronic renal failure showed that aluminium was eluted as a single peak which was coincidental with the transferrin peak (Figure 2).

When gel filtration was performed with 2.5mg/ml and 5mg/ml of commercially available human transferrin (Behringwerke) before and after spiking with 200μg/L aluminium, the aluminium was eluted from the column as a prominent peak coincident with transferrin. However, when similar gel filtration studies were carried out with 5mg/ml and 50mg/ml of human albumin (Hoechst), no aluminium peak was detected in the column fractions coincident with albumin peak.

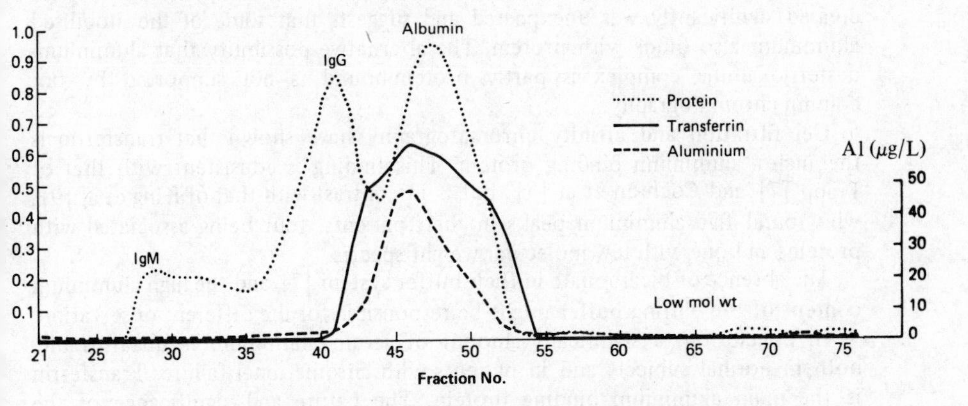

Figure 2. Gel filtration of serum from patients with chronic renal failure

Gel filtration studies after desferrioxamine infusion showed that aluminium was eluted as two distinct peaks, the first coincident with transferrin peak and the second coincident with the aluminium-desferrioxamine marker peak on the column. During gel filtration, it was observed that the gel matrix absorbs aluminium from the buffer and serum ultrafiltrate. This aluminium can subsequently be taken up by the serum or desferrioxamine passing through the column. For this reason gel filtration could not be used for quantitative studies.

Affinity chromatography in five patients with chronic renal failure showed that about 35 per cent of the serum aluminium was eluted in the fractions from the washing cycle and the remaining 65 per cent was eluted in the dissociation cycle with the transferrin.

Discussion

In normal subjects 46 per cent (range 30–58%) of the serum aluminium is ultrafiltrable. This finding agrees with that of Lundin et al [3], but is in contrast with the finding of Elliott et al [4], who could not detect significant amounts of ultrafiltrable aluminium in normal subjects. In patients with chronic renal failure, 33 per cent (range 20–47%) of the serum aluminium is ultrafiltrable which is consistent with the finding of Elliott et al [4], but is higher than the 20 per cent ultrafiltrable aluminium reported by Graf et al [5]. The difference in the membranes used and conditions for the ultrafiltration studies between the workers [4,5] may explain this discrepancy. Our observation of an inverse correlation between the percentage ultrafiltrable aluminium and total serum aluminium agrees with the finding of Hosakawa et al [6].

After desferrioxamine infusion, the serum, ultrafiltrable and protein-bound aluminium showed a remarkable increase. The percentage of ultrafiltrable aluminium increased significantly. This finding is in agreement with that reported by Graf et al [5] and suggests that desferrioxamine mobilises aluminium from the body tissues. The observation that the protein bound aluminium also increased significantly was unexpected and suggests that some of the mobilised aluminium also binds with protein. The alternative possibility that aluminium-desferrioxamine complex is partly protein-bound is not supported by our column chromatography.

Gel filtration and affinity chromatography have shown that transferrin is the major aluminium binding protein. This finding is consistent with that of Trapp [7] and Cochran et al [8], but is in contrast with that of King et al [9], who found five aluminium peaks in their patients, four being associated with proteins and one with low molecular weight species.

The absence of bicarbonate in their buffer system [7] and the high aluminium content of the eluting buffer might be responsible for the different observation.

In conclusion, a significant amount of serum aluminium is ultrafiltrable, both in normal subjects and in patients with chronic renal failure. Transferrin is the main aluminium binding protein. The nature and significance of the ultrafiltrable and protein bound aluminium in health and in patients with chronic renal failure need further study.

Acknowledgments

This study was supported by grants from the Association of Commonwealth Universities and Northern Counties Kidney Research Fund.

References

1 Kaehny WD, Alfrey AC, Holman RE, Shorr WJ. *Kidney Int 1977; 12:* 361
2 Ackrill P, Day JP, Garstang FM et al. *Proc EDTA 1982; 19:* 203
3 Lundin AP, Carusco C, Sass M, Berlyne GM. *Clin Res 1978; 26:* 636A
4 Elliott HL, MacDougall AI, Fell GS, Gardiner PHE. *Lancet 1978; ii:* 1255
5 Graf H, Stummvoll HK, Meinsinger V. *Proc EDTA 1981; 18:* 674
6 Hosakawa S, Imai T, Okumura T et al. *Medicine 1984; 1(2):* 59
7 Trapp GA. *Life Science 1983; 33:* 311
8 Cochran M, Neoh S, Stephens E. *Clin Chim Acta 1983; 132:* 199
9 King SW, Savory J, Wills MR. *Ann Clin Lab Sci 1982; 12:* 143

Open Discussion

VAN HERWEGHAM (Brussels) Do you have any information on the possible competition between iron and aluminium for transferrin and for desferrioxamine?

RAHMAN That is a very interesting question. We have no data on this particular point. It is however important as it might explain the anaemia associated with dialysis dementia, but this is hypothetical.

KERR (Chairman) I would like to comment that data presented at the Ankara meeting suggests that transferrin has about four orders of magnitude tighter binding for iron than for aluminium.

BRANCACCIO (Milan) Do you believe that patients affected by aluminium induced encephalopathy have different protein binding of aluminium in comparison to other patients on chronic dialysis?

RAHMAN We have not found any difference. There is one study showing that in patients with dialysis dementia most of the aluminium is bound to a low molecular weight protein of about 10,000 daltons. In addition the plasma concentration of this protein is higher in dementia patients than others.

KERR There are a number of papers indicating that aluminium appears to be bound to protein.

CANNATA Have you done any study on transferrin arising from the gastro-intestinal mucosa?

RAHMAN I think that iron and aluminium shares the same transport mechanism in the intestinal wall. The competition between these two substances for absorption may explain some of your findings.

Proc EDTA–ERA (1984) Vol 21

DESFERRIOXAMINE INDUCED ALUMINIUM REMOVAL IN HAEMODIALYSIS

J Bonal, **J Montoliu**, J Löpez Pedret, E Bergadä, L Andrew, M Bachs, L Revert

Hospital Clinic, Barcelona, Catalunya, Spain

Summary

This study evaluates if the use of high flux membranes and the type of dialysate influences aluminium removal in haemodialysis. Aluminium kinetics and dialysance were determined in baseline conditions and after infusion of desferrioxamine. The free diffusible fraction of plasma aluminium correlated significantly with the plasma aluminium post desferrioxamine, independently of the type of membrane or dialysate used. Aluminium removal therefore depends on the plasma concentration reached and the dialysate concentration. High flux membranes do not improve aluminium removal in vivo.

Introduction

Aluminium (Al) accumulates in patients on chronic haemodialysis [1], and causes dialysis encephalopathy, Vitamin D resistant osteomalacia and a non-specific toxicity syndrome [2]. It has proved possible to remove significant quantities of aluminium by haemodialysis following desferrioxamine (DFO) chelation [3,4], but only a very small number of studies have reported dialysance of aluminium post DFO. Since acrylonitrile membranes have demonstrated far superior dialysance of the chelate ferrioxamine [5] and that the difference in the pH of dialysis fluid can increase the solubility of aluminium, we have carried out studies to investigate aluminium removal post DFO in acetate and bicarbonate haemodialysis using cuprophan and acrylonitrile membranes.

Methods

After informed consent, ten patients on chronic haemodialysis (HD) for four hours thrice weekly, were studied. Mean duration of dialysis was 69 months. All patients had taken oral Al-containing phosphate binders and were dialysed with deionised tap water. None had evidence of aluminium toxicity. Throughout the study, oral Al-hydroxide was withheld.

Aluminium was determined by atomic absorption spectrophotometry using graphite furnace (Instrumentation Laboratory, IL-551). Samples were collected under acid conditions.

We evaluated aluminium kinetics (plasma Al, free diffusible plasma Al fraction, dialysate Al) as published elsewhere [6]. Dialysance (ml/min) was calculated by:

$$D = Q_D \frac{C_D - C_{DI},}{C_A}$$

where Q_D is dialysate flow rate (ml/min, C_D is dialysate aluminium efflux (μg/L), C_{DI} is dialysate aluminium influx (μg/L), C_A equals aluminium concentration in blood and D equals dialysance.

Aluminium kinetics were studied before and 48 hours after an intravenous infusion of 2g desferrioxamine (DFO). DFO was administered to all patients after a baseline haemodialysis.

Measurements were performed at dialysate flow rates of 500ml/min in acetate and bicarbonate single pass dialyses. Half of the patients were dialysed with acetate dialysate and the other half with bicarbonate. Two membranes were evaluated: cuprophan (CF 15.11, Travenol) and acrylonitrile AN-69 (H-12.10, Hospal).

Student's test for paired data and linear regression analysis were employed for statistical analysis. Data are expressed as average values and the standard error of the mean.

Results

Basal plasma aluminium was within 'safe' limits in all patients (mean plasma Al = 70 ± 7μg/L).

During baseline dialysis, a significant increase in plasma aluminium was observed and can be accounted for by plasma concentration by ultrafiltration. Baseline aluminium dialysance was 0ml/min (Table I and Figure 1).

TABLE I. Value of measurements before and after desferrioxamine (DFO) haemodialysis (HD)

	Plasma aluminium μg/L	Free diffusible plasma aluminium % (μg/L)	Dialysate aluminium influx μg/L	Dialysate aluminium efflux μg/L	Aluminium dialysance ml/min
			Baseline HD		
start	70±7	19 (13.3±1.3)	13.1±1.2	13.1±1.8	0
end	92±9				
			Post-DFO HD		
start	143±34*	37.5 (53.7±26.8)*	14.3±1.5	14.9±1.5	2.1*
end	90±7				

* p<0.001 in relation to initial values

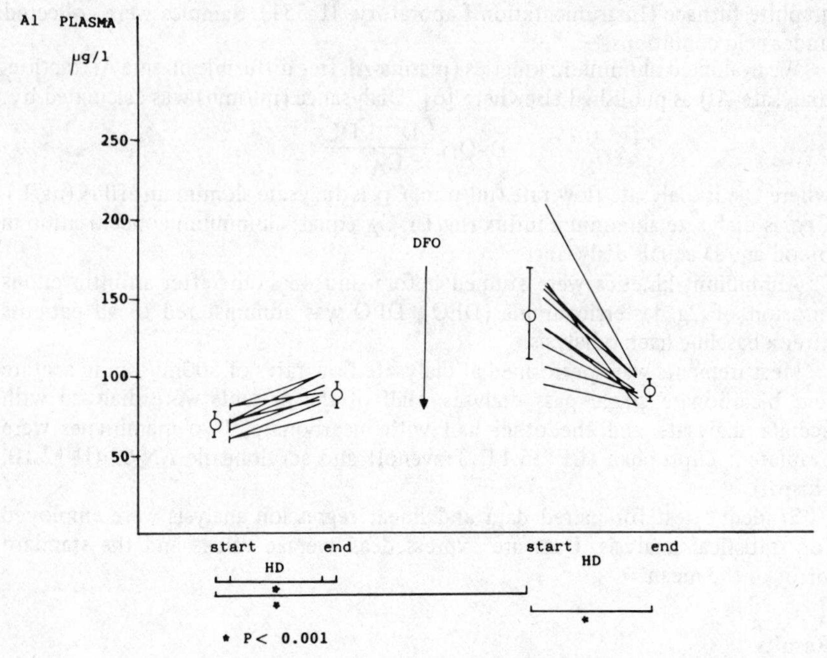

Figure 1. Changes in plasma aluminium (Al) after baseline haemodialysis (HD) (left side) and following desferrioxamine (DFO) administration

The significant decrease in plasma aluminium during post DFO dialysis demonstrates aluminium removal (post DFO Al dialysance = 2.1ml/min) (Table I).

The plasma aluminium reached at the end of haemodialysis was similar in baseline and post DFO dialysis, and depends on the concentration of aluminium in the dialysate (Figure 1).

The free diffusible plasma aluminium fraction (which is the Al content of in vivo obtained ultrafiltrate) was 19 per cent of total plasma aluminium in baseline haemodialysis with both cuprophan and acrylonitrile membranes, and rose to (24–53%) (mean 37.5%) of total plasma aluminium post DFO (Table I). The free diffusible plasma aluminium fraction correlated significantly with the plasma aluminium reached post DFO ($r = 0.79$, $p<0.01$), independently of the type of membrane used (Figure 2).

There was no difference in any of the measurements made between the patients dialysed with acetate or bicarbonate.

Discussion

In all patients with a safe plasma aluminium administration of 2g DFO caused a 100 per cent increase in plasma aluminium concentration due to tissue mobilisation. This was accompanied by an increase in the free diffusible plasma

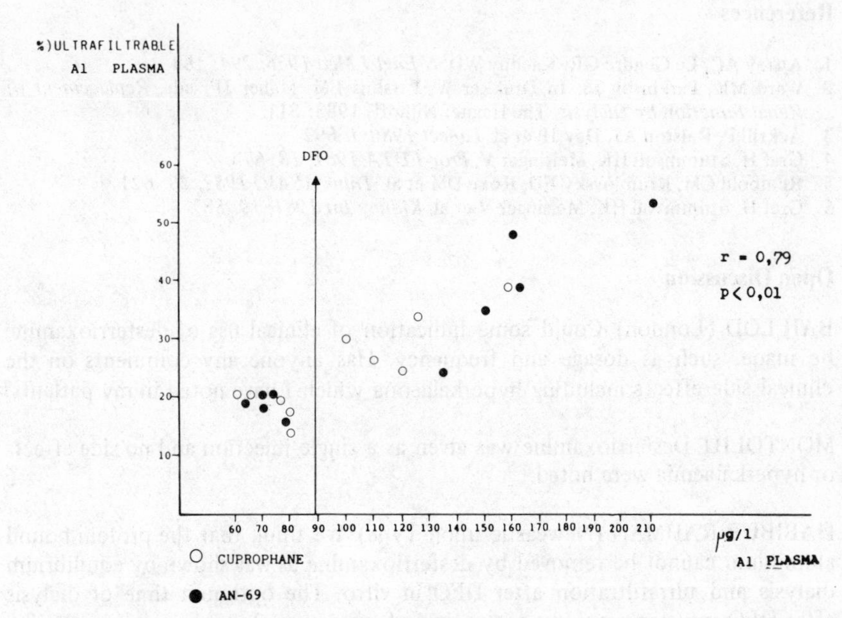

Figure 2. Correlation between per cent ultrafiltrable plasma aluminium (Al) and plasma Al in baseline haemodialysis (HD) and post desferrioxamine (DFO) (HD), using either cuprophan or acrylonitrile (AN-69) membranes

aluminium fraction, raising the effective concentration gradient between dialysate aluminium and the free diffusible aluminium. Therefore aluminium dialysance which was zero in baseline conditions rose to a mean of 2.1ml/min post DFO.

Using standard dialysis baths, with a constant dialysate aluminium there was no significant difference in aluminium kinetics between acetate or bicarbonate.

The free diffusible aluminium fraction correlated significantly with plasma aluminium reached post DFO. There was no significant difference between cuprophan or acrylonitrile membranes. In consequence, aluminium kinetics post DFO suggest that aluminium removal is more dependent on the free plasma aluminium fraction than on DFO-Al chelates.

We conclude that the main factors that influence aluminium removal in haemodialysis are dialysate aluminium and the plasma aluminium reached post DFO and are independent of the bath or type of membrane used. The main goal should be to obtain a high effective concentration gradient.

Acknowledgments

We thank the staff of the Toxicology Laboratory (Professor J Corbella) of the Hospital Clinic, for performing aluminium measurements.

References

1 Alfrey AC, Le Gendre GR, Kaehny WD. *N Engl J Med 1976; 294:* 184
2 Ward MK, Parkinson IS. In Drukker W, Parsons FM, Maher JF, eds. *Replacement of Renal Function by Dialysis.* The Hague: Nijhoff. 1983: 811
3 Ackrill P, Ralston AJ, Day JP et al. *Lancet 1980; i:* 692
4 Graf H, Stummsoll HK, Meisinger V. *Proc EDTA 1981; 18:* 674
5 Rembold CM, Krumlovsky FD, Roxe DM et al. *Trans ASAIO 1982; 28:* 621
6 Graf H, Stummvoll HK, Meisinger V et al. *Kidney Int 1981; 19:* 587

Open Discussion

BAILLOD (London) Could some indication of clinical use of desferrioxamine be made, such as dosage and frequency. Has anyone any comments on the clinical side effects including hyperkalaemia which I have noted in my patients?

MONTOLIU Desferrioxamine was given as a single injection and no side effects or hyperkalaemia were noted.

HABIBUR-RAHMAN (Newcastle upon Tyne) We think that the protein bound aluminium cannot be removed by desferrioxamine as was shown by equilibrium dialysis and ultrafiltration after DFO in vitro. The optimum time of dialysis after DFO infusion would be at the time of maximum aluminium concentration. The serum aluminium peaks after 48 hours and remains the same for 72 hours. So, dialysis after 48 hours of DFO infusion seems to be logical.

MONTOLIU I agree with your first comment. Our experimental protocol was designed to reproduce a kinetic model and not to obtain maximum aluminium estimation.

BRANCACCIO (Milan) I have promoted a protocol similar to yours and I want to tell you that I have obtained aluminium clearances higher than those presented by you. The reason is likely to be due to the fact that the aluminium in our dialysate was always low, about 5μg/L.

MONTOLIU As previously stated our dialysate aluminium was approximately 13μg/L and this probably explains the difference between your results and mine.

Proc EDTA–ERA (1984) Vol 21

THE DESFERRIOXAMINE TEST PREDICTS BONE ALUMINIUM BURDEN INDUCED BY $Al(OH)_3$ IN URAEMIC PATIENTS BUT NOT MILD HISTOLOGICAL OSTEOMALACIA

A Fournier, *P Fohrer, P Leflon, P Moriniere, **M Tolani**, *G Lambrey, R Demontis, JL Sebert, †F Van Devyver, †M Debroe

*Hôpital Nord, Amiens, *Centre Hospitalier, Beauvais, France, †Akadem Zirkenhuis, Antwerp, Belgium*

Summary

Desferrioxamine (DFO), a chelating agent of aluminium was administered to 27 uraemic patients on chronic haemodialysis or haemofiltration with a minimal parenteral exposure to aluminium but taking various amounts of $Al(OH)_3$ for about two years. All these patients had a double bone biopsy for measurement of their aluminium content and histomorphometric evaluation.

Bone aluminium of our patients were 10 times greater than in our uraemic controls. Plasma aluminium increase (△ Al) induced by DFO correlated better than basal plasma aluminium with bone aluminium and cumulative dose of $Al(OH)_3$ correlated with bone aluminium and △ Al DFO. None of the patients had florid osteomalacia and only two had traces of aluminium staining. However 16 had mild mineralisation defect as demonstrated by low mineral appositional rate. The aluminium parameters were not different between the two groups of patients with or without mild mineralisation defect.

It is concluded that the DFO test predicts bone aluminium but not mild histological osteomalacia in uraemic patients moderately aluminium over-loaded with phosphate binders.

Introduction

Although Winney et al [1] have shown that basal plasma aluminium was a good predictor of clinical bone toxicity (when $>200 \mu g/L$) and even in that regard a better predictor than total bone aluminium. Milliner et al [2] proposed the desferrioxamine test (DFO test) as a non-invasive test for diagnosis of histological aluminium osteomalacia. In their experience basal aluminium was $<100 \mu g/L$ in six of 28 cases of aluminium bone disease whereas the plasma aluminium increase induced by desferrioxamine (△Al DFO) was $>180 \mu g/L$ in 28 of 28 patients with aluminium osteomalacia and $<180 \mu g/L$ in three patients without this disease.

Berland et al [3] have confirmed the value of the DFO test to distinguish between patients with and without aluminium stainable deposits, regardless of the histological presence of osteomalacia or hyperparathyroidism [2]. Simon et al [4] have suggested that the DFO test is a better predictor than basal plasma aluminium for the evaluation of total body overload with aluminium since in patients exposed mainly to oral aluminium overload, they found a correlation between the cumulative dose of $Al(OH)_3$ and $\triangle$Al DFO but not with basal plasma aluminium.

Malluche et al [5] on the other hand have concluded that the DFO test has no value for the diagnosis of aluminium bone accumulation since between 12 patients having three criteria of aluminium deposition (positive Aluminon staining, increased bone aluminium content measured by atomic absorption and abnormal peak of aluminium by X-ray analysis) and 10 patients having not all these criteria, they did not find significant difference in $\triangle$Al DFO.

Because of these opinions on the diagnostic value of the desferrioxamine test, we performed the following study in 27 uraemic patients exposed essentially to oral aluminium overload to assess the value of the DFO test in the measurement of bone aluminium content and in the diagnosis of aluminium osteomalacia as well as to assess its link with the cumulative dose of $Al(OH)_3$.

Patients and methods

Patients

We studied 15 patients maintained on chronic haemodialysis for 25 ± 3 months with a dialysate aluminium always <0.3μmol/L due to reverse osmosis water treatment. They were eight men and seven women of a mean age of 54 ± 4 years. Their cumulative dose of $Al(OH)_3$ was 1,600 ± 345g.

Twelve patients on haemofiltration for 24±4 months were also studied. The aluminium concentration of their substitution fluid always ranged 0.15–0.6μmol/L. They were six men and six women aged 67±3 years. Their cumulative dose of $Al(OH)_3$ was 2,740±1,000g. No difference between haemodialysis and haemofiltration was noted for any of these clinical data.

Methods

All patients accepted having two iliac bone biopsies after double labelling by tetracycline.

One biopsy was used for histomorphometry evaluation [6] and staining of aluminium by Aluminon [7]. The other biopsy was divided into two pieces after muscle particle removal for independent measurement of aluminium content, after being precisely weighed: one determination performed in Amiens by inductively coupled plasma emission spectrometry (ICPES) [8]. The other performed in Antwerp by electrothermal atomic absorption spectrometry with a graphite furnace (EAAS) [9].

The normal mean ± SD in seven non-uraemic cadavers is 0.07 ± 0.04μmol/g of fresh bone or 1.8 ± 1.1ppm with ICPES.

The desferrioxamine test was performed by injection of 2g of DFO during the last hour of a dialysis and measurement of plasma aluminium by ICPES before this dialysis and 44 hours later before the subsequent dialysis. Normal non uraemic plasma aluminium is less than 0.3μmol/L.

Results

Comparison of the measurements of bone aluminium by ICPES and EAAS

The correlation between the results of the two methods obtained in 22 cases is good since the correlation coefficient is 0.91, the slope of the regression line 1.075 and the origin ordinate 2.9. This relation shows that the measurement by ICPES usually overestimates the aluminium concentration in the bone. This can be explained by the fact that bone calcium content is high and that there is a known overlap of the emission rays of aluminium and calcium, a phenomenon which does not occur with atomic absorption.

Bone aluminium content

Mean bone aluminium (±SEM) is 16.5 ± 4ppm by ICPES and 12 ± 14 by EAAS. These figures are considerably higher than in bone from non uraemic cadavers (1.8 ± 1.1ppm measured by ICPES).

Nevertheless the aluminium staining of undecalcified sections showed only traces of aluminium in two patients.

Correlation of the △Al DFO with bone aluminium content

The correlation coefficient r for the regression of bone aluminium versus △Al DFO was 0.78 ($p<0.001$) when bone aluminium was measured by ICPES and 0.83 when it was measured by EAAS (Figures 1 and 2). These correlations are better than those with basal plasma aluminium, the r coefficient being respectively 0.49 ($p<0.05$) and 0.51 ($p<0.05$). We suggest therefore that bone aluminium (in ppm) can be evaluated indirectly by the equation obtained with EAAS data: 3.35 △Al DFO μmol/L ± 3.4

Correlation of bone aluminium and △Al DFO with cumulative dose of $Al(OH)_3$

Significant correlations were found between the cumulative dose of $Al(OH)_3$ and bone aluminium content ($r = 0.81$ for EAAS data, $p<0.001$ and $r = 0.64$ for ICPES data, $p<0.01$ or △Al DFO ($r = 0.83$ for EAAS data and $r = 0.81$, $p<0.001$ for ICPES data).

Histomorphometrical evaluation according to osteomalacia criteria

None of our patients had florid osteomalacia with increased osteoid thickness coexistent with decreased mineral appositional rate or decreased mineralisation front. All had the osteoid thickness index of Meunier in the normal range 14–22 [6]. However 16 patients had either decreased appositional rate (<0.48μg/day) and/or decreased mineralisation front (<54 % of the osteoid surface) suggesting

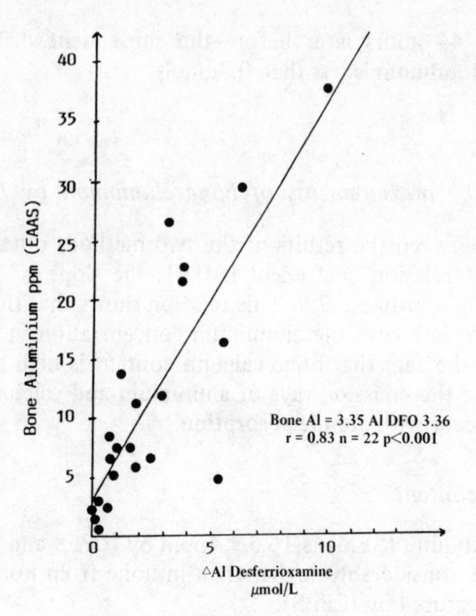

Figure 1

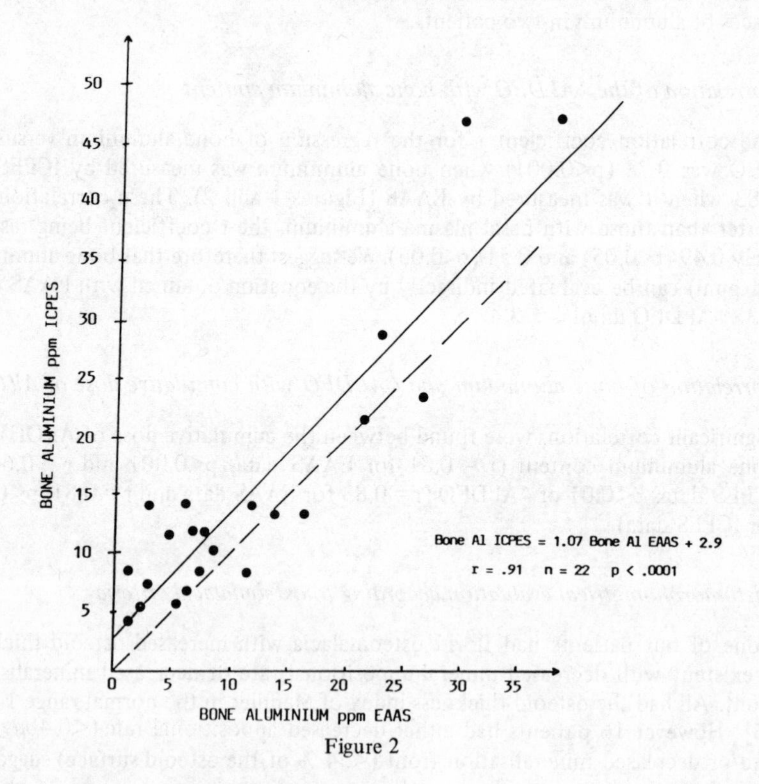

Figure 2

a mild mineralisation defect considered by Evans as osteomalacia of type II. In the 11 other patients no such mineralisation defect was present.

Comparison of the aluminium parameters between these two groups did not show any significant difference (Table I).

TABLE I. Comparison of the aluminium parameters in patients with and without mineralisation defect

Aluminium parameters mean ± SEM	Mineralisation Defect present n = 16	absent n = 11
Basal plasma aluminium μmol/L	1.4 ± 0.4	2.2 ± 0.5
△Al DFO μmol/L	2 ± 0.3	2.7 ± 0.5
Bone aluminium ppm		
ICPES method	14.8 ± 2.5	17.6 ± 5
EAAS method	10.9 ± 3	13.4 ± 4

There is no significant difference between the two groups of patients:
ICPES = Inductively Coupled Plasma Emission Spectrometry
EAAS = Electrothermal Atomic Absorption Spectrometry

Discussion

The bone aluminium content of our patients being much higher (10 times) than that of non uraemic controls, our patients are aluminium over-loaded although much less so than the patients of Malluche or Winney (82 and 79ppm of aluminium from dry bone). Since the cumulative doses of $Al(OH)_3$ correlated with the bone aluminium content and the △Al DFO the over-load of our patients appears to be mainly due to the intestinal absorption of aluminium from their phosphate binders. In fact, these patients were always exposed to low aluminium concentrations in the dialysate and the substitution fluid.

Histomorphometrical studies of their bone showed that none of our patients had florid osteomalacia since none had increased osteoid thickness and only two had traces of histological stainings for aluminium. Since in none was △Al DFO greater than 180μg/L (the highest was 162μg/L) our data agree with those of Milliner [2]. However this test does not discriminate between patients with or without mild alteration of the mineralisation process, the presence of which being demonstrated by decreased mineral appositional rate and/or decreased mineralisation front without increased osteoid thickness. Bone aluminium content did not discriminate between these two groups of patients. This lack of difference of the aluminium parameters between the two groups with and without mineralisation defect, does not mean however that their aluminium overload does not have a deleterious effect on their bone. As shown in another paper of this EDTA issue [10] when the influence of the other potentially pathogenetic factors such as plasma concentrations of PTH, $1.25(OH)_2 D_3$, $24.25(OH)_2 D_3$, Mg, Ca and PO_4 are excluded by the means of a multidimensional analysis, aluminium parameters (bone aluminium, basal plasma and to a lesser degree △Al DFO are negatively correlated to osteoblastic surfaces, mineral appositional rate and bone formation rate, suggesting that their aluminium over-load decreases their bone formation.

Conclusion

In uraemic patients with mild aluminium over-load induced mainly by $Al(OH)_3$, the DFO test can predict bone aluminium content but not the presence of mild mineralisation defect.

References

1 Winney RJ, Cowie JF, Robson JS. *Kidney Int 1984; Supplement. Symposium of New Port Beach on Aluminium Related Diseases*
2 Milliner DS, Nebeker HG, Ott SA et al. *Kidney Int 1984; 25:* 149
3 Berland Y, Olmer M, Charlton S et al. *Abstract Book 9th International Congress Nephrology.* Los Angeles. 1984: 150A
4 Simon P, Allain P, Ang KS et al. *Actualités Néphrologiques de l'Hôpital Necker.* Paris: Flammarion. 1984: 383–414
5 Malluche HH, Smith AJ, Abreo et al. *N Engl J Med 1984; 311:* 140
6 Fournier A, Boudailliez B, Tolani M et al. In Fournier A, Garabédian M, Sebert JL, Meunier PJ. *Vitamine D et Maladies des Os et du Métabolisme Minéral.* Paris: Masson. 1984: 186–189
7 Maloney NA, Ott SM, Alfrey AC et al. *J Lab Clin Med 1982; 99:* 206
8 Allain P, Mauras Y. *Anal Chem 1979; 51:* 2089
9 De Broe ME, Van De Vyver FL, Bekaert AB. *Contre Nephrol 1984; 38:* 37
10 Fournier A, Sebert JL, Fohrer P et al. *Proc EDTA–ERA 1984: 21:* this volume

Open Discussion

KERR (Chairman) Can you tell us how marked is the reduction in ossification in your patients compared with what you find in pre-dialysis patients?

TOLANI I am afraid I do not have the data to hand but in pre-dialysis patients there is a low appositional rate. In our multidimensional analysis the appositional rate was mainly dependent upon vitamin D metabolites. We could detect a link, let us not say an effect, between vitamin D metabolites and apposition.

Proc EDTA–ERA (1984) Vol 21

IRON REMOVAL BY DESFERRIOXAMINE DURING HAEMODIALYSIS: IN VITRO STUDIES

D E Müller-Wiefel, U Vorderbrügge, K Schärer

University Children's Hospital, Heidelberg, FRG

Summary

To find the best way of using desferrioxamine (DFO) in iron (Fe) overloaded patients on regular haemodialysis (HD) treatment, we investigated Fe removal in a standard in vitro haemodialysis model by measuring ^{59}Fe activity in plasma and dialysate after the addition of DFO as a constant infusion during or as a bolus at different time intervals before haemodialysis. Data were compared with those obtained without DFO at normal or high plasma Fe concentrations and by haemofiltration (HF). In summary HF was superior to haemodialysis for Fe removal. No DFO was needed to dialyse high plasma Fe concentrations whereas the efficiency of removal of normal plasma Fe content increased with equilibration time of DFO before haemodialysis. For treatment of Fe overload DFO is therefore recommended to be given after each haemodialysis treatment.

Introduction

Iron overload is a common clinical problem in children on regular haemodialysis treatment – in contrast to aluminium overload. It is the consequence of a more pronounced anaemia compared with adults, making blood transfusion necessary at regular intervals [1]. Serum ferritin (SF) as the most reliable biochemical indicator of Fe stores was found to reflect Fe overload in three-quarters of haemodialysed children [2] and to be dependent not only on the number of blood transfusions but also on genetic conditions. In order to prevent its harmful consequences on heart, liver, endocrine system, muscle, blood and probably bone, severe hemosiderosis (SF >2,000μg/L) which occurs in about 25 per cent of children after 12 months on haemodialysis, has to be treated with the Fe chelating agent DFO. An increasing number of reports in adult patients on haemodialysis deals with the effect of DFO treatment [3–8], although no uniform mode of therapy is practised. To find out the most adequate schedule of DFO therapy in haemodialysis the following in vitro studies were performed. From ethical reasons we did not perform in vivo studies with radioactivity in children.

Methods

Imitating in vivo paediatric conditions pooled human plasma (1l) was haemodialysed for four hours by a cuprophan hollow fibre dialyser (surface area $1.0m^2$, Hemoflow MTS C, Fresenius) against a routine acetate containing dialysate (flow 500ml/min) without transmembrane pressure and a flow of 100ml/min by means of a Drake Willock monitor. The circulating volume was kept constant by a saline infusion. The concentration of plasma Fe was either unphysiologically high (2,500mmol/L) or normal (18mmol/L) with a constant transferrin of 250mg/dl. DFO of 1.0g was given either by an infusion during haemodialysis (3 hours) or as a bolus at the start of, or 30 minutes, two hours, or 24 hours before haemodialysis. At the time of DFO administration ^{59}Fe (0.1mCi) was additionally added to the plasma. Fe elimination was measured by the loss of radioactivity in plasma and its increase in the dialysate. Three plasma samples were taken at 0, 3, 15, 30, 60, 120, 180 and 240 minutes and three dialysate samples at 1, 2, 3, 8, 15, 30, 60, 120, 180 and 240 minutes respectively after start of dialysis. Radioactivity was measured three times per sample in a β-counter and plasma Fe loss expressed as per cent of initial (100%) activity = $\triangle$. All data are given as mean values of simultaneous measurements. Additionally Fe concentrations were measured under the same conditions by atomic absorption spectrophotometry in two studies. Moreover modification of Fe removal was attempted by increasing the plasma transferrin content by 1,000mg/L addition of ascorbic acid (0.5g/L) or by using post dilution haemofiltration (cellulose triacetate filter, SM40041, Sartorius).

Results

At toxic plasma Fe consentrations (2,500μmol/L) Fe was eliminated almost completely within the first hour of haemodialysis both without (Figure 1A) and with DFO ($\triangle$ 97.1 and 99.1%). In contrast, at normal in vivo plasma Fe concentrations (18μmol/L) nearly no Fe could be removed without DFO by haemodialysis (Figure 1B), although about half the plasma content could be removed by HF ($\triangle$ 52%) (Figure 1C). By DFO infusion Fe removal by haemodialysis could only be increased to a minor extent ($\triangle$ 4%) (Figure 1D). Giving DFO as a bolus at the start of haemodialysis Fe removal, however, could be further augmented to 15.8 per cent (Figure 1E) and at 30 minutes before start to 48 per cent (Figure 1F). With this timing of DFO administration Fe removal could be further increased by HF ($\triangle$ 72%, Figure 1G), but not significantly by the addition of either transferrin ($\triangle$51.1%, Figure 1H) or ascorbic acid ($\triangle$ 51.0%, Figure 1I). By increasing the interval between the application of DFO and start of haemodialysis a progressive increase of Fe removal could be detected which amounted to 82.4 per cent at an interval of two hours and to 99.7 per cent at 24 hours (Figures 1K, 1L).

The atomic absorption method revealed comparable results of Fe elimination by haemodialysis without DFO ($\triangle$ + 2.0%) as well as with DFO 24 hours before haemodialysis ($\triangle$99.5%) (Figure 1M).

Radioactive counts of the dialysate/filtrate gave an initial peak three minutes

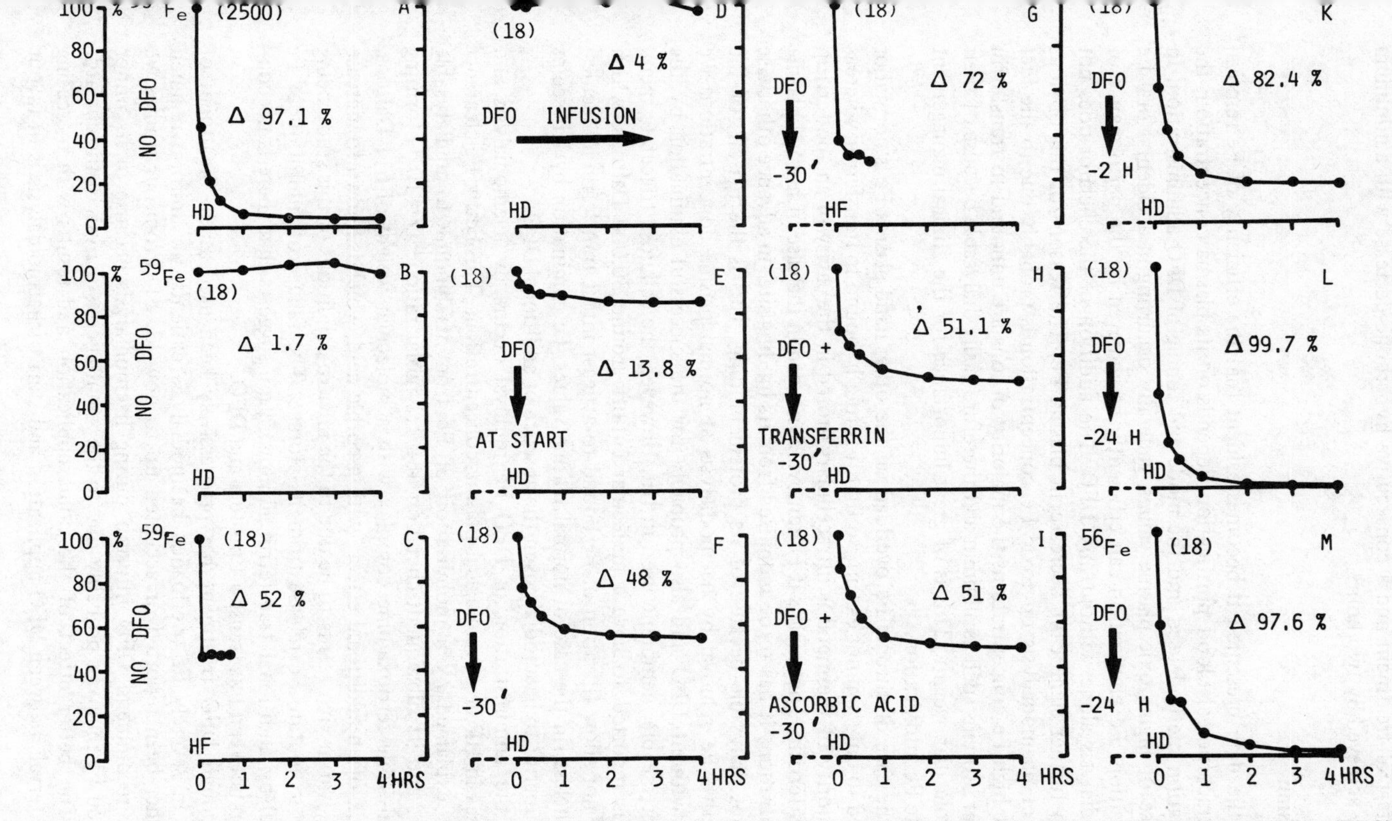

Figure 1. Kinetics of plasma iron (Fe) removal by haemodialysis (HD) or haemofiltration (HF) at different modes of desferrioxamine (DFO) application. Final removal is given by Δ of initial ^{59}Fe radioactive counts or of total Fe as measured by atomic absorption (M). The pre-treatment plasma Fe concentration is presented in brackets. For details see text. ● represent mean values

after start of treatment and a consecutive hyperbolic decrease with a minimum usually achieved by two hours.

Discussion

From the data presented it becomes evident that the efficiency of Fe removal by extracorporeal blood purification depends on its plasma concentration, the purification method used and the mode and timing of DFO administration. In the case of high, toxic plasma values, generally occurring in accidental oral Fe intoxication, Fe elimination can efficiently be achieved by the performance of haemodialysis alone without any DFO. The addition of the chelator does not lead to further significant decrease in plasma Fe. The reason for this obvious effect of haemodialysis on toxic Fe concentration is probably due to the fact that at high Fe concentrations the majority of ions are unbound to transferrin or other proteins such as albumin and therefore easily dialysable because of their low molecular weight. This is of great importance for the clinical management of acute Fe intoxication.

In the case of chronic Fe overload on the other hand plasma Fe is normal or elevated to only a minor extent, so that it is totally bound to transferrin, whereas the majority of the metal is intracellulary stored in the cavity of an apo-ferritin shell as ferritin. That means if Fe removal is needed in the case of normal plasma concentrations it has to be mobilised from its binding sites in advance otherwise it is not dialysable because of its protein binding. This is the reason for the ineffective Fe removal by haemodialysis at normal plasma concentrations we found without DFO and also probably for the successful elimination by the more permeable membrane used in HF. However, the cellulose triacetate membrane is reported to have a molecular cut-off around 30,000 Daltons, which should not allow filtration of Fe bound to at least intact transferrin molecules. Nevertheless, in the case of normal plasma values, Fe elimination by HF seems to be superior to haemodialysis in the absence of additional DFO.

By the administration of DFO Fe removal occurs by haemodialysis and increases in HF at physiological plasma concentrations. This better Fe clearance can be explained by a mobilisation of Fe from its binding to undialysable proteins and its subsequent chelation in a 1:1 molar ratio ($^{K}D = 10^{30}$) forming the dialysable ferrioxamine complex with a molecular weight of 611 Daltons. The lack of an additional effect with ascorbic acid, which is known to increase Fe removal in vivo seems to be due to the in vitro conditions of this study being unable to imitate Fe release from its storage. The failure of inhibition of Fe removal by additional transferrin administration suggests that there is no competition between this transport protein and DFO.

Although DFO treatment has progressively been used as the treatment of choice in severely Fe overloaded haemodialysis patients, no uniform regimen has either been reported or accepted up to now [3–8]. From our studies we cannot recommend DFO infusion during haemodialysis because of its minor effect on Fe removal, probably due to the simultaneous dialysis of the chelator itself. The better effect of the bolus application and results with increasing time intervals between DFO application and start of haemodialysis is in favour of an Fe mobilising and chelating reaction which is time dependent. Although in vitro studies of DFO binding are not totally representative of in vivo conditions because of their multiple compartment character it seems reasonable from the practical clinical point of view to apply DFO at the end of each haemodialysis session. By this procedure there is time enough for DFO to react with plasma and certainly storage Fe to lead to an efficient Fe removal via the dialysis membrane. This therapeutic regimen is in accordance with the subcutaneous DFO infusion used successfully in hemosiderotic children suffering from homozygous β-thalassaemia. As the elimination depends on membrane permeability an effect is achieved more quickly and better not only by HF as demonstrated by us and other investigators [9] but also by additional haemoperfusion [10]. As Fe removal could not be optimised after DFO application 24 hours before haemodialysis we did not investigate the effect of other haemodialysis membranes [5] with this regimen.

The plasma ^{59}Fe removal accorded with the elimination kinetics found by dialysate counting. The data obtained are not artificially influenced by radioactive Fe because the results by atomic spectrophotometry were similar.

References

1 Müller-Wiefel DE. *Renale Anämie im Kindesalter – Untersuchungen zur Pathogenese und Kompensation.* Stuttgart and New York: Thieme. 1982
2 Müller-Wiefel DE, Waldherr R, Feist D et al. *Contr Nephrol 1984; 38:* 141
3 Baker LRI, Barnett MD, Brozovich B et al. *Clin Nephrol 1976; 1:* 326
4 Gokal R, Millard PR, Weatherhall DJ et al. *Q J Med 1979; 191:* 369
5 Rembold CM, Krumlovsky FA, Roxe DM et al. *Trans ASAIO 1982; 28:* 621
6 Simon P, Bonn F, Guezennec M et al. *Néphrologie 1981; 2:* 165
7 Hilfenhaus M, Koch KM, Bechstein PB et al. *Contr Nephrol 1984; 38:* 167
8 McGonigle RJS, Keogh AM, Weston MJ et al. *Dial Transpl 1984; 4:* 214
9 McCarthy JT, Libertin CR, Mitchell JC et al. *Mayo Clin Proc 1982; 57:* 439
10 Chang TMS, Barre P. *Lancet 1983; ii:* 1051

Open Discussion

CAROZZI (Genoa) Did you observe in your study a clinical improvement in the anaemia or in the bone marrow iron incorporation?

MÜLLER-WIEFEL We only have the clinical impression that in some patients who are on long-term desferrioxamine there is some improvement in anaemia. We have not routinely looked at the bone marrow of these patients and so I can't comment on marrow iron.

KERR (Chairman) How do you explain the removal of half of your serum iron by haemofiltration, even without desferrioxamine, when you get almost no removal by haemodialysis suggesting that it is all protein bound. Have you looked at the iron content of the filtrate?

MÜLLER-WIEFEL By our methods we were unable to differentiate between bound and unbound iron. It has been shown by other investigators that the removal of iron increases with the permeability of the membrane.

Proc EDTNA–ERA (1984) Vol 21

DESFERRIOXAMINE TREATMENT FOR ALUMINIUM AND IRON OVERLOAD IN URAEMIC PATIENTS BY HAEMODIALYSIS OR HAEMOFILTRATION

C A Baldamus, H Schmidt, E-H Scheuermann, *E Werner,
J P Kaltwasser, W Schoeppe
*University Hospital, *Gesellschaft für Strahlen- und Umweltforschung, Frankfurt, FRG*

Summary

In an attempt to quantify iron and aluminium removal in patients following desferrioxamine therapy, a comparative study of haemodialysis and haemofiltration was performed in five patients. Haemofiltration was found to be the more efficient treatment especially if iron and aluminium clearances were related to urea clearances. The distribution volumes of chelated iron and aluminium at the low clearance rates obtained, behave as a single compartment which amounts to about 40 per cent of body weight.

Introduction

Desferrioxamine (DFO) treatment is increasingly used in end-stage renal disease to eliminate iron (Fe) and/or aluminium (Al) during haemodialysis (HD) [1–4]. The rationale is that DFO chelates Fe and Al, deposited in body stores, and that this soluble complex is then removed by diffusion through the dialysis membrane. This therapeutic principle is now frequently used in patients needing frequent blood transfusions over a prolonged period of time and suffering from clinical signs of haemosiderosis. The other major indication for DFO therapy is in chronic Al intoxication leading to specific osteopathy and/or encephalopathy [3,4]. Although this therapy is widely used only limited data is available about efficiency of removal in different forms of ESRD treatment. Therefore it is the purpose of this study to quantify Fe and Al removal during haemodialysis and haemofiltration (HF) and to relate it to urea clearances and to try to establish the distribution volumes of chelated Fe and Al.

Patients and methods

Five RDT patients (3 females, 2 males, age 25–64 years) who had been on haemodialysis for a mean of 5.9 years (2.8–8.5 years) entered the study. They all suffered from haemosiderosis. During the last 12 months a mean of 20.4

units of blood per patient were transfused because of different clinical requirements. In all cases plasma ferritin was elevated to a mean of 4,000μg/L (2,300–8,000μg/L). DFO (30mg/kg/BW) was given as infusion (IV) 48 hours before treatment. Then the same patients were treated with either haemodialysis or HF in a chance sequence of a one week time interval. Haemodialysis was performed at a QB of about 200ml/min and a QD of about 500ml/min for four hours using a 1.2m^2 cuprophan hollow fibre dialyser with a wall thickness of 11μm (GF 120M; Gambro). A four hours HF with a 1.6m^2 polyamide hollow fibre haemofilter (HF 202; Gambro) was performed at a maximal QB (>300ml/min) resulting in ultrafiltration rates of about 120ml/min. In addition to plasma, ultrafiltrate was harvested immediately before and after the dialyser by installing small ultrafilters with a polyamide membrane (UF1700; Gambro). During HF, ultrafiltrate was directly sampled from the haemofilter. Plasma and filtrate was taken every 30 minutes during the test treatments. The plasma, filtrate, and dialysate concentrations of urea, Fe and Al were measured, urea in a Beckman autoanalyser by the enzymatic conductivity rate method whereas Fe and Al were analysed in a flameless atomic absorption spectrometer (Perkin Elmer). Then from these concentrations, measured QB and QD, HCT and plasma protein concentration, whole blood and plasma water clearances were calculated. Distribution volumes (V) of urea and chelated Fe and Al were calculated:

$$V = \frac{UF}{BW \cdot \ln \frac{Ct}{Co}}$$

UF is total ultrafiltrate volume during HF, BW is body weight and C represents plasma water concentrations at start (Co) and end of treatment (Ct).

Results

Forty-eight hours after infusion of DFO a significant increase in plasma concentration of Fe and Al was observed: Fe increased from a mean of 166 ± 75 to 307 ± 88μg/dl and Al from 99 to 237μg/dl. The total amount of Fe removed during haemodialysis and HF varied from patient to patient dependent on the initial plasma concentration. In the patients of this study amounts ranged from 2–17mg of Fe and between 1 and 4mg of Al per treatment. Average plasma Fe concentrations fell during haemodialysis by a mean of 16.5 per cent compared to 21.5 per cent during HF. The corresponding values for Al were 47.6 per cent for haemodialysis and 50.7 per cent during HF (Figure 1). Plasma urea concentrations decreased 65 per cent during haemodialysis and 57 per cent during HF. During the same trial runs plasma water concentrations of Fe fell during HD by 43 per cent and during HF by 59 per cent and for Al 43 per cent during haemodialysis but 64 per cent during HF (Figure 1). Plasma water urea concentrations fell by the same percentage as plasma concentration, 65 per cent during haemodialysis and 57 per cent during HF. From QB and plasma water concentration clearances during haemodialysis were calculated to be for urea 60 ± 17ml/min, Fe 58 ± 26ml/min and Al 80 ± 20ml/min. Knowing from in vitro studies that water soluble urea, chelated Fe and Al were filtrable with a sieving coefficient of one the mean filtration rate was identical to clearance

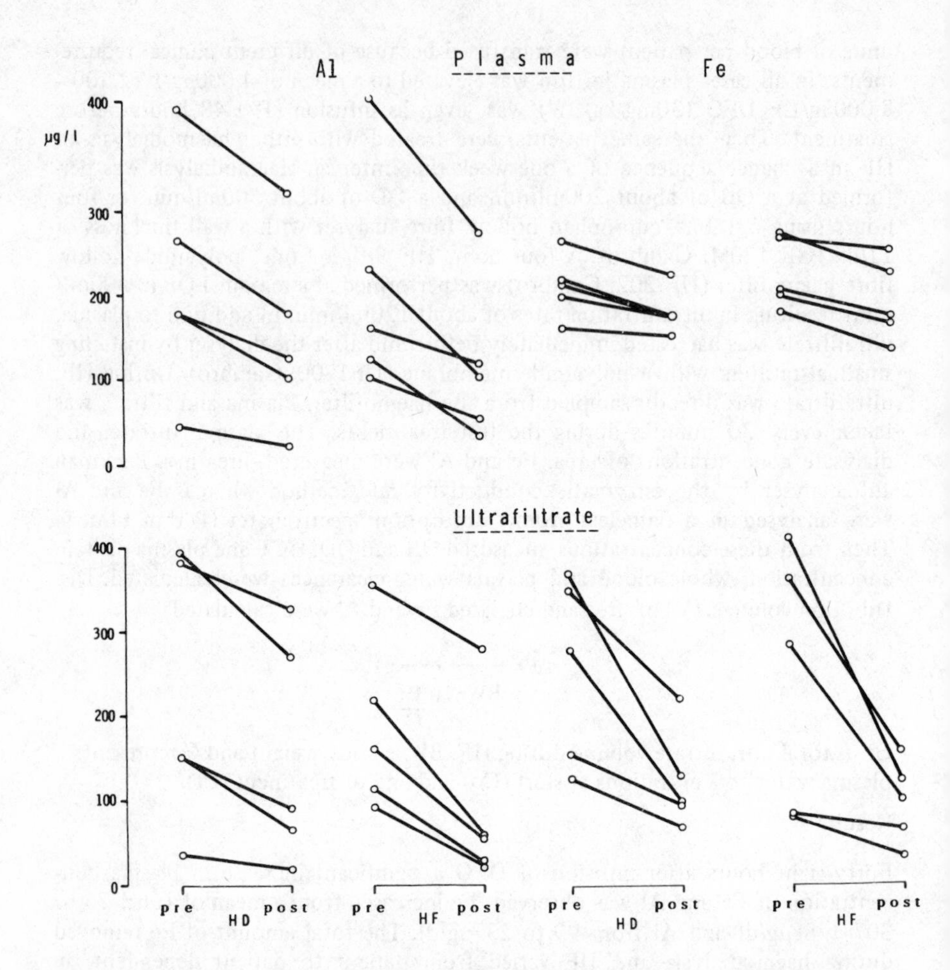

Figure 1. Concentrations of aluminium (Al) and iron (Fe) in plasma and arterial ultrafiltrate at start (pre) and end (post) of treatment, either haemodialysis (HD) or haemofiltration (HF)

rate for all three solutes (105 ± 21ml/min). Although the absolute clearances for Fe and Al during HF were superior to those during haemodialysis, for practical purposes it seems to be logical to correlate it to the particular urea clearances (Cl_{Fe} or Cl_{Al}/Cl_{U}). Calculating this for haemodialysis, the ratio is only 36 per cent for Fe and 50 per cent for Al in contrast to HF where both figures reach 100 per cent indicating that for a comparable urea clearance Fe and Al are far more efficiently removed by HF than by haemodialysis.

The intra-treatment concentration profiles for plasma water urea, Fe and Al concentrations fitted very well a semilogarithmic curve. From the high correlation coefficient for this semilogarithmic decay of 0.9994 for urea, 0.912 for Fe and 0.917 for Al it can be assumed that at least at these relatively low clearance rates distribution volumes for all three solutes behaved like a single compartment. Distribution volume of urea was 60 per cent of BW, that of Fe 44 per cent and of Al 38 per cent of BW.

Discussion

Treatment of choice of clinically evident haemosiderosis in ESRD patients should either be discontinuation of further Fe substitution or renal transplantation. Iron prescription can be stopped immediately but usually regular blood transfusions have to be continued because they are clinically mandatory to keep the patient alive. In these cases renal transplantation offers the best chance of curing haemosiderosis. Iron stores can eventually become depleted by repetitive plebotomy since patient's haemopoetic function is normal after transplantation and high haematocrits are usual. Unfortunately only a minority of patients needing regular blood transfusion can be transplanted with an acceptable risk, because these patients are either older or they suffer not only from renal failure but also from secondary diseases, interfering with renal transplantation. For these patients DFO treatment has become the treatment of choice in clinically evident haemosiderosis [2]. In ESRD patients chelated Fe is either excreted into the faeces in rather limited amounts or it passes across the dialysis membrane during treatment ·[2]. Quantititative differences exist between the different membranes used in commercially available dialysers. In in vitro studies polyacrylonitrile membranes show a higher efficiency than cellulosic membranes and it can be expected the clearance will become even higher with membranes

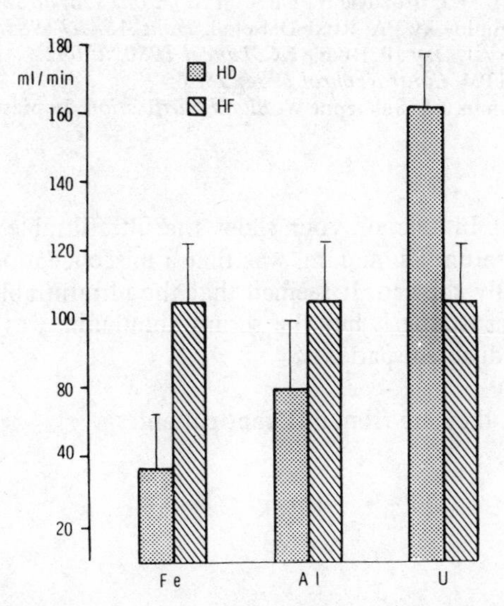

Figure 2. Comparative clearances of iron (Fe), aluminium (Al) and urea (U) during haemodialysis (HD) and haemofiltration (HF). For operating conditions and explanations see text

of high hydraulic permeability [5]. From a therapeutic point of view convective treatment modes should be superior to diffusive modalities since chelated Fe and Al are filtered without restraint and in this regard behave like urea. In contrast in haemodialysis the greater size of chelated Fe compared to urea explains the differences in clearances. These theoretical considerations are confirmed by results of this study (Figure 2).

Aluminium intoxication resulting in encephalopathy and/or osteomalacia is also preferably treated by renal transplantation, but if this is not possible DFO therapy is the treatment of choice [3,4]. The arguments used above for Fe in regard to removal during diffusive or convective treatment modes also hold for chelated Al as seen from Figure 2. As the amount removed during treatment is determined by initial concentration, clearance rate and distribution volume we tried to determine the distribution volume of chelated Fe and Al and to compare it with urea distribution volume, an established measure for total body water. From concentration decay curves it was estimated to about 40 per cent BW for chelated Al. Whether it is possible to imagine the size of the Fe and Al pool by measuring the increase in plasma water concentration following DFO administration needs to be investigated. The increase in plasma water concentration, however, could be a more relevant clinical parameter if iron and aluminium overload than the increase in plasma concentrations. The increases in plasma water concentration might also help to find the correct DFO dose to mobilise the optimal amount of Fe and Al from body stores.

References

1 Baker LRI, Barnett MD, Brozovic B et al. *Clin Nephrol 1976; 6:* 326
2 Rembold CM, Krumlovsky FA, Roxe DM et al. *Trans ASAIO 1982; 28:* 621
3 Ackrill P, Ralston AJ, Day JP, Hodge KC. *Lancet 1980; ii:* 692
4 Pierides AM, Myli PM. *Contr Nephrol 1984; 38:* 65
5 Schmidt M, Baldamus CA, Schoeppe W. *Blood Purification.* In press

Open Discussion

KERR (Chairman) In one of your slides the ultrafiltrable aluminium was as high as the total serum aluminium, was that a misconception on my part or is that what you really showed. It seemed that the ultrafiltrable aluminium in the highest patient was 400μg/L but the serum aluminium was also 400μg/L. Are these results from different patients?

BALDAMUS Yes, they are from different patients.

Proc EDTA–ERA (1984) Vol 21

INCORPORATION OF ALUMINIUM AND EFFECT OF REMOVAL IN EXPERIMENTAL OSTEOMALACIA AND FIBRO-OSTEOCLASIA

T Hess, K Gautschi, H Jungbluth, U Binswanger

University of Zurich, Zurich, Switzerland

Summary

Osteomalacia (OM) and fibro-osteoclasia (FO) was induced in growing rats by feeding a low calcium, low phosphorus, vitamin D poor diet or a low calcium diet respectively. After two weeks, Al Cl_3 was supplemented up to a content of 0.1% of elemental Al and feeding continued for another two weeks, when half of the animals were sacrificed and the rest treated with desferrioxamine. In spite of similar blood aluminium concentrations, OM rats retained more Al in bone than FO rats. Desferrioxamine (DFO) treatment resulted in a significant decrease of Al in OM rats and tended to decrease Al in FO rats. Simultaneously the per cent mineral weight of the bone increased. We conclude that the pre-existing bone pathology (OM or FO) is related to aluminium uptake and removal by DFO, which results in rapid mineral uptake.

Introduction

Aluminium (Al) retention and incorporation into bone is recognised to occur in renal insufficiency before and during haemodialysis treatment [1]. Deposition of Al in the mineralising front of osteomalacic bone suggests that mineralisation is impaired due to the uptake of the element [2]. Al comes from the prescribed oral Al $(OH)_3$ as a phosphate binder [3] or from contamination of the dialysate [4]. Since phosphate binders are prescribed at a creatinine clearance of 25ml/min when renal osteodystrophy already exists [5], we devised an experiment to study the relationship between pre-existing bone pathology and Al uptake as well as its removal by desferrioxamine (DFO) [6].

Methods

Thirteen rats each weighing 90g were raised on a vitamin D deficient rachitogenic diet (calcium 0.03%, magnesium 0.18%, phosphorus 0.15%) and low calcium diet (calcium 0.03%, magnesium 0.1%, phosphorus 0.7%) respectively. After two weeks the diet was supplemented with Al Cl_3, 0.1%. At four weeks half of the animals were sacrificed. The rest received 40mg/kg/BW of desferrioxamine 24 and 12 hours before sacrifice. Serum analysis was performed by atomic absorption spectrometry (CV<3%). Bones were cleaned, weighed and ashed; ash analyses were obtained by the same chemical analytical procedures. Statistical analysis was performed using Student's 't' test. Data are expressed as mean ± SE.

Results (Table I)

Due to the diet, OM rats exhibit hypercalcaemia and hypophosphataemia. In spite of a higher diet content, the serum magnesium concentration is lower in OM than in FO rats. Due to a large spread, the serum Al concentration is similar in both groups of animals. The chemical ash analysis reveals a higher calcium and aluminium, but a lower magnesium content in OM as compared to FO rats. Weight percentage of mineral and matrix are in accord with the expected bone changes (differences not significant).

TABLE I. Effect of feeding rachitogenic and low calcium diet supplemented with Al Cl_3 (mean ± SE)

Serum	Osteomalacia n=6	Fibro-osteoclasia n=6
Ca (mmol/L)	2.90 ± 0.03	2.24 ± 0.07 **
Mg (mmol/L)	0.79 ± 0.03	0.97 ± 0.03 **
P (mmol/L)	0.43 ± 0.05	2.66 ± 0.10 **
Al (μmol/L)	2.52 ± 0.41	2.20 ± 0.28 NS
Bone ash		
Ca (mmol/g)	10.23 ± 0.24	9.70 ± 0.11 *
Mg (mmol/g)	0.29 ± 0.01	0.36 ± 0.005**
P (mmol/g)	6.36 ± 0.13	6.53 ± 0.09 NS
Al (μmol/g)	1.90 ± 0.12	1.30 ± 0.16 *
Bone weight		
Mineral %	44.9 ± 0.73	46.5 ± 1.02 NS
Matrix %	55.1 ± 0.73	53.5 ± 1.02 NS

OM vs FO: *p<0.05; **p<0.001

The DFO treatment resulted in a decrease of the serum Al concentration from 2.5±0.4 to 1.6±0.1μmol/L ($p<0.05$) and from 2.2±0.3 to 2.0±0.1μmol/L (NS) in OM and FO animals respectively. Correspondingly, bone ash Al changed from 1.9±0.1 to 1.5±0.1μmol/g ($p<0.01$) and 1.3±0.2 to 1.2±0.1μmol/g (NS).

Concomitantly percentage mineral weight rose from 44.9±0.7 to 49.5±0.6 ($p<0.005$) in OM rats and from 46.5±1.0 to 49.4±0.6 in FO rats ($p<0.05$).

Discussion

The data presented indicate that osteomalacic bone accumulates more aluminium than fibro-osteoclastic bone during similar exposure and at low serum values. The extraction of Al by DFO is followed by a rapid increase of the percentage mineral weight. This fact is in accord with the assumption of a mineralisation blockade by Al which is rapidly reversible. Whether this block is located in the mineralising front and/or more generalised at the crystal surface of bone mineral [2] remains to be elucidated.

References

1 Alfrey AC, Hegg A, Craswell P. *Am J Clin Nutr 1980; 33:* 1509
2 Maloney NA, Ott SM, Alfrey AC et al. *J Lab Clin Med 1982; 99:* 206
3 Fleming LW, Stewart WK, Fell GS et al. *Clin Nephrol 1982; 17:* 222
4 Walker GS, Aaron JE, Peacock M et al. *Kidney Int 1982; 21:* 411
5 Malluche HH, Ritz E, Lange HP et al. *Kidney Int 1976; 9:* 335
6 Ackrill P, Ralston AJ, Day JP et al. *Lancet 1980; ii:* 692

Proc EDTA–ERA (1984) Vol 21

1αOH VITAMIN D_3 INCREASES PLASMA ALUMINIUM IN HAEMODIALYSED PATIENTS TAKING Al $(OH)_3$

A Fournier, R Demontis, Y Tahiri, *M Herve, **Ph Moriniere**, †A Leflon, Z Abdull-Massih, H Atik, S Benelmouffok

*Hôpital Nord, *Société de Mathématiques appliquées à la Recherche Médicale et Biométrique, †Laboratoire de Biochimie, Amiens, France*

Summary

1αOH vitamin D_3 at the dose of 6μg per week was given for four weeks to 16 stable patients on chronic haemodialysis with a low dialysate aluminium while taking a constant dose of $Al(OH)_3$. A significant increase of their plasma aluminium was observed while on $1\alpha(OH)D_3$ therapy and during the six weeks following. This increase correlated with the cumulative dose of $Al(OH)_3$ and duration on dialysis but not with the recent dose of $Al(OH)_3$. The increase in plasma aluminium observed with $1\alpha(OH)D_3$ and after its discontinuation is more likely to be due to aluminium redistribution than to increased intestinal aluminium absorption. This effect indicates the need for close monitoring of plasma aluminium in uraemic patients treated with $1\alpha(OH)D_3$.

Introduction

Prevention of hyperparathyroidism is based on the correction of hyperphosphataemia by aluminium containing phosphate binders and the correction of persistent hypocalcaemia by administration of calcium supplements and pharmacological doses of vitamin D or physiological doses of 1αhydroxylated metabolites of vitamin D [1]. Hyperaluminaemia is frequent in this population and, even in the absence of high water concentrations, may lead to clinically significant bone disease or encephalopathy [1]. Vitamin D metabolites may increase plasma aluminium indirectly by increasing the need of $Al(OH)_3$ in order to control hyperphosphataemia [1]. Whether or not vitamin D metabolites may directly increase plasma aluminium has not been adequately studied. Experimentally Drueke et al [2] have observed higher plasma aluminium contrasting with lower liver aluminium content in uraemic rats intoxicated by oral $Al(OH)_3$ when they were taking $1,25(OH)_2D_3$. Clinically de Vernejoul has reported an increase of plasma aluminium in four of five uraemic patients taking pharmacological doses of $25(OH)D_3$ [3].

We have studied the effect of 1αOH vitamin D_3 on plasma aluminium in uraemic patients taking a constant dose of $Al(OH)_3$ but not exposed to high aluminium dialysate.

Patients and methods

Patients

Sixteen patients (9 men and 7 women: 42–72 years) on chronic haemodialysis were selected because they needed a constant dose of $Al(OH)_3$ (0.5–4g; mean 2g per day) for control of their plasma phosphate between 1–2mmol/L, while their plasma calcium was <2.5mmol/L. They had been on haemodialysis for 2–108 months (mean = 43 months) and their dialysate aluminium had always been <0.3μmol/L due to reverse osmosis treatment of the water.

Treatment protocol

After three weeks observation 1α(OH)D_3 was given for four weeks at the dose of 6μg weekly, the drug being given at the end of each dialysis. After 1α(OH)D_3 discontinuation follow-up was continued for two weeks in the 16 patients and for eight weeks in six patients. Throughout this study, the dose of $Al(OH)_3$ has been kept constant.

Analytical methods

Before the first dialysis of each week the following plasma concentrations were measured throughout the study: Calcium, proteins and phosphate by auto-analyser technique. Aluminium by inductively coupled plasma emission spectrometry (normal range 0.46±0.15μmol/L: detection limit 0.15μmol/L) [4].

Plasma PTH was measured in only six patients with an antibody specific to the middle region of the molecule (normal range 80–220pg/ml) [5], at the end of the control period and of the 1α(OH)D_3 administration, then four to eight weeks after 1α(OH)D_3 discontinuation.

Statistical methods

Significance of the changes of plasma parameters compared to the control period was assessed by analysis of variance, pooling the data of the control period and then the data of successive two week periods. Association between plasma aluminium increase and parametric variables was looked for with simple regression and covariance analysis.

Results

The individual data of this study are reported in the thesis of Demontis [6].

Table I shows that plasma concentration of aluminium (P Al) increases significantly after two to four weeks of 1αOH vitamin D_3 administration

TABLE I. Effect of $1\alpha(OH)D_3$ on plasma concentration of aluminium, calcium, phosphate and PTH 44–48 (Mean + SEM; percent increase versus control period in parenthesis)

Plasma Concentration (Numbers of patients)	Control Period	Weeks on $1\alpha(OH)D_3$		Weeks after $1\alpha(OH)D_3$ discontinuation			
		1 + 2	3 + 4	1 + 2	3 + 4	5 + 6	7 + 8
	(16)	(16)	(16)	(16)	(6)	(6)	(6)
Aluminium (μmol/L)	1.20 ± 0.25	1.51 ± 0.3*	1.69 ± 0.35**	1.71 ± 0.3**	1.52 ± 0.4*	1.72 ± 0.6*	1.05 ± 0.50
Calcium (mmol/L)	2.28 ± 0.06	2.48 ± 0.07	2.54 ± 0.09*	2.38 ± 0.06	2.31 ± 0.06	2.28 ± 0.08	2.18 ± 0.06
Phosphate (mmol/L)	1.72 ± 0.13	1.86 ± 0.12	1.90 ± 0.11	1.81 ± 0.12	1.31 ± 0.20	1.33 ± 0.15	1.41 ± 0.10
PTH 44–68 (pg/ml) (only 6 pts)	1375 ± 300	–	654 ± 200**	–	1173 ± 300*	–	911 ± 150*

Comparison versus control period: $*p<0.05$; $**p<0.01$

(28 and 43%) and that the increase is still observed during the six weeks following the discontinuation of $1\alpha(OH)D_3$. After seven to eight weeks plasma aluminium reaches control values.

Plasma calcium increases significantly during $1\alpha(OH)D_3$ administration but the increase is no longer significant 15 days following $1\alpha(OH)D_3$ discontinuation. No significant change in plasma phosphate was observed. Plasma PTH measured in only six patients decreased significantly (−50%; $p<0.01$) at the end of $1\alpha(OH)D_3$ administration and the decrease is still present during the eight weeks after $1\alpha(OH)D_3$ discontinuation.

The greater plasma aluminium increase (measured during the two weeks following $1\alpha(OH)D_3$ discontinuation correlates to the total previously prescribed dose of $Al(OH)_3$ and to the duration on dialysis ($p<0.01$ and <0.02 respectively by covariance analysis) but not to the recent dose of $Al(OH)_3$ nor to age, sex or the nature of nephropathy.

Variations of plasma aluminium (versus control period) positively correlate with plasma calcium variations during $1\alpha(OH)D_3$ administration and during the four weeks following $1\alpha(OH)D_3$ discontinuation at a borderline significance ($p = 0.08$). No correlation between plasma aluminium variation and plasma phosphate variation is observed. Significant negative correlations are however observed between variations of plasma aluminium and variations of plasma PTH only during the period of $1\alpha(OH)D_3$ administration.

Discussion

$1\alpha OH$ vitamin D_3 induces in stable patients on chronic haemodialysis, not exposed to parenteral aluminium intoxication but taking a constant dose of $Al(OH)_3$, an increase in their plasma aluminium concentration which lasts six weeks after the drug discontinuation while the effects on plasma calcium resolve within 15 days.

These findings may be explained either by an increase in intestinal absorption of aluminium or by a decreased capacity of tissue storage of aluminium. The fact that plasma aluminium increase is correlated to duration on dialysis and to the cumulative dose of $Al(OH)_3$ but not to the recent dose of $Al(OH)_3$ favours the hypothesis of a redistribution of total body aluminium. The experimental data of Drueke et al [2] showing that uraemic rats treated with $1,25(OH)_2D_3$ accumulate less aluminium in their liver than paired controls when exposed to an oral aluminium load but have higher plasma aluminium values, are consistent with our observation. To determine between the two hypotheses the effect of $1\alpha(OH)D_3$ on the plasma aluminium of uraemic patients previously loaded with aluminium but no longer taking aluminium hydroxide, will have to be studied. Which of the hypotheses is eventually proven, the fact that plasma aluminium increases with $1\alpha OH$ vitamin D_3 should lead to a close monitoring of plasma aluminium when 1α hydroxylated vitamin D metabolites are given to patients previously loaded with aluminium or still taking aluminium containing phosphate binders.

References

1 Fournier A, Moriniere Ph, Sebert JL et al. In Robinson R, ed. *Proceedings of the 9th International Congress of Nephrology, Los Angeles.* New York: Springer Verlag. 1984
2 Drueke T, Lacour B, Basile C et al. *Abstract 8th International Congress of Nephrology, Athens; 1981:* 357
3 de Vernejoul MC. *Calcif Tissue Int 1983; 35:* A49
4 Allain P, Mauras Y. *Anal Chem 1979; 51:* 2089
5 Gueris J, Ferriere C. *Pathol Biol 1975; 23:* 821
6 Demontis R. *Effet de la 1α(OH) vitamine D_3 l'aluminémie des hémodialysés. Thèse Doctorat en Médecine.* Amiens. Septembre 1984

Proc EDTA–ERA (1984) Vol 21

RED BLOOD CELLS INDICES AND ALUMINIUM TOXICITY IN HAEMODIALYSIS PATIENTS

C Tielemans, L Kalima, F Collart, R Wens, *J Smeyers-Verbeke, †D Verbeelen, M Dratwa (Introduced by P Vereerstraeten)

*Hôpital Brugmann, *Farmaceutisch Instituut, †Akademisch Ziekenhuis, Brussels, Belgium*

Summary

Microcytic, hypochromic anaemia is a feature of aluminium toxicity. To detect the possible influence of aluminium on erythropoiesis in a general haemodialysis population we studied the evolution of red blood cell parameters and aluminium status in 30 patients (27 without aluminium toxicity symptoms). Aluminium status was assessed by serum aluminium measurements before (BAl) and after (PAl) a desferrioxamine infusion. The evolution with time (△) of PAl and DAl (= PAl – BAl) during the prospective study inversely correlated with △ mean corpuscular volume ($2\alpha<0.01$) and △ mean corpuscular haemoglobin ($2\alpha<0.001$). Patients with DAl > 180μg/L had lower mean corpuscular haemoglobin values ($p<0.05$). These findings suggest that aluminium inhibits haemoglobin synthesis even in haemodialysis patients free of aluminium toxicity symptoms.

Introduction

Aluminium toxicity has become one of the most preoccupying problems in the long-term management of haemodialysis patients. Besides the neurological manifestations [1] and vitamin D-resistant osteomalacia [2], severe aluminium intoxication can be responsible for a microcytic hypochromic anaemia [3–5].

The present prospective study was conducted to detect a possible influence of exposure to aluminium on erythropoiesis in a general haemodialysis population.

Patients and methods

We prospectively studied 30 haemodialysis patients (18 men, 12 women) whose mean age was 49.8 years (range: 23–85). They had been on regular haemodialysis for a mean period of 63.1 months (range: 5–132). They were dialysed nine to 12 hours weekly, with hollow fibre cuprophan dialysers. Dialysate was prepared from softened water; its aluminium content had been randomly assessed during the last three years and never exceeded 25μg/L. Aluminium containing phosphate binders had been prescribed to each patient at doses varying with time of 0 to 4g/day.

Diet, dialysis schedules, vitamin and drugs regimens remained unchanged throughout the study as did our transfusion policy: one unit packed red blood cell was transfused when the weekly-measured haematocrit fell below 25 per cent. The individual transfusion rate ranged from 0 to 3 units per month.

Patients who developed systemic infection, hepatitis, overt bleeding, who had iron, vitamins or nutritional deficiencies or underwent surgery during the study period or the six preceding months were excluded. Among the 30 patients retained for the study, three had been previously identified with aluminium-induced bone disease [2] and were treated with desferrioxamine [6] during the study (2g at the end of each dialysis). The 27 others remained free of any symptoms of aluminium toxicity.

Each patient was evaluated at the start of the study (time 0) and six to 12 months later (time 1) for erythropoiesis and aluminium status. The latter was assessed by a 'desferrioxamine test' [7]: serum aluminium was measured by flameless atomic absorption spectro-photometry [8] before (basal aluminium, BAl) and 48 hours after (peak aluminium, PAl) a single 50mg/kg desferrioxamine infusion, and the difference (DAl) was calculated. Several red blood cell parameters (assessed by an automatic counting procedure) were determined at times 0 and 1 by calculating the mean of the values available during the two months preceding the 'desferrioxamine test': haematocrit, haemoglobin, number of red blood cells and their indices (mean corpuscular volume, mean corpuscular haemoglobin and mean corpuscular haemoglobin concentration). The evolution with time (△) of serum aluminium and red blood cell indices was expressed by the difference of their values at the start (time 0) and the end (time 1) of the prospective study.

Results were expressed as means ± SEM. Statistical analysis was performed using Student's 't' test for unpaired data and correlation test.

Results

Some prominent characteristics of the studied population at the start of the study are shown in Table I. Among the studied patients, only one disclosed microcytosis (mean corpuscular volume < 80fl) and six had hypochromia (mean corpuscular haemoglobin < 27μg). Twenty-five had a DAl over 180μg/L. Aluminium (BAl, PAl, and DAl) were not correlated with haematocrit, haemoglobin, mean corpuscular volume, mean corpuscular haemoglobin or mean corpuscular haemoglobin concentration. On the contrary, △PAl and △DAl during the prospective study were negatively correlated with △mean corpuscular volume ($r = -0.505$, $2\alpha<0.01$ and $r = -0.521$, $2\alpha<0.01$, respectively) and △ mean corpuscular haemoglobin ($r = -0.629$, $2\alpha<0.001$ and $r = -0.633$, $2\alpha<0.001$, respectively). △mean corpuscular volume and △mean corpuscular haemoglobin were also correlated together ($r = 0.88$, $2\alpha<0.001$). △mean corpuscular haemoglobin concentration was not correlated with △mean corpuscular volume, △mean corpuscular haemoglobin, or △aluminium. Moreover, patients with DAl over 180μg/L also had a significantly lower mean corpuscular haemoglobin (28.2 ± 0.5 vs 30.4 ± 0.6, $p<0.05$); mean corpuscular volume was also lower in that subpopulation but the difference failed to reach significance (87.4 ± 2.3 vs 91.9 ± 1.5, NS).

TABLE I. Characteristics of the studied patients at time 0

	mean	± SEM	range
BAl (μg/L)	78.7	11.7	7.2–275.1
PAl (μg/L)	407.4	46.8	90.1–1037.0
DAl (μg/L)	328.8	41.5	77.8–1002.8
Hct (%)	27.2	0.5	23.6–32.4
Hb (g/dl)	8.7	0.1	6.8–10.2
MCV (fl)	88.8	1.0	78.0–96.8
MCH (pg)	29.0	0.4	25.2–32.6
MCHC (%)	32.1	0.2	28.6–33.6
Tx*		– –	0–3

*number of packed red blood cell units transfused monthly to maintain Hct at 25 per cent
Hct = haematocrit; Hb = haemoglobin; MCV = mean corpuscular volume; MCH = mean corpuscular haemoglobin; MCHC = mean corpuscular haemoglobin concentration

Discussion

Severe aluminium intoxication has been demonstrated to cause a non-iron deficient microcytic hypochromic anaemia in haemodialysis patients [3–5]. That conclusion was also supported by more recent experimental data in aluminium-loaded uraemic rats [9].

The haemodialysis population presented in this work was free of manifestations of aluminium toxicity and the mean corpuscular volume and mean corpuscular haemoglobin remained within normal limits. However, it seems evident that most patients were aluminium-overloaded on the basis of the 'desferrioxamine test'. Indeed, in a study by Milliner et al [7] a DAl over 180μg/L was always associated with increased aluminium tissue stores, as demonstrated by specific staining of bone tissue specimens.

In one aluminium-intoxicated patient, Touam et al [9] were able to demonstrate an inverse correlation between mean corpuscular volume and aluminium. In our population we did not observe any correlation between serum aluminium and red blood cell indices. However, this fact is not surprising since the determinism of red blood cell indices such as mean corpuscular volume and mean corpuscular haemoglobin must be multifactorial (role of genetic factors, nutritional and iron status, mixing with transfused red blood cells, possible influence of aluminium exposure and so on). Thus, in looking for a possible influence of aluminium on those parameters it seemed more appropriate to study their evolution with time in a given population. This allowed us to demonstrate a high inverse correlation between the changes in PAl and DAl (which are good indicators of total body aluminium [7] and those in mean corpuscular volume and mean corpuscular haemoglobin. Moreover in patients with significant aluminium overload (defined by DAl>180μg/L) mean corpuscular haemoglobin was significantly lower, while remaining within normal limits.

In conclusion, although microcytosis and hypochromia seem to be hallmarks of severe aluminium intoxication, the present findings suggest that milder degrees of aluminium exposure could inhibit haemoglobin synthesis and aggravate anaemia in haemodialysis patients, even in the absence of overt aluminium toxicity symptoms, microcytosis or hypochromia. Also, in the absence of iron deficiency a fall in mean corpuscular volume and mean corpuscular haemoglobin with time in a given patient must raise the possibility of aluminium overload.

Acknowledgments

This work was encouraged by a grant from the Fonds Lekime-Ropsy (Université Libre de Bruxelles) and grant number 3.0033.82 from the Belgian Medical Research Council (FGWO).

References

1 Alfrey AC, Legendre GR, Kaehny WD. *N Engl J Med 1976; 294:* 184
2 Pierides AM, Edwards WG Jr, Cullum UX Jr et al. *Kidney Int 1980; 18:* 115
3 Elliott HL, MacDougall AJ, Fell GS. *Lancet 1978; i:* 1203
4 O'Hare JA, Murnaghan DJ. *N Engl J Med 1982; 306:* 654
5 Short AI, Winney AJ, Robson JS. *Proc EDTA 1980; 17:* 226
6 Arze RS, Parkinson JS, Cartlidge WEF et al. *Lancet 1981; ii:* 1116
7 Milliner DS, Nebeker HG, Ott SA et al. *Abstracts 16th Annual Meeting of the American Society of Nephrology 1983; 12A*
8 Smeyers-Verbeke J, Verbeelen D, Massart DL. *Clin Chem Acta 1980; 108:* 67
9 Touam M, Martinez F, Lacour B et al. *Clin Nephrol 1983; 19:* 295

Proc EDTA–ERA (1984) Vol 21

BONE SCAN IN HAEMODIALYSIS PATIENTS: RELATION TO HYPERPARATHYROIDISM AND ALUMINIUM TOXICITY

C Tielemans, R Wens, F Collart, M Dratwa, *J Smeyers-Verbeke, I Van Hooff, G De Roy, P Bergmann, †D Verbeelen (Introduced by P Vereerstraeten)

*Hôpital Brugmann, *Farmaceutisch Instituut,*
†Akademisch Ziekenhuis, Brussels, Belgium

Summary

The influence of hyperparathyroidism and aluminium toxicity on bone scan scores (FS = Fogelman's score) was studied in 37 haemodialysis patients (group 1), of whom 24 had aluminium toxicity (group 2). FS, parathormone (iPTH) and aluminium status were assessed simultaneously, the latter by measuring serum aluminium before (BAl) and 48 hours after (PAl) a desferrioxamine infusion. FS correlated directly with iPTH in all groups, inversely with PAl and DAl (= PAl – BAl) in group 1; FS $\leqslant$2 was found in group 2 only ($2\alpha < 0.05$). The time course of FS and aluminium concentration was also studied in 24 patients: a PAl increase was significantly associated with a FS decrease ($2\alpha < 0.001$). In conclusion, hyperparathyroidism increases and aluminium toxicity decreases FS; FS$\leqslant$2 is very suggestive of aluminium toxicity.

Introduction

Bone scintigraphy often shows increased isotope uptake in patients with chronic renal failure [1,2]; however, its exact significance remains debated. To better understand the scintigraphic patterns of renal osteodystrophy we studied the respective influence of hyperparathyroidism and aluminium status on bone scintigrams in 37 haemodialysis patients.

Patients and methods

Thirty-seven haemodialysis patients (20 men, 17 women) were selected for this study (group 1). Their mean age was 51.1 years (range: 23–85) and they had been on regular dialysis for a mean period of 54.8 months (range: 9 to 127). They were dialysed nine to 12 hours weekly with hollow fibre cuprophan dialysers using dialysate prepared with softened water and containing 7.5mg/dL of calcium. The dialysate aluminium content, measured randomly during the last three years, ranged from 10 to 25µg/L. Most of the patients had been prescribed vitamin D analogues, calcium carbonate and aluminium containing phosphate-binders at varying doses to control hyperparathyroidism.

Aluminium was measured in serum and dialysate samples by flameless absorption spectrophotometry [3].

Aluminium status was evaluated in each patient by a 'desferrioxamine test': serum aluminium was measured before (BAl) and 48 hours after (PAl) a 50mg/kg desferrioxamine infusion [4] and the difference (DAl) was calculated. On the basis of this desferrioxamine test we selected for the study 24 patients with asymptomatic aluminium intoxication [4] (group 2, DAl ⩾220µg/L) and 13 patients without aluminium intoxication (group 3, DAl <150µg/L). This classification was confirmed by specific staining of bone tissue specimens for aluminium [5] in the 16 patients of group 2 and the 5 of group 3 in whom it could be performed.

Serum alkaline phosphatase (SAP) and immunoreactive carboxyl-terminal parathyroid hormone (iPTH) [6] were measured at the same time. A ^{99m}-Tc pyrophosphate BS was performed in each patient during the four weeks preceding or following the desferrioxamine test; it was scored semi-quantitatively according to Fogelman [1] (Fogelman's score, FS, normal ⩽4, (Table I)) by one of the authors who was not informed of the results of iPTH, SAP and desferrioxamine tests.

TABLE I. Scoring of bone scan: FS ranging from 0 to 12, normal ⩽4 [1]

Each sign was scored 0, 1 or 2 as follows;
hyperactivity of the cranial vault and face
hyperactivity of the axial skeleton
hyperactivity of the sternal edges
hyperactivity of the chondro-costal joints
periarticular activity
hyperactivity of the long bones diaphysis

In 24 patients, the BS, desferrioxamine test, iPTH and SAP measurements were repeated at a one-year interval. The evolution with time (△ = 1st-2nd measurement) of those parameters was also studied.

Results were expressed as means ± SEM. Statistical analysis of the data was performed using Wilcoxon's test (for comparison of FS), Student's 't' test for unpaired data (for iPTH, SAP, aluminium), correlation and chi-square tests.

Results

Comparisons of groups 2 and 3 is shown in Table II. The iPTH, SAP and FS were not different in the two groups but low FS ⩽2) were seen in group 2 only ($2\alpha<0.05$).

FS was directly correlated with PTH in each group ($2\alpha<0.001$ in 1 and 2; $2\alpha<0.05$ in 3); moreover in each group, patients with FS >4 had higher iPTH

TABLE II. Comparison of groups 2 and 3

Parameter	Normal	Group 2 (n = 24)	Group 3 (n = 13)	2 vs 3
FS	$\leqslant 4$	5.7 (range: 1–12)	65 (3–12)	NS
Number of patients with FS $\leqslant 2$	–	6	0	$2\alpha < 0.05$
BAl (μg/L)	$\leqslant 1$	88.4 ± 20.5	39.4 ± 12.9	$p < 0.001$
PAl (μg/L)	–	486.6 ± 41.3	138.8 ± 20.5	$p < 0.001$
DAl (μg/L)	–	380.2 ± 36.2	99.4 ± 12.7	$p < 0.001$
iPTH (pg/ml)	65–630	1258 ± 107	1243 ± 127	NS
SAP (IU/L)	65–230	453.3 ± 86.7	350.5 ± 91.2	NS

than those with FS <4 (group 1: 1441 ± 108 vs 976 ± 86pg/ml, $p<0.005$; group 2: 1506 ± 141 vs 964 ± 115pg/ml, $p<0.01$; group 3: 1348 ± 171 vs 1009 ± 81 pg/ml, NS). FS was inversely correlated with PAl and DAl in group 1 ($2\alpha<0.05$). Plasma aluminium did not correlate with iPTH or SAP, nor did SAP with FS.

In the group of 24 patients studied twice at a one-year interval we found an inverse correlation between ΔFS and ΔPAl ($2\alpha<0.01$) and ΔDAl ($2\alpha<0.01$). Furthermore, a rise of PAl was significantly associated with a decrease of FS ($2\alpha<0.001$). ΔFS was not correlated with ΔPTH, ΔSAP nor ΔBAl.

Discussion

Bone scintigraphy has been advocated for detection of renal osteodystrophy, follow-up of its evolution [1,2] and selection of patients who can benefit from vitamin D therapy [7].

The present work was undertaken to assess the respective influence of hyperparathyroidism and aluminium toxicity on bone scintigraphy. Therefore we selected in our haemodialysis population a group of patients with aluminium intoxication and another free of it. This selection was based on the 'desferrioxamine test', which has been shown to correlate with bone aluminium content [4]; in the study by Milliner et al [5] aluminium overload was associated with a DAl $>180\mu$g/L 36 hours after a 40mg/kg desferrioxamine infusion while patients with a DAl $<180\mu$g/L were free of it. Based on these findings, we selected the patients with DAl $\geqslant 220\mu$g/L as aluminium-intoxicated subjects (group 2) and those with DAl $<150\mu$g/L (group 3) as non-intoxicated. This selection was confirmed by the staining for aluminium in each of the 21 patients in whom bone biopsy was performed.

Groups 2 and 3 could not be differentiated by FS, PTH or SAP. However, our results in each group show that FS increased with increasing iPTH. On the contrary, in the whole population under study, FS was lower when aluminium overload (assessed by PAl and DAl) was more severe, and in the 24 patients studied prospectively a worsening of aluminium toxicity resulted in a decrease of FS and vice versa. FS ⩽2 was characteristic of aluminium bone disease. It is noteworthy that already in 1981, we [7] demonstrated that low FS (<5) could serve as a criterion to predict which patients were at risk of hypercalcaemia when given vitamin D analogues, which has since been recognised as a feature of aluminium bone disease [8].

We conclude that aluminium intoxication decreases FS and that very few low scores (FS ⩽2) are very suggestive of toxicity in haemodialysis patients. In a given patient an aggravation of aluminium overload is associated with a lowering of FS and vice versa. On the other hand, iPTH increases the skeletal isotope uptake whatever the aluminium status. Thus, bone scintigraphy appears a valuable tool in the evaluation and follow-up of osteodystrophy in haemodialysis patients.

References

1 Fogelman I, Citrin DL, Turner JG et al. *Eur J Nucl Med 1979; 4:* 287
2 Olgaard K, Heesfordt J, Madsen S. *Nephron 1976; 17:* 325
3 Smeyers-Verbeke J, Verbeelen D, Massart DL. *Clin Chem Acta 1980; 108:* 67
4 Milliner DS, Nebecker HG, Ott SA et al. *Abstracts 16th Annual Meeting of the American Society of Nephrology 1983; 12A*
5 Buchanan MRC, Ihle BV, Dunn CM. *J Clin Pathol 1981; 34:* 1352
6 Heynen G, Reuter A, Vriendts Y et al. *IRE Bull 1979; 9:* 4
7 Van Herweghem JL, Dhaene M, Tielemans C et al. *Proc EDTA 1981; 18:* 648
8 Ellis HA, Pierides AM, Feest TG et al. *Clin Endocrinol 1977; 7 (Suppl):* 30S

Proc EDTA–ERA (1984) Vol 21

THE BONE SCAN IN PATIENTS WITH ALUMINIUM-ASSOCIATED BONE DISEASE

J Botella, J L Gallego, J Fernandez-Fernandez, D Sanz-Guajardo, A de Miguel, J Ramos, P Franco, R Enriques, C Sanz-Moreno

Universidad Autónoma de Madrid, Clinica Puerta de Hierro, Madrid, Spain

Summary

The bone scans of 32 patients on regular dialysis who received desferrioxamine therapy for fracturing osteomalacia secondary to aluminium intoxication are reviewed. All scans show the same pattern, with lack of tracer deposition in bone and deposition in soft tissues. Therapy with desferrioxamine controlled the aluminium intoxication in all cases, and in 21 patients the bone scan reverted to normal or showed a pattern typical of hyperparathyroidism.

Introduction

The value of radioisotope bone scans for the diagnosis of renal osteodystrophy was first described in 1976 by Olgaard [1]. The few reports which have appeared since that time [2] maintain the quantitative and not the qualitative interpretation of this diagnostic technique. Almost simultaneously, Drüeke [3] and ourselves [4] have supported the usefulness of this technique for differentiating between fracturing osteomalacia secondary to aluminium intoxication and osteitis fibrosa due to the secondary hyperparathyroidism of regular dialysis patients.

In secondary hyperparathyroidism there is excessive deposition of the tracer in bone, both in long and short bones. There is no deposition outside bony structures (Figure 1A).

In osteomalacia, the scan pattern has completely opposite characteristics, with almost no deposition of the tracer in the bones, which are difficult to see and show an 'erased' appearance, and heavy deposits in non-bony tissues both in the thorax and in the abdomen, so that the full silhouette of the body with the legs and calves becomes visible (Figure 1C).

Between these two extreme patterns, A and C, there is an intermediate one in which excessive deposits of the tracer in the facial bones and the cranial vault coexist with deposits in the thorax and abdomen. This pattern is found in patients who have high serum values of both aluminium and PTH.

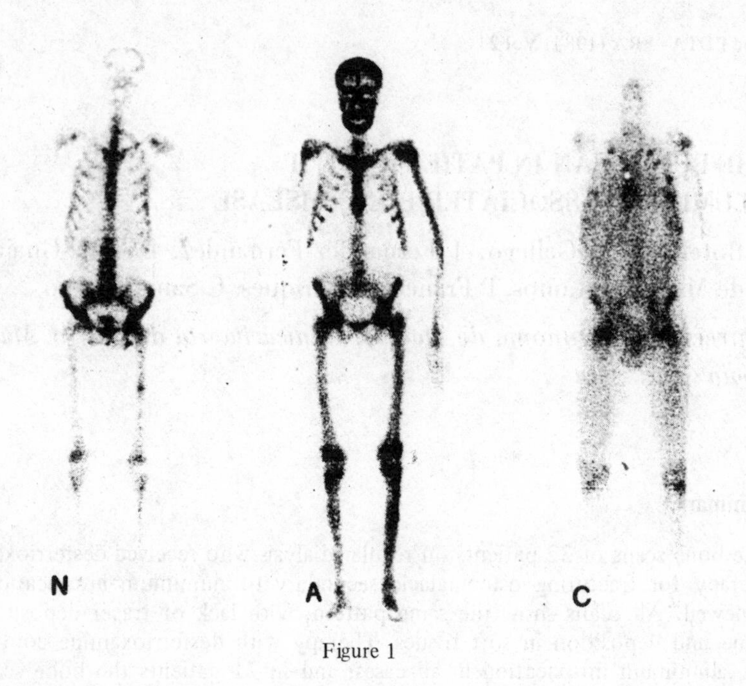

Figure 1

Pattern N in Figure 1 is a normal bone scan in a haemodialysis patient. The present study evaluates the clinical, analytical and X-ray data of patients with a type C bone scan, corresponding to aluminium intoxication, before and after therapy with desferrioxamine.

Patients

The study group comprised 60 patients under regular dialysis treatment at two different dialysis Units. The first Unit used a softener from 1968 to 1980, at which time a mixed bed ionic resin water processing system was installed; the second Unit used a softener from 1972 to March 1982, then installing a reverse osmosis system.

Thirty-nine patients had a type C bone scan, but only 32 received desferrioxamine therapy, as only these patients had clinical osteomalacia; 16 of them were being treated at Unit 1, and 16 at Unit 2. the remaining, 28 patients, nine from Unit 1 and 19 from Unit 2, formed the control group.

All patients treated with desferrioxamine presented bone pain, more or less marked proximal myopathy and type C bone scan. Eleven of them had costal fractures, three had femoral neck fractures, and two dialysis dementia.

Methods

The serum aluminium was assayed in a flameless atomic absorption spectrophotometer with a graphite furnace; the PTH was determined by a C-terminal-fragment-sensitive radioimmunoassay. The bone scans were carried out with a gamma camera after intravenous injection of 10–15mCi of ^{99m}Tc labelled diphosphonate (MDP). The statistical analysis was performed by non-parametric methods, using the Mann-Whitney 'U' test and the Wilcoxon 'T' test; the chi-square method was also used for some data.

Desferrioxamine was given at a dosage of 1g, dissolved in 100ml normal saline and infused via the arterial line over the first two hours of each haemodialysis session, three times weekly, for six consecutive months; this treatment was started in January 1983.

Aluminium and PTH values and bone scans were recorded immediately before starting desferrioxamine treatment, and again one month after this therapy was stopped.

Results

Prior to the institution of desferrioxamine therapy, the comparison of the data of the 39 patients with type C bone scan and the 21 patients in the control group yielded the following differences: the patients with type C bone scans had higher serum aluminium values, 88.63 ± 74.74 versus 39.58 ± 27.85μg/L ($p<0.001$); they had lower PTH values, 4.72 ± 3.00 versus 15.09 ± 10.54ng/ml ($p<0.001$), and they had been undergoing dialysis treatment for longer, 49.33 ± 36.07 versus 30.24 ± 35.18 months ($p<0.05$).

Under desferrioxamine therapy the clinical symptoms of osteomalacia and dialysis dementia disappeared, all fractures healed, and the EEG studies showed considerable improvement in both patients.

Figure 2 shows the changes in serum aluminium both in the group treated with desferrioxamine and in the one without. In the treatment group the initial

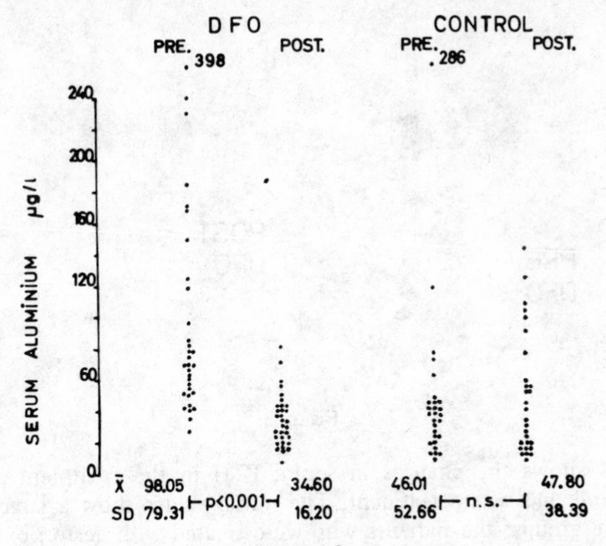

Figure 2

values are very high, with a mean value of 98.04μg/L and a large standard deviation, as there are individual values of up to 398μg/L. After treatment the mean value was 34.60μg/L with a small standard deviation, and of course with a statistically highly significant difference. The control group had practically unchanged values, 46.01μg/L at the beginning of the study and 47.80μg/L at the end.

All patients in the treatment group had a type C bone scan, whereas only seven patients in the control group had this pattern. After treatment, only 11 patients still have the type C pattern ($p<0.01$), which persisted in five patients of the control group (p NS) (Figure 3).

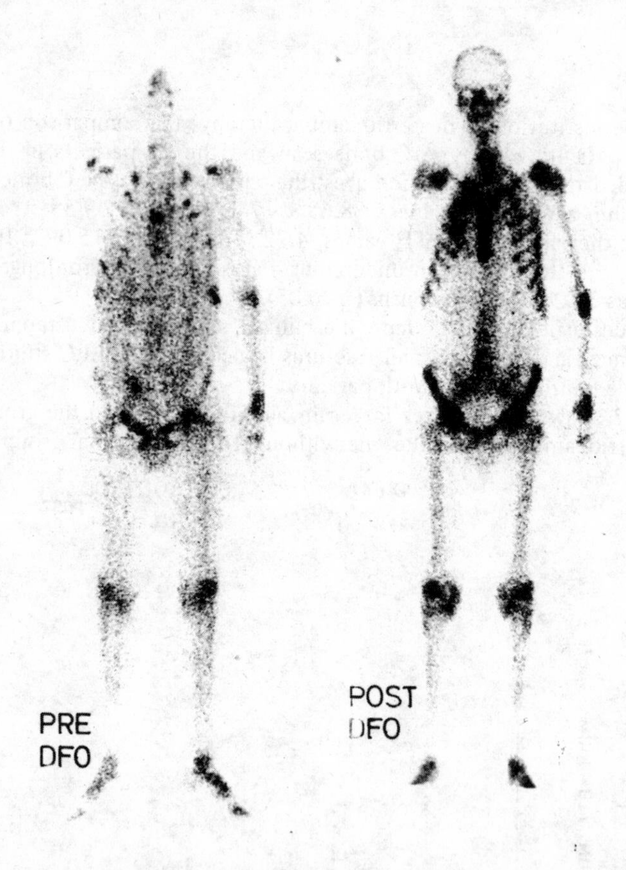

Figure 3

Figure 4 shows the changes in serum PTH in the treatment and control groups before and after treatment. The initial values show a large difference between the groups; the patients who were treated with desferrioxamine were near normal, 4.30 ± 2.79 versus 12.97 ± 17.30ng/ml ($p<0.001$). Six months later, the control group showed no significant changes, while those of the treatment group have risen to 6.20 ± 4.47ng/ml ($p<0.05$).

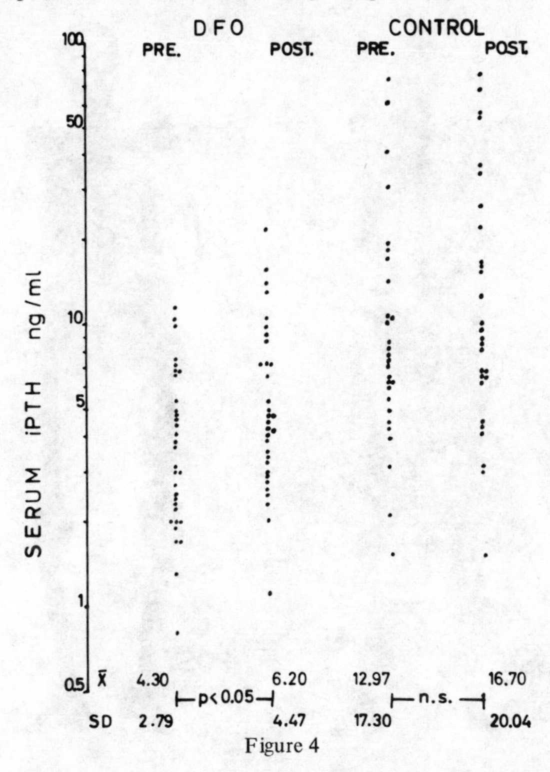

Figure 4

This change in PTH explains why, in some cases, the bone scan did not revert towards normal but to type A, characteristic of hyperparathyroidism (Figure 5).

Discussion

The bone scan with ^{99m}Tc MDP reflects the calcium turnover in bone [5]; it is therefore logical that in situations of secondary hyperparathyroidism there is excessive uptake of the tracer by the bones [1] which results in the type A pattern. In osteomalacia the situation is fully reversed, the deposition of calcium salts in the bone is slowed and therefore the binding of MDP will be less and slower. The skeleton of the patients is barely visible or not at all, giving rise to the type C pattern. In this pattern, the deposition of the tracer delineating the soft tissues, or as intense deposits in the thorax and abdomen, may be due to two causes: either it shows simply the presence of the tracer in the circulating blood, or it really reflects an excessive deposition of calcium in the viscera and soft tissues.

Akrill, in 1980, first reported the treatment of aluminium intoxication with

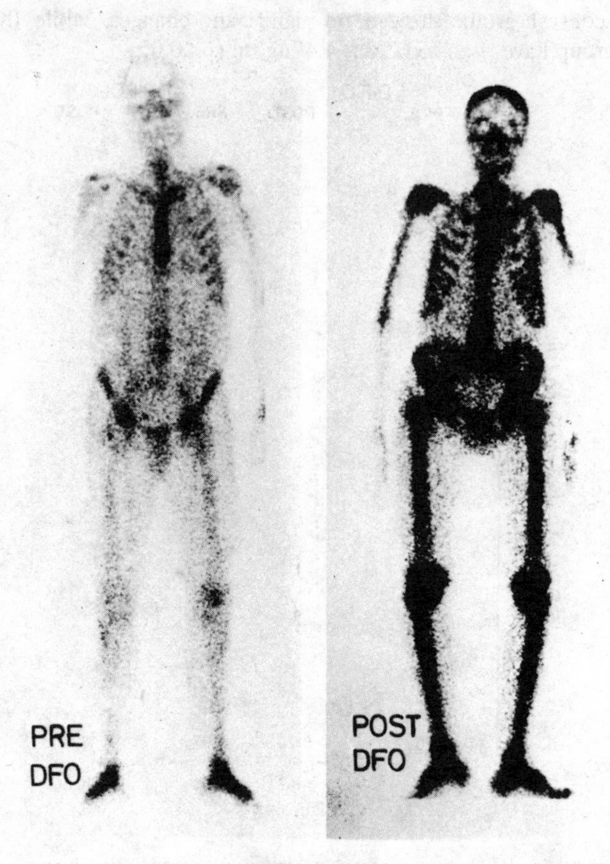

Figure 5

desferrioxamine [6] ; other authors have later confirmed it [7] and also point out the appearance of increased plasma PTH as the aluminium is eliminated from the body. In our patients, therapy with desferrioxamine improves the clinical, biochemical and X-ray findings of aluminium intoxication, but at the same time induces higher serum PTH. Coincidentally, the bone scans change from the type C pattern (osteomalacia) either to type N (normal) or to type A (hyperparathyroidism). In our opinion, all these data prove that the type C pattern is caused by aluminium intoxication, although it could be a reflection of the low calcium turnover in bone and might therefore exist in any other form of osteomalacia in patients under regular dialysis treatment.

References

1 Olgaard K, Heerfordt J, Madsen S. *Nephron 1976; 17:* 325
2 Kessler M, Ferry JP, Naoun A et al. *Néphrologie 1981; 2:* 171
3 Karsenty G, Vigneron N, Jorgetti V et al. *Abstracts 9th Int Congress Nephrol 1984:* 169A

4 Botella J, Lauzurica R, Fernandez-Fernandez J et al. *Kidney Int 1983; 24:* 126
5 Jones AG, Francis MD, Davis MA. *Seminars in Nuclear Medicine 1976; 6:* 3
6 Ackrill P, Ralston AJ, Day JP et al. *Lancet 1980; i:* 692
7 Ihle BU, Buchanan MRC, Stevens B et al. *Proc EDTA 1982; 19:* 195

Proc EDTA–ERA (1984) Vol 21

ASSESSING THE BENEFIT OF CHANGING ALUMINIUM HYDROXIDE SCHEDULE ON ANAEMIA AND SERUM PHOSPHORUS CONTROL

J B Cannata, C Suarez Suarez, C Rogrigues Suarez, V Cuesta, A Sanz-Medel, V Peral, J Herrera

Hospital General de Asturias, Oviedo, Spain

Summary

Convincing evidence exists concerning aluminium hydroxide (Al $(OH)_3$) absorption and risk of toxicity. Over recent years our aim has been to reduce exposure to this risk. In this study we evaluated the effect of changing our Al $(OH)_3$ prescription policy, reducing its intake by stopping the breakfast dose, separating the iron intake from the binder's influence, and tailoring the Al $(OH)_3$ dose according to the protein intake patterns. The change was done gradually, initially in a pilot group and then in the whole unit.

The results from the pilot group, who completed two years follow-up and from the whole unit, when more patients adhered to the new scheme, were similar. After the Al $(OH)_3$ reduction serum phosphorus did not change, haemoglobin increased and the blood transfusion requirements decreased. These results support our preliminary findings that Al $(OH)_3$ might interfere with erythropoiesis and stress the necessity of reassessing the prescription of binders thoroughly, aiming to give adequate individual doses according to the different protein intake patterns.

Introduction

There is increasing evidence that aluminium (Al) can induce anaemia. Most data comes from studies with high-aluminium exposure either in haemodialysis patients dialysed with high-aluminium dialysate or in experimental works with aluminium-intoxicated animals [1–5]. Nevertheless, little has been published about the likely toxic effect of aluminium hydroxide (Al $(OH)_3$) on erythropoiesis.

In previous studies we drew attention to the need for reassessing phosphate binders in chronic renal failure as in many countries the daily phosphorus (P) intake is divided between two rather than three meals. We also demonstrated benefits achieved by separating the oral iron intake from the influence of the binder [6].

In this study we progressively evaluated over a six year period the binder efficacy of changing the Al $(OH)_3$ prescription policy, together with influence on the control of anaemia in an increasing number of patients on haemodialysis.

Patients and methods

We have studied a total number of 57 patients on haemodialysis. From years one to six (Table I) dialysis was thrice weekly (13.5hr/week) employing de-ionised water with an aluminium content ranging between 0.1 to 0.6μmol/L and using dialysers 1.0 to 1.3m^2. The daily protein intake was between 1.2 to 1.5g/kg/day, and all patients received either oral or intravenous water soluble vitamin supplements and oral ferrous sulphate in a single fasting dose of 150–300mg/day. Intravenous iron was used as has been previously described [6] for short periods without stopping the oral iron in those patients who did not respond to the oral supplements.

Throughout the study we had a very restrictive policy about transfusions using blood or packed red cells only when symptoms of anaemia became unacceptable.

TABLE I. Patients included in the study. The cumulative number (No.) and percentage refer to patients on haemodialysis (HD) who have completed at least 12 months with the new Al $(OH)_3$ policy

Years	1	2	3	4	5	6
No. of patients in the HD unit	27	30	39	42	45	57
No. and percentage of patients included in the new Al $(OH)_3$ policy	0	0	9 (23%)	18 (43%)	27 (60%)	57 (100%)
			↑ Beginning of the pilot group			

The change in the Al $(OH)_3$ prescription policy is shown in Figure 1. We reduced the Al $(OH)_3$ intake by stopping the breakfast dose giving it twice or occasionally once, with main meals and tailoring the dose according to each patient's protein intake pattern.

The above described change in the Al $(OH)_3$ intake policy was deliberately introduced gradually throughout years two to six (Table I) with the aim of diminishing the almost inevitable error implicit in comparing pre and post values over periods longer than one year and simply attributing the better performance achieved to technical improvements.

The preliminary results from the first 27 patients included in this study (pilot group) have been previously reported [6] after being treated for one year with the new policy. Twenty-two of them completed two years follow-up and they will be briefly analysed here again, as a separate group. As the change was gradually introduced approximately one-third of the patients had completed the two years follow-up in year four, the second third in year five and the rest in year six (Table I).

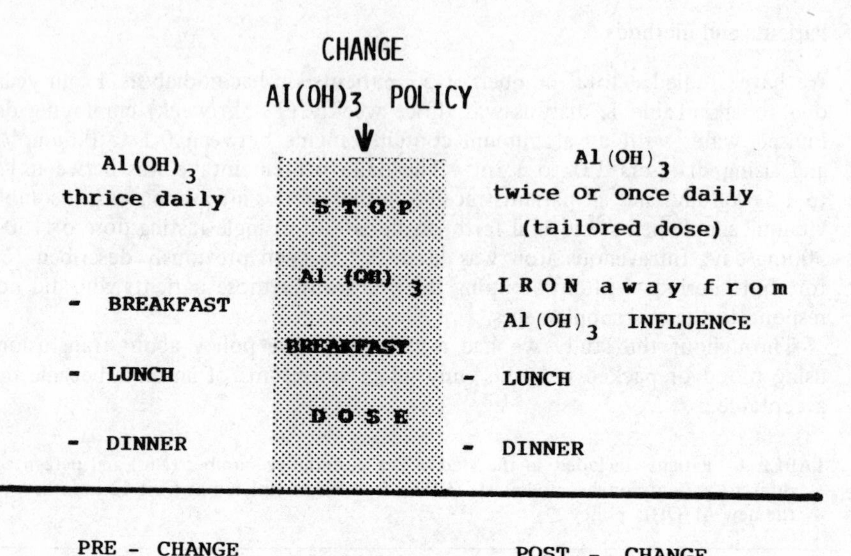

Figure 1. Change in the Al $(OH)_3$ prescription policy

Statistical analysis

All values are expressed as a mean ± SD and they were compared using paired and unpaired 't' tests.

Results

The main results from the whole unit are displayed in Table II showing the benefits achieved according to the progressive adherence of patients to the new policy.

Despite the Al $(OH)_3$ intake reduction the serum P did not change, the haemoglobin (Hb) increased, and the number of units of blood transfused decreased significantly ($p<0.05$). In year six we started measuring serum aluminium levels finding at the end of this year a serum mean aluminium level of 2.75 ± 1.8μmol/L.

The pilot group showed similar results. The significant Al $(OH)_3$ reduction did not modify the serum P control (1.6 ± 0.3 to 1.6 ± 0.4mmol/L) nevertheless, the Hb increased from 8.0 ± 2 to 9.4 ± 3 remaining stable throughout the second year and the number of blood transfusion requirements decreased from 3.19 to 1.22 units per patient year. At the end of this period only three out of 11 checked patients had serum aluminium values higher than 2.5μmol/L and the mean corpuscular volume was 84 ± $4\mu^3$.

TABLE II. Main results from the whole haemodialysis unit throughout years 1–6

Years	1	2	3	4	5	6
$Al(OH)_3$ Intake (g/day)	2.1 ± 1.6	2.6 ± 1.8	1.6 ± 1.1	1.0 ± 0.7 *	1.1 ± 0.9 *	1.2 ± 1.1 *
Serum P (mmol/L)	1.7 ± 0.4	1.6 ± 0.4	1.6 ± 0.4	1.7 ± 0.5	1.7 ± 0.4	1.7 ± 0.5
Haemoglobin (g/dl)	7.4 ± 2	7.4 ± 2	7.8 ± 2	8.6 ± 3 *	8.4 ± 3 **	8.6 ± 3 *
Blood Transfusion (units/patient year)	–	2.6	2.8	1.8	1.7	1.2 *

* $p<0.02$, ** $p<0.05$, compared with values from years 1 and 2

Discussion

For some years the main source of aluminium in haemodialysis has been the dialysate prepared using untreated tap water. However, currently most haemodialysis units use adequate water treatment systems and thus $Al(OH)_3$ ingestion has become proportionately more important as an aluminium source and it merits particular attention.

Nowadays, there is no doubt about the risk of aluminium absorption from aluminium containing phosphate binders, and recently some authors have demonstrated elevated serum aluminium levels and bone toxicity in renal patients receiving $Al(OH)_3$ who were not undergoing dialysis [7,8]. As aluminium accumulates in bone inducing a form of osteomalacia it might also act in the bone marrow interfering with red cell production [4]. Although this mechanism can be valid for partially explaining the aluminium-induced microcytic anaemia, we have also proposed that $Al(OH)_3$ could interact with iron absorption either through gastroduodenal pH elevations or by its action as an iron binder in the gastrointestinal tract [6] providing another factor impairing erythropoiesis.

Even though mean serum ferritin levels have been reported as normal in dialysis patients [3,4] in our experience there are a number of haemodialysis patients with low or 'low-normal' serum ferritin levels which we believe should be interpreted in a different way than those similar results in non-renal patients. Recently with the advent and use of better developed dialysers (low-priming volume, low residual-blood volume, better washback, etc) blood loss has decreased. Nevertheless, most haemodialysis patients including ours received similar oral ferrous sulphate supplements. Therefore, if the iron were adequately absorbed from the gastrointestinal tract and providing there are no other blood losses, one would expect most of them to have a tendency for impaired erythropoiesis and consequently a low-iron utilisation.

In fact, in one recent study seeking a likely effect of serum ferritin on gastrointestinal aluminium absorption, we found 43 per cent of patients to have low or 'low-normal' serum ferritin levels [9].

The significant haemoglobin improvement and the striking reduction in blood transfusion requirements after reducing the Al $(OH)_3$ intake and separating the iron intake from the binder's influence in both the pilot group and the whole unit, suggest that Al $(OH)_3$ might interfere with erythropoiesis. Unfortunately, this study does not contain enough data to support the previously mentioned hypothesis. Nevertheless, bearing in mind the low mean serum aluminium levels, the results obtained allow us to think that in some cases not only direct aluminium toxicity is of importance. Al $(OH)_3$ could also be implicated through its interaction with iron in the gastrointestinal tract and impairing its absorption.

On the other hand, Al $(OH)_3$ has proved its efficacy in lowering serum P and although new phosphate aluminium-free binders are being developed [10] Al $(OH)_3$ is, so far, the most used phosphate binder and therefore all the recent warnings about its hazards should be considered thoroughly, and the dose tailored individually according to different protein intake patterns.

References

1 Elliot HL, Dryburgh F, Fell GS et al. *Br Med J 1978; 1:* 1101
2 Short AIK, Winney RJ, Robson JS. *Proc EDTA 1980; 17:* 226
3 O'Hare JA, Murnagham DJ. *N Engl J Med 1982; 306:* 654
4 Touam M, Martinez F, Lacour B et al. *Clin Nephrol 1983; 19:* 295
5 Wills MR, Savoury J. *Lancet 1983; ii:* 29
6 Cannata JB, Ruiz-Alegría P, Cuesta MV et al. *Proc EDTA 1983; 20:* 719
7 Kaye M. *Clin Nephrol 1983; 20:* 208
8 Andreoli SP, Bergstein JM, Sherrard DJ. *N Engl J Med 1984; 310:* 1079
9 Cannata JB, Saurez Suarez C, Cuesta MV et al. *Proc EDTA 1984; 21:* this volume
10 Schneider H, Kulbe KD, Weber H, Streicher E. *Proc EDTA 1983; 20:* 725

Proc EDTA–ERA (1984) Vol 21

INFLUENCE OF BIOCHEMICAL AND HORMONAL FACTORS ON THE BONE HISTOMORPHOMETRIC FEATURES OF URAEMIC PATIENTS

A Fournier, **J L Sebert**, *P Fohrer, Ph Moriniere, *G Lambrey, †M A Herve, P Leflon, **J Gueris, ††M Garabedian

*Hôpital Nord, Amiens, *Centre Hospitalier, Beauvais, †Société de Mathématiques appliquées à la recherche médicale et biométrique, Amiens, **Hôpital Lariboisière, Paris, ††Hôpital des enfants malades, Paris, France*

Summary

A multidimensional analysis was used to evaluate, the influence on bone histology of various biochemical and hormonal factors in 20 uraemic patients on chronic haemodialysis or haemofiltration.

A positive relationship ($p<0.1$) was found between PTH and osteoclastic and osteoblastic surfaces but not with mineral apposition and bone formation rates. The mineral appositional rate which reflects the cellular activity of osteoblasts was positively related to D metabolites $25(OH)D_3$ and $1,25(OH)_2D_3$ and to phosphate ($p<0.1$). Mineral appositional rate and bone formation rate were negatively related to bone aluminium ($p<0.05$). These data indicate that: 1) PTH simulates bone turnover but has no direct effect on the bone cellular activity of osteoblasts which is mainly dependent on D metabolites and phosphate; 2) mild aluminium overload not severe enough to cause osteomalacia decreases bone formation in uraemic patients.

This study evaluates the role of various simultaneously measured biochemical and hormonal factors on bone histological parameters in uraemic patients.

Patients and methods

Twenty uraemic patients (14 males, 6 females), aged 52 ± 15 (SD) years underwent iliac bone biopsy after double tetracycline labelling. They had been on chronic haemodialysis (12 patients) or haemofiltration (8 patients), three times weekly, for 28 ± 15 months and had no symptoms of bone disease. Aluminium concentration has always been less than 0.3μmol/L in the dialysate (due to double step reserve osmosis) and ranged from 0.15 to 0.60μmol/L in the substitution fluid for haemofiltration. Calcium carbonate (0.5–7g/day) was given in 17 patients. Furthermore, to control plasma phosphate, aluminium hydroxide was given to 17 patients at a total cumulative dose ranging from 100 to 6,400g. Six patients were treated with $25(OH)D_3$ (5–30μg/day) and two were treated with $1\alpha(OH)D_3$, 1μg/day.

Quantitative bone histomorphometry

Quantitative bone histomorphometry was carried out on undecalcified, 5μ thick sections, using Zeiss integrated eye-pieces. The following parameters were measured on Goldner stained sections: osteoid volume (OV) expressed in per cent trabecular bone volume, osteoid surface (OS) in per cent trabecular surfaces, osteoid thickness index (OTI) calculated as the ratio $\frac{OV}{OS} \times 100$ according to Meunier [1] osteoblastic surface (OBL.S) defined as osteoid surface covered by plump osteoblasts, and active resorption surface (ARS) defined as the per cent trabecular bone surface occupied by Howship lacunae with close osteoclasts. The following dynamic parameters of bone remodelling were derived from fluorescence study of unstained sections: Tetracycline – labelled surface (Lab. S) defined as the per cent extent of single and double labelled surfaces, mineral appositional rate (App. Rate), bone formation rate at tissue level (BFR) calculated as App. Rate x Lab. S.

Aluminium content of bone and plasma

The bone concentration of aluminium was determined on a second bone sample taken during the same biopsy procedure. Aluminium was measured in plasma and bone by inductively coupled plasma emission spectrometry using a JOBIN YVON Elemental Analyser JY 38 P. [2]. The upper limit of the normal range of plasma aluminium concentration was 0.3μmol/L with a detection limit of 0.15 μmol/L. Normal values for aluminium bone concentration, obtained from seven non-uraemic corpses were 0.068 ± (SD) 0.36μmol/g of fresh bone tissue.

Biochemical and hormonal parameters

Plasma parathyroid hormone (PTH) was measured by radioimmunoassay using an antiserum specific for the mid-region of the molecule [3]. Normal range was 80–220pg/ml. Vitamin D metabolites were measured by radiocompetition [4,5] after previous lipidic extraction of the plasma and purification by chromatography on a sephadex LH 20 column followed by a second high pressure liquid chromatography. The normal range of D metabolites was: 6–30ng/ml for $25(OH)D_3$, 1–3ng/ml for $24,25(OH)_2D_3$ and 20–60pg/ml for $1,25(OH)_2D_3$.

Statistical methods

To study the potential link between the various bone histomorphometric parameters and the biochemical and hormonal factors, two statistical methods were successfully used:

1. the least squares simple linear regression analysis.
2. a multidimensional analysis combining polynomial evaluation and matricial discriminant analysis [6].

The linear regression analysis shows the apparent link between two variables but the concomitant influence of the other parameters has not been eliminated. The combination of polynomial evaluation and matricial discriminant analysis allows identification of the specific link between an explained variable (here one of the histomorphometric parameters) and each of the explanatory variables (here aluminium and the other biochemical parameters) independently of the influence of the others. The multidimensional analysis was performed on a COMMODORE 8000 computer using programs from the University of Paris VII and of the Institute of Technology of Compiegne. This multidimensional analysis leads to the determination of the D^2 coefficient of Mahalanobis which represents the proper influence of each explanatory variable on a given explained one. This coefficient ranges from -1 to $+1$ according to the sign of the relation.

Results

Bone histology

Eight patients had increased bone resorption coupled with decreased mineral appositional rate. These patients had hyperparathyroidism but low bone formation rates. The remaining four patients had low-normal bone resorption but had decreased labelled surfaces and mineral appositional rates in spite of increased osteoid surfaces. These patients had low bone turnover with low bone formation rate. No patient had true osteomalacia since in no case was the osteoid thickness index increased.

As shown in Figure 1 osteoblastic surfaces were strongly correlated with active resorption surfaces ($r=0.86$), labelled surfaces ($r=0.82$), mineral appositional rate ($r=0.79$) and bone formation rate ($r=0.87$).

Biochemistry

Bone aluminium was increased in all patients with a mean value of $0.59 \pm 0.44 \mu mol/g$ about 10 times higher than the mean value of $0.068 \pm 0.036 \mu mol/g$ obtained in non-uraemic controls. Plasma aluminium was increased in 17/20 patients with a mean value of $1.94 \pm 0.44 \mu mol/L$. The three patients with normal plasma levels were those who did not receive oral aluminium hydroxide. Plasma and bone aluminium concentrations were significantly correlated ($r=0.59$, $p<0.02$).

The mean plasma level of $25(OH)D_3$ was $14.3 \pm 11.8 ng/ml$ and only 5/20 patients had values under the lower limit of the normal range.

The two patients who were taking $1\alpha(OH)D_3$ had increased levels of $1,25(OH)_2D_3$. When these two patients were excluded, the mean plasma level of $1,25(OH)_2D_3$ was $24.2 \pm 17 pg/ml$, a value located in the lower part of the normal range. Plasma $24,25(OH)_2D_3$ was in the normal range in all cases and correlated with plasma $25(OH)D_3$ ($r=0.52$).

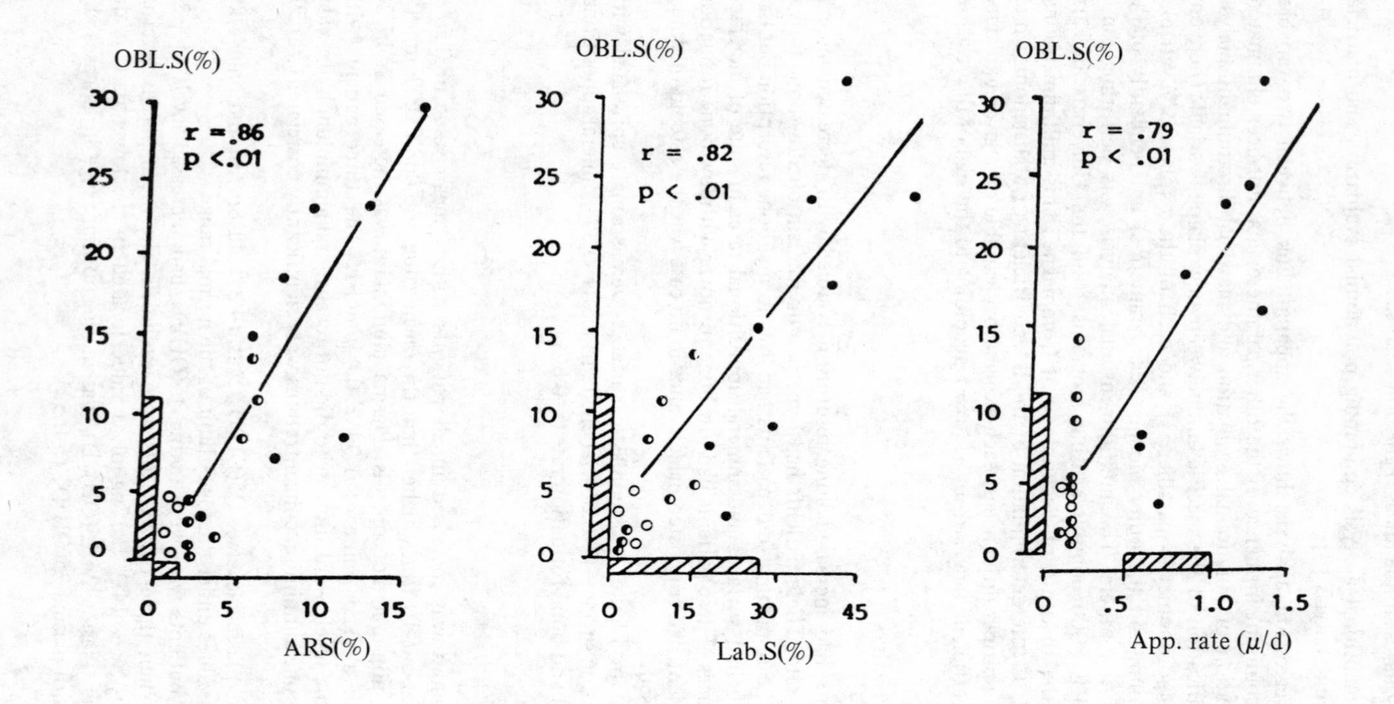

Figure 1. Correlations between osteoblastic surfaces (OBL.S) and: *left panel,* active resorption surfaces (ARS), *middle panel,* extent of labelled surfaces (Lab.S), *right panel,* mineral appositional rate (app. Rate).
Symbols indicate: (●) patients with pure hyperparathyroidism, (○) patients with low mineral appositional rate and normal resorption, (◐) patients with both hyper-resorption and low mineral appositional rate. Hatched area represents the normal mean ± 2 SD

Plasma PTH was increased in all but one patient. Plasma PTH did not correlate with plasma calcium (r=0.15) but there was a trend for a positive correlation between PTH and phosphate (r=0.36, $p<0.10$). Six patients had increased alkaline phosphatase activity.

As shown in Table I, PTH and alkaline phosphatase correlated with many parameters of bone remodelling. In contrast, there was no significant correlation between bone or plasma aluminium and any of the histological parameters.

TABLE I. Linear correlations between biochemical and histological data

	PTH		Alkaline phosphatase	
	r	p	r	p
Osteoid Volume	0.33	NS	0.42	0.05
Osteoid Surface	0.12	NS	0.33	NS
Osteoid thickness index	0.43	0.05	0.34	NS
Osteoblastic Surface	0.69	<0.01	0.75	<0.01
Active resorption surface	0.65	<0.01	0.60	<0.01
Osteoclast number	0.61	<0.01	0.64	<0.01
Labelled Surface	0.60	<0.01	0.53	<0.05
Apposition rate	0.58	<0.01	0.61	<0.01
Bone Formation Rate	0.61	<0.01	0.66	<0.01

TABLE II. Results of the multidimensional analysis

	value of D^2 for					
	Bone Al	Plasma Al	PTH	25(OH)D_3	1,25(OH)$_2$ D_3	Phosphate
Osteoid Volume	0.44*	0.51*	0.09	0.18	0.19	0.50**
Osteoid Surface	0.47*	0.63**	0.15	0.09	0.20	0.40*
Osteoid Thickness Index of Meunier	0.13	0.28	0.50**	0.12	0.35	0.72**
Osteoblast Surface	−0.47*	−0.62**	0.82**	0.12	0.49*	0.73**
Active resorption surface	0.27	0.29	0.80**	0.11	0.11	0.23
Labelled Surface	−0.07	−0.10	0.23	0.67**	0.82**	0.38
Apposition Rate	−0.39*	−0.31	0.19	0.53**	0.80**	0.49**
Bone Formation Rate	−0.39*	−0.22	0.20	0.29	0.41*	0.40*

D^2 is the D^2 coefficient of Mahanalobis which measures the proper influence on histological parameters of each biochemical factor, independently of the others. The p value of this D^2 coefficient is: *$p<0.05$, **$p<0.01$

Results of the multidimensional analysis are presented in Table II.

When the influence of all the other simultaneously measured biochemical factors was eliminated, negative relations were found between either bone or plasma aluminium and the histological parameters of bone formation and mineralisation including osteoblastic surfaces, mineral appositional rate and bone formation rate. PTH had a positive link with osteoblastic surfaces and active reabsorption surfaces but had no link with appositional rate. In contrast D metabolites $25(OH)D_3$ and $1,25(OH)_2 D_3$ and phosphate had a positive link with appositional rate and bone formation rate.

Conclusions

These data suggest that in uraemic patients with mild aluminium overload mainly induced by $Al(OH)_3$, 1) aluminium intoxication is not severe enough to induce true osteomalacia but decreases bone formation. 2) PTH stimulates bone turnover but has no direct effect on osteoblast activity which is mainly dependent on D metabolites (25 and $1,25(OH)_2 D_3$) and on phosphate.

References

1 Meunier P, Edouard C, Richard D, Laurent J. In Meunier P, ed. *Bone Histomorphometry.* Paris: Armour-Montagu. 1977: 249

2 Allain P, Mauras Y. *Anal Chem 1979; 51:* 2089

3 Gueris J, Ferriere C. *Pathol Biol 1975; 23:* 821

4 Preece MA, O'Riordan JLH, Lawson DEM, Kodicek E. *Clin Chem Acta 1974; 54:* 235

5 Shepard RM, Horst RL, Hamstra AJ, De Luca HF. *Biochem J 1979; 182:* 55

6 Lefebvre J. *Introduction aux Analyses Statistiques Multidimensionnelles.* Paris: Masson. 1983

Proc EDTA–ERA (1984) Vol 21

FLUORIDE METABOLISM AND RENAL OSTEODYSTROPHY IN REGULAR DIALYSIS TREATMENT

J Erben, B Hájaková, M Pantůček, L Kubes

University Hospital, Hradec Králové, Czechoslovakia

Summary

Fluoride in plasma, urine and bone tissue ash were estimated using a fluoride-ion electrode in 20 control persons (CP), 32 patients with compensated chronic renal failure (CRFP) and 59 patients in RDT (RDTP). The increase in plasma fluoride (CP: 2.4 ± 1.4, CRFP: 6.5 ± 2.2, RDTP: 12.3 ± 4.5μmol/L) and bone fluoride (CP: 55.1 ± 31, CRFP: 99.9 ± 31.2, RDTP: 339.1 ± 150.6μmol/g) significantly correlated with the decrease of residual glomerular filtration rate (RGFR), in RDT with the number of haemodialyses, so that the maximum increase in fluoride was found in completely anuric patients (379.7 ± 153μmol/g). The increase in fluoride retention was intensified by body retention of considerable amounts of fluoride each dialysis (the fluoridated dialysate increased the post-dialysis plasma fluoride by 195%). Development of bone fluorose known to develop with a bone fluoride greater than 180μmol/g was not found in any of 40 iliac crest trephine bone biopsy specimens. No correlation was found between laboratory and histological findings of renal osteodystrophy and plasma or bone fluoride. No patient developed spontaneous fractures even after 11 years of using fluoridated dialysate. In conclusion, this report indicates that fluoride might have a protective effect against the progression of renal osteodystrophy in patients with high retention values. The longer the exposure of RDT patients to the fluoridated dialysate, the greater the bone fluoride concentration.

Introduction

Uraemic patients on maintenance RDT employing fluoridated water for dialysate preparation absorb and retain considerable amounts of fluoride each dialysis with subsequent long-term storage in bone tissue [1–3]. The incorporation of fluoride in bone results in the formation of fluorapatite with improved mineral crystallisation leading to reduced chemical reactivity and reduced solubility of the bone mineral [4] following reduction in calcium. The hypocalcaemia results in secondary hyperparathyroidism [5,6] associated with the development of osteomalacia because of poor mineralisation [7], in which increased bone resorption is blocked and only the stimulation of new bone formation is manifest [4]. The high fluoride accumulation in bone induces the activated osteoblasts to produce excessive osteoid in which the collagen fibrils are disarrayed [8]. A somewhat analogous activating effect on the osteoblasts occurs also in chronic renal failure itself.

We have studied 91 chronic renal failure patients before and during RDT to determine quantitatively the bone uptake of fluoride and its influence on bone changes in patients maintained for many years on RDT.

Methods

Twenty control persons and ninety-one uraemic patients aged 21–54 years (48 males and 43 females) were included in this study. The mean duration of RDT was between two (patients with RGFR) and five (patients with anuria) years but in some patients the duration exceeded even 11 years. Forty patients had chronic glomerulonephritis, 21 chronic pyelonephritis, 12 polycystic kidney disease, and 18 other causes of CRF. All patients were receiving aluminium hydroxide when serum phosphate exceeded 2.5mmol/L and/or Ca.P product was over six; oral calcium and/or dihydrotachysterol when serum calcium fell below 2.2mmol/L. Dialysis schedules were individually adapted to maintain a stable clinical status and pre-dialysis blood values below 30mmol/L for urea. On average the patients dialysed three times weekly for six hours. The preparation of the dialysate was performed by means of special decalcification-cation-exchange filter. The dialysate calcium concentration was 1.75mmol/L.

The biochemical determinations were performed by commonly used methods. Parathormone (PTH) was measured by RIA. Fluoride in plasma, urine and bone tissue was estimated using fluoride-selective-electrode [9]. The double iliac crest trephine bone biopsy specimens were obtained from 40 patients with CRF who were dialysed with fluoridated dialysate (14 with RGFR:RDTDP, 26 with anuria:RDTAP), 15 patients waiting for RDT (:CRFT) and 20 control orthopaedic patients (CP) in order: a) to determine tissue fluoride concentration in the bone ash after burning in 400°C and b) to establish the histological criteria of renal osteodystrophy. Biopsies were classified as normal or as showing hyperparathyroid bone disease, osteomalacia, or combined disease. The following parameters were evaluated in four stages in trabecular bone: total bone volume, relative osteoid surfaces and volume, mean width of osteoid seams, trabecular osteoclastic resorption surface, mean size of periosteocytic lacunae, osteoblastic surface, and area of fibrosis.

Results and discussion

Absorbed fluoride is quickly distributed through body fluids. The amount of fluoride in the plasma is determined mainly by the rate of absorption from the gut, excretion by the kidney and the uptake by calcified tissue [4]. Regulation of plasma fluoride content is effected mainly by the skeletal uptake of fluoride. The skeletal regulation of the body fluoride is faster and quantitatively more significant than the excretion of fluoride by the kidney [10]. These homeostatic mechanisms may be overwhelmed by very large fluoride uptake occurring when patients are dialysed with fluoridated dialysate. Excessive osteoid may be produced and severe osteomalacia may develop [8].

The fluoride concentration of fluoridated tap water and also of our dialysate fluctuated between 20–50μmol/L. The influence of single haemodialyses on the plasma fluoride of dialysed patients was studied in 22 patients. Mean ± SEM values were 12.9 ± 5.1μmol/L before and 38.1 ± 5.7μmol/L after haemodialysis ($p<0.0001$), i.e. a 195 per cent increase in plasma fluoride after dialysis.

Table I demonstrates mean ± SEM values of fluorides in single patient groups with significant differences between plasma concentrations of CP and CRFP, RDTDP and RDTAP, and between bone tissue ash concentrations of all groups of patients.

TABLE I. Fluoride concentration in plasma and in bone tissue ash in single groups of patients: control persons (CP), patients waiting for RDT in chronic compensated renal failure (CRFP), RDT patients with RGFR (RDTDP), and RDT patients with anuria (RDTAP)

Group of patients	n	Plasma fluoride in μmol/L:	't' test	p<	n	Bone tissue fluoride μmol/g	't' test	p<
CP	20	2.40 ± 1.40			20	55.1 ± 31.0		
			4.91	0.001			4.88	0.0001
CRF	32	6.47 ± 2.2			15	99.9 ± 31.2		
							5.05	0.0001
RDTDP	34	10.8 ± 4.5			14	263.5 ± 117.54		
			4.25	0.001			2.68	0.01
RDTAP	25	15.6 ± 2.99			26	379.7 ± 153		

Figure 1 shows a significant correlation between fluoride and creatinine clearance in 31 RDTDP ($r=0.868$, $p<0.0001$), but a non-significant correlation was found in 28 CRFP ($r=0.275$, $p>0.05$). The fluoridated drinking water and/or fluoridated dialysate influence body fluoride concentration of all persons depending on their RGFR. Fluoride absorption from the alimentary tract depends on the solubility of the compound and/or dietary factors, i.e. dietary fat increases their skeletal storage. The fluoride intake from domestic water supplies is too small in comparison with the intake from the fluoridated dialysate during haemodialysis. It seems that when plasma fluoride concentrations are low they are influenced mainly by bone clearance [10] and when in high concentrations by renal clearance. A significant correlation between the fluoride in plasma and in bone tissue in 55 renal insufficiency patients ($r=0.97$, $p<0.05$) suggests that the bone flow is the major determinant of fluoride transfer from blood to bone where the fluoride is stored.

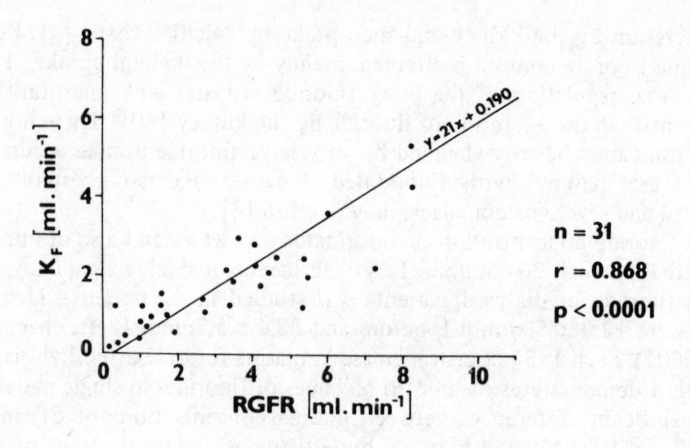

Figure 1. Correlation between fluoride clearance (K_F) and RGFR (creatinine clearance) in RDT patients

Fluoride storage in the bone depends on the duration of RDT. Significant correlations between bone fluoride concentration and the number of haemodialyses in RDT (n=40, r=0.75, p<0.01), or the number of RDT-months (r=0.81, p<0.01), indicate that the longer the exposure of RDT patients to the fluoridated dialysate, the greater bone fluoride storage. The influence of the exposure to uraemia and RDT with fluoridated dialysate play their own part. There are close morphological resemblances between uraemic osteodystrophy and chronic skeletal fluorosis [4], but this most probably is a measure of the involvement of the parathyroid glands in fluorosis. The histological features in bone biopsies are similar in hyperparathyroidism and in osteoporotic patients treated with fluoride [6]. The demonstrable hyperactivity of the parathyroid glands, may be a compensatory phenomenon maintaining the serum calcium at a constant value [1] in fluorosis and also in chronic renal failure. Our results demonstrate a non-significant correlation between plasma and/or bone fluoride and alkaline phosphatase (r=0.32), bone isoenzyme of alkaline phosphatase (r=−0.01), serum calcium (r=−0.06), serum PTH (r=0.162), and histological and radiological evidence of renal osteodystrophy. We have never found any spontaneous fractures during RDT for as long as 11 years using fluoridated dialysate, but there is a borderline significant correlation between bone fluoride and histological stages of secondary hyperparathyroidism (r=0.39, p<0.05).

In conclusion, our results show that 1) dialysate fluoridation was not associated with the development of osteodystrophy of the advanced degree found in fluorosis, 2) the longer the exposure of RDT patients to the fluoridated dialysate, the higher is the bone fluoride storage concentration, 3) plasma fluoride is influenced by bone clearance (fluoride sequestration) in low concentrations and largely by RGFR at higher concentrations, 4) our results suggest a certain protective effect of bone fluoride storage against the progression of renal osteodystrophy. It remains to be established whether there exists an upper limit of bone fluoride concentration beyond which the bone would be damaged rather than strengthened.

Acknowledgment

The authors wish to express their appreciation to the laboratory team of Radiobiology in Prague led by Ing Marie Zichova for the determination of serum parathormone.

References

1 Nielsen E, Solomon N, Goodwin NJ et al. *Trans ASAIO 1973; 19:* 450
2 Taves D, Freeman RB, Tamm DE et al. *Trans ASAIO 1968; 14:* 412
3 Siddiqui JY, Sampson SW, Ellis HE et al. *Proc EDTA 1971; 8:* 149
4 Faccini JM. *Calc Tiss Res 1969; 3:* 1
5 Ream LS, Principato R. *Am J Anat 1981; 162:* 233
6 Baylink DJ, Bernstein DS. *Clin Orthoped 1967; 55:* 51
7 Jowsey J, Riggs BL, Keller PJ, Hoffman DL. *Am J Med 1972; 53:* 43
8 Lough J, Noonan R, Gagnon R, Kaye M. *Arch Pathol 1975; 99:* 484
9 Pantůček MB. *Cs Hygiene 1978; 23:* 187
10 Yeh MC, Singer L, Armstrong WD. *Proc Soc Exp Biol Med 1970; 135:* 421

Proc EDTA–ERA (1984) Vol 21

EVALUATION OF THE INTACT HORMONE ASSAY IN THE STUDY OF PARATHYROID AUTOGRAFT FUNCTION

E C R Wijeyesinghe, A J Parnham, *J R Farndon, R Wilkinson

*Freeman Hospital, *Royal Victoria Infirmary, Newcastle upon Tyne, United Kingdom*

Summary

Parathyroid autograft function was studied using radioimmunoassay measuring intact parathormone (iPTH) and C-terminal PTH (C-PTH). Blood samples were taken from graft and non-graft arms pre- and post-dialysis. The intact assay showed an appreciably greater differential gradient pre- and post-dialysis than the C-terminal assay. We conclude that particularly in patients with chronic renal failure the intact assay is of greater value in monitoring graft function than the commonly used C-terminal assay. It may also be of value in distinguishing between graft overactivity and ectopic cervical parathyroid tissue in cases of recurrent hyperparathyroidism following apparently total parathyroidectomy and autotransplantation.

Introduction

Secondary hyperparathyroidism is a common problem in the long-term dialysis population. Uncontrolled, this leads to the well known complications of bone pain, fractures, and soft tissue calçification. Although in many, conservative management may halt and/or improve the process, a proportion need to be subjected to surgery.

Following the pioneering experimental work of Alveryd [1], Wells et al [2] described the technique of total parathyroidectomy with forearm grafting in 1973. Due to its many advantages this method is now widely accepted as the procedure of choice in many centres including our own.

Parathyroid hormone (PTH) is secreted mainly in the intact form consisting of 84 amino acids. Peripheral cleavage occurs mainly in the liver and kidney to form a biologically active aminoterminal fragment (N-PTH) and an inactive carboxyterminal fragment (C-PTH) [3]. The two principal forms of PTH in peripheral blood are intact PTH (iPTH) and (C-PTH) [4]. The half lives of the two types are quite different, that of the intact molecule being less than five minutes while the C-terminal fragment has a half life of about 40 minutes. In chronic renal failure the half lives of both peptides are prolonged, that of the intact molecule to 30 minutes and of the C-terminal fragment to 24–36 hours. This is because the kidney while being a major organ responsible for clearance of the intact hormone is the only organ responsible for clearing the C-terminal fragment. This leads to a high concentration of C-PTH in chronic renal failure with only 20 per cent of the concentration present representing true hypersecretion [5].

Radioimmunoassays (RIAs) for PTH have commonly used antisera directed at the C-terminal end of the molecule but recently a convenient sensitive RIA has become available for the measurement of 'intact PTH' (peptides containing N-terminal, mid region and substantial portions of C-terminal determinants within the same molecule) [6]. In previous work the response of transplanted parathyroid tissue to changes in plasma calcium has been demonstrated using a PTH assay measuring predominantly the carboxyl determinants [7].

We have compared intact PTH measurements with C-terminal measurements in monitoring PTH secretion from parathyroid autografts in patients on haemodialysis.

Patients and methods

Six patients on chronic intermittent in-centre haemodialysis were studied. The standard dialysate used in all cases contained 1.55mmol/L of ionised calcium. Samples for PTH analysis were withdrawn simultaneously from an arm vein draining the graft and the contralateral arm (fistula arm) via the arterial line. Samples for RIA were collected into EDTA tubes, separated and frozen quickly after withdrawal. They were stored at −20°C until assayed and the analyses for intact and C-terminal PTH were performed in parallel. The C-terminal PTH immunoassay was manufactured by Immunonuclear Corporation (Minnesota, USA). The method is based on an antibody prepared against intact beef PTH and reactive to the 65–84 sequence of human PTH. The tracer used is an iodinated human PTH. The synthetic standard PTH 65–84 has been calibrated against WHO preparation 79/500. The intact PTH kit is also manufactured by Immunonuclear Corporation [6].

The upper limit of normal for the intact and C-terminal PTH assays are 10pmol/L and 88pmol/L respectively.

Results (Table I)

Figures 1 and 2 illustrate the percentage differences between graft and non-graft arms pre- and post-dialysis using the two different assays. Using the intact assay, the pre-dialysis samples on five of the six patients showed a differential gradient of at least 59 per cent, the sixth patient showing a value of less than 20 per cent. The same samples analysed using the C-terminal assay showed a gradient of less than 30 per cent. The post-dialysis samples when analysed using the intact assay showed a minimum difference of 43 per cent in the same five patients with high gradients pre-dialysis. Only one of the six showed a gradient greater than 20 per cent when analysed using the C-terminal assay.

TABLE I

Patient	Arm	Pre-dialysis Intact	Pre-dialysis C-terminal	Post-dialysis Intact	Post-dialysis C-terminal
1	G	26	100	6	100
	NG	5	90	3	90
2	G	78	420	48	440
	NG	10	360	10	390
3	G	23	180	36	230
	NG	9	150	7	190
4	G	60	200	>125	900
	NG	10	170	4	160
5	G	37	180	14	140
	NG	3	130	8	140
6	G	115	>1000	118	>1000
	NG	98	>1000	125	>1000

G = Graft, NG = Non Graft, all values in pmol/L

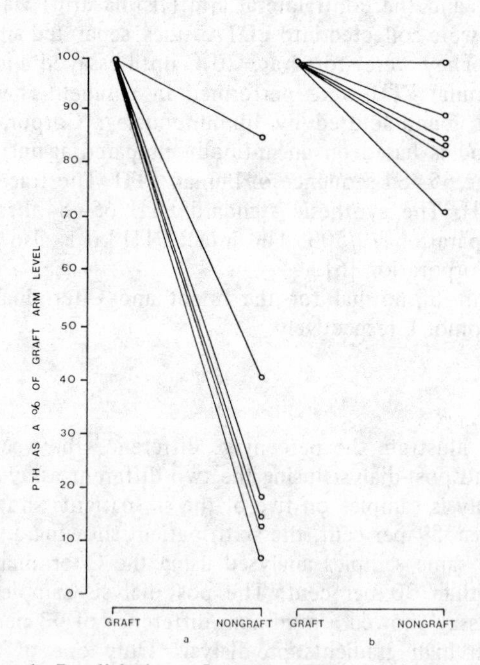

Figure 1. Pre-dialysis. a = Intact Assay, b = C-terminal Assay

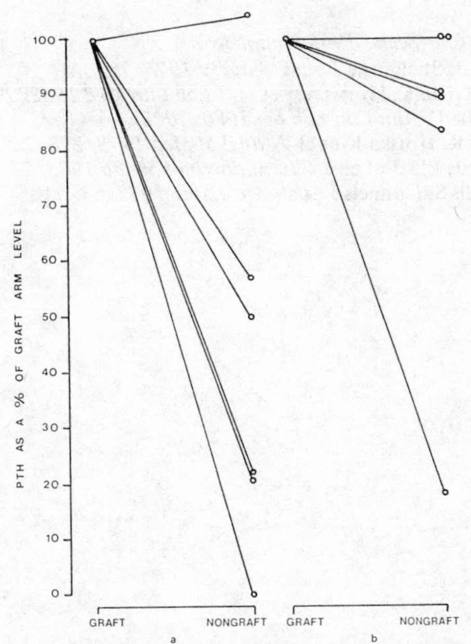

Figure 2. Post-dialysis. a = Intact Assay, b = C-terminal Assay

In fact, using the C-terminal assay, the differential gradient in four of the six patients lay just outside the within-batch assay variability (a differential gradient of 15 per cent or less pre- and post-dialysis with a within-batch assay variability of 6 per cent at 690pmol/L). The within-batch assay variability for the intact assay was 11 per cent at 20pmol/L.

Discussion

These results suggest that although both assays are of some value in confirming autograft function, the intact PTH assay enables us to do this with greater confidence. In fact, in the sixth patient who probably has residual parathyroid tissue in his neck despite three cervico-mediastinal parathyroidectomies, the C-terminal assay was unable to detect a differential gradient indicative of a functioning autograft, while the intact assay showed a difference of 15 per cent.

This study suggests that particularly in patients with chronic renal failure, the RIA for intact PTH is of greater value in the monitoring of autograft function than the C-terminal assay. It may also be of value in distinguishing between over-activity of the graft and residual parathyroid tissue in the neck when hyperparathyroidism persists after total parathyroidectomy and autotransplantation.

References

1 Alveryd A. *Acta Chir Scand 1968; Suppl 389:* 1
2 Wells S, Gunnels J, Shelburne J et al. *Surgery 1975; 78:* 34
3 Slatopolsky E, Martin K, Morrisey J et al. *J Lab Clin Med 1982; 99:* 309
4 Hawker G, DiBella F. *Ann Clin Lab Sci 1980; 10:* 76
5 Freitag J, Martin K, Hruska K et al. *N Engl M Med 1978; 298:* 29
6 Lindall A, Elting J, Ells J et al. *J Clin Endocrinol Metab 1983; 57:* 1007
7 Schneider A, Wells S, Gunnels J et al. *Am J Med 1977; 63:* 710

Proc EDTA-ERA (1984) Vol 21

USEFULNESS OF ^{99m}Tc PYROPHOSPHATE BONE SCINTIGRAPHY IN THE SURVEY OF DIALYSIS OSTEODYSTROPHY

J L Vanherweghem, A Schoutens, P Bergman, M Dhaene, M Goldman, M Fuss, P Kinnaert

Erasme and Brugmann Hospitals, University of Brussels, Brussels, Belgium

Summary

Fogelman's score (FS) was used to determine the usefulness of ^{99m}Tc pyrophosphate (Tc-PP) bone scintigraphy in the evaluation of dialysis osteodystrophy. FS correlated well with bone ^{47}Ca accretion rate. It remained stable after six months in patients treated with $1\alpha(OH)D_3$ and increased significantly in a randomised group of untreated patients. It decreased after two years of $1\alpha(OH)D_3$ therapy while serum calcium increased and iPTH and alkaline phosphatases decreased. Patients with low FS, treated by $1\alpha(OH)D_3$, rapidly developed hypercalcaemia. In cases of spontaneous hypercalcaemia, parathyroidectomy did not normalise serum calcium in patients with low FS despite a significant decrease in serum iPTH. Lower FS were associated with a higher increase in serum aluminium after desferrioxamine (DFO) administration and in two cases of proven aluminium osteomalacia, DFO therapy was followed by a dramatic increase in FS.

Introduction

After five years of dialysis, about 15 per cent of patients present with a severe disabling bone disease [1]. For early diagnosis, X-rays are disappointing since the radiological signs are often late. Histological examination is certainly the most accurate method but bone biopsies may obviously not be done repeatedly in asymptomatic patients. As bone scintigraphy was shown to be a sensitive method for revealing renal osteodystrophy [2–5] we decided to evaluate the systematic use of this technique in a survey of dialysis patients.

Methods

Bone scintigraphy was performed four hours after an IV injection of ^{99m}Tc pyrophosphate (Tc-PP). A dialysis session was systematically begun 30 minutes after the injection of the tracer and ended before scan imaging [5]. Bone uptake of Tc-PP was estimated semi-quantitatively according to Fogelman [6]. Six areas of interest were examined and Tc-PP uptake of each area was scored from 0 to 2, giving a global Fogelman's score (FS) from 0 to 12. Very poor uptake by bone scored 0. Bone scintigraphy was routinely done every six months in all dialysed patients. The following data were retrospectively recorded:

1. In 11 patients, FS was correlated with the bone calcium accretion rate calculated from the radiocalcium retention and the plasma radioactivity curve recorded during seven days following IV injection of $^{47}CaCl_2$ [7].
2. The significance of FS as an index of bone disease activity was assessed
   a) by comparing FS of 12 patients treated by $1\alpha(OH)D_3$ ($1\mu g/d$) to FS of nine patients treated by placebo, in correlation with biological parameters of bone disease
   b) by correlating the evolution of FS in 14 patients after two years of $1\alpha(OH)D_3$ therapy to that of the biological parameters of bone disease.
3. The predictive value of Tc-PP bone scintigraphy was analysed
   a) by comparing the FS of 10 patients who developed an increase in serum calcium above 11mg/dl in the course of the first month of $1\alpha(OH)D_3$ therapy ($1\mu g/d$) to the FS of 10 patients who tolerated the treatment
   b) by comparing the FS of 13 patients whose serum calcium fell after parathyroidectomy performed according to Wells [8] to the FS of two hypercalcaemic patients who did not respond to parathyroidectomy.
4. The value of Tc-PP bone scintigraphy in aluminium bone disease was examined
   a) by correlating FS and the increase in serum aluminium after the IV administration of 1g desferrioxamine (DFO)
   b) by observing the changes in Tc-PP bone uptake during DFO therapy in two cases of histologically proven aluminium bone disease.

Results

1. In 11 dialysed patients, there was a good correlation between FS and bone ^{47}calcium accretion rate ($r = 0.80$, $p < 0.01$).
2. a) Table I shows the significant difference in the evolution of FS, iPTH, serum calcium and alkaline phosphatases in two groups of dialysed patients (a group treated by $1\alpha(OH)D_3$ and a control group treated by placebo). After six months, FS remained stable in 12 patients treated by $1\alpha(OH)D_3$ while it significantly increased in nine control patients treated by placebo.
   b) In 14 patients treated with $1\alpha(OH)D_3$ for two years, FS decreased from 8.4 ± 0.6 to 5.3 ± 0.4 ($p < 0.001$) while serum calcium rose from 8.5 ± 0.2 to 10.3 ± 0.1mg/dl ($p < 0.001$) with a concomitant decrease in iPTH (from 84 ± 6 to $59 \pm 4\mu Eq/ml$, $p<0.001$) and in alkaline phosphatases (from 84 ± 12 to 41 ± 3UI/ml, $p<0.01$).

TABLE I. Evolution of the signs of metabolic bone disease activity in dialysed patients treated by $1\alpha(OH)D_3$ as compared to a placebo group (after six months)

	$1\alpha(OH)D_3$ (n = 12)		placebo (n = 9)
Variations expressed in percent of basal values:			
iPTH	– 34 ± 9	p<0.01	+ 24 ± 10
Alkaline phosphatases	– 45 ± 21	p<0.01	+ 20 ± 11
Serum calcium	+ 24 ± 5	p<0.001	– 8 ± 6
Fogelman's score	– 8 ± 8	p±0.05	+ 27 ± 10

3. a) Thirty-eight dialysed patients received oral calcium supplements (calcium carbonate 1.8g/d) and $1\alpha(OH)D_3$ (1μg/d). The calcium dialysate content was 7.5mg/dl. Ten of these patients developed hypercalcaemia (serum calcium above 11.0mg/dl) in the course of the first month of therapy while the 23 other patients tolerated the treatment well. The former had lower FS (3.6 ± 0.8) as compared to the latter (8.0 ± 0.4) ($p < 0.001$).

   b) In 13 patients with high FS (9.2 ± 0.5), parathyroidectomy was followed by a prompt decrease in serum calcium from 10.7 ± 0.3 to 8.0 ± 0.4mg/dl. On the opposite, in two patients with low FS (1.0 ± 0.0), parathyroidectomy did not correct hypercalcaemia (11.0 ± 0.5mg/dl before and after surgery) despite a significant reduction of iPTH. Further investigations demonstrated that these two patients had bone aluminium intoxication.

4. a) After administration of DFO, the increase in serum aluminium was significantly ($p < 0.001$) higher (△ = 100.9 ± 18.5μg/L) in seven patients with low FS (3.3 ± 0.3) than in 17 patients (△ = 42.1 ± 9.2μg/L) with greater FS (6.8 ± 0.3, n = 17).

   b) In two patients with histologically proven aluminium bone disease the Tc-PP bone uptake was very poor. After some weeks of DFO therapy, FS increased to 8–9 while the clinical condition improved dramatically.

Discussion

Tc-PP bone scintigraphy was suggested to be a sensitive method for detecting early stages of metabolic bone disease in renal failure [2–5]. Tc-PP bone uptake seems indeed related to the calcium bone turnover as suggested by our data. FS correlated with radio calcium accretion rate as well as with biological parameters of metabolic bone disease activity. Moreover, treatment with an active vit D metabolite which is supposed to decrease the activity of the disease, reduced Tc-PP bone uptake.

If FS correlates with calcium accretion rate, the patients with low FS are expected to develop rapidly hypercalcaemia during an exogenous calcium load. Indeed, our previous suggestion [9] that Tc-PP bone scintigraphy could be useful for detecting uraemic patients at risk for vitamin D intoxication is confirmed by the present data in a larger group of patients. On the other hand, the effect of parathyroidectomy in patients with spontaneous hypercalcaemia can be predicted by FS. A low FS suggests that parathyroidectomy will fail to decrease serum calcium. Finally, a poor Tc-PP bone uptake may also suggest a blockade in bone mineralisation due, for instance, to aluminium intoxication [10].

In conclusion, Tc-PP bone scintigraphy is an easy non-invasive technique that may be useful 1) to diagnose early bone diseases in uraemic patients, 2) to evaluate the degree of disease activity and the effects of a treatment, 3) to detect the patients at risk for vitamin D intoxication, 4) to discuss the indications of parathyroidectomy especially in cases of spontaneous hypercalcaemia and 5) to detect the patients with a blockade in bone mineralisation such as aluminium intoxication.

References

1 Brunner FR, Brynger H, Chantler C et al. *Proc EDTA 1979; 16:* 4
2 Sy W, Mittal AK. *Br J Radiol 1975; 48:* 878
3 Ølgaard K, Heerfordt J, Madsen S. *Nephron 1976; 17:* 325
4 Ritz E, Clorius JH. *Nephron 1976; 17:* 321
5 Degraaf P, Schicht IM, Pauwels EKJ et al. *J Nucl Med 1978; 19:* 1289
6 Fogelman I, Citrin DL, Turner JG et al. *Eur J Nucl Med 1979; 4:* 287
7 Bergman P, Paternot J, Schoutens A. *Calcif Tissue Int 1983; 35:* 21
8 Wells SA, Ellis GI, Gunnels JC. *N Engl J Med 1976; 295:* 57
9 Vanherweghem JL, Dhaene M, Tielemans C et al. *Proc EDTA 1981; 18:* 648
10 Vanherweghem JL, Schoutens A, Bergman P et al. *Trace Elements Med 1984; 2:* 80

Proc EDTA–ERA (1984) Vol 21

BONE MINERAL LOSS DURING CENTRE AND HOME HAEMODIALYSIS. INFLUENCE OF VITAMIN D METABOLITES AND SERUM PARATHYROID HORMONE

H Rickers, *M Christensen, †C Christiansen, *P Rødbro

*Aarhus Kommunehospital, Aarhus, *Aalborg Hospital, Aalborg, Glostrup Hospital, Glostrup, Denmark*

Summary

Twenty-seven patients on chronic haemodialysis were investigated for a mean of 4.8 years (3.0–6.5 years). Mean bone mineral content fell constantly and similarly at a rate of three to four per cent per year in both centre (n=14) and home (n=13) haemodialysis patients. Mean serum values of $25(OH)D_3$ (normal), $24,25(OH)_2D_3$ (decreased to half the normal level and $1,25(OH)_2D_3$ (severely decreased and almost non-detectable) were similar in patients with a rapid bone mineral content loss (>10%/3 years) and a slow bone mineral loss (<10%/ 3 years). Mean serum parathyroid hormone was markedly elevated, but significantly higher (about twice the level) in the 'rapid losers' than in the 'slow losers'; whereas the two groups did not differ with regard to mean serum concentrations of calcium, phosphate and alkaline phosphatase.

Introduction

Metabolic bone disease is a well-known complication of chronic renal failure and maintenance haemodialysis. The nature of this bone disease, known as uraemic osteodystrophy, is complex, but secondary hyperparathyroidism [1] and disturbances in vitamin D metabolism [2] seem to be of major importance in the pathogenesis. However, neither parathyroidectomy nor treatment with high potency vitamin D analogues, such as 1,25-dihydroxy vitamin D ($1,25(OH)_2D_3$), invariably cure bone lesions despite normalisation of biochemical abnormalities.

Longitudinal studies of the bone mineral content have shown marked differences in the natural history of bone mineral content loss during haemodialysis, indicating that some haemodialysis patients are rapid losers of bone calcium [3]. In the present longitudinal bone mineral content study of haemodialysis patients, the influence of the serum levels of parathyroid hormone (iPTH) and the vitamin D metabolites ($25(OH)D_3$, $24,25(OH)_2D_3$ and $1,25(OH)_2D_3$) on the rate of bone mineral content loss during haemodialysis was assessed. Moreover, a comparison was made between patients on centre and home haemodialysis as regards bone mineral content loss.

Patients and methods

Twenty-seven consecutive patients (12 females, 15 males; mean age 42 years (20–68 years)) entering chronic haemodialysis in the period 1977–1979 were studied. Fourteen of them were on centre haemodialysis, 13 on home haemodialysis. Excluded were patients with a previous renal transplant, bilateral nephrectomy, treatment with anticonvulsants, vitamin D derivatives and corticosteroids. They were haemodialysed four to six hours thrice weekly on a C-DAK 1.8m^2 or 2.5m^2 kidney. The dialysate was aluminium-free deionised tap water with a calcium concentration of 3.0mEq/L and a magnesium concentration of 1.0mEq/L. Heparin was routinely given, and all patients took 3g aluminium aminoacetate daily as a phosphate binder. They had received no calcium supplements and were on an unrestricted diet.

Bone mineral content was measured by two dimensional scanning photon absorptiometry (^{125}I) on both forearms as the mean of 12 scans. Measurements were made at the start of haemodialysis and usually at six month intervals for a mean of 4.8 years (range 3.0–6.5 years). Serial measurements were corrected for the physiological bone mineral content loss with age [4]. Long-term reproducibility of bone mineral content measurements is 1.2 per cent in normals and about two per cent in osteopenic patients [5]. Bone mineral content values were calculated as: 1) per cent of published reference values [6], 2) per cent of individual start value (=100%) in serial measurements, and 3) individual bone mineral content slope (mean percentage bone mineral content change/year) by linear regression in each patient in order to compare patients with differing periods of observation.

Serum values of immunoreactive parathyroid hormone (iPTH) (measured by a C-terminal assay), calcium (albumin corrected), phosphate and alkaline phosphatase were measured at three month intervals and mean values for the period of haemodialysis were calculated for each patient.

Serum values of 25(OH)D_3, 24,25(OH)$_2$ D_3 and 1,25(OH)$_2$ D_3 were measured by previously described methods [7]. Interassay variation in our laboratory for the three metabolites was: 11.5 per cent, 10.5 per cent and 9.8 per cent, respectively. All blood samples were drawn in October to eliminate the seasonal variation in 25(OH)D_3 and 24,25(OH)$_2$ D_3.

Results

At the start of haemodialysis the mean bone mineral content (89.0% ± 18.2%) was significantly lower than in matched controls ($p<0.05$). There was no significant difference between males (92.9% ± 16.2%, SD)) and females (84.0% ± 19.8%) or between centre (85.9% ± 20.1%) and home (92.2% ± 16.2%) haemodialysis patients. Figure 1 shows individual percentage slopes during haemodialysis

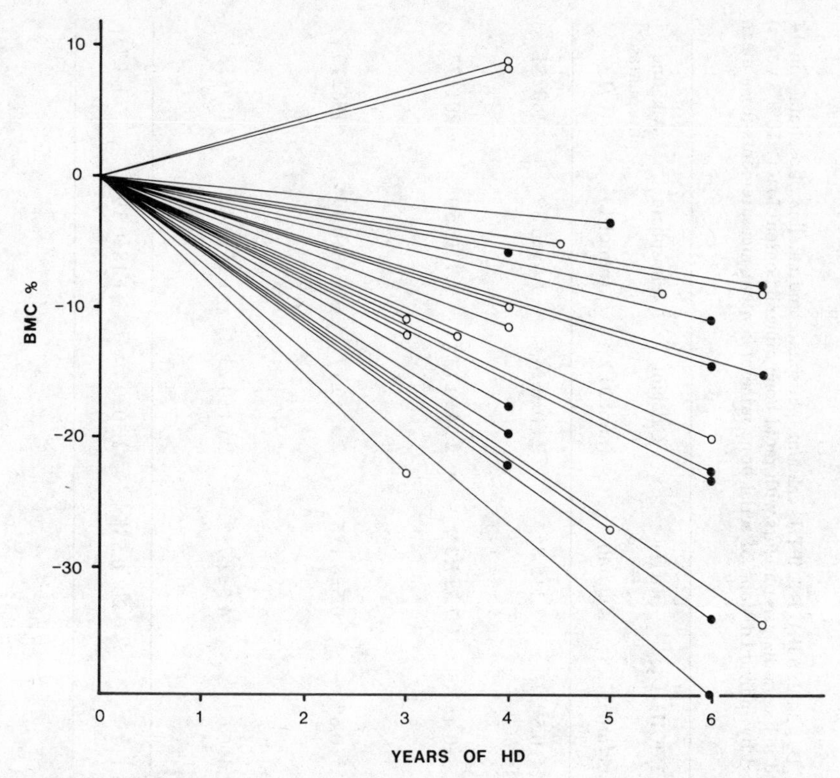

Figure 1. Individual bone mineral content changes during haemodialysis in 27 patients (○= centre haemodialysis, ●=home haemodialysis). The bone mineral content slope (=mean percentage bone mineral content change/year) was calculated from serial measurements by analysis of regression

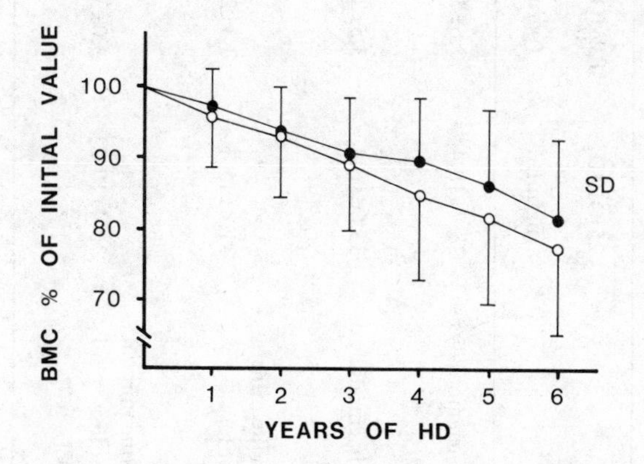

Figure 2. Mean annual bone mineral values (% of initial value) in patients on centre (○) and home (●) haemodialysis

TABLE I. Serum values (mean ± SD) of $25(OH)D_3$, $1,25(OH)_2D_3$, $24,25(OH)_2D_3$, iPTH, calcium, phosphate and alkaline phosphatase in 14 patients on centre and 13 patients on home haemodialysis, and in haemodialysis patients with rapid bone mineral content loss (>10%/3 years) and slow bone mineral content loss (<10%/3 years) estimated by individual slope of serial bone mineral content measurements for a mean of 4.8 years (range 3.0–6.5 years) of haemodialysis

	$25(OH)D_3$ (ng/ml)	$1,25(OH)_2D_3$ (pg/ml)	$24,25(OH)_2D_3$ (ng/ml)	iPTH (μg/ml)	Calcium (mmol/L)	Phosphate (mmol/L)	Alkaline phosphatase (U/L)
Centre haemodialysis (n=14)	28.7±18.4	2.6±3.2**	1.4±0.5*	7.1±5.8**	2.60±0.24	1.69±0.39*	180±51
Home haemodialysis (n=13)	33.6±13.8	3.5±2.8**	1.6±0.4*	7.7±7.3**	2.54±0.13	1.92±0.50*	201±77
rapid bone mineral content loss (n=13)	31.8±15.8	3.1±2.7**	1.5±0.6*	10.5±8.1**	2.55±0.17	1.76±0.49*	196±73
slow bone mineral content loss (n=14)	30.4±17.2	3.0±3.3**	1.4±0.6*	4.6±2.7**	2.61±0.23	1.86±0.43*	184±58
Controls	26.1±11.3	28.6±10.8	3.1±2.7	0.20–0.55†	2.50±0.10	1.18±0.19	70–275†

Difference from control value: *=$p<0.05$; **=$p<0.01$; †=range calculated from serial measurements and Figure 2 mean annual bone mineral content values (per cent of starting value) in centre and home haemodialysis patients; the mean bone mineral content loss was similar in the two groups, and amounted to three to four per cent per year of haemodialysis.

Table I lists the mean serum values in patients on centre and home haemodialysis, and in patients with rapid bone mineral content loss ($>10\%/3$ years) and slow bone mineral content loss ($<10\%/3$ years), respectively. In the total group mean serum $25(OH)D_3$ was normal, $24{,}25(OH)_2D_3$ reduced to half the normal level, and $1{,}25(OH)_2D_3$ severely reduced to almost non-detectable levels. Mean serum iPTH was markedly elevated (about 15 times upper normal limit), and was significantly higher ($p<0.05$) in patients with rapid than in those with slow bone mineral content loss; whereas the vitamin D metabolites were similar in the two groups. Mean serum calcium was normal, whereas both serum phosphate and alkaline phosphatase were significantly elevated. Serum iPTH did not correlate significantly to *individual* bone mineral content loss ($r=0.22$) nor to any of the biochemical parameters. Patients on centre and home haemodialysis did not differ significantly with regard to any of the biochemical parameters.

Discussion

The bone mineral content results suggest the presence of osteopenia at the start of haemodialysis and progressive bone loss during haemodialysis in both centre and home haemodialysis patients. A less pronounced bone mineral content loss in home haemodialysis patients might have been expected since they are usually in better general health and less immobilised than centre patients. However, normal serum $25(OH)D_3$ in the two groups suggest a similar nutritional status; moreover, they did not differ in either degree of hyperparathyroidism or in serum calcium or phosphate. Therefore, it must be concluded that home haemodialysis as such does not prevent development of uraemic osteodystrophy.

Besides measuring the intact C-84 molecule and the C-34 fragment, the C-terminal iPTH assay also detects biologically inactive polypeptides which are normally cleared by the kidney. This is probably the main reason for the lack of correlation between serum iPTH and serum calcium and phosphate, which are kept in a narrow range due to haemodialysis and treatment with phosphate binders; however, also a shift in 'set point' or autonomy of the parathyroid gland is likely. Despite difference in iPTH between the rapid and slow losers of bone calcium (the former group about twice as high), individual iPTH values correlated poorly to rate of bone mineral content loss mainly due to a considerable variation and overlapping of iPTH values between the two groups. Thus the iPTH was of no value in predicting bone mineral content loss in the individual patient.

The findings in serum of normal $25(OH)D_3$ and severely reduced or non-detectable $1{,}25(OH)_2D_3$ values are in accordance with most other reports and confirm that there is no impairment of liver C-25 hydroxylation of vitamin D in patients with chronic renal failure whereas the C-1α hydroxylation is severely reduced due to progressive destruction of the nephron mass. No extra-renal C-1α hydroxylation of $25(OH)D_3$ seems to take place. In contrast, our findings of only moderately reduced $24,25(OH)_2D_3$ values, although at variance with some reports of non-detectable serum $24,25(OH)_2D_3$ values in anephric patients [8], indicate an extra-renal C-24 hydroxylase activity, which may be of significance in the development of renal osteodystrophy. Some reports point to a specific effect of $24,25(OH)_2D_3$ on bone mineralisation [9].

However, the present results showing virtually identical serum concentrations of vitamin D metabolites in haemodialysis patients with rapid and slow bone mineral content loss point to the importance of factors other than impaired vitamin D metabolism in the pathogenesis of renal osteodystrophy. The possible role of $24,25(OH)_2D_3$ in the development of bone disease is particularly difficult to assess.

References

1 Massry SG, Dua S, Garty J, Friedler RM. *Min Elect Metab 1978; 1:* 172
2 Stanbury SW. *Clin Endocr 1977; suppl 7:* 25
3 Rickers H, Christensen M, Rødbro P. *Clin Nephrol 1983; 20:* 302
4 Smith DM, Khairi MRA, Johnston CC. *J Clin Invest 1975; 50:* 311
5 Christiansen C, Rødbro P. *Scand J Clin Lab Invest 1977; 37:* 321
6 Christiansen C, Rødbro P. *Scand J Clin Lab Invest 1975; 35:* 425
7 Shepard M, Horst RL, Hamstra AJ, Deluca HF. *Biochem J 1979; 182:* 55
8 Horst RL, Littledike ET, Gray RW, Napoli JL. *J Clin Invest 1981; 67:* 274
9 Coburn JW, Wong EGC, Sherrard DJ et al. *Clin Res 1980; 28:* 532A

Proc EDTA–ERA (1984) Vol 21

HAEMOFILTRATION WITHOUT SUBSTITUTION FLUID

G Civati, C Guastoni, U Teatini, A Perego, D Scorza, L Minetti

Niguarda Ca' Granda Hospital, Milan, Italy

Summary

Haemofiltration at its best is able to give excellent clearances of solutes in a wide molecular weight range, but the need of large amounts of reliable substitution fluid makes this technique too expensive for more widespread application.

In order to give an 'adequate' treatment by a safe, well tolerated, effective and comparatively cheaper method, haemofiltration without substitution fluid (HWSF) has been carried out. This method, consisting of the regeneration of appropriate amounts of pure convected plasma water by highly permeable, high surface area dialysis membranes, has been utilised for several months in four patients, with encouraging clinical results.

Introduction

In 1975, according to the middle molecule hypothesis, Babb [1] and Funck-Brentano [2] claimed that a vitamin B_{12} dialysis clearance of 30L/week (2.9ml/min) and a urea clearance of 69L/week (6.8ml/min) was sufficient to prevent or cure uraemia.

In 1979 Savazzi [3] demonstrated that signs of polyneuropathy were absent in conservative patients whose residual glomerular filtration rate (GFR) was equal to or greater than 13ml/ min, while polyneuropathy was still seen in all dialysis patients despite a dialysis index equal to or greater than one. Teschan [4] has recently (1983) shown the disappearance even of the mildest 'uraemic' symptoms (neurobehavioural changes) in his patients receiving a dialysis prescription of 3000ml/week/L body water as total urea clearance, which corresponds to about 10 per cent of normal GFR.

High efficiency haemofiltration (HEHF) was proposed in 1983 by Civati [5] as the best method to clear solutes in the widest molecular weight range: by this prescription (3.3L/L body water/week) it was possible to achieve a urea clearance of 115.9L/week (11.5ml/min) and an inulin clearance of 86.6L/week

(8.6ml/min). Using this technique Minetti [6] found a marked improvement of the peripheral uraemic polyneuropathy. The aim of our study was to test a new model of haemofiltration without substitution fluid (HWSF) evaluating the removal of various sized solutes, the haemodynamic stability, the acid-base equilibrium, the easy feasibility, safety and costs.

Patients, materials and methods

Four patients, all women aged 22 to 65 years (mean 45.7), body weight 47.5 to 82.5kg (mean 59.2 ± 15.7) have been treated thrice weekly for four hours by HWSF for a period of three to nine months (mean 6.1 ± 3.2). Maximum weight loss was between 2.2 per cent and 6.5 per cent of total body weight (mean 2.6 ± 1.06%). An average pure convective ultrafiltration rate of 180–200ml/min was obtained by an appropriate haemofilter (ASAHI PAN 250) with a blood flux of 500ml/min or more, transmembrane pressure of 400 ± 50mmHg, the mean Ht being 20.3 ± 3.4 per cent and total protein concentration 6.9 ± 0.9g/dl. This acellular and aproteic ultrafiltrate the composition of which is nearly identical to the plasma water, was then passed into a high surface area, highly permeable dialyser (2 TORAY B1L = 4.2m² PMMA) through which 600ml/min of dialysate (Na^+ 140; K^+ 1.5; Mg^{++} 1.5; Ca^{++} 4; Cl^- 107; Acetate 40mEq/L; Glucose 1g/L; mOsm 300/L) were flowing (Figure 1).

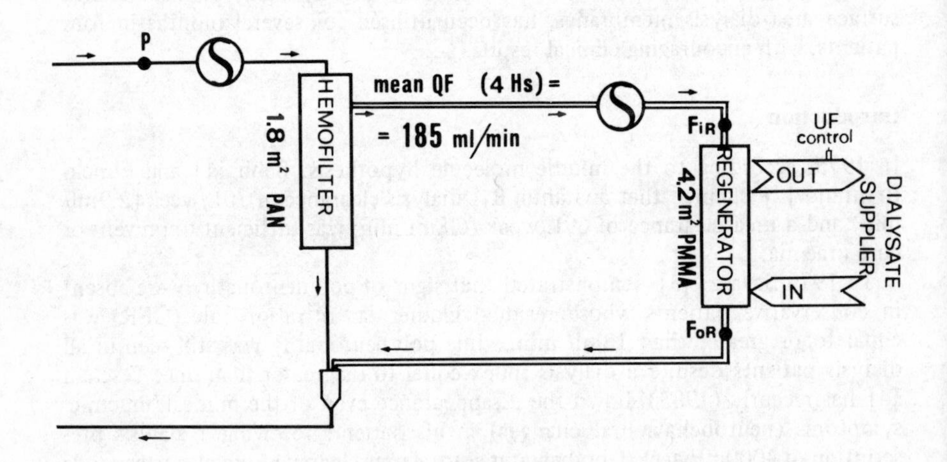

Figure 1. Haemofiltration without substitution fluid (HWSF). Clearances calculated as: $Cl = \frac{QF\ (F_{iR} - F_{oR})}{P}$ where QF = ultrafiltration rate (ml/min) F_{iR} = solute concentration at regenerator inlet; F_{oR} = solute concentration at regenerator outlet; P = plasma solute concentration

Results

Plasma clearances

All data were calculated as the mean of three values, each of them being the average of duplicate samples at 15, 120, 240min from the start, for three consecutive runs. The mean ultrafiltration rate (QF) was 187.1 ± 15.3ml/min. Average urea clearance was 188.8 ± 3.7ml/min and creatinine clearance 182.6 ± 3.3ml/min, while inulin (MW 5200) and β_2-microglobulin (MW 11800) clearances were 70.2 ± 3.8ml/min and 11.9 ± 1.4ml/min respectively.

Plasma solute values and balances

The data are shown in Table I.

TABLE I. Average plasma values before and after three HWSF sessions in four patients and solutes balance. All the effluent dialysate was collected in a large tank

	Before	After	Balance
Urea g/L	1.5 ± 0.3	0.6 ± 0.2	−33.7 ± 7.2g*
Creatinine mg/dl	8.7 ± 1.9	3.7 ± 1.1	−1.1 ± 0.1g*
Phosphate mg/dl	4.6 ± 0.5	2.8 ± 0.2	−0.6 ± 0.05*
Calcium mg/dl	9.1 ± 0.2	11.1 ± 0.5	+0.3 ± 0.06g†
Potassium mEq/L	4.9 ± 1.1	3.3 ± 0.3	−130 ± 36mEq†

* Solutes amount collected in the tank

† Solutes supplied – solutes collected in the tank

Blood gas equilibrium

PaO_2 and $PaCO_2$ were quite similar in HWSF and in a pure convective method such as high efficiency haemofiltration (Figure 2), while the same did not happen in acetate dialysis where a fall of PaO_2 and $PaCO_2$ was seen especially with large surface area dialysers.

Acetate and bicarbonate equilibrium

In HWSF plasma acetate at 120min from the start (5.2 ± 1.5mMol/L) was lower than that observed during a haemodialysis session with the same surface area and membrane type (6.4 ± 1mMol/L). HWSF was well tolerated while sudden hypotension with a fall in PaO_2 and $PaCO_2$ occurred during haemodialysis. During HWSF plasma bicarbonate never fell below starting values during the whole session, while they usually fell during haemodialysis. During HWSF, but also in a pure convective way, there is a characteristic pattern of acetate and bicarbonate (Figure 3).

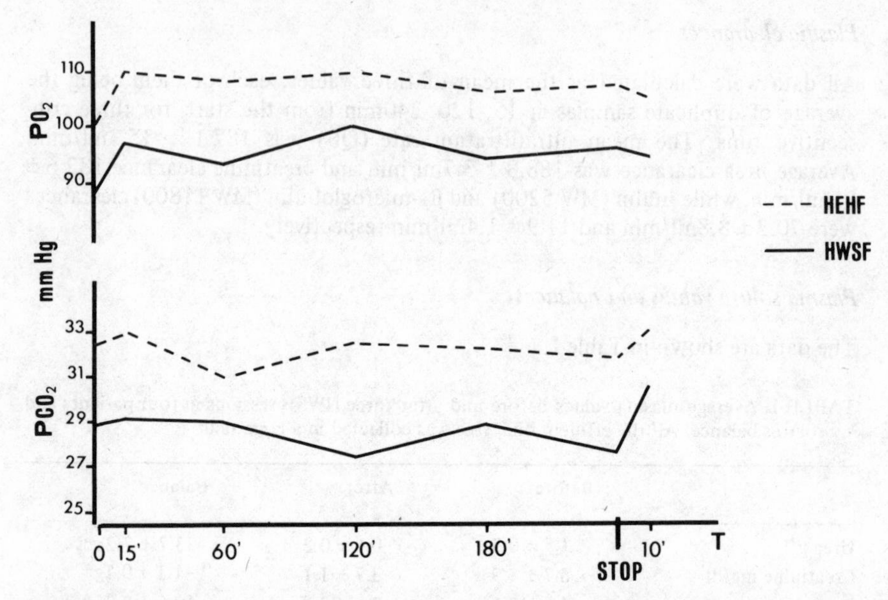

Figure 2. Blood gas behaviour in HWSF and in HEHF (mean values of three consecutive sessions in four patients)

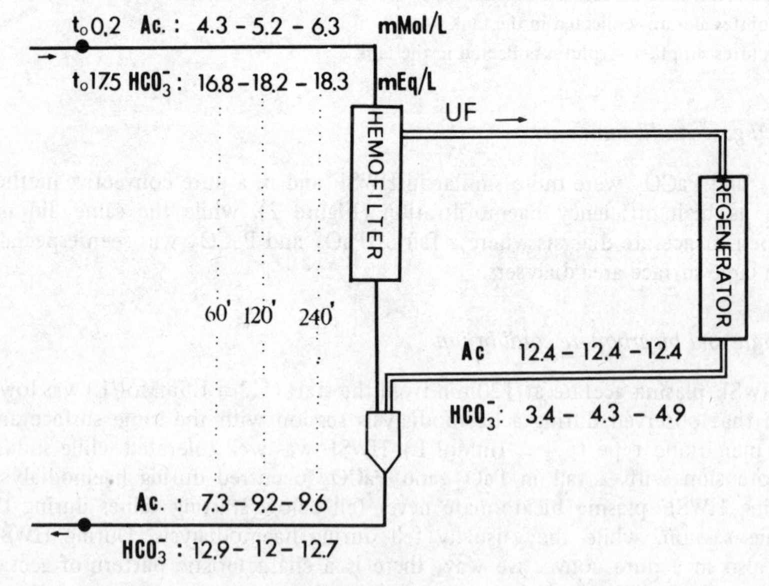

Figure 3. Behaviour of acetate (mMol/L) and bicarbonate (mEq/L) in HWSF at 0, 60, 120 and 240 minutes. Mean values of four experiments

Discussion

HWSF is able to achieve very good clearances of small solutes, Average urea (MW 60) clearances (13.2ml/min = 132L/week) were identical to those obtained by HEHF, but creatinine (MW 113) clearances were somewhat lower (12.8ml/min = 129.6L/week). These data are as good as those obtained by conventional diffusive or convective methods.

According to Savazzi, Teshan and ourselves, this regimen could be 'adequate' and effective in preventing and curing neurobehavioural changes and peripheral polyneuropathy. The role of middle and larger molecules in the pathogenesis of uraemia is less clearly defined; but good removal of some of them (i.e. PTH and its fragments) must be of value. HWSF obtains a mean inulin (MW 5200) clearance of 5ml/min = 50L/week and β_2-microglobulin (MW 11800) clearance of 0.8ml/min = 8.6L/week. From this point of view HEHF seems still superior to HWSF, nevertheless the need for removal of such large solutes is not yet completely understood. On the other hand small clearances of β_2-microglobulin may not be disadvantageous because normally the kidney retains microproteins and free light-chains. Anyway neither a deficiency syndrome, nor a loss of significant amounts of albumin and globulins were seen in our patients.

In our experience HWSF was also a powerful tool for removing phosphates (mean -625mg/session) and potassium (mean 130mEq/session), allowing patients to have a completely free diet. Calcium balance studies showed in our patients a mean gain of 310mg/session. These two latter features, together with PTH removal, could possibly slow the evolution of the uraemic osteodystrophy. Sodium balance of each HWSF session is precisely defined by the sodium concentration in the effluent dialysate per volume of weight loss, no other sodium escaping the system. Several factors may play a role in cardiovascular stability during HWSF sessions, where no significant change between pre and post mean arterial pressure (MAP) was recorded. First of all there is a small decrease of plasma osmolality, because the sodium saving effect of haemofiltration is fully maintained: no disequilibrium syndromes were seen in spite of the rapid decrease in urea even to one-third of the starting values. Moreover HWSF involves a bicarbonate saving effect and ensures blood gas stability and moderate acetate at the reinfusion site.

HWSF is a simple renal replacement therapy with no need of special equipment. We used commercial bags of pure salt, aluminium free (0.02 PPM) concentrate with deionised water. A Hospal Monitral machine with reliable control of weight loss at the effluent dialysate site and with a supplementary pump for convected ultrafiltrate was regularly utilised. No pyrogenic reactions were recorded; limulus amebocyta tests in affluent and effluent dialysate and in reinfused dialysed ultrafiltrate were always negative as well as cultures in the same samples and in blood. At least HWSF is cheaper than HEHF because no substitution fluid is required. Furthermore, regenerators can be utilised dozens of times without any problem at a very low price per session. Thus HWSF may be introduced as a routine procedure for 'adequate' and well tolerated treatment in end-stage renal failure.

Acknowledgments

We would like to thank Mrs Daniela Bongiorno for her technical assistance.

References

1 Babb AL, Strand MJ, Uvelli DA et al. *Kidney Int 1975; Suppl 2:* S23
2 Funck-Brentano JL, Man NK, Sausse A. *Kidney Int 1975; Suppl 2:* S52
3 Savazzi GM, Migone L, Cambi V. *Clin Nephrol 1980; 13:* 74
4 Teschan PE, Ginn E, Bourne JR et al. *Asaio J 1983; 6:* 108
5 Civati G, Guastoni C, Perego A et al. *Blood Purification 1983; 1:* 184
6 Minetti L, Civati G, Guastoni C et al. *Blood Purification 1983; 1:* 170

Open Discussion

ALBERTINI (Los Angeles) I think this is an elegant way of helping haemofiltration a little further. At the same time it seems to me that you are taking away one crutch to justify the treatment, which is the high molecular solute removal component, which in dialysis uses a purely diffusive transport mechanism. I also want to bring to your attention, as you will see in the next paper, if in your set up you left out the first step, which is the needles, you would double the efficiency of the whole treatment. You would have a blood flow of 500ml/min and you would get a clearance of about 440ml/min for urea.

CIVITI Yes, our philosophy in this kind of treatment was mainly to achieve a very good clearance of small and larger solutes. It could be done better by high efficiency haemofiltration with substitution fluid. We wanted to reduce the cost but maintaining the results in terms of stability and of acid base and bicarbonate equilibrium. This is the first clinical experiment in this field of regenerating ultrafiltrate. I think that we will improve this system when we can adapt as regenerators more compact membranes with higher cut off. This is a bioengineering problem.

SHALDON (Chairman) I think your goal of trying to achieve haemofiltration without fluid replacement is obviously the right way to go if one ever wants to achieve a transportable, portable or inplantable artificial kidney. It seems to me that you have to produce 600ml per minute of dialysis fluid to achieve this goal. Have you considered looking at the sorbent field? A few years ago we published on the problems with aluminium but Dr Shapiro in New York has had quite encouraging results recently using a sorbent based system. What is your opinion on that?

CIVITI I think that the way of the sorbent is very hard. It is certainly harder than our method and I think is more expensive. I know the experiments you mention but I think that the method we have described is superior.

Proc EDTA–ERA (1984) Vol 21

PERFORMANCE CHARACTERISTICS OF HIGH FLUX HAEMODIAFILTRATION

B von Albertini, J H Miller, P W Gardner, J H Shinaberger

Wadsworth Veterans Administration Wadsworth Medical Center, Los Angeles, California, USA

Summary

With the described technique of high diffusive and convective solute transport an almost threefold increase in efficiency over conventional dialysis was clinically demonstrated. Coupled with the better tolerance to high solute and weight removal rates, this approach permits drastic reduction of treatment time, without sacrificing treatment adequacy.

Introduction

The goal of any renal replacement therapy for end-stage renal disease should be to maintain the patient's solutes within acceptable ranges, remove interdialytic weight gain and, for obvious socio-economic reasons, do both in the shortest possible time. It follows that treatment time must not be reduced unless it is accompanied by an at least commensurate increase in solute and weight removal rates. Efforts to augment efficiency in conventional acetate haemodialysis are effectively curtailed by clinical intolerance to high solute transfer and ultrafiltration rates [1,2]. The standard technique, as it has emerged over the years, mirrors these limitations and consists of comparatively low blood and dialysate flow rates as well as membranes with low hydraulic permeability and surface area. Previously thought to be absolute, this barrier to efficiency was challenged in recent years by observations of better treatment tolerance with bicarbonate dialysis, haemofiltration and haemodiafiltration [3–6]. The key elements of these techniques, a greater convective solute transport component and bicarbonate dialysate, were recently combined in a new modality termed high flux haemodiafiltration. By augmenting simultaneously blood and dialysate flow rates as well as membrane surface area and permeability a more than twofold increase in efficiency over conventional haemodialysis was achieved with this technique [7]. Clinical tolerance to high solute and weight removal rates permitted the reduction of treatment time to under six hours weekly, without sacrificing treatment adequacy [8].

The present clinical study explores the gains in efficiency to be made by further augmenting blood flow rates and evaluates the performance of different membranes with greater permeability in high flux haemodiafiltration.

Materials and methods

The study was performed with four stable end-stage renal disease patients who had mature forearm Cimino-Brescia fistulae. Cannulation was by 14-gauge, 1 inch needles (Lifepath®, Drake Willock). Haemodiafiltration was performed in a new mode, described in greater detail elsewhere [7]. Three different hollow fibre dialysers were evaluated, made of a new anisotropic polysulfone membrane ($1.25m^2$, Hemoflow F-60, Fresenius AG), polymethylmethacrylate ($2.1m^2$, Filtryzer B1L, Toray Industries) and an older model of cellulose acetate ($1.8m^2$, C-DAK DuoFlux, Cordis Dow). The listed dialysers were used in pairs in a serial configuration in the extracorporeal circuit. Sterile, pyrogen-free dialysate was delivered at 1000 ± 22ml/min in countercurrent mode by an automated system providing volumetric control of net ultrafiltration. By adjusting differential pressures, maximal ultrafiltration was obtained in the first device, while volume replacement by backfiltration of dialysate occurred in the second device. In this serial configuration the entire surface area of both devices was available for diffusion. Dialysate composition was Na 140, K 2, Ca 4.5, Mg 0.7, Cl 108.2, HCO_3 35, Acetate 4mEq/L and glucose 100mg/100ml.

During the clinical treatment the performance was routinely evaluated by whole blood clearances obtained from blood and dialysate [7]. Blood flows were measured by using occlusive blood pumps calibrated at the same input and output pressures as encountered clinically. Dialysate flow rates were measured by timing the strokes of the proportioning cylinders, whose volume had previously been calibrated. Net weight loss was determined from the slope of a continuous recording of the weight of the bed scale plus the patient. Ultrafiltration rate in the first device was measured by a ball-in-tube flow meter (F1300, Gilmont Instruments) during brief periods when the dialysate flow to this dialyser had been bypassed, but pressures maintained at operating levels. Pressures were automatically measured using differential transducers (140 PC, Microswitch) and continuously recorded by a microcomputer (TRS 80®). Solute concentrations were determined by the Autoanalyser technique. The erythrocytes were allowed to equilibrate with plasma for at least one hour before separation. Inulin (American Critical Care) was injected in bolus of 2g intravenously at least 10 minutes before clearance periods. Plasma inulin was analysed by the Anthrone method for Autoanalyser, and clearances were calculated using plasma flows.

In order to exclude possible measurement errors, all clearance periods were analysed for mass balance error of urea [9]. Only periods with a urea mass balance error of less than 10 per cent were used for analysis. The reported whole blood clearances are means of values calculated from both blood and dialysate concentrations.

Results

All treatments, where the high flows and clearances were maintained throughout, were well tolerated by the patients. Treatment times were less than two hours, three times weekly.

Fluid movements across the membrane in the described serial configuration are illustrated in Figure 1. The upper panel depicts ultrafiltration from blood to

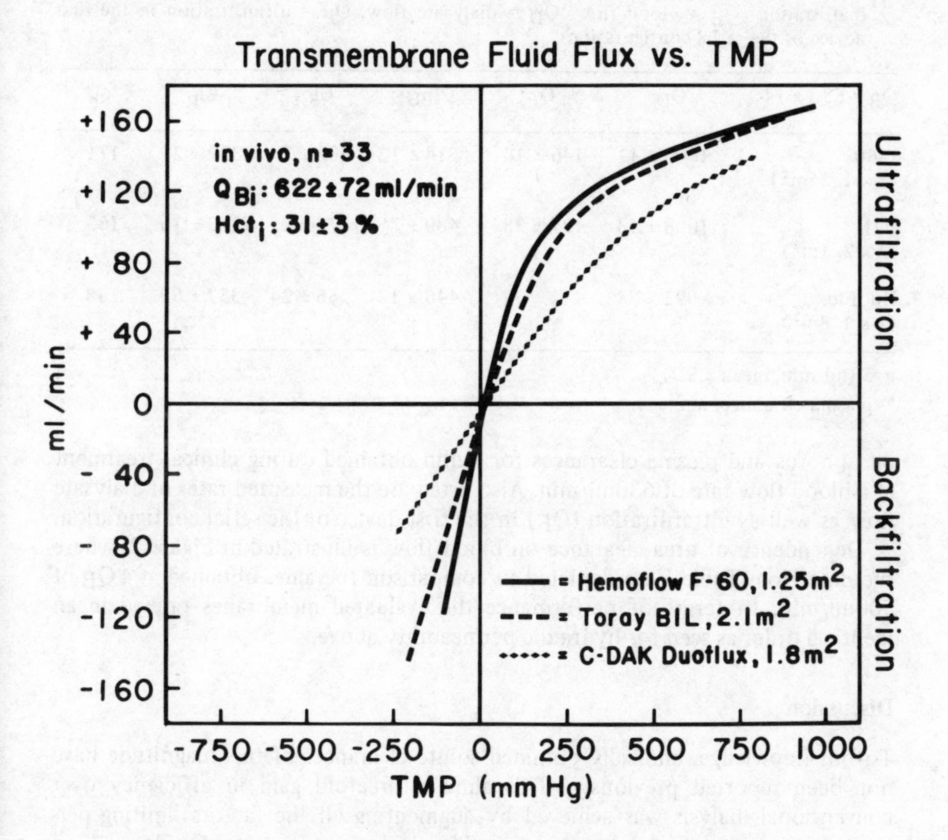

Figure 1. Transmembrane fluid fluxes in response to transmembrane pressures in high flux haemodiafiltration (see text)

dialysate occurring in the first device of the serial configuration in response to a range of positive or forward transmembrane pressure gradients. Back filtration from dialysate to blood, occurring simultaneously in the second device in response to automatically adjusted negative or reverse transmembrane pressure gradients, is shown in the lower panel. Note that it takes less negative or reverse transmembrane pressure to transfer back filtrate into blood than it takes to ultrafiltrate the same volume from the blood.

Compared to conventional haemodialysers, all of the evaluated membranes demonstrated substantially greater hydraulic permeabilities. If surface area is taken into account, the new polysulfone membrane performed best followed by polymethylmethacrylate and cellulose acetate.

The overall performance for solute removal of the different devices in series is summarised in Table I. Listed are whole blood clearances for urea, creatinine,

TABLE I. Overall clearances for urea, creatinine, phosphorus and inulin in high flux haemodiafiltration (Q_B = blood flow, Q_D = dialysate flow, Q_F = ultrafiltration in the first device of the serial configuration)

Q_B = 630 ± 14	Q_D	Q_F	Cl_{BUN}	Cl_{Cr}	Cl_P	Cl_{In}*
F-60 (2 x 1. 25m²)	1006 ± 11	146 ± 18	514 ± 12	432 ± 7	399 ± 29	171
B1L (2 x 2. 1m²)	1008 ± 13	127 ± 38	480 ± 7	377 ± 9	373 ± 16	163
DuoFlux (2 x 1. 8m²)	992 ± 34	111 ± 39	446 ± 17	366 ± 24	357 ± 69	148

n=5 (ml/min, mean ± SD)
*=plasma clearance, n=2

phosphorus and plasma clearances for inulin obtained during clinical treatment at a blood flow rate of 630ml/min. Also listed are the measured rates of dialysate flow as well as ultrafiltration (Q_F) in the first device of the serial configuration.

Dependence of urea clearance on blood flow is illustrated in Figure 2, where the data from Table I are depicted in comparison to values obtained at a Q_B of 500ml/min. In terms of performance the evaluated membranes ranked in an identical order as seen for hydraulic permeability above.

Discussion

To our knowledge, clinically obtained solute clearances of this magnitude have not been reported previously. This almost threefold gain in efficiency over conventional dialysis was achieved by augmenting all the factors limiting performance in the extracorporeal circuit. The single most important under-utilised resource to increase efficiency is the blood flow delivering the solute to the membranes. Actual blood flows in mature vascular accesses substantially exceed the flow rates used in conventional dialysis and may be up to 2L/min in many patients [10]. Given an optimal technique, bypassing a larger fraction through the extracorporeal circuit in itself poses no additional risk for the patient.

In order to fully exploit higher blood flow rates, every other factor contributing to mass transfer must be optimised. To maintain solute gradients, dialysate flow must be proportionally increased. The doubling of surface area in our serial configuration maintains flow dependence of clearance even in the range illustrated in Figure 2. The effect of high surface area and simultaneous diffusion and convection on solute transport is apparent from the magnitude of the

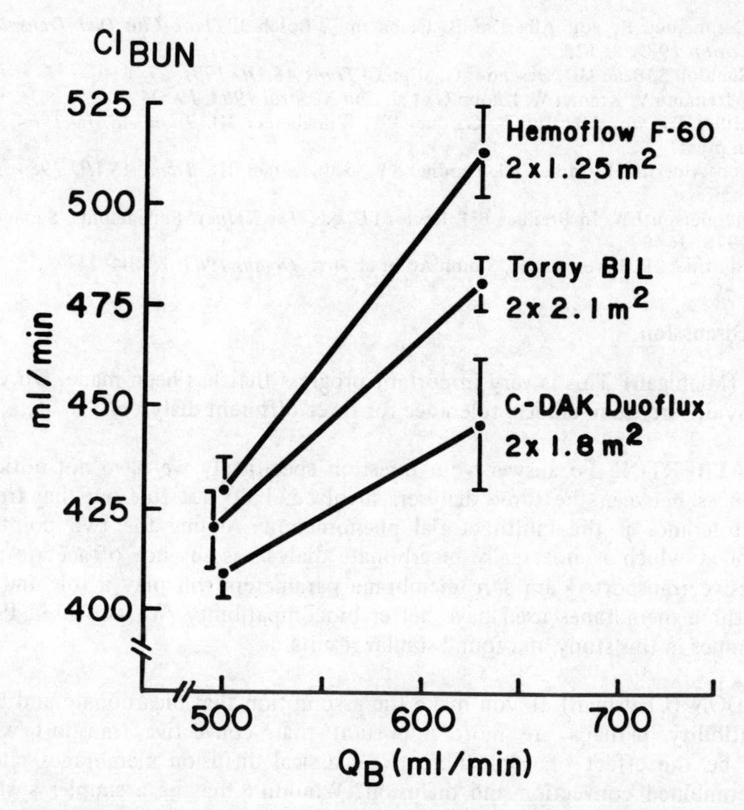

Figure 2. Effect of blood flow on urea clearance in high flux haemodiafiltration (n=5, Q_D= 1000 ± 22ml/min)

clearances obtained for small as well as large solutes. The comparison between the different membranes in our study indicates that diffusive permeability correlates with hydraulic permeability.

The observed clinical tolerance to high solute and weight removal rates differs from the experience with rapid conventional haemodialysis and must be related to the only two fundamentally different features of our approach: bicarbonate dialysate and a high convective solute transport component.

The clinically demonstrated efficiency in this study opens new horizons to shorten treatment time. Treatments of well under two hours, equivalent in terms of total clearance of small and large solutes to conventional haemodialysis, are within reach.

References

1 Maher JF, Schreiner GE. *N Engl J Med 1965; 273:* 370
2 Arieff AI, Massry SG, Barrientos A et al. *Kidney Int 1973; 4:* 177
3 Graefe V, Follette WC, Vizzo JE et al. *Proc Clin Dial Transplant Forum 1976; 6:* 203

4 Geronemus R, von Albertini B, Glabman S, Bosch JP. *Proc Clin Dial Transplant Forum 1979; 9:* 125
5 Shaldon S, Beau MC, Deschodt G, Mion C. *Trans ASAIO 1981; 27:* 610
6 Wizemann V, Kramer W, Knopp G et al. *Clin Nephrol 1983; 19:* 24
7 Miller JH, von Albertini B, Gardner PW, Shinaberger JH. *Trans ASAIO 1984; 30:* in press
8 von Albertini B, Miller JH, Gardner PW, Shinaberger JH. *Trans ASAIO 1984; 30:* in press
9 Henderson LW. In Brenner BM, Rector FC, eds. *The Kidney.* Philadelphia: Saunders. 1976: 1649
10 Bouthier JD, Levenson JA, Simon AC et al. *Artif Organs 1983; 7:* 404

Open Discussion

PORT (Michigan) This is very important progress that has been made. Do you find any difference in clinical tolerance for three different dialysers?

VON ALBERTINI To answer your question specifically we have not noticed differences between the three dialysers involved. I do not rule out that treatment tolerance is the multifactorial phenomenon. Among the two points I mentioned which is not really bicarbonate dialysis is absence of acetate and convective transport. I am sure membrane parameters will play a role and all these three membranes used have better biocompatibility. We also used PAN membranes in this study and found similar results.

SHALDON (Chairman) If you make the assumption that bicarbonate and biocompatibility perhaps are more important than convective transport what would be the effect of using those two classical diffusion membranes rather than combined convection and diffusion? Wouldn't that be a simpler system with less potential risk to the patient?

VON ALBERTINI Your comment is very well taken. The point I would like to make is using these very highly permeable membranes, thinner membranes in zero configuration, will always entail the necessary consequences of the flow resistance and transmembrane fluxes in both directions. Our configuration allows us to adequately control this, maximise ultrafiltration and at the same time provide a safe way of working with sterile dialysate with necessarily occurring back filtration.

SHALDON (Chairman) That just leads me to one very quick question. Your experience is limited in terms of numbers so obviously you have not had pyrogen reactions. We have been pouring dialysate into people's veins for years through filters, and I firstly would be concerned with using only one Amicon filter between the patient and your crude dialysate system preparing the fluid. Do you have any view on that?

VON ALBERTINI Well, my observation number may be limited for haemodiafiltration but I happen to have made about 80,000 litres of haemodiafiltration fluid which I have instilled into patients. The point is that you have double security because the membrane you are using for the treatment itself acts as the last filter and, in our experience, it is sufficiently safe to use an Amicon in the dialysate part and then use the final filter as a second safety measure. We have, by the way, always checked with limulus and cultures and found that it is negative.

KESHAVIAH (Minneapolis) Do you have any experience with chronic exposure using this form of therapy? Secondly, what was the needle you used? You mentioned 14 gauge one inch, was it an Avon needle?

VON ALBERTINI No it was Drake Willock. I would love to have the needle you have which is shorter but we have not been able to get it. Obviously the only limit to increase efficiency is, in fact, the needle. I told you that we have made flow measurement in accesses and it exceeds by far the rate that we are using here. As far as chronic studies are concerned we have now six patients who have undergone at least one month of this therapy. We have one patient who has undergone eight weeks therapy.

Proc EDTA–ERA (1984) Vol 21

STANDARD METHODS FOR THE MICROBIOLOGICAL ASSESSMENT OF ELECTROLYTE SOLUTION PREPARED ON LINE FOR HAEMOFILTRATION

H U Mayr, *F Stec, *C M Mion

*Hospital St Josef, Regensburg, FRG, *University Hospital of Montpellier, Montpellier, France*

Summary

The preparation of large volumes of intravenous solution at the bedside (on-line preparation) requires bacteriological monitoring and pyrogen control before final sterilisation. We tested the sensitivity of microbiological methods and their applicability in routine clinical conditions during haemofiltration. In vitro, the 0.22μm membrane filter technique (MF) showed a better or equal bacterial recovery with two test organisms. Parallel tests of 0.22 and 0.45μm MF under simulated haemofiltration conditions showed no significant difference in the number of detected bacteria. Routine MF with 0.22μm membranes offers a simple and reliable way to monitor the bacterial content of on-line prepared electrolyte solution in clinical conditions.

Introduction

The bedside preparation of electrolyte solution for intravenous injection is a cost-reducing approach in the development of haemofiltration [1] and long-term experience has shown successful results [2]. To increase the safety of this technique we felt that it was essential to monitor each prepared batch of electrolyte solution for its bacterial content. Such a rigorous quality control should detect any degradation in the preparation procedure. Standard recommendations for the bacteriological assessment of water are based on the collection of small samples (i.e. a few millilitres of water) to detect coliform and faecal organisms [3].

The purpose of the present study was to define the most sensitive system for quantitative bacterial testing of large volumes (i.e. several litres) of bedside prepared electrolyte solution and to assess the microbiological monitoring including pyrogen assay of electrolyte solution in clinical routine conditions during haemofiltration.

Materials and methods

Equipment for preparation of electrolyte solution and for haemofiltration

The water treatment consisted of the sequential system: raw water supply, pressure reducing valve, 5μm depth filter, water softener, activated charcoal filter, 1μm depth filter, heater, flow valve and reverse osmosis (RO, Milli RO 120, Millipore Corporation, Bedford, MA, USA). A system of ultrafilters (UF, polyamid haemofilters FH 202, Gambro AB, Lund, Sweden) surrounded the reverse osmosis membrane to protect it against bacterial contamination. These ultrafiltrators were mounted in a special plastic housing (Lopez SA, Toulouse, France).

For the preparation of electrolyte solution, an acetate containing dialysate concentrate was infused in the flow path of reverse osmosis product water by means of a roller pump. The mixed solution passed through an ultrafiltrator (presterilising ultrafiltration) and a conductivity meter, the latter serving as feedback control for the speed of the pump. The fluid was then stored in a clean 35L PVC bag in a rigid container. During the haemofiltration session the stored fluid passed through two ultrafiltrators placed in series and was then infused intravenously.

Haemofiltration was carried out with haemofilters FH 202 or FH 303 (Gambro) and the Gambro system for haemofiltration using gravimetric balancing of the prepared fluid and patient's filtered plasma water. The electrolyte solution had a temperature of 20°C in the batch and 37°C before intravenous injection. Its glucose concentration was 11.1mMol/L.

Formalinisation with 4% formaldehyde (final concentration) was performed daily for all ultrafiltrators and the reverse osmosis, and weekly for the water softener. Depth filters and charcoal were replaced weekly, ultrafiltrators once a month. The equipment was rinsed for three hours before starting the preparation of electrolyte solution.

Microbiological methods

The culture medium for all bacteriological studies was standard agar (Tryptone glucose yeast agar). Results were expressed as colony forming units per volume of sample after 72 hours of incubation at 20°C. All manipulations were carried out under aseptic conditions.

The agar pour plate technique was performed in accordance with standard methods, section 406 [3]. One millilitre of the sample or of a serial dilution was placed in an empty Petri dish before the liquid medium was added at 45°C and carefully mixed. The membrane filter technique was performed either in the laboratory or in the dialysis unit using membranes of mixed esters of cellulose acetate with retention pore sizes of 0.22 and 0.45μm (GVWP 04700 and HAWP 04700, Millipore), sterilised by autoclaving prior to use. In the laboratory, a 100ml portion of inoculated sample was passed through the membrane placed in a filter-holding assembly, and processed according to standard methods, section 408 [3] by a laboratory technician.

In the dialysis unit, membrane filtration of large volumes (12 to 30L) of electrolyte solution was performed using a Swinnex filter holder (47mm diameter, Millipore). This filter holder, heat sterilised with the membrane before use, was easy to insert in the PVC tubings used for the delivery of electrolyte solution, allowing MF to be done in situ by a nephrologist. At the end of each filtration procedure, membranes were immediately cultured: the membrane, removed from its filter holder was placed onto the plate containing the solid culture medium. Sterile membranes were used as controls for each series of assays.

The endotoxin test used was the LAL Pyrogent test (Mallinckrodt Inc, St Louis, MO, USA). The sample was mixed with the test reagent and incubated in a water bath at 37°C for 60 minutes. The sensitivity of this test varies between 0.006 and 0.25ng of standard endotoxin (*E coli*) per millilitre.

Study design

In vitro studies were made to determine the most sensitive filter membrane for bacterial testing of electrolyte solution during haemofiltration. In the laboratory, a comparison of the pour plate technique versus membrane filtration with 0.22 and 0.45μm pore sized membranes was conducted using *Pseudomonas aeruginosa (P aerug)* and *Staphylococcus coagulase negative (Staph coag neg)* as test organisms. Five bacterial suspensions of each test organism containing about 10^6 bacteria/L were examined by the three methods. Filtration was performed with portions of a 10^3- and 10^4-fold diluted preparation in replicates of four. In the dialysis unit, we performed a parallel testing of 0.22 and 0.45μm MF under simulated haemofiltration conditions. Thirty litres of electrolyte solution were prepared and stored in a PVC bag. The fluid was then pumped at similar flow rates of about 150ml/min through two Swinnex with 0.22 and 0.45μm membranes respectively, placed in parallel downstream of two roller pumps. The similar flow rates were achieved by manual adjustment of the pumps after repeated flow rate measurements with a graduated cylinder and stop watch.

In vivo studies were made during the routine haemofiltration treatment of five stable patients (2 females, 3 males, mean age 54.6 ± 17.9 years). The patient's temperature was measured just before and after completion of the treatment. All treatments were carefully followed to observe any fever or shivering reaction. Endotoxin tests were performed from the stored electrolyte solution after the treatment. The bacterial contamination of the electrolyte solution was assessed with MF using a Swinnex filter holder. Before starting the haemofiltration session, the Swinnex was fitted in the flow path of the electrolyte solution between the infusate pump of the haemofiltration apparatus and the sterilising ultrafiltrator. The total volume of electrolyte solution was sampled during each session.

Statistical methods

Results were expressed as mean ± standard deviation (SD). Statistical analyses were performed with the Wilcoxon matched-pair signed-rank test.

Results

The results obtained by the in vitro comparison (Table I) showed an underestimation of the number of the two test organisms by membrane filtration in comparison with the number obtained by the agar pour plate technique. For

TABLE I. In vitro comparison of agar plate count versus membrane filtration with 0.22 and 0.45μm pore-sized membranes using *Pseudomonas aeruginosa* and *Staphylococcus coagulase negative* as test organisms

		Plate count	Membrane filtration 0.22μm	Membrane filtration 0.45μm
Pseudomonas aeruginosa				
CFU* x 10^6/L series	1)	6.3	3.4	3.1
	2)	5.6	3.1	3.2
	3)	0.9	0.6	0.6
	4)	1.4	0.9	0.9
	5)	1.0	1.1	1.0
Mean ± SD		3.0 ± 2.7	1.8 ± 1.3	1.8 ± 1.3
2-tailed probability†		p=0.080 (plate count vs 0.22μm)		
		p=0.068 (plate count vs 0.45μm)		
			p=0.593 (0.22μm vs 0.45μm)	
Staphylococcus coagulase negative				
CFU* x 10^6/L series	1)	3.8	2.9	1.9
	2)	6.0	4.8	3.0
	3)	3.0	2.2	1.8
	4)	1.9	2.0	1.7
	5)	3.0	2.0	1.7
Mean ± SD		3.5 ± 1.5	2.8 ± 1.2	2.0 ± 0.6
2-tailed probability†		p=0.080 (plate count vs 0.22μm)		
		p=0.043 (plate count vs 0.45μm)		
			p=0.043 (0.22μm vs 0.45μm)	

* colony forming units
† using the Wilcoxon matched-pair signed-rank test

Staph coag neg the 0.22μm MF showed a better sensitivity than the 0.45μm (p=0.043), for *P aerug* equal results were obtained.

The parallel test of 0.22 and 0.45μm MF during in vitro haemofiltration (n=44 sample pairs, flow rate 130–150ml/min, throughput 12L/membrane) showed mean values of 57 ± 72 colony forming units for the 0.22μm and 63 ± 76 colony forming units for the 0.45μm MF (four culture pairs were too numerous to count and excluded); the median values were 27.5 and 28 colony forming units respectively, p=0.116, not significant. The results of the 29 values ranging from 10 to 300 colony forming units/12L are shown in Figure 1.

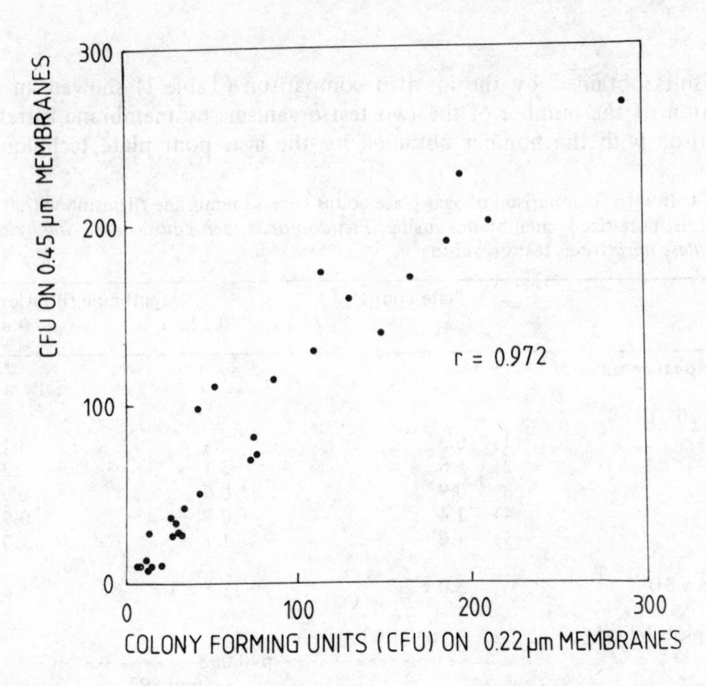

Figure 1. Colony forming units on 0.22 and 0.45μm pore-sized membranes, obtained under simulated haemofiltration conditions after parallel membrane filtration of 12L of batch prepared electrolyte solution per membrane

In vivo studies were performed during 264 routine haemofiltration treatments from April to July 1983 Patients did not have any fever or shivering reaction. The body temperature showed a mean of 36±0.2°C before the treatment and a mean pre-/post-haemofiltration increase of 0.4±0.3°C (range 0–1.3°C). LAL Pyrogent tests in the batches of electrolyte solution were performed after a temperature rise of 1°C or more and showed negative reaction in all cases (n=11).

The bacterial assessment of electrolyte solution was made using exclusively 0.22μm MF in order to combine sampling and sterilising of the fluid. The results of cultures showed 0–9 CFU in 67.4 per cent, 10–99 CFU in 20.1 per cent and more than 100 CFU in 8.5 per cent of samples; four per cent of samples showed confluent or excessive CFU. The mean volume of throughput was 25.7±2.1L. Controls made of sterile unused membranes showed no growth in all cases. On only one occasion the filter membrane was ruptured. The time for handling the Swinnex including culturing was less than five minutes for each treatment.

Discussion

Standard microbiological methods for field testing of water are mainly adapted to detection of coliform bacteria [3,4]. The membrane filtration, able to concentrate bacteria from large sample volumes, is usually done in the laboratory using 0.45μm pore sized membranes.

The conditions in clinical haemofiltration, where 20–30L of ultrapure electrolyte solution are infused after final sterilisation during a treatment session, allow the in situ MF of the electrolyte solution, unifying sampling and filtering in a single step.

The results obtained in this study demonstrate that the sterilising 0.22μm filter membrane can replace the 0.45μm membrane as test membrane for bedside prepared electrolyte solution in haemofiltration, as it offers similar sensitivity for quantitative bacterial assessment.

The in vitro tests showed a better or equal recovery on 0.22μm membranes with two test organisms, commonly found in the stored bedside prepared electrolyte solution. The lower retention of *Staph coag neg* on 0.45μm membranes may be explained by the small size of this organism (0.5–0.8μm).

Under haemofiltration conditions with an undefined 'natural' bacterial flora in the stored electrolyte solution and similar flow rates the results of the parallel testing of 0.22 and 0.45μm MF showed no significant difference.

The in vivo study demonstrated that continuous membrane filtration of batch prepared electrolyte solution is practicable under routine conditions, because it is easy to perform and not time-consuming. Membranes can easily and reproducibly be cultured by the staff of the dialysis unit. This 'on the spot' culture procedure suppresses any delay for culturing the membrane at the end of the HF session. In the presence of positive membrane cultures, the plate may then be sent to the microbiological laboratory for identification of the contaminating organism.

The reason for the constant temperature rise during haemofiltration treatment may lie in the net heat transfer or in small amounts of exogenous pyrogens not detectable by the LAL test used. However, it is important to note that neither fever nor shivering occurred during or after the treatment. Further investigations using a more sensitive pyrogen assay could help to elucidate this problem [5].

In conclusion, routine MF with 0.22μm membranes offers a simple and reliable way to monitor the bacterial content of batch-prepared electrolyte solution. Such monitoring seems to us the only way to maintain a high degree of awareness among staff members of the dialysis unit concerning the utmost care which should be used in the bedside preparation of sterile apyrogenic solution for intravenous infusion.

Acknowledgments

We wish to express our thanks to Mrs Francine Sinègre, Maître-ès-sciences, Head Laboratory of Microbiology, Institut Bouisson-Bertrand, Montpellier, France, for her help with the in vitro microbiological studies, and to M Günther Hinderer, Dipl Volkswirt, Assistant Computer Center of the University of Regensburg, West Germany, for the statistical analyses.

This work has been partially supported by a grant from Gambro AB, Lund, Sweden.

References

1 Henderson LW, Sanfelippo ML, Beans E. *Trans ASAIO 1978; 24:* 465
2 Shaldon S, Beau MC, Deschodt G et al. In Schaefer K, ed. *Contributions to Nephrology; 32.* Basel: Karger. 1982: 161–164
3 American Public Health Association. In *Standard Methods for the Examination of Water and Wastewater; 13.* New York: American Public Health Association Inc. 1971: 635–685
4 Sladek KJ, Suslavich RV, Sohn BI et al. *Appl Microbiol 1975; 30:* 685
5 Dinarello CA. *Blood Purification 1983; 1:* 197

Open Discussion

PORT (Michigan) You seem to suggest that when the temperature rises more than 1°C endotoxin must have been present in the infusate. I was wondering how you substantiate that conclusion and whether some other factors such as temperature of the infusate and perhaps endogenous pyrogens could not also be responsible?

MAYR Yes, we could not conclude that the temperature rise of 1°C relates only to the amount of pyrogen in the blood. I think that the temperature rise may be due to the net heat transfer because the injection fluid was 37°C and there may be low amounts of pyrogen not detectable by the LAL test used.

Proc EDTA–ERA (1984) Vol 21

PHOSPHORUS KINETICS DURING HAEMODIALYSIS AND HAEMOFILTRATION

H Pogglitsch, W Petek, E Ziak, F Sterz, H Holzer

University of Graz, Graz, Austria

Summary

Phosphorus excretion during haemodialysis positively correlates with the plasma inorganic phosphorus (P_i) concentration and the dialyser P_i clearance. Therefore, by using highly efficient dialysers or haemofilters, disciplined patients may achieve a well regulated P-balance.

The plasma P_i concentration time curve during haemodialysis or haemofiltration must be seen as the result of passive diffusion combined with an active mobilisation of P_i from a rapidly exchangeable P-pool.

The plasma P_i concentration hardly falls below the normal range in dialysis patients, regardless of the quantity of P_i removed by haemodialysis or haemofiltration. The stability of plasma P_i demonstrates the existence of a mechanism for phosphate regulation in body fluids, independent of the calcium homeostasis

Introduction

Our investigations were originally designed to determine whether or not high efficiency dialysers could effect a significant increase in phosphate elimination, resulting in a reduced oral phosphate binder requirement. In the course of these investigations, several conspicuous peculiarities in the phosphate kinetics became apparent. Owing to their basic significance, they are reported in this paper.

Patients and methods

The following parameters were determined in dialysis patients with no residual renal excretory function:

1. The behaviour of the plasma concentrations of urea, creatinine, and inorganic phosphorus (P_i) during a four or eight hour haemodialysis or haemofiltration.
2. The concentration and total excretion of these metabolites in the dialysate (or haemofiltrate).

3. The clearance of these substances, whereby we distinguished between:

a) the in vivo blood clearance: $K_{Bm} = \frac{Q_{Bi}(C_{Pi} - C_{Po})}{C_{Pi}} + \frac{Q_F(C_{Pi})}{C_{Pi}}$

b) the blood side plasma clearance: $K_{p\text{-}blood} = \frac{Q_{Bi}(1\text{-}H)C_{Pi} - [Q_{Bi}(1\text{-}H) - Q_F]C_{Po}}{C_{Pi}}$

c) the dialysate side plasma clearance: $K_{p\text{-}dial} = \frac{Q_{Do}\,C_{Do}}{C_{Pi}}$

(Q_{Bi} and Q_{Bo}: incoming or outgoing blood flow (cm^3/min); C_{Do}: solute concentration in the dialysate outlet flow (cm^3/min); Q_{Do}: dialysate outlet flow (cm^3/min); C_{Pi} and C_{Po}: plasma concentrations in the incoming or outgoing blood stream; Q_F: ultrafiltrate (cm^3); H: haematocrit).

The plasma concentrations of urea, creatinine, uric acid, and P_i were determined on a SMA II Technicon®, the concentrations in the dialysate or haemofiltrate on a GSA 300 Greiner®. The method of Itaya et al [1] was used with minor modifications to determine the P_i concentration in the dialysate.

Blood flow was measured by the bubble-flow method, the dialysate outflow was collected and measured.

Either a GF 180M hollow fibre or a GL 1.36 plate dialyser was used for phosphorus kinetic studies during haemodialysis; FH 303 for haemofiltration.

Results

Clearance and total excretion of urea, creatinine and P_i

The values of the in vivo blood clearance (K_{Bm}) disagree with those of the blood side and dialysate side plasma clearance considerably (Table I). The K_{Bm} is useful only in comparing the efficiencies of different dialysers. Conclusions as to the actual mass transfer during haemodialysis are not possible, however, since the K_{Bm} is influenced by haematocrit, active and passive transport of substances between plasma and erythrocyte ICF, and by changes in protein binding during passage through the dialyser. During dialyser passage, urea diffuses not only out of the plasma, but also out of the erythrocytes. This easily explains the higher $K_{p\text{-}dial}$ compared with the $K_{p\text{-}blood}$ values. The differences between $K_{p\text{-}dial}$ and $K_{p\text{-}blood}$ for creatinine and P_i undoubtedly also result from their different concentrations in plasma and erythrocyte ICF and their diffusion disequilibrium during the blood passage. Disregarding methodological error the $K_{p\text{-}dial}$ proved best for kinetic calculations.

The total quantity of P_i excretion correlates with the plasma P_i concentration during haemodialysis or haemofiltration, and with the P_i clearance of the dialyser or the ultrafiltrate efficiency of the haemofilter (comparisons in Table I).

TABLE I

	Urea (mmol/L)			Creatinine (μmol/L)			Inorganic phosphorus (mmol/L)		
	GF 180 M	Gl 1.36	HF 303	GF 180 M	Gl 1.36	HF 303	GF 180 M	Gl 1.36	HF 303
Initial concentration	27.47±1.93	25.26±4.24	32.50±8.52	1072±38.1	933±296	1042±287	2.16±0.55	1.87±0.12	1.96±0.56
Removal (g)	45.5 ±7.84	31.92±9.71	20.44±6.44	2.18±0.1	1.54±0.57	1.65±0.82	1.19±0.17	0.68±0.08	0.75±0.22
K_{Bm} (cm^3/min)	192±3.8	173±4.4		171±5.0	147±8.6		159±5.1	137±8.8	
$K_{p\text{-}dial}$ (cm^3/min)	163±17.0	150±4.0		105±9.6	96±6.3		104±11.8	86.0±7.5	
$K_{p\text{-}blood}$ (cm^3/min)	142±10.9	137±29.5		127±10.6	103±7.3		119±11.5	97±5.2	

Passive diffusion and active mobilisation of P_i during haemodialysis and haemofiltration. Evidence for an independent P_i regulatory mechanism

The decrease in concentration of a solute which is uniformly distributed throughout the total body fluid and which can diffuse from ICF to ECF with negligible resistance and thus may diffuse freely through a dialyser membrane, demonstrates first order kinetics:
$C_t = C_o.e^{-K.T/V}$ = (where C_t=concentration at time t, C_o=initial concentration, T=duration of haemodialysis in minutes, $K=K_{p\text{-}dial}$, and V-effective volume of solute distribution, e.g. 60 per cent of body water).

Urea is undoubtedly the only substance for which these assumptions obtain. Therefore, the experimentally determined fall in urea concentration corresponds well to the calculated value (Figure 1). Creatinine and uric acid behave similarly,

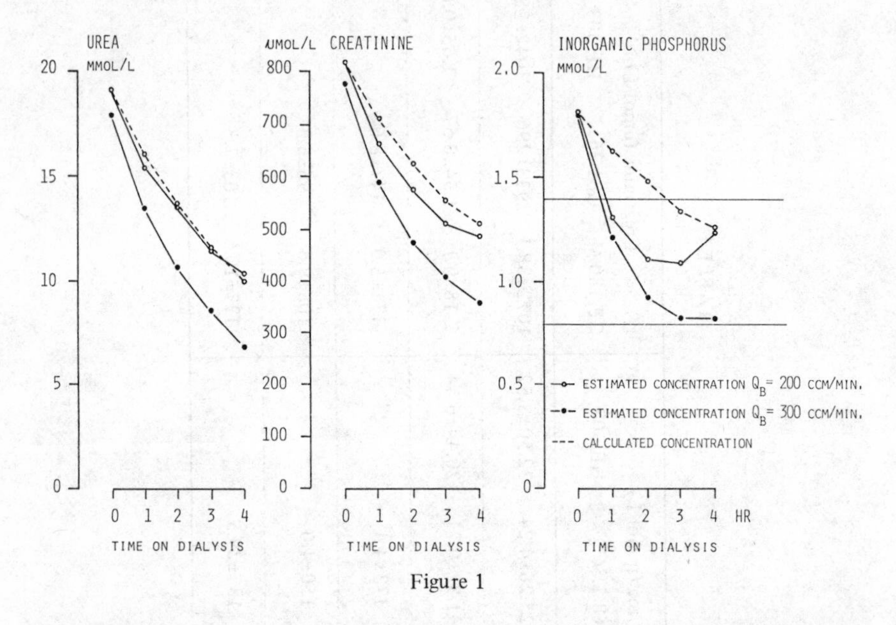

Figure 1

although not identically. The behaviour of P_i is different: with initial P_i above normal, the concentration falls sharply within the first two hours of haemodialysis or haemofiltration (Figure 2). As soon as the normal P_i range (0.8–1.4mmol/L) is reached, the concentration remains constant or varies within the normal range, regardless of haemodialysis blood flow (Q_B=300 or 200cm^3/min). Even after eight hours of haemodialysis values do not fall below the normal range, unless the patient is suffering from a wasting disease (sepsis, cachexia, etc). Theoretically, the P_i concentration should be reduced to considerably lower levels after eight hours of haemodialysis and presuming constant diffusion from the ICF. If the P_i level is normal before haemodialysis, it remains in the normal range in spite of continuous removal of P_i.

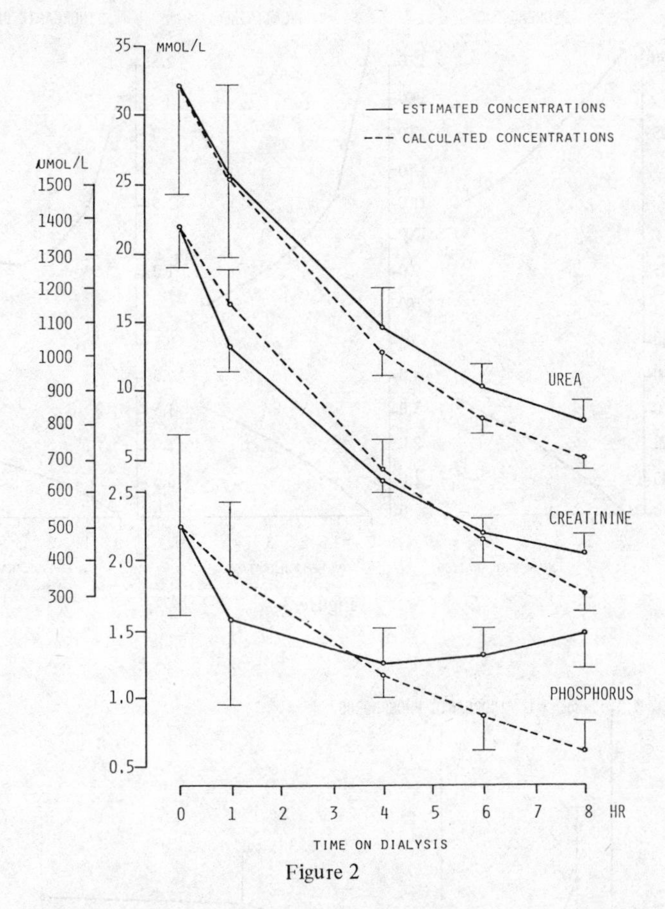

Figure 2

Since the clearance rates of P_i during haemofiltration are of the same order of magnitude as the $K_{p\text{-dial}}$ for dialysers of average efficiency, the P_i concentration behaves the same during haemofiltration as in haemodialysis (Figure 3).

Upon termination of haemodialysis, a rebound in P_i occurs in a regular fashion. In the first hour following a four hour haemodialysis, a relatively steep rebound is observed in all patients. It depends on the level of the initial concentration as well as on the rate of removal. After the third hour following haemodialysis, P_i continues to rise only in patients whose predialysis concentrations were above normal. In these patients, P_i values reach approximately 80 per cent of their original value after 12 hours (Figure 4).

Discussion

The behaviour of the P_i concentration during and after haemodialysis or haemofiltration cannot be explained solely on the basis of passive diffusion of P_i out of

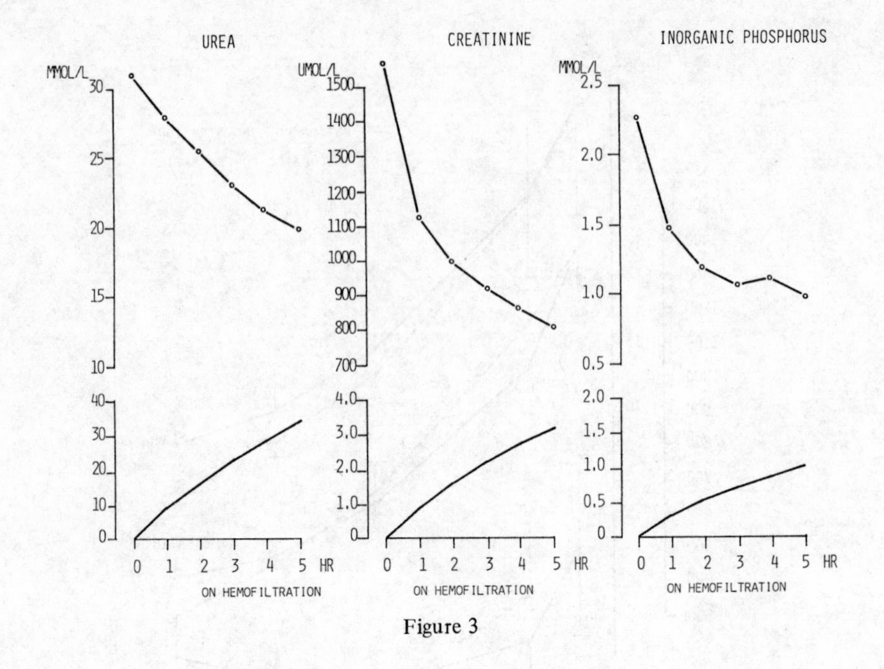

Figure 3

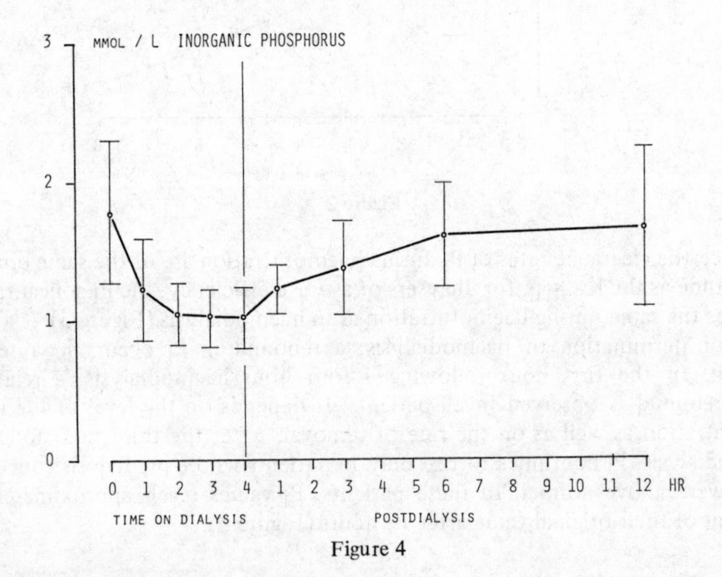

Figure 4

diverse compartments. During haemodialysis or haemofiltration, the P_i concentration should drop continuously, even assuming that P_i is replaced by passive diffusion from a large reservoir. However no tendency to a continuous decrease in P_i concentration was observed in any patient. Similar observations have been reported by Sugisaki et al [2].

To explain the P_i kinetics during haemodialysis or haemofiltration, it must be seen as the result of a combination of two processes: passive diffusion out of a fluid space somewhat larger than the ECF, and active mobilisation of P_i. Apparently, P_i in concentrations exceeding 1.4mmol/L represents a waste product, which diffuses freely during haemodialysis, much like urea. Active mobilisation begins when the P_i value reaches the normal range. Possibly, P_i is mobilised from a rapidly exchangeable phosphate pool, whose existence has long been postulated in nuclear medicine [3], but has not yet been proven.

Active P_i mobilisation during haemodialysis and haemofiltration to stabilise the plasma P_i in the normal range implies the existence of a regulatory mechanism for P_i homeostasis. This mechanism still functions in kidney failure and is also independent of calcium homeostasis. During haemodialysis, the total and ionised calcium concentrations increase, and the secretion of parathyroid hormone often decreases. Therefore, the well-known physiological dependence of the P_i concentration on calcium homeostasis plays no part in this regulatory mechanism.

The obvious tendency to protect itself from a drop in P_i level below a certain value by means of an independent regulatory system poses a series of new questions. The nature of the regulatory mechanism, the capacity of the rapidly exchangeable P-pool, and the consequences of acute hypophosphataemia as well as chronic phosphate depletion which might be masked by normal pre- or post-dialytic P_i values, are subjects which will merit future interest.

References

1 Itaya K et al. *Clin Chim Acta 1966; 14:* 361
2 Sugisaki H et al. *Trans ASAIO 1982; 28:* 302
3 Fueger GF. In *Ergebn Med Radiol.* Stuttgart: Georg Thieme. 1973

Open Discussion

BAZZATO (Mestre) In the last slide you postulate the removal of phosphorus from the intracellular space. Do you have any data concerning the removal using acetate or bicarbonate buffers? Mastrangelo* has shown a better reduction in plasma phosphate during bicarbonate dialysis.

POGGLITSCH I heard your interesting paper but we do not have any experience with the bicarbonate or acetate dialysis to differentiate the relative efficiencies. I think that pH must have an influence on the phosphate pool but not on the principle behaviour.

VON ALBERTINI (Los Angeles) We were particularly concerned with our method, which was shown previously†, whether we would be able to remove

*Mastrangelo F, Rizzelli S, De Blasi V et al. *Proc EDTA–ERA 1984: 21:* this volume
†von Albertini B, Shinaberger J, Gardner P, Miller J. *Proc EDTA–ERA 1984; 21:* this volume

the same amount of phosphorus in a short time. Indeed, we have been able to achieve it and we have confirmed this independently by measurements in the dialysate effluent. The pattern of behaviour is indeed such that plasma phosphate falls initially and very often goes up again in the second half of dialysis. Studies in the first half of dialysis indicates that phosphate seems to be removed adequately.

Proc EDTA–ERA (1984) Vol 21

CONCOMITANT REMOVAL OF ALUMINIUM AND IRON BY HAEMODIALYSIS AND HAEMOFILTRATION AFTER DESFERRIOXAMINE INTRAVENOUS INFUSION

C Ciancioni, J L Poignet, C Naret, S Delons, *Y Mauras, *P Allain, †N K Man

*Centre Médical Rist, Paris, †Hôpital Necker, Paris, *Pharmacologie, CHU, Angers, France*

Summary

In six anuric haemodialysed patients, aluminium and iron mass transfer were determined 48 hours after 40 and 80mg/kg body weight desferrioxamine intravenous infusion.

All patients were aluminium overloaded (mean ± SEM: 2.91 ± 1.05μmol/g wet tissue bone) and two had high plasma ferritin. Haemodialysis and haemofiltration were performed using a highly permeable membrane.

The adequate dose of desferrioxamine for aluminium removal is 40mg/kg, since aluminium mass transfer induced by haemodialysis and haemofiltration (47.4 and 40μmol/session) are not significantly different from that obtained with 80mg/kg.

Iron removal is dose related in high plasma ferritin concentration patients: 50 and 100μmol/session with haemodialysis and 29 and 175μmol/session with haemofiltration after 40 and 80mg/kg body weight respectively.

Introduction

Desferrioxamine (DFO), a chelating agent of Iron (Fe) [1], is now currently used in the treatment of aluminium (Al) overload [2–4], but, as yet, no clinical studies have determined concomitant removal of Al and Fe by dialysis in Al overloaded haemodialysed patients. We have, therefore, carried out a study of the concomitant removal of Al and Fe by haemodialysis and haemofiltration 48 hours after DFO intravenous infusion with different doses (40 and 80mg/kg body weight) in six Al overloaded haemodialysis patients.

Subjects and methods

Six patients (2 males, 4 females), mean age (± SEM) 46.7 ± 10.8 years (range 30–57 years) and mean weight 50.5 ± 7.7kg (range 38.5–60kg) with chronic renal failure, on regular haemodialysis for 120 ± 14.3 months (range 77–156 months) were studied.

Aluminium hydroxide has been used in all patients for many years but was stopped three months before the study. Five patients had histological evidence of osteomalacia and one suffered from dialysis encephalopathy.

Mean Al content in bone biopsy was 2.91 ± 1.05μmol/g wet tissue (range 1.74–4.26μmol/g wet tissue).

Mean plasma ferritin concentration was 750 ± 384ng/ml (range 42–2100ng/ml), but only two patients had high plasma ferritin (1810 and 2100ng/ml respectively).

Haemodialysis was performed with a 150L dialysate batch and haemofiltration with a substitution fluid volume corresponding to one-third body weight, using in both cases a highly permeable membrane (Bio-Hospal 3000 S).

Dialysate Al concentration was less than 0.18μmol/L and dialysate Fe concentration was less than 0.35μmol/L in fresh dialysate and haemofiltration substitution fluids.

Desferrioxamine (Desferal-Ciba) was administered intravenously at different doses: 40 and 80mg/kg body weight, during 30 minutes after the end of a dialysis session, 48 hours prior to aluminium and iron removal.

Aluminium and iron were measured by inductively coupled plasma emission spectrometry [5] in plasma before (Pi) and after (Pf) each session, fresh (Di) and used dialysate (Df) substitution fluid (Sf) and haemofiltrate (Uf). Mass transfer (N), extraction ratio (ER) and integrated clearance (ICI) were calculated according to the following formulae (V=volume, L/session):

1. Mass transfer: N (μmol/session)

Haemodialysis	N_{Al}:	$[(Al_{Df}\, V_{Df}) - (Al_{Di}\, V_{Di})]$
	N_{Fe}:	$[(Fe_{Df}\, V_{Df}) - (Fe_{Di}\, V_{Di})]$
Haemofiltration	N_{Al}:	$[(Al_{Uf}\, V_{Uf}) - (Al_{Sf}\, V_{Sf})]$
	N_{Fe}:	$[(Fe_{Uf}\, V_{Uf}) - (Fe_{Sf}\, V_{Sf})]$

2. Extraction ratio: ER (per cent)

$$ER = 100\left(1 - \frac{(Al)_{Pf}}{(Al)_{Pi}}\right)$$

3. Integrated clearance: ICI (ml/min)

$$ICI = \frac{N}{\dfrac{Al_{Pi} - Al_{Pf}}{Ln\, Al_{Pi} - Ln\, Al_{Pf}}\, T}$$

Student's 't' test was used in all statistical evaluations.

Results

Figure 1 shows mean Al mass transfer, extraction ratio and integrated clearance in the six patients, for haemodialysis and haemofiltration, 48 hours after intravenous infusion of 40 and 80mg/kg body weight.

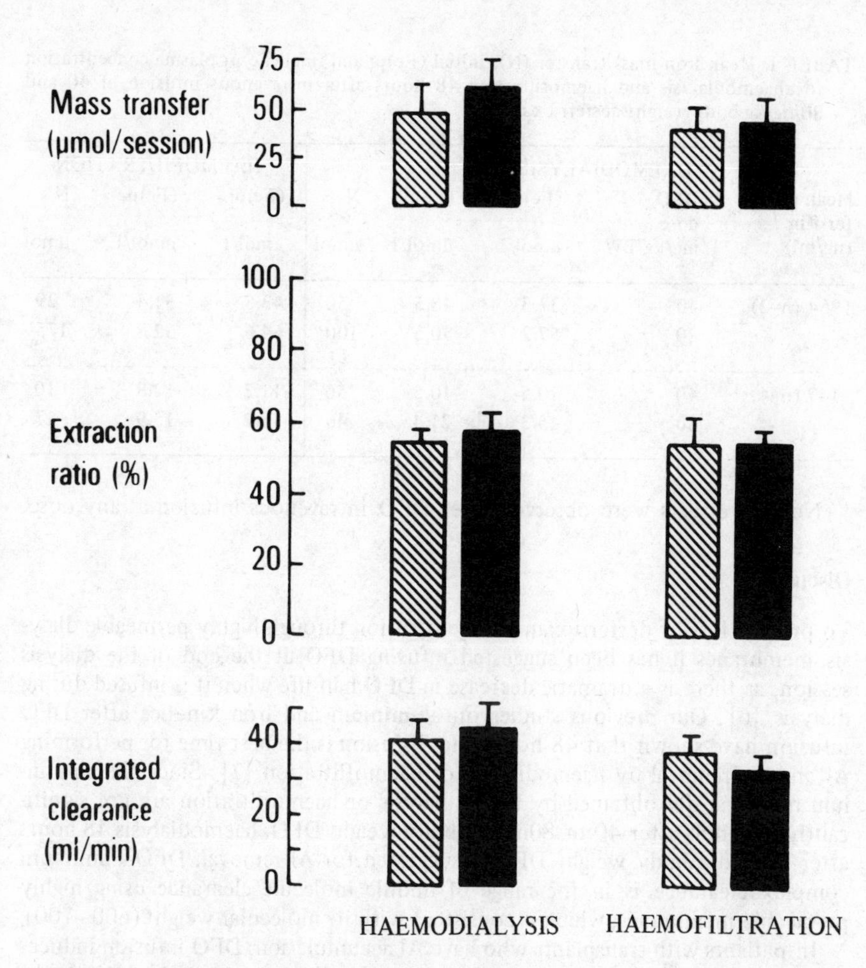

Figure 1. Mean aluminium mass transfer, extraction and integrated clearance for haemodialysis and haemofiltration 48 hours after intravenous infusion of 40 (hatched column) and 80 (solid column) mg/kg body weight desferrioxamine

Mean N(Al), ER(Al) and ICl(Al) are of the same order of magnitude and not significantly different for all DFO doses after haemodialysis or haemofiltration.

Table I lists mean iron mass transfer, initial plasma iron concentration and final plasma iron concentration in the six patients, for haemodialysis and haemofiltration, 48 hours after intravenous infusion of 40 and 80mg/kg body weight DFO.

Mean (Fe)Pi and Fe(Pf) are not significantly different for haemodialysis and haemofiltration in the high plasma ferritin group (1954ng/ml) and in the low plasma ferritin group (147ng/ml).

In addition, N(Fe) is dose related in high plasma ferritin concentration patients: 50 and 100μmol/session with haemodialysis and 29 and 175μmol/session with haemofiltration after 40 and 80mg/kg body weight respectively.

TABLE I. Mean iron mass transfer (N), initial $(Fe)_{Pi}$ and final $(Fe)_{Pf}$ plasma concentration for haemodialysis and haemofiltration 48 hours after intravenous infusion of 40 and 80mg/kg body weight desferrioxamine

	HAEMODIALYSIS				HAEMOFILTRATION		
Mean plasma ferritin (ng/ml)	DFO dose mg/kg/BW	$(Fe)_{Pi}$ μmol/L	$(Fe)_{Pf}$ μmol/L	N μmol	$(Fe)_{Pi}$ μmol/L	$(Fe)_{Pf}$ μmol/L	N μmol
1954 (n=2)	40	37.3	48.5	50	43.6	31.4	29
	80	57.2	50.3	100	64.6	62.8	175
147 (n=4)	40	10.5	10.5	50	11.2	12.9	10
	80	15.3	21.4	46	18.2	15.9	17

No side effects were observed after DFO intravenous infusion at any dose.

Discussion

To prevent loss of desferrioxamine by diffusion through highly permeable dialysis membranes it has been suggested infusing DFO at the end of the dialysis session, as there is a dramatic decrease in DFO half-life when it is infused during dialysis [6]. Our previous studies on aluminium and iron kinetics after DFO infusion have shown that 48 hours after infusion is the best time for performing Al and Fe removal by haemodialysis or haemofiltration [7]. Since the aluminium mass transfer obtained by haemodialysis or haemofiltration are not significantly different after 40 or 80mg/kg body weight DFO, haemodialysis 48 hours after 40mg/kg body weight DFO is suggested for Al removal. DFO-aluminium complex clearance is in the range of middle molecule clearance using highly permeable membranes, which is consistent with its molecular weight (600–700).

In patients with transplants who have Al accumulation, DFO infusion induces urinary excretion around 150μmol a day [4], which is three times the mass transfer obtained by a dialysis session. However, Keberle has reported in dog experiments that 75 per cent of injected DFO was excreted into the bile [8]. One could expect that between 300–500μmol of aluminium are excreted via the intestinal route after each DFO infusion.

Desferrioxamine infusion is an effective treatment for iron removal in patients with chronic anaemia who required repeated blood transfusions with the risk of long-term iron poisoning: average weekly iron excretion is 0.3–0.9mmol in urine and 2–4mmol in faeces [9,10].

Our study shows that after 80mg/kg body weight DFO intravenous infusion, 0.3–0.5mmol of iron could be removed weekly in high plasma ferritin patients which is similar to the weekly urinary iron excretion in subjects with normal renal function. Since desferrioxamine and its metabolites are mostly excreted via the intestinal route, quantification of Al and Fe output must take faecal excretion into account in aluminium and iron overloaded patients.

References

1 Bickel M, Gaumann E, Keller-Schierlein W et al. *Experentia 1960; 16:* 128
2 Ackrill P, Ralston AJ, Day JP, Hodge KC. *Lancet 1980; ii:* 692
3 Wills MR, Savory J. *Lancet 1983; ii:* 29
4 Malluche HH, Smith AJ, Abreo K, Faugere MC. *N Engl J Med 1984; 311:* 140
5 Allain P, Mauras Y. *Anal Chem 1979; 51:* 2089
6 Fosburg M, Hakim RM, Schulman G et al. *Kidney Int 1984; 25:* 183
7 Ciancioni C, Poignet JL, Mauras Y et al. *Trans ASAIO 1984:* in press
8 Keberle H. *Ann NY Acad Sci 1964; 119:* 2758
9 Blume KG, Beutler E, Chillar RK et al. *JAMA 1978; 239:* 2149
10 Pippard MJ, Callender ST, Finch CA. *Blood 1982; 60:* 288

Open Discussion

SHALDON (Chairman) Your observation on the faecal balance completing the total amount removed is interesting. Do you think there is a large variation from individual to individual in the amount of desferrioxamine that comes out via the gut as opposed to across the membrane?

CIANCIONI Yes. I think there is a large variation between patients with iron stores but the question is to evaluate iron stores in patients.

SHALDON If you have a normal iron store some people have recommended giving supplementary iron if you are using desferrioxamine for aluminium removal. Do you have any opinions on that?

CIANCIONI In our long-term experience of desferrioxamine we think that it is necessary to give iron to patients who are treated only for aluminium overload.

SHALDON So you would recommend in patients who are not iron overloaded routine intravenous iron supplements if on a long course of desferrioxamine?

CIANCIONI Yes, I think that is quite important.

PART XI

SYMPOSIUM

OPTIMUM DIALYSIS TREATMENT FOR PATIENTS OVER 60 WITH PRIMARY RENAL DISEASE. SURVIVAL, CLINICAL RESULTS, QUALITY OF LIFE

Chairmen: H Klinkmann
L Minetti

Proc EDTA–ERA (1984) Vol 21

MAINTENANCE HAEMODIALYSIS TREATMENT IN PATIENTS AGED OVER 60 YEARS. DEMOGRAPHIC PROFILE, CLINICAL ASPECTS AND OUTCOME

C Jacobs, A Diallo, *E A Bàlàs, †M Nectoux, †S Etienne

*Centre Pasteur-Vallery-Radot, Paris, France, *EDTA Registry, London, United Kingdom, †CITI2, Paris, France*

Introduction

Over the last decade an increasing number of elderly patients with end-stage renal failure have been accepted for renal replacement therapy in most of the economically developed countries. This more liberal access to a life-saving treatment was made possible by a continuous expansion and also diversification of the therapeutic procedures made available to patients with end-stage renal failure. Kidney transplantation is not considered as the preferred mode of treatment for elderly patients. Nonetheless, the results reported in small series of patients demonstrate that the widely advocated upper age limit of 50 years should not be considered as absolute for performing renal transplantation [1,2]. Chronic peritoneal dialysis, mainly continuous ambulatory peritoneal dialysis (CAPD), has become an established method of treatment of end-stage renal failure and is often considered particularly suitable for elderly patients as it is performed at home by the patients themselves [3–5]. The excellent haemodynamic tolerance and high efficiency of haemofiltration render this technique quite attractive for elderly patients, particularly those with unstable cardiovascular conditions [6]. Nonetheless, to this date, the vast majority of elderly patients accepted for renal replacement therapy have been treated by maintenance haemodialysis. Most of the reports published in the literature in this category of uraemic patients are, however, based on limited series of cases often comprising a high proportion of patients in the 50 to 60 year age range at start of treatment [7–11]. For the purpose of this Symposium, we conducted a study based on information made available by consulting two Computerised Dialysis Registries with large numbers of elderly patients treated by maintenance haemodialysis on their files.

Populations and methods

The demographic and survival data concerning the patients treated in the various European countries were provided by the EDTA Registry. The analyses were carried out for 11,507 patients with non-systemic renal disease listed on the Registry's file by December, 1982, who were aged 60 years or more when they started Centre/Home haemodialysis and were never transferred subsequently to any other mode of renal replacement therapy (Figure 1).

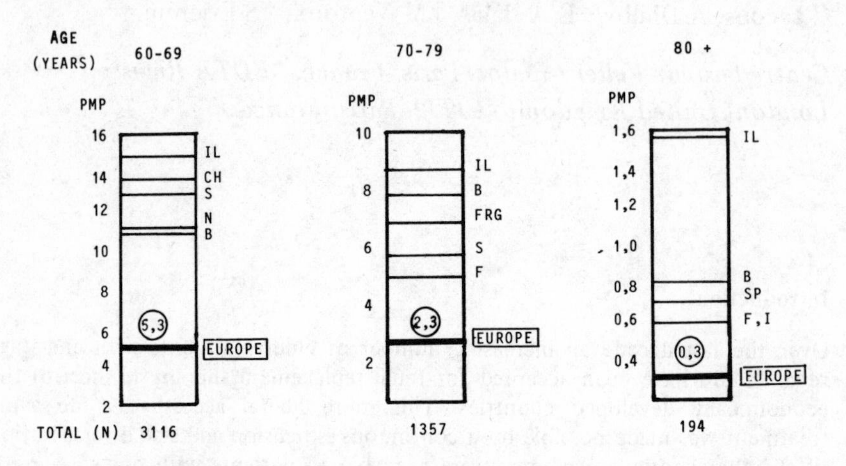

Figure 1. New patients aged over 60 years per million population taken onto renal replacement therapy in 1982 in countries reporting to the EDTA Registry

A comparative study of clinical results and outcome recorded in patients treated exclusively by maintenance haemodialysis has been performed for 674 patients under 60 years and 348 patients older than that age with non-systemic renal disease who were treated during the calendar years 1982–1983 in 33 French Dialysis Centres which participate in the French Diaphane Dialysis Registry. Actuarial survival rates were determined for the seven year period 1977–1983 in patients aged under and over 60 years at start of haemodialysis and who were treated exclusively by that mode of therapy.

The methods used by each of the two Registries for their data collection and processing have been described extensively elsewhere [12,13]. Statistical comparisons of the results provided by the Diaphane Registry were made using the Student's 't' or chi-square tests as appropriate, the statistical level of significance being set at five per cent. Cumulative survival rates were determined according to life table methods, statistical differences between survival curves were determined with the Mantel-Cox test. The analyses carried out for the patients on the Diaphane Dialysis Registry were performed with the BMDP program (University of California, Los Angeles) adapted for the Honeywell Computer series 66 (CITI2, Paris).

Results

Demographic data

As of December 1982, there were 64,914 patients with non-systemic renal disease treated exclusively by Centre/Home haemodialysis alive in 17 countries reporting to the EDTA Registry where more than 80 patients per million population (PMP) were being treated by all modes of renal replacement therapy. Of these patients 17.7 per cent were older than 60 years at start of their haemodialysis treatment. The male/female sex ratio was 1.32 per cent. Of the 13.7 per cent of patients aged 60–69 years at start of haemodialysis, 3.9 per cent were in their seventh decade of life and 0.1 per cent were older than 80 years. The proportion of elderly patients reached 30.9 per cent of all patients with non-systemic renal disease alive on haemodialysis in Sweden; it was between 20 and 25 per cent in the Netherlands, Switzerland, the Federal Republic of Germany, Belgium, Israel, Italy and France, whereas it was only 5.8 per cent in the United Kingdom and lower than 5 per cent in all Eastern European Countries, except Yugoslavia (7.1%).

The overwhelming majority of elderly patients were treated by Centre haemodialysis: of the 7,274 patients alive in December 1982 on home dialysis in the five European countries with more than 200 home haemodialysis patients, only 295 (4.1%) were over 60 years at start of treatment. Even in the United Kingdom, where home haemodialysis is a common method of treatment of end-stage renal disease, the patients older than 60 years at start of treatment accounted for only 3.2 per cent of the total population treated exclusively by home haemodialysis. Among the 1,022 patients fulfilling the above mentioned selection criteria listed on the Diaphane Registry, 34.1 per cent were aged over 60 years when treated by haemodialysis during the calendar years 1982–1983. Their mean age at start of treatment was 67 ± 5 years, whereas it was 44 ± 11 years for those younger than 60 years. The male/female sex ratio was similar for the younger and older age group, 1.19 and 1.23 respectively.

The age distribution of the new patients taken on renal replacement therapy in the years 1980–1982 in the countries reporting to the EDTA Registry shows a sustained trend towards an increase in the proportion of patients aged over 60 years at start of treatment, which was higher by 3.4 per cent in 1982 (25.4%) than two years earlier (Table I). The acceptance rate of elderly patients to renal replacement therapy in 1982 in the countries reporting to the EDTA Registry is depicted in Figure 1: overall, 7.9 patients with non-systemic renal disease older than 60 years PMP were accepted onto renal replacement therapy, of whom 5.3 PMP were aged 60 to 69 years, 2.3 PMP were in their seventh decade of life and 0.3 PMP were over 80 years. The countries with the highest acceptance rates of new elderly patients were Israel, Sweden, Belgium, Switzerland and Norway with respectively 25, 23, 20, 15 and 11 new elderly uraemic patients taken onto renal replacement therapy per million population. Renal vascular disease and primary interstitial nephritis were more frequently recorded diseases among elderly patients accepted onto renal replacement therapy, whereas primary glomerulonephritis and cystic renal diseases are decreasing with age (Table II).

TABLE I. Age distribution of new patients accepted for renal replacement therapy (all countries on EDTA Registry)

Age (years)	1980 All patients %	1980 Patients with NSRD* %	1981 All patients %	1981 Patients with NSRD %	1982 All patients %	1982 Patients with NSRD %
80+	0.6	0.4	0.7	0.4	1.1	0.7
70–79	5.9	5.6	7.1	6.4	7.9	7.5
60–69	16.7	16.0	17.2	15.8	18.1	17.2
0.1–59	76.8	78.0	75.0	77.4	72.9	74.6
Patients (N)	14,886	10,328	16,394	11,433	17,242	11,632

*NSRD=non-systemic renal diseases

TABLE II. Distribution according to type of renal disease in new patients taken onto renal replacement therapy in 1982 (all countries on EDTA Registry)

Primary renal diseases	All ages combined %	0.1–59 years %	60–69 years %	70–79 years %	80+ years %
Primary glomerulonephritis	39.6	44.7	26.1	20.6	25.3
Primary interstitial nephritis	34.3	30.8	43.5	48.5	36.7
Cystic kidney diseases	13.1	13.8	12.0	8.8	8.9
Renal vascular diseases	13.0	10.7	18.4	22.1	29.1
Total (N)	11,632	8,681	1,999	873	79

Clinical aspects

The primary renal diseases recorded in the patients aged less and more than 60 years who were treated exclusively by haemodialysis during the calendar years 1982–1983 in the Centres reporting to the Diaphane Registry are shown in Table III. The proportions of patients with renal vascular disease or nephropathy of undetermined origin are markedly increasing in the more advanced age groups. The average duration of the dialysis sessions performed in young and elderly patients was similar (249 ± 91 and 254 ± 69 minutes respectively, p=NS). An artiovenous fistula was used as vascular access in 85.6 per cent of the dialysis sessions carried out in elderly patients. No significant difference was seen in the overall distribution of the various types of vascular access used in the two

TABLE III. Distribution of primary renal diseases in patients treated exclusively by maintenance haemodialysis (Diaphane Registry)

Age range (years)	20–59	60–69	70–79	80+
	%	%	%	N
Primary renal diseases				
Chronic glomerulonephritis	30.0	20.7	18.0	1
Chronic interstitial nephritis	16.2	21.5	18.0	0
Cystic renal diseases	15.6	15.3	11.0	1
Renal vascular diseases	14.1	14.9	19.0	1
Other identified renal diseases	16.4	15.6	12.0	1
Undetermined renal diseases	7.7	12.0	22.0	2
Total (N)	674	242	100	6

age groups of patients, whilst single needle technique is used more frequently in elderly patients (4% versus 2%). Among pre- and post-dialysis clinical data, only the average systolic blood pressure was found higher in the older patients: 147 ± 17 versus 143 ± 18mmHg pre-dialysis and 140 ± 19 versus 132 ± 19mmHg post-dialysis ($p<0.01$ and <0.001 respectively). Hypotensive episodes requiring intravenous infusion of solutes or plasma volume expanders were recorded in a similar proportion of younger and elderly patients, 39.7 per cent and 46.5 per cent respectively (p=NS). The pre-dialysis blood urea, serum creatinine, potassium and phosphate were significantly lower in elderly patients as well as post-dialysis values of serum creatinine, calcium and phosphate (Table IV). The

TABLE IV. Pre- and post-dialysis blood chemistries in patients treated by maintenance haemodialysis (Diaphane Registry)

	Pre-dialysis			Post-dialysis		
	Patients <60 years (n=674)	p*	Patients ⩾60 years (n=348)	Patients <60 years (n=674)	p*	Patients ⩾60 years (n=348)
Blood urea (mmol/L)	29.4±6.5	0.02	28.4±6.6	11.3±6.2	NS	10.9±4.4
Creatinine (μmol/L)	1973±239	<0.001	937±207	473±148	<0.001	437±127
Potassium (mmol/L)	5.0±0.6	<0.001	4.9±0.5	3.4±0.5	NS	3.4±0.5
Calcium (mmol/L)	2.3±0.1	NS	2.3±0.1	2.8±0.2	<0.01	2.7±0.2
Phosphate (mmol/L)	1.9±0.4	<0.001	1.7±0.4	1.1±0.3	0.01	1.0±0.3
Haematocrit (%)	26.5±5.5	<0.001	27.5±4.5	–		

*Student's 't' test

average pre-dialysis haematocrit was higher in elderly patients, but the respective number of blood transfusions administered to each of the two groups of patients is unknown. About 10 per cent of the elderly patients were reported as taking digitalis or anti-arrhythmic drugs, either temporarily or permanently versus three per cent of the younger patients ($p<0.001$). Twenty per cent of the elderly patients were given oral anticoagulants between dialysis sessions, at least at some time. No significant difference is recorded between younger and older patients regarding the intake of anti-hypertensive drugs, ion resins, calcium supplements and vitamin D derivatives or aluminium phosphate binders. The number and reasons for blood transfusions have been carefully documented for the 77 patients who have been treated by haemodialysis in our own Unit during the calendar year 1982 (Table V). There was no significant difference

TABLE V. Blood transfusions in patients treated by maintenance haemodialysis in 1982 (Centre Pasteur-Vallery-Radot, Paris)

	Patients <60 years (n=53)	p*	Patients 60–69 years (n=17)	p*	Patients 70–85 years (n=7)
Haematocrit level in 1982	26.3±2	NS	27.3±5.2	<0.001	23.7±3.2
($\bar{m}$ ± SD)				26.6±5	
Number of blood units transfused	2.1±0.4	NS	2.1±5.2	NS	2.1±0.2
Reasons for transfusion					
Medical/surgical/technical %	27		79		36
Alleviation of fatigue or improvement of overall comfort %	73		21		64

*Student's 't' test

between the average number of blood units received by the patients in the different age groups. Meanwhile, the average haematocrit was not significantly different in patients younger or older than 60 years ($t=0.88$, p=NS). However, patients in their seventh decade of life had a significantly lower haematocrit than those aged 60–69 years ($t=6.18$, $p<0.001$). With blood transfusions administered for pre-transplantation protocols being excluded from the calculations, 73 per cent of the blood units transfused to patients under 60 years were given to alleviate fatigue or relieve discomfort, whereas 79 per cent of those given to patients who were in their sixth decade of life were ordered for medical/technical reasons. Improvement of well-being is again the most frequent ground for blood transfusions in patients older than 70 years.

Cardiovascular complications as well as clinical symptoms of peripheral arteriopathy and neuropathy were reported in a significantly higher proportion of elderly patients in the Diaphane Registry, whereas no significant difference emerged between younger and older patients as regards the occurrence of cerebrovascular accidents or clinical manifestations of renal osteodystrophy (Table VI). Twenty-five per cent of the elderly patients treated in our Unit did not need any hospitalisation during the two calendar years 1982 and 1983 whereas this proportion reached 53.6 per cent for the patients under 60 years.

TABLE VI. Complications in patients with non-systemic renal diseases treated by maintenance haemodialysis (Diaphane Registry)

Complications		Patients <60 years (n=256)	Patients $\geqslant 60$ years (n=245)	χ^2 p
Cardiac insufficiency	%	9.1	20.4	<0.0001
Ischaemic heart disease	%	8.5	25.3	<0.0001
Cardiac arrhythmias	%	6.7	16.7	<0.0001
Peripheral arteriopathy	%	1.0	6.1	<0.0001
Peripheral neuropathy	%	9.1	14.7	0.02
Malignancy	%	1.7	5.7	0.002
Cerebrovascular accident	%	1.3	2.4	NS
Clinical hepatitis	%	3.5	5.3	NS
Peptic ulcer	%	2.7	2.4	NS
Clinical osteodystrophy	%	26.0	30.6	NS

One fourth of the elderly patients were hospitalised for 31 to 60 days during this period for medical or social reasons.

Survival data

The cumulative actuarial survival rates determined for the different age groups of Centre and home haemodialysis patients listed on the EDTA Registry are given in Table VII. Five years after the start of haemodialysis the survival rates were 41.2 per cent and 25.2 per cent in patients aged 60–69 and 70–79 years at start of treatment, respectively. In both age groups they were better in female than in male patients, 42.8 versus 40 per cent and 29 versus 23 per cent respectively. Survival rates of home haemodialysis patients were higher than in centre patients at all time intervals up to five years for those in the 60–69 year age group. Female patients also fared slightly better than males with survival rates at five years being 57.8 and 54.7 per cent respectively.

The differences observed in survival rates of patients belonging to various age groups according to the systemic or non-systemic nature of their underlying renal disease have been evaluated for the patients listed on the Diaphane Registry. A significantly lower survival rate is recorded up to five years in patients with systemic renal diseases aged under 60 years at start of haemodialysis as compared with patients with non-systemic renal disease of the same age group (Mantel-Cox test: $p<0.001$), whereas no such difference is observed in patients aged over 60 years at start of treatment (Figure 2). The analysis of the distribution of the underlying systemic renal diseases among the two age groups of patients shows, however, that chronic glomerular nephropathies secondary to diabetes mellitus, lupus, and amyloidosis are recorded in 77.6 per cent of the younger patients and only in 59 per cent of the elderly patients, whereas

TABLE VII. Actuarial survival rates of patients with non–systemic renal diseases treated exclusively by maintenance haemodialysis (EDTA Registry)

Age groups (years)	Months on haemodialysis (per cent survival and number (N) at risk)											
	6		12		24		36		48		60	
	%	(N)	%	(N)	%	(N)	%	(N)	%	(N)	%	(N)
Centre haemodialysis												
60–69	91.5	(11,315)	84.4	(9,457)	70.9	(7,528)	59.4	(4,914)	49.6	(3,078)	41.2	(1,847)
70–79	87.4	(3,784)	77.6	(2,927)	59.9	(2,127)	46.0	(1,162.5)	33.9	(611)	25.3	(286.5)
80–89	80.2	(76)	66.4	(46.5)	+		+		+		+	
Home haemodialysis												
60–69	95.6	(574)	90.8	(517.5)	82.3	(446)	72.6	(350)	63.0	(263.5)	55.5	(184)
70–79	91.4	(58)	75.4	(51.5)	61.3	(37.5)	+		+		+	
80–89	+		+		+		+		+		+	

+ = number at risk less than 30

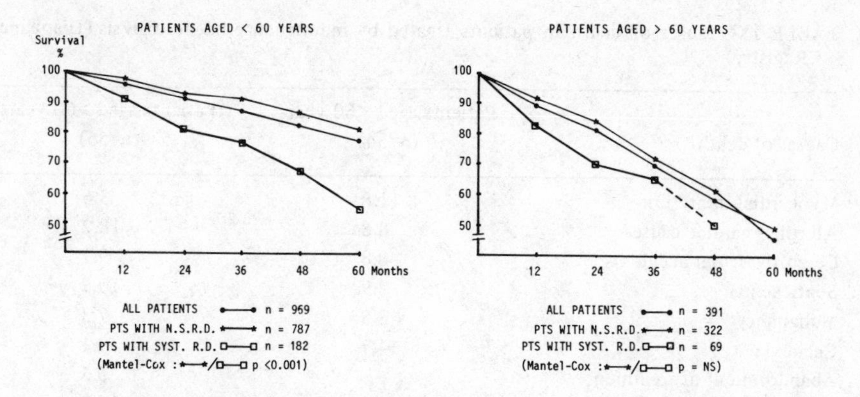

Figure 2. Actuarial survival rates of patients treated exclusively by maintenance haemodialysis according to age and underlying renal disease (Diaphane Dialysis Registry). NSRD=non-systemic renal disease

interstitial nephropathies of metabolic origin account for 10.7 per cent of renal diseases in younger patients and 27.4 per cent in the elderly.

Cardiac and vascular causes account for 43 per cent of all deaths recorded in patients aged over 60 years at start of haemodialysis who died in the years 1980–1982 in the countries reporting to the EDTA Registry (Table VIII).

TABLE VIII. Causes of death in patients with non-systemic renal diseases aged over 60 years at start of maintenance haemodialysis (EDTA Registry)

Causes of death	All patients >60 years %	Patients 60–69 years %	Patients 70–79 years %	Patients >80 years %
Myocardial infarction	13.9	15.0	11.0	8.2
All other cardiac and vascular causes	27.7	27.7	27.7	31.5
Cerebrovascular accident	11.5	11.4	11.8	11.0
Infectious causes	10.0	10.1	9.5	13.6
Malignancy not induced by immunosuppressive drugs	6.6	6.6	7.0	1.4
Abandonment of treatment	3.1	3.0	3.3	2.7
All other causes	27.2	26.2	29.7	31.6
Total (N)	3,848	2,745	1,030	73

Myocardial infarction accounts for a smaller proportion of deaths in patients aged over 70 years at start of treatment. Fatal cerebrovascular accidents and deaths related to abandonment of treatment were of similar frequency in all subgroups of elderly patients. No significant difference is seen regarding the

TABLE IX. Causes of death in patients treated by maintenance haemodialysis (Diaphane Registry)

Causes of death		Patients aged <60 years (n=34)	Patients aged ≥60 years (n=55)
Myocardial infarction	%	11.8	3.6
All other cardiac causes	%	8.8	18.2
Cerebrovascular accidents	%	8.8	9.1
Septicaemia	%	2.9	10.9
Malignancy	%	8.8	7.3
Cachexia	%	14.7	9.1
Abandonment of treatment	%	–	5.5
All other causes	%	44.2	36.3

X^2 =NS

overall distribution of the causes of death in patients aged less and more than 60 years whose death was reported to the Diaphane Registry during the calendar years 1982–1983 (Table IX).

Discussion

A sustained, albeit moderate, trend indicating a more liberal acceptance of elderly patients onto renal replacement therapy in the years 1980–1982 can be observed from the data collected by the EDTA Registry. In 1982, 2,951 new patients older than 60 years with non-systemic renal disease were taken on renal replacement therapy in the countries reporting to the Registry compared to only 2,268 two years earlier (+30.1%). During the same time interval, the number of new patients belonging to the same age group with underlying renal diseases due to systemic illnesses rose from 1,192 to 1,716 (+44%). The relatively lower proportion of patients with non-systemic renal disease who started renal replacement therapy for end-stage renal failure in recent years can be accounted for by the increasing number of patients with end-stage diabetic nephropathy accepted for treatment in many countries [14]. This improvement in the acceptance rate of elderly patients may also be due to the widespread development of CAPD in several countries [12]. Large differences in acceptance rates of elderly patients are, however, persistent among the various countries. The vast majority of these patients being currently treated by Centre haemodialysis (and to a lesser extent, CAPD), it is therefore obvious that access to treatment is severely limited in those countries where Centre dialysis facilities are particularly scarce, e.g. in the United Kingdom) or devoted primarily to patients awaiting renal transplantation, as is the case in some Eastern European countries. In contrast, acceptance of elderly patients is facilitated in the countries that have developed a well integrated programme for the treatment of end-stage renal failure, with all modes of dialysis procedures as well as renal transplantation being available, such as in Scandinavia, Israel, Switzerland and Belgium. The opportunity for elderly patients with end-stage renal failure of getting access to some mode of renal replacement therapy is thus not exclusively related to the economic status prevailing in a given country, but also to the degree of availability of alternative methods to Centre haemodialysis, whose implementation depends largely on the activity and underlying spirit of its medical community.

No striking differences were recorded for elderly patients regarding pre- and post-dialysis clinical and biological data when compared to those observed in younger patients. No significantly more elderly patients experienced acute falls in blood pressure during the dialysis sessions than younger ones, but this finding would need to be substantiated by detailed information concerning the dialysis procedures used for each group of patients such as distribution of dialysis machines equipped with ultrafiltration control and use of bicarbonate dialysate etc.

A significantly higher incidence of cardiac complications was recorded in the populations of elderly patients listed on the two Registries, whereas no such difference was shown concerning cerebrovascular accidents. Myocardial infarction becomes, however, a decreasing cause of death with advancing age in patients over 60 years at start of haemodialysis.

A significantly lower survival rate at all time intervals up to five years after the start of haemodialysis is found for patients older than 60 years listed on the Diaphane Registry when compared to the figures recorded in those under that age (48.2% vs 82.9% at five years). In contrast, no significant difference in survival rates up to three years after the start of therapy is shown between patients aged 60–69 years and 70–79 years at start of treatment. This may be due, among other factors, to a more rigid selection process applied to patients over 70 years of age. In the much larger cohort of elderly patients on the EDTA Registry, the actuarial survival rates at five years after the start of treatment were 41 per cent for patients aged 60–69 years treated by Centre haemodialysis and 55 per cent for those who performed home haemodialysis. The patients' selection criteria applied for this study, together with considerable differences in sample sizes, render a comparison of the survival rates determined in this survey with the ones reported in the small series previously published in the literature difficult, all the more so as many papers refer to patients who commenced haemodialysis more than 10 years ago [10–13]. The three and five year actuarial survival rates determined for 141 non-diabetic patients aged over 61 years and treated in the years 1966–1975 at the Hennepin County General Hospital in Minneapolis were around 40 per cent and 25 per cent respectively. They rose to 55 per cent and 40 per cent at similar time intervals for the 310 non-diabetic patients in the same age range taken onto haemodialysis in the years 1976–1982 [15]. The latter results are well in accordance with those found for the patients listed on the Diaphane Registry whose survival rate at five years after the start of haemodialysis is 48 per cent. The risk factors that influence the prognosis in patients treated by various modes of renal replacement therapy have been investigated in several recent studies by the means of sophisticated statistical methods, mainly the Cox Proportional Hazards Model [15,16]. Age, diabetes mellitus, arteriosclerotic heart disease, cerebrovascular accident, chronic obstructive pulmonary disease and non-skin malignancy were individualised as the prominent factors that influence prognosis in patients submitted to maintenance haemodialysis [15]. Male sex, elevated systolic and diastolic blood pressure were also reported as significant risk factors in a study conducted some years ago on patients on the Diaphane Registry [13]. Whereas the mortality rate was not significantly different amongst patients with primary glomerulonephritis, interstitial nephritis or polycystic kidney disease, it was found markedly higher in those with renal vascular diseases, whose proportion, as evidenced in the present study, increases steadily with advancing age.

Overall, 40 to 50 per cent of the patients with non-systemic renal disease aged over 60 years when taken onto haemodialysis are still alive five years later. Poorer prognosis and a deteriorated quality of life are the arguments most frequently set forward for limiting or denying dialysis therapy to elderly patients. The survival rates of elderly dialysis patients reported in this and other studies compare very favourably indeed with those obtained for many patients with malignant tumours of the lung, colon, bladder etc, or alcoholic cirrhosis of the liver whose aggressive and often very costly therapeutic regimens suffer currently little or no criticism in the economically developed countries. The quality of life experienced by an individual being actually what this individual thinks it to be [17], a reliable and valid measurement of such subjective a criterion through the various scoring systems devised by many authors appears as very difficult to be achieved for each single patients. Several factors considered to be of major importance for a good quality of life in young patients submitted to maintenance haemodialysis (e.g. occupational rehabilitation, sexual activity, or ability to travel to remote places) are deemed less essential for elderly patients. A state of physical fitness and well-being allowing elderly patients to pursue their customary activities in their usual environment without requiring assistance from spouse, relatives or third parties, together with the preservation of a rewarding social life are the prominent criteria for an acceptable quality of life that are actually fulfilled for most cases. Many authors who reported on the subject share the opinion that the quality of life achieved by elderly patients submitted to maintenance haemodialysis compares very favourably with the one obtained in young patients, and that age alone should not be considered as a contraindication to this mode of therapy [7–11]. Clearly, dialysis in itself is ineffective for mitigating the extra-renal deficiencies related to the ageing process and the sometimes disastrous consequences of physical dependence or isolation that are often the plight of elderly persons. The modern advances made in dialysis techniques are likely to reduce the additional inconveniences and discomfort suffered by elderly patients who undergo haemodialysis to an acceptable minimum level. Closer clinical surveillance often required during the dialysis sessions, together with frequent poor ability for managing properly home haemodialysis or CAPD, make hitherto Centre dialysis the most widespread mode of treatment for elderly patients with end-stage renal failure. The increasing constraints put on financial resources required for expanding Centre dialysis facilities account, therefore, for the major limiting factor that may preclude a wider acceptance of elderly patients onto renal replacement therapy [18]. Even to a larger extent than for young patients, provision of care for *all* elderly patients with end-stage renal failure can only be considered in the future through an optimal application to this group of patients of all other methods alternative to Centre haemodialysis therapy.

References

1 Bailey GL, Mocelin AJ, Griffiths AJL et al. *J Am Geriatrics Soc 1972; 20:* 421
2 Kjellstrand CM, Shideman JR, Lynch RE et al. *Geriatrics 1976; 31:* 65
3 Kaye M, Pajel PA, Somerville PJ. *Perit Dial Bull 1983; 3:* 17
4 Nicholls AJ, Waldek S, Platts MM et al. *Br Med J 1984; 288:* 18
5 Marai A, Rathaus M, Gibor Y et al. *Perit Dial Bull 1983; 3:* 183
6 Baldamus CA, Ernst W, Kachel HG et al. *Contr Nephrol 1982; 32:* 56
7 Garini G, Cambi V, Arisi L et al. *Minerva Nefrologica 1976; 23:* 371
8 Walker PHW, Ginn HE, Johnson HK et al. *Geriatrics 1976; 31:* 55
9 Rathaus M, Bernheim J. *Geriatrics 1978; 33:* 56
10 Chester AC, Rakowski TA, Argy WP Jr et al. *Arch Int Med 1979; 9:* 1000
11 Taube DH, Winder EA, Ogg CHS et al. *Br Med J 1983; 286:* 2018
12 Wing AJ, Broyer FP, Brynger H et al. *Proc EDTA 1983; 20:* 2
13 Degoulet P, Legrain M, Reach I et al. *Nephron 1982; 31:* 103
14 Jacobs C, Brunner FP, Brynger H et al. *Diabetic Nephrop 1983; 2:* 12
15 Shapiro FL, Umen A. *ASAIO J 1983; 6:* 176
16 Hutchinson TA. Thomas DC, MacGibbon B. *Ann Int Med 1982; 96:* 417
17 Churchill DN, Morgan J, Torrance GW. *Perit Dial Bulletin 1984; 3:* 20
18 Berlyne GM. *Nephron 1982; 31:* 189

Proc EDTA–ERA (1984) Vol 21

MAINTENANCE DIALYSIS IN THE ELDERLY. A REVIEW OF 15 YEARS' EXPERIENCE IN LANGUEDOC-ROUSSILLON

C Mion, *R Oulès, B Canaud, G Mourad, †A Slingeneyer, *B Branger, *C Granolleras, *B Al Sabadani, **P Florence, **R Chouzenoux, **F Maurice, †R Issautier, †J L Flavier, C Polito, F Saunier, ††L Marty, ††P Fontanier, †C Emond, ††H Ramtoolah, ††F de Cornelissen, ¶ G Huchard, ¶ H Fitte, ¶ R Boudet

*Montpellier University Hospital, *Nîmes University Hospital, †Association pour l'Installation à Domicile des Epurations Rénales, Montpellier, **Centre d'Hémodialyse du Languedoc méditerranéen, Montpellier, ††Centre Hospitalier Général, Carcassonne, ¶ Centre Hospitalier Général, Perpignan, France*

Introduction

Since its inception in 1965 at Montpellier University Hospital, one of the main goals of our end-stage renal disease treatment programme was to avoid the need for selecting patients on the basis of an arbitrary age limit.

Our first experience with an end-stage renal disease patient over 60 occurred in one isolated case in 1967. However, it was not until 1969, when an increasing number of uraemic patients were referred to us by general practitioners who became aware of the long-term efficacy of maintenance haemodialysis. This increased referral of older patients, together with the remarkable results obtained in some of them with home haemodialysis created a situation in which flat refusal of treatment because of advanced age became progressively an untenable attitude. Over the years, new methods of treatment were implemented and old age was definitively eliminated in practice as a selection criteria about 1976–1977.

This report is a description of the pragmatic approach used by nephrologists in the six dialysis units of Languedoc-Roussillon to face the problem of maintenance dialysis in patients over 60. Admittedly, the data presented here have all the limitations of any historical series [1]. To interpret our results, it is essential to realise that two fundamental postulates have oriented the technical choices offered to a majority of elderly. Firstly, every effort was made to treat at home older patients in an attempt to avoid the rapid 'crowding' of dialysis facilities by patients over 60. Indeed, epidemiological data indicate an exponential increase with age in the number of deaths due to uraemia [2] ; such information convinced us that admitting to hospital haemodialysis all end-stage renal disease patients, whatever their age, would quickly overwhelm the capacity of the existing dialysis centres with elderly patients and later impede the admission of younger subjects. Secondly, it was felt that home haemodialysis, a rather complex technique, could not be successful in the majority of old patients because many of them would not have the skill, learning ability and emotional stability to master home haemodialysis procedures, and also because of the cardiovascular instability frequently observed during haemodialysis among older patients [3] . As a consequence, when difficulties with home haemodialysis were expected (for medical or social reasons) maintenance peritoneal dialysis in the home was the preferred technique, initially as intermittent peritoneal dialysis and later continuous ambulatory peritoneal dialysis (CAPD). In fact, this home dialysis oriented policy was mainly practised by the Divisions of Nephrology of both Montpellier and Nîmes University Hospitals, in conjunction with the Association pour l'Installation à Domicile des Epurations Rénales; the three other dialysis units admitted a greater number of elderly uraemic patients to hospital haemodialysis.

In spite of these main biases and of the inherent limitations in interpreting the data presented here, we will attempt to address the problem of optimum dialysis in end-stage renal disease patients over 60 with primary renal disease in the light of our 15 years' experience. Special emphasis will be given to rehabilitation and morbidity, causes of death and survival, according to the mode of treatment. An attempt will also be made to evaulate the effect on survival of age and cardiovascular status at start of treatment. Finally, the relative efficacy of haemodialysis compared to peritoneal dialysis in the development of home dialysis in the elderly will also be discussed.

Patients and methods

Patients

As shown in Figure 1, the admission in our end-stage renal disease treatment programme of patients aged 60 or older, started in 1969. The number of patients in this age group increased steadily up to 1977 and, since then, represented 35.8 to 45.9 per cent of the flow of new patients accepted each year on dialysis. From 1st January 1969 to 31st December 1983, 480 patients 60 years or older entered our programme: they represent 31.6 per cent of the 1,516 patients taken on dialysis since the inception of our programme in 1965.

As shown in Table I, there were 390 patients with primary renal diseases; among these, 21 patients were excluded from the study: one patient treated in 1967 and 20 patients with multiple cross-overs. In Table II, the aetiologies of primary renal diseases of the 369 patients under study are presented.

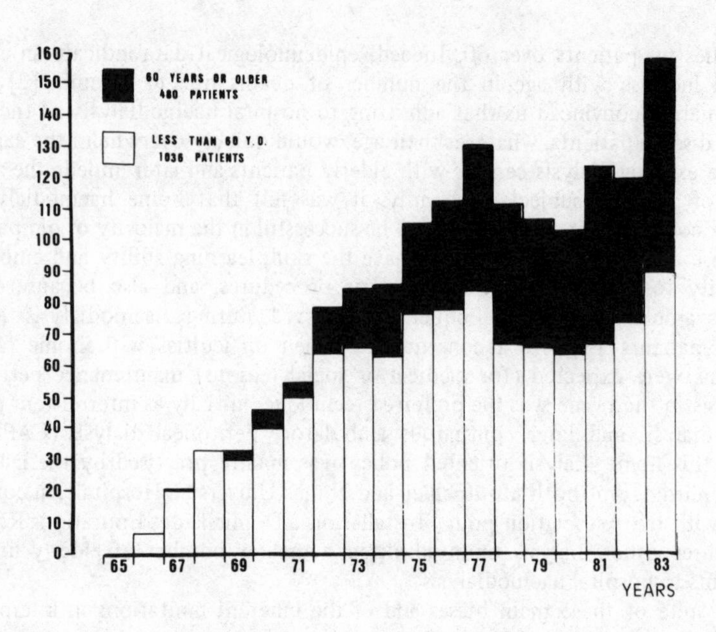

Figure 1. Number of new patients admitted yearly in Languedoc-Roussillon from 31st March 1965 to 31st December 1983

TABLE I. Population of end-stage renal disease patients 60 years or older, admitted to maintenance dialysis in Languedoc-Roussillon from 31st March 1965 to 31st December 1983

	Number of patients		Percent total
	Total	Subgroups	
Total population	480		100
Primary renal disease	390		81.25
included in study		369	
excluded: treated in 1967		1	
multiple therapeutic cross-overs		20	
Secondary renal diseases	90		18.75
diabetes mellitus		54	
myeloma		9	
visceral cancer		9	
amyloidosis		6	
SLE, vasculitis		6	
tuberculosis		4	
primary biliary cirrhosis		2	

TABLE II. Aetiology of primary renal diseases in 369 end-stage renal disease patients included in the present study

	n	%
Nephroangiosclerosis	108	29.3
Chronic glomerulonephritis	71	19.2
Chronic pyelonephritis	65	17.2
Adult polycystic kidneys	40	10.8
Gout	11	3.0
Analgesic nephropathy	3	0.8
Miscellaneous	3	0.8
Unknown	68	18.5
Total	369	

Methods

As described elsewhere [4] several modes of end-stage renal disease treatments were developed progressively: our programme started at first with haemodialysis on 31st March 1965. Hospital intermittent peritoneal dialysis with the repeated puncture technique [5] was introduced in 1966. A home haemodialysis programme was initiated in 1968 [6] while a home intermittent peritoneal dialysis programme with the Tenckhoff catheter and a cycler was started in 1974 [7]. Finally, CAPD completed our therapeutic armamentarium in 1978 [8]. Postdilutional haemofiltration was also used in a small group of patients [9] but older patients receiving this mode of treatment form too small a group to be analysed separately from those receiving haemodialysis.

According to the available dialysis techniques, three successive five year periods were schematically delineated concerning the admission to dialysis of elderly patients:

Period I: 1st January 1969 to 31st December 1973, 43 patients;
Period II: 1st January 1974 to 31st December 1978, 144 patients;
Period III: 1st January 1979 to 31st December 1983, 182 patients.

During Period I, intermittent peritoneal dialysis with the repeated puncture technique and haemodialysis (mainly in the home) were the two dialytic techniques available. During Period II home and hospital haemodialysis as well as home intermittent peritoneal dialysis with the Tenckhoff catheter were the two main modes of dialysis used and intermittent peritoneal dialysis with the repeated puncture technique was progressively abandoned. Finally, during Period III, haemodialysis and intermittent peritoneal dialysis with the Tenckhoff catheter (home or hospital) and CAPD were the methods of treatment for the older patient.

The technical aspects of the dialytic methods have been described elsewhere [4–8] and will not be repeated here.

Table III presents the number of patients admitted to dialysis during each period, as well as their first mode of treatment. As shown in Table IV, 70

TABLE III. Identification of patients' subgroups according to the period of treatment and the first mode of dialysis

First treatment mode	PERIOD I 1969–1973	PERIOD II 1974–1978	PERIOD III 1979–1983
Number of patients	43	144	182
Sex M:F	24:19	88:56	120:62
Hospital IPD-RPT*	40	11	–
Haemodialysis	3	36	64
Hospital IPD-TK**	–	20	26
Home IPD-TK**	–	77	45
CAPD***	–	–	47

* IPD-RPT = Intermittent peritoneal dialysis with repeated puncture technique

** IPD-TK = Intermittent peritoneal dialysis with a permanent peritoneal access and automatic delivery of dialysate (cycler or RO machine)

*** Continuous ambulatory peritoneal dialysis

TABLE IV. Number of patients transferred from one therapeutic mode to another (single transfer

	n	percent
Total population	369	100
Transferred patients	70	19
Type of transfer*		
IPD-RPT to HD	22	
IPD-TK to HD	26	
IPD-TK to CAPD	9	
CAPD to IPD-TK	9	
CAPD to HD	3	
HD to HF	2	

*HD = haemodialysis; HF = haemofiltration. For other abbreviations see Table III

patients representing 19 per cent of the population under study were transferred from one treatment mode to another. The most frequent transfers occurred from intermittent peritoneal dialysis with repeated puncture technique to haemodialysis and from intermittent peritoneal dialysis with a permanent peritoneal access and automatic delivery of dialysate to haemodialysis. This reflects a common trend to use intermittent peritoneal dialysis as a waiting procedure in the early stage of treatment, until admitting the patients to haemodialysis. As already mentioned, 20 patients who experienced several successive transfers during their treatment were excluded from this study, due to the difficulty in interpreting the data obtained in such complex cases.

Rehabilitation

Most patients in our series being retired, it was not possible to use the EDTA Registry scoring system [10] to assess the degree of rehabiliation and quality of life. The patients were therefore graded according to the following scale: working full time; working part time; retired and active, i.e. patients having a good physical and social activity (e.g. sport, gardening, regular participation in the activities of a club, taking care of the household, etc); retired and inactive (e.g. little physical activity, few social contacts, need for some help in shopping or cooking); dependent (e.g. requiring help to wash and dress and in other daily life activities); hospitalised. Rehabilitation was studied in patients treated during Period III exclusively, and followed up for at least one year. The degree of rehabilitation was graded yearly in patients followed for several years consecutively to detect the possible decrease in the quality of life that might occur with time.

Morbidity

A survey was conducted among 118 patients receiving dialysis therapy at any one time during the calendar year 1983. Morbidity was evaluated as the number of days spent in hospital and the causes of hospitalisation were recorded. Complete information was available for 102 patients (86.4% of patients on treatment during 1983).

Mortality

The number and causes of death were analysed among 326 patients entering dialysis during Periods II and III. The main treatment mode, defined as the dialysis technique in a given patient for more than 90 per cent of his total treatment duration was used to identify two subgroups: haemodialysis – 128 patients and peritoneal dialysis – 198 patients. The number of deaths per 1000 patient treatment months [11] was computed for each subgroup.

Assessment of initial cardiovascular status

The cardiovascular status was assessed at the start of treatment among 182 patients admitted on dialysis (71 haemodialysis and 111 on peritoneal dialysis) from 1st January 1979 to 31st December 1983 and followed up to 30th June 1984. Patients were classified as 'high risk' or 'low risk' according to the presence or the absence of the following criteria: prior myocardial infarction and/or cerebrovascular accident; clinical signs and symptoms of carotid, coronary, aortic, iliac or femoral arteriosclerotic obstructive disease; electrocardiographic abnormalities, including left ventricular hypertrophy, bundle branch blocks, ventricular ectopics and other arrhythmias; abnormal echocardiographic pattern. A complete set of data was obtained in 57 patients on haemodialysis and 100 patients on peritoneal dialysis.

Cardiovascular stability during haemodialysis sessions

In 38 patients receiving hospital haemodialysis, the number of sessions with hypotensive episodes was evaluated during the period 1st October 1983 to 31st December 1983. Overall, 1,336 sessions were analysed. The mean session duration was 4.1 ± 0.7 hours. The dialysate sodium concentration was 140.6± 1.8mmol/L. One thousand one hundred and sixty-five sessions were with a dialysis fluid containing acetate, and 171 with a bicarbonate dialysate. In each session, symptomatic hypotension (i.e. fall in blood pressure accompanied with signs and symptoms requiring an intervention from the nurse in charge) was recorded. The patients were then classified according to the number of sessions with hypotensive episodes: Group I, none; Group II, less than 10 per cent; Group III, 10 to 50 per cent; Group IV, 50 per cent or more sessions with symptomatic hypotension.

Survival

Actuarial survival was computed according to the life table method [12]. In an attempt to reduce the over-estimation of survival that might result from actuarial computation when a patient is transferred from one mode of treatment to another (i.e. counted as lost for follow-up) [4], actuarial survival was calculated in homogenous subgroups of patients in whom one dialytic mode was almost exclusively used; furthermore, survival curves were computed by considering the complete period of observation ending with the continuation of treatment on 30th June 1984 or until patient's death.

The effects on actuarial survival of several variables (e.g. age and/or cardiovascular status at start of treatment, dialysis technique) were assessed in subgroups of patients admitted to treatment during Periods I, II and/or III.

Results

The degree of rehabilitation could be assessed in 142 patients who received dialysis during period III. Their mean age was 65.1±4.7 on haemodialysis and 70.1±5.5 on intermittent peritoneal dialysis/CAPD. As shown in Table V, the

TABLE V. Degree of rehabilitation in 142 patients receiving maintenance dialysis during Period III (1st January 1979 to 31st December 1983) and followed up for at least one year

		Haemodialysis					IPD/CAPD			
Year on treatment		1st	2nd	3rd	4th	5th	1st	2nd	3rd	4th
Number of patients		52	37	22	9	4	90	60	25	11
Mean age (years ± SD)*		65.1 ± 4.7					70.1 ± 5.5			
Working†	full time	–	–	–	–	–	1.5	–	–	–
	part time	2	–	–	–	–	1.5	–	–	–
Retired†	active	65	60	68	66	100	41.1	49	56	54.5
	inactive	27	32	23	33	–	32.2	36	20	27.3
	dependent	4	8	9	–	–	15.5	150	24	18.2
Hospitalised†		–	–	–	–	–	8.2	–	–	–

*= mean age on 1st year of treatment; †=percent patients in each class

percentage of working patients was two per cent or less during the first year of treatment. The percentage of patients classified as 'retired active' ranged from 60 to 68 per cent with haemodialysis and from 41 to 56 per cent with intermittent peritoneal dialysis and CAPD. There was an increase in the percentage of dependent patients from the first to the third year of dialysis with both modes of treatment.

The number and the main characteristics of the population entering the survey on morbidity are shown in Table VI: the number of days spent in the

TABLE VI. Survey on morbidity among 102 patients receiving maintenance dialysis during the calendar year 1983

	Haemodialysis	IPD/CAPD
Number of patients	48	54
Sex, M:F	33:15	32:22
Mean age (years±SD)	67.4 ± 4.9	71.2 ± 5.7
Treatment duration (months)		
Cumulative	478	482
mean/patient	9.9 ± 3.4	8.1 ± 3.2
Patients hospitalised in 1983		
number (%)	27 (56)	32 (59)
mean age (years± SD)	68.2 ± 5.2	70.7 ± 5.9
Number of days in hospital		
cumulative	579	1,336
range	1 to 300	1 to 133
per patient per year	14.6	36.6

TABLE VII. Main causes of hospitalisation among 102 patients treated by haemodialysis, intermittent peritoneal dialysis or CAPD during the calendar year 1983

	Haemodialysis		IPD/CAPD	
	n days	(%)*	n days	(%)*
1. Complications directly related to dialysis				
Connection	64	(11.0)	30	(2.2)
ECV overload**	14	(2.9)	31	(2.3)
Pericarditis	4	(0.7)	12	(0.9)
Septicaemia	4	(0.7)	–	
Peritonitis	–		403	(30.2)
Loss of UF†	–		154	(11.5)
2. Other causes				
Medical problems	445	(76.8)	354	(30.9)
Surgery††	48	(8.3)	59	(4.4)
Technical problems	–		57	(4.3)
Social problems	–		236	(17.6)

* Percent of total number of days in hospital
** Extracellular volume overload
† Loss of the capacity of the peritoneum to ultrafiltrate
†† Except operations for blood access or placement of peritoneal catheter (included under 'connection')

hospital per patient per year was 14.6 in patients receiving haemodialysis and 36.6 days in those treated with intermittent peritoneal dialysis and/or CAPD. The main causes of hospitalisations are presented in Table VII: in haemodialysis patients, miscellaneous medical problems (76.8%), blood access problems (i.e. fistula thrombosis and blood access surgery) (11%) and abdominal surgery (8.3%) were the major reasons for patients' admission. In patients receiving peritoneal dialysis, peritoneal infection (30.2%), miscellaneous medical diseases (30.9%), loss of the capacity of the peritoneum to ultrafiltrate (loss of UF) were the three leading causes of hospitalisation. However, 17.6 per cent of hospital days were due to difficulties in maintaining in their homes dependent patients on intermittent peritoneal dialysis and/or CAPD, as a result of a temporary unavailability of the helper (spouse or other third party).

TABLE VIII. Number of deaths observed among 326 patients admitted on haemodialysis or peritoneal dialysis during Periods II and III (see text for definition of the two subgroups haemodialysis or peritoneal dialysis)

	Haemodialysis	IPD/CAPD
Total number of patients	128	198
Age, years (mean± SD)	65.1 ± 3.9	67.5 ± 4.7
Number of deaths	35	135
Percent of deaths	27.3	68.2
Cumulative treatment duration (months)	5,267	4,474
Number of deaths per 1,000 patient treatment months	6.64	30.17

TABLE IX. Causes of death among 326 patients receiving haemodialysis or peritoneal dialysis during Periods II and III

	Haemodialysis		IPD/CAPD	
	n	%	n	%
Total number of deaths	35	100	135	100
Cardiovascular deaths	12	34.3	66	48.8
Cerebrovascular accident	3		21	
Congestive cardiac failure	2		15	
Sudden death	4		11	
Myocardial infarction	–		9	
Arrhythmias	2		6	
Hyperkalaemia	1		1	
Other	–		3	
Infections	9	25.7	30	22.2
Septicaemia	6		–	
Pneumonia	1		4	
Peritonitis	–		24	
Tuberculosis	2		2	
Miscellaneous	14	40.0	40	29.0
End-stage uraemia	6		6	
Treatment discontinued*	1		15	
Visceral cancer	2		2	
Other	5		17	

* Cessation of therapy at the patient's or family's request

Table VIII summarises the number of deaths observed among 326 patients treated by dialysis during Periods II and III (1st January 1973 to 31st December 1983). The main causes of death are given in Table IX. There were 35 and 135 deaths respectively among haemodialysis and peritoneal dialysis patients, corresponding to 27.3 and 68.2 per cent of deaths in each subgroup, and to a death rate per 1,000 treatment months of 6.64 with haemodialysis versus 30.17 with peritoneal dialysis. Cardiovascular deaths were more frequently observed in intermittent peritoneal dialysis/CAPD than in haemodialysis patients, representing 48.8 per cent versus 34.3 per cent of all deaths in each group respectively. The percentage of deaths due to infection was similar in both techniques (25.7% in haemodialysis versus 22.2% in intermittent peritoneal dialysis/CAPD); however, the most common infections were septicaemia in haemodialysis and peritonitis in peritoneal dialysis. It should be stressed that seven deaths due to a complication of peritonitis (e.g. cerebrovascular accident, pulmonary embolism, cachexia) were considered as deaths from peritonitis even though the peritoneal infection was cured at the time of death.

Actuarial survival curves are presented in Figures 2 to 7. During Period I, the survival of older patients treated with intermittent peritoneal dialysis with repeated puncture technique was very poor as all the patients had died before the third year of treatment; on the other hand, 22 patients who were transferred to haemodialysis had a 48 per cent survival at seven years (Figure 2). The overall

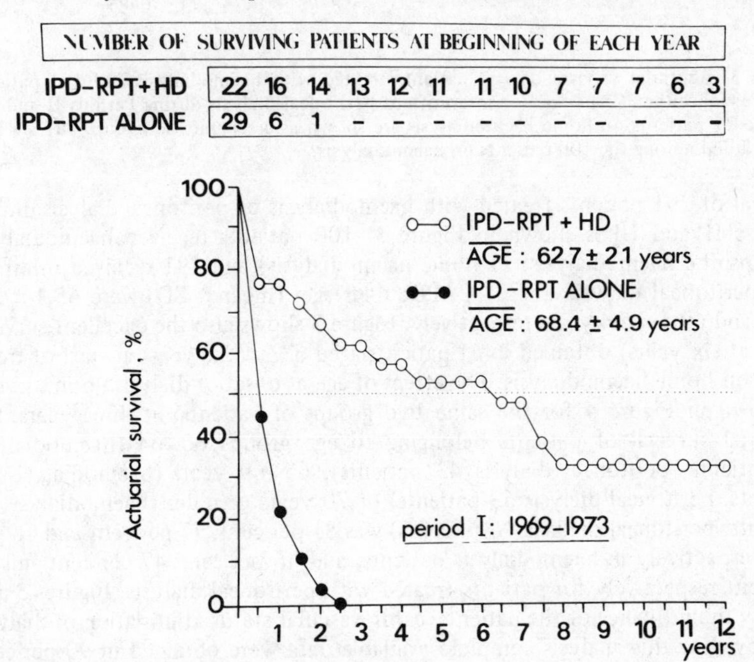

NUMBER OF SURVIVING PATIENTS AT BEGINNING OF EACH YEAR													
IPD-RPT+HD	22	16	14	13	12	11	11	10	7	7	7	6	3
IPD-RPT ALONE	29	6	1	-	-	-	-	-	-	-	-	-	-

Figure 2. Actuarial survival of patients dialysed during Period I and receiving either intermittent peritoneal dialysis with repeated puncture technique (IPD-RPT) alone or IPD-RPT followed by haemodialysis (HD)

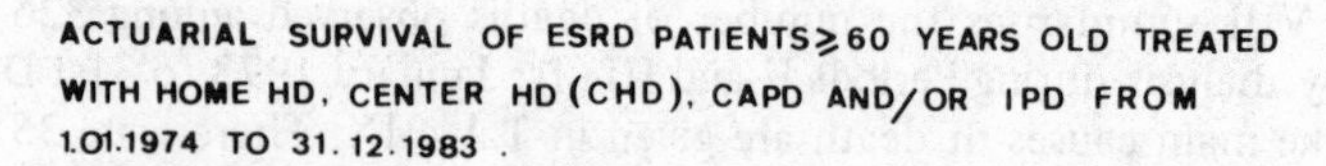

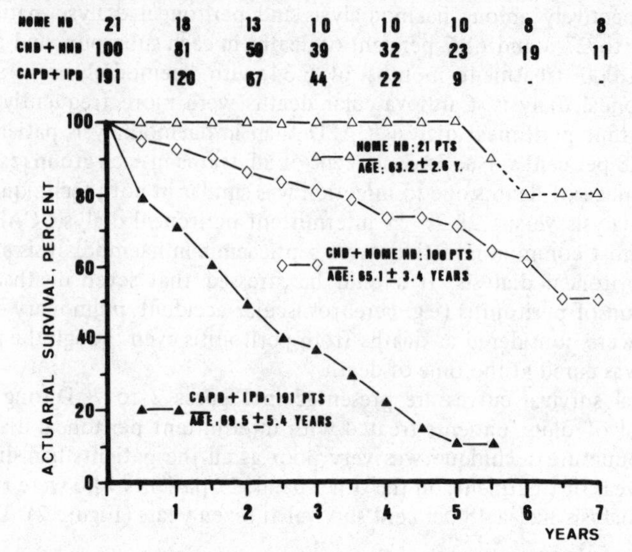

Figure 3. Actuarial survival of 100 patients treated with haemodialysis and 191 patients receiving either CAPD and/or intermittent peritoneal dialysis during Periods II and III. The 21 patients on home haemodialysis are shown as a separate subgroup, but are also included among the 100 patients on haemodialysis

survival of 291 patients treated with haemodialysis or peritoneal dialysis during Periods II and III is shown in Figure 3: 100 patients received haemodialysis (79 hospital haemodialysis, 21 home haemodialysis) and 191 received intermittent peritoneal dialysis and/or CAPD; their ages (mean ± SD) were 65.1 ± 3.4 years and 69.5 ± 5.4 years respectively. Figure 3 shows also the excellent survival (80% at six years) obtained in 21 patients aged 63.2 ± 2.6 years at start of treatment on home haemodialysis. The effect of age at onset of dialysis upon survival is shown in Figure 4 for the same two groups of patients: at three years, the actuarial survival of patients belonging to age groups 60–64 (haemodialysis 51 patients, peritoneal dialysis 43 patients), 65–69 years (haemodialysis 33 patients, peritoneal dialysis 53 patients) or 70 years or older (haemodialysis 16 patients, peritoneal dialysis 96 patients) was 81 per cent, 77 per cent and 75 per cent respectively in haemodialysis patients, and 46 per cent, 47 per cent and 11 per cent respectively for patients treated with peritoneal dialysis. Figures 5 and 6 show the influence of the patients' cardiovascular status at initiation of dialysis therapy. For this analysis complete available data were obtained in 90 per cent of the patients admitted to peritoneal dialysis or haemodialysis during Period III; the characteristics of this population are presented in Table X, showing a greater percentage of patients with clinical, ECG and echocardiographic evidence

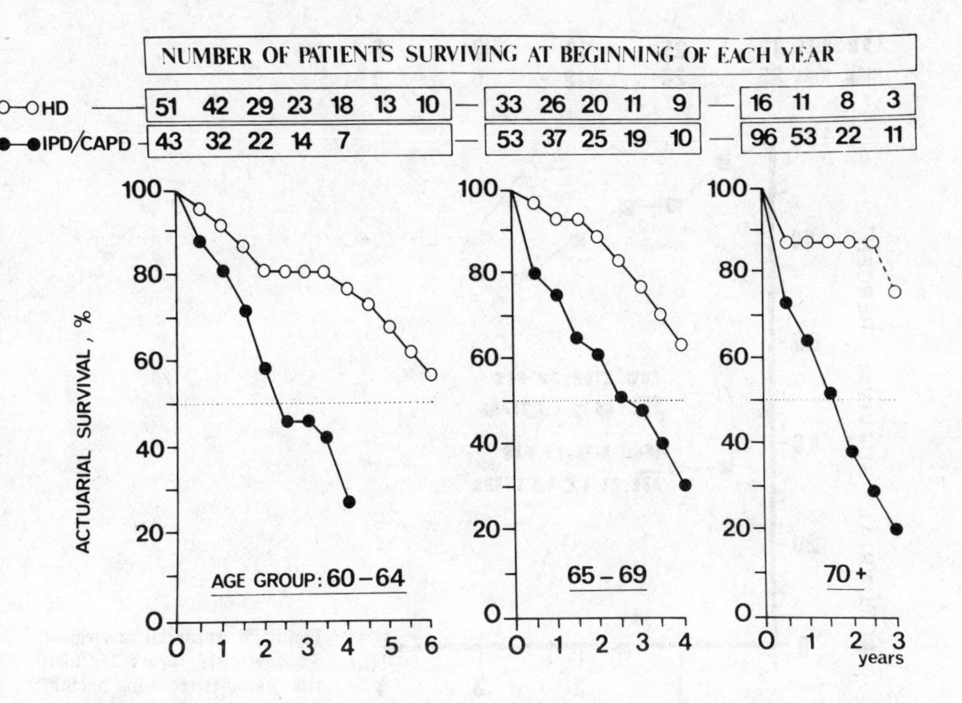

Figure 4. Actuarial survival on haemodialysis or peritoneal dialysis of patients in three different age groups

TABLE X. Cardiovascular status of 157 patients admitted on dialysis during Period III (see text for definition of 'high risk' and 'low risk')

	Haemodialysis	IPD/CAPD
Total patients admitted during Period III	64	111
Patients with available data for definition of risk (%)*	57 (87)	100 (90)
'Low risk' (%)†	32 (56.2)	23 (23)
Age, years (mean±SD)	65.1 ± 4.7	70.1 ± 6.5
'High risk' (%)†	25 (43.8)	77 (77)
Age, years (mean±SD)	65.1 ± 4.6	70.0 ± 4.9

* Percent from total patients admitted on treatment

† Percent from patients in each subgroup

of atherosclerotic cardiovascular diseases among peritoneal dialysis patients (77%) than in haemodialysis patients (43.8%). It should also be noted that patients on peritoneal dialysis were five years older, on average, than patients on haemodialysis. The survival of 'high risk' patients on haemodialysis was clearly poorer than the 'low risk' group (75% versus 95% at three years). In patients treated with peritoneal dialysis (Figure 6), 'high risk' and 'low risk'

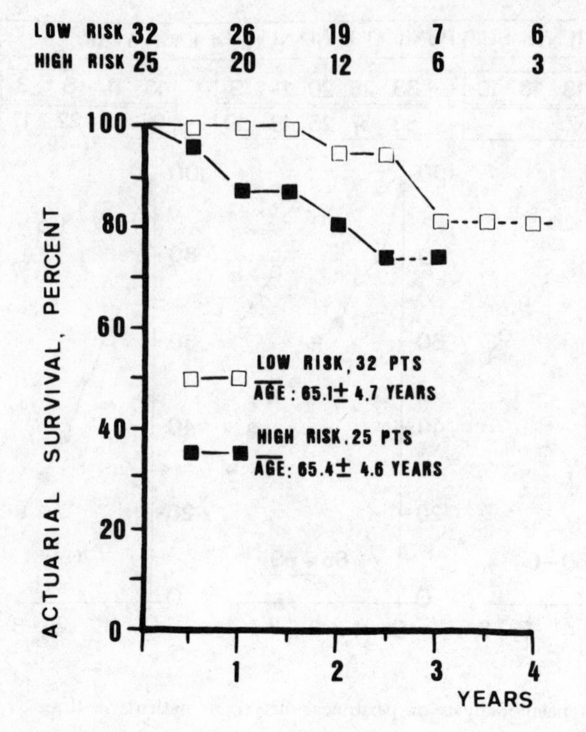

Figure 5. Actuarial survival of 32 'low risk' versus 25 'high risk' patients receiving haemodialysis during Period III

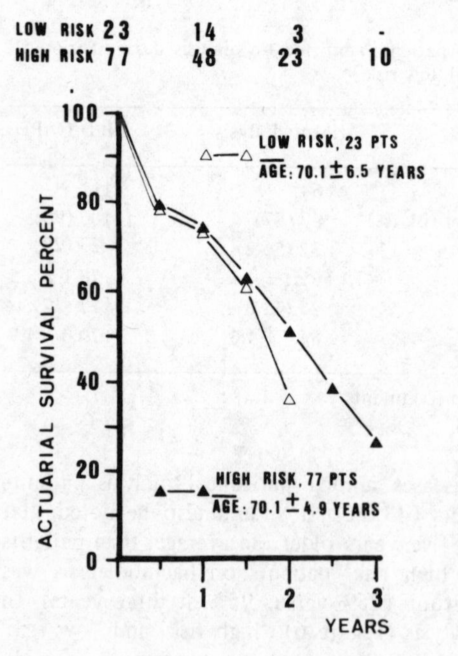

Figure 6. Compared actuarial survival of 23 'low risk' versus 77 'high risk' patients treated with peritoneal dialysis (CAPD and/or intermittent peritoneal dialysis) during Period III

patients had similar survivals (60% at 18 months). An analysis of the causes of deaths in 'high risk' and 'low risk' patients suggests that the lack of a difference in survival between these two subgroups receiving peritoneal dialysis was the consequence of an excess mortality from peritonitis in 'low risk' patients as shown in Table XI. Finally, actuarial survival of patients treated either with

TABLE XI. Causes of death in 23 'low risk' and 77 'high risk' patients receiving intermittent peritoneal dialysis or CAPD from 1st January 1979 to 31st December 1983

	Low risk	High risk
Number of patients	23	77
Age, years (mean±SD)	70.1 ± 6.5	70.1 ± 4.9
Treatment duration (years)		
cumulative	28.7	98.9
mean/patient	1.2 ± 1.0	1.3 ± 1.0
range	0.02 – 3.9	1 day – 4.7 years
Number of deaths	15	46
Causes of death (%)*		
Cardiovascular	4 (26.6)	30 (65.2)
Peritonitis	4 (26.6)	7 (15.2)
Discontinuation of peritoneal dialysis	1 (6.6)	4 (8.7)
Cancer	1 (6.6)	–
Miscellaneous	5 (33.3)	5 (10.9)
Number of deaths per 1,000 treatment months		
All causes	43.5	37.7
From peritonitis	11.6	5.7

* Percent deaths in each subgroup

intermittent peritoneal dialysis or with CAPD beyond the first month of dialysis during Period III are compared in Figure 7; to allow a fair comparison between intermittent peritoneal dialysis with the Tenckhoff catheter and cycler with CAPD, exclusion of early deaths was deemed necessary as patients were transferred to the CAPD training centre only when clinical improvement was obtained with hospital intermittent peritoneal dialysis with the Tenckhoff catheter and cycler. Patients on intermittent peritoneal dialysis were 3.3 years older on average than those treated with CAPD, and yet the survival was absolutely identical with both peritoneal dialysis techniques during the first two years of treatment.

Table XII presents the data on cardiovascular tolerance during hospital haemodialysis obtained in 38 patients: 23 of them (60.4%) were in Groups I and II (less than 10% of sessions with SH); Groups III and IV included respectively 9 (23.6%) and 6 (16.0%) of the patients.

The impact of haemodialysis versus peritoneal dialysis on the development of home dialysis in our area is analysed in Table XIII: among 126 patients accepted on haemodialysis from 1st January 1974 to 31st December 1983, 37 (30.9%) were admitted at the training centre, 27 of whom were eventually installed at

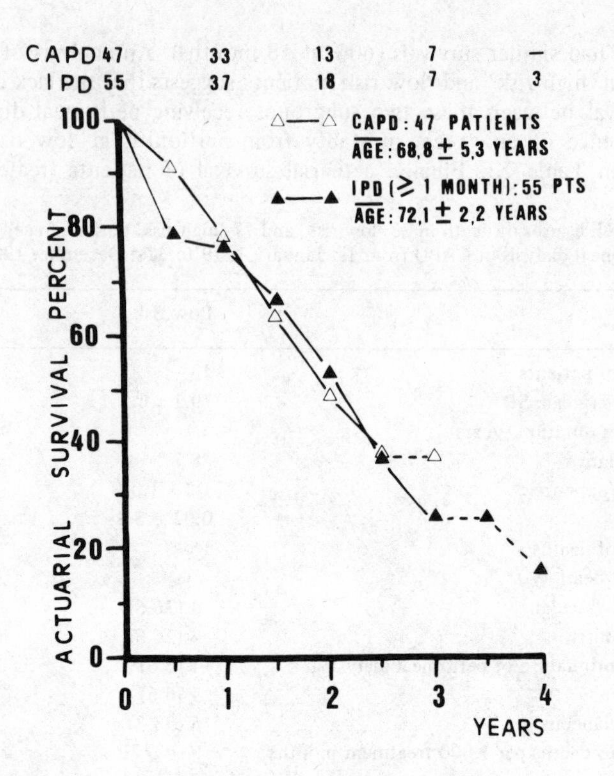

Figure 7. Survival of patients treated with peritoneal dialysis for one month or more during Period III: a comparison of CAPD versus IPD

TABLE XII. Hypotensive episodes during haemodialysis. Results of a three month survey during 1,336 sessions among 38 patients on hospital haemodialysis from 1st October to 31st December 1984

	Patients subgroups			
	I	II	III	IV
Number of patients M : F	8 : 1	9 : 5	7 : 2	4 : 2
Number of haemodialysis sessions	327	472	324	213
% haemodialysis sessions with HCO_3^-	28.4	16.5	0	0
Haemodialysis sessions with hypotensive episodes	0	35	103	137
% sessions with hypotension	0	7.4	31.7	64.3

home (21.4% of overall success). By contrast, among the 194 patients accepted on peritoneal dialysis during the same period of time, 148 (76.4%) were submitted to a period of training that resulted in home intermittent peritoneal

TABLE XIII. Home versus hospital dialysis: a comparison of haemodialysis versus intermittent peritoneal dialysis or CAPD during Periods II and III

	Haemodialysis		Intermittent peritoneal dialysis		CAPD	
	n	%	n	%	n	%
Total number of patients	126	100	136	100	46	100
Hospital dialysis*	90	71.4	46†	33.8	–	
Admitted at training centre	36	28.6	90	66.2	46	100
Installed at home	27	21.4	85	62.5	46	100
Not installed	9	7.1	5	3.7	–	

* Hospital dialysis without prior attempt to train the patient for home dialysis;
† Including 20 patients who died in less than one month

TABLE XIV. Duration of treatment of 46 patients receiving hospital intermittent peritoneal dialysis with the Tenckhoff catheter and a cycler during Periods II and III

	Duration of treatment (days) (as on 30th June 1984)		
	1–30	30–120	> 120
Total patients	19	9	18
Surviving patients	0	0	4
Treatment duration (days)			
Range	1–27	32–103	136–1702
Mean/patients (±SD)			
All patients	11.2 ± 7.5	59.6 ± 19.7	557.7 ± 552.7
Surviving alone	–	–	385.0 ± 147.1

dialysis or CAPD in 143 of them, a success rate of 73.7 per cent. In fact, the success rate of home peritoneal dialysis was even greater, if one takes into account that early deaths on peritoneal dialysis, in most instances due to severe uncontrolled uraemia, were included in the hospital peritoneal dialysis group, as shown in Table XIV.

Discussion

During the last 15 years, an increasing number of older uraemic patients were accepted on maintenance dialysis therapy in Languedoc-Roussillon. Our pragmatic approach integrated progressively newer modes of treatment as they became available [4] and gave a high priority to the development of home dialysis [6,8]. This approach permitted us to eliminate completely the arbitrary selection of end-stage renal disease patients on non-medical criteria such as age;

senile dementia and metastatic cancer remained two contraindications to end-stage renal disease therapy. Thus one of the main goals of our regional end-stage renal disease treatment programme was achieved.

As expected, the retrospective analysis presented here did not provide a clearcut answer to the question of optimum dialysis in the elderly. Two facts explain our uncertainties about the best therapeutic choice to offer elderly patients: 1) intermittent peritoneal dialysis and/or CAPD were used as the first mode of treatment, mainly to promote home dialysis, as these methods are simpler and do not induce cardiovascular intolerance; 2) due to a negative selection bias, patients on peritoneal dialysis were five years older on average than those receiving haemodialysis and had a higher frequency of arteriosclerotic cardiovascular disease at start of treatment. Therefore, our data did not permit an objective comparison between the results obtained with haemodialysis versus peritoneal dialysis, as older age and arteriosclerotic cardiovascular disease have been identified as factors affecting negatively the prognosis in dialysed uraemic patients [13,14].

However, our data arouse a high degree of suspicion as to the superiority of haemodialysis over peritoneal dialysis. First, actuarial survival curves were constantly better with haemodialysis even when the patients were age-matched.

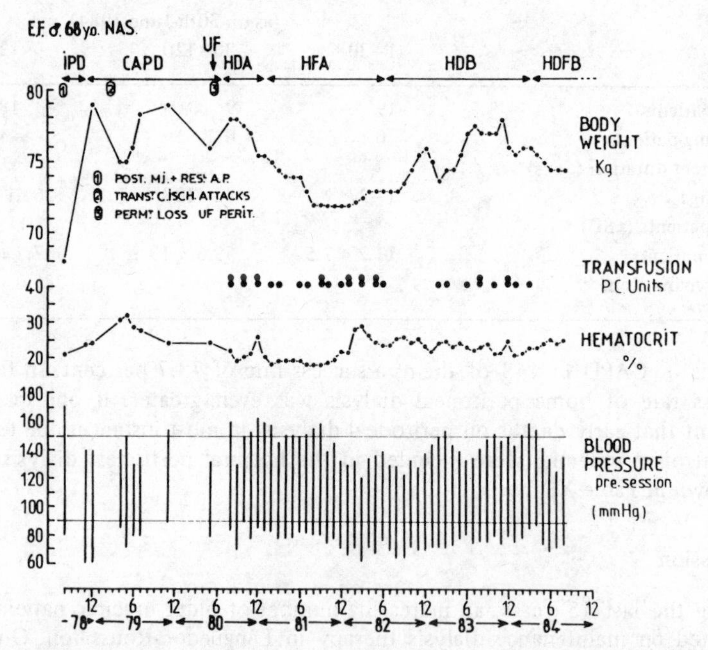

Figure 8. Clinical course of patient E.F., 68-year old male, with nephroangiosclerosis treated successively with intermittent peritoneal dialysis, CAPD, haemodialysis with acetate dialysate, haemofiltration with acetate containing substitution fluid, haemodialysis with bicarbonate dialysate and haemodiafiltration with bicarbonate dialysate and substitution fluid

Second, when comparing 'low risk' and 'high risk' patients, it is worth noting that 'high risk' patients on haemodialysis did not have as good a survival as the 'low risk' ones, whereas there was no difference between the two subgroups of patients receiving peritoneal dialysis. The reason for the lack of a better survival of 'low risk' patients on peritoneal dialysis was an excess mortality from peritonitis. This observation is in agreement with the well known fact that mortality from peritoneal infection rises with older age [15,16] and suggests that the risk of dying from peritonitis may well cancel the potential benefit of the absence of arteriosclerotic cardiovascular disease on improved survival of 'low risk' patients receiving peritoneal dialysis.

The suspicion that haemodialysis may be superior to peritoneal dialysis is further illustrated by the case of the 'high risk' patient E.F., a 60-year old male

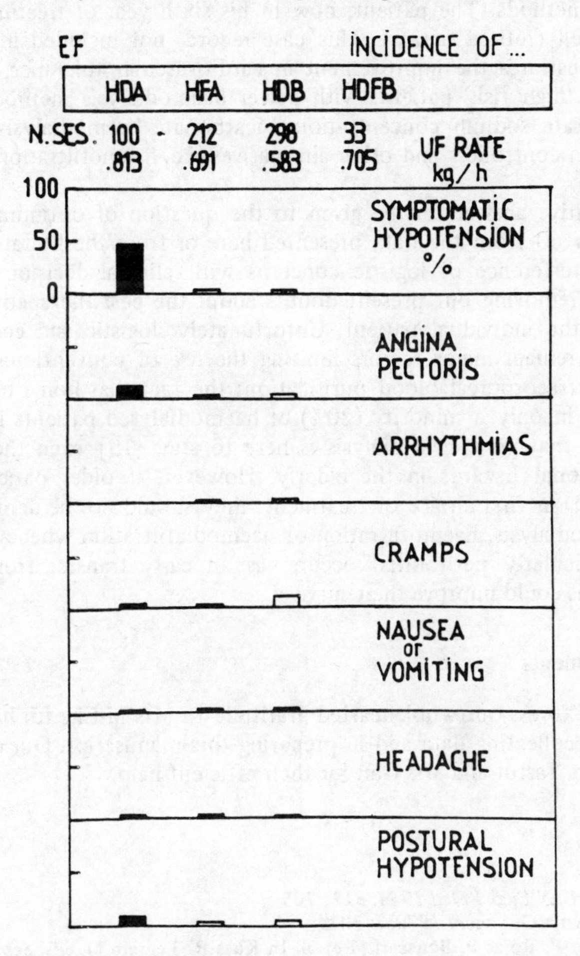

Figure 9. A comparison of ultrafiltration rates and symptoms occurring during sessions using various extracorporeal dialysis techniques in patient E.F.

whose clinical course is summarised in Figures 8 and 9. This patient had posterior myocardial infarction when intermittent peritoneal dialysis was initiated in September 1978 and a transient cardiovascular accident from hypotension six months later while on CAPD; he lost his peritoneal cavity from sclerosing peritonitis after two years on peritoneal dialysis and was transferred to extracorporeal dialysis in August 1980. He used successively haemodialysis with acetate dialysate; haemofiltration using a substitution fluid containing acetate; haemodialysis with bicarbonate in the dialysate; haemofiltration with bicarbonate dialysate. The compared tolerance to dialysis sessions with these various dialysis methods is shown in Figure 9: symptomatic hypotension and angina pectoris occurred respectively in 50 per cent and 10 per cent of sessions during haemodialysis with acetate dialysate and were observed in less than two per cent of sessions with other methods. The patient, now in his sixth year of treatment, remains alive and well (retired active). This case record, not included in the present series, demonstrates the improvement in cardiovascular tolerance that may be obtained in 'high risk' patients with newer haemodialysis methods, including higher dialysate sodium concentration, bicarbonate haemodialysis, the use of ultrafiltration controllers and other alternatives (i.e. haemofiltration and haemodiafiltration).

No definitive answer can be given to the question of optimum dialysis of patients over 60 from the data presented here or from the literature [17,18]. Persisting interference of logistic concerns with clinical decision making will not help in removing our present doubts about the best therapeutic choice to propose to the individual patient. Unfortunately, logistic and economic constraints will remain major factors limiting the use of conventional and newer modes of extracorporeal blood purification; the fact that home haemodialysis was feasible in only a minority (20%) of haemodialysed patients in our series, convince us that peritoneal dialysis is here to stay [19] even though it may entail additional hazards in the elderly. However, if older patients are still offered CAPD as first choice of treatment, they should not be denied the benefits of haemodialysis, haemofiltration or haemodiafiltration whenever a complication (particularly peritonitis) occurs, as an early transfer from CAPD to haemodialysis could improve their survival.

Acknowledgments

We wish to express our wholehearted gratitude to Mrs M Elie for her invaluable assistance in collecting data and in preparing this manuscript. Our many thanks go also to Mrs Tastut and Mrs Gau for their efficient help.

References

1 Moses LE. *N Engl J Med 1984; 311:* 705
2 McGeown MG. *Lancet 1972; i:* 307
3 Degoulet P, Rojas P, Bouscari M et al. In Küss R, Legrain M, eds. *Séminaires d'Uro-Néphrologie, 4e Série.* Paris: Masson. 1978: 164
4 Mion CM, Mourad G, Canaud C et al. *Asaio J 1983; 6:* 205

5 Mirouze J, Mion C, Jullien C. *Proc EDTA 1967; 4:* 156
6 Mion CM, Issautier R, Polito C et al. In Hamburger J, Crosnier J, Maxwell H, eds. *Advances in Nephrology, Volume 3.* Chicago: Year Book Medical Publishers. 1974: 311
7 Mion CM. *Proc EDTA 1975; 12:* 140
8 Mion CM. *Proc EDTA 1981; 18:* 91
9 Shaldon S, Beau MC, Claret G et al. *Artif Organs 1978; 2:* 323
10 Gurland HJ, Brunner FP, Chantler C et al. *Proc EDTA 1976; 13:* 49
11 Comty CM, Kjellstrand D, Shapiro FL. *Trans ASAIO 1976; 22:* 404
12 Kaplan EL, Meier P. *J Am Stat Assoc 1958; 53:* 457
13 Hutchinson TM, Thomas DC, MacGibbon B. *Ann Int Med 1982; 96:* 417
14 Shapiro FL, Umen A. *Asaio J 1983; 6:* 176
15 Hau T, Ahrenholz DH, Simmons RL. In *Current Problems in Surgery.* Chicago: Year Book Medical Publishers. 1979: 1
16 Fenton SSA. *Perit Dial Bull 1983; 3:* S9
17 Rao TK, Nathanson G, Avram M et al. *Proc Dial Transpl Forum 1976; 6:* 62
18 Mion C, Slingeneyer A, Huchard G et al. In Hamburger J, Crosnier J, Funck-Brentano JL, eds. *Actualités Néphrologiques de l'Hôpital Necker.* Paris: Flammarion. 1979: 71
19 Oreopoulos DG. *Nephron 1979; 24:* 7

Proc EDTA–ERA (1984) Vol 21

OPTIMUM DIALYSIS TREATMENT FOR PATIENTS OVER 60 YEARS WITH PRIMARY RENAL DISEASE. SURVIVAL DATA AND CLINICAL RESULTS FROM 242 PATIENTS TREATED EITHER BY HAEMODIALYSIS OR HAEMOFILTRATION

K Schaefer, G Asmus, *E Quellhorst, A Pauls, D von Herrath, J Jahnke

*St Joseph-Krankenhaus I, Berlin, *Nephrologisches Zentrum Niedersachsen, Muenden, FRG*

Summary

An analysis of the data of 180 haemodialysis patients and 62 haemofiltration patients over 60 years of age when commencing treatment, clearly shows that this age group of patients (when suffering from primary renal disease) has a very good chance of surviving many years when treated with either haemodialysis or haemofiltration. This refers also to patients being older than 75 or 80 years, who have survival rates of 50 per cent after five years and three years respectively. The presented data further indicate that chronic haemofiltration seems to be the superior treatment when compared with acetate haemodialysis for the treatment of elderly renal patients, as the survival rates are at any chosen time interval higher with haemofiltration than with haemodialysis.

Introduction

Only a few reports are available concerning treatment experiences with haemodialysis, haemofiltration, or chronic ambulatory peritoneal dialysis (CAPD) in patients older than 60 years [1–4].

The present report is an attempt to summarise our experiences with this age group. It is of note, that the majority of the elderly patients considered here has even surpassed the age of 70 years. In addition, we will compare the influence of two different treatment strategies (haemodialysis versus haemofiltration) on the survival rate of elderly patients and to analyse the causes for hospitalisation and mortality. Finally, data will be reported concerning the use of different therapeutic modes in these elderly patients compared to those used in patients younger than 60 years.

Patients and methods

Haemodialysis patients

Data are derived from a total of 180 haemodialysis patients, all being older than 60 years when starting haemodialysis (mean age 72.7 ± 5.8 years, female patients n=111, mean age 72.0 ± 5.7 years, male patients n=69, mean age 73.9 ± 5.6 years). All the patients were treated in the haemodialysis facilities of the St Joseph Hospital (SJK) in Berlin since 1975. Only patients with primary renal diseases were included, i.e. patients suffering from diabetes mellitus, amyloidosis myeloma, immunological diseases or other systemic diseases were not considered. The records of the 180 patients were analysed for causes of hospitalisation, duration of the hospital admission, and, in the case of death, for the reason of mortality. Survival data were calculated according to Cutler and Ederer [5]. In addition data are reported of a recently conducted survey in five different German haemodialysis centres (Hann. Muenden, Heidelberg, Darmstadt, Muenchen, Berlin), which was initiated to evaluate the kind and frequency of drugs being prescribed to haemodialysis patients. In this context it seemed of special interest to compare the therapeutic regimen of patients under (n=453) and over (n=301) 60 years of age.

Haemofiltration patients

Data are reported from a total of 62 patients (mean age 73.8 years) who were treated either since 1972 in Hann. Muenden (n=45) or since 1976 in Berlin (n=17); (mean age: female patients 72.1 ± 5.7 years, male patients 74.9 ± 5.4 years). The same exclusion criteria for haemofiltration patients were applied as for haemodialysis.

The data analysis was performed accordingly, with the exception that no special inquiries concerning the medical therapy were performed in this patient group.

Haemodialysis and haemofiltration techniques

Haemodialysis was performed thrice weekly always using acetate as buffer. For seven years a sodium concentration of 150mEq/L was regularly applied. The duration of each haemodialysis session varied between four to six hours. Haemofiltration was always performed in the post-dilution mode. During a normal treatment session, undertaken thrice weekly, 20 to 35 litres were exchanged, the volume of exchange depending mainly upon the general feeling of the patients, their blood urea nitrogen and the fluid, which had to be removed during one session, respectively. The dialysers and haemofilters, which were used differed according to the market situation. This also applied to the haemofiltration equipment which was changed according to the technical progress achieved in the last years in this field.

Results

Haemodialysis

Table I depicts the different diseases which caused the end-stage renal failure in the elderly patients. The most frequent cause was chronic glomerulonephritis, which was in most cases diagnosed from clinical criteria, i.e. the absence of a

TABLE I. Renal diseases of haemodialysis/haemofiltration patients being older than 60 years (SJK, Berlin)

1. Glomerulonephritis	38.8%
2. Interstitial nephritis	29.8%
3. Nephrosclerosis	8.5%
4. Cystic kidney disease	8.0%
5. Acute renal failure	1.6%
6. Nephrolithiasis	0.5%
7. Unknown	12.8%

special history indicating another disease, or the presence of hypertension, haematuria and proteinuria. Most of the cases of interstitial nephritis were associated with an abuse of analgesics and could therefore presumably be classified as analgesic nephropathies.

TABLE II. Overview about diseases which were frequently the cause for hospital admission (SJK, Berlin)

1. Shunt problems	158 admissions	24.5%
2. Cardiovascular diseases	103 admissions	15.6%
3. Gastrointestinal diseases	94 admissions	14.6%
4. Sepsis	40 admissions	6.2%
5. Overhydration	32 admissions	5.0%
6. Hypertensive episodes	26 admissions	4.0%
7. Miscellaneous (n=47)	193 admissions	31.1%
53 different diseases	646 admissions	100%

Table II shows the diseases which were mainly responsible for hospitalisation. It can be seen that shunt problems and cardiovascular and gastrointestinal (mainly bleeding) disorders were the most frequent reasons for admission. However, it is of note that the duration of the hospital stay was especially long when it was due to sepsis, whereas the other complications resulted in comparable lengths of hospital stay (Table III). The average time of a hospital stay for haemodialysis patients older than 60 years was 2.4 days per month.

TABLE III. Duration of hospital stay for diseases which most frequently caused hospital admission (SJK, Berlin)

1. Sepsis	41.0 days
2. Shunt problems	17.7 days
3. Cardiovascular diseases	15.1 days
4. Gastrointestinal diseases	15.1 days
5. Hypertensive episodes	13.6 days
6. Overhydration	11.9 days

TABLE IV. Causes of death in 72 haemodialysis/haemofiltration patients being older than 60 years (SJK, Berlin)

1.	Cardiovascular diseases	27.8%
2.	Cerebrovascular diseases	11.1%
3.	Sepsis	9.7%
4.	Gastrointestinal bleeding	8.3%
5.	Hyperkalaemia	6.9%
6.	Chronic obstructive lung disease	5.6%
7.	Malignancy	2.8%
8.	Liver coma	2.8%
9.	Peritonitis	2.8%
10.	Suicide	2.8%
11.	Hypertension, uraemia, hepatitis, undernutrition, leukaemia, pulmonary embolism	1.4%
12.	Unknown	11.0%

Table IV shows the causes of death in 72 haemodialysis/haemofiltration patients who died in the St Joseph Krankenhaus since 1975. Vascular disorders localised either in the heart or cerebrum were the cause of death in almost 40 per cent in elderly renal patients. Almost 10 per cent of this age group died because of sepsis or gastrointestinal bleeding. In almost all cases the listed diagnoses are based on findings obtained at autopsy.

Figure 1 shows the seven year survival rate of 180 haemodialysis patients and 62 haemofiltration patients. It is evident that at all time intervals patients being treated by haemodialysis have a lower survival rate than haemofiltration patients. The difference is especially pronounced after the fourth year.

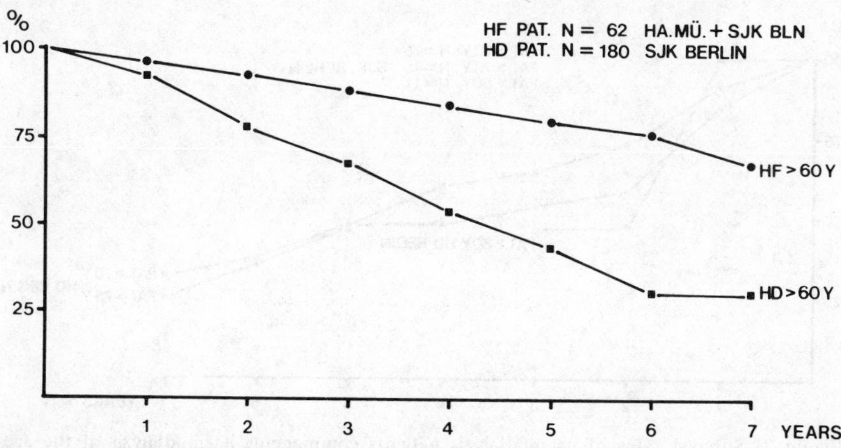

Figure 1. Survival rates of haemodialysis and haemofiltration patients up to seven years

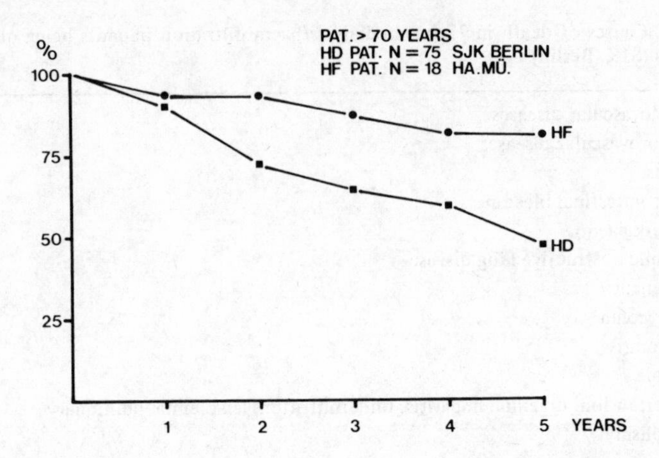

Figure 2. Survival rates of haemodialysis and haemofiltration patients being older than 70 years

A similar gap between both treatment methods could be demonstrated when the survival rate was calculated only for patients being older than 70 years when they started receiving regular haemodialysis (Figure 2).

Figure 3 shows the survival rates for three different age groups. It is remarkable, that after five years no difference could be demonstrated for patients being 70 years or 75 years when commencing haemodialysis. Even patients being older than 80 years at the start of regular haemodialysis have an almost 50 per cent chance of surviving for three years.

Table V summarises the 12 therapeutic agents which were most frequently prescribed in elderly haemodialysis patients. Major differences with regard to the patients being younger than 60 years refer especially to the prescription

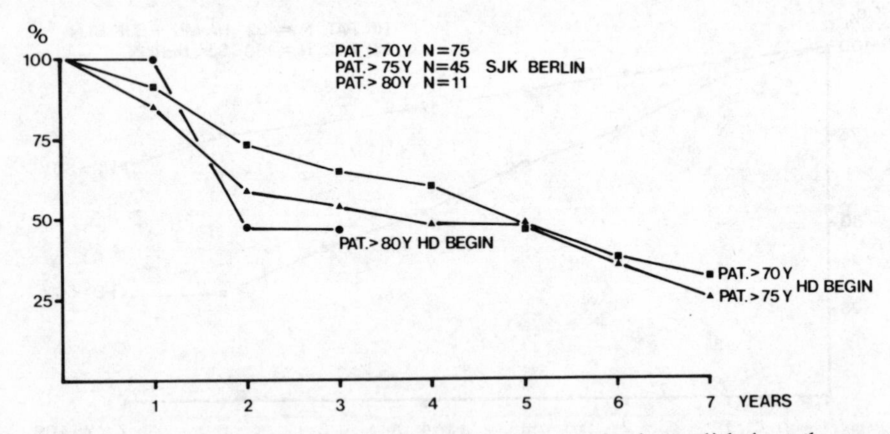

Figure 3. Survival rates of haemodialysis patients commencing haemodialysis at the age of 70, 75 or 80 years, respectively

TABLE V. Frequency of use of individual therapeutic agents in patients being younger (n=453) and older (n=301) than 60 years (% of all patients)

Drug		<69 years	>60 years	Difference
1.	Phosphorus binder	96.5	90.7	– 5.8
2.	Vitamin preparations	77.7	75.1	– 2.6
3.	Antihypertensive agents	48.3	52.2	+ 3.9
4.	Digoxin/Digitoxin	32.2	61.1	+ 28.8 (!)
5.	Iron preparations	30.5	18.6	– 11.9
6.	Vitamin D preparations	30.0	16.3	– 13.7
7.	Sedatives/Hypnotics	14.3	14.3	0
8.	Calcium salts	13.9	8.6	– 5.3
9.	Antiplatelet agents	13.7	23.3	+ 9.6
10.	Laxatives	9.7	12.0	+ 2.3
11.	Nitrates	9.1	30.2	+ 21.1 (!)
12.	Analgesics	8.4	17.6	+ 9.2

of digoxin/digitoxin, which was substantially increased in the elderly group. This is also true for nitrates which also increase by more than 20 per cent.

The actual number of drugs which is prescribed does not differ between the two age groups (Figure 4). Most of the patients being either younger or older than 60 years are taking five different therapeutic agents daily. There are also no differences demonstrable for patients of each group taking either more or fewer drugs per day (Figure 4).

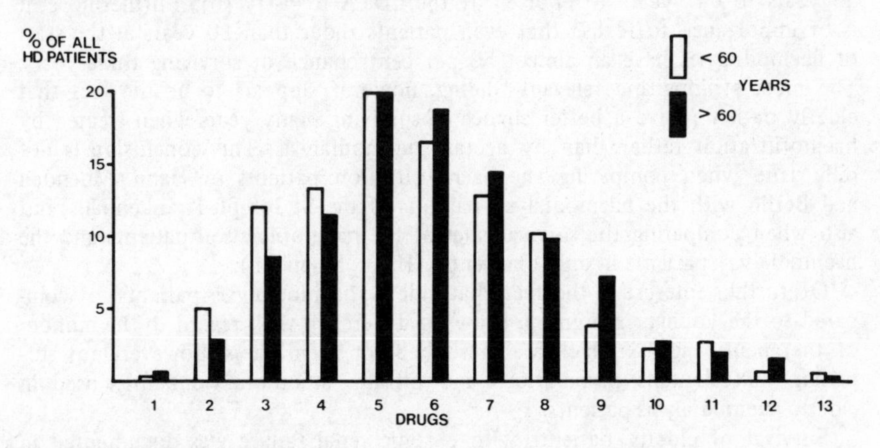

Figure 4. Number of different therapeutic agents prescribed to haemodialysis patients being younger or older than 60 years

Haemofiltration

The diseases which most frequently caused hospitalisation in haemofiltration patients (Hann. Muenden) were myocardial insufficiency, myocardial infarction and shunt problems. The average duration of the hospital stay was slightly shorter than for haemodialysis patients, although it is difficult to establish any significant differences, as the haemodialysis patients from the St Joseph Krankenhaus were compared with the haemofiltration patients from the Hann. Muenden hospital.

The causes of death of the haemofiltration patients treated in Hann. Muenden were comparable to haemodialysis patients. In haemofiltration patients, too, vascular complications of the heart and cerebrum were the leading reasons for mortality.

The survival rates for haemofiltration patients compared to haemodialysis patients are, as already indicated, depicted in Figures 1 and 2.

Discussion

The results presented show very convincingly that it is extremely rewarding to treat elderly patients with haemodialysis or haemofiltration, especially when they are suffering from a primary renal disease. When comparing our data obtained in haemodialysis patients with those of the EDTA Registry it becomes evident that the St Joseph Krankenhaus haemodialysis patients with an age over 60 years have a better chance of surviving five years than patients of a comparable age group reported in the EDTA Registry, who are subjected either to haemodialysis, peritoneal dialysis or transplantation, respectively [6]. Even our patients being older than 75 years when commencing haemodialysis, have a significantly better five-year survival chance than the age group ranging from 55 years to 64 years as reported by the EDTA Registry [6]. Furthermore, it is of importance to realise that even patients older than 80 years at the start of haemodialysis have an almost 50 per cent chance of surviving three years. The most striking and relevant finding, however, appears to be the fact that elderly patients have a better chance of surviving many years when treated by haemofiltration rather than by acetate haemodialysis. This conclusion is not only true when comparing the haemofiltration patients of Hann. Muenden and Berlin with the haemodialysis patients of the St Joseph Krankenhaus, but also when comparing the survival rates of the haemofiltration patients and the haemodialysis patients in only one centre (Hann. Muenden).

Of further interest is the fact that elderly haemodialysis patients, as compared to the younger age group, show no difference with regard to the number of therapeutic agents which are prescribed for them. It is, however, not surprising that digitalis preparations and nitrates are more frequently used in elderly haemodialysis patients.

Survival of elderly patients with chronic renal failure was documented in the recent past by different groups. Taube et al reported data from the United Kingdom, which showed that patients with an age range from 55 to 72 years could be successfully treated by renal replacement therapy. The five-year survival was a little higher than in our haemodialysis patients; however, the mean age was only 59.6 years, and therefore lower by 13.1 years [2]. Of interest in this context are the findings by Chester et al [1] who reported on 45 patients starting haemodialysis after the age of 70 years (mean 75 years). The three-year survival of their patients was 23 per cent, which is markedly lower than in our elderly haemodialysis patients with a starting age of 70 years, who show a three-year survival of 65 per cent. Nine of their elderly patients were over 80 years old and had a similar two-year survival as compared to our patients over 80 years (41% versus 47%). However, none of their patients survived more than three years, a finding which is significantly different from our data. In 1984 Westlie et al reported a five-year survival of 22 per cent in their patients being older than 70 years , which again is lower than in our patients [4] . More recently Mallinson et al reported that elderly patients requiring haemodialysis had a significantly poorer survival than non-dialysed patients [3]. However, when comparing the actuarial survival rate of the non-dialysed patients with our dialysed patients it is evident that only less than 35 per cent of their patients have survived four years, whereas 60 per cent of our haemodialysis patients are still alive after that time interval. The superiority of haemofiltration, especially when treating elderly patients or so-called poor-risk patients, is well documented [7]. According to our knowledge, the data reported above permit for the first time a direct comparison between haemodialysis and haemofiltration on a large scale. The reasons for the superiority of haemofiltration for elderly patients cannot be discussed in this context, especially as it is presently not clear whether differences in the solute transport (convection versus diffusion), membrane properties or other factors are of major importance, for the difference between these two blood purification treatments.

In conclusion, data obtained in 242 patients older than 60 years at the start of replacement therapy clearly demonstrate that it is very rewarding to start treatment in this age group. Good survival rates could also be documented for patients older than 75 years. According to the experiences of two large haemodialysis centres it appears that post-dilution haemofiltration seems to be a better method than acetate haemodialysis. However, it remains to be seen whether or not CAPD could compete with both methods.

Acknowledgments

We thank D Kutschera, D Weisselberg and S Schaefer for their help in compiling and typing the data.

References

1 Chester AC, Rakowski TA, Argy WP et al. *Arch Int Med 1979; 139:* 1001
2 Taube DH, Winder EA, Ogg CS. *Br Med J 1983; 286:* 2018
3 Mallinson WJM, Fleming SJ, Shaw JEH et al. *Q J Med 1984; New Series LIII:* 301
4 Westlie L, Umen A, Nestrud S et al. *Abstr IXth Int Congr Nephrology 1984:* 197A
5 Cutler SJ, Ederer F. *J Chron Dis 1958; 8:* 699
6 Jacobs C, Broyer M, Brunner FP et al. *Proc EDTA 1981; 18:* 4
7 Quellhorst E, Schuenemann B, Hildebrand U. *Blood Purif 1983; 1:* 70

Open Discussion

KLINKMANN (Chairman) I was looking forward in the presentations to an answer to a question, but probably the question was too difficult because almost no speaker touched on it. How do the elderly patients compare to any other chronic disease social-wise, economic-wise, as well as from the medical survival rates. You all compared and showed younger patients, but none of you have, for example, compared data from chronic patients with carcinoma, cardiovascular disease and so on. I do not know if any of you are prepared to answer the question, but I would like, as a start to the discussion, to have a statement from all of you.

MION Your question is very difficult to answer because we do not have the experience. However, I would say that those we call 'retired active' who are 60 per cent of the haemodialysis population and 50 per cent of the intermittent peritoneal dialysis population, do as well, if not better, than any other chronic disease.

KLINKMANN But there is no hard data available at present from your Centre?
Dr Schaefer, do you have any such data on patients treated by haemofiltration?

SCHAEFER I am sorry that I also have no data on that, but I know that Dr Jacobs will present some examples of different diseases.

JACOBS Well, I anticipated that this question would come up, so I enquired from one of my colleagues. We looked at the reports of cancer patients from the National Institute of Health in the United States issued in 1976 and reprinted in 1981. In male patients in the 55–74 year age range, the age range which was considered in our presentations, for five very common cancers. The five-year survival in patients with carcinoma of the prostate in the 55–64 year age range was 67 per cent, and in the 65–74 year age range 45 per cent; bladder carcinoma is exactly the same, but for cancer of the stomach, bronchus and lung the five-year survival is less than 10 per cent. In females the five-year survival in the 55–64 year age range is 60 per cent for carcinoma of the breast and 55 per cent in the 65–74 year age range; cancer of the cervix uterus is 48 per cent at five years in those aged 55–64 and only 37 per cent in the 65–74 year age range. I do not want to get you baffled with all these figures: the only conclusion that can be drawn is that the results which are obtained in patients with life-threatening diseases like end-stage renal failure, which kills people within a five day interval if not treated, compares very favourably indeed with the results of many malignant tumours whose treatment is not really disputed or criticised strongly in the developed countries.

KLINKMANN Thank you Dr Jacobs: we might come back to this during the discussion.

IAINA (Israel) Do you think that the elderly patient needs less or more hours of dialysis each week and do you have any data on the deterioration of residual kidney function during dialysis in these patients?

MION I have no information on the residual renal function, although I know some patients maintain kidney function: they need as much dialysis as younger people, if you take into account their body size and residual kidney function.

SCHAEFER I have no data on residual renal function and the duration of haemodialysis as compared to the younger patients. In our patients it ranges between four to six hours per session.

JACOBS I have no data on the evolution of renal function. In the retrospective study, the clinical and biological results that I have shown were obtained with a duration of dialysis sessions of approximately the same time.

KLINKMANN Clearly this is a matter to be examined at future congresses.

BERNHEIM (Israel) I want to give my experience because over 60 per cent of my patients are older than 60 and 25 to 30 per cent of my patients are over 70. It is very difficult to compare the results between haemodialysis and peritoneal dialysis because in patients aged 65 years and 75 years without dialysis have a different life expectancy. In our patients there is no difference between haemodialysis and peritoneal dialysis. I do not think that you can make a comparison between peritoneal dialysis and haemodialysis because the same patient is on either haemodialysis or peritoneal dialysis. I do not think there is a particularly good treatment as I think that all the treatments are good for these patients.

KLINKMANN So you do not believe that we need any specific mode of treatment?

MION I would agree generally with that statement, except in our experience in the age groups 60–64, 65–69 and 70 plus, haemodialysis always gives better results than peritoneal dialysis. However, we have a negative bias selection because the worst patients are put on peritoneal dialysis. Therefore it is very difficult to compare patient groups and so I agree with you completely.

BERNHEIM Exactly, that is the problem because you are taking the bad patients onto peritoneal dialysis.

KLINKMANN Well, it should not be that way and I do not think it can be stated generally.

ANDREUCCI (Naples) First of all let me congratulate you for your excellent presentations. It is very exciting to hear that we are so good in giving a good life to older patients. Professor Mion, you are one of the most expert in CAPD, what is your present policy in deciding which kind of treatment to give these older patients? I would like to ask the same question of Professor Schaefer. Could Professor Jacobs tell us of the incidence of side effects, such as hypotension, during or after haemodialysis in these older patients? Finally, what was the cost of haemofiltration?

MION I presented the original experience from several centres. In Montpellier we try to take all patients over 65 on peritoneal dialysis and let them try both intermittent peritoneal dialysis and continuous ambulatory peritoneal dialysis and then select what is most convenient. The trend is that the older patients, who are usually more dependent upon their spouses or helpers, will have home intermittent peritoneal dialysis, whereas those who are fitter and more independent will have CAPD because they can do that themselves. In the age group 60–65 there is more versatility and if the patient is fitter we will treat by haemodialysis. Some of them will prefer peritoneal dialysis because they do not tolerate haemodialysis. There are some other centres like Perpignan, for instance, or Carcassone where there is a trend to take more patients into hospital haemodialysis whatever the age group.

SCHAEFER I would presently consider haemofiltration as a superior method for elderly patients when comparing survival rates. However, what is necessary for these patients is that they have very good vascular access. It could well be that patients who have poor vascular access are also the patients who have a poor survival rate after transfer to haemodialysis. However, I would like to emphasise that it is very rewarding to treat these elderly patients with haemodialysis. Therefore I would start patients with good vascular access on haemofiltration and if they have poor vascular access I would transfer them to haemodialysis. As far as costs are concerned I think there is presently no doubt that haemofiltration is much more expensive than haemodialysis, but as you may know there are some advances in reducing the costs of the substitution fluid.

JACOBS I would like to answer the question about the side effects during dialysis sessions in elderly patients. There are three main side effects: hypotensive episodes, angina and arrhythmias. If you want to compare the incidence and the severity of the side effects you have to compare similar groups of patients under and over 60 years with, for instance, bicarbonate dialysate or ultrafiltration controls and so forth. Secondly, you must compare the groups with respect to haematocrit and hypotensive medications and so on. It is very difficult to give an overall figure that is meaningful about the relative proportions of side effects in younger and elderly patients if you have not very homogenous groups.

KERR (London) I was surprised to hear from Professor Jacobs that osteodystrophy is as common in the elderly as in the young, which is contrary to a lot of previous experience. I wonder whether osteoarthritis in the elderly is being misinterpreted as osteodystrophy or whether this is confirmed radiologically or by histology?

JACOBS The question on the form regards the clinical manifestations of renal osteodystrophy, which may mimic the manifestations due to arthrosis from ageing of joints and bones. This item must be subject to careful interpretation. I do agree with you.

PORT (Michigan) Could the panel please comment on the withdrawal from therapy as a cause of death. Is this a significant proportion that is hidden in causes of death, such as cerebrovascular accident or other cardiovascular causes? Secondly, are patients allowed to withdraw from therapy in this elderly age group when they have major complications?

MION Withdrawal represented about two per cent of deaths on haemodialysis and about 8–10 per cent on peritoneal dialysis. The patients were allowed to withdraw if the family decided they should not continue the treatment when the complications are particularly severe, such as cerebrovascular accident.

SCHAEFER In my experience it is about two per cent and I have listed it as suicide.

JACOBS In a large series of patients suicide and abandonment of treatment have been combined and represented about three per cent of all deaths.

KLINKMANN Is the suicide rate in these patients higher than in the general population?

JACOBS I cannot answer that question.

BÁRTOVÁ (Prague) Dr Jacobs mentioned that patients were taking anticoagulant drugs and I wondered if it was for ischaemic or cardiovascular disease or for fistula problems?

JACOBS It was not specified on the questionnaire so I cannot answer that question.

BÁRTOVÁ Professor Schaefer, is the hyperkalaemia the cause of death, the result of clinical complications, such as the catabolic state of the patient or gastrointestinal bleeding? If it was just a high potassium intake in the diet you can take it as a mode of suicide.

SCHAEFER Yes, it depends on whether it was done on purpose.

SHALDON (Montpellier) This is a comment and a question. I do not accept that there are no differences between these forms of treatment. We should be concerned that our methodology of retrospective analysis is being analysed perhaps less sensitively than it should be. A consequence of this is that in the small numbers of patients the actuarial survival rates are probably not the appropriate way to determine survival. The second point which I would like the panel's opinion on is that perhaps our picture in Europe is too rosy for the real geriatric picture, in that already from 1982–1983 the mortality rate versus the new intake rate has shown a considerable increase of 52 per cent from 42 per cent in one year. Do you really think that your survival rates in your elderly patients is merely reflecting the use of geriatric dialysis and that you are really just showing a positive selection process?

JACOBS Well, I think it is always a problem in showing the results from one Unit or two Units where they have their own selection criteria as compared to very large series where the selection criteria may cancel out. One particular experience is valid for one group and it is probably not a general view for every case. If you have a very large series you may also restrict the validity of the results due to the fact that you have a mixture, so there is no totally satisfactory answer to your question.

MION I fully agree on the limitations of retrospective analysis. It is very difficult to get hard data and to show up comparative results. My feeling is that peritoneal dialysis is inferior to haemodialysis but I cannot prove it because it is not randomised and prospective.

RITZ (Heidelberg) We have heard very sophisticated and detailed analysis, but I would appreciate to point a question directly to Professors Mion and Schaefer. God be graceful to you, but if you were aged 70 or 80 what kind of treatment would you prefer for yourself?

MION That is a very important question which I have asked myself when I am doing this kind of prospective study. Firstly, I wonder whether beyond a certain age, such as 70 or 80, I would like to have a life of peritoneal dialysis or whether I would prefer to die. I cannot tell you because the courage to die is not given to everyone. If I had to choose I would first like to try some kind of haemodialysis or haemofiltration before going to peritoneal dialysis.

SCHAEFER I think the answer would depend on my vascular status. I think it would be fair to put to the golden-standard Claude Jacobs the same question.

JACOBS Well, I think it is one of the most difficult questions that can be asked because it relates to the quality of life that one wishes to have for his or her last years of life. As somebody put in the literature very clearly, the quality of life of an individual is actually what the individual thinks it to be and so this is a very personal question. On the one hand, centre haemodialysis treatment is very well appreciated by elderly patients because they have created another family; this is particularly true for patients who are single because the husband or wife has died. On the other hand, at the other extreme, retired people with nothing much to do and having CAPD at home have found another profession. They are CAPD people. This is to show that each individual is different and what is excellent for one is very bad for the other. I think that there is no general way of dealing with this topic.

ROTELLAR (Barcelona) Have you compared people over 60 years in the general population with patients over 60 on dialysis with respect to the mortality rate?

JACOBS I would not like to answer this question because I have not examined this.

PART XII

GUEST LECTURES ON NEPHROLOGY I

Chairmen: L Morel Maroger Striker
G Remuzzi

J Bergström
R Maiorca

Proc EDTA–ERA (1984) Vol 21

GLOMERULONEPHRITIS IN ESSENTIAL MIXED CRYOGLOBULINAEMIA

G D'Amico, F Ferrario, G Colasanti, A Bucci

S Carlo Borromeo Hospital, Milano, Italy

Introduction

Cryoglobulins are immunoglobulins that have the property of reversible precipitation in the cold. Immunochemical analysis identifies three types of cryoglobulins [1], type 1 are single monoclonal immunoglobulins, mainly myeloma proteins and macroglobulins that, by virtue of some as yet poorly understood structural features, possess the property of cryoprecipitability [2,3]. Type 2 and type 3 are mixed cryoglobulins, composed of at least two immunoglobulins. In both a polyclonal IgG is bound to another immunoglobulin, which acts as anti-IgG rheumatoid factor [4]. This antiglobulin component, which is usually of the IgM class, but occasionally also of the IgG of IgA class [5], is monoclonal in type 2 and polyclonal in type 3. It functions as an antibody with affinity for determinants present primarily in the Fc portion of polyclonal IgG, although a concomitant reaction between the IgG and the IgM variable regions, an anti-idiotype interaction, has been recently suggested for the type 2 IgG-IgM cryoglobulins [6].

As is typical for some rheumatoid factors, the monoclonal or polyclonal IgM reacts best with an IgG when the latter is complex to an antigen [7], probably because this complexing modifies the Fc portion of the IgG molecule so that it becomes antigenic. In fact, careful analysis of mixed cryoglobulins taken from patients with SLE or with hepatitis B virus infection demonstrate the presence of an initiating antigen: DNA and anti-DNA antibodies have been demonstrated in the isolated cryoglobulins in the former [8], HBsAg in the latter [9,10].

In the last 20 years, cryoglobulinaemia has been found in an increasing variety of diseases [1]. Cryoglobulins may be a marker for lymphoproliferative disorders or may be immune complexes that suggest the presence of some chronic inflammatory or infectious disease.

A great percentage of the mixed cryoglobulins, which constitute 60–75 per cent of all cryoglobulins described in the two largest reviews [1,11], are found in connective tissue diseases, in infectious or lymphoproliferative disorders or in immunologically mediated glomerulonephritis [12].

However, in approximately 30 per cent of all mixed cryoglobulins, the aetiology is not clear-cut and the cryoglobulinaemia is called 'essential' (Table I). Since the main clinical symptoms were accurately described by Meltzer et al [19], it has been called the 'purpura-weakness-arthralgia syndrome'. The same investigators described renal damage and subsequently this has been found to be common in essential mixed cryoglobulinaemias, being reported in 20–60 per cent of cases (Table I).

TABLE I. Some characteristics of cases of essential mixed cryoglobulinaemia (EMC) in different reports from the literature

	Number of patients with EMC*	% of female patients	% of cases with type II cryoglobulinaemia
Series of unselected cases			
Brouet et al [1]	29 (34%)	–	7.0%
Invernizzi et al [13]	54 (53%)	71.0%	73.0%
Gorevic et al [11]	40 (32%)	67.5%	32.5%
Popp et al [14]	12 (35%)	92.0%	–
Realdi et al [10]	25 –	84.0%	20.0%
Migliorini et al [15]	36 –	83.3%	–
Monteverde et al [16]	11 (14%)	81.8%	72.7%
Series of cases selected for the presence of renal involvement			
Cordonnier et al [17]	18	44.5%	100%
Tarantino et al [18]	44	61.4%	–
Our cases	29	37.9%	100%

* Numbers in brackets indicate percentage of all observed cases of cryoglobulinaemia

While women outnumber men in the largest unselected populations of patients with essential mixed cryoglobulinaemia, no such prevalence was evident in the three largest populations selected for the presence of renal disease, including ours (Table I).

Characterisation of the immunoglobulin which compose the cryoglobulins in these patients with renal disease demonstrate the exclusive monoclonal nature of the IgM, with a large prevalence of the Ig Mk (Table I). Since in the largest rheumatological surveys of patients with essential mixed cryoglobulinaemia the percentage of those with monoclonal igM nearly equals that of patients with polyclonal IgM, we must conclude that renal involvement is more frequent, for unknown reasons, in type II mixed cryoglobulinaemia, in which Ig Mk is the monoclonal component. In our population of 29 biopsied patients with essential mixed cryoglobulinaemia and renal manifestations, all of whom had an IgG-IgM cryoglobulinaemia (with some additional IgA in 5 cases) with rheumatoid factor activity, monoclonal Ig Mk was documented in all 24 patients for whom it was possible to carry out an accurate characterisation of the immunoglobulins.

Renal pathology

Some specific features enable us to differentiate the renal involvement of essential mixed cryoglobulinaemia from both the glomerular lesions found in the various types of idiopathic glomerulonephritis and from the kidney lesions found in simple cryoglobulinaemia (which do not differ significantly from those of myeloma and macroglobulinaemia). Four main lesions can be identified histologically and immunohistologically in kidney biopsy specimens:

1. *Deposits* of variable intensity, either on the subendothelial aspect of the glomerular capillary walls and/or in the capillary lumen, filling it completely as 'intraluminal thrombi'.
2. *Infiltration of the capillary lumen by monocytes-macrophages*, migrated from the bloodstream (exudative glomerulonephritis).
3. *Thickening of the glomerular basement membrane, with a double-contoured appearance* not due to the peripheral interposition of mesangium.
4. *Renal arteriolitis*, with infiltration of monocytes in the vessel walls.

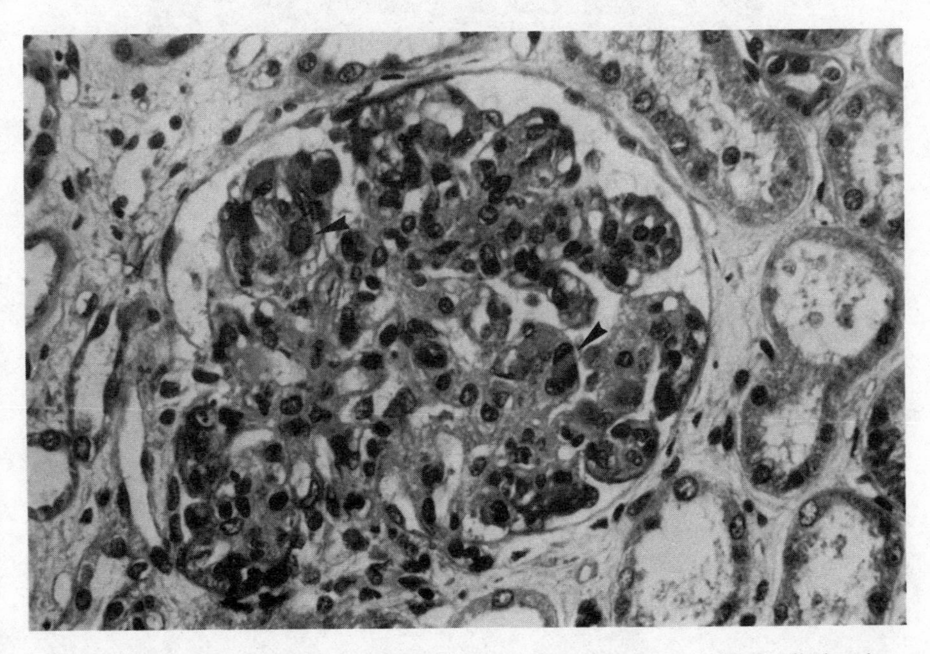

Figure 1. Diffuse endocapillary proliferation, and infiltration of mononuclear cells (exudation. Intraluminal thrombi (arrow). Thickening of glomerular basement membrane. Trichrome stain (250 x)

These four lesions can be variably associated in the different patients. They are also irregularly distributed in different glomeruli in the same patient and even in different loops.

The most common histological pattern by light microscopy is that of an endocapillary proliferative-exudative glomerulonephritis, which is characterised by an hypercellularity not only due to the proliferation of resident cells (more endothelial than mesangial cells), but also to the infiltration of migrating cells, mainly monocytes-macrophages (Figure 1).

Large, amorphous, eosinophilic, PAS-positive, congo-red negative deposits lying on the inner side of the glomerular capillary wall and filling the capillary lumen, the so-called 'intraluminal thrombi', are seen in about one-third of cases, especially in those presenting with a more acute renal disease (Figure 1).

Thickening of glomerular basement membrane, with a widespread double-contoured appearance, is a very common feature of this proliferative-exudative glomerulonephritis, and gives a picture of 'membrano-proliferative' glomerulonephritis, which can not be defined as 'mesangiocapillary'. In fact, peripheral interposition of mesangial matrix and cells is not as prominent as in idiopathic mesangiocapillary glomerulonephritis, and insufficient to account for the double-

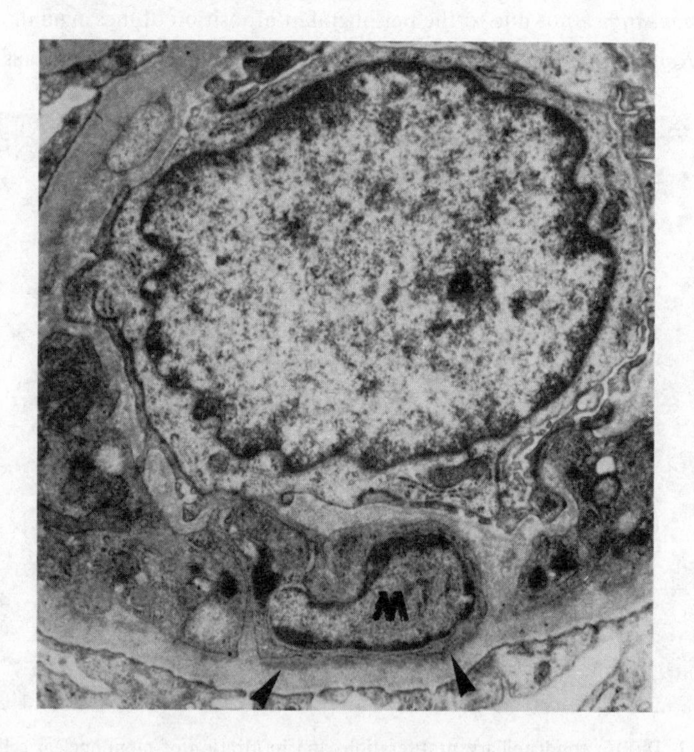

Figure 2. A monocyte (M), in close contact with subendothelial deposits (arrows), in the double-contoured glomerular basement membrane (10,800 x)

contoured appearance, while monocytes and subendothelial deposits more commonly fill the space between the glomerular basement membrane and the newly formed basement membrane-like material lying on its inner side (Figure 2).

Mesangial activation and peripheral circumferential expansion are prominent phenomena in only a few patients, in whom centrolobular sclerosis and a picture typical of lobular mesangiocapillary glomerularnephritis are found.

Staining with non-specific esterase, which is rather specific for monocytes-macrophages, shows that infiltration of monocytes may be massive: up to 80 cells per glomerulus can be counted, and they are in close contact with the parietal and intraluminal deposits (Figure 3). By electron microscopy these cells seem to be engaged in the phagocytosis of the deposits.

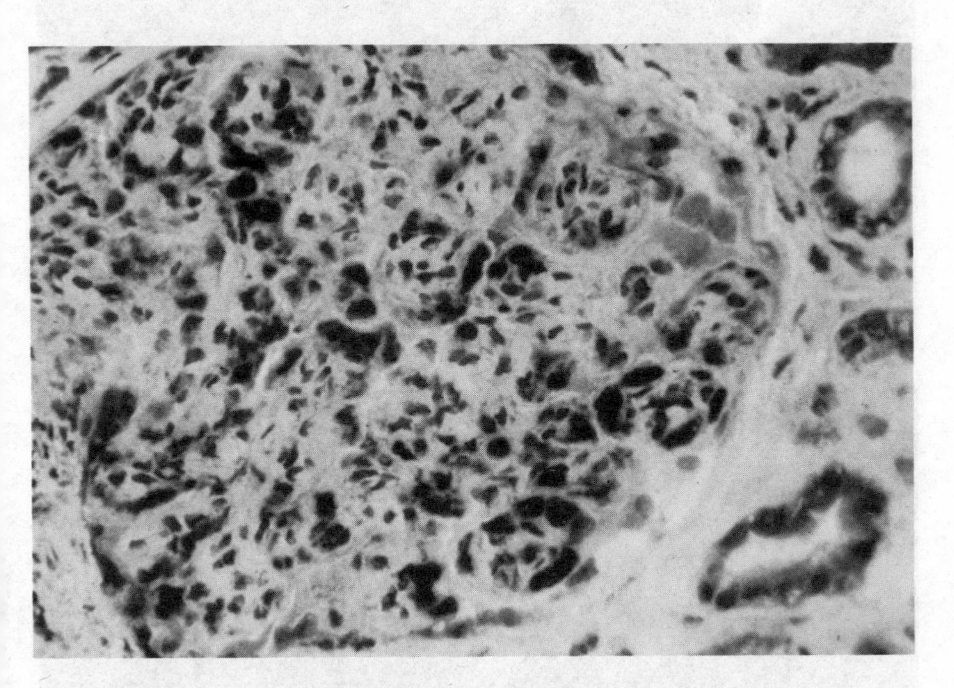

Figure 3. Numerous monocytes (dark cells) infiltrating the glomerular capillary lumen. Non-specific esterase stain (250 x)

Renal vasculitis, mainly involving small arteries, is present in about one-third of cases. It is often more frequent when massive intraluminal thrombi are also present, and it is frequently associated with other signs of systemic vasculitis. Histologically, it is characterised by an infiltration of monocytes-macrophages in the wall of vessels.

Three patterns of deposition can be seen by immunofluorescence [20–22]:

1. Intense massive staining of huge deposits filling the capillary lumen (intraluminal thrombi), usually associated with faint irregular segmental parietal staining of some peripheral loops, in a subendothelial position (Figure 4a).

Figure 4 (a–c). Three common patterns of immunofluorescence staining. Anti-IgM antiserum. See text for description

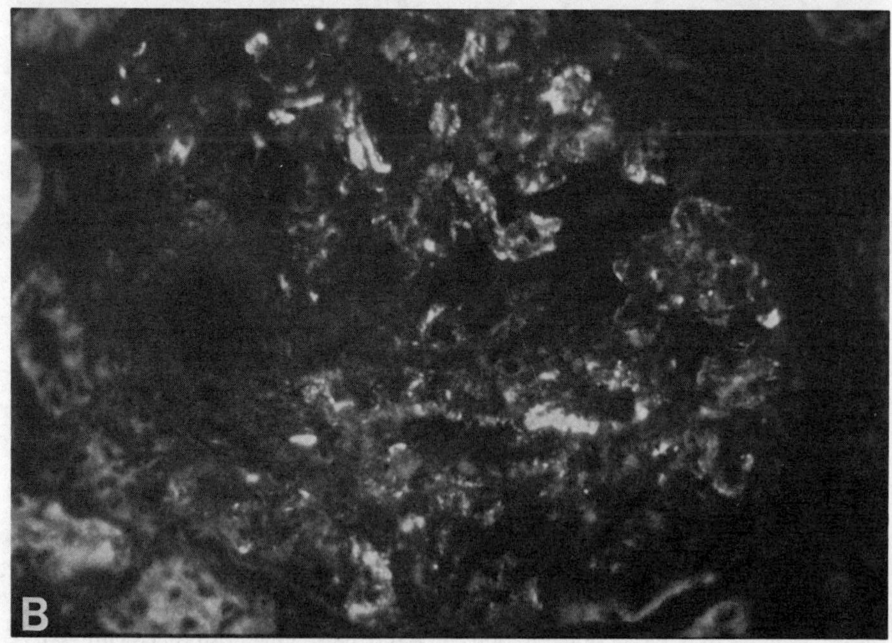

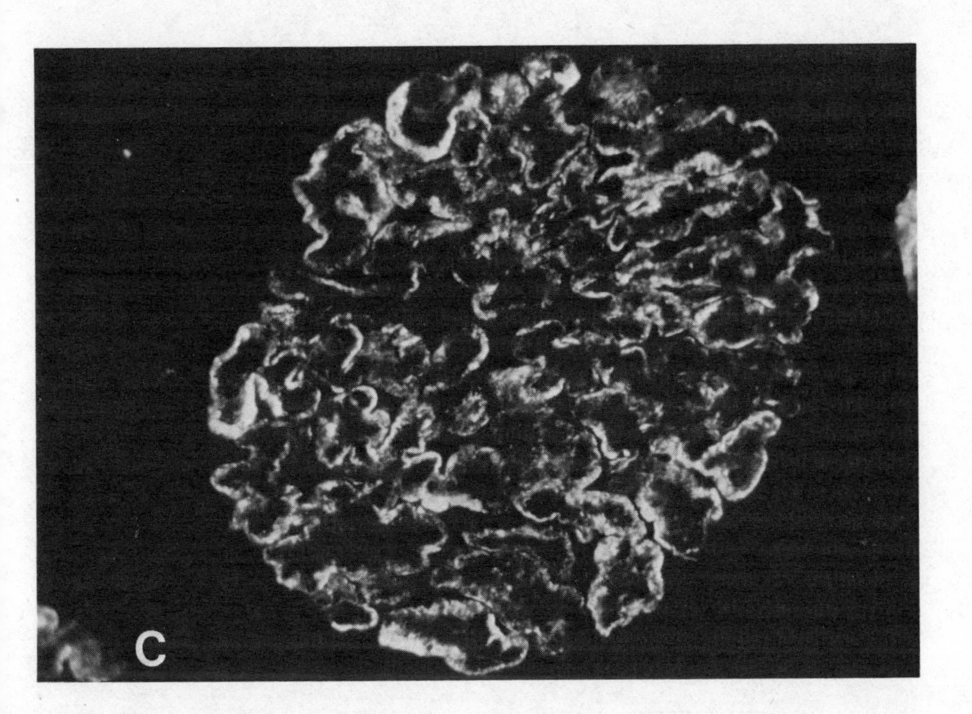

2. Faint irregular segmental parietal staining of some peripheral loops, in a subendothelial position, without staining of intraluminal thrombi (Figure 4b).
3. Intense granular diffuse staining of peripheral loops, with a subendothelial pattern similar to that of idiopathic membrano-proliferative glomerulonephritis (Figure 4c), sometimes with a lobular pattern.

The deposits mainly stain with antisera to IgM and IgG, usually associated, especially in the subendothelial deposits, with C_3 (92% of our 29 cases), Clq and fibrinogen (56% in our cases), and C_4 (32% in our cases). There is a good correlation between the immunoglobulin composition of the circulating cryoglobulins and the immunoglobulin classes in the glomerular deposits. This close similarity, which suggests that the deposits are locally trapped or precipitated cryoglobulins, is confirmed by our demonstration that glomerular immunodeposits display antiglobulin activity similar to that of serum cryo-IgM [23]. It is also confirmed by electron microscopy findings: in addition to classical immune-complex-like deposits, one can observe a peculiar fibrillar or crystalloid material, identical to that seen in the in vitro cryoprecipitate of the same patients [24,25], which in cross-sectional views appears to be made up of tubular units, while longitudinal sections reveal parallel fibrils (Figure 5).

Twenty-one of our 28 patients, for whom sufficient material for light microscopy examination of the renal biopsy specimen was obtained, showed the proliferative-exudative pattern, while in five a typical pattern of lobular mesangiocapillary glomerulonephritis was found, and in the remaining two patients a

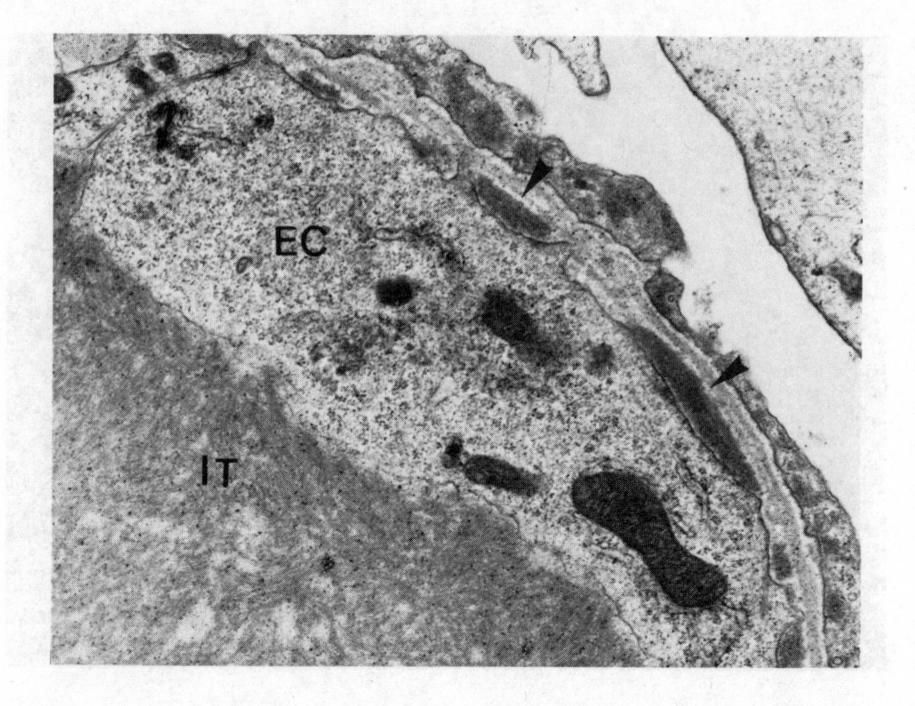

Figure 5. Intraluminal thrombus (IT) showing a fibrillar structure. Subendothelial deposits (arrows) (20,000 x)

very mild mesangial proliferative glomerulonephritis was found (Table II). Three subgroups could be identified among the 21 patients with endocapillary proliferative-exudative pattern, according to the three different patterns of deposition by immunofluorescence. They showed different clinical features.

Clinical manifestations and clinico-pathological correlations

Renal symptoms are usually a late manifestation, which appears some years after the onset of the extrarenal symptoms (4 years on the average, according to Gorevic et al [11] ; 77 months according to Tarantino et al [19]). However, renal disease may be the initiating clinical manifestation of essential mixed cryoglobulinaemia. In 23 of our 29 patients, renal signs appeared at an average of 55.2 months after the onset of the disease, whereas in the other six they were the presenting symptom. Any of the three main clinical syndromes typical of glomerular diseases may characterise at presentation the renal disease which accompanies essential mixed cryoglobulinaemia [11,19] :

1. An acute nephritic syndrome, defined by severe proteinuria, haematuria usually macroscopic, hypertension and sudden rise in serum creatinine is found in 40–50 per cent of cases (55% in our population); it is sometimes (about 5% of cases) complicated by an acute oliguric renal failure.

TABLE II. Clinico-histological correlations for 28 cases of type 2 IgG–IgM essential mixed cryoglobulinaemia

CLINICAL AND HISTOLOGICAL FEATURES AT THE TIME OF BIOPSY								Duration of post-biopsy follow-up (months)	Laboratory data at the last examination	
No. of patients	Pattern of deposition by IF	Features by light microscopy	Clinical syndrome	Interval from onset (months)	Laboratory data				Serum creatinine (mg/24hrs)	Proteinuria (g/24hrs)
					Cryocrit (%)	Serum creatinine (mg/100ml)	Proteinuria (g/24hrs)			
9	Massive thrombi + faint segmental subendothelial	Severe proliferation + massive thrombi M/G=50 vasculitis in 6 cases	Acute nephritic (9 cases)	1.8±1.3	13.8±9.3	2.8±2.5	3.6±5.5	37.1±29.3	1.5±0.5	0.47±0.8
5	Faint segmental subendothelial without thrombi	Mild proliferation without massive thrombi M/G=20 vasculitis in 1 case	Acute nephritic (4 cases)	3.6±3.0	9.8±4.5	1.5±0.8	0.4±0.5	15.4±14.3	1.3±0.5	0.60±0.8
7	Intense diffuse subendothelial	Severe proliferation without thrombi M/G=35 vasculitis in 1 case	Nephrotic (3 cases) urinary abnormalities (3 cases)	25.0±26.0	21.3±19.5	1.6±0.5	4.0±2.5	23.3±22.1	1.4±0.6	0.9 ±0.6
5	Intense diffuse subendothelial with lobular pattern	Lobular membrano-proliferative GN M/G=16 vasculitis in 2 cases	Variable (mainly nephrotic)	55.4±62.6	25.2±21.2	2.0±0.6	1.8±1.2	33.6±22.1	2.2±1.6	1.1±1.3
2	Faint segmental subendothelial	Mild mesangial GN M/G=0 No vasculitis	Urinary abnormalities	1–24	5.5–10.0	0.8–1.5	0.3–0.5	25–66	1.1–1.4	0.2–0.3

M/G=average number of monocytes per glomerulus; GN=glomerulonephritis

2. A nephrotic syndrome, sometimes more humoral than clinical, is found in 20 per cent of cases (17% in our population).

3. An isolated proteinuria and haematuria is found in the remaining patients (28% in our population).

The first two syndromes present with overlapping features in some cases, when an acute exacerbation is heralded by an abrupt increase in proteinuria and a moderate increase in serum creatinine, without macroscopic haematuria. Table II shows that an acute nephritic syndrome was reported by all the nine patients with severe endocapillary proliferation and monocyte infiltration, massive intraluminal thrombi and an immunofluorescent pattern of faint segmental parietal deposition, and also by four of the five patients with the same pattern of parietal deposition, but without thrombi and with a less severe proliferation and accumulation of monocytes.

Patients of both subgroups were biopsied only few months after the onset of renal symptoms, but those without thrombi and less severe exudative lesions were biopsied later (on average, 3.6 months instead of 1.8 months). It is possible that even in the latter subgroup of patients the nephritic syndrome was due to the sudden accumulation of cryoglobulins in the capillary lumen, but intraluminal thrombi were not evident any more because they had already been eliminated by the accumulated macrophages, or because they had decreased and were not represented in the biopsy specimen.

On the contrary, proteinuria, often in the nephrotic range, was the presenting renal sign in the majority of the seven patients with a severe proliferative-exudative pattern but without thrombi and in the five patients with lobular mesangio-capillary glomerulonephritis; in both subgroups diffuse parietal staining was seen by immunofluorescence and renal symptoms had been present for a long time when biopsy was carried out (Table II).

We suggest that an acutely presenting renal disease, probably due to massive precipitation of cryoglobulins in the capillary lumen, without diffuse subendothelial trapping of circulating cryoglobulins, develops in an early phase of renal involvement in some patients with essential mixed cryoglobulinaemia, while a more chronic clinical syndrome sluggishly develops in some other patients, or in the same patient in a later phase. The latter one is associated with a more marked trapping of circulating cryoglobulins on the inner aspect of the glomerular basement membrane, similar to that which results from local trapping of circulating immunocomplexes in diffuse proliferative lupus nephritis.

The kidney disease in essential mixed cryoglobulinaemia has a variable course. The data in the literature, recently reviewed by Tarantino and Ponticelli [29], show that about 30 per cent of cases have complete or partial remission of renal symptoms. In another 30 per cent the course of the disease is rather indolent, and does not progress to renal insufficiency for several years, in spite of the persistence of urinary abnormalities. About 20 per cent of patients experience reversible clinical exacerbations, such as nephrotic or acute nephritic syndromes, during the follow-up period. Table III shows that changes in the histological and immunohistological features may correspond with these changes in the clinical manifestations. In patient L.E., who died because of massive

TABLE III. Changes in the histological features in six patients with type 2 essential mixed cryoglobulinaemia (re-biopsy or post-mortem examination)

Name		Interval (months)	Histological pattern	Immunohistological pattern
L.E.	1st biopsy		SDPE + T + V	Segm. subendothelial + T
	Autopsy	1	SDPE + V	–
A.M.	1st biopsy		SDPE + T + V	Segm. subendothelial + T
	Autopsy	5	MDPE + V	–
C.M.	1st biopsy		SDPE + T	Segm. subendothelial + T
	2nd biopsy	10	Mesangial	Segm. subendothelial
M.P.	1st biopsy		SDPE + T + V	Segm. subendothelial + T
	2nd biopsy	30	SDPE + V	Intense diff. subendothelial
A.A.	1st biopsy		SDPE + V	Intense diff. subendothelial
	Autopsy	3	Mesangial + V	–
P.L.	1st biopsy		MDPE	Negative
	2nd biopsy	10	SDPE + V	Intense diff. subendothelial

SDPE=severe diffuse proliferative-exudative glomerulonephritis; MDPE=mild diffuse proliferative-exudative glomerulonephritis; T=thrombi; V=vasculitis

gastrointestinal bleeding complicating a systemic vasculitis only one month after the first biopsy, post-mortem examination showed complete disappearance of the intraluminal thrombi, without any change in the extent of the proliferative-exudative and of the vasculitic lesions. In three other patients (A.M., C.M. and M. P.) second biopsies or post-mortem examinations, 5, 10 or 30 months after the first biopsy, also demonstrated the complete disappearance of intraluminal thrombi. In one of them (M.P.), proliferative-exudative changes were still very severe after 30 months, but the pattern of subendothelial immunofluorescent staining had shifted from faint segmental to intense diffuse. In another case (A.A.), disappearance of exudative features was total after 10 months, leaving only a mild mesangial proliferation. Finally, in patient P.L., aggravation of the histological and immunohistological lesions was documented in a second biopsy, performed because of clinical exacerbation with a sudden increase of proteinuria to nephrotic levels.

A moderate degree of renal insufficiency is rather frequent long after the onset of the renal disease in essential mixed cryoglobulinaemia. However, severe chronic renal failure is a less common complication than was believed in the past. Only 12 of the 102 patients reviewed by Tarantino and Ponticelli [29] developed chronic uraemia, usually several years after the onset of renal symptoms, and only three of them required maintenance dialysis.

Our data (Table II and Figure 6) confirm this. Recurrence of an acute nephritic syndrome or appearance of a nephrotic syndrome was observed in five patients during the post-biopsy follow-up. After an average total clinical follow-up of 50.4 ± 44.5 months (2–200) from the onset of renal symptoms, 13 of the 29

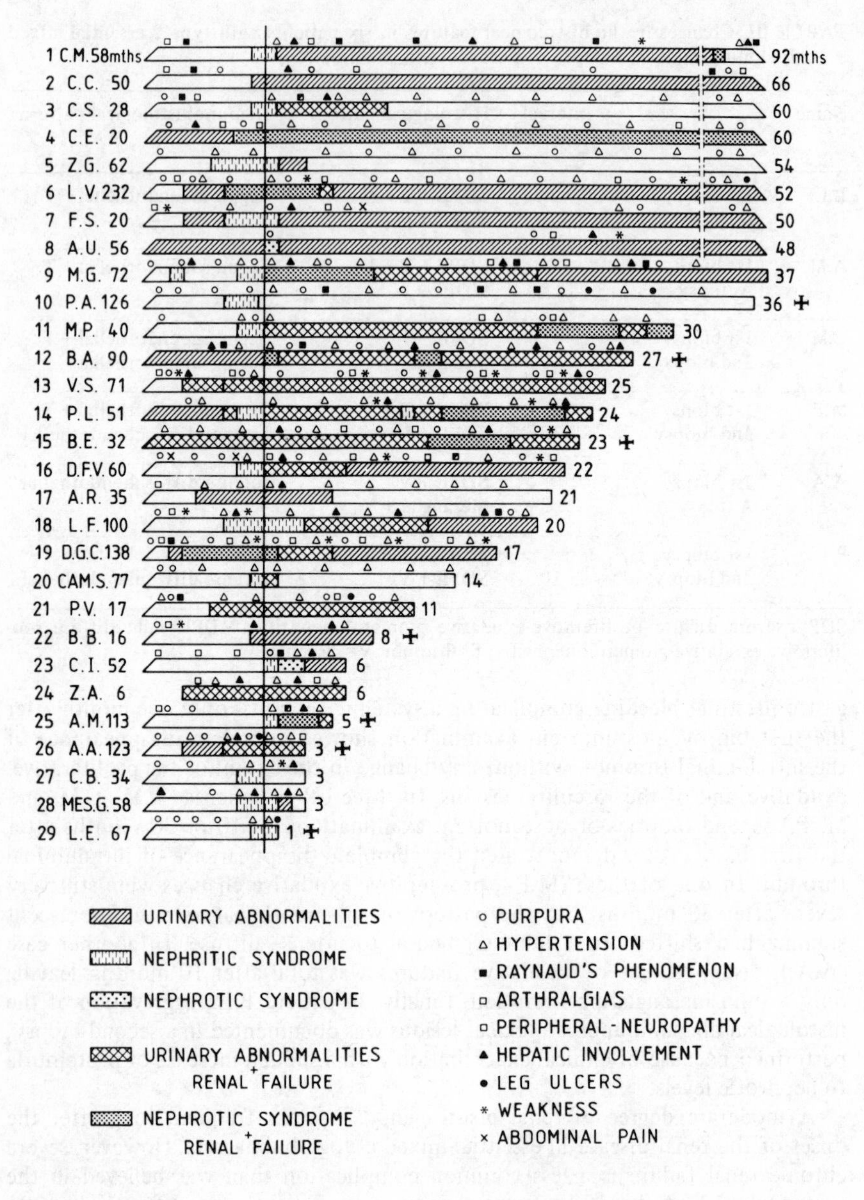

Figure 6. Clinical course of renal disease, and occurrence of extrarenal symptoms during the follow-up, in 29 patients

patients have abnormally increased serum creatinine, quite moderate (<2mg/100ml/ in nine patients, between 2 and 3mg/100ml in three patients, and >3mg/100ml in only one patient, who increased slowly from 3mg/100ml to 5mg/100ml in 27 months, at which time he died for unknown reasons. No

TABLE IV. Clinical and serological features of essential mixed cryoglobulinaemia in the different series reported in the literature

	Series of unselected cases					Series selected for the presence of renal involvement		
	Invernizzi et al [13]	Gorevic et al [11]	Popp et al [14]	Migliorini et al [15]	Realdi et al [10]	Cordonnier et al [17]	Tarantino et al [18] *	Our results
Number of patients	54	40	12	36	25	18	44	29
Renal disease %	20	55	58	33	40	100	100	100
Purpura %	100	100	75	83	80	61	59	89
Leg ulcers %	9	30	–	–	–	–	5	17
Raynaud's phenomenon %	14	25	–	–	–	33	7	34
Arthralgias %	–	73	50	97	84	50	57	72
Sjogren's syndrome %	–	15	–	30	–	–	2	0
Peripheral neuropathy %	2	20	–	50	36	11	–	10
Hepatic involvement %	–	70	16	64	100	28	52	72
Chronic hepatitis %	6	32	8	39	84	11	14	17
Serum HBsAg positivity %	–	10	0	6	12	–	7	0
Serum HBsAb positivity %	–	41	0	28	20	–	–	30

* At first admission

patient has developed severe chronic uraemia requiring maintenance dialysis. Seven patients have died. In three of them death was from clinical complications due to severe systemic vasculitis. Myocardial infarction, cancer of the larynx and acute peritonitis were responsible for another three deaths. The cause of death for the remaining case is unknown.

Arterial hypertension is an early and very common clinical feature [11,18]. It was present in 27 of our 29 patients. It is often severe and difficult to control, even with vigorous therapy, and may have an accelerated course in some patients. No difference can be found in the incidence of the most typical extrarenal symptoms, between populations of patients with renal disease and non-selected populations of patients with essential mixed cryoglobulinaemia (Table IV). Liver involvement is rather common, even in patients without signs of previous viral hepatitis. The positive markers for hepatitis and in particular for HBsAg have variable incidences in the different surveys. In three recent series [9,30,31] a high prevalence of HBs viral antigens and anti-HBs antibodies was observed in both serum and cryoprecipitates of patients with essential cryoglobulinaemia. Moreover, HBs particles were seen on electronmicroscopy of some cryoprecipitates [9]. Staining with anti-HBsAg antisera has also been found by immunofluorescence in the glomeruli of some patients, but we demonstrated that this is a non-specific staining, due to some interference with the antiglobulin IgM deposits present on the capillary walls, and that it disappears after the inactivation of rheumatoid factor [32]. Many other series, including ours, do not confirm such a high prevalence of the markers of viral hepatitis (Table IV). We think that these differences are due to the criteria used to recruit the patients with essential mixed cryoglobulinaemia and to differences in the incidence of subclinical infection in the different geographical areas. The concentration of circulating cryoglobulin is usually very high ($>$1g/L), with cryocrits always $>$2 per cent and up to 60–70 per cent. This varies very much in a given patient during the course of the disease. No correlation can be found between cryocrit and degree of activity of the disease, although a decrease in cryoglobulin concentration has been claimed to have corresponded with clinical improvement in some patients [33,34]. Serum titres of IgM rheumatoid factor are also frequently elevated, without any correlation with the severity of the renal disease.

The serum complement pattern is rather peculiar in essential mixed cryoglobulinaemia-glomerulonephritis (Table V). Early complement components (C_{1q}, C_4) and CH_{50} are usually strongly depressed, while C_3 is slightly but significantly reduced and later components (C_5 and C_9), C_3PA and C_1INH tend to be higher than in normal controls [18,35,36]. The C_3 breakdown product C_{3d} is often increased. In other words, there is activation of complement through the classical pathway, with further activation of complement factor C_3, but not sufficient to activate the membrane attack complex and to reduce the serum concentration of the late components.

This pattern does not change very much with changes in clinical activity of the disease. Early complement components are persistently low, whatever the degree of activity of the systemic and renal disease. This is in contrast with the trend to normalisation of the complement profile during the phases of clinical inactivity which is typical of the majority of hypocomplementaemic renal diseases.

TABLE V. Repeated determinations of serum complement components in 29 patients with type 2 IgG-IgM essential mixed cryoglobulinaemia and in normal subjects

	Patients with essential		Normal controls			
	No. of samples	Mean ± SD	No. of samples	Mean ± SD	Range	p
CH_{50}	220	11.6±7.1U/ml	54	22.0±3.7U/ml	14.5–29.5U/ml	<0.001
C_1q	314	46.4±32.1%†	54	92.7±14.0%†	64–120%	<0.001
C_4	314	23.4±17.5%†	54	99.5±28.4%†	43–156%	<0.001
C_3	314	78.3±24.7%†	54	102.5±18.8%†	65–140%	<0.001
C_3d	90	48.7±18.9%*	32	19.0±4.5%*	10–28%	<0.001
C_5	90	110.0±36.4%†	32	98.5±15.2%†	68–129%	NS
C_9	90	156.3±52.8%	32	97.5±30.7%†	36–159%	<0.001
C_3PA	90	92.7±40.7%†	32	100.0±18.5%†	63–137%	NS
C_1INH	90	127.2±35.5%†	32	91.5±12.7%†	66–117%	<0.001

† % of a normal serum pool

* % of a calibrated serum pool previously activated

This could be explained by assuming reduced synthesis of the early complement components [37] instead of their increased consumption. Alternatively, a block of the classical pathway C_3 convertase could explain it, which has recently been suggested by Gigli's groups [38]. These investigators demonstrated that purified cryoglobulins were able to activate C_4 and C_2 and to initiate the formation of C_{42} enzyme in normal human serum but were unable to cause the subsequent C_3–C_9 activation, because of their effect on the C_4-binding protein, a control protein which is important for the modulation of the classical pathway C_3 convertase [39]. Cryoglobulins would impair the action of the classical pathway C_3 convertase just as the immunoglobulin called C_3NeF impair the action of the alternate pathway C_3 convertase in idiopathic membrano-proliferative glomerulonephritis.

Treatment

It is still a matter of controversy whether or not, and if so, when and how to treat the renal and extrarenal manifestations of essential mixed cryoglobulinaemia. Two types of therapeutic approach can be justified theoretically:

1. Use of *corticosteroids* and *cytotoxic drugs*, to inhibit the activated clones of B lymphocytes which produce the antiglobulin antibodies and to block the mediators of inflammation.
2. Unselective *(plasma exchange)* or selective *(cryopheresis)* removal of circulating cryoglobulins.

Corticoids and cytotoxic agents (in particular cyclophosphamide or chlorambucil) have been extensively used, especially for treatment of acute renal and extrarenal exacerbations [40–45], with allegedly good results in some cases

[40,41], but none has been thoroughly evaluated in controlled studies. In the most acute phases of the disease, especially when signs of systemic vasculitis and/or acute oliguric renal failure are present, short-term *high dose methylprednisolone pulses* have also been used [18], followed by a maintenance regimen of small doses of steroids or cytotoxic agents for a few months. According to our experience clinical symptoms reverse and renal function improves in most patients a few days after the administration of cytotoxic drugs combined with steroids, which we usually also combine with plasma exchange. Protracted immunosuppression with maintenance doses of cyclophosphamide is used by us with caution, since patients with essential cryoglobulinaemia are prone to infection.

Partial removal of circulating cryoglobulins through plasma exchange is theoretically a useful therapeutic approach for many reasons:

1. It has been demonstrated [34] that the concentration of circulating cryoglobulins raises the temperature at which cryoprecipitation may occur in the small visceral vessels, including renal arterioles and glomerular capillaries.

2. It is now well established that this local precipitation is responsible for the renal damage in essential mixed cryoglobulinaemia.

3. High levels of circulating cryoglobulins, probably because of a chronic overload of the intake mechanisms, induce saturation of reticulo-endothelial function [46]. Partial removal with plasma exchange, by reducing this overload, can restore reticulo-endothelial function [47], with consequent temporarily increased removal of the remaining circulating cryoglobulins.

Many investigators have reported good results with these procedures, which are becoming accepted as routine treatment of acute clinical manifestations of essential mixed cryoglobulinaemia–glomerulonephritis, usually as an adjunct to the pharmacological therapy with steroids and/or cytotoxic drugs [47–53]. However, their efficacy is difficult to prove and is still controversial, for the following reasons:

1. Acute exacerbation of renal damage in essential mixed cryoglobulinaemia is a totally or partially reversible phenomenon, even without any therapeutic regimen.

2. Plasma exchange on cryopheresis is usually combined with cytotoxic drugs, to avoid the rebound increase in circulating cryoglobulins after their removal [54] and it is therefore very difficult to know which part of the treatment was responsible for any improvement.

3. Measurement of circulating cryoglobulins is unreliable for monitoring the efficacy of plasma exchange since there is only an approximate correlation between the serum cryoglobulin level and the presence and severity of the signs of the renal disease. One can only assume that reduction of these circulating immune complexes in symptomatic patients will reduce their local damaging effects.

4. There are still no controlled trials on the efficacy of plasma exchange in essential mixed cryoglobulinaemia–glomerulonephritis.

We treated 15 patients with essential mixed cryoglobulinaemia–glomerulonephritis by plasma exchange, because of clinical flare-ups characterised in all cases by acute exacerbation of extrarenal signs and deterioration of renal function of rather rapid appearance, associated with acute nephritic syndrome or acute renal failure in 10 cases and with proteinuria which increased to more than 2g/24 hours in five and reached the nephrotic range in two. Clinical signs of systemic vasculitis were also present in 10 cases. The cryocrits were highly variable at the beginning of treatment. Serum C_4 levels were very low in the majority of cases, while C_3 levels were abnormally low in only six of them.

Severe diffuse proliferative-exudative lesions were documented in 10 cases, with massive intraluminal thrombi in seven, while mild proliferative-exudative glomerulonephritis without thrombi was present in two cases and a lobular pattern in the remaining three cases. Renal vasculitis, always associated with clinical signs of systemic vasculitis, was also present in eight patients. Steroids were also given to 14 of the 15 patients and were administered at the beginning as intravenous pulses (3 in the first 6 days) to three of them. Cyclophosphamide was also administered to all cases at an average dose of 3mg/kg body weight. Both intermittent flow cell separator and surface plasma filters were used for the plasma exchange, as already described [55]. The number of plasma exchanges in the acute stage ranged from 4 to 39, and was 13 as the average.

TABLE VI. Changes in some laboratory parameters after combined plasmapheretic and pharmacological treatment of 15 patients with type 2 IgG-IgM essential mixed cryoglobulinaemia

	Before treatment	After treatment	p
Cryocrit (%)	23.5±14.0	7.0±4.7	<0.001
Serum creatinine (mg/100ml)	2.86±1.94	1.66±0.47	<0.01
Proteinuria (g/24hrs)	5.5±5.3	1.7±2.6	<0.01
Serum C_3 (% of normal human serum pool)	74.2±26.1	85.3±33.8	NS
Serum C_4 (% of normal human serum pool)	9.4±8.6	12.3±12.0	NS

As Table VI shows, the mean cryocrit, serum creatinine and proteinuria decreased significantly.

On the contrary, in accordance with the observation of others [46,51], mean C_3 and C_4 levels did not change significantly. It is difficult to say whether this clinical improvement was influenced by the plasma exchange. It is our experience, confirmed by others, that regression of the signs of clinical exacerbation, with reduction of circulating cryoglobulins, serum creatinine and proteinuria, may take place in patients treated only with cytotoxic drugs and steroids (especially when methylprednisolone pulse therapy is used), and even in patients not treated at all. A controlled trial must be carried out, but until the results of it are available, we think plasma exchange should be added to pharmacological treatment when acute exacerbation of extrarenal and renal signs is accompanied by an abrupt rise in serum creatinine and/or proteinuria, especially when signs of systemic vasculitis are present.

Another controversial application of plasmapheresis in essential mixed cryoglobulinaemia is to use periodical intermittent plasma exchange as a long-term treatment, to prevent acute exacerbations of renal as well as of other visceral and cutaneous signs of the disease [52,56]. Since it has been demonstrated that cryoglobulin concentration increases to the pre-plasma exchange levels over a period of time which, although variable, is often rather long, this therapeutic approach is theoretically correct. It has been widely seen, also by us, that weekly or bi-monthly removal of circulating cryoglobulins through plasma exchange, especially for patients also treated with a maintenance dose of an immunodepressive drug to slow down the rebound of these substances, controls the systemic signs of the disease for long periods of time quite well. However, all investigators with a large experience in this field know very well that such a long-lasting, symptomless, sluggish clinical history is also rather common for essential mixed cryoglobulinaemia patients not treated at all, independently of the levels, sometimes very high, of circulating cryoglobulins. A controlled trial is probably warranted. In the absence of significant controlled data, and in view of the elevated cost and the technological complexity of plasma exchange, we think that this treatment should be used on a long-term basis only when a prolonged period of clinical activity has been incompletely controlled with the usual time-limited treatment of the acute phase which combines the pharmacological and the plasmapheretic approaches.

Pathophysiology of the renal damage

The aetiology of the pathogenesis of cryoglobulin formation and precipitation, as well as the pathogenetic mechanisms of tissue damage in essential mixed cryoglobulinaemia still remain unclear [3,57].

Type II cryoglobulins, which cause the renal lesions that we have described, are probably produced as a consequence of an immunoproliferative disorder of IgM producing B-lymphocytes that favours increased production, by selected clones, of monoclonal IgMk with rheumatoid factor antibody activity versus polyclonal IgG, without signs of lymphoplasmacytic neoplasms.

It is still debated whether or not this IgG is always bound to as yet unidentified antigens, probably viral antigens, to form immune complexes, so that monoclonal IgM acts as an anti-immune complex antibody. As we have said before, HBs could be such an antigen in a minority of patients with essential mixed cryoglobulinaemia. Antigens related to coccidiodomycosis were suspected to be the initiating antigens in a single case [43].

It is also a matter of speculation whether or not the renal lesions are produced in all cases because of an immunologically mediated mechanism, with cryoglobulins acting at the glomerular level as trapped immune complexes that bind complement. Alternatively, it has been postulated that local precipitation can take place because of less specific physico-chemical mechanisms, as a consequence of higher endocapillary cryoglobulin concentration induced by the filtration process at the glomerular level.

As we said, this second mechanism seems to be at work in cases in whom massive intraluminal accumulation of cryoglobulins is found histologically, together with an immunohistological pattern of almost exclusively intraluminal staining of thrombi. On the contrary, the immunohistological pattern of intense and diffuse subendothelial staining with anti-IgG, anti-IgM and anti-C_3 antisera suggests an immunological mechanism of trapping similar to that which is characteristic of lupus glomerulonephritis. The involvement in this local subendothelial trapping of circulating immune complexes other than circulating IgG-IgMk cryoglobulins can not be excluded. Such circulating immune complexes have been found by some investigators [52,58].

However, whatever is the prevailing mechanism for their deposition in the glomerulus, locally trapped cryoglobulins trigger endocapillary accumulation of activated monocytes from the blood stream. These cells act as scavenger cells, but probably also as mediators of the local damage, through liberation of lytic enzymes and other humoral factors able to induce proliferation of resident cells [59–61].

The renal disease in essential mixed cryoglobulinaemia may be characterised by time-limited, sometimes recurrent bouts of intense cryoglobulin precipitation in the glomerulus, which give rise to reversible acute clinical exacerbations (acute nephritic syndrome or sudden increase of proteinuria), sometimes with temporary deterioration of renal function, and the histological features of an acute endocapillary proliferative-exudative glomerulonephritis, which is also reversible when the accumulated monocytes have been disposed of. For as yet unknown reasons, the more chronic discrete deposition of cryoglobulins (and possibly of other circulating immune complexes) on the subendothelial aspect of the glomerular basement membrane does not induce the marked chronic activation and peripheral circumferential expansion of mesangial cells and matrix which are typical of idiopathic membrano-proliferative glomerulonephritis. Mesangial sclerosis and progressive irreversible impairment of renal function are indolent inconstant phenomena in essential mixed cryoglobulinaemia –glomerulonephritis, even when no treatment is given.

The damage induced by the precipitation of cryoglobulins in the arterioles, leading to a vasculitis usually systemic, frequently has a poorer prognosis. However, even in this circumstance severe impairment of renal function is not frequent, and the causes of death are more often extrarenal complications.

Acknowledgments

The study was supported by Consiglio Nazionale delle Ricerche Grant No. 83.02.898.04. We thank Professor C Schmid and Professor M Bestetti Bosisio of the Department of Pathology of the S Carlo Hospital, for their valuable advice.

References

1 Brouet JC, Clauvel JP, Danon F et al. *Am J Med 1974; 57:* 775
2 Meltzer M, Franklin EC. *Am J Med 1966; 40:* 828

3 Grey HH, Kohler PF. *Semin Hemathol 1973; 10:* 87
4 Franklin EC, Frangione B. *J Immunol 1971; 107:* 1527
5 Wager O, Mustakallio KK, Rasanen JA. *Am J Med 1968; 44:* 179
6 Goldman M, Renversez JC, Lambert PH. *Springer Semin Immunopathol 1983; 6:* 33
7 Warner NL, Mackenzie MR, Fudenberg HH. *Proc Nat Acad Sci USA 1971; 68:* 2846
8 Winfield JB, Koffler D, Kunkel HG. *J Clin Invest 1975; 56:* 563
9 Levo Y, Gorevic PD, Kassab HI et al. *N Engl J Med 1977; 296:* 1501
10 Realdi G, Bortolotti F, Alberti A et al. In Chiandussi L, Bartoli E, Sherlock S, eds. *Systemic Effects of HbsAg Immune Complexes.* Padova: Piccin. 1980: 242
11 Gorevic PD, Kassab HJ, Levo Y et al. *Am J Med 1980; 60:* 287
12 Gorevic PD, Franklin EC. In Zabiskie JB, Fillit H, Villareal H Jr, Lovell Becker, E, eds. *Clinical Immunology of the Kidney.* New York: J Wiley. 1982: 317
13 Invernizzi F, Pioltelli P, Cattaneo R et al. *Acta Haematol 1979; 61:* 93
14 Popp JW Jr, Dienstag JL, Wands JR, Bloch KJ. *Ann Int Med 1980; 92:* 379
15 Migliorini P, Bombardieri S, Castellani A, Ferrara GB. *Arthritis Rheum 1981; 24:* 932
16 Monteverde A. *Le crioglobulinemie miste. Atti dell.84° Congr Soc It di Medicina Interna.* Roma: Pozzi Ed. 1983: 357
17 Cordonnier D, Vialtel P, Renversez JC et al. In Hamburger J, Crosnier J, Funck Brentano JL, eds. *Actualités nephrologiques de l'Hôpital Necker.* Paris: Flammarion. 1982: 49
18 Tarantino A, De Vecchi A, Montagnino G et al. *Q J Med 1981; 50:* 1
19 Meltzer M, Franklin EC, Elias K et al. *Am J Med 1966; 40:* 837
20 Morel-Maroger L, Mery JP. In *Proc Fifth Congr Nephrol.* Basel: Karger. 1974: 173
21 D'Amico G, Ferrario F, Colasanti G et al. *La Ricerca Clin Lab 1980; 10:* 59
22 Barbiano di Belgioioso G, Tarantino A, Colasanti G et al. In Leaf A, Giebisch G, Bolis L, Gorini S, eds. *Renal Pathophysiology. Recent Advances.* New York: Raven. 1980: 245
23 Maggiore Q, Bartolomeo F, L'Abbate A et al. *Kidney Int 1982; 21:* 387
24 Cordonnier D, Martin H, Groslambert P et al. *Am J Med 1975; 59:* 867
25 Feiner H, Gallo G. *Am J Pathol 1977; 88:* 145
26 Monga G, Mazzucco G, Coppo R et al. *Virchows Archiv für Zell-pathologie 1976; 20:* 185
27 Monga G, Mazzucco G, Barbiano di Belgioioso G, Busnach G. *Am J Pathol 1979; 94:* 271
28 Ferrario F, Castiglione A, Colasanti G et al. *Contr Nephrol 1984; 45:* in press
29 Tarantino A, Ponticelli C. In Bacon PA, Hadler NM, eds. *The Kidney and Rheumatic Diseases.* London: Butterworth. 1982: 128
30 Bombardieri S, Paoletti P, Ferri C et al. *Am J Med 1979; 66:* 748
31 Garcia-Bragado F, Vilardell M, Fonollosa V et al. *Nouv Presse Med 1981; 10:* 2955
32 Maggiore Q, L'Abbate A, Bartolomeo F et al. In Chiandussi L, Bartoli E, Sherlock S, eds. *Systemic Effects of HbsAg Immune Complexes.* Padova: Piccin. 1980: 242
33 Goldberg LS, Barnett EV. *Arch Int Med 1970; 125:* 145
34 Lockwood CM. *Kidney Int 1979; 16:* 522
35 Linscott WD, Kane JP. *Clin Exp Immunol 1975; 21:* 510
36 Tarantino A, Anelli A, Costantino A et al. *Clin Exp Immunol 1978; 32:* 77
37 Ruddy S, Carpenter GB, Chin KW et al. *Med (Baltimore) 1975; 54:* 165
38 Haydey RP, Patarroyo de Royas M, Gigli I. *J Invest Derm 1980; 74:* 328
39 Gigli I, Fujita T, Nussenzweig V. *Proc Nat Acad Sci USA 1979; 76:* 6596
40 Brody JI, Samitz MH. *Am J Med 1973; 55:* 211
41 Ristow SC, Griner PF, Abraham GN, Shoulson I. *Arch Int Med 1976; 136:* 467
42 Skrifvars B, Tallovist G, Törnröth T. *Acta Med Scand 1973; 194:* 229
43 Gamble CN, Ruggles SW. *N Engl J Med 1978; 299:* 81
44 Zlotnick A, Slavin S, Eliamkin M. *Isr J Medical Sci 1972; 8:* 1968
45 Reza MJ, Roth BE, Popps MA, Goldberg LS. *Ann Int Med 1974; 81:* 632
46 Hamburger MJ, Gorevic PD, Lawley TJ et al. *Trans Ass Am Physic 1979; 62:* 104
47 Pusey CD, Schifferli JA, Lockwood CM, Peters DK. *Artif Organs 1982; 5 (suppl):* 183

48 Houwert DA, Hené RJ, Struyvenberg A, Kater L. *Proc EDTA 1980; 17:* 650
49 Berkman EM, Orlin JB. *Transfusion 1980; 20:* 171
50 McLeod BC, Sassetti RJ. *Blood 1980; 55:* 866
51 Geltner D, Kohn RW, Gorevic O et al. *Arthritis Rheum 1981; 24:* 1121
52 Bombardieri S, Maggiore Q, L'Abbate A et al. *Plasma Ther 1981; 2:* 101
53 Maggiore Q, L'Abbate A, Caccamo A et al. *Artif Organs 1981; 5 (suppl):* 126
54 L'Abbate A, Maggiore Q, Caccamo A et al. *Int J Artif Organs 1983; 6 (suppl):* 51
55 Sinico R, Fornasieri A, Fiorini G et al. *Int J Artif Organs 1983; 6 (suppl):* 21
56 Bombardieri S, Ferri C, Paleologo G et al. *Int J Artif Organs 1983; 6 (suppl):* 47
57 Tarantino A, Montagnino G. In Bertani T, Remuzzi G, eds. *Glomerular Injury 300 Years after Morgagni.* Milano: Wichtig. 1983: 245
58 Lawley TJ, Gorevic PD, Hamburger MI et al. *J Invest Derm 1980; 75:* 297
59 Unanue ER. *Am J Pathol 1976; 83:* 396
60 Nathan CF, Murray HW, Cohen ZA. *N Engl J Med 1980; 803:* 622
61 Dubois CH, Foidart JB, Hautier MB et al. *Eur J Clin Invest 1981; 11:* 91

Open Discussion

CAMERON (London) I would like to ask three related questions about hypertension. First there seems a tremendous contrast between the incidence of severe hypertension in mixed cryoglobulinaemia compared with lupus and with polyarteritis. Have you any idea what is the cause of the hypertension in this disease? Secondly you mention that it seems that renal failure is perhaps less common than it used to be. Is this perhaps the result of hypertension? What role do you think hypertension plays in the genesis of the renal failure? Finally, a very interesting observation by Kincaid-Smith some years ago was that in severe hypertension complicating nephritis, plasmapheresis itself sometimes renders the hypertension amenable to treatment. I wondered if you had ever seen anything like this in your patients with mixed cryoglobulinaemia?

D'AMICO I do not have a precise idea why hypertension is so common in this disease. I do not think it is a problem of glomerulosclerosis. What we find more frequently than in other diseases are interstitial lesions (because the phenomenon of monocyte infiltration is also at the interstitial level sometimes) and the vascular lesions. Nephrosclerosis was of the vascular type and our idea is that the cause of hypertension and renal insufficiency, although it is just a suggestion, can be the combined effect of the vascular and of the interstitial lesions. Vascular lesions of the vasculitic type, as I said, are very frequent in this disease in the small and average size arterioles and can also be the cause of some sclerosis. What we want to stress is that we do not find mesangial sclerosis as commonly as in other diseases of the same type due to trapping of immune complexes. As for the third question, I do not think we did notice this phenomenon with plasma exchange, but probably we did not check: I think we have to look at the behaviour of blood pressure after plasma exchange.

MAGGIORE (Reggio Calabria) What do you think is the mechanism for cryoglobulin precipitation in the kidney? Do you consider this a physical problem or an immunological problem?

D'AMICO I think that an immunological phenomenon takes place in the blood in any case of cryoglobulinaemia. We have demonstrated recently that the procoagulant activity of monocytes is increased in cryoglobulinaemic patients, which means that something is going on in the blood. What I want to say is that apart from the chronic precipitation of this cryoglobulin immune complexes a kind of precipitation due to physical reasons may take place acutely. The more massive precipitation triggers an immunological mechanism. Some physical mechanisms may give more acute damage to the kidney. The reasons by which this more acute precipitation in the kidney takes place are just a matter of speculation. Lockwood states that the temperature at which the precipitation takes place increases when the concentration increases,* and maybe at the glomerular level there is more concentration of protein in general for physical reasons (ultrafiltration).

SIMOES (Lisbon) Professor D'Amico, my first question is did your cases showing thrombi in the glomeruli have low platelet levels or did they have signs of disseminated intravascular coagulation? Secondly, have not some of these patients been exploited for the diagnosis of so-called mixed connective tissue disease?

D'AMICO As for the second question, the answer is no. In no case did we find this kind of association: we just selected cases where we were able during the follow-up to exclude that other diseases could be the cause of cryoglobulinaemia formation. For the first question, we saw the mechanism which could be considered intravascular coagulation in only one case.

PONTECHELLI (Milan) In still unpublished work we demonstrated that in mixed cryoglobulinaemia there is an acquired immune storage disease. This disorder seems to be related in some way to the exacerbations. Could you plan some trials about the possible role of anti-platelet agents in mixed cryoglobulinaemia and in particular about the possible value of prostacyclin?

D'AMICO I do not know. I do not think there is enough evidence at the moment, but I do not know of this new data of yours.

*Lockwood CM. *Kidney Int 1979; 16:* 522

Proc EDTA–ERA (1984) Vol 21

EFFECTS OF VARIOUS DIETS ON THE PROGRESSION OF HUMAN AND EXPERIMENTAL URAEMIA

C Giordano, G Capodicasa, N G De Santo

1st Faculty of Medicine and Surgery, University of Naples, Naples, Italy

Introduction

Recently there has been a renewed interest in the progression of renal diseases to renal failure. Among many factors this is due to the socioeconomic impact of uraemia therapy and to the scientific failure to develop proper preventive strategies during the last 25 years. The new interest arises from data gained in the field of clinical nutrition in uraemic patients [1–4] and in experimental uraemia [6–7].

In Brenner's view the mechanism by which a high protein intake accelerates the development and the severity of glomerular sclerosis has been related to increased glomerular perfusion and hyperfiltration [5,6]. Studies of fractional clearance of charged molecules demonstrates an alteration and a reduction in charge selective properties of glomerular capillaries in remnant kidneys. Since albumin is a polyanionic protein, defects in the charge selectivity will contribute to the development of albuminuria [7–9].

It is possible that changes in dietary proteins may affect plasma peptides, poly-L-aminoacids charged positively (polycations), and therefore, may contribute to the progression of renal disease by the neutralisation of anionic sites producing alterations in charge dependent permselectivity. The increased permeability of albumin may itself cause glomerular damage by mesangial overload [10–11]. There is therefore a need to reconsider nutritional therapy for chronic renal failure and to come to a definite conclusion about the possibility of taking advantage of Brenner's data in terms of nutrition.

The goal of this study is twofold: 1) to evaluate the effects of early low protein nutrition in the time course progression of renal failure, 2) to evaluate the effects of dietary proteins and urea or polyaminoacid (polycations) infusion on increased albumin excretion. Data will be presented indicating the feasibility of a strategy for the prevention of nephron loss in man since in experimental uraemia positively charged loads of polyaminoacids (homo copolymers) accelerate renal damage by albuminuria.

Methods

Early low protein nutrition

A group of 20 patients (20 males) with initial serum creatinine concentrations of 1.35–2.77μmol/L were studied. Impairment of renal function was due either to glomerulonephritis (11) cases, pyelonephritis (8 cases) or to type II diabetes mellitus (1 case). They underwent a nutrition protocol [1–4] providing a mean protein intake of 0.6g/kg associated with 160KJ/kg of energy, in so doing the diet was relatively high in phosphorus. Progression of renal failure was assessed by the method of Rutherford et al [12] correlating the time with the log 1/creatinine and evaluating each individual case by the tangent of the angle between the regression line of log 1/creatinine and the abscissa [4]. In this way we could calculate the creatinine doubling time (in years), that is the time corresponding to a 50 per cent destruction of nephron population available at the time treatment with low protein nutrition was started.

Experimental uraemia

All studies were performed on Morini Wistar rats weighing 200–250g. The rats were anaesthetised with sodium petobarbital (20–50mg/kg). An IV polyethylene catheter (PE 50) was introduced into a femoral vein. The bladder was catheterised with polyethylene tubing (PE 50). Glomerular filtration was measured by inulin clearance. The animals received an appropriate priming dose followed by a constant infusion in a solution containing 51mM of potassium chloride at a rate of 1.2ml/hr. Blood samples for clearance were drawn at the mid point of each collection period. Inulin was measured by the anthrone method [13]. Albuminuria was evaluated by standard quantitative radial immunodiffusion methods in normal, uninephrectomised and rats made uraemic by means of severe renal ablation, under different conditions: a) Switching from a standard diet (20% casein) to a low protein diet (5% casein), b) Under various systemic infusions in 0.9 per cent saline urea loads (from 5mg/g to 15mg/g) in a group of rats on low protein diet (5% casein), c) Under systemic intravenous infusions of Poly-L-Aminoacid homopolymers (Sigma Chem.) of different M.W at a dosage of 0.6mg/g BW and of Poly-L-Aminoacid copolymers (Sigma Chem.) at a dose of 500μg IV in a different group of rats on low protein diet (5% casein). Data are expressed as the means ± SD for each group and Student's 't' test for unpaired data was used for statistical analysis.

Results

Early low protein nutrition

Data in Table I show that the creatinine doubling time for patients 1–10 ranged between two and 19 years. In patients 11–13 the doubling time was eight, 13 and 14 years respectively. In patients 14–20 a remission is observed which persists for two to nine years.

TABLE I. Serum creatinine doubling time

Patient Number	Creatinine at start μmol/L	Doubling Time years
1	189	3
2	225	3
3	300	4
4	288	5
5	198	6
6	135	7
7	198	9
8	144	11
9	171	11
10	162	19
11	216	not yet after 8 years
12	235	not yet after 13 years
13	189	not yet after 14 years
14	216	remission lasting 2 years
15	270	remission lasting 2 years
16	252	remission lasting 3 years
17	252	remission lasting 4 years
18	243	remission lasting 4 years
19	198	remission lasting 7 years
20	171	remission lasting 9 years

Experimental uraemia

When renal mass is reduced GFR is not affected by a low protein diet while it is affected in the control animal ($p = 0.01$) indicating that at the end of the second week hypertrophy was progressing in the presence of a reduced renal mass (Table II). In control rats a low protien diet reduces albuminuria significantly ($p = 0.01$). This protective effect was better appreciated in the presence a reduced renal mass (Table II). In other words, in rats with reduced residual renal mass (uraemic and uninephrectomised) there is a significant change in the rate of albuminuria (from $34.7 \pm 5.7 \mu g/min$ to $6.3 \pm 1.4 \mu g/min$ in uraemic rats, $p = 0.001$; and from $19.4 \pm 9.0 \mu g/min$ to $4.8 \pm 1.6 \mu g/min$ in uninephrectomised rats, $p = 0.001$).

Data also indicate that albumin excretion remained unchanged in control animals as well as in uninephrectomised rats during infusions of 5mg/g urea (Table III).

In contrast to the data on urea loads, when we infused polycations such as polyaminoacid homopolymers (Poly-L-Histidine and Poly-L-Arginine) of different molecular weights (Table IV) the albumin excretion doubled in the control rats ($p = 0.001$) and increased 10 times or more ($p = 0.001$) in the uninephrec-

TABLE II. Inulin clearance and albumin excretion in experimental groups of rats by changing dietary protein intake (standard diet versus low protein diet). Data are mean ± SD. U_{ALB}, μg/min; C_{IN}, ml/min

		Experimental Periods					
		Standard Diet (20%) 1st week		Low Protein Diet (5%) 2nd week		P	
Groups	n	C_{IN}	U_{ALB}	C_{IN}	U_{ALB}	C_{IN}	U_{ALB}
Control	20	1.50±0.26	6.9±1.8	0.98±0.30	3.4±1.2	0.01	0.01
Uraemic	20	0.38±0.10	34.7±5.7	0.36±0.1	6.3±1.4	n.s	0.001
Uninephrectomy	20	0.60±0.24	19.4±9.0	0.50±0.20	4.8±1.6	n.s.	0.001

TABLE III. Effects of urea infusion on albumin excretion. Data are Mean ± SD. The number of animals participating in the study is given in parenthesis

Treatments	Albumin Excretion μg/min	
	LPD	Urea Infusion (5mg/g)
Control (20)	5.01 ± 1.9 (10)	5.8 ± 1.6 (9)
Uraemic (20)	7.2 ± 1.2 (9)	8.4 ± 1.4 (10)
Uninephrectomy (20)	6.2 ± 0.8 (8)	6.8 ± 1.2 (9)

TABLE IV. Effects of Poly-L-Aminoacids (homopolymers) infusion on albumin excretion. Data are mean ± SD. Poly-L-Aminoacids (homopolymers) were given at a dose of 0.6mg/g BW intravenously. Poly-L-Histidine MW 5000–15000. Poly-L-Arginine MW 15000–70000

		Albumin Excretion, μg/min					
Treatment	n	LPD (A)	Poly-L-Histidine (B)	Poly-L-Arginine	A vs B	B vs C	B vs C
Control	20	4.5±1.4	8.4±1.2	7.8±1.5	0.01	0.01	
Uraemic	20	6.8±1.3	77.2±5.7	52.4±4.7	0.001	0.001	0.01
Uninephrectomy	20	5.6±1.7	46.5±5.9	32.4±3.9	0.001	0.001	0.01

tomised and uraemic rats. This effect was greater the lower the molecular weight of the infused peptide. The same effect on albuminuria was obtained with a polycationic (copolymer) aminoacid (Table V).

Discussion

Early low protein nutrition represents a powerful tool in a preventive strategy for end-stage renal disease, a goal which has been missed during the last 20 years. Our previous experience [1,2] indicated that working with two groups of

TABLE V. Effects of Poly-L-Aminoacids (copolymer) infusion on albumin excretion. Data are means ± SD. Poly(Lys Hbr,Ala) 2:1 M.W 20000–50000 given at a dosage of 500μg IV

Treatment	LPD	Poly(Lys HBr,Ala) 2:1	p
Control	3.8 ± 2.0	9.2 ± 1.6	0.01
Uraemic	6.5 ± 1.4	70.8 ± 4.5	0.001
Uninephrectomy	5.2 ± 1.7	42.3 ± 3.6	0.001

uraemic patients matched for age, sex, aetiology and hypertension with a mean plasma creatinine of 225μM/L progression to terminal renal failure was complete in 16 months in patients on free diets while it took 7.6 years for patients on low protein nutrition.

The present data which was collected in a group of adult patients, excluding polycystic renal disease, are important for three reasons: 1) they indicate that a low protein nutrition is a preventive approach for the progression to end-stage renal disease when started at plasma creatinine concentrations greater than 180μM/L, 2) some patients enter a stage of spontaneous remission of renal damage lasting for years during low protein nutrition and 3) the effect of low protein nutrition is not linked to phosphate intake so that doubts are cast on the validity of data collected in uraemic rats [14] and man [15] treated with low phosphate diets. In other words working with diets containing a mean daily protein intake of 0.6g/kg it seems that it does not matter how much phosphate is taken with the protein suggesting a primary role for protein and not for phosphate intake.

Data in experimental uraemia allows three major conclusions 1) low protein diet reduces albuminuria in uraemic rats, uninephrectomised rats and in control rats, 2) a urea load infusion, at rates comparable to that generated by high protein feeding, does not affect albumin excretion [16], and 3) positively charged load of polyaminoacids (homo copolymers) increases protein excretion significantly. The role of sustained increments in glomerular pressure ($\triangle P$) and flows (Q_A) in the initiation and progression of glomerular sclerosis on the basis of Brenner's scheme [6] is now universally accepted because such haemodynamic factors are important. But this approach fails to fully explain many clinical and experimental findings including ours. In fact proteinuria is the determinant of glomerular sclerosis in oligomeganephronia [17] as well as of the mesangial overload in stages I to III of patients with diabetic nephropathy [18]. In addition our data on polyaminoacids are in keeping with results obtained under lysine infusion in normal man [19] showing dramatic increase of albuminuria without changes in GFR, RPF, $\triangle P$ and Kf. In these experiments the authors did not take into consideration the possibility that a lysine load results in a load of positive charges as was evident by the reduction of glomerular wall fixed negative charge density for albumin. Further to this, in experimental uraemia, proteinuria precedes the development of histological changes that are indistinguishable from changes seen in human focal glomerulosclerosis and the degree of proteinuria seems to correlate with the severity of glomerular sclerosis [20]. Finally, Kaysen and Kropp were able to show that a tryptophan enriched diet reduces albuminuria and proteinuria in seven of eight nephrectomised rats. The relevant aspect of this experiment had to be identified in the fact that the addition of tryptophan (neutral aminoacid) to a diet basically constituted of neutral aminoacids was associated with a reduction of positively charged aminoacids (basic aminoacids) such as arginine, histidine, lysine and hydroxylysine.

On these grounds it is evident that Brenner's hypothesis, in terms of permselectivity, gives relatively little space for other factors which also may be involved in explaining albuminuria. For this reason we take the liberty of redrawing Brenner's hypothesis by taking into account the data obtained from the present study (Figure 1) on the effect of protein feeding, inappropriate for

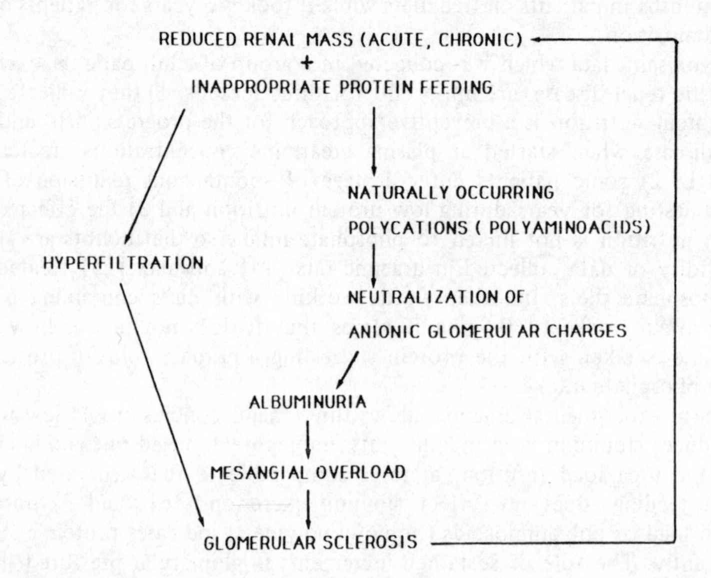

Figure 1. The role of neutralisation of anion glomerular sites by Poly-L-Aminoacids (polycation) in the progression of glomerularsclerosis by albuminuria

reduced renal mass, on negatively charged sites. Of course this approach may be considered complimentary to that studied in Brenner's laboratory. Our approach is confirmed by the data on urea load. When urea was infused at a rate equivalent to a protein intake similar to that obtained in switching from a low protein diet to a standard diet, albumin excretion was not affected significantly in all experimental groups. As a matter of fact albumin excretion did not change significantly even when large amounts of urea were given (up to 15mg/BW, a quantity of urea corresponding to that derived from a high protein diet (50% protein). These data cast some doubts on high protein diet driven haemodynamic effects.

Since polycations have been shown to induce albuminuria by blocking negatively charged sites it is possible to suspect that progression of renal disease might be linked to naturally occurring polycations generated in the digestion of protein diets if in large quantities and inappropriate for residual renal mass. Having indicated that positively charged polypeptides exert undue effects and may solicit damage to renal structure it would be very exciting to see whether negatively charged polyaminoacids may consistently show protective effects on the renal parenchyma. This aspect is currently under evaluation in our laboratory and looks most promising for the prevention of renal damage.

Acknowledgments

This work was financed by the Ministero della Pubblica Istruzione through funds made available to Programs of National and Local Interest.

References

1 Giordano C. In Giordano C, Friedman EA, eds. *Uremia, Pathobiology of Patients Treated for 10 Years or More.* Milan: Wichtig. 1981
2 Giordano C. *Proceedings of the 8th International Congress of Nephrology, Athens.* Basel: Karger. 1981: 71–82
3 Giordano C. *Kidney Int 1982; 22:* 401
4 Giordano C. *Kidney Int.* Submitted for publication
5 Brenner BM, Meyer TW, Hostetter TH. *N Engl J Med 1982; 307:* 652
6 Brenner BM. *Kidney Int 1983; 23:* 647
7 Chang RLS, Deen WM, Robertson CR, Brenner BM. *Kidney Int 1975; 8:* 212
8 Renke HG, Matel Y, Venkatachalam MA. *Kidney Int 1978; 13:* 324
9 Renke HG, Venkatachalam MA. *Kidney Int 1977; 11:* 44
10 Glasser RJ, Velosa JA, Michael AF. *Lab Invest 1977; 36:* 519
11 Velosa JA, Glasser RJ, Nevins TE, Michael AF. *Lab Invest 1977; 36:* 527
12 Rutherford WJ, Blondin J, Miller JP et al. *Kidney Int 1977; 11:* 62
13 White RP, Samson FC. *J Lab Clin Med 1954; 43:* 475
14 Ibels LS, Alfrey AC, Haut L, Huffer WE. *N Engl J Med 1978; 298:* 122
15 Maschio G, Oldrizzi L, Tessitore N et al. *Kidney Int 1982; 23:* 371
16 Osborne TB, Mendell LB, Parj EA, Winternitz MC. *J Biol Chem 1927; 71:* 317
17 McGraw M, Poucell S, Sweet J, Baumal R. *Int J Pediatr Nephrol 1984; 5:* 67
18 Royer P, Habib R, Le Clerc F. *Proceedings 3rd International Congress of Nephrology (Washington 1966).* Basel: Karger. 1967: 251–275
19 Bridges CR, Myears BD, Brenner BM, Deen WM. *Kidney Int 1982; 22:* 677
20 Shimamura J, Morrison AB. *Am J Pathol 1975; 79:* 97

Open Discussion

BERGSTROM (Chairman) Thank you very much for your very interesting paper. I would like to add to what Dr Maiorca said in the beginning that Carmelo Giordano belongs to the traditionalists of modern medicine. Now we have been presented with a new and very challenging hypothesis which adds another flavour to the Brenner hypothesis.

ABORAS (Jeddah) Thank you Professor Giordano for this very interesting lecture you have delivered. I would like to ask something about what you have named 'spontaneously hypertensive rats'. Do you mean essential hypertension in rats? Secondly what about the rapidly progressive glomerulonephritis?

GIORDANO I am not referring to the Dale breed of rats but to spontaneous hypertension in rats. About progressive glomerulonephritis, the timing of the destruction of renal function has no comparison with our data. I should state that all the patients forming the group of 22 patients shown to you were all hypertensive so at least they had in the beginning some of the stigmata not seen in long progressive nephropathies.

CASSIDY (Cape Town) What is your clinical approach to patients with heavy proteinuria and mild renal impairment with regard to low protein diet?

GIORDANO I slipped over the methodology of low protein intake which we like to call alimentation rather than protein diet. The diet should not be very strict in a patient with a serum creatinine of 2mg/dl. The concept is that the alimentation should be kept as low as possible. In those patients with heavy proteinuria we should add for each gram of protein lost daily 0.6g of protein in excess of the diet.

AVIRAM Could you give us some examples of food that contains good proteins and bad proteins?

GIORDANO In terms of biological values we should put in the first place the egg proteins, subsequently casein and milk derivatives, rice proteins will come before meat. You have enough data to establish that low protein intake containing eggs and milk will not be a low phosphorus intake, but will be a relatively high phosphorus intake. Nevertheless, the results are those that I have shown.

ONUZO (Nigeria) In view of the fact that you postulated that negative charge is significant in the injury to the glomerular membrane by the protein and in view of the fact that a low protein diet leads to tissue catabolism and progressive weight loss in this patient, would you recommend that counteracting the negative charges is the primary aim? If you anticipate the progressive renal disease you should probably give a higher protein diet with heparin, as you suggested, rather than reducing the protein diet in this patient's regime.

GIORDANO This is a very good point. Of course, perhaps a new horizon is widening in front of us in terms of the possibility to slow or to arrest the progression of renal impairment by using molecules such as negatively charged molecules. When I say polycations I mean basic compounds. When I say polyanions, I mean acidic compounds. So heparin may be one of these. Who does not remember 15 to 20 years ago patients being transplanted receiving heparin when they had rejection?

BERGSTROM I would just like to ask you one thing. We are all aware that blood pressure control is also very important in slowing the progression of renal failure. What is your attitude to this? Also if you reduce the protein intake of the patient is there the possibility that it will also provide them with less sodium and thereby better blood pressure control?

GIORDANO This is a very good point. I have no data to answer this question properly. I may imply that there are some experiments pointing to the possibilities that blood pressure is influenced by protein intake. We know that casein given to rats has very little sodium, but of course, in all experiments the change from low protein to high protein intake may be so marked that it is far from clinical implication. I have no data to answer your question properly.

Proc EDTA–ERA (1984) Vol 21

INFLUENCE OF PHOSPHATE RESTRICTION, KETO-ACIDS AND VITAMIN D ON THE PROGRESSION OF CHRONIC RENAL FAILURE

P T Fröhling, *R Schmicker, **F Kokot, †K Vetter, †J Kaschube, ††K-H Götz, M Jacopian, ¶K Lindenau

*St Josefs-Hospital, Potsdam, *Wilhelm Pieck University, Rostock, GDR, **Silesian School of Medicine, Katowice, Poland, †Research Clinic of Nutrition, Potsdam Rehbrücke, ††Clinic of Internal Medicine, Neuruppin, ¶Humboldt University, Berlin, GDR*

Summary

Progression of renal failure was investigated in a prospective follow-up study of 60 patients with mild and 100 patients with advanced renal failure. Phosphate restriction in mild and protein restriction in advanced renal failure can delay the progression rate of patients with chronic renal insufficiency. Simultaneous administration of Vitamin D does not effect the progression. Administration of keto-acids to the low protein diet has a strong delaying effect on the progression rate.

Introduction

Experimental renal insufficiency in rats revealed that protein restriction can delay the progression of chronic renal failure [1]. Some clinical studies on uraemic patients in the pre-dialysis period have also shown the positive effect which low protein diets (LPD) have on the progression of chronic renal failure [2–5]. Walser found that the administration of keto-acids (KA) had a most remarkable effect [4]. By phosphate restriction alone Mashio [6] achieved a successful delay of progression in early renal failure. Administration of vitamin D or its metabolites has been described as a risk towards deterioration of renal insufficiency [7]. Our contribution informs about the investigation of the influence of protein and phosphate restriction and vitamin D and keto-acids in a prospective study of patients with chronic renal failure.

Patients and methods

Thirty patients in mild renal failure were treated with phosphate restriction and pharmacological doses of vitamin D (as a rule between 20 and 400,000 units daily). Thirty patients without treatment served as control group.

Twenty patients in advanced renal failure received protein and phosphate restriction only (0.4–0.6g of protein per kg body weight), 30 patients under the same dietary regime received pharmacological doses of vitamin D, and 30 vitamin D and a mixture of amino acids and keto-acids. Twenty patients without dietary treatment or vitamin D administration served as control group. Several patients in the group of dietary treatment without keto-acid administration received a mixture of essential amino acids of the same composition. The characteristics of the patient-groups are documented in Table I.

TABLE I. Distribution of diagnosis, incidence of hypertension, initial values of creatinine and the observation time in the investigation groups

	Diagnosis GN	PN	Cy	n	Hypertension	Creatinine (mg/dl)	Observation time (Months)
Mild renal failure without treatment	4	23	3	30	13	2.3 ± 0.6	21.7 ± 6.4
Mild renal failure low phosphorus diet + vitamin D	9	19	2	30	17	2.1 ± 0.7	42.3 ± 18.9
Control group	6	10	4	20	12	6.0 ± 1.4	14.3 ± 7.3
Low protein diet	5	11	4	20	12	5.9 ± 1.5	18.7 ± 7.2
Low protein diet + vitamin D	3	17	10	30	20	7.4 ± 1.8	20.8 ± 8.5
Low protein diet + vitamin D + KA	5	17	8	30	18	9.2 ± 2.5	25.5 ± 12.5

Progression was estimated as the regression of mean reciprocal plasma creatinine concentration over time.

Creatinine, calcium and phosphate were determined monthly, and PTH every six months, measured by radioimmunoassay which has been described in detail elsewhere [8]. Muscle wasting as a cause of the decline of creatinine was excluded by anthropometric measurements every three months in patients with advanced renal failure. Patients with a rapid progression based on the activity of the underlying disease and with uncontrolled hypertension were not accepted for this study.

Results and discussion

Figure 1 gives the influence on progression of vitamin D administration in mild and advanced renal failure.

As can be seen a deterioration of renal function under the vitamin D treatment is not detectable in mild and in advanced renal failure.

This result confirms the observation by Mazur [9] that effective $25(OH)D_3$

levels for the prevention of bone diseases does not cause a deterioration in renal function. A marked delaying effect on the progression due to phosphate restriction, however, was observed in mild renal failure.

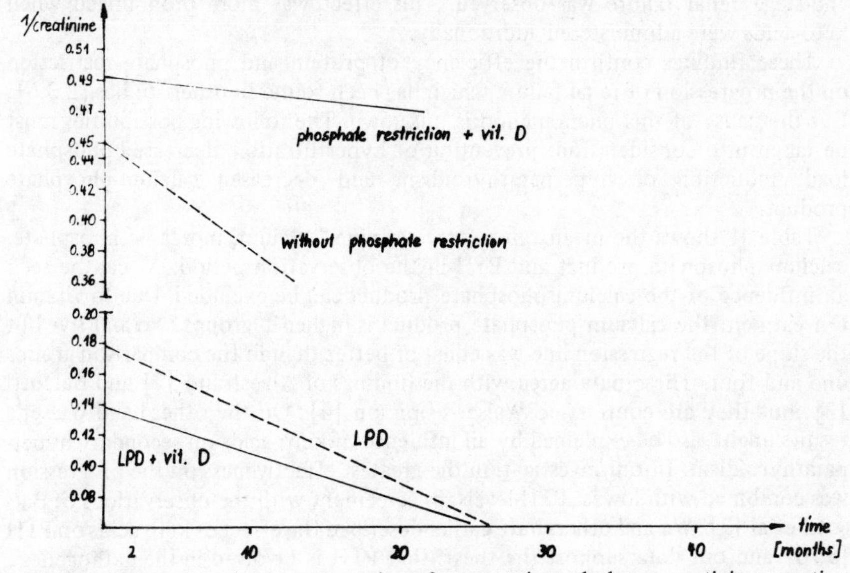

Figure 1. Comparison between regression of mean reciprocal plasma creatinine over time in mild advanced renal failure with and without vitamin D administration

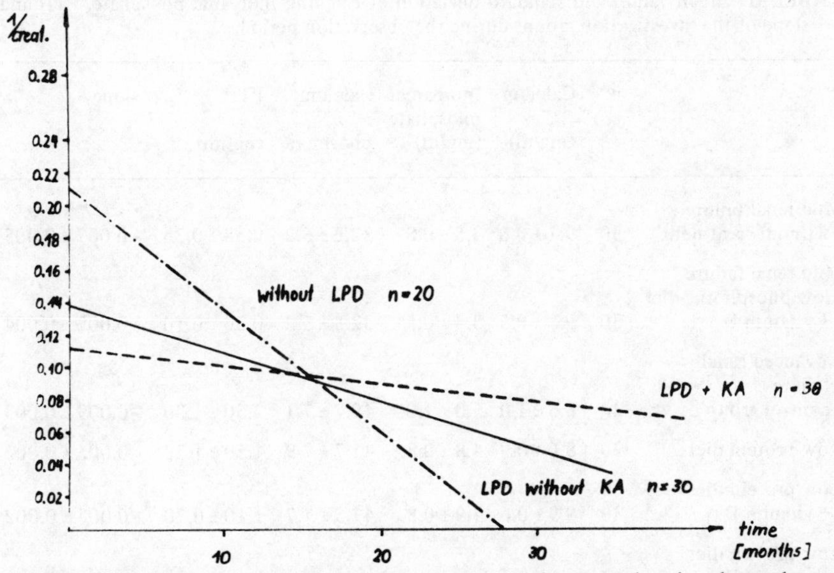

Figure 2. Regression of mean reciprocal plasma creatinine over time in advanced renal failure under LPD treatment with and without KA administration compared with a control group

Figure 2 shows the influence of low protein diet and keto-acids on the course of renal failure. A strong negative correlation between reciprocal creatinine and time was found in patients without LPD. During LPD a marked delay of the end-stage renal failure was observed. This effect was more pronounced when keto-acids were administered additionally.

These findings confirm the efficiency of protein and phosphate restriction on the progression of renal failure which has been found in other studies [2,3,6]. But the cause of this phenomenon is unknown. The following possibilities must be taken into consideration: prevention of hyperfiltration, decreased phosphate load, reduction of hyperparathyroidism and decreased calcium-phosphate product.

Table II shows the mean values of creatinine, calcium, inorganic phosphate, calcium phosphate product and PTH in the observation period. As can be seen an influence of the calcium phosphate product can be excluded. Due to vitamin D treatment the calcium phosphate product is higher in groups two and five but the slope of the regression line was equal or better than in the comparison groups one and four. These data agree with the findings of Alvestrand [2] and Barsotti [3], but they are contrary to Walser's opinion [4]. On the other hand Walser's results might also be explained by an influence of keto acids on secondary hyperparathyroidism. In our investigation the greatest effectiveness on the progression was combined with lowest PTH-levels, in agreement with the observations of Barsotti et al [3]. We and others have earlier described the effect of keto acids on PTH [5,10] and our data support the thesis that PTH is involved in the pathogenesis

TABLE II. Mean values and standard deviation of calcium, inorganic phosphate, PTH and slope of the investigation groups during the observation period

	n	Calcium (mg/dl)	Inorganic phosphate (mg/dl)	Calcium x phosphate	PTH (ng/ml)	slope
Mild renal failure without treatment	30	9.1 ± 0.6	3.5 ± 0.8	32.2 ± 6.2	0.68 ± 0.38	−0.007 ± 0.005
Mild renal failure low phosphorus diet + vitamin D	30	9.5 ± 0.5	3.4 ± 0.5	32.5 ± 5.2	0.30 ± 0.10	−0.002 ± 0.004
Advanced renal failure control group	20	8.6 ± 1.0	6.0 ± 1.0	49.7 ± 7.1	2.50 ± 1.30	−0.009 ± 0.003
Low protein diet	20	8.6 ± 0.9	4.8 ± 0.8	41.7 ± 7.9	1.50 ± 1.30	−0.005 ± 0.002
Low protein diet + vitamin D	30	9.9 ± 0.6	4.9 ± 0.8	47.5 ± 7.7	1.10 ± 0.70	−0.003 ± 0.002
Low protein diet + vitamin D + KA	30	9.9 ± 0.7	4.6 ± 0.7	45.5 ± 6.8	0.70 ± 0.30	−0.001 ± 0.001

of the progression of chronic renal insufficiency. Lower PTH-levels can reduce the phosphate load of the remaining nephrons and the calcium phosphate deposition in the kidney. It may be possible that the hyperfiltration of the remaining nephrons can also be prevented in this way.

Conclusions

In mild renal failure phosphate restriction should be initiated as early as possible. Effective prophylaxis to prevent renal osteodystrophy with vitamin D does not cause a deterioration in renal function. Protein and phosphate restriction can delay the progression of renal failure in many patients. PTH is involved in the mechanism of progression in chronic renal failure. In our investigation the keto-acid substituted diet was the most effective form of LPD on the progression. This can be explained by the better control of PTH in this group.

References

1 Meyer TW, Lawrence WE, Brenner BM. *Kidney Int 1983; 16 (Suppl):* 243
2 Alvestrand A , Ahlberg M, Bergström J. *Kidney Int 1983; 16 (Suppl):* 268
3 Barsotti G, Guiducci A, Ciardella F et al. *Nephron 1981; 27:* 113
4 Walser M, Mitch WE, Collier VU. *Contr Nephrol 1980; 20:* 92
5 Barsotti G, Morelli E, Guiducci A et al. *Proc EDTA 1983; 19:* 773
6 Mashio G, Oldrizzi L, Tessitore N et al. *Kidney Int 1983; 16 (Suppl):* 273
7 Nielsen HE, Christensen MS, Romer FK et al. *Lancet 1978; ii:* 1259
8 Kokot F, Stupnicki R. *State Medical Publisher.* Warsaw. 1979: 161
9 Mazur AT, Norman ME. *Proc V Workshop on Vitamin D 1982;* 865
10 Fröhling PT, Kokot P, Schmicker R. *Clin Nephrol 1983; 4:* 212

Open Discussion

ZOCCALI (Reggio Calabria) Your two groups were followed for very different periods, some for twice as long as others. How can you compare slopes obtained from data points referring to different ranges. I also noticed that you employed non-parametric tests for statistical analysis and in addition you compared several parameters in the study groups. Did you use multiple regression analysis because if not I suspect that some of your data will not be significant after appropriate statistical tests are applied.

FRÖHLING The observation time was different in patients with mild renal failure because we must give vitamin D after the first control biopsy. If we observe a deterioration in the bones we must restrict phosphate and give vitamin D. I think a slope of more than 12 months can be compared with a non-parametric test.

BONE (Liverpool) Your final hypothesis was that an increase in parathyroid hormone is more nephrotoxic than a rise in the calcium phosphate product: can you explain the commonly observed phenomenon that the treatment of patients in severe renal failure with vitamin D metabolites $1\alpha(OH)D_3$ and $1,25(OH)_2D_3$ leads to a disproportionate loss of renal function compared to the rise in the calcium phosphate products at a time when renal osteodystrophy is healing and parathyroid hormone values are falling.

FRÖHLING We have not seen a deterioration of renal function after administration of vitamin D and the same was found by Dr Mazur of Philadelphia after the administration of $25(OH)D_3$.* We have no experience of $1,25(OH)_2D_3$ and it is my opinion that the adverse effect on renal function is seen only after the use of the dihydroxy metabolites.

*Mazur AT, Norman ME. *Proc V Workshop on Vitamin D 1982:* 865

Proc EDTA–ERA (1984) Vol 21

EARLY PROTEIN RESTRICTION IN CHRONIC RENAL FAILURE

J B Rosman, P M ter Wee, G Ph M Piers-Becht, W J Sluiter,
F J van der Woude, S Meijer, A J M Donker

University Hospital of Groningen, Groningen, The Netherlands

Summary

We performed a prospective randomised trial in 199 patients with various stages of renal failure. Stratified for sex, age and renal insufficiency, 105 patients were assigned to a protein (Pr)-restricted group (0.4–0.6g/kg/BW), and 94 to a control group. Pr-restriction led to a significant reduction in the excretion of urea, phosphate and protein. Survival of renal function was significantly better in Pr-restricted patients. Median serum creatinine concentration increased in the control group ($p<0.05$), but remained stable in Pr-restricted patients. We conclude that Pr-restriction retards, or even halts the progression of chronic renal failure.

Introduction

Once chronic renal insufficiency has been established, progression to end-stage renal disease seems inevitable. Since the beginning of this century, it has been known that protein-restricted diets are useful in uraemic patients [1].

In animals, especially in the so-called remnant kidney model, it has been shown that protein restriction retards or even halts the progression of renal failure, apparently by means of reducing glomerular hyperfiltration and thereby preventing histological lesions in remnant glomeruli [2].

However, does an early started low protein diet influence the progression of renal failure in man? Several retrospective studies concerning this topic have been published since 1976 [3–7]. Definitive proof, however, may only be drawn from a prospective randomised trial, which is still lacking. We decided to initiate such a study in December, 1981.

Patients and methods

One hundred and ninety-nine patients visiting our nephrologic outpatient department between 1 January 1982 and 1 January 1984, with creatinine clearances between 10 and 60ml/min/1.73m^2 entered the trial. Patients with SLE, PAN, Wegeners' granulomatosis and potentially lethal diseases (e.g. cancer) and those on non-steroidal anti-inflammatory drugs were excluded in advance. After stratification for sex, age and renal failure, the patients were randomly allocated to a protein (Pr)-restricted and a control group.

Altogether, 105 patients were randomly assigned to the Pr-restricted group (0.4–0.6g/kg/BW) and 94 patients to the control group. Further differentiation produced four groups:

A1 (n=57): creatinine clearance 31–60ml/min/1.73m^2 ; no protein restriction,

B (n=57): creatinine clearance 31–60ml/min/1.73m^2 ; 0.6g protein per kg/BW,

A2 (n=37): creatinine clearance 10–30ml/min/1.73m^2 ; no protein restriction,

C (n=48): creatinine clearance 10–30ml/min/1.73m^2 ; 0.4g protein per kg/BW.

At entry all patients visited the dietician for dietary history. A computerised data base of 950 variables was established. Every three months it was updated with the newest results. Patients adhering to the Pr-restricted diets consulted the dietician every three months, patients in the control groups only on indication. Non-restricted patients were advised to reduce protein intake if their serum urea exceeded 25mmol/L. All patients received supplemental vitamins and trace-elements; patients in Pr-restricted groups also received methionine. Blood pressure, serum calcium and serum phosphate were, if possible, kept within normal limits in all patients. Some patients with creatinine clearance <10ml/min/1.73m^2 received a cadaveric kidney transplant during the follow-up. Dialysis was initiated when creatinine clearance dropped below 4ml/min/1.73m^2.

Statistical analysis was performed by means of survival computation. A persistent 10 per cent increase in serum creatinine was chosen as the non-survival criterion. Increase of median serum creatinine was studied in all four groups separately by linear regression analysis. Data comparison between the groups was analysed with Mann Whitney U tests, and within the groups by Wilcoxon matched pairs signed ranks tests.

Results

In the first instance we created two groups: A1+A2 being the non-restricted, and B+C being the Pr-restricted patients. At the moment of randomisation we scanned all relevant variables for statistical differences between these two main groups. No difference could be established for diagnosis (see Table I), serum phosphate, serum calcium, serum alkaline phosphatase, haematocrit, body weight, blood pressure, serum urea, serum creatinine, and the 24-hour excretion values of sodium, urea, creatinine and protein.

Reliable conclusions can only be drawn if dietary compliance can be measured and we used the 24-hour excretion of urea. At randomisation, it amounted to 288mmol in the combined Pr-restricted group, against 280mmol in the combined control group (Figure 1). After three months (and persistently thereafter), a decreased urea excretion in the Pr-restricted patients proved that they actually adhered to their diets. At 18 months the values were 180 and 290mmol/24hr,

TABLE I. Distribution of the various renal diseases over the protein-restricted and the control group. χ^2_{16} =15.7, not significant

Diagnosis (n=199)	A1+A2	B+C	%
1. unknown	2	5	3.5
2. membrano-proliferative glomerulonephritis	2	6	4
3. focal glomerulosclerosis	9	14	11.6
4. IgA-glomerulopathy	6	3	4.5
5. glomerulonephritis (other types)	6	10	8
6. adult polycystic kidney disease	4	10	7
7. unilateral agenesis	2	2	2
8. uninephrectomy	5	7	6
9. hypoplasia/dysplasia	1	2	1.5
10. pyelonephritis	5	5	5
11. Alport's syndrome	1	1	1
12. nephrosclerosis	16	9	12.6
13. analgesic nephropathy	14	9	11.6
14. interstitial nephritis	3	3	3
15. reflux nephropathy	7	5	6
16. other urological diseases	6	3	4.5
17. miscellaneous (amyloidosis, diabetes, etc)	5	11	8

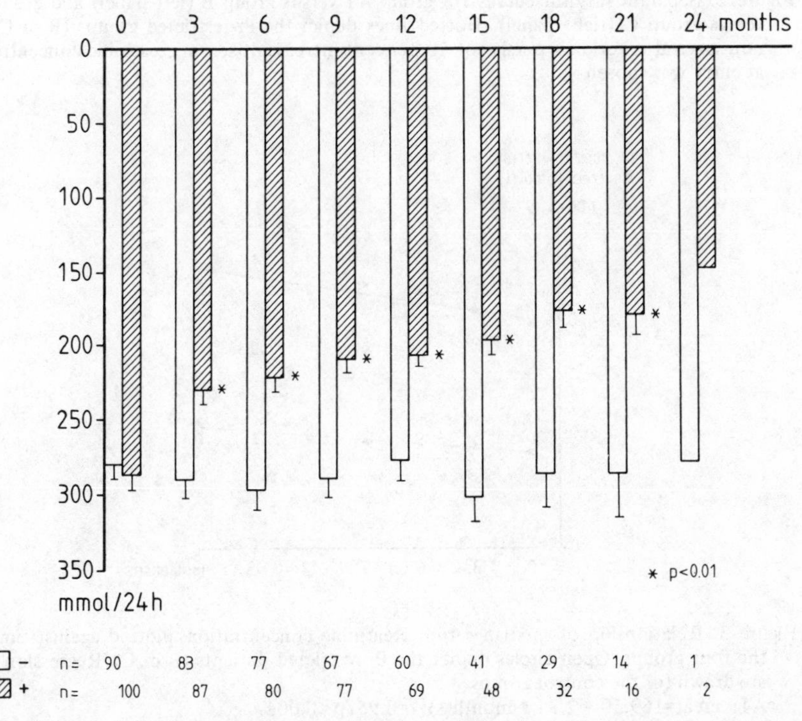

Figure 1. Mean 24-hour urea excretion. Open bars depict the control group, hatched bars the Pr-restricted group. The significance signs represent significant differences between the two groups. The number of patients at any moment is shown as well

respectively (p<0.001). Roughly, the mean decrease in urea excretion in the Pr-restricted groups equals 20g of protein per day.

In the Pr-restricted group, phosphate- and protein excretion also decreased; no change occurred in the control group. Serum phosphate fell significantly in

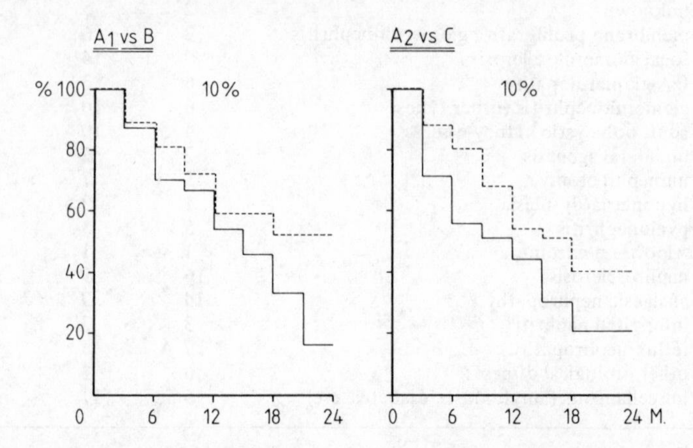

Figure 2. Separate survival curves for group A1 versus group B (left panel) and group A2 versus group C (right panel). Dotted lines depict the Pr-restricted groups (B or C). As non-survival criterion a persistent 10 per cent increase of serum creatinine concentration at entry was chosen

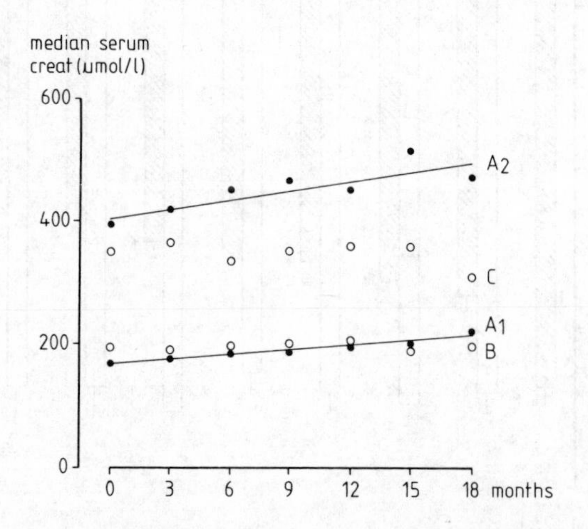

Figure 3. Relationship of median serum creatinine concentration, plotted against time, in the four groups. Open circles depict the Pr-restricted patients (B or C). Regression lines are drawn for the control groups.

A1: creat=169.50 + 2.31 * month(s); r=0.95; p<0.001.
B: creat=193.32 + 0.25 * month(s); r=0.30; not significant.
A2: creat=405.82 + 5.08 * month(s); r=0.84; p<0.05.
C: creat=357.93 – 1.36 * month(s); r=0.48; not significant.

the Pr-restricted group, and patients in this group ultimately used less aluminium hydroxide. Serum urea fell immediately after institution of the Pr-restricted diet. Thereafter it remained below the starting value. In the controls, a moderate, but consistent increase was noticed. During follow-up no differences in blood pressure, haematocrit, serum alkaline phosphatase, serum albumin and 24-hour excretion of sodium developed within and between the groups.

For analysis of serum creatinine changes we used the four group divisions (A1 versus B, A2 versus C). Survival curves revealed a better survival of Pr-restricted patients (Figure 2). Moreover, it appeared that median serum creatinine concentration in the control group increased every month (Figure 3). In the Pr-restricted patients, median serum creatinine remained constant during the whole observation period.

Discussion

In this study we have shown in a prospective randomised way that early dietary protein restriction can be beneficial in patients with chronic renal failure.

From the urea excretion we concluded that the dietary compliance was sufficient, for some patients it was difficult to adhere to the low protein diet. It is our strong belief that frequent visits to a dietician are a necessary tool for maintaining patients on these diets. This idea is reinforced by the fact that the urea excretion in the combined Pr-restricted group was still decreasing after 12 months. From the dietary interviews we learned that our patients became used to their diets. There were no signs of malnutrition.

The finding of retarding progression of renal failure by low protein diets is in accordance with the hyperfiltration theory of Brenner's group. As recently stated by Alvestrand and Bergström, amino acids (and proteins) might trigger the secretion of a liver hormone, called glomerulopressin, that reduces the tone of the afferent arteriole [8]. The resulting hyperperfusion and hyperfiltration will lead to structural changes in remnant glomeruli. The histological substrate is glomerulosclerosis.

Not all patients benefit from these diets. There are groups, for example polycystic kidney disease and nephrosclerosis, that show a steady progression to end-stage renal disease despite all efforts to retard it [9,10]. In the case of nephrosclerosis one should realise that a situation exists in which glomeruli hypofiltrate because of afferent vascular abnormalities. The same might hold true for polycystic kidney disease since this disorder is also not characterised by proteinuria. Further analysis of our data might reveal which diagnoses benefit most from the diet.

Thus, low protein diets will have a strong impact on the treatment of patients with chronic renal disease in the future.

References

1 Ambard L. *Physiologie Normale et Pathologique des Reins.* Paris: Masson et Cie. 1920: 264

2 Brenner BM, Meyer TW, Hostetter TH. *N Engl J Med 1982; 307:* 652

3 Giordano C. In Zurukzoglu W, Papadimitriou M, Pyrpasopoulos M, Sion M, Zamboulis C, eds. *Proceedings Eighth International Congress of Nephrology, Athens, June 1981*. Basel: Karger. 1981: 71–81
4 Sitprija V, Suvanpha R. *Br Med J 1983; 287:* 469
5 Alvestrand A, Ahlberg M, Bergström J. *Kidney Int 1983; 24:* S268
6 Bennett SE, Russell GI, Walls J. *Br Med J 1983; 287:* 1344
7 Maschio G, Oldrizzi L, Tessitore N et al. *Kidney Int 1982; 22:* 371
8 Alvestrand A, Bergström J. *Lancet 1984; i:* 195
9 Maschio G, Oldrizzi L, Rugiu C. *Kidney Int 1984; 25:* 112A
10 El-Nahas AM, Masters-Thomas A, Moorhead JF. *Kidney Int 1984; 25:* 106A

Open Discussion

BROYER (Paris) What was the change in weight in the control group compared to that of the group treated by the low protein diet?

ROSMAN There was no statistical difference between the groups.

ZAKAR (Hungary) Have you seen an increase in the creatinine clearances in patients with an initial clearance of 50–60ml/min when they are given protein restriction?

ROSMAN There is an improvement in renal function in some patients in this group but others remain stable or deteriorate. Further analysis of our data will be needed to determine if this is dependent on the underlying cause of the renal failure.

ZAKAR Have you had patients with the nephrotic syndrome and chronic renal failure and what was the course of the proteinuria in such patients?

ROSMAN Unfortunately we have not looked specifically at this group.

ZAKAR In my experience a certain decrease in proteinuria is sometimes observed but not in patients with the nephrotic syndrome. It is much more commonly observed in patients with pyelonephritis. In addition in some patients there is an increase in creatinine clearance but this never exceeds 10 per cent of the starting value.

ROSMAN We believe that the reduction in proteinuria we have seen supports Brenner's hypothesis on hyperfiltration.*

BONE (Liverpool) You mentioned that there was a reduction in the phosphate excretion rate and also in the aluminium hydroxide consumption of your patients on protein restriction. Could you tell us of the magnitude of these reductions?

*Brenner BM, Meyer TW, Hostetter TM. *N Engl J Med 1982; 307:* 625

ROSMAN The 24 hour phosphate excretion shows significant differences between the groups.

BONE Was there a concomitant reduction in plasma phosphorus concentrations?

ROSMAN The serum phosphate decreased in protein restricted patients even though these patients used less aluminium hydroxide.

Proc EDTA–ERA (1984) Vol 21

LONG-TERM OUTCOME OF RENAL FUNCTION AND PROTEINURIA IN KIDNEY TRANSPLANT DONORS

O Mathillas, P-O Attman, M Aurell, I Blohmé, H Brynger, G Granerus, G Westberg

Sahlgrenska sjukhuset, University of Göteborg, Sweden

Summary

Ten to eighteen years after donor uninephrectomy (UN) there was no evidence for deterioration of renal function. The mean urinary albumin excretion was slightly increased. There was a positive correlation between the assessed protein intake at investigation and the glomerular filtration rate (GFR).

Introduction

Living donors are important sources of transplant kidneys in many centres. Donor uninephrectomy (UN) has been considered in both short and long-terms to carry little risk as reported by us and others [1–4]. The recent suggestion that donor UN may cause hypertension and proteinuria and eventually progressive renal failure has caused considerable concern [5]. It prompted us to re-examine the living donors in Göteborg more than ten years after donation with special reference to glomerular filtration rate (GFR), hypertension and proteinuria. These data have been correlated to the actual protein intake in these subjects as this factor has been incriminated in causing long-term deterioration of renal function.

Subjects

During the years 1965–1973 sixty-four living donor UN were performed at our transplant centre in Göteborg, Sweden. Thirty-eight donors were men with a mean age at donation of 42 (23–63) years while 26 were women with a mean age of 47 (23–69) years. Ten donors have died during the observation period. Thirty-eight donors have been investigated 10–18 years after donation. Sixteen donors have not yet been investigated but they have been interviewed and their case records have been examined.

Methods

Glomerular filtration rate (GFR) was measured as renal clearance of inulin and the plasma clearance of ^{51}CrEDTA using the single injection technique. The results of these methods are almost identical and expressed by the equation:

$$\text{Plasma } ^{51}\text{CrEDTA clearance} = \frac{\text{renal inulin clearance}}{1.10} + 3{,}7$$

The reference values of Granerus and Aurell [6] for various age groups were used.

Urinary excretion of total protein, albumin, beta$_2$ microglobulin and nitrogen was measured in three consecutive 24hr samples in each individual. Protein intake was calculated as 6.25 x (urinary nitrogen + 2) g/24hr according to Isaksson [7]. Reference values were obtained in a group of 14 healthy 55-year-old men. Glomerular and tubular proteinuria were defined according to Petersson, Evrin and Berggard [8].

Student's 't' test, Wilcoxon rank sum test, Spearman's rank correlation and multiple linear regression analysis have been used for statistical evaluation of the results.

Results

Renal function

GFR 10–18 years after donor UN is shown in Figure 1. Mean GFR for the whole group of donors was 75.8 per cent (50.0–103.9%) of pre-UN GFR for two kidneys. Only seven donors had reduced GFR. Their GFR was 62.5 per cent (50.0–80.0%) of their pre-UN value.

GFR was also measured one year after kidney donation in 10 donors and after five years in 11 donors. As shown in Table I there was a significant increase in GFR from one year to ten years after donation but not between five years and 10 years after donation.

Protein excretion is given in Table II. Only albumin excretion was significantly increased in the donor UN group but a total protein excretion of more than 0.5g/24hr was found in four donors.

Four donors had glomerular proteinuria and two had tubular proteinuria. Mean GFR for these six subjects was 70 (58–84) ml/min1.73m^2 (Figure 1).

Nitrogen excretion

The nitrogen excretion was 11.4 (7.1–18.4) g/24hr corresponding to 79.2 (54.6–113.7) g protein intake/24hr/1.73m^2. As seen in Figure 2 there was a positive correlation between protein intake and GFR after donor UN ($p<0.02$). Age accounted only partly for this significant correlation.

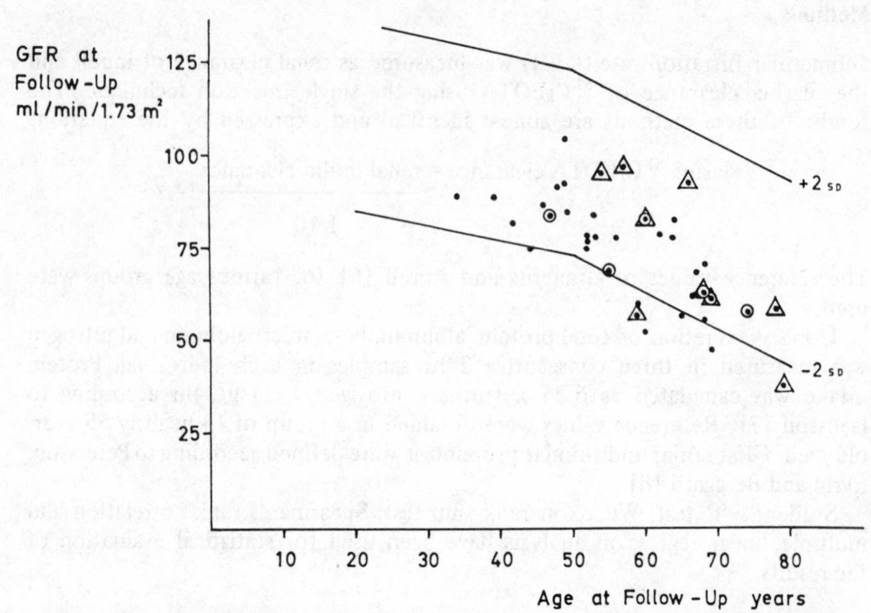

Figure 1. Glomerular filtration rate (GFR) at follow up 10–18 years after donor uninephrectomy (UN) in 38 subjects: • normotensive subjects without proteinuria; ⊙ subjects with proteinuria; △ subjects with hypertension; ⚠ subjects with both hypertension and proteinuria

TABLE I. Mean glomerular filtration rate (GFR) before, 1, 5, and 10–18 years after uninephrectomy (UN) in 32 living kidney transplant donors

GFR (ml/min/ 1.73m²)	pre-UN	1 year	5 years	10–18 years
n = 10	90.6	62.6	–	69.5*
n = 11	101.2	–	77.0	78.6
n = 32	99.2	–	–	75.8

*$p<0.05$ vs 1 year value

Morbidity

Ten donors have died since nephrectomy. The causes of death were malignant tumours (3), myocardial infarction (3), alcoholism (2), cerebrovascular lesion (1), and trauma (1). UN may have had some importance in only one of these cases, a 58-year-old woman who developed hypertension three years after the nephrectomy. The hypertension was well controlled but she died five years after nephrectomy from a cerebrovascular lesion.

TABLE II. Mean urinary excretion of total protein, albumin, and β_2 microglobulin 10–18 years after donor uninephrectomy (UN) and in control subjects. Three consecutive 24 hour urinary samples were collected in each individual

	Donor UN (n = 31)	Control (n = 14)	Significance
Total protein (mg/24hr)	306 (92–1217)	210 (65–326)	NS
Albumin (mg/24hr)	69.5 (5–620)	11.4 (8–23)	p<0.01
β_2 microglobulin (μg/24hr)	1039 (49–28400)	157 (61–420)	NS

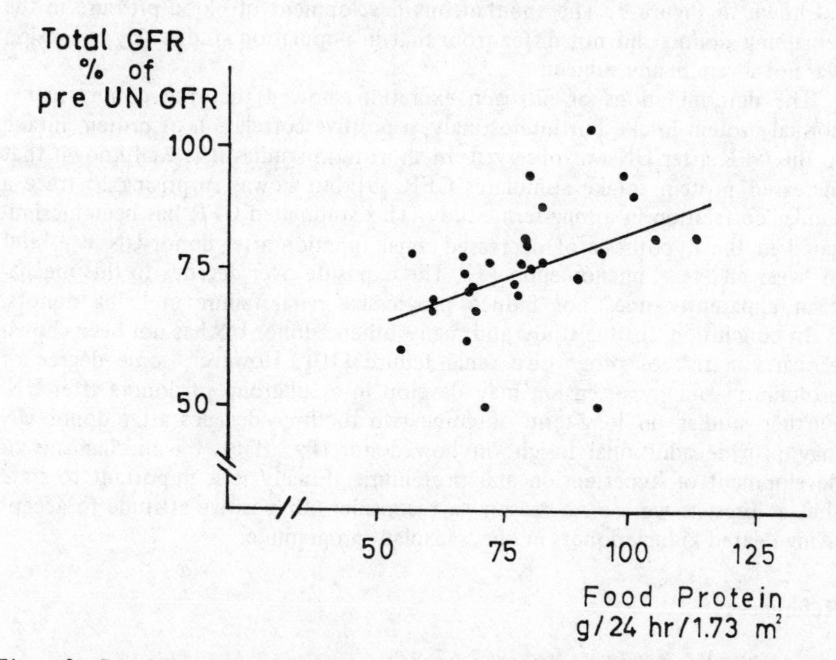

Figure 2. Correlation between food protein and total GFR in per cent of pre-UN GFR in 29 donors 10–18 years after donor UN

No donor was treated for hypertension at the time of UN. At follow-up nine donors (seven men and two women) were treated with one to three drugs. No one had malignant or treatment resistant hypertension. For the remaining subjects the mean systolic blood pressure increased from 138 to 149mmHg corresponding to 0.72mm/year and mean diastolic pressure from 85 to 87mm corresponding to 0.10mm/year.

Urinary sediments were normal in all subjects. Four subjects developed diabetes mellitus after UN. GFR in these patients was not reduced compared to the other donors and they did not have albuminuria. Nine women had had one or more urinary tract infections after the UN but their GFR was not reduced;

Discussion

Renal function 10–18 years after donor UN is well maintained and only seven donors were found to have a slightly decreased GFR compared to normal values for healthy subjects with two kidneys. Repeated determinations during the observation time did not indicate progressive decrease of renal function in any subject.

Some donors, however, developed hypertension and/or proteinuria. Twenty-four per cent of the donors were treated for hypertension. This is a high prevalence but hypertension only developed in the upper age groups (54–78 years) as shown in Figure 1. The spontaneous development of blood pressure in the remaining donors did not differ from that in population studies. Hypertension was not severe in any subject.

The determinations of nitrogen excretion showed that the donors had a normal protein intake but interestingly, a positive correlation of protein intake to the GFR after UN was observed. In short-term studies it is well known that increased protein intake stimulates GFR [9] but it was surprising to trace a similar correlation in a long-term study. This stimulated GFR has been incriminated in the hypothesis of decreased renal function after donor UN as a kind of 'wear and tear' phenomenon [5]. The exposure over decades to this mechanism apparently does not induce progressive renal failure in living donors.

In conclusion, in this study and many others, donor UN has not been shown to carry a risk of progressive renal failure [10]. However, some degree of proteinuria and hypertension may develop in a subgroup of donors after UN. Further studies on long-term outcome two to three decades after donor UN may provide additional insight in how donor UN affects the mechanisms of development of hypertension and proteinuria. Finally it is important to state that at present we see no reason to reconsider our positive attitude to accept living related kidney donors in our transplant programme.

References

1. Ogden DA. *Ann Intern Med 1967; 67:* 998
2. Boner G, Shelp D, Newton M et al. *Am J Med 1973; 55:* 169
3. Aurell M, Ewald J. *Scand J Urol Nephrol 1981; 64 (Suppl):* 137
4. Blohmé I, Gäbel H, Brynger H. *Scand J Urol Nephrol 1981; 64 (Suppl):* 143
5. Brenner BM, Meyer WT, Hostetter TH. *N Engl J Med 1982; 216:* 652
6. Granerus G, Aurell M. *Scand J Clin Lab Invest 1981; 41:* 611
7. Isaksson B. *Am J Clin Nutr 1980; 33:* 4
8. Petersson P, Evrin PE, Berggard I. *J Clin Invest, 1969; 48:* 1189
9. Pullman TN, Alving AS, Dern RJ et al. *J Lab Clin Med 1954; 44:* 320
10. Hakim R, Goldszer R, Brenner BM. *Kidney Int 1984; 25:* 930

Open Discussion

MAGGIORE (Reggio Cal.) If you look at the relatives of an unselected population with kidney disease you would expect a greater incidence of hypertension and proteinuria in the relatives of patients with polycystic kidney disease, nephrosclerosis, and hereditary nephropathies. Would it not be more reasonable to choose your control population from among the relatives of uraemic patients rather than the general population?

MATHILLAS I think that will be the idea in a future donor study following the relatives of the donor.

RENAL FUNCTION AND BLOOD PRESSURE AFTER DONOR NEPHRECTOMY

J S Tapson, S M Marshall, S R Tisdall, R Wilkinson, M K Ward, *D N S Kerr

*Royal Victoria Infirmary, Newcastle upon Tyne, *Hammersmith Hospital, London, United Kingdom*

Summary

Pre- and post-nephrectomy renal function and blood pressure were compared in 75 subjects who had donated a kidney for transplantation during the past 20 years. The function of the remaining kidney was not adversely affected by prolonged compensatory hyperfiltration. However, an increased prevalence of hypertension was found in 'long-term' kidney donors.

Introduction

The advantage of living related kidney transplantation to the recipient are well recognised. Any benefits to the donor, however, are limited to the possible detection of remedial disorders during pre-nephrectomy evaluation and to the positive psychological effects of giving. Donor nephrectomy is a major procedure and is associated with significant morbidity, and a small risk of mortality [1]. For these reasons, many European centres consider living related kidney transplantation to be 'ethically unjustified' [2].

Recent experimental work also questions the long-term safety of donor nephrectomy. In animals, destruction of a critical mass of renal tissue is followed by the development of proteinuria, hypertension and progressive loss of function of the remaining nephrons, and it has been suggested that compensatory hyperfiltration of remnant nephrons is responsible for these developments [3]. As hyperfiltration occurs consistently following unilateral nephrectomy in man, kidney donors would seem to be at risk. In support of this concern is the observation by Kiprov that patients with unilateral renal agenesis may develop focal glomerulosclerosis in their single kidney [4].

Studies of kidney donors performed during the first six years after nephrectomy have all demonstrated compensatory hyperfiltration but have not shown any progressive deterioration of function in the remaining kidney [5,6]. However, there is no information available from these studies on urinary protein excretion or blood pressure, and until recently, long-term follow-up of kidney donors has not been reported. This paper examines glomerular filtration rate (GFR), urinary albumin excretion and blood pressure in subjects who have donated a kidney during the past 20 years in Newcastle.

Methods and materials

Seventy-five subjects (38 males, 37 females) who had undergone donor nephrectomy during the period of March 1963 to June 1982, participated in the study. Patients who had donated a kidney less than one year before the study were excluded.

All donors had been investigated before nephrectomy to exclude renal or systemic disease. Their case notes were reviewed and all subjects were interviewed and examined. Blood pressure was measured after resting on three separate occasions, each in supine, sitting and erect positions. Diastolic pressure was taken at the fifth phase of Korotkoff.

Blood was drawn for electrolytes, urea, creatinine, calcium, phosphate, magnesium, albumin and haemoglobin estimations. A 24 hour urine collection was provided for protein and creatinine measurement. On a separate occasion, a timed overnight urine collection was performed for measurement of albumin excretion. All patients gave a freshly voided mid-stream urine specimen for culture and microscopic examination.

A 'single-shot' Cr^{51}-EDTA clearance test was performed on all subjects.

Biochemical determinations were performed using routine laboratory methods. Urinary albumin was measured using a single antibody radioimmunoassay. The antibody was specific for human albumin, and separation was by polyethylene glycol precipitation (intra-assay coefficient of variance, 2.5%; inter-assay coefficient of variance, 5.2%).

Sixty-one of the 75 donors completed a week weighed dietary record for assessment of daily sodium, protein and phosphate intake.

For statistical evaluation, the Student's 't' test, the Student's paired 't' test, and the test for the Spearman correlation coefficient were used. All means are expressed as the mean ± standard error. P values of less than 0.05 are considered statistically significant.

Results

As no long-term prospective longitudinal study of renal function in donors had been undertaken, we examined the effect of prolonged hyperfiltration by comparing 38 'short-term' donors, studied between 1.4 and 9.9 (mean 4.7 ± 0.3) years after nephrectomy with 37 'long-term' donors who had given a kidney between 10.4 and 20.8 (mean 13.5 ± 0.4) years earlier. There were 19 males and 19 females in the 'short-term' group and 19 males and 18 females in the 'long-term' group. There was no significant difference between the current age of the 'short-term' donors (51.6 ± 1.6 years) and the 'long-term' group (53.4 ± 1.6 years). We have also analysed the relationship between renal function and blood pressure, and time since nephrectomy for the group as a whole.

Renal function

The current renal function of the 'short-term' donors is compared with pre-donation figures in Table I.

TABLE I. Renal function before and up to 10 years after nephrectomy ('short-term' donors)

	Pre-Nx		Post-Nx
Urea (mmol/L)	4.8 ± 0.2	$p<0.01$	5.4 ± 0.2
Serum creatinine (μmol/L)	88.4 ± 2.7	$p<0.001$	104.0 ± 2.8
Creatinine clearance			
(mls/min)	105.1 ± 3.9	$p<0.001$	78.3 ± 3.0
(mls/min/1.73m²)	-	-	74.9 ± 2.8
Cr^{51}-EDTA clearance (mls/min/1.73m²)	-	-	66.9 ± 2.6

Nx = Nephrectomy

Only one of the 38 'short-term' donors had pathological proteinuria (700mg/24hrs). The mean albumin excretion rate of this groups was 7.6 ± 3.2μg/min (reference range for age-sex matched non-nephrectomised controls: 0–10.4μg/min).

The pre- and post-nephrectomy renal function of the 'long-term' donors is compared in Table II.

TABLE II. Renal function before and between 10 and 20 years after nephrectomy ('long-term' donors)

	Pre-Nx		Post-Nx
Urea (mmol/L)	4.6 ± 0.2	$p<0.001$	5.3 ± 0.2
Serum creatinine (μmol/L)	87.2 ± 3.6	$p<0.01$	98.6 ± 2.1
Creatinine clearance			
(mls/min)	101.4 ± 5.3	$p<0.01$	83.3 ± 3.7
(mls/min/1.73m²)	-	-	81.5 ± 2.7
Cr^{51}-EDTA clearance (mls/min/1.73m²)	-	-	76.9 ± 2.3

Nx = Nephrectomy

No patient in the 'long-term' donor group had pathological proteinuria and the mean albumin excretion rate was 8.6 ± 2.6μg/min.

None of the 75 donors had an abnormal urine sediment on microscopy to suggest parenchymal renal disease. Three women in each group had asymptomatic urinary tract infections.

No significant difference was found between the pre- and post-nephrectomy values of calcium, magnesium, phosphate, albumin and haemoglobin in either group of donors.

When the two groups of donors were combined, there was a positive correlation between time since nephrectomy and current creatinine clearance (r = +0.26) and current Cr^{51}-EDTA clearance (r = +0.25), neither of which reached significance.

Blood pressure

The mean systolic blood pressure of the 'short-term' donors rose from 128.4 ± 2.1 pre-nephrectomy to 132.8 ± 2.6 post-nephrectomy ($p<0.05$), and the mean diastolics from 78.0 ±1.2 to 83.8 ± 1.4 ($p<0.01$). No patient in this group was hypertensive prior to nephrectomy, but at follow-up, four had systolic hypertension (>160mmHg) and five had diastolic hypertension (>95mmHg).

In the long-term group, systolic blood pressure rose from 129.9 ± 3.9 pre-nephrectomy to 144.1 ± 4.8 post-nephrectomy ($p<0.01$) and the diastolics from 80.1 ± 1.3 to 89.7 ± 2.1 ($p<0.01$). Eleven donors in this group had developed de novo systolic hypertension and 14 diastolic hypertension.

All donors found to be hypertensive were subsequently reviewed by their general practitioner and in all cases hypertension was confirmed and appropriate therapy commenced.

When the 'short-term' and 'long-term' groups were combined, there was a positive correlation between time since nephrectomy and systolic (r = +0.21; p = NS), diastolic (r = +0.26; $p<0.05$) and mean arterial (r = 0.25; $p<0.05$) blood pressure.

Effect of diet on renal function and blood pressure

No correlation was found between current renal function (or the percentage change in renal function after nephrectomy) and protein or phosphate intake. Similarly, there was no significant correlation between salt intake and current blood pressure.

Discussion

This study demonstrates a remarkable preservation of GFR in 'long-term' kidney donors. We found no significant difference between the post-nephrectomy creatinine clearances of our 'long-term' donors compared to the 'short-term' group. Cr^{51}-EDTA clearance was significantly greater in the 'long-term' donors ($p<0.01$). Both of these findings, considered with the positive correlations bound between both current creatinine clearance and Cr^{51}-EDTA clearance and time since nephrectomy for the group as a whole, suggest that over this time course at least hyperfiltration had not resulted in glomerular damage. We did not assess renal function in the early post-operative period, but our late measurements are similar to those reported by others at one week after nephrectomy [5], again suggesting stable function. It seems unlikely that the greater GFR in the long-term donors could be accounted for by their slightly younger age at donation, since although compensatory hypertrophy of the remaining kidney is greater when nephrectomy is performed in childhood than in later life, the mean difference in age at donation in our patients was only seven years (39.9 ± 1.1 years v 46.9 ± 1.5 years).

The second marker of hyperfiltration-induced renal damage is proteinuria. Only one of our 75 donors had pathological proteinuria. There was also no significant difference between the urinary albumin excretion rates of our 'short-term' and 'long-term' donors. Both figures were within the normal range for non-nephrectomy controls in our laboratory. These findings are in agreement with the study of long-term kidney donors reported by Vincent et al [7] and again argue against any progressive glomerular damage in this group of patients.

Both mean systolic ($p<0.05$) and diastolic ($p<0.05$) blood pressure were greater in the 'long-term' donors compared with the 'short-term' group. Our data suggest that there is an increased prevalence of hypertension in 'long-term' kidney donors. Thirty-eight per cent of our 'long-term' group had developed de novo diastolic hypertension, a figure far in excess of that observed in the otherwise similar group of 'short-term' donors (13%), and more than would be expected in the general population of this age (15%) [8]. In fact, a figure less than that for the general population would be expected as all donors were selected in being normotensive at the time of nephrectomy. The younger age of the 'long-term' donors at nephrectomy gives a possible explanation of their greater proneness to hypertension in that essential hypertension was less likely to have manifest itself at the time of pre-nephrectomy assessment than in the older 'short-term' donors and may have subsequently emerged. However, the age difference between the groups at the time of nephrectomy (7 years) does not seem sufficient to explain the observed difference in prevalence of hypertension.

After partial renal ablation in animals, the perfusion of remnant nephrons is augmented by a high protein diet and it is suggested that this contributes to the eventual destruction of these nephrons [9]. In addition, preservation of renal function by phosphate restriction has been demonstrated in the renal ablation model [10]. Using an assessment of current dietary intake as an index of eating habits since nephrectomy, we found no correlation between protein or phosphate intake, and either present renal function or the percentage change in creatinine clearance since surgery. However, the range of dietary intake of protein and phosphate and of change of renal function in our patients was small so that any effect of diet may have been missed.

We would conclude from our observations that unilateral nephrectomy does not result in progressive deterioration in renal function, nor in the development of albuminuria. However, nephrectomised patients do appear to be at increased risk of developing systemic hypertension, the mechanism of which is not clear.

References

1 Ogden DA. *Am J Kidney Diseases 1983; 2:* 501
2 Jacobs C, Broyer M, Brunner FP et al. *Proc EDTA 1981; 18:* 20
3 Hostetter TH, Olson JL, Rennke HG et al. *Am J Physiol 1981; 241:* F85
4 Kiprov DD, Colvin RB, McCluskey RT. *Lab Invest 1982; 46:* 275
5 Donadio JV, Farmer CD, Hunt JC et al. *Ann Intern Med 1976; 66:* 105
6 Davison JM, Uldall PR, Walls J. *Br Med J 1976; 1:* 1050
7 Vincenti F, Amend WJ, Kayson G et al. *Transplantation 1983; 36:* 626
8 *National Health Examination Survey United States 1960–1962.* National Centre for Health Statistics, Series 11. Washington DC: DHEW. 1984
9 Brenner BM, Meyer TW, Hostetter TH. *N Engl J Med 1982; 307:* 652
10 Ibels LS, Alfrey AC, Haut L et al. *N Engl J Med 1978; 298:* 122

Open Discussion

ALBERTAZZI (Chairman) Dr Tapson can you tell us some detailed information concerning the number of patients in your series having serum creatinine values of more than 1.5mg/dl and also having a glomerular filtration rate of less than 60ml/min.

TAPSON Yes, well the numbers were very small but I can't remember the actual numbers for you.

HABERAL (Ankara) How do you explain why some patients develop hypertension?

TAPSON I have no idea why some patients should become hypertensive and others should not become hypertensive. I think from our study, being of relatively small numbers, it is difficult to come to firm conclusions. Other workers have also found this increased incidence of hypertension in kidney donors. Brenner [1] found that in 25 long-term donors 10 became hypertensive, Friedman in New York [2] found that 60 per cent of their donors had become hypertensive and in a study by Miller [3] that was reported as an abstract at the American Society of Nephrology meeting in 1982 it was found that 33 per cent of their donors had also become hypertensive. I have no explanation as to why some should and some should not become hypertensive.

HABERAL Did you biopsy these patients?

TAPSON No, they only have one kidney and so it would be wrong to biopsy.

SOBH (Mansoura, Egypt) You have one donor who developed end-stage renal failure needing haemodialysis and some other patients who died for other reasons. What is the fate of the kidney donated from people who have subsequently died?

TAPSON As far as the patient whose function deteriorated after he donated a kidney to his brother so that he required haemodialysis, the donated kidney is still functioning extremely well. This patient developed three years after donor nephrectomy severe malignant hypertension associated with nephrotic range proteinuria and there was subsequently a period of poor blood pressure control and gradual deterioration of his function.

SOBH Do you know what was the cause of death in the donor who died?

TAPSON In one of the four subjects who died I have no information. In the other three all died of malignancies, one died of carcinoma of the bladder, one died of carcinoma of the lung and one died of skin lymphoma. Incidentally since reporting this work one of the donors that we studied has also died, an unfortunate young 39 year old man, who as a matter of interest had the best function of all 75 donors, dropped dead with a myocardial infarction.

BANKS (Bristol) I take your point that the 30 per cent or more in your long-term follow-up of hypertensive patients is more than you would expect in the normal population although it depends on what age your patients ended up at. The trend towards an increased blood pressure surely is an age related phenomenon.

TAPSON I think a lot depends on whether you feel that blood pressure increases with age in the individual. If you do feel that way then perhaps long-term donors who were on average seven years younger than the short-term donors at the time of nephrectomy may not yet have had time for their essential hypertension to become manifest. Cross sectional studies of normal populations have certainly shown that systolic blood pressure increases with age but diastolic blood pressure increases to a much lesser extent with time. This is one of the reasons why I reported the diastolic trends. Probably what is more relevant to our study is the longitudinal studies, one in 1967 of Welsh miners [4] and another in 1973 of a group of American fighter pilots [5], where these healthy men were studied longitudinally for 30 years. The conclusions were that the rate of rise of the blood pressure had no correlation whatsoever with the original age of the patient at the time of entry into the study. It was correlated with the level of blood pressure entry into the study, i.e. those patients with a higher blood pressure at entry developed a greater rate of rise, so I think for these reasons our point is valid.

AVIRAM (Tel Aviv) Your data seems to disprove Brenner's theory [6]. Do you support de Wardener's theory [7] about the natriuretic factor being the agent for hypertension?

TAPSON I think there is little in our work to support Brenner's theory but one presumes that reduction of the nephron mass by 50 per cent isn't enough. The other point of course being Brenner's work was done on rats and rats tend to get proteinuria with advancing age and so I don't think there is very much in our work that supports Brenner's theory. As far as de Wardener's natriuretic hormone is concerned I can't answer that, although we are studying our hypertensive kidney donors.

MAGGIORE (Reggio Calabria) Is the hypertension and proteinuria of some of your donors due to them being the relatives of patients with inherited renal diseases?

ONUZO (Lagos) I wonder what observations you have made on the family history of hypertension in the relatives of the patients to whom the kidneys were donated?

TAPSON We did not study the family history further than the patients to whom the kidney was donated so I'm afraid I can't answer that.

References

1 Goldszer RC, Hakim RM, Brenner BM. *Kidney Int 1983; 23:* A124
2 Delano BG, Lazar IL, Friedman EA. *Kidney Int 1983; 23:* A168
3 Miller I, Riggio RR, Suthanthiran M et al. *Kidney Int 1984; 25:* A346
4 Miall WE, Lovell HG. *Br Med J 1967; 2:* 660
5 Harlan WR, Oberman A, Mitchell RE, Graybiel A. In Onesti G, Kim KE, Moyer JH, eds. *Hypertension: Mechanisms and Management.* New York: Grune and Stratton. 1973: 85
6 Brenner BM, Meyer TW, Hostetter TM. *N Engl J Med 1982; 307:* 625
7 de Wardener HE, Mills IH, Clapham WF, Haytor CJ. *Clin Sci 1961; 21:* 249

ADULT-ONSET NEPHROTIC SYNDROME WITH MINIMAL CHANGES: RESPONSE TO CORTICOSTEROIDS AND CYCLOPHOSPHAMIDE

***F Nolasco,** J Stewart Cameron, J Hicks, C S Ogg, D G Williams

**Hospital Curry Cabral, Lisbon, Portugal, Guy's Hospital, London, United Kingdom*

Summary

Eighty-nine patients with onset of nephrotic syndrome over the age of 15 years and minimal changes on renal biopsy have been studied. Seventy-five patients were given a course of prednisone in an initial dosage of 60mg/24hr, tapering over the following 8–16 weeks. Only 45 were in remission after eight week's treatment, 55 after 16 weeks; eventually, a total of 58 lost their proteinuria completely. Of these, 24 per cent never relapsed, 56 per cent relapsed on a single occasion or infrequently, and only 21 per cent were frequent relapsers. Cyclophosphamide, used in 36 patients, had a similar time to response. Stability of remission was better than in similar children, 66 per cent being in remission at five years, after which no further relapses were seen.

Introduction

About one-quarter to one-third of patients with an onset of a nephrotic syndrome over the age of 15 years have, on renal biopsy, a pattern usually described as 'minimal change disease'; that is, no obvious abnormality of the glomeruli on light microscopy. Although there are many good descriptions of the short- and long-term outcome in children with this appearance [1], the literature on similar adults is scanty [2]. Until recently, the practice of renal biopsy in all adult onset nephrotics has been unchallenged, whilst in children it has been generally accepted that biopsy is only performed after the demonstration of resistance to corticosteroid treatment at four to eight weeks. If such a policy is to be applied to adults [3,4], then data are needed for adults on just how rapidly they will respond to corticosteroid treatment. In addition, there are no detailed long-term descriptions in the literature of the results of cyclophosphamide treatment of adult-onset nephrotics with minimal change lesions on renal biopsy; this study set out to fill these gaps in our knowledge of the short- and long-term behaviour of adult-onset nephrotics with minimal changes on biopsy.

Patients, materials and methods

The *nephrotic syndrome* was defined as oedema, proteinuria in excess of 3g/24hr, and a serum albumin of less than 30g/L. *Remission* was defined as return of proteinuria to normal limits ($p<0.3g/24hr$), *relapse* as return of a nephrotic syndrome, defined as above. The pattern of relapses was described as: *frequent relapse* if two or more relapses occurred within the first six months of stopping corticosteroid treatment, or four or more relapses within the first year of disease. *Steroid dependence* was defined as two consecutive relapses appearing when the steroid dosage was reduced, or immediately upon stopping. *Infrequent relapse* was defined as the return of a nephrotic syndrome not satisfying the definitions just given.

During the period 1964–1982, 91 patients satisfying the criteria given above were biopsied at Guy's Hospital, London. Two patients did not have adequate follow-up and were excluded from analysis. Protein excretion was measured by the biuret method, serum albumin by AutoAnalyser®, as was plasma creatinine and urine creatinine.

The mean age of the patients at presentation was 42 ± 19 years (SD) (range 15–82 years) with a slight preponderance of males (50 : 39). Mean serum albumin was 19.7g/L (range 2–28g/L), and mean proteinuria was 10.2g/L, range 1.2 (in acute renal failure) to 42g/24hr.

Initial treatment

Eight patients were not treated specifically; six of these eight eventually went into remission. Of the remaining 81, 75 were treated with prednisone alone, four (in 1966–1967) with prednisone and azathioprine, and two with cyclophosphamide alone. Sixty-four of the 75 patients who were treated with prednisone alone were given a mean of 13 ± 5 weeks of treatment, range eight to 20 weeks, but 11 patients received long courses of prednisone, under the care of local physicians, of 25–32 weeks.

Secondary treatment

In general, eight weeks of cyclophosphamide 3mg/kg/24hr was given. Three patients received longer courses of 16, 40 and 70 weeks each.

Results

The rate of entry into remission is shown in Figure 1; by 16 weeks, none of the eight untreated patients had gone into remission ($p<0.005$ compared with prednisone-treated). Twenty patients still had proteinuria after 16 weeks' treatment, but three went into remission subsequently; of the remainder, 10 had persistent proteinuria and seven a persistent nephrotic syndrome.

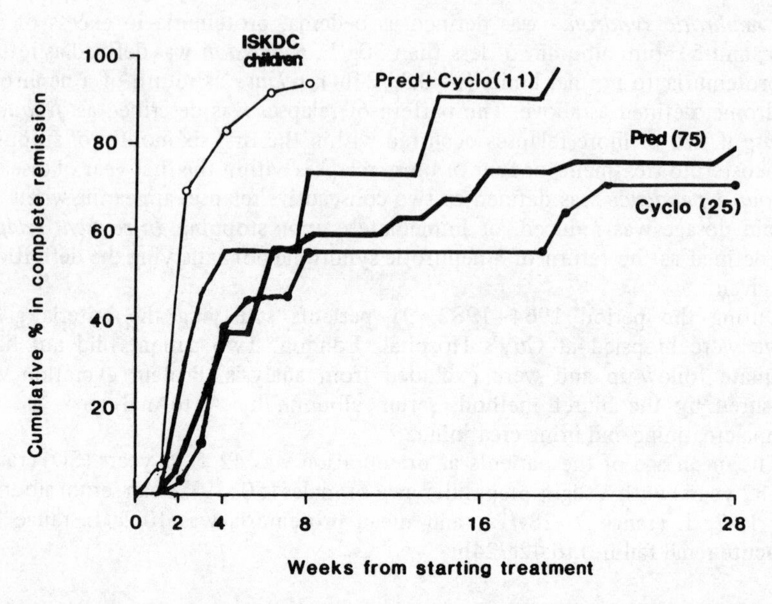

Figure 1. Time course to loss of proteinuria in adult onset nephrotic patients with minimal changes calculated by actuarial methods. Patients were treated with prednisone alone (75 patients), or cyclophosphamide (36 patients). Eleven patients were given both drugs and 25 cyclophosphamide alone. When tested by the logrank method there was no statistical difference between the curves. Corresponding data for children treated with prednisone alone are shown for comparison [1]

Relapses following corticosteroid treatment

Duration of remission in the 55 patients who achieved remission after corticosteroid treatment is shown in Figure 2: only 34 per cent remained in remission after two years. Eventually all but nine relapsed. The majority of relapses were within the first year of follow-up, and were more frequent in younger patients, relapse being infrequent in those over 50 years of age. Only six patients were steroid-dependent, as defined above.

Treatment with cyclophosphamide

The rate of remission with cyclophosphamide was similar to that achieved with corticosteroids in the 36 patients treated (Figure 1); 69 per cent lost proteinuria within 16 weeks of starting treatment. Those with rapid induction of remission with corticosteroids, also showed rapid loss of proteinuria with cyclophosphamide, but the age at presentation had no effect upon the results. Two-thirds of the patients treated with cyclophosphamide were still in remission after four years (Figure 2); of the 16 patients followed beyond this point, only one relapsed.

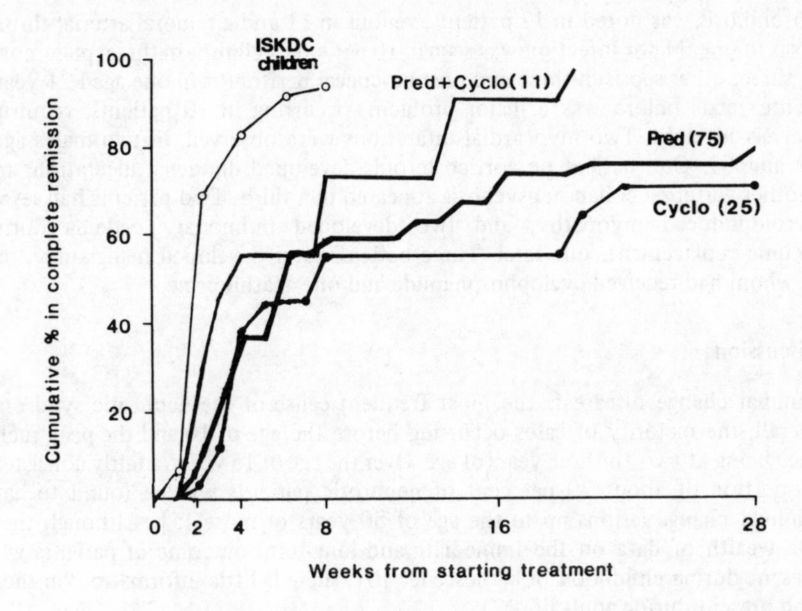

Figure 2. Long-term stability of remission in adult onset minimal change nephrotics following treatment with prednisone (broken line, 58 patients) and cyclophosphamide (continuous line, 28 patients). Remission is more prolonged after treatment with cyclophosphamide, and is superior to results obtained in similar children, even though relapse after prednisone is more frequent

Status in last follow-up

The median follow-up period was 91 ± 63 (SD) months, range two years to 24 years potential follow-up; 30 patients had been followed for more than 10 years. Fifteen patients had died (17%); all the patients who died were aged more than 42 years at onset, and the majority were aged over 60 years. Eleven of the 15 deaths occurred within three years of presentation, but only one was in chronic uraemia. Another four deaths related to complications of the nephrotic syndrome: one from pulmonary oedema following an albumin infusion whilst anuric, one from a pulmonary embolus, another during a prolonged acute renal failure during which it was discovered that she had extensive cerebral atrophy, and another from myocardial infarction. The other deaths were from incidental causes all malignant, and vascular disease in older patients.

Of the surviving patients, 59 (80%) were in complete remission, 10 had persisting proteinuria and only five (7%) had a persisting nephrotic syndrome. Surprisingly even when allowance was made for age, 14 per cent of survivors had some evidence of renal function impairment, and 21 per cent hypertension. All the patients with renal functional impairment were over 45 years of age, and four had suffered episodes of acute renal failure.

Complications

Thrombosis was noted in 12 patients, venous in 11 and a femoral arterial thrombosis in one. Major infection was seen in 10 patients, cellulitis in three, pneumonia in three, other sepsis in three and pneumococcal peritonitis in one aged 21 years. Acute renal failure was a major problem, occurring in 10 patients, requiring dialysis in eight. Two myocardial infarctions were observed, in two males aged 49 and 69. One patient on corticosteroids developed duodenal ulceration, and another vertebral collapse; psychosis appeared in a third. Two patients had severe steroid-induced myopathy, and two developed pulmonary oedema during volume replacement, one fatal. Three patients later developed malignancy, one of whom had received cyclophosphamide and one azathioprine.

Discussion

Minimal change disease is the most frequent cause of the nephrotic syndrome overall, the majority of cases occurring before the age of 10 and the peak incidence being at two to three years of age. Over the age of 15 years, a fairly consistent proportion of about 20 per cent of nephrotic patients will be found to have minimal change lesions up to the age of 80 years or more [5]. Although there is a wealth of data on the immediate and long-term outcome of patients who present during childhood or adolescence [6], there is little information on those who present during adult life [2].

It is clear that the spontaneous evolution of the minimal change nephrotic syndrome in adults is, as in children, towards remission [2]. However, these data, and our own, indicate that this spontaneous remission in general takes one to three years. During this period the patient is at risk from the many complications of the nephrotic syndrome, often in its most severe form, since the patients have relatively large protein losses through a well-preserved glomerular filtering surface. Controlled trials have shown that minimal change adult nephrotics will enter remission much more rapidly than those not treated [2,7], but the rate of remission is slower than in comparable children: only 60 per cent of our adults were in remission at eight week's treatment compared with 93 per cent of children. Again, of those 30 patients not in remission by eight weeks, only 43 per cent subsequently lost their proteinuria, compared with 60 per cent of comparable children. This may relate to relatively lower doses of prednisone given to adults than children, on a body weight or surface area basis, but response to identical body weight doses of cyclophosphamide was also slower (Figure 1).

Relapse after steroid treatment was as frequent in our adult population as in the reported children (Figure 1), but the total number of relapses per patient was fewer. In particular, steroid dependence and a frequent relapsing course were less often seen. This phenomenon was age-dependent in the adult group, in that relapse became progressively less common with age.

Treatment with cyclophosphamide resulted in an almost identical response to that observed with corticosteroids (Figure 1), even though patients with complete or relative corticosteroid resistance were included. Eight weeks of 3mg/kg/24hr of cyclophosphamide produced a stable remission in the majority of adults, two-thirds remaining in remission five years from treatment (Figure 2). This is considerably better than comparable children, but not better if only those with infrequent relapses are considered [8]. Thus, the general impression of a more mild disease with increasing age of onset is sustained.

However, outlook for the older patients was poor, mainly from complications of the nephrotic syndrome and/or its treatment; only one patient (who was not biopsied again) went into renal failure and had been corticosteroid-resistant from the outset. Only five of the 15 deaths in the older patients could be directly related to the nephrotic syndrome, or its treatment, but thrombosis and sepsis were, in this group, as overall in the nephrotic syndrome, the major problem [9].

Thus adults respond more slowly and less completely to treatment with corticosteroids than comparable children with minimal change nephrotic syndrome, but adults show greater long-term stability of remission after cyclophosphamide, which is concordant with their less frequent relapses. If adults are to be treated blind without biopsy, as some have advocated [3,4], then at least 16 weeks' treatment with its attendant toxicity will be necessary to establish 'steroid resistance'.

References

1 International Study of Kidney Disease in Children. *J Pediatr 1981; 98:* 560
2 Coggins CH. In Zurukzoglu W et al, eds. *Proceedings of the 8th International Congress of Nephrology.* Basel: Karger. 1981: 336–344
3 Hatley MA. *Lancet 1982; ii:* 1264
4 Kassirer J. *Kidney Int 1983; 24:* 561
5 Cameron JS, Glassock RJ. In Cameron JS, Glassock RJ, eds. *The Nephrotic Syndrome.* New York: Marcel Dekker. 1985: In press
6 Barnett H, Schoenman M, Bernstein J, Edelmann CM. In Edelmann CM, ed. *Pediatric Nephrology.* Boston: Little Brown. 1978: 695–711
7 Black DAK, Rose G, Brewer DB. *Br Med J 1970; 3:* 421
8 Garin EH, Pryor ND, Fennell RS, Richard GA. *J Pediatr 1978; 92:* 304
9 Cameron JS, Ogg CS, Wass V. In Cameron JS, Glassock RJ, eds. *The Nephrotic Syndrome.* New York: Marcel Dekker. 1985: In press

FOLLOW-UP PREDNISOLONE DOSAGE IN RAPIDLY PROGRESSIVE CRESCENTIC GLOMERULONEPHRITIS SUCCESSFULLY TREATED WITH PULSE METHYLPREDNISOLONE OR PLASMA EXCHANGE

M E Stevens, J M Bone

Royal Liverpool Hospital, Liverpool, United Kingdom

Summary

Of nineteen patients with RPCGN who responded promptly to initial treatment with PMP or PX, and who were subsequently maintained on oral immunosuppression with prednisolone (reducing dosage from 30mg/day) and azathioprine/cyclophosphamide (1–3mg/kg/day), five showed progressive loss of renal function within one year of responding to treatment. Both the daily dose at four weeks and the cumulative dose of prednisolone at six months were significantly lower ($p<0.01$) in the group whose renal function deteriorated. We suggest that the follow-up dosage of prednisolone may be critical in maintaining continued stable renal function in the first few months after starting PMP or PX.

Introduction

The poor immediate prognosis usually associated with rapidly progressive crescentic glomerulonephritis (RPCGN) has been shown in an earlier communication to be reversible in >70 per cent of cases by early aggressive therapy with pulse methylprednisolone (PMP) or plasma exchange (PX) [1]. The long-term outcome, however, is uncertain, with some patients showing progressive loss of renal function within one year of responding to initial treatment. The factors associated with relapse following initial response to PMP or PX were investigated and are reported in this paper.

Patients and methods

Twenty-six patients with RPCGN were treated with PMP or PX between 1976–1983 at the Regional Renal Unit, Liverpool, United Kingdom as previously described [1]. RPCGN was diagnosed when rapidly progressive renal failure was associated with normal sized kidneys and histologically confirmed severe crescentic glomerulonephritis (>50% crescents). Patients with lupus nephritis, post

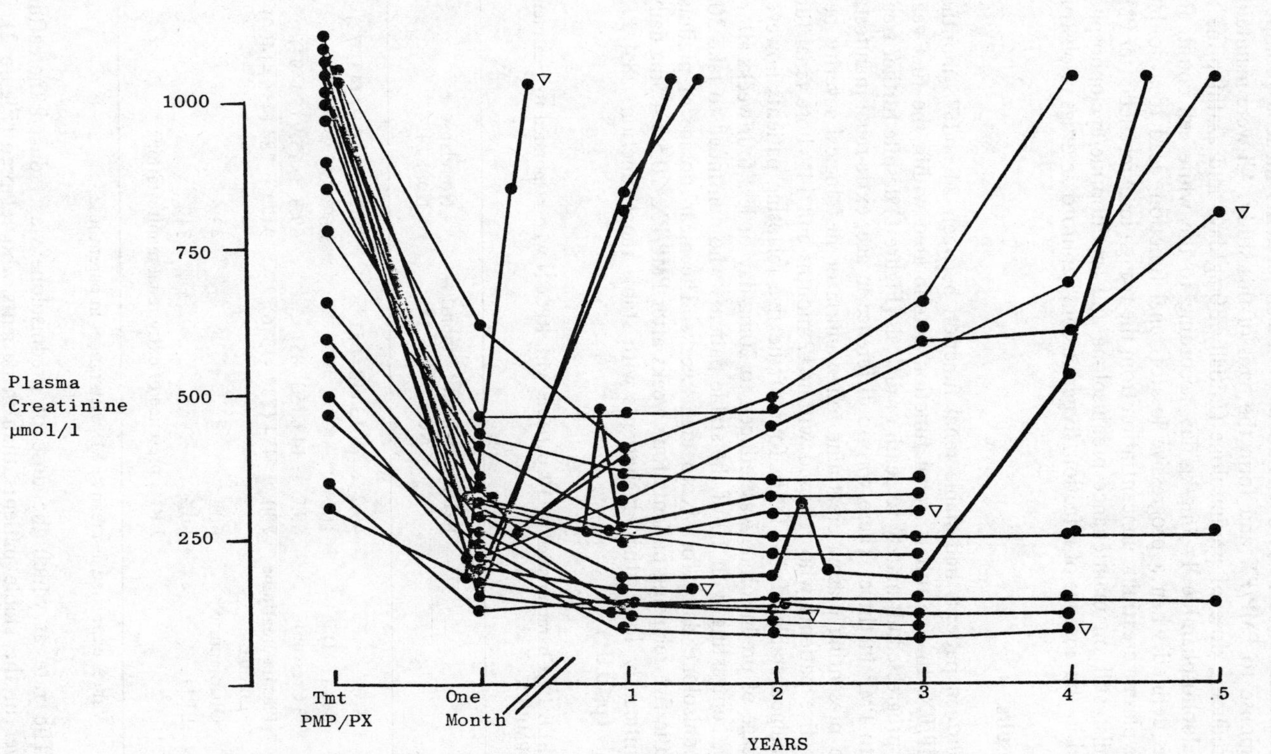

Figure 1. Follow-up renal function of 19 patients with RPCGN who responded to treatment with PMP/PX

streptococcal disease, malignant hypertension and anti-glomerular basement membrane disease were excluded. Nineteen patients made a prompt initial response to PMP/PX and form the basis of this study. All were maintained on a reducing dose of prednisolone (initially 30mg/day) and azathioprine and/or cyclophosphamide 1–3mg/kg/day according to the white cell count. Relapse was defined when a progressive loss of renal function lead to a doubling of the plasma creatinine concentration from the trough level reached after PMP/PX, whilst still on maintenance prednisolone and azathioprine/cyclophosphamide, and in the absence of infection, hypertension, nephrotoxic drugs or obstruction.

Results

Fourteen patients had stable renal function between 24 and 78 months after PMP/PX. In five others, renal function deteriorated within the first year, and all progressed ultimately to death or dialysis (Figure 1) despite further treatment with PMP in three. There was no difference in age, extra-renal manifestations, the presenting plasma creatinine, oligo-anuria or histological severity between the five patients who relapsed within 12 months and 14 whose renal function remained stable (Table I). In four of the five relapsing patients however, the dosage of prednisolone was reduced to 20mg/day or less four weeks after PMP/PX, contrasting with 10 of the stable patients who continued to take 30mg of prednisolone daily for six to eight weeks. The mean dosage of prednisolone in the five relapsing patients four weeks after PMP/PX, 20.5 ± 4.5mg daily, was significantly lower than in patients with stable renal function, 28.4 ± 3.6mg daily ($p<0.01$).

TABLE I. Characteristics of 19 patients with RPCGN who responded to treatment with PMP/PX

	Relapsed within 12 months No. : 5	No relapse No. : 14
Age	49.2 ± 18.13 (18–65)	58.79 ± 10.97 (43–72)
Sex (%M)	100%	86%
% Crescents	69.2 ± 14.6 (53–85)	69.3 ± 15.9 (50–91)
Plasma creatinine μmol/L	730 ± 454 (237–1376)	917 ± 284 (456–1400)
Oligo-anuric	60%	43%
ERM	60%	71%

ERM: presence of extra renal manifestations

Values expressed as means ± SD with ranges in parentheses

The rate at which the dose of prednisolone was reduced thereafter was lower in the stable patients than in patients who relapsed (Figure 2). Three months after PMP/PX, four of the five patients were taking 10mg or less of

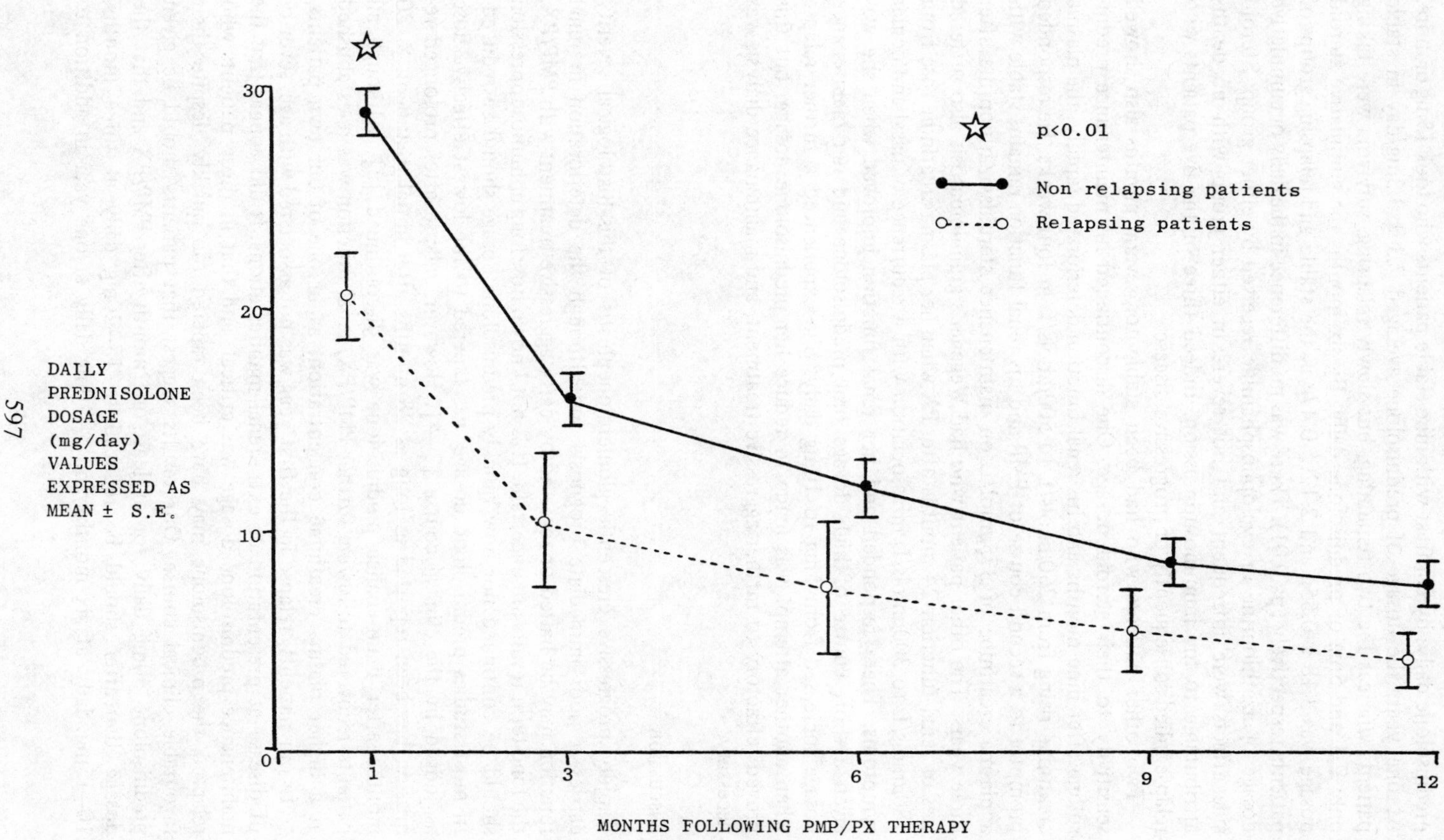

p<0.01
Non relapsing patients
Relapsing patients
DAILY PREDNISOLONE DOSAGE (mg/day) VALUES EXPRESSED AS MEAN ± S.E.
30
20
10
0
1
3
6
9
12
MONTHS FOLLOWING PMP/PX THERAPY

Figure 2

prednisolone daily, by contrast with nine stable patients who took 15mg or more. At one year, the dosage of prednisolone averaged 7.3 ± 3.3mg/day in stable patients whereas the two remaining, but slowly relapsing, survivors were taking only 2.5 and 5mg of prednisolone daily. By six months the cumulative steroid dosage was 3.18 ± 0.55g and 2.12 ± 0.47g in the stable and relapsing groups of patients respectively ($p<0.01$). There was no difference in the daily or cumulative dosage of azathioprine or cyclophosphamide received by either group. Steroid side effects were infrequent and not severe in either group with no deaths attributable to the immunosuppression. Indeed three of the five patients who relapsed died subsequently of progressive disease.

Two other patients who had been stable for several months also showed sensitivity to their steroid dosage. One discontinued his maintenance prednisolone after nine months and his renal function deteriorated rapidly, the plasma creatinine rising from 260μmols/L to 500μmols/L in four weeks. He responded promptly to a second course of PMP and his renal function remains stable with a plasma creatinine of 375μmol/L on maintenance steroids/azathioprine after three years. The other patient who had Wegener's granulomatosis also suffered loss of renal function 24 months after PX when her plasma creatinine rose from 180μmols/L to 300μmols/L in association with a recurrence of nasal and ocular symptoms. These responded and her renal function improved when she was given a short course of high dosage oral prednisolone and her maintenance dosage increased from 5mg to 10mg daily. She experienced a further relapse when another attempt was made to reduce her prednisolone dosage, but this proved refractory to further aggressive treatment, and maintenance dialysis was necessary.

Discussion

Rapidly progressive crescentic glomerulonephritis of this histological severity carries a poor immediate prognosis and although the deterioration in renal function may be halted or reversed by early aggressive treatment with PMP/PX, the long-term results are uncertain. It is not known how long immunosuppression should be continued nor how quickly prednisolone dosage should be reduced. In most studies prednisolone dosage was tapered to the lowest effective dose as judged by the clinical course [2–4]. However, relapse after prolonged remission has been reported as long as 28 months after initial treatment, 8–20 months after maintenance prednisolone was discontinued [5]. Although all patients recovered following further PMP/PX, renal function was re-established at a higher plasma creatinine concentration, as in one of our own patients.

In our patients, relapse in the first year was not associated with age, severity of disease at presentation or extra-renal manifestations. It did appear that the maintenance prednisolone dosage was critical, and that in those patients who relapsed the prednisolone may have been tapered too quickly despite their favourable clinical course. Our results suggest that patients should be given prednisolone 30mg daily for at least one month after PMP/PX and that the dosage thereafter should be reduced to 15–20mg daily at three months, 10–15mg daily at six months and 5–10mg daily at one year in addition to continuing azathioprine and/or cyclophosphamide. Some patients may require maintenance prednisolone indefinitely even in the absence of associated disease such as polyarteritis nodosa or Wegener's granulomatosis.

References

1 Stevens ME, McConnell M, Bone JM. *Proc EDTA 1982; 19:* 724
2 Kline Bolton W, Couser WG. *Am J Med 1979; 66:* 495
3 O'Neill WM, Etheridge WB, Bloomer HA. *Arch Intern Med 1979; 139:* 514
4 Lockwood CM, Rees AJ, Pinching AJ et al. *Lancet 1977; i:* 63
5 Adler S, Bruns FJ, Fraley DS et al. *Arch Intern Med 1981; 141:* 852

Proc EDTA–ERA (1984) Vol 21

EFFECTS OF DIETARY PROTEIN RESTRICTION ON THE COURSE OF EARLY RENAL FAILURE

G Zakar (Introduced by I Taraba)

B.A.Z. County Hospital, Miskolc, Hungary

Summary

Two groups of patients with histologically proven glomerulonephritis and chronic renal failure were studied. Group 1 comprised 28 patients with an endogenous creatinine clearance of 66ml/min, daily protein excretion of 1.7g. These patients were maintained for eight to 18 months on a diet containing about 40Kcal/kg and 1g/kg of protein. Phosphorous and calcium intake were unrestricted. Group 2 had 32 patients with a mean creatinine clearance of 60ml/min and daily protein excretion of 1.8g. They followed no specific dietary regimen for six to 14 months. Their dietary calorie and protein intakes averaged 30Kcal/kg and 70g respectively. At the end of the observation period creatinine clearance values of Group 1 averaged 74ml/min, the daily protein excretion amounted to 1.1g. Values for Group 2 averaged 53ml/min and 1.9g respectively.

Moderate dietary restriction of protein alone results in improving creatinine clearance values and a decreased protein excretion in the early stages of chronic renal failure in patients with glomerulonephritis.

Introduction

Severe reduction of renal mass leads to substantial proteinuria and the magnitude of this proteinuria is markedly lessened in rats by the low protein diet [1]. Protein restriction has been shown to limit the progression of nephrotoxic serum nephritis in rats [2] and of the renal disease of the NZB/NZW mouse, an animal model of lupus erythematosus [3].

Several observations indicate that in humans a high intake of protein will increase renal blood flow and glomerular filtration rate, thus producing hyperfusion and hyperfiltration, findings which are similar to those obtained in experimental animals [4,5]. Circulating hormones may be responsible for renal vasodilatation and increased glomerular filtration induced by protein ingestion [6].

TABLE I. Histological diagnoses of glomerulonephritis in the two groups of chronic renal failure patients

Histological diagnosis*	Group 1 n	Group 2 n
Glomerular minimal change	6	4
Epimembranous GN	3	2
Mesangial proliferative GN	13	18
Membranoproliferative GN	2	1
Focal-segmental GN	4	7
*(Zollinger/Michatsch)	(n = 28)	(n = 32)

GN = glomerulonephritis

TABLE II. Laboratory data and body weight in Group I chronic renal failure patients (protein restricted diet)

	Before PRD*	After PRD*
CCR (ml/min)	66.0 ± 7.6	74.0 ± 10.4
DPE (g/day)	1.7 ± 1.2	1.1 ± 1.8
BUN (mg/dl)	22.4 ± 5.3	19.0 ± 4.2
SCR (mg/dl)	1.6 ± 0.6	1.4 ± 0.4
S-TPROT (g/dl)	7.2 ± 0.4	7.0 ± 0.3
BW (kg)	72.9 ± 18.5	72.4 ± 16.4

*(All mean values ± SE)

CCR = endogenous creatinine clearance; DPE = daily protein excretion; BUN = blood urea nitrogen; SCR = serum creatinine; S-TPROT = total serum protein; BW = body weight

TABLE III. Laboratory data and body weight in Group 2 chronic renal failure patients (no diet)

	Initial values*	Final values*
CCR (ml/min)	60.0 ± 12.4	53.0 ± 8.9
DPE (g/day)	1.8 ± 0.8	1.9 ± 1.3
BUN (mg/dl)	16.8 ± 6.2	18.9 ± 5.4
SCR (mg/dl)	1.5 ± 0.8	1.6 ± 0.7
S-TPROT (g/dl)	6.9 ± 0.6	6.7 ± 0.8
BW (kg)	73.8 ± 18.2	74.2 ± 16.3

*(All values mean ± SE)

CCR = endogenous creatinine clearance; DPE = daily protein excretion; BUN = blood urea nitrogen; SCR = serum creatinine; S-TPROT = total serum protein; BW = body weight

We require treatment aimed at preventing excessive glomerular pressures and flows to check the relentless progression of clinical renal disease. An obvious first step might be a reduction of protein intake early in the course of intrinsic renal disease [7].

Methods

Two groups of patients with histologically proven glomerulonephritis and chronic renal failure were studied. *Group 1* comprised 28 patients (male 21, female 7, mean age 46 years). These patients were maintained for eight to 18 months on a diet containing about 40Kcal/kg and 1g/kg of protein (protein restricted diet) phosphorous and calcium intake were not restricted. *Group 2* had 32 patients (male 19, female 13, mean age 42 years) who had followed no specific dietary regimen for six to 14 months. Data and histological diagnoses of glomerulonephritis in both groups are summarised in Tables I, II and III. Allocation of patients to both groups was prospective and selected. All patients attended the nephrological clinic regularly, for biochemical monitoring, body weight and dietary compliance checked in form of dietary interviews. Protein and calorie intake of Group 2 patients were calculated on the basis of written dietary diaries.

Sixteen of 28 patients in Group 1, and 18 of 32 patients in Group 2 were hypertensive, good control of blood pressure could be achieved in nearly all patients.

Through the observation period no steroid or cytostatic treatment had been given. Ten of 28 patients in Group 1, and eight of 32 patients in Group 2 had been treated with cyclophosphamide and/or steroids previously.

Seven patients in Group 1, and 10 patients in Group 2 had a nephrotic range proteinuria at the time of the study.

Comments and conclusions

1. Mean values of creatinine clearance in protein restricted diet patients showed slight improvement through the observation period and daily protein excretion decreased, whereas creatinine clearance values in patients with no dietary restriction deteriorated and daily protein excretion did not change.
2. The mechanisms responsible for the observed increase of creatinine clearance values in protein restricted diet treated patients remain speculative. One can assume that a reduction in nitrogen load diminishes hyperfiltration in the remnant nephrons, which may exert a beneficial effect on the affected ones (probably via intrarenal hormones) thus increasing total filtration of the kidney.
3. The patients adherence to the diet was good. In most patients serum urea was consistently low relative to the serum creatinine, serum protein levels and body weight remained stable indicating compliance with the prescribed diet.

4. The diet caused no protein depletion or malnutrition in the seven nephrotic patients included in the protein restricted diet treatment. The practice of high protein feeding of nephrotic patients with chronic renal failure seems to be questionable on the basis of our findings.

References

1 Olson JL, Hostetter THH, Rennke HG. *Kidney Int 1982; 22:* 112
2 Farr LE, Smadel JE. *J Exp Med 1939; 70:* 615
3 Friend PS, Fernandes G, Good RA. *Lab Invest 1978; 38:* 629
4 Klahr S, Buerkert J, Purkerson ML. *Kidney Int 1983; 24:* 579
5 Pullman TN, Alving AS, Dern RJ. *J Lab Clin Med 1954; 44:* 320
6 Meyer TW, Lawrence WE, Brenner BM. *Kidney Int 1983; 24:* S243
7 Brenner BM. *Kidney Int 1983; 23:* 647

Proc EDTA–ERA (1984) Vol 21

ALTERATION OF THE COURSE OF CHRONIC RENAL FAILURE BY DIETARY PROTEIN RESTRICTION

A J Williams, S E Bennett, G I Russell, J Walls

Leicester General Hospital, Leicester, United Kingdom

Summary

The role of dietary protein restriction in the progression of chronic renal failure has been investigated in 47 patients suffering with moderate and severe renal impairment. Prescription of a diet containing 0.6g/kg body weight protein resulted in 87 per cent of the patients exhibiting a diminution in the decline of renal function, as assessed by a reciprocal creatinine plot. Those patients with the most rapidly declining renal function showed the greatest response, but no correlation with initial renal function, age, sex or drug therapy could be found. No adverse nutritional effects, within the parameters. examined, was found.

Introduction

Dietary protein restriction in the management of chronic renal failure has been previously employed to alleviate or reduce uraemic symptoms in patients with poor renal function. However, it has been shown in experimental animals made uraemic by a variety of methods that a low protein diet will prolong survival and lessen structural histological damage [1–3].

Recently a study concerned with the prevention of renal osteodystrophy in patients with chronic renal failure [4] found that a diet restricted in phosphorous and protein would reduce the rate of decline in renal function, whilst producing no adverse nutritional effects.

The present study reviews the effects of dietary protein restriction in patients with chronic renal failure over the period 1977 to 1983, with reference to change in renal function and attempts made to define the indices of those patients who demonstrate a beneficial effect.

Patients and method

Forty-seven patients, 29 male, 18 female, suffering with 'non-nephrotic' chronic renal failure of varied aetiology were studied over a period of 17.4 ± 2.24

months (mean ± SEM) (range 3 to 62 months) on a protein restricted diet. The mean age was 44.5 ± 2.1 years (mean ± SEM) (range 17 to 69 years) and serum creatinine at the start of dietary protein restriction was 780.3 ± 37.8μmol/L (range 399 to 1570μmol/L).

A diet consisting of approximately 0.6g/kg body weight protein and 20mmol phosphorous per day was prescribed, and dietary adherence was monitored by regular dietetic assessment and estimation of the urea:creatinine ratio. Blood pressure control and the patients' general care were regularly reviewed, and the serum phosphate controlled with the use of oral aluminium phosphate binding agents when necessary.

The rate of decline in renal function was assessed by the slope of the reciprocal creatinine with time [5], and a response to the diet was deemed to have occurred if a diminution in the slope of the reciprocal creatinine plot took place after the institution of protein restriction.

Results

A significant change ($p<0.001$) was seen in the slope of the reciprocal creatinine plot after the institution of the diet in 41 patients (87%) together with a significant reduction in the urea: creatinine ratio ($p<0.001$) (Table I). Blood pressure,

TABLE I. 1/Creatinine plots and urea: creatinine ratios, serum phosphate, proteins, transferrin, body weight before and after protein restriction (values expressed as mean±SEM)

	Pre Diet	Diet	p value
1/Creatinine slope	-1.1667×10^{-4} $\pm 0.1969 \times 10^{-4}$	-0.3900×10^{-4} $\pm 0.0757 \times 10^{-4}$	<0.001
Urea: creatinine ratio	0.0475 ±0.0019	0.0398 ±0.0017	<0.001
Serum phosphate ($\mu mol\ L^{-1}$)	2.063 ±0.093	1.919 ±0.060	n.s
Total proteins ($g\ L^{-1}$)	65.32 ±1.38	66.18 ±0.92	n.s
Serum Albumin ($g\ L^{-1}$)	39.18 ±0.75	39.63 ±0.68	n.s
Serum transferrin ($g\ L^{-1}$)	Not available	2.68 ±0.113	
Body weight (kg)	63.71 ±1.98	62.23 ±1.84	<0.05

serum phosphate and serum proteins remained stable throughout the period of review, and values for serum transferrin obtained were within the normal range (Table I). A small decrease in body weight was noted.

If the magnitude of change in the slope of the reciprocal creatinine is expressed as a function of the initial pre diet slope, then it was found that the greatest improvement occurred in those patients with the most rapidly declining renal function (re = -0.7504, p<0.001). If this is further examined in the aetiological groups of renal disease, the greatest change occurred in those patients with glomerulonephritis, the group with the highest value for the pre diet slope (Table II). This correlation could not be found for patients with hypertensive or polycystic renal disease. The magnitude of change in slope was not dependent upon the initial serum creatinine, age, sex, blood pressure or drug therapy. Examples of response to dietary protein restriction are shown in Figures 1 and 2.

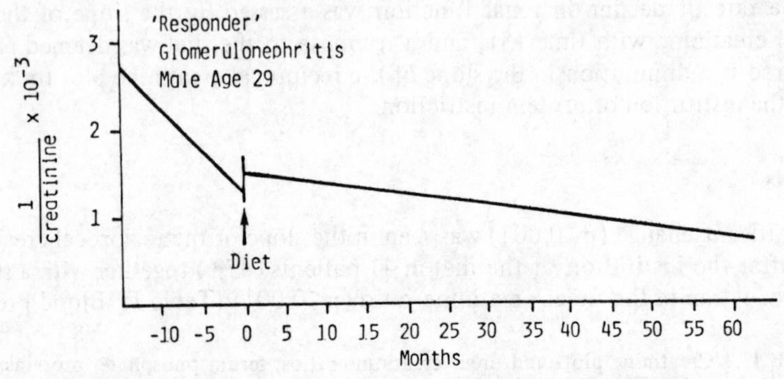

Figure 1. The effect of protein restriction on reciprocal creatinine plots in a patient with chronic glomerulonephritis

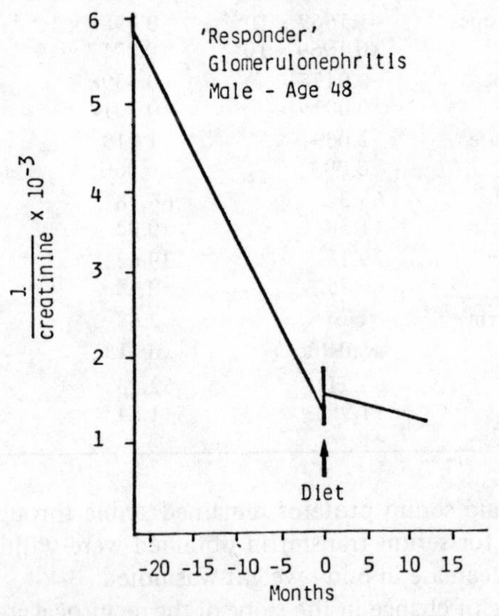

Figure 2. The effect of protein restriction on reciprocal creatinine plots in a patient with chronic glomerulonephritis

TABLE II

Aetiology of renal disease	Value of pre diet slope	Magnitude of change in slope	Correlation coefficient	p value
Glomerulonephritis n = 10	−1.511 ±0.322 x 10^{-4}	0.9939 ±0.283 x 10^{-4}	−0.762	<0.01
Chronic pyelonephritis n = 13	−0.473 ±0.066 x 10^{-4}	0.2911 ±0.1044 x 10^{-4}	−0.676	<0.01
Hypertensive nephropathy n = 7	−0.755 ±0.143 x 10^{-4}	0.442 ±0.1485 x 10^{-4}	−0.403	n.s
Polycystic kidney n = 5	−0.943 ±0.245 x 10^{-4}	0.205 ±0.113 x 10^{-4}	−0.036	n.s
Others n = 12	−1.340 ±0.387 x 10^{-4}	0.989 ±0.326 x 10^{-4}	−0.657	<0.05

All values expressed as mean ± SEM

Comment

This study shows that a protein restricted diet will favourably alter the decline in renal function in the majority of patients with chronic renal failure, thereby enabling the initiation of renal replacement therapy to be deferred. In patients with glomerulonephritis and chronic pyelonephritis, it is demonstrated that those with the most rapid decline in renal function exhibited the greatest magnitude of change in the reciprocal creatinine slope. This relationship did not exist for the patients suffering with hypertensive nephropathy or polycystic renal disease, but the numbers reviewed in these groups were small.

From the parameters examined it appeared that the protein restricted diet produced no signs of malnutrition, however a small decline in body weight was observed, but this is a common finding during the progression of the disease process [6].

We could demonstrate no correlation between the magnitude of response to the diet and the degree of renal function at which the diet was commenced, so the question still remains as to when dietary protein restriction should be commenced to obtain maximum benefit, with no detriment to the patient.

No attempt has been made to suggest the mechanism of action of the diet, however a reduction in hyperfiltration of remnant nephrons has been demonstrated in experimental animals fed a protein restricted diet [2], and a similar mechanism may be in operation in the patients reviewed.

References

1 Kleinknecht C, Salusky I, Broyer M, Gubler MC. *Kidney Int 1979; 15:* 534
2 Hostetter TH, Olson JL, Rennke HG et al. *Am J Physiol 1981; 241:* F85
3 Friend PS, Fernandes G, Good RA et al. *Lab Invest 1978; 38:* 629
4 Maschio G, Oldrizzi L, Tessitore N et al. *Kidney Int 1982; 22:* 371
5 Mitch WE, Walser M, Buffington GA, Lemann J et al. *Lancet 1976; ii:* 1326
6 Blumenkrantz MK, Kopple JD, Gutman RA et al. *Am J Clin Nutr 1980; 33:* 1567

Proc EDTA–ERA (1984) Vol 21

EFFECTS OF NALOXONE INFUSION ON GONADOTROPHIN AND PROLACTIN CONCENTRATIONS IN PATIENTS WITH CHRONIC RENAL FAILURE

M Cossu, G B Sorba, *M Maioli, *P Tomasi, *G Delitala

*Ospedali Riuniti, *Università di Sassari, Sassari, Italy*

Summary

Serum prolactin, luteinising hormone and follicle-stimulating hormone, determined by radioimmunoassay were measured during the infusion of 10mg naloxone or saline in eight male patients with chronic renal failure on regular dialysis and in seven normal controls. Neither saline nor naloxone caused any significant change in luteinising hormone, follicle-stimulating hormone or prolactin in patients with chronic renal failure. On the contrary, luteinising hormone secretion was significantly stimulated by naloxone in normal controls. Since naloxone is a specific antagonist of opiate receptors, the results would suggest a reduced hypothalamic opiate tone in patients with chronic renal failure. The data does not support the concept that the high circulating met-encephalin reported in chronic renal failure represents the pathogenetic cause of hyperprolactinaemia in chronic renal failure.

Introduction

It has been established for some time that morphine and its derivatives suppress the hypothalamic-pituitary-gonadal-axis in mammals. In particular, opiate alkaloids and endogenous opioid peptides have been shown to stimulate prolactin secretion while reducing gonadotrophin release through hypothalamic mechanisms [1]. Moreover, the continuous infusion of the opiate antagonist naloxone significantly stimulates gonadotrophin and testosterone secretion in adult males [2]. This would suggest that the hypothalamic-pituitary-gonadal-axis is tonically inhibited by an endogenous opiate tone.

Several endocrine distrubances have been described in patients with chronic renal failure. In particular, impotence, hyperprolactinaemia and reduced plasma androgen are not uncommon findings in chronic renal failure patients on long-term haemodialysis [3]. However, the mechanisms of these endocrine disfunctions are not completely clear. More recently, high values of circulating endogenous opioid peptides have been reported in uraemic patients [4]. This finding might suggest that endogenous opioid peptides are involved in the pathogenesis of hypogonadism in chronic renal failure. To test this hypothesis, the effects of the opiate antagonist naloxone on prolactin and gonadotrophin secretions was evaluated in a group of patients with chronic renal failure and in normal controls.

Materials and methods

Eight men (mean age 31.2: range 19–49 years) participated in this study for which ethical committee approval was obtained. All had chronic renal failure (creatinine clearance less than 15ml/min) due to chronic glomerulonephritis. No patient was taking digoxin, spironolactone, dopamine antagonists or any other medication with known androgenic or oestrogenic side effects. None had ever been treated with cytotoxic drugs or irradiation. Haemodialysis was performed three times weekly for an average 12–14 hours per week. Seven normal men of comparable age were studied as controls.

All studies were performed in the morning on fasting subjects. Forearm vein cannulation was performed at 9.00–9.30 hours and samples were taken -30, 0, 15, 30, 45, 60, 90, 120, 150 and 180 minutes before and following 10mg naloxone (Narcan) or normal saline. All serum samples were stored at -20°C until the individual samples from a subject could be assayed in duplicate in the same assay. Serum luteinising hormone, follicle-stimulating hormone and prolactin were estimated by radioimmunoassay using commercial kits (Biodata, Milan) with MRC standard 21/222 for prolactin, 68/40 for luteinising hormone and 69/104 for follicle-stimulating hormone. The intra- and interassay coefficients of variation of these assays were below 10 per cent. Serum testosterone was measured on extracted samples by specific radioimmunoassay (Biodata, Milan). Results are reported as mean ± SEM. Two-tailed paired 't' test and Student's 't' test were used for statistical analyses.

Results

The effects of naloxone or saline infusion on prolactin, luteinising hormone and follicle-stimulating hormone secretion are reported in Tables I and II. Naloxone had no appreciable effect on prolactin release both in uraemic and normal subjects when compared to the control infusion ($p>0.05$). Moreover, the infusion of naloxone did not suppress prolactin secretion in four uraemic patients with hyperprolactinaemia (i.e., prolactin >15ng/ml: range 22–31ng/ml). On the contrary, the opiate antagonist significantly stimulated luteinising hormone release when compared to the saline control in normal subjects (Table II). The naloxone and saline response areas also were significantly different for luteinising hormone ($p<0.01$). No appreciable luteinising hormone and follicle-stimulating hormone change was observed in uraemic patients following naloxone administration ($p>0.05$). Values of plasma testosterone in the uraemic patients were 289 ± 41ng/100ml. Normal control values were 560 ± 60ng/100ml ($p<0.05$).

The infusion of naloxone did not cause any side effects in patients or controls.

TABLE I. Serum prolactin (ng/ml) following 10mg naloxone or saline infusion in patients with chronic renal failure and in controls. Mean ± SEM is reported

		−30	0	+15	+30	+45	+60	+90	+120	+150	+180 min
Patients	saline	17.3 3.5	20.7 4.8	18.7 3.6	18.4 3.8	18.2 3.9	18.4 3.5	18.0 3.1	17.5 4.6	18.9 4.7	18.5 3.9
	naloxone	19.0 3.8 n.s	17.5 3.6 n.s	18.7 3.8 n.s	17.3 3.9 n.s	18.4 3.5 n.s	18.5 3.7 n.s	18.3 3.6 n.s	17.5 3.7 n.s	17.4 3.8 n.s	17.7 4.0 n.s
Controls	saline	6.8 1.9	7.1 1.9	5.1 1.4	5.2 1.2	5.1 1.1	5.4 1.2	5.3 1.5	6.3 1.8	6.2 2.1	5.8 1.5
	naloxone	6.1 2.0 n.s	6.3 1.7 n.s	6.8 1.9 n.s	6.0 2.0 n.s	5.7 1.5 n.s	5.9 1.5 n.s	5.5 1.4 n.s	5.7 1.4 n.s	5.8 1.6 n.s	6.5 1.7 n.s

TABLE II. Serum luteinising hormone (mIU/ml) and follicle-stimulating hormone (mIU/ml) (mean ± SEM) following saline or naloxone infusion in patients with chronic renal failure and in controls

		−30	0	+15	+30	+45	+60	+90	+120	+150	+180 min
						Luteinising hormone					
	saline	50.8	50.0	49.9	66.5	58.0	47.7	60.8	55.5	55.7	60.6
		25.6	23.4	24.5	28.9	25.8	26.5	24.6	25.7	29.5	27.6
Patients											
	naloxone	52.1	50.6	52.3	54.3	54.0	63.4	61.0	53.5	53.4	61.7
		17.5	24.3	28.1	28.5	27.6	29.5	27.4	24.5	23.1	26.4
		n.s	n.s	n.s	n.s	n.s	n.s	n.s	n.s	n.s	n.s
	saline	6.3	6.2	6.0	5.7	4.9	4.8	6.8	6.9	6.0	6.3
		1.0	1.2	1.5	1.7	1.2	1.8	1.7	1.6	1.4	1.3
Controls											
	naloxone	5.5	6.2	9.0	9.2	9.1	9.3	7.9	7.9	8.8	7.9
		1.7	0.8	2.6	2.9	1.0	0.9	1.2	1.1	1.7	1.8
		n.s	n.s	n.s	n.s	0.05	0.02	n.s	n.s	0.02	n.s
						Follicle-stimulating hormone					
	saline	21.7	24.0	24.3	27.0	22.3	18.4	23.2	22.0	21.8	22.6
		12.1	15.2	15.1	14.3	14.3	11.9	12.0	11.7	12.1	10.0
Patients											
	naloxone	18.7	19.1	20.7	22.1	21.0	21.4	22.5	23.1	21.5	21.7
		10.1	12.3	12.4	11.5	11.0	10.7	11.3	10.5	9.9	8.7
		n.s	n.s	n.s	n.s	n.s	n.s	n.s	n.s	n.s	n.s
	saline	7.0	7.5	7.1	6.8	6.5	6.5	6.6	7.0	7.2	7.3
		1.5	1.2	1.8	1.0	1.4	1.6	1.7	1.8	1.6	1.8
Controls											
	naloxone	6.5	7.1	7.3	7.0	7.1	7.0	7.2	7.5	7.6	7.8
		2.0	1.5	1.6	1.1	1.6	1.6	1.7	1.8	1.7	2.0
		n.s	n.s	n.s	n.s	n.s	n.s	n.s	n.s	n.s	n.s

Discussion

Endogenous opioid peptides are putative neurotransmitters or neuromodulators in the brain and recent data suggest an important role of opioids and their receptors in the regulation of anterior pituitary hormone secretion in man [5]. In particular, the stimulation of opiate receptors reduces gonadal steroid and gonadotrophin secretion while stimulating prolactin release both in men and women [6]. The endocrine effects induced by opioids likely represent the pathogenetic mechanism of amenorrhoea and impotence in chronic heroin addicts. It is commonly accepted that opiates suppress gonadotrophin and stimulate prolactin secretion at the hypothalamic level whereby they alter the activity of hypothalamic neurotransmitters controlling anterior pituitary hormone secretion. Experimental evidence also suggests that narcotics depress the activity of hypothalamic luteinising hormone-releasing hormone producing neurons and reduce the amount of dopamine released from the tubero-infundibular neurone terminals, thereby allowing prolactin to increase [1]. Since naloxone is a specific opiate receptor antagonist, alteration of endogenous hormones following opiate receptor blockade is interpreted as indicating that the basal secretion of these hormones is tonically regulated by an endogenous opioid tone. The results of the present experiment show that the infusion of naloxone clearly stimulated luteinising hormone secretion in normal adults confirming previously published data [7]. Moreover, the administration of naloxone failed to modify gonadotrophin secretion in patients with chronic renal failure. Finally, the results show that the opiate antagonist did not reduce serum prolactin both in normo- and hyperprolactinaemic uraemic patients as well as in normal controls.

Alterations in hypothalamic-pituitary-gonadal function in ureamia are common clinical findings. These include reduced libido, impotence, varying degrees of spermatogenic arrest resulting in oligospermia or azospermia, low values of gonadal steroids with high normal or modestly elevated gonadotrophins. Since no correlation between serum testosterone and serum luteinising hormone may be found in uraemic patients, some authors suggested that a defect in hypothalamic responsiveness to feed-back control may be present in these patients [8,9]. Our additional data show a lack of luteinising hormone responsiveness to naloxone in patients with chronic renal failure. Since modulations of frequency and amplitude of luteinising hormone-releasing hormone secretory activity appear to be mediated through an inhibitory action of endogenous opioids, the lack of luteinising hormone response to opiate receptor blockade may be an expression of markedly reduced endogenous opiate receptor activity in uraemic patients. If so, this aspect may represent another mechanism of the deranged hypothalamic-pituitary-gonadal status in uraemia. The prolonged half-life of luteinising hormone as well as reduced testosterone in chronic renal failure may partly contribute to the explanation of our findings, since experimental data have recently shown that the opiate control of luteinising hormone secretion is eliminated by orchidectomy [10]. However, we could not find any effect of naloxone on luteinising hormone secretion in some patients with normal serum testosterone. Moreover, the anti-oestrogenic drug clomiphene is still able to stimulate luteinising hormone secretion in chronic renal failure patients, suggesting the existence of still adequate blood sex steroids and/or the hypothalamic-pituitary-unit of these patients. Therefore, the reduced testosterone in patients with chronic renal failure do not seem to represent the main cause of the absent luteinising hormone response to naloxone observed in the present experiment.

Finally, naloxone had no significant effect on prolactin both in normal and uraemic patients. There is thus no evidence to support the view that endogenous opiates are physiologically important in human prolactin secretion, at least in basal conditions. In particular, the lack of naloxone effect on prolactin in hyperprolactinaemic uraemic patients does not suggest that the high circulating opioid peptides reported in uraemic patients represent the pathogenetic cause of hyperprolactinaemia in chronic renal failure.

Acknowledgments

This work was supported by a grant from the Ministero della Pubblica Istruzione to G Delitala.

References

1 Meites J, Bruni JF, Van Vugt DA, Smith AF. *Life Sci 1977; 24:* 1325
2 Delitala G, Giusti M, Mazzocchi G et al. *J Clin Endocrinol Metab 1983; 57:* 1277
3 Nagel TC, Freinkel N, Bell RH et al. *J Clin Endocrinol Metab 1973; 36:* 428
4 Smith R, Grossman A, Gaillard R et al. *Clin Endocrinol 1981; 15:* 291
5 Grossman A, Rees LH. *Br Med Bull 1983; 39:* 83
6 Delitala G, Grossman A, Besser GM. *Neuroendocrinology 1983; 37:* 275
7 Grossman A, Moult P, Gaillard R et al. *Clin Endocrinol 1981; 14:* 41
8 Krølner B. *Acta Med Scand 1979; 205:* 623
9 Lim VS, Fang VS. *Am J Med 1975; 58:* 655
10 Bhanot R, Wilkinson M. *Endocrinology 1983; 112:* 399

Proc EDTA–ERA (1984) Vol 21

HUMAN PANCREATIC POLYPEPTIDE AND SOMATOSTATIN IN CHRONIC RENAL FAILURE

R Lugari, **S David,** P Dall'Argine, V Nicolotti, A Parmeggiani, A Gnudi, *A Luciani, *S Toscani, *R Zandomeneghi

*Università di Parma, Parma, *Universita di Modena, Modena, Italy*

Summary

Human pancreatic polypeptide is the only hormone so far reported which clearly suppresses somatostatin release, suggesting that this peptide may have a role in controlling somatostatin secretion from the gut and pancreas.

In this study endogenous high circulating human pancreatic polypeptide concentrations in patients with chronic renal failure do not decrease somatostatin circulating levels.

The reduced clearance rate of somatostatin in chronic renal failure may partially account for the normal circulating levels of somatostatin observed in our patients with respect to controls.

Renal insufficiency may, itself, induce an increase in some gastrointestinal peptides capable of stimulating somatostatin secretion.

Introduction

The kidneys play a central role in the turnover and regulation of human pancreatic polypeptide [1–3].

Elevated plasma human pancreatic polypeptide concentrations have been found in patients with chronic renal failure [1–4], suggesting the possibility of human pancreatic polypeptide being involved in the development of uraemic gastrointestinal symptoms.

Acute intravenous administration of synthetic human pancreatic polypeptide clearly decreases immunoreactive somatostatin in the rat [5]. It has also been reported that in a patient with a human pancreatic polypeptide cell-tumour who showed extremely high circulating human pancreatic polypeptide levels, plasma somatostatin was undetectable, but rose to a normal level after removal of the tumour [5].

The aim of this study was to evaluate the effect of chronic high endogenous human pancreatic polypeptide plasma levels on circulating somatostatin in man.

The effect of haemodialysis on human pancreatic polypeptide and somatostatin concentrations was also studied.

Materials and methods

Ten patients with stable chronic renal failure (mean creatinine = 2–6mg/dl) were studied (Group A). After an overnight fast, two baseline blood samples were collected for human pancreatic polypeptide and somatostatin determination.

Twelve age-matched healthy volunteers were examined as controls. In another 12 overnight fasted chronic renal failure patients undergoing chronic dialysis (Group B), blood samples for human pancreatic polypeptide and somatostatin determination were collected before and after haemodialysis.

Results are expressed as mean ± SD. Statistical evaluation was performed using Student's 't' test.

Determination of human pancreatic polypeptide

Immunoreactive human pancreatic polypeptide was measured by a sensitive (4pg/tube) and validated by radioimmunoassay [6]. After centrifugation, 1ml of plasma was precipitated with 2ml of ethanol and supernatant was stored at $-20°C$ until the radioimmunoassay. Highly purified human pancreatic polypeptide (from Dr Ronald Change) was used both for iodination and standard. A standard curve was made in hormone-free plasma and treated as above before use in the assay.

Iodination of human pancreatic polypeptide was performed with ^{125}I using chloramine T as the oxidising agent. The reagents were added in the following sequence and amounts: 100μl of 100μg/ml solution of human pancreatic polypeptide (10μg) in 2M phosphate buffer pH 7.4, 10μl of a 100mCi/ml solution of Na ^{125}I in 4M sodium phosphate buffer (1mCi). The reaction was initiated by adding 10μl of a 2g/l solution of chloramine T (20μg) and was allowed to continue for 15 seconds at room temperature; it was stopped by adding 100μl of a 5g/l solution of metabisulphite (50μg).

Purification of ^{125}*IPP* the iodinated material was chromatographed on a Sephadex C-25 column (0.9 x 30cm) and eluted with phosphate buffer 40mM pH 7.4 with 0.2 per cent human serum albumin at 8ml/hr speed. Fractions of 1ml were collected and the radioactivity in each fraction was determined in a gamma counter. The ^{125}IPP peak arrived between fractions 60–80. Rabbit anti human pancreatic polypeptide serum (lot 204, generous gift from Professor KD Buchanan, University of Belfast) was used at final dilution of 1:48,000.

No crossreactivity was found with other gastrointestinal hormones.

Conditions for incubation 0.1ml ethanol plasma extract and 0.1ml antiserum were kept at 4°C for 48 hours. Then 0.1ml ^{125}IPP (5,000cpm/tube) was added and the incubation continued at 4°C for a further 48 hours.

Separation of bound from free ^{125}IPP was performed with 1ml charcoal-dextran at one per cent (10/1).

Accuracy and reproducibility the intra-assay variation of the assay (a measure of accuracy), determined from 25 identical tubes, was 10.2 per cent; the inter-assay variation among 10 assays (a measure of reproducibility) was 16.3 per cent.

Sensitivity the smallest amount that can be detected is about 4pg/tube.

Determination of somatostatin

Immunoreactive somatostatin was measured by a sensitive and specific radioimmunoassay. After centrifugation 1ml of plasma was precipitated with 2ml of ethanol and supernatant was stored at -20°C until radioimmunoassay. Synthetic somatostatin was used as standard; standard curve was performed in hormone-free plasma and treated as the samples before use in the assay. Iodinated somatostatin was obtained by NEN. Rabbit anti somatostatin (gift from Professor KD Buchanan, University of Belfast) was used.

Conditions for incubation 0.1ml ethanol plasma extract and 0.1ml antiserum were maintained at 4°C for 24–48 hours. Then 0.1ml ^{125}somatostatin was added and the incubation continued at 4°C for a further 48 hours.

Separation of bound from free ^{125}somatostatin was performed using 1ml charcoal-dextran at one per cent (10/1).

Accuracy and reproducibility the intra-assay variation determined from 30 identical tubes was 11.3 per cent; the inter-assay among 15 assays was 14.9 per cent.

Sensitivity the smallest amount that can be detected is about 5pg/ml.

Results

Basal plasma PP concentrations in Group A (180 ± 50pg/ml) were significantly higher ($p<0.001$) than those in normal subjects (41 ± 8pg/ml). Mean fasting human pancreatic polypeptide in Group B (725 ± 517pg/ml) did not change after haemodialysis (765 ± 616pg/ml). The degree of human pancreatic polypeptide elevation in patients with chronic renal failure well correlated with the degree of renal insufficiency ($p <0.001$).

No significant variation in somatostatin was observed between controls: 55 ± 12pg/ml and Group A: 48 ± 18pg/ml.

Group B showed similar basal somatostatin values (52 ± 15pg/ml) which were not significantly modified by haemodialysis (74 ± 22pg/ml).

Discussion

Our findings confirm that the mean basal plasma human pancreatic polypeptide is clearly increased in patients with stable chronic renal failure. In the present study a close correlation between the degree of renal insufficiency and plasma human pancreatic polypeptide increase is further shown. Haemodialysis has no effect on the human pancreatic polypeptide concentration.

Previous studies on the physiological function of human pancreatic polypeptide suggest that somatostatin producing D-cells and human pancreatic polypeptide cells control the secretory activity of each other in a paracrine fashion [5]. It is known that somatostatin exerts an inhibitory effect on human pancreatic polypeptide in man [7]. The suppression of somatostatin by high human pancreatic polypeptide plasma levels has been shown in the rat after acute intravenous injection of the peptide and in a single case of a patient with a human pancreatic polypeptide-cell tumour [5].

In our study mean basal plasma somatostatin value in subjects with chronic renal failure is not reduced with respect to controls, in spite of high circulating levels of human pancreatic polypeptide.

The immediate plasma somatostatin fall observed in the rat following an acute stimulus may be the consequence of a pharmacological treatment and thus the human pancreatic polypeptide inhibitory effect could not necessarily be of physiological significance. The extremely high circulating human pancreatic polypeptide levels in the presence of a human pancreatic polypeptide-cell tumour are not documented as absolute values [5]. Presumably the tumour is able to produce higher plasma human pancreatic polypeptide concentrations than stable chronic renal failure.

In both conditions, however, the renal function was normal. It is known that somatostatin clearance rate is reduced in chronic renal failure, suggesting that the kidney may play a role in the metabolism of the hormone [8,9]. It could partially account for the human pancreatic polypeptide failure to reduce somatostatin levels in the patients studied.

Finally, since various gut hormones and pancreatic glucagon stimulate release of somatostatin [10], stable chronic renal failure could induce an increase of circulating peptides able to maintain a normal plasma somatostatin.

Acknowledgment

This work was supported by 'CNR- Contratto Finalizzato, Medicina Preventiva e Riabilitativa' SP4, Obiettivo 47b, Rome, Italy.

References

1 Boden G, Master RW, Owen OE et al. *J Clin Endocrinol Metab 1980; 51:* 573
2 Lamers CBHW, Diemel CM, Van Leer E et al. *J Clin Endocrinol Metab 1982; 55:* 922
3 Boden G, Master RW, Rudnick MR et al. *Gastroenterology 1979; 76(5):* 1104
4 Hallgren R, Lunquist G, Change RE. *Scand J Gastroent 1977; 12:* 923
5 Arimura A, Meyers CA, Case WL et al. *Biochem Biophys Res Comm 1979; 89:* 913
6 Zandomeneghi R, Lugari R, Parmeggiani A et al. *J Endocrinol Invest 1984; 7 (Suppl 1):* 99
7 Marco J, Hedo JA, Villaneuva ML. *Life Sci 1977; 21:* 789
8 Sheppard M, Shapiro B, Pimstone B et al. *J Clin Endocrinol Metab 1979; 48:* 50
9 Polonsky KS, Jaspan B, Berelowitz DS et al. *J Clin Invest 1981; 68:* 1149
10 Unger RH, Dobbs RE, Orci L. *Ann Rev Physiol 1978; 40:* 307

PART XV

GUEST LECTURE ON DIABETIC NEPHROPATHY

Chairmen: L Migone
S Giovannetti

Proc EDTA–ERA (1984) Vol 21

DIABETIC NEPHROPATHY: A PREVENTABLE COMPLICATION?

C G Viberti, M J Wiseman, J J Bending

Guy's Hospital Medical School, London, United Kingdom

Renal failure is a complication of diabetes mellitus that occurs in approximately 45 per cent of insulin-dependent patients [1]. Its onset is heralded by the development of dipstix positive proteinuria (i.e. total urinary protein excretion >0.5g/24 hours) in a patient with usually 10 years or more of diabetes who has concomitant retinopathy and rising arterial pressure, but is free of other renal disease, urinary tract infection and cardiac failure. The incidence of diabetic nephropathy peaks at about 16 years after the onset of the disease and once manifest the condition progresses relentlessly to end-stage renal failure. If left untreated 50 per cent of the patients die within seven years [1]. With the appearance of persistent proteinuria glomerular filtration declines linearly with time at an average rate of 1.2ml/min/month, ranging widely between different individuals from 0.5 to 2.4ml/min/month [2]. The reasons for the different rates of decline are largely unknown but different degrees of blood pressure rise may play a role. Indeed arterial pressure starts to climb at a very early stage of renal involvement well before it was previously thought [3]. With the fall in glomerular filtration rate the proteinuria becomes heavier and the clearances of albumin and IgG increase [4]. There is a passage from a proteinuria highly selective for albumin in the early stages when the glomerular filtration rate is still normal or only moderately reduced to a low selectivity proteinuria with proportionally more IgG being cleared when the filtration rate is markedly reduced (Figure 1). One possible interpretation of this phenomenon is that size selectivity defects of the glomerular membrane develop in late nephropathy accounting for the proportionately increased transit of IgG [5], the early selective albuminuria being more likely explained by a change in the charge selectivity properties of the glomerular membrane [6]. A compensatory intra-glomerular pressure increase in the surviving glomeruli would almost certainly accompany these membrane injuries and perhaps perpetuate them [7].

Management of established diabetic nephropathy

Therapeutic attempts of different kinds have been made in order to arrest progression of the disease. An obvious possibility was that improvement of

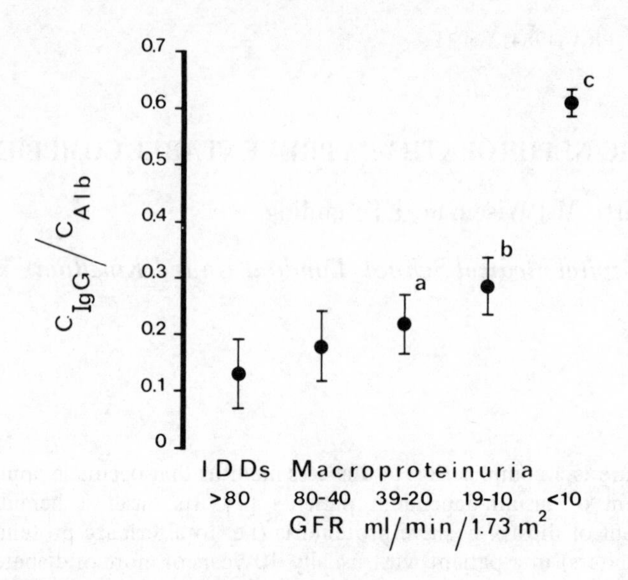

Figure 1. Selectivity index (SI=clearance of IgG/clearance of albumin) in proteinuric diabetic patients with different degrees of glomerular failure. As the glomerular filtration rate (GFR) decreases the SI increases from 0.13 for GFRs >80ml/min/1.73m² to 0.62 for GFRs >80ml/min/1.73m². This is caused by a greater filtration of IgG relative to albumin in advanced diabetic nephropathy [6]

glycaemic control would affect the course of this condition. Results with strict diabetic control have been by and large disappointing and controlled studies have failed to show a significant effect of correction of hyperglycaemia [8]. More promising were the findings of Parving et al [9] showing that hypotensive treatment, started early in the course of diabetic renal failure, is capable of slowing deterioration of renal function. The effect of low protein diet has been little explored in diabetic renal failure, but is known to retard progression in other renal disease [10]. Figure 2 shows two patients with early and moderate diabetic renal failure followed for approximately five years. The blood pressure in these patients was kept stable throughout. Correction of hyperglycaemia with insulin pump treatment had little impact on the progressive linear deterioration of their renal function while on a normal protein intake of 80g of protein per day. Changing to low protein diet (40g/day) over the last year had a dramatic effect on their renal function which appears to have stabilised. This was so both in the patient who continued pump therapy (WP) and in the one who stopped it (RW). A longer follow-up is clearly needed to substantiate these findings, which nevertheless look promising. Except for the treatment of systemic arterial hypertension no pharmacological intervention has thus far been tried in order to remove some of the severe haemodynamic and chemical-physical disturbances that occur in the surviving glomeruli and contribute to renal disease, and perhaps future research should concentrate more on these aspects. The overall conclusion is that therapeutic manoeuvres in overt diabetic nephropathy can at best slow progression, but no arrest of reversal of the disease is possible at present.

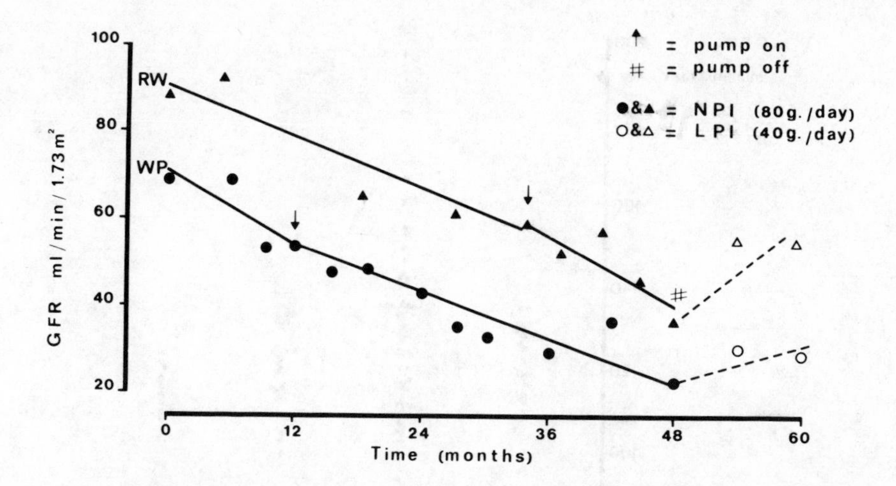

Figure 2. Glomerular filtration rate (GFR) changes over time in diabetic patients with clinical nephropathy. Institution of continuous subcutaneous insulin infusion (pump=↓) with marked amelioration of glycaemic control did not affect significantly the rate of fall of GFR. Reduction of protein intake to 40g/day (open symbols; broken line) for up to one year appears to have stopped the decline in GFR

Early diabetic renal disturbances: a new story

The state of affairs in established nephropathy has prompted a number of investigations aimed at characterising the early disturbances of renal function in diabetes before clinical proteinuria develops. A number of abnormalities have been demonstrated. It was found that both in newly diagnosed and in patients of some years standing the glomerular filtration rate is often elevated [11,12]. This seems to occur in approximately 20–30 per cent of the patients (Figure 3) and to be induced by moderate degrees of hyperglycaemia [12] as well as elevation of glucoregulatory hormones [13,14]. The determinants of this glomerular hyperfunction have been explored with micropuncture techniques in diabetic rats and have been shown to consist mainly of profound renal vasodilatation, more marked in afferent than efferent glomerular arterioles and so resulting in both increased glomerular plasma flow rate, and mean transglomerular hydraulic pressure gradient [15]. The high glomerular filtration rate in humans has been shown to be strongly associated with a large kidney volume [16,17]: so much so that although a large kidney volume can be associated with a normal glomerular filtration rate, a high glomerular filtration rate cannot occur in a normal size kidney (Figure 4). Large kidneys reflect primarily an increase in the tubular mass but large glomeruli are also present. The latter would increase the surface area available to filtration and a strong correlation has been described between glomerular filtration rate and glomerular filtering area in diabetic patients [18]. It is therefore possible that in man the elevation of glomerular filtration rate is a combination of increases not only in flow and pressure but also in the surface area available for filtration.

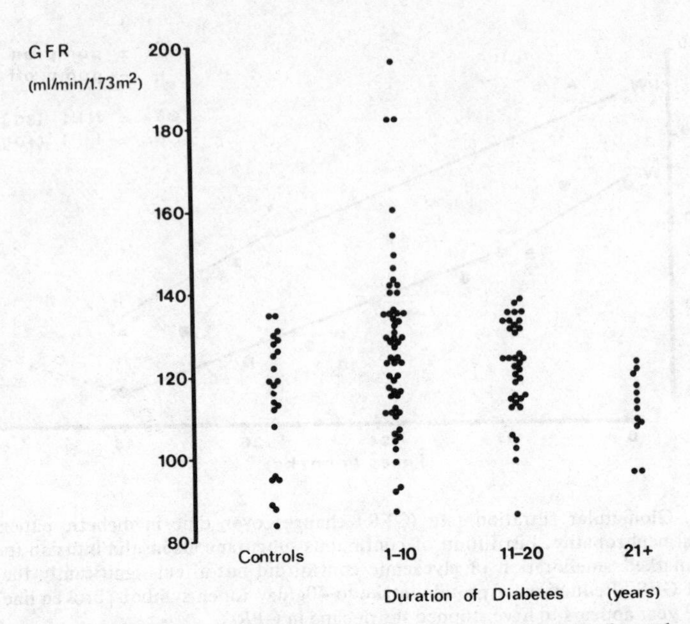

Figure 3. Glomerular filtration rate (GFR) in 118 non-proteinuric insulin-dependent diabetic patients divided by duration of diabetes. Supranormal GFR is present in approximately 20 per cent of the patients, the higher proportion in the younger group with diabetes duration between 1–10 years

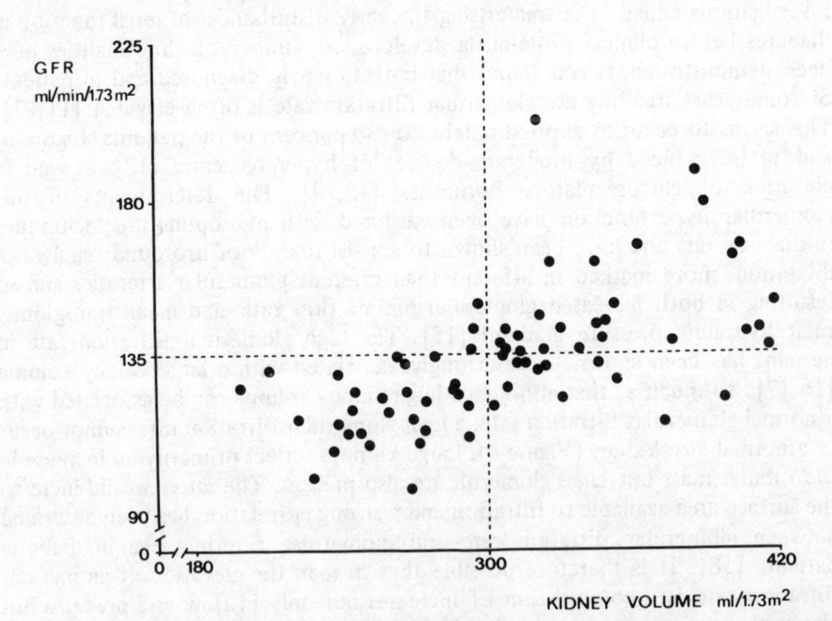

Figure 4. Relationship between 69 paired measurements of glomerular filtration rate (GFR) and kidney volume (KV) in 35 non-proteinuric insulin-dependent diabetics. Broken lines indicate the upper limit of normal for GFR and KV respectively. A large kidney can be associated with high or normal GFR, but a high GFR occurs only in large kidneys

Another abnormality that can be encountered in the kidney in early diabetes is that of an increased urinary excretion of albumin [19]. This is below the limit of detection of conventional urine tests for albumin (e.g. Albustix test) but above the normal values that in our laboratory range between 2.0 and 25mg/24hr. Albumin excretion rates above 25mg/24hrs but usually below 250mg/24hr have been defined as microalbuminuria. With excretion rates in excess of 250mg/24hr the Albustix test becomes positive and a different phase of macroalbuminuria and clinical proteinuria is entered (Table I). The microalbuminuria of diabetes is accompanied by increases in the urinary excretion

TABLE I. Definition of different levels of albumin excretion rate in healthy controls and in diabetic patients

	Albumin excretion rate (mg/24hr)
Normal subjects (also ca 60% of insulin dependent diabetics)	⩽25
Microalbuminuric diabetics (ca 40% of insulin dependent diabetics)	25–250
Macroalbuminuric diabetics (up to 45% of insulin dependent diabetics eventually)	>250

of IgG but not by changes in that of β_2-microglobulin [6]. These findings, together with other theoretical and in vitro experimental data [20,21] suggest that the microalbuminuria of diabetes is glomerular in origin. Interestingly the progressive increase of albumin excretion rates within the microalbuminuric range is not followed by a progressive proportional increase in the IgG excretion [6]. This leads to a fall in the selectivity index and to the appearance of a high selectivity proteinuria (for albumin), a condition encountered in early clinical nephropathy (Table II). Different studies have now shown that microalbuminuria

TABLE II. Urinary excretion of albumin and IgG and selectivity index (SI) in diabetic patients with normo- and microalbuminuria

	Normoalbuminuria		Microalbuminuria		
Albumin (mg/24hr)	3.5	18	45	90	200
IgG (mg/24hr)	0.7	3.5	9	9	10
SI	0.56	0.55	0.56	0.28	0.14

is associated with poorer glycaemic control and independently with higher blood pressure [22]. The arterial pressure need not be in the hypertensive range, but is found to be significantly higher than that of a matched normoalbuminuric group, although within the so-called normal range. These early physiological changes are also accompanied by histological renal changes which have been better characterised in the diabetic animal model [23]. However, these are not meant to be part of this presentation and will not be discussed further.

Markers of diabetic nephropathy

The significance of these early diabetic renal abnormalities has remained obscure for several years. Only recently, long-term prospective studies [24–26] have indicated that certain levels of albumin excretion rate in the microalbuminuric range are strongly predictive of late diabetic nephropathy. Although no clear agreement has been reached on the discriminating level of albumin excretion rate that identifies an at risk patient, probably because of different study protocols, it would appear that values in excess of 45–50mg/24hr carry a risk of late nephropathy which is 20 times higher than that of patients with lower albumin excretion rates. Moreover, recent evidence suggests that diabetic patients with extreme hyperfiltration (i.e. glomerular filtration rate $>$150ml/min/1.73^2) and microalbuminuria may lose glomerular function at a greater rate than normofiltering diabetics [26]. Although these data are preliminary and need confirmation they would support the hypothesis, generated by animal evidence, that glomerular hyperfunction and hypertension is, in the long-run, detrimental to the kidney, leading to glomerular destruction [7]. The glomerular hyperfiltration has also been claimed to be associated with marginal increases in arterial pressure by some workers [26] but this seems to be mediated only through an increase of albumin excretion rate [25,26]. Thus at least two early markers of diabetic renal disease have been identified: microalbuminuria (of a certain degree) and hyperfiltration (of a certain level).

Management of the early renal anomalies and potential for prevention of diabetic nephropathy

In sharp contrast to the relative insensitivity to treatment of established diabetic nephropathy the early phase abnormalities respond to therapeutic intervention. Both the microalbuminuria and the hyperfiltration are significantly reduced and often normalised by intensified insulin treatment and strict glycaemic control [27,28]. Moreover pharmacological intervention with a thromboxane synthetase inhibitor has been shown to reduce microalbuminuria independently of blood glucose changes [29].

Preliminary evidence in the diabetic rat would suggest that at similar levels of hyperglycaemia a low protein diet protects the kidney from hyperfiltration, albuminuria and the consequent histological lesions [30]. Interestingly recent observations in humans with different protein intakes indicate that vegans, who eat less protein, and all of vegetable origin, have significantly lower glomerular filtration rates than omnivores who eat more protein, largely of animal origin (Table III).

Thus the early abnormalities of the diabetic kidney are reversible, at least in part, by a variety of therapeutic manoeuvres. Correcting an early marker of

TABLE III. Mean (± SD) glomerular filtration rate (GFR) and daily protein intake (PI) in omnivore and vegan subjects

	GFR (ml/min/1.73m²)	PI (g/24hr)	Animal protein (g/24hr)	Vegetable protein (g/24hr)
Omnivores (n=17)	115±16	74±18	52±16	22±6
Vegans (n=14)	100±13**	59±22*	0	59±22

**=p<0.02; *=p<0.05

disease does not necessarily imply that the disease will be abolished; however, the possibility now exists to test, in controlled clinical trials of diabetic patients at risk of nephropathy, whether prevention of diabetic kidney failure is attainable.

Acknowledgments

We thank all subjects who helped us in these studies, Professor H Keen for unfailing support and inspiration and Mrs J Gray for typing the manuscript.

References

1 Andersen AR, Christiansen JS, Andersen JK et al. *Diabetologia 1983; 25:* 496
2 Viberti GC, Bilous RW, Mackintosh D, Keen H. *Am J Med 1983; 74:* 256
3 Parving HH, Smidt UM, Friisberg B et al. *Diabetologia 1981; 20:* 457
4 Viberti GC, Mackintosh D, Bilous RW et al. *Kidney Int 1982; 21:* 714
5 Myers BD, Winetz JA, Chui F, Michaels AS. *Kidney Int 1982; 21:* 633
6 Viberti GC, Keen H. *Diabetes 1984; 33:* 686
7 Hostetter TH, Rennke HG, Brenner BM. *Am J Med 1982; 72:* 375
8 Viberti GC, Bilous RW, Mackintosh D et al. *Br Med J 1983; 286:* 598
9 Parving HH, Anderson AR, Smidt UM, Svendsen PA. *Lancet 1983; i:* 1175
10 Maschio G, Oldrizzi L, Tessitore N et al. *Kidney Int 1982; 22:* 371
11 Mogensen CE. *Diabetes 1976; 25:* 872
12 Wiseman MJ, Viberti GC, Keen H. *Nephron 1984:* in press
13 Parving H-H, Christiansen JS, Noer I et al. *Diabetologia 1980; 19:* 350
14 Christiansen JS, Gammelgaard J, Frandsen M et al. *Diabetologia 1982; 22:* 333
15 Hostetter TH, Troy JL, Brenner BM. *Kidney Int 1981; 19:* 410
16 Mogensen C, Andersen A. *Diabetes 1973; 22:* 706
17 Wiseman MJ, Viberti GC. *Diabetologia 1983; 25:* 530
18 Hirose K, Tsuchida H, Østerby R, Gundersen HJG. *Lab Invest 1980; 43:* 434
19 Viberti GC, Pickup JC, Jarrett RJ, Keen H. *N Engl J Med 1979; 300:* 638
20 Deen WM, Satvat B. *Am J Physiol 1981; 241:* F162
21 Park CH, Maak T. *J Clin Invest 1984; 73:* 767
22 Wiseman MJ, Viberti GC, Jarrett RJ, Keen H. *Diabetologia 1984; 26:* 401
23 Brown DM, Andres GA, Hostetter TH et al. *Diabetes 1982; 31 (Suppl 1):* 71
24 Viberti GC, Hill RD, Jarrett RJ et al. *Lancet 1982; i:* 1430
25 Mathiesen ER, Oxenboll D, Johansen K et al. *Diabetologia 1984; 26:* 406
26 Mogensen CE, Christensen CK. *N Engl J Med 1984; 311:* 89

27 KROC Collaborative Study Group. *N Engl J Med 1984; 311:* 365
28 Wiseman MJ, Viberti GC, Saunders A, Keen H. *Diabetologia 1983; 25:* 205
29 Barnett AH, Wakelin K, Leatherdale BA et al. *Lancet 1984; i:* 1322
30 Brenner BM. *Proc IXth International Congress of Nephrology, Los Angeles.* 1984: 6

Open Discussion

CHAIRMAN Thank you very much Dr Viberti. I want to congratulate you on your lecture.

SOBH (Egypt) The prostaglandin synthetase inhibitor indomethacin decreases glomerular filtration rate and proteinuria. Do you think it has a role in decreasing the rate of deterioration in kidney function in diabetic patients?

VIBERTI At what stage?

SOBH In the early stages to prevent the mesangial expansion by decreasing the glomerular filtration rate and protein filtration.

VIBERTI There have been a few studies done to test this hypothesis. They have not been published because the results have been negative. People have been giving indomethacin to patients with hyperfiltration over different periods of time – three days, one week – and they did not see any change.

SOBH In spite of a reduction in the glomerular filtration rate?

VIBERTI The glomerular filtration rate was not reduced. I think the results of this study will eventually be published but it is always more complex than we think.

HAAPANEN (Helsinki) I just wonder if it would be possible to influence the prognosis by treating the hypertension by a low sodium diet?

VIBERTI That is a good suggestion. There are a number of studies that are directed at correcting this marginal increase of blood pressure by different manoeuvres such as sodium restriction. People are using drugs and those who are using drugs find it very difficult to perform the study because of compliance. Certainly it is a good suggestion.

ANDREUCCI (Naples) Congratulations for your excellent presentation. Since we have so early in diabetic nephropathy, microalbuminuria, microhypertension and micro increase in renal perfusion, we should try to microconstrict the afferent arterioles to reduce the hyperfiltration. This will decrease the perfusion and the capillary hypertension and the transcapillary pressure gradient.

VIBERTI That is something that is obviously worth thinking about. What we have to do is to modify the balance between afferent and efferent arterioles, which are both vasodilated in diabetes. The afferent is more vasodilated than the efferent so we will have to find a drug that refines the balance to reduce the glomerular hypertension. Maybe this will delay progression. However, in man there are other components, in addition to ΔP and QA, and I am convinced that the filtration surface area plays a role in the hyperfiltration. Whether that is of any importance in the progression of the condition I do not know.

PART XVI

DIABETIC NEPHROPATHY

Chairmen: L Migone
S Giovannetti

Proc EDTA–ERA (1984) Vol 21

URINARY EXCRETION OF ALBUMIN WITH AN ALTERED THREE-DIMENSIONAL CONFORMATION IN DIABETIC FUNCTIONAL NEPHROPATHY. EVIDENCE FOR A PATHOGENETIC ROLE

G M Ghiggeri, G Candiano, G Delfino, C Querirolo

Hospital of Lavagna, Lavagna, Italy

Summary

The free sulfhydryl group content of serum and urinary albumin has been evaluated in eight normal and 23 diabetic patients with various grades of urinary albumin excretion rates. While in normal subjects and in diabetics with either normal albuminuria or functional nephropathy, urinary albumin showed a statistically higher content of free sulfhydryl groups compared to homologous serum, diabetic patients with clinical nephropathy showed no difference. These results indicate that an increased urinary excretion of albumin altered in its conformational status is the main feature of diabetic functional nephropathy and suggest that a molecular mechanism determines the glomerular accumulation of albumin.

Introduction

Very early in the course of diabetes mellitus the urinary excretion of albumin increases to values detectable only by sophisticated techniques and characterises the so-called functional phase of diabetic nephropathy [1]. While a strict control of metabolic parameters can return this early dysfunction to normal [1], the chronic renal leakage of albumin leads to its accumulation in the mesangium, which is perhaps the decisive stimulus towards glomerulosclerosis and the consequent decline of the renal function [2]. Once urinary albumin has become detectable by common laboratory methods (0.25–0.5g/24hr), any intervention aimed at decreasing pressure values only succeeds in delaying but not reversing the deteriorating course of the renal disease [3]. On the basis of animal studies, it has been suggested that the lesions of diabetic nephropathy develop as a consequence of altered haemodynamics [4]; however an unequivocal proof of the influence of haemodynamic factors as unique determinants of diabetic nephropathy is still lacking. Furthermore other studies have shown that diabetes is a necessary prerequisite to the development of renal lesions in diabetic rats [5]. Recent evidence indicates that glycosyl albumin (a posttranslational derivate of albumin, the serum concentration of which is increased in diabetes mellitus) has an accelerated renal handling compared to normal homologous albumin, supporting the hypothesis of a molecular origin of diabetic microalbuminuria [6]. A clear understanding of the mechanisms governing renal filtration of albumin and glycosyl albumin in vivo could contribute to the development of a preventional strategy as a logical approach to an otherwise malignant disease.

Materials and methods

Eight normal controls (3 males, 5 females) ranging from 21 to 49 years (mean 32.7) and 23 patients with type I diabetes mellitus (14 males, 9 females) ranging in age from 6 to 60 years (mean 41.6) were studied. Diabetic patients were subdivided into three groups according to their urinary excretion rates of albumin (U_{alb}): (A) eight with U_{alb} <10μg/min; (B) eight with U_{alb} from 10 to 100μg/min, and (C) seven with U_{alb} >100μg/min. All patients in groups A and B were normotensive, whereas four in group C had been previously hypertensive and were receiving therapy with diuretics and β-blockers to keep their blood pressure normal. Twenty-four hour urine collections were performed at home during a period of normal physical activity using thymol as preservative; only urines without signs of bacteriuria were examined. Blood samples were obtained in the morning after an overnight fast and centrifuged at 3000g x 30 min; cells were then discarded and sera kept at −20°C. Albumin was purified from other serum and urinary proteins by pseudo-ligand chromatography on Affi-Gel Blue [7] (Bio-Rad, Richmond, California, USA) previously regenerated according to the supplier's instructions. One millilitre of serum and 300ml of urine were applied to 1.5 x 7cm columns washing out other proteins with 0.05M Tris-HCl, 0.1M KCl pH 7 with about 600ml of buffer until the concentration of protein in the effluent was less than 0.01 absorbance at 280nm. Albumin was eluted with 0.05M Tris-HCl, 1.5M KCl pH 7 with about 400ml of buffer. The fractions containing albumin were then dialysed and ultrafiltrated (Amicon PM-30 membranes). In all cases albumin purity was tested by rocket immunoelectrophoresis against a two layer gel containing either anti-albumin (first layer) or anti-total serum protein (second layer). Free sulfhydryl groups (SH) were measured by a modification of Ellman procedure as previously described [8] and results are given as nmol SH/nmol albumin, protein concentration was evaluated by the Coomassie Dye binding assay. Serum and urinary albumin was quantitated by nephelometry (Beckman Immunochemistry Analyzer). β_2-microglobulin by a radioimmunoassay (Pharmacia, Uppsala, Sweden).

Statistical significance in Table I was calculated using Student's 't' test for paired groups. The least squares method was used for the calculation of the correlation coefficient.

Results

Urinary excretion rates of albumin and total proteins but not of β_2-microglobulin were higher in group B compared to group A and to normal controls, values

TABLE I. Free sulfhydryl groups of serum and urinary albumin in eight normal and 23 type I diabetic patients

	No.	U_{alb} (µg/min)	SH albumin (mol SH/mol alb) Serum	Urine	Urine/Serum
Normals	8	3.8 (2.6–7.5)	0.49±0.02	1.67±0.37	2.95±0.54*
Group A U_{alb} <10µg/min	8	4.7 (2.4–10)	0.43±0.02	2.03±0.52	4.65±1.17*
Group B U_{alb} 10–100µg/min	8	32.9 (15.1–99.5)	0.47±0.05	2.01±0.62	4.09±0.97*
Group C U_{alb} >100µg/min	7	1648 (199–3800)	0.49±0.05	0.32±0.07	0.66±0.14†

* $p<0.02$ compared to serum
† $p<0.001$ compared to urinary concentration of other groups
Abbreviations: U_{alb}=urinary excretion rate of albumin; SH=sulfhydryl groups

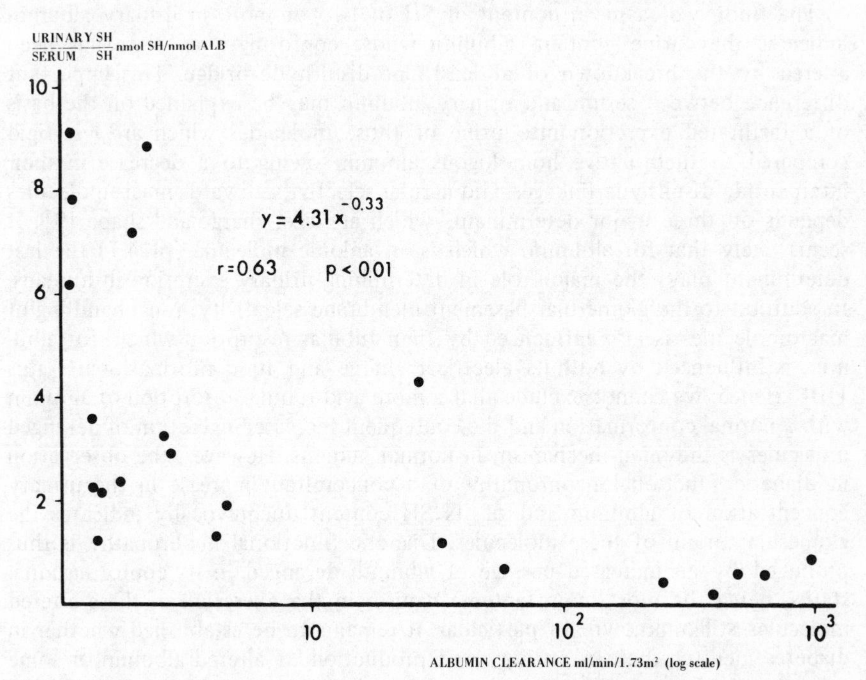

Figure 1. Relationship between the urinary/serum ratio of albumin SH and the clearance of albumin in 23 type I diabetic patients. $y=4.31x^{-0.33}$ is the curve giving the best correlation coefficient

fitting the diagnosis of functional nephropathy. Five patients of group C had proteinuria characteristic of the nephrotic syndrome, showing at the same time a depressed creatinine clearance; by current criteria this last pattern is defined as clinical nephropathy. SH content of serum albumin was the same in all diabetic groups and in normal subjects with a mean value of 0.46±0.04nmol SH/nmol albumin. Urinary albumin showed a statistically higher content of SH compared to homologous serum albumin in normals in group A and group B, but it was significantly depressed in diabetics with overt proteinuria (Group C). Table I gives the detailed values for each group. The relationship between the urinary/serum ratio of albumin SH and the clearance of albumin is depicted in Figure 1.

Discussion

As for any other globular protein, the three-dimensional structure of albumin is determined by the presence in its amino acid sequence of cysteines. In albumin there are 35 molecules of cysteine, 34 of which are linked together by means of the oxidative reduction of their SH to form 17 disulfhyde bridges. The remaining one SH may be masked by a link with circulating cysteine or glutathione in about 50 per cent of the molecules, so that each mole of human serum albumin contains about 0.4–0.5 moles of SH.

The finding of a mean content of SH of two or more in urinary albumin indicates that urine contains albumin whose conformational state has been altered by the breakdown of at least one disulfhyde bridge. This important difference between serum and urinary albumin may be explained on the basis of a facilitated excretion into urine of those molecules which are less rigid compared to their native homologous albumin owing to a decrease in their interpeptide disulfhyde linkages. Glomerular selectivity towards macromolecules depends on three major determinants which are size, charge and shape [9]. It seems likely that for albumin, which is an anionic molecule (pI 4.7) the last determinant plays the major role in determining urinary excretion in humans. In addition to the glomerular basement membrane selectivity, renal handling of macromolecules is also influenced by their tubular resorption which, for albumin, is influenced by both its electrical charge and its conformational status [10]. Hence, we cannot exclude that a more avid tubular resorption of albumin with a normal conformation and the consequent increased excretion of deranged molecules is the main mechanism in normal humans. However, the observation in diabetic functional nephropathy of a concomitant increase in the urinary concentration of albumin and of its SH content unequivocally indicates the glomerular origin of these molecules. Diabetic functional nephropathy is thus produced by an increased passage of albumin deranged in its conformational status, being the most important mechanism in the excretion of these altered molecules still unknown. In particular, it remains to be established whether in diabetes mellitus there is an increased production of altered albumin or some other mechanism, such as altered renal haemodynamics playing a central role. It seems likely that the handling of these molecules may be influenced by high pressure values in the renal microcirculation.

The chronic increase in filtration of altered albumin seems able to produce mesangial deposition and the induction of lesions characteristic of the late phase of the diabetic nephropathy. A knowledge of the intimate mechanisms of the increased production and/or excretion of altered albumin could lead to an understanding of the causes of diabetic nephropathy and eventually achieve specific prevention.

References

1 Viberti GC, McKintosh D, Bilous RW. *Kidney Int 1982; 21:* 714
2 Hostetter TH, Rennke HG, Brenner BM. *Am J Med 1982; 72:* 375
3 Mogensen CE. *Acta Endocrinol 1980 94 (Suppl 238):* 103
4 Hostetter TH, Troy JL, Brenner BM. *Kidney Int 1981; 19:* 410
5 Mauer SH, Steffes MW, Azar S et al. *Diabetes 1978; 27:* 738
6 Ghiggeri GM, Candiano G, Delfino G et al. *Kidney Int 1984; 25:* 565
7 Travis J, Pennel R. *Clin Chim Acta 1973; 49:* 49
8 Ghiggeri GM, Candiano G, Delfino G et al. *Clin Chim Acta 1983; 130:* 257
9 Deen WM, Bohrer MP, Brenner BM. *Kidney Int 1979; 16:* 353
10 Christensen E, Rennke HG, Carone FA. *Am J Phys 1983; 244:* F436

Open Discussion

MIGONE (Chairman) Can you comment on your finding of a serum sulfhydryl albumin content even in your normal control groups?

GHIGGERI Yes, but I said that the mean content per mole of albumin is the same in normal and diabetic groups. The actual content, obtained by multiplying the excretion rates of the albumin by SH content per mole of albumin, is increased because of increased albuminuria. The actual content of the uraemia and diabetic groups is increased by three to four times.

Proc EDTA–ERA (1984) Vol 21

LOW MOLECULAR WEIGHT PROTEINURIA IN DIABETIC CHILDREN – A MARKER OF EARLY DIABETIC NEPHROPATHY?

R Wartha, *D Nebinger, *D Gekle

*University Children's Hospital, Heidelberg, *University Children's Hospital, Wuerzburg, FRG*

Summary

Twenty-four hour urine specimens of 67 diabetic children aged 1–17 years without any renal manifestations were examined by SDS-polyacrylamide gel electrophoresis (SDS-PAGE). The excretion of high molecular weight, i.e. glomerular proteins was compared to that of low molecular weight, i.e. tubular proteins corresponding to more or less than 68,000 daltons. The glomerulo-tubular protein ratio (GTPR) obtained was significantly lower in diabetic patients compared with 30 healthy children of the same age and showed a linear decrease with longer duration of diabetes.

Introduction

About one-third of young insulin dependent diabetics develop diabetic nephropathy leading finally to terminal renal failure. The characteristic histological lesion, glomerulosclerosis, is usually accompanied by proteinuria starting 10–15 years after the first manifestation of diabetes. Proteinuria rapidly increases during the following years concomitant with the progressive damage of the glomerular capillaries. Conventional clinical parameters such as measurement of total proteinuria have failed to recognise renal involvement in diabetes before advanced morphological lesions become evident. Measurement of microalbuminuria determined overnight and after exercise seems to be a better indicator of impending kidney dysfunction [1,2]. Exercise-induced microalbuminuria increased significantly in diabetic children even with a duration of diabetes of only more than five years [3]. In recent years SDS-PAGE has proven to be a sensitive method of characterising proteinuria according to the molecular size of proteins excreted. This enables SDS-PAGE to differentiate glomerular, i.e. high molecular weight (HMW) proteinuria and tubular, i.e. low molecular weight (LMW) proteinuria. It has been applied to adult patients at different stages of diabetes, who often showed an unselective glomerular proteinuria. We have used SDS-PAGE in a group of paediatric patients without clinical manifestations of renal disease, in order to recognise disturbed handling of protein excretion at a subclinical stage.

Patients and methods

Sixty-seven children aged 1–17 years with type I diabetes known for one month to 12 years were examined. All patients were treated by insulin and diet on an ambulatory basis. No patient was known to have renal disease as demonstrated by sterile urine with normal cell counts. Urinary protein excretion was negative to Albustix and sulfosalicylic acid. Serum creatinine and blood pressure were within normal limits for age. In 34 patients blood Haemoglobin A1c were determined. All patients were in good metabolic control. Thirty healthy children aged 2–15 years served as controls.

Twenty-four hour urine samples were collected after addition of 0.05% sodium azide. Protein concentration was measured by a modified Lowry technique. Urine samples were concentrated and dialysed at 4°C against 20% polyethyleneglycol using flexible 'Visking' tubes (Serva, Heidelberg). SDS-PAGE was performed on slab gel units (Hoelzel, Munich) using the Laemmli buffer system [4]. The acrylamide concentrations used were %T=10, %C=2.7 [5]. The gels were stained with Coomassie Blue G 250. For densitometry and integration of the protein peaks Desaga Quick Scan and Desaga Quick Quant units were used (Desaga, Heidelberg). All chemical reagents were obtained from Serva (Heidelberg) and Merck (Darmstadt). GTPR was calculated by dividing the total excretion of HMW proteins with molecular weight (MW) above 68,000 daltons by that of LMW proteins with MW below 68,000 daltons. The albumin peak was excluded from the calculation.

Results

In no patient total proteinuria exceeded 100mg/24hr/m^2 body surface area. Glomerulo-tubular protein ratio (GTPR) values in diabetics and healthy children are compared in Figure 1. In the diabetic group GTPR ranged between 0.2 and

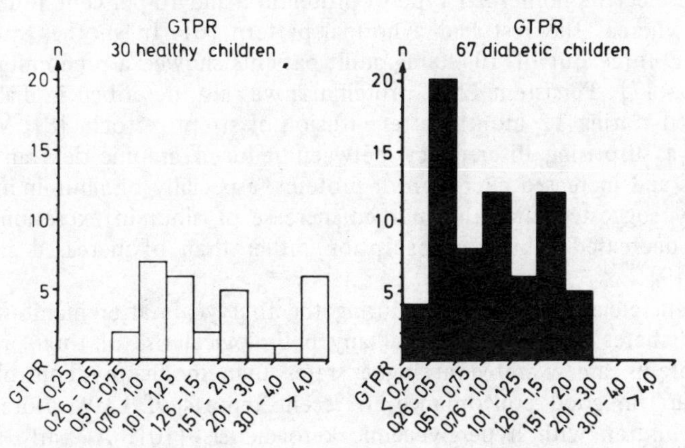

Figure 1. GTPR-values (GTPR=glomerulo-tubular protein ratio) of 30 healthy and 67 diabetic children

1.9 (mean 0.8, median 0.8). In the control group the range was between 0.7 and 5.0 (mean 2.0, median 1.3). This difference was significant with a confidence interval of 99 per cent in the Wilcoxon test. About 35 per cent of the diabetic patients had GTPR values lower than 0.7 which was the lowest value seen in the control group. The LMW proteinuria and the corresponding GTPR values in the diabetics did not correlate with age of patients, urine volume per 24 hours, total proteinuria and blood Haemoglobin A1c, but the GTPR values decreased with longer duration of diabetes (coefficient of correlation 0.32, $p<0.01$, 't' test), as shown in Figure 2.

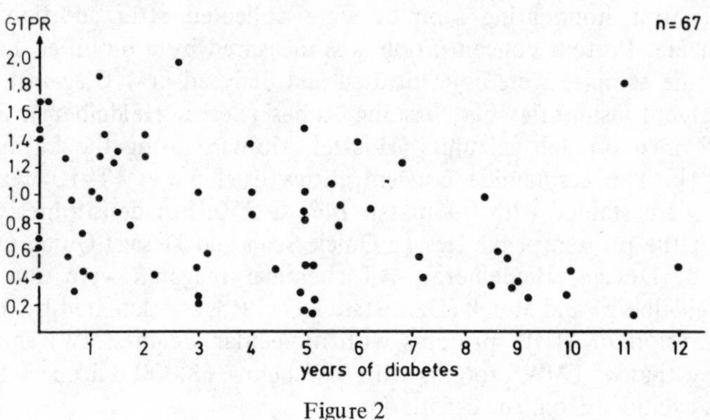

Figure 2

Discussion

There is a general belief that proteinuria in manifest diabetic nephropathy is of glomerular, i.e. of HMW, origin, but some recent data indicate a very early involvement of tubular protein handling in diabetics. In a study of 200 adult diabetics Boesken et al, using SDS-PAGE, found that 50–60 per cent of patients had an unselective glomerular type of proteinuria and 10 per cent a tubular proteinuria, whereas the rest had a normal pattern [6]. In another study using SDS-PAGE three out of 10 young adult patients showed a predominant LMW proteinuria [7]. Persistent LMW proteinuria was also described in diabetic rats investigated during 12 months after infusion of streptozotocin [8]. Michels et al found a surprising discrepancy between reduced anionic dextran sulphate clearances and increased excretion of proteins, especially of albumin in diabetic rats. They suggested that the marked increase of albumin excretion was the result of decreased tubular reabsorption rather than of increased glomerular filtration [9].

Our data clearly express that during the first years after manifestation of juvenile diabetes in the absence of any pathological rise of total proteinuria LMW proteins are excreted at higher rates than in healthy controls. These findings are in good concordance to recent reports of LMW proteinuria in diabetic children with hyperglycaemic ketoacidosis [10]. At early stages of insulin-treated diabetes glomerular filtration rate is increased by about 10–25 per cent. Therefore, it is possible that an increased filtered load of LMW proteins is responsible for the LMW proteinuria observed. Even if the limited capacity of the tubular system to handle a LMW overflow proteinuria is considered it seems unlikely that an increased delivery of these proteins at the rates described above could cause the degree of excretion observed. Moreover proteinuria induced by increased glomerular filtration rate could not explain the reduction in GTPR values with the duration of diabetes. For this reason we would suggest that the low molecular weight proteinuria in diabetic children is the consequence of beginning alteration of renal protein handling by the tubular epithelium due to, for example, early changes in the peritubular vasculature.

References

1 Mogensen CE, Christensen CK. *N Engl J Med 1984; 311:* 89
2 Viberti GC, Jarrett RJ, McCartney M, Keen H. *Diabetologia 1978; 14:* 293
3 Dahlquist G, Aperia A, Carlsson L et al. *Acta Paedtr Scand 1983; 72:* 895
4 Laemmli UK. *Nature 1970; 227:* 680
5 Thanner F, Wartha R, Gekle D. *Klin Wochenschr 1979; 57:* 285
6 Boesken WH, Schneider G, Reuscher A. *Verh Dt Ges Inn Med 1976; 82:* 785
7 Lopes-Virella MF, Virella G, Rosebrock G et al. *Diabetologia 1979; 16:* 165
8 Pennell JP, Meinking TL. *Kidney Int 1982; 21:* 709
9 Michels LD, Davidman M, Keane WF. *Kidney Int 1982; 21:* 699
10 Miltenyi M, Koerner A, Dobos M et al. *Int J Ped Nephrol 1983; 4:* 247

Open Discussion

MIGONE (Chairman) Have you considered the possibility of a relationship between low molecular weight proteinuria and the level of serum β_2 microglobulin which has been shown to be increased in diabetic patients?

WARTHA The problem is there are very few studies on children. There is one from Belgium reporting that β_2 microglobulin is increased in diabetic children and the excretion of β_2 microglobulin then decreases with a decrease in GFR in older patients. At this time serum β_2 microglobulin seems to increase.

MIGONE What happens to the microproteinuria in your patients with decreasing glomerular filtration rate?

WARTHA I think that tubular proteinuria in diabetic patients is still manifest in adults, but you won't detect it because it will be masked by the increased glomerular proteins that are excreted in consequence of the progressive damage of the glomerular filter. Our patients were only followed in childhood.

Proc EDTA-ERA (1984) Vol 21

RADIONUCLEAR DETERMINATION OF GLOMERULAR FILTRATION RATE AND RENAL PLASMA FLOW TO DETECT EARLY DECREASE OF RENAL FUNCTION IN INSULIN DEPENDENT DIABETES

C Cascone, G Beltramello, N Borsato, F Cavallin, P Calzavara, M De Luca, F Susanna, D Madia, P Zanco, B Saitta, M Camerani, G Ferlin

General Hospital, Castelfranco Veneto, Italy

Summary

To evaluate the role of renal haemodynamic factors in the pathophysiology of diabetic nephropathy, we determined by radionuclear techniques glomerular filtration rate (GFR) and renal plasma flow (RPF) in 18 patients affected by insulin dependent diabetes mellitus (IDDM) in good metabolic control, with normal blood pressure and plasma creatinine. GFR and RPF measured in the same patients after ten months correlated with proteinuria and duration of diabetes.

Our finding of a significant correlation between the decline of RPF and duration of diabetes may support the haemodynamic hypothesis of progression of diabetic nephropathy.

Introduction

In spite of the noteworthy contributions given by Danish authors [1–4] and Viberti [5–7] in the study of the pathophysiology of diabetic nephropathy, we cannot explain the reasons why more than 50 per cent of insulin-dependent diabetics are protected against the development of clinical nephropathy. Viberti [5] has emphasised the role played by microproteinuria, Hostetter [8], on the basis of his studies on diabetic rats, pointed out the hypothetical role of glomerular hyperfiltration in the initiation and progression of diabetic nephropathy.

The aim of the present study was to verify this haemodynamic hypothesis in patients affected by IDDM.

Patients and methods

Eighteen patients, ten males and eight females, 13–58 years old (mean 38), affected by IDDM from 2–34 years (mean 12) were studied.

All patients had a normal plasma creatinine (less than 1.2mg/100ml in our laboratory) and blood pressure and good metabolic control in the preceding year. We evaluated twice (the second time after 10 months) the following parameters:

1. GFR by a single injection (5mCi) of $Tc^{99}m$-DTPA according to a previously described technique [9].
2. RPF by a single injection (800μCi) of I^{123}-Orthoiodo-hippurate (OIH) and calculation made as previously described.
3. Daily proteinuria by biuret method and radial immunodiffusion (LC-Partigen Behringwerke AG, Marburg, WG).
4. Urinary excretion of lysozyme (Quantiplate. Lysozyme Test Kit. Kallestad, Chaska, Mn, USA) in the first study, of Beta 2-Microglobulin (Phadezym Beta 2-micro Test. Pharmacia Diagnostics, Uppsala, Sweden) in the second, in order to exclude tubular dysfunction responsible for proteinuria.

The correlation between these parameters and the duration of the disease were determined by linear regression analysis, statistical significance by Student's 't' test and Fisher's test.

Results

Results are summarised in Table I.

TABLE I. Renal haemodynamic factors and proteinuria

		1st Study	2nd Study	Difference
GFR	ml/min	67–154 (Mean 103±23 SD)	54–154 (Mean 107±29 SD)	NS
RPF	ml/min	450–770 (Mean 611±95 SD)	390–894 (Mean 621±149SD)	NS
Proteinuria	mg/day	60–1295 (Mean 334±321SD)	30–1030 (Mean 231±284SD)	NS

Lysozyme (1st study): small traces in the urine of two patients.

Beta 2-microglobulin: urinary excretion in the normal range in all patients.

GFR was below the normal limit in three patients and increased in one (Figure 1). RPF was at the upper normal limit in four patients and decreased in five (Figure 2). A poor statistical correlation was found between proteinuria and duration of diabetes (Figure 3).

A significant negative correlation was observed (Figure 2) between the duration of diabetes and RPF, while duration of the disease does not correlate with GFR (Figure 1). The values of the second study are not significantly different from the first findings, but the evaluation of the single cases gives a

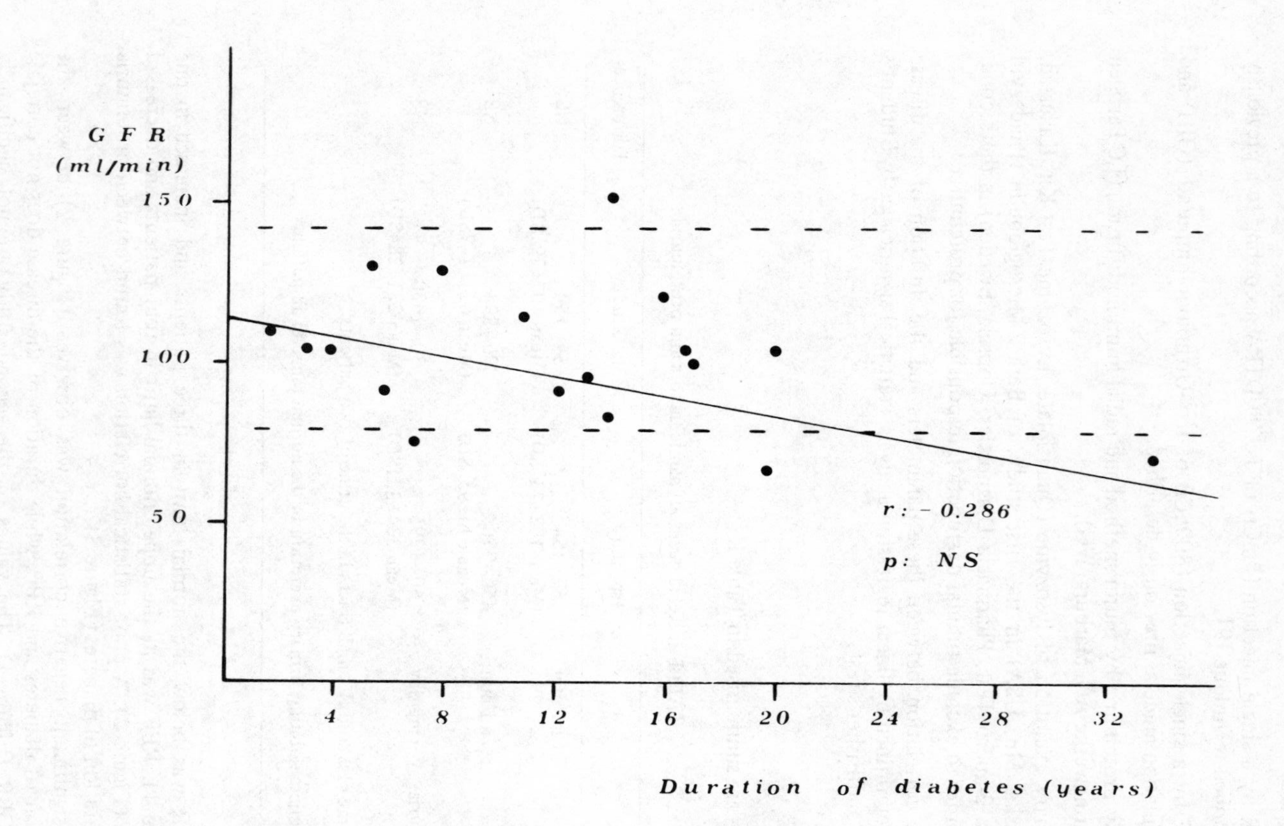

Figure 1. Poor statistical correlation between duration of diabetes (abscissa) and GFR. Only three patients with normal plasma creatinine are just below the lower normal limit (dotted line)

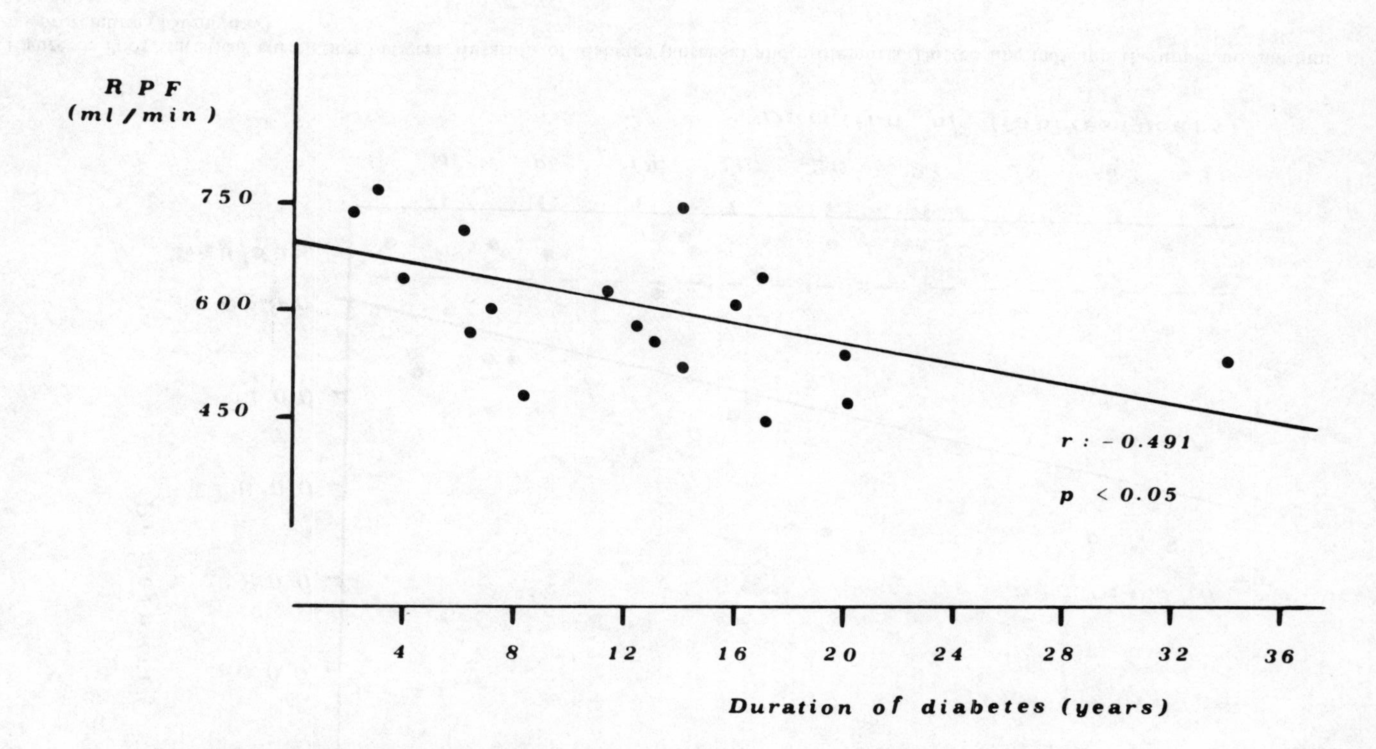

Figure 2. Correlation between duration of diabetes (abscissa) and RPF. The comparison with GFR (Figure 1) supports the hypothesis that other factors (transcapillary hydraulic pressure differences and glomerular capillary ultrafiltration coefficient) may determine variation of GFR in diabetics without clinical nephropathy

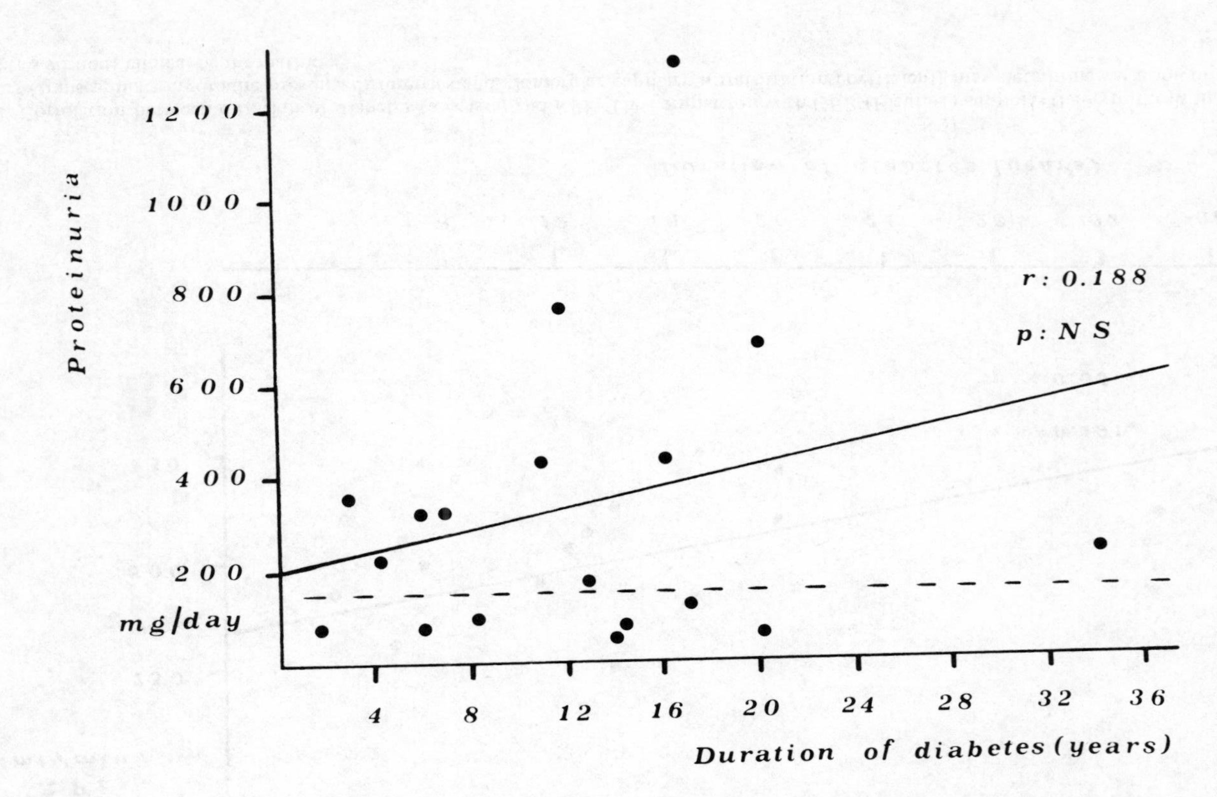

Figure 3. Poor statistical correlation between duration of diabetes (abscissa) and proteinuria. Dotted line indicates the upper normal limit of proteinuria (150mg/day)

partial confirmation about the predictive value of heavy proteinuria. In fact two of the four patients with daily proteinuria greater than 0.5g showed significant decline of their GFR and RPF after the short period of 10 months. Such a decline was not found in patients with microproteinuria.

Discussion

A general agreement exists among the investigators about two features of diabetic nephropathy; the increase of GFR and RPF in early diabetes and the progressive decline in renal function in diabetics with constant gross proteinuria. Some conflicting opinions have been expressed about the importance of microproteinuria in the initiation and progression of clinical nephropathy. An obvious question arises – is proteinuria only a 'marker' of a functional and reversible derangement or does it indicate an irreversible structural damage?

The hypothesis of Hostetter et al [8] is fascinating, since it postulates that haemodynamic factors may be responsible for the glomerular structural damage and consequently for proteinuria. This hypothesis needs confirmation from clinical studies, because of the great number of variables which are difficult to quantify such as physical activity and metabolic control. Although we were conscious of these difficulties, we thought it interesting to evaluate renal haemodynamic factors before the initiation of overt clinical nephropathy and to subsequently restudy the same patients to detect some derangements responsible for clinical manifestations. Our data, documenting a slight but statistically significant decline in RPF in the course of IDDM, without a parallel change of GFR, may indicate that other factors, other than RPF, contribute to determine GFR. Indeed we do not know the role played by the transcapillary hydraulic pressure gradient and the glomerular capillary ultrafiltration coefficient.

However our finding of a parallel decline in GFR and RPF in patients with constant gross proteinuria may indicate that in the evolutive phase of diabetic nephropathy RPF is the main determinant of GFR.

Finally, if the role of microproteinuria postulated by Viberti is confirmed, the question posed by Deckert and Poulsen [5] could be reviewed with new dramatic significance since strict metabolic control capable of reversing microproteinuria could be obtained only by an artificial pancreas.

References

1. Sandahl Christiansen J, Gammelgaard J, Frandsen M et al. *Diabetologia 1981; 20:* 451
2. Mogensen CE, Osterby R, Gundersen HJG. *Diabetologia 1979; 17:* 71
3. Sandahl Christiansen J, Frandsen M, Parving HH. *Diabetologia 1981; 21:* 368
4. Deckert T, Poulsen JE. *Diabetologia 1981; 21:* 178
5. Viberti GC, Jarrett RJ, Mahmud U et al. *Lancet 1982; i:* 1430
6. Viberti GC, Bilous RW, Mackintosh D et al. *Am J Med 1983; 74:* 256
7. Viberti GC, Wiseman MJ. *Diabetic Nephropathy 1983; 2:* 22
8. Hostetter TH, Rennke HG, Brenner BM. *Am J Med 1982; 72:* 375
9. Piepsz A, Dobbeleir A, Erbsmann F. *Eur J Nucl Med 1977; 2:* 173
10. Tauxe WN, Dubovsky EV, Kidd T et al. *Eur J Nucl Med 1982; 7:* 51

Open Discussion

SCHARER (Heidelberg) I wonder if you corrected your figures of proteinuria, GFR and plasma flow for the body surface area because you included children in your study?

CASCONE All the data was corrected to surface area of $1.73m^2$, but we had only one child of 13 years and all the other patients were more than 20 years of age.

MIGONE (Chairman) Did you perform renal biopsy to look for other parameters in your patients?

CASCONE No, we followed the general concept that biopsy is useful in diabetic patients only if we can determine some new element for prognosis and therapy.

Proc EDTA–ERA (1984) Vol 21

GLYCOSYLATED PROTEINS IN DIABETIC NEPHROPATHY

M Ramzy, M Habib, N El-Sheemy, *M Dunn, I Iskander, S Samney, S H Doss

*Faculty of Medicine, Cairo University, *NAMRU 3, Cairo, Egypt*

Summary

This study was carried out on 55 diabetic patients, 20 of whom had diabetic nephropathy, and 10 controls. Glycosylated haemoglobin, glycosylated serum protein, glucoprotein, serum protein electrophoresis, blood urea, serum creatinine and β_2-microglobulin were measured.

A significant increase of glucoprotein was observed in patients with diabetic nephropathy. No correlation was found between glycosylated serum protein and glycosylated haemoglobin and duration of diabetes. Glycosylated serum protein showed a positive correlation with β_2-microglobulin, indicating a link between renal involvement and the rise in glycosylated serum protein. Whether there is a pathogenic relation between glycosylated serum protein and the development of nephropathy awaits further evidence.

Introduction

Renal disease is a major cause of morbidity in people with type I diabetes mellitus. About half of the patients with insulin dependent diabetes mellitus develop renal failure by a mean of 20 years after the apparent onset of the disease. The pathogenesis of diabetic glomerulopathy, which is characterised by increased thickness of the capillary basement membranes and accumulation of proteinatious material in the mesangium is still speculative [1]. This proteinatious material was previously described as glycoproteins and on other occasions as immunoglobulins with various complement components and fibrin. Currently it is assumed to be glycosylated proteins, especially albumin.

Tischer [1] concluded that beside the metabolic effect of hyperglycaemia, there are additional factors, such as hypertension, growth hormone, platelet aggregation and abnormalities in the coagulation and fibrinolytic systems. Other factors include abnormalities in metabolism of proteoglycans and glycosaminoglycan and genetic factors.

In this study we investigated the possible pathogenic role of glycosylated proteins in diabetic nephropathy.

Materials and methods

This study included 55 patients with insulin dependent diabetes mellitus, 10 of which were without clinically evident complications (6 females and 4 males) with a mean age of 44.9±14.1 years. Twenty patients had diabetic nephropathy (13 females and 7 males) with a mean age of 51.2±9.2 years. Fifteen patients had ischaemic heart disease (10 males and 5 females) with a mean age of 52.5±10.8 years. The remaining 10 patients had diabetic retinopathy (6 males and 4 females) with a mean age of 54.4±7.4 years.

Ten healthy persons with no history of family predisposition to diabetes, of matching age and sex were chosen as normal controls (5 males and 5 females with a mean age of 40.8±16.8 years).

Our patients were classified according to the main clinical presentation as most of the cases had combined complications.

For the classification of our material all cases were subjected to full clinical assessment, funduscopic examination, and the following investigations: electrocardiogram; blood sugar, both fasting and two hours post-prandial; full urine analysis including quantitative urinary proteins per 24 hours; urine culture and bacterial count (cases with upper urinary tract infection were excluded); blood picture; serum creatinine and blood urea; serum proteins (total and albumin): diabetics with recent febrile illness or serious overt infections were excluded to avoid any change in plasma proteins secondary to increased acute phase of protein production; serum glucoproteins electrophoresis; glycosylated haemoglobin; glycosylated serum proteins; urinary tract plain X-ray and intravenous pyelogram or sonogram (cases with any suspicion of obstructive uropathy were excluded); renal biopsy for cases with proteinuria more than 1g/24 hours, and serum β_2-microglobulin as an index of glomerular filtration rate (cases with any possible tubular injury as infection, obstruction or receiving nephrotoxic drugs were excluded).

Results

Blood sugar, serum creatinine and blood urea are shown in Table I.

β_2-microglobulin ($\beta_2M\mu g/L$) In the controls the mean was 1656.2±863.8 (mean ± SD), in uncomplicated diabetic patients 1693.3±789.7 (NS), patients with diabetic nephropathy 9824.2±1250.7 ($p<0.01$), those with diabetic retinopathy 3731.8±243.4 ($p<0.05$) and those with ischaemic heart disease 3244.5±132 ($p<0.05$).

Plasma proteins: (a) total serum proteins (mg%) Total serum proteins of uncomplicated diabetic patients were 7.3±0.5 which is insignificantly different from control values (7.6±0.6). The mean values in the complicated subgroups were 6.7±0.8 for diabetic nephropathy, 6.9±0.6 for diabetic retinopathy and 6.5±0.5 for ischaemic heart disease.

TABLE I

Group	Age (years)	Duration of diabetes	Fasting glucose mg%	Two hours post-prandial mg%	Blood urea mg%	Serum creatinine mg%	β_2 M μg/L	Proteinuria g/24hr	GHB%	GSP mmol HMF/ mg prot.
Control n=10	40.8 ±16.8	–	79.9 ±21.8	125.5 ±15.5	30.8 ±10.3	0.8 ±0.2	16562 ±863.6	–	7.9 ±0.6	0.04 ±0.04
Uncomplicated diabetics n=10	44.9 ±14.1	3.2 ±3.1	170.5** ±94.1	195.5* ±96.5	35 ±4.8	0.8 ±0.3	1693.3 ±789.7	–	11.4** ±1.5	0.133 ±0.146
Diabetic nephropathy n=20	51.2 ±9.2	9.4 ±7.9	263.6** ±105	317.4** ±125.2	139.3 ±34.1	5.7 ±3.9	9824.5** ±1250.7	4.2 ±1.9	12.9** ±2.5	0.59** ±0.23
Diabetic retinopathy n=10	54.4 ±7.4	7.7 ±4.5	234.1** ±132	321** ±134.2	43 ±18	1.9 ±0.2	3731.8 ±243.4	0.9 ±0.2	13.6** ±2.2	0.29** ±0.17
Ischaemic heart disease n=15	52.5 ±10.8	12.7 ±9.8	283.8** ±93.1	384.7** ±114.3	41.9 ±16	1.2 ±0.1	3244.5** ±132.0	0.3 ±0.01	14.4** ±2.2	0.38** ±0.24

** $p<0.01$
* $p<0.05$
β_2 M = β_2-microglobulin
GHB = Glycosylated haemoglobin
GSP = Glycosylated serum protein
HMF = 5-hydroxy methylfurfural

(b) serum albumin (mg%) in uncomplicated diabetic patients was 4.1±0.7 which is insignificantly different from controls (4.5±0.3). In complicated subgroups it was 3.3±0.7 in diabetic nephropathy, 3.6±0.7 in diabetic retinopathy and 4.1± 0.8 in ischaemic heart disease, a significant reduction in both diabetic nephropathy and diabetic retinopathy groups.

Glucoproteins (GP%) $alpha_1$, $alpha_2$ and beta-glucoproteins in uncomplicated diabetic patients was 11.46±1.42, 35.62±3.6 and 24.28±4.51 respectively. The corresponding control figures were 12.76±2.05, 36.56±2.75 and 24.19±4.45 respectively. The mean value of $alpha_1$-glucoproteins was 13.0±2.7 in diabetic nephropathy, 9.8±4.2 in diabetic retinopathy and 9.6±3.4 for ischaemic heart disease. The corresponding figures for $alpha_2$-glucoproteins were 42.03±5.9, 36.6±4.7 and 38.5±6.2 respectively. A highly significant increase was found in the renal group compared to controls ($p<0.01$) and to uncomplicated diabetic patients ($p<0.05$). While beta-glucoprotein was 22.9±4.9, 23.6±3.2 and 23.8±4.1 respectively. Beta-glucoproteins showed insignificant differences between all groups.

Glycosylated haemoglobin (GHB %) in uncomplicated diabetic patients was 11.36±1.5 which is significantly higher than the control value (7.89±0.6) ($p<0.01$). In diabetic nephropathy it was 12.95±2.5, in diabetic retinopathy 13.6±2.2 and 14.4±2.2 for ischaemic heart disease. Glycosylated haemoglobin was significantly high in all complicated subgroups compared to controls ($p<0.01$), while compared to uncomplicated diabetic patients significantly higher values were found in both ischaemic heart disease ($p<0.01$) and diabetic retinopathy ($p<0.05$). No correlation could be found between glycosylated haemoglobin and age, sex or duration of diabetes.

Glycosylated serum proteins (GSP, mmol HMF/mg protein) in uncomplicated diabetics it was 0.133±0.14 which was insignificantly higher than controls (0.036±0.04). The mean value in diabetic nephropathy was 0.59±0.23, in diabetic retinopathy was 0.29±0.17 and 0.35±0.24 in ischaemic heart disease.

Comparing these later subgroups to uncomplicated diabetic patients and comparing the different complicated subgroups to each other, diabetic nephropathy was the only group that showed significantly higher values throughout. No correlation was found between glycosylated serum proteins and fasting blood sugar. No correlation could be found between glycosylated serum proteins and glycosylated haemoglobin, total serum protein, serum albumin, serum creatinine or blood urea. A positive correlation was found between glycosylated serum proteins and β_2-microglobulin in the nephropathy group (pr=0.8637, $p<0.001$).

Discussion

In our study, a highly significant increase in $alpha_2$-glucoproteins in the nephropathy group was found whether compared with the control group or the uncomplicated diabetic patients. This agrees with the studies of Levin et al [2], who reported abnormal glucoproteins in the glomerular basement membrane resulting in diabetic glomerulosclerosis, and later in 1982 Tischer [1] described diffuse accumulation of glucoproteins in intercapillary glomerulosclerosis.

The percentile value of beta-glucoproteins did not show any statistically significant difference between controls and diabetic patients, whether complicated or uncomplicated. A significant increase of glycosylated haemoglobin was found in all diabetic groups (complicated and uncomplicated) compared to controls, while a significant rise was found only in the retinopathy and the coronary groups on comparing the complicated with the uncomplicated diabetic patients. The relatively low glycosylated haemoglobin in the renal cases compared to other diabetic groups could be explained by the shortened lif span of erythrocytes in renal failure [3] and frequent blood transfusions in these cases.

A positive correlation could be found between glycosylated haemoglobin and fasting blood sugar in all diabetic groups. This is in agreement with the findings of Graf et al [4]. Glycosylated haemoglobin was not related to age, sex or duration of diabetes. This disagrees with Goldstein et al [5], who found a significant relationship between glycosylated haemoglobin and the duration of diabetes.

In our study, glycosylated serum protein values were similar to the figures found by Kinnedy et al [6], but contrary to the same author, we found no correlation between fasting blood sugar and glycosylated serum protein. A significant increase in glycosylated serum protein was found in all complicated groups while an insignificant rise was found in uncomplicated diabetic patients. Patients with diabetic nephropathy were the only group that showed a significant rise in glycosylated serum proteins on comparing different complicated subgroups to each other.

These data are suggestive of a possible pathogenic role of glycosylated serum proteins in the pathogenesis of diabetic renal disease. This is in agreement with the experimental study of Cohn [7], who demonstrated glycosylation of rat glomerular basement membrane after incubation with glucose. Also, it has been found that incorporation of glycosylated serum proteins in renal basement membrane may explain the pathological changes seen in diabetic nephropathy.

No correlation could be found between glycosylated serum proteins and other variables, such as glycosylated haemoglobin, total serum proteins, serum albumin, serum creatinine or blood urea. There was also no correlation between glycosylated serum proteins and age, sex or the duration of the disease. This is in agreement with Kinnedy et al [6].

A statistically significant correlation could be found between glycosylated serum proteins and β_2-microglobulins as it is a sensitive index of renal function in diabetic patients [8]. The accumulating evidence points to changes in the plasma proteins of diabetic patients playing a role in accelerating the progression rate of diabetic microangiopathy. A disturbance in the average molecular shape of plasma proteins due to raised glycosylated serum proteins directly increases both plasma viscosity and erythrocyte aggregation leading to impairment of the microcirculation [9]. These findings can explain the positive correlation between glycosylated serum proteins and β_2-microglobulins, especially in the late second junctional and early third phase of diabetes.

Therefore we can conclude that elevated glycosylated serum protein, together with other factors, have a pathogenic role in the evolution of diabetic renal disease.

References

1 Tischer CC. In Wyngaarden JB, Smith LH Jr, eds. *Cecil Textbook of Medicine, 16th edition.* Philadelphia: Saunders. 1982: 569
2 Levin NW, Cortes P, Silveira E, Rubenstein AH. *Lancet 1975; i:* 1120
3 Horton BF, Huisman THJ. *Br J Haematol 1965; 11:* 196
4 Graf RJ, Halter JB, Porte D. *Diabetes 1978; 27:* 834
5 Goldstein DE et al. *Diabetes 1982; 31 (suppl 3):* 70
6 Kinnedy L, Mehl TD, Rieley WJ, Merimee TJ. *Diabetologiea 1981; 21:* 94
7 Cohen MP et al. *Diabetes 1981; 30:* 367
8 Viberti GC, Keen H, Mackintosh D. *Br Med J 1981; 282:* 95
9 McMillan DE. *Diabetes 1976; 25 (suppl 2):* 858

Open Discussion

MIGONE (Chairman) May I ask you if you related the microproteinuria to the increased β_2-microglobulin you have measured in the serum?

RAMZY Yes, the serum β_2-microglobulin is an index of glomerular filtration rate.

MIGONE Was there a parallel increase in both serum and urine?

RAMZY No, the increase was between the serum β_2 and the glycosylated serum proteins.

MIGONE You did not find any increase of microproteinuria in general without looking for the β_2-microglobulin – I mean the microproteinura, the so-called tubular proteinuria?

RAMZY No, in some cases there is increased β_2-microglobulin but I have excluded them because I tried to concentrate on the glomerulus. That is why if we have increased β_2 it is suggestive of tubular injury of unknown cause. That is why these cases had been excluded because they may have infection or something else.

MIGONE In any case you are dealing with β_2-microproteinuria?

RAMZY Yes.

Proc EDTA–ERA (1984) Vol 21

RESULTS OF TRANSPLANTATION OF KIDNEYS FROM DIABETIC DONORS

H van Goor, M J H Slooff, G D Kremer, A M Tegzess

University Hospital, Groningen, The Netherlands

Summary

Diabetic donors are still reluctantly accepted as potential organ donors because of supposed poor graft function caused by diabetic lesions. The results of transplantation of six kidneys from three donors with insulin dependent diabetes mellitus are reported. All three donors had a normal creatinine clearance and absence of proteinuria. Renal biopsies were taken.

Five grafts are still functioning, six months to two years after transplantation with a mean creatinine clearance of 69ml/min (range 51–95).

Three of five biopsies taken six months after transplantation showed marked decrease of the diabetic lesions. On the basis of these findings it seems justified to accept kidneys from diabetic donors for transplantation.

Introduction

Diabetic nephropathy is a well known complication of long standing insulin dependent diabetes. This complication usually leads to progressive renal failure requiring dialysis or transplantation. For this reason kidneys of donors with known insulin dependent diabetes have been reluctantly accepted for transplantation.

However ample experimental evidence exists on the prevention and reversibility of diabetic glomerular changes when diabetic kidneys are placed in a normoglycaemic environment. In view of the shortage of suitable organ donors it was thought worthwhile to investigate the utilisation of kidneys from diabetic donors for transplantation.

Methods

From June 1982 until February 1984 three potential kidney donors with insulin dependent diabetes mellitus came available within the region of Eurotransplant.

Donor ages at referral were twenty-one, thirty-seven and fifty-nine years. Brain death was the result of cerebral injury in two cases and thrombosis of the basillar artery with brain stem infarction in the other case. The donors suffered from diabetes mellitus for ten, seventeen and eighteen years respectively. All donors had stable blood pressure with small amounts of vasopressors, abundant diuresis, a normal creatinine clearance and absence of proteinuria. Consent of next of kin was obtained before the donor operation was started. In all cases no gross anatomical abnormalities were observed and the renal arteries were without signs of arteriosclerotic changes. The first warm ischaemic time was within the usual range. Four kidneys flushed excellently with Eurocollins and afterwards preserved by cold storage at 4°C.

Two kidneys of the same donor were placed on the Gambro preservation machine and had good perfusion characteristics. Prior to transplantation wedge biopsies were taken from each kidney for histological examination.

Results

Kidneys were transplanted after a cold ischaemic time ranging between twenty-six and fifty-two hours. All showed immediate diuresis.

One graft failed after three days due to rejection. The remaining five grafts are still functioning between six months and two years after transplantation. None of the patients died during their follow-up period. All kidneys were biopsied prior to and/or one hour after transplantation. Two biopsies of kidneys from the same donor showed no diabetic lesions. Light microscopic examination of the remaining four renal biopsies showed diffuse glomerulosclerosis with obliteration of capillary lumina in most glomeruli, basement membrane thickening with PAS positive material, arteriolosclerosis and insudative lesions in the form of capsular drops (Figure 1). Immunofluorescence examination for IgA, IgG, IgM and C3 were positive in two biopsies.

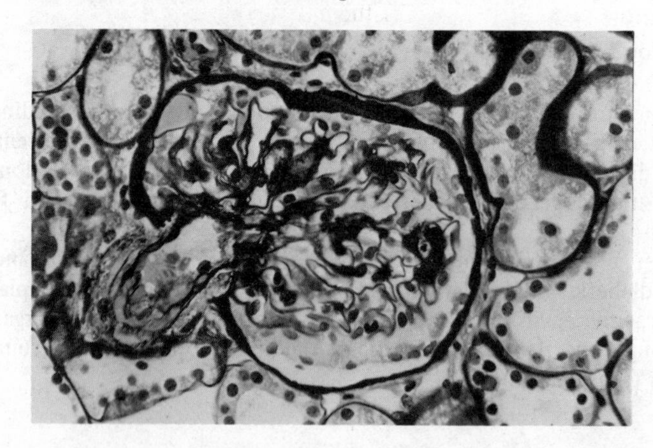

Figure 1. Glomerulus of kidney from a 21-year old donor with a 10 year history of insulin dependent diabetes. Biopsy taken one hour after transplantation: mild diffuse glomerulosclerosis, capsular drop and arteriosclerosis. Methanamin-silver, magnification x100

In three out of four biopsies taken at six months after transplantation a marked decrease of these lesions was observed compared to the pre-transplant observed lesions. One biopsy taken at two months after transplantation showed a slight resolution of glomerular changes although there was still basement membrane thickening and arteriolosclerosis (Figure 2).

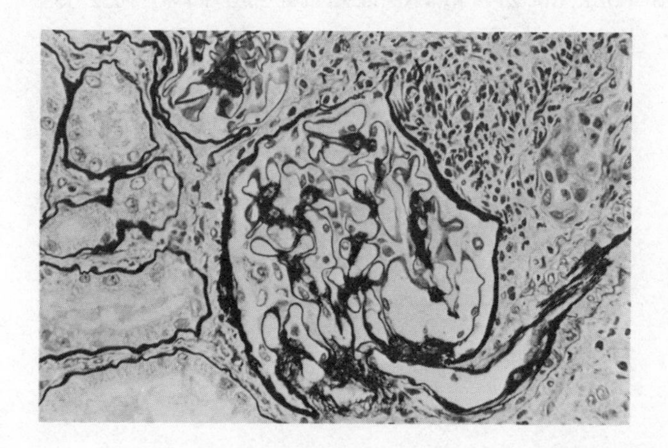

Figure 2. Glomerulus of same kidney as in Figure 1. Biopsy taken two months after transplantation. Slight decrease in diabetic glomerular lesions. Methanamin-silver, magnification x 100

Discussion

Clinical features of renal involvement in diabetic patients appear at least ten years after the diagnosis of diabetes mellitus [1,2]. Albustix positive proteinuria is often the first sign of diabetic nephropathy. In contrast to clinical signs morphological changes of the glomerulus can be seen in an earlier phase. In the late phase, when overt proteinuria develops, these morphological changes have severely progressed. From experimental studies it is reported that glomerular changes occurring in the early stage of diabetic renal involvement can be prevented or reversed by accurate blood sugar control in contrast with glomerular changes in the late phase [3,4].

In man, regression of clinical diabetic nephropathy after control of hyperglycaemia is still a disputed matter although regression of light microscopic diabetic lesions in a transplanted kidney after a successful segmental pancreatic grafing was reported by Sutherland et al [5,6].

From this report it seems evident that early diabetic glomerulopathy is reversible in a normoglycaemic situation. These findings have important implications for the potential donor pool because they justify the use of kidneys from insulin dependent diabetic donors with a normal creatinine clearance and absence of proteinuria for transplantation.

References

1 Goldstein DA, Massry SG. *Nephron 1978; 20:* 286
2 Kussmann MJ, Goldstein HH, Gleason RE. *JAMA 1976; 236:* 1861
3 Rasch R. *Acta Endocrinol 1981; 97 (Suppl 242):* 43
4 Lee CS, Mauer M, Brown DM et al. *J Exp Med 1974; 139:* 793
5 Viberti GC, Bilous RW, Mackintosh D et al. *Br Med J 1983; 286:* 598
6 Sutherland DER, Goetz FC, Rywasternicz J et al. *Surgery 1981; 90, 2:* 159

PART XVII

NEPHROLOGY: GLOMERULAR DISEASE I 661

Chairmen: P Verroust
G Piccoli

PART XVIII

NEPHROLOGY: GLOMERULAR DISEASE II 693

Chairmen: A M Davison
F P Schena

PART XIX

NEPHROLOGY POSTERS II 722

Chairmen: J Stewart Cameron
A Vangelista

Proc EDTA–ERA (1984) Vol 21

MACROPHAGE FUNCTION IN PRIMARY AND SECONDARY GLOMERULONEPHRITIS

D Roccatello, R Coppo, G Martina, C Rollino, B Basolo, G Picciotto, *F Bayle, †P Bajardi, +A Amoroso, *D Cordonnier, G Piccoli

*Ospedale S, Giovanni di Torino, Italy, *Centre Hospitalier Regional et Universitaire de Grenoble, France, †Ospedale degli infermi, Biella, +CNR Centro Immunogenetica ed Istocompatibilità, Torino, Italy Italy*

Summary

The function of the mononuclear phagocyte system was assessed in vivo in 85 patients with primary and secondary glomerulonephritis, by measuring the clearance of IgG – sensitised ^{51}Cr-labelled autologous erythrocytes.

Eleven per cent of patients in clinical remission were found to have a delayed clearance, whereas impaired macrophage function was present in 62.5 per cent of the patients with major urinary abnormalities.

Blockade of mononuclear phagocyte system, induced at least in part by unidentified factors, might have a role in development and perpetuation of glomerular injury.

Introduction

Experimental models and immuno-histochemical data suggest that immune complexes (IC) might play a role in the pathogenesis of many glomerular diseases [1]. This concept is supported by the detection of IC-like material in both the circulation and the affected tissues, where deposition might result in immunologically mediated damage.

The mononuclear phagocyte system (MPS) is thought to remove IC from the bloodstream in animals [2]. Blockade of the MPS might lead to decreased clearance of IC and their enhanced deposition in the kidney.

We studied the functional activity of the MPS in 85 patients affected by primary or secondary glomerulonephritis to investigate whether an impaired removal of immunologically active substances is present in these nephropathies. In addition the relationship between the presence of IC in serum and the Fc-receptor function of MPS was investigated.

Patients and Methods

Patients

Patients included 19 cases of membranous glomerulonephritis (MGN) (16 males, mean age 41.5, 20–65 years), 17 primary IgAGN (16 males, mean age 37.1,

11–55 years), eight membranoproliferative GN (5 males, mean age 35.6, 18–50 years), nine focal glomerulosclerosis (FGS) (6 males, mean age 37.2, 19–60 years), two GN with mesangial IgM deposits (IgMGN) (50 and 61 year old women), 13 lupus nephritis (LN) (all females, mean age 41.8, 27–55 years), three polyarteritis (P) (1 male, mean age 49, 27–65 years), four cryoglobulinaemias (C) (all females, mean age 56.2, 45–65 years), five Henoch-Schönlein purpura nephritis (HS) (3 males, mean age 31.6, 19–50 years). Five cirrhosis – associated GN (C GN) (all males, mean age 57, 50–65 years).

The diagnosis in the patients with systemic disease was based on both clinical and histological data (renal biopsy having been performed in every case). Twenty-nine patients (3MGN, 6 IgAGN, 2 MPGN, 5 FGS, 5LN, 2 P, 3 C, 1 HS, 2 C GN) were studied during a clinical stage of spontaneous or therapy-induced remission (as defined by the absence of urinary abnormalities or the presence of trace amounts of haematuria or proteinuria). Ten patients (4 MGN, 1 FGS, 3 IgAGN, 2 HS) presented with non-nephrotic proteinuria and/or persistent moderate haematuria. The others had nephrotic-range proteinuria and/or severe haematuria (>25 red blood cells (RBC)/high power microscopic field), with signs of systemic vasculitis (fever or arthralgia or cutaneous manifestation or gastrointestinal symptoms) in the cases of secondary GN.

Healthy subjects

Twenty normal volunteers acted as controls for the studies of macrophage function. Sera from 30 healthy subjects were analysed for the detection of IC.

Fc-receptor function of macrophages

The procedure described by Crome [3] recently developed by Frank [4], was employed with slight modifications [5]. Three times washed autologous RBC were labelled with ^{51}Cr and sensitised with an amount of IgG anti-Rh (D) (CTS Lyon) selected to obtain a clearance of half-time (T 1/2) in normals of about 35 minutes.

IC detection

The detection of IgG containing IC (IgGIC) was performed by employing a modified Clq solid phase assay as previously described [6]. The conglutinin solid phase assay was used for measuring IgAIC levels [7]. Results were expressed here in optical density (OD) units.

Typing for HLA and D-locus-related antigens

Tissue typing was performed on peripheral blood leucocyte by the Medical Genetics Institute of the University of Turin.

Statistical analyses employed the Mann-Whitney U-Test and the r correlation.

Results

Fc-receptor function of macrophages

The mean values of T 1/2 in patients with MPGN, IgAGN and LN were significantly higher than those in healthy people (Table I). Half-lives exceeding the upper 95 per cent confidence limit (95th percentile) in healthy people were considered as delayed. Prolonged T 1/2 were observed only in three of 29 patients (1 MPGN and 2 LN) in clinical remission as defined above. Conversely, in the presence of urinary abnormalities, seven of 16 cases of MGN, nine of 11 IgAGN, four of 6 MPGN, one of four FGS, one of two IgMGN, seven of eight LN, one of one P, one of one C, two of four HS, two of three C GN had delayed RBC clearance.

TABLE I. Fc-receptor function of macrophages in 85 patients with primary and secondary glomerulonephritis

	N	T 1/2 min.
Healthy people	20	32.9 (27–46)
Membranous GN	19	47.3 (15–136)
Primary IgAGN	17	59.5* (22–180)
Membranoproliferative GN	8	46.5* (26–66)
Focal glomerulosclerosis	9	33.2 (25–48)
GN with mesangial IgM deposits	2	33 (29–47)
Lupus nephritis	13	59** (25–92)
Polyarteritis	3	45 (30–69)
Cryoglobulinaemia	4	37.3 (23–70.5)
Henoch-Schönlein purpura nephritis	5	49.8 (21–92)
Cirrhosis-associated GN	5	45 (11–81)

*p<0.01), **p<0.001, N = number of subjects

IC levels

IgGIC levels were significantly higher than normal control values only in LN (mean value 21μg Agg IgG Eq/ml, range 0–94 versus 2.2, range 0–16, p<0.02). In this subset IgGIC levels were found to exceed the 90th percentile in normals of 8μg Agg IgG Eq/ml in each patient with urinary abnormalities. However the correlation with T 1/2 values did not reach significant levels (r = 0.4). Low amounts of IgGIC were detected in variable percentages in other nephropathies, Sixteen point six per cent in MGN, 25 per cent in IgAGN and MPGN, 40 per cent in HS.

The mean levels of IgAIC in patients with IgAGN (0.97 OD 400nm, range 0.3–3), HS (0.9 OD 400nm, range 0.23–1.83) and C GN (0.8 OD, 400nm, range 0.55–1.33) were significantly higher (p<0.01) than those in healthy controls (0.25 OD 400nm, range 0.17–0.61). As reported in a previous study [5] a good correlation was confirmed in primary and secondary IgA nephropathies, between IgAIC levels and T 1/2 values.

Typing for HLA and D-locus-related antigens

The frequency of DR3 positive subjects in MGN was about 63 per cent (36.6% B8 DR3). Fifty-seven per cent of DR3 positive MGN patients had normal RBC clearance. No difference in clinical status was found by comparing this group with the one of DR3 positive subjects with prolonged RBC half-life. The mean T 1/2 values of DR3 positive MGN patients were not different from the negative subjects with the same nephropathy. No HLA- A, B, DR types was prevalent in IgAGN patients who had Fc-receptor dysfunction. Five out of eight MPGN patients were typed for HLA- A, B DR antigens: four of five were DR3-positive and three of them had a prolonged RBC clearance.

Discussion

Defective macrophage function affected a conspicuous number of GN patients with urinary abnormalities, whereas this defect was rarely found in patients in remission. The mechanism responsible for the macrophage impairment is unclear. In primary and secondary IgA-nephropathies, a role – as blocking factors – might be played by circulating IgA-containing materials. Indeed the levels if IgAIC correlated significantly with the magnitude of the clearance defect. In LN high levels of IgGIC were present in all patients with nephrotic proteinuria and/or severe haematuria. These cases were also characterised by a delayed immune clearance. However the correlation between these two parameters did not attain statistical significance, suggesting that the dynamics of IC production and macrophage saturability could be non-consensual or even individual. Impaired macrophage function was present, to variable degrees in all patient groups. However only one of nine patients with FGS had a slightly prolonged clearance half-time. In the other cases of primary or secondary GN a clearance defect affected almost exclusively the patients showing urinary abnormalities, but IC were detected, at low levels, in a few. Recently it was suggested that Fc-receptor dysfunction might be linked to some HLA-, A, B, DR types [8]. In the present study an unsuspected prevalence, although in a quite small group of patients, of DR3 positive subjects seemed to characterise MPGN patients with MPS dysfunction. However, in MGN patients – one of the most extensively studied group – in which a high frequency of DR3 positive subjects was observed, the attempt to relate macrophage dysfunction with the presence of this antigen failed. Therefore, although an intrinsic intracellular defect – perhaps related to an active phase of the disease – could not be excluded, these data might also suggest that still undetectable immunologically active substances might account for the macrophage dysfunction in some human GN. Whatever the mechanism involved, the frequent finding of macrophage dysfunction in patients with urinary abnormalities suggest a role of this defect in the development of glomerular injury.

References

1 Dixon FJ. *Am J Med 1968; 44:* 493
2 Mannik M, Arend WP, Hall AP et al. *J Exp Med 1971; 133:* 713
3 Crome P, Mollison PL. *Br J Haemat 1964; 10:* 137
4 Frank MM, Hamburger MI, Lawley TJ et al. *N Engl J Med 1979; 300:* 518
5 Roccatello D, Coppo R, Basolo B et al. *Proc EDTA 1983; 20:* 610
6 Coppo R, Bosticardo GM, Basolo B et al. *Nephron 1982; 32:* 320
7 Coppo R, Basolo B, Martina G et al. *Clin Nephrol 1982; 5:* 230
8 Lawley TJ, Hall RP, Fauci AS et al. *N Engl J Med 1981; 304:* 185
9 Lawley TJ, Hall RP, Fauci AS et al. *N Engl J Med 1981; 304:* 185

Open Discussion

VALLENTINE (Leiden) I am impressed by the high number of patients that you have with decreased Fc clearance. The number of patients that you have with decreased function is remarkably higher than most other studies that have been described in glomerulonephritis. As you know one of the most important factors of studying Fc clearance is the amount of IgG molecules that you have coated on the erythrocytes. The amount of IgG molecules will determine how long the clearance studies will be. What kind of studies did you do to standardise the amount of IgG of your erythrocytes to show that your assay is standardised?

ROCCATELLO As I told you before, we incubated sensitised erythrocytes with ^{125}I-labelled anti-IgG to evaluate the number of coated sites. To standardise our test, each time we changed our batch of anti-D immunoglobulins we verified, by using this method, that we obtained the same degree of sensitisation in normal subjects and in pathological ones.

Proc EDTA–ERA (1984) Vol 21

INTERSTITIAL FOAM CELLS IN THE NEPHROTIC SYNDROME BELONG TO THE MONOCYTE/MACROPHAGE LINEAGE

F Nolasco, * J S Cameron, *R Reuben, *B Hartley, F Carvalho, R A Coelho

*Hospital Curry Cabral, Lisbon, Portugal, *Guy's Hospital, London, United Kingdom*

Summary

Interstitial foam cells are occasionally seen in patients with the nephrotic syndrome. In a group of patients with the nephrotic syndrome we were able to demonstrate that these cells express markers characteristic of the monocyte/macrophage lineage. Their presence was related to the previous duration of proteinuria, but they had no apparent influence on the subsequent evolution of renal function. The mechanisms leading to their presence are unknown.

Introduction

Lipid-containing cells with a foamy appearance have long been described in the renal interstitium of patients with the nephrotic syndrome [1]. Such cells seem to be more frequent in the nephrotic syndrome due to mesangio-capillary or focal and segmental sclerosing lesions and in certain cases they are also noted in glomeruli.

However, the origin of these interstitial cells is disputed. Some believe they represent tubular cells [2], which have acquired lipids or lipoproteins that have escaped the glomerular filter. Others suggest they represent lipid-containing macrophages [3], similar to those observed in atheroma lesions, in which extensive data suggest their macrophage origin [4]. They have not been correlated with clinical findings or prognosis of the underlying renal disease.

We describe our findings in patients with the nephrotic syndrome of various aetiology, of large lipid-loaded interstitial cells with a foamy appearance, strongly HLA-DR positive and expressing other markers characteristic of cells of the monocyte/macrophage lineage.

Patients and methods

Twenty-nine adult patients with the nephrotic syndrome were studied. Systemic lupus erythematosus syndrome was present in four cases, Henoch-Schönlein purpura in one, and a membranous nephropathy associated with penicillamine in another patient. The remaining 23 had the idiopathic types of glomerulonephritis (membranous 15, mesangio-capillary 3, focal and segmental sclerosing lesions 2, minimal change 3).

Renal biopsies were performed in 19 patients within six months of apparent onset of the nephrotic syndrome. The remaining biopsies were performed later in the course, two of them being second biopsies obtained at 18 and 69 months from onset respectively because of deteriorating renal function.

Renal tissue was obtained in all cases by percutaneous renal biopsy and processed for conventional optical and immunoperoxidase studies.

A panel of previously well-characterised monoclonal antibodies was applied at appropriate dilutions to sequential cryostat sections, and revealed using an indirect immunoperoxidase method. Sections were counterstained with Mayer's haemalum. We employed monoclonal antibodies recognising epitopes expressed in all leucocytes (2D1), B cells (TO 15), monocyte/macrophages (FMC 32), natural killer cells (Leu 7), Pan T (UCHT 1), T helper/inducer (Leu 3a) and T cytotoxic/suppressor lymphocytes (UCHT 4), and an anti-C3 b receptor. Anti-HLS-DR monoclonal antibodies were also used, as well as M704 and DH 2. A positive control (tonsil section) was used with each monoclonal antibody, as well as a negative control in each biopsy.

Double staining with immunoperoxidase-lipid (Sudan B) was additionally used in two cases.

Results

In all biopsies studied, HLA-DR was expressed on capillaries (either peritubular or glomerular), on interstitial round or stellate cells, and with a variable intensity on the cytoplasm of epithelial tubular cells. These epithelial cells also reacted with the anti-NK cell antibody (Leu 7), but not with the remaining monoclonal antibodies.

TABLE I. The nephrotic syndrome patients studied: type of glomerular lesion and presence or absence of HLA DR +ve foam cells (see text)

Histological diagnosis	Foam cells (HLA DR +)	No foam cells	Total
Membranous	5	11*	16
Mesangio-capillary	2	1	3
Focal and segmental sclerosing lesions	2	–	2
Minimal change	–	3	3
SLE: Membranous	1	–	1
SLE: Crescentic	–	3	3
Henoch-Schonlein Purpura	–	1	1
Total	10	19	29

*one patient penicillamine-induced

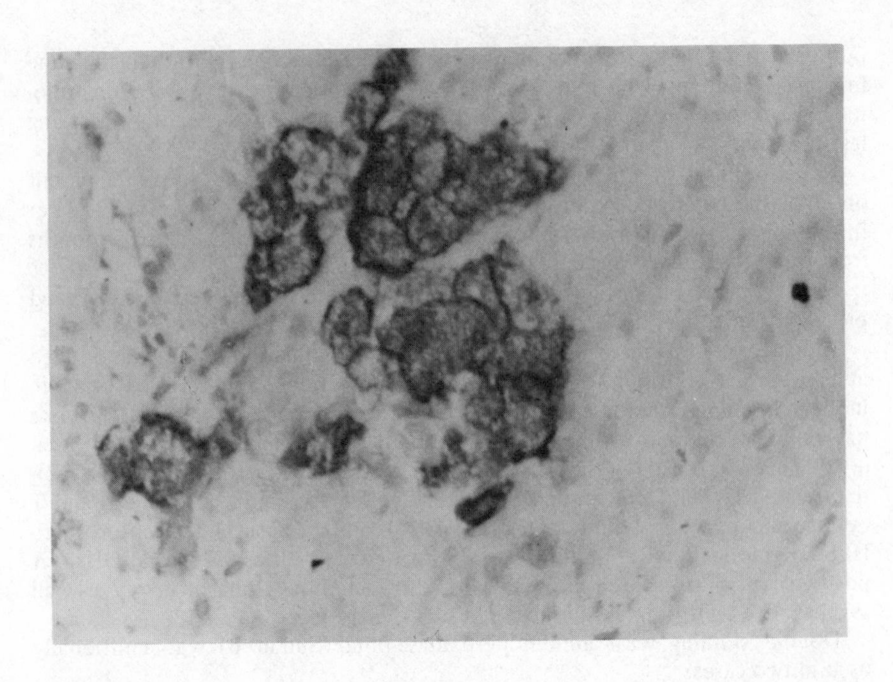

Figure 1. Immunoperoxidase study of a renal biopsy from a 17-year old girl with type I mesangio-capillary nephritis and a six months' duration of nephrotic syndrome. A number of large HLA DR-positive interstitial foam cells are present. Cryostat section counterstained with haematoxylin X 250

In 10 biopsies (Table I) variable numbers of interstitial cells showed a strong and characteristic pattern with the anti HLA-DR monoclonal antibodies (Figure 1). They were large cells expressing both membrane and cytoplasmic HLA-DR, in the latter case in a reticulated pattern suggesting a foamy appearance. Identical results were obtained with both anti-HLA-DR monoclonal antibodies.

A similar characteristic pattern was consistently obtained with the monoclonal antibodies FMC 32 (Figure 2), recognising respectively epitopes present on monocyte/macrophages and all leucocytes. With the anti-C3b (TO 5) and anti-helper/inducer lymphocyte (Leu 3a) monoclonal antibodies only a faint, weaker reaction was detected. In contrast, no reaction was noted with the remaining monoclonal antibodies.

The use of the double-staining peroxidase-lipid revealed that the positive HLA-DR foamy cells were also strongly Sudan B positive, identifying them as lipid-containing foam cells.

Review of the light microscopy sections in these 10 cases revealed in all of them variable numbers of interstitial foam cells.

Table I shows the types of nephrotic syndrome in the patients studied. The interstitial foamy macrophages were noted in 33 per cent of the idiopathic membranous, 66 per cent of the mesangio-capillary and in the two cases with focal and segmental sclerosing lesions.

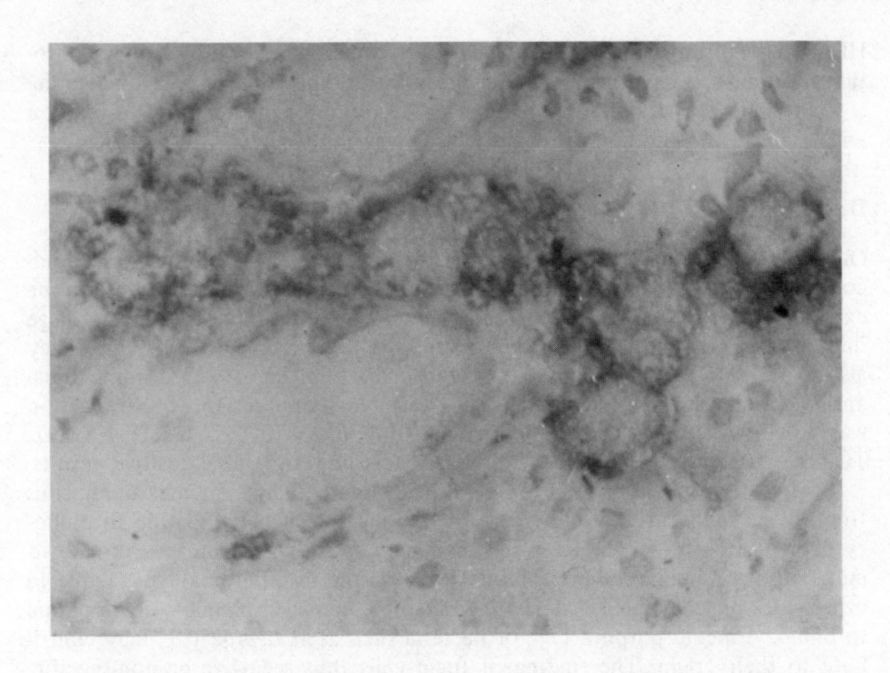

Figure 2. Immunoperoxidase study of a renal biopsy from a 48-year old female patient with a long-standing nephrotic syndrome (69 months). This was a second renal biopsy, done because of deteriorating renal function. A large number of interstitial foamy macrophages (FMC 32-positive) are present. Cryostat section counterstained with haematoxylin X 400

TABLE II. Clinical findings in the patients with idiopathic membranous glomerulonephritis, divided according to the presence of HLA-DR foam cells (see text)

	Foam cells (HLA DR+) n=5	p value	No foam cells n=10
Classical findings when biopsied			
Time from apparent onset	32±27*	<0.005	5.4±6.4*
Range	3–69		0.5–24 months
Decreased renal function	2/5	NS	2/10
Proteinuria	6.9±2.7*	NS	8±5.1*
Follow-up			
Deteriorating renal function	2/5	NS	5/9

(Unpaired student 't' test or X^2 test where appropriate)
*Mean ± SD

Table II compares the clinical findings in the group of patients with idiopathic membranous glomerulonephritis (where enough patients were available for comparison), divided according to the presence or absence of the interstitial

HLA-DR positive foam cells. Only the time from apparent onset of the nephrotic syndrome was significantly increased in the group with foam cells, indicating a relation between their presence and the duration of proteinuria. No relation with subsequent evolution of renal function was observed.

Discussion

Our results using monoclonal antibodies demonstrate that the interstitial lipid-containing foam cells observed in a number of patients with nephrotic syndrome express markers characteristic of cells belonging to the monocyte/macrophage lineage (positivity with monoclonal antibodies 2D1, HLA-DR, FMC 32). They also express a C3b receptor and react with an anti-helper/induced monoclonal antibody, findings described in monocytes or macrophages [5,6]. No reaction was noted with the remaining monoclonal antibodies (UCHT 1, UCHT 4, Leu 7, TO 15), or with the negative controls, excluding any false positive results.

Foamy macrophages have been described in a number of diseases and in sites from vessel walls in atherosclerosis [4] to the spleen in idiopathic thrombocytopenia [7]. In some cases they can be related with increased uptake or metabolism of lipids or lipoproteins (for example in atheroma) [8], while in other cases phagocytosed material (for example damaged platelets in idiopathic thrombocytopenic purpura [7], or bacteria such as *M Leprae* [9], may contribute to their origin. The finding of foam cells thus seems to be non-specific, possibly indicating only exhaustion of the cell's capacity to process all the engulfed material [7].

In the kidney, they are observed most often in the interstitium, as we have shown, but are also seen in glomeruli on occasion [10]. The interstitial cells could acquire their lipid inclusions in the tubules, or in the glomeruli, subsequently migrating into the interstitium. Alternatively lipoproteins escaping through the glomerulus in proteinuric states and taken up by tubular cells could subsequently be transferred into phagocytic cells, since monocytes possess receptors for very low-density lipoproteins.

However, it is possible that the lipid inclusions present in the foamy macrophages are unrelated to the lipoproteinuria, being related instead to other factors such as intra-macrophage metabolic changes or increased intra-renal platelet destruction.

Whatever the reason for the appearance of interstitial foam cells, they seem to correlate with the duration of the proteinuria, but not with subsequent changes in renal function; apparently they have no direct prognostic significance, at least in our small series of patients.

Acknowledgment

Monoclonal antibodies FMC 32, 2D1, UCHT 1, UCHT 4, TO 15 were gifts from Doctors H Zola, P Beverley and D Mason respectively.

References

1 Heptinstall RH. *Pathology of the Kidney, 3rd Edition.* Boston: Little, Brown & Company. 1983
2 Zollinger HU, Mihatsch MJ. *Renal Pathology in Biopsy.* Berlin: Springer. 1978: 142
3 Bohman SO. In Cotran R, ed. *Tubulo Interstitial Nephropathies.* London: Churchill Livingstone. 1982: 15
4 Schaffner T, Taylor K, Bartucci EJ et al. *Am J Pathol 1980; 100:* 57
5 Fearon DT. *Immunol Today 1984; 5:* 105
6 Wood G et al. *J Immunol 1983; 131:* 212
7 Ishihara T, Akizuki S, Yokota T et al. *Am J Pathol 1984; 114:* 1
8 Clevidence BA, Morton RE, West G et al. *Arteriosclerosis 1984; 4:* 196
9 Andrade ZA, Reed SG, Roters SB et al. *Am J Pathol 1984; 114:* 137
10 Zollinger HU, Mihatsch MJ. *Renal Pathology in Biopsy.* Berlin: Springer. 1978: 90

Open Discussion

CHAIRMAN Thank you very much for this very clear and interesting presentation.

AHLMEN (Gottenburg) I would very much like to hear your opinion on the foam cells in patients with Alport's syndrome, who most often have mild proteinuria.

CAMERON Well, this comes back to the point that we found no correlation between the amount of proteinuria in these nephrotic patients and the appearance of the foam cells, and we think they are not related to proteinuria. I would have liked to have shown you some pictures from Alport's syndrome, and we intended to do this, but the problem was that we have used up all our Alport's material on other studies. Clearly we need to get hold of some biopsies prospectively and do this. I would suspect that they are going to turn out to be the same cells and we will find them later in the course of the Alport's evolution. Certainly in our own experience of children and adults with Alport's syndrome they are usually visible in the later course of the disease, especially in uraemic patients. For example, in a five year old with isolated haematuria you virtually never see the foam cells.

BONE (Liverpool) What are the chances of finding foam cells in peripheral blood?

CAMERON We have not looked in this study recently. We were looking very actively at monocytic cells in peripheral blood in our transplant patients and incidentally you might find some foam cells in such patients. You can, of course, see them in the plasma of some lipid storage diseases as well as in the tissues.

Proc EDTA–ERA (1984) Vol 21

AETIOLOGY AND PROGNOSIS OF DE NOVO GRAFT MEMBRANOUS NEPHROPATHY

Y Pirson, J Ghysen, J P Cosyns, J P Squifflet, G P J Alexandre, C van Ypersele de Strihou

Cliniques Universitaires St-Luc, Brussels, Belgium

Summary

In order to investigate the aetiology and prognosis of de novo graft membranous nephropathy (DNGMN), we review 25 such cases observed among 1258 grafts. Coexistence of chronic rejection lesions and their parallel progression with DNGMN suggest that DNGMN may be part of the rejection process. DNGMN developed in 12 per cent of HLA-identical living donor recipients vs only two per cent of both haplo-identical and cadaver donor recipients; in the latter group, all DNGMN patients had ⩽2 HLA-AB mismatches. Graft survival after diagnosis of DNGMN is only 49 per cent at five years. We conclude that DNGMN is associated with chronic rejection, develops preferentially in well-matched grafts and carries a rather poor prognosis.

Introduction

De novo graft membranous nephropathy (DNGMN) is an increasingly recognised distinct clinical entity. In 1982 we described nine cases of DNGMN [1]. In a French collaborative study analysing 19 other cases, a review of the literature recorded 42 published cases [2].

In none of these reports was the aetiology of DNGMN elucidated. The prognosis appeared to be poor in our limited previous study [1], while others indicated that DNGMN did not adversely affect graft function [2] and survival [3].

In order to assess the aetiology and prognosis of DNGMN, we review 25 such cases observed in our centre.

Methods

Between 1963 and December 31st, 1983, we performed 1258 renal grafts. Detailed histocompatibility data are available since 1973. Since 1976, all graft specimens were routinely processed for light and immunofluorescence microscopy using standard fluorescent antisera; electron microscopy is performed only in selected cases. The present study is based on the 450 grafts subjected to pathological examination (biopsy or transplant nephrectomy) from 1976 to July 31st, 1984. Our indications for biopsy are acute rejection resistant to therapy, unexplained graft dysfunction and persistent proteinuria.

Patients were treated with a conventional immunosuppressive regimen (azathioprine and steroids) except for a few receiving cyclosporine; most of the patients transplanted in the last seven years were also given antilymphocyte or antithymocyte serum.

The criteria for the diagnosis of membranous graft nephropathy are detailed in our previous report [1]. The diagnosis of DNGMN is certain when sufficient clinical and/or pathological data allow the exclusion of membranous nephropathy as the cause of the original renal disease. DNGMN is stated to be probable when available pre-transplant clinical data indicate a diagnosis other than membranous nephropathy without definitive histological proof.

The patients' notes were reviewed up to July 31st, 1984. Nephrotic syndrome is defined by proteinuria >5g/L or >3g/24hr or proteinuria with albuminaemia <3g/dl; improvement of nephrotic syndrome by a regression of proteinuria <1g/24hr; remission by reduction of proteinuria <350mg/24hr or absence on morning urine sample. Deterioration of graft function is defined as a doubling of the serum creatinine.

Results

DNGMN was observed in 25 patients: de novo characteristics were certain in 18 cases and probable in the seven others. There were 14 men and 11 women, aged from seven to 45 (mean: 28) years at the time of transplantation. In 18 of them, quantifiable proteinuria was noted three to 48 (mean: 13) months after transplantation and was the main indication for biopsy demonstrating DNGMN 3 to 30 (mean: 13) months later; in five cases, DNGMN was discovered on a biopsy specimen performed six to 54 (mean: 24) months after transplantation to elucidate graft dysfunction; in the last two cases, DNGMN was found in a graft removed 24 to 39 months after transplantation and was associated with other glomerular lesions. All patients were on conventional immunosuppressive regimens.

The overall incidence of DNGMN is two per cent (25/1258) in the whole population and 5.6 per cent (25/450) among histologically documented grafts.

Mild to severe chronic rejection lesions are present together with DNGMN in all patients. Serial histological data are available in six patients. In three of them appearance of DNGMN, absent on a previous biopsy 12 to 31 months earlier, was associated with progression of chronic rejection lesions. In the three others a subsequent pathological examination two to 46 months after the discovery of DNGMN showed a parallel progression of DNGMN and chronic rejection lesions.

If we restrict the analysis to grafts performed since 1973, it appears that incidence of DNGMN is significantly higher in HLA-identical living donor groups (2/17) compared to that of the HLA-haploidentical living donor group

(3/141) and cadaver group (18/847) ($p<0.05$). This is not due to more frequent pathological examination in the first group since only 5/17 grafts were histologically documented in this group, vs 55/141 in the second and 377/847 in the third group respectively. In the 18 cadaver graft recipients with DNGMN, histocompatibility was particularly good: five had none, five had only one, and eight had two HLA-A,B mismatches with their donors.

Among the 23 patients in whom DNGMN was discovered on biopsy, graft survival was 91, 86, 79, 63 and 49 per cent respectively 1, 2, 3, 4 and 5 years after diagnosis. The clinical course was assessed in 19 patients with a potential follow-up of at least one year after demonstration of DNGMN (range: 5–85, mean: 38 months). One patient with minimal proteinuria at the time of diagnosis has normal graft function and no proteinuria 46 months later. Two patients with proteinuria <3g/L at the time of diagnosis are in remission with good graft function 60 to 85 months later. In one patient, the nephrotic syndrome is markedly improved and renal function remains normal 66 months after diagnosis. In five patients, proteinuria remains unchanged with stable graft function, 25 to 48 months after diagnosis. In the last 10 patients, proteinuria did not remit and graft function deteriorated two to 40 months after diagnosis, leading to return to haemodialysis in seven of them five to 50 months after diagnosis. Deterioration of graft function was more frequent (although not significantly) in cadaver grafts (10/16) than in living donor grafts (0/3), when serum creatinine was above 1.7mg/dl (4/5) rather than below this level (6/14) and when nephrotic syndrome was present (7/11) rather than absent (3/8) at the time of biopsy.

Two patients received a second graft: recurrence of the DNGMN leading to graft loss was observed in one patient previously reported [4], while the other has normal graft function and no proteinuria nine months after the second graft.

Discussion

DNGMN is more frequent than previously assumed. Its occurrence in our population is at least two per cent. If we restrict the analysis to histologically documented grafts, the prevalence of DNGMN reaches 5.8 per cent, compared to a two per cent rate reported by others [2,5]: variable incidence among series might simply reflect differences in indications for graft biopsy. Thus, Cheigh [6] found DNGMN in 7.8 per cent of grafts which developed nephrotic syndrome.

Pathogenesis of DNGMN and particularly identification of offending antigen(s) are not yet elucidated. Various exogenous antigens such as hepatitis B virus and antilymphocyte globulins have been suspected but never demonstrated as responsible [1]. Another possible source of antigen is the graft itself: histoincompatible alloantigens might elicit antibody production leading to immune complex glomerulonephritis. In other words, DNGMN could in some way be part of the rejection process. Some of our observations suggest this hypothesis. DNGMN coexists with chronic rejection lesions of variable degree in all of our patients. In two of them DNGMN was superimposed upon other severe and complex lesions of transplant glomerulopathy found in graft nephrectomy specimens. Repeated pathological examinations in six cases show a striking parallelism between the appearance or evolution of DNGMN lesions and those of chronic rejection. Association of DNGMN and chronic rejection damage has already been described by us [1] and by Dische [7] and can also be found by reviewing pathological data provided in other reports [2,5,8].

If alloantigens are responsible for the development of DNGMN it might seem paradoxical that in our series the disease appears in well-matched grafts and is even more frequent among HLA-identical living donor grafts than in cadaver grafts. It must be added that three other cases of DNGMN occurring in HLA-identical living donor grafts were reported [1,5,8]. In fact, Thoene's experimental studies support the suggestion that the development of DNGMN is enhanced by a good histocompatibility: in the rat, immune complex glomerulonephritis develops in grafts coming from MHC-identical donors and is thought to result from immunisation against minor renal alloantigens independent of MHC [9]. Other factors that may play a role include the immunosuppressive treatment, as discussed elsewhere [1] and individual susceptibility, as suggested by our case of DNGMN recurrence [4].

In our limited previous report, the prognosis of DNGMN appeared poorer than that of idiopathic membranous nephropathy [1]. This is confirmed by the present study, now extended to 25 patients: five year graft survival is only 49 per cent, compared to a five year kidney survival of 88 per cent in a large series of untreated idiopathic membranous nephropathy [10]. Others have claimed that DNGMN has a benign course [6] and does not adversely affect graft function [2] and survival [3]; however two of these reports involve only three [6] and 6 [3] patients. The discrepancy between our results and those of Charpentier [2] may be explained by a greater proportion of DNGMN patients with abnormal renal function and nephrotic syndrome at the time of diagnosis in our series, perhaps resulting from a more severe form or a more advanced stage of the disease. Extended follow-up of such patients in large series will give a more precise assessment of their long-term course.

References

1 Cosyns JP, Pirson Y, Squifflet JP et al. *Kidney Int 1982; 22:* 177
2 Charpentier B, Lévy M. *Néphrologie 1982; 3:* 158
3 Berger BE, Vincenti F, Biava C et al. *Transplantation 1983; 35:* 315
4 Cosyns JP, Pirson Y, van Ypersele de Strihou C et al. *Nephron 1981; 29:* 142
5 Honkanen E, Törnroth T, Pettersson E, Kuhlback B. *Clin Nephrol 1984; 21:* 210
6 Cheigh JS, Mouradian J, Susin M et al. *Kidney Int 1980; 18:* 358
7 Dische FE, Herbertson BM, Melcher DH, Morley AR. *Clin Nephrol 1981; 15:* 154
8 Case records of the Massachusetts General Hospital (Case 45-1979). *N Engl J Med 1979; 301:* 1052
9 Thoenes GH. *Transplant Proc 1981; 13:* 1197
10 Noel LH, Zanetti M, Droz D, Barbanel C. *Am J Med 1979; 66:* 82

Open Discussion

CAMERON (LONDON) May I make one comment on your very nice presentation and ask a question. The comment is that it is interesting that different units in different areas seem to be finding very differing incidences. For example, out of 1,200 grafts we have just one case and we biopsy every nephrotic patient and very many others. When I was looking at the literature, before the publication of your series, I noticed that there seemed to be a deficit of patients with other forms of glomerulonephritis as an aetiology of the renal failure in patients who developed subsequently de novo membranous. Clearly if the initial disease is membranous and they develop a membranous lesion in the graft it is considered a recurrence. It is striking that in the reported cases there is cystinosis, reflux nephropathy, and very curiously three cases of secondary amyloid, which was a grossly distorted pattern. Did you also see this absence of other forms of primary glomerulonephritis in these patients?

PIRSON The cause of the initial renal disease leading to de novo graft nephropathy in 18 cases where the diagnosis was certain were, five cases of glomerulonephritis, one focal glomerulosclerosis, one extracapillary glomerulonephritis, one membranous glomerulonephritis and two other cases which could not be specified but with sufficient data allowing the exclusion of membranous glomerulonephritis and eight chronic interstitial nephritis, two haemolytic uraemic syndrome, two polycystic renal disease, and one oxalosis. There were no cases of amyloidosis. I would add that this prevalence is more than five per cent if we add all the histology we documented.

EDTA–ERA (1984) Vol 21

RENAL INVOLVEMENT IN A SYNDROME OF VASCULITIS COMPLICATING HBsAg NEGATIVE CIRRHOSIS OF THE LIVER

J Montoliu, A Darnell, J Mª Grau, A Torras, L Revert

Hospital Clinic, Barcelona, Catalunya, Spain

Summary

Six HBsAg negative patients with cirrhosis of the liver (CL) presented with recurrent bouts of palpable purpura in the legs due to small vessel leucocytoclastic vasculitis. In addition, all patients had renal failure, proteinuria and microhaematuria.

Renal biopsy disclosed either diffuse proliferative (3 cases) or focal necrotising glomerulonephritis with crescents (2 cases). One patient had IgM-IgG mixed cryoglobulinaemia (type II).

Four patients died of complications of their CL. Hepatocellular carcinoma was found in 1 case. In the patient without renal biopsy renal function improved following steroids and cyclophosphamide. The pathogenesis of this syndrome of cutaneous vasculitis with severe glomerular involvement in CL is unknown but could be immune-complex mediated.

Introduction

Hypersensitivity vasculitis is characterised pathologically by inflammation of small vessels and includes a heterogeneous group of clinical disorders. In recent years we have seen 6 HBsAg negative patients with cirrhosis of the liver who developed a syndrome of cutaneous vasculitis accompanied by severe renal involvement. This report describes the features of the syndrome.

Methods

The patients are five males and one female, aged 50 to 74 years. All patients were HBsAg negative and had histologically documented cirrhosis of the liver, of either alcoholic (4/6) or cryptogenic (2/6) origin.

In addition to the features of liver disease, the clinical presentation of the syndrome was similar in all cases and consisted of the appearance of recurrent bouts of palpable purpura in the lower extremities closely associated in time with the development of renal failure, proteinuria and microhaematuria. To further investigate these abnormalities five patients underwent skin biopsy and renal biopsies were also performed in five.

Most patients had advanced liver disease. Their serum albumin ranged from 16 to 32g/L (mean 24.8 ± 5.5g/L, ± = SD) and their serum bilirubin from 0.4 to 7.2mg/dl (mean 2.5 ± 2.6). Initially, only two patients (cases 2 and 3) had moderately prolonged prothrombin times with respect to controls (3 and 5 seconds respectively). Platelet counts were greater than 110,000mm^3 in every instance and averaged 181,000 platelets/mm^3.

Results

All cases but patient Number 1 had skin biopsies of purpuric lesions and they uniformly showed small vessel leucocytoclastic vasculitis with subepidermal haemorrhage (Figure 1). No drug or offending antigen could be identified in any case.

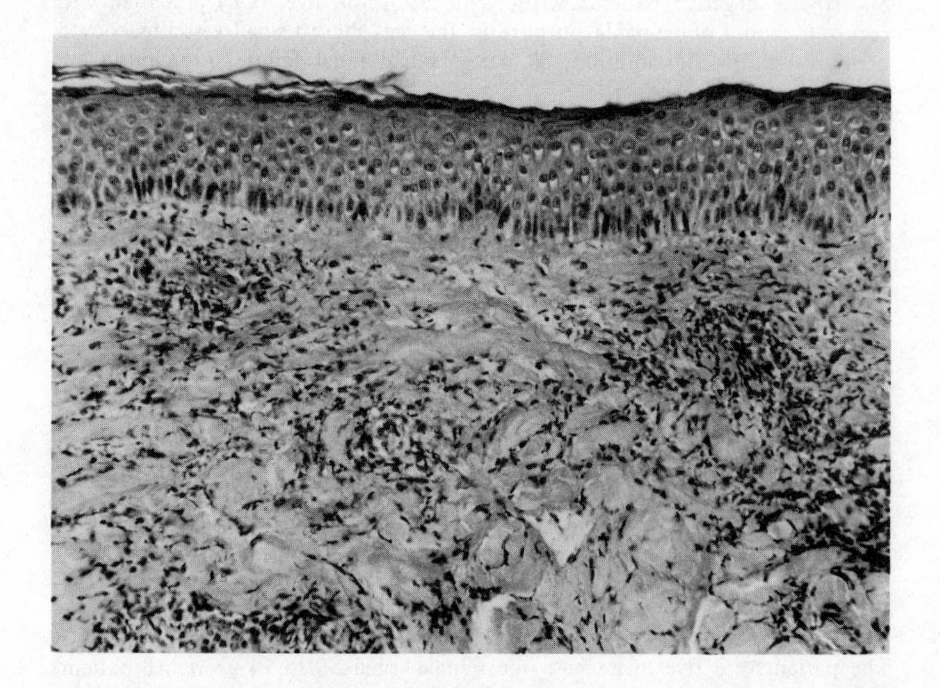

Figure 1. Skin biopsy showing inflammatory perivascular infiltrate, thrombosis of dermal vessels and subepidermal haemorrhage (H&E x 100 – reduced for publication)

Coincident in time with the cutaneous vasculitis, the patients also developed significant renal failure associated with proteinuria and microscopic haematuria

(Table I). Two patients (cases 3 and 4) required dialysis. Cases 2 and 3 had transient episodes of arthritis and otherwise vasculitic involvement of other organs was lacking.

TABLE I. Summary of the main features of the syndrome

Patient number	Age Sex	Skin biopsy	Peak serum creatinine (mg/dl)	Proteinuria (g/24hr)	Microscopic Haematuria	Renal biopsy
1	65 F	NP	8.5	3.2	Yes	Diffuse proliferative GN
2	50 M	Vasculitis	2.8	0.8	Yes	Diffuse proliferative GN
3	74 M	Vasculitis	9.4	1.5	Yes	Focal necrotising GN with 60% crescents
4	73 M	Vasculitis	9.6	2.4	Yes	Focal necrotising GN with 75% crescents
5	61 M	Vasculitis	2.7	3.4	Yes	Diffuse proliferative GN
6	56 M	Vasculitis	8.8	1.2	Yes	NP

M = Male, F = Female, NP = Not performed, GN = Glomerulonephritis

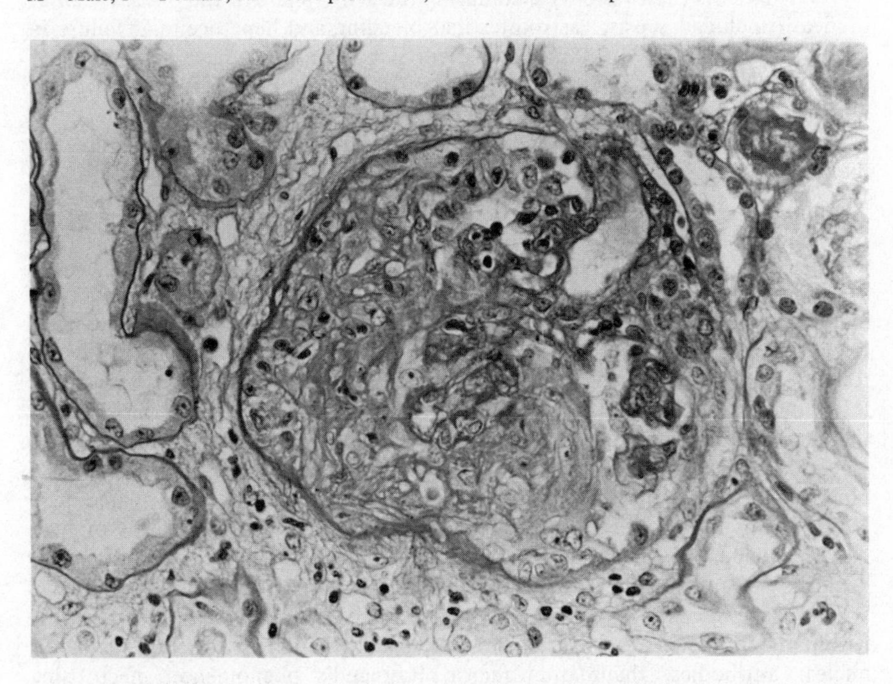

Figure 2. Renal biopsy: glomerulus showing focal necrotising glomerulonephritis and extracapillary proliferation (Masson's trichrome x 1000 – reduced for publication)

Renal biopsy results were as follows: case 1 had a diffuse endocapillary proliferative glomerulonephritis with mesangial IgA and C3 deposits; case 2 had a mild diffuse proliferative glomerulonephritis with negative immunofluorescence, and cases 3 and 4 both had a severe focal necrotising glomerulitis (Figure 2) with extracapillary proliferation in 60 per cent and 75 per cent of glomeruli respectively. Immunofluorescence in cases 3 and 4 showed only fibrinogen within the crescents. Finally, patient Number 5 had endocapillary diffuse proliferative glomerulonephritis with granular capillary C3 deposits and type II mixed essential cryoglobulinaemia (monoclonal IgM-Kappa against polyclonal IgG). No inflammatory lesions of the renal arteries or arterioles were seen in any case.

Cryoglobulins were negative in all patients but case 5, and antinuclear antibodies were absent in all cases. Serum C3 was low in patients 1, 2, 4 and 5, normal in case 6 and not performed in case 3. Alpha 1 fetoprotein was raised in patient 2, later found to have hepatocellular carcinoma, and normal in the remaining cases. Unfortunately, serum IgA and circulating immune complexes were not measured in a meaningful number of patients.

In patient 6 no renal biopsy was performed, but he was treated with steroids and cyclophosphamide and his serum creatinine came down within two months of initiating treatment. After a follow-up of three years he is now off therapy, without skin lesions and with a serum creatinine of 2.1mg/dl.

Four patients (cases 1to 4) died during the acute phase of the disease. Causes of death included sepsis, gastrointestinal bleeding and hepatocellular failure in various combinations. In addition, patient 2 was found to have hepatocellular carcinoma at autopsy. Of the patients who died only case 3 had been treated with steroids, cyclophosphamide and a short course of plasmapheresis.

Patient 5 was not treated and is now clinically stable with a serum creatinine of 2.2mg/dl after six months of follow-up.

Discussion

The present work identifies a syndrome of cutaneous vasculitis with glomerular involvement developing in HBsAg negative patients with cirrhosis of the liver. Both the types of glomerular disease encountered, diffuse proliferative and focal necrotising glomerulonephritis with crescents are consistent with vasculitic inflammation of the glomerular capillaries [1]. To our knowledge this report is the first to describe serious renal disease in this syndrome. However, Grau Junyent et al have recently described a predominantly cutaneous vasculitis in nine cirrhotic patients, eight of whom had their visceral organs spared [2]. Hypersensitivity vasculitis has been reported to occur in other forms of liver disease, such as primary biliary cirrhosis [3] acute viral hepatitis, chronic active hepatitis [4] and alpha 1 antitrypsin deficiency with liver disease [5].

The development of an immunologically mediated disease like vasculitis within the framework of liver cirrhosis is not particularly surprising since antinuclear antibodies, rheumatoid factor, Raynaud's phenomenon, necrotising pulmonary vasculitis and other clinical signs of systemic disease have been noted in post-necrotic cirrhosis of the liver [6].

Apart from the patient with cryoglobulinaemia, no other aetiological agent for vasculitis could be identified. The relation of cryoglobulinaemia with liver disease is well known [7].

It is generally believed that hypersensitivity vasculitis represents an immunological response to antigenic material which leads to antibody formation and immune complex deposition in the vessel wall [8]. The nature of the antigenic material in this group of patients is unclear, but we could speculate firstly, the patients with cirrhosis of non alcoholic origin might be suffering from chronic undetected viraemia, perhaps due to the non A non B hepatitis virus. Secondly, impaired clearance by the diseased liver of absorbed intestinal antigens could be an alternative explanation. Finally, release of tumoral antigens [9] may provide the antigenic stimulus in those cases with superimposed hepatocellular carcinoma, such as our patient Number 2.

References

1 Glassock RJ, Cohen AH. In Brenner BM, Rector FC, eds. *The Kidney.* Philadelphia: WB Saunders. 1981: 1520
2 Grau Junyent JM, Urbano Marquez A, Rozman C et al. *Med Clin (Barc).* In press
3 Gilliam JN, Smiley JD. *Ann Allergy 1976; 37:* 328
4 Popp JW, Harrist TJ, Dienstag JL et al. *Arch Intern Med 1981; 141:* 623
5 Bandrup F, Ostergaard PA. *Arch Dermatol 1978; 114:* 921
6 Morrison EB, Gaffney FA, Eigenbrodt EH et al. *Am J Med 1980; 69:* 513
7 Druet P, Letonturier P, Contet A, Mandet C. *Clin Exp Immunol 1973; 15:* 483
8 Cupps TR, Fauci AS. *The Vasculitides.* Philadelphia: WB Saunders. 1981: 6–19
9 ibid. Pages 116–122

Open Discussion

SIMOES (LISBON) A very interesting paper. Did you exclude the action of drugs in these patients? Some of these histological reactions, vasculitis in the skin and lesions in the kidney, I assume could be produced by hypersensitivity?

MONTOLIU As I said before in the three months immediately preceding admission, these patients were not receiving any drugs and specifically were not taking diuretics. We were also unable to identify any specific infection or antigenic stimuli responsible for the development of vasculitis.

DAL CANTON (Naples) Did you carry out immunofluorescence studies in your renal biopsies? I enquire about that because we have recently observed a quite similar case and we have found diffuse deposits of IgA.

MONTOLIU Yes, all renal biopsies were studied by immunofluorescence. Of the patients with diffuse proliferative glomerulonephritis, one had mesangial IgA deposits, another one negative immunofluorescence and the patient with cryoglobulinaemia had isolated capillary C_3 deposits. The two patients with focal necrotising glomerulonephritis had negative immunofluorescence except for fibrinogen in the necrotic areas and within the crescent. I might add that in our experience IgA deposition in glomerular disease associated with cirrhosis of the liver is not as common as generally believed.

VERROUST (Chairman) Perhaps I could ask you if you have any information on the serum IgA of these patients?

MONTOLIU As I said before serum IgA levels were measured in few patients, exactly three. They were raised in two patients, one of them with glomerular IgA deposits. This was a retrospective study and we could not obtain more data.

VERROUST Can you exclude also, apart from the one with mixed cryoglobulinaemia, that other patients did not have monoclonal bands in their serum?

MONTOLIU All I can say is that cryoglobulins were negative in all patients but one and that the admission serum electrophoresis did not show a monoclonal component in any patient.

Proc EDTA–ERA (1984) Vol 21

FAMILIAL NEPHROPATHIC AMYLOIDOSIS ASSOCIATED WITH INDOMETHACIN RESPONSIVE FEVER

T G Feest, J P Wallis

Royal Devon and Exeter Kidney Unit, Exeter, United Kingdom

Introduction

We describe a patient with familial nephropathic amyloidosis who also has recurrent fever and polyserositis which is responsive to indomethacin. The association within families of periodic fever and nephropathic amyloid is most commonly seen in familial Mediterranean fever, but in that condition the mode of inheritance and the amyloid fibrils are different from the family described here. The family described seems to be different from other forms of familial nephropathic amyloidosis.

Case history

A 45 year old woman presented with chronic renal failure due to renal amyloid. She has a family history of autosomal dominant nephropathic amyloidosis, affecting six members of two successive generations, all descended from the patient's apparently unaffected maternal grandmother. The family has been described in detail elsewhere [1]. Continuous ambulatory peritoneal dialysis (CAPD) was started in March 1982. During admission for training for CAPD she was noted to have recurrent episodes of fever, usually symptomless, although on one occasion there was pleuritic pain. No infection or pulmonary embolus could be demonstrated. The fevers settled after six weeks. No treatment was given. In the next nine months she had three episodes of fever and malaise each of which settled spontaneously after two weeks. There were also four episodes of apparent sterile peritonitis with abdominal pain and fever, and white cells in the peritoneal dialysate, but no organisms were grown from peritoneal fluid. In February 1983 she developed candida peritonitis for which the peritoneal catheter had to be removed and amphotericin and ketoconazole were given. There was quick recovery with no subsequent evidence of peritoneal infection.

During five subsequent months of haemodialysis she had recurrent fever (Figure 1) with intermittent pleuritic pain and flitting pleural rubs. Chest X-rays

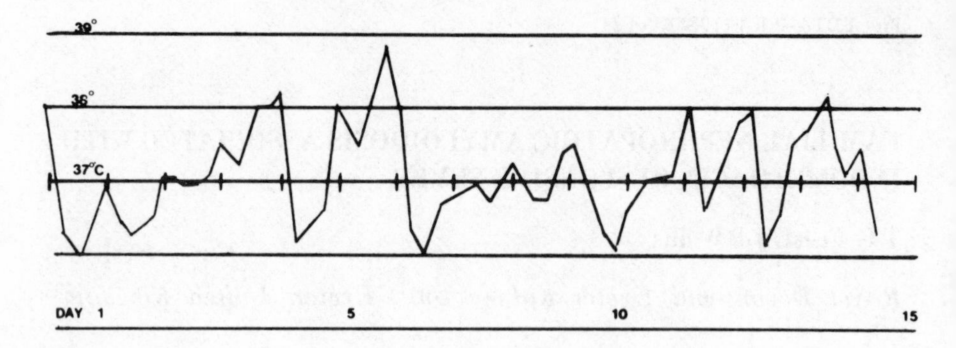

Figure 1. Example of recurrent fever

showed small transient pleural effusions but no other lesions. Perfusion lung scan was normal. Cultures of blood, urine, and throat swabs were negative for tuberculosis, bacteria, and fungi. Viral antibody screening was negative. Rheumatoid factor and autoantibody screens were negative. Ultrasound and gallium scan of the abdomen showed no evidence of intra-abdominal abscess. Antibiotics had no effect upon the fever. She then developed pericardial pain and friction with fever, despite excellent dialysis. Pericardial ultrasound was normal. The pericardial pain was treated with indomethacin, and immediately the fever resolved for the first time for six months. After two weeks indomethacin was stopped and fever returned. On restarting indomethacin the fever subsided, but when the dose was reduced from 50mg tds to 25mg tds the fever returned, settling again with increased dosage. Indomethacin was eventually withdrawn after five months. The patient was subsequently restarted on CAPD without recurrence of fever.

The amyloid deposits in this patient showed permangante sensitive congophilia. This is suggestive of amyloid A, as seen in reactive systemic amyloidosis and familial Mediterranean fever, but the amyloid in this family is distinct from those conditions, as there is no staining with anti-amyloid A or anti-prealbumin sera [1].

Discussion

This patient has autosomal dominant familial nephropathic amyloidosis associated with recurrent fever and polyserositis. We found no evidence of an infective cause for this, and the response to indomethacin suggests a non-infective inflammatory mechanism.

Familial nephropathic amyloidosis is rare. Ostertag [2] described a syndrome of familial nephropathic non-neuropathic amyloidosis with autosomal dominant inheritance which leads to renal failure. There was also massive hepatosplenomegaly. A second family [3] was considered to have the same syndrome although in this family there was only minimal hepatic involvement confined to vessel walls. Our patient is a member of the third family assigned to familial amyloidosis of Ostertag [2] although in this family there is also little hepatic or splenic involvement and histologically hepatic involvement is limited to small deposits in middle sized arterioles. We propose that this syndrome is distinct from that described by Ostertag. The condition presents with severe nephropathic amyloidosis, there is polyserositis and indomethacin responsive fever, and relatively little early clinical involvement of other organs.

There are other associations of familial nephropathic amyloidosis and periodic fever. In familial Mediterranean fever nephropathic amyloidosis frequently develops, but inheritance is autosomal recessive, and the amyloid fibrils are different from those seen in this family [1]. In the Muckle-Wells syndrome [4] autosomal dominant familial nephropathic amyloidosis is associated with nerve deafness and recurrent urticaria and fever. There is no record of deafness or urticaria in our family. Of six patients recently described with a syndrome of periodic fever and hyperimmunoglobulinaemia D [5] three had a positive family history, and in one family two members in successive generations developed renal amyloidosis. The case of Fox and Morrelli [6] probably had a similar condition. As in familial Mediterranean fever, fever of this condition is responsive to colchicine. Nilsson and Floderus also described an autosomal dominant condition in which renal amyloid occurred in association with recurrent abdominal pain and fever [7].

In familial Mediterranean fever treatment with colchicine appears to prevent recurrent fever and to slow or stop the deposition of amyloid [8,9]. It is interesting to speculate, should apparently unaffected members of this family develop recurrent fever, whether indomethacin may be of value not only in treating the fever but in preventing amyloidosis. It might also be worthwhile to study the effects of indomethacin or other prostaglandin synthetase inhibitors on fever and amyloidosis in familial Mediterranean fever or on periodic fever associated with hyperimmunoglobulinaemia D.

References

1 Lanham JG, Meltzer ML, De Beer FC et al. *Q J Med 1982; 201:* 25
2 Ostertag B. *Z Menschl Vererb Konstit 1950; 30:* 105
3 Weiss SW, Page DL. *Am J Pathol 1973; 72:* 447
4 Muckle TJ, Wells M. *Q J Med 1962; 122:* 235
5 Van Der Meer JWM, Vossen JM, Radl J et al. *Lancet 1984; i:* 1087
6 Fox M, Morrelli H. *N Engl J Med 1960; 263:* 669
7 Nilsson SE, Floderus S. *Acta Med Scand 1964; 175:* 341
8 Dinarello CA, Wolff SM, Goldfinger SE et al. *N Engl J Med 1974; 291:* 934
9 Goldstein RC, Schwabe AD. *Ann Intern Med 1974; 81:* 792

Proc EDTA–ERA (1984) Vol 21

IMMUNE FUNCTION AND RENAL TRANSPLANTATION IN FABRY'S DISEASE

D Donati, *M G Sabbadini, *F Capsoni, L Baratelli, D Cassani, A De Maio, G Frattini, M Martegani, L Gastaldi

Ospedale Regionale, Varese, Clinica Medica II, Università degli Studi, Milan, Italy

Summary

A deficient leucocyte immunological function could cause the reported high rate of lethal infections following renal transplantation in patients affected by Fabry's disease. We have studied humoral immunity, peripheral lymphocyte subsets, mitogenic lymphocyte response in vitro and granulocyte function in three patients with Fabry's disease.

The immunological state appears to be quite similar to that of the uraemic population in general, not showing any specific impairment.

Introduction

Renal transplantation has been regarded as the first choice treatment of uraemia due to Fabry's disease providing a correction of both uraemia and the enzymatic defect [1,2] but recently a high rate of life threatening infections has been reported after transplantation so that this treatment can no longer be recommended [3,4].

It has been suggested that the ceramide-trihexoside storage in leucocytes could alter their host defence function exposing the patients, when transplanted, to serious infectious complications [3].

In order to verify the hypothesis of a specific immunological defect, we performed a study of the immune function in these patients.

Patients and methods

Patients

Two male uraemic patients and one female without clinical nephropathy were studied. The diagnosis of Fabry's disease was both clinical and biochemical in all. Their age ranged from 45–50 years and the two uraemic patients had undergone maintenance haemodialysis for 49 and 52 months respectively.

Control blood samples were obtained from six haemodialysis patients (5 male, 1 female) matched for age and length of treatment. All the patients were free from infectious episodes and had not received drugs affecting the immune system for at least two years.

Methods

Immunoglobulins A, G, M, and serum Complement fractions (C_3, C_4) were determined by immunonephelometry.

The total lymphocyte counts were determined from blood cells count and differential counts.

Peripheral mononuclear cells were prepared from heparinised blood by gradient (lymphoprep), washed three times and suspended at optimal concentrations. The T cell subsets were analysed by immunofluorescence staining with monoclonal antibodies (Ortho Diagnostic System). The total T cell population was quantified with OKT_3 antibody, The T helper cells with OKT_4, the suppressor/cytotoxic T cells with OKT_8 [5,6]. The B cells were estimated with surface Ig immunofluorescence and the Natural Killer cells with monoclonal antibody Leu 7 (Becton-Dickinson, Torino, Italy). At least 2,000 lymphocytes were scored for each determination.

Mitogenic lymphocyte response in vitro to phytohaemoagglutinin A, Concanavalin A, and pokeweed mitogen was measured. Lymphocytes were incubated in culture (RPMI-Gibco) supplemented with 10 per cent fetal calf serum. Mitogenic stimulations were carried out in triplicate in 96 well microtitre plates (Sterilin); 10^5 lymphocytes per well with the medium alone and the mitogens at different concentrations were incubated at 37°C in humidified and five per cent CO_2 supplemented air. After 48 hours (phytohaemoagglutinin A and Concanavalin A) and 72 hours (pokeweed mitogen) 1μCi of ^{3}H-metil-thymidine was added at each well then after a further 24 hours cultures were harvested (Sacrificator) and the incorporation of ^{3}H-thymidine measured as counts per minute (β-counter Packard Tricarb).

Autologous E rosette formation was estimated with 5 x 10^6 lymphocytes, previously incubated with RPMI medium supplemented with 20 per cent autologous serum to which 5μl of autologous erythrocytes at concentration of 280 x 10^6/ml were added. After centrifugation and incubation for 24 hours, 200 elements were scored. Each test was carried out in duplicate.

Polymorphonuclear cells were prepared from heparinised blood by lymphoprep gradient at a concentration of 2.5 x 10^6/ml (chemotaxis) and 10^7/ml (phagocytosis). Chemotactic activity was studied with migration wells for chemotaxis through porous filters ('blind well' model). As chemotactic factor the synthetic peptide formil-metionil-leucil-phenil-alanine at concentration of 10^{-8}M was used. Chemotactic activity was assessed as the number of cells per microscopic field (400x) reaching the last floor of the filter. Five fields were scored.

The morphological test for Candida albicans phagocytosis was performed. Candida, previously killed by heat, were brought at concentration of 10^7/ml. Polymorphonuclear cells were incubated in autologous fresh serum then Candida was added. The phagocytic properties were assessed as: a) number of cells phagocyting at least one Candida (%Ph) and b) number of phagocytosed Candida per cell (Phagocytic index) per microscopic field (1000x). Both tests were carried out in duplicate. The polymorphonuclear cell tests were also performed on healthy volunteers to which the normal values mentioned in the text refer.

Results

Serum immunoglobulin and Complement fractions were in the normal range in all the patients.

The number of white blood cells (WBC), total peripheral lymphocytes, T_3, T_4, T_8, B lymphocytes and NK cells are shown in Table I. It is apparent that although the number of WBC varies from reduced to normal values, all the patients of both groups show absolute lymphocytopenia. The absolute number of T_3 and T_4 cells is also low when compared to the values usually obtained from a healthy population matched for age and sex but Fabry's patients do not show any notable difference from the control group. In particular T_4 cells are not relatively reduced and the T_4/T_8 ratio was over 1.7 in Fabry's patients. The absolute number of peripheral B lymphocytes and Natural Killer cells of patients affected by Fabry's disease are not depressed.

TABLE I. Absolute number of white blood cells (WBC), peripheral lymphocytes (L), T cell subsets, Natural Killers (NK), B lymphocytes and percentage of autologous E rosette (R) in Fabry's (F) and control groups (C)

Patient	WBC	L	T_3	T_4	T_8	NK	B	R
F1	6000	1440	1008	734	274	158	14	21
F2	3600	1116	636	234	134	56	134	21
F3	6200	1612	967	500	177	338	81	15
C1	7400	1332	653	573	147	66	13	36
C2	7000	1190	797	512	226	131	12	17
C3	4800	1344	672	417	148	67	67	28
C4	4500	1125	652	394	112	123	11	5
C5	4900	1617	1100	728	291	194	81	3
C6	6900	1242	534	149	149	62	62	11

Mitogenic lymphocyte response.

In Figure 1 the mitogenic lymphocyte responses in vitro to Concanavalin A (Con A), pokeweed mitogen (PWM), phytohaemoagglutinin A (PHA) are shown.

Con A although a large variability in the responses among individual patients exists, Fabry's patients response to Con A (T cell and T suppressor cells mitogen) is not lower than in the control group.

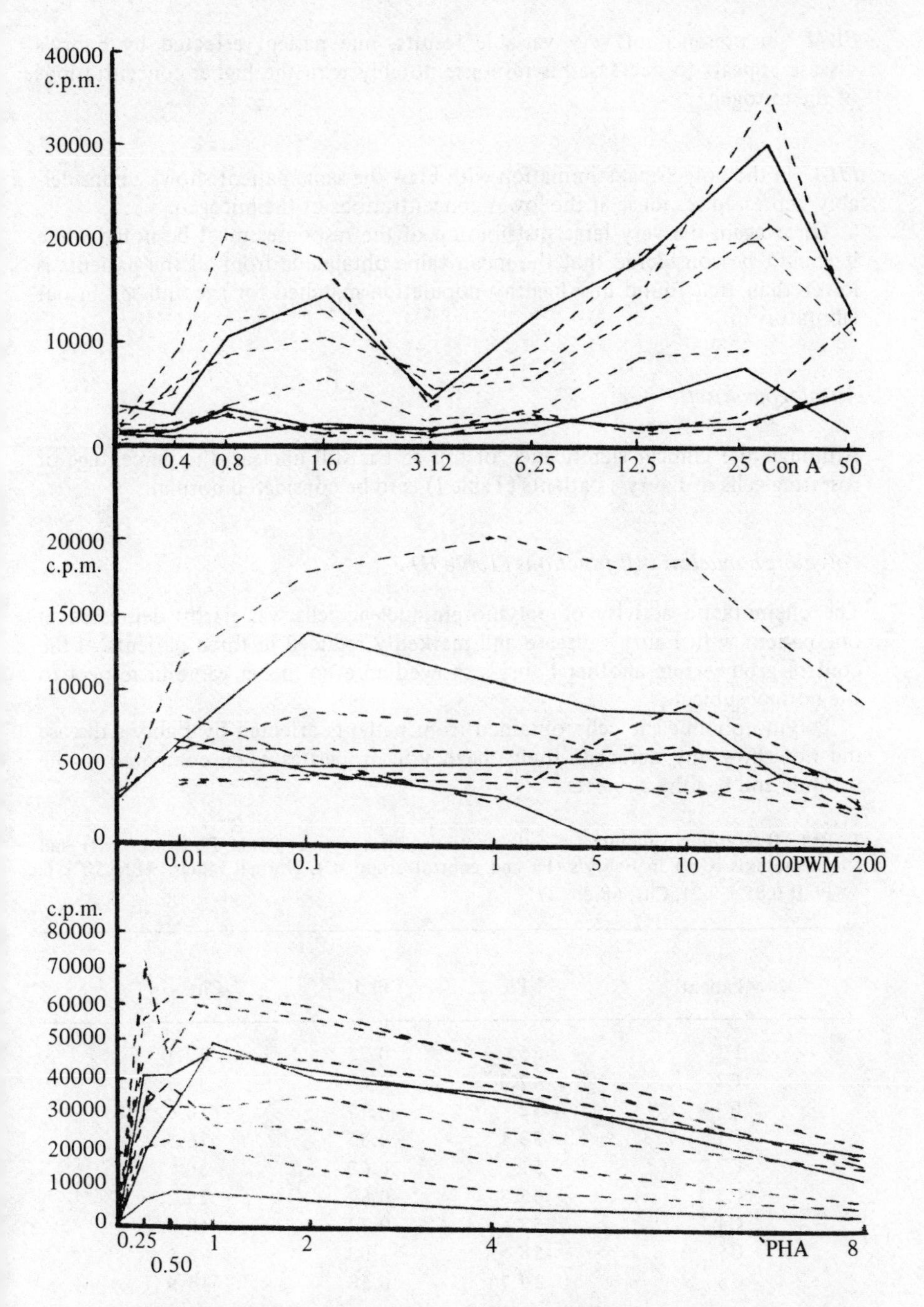

Figure 1. Mitogenic lymphocyte response in vitro to increasing concentrations of Con A, PWM, PHA in Fabry's (——) and control group (- - -). On the zero points of the abscissas the spontaneous stimulation (when determined) with the medium alone is represented

PWM in presence of very variable results, one patient affected by Fabry's disease appears to decrease his response notably with the higher concentrations of the mitogen.

PHA in the polyclonal stimulation with PHA the same patient shows a considerably depressed response at the lower concentrations of the mitogen.

Once again the very large distribution of the responses must be noted while it should be considered that the mean value obtainable from all the patients is lower than that found in a healthy population matched for age and sex in our laboratory.

Autologous rosette

Although the clinical significance of this test is still unclear, the percentage of rosetting cells of Fabry's patients (Table I) is to be considered normal.

Polymorphonuclear cell functions (Table II)

The chemotactic activity of polymorphonuclear cells was partly depressed in one patient with Fabry's disease and markedly reduced in three patients of the control group while another Fabry's showed an even higher value in respect to the normal subjects.

Polymorphonuclear cells obtained from patients affected by Fabry's disease did not show any defect in their phagocytic properties when compared to the uraemic and healthy subjects.

TABLE II. Polymorphonuclear cell functions, Phagocytosis (%Ph, Ph.I, see text) and chemotaxis (Cht) in Fabry's (F) and control group (C). Normal values: %Ph: 50 ± 6, Ph.I: 0.85 ± 0.21, Cht: 68.2 ± 27

Patient	%Ph	Ph.I	Cht
F1	55	0.75	62.1
F2	47.3	0.67	34.7
F3	52	0.77	118.6
C1	58.1	0.85	76.3
C2	51	0.69	30
C3	45.3	0.62	29.2
C4	53.5	0.73	48.6
C5	58	1.03	63.5
C6	39.7	0.56	15.9

Discussion

Uraemia is associated with impairment of the immune potential and our data are in agreement with those of others [7–9]. Patients affected by Fabry's disease do not seem to differ from the uraemic pattern in our study. Lymphocytopenia [7], reduced the number of total T and helper T cells [7,9] and low mitogenic lymphocyte response in vitro [7,8] are the immunological defects which are frequently observed in uraemic patients undergoing long-term maintenance haemodialysis.

Humoral immunity as well as the polymorphonuclear cell functions appear to be preserved in Fabry's disease. Our data do not show any specific immunological impairment in these patients which agrees with our previously reported clinical review of eight Fabry's patients receiving 10 renal transplantations [10]. We recorded an incidence of 11 major infectious episodes none of which was lethal and all responded to medical treatment and no transplant nephrectomy was required.

The reason for the high rate of lethal septic complications [3] does not appear related to the Fabry's disease. However, in spite of a lack of a qualitative immune function defect, one of our patients showed a more marked depressed response to stimulation with pokeweed mitogen and phytohaemoagglutinin A. Although it can hardly be ascribed to the basic disease, it could lead us to suppose that, in some subjects, Fabry's disease may make the uraemia induced immunodeficiency worse, probably related to a more severe involvement of the bone marrow and other organs of the immune system in this unusual systemic disease.

References

1 Philippart M, Franklin SS, Gordon A. *Ann Intern Med 1972; 77:* 195
2 Clarke JTR, Guttmann RD et al. *N Engl J Med 1972; 287:* 1215
3 Maizel SE, Simmons RL, Kjellstrand CM, Fryd DS. *Transplant Proc 1981; 13:* 57
4 Sutherland DER, Fryd DS, Morrow CE. In Brynger H, ed. *Clinical Kidney Transplantation.* Gothenburg: Grune and Stratton. 1982: 19–27
5 Kung PC, Talle MA, DeMaria ME et al. *Transplant Proc 1980; 12 (Suppl 1):* 141
6 Reinherz EL, Kung PC, Goldstein G. *Proc Natl Acad Sci USA 1979; 76:* 4061
7 Raska K Jr, Raskova J, Shea SM et al. *Am J Med 1983; 75:* 734
8 Kunori T, Fehrman I, Ringdèn O et al. *Nephron 1980; 26:* 234
9 Lortan JE, Kiepiela P, Coovadia HM et al. *Kidney Int 1982; 22:* 192
10 Donati D, Baratelli L, Cassani D et al. In *Nefrologia, Dialisi, Trapianto.* Milan: Witchig. 1984: In press

Open Discussion

VERROUST (Chairman) Do you think that there is immune deficiency in the disease? You first said that there was not but a later slide seemed to contradict this.

DONATI Yes, we have found a decreased response to stimulation with antigens in one patient. I think that we can hardly ascribe this to the patient's disease. I think that one of the rarest complications of this disease is aplastic anaemia.

Probably there will be more severe involvement of the bone marrow and the other organs in some subjects. The best way is to evaluate the medullary reserve before subjecting the patients to transplantation. I don't think that clinically the Fabry's disease patients have any specific immunological failure. Maybe their basic disease makes the renal induced immune deficiency state worse probably depending on the degree of involvment of the bone marrow.

Proc EDTA–ERA (1984) Vol 21

HYPERTENSION IN MESANGIAL IgA GLOMERULONEPHRITIS

M Rambausek, R Waldherr, K Andrassy, E Ritz

University of Heidelberg, Heidelberg, FRG

Summary

Blood pressure in 75 patients with IgA nephropathy (IgA-GN), confirmed by renal biopsy, was related to clinical, immunological and morphological findings. The findings were compared with an age-matched control group of patients with non-IgA-GN. Overall prevalence of hypertension (HT) was similar in IgA-GN and non-IgA-GN (38.7% vs 38.2%). The presence of HT in IgA-GN was related to age, renal function, immunohistological pattern and degree of glomerular sclerosis or vascular lesions respectively. No correlation was found between HT and elevated serum IgA, circulating IgA immune complexes and IgA skin deposits. The current observations underline the value of hypertension for predicting development of renal failure. Vascular lesions are not only strongly correlated with, but may even precede development of, hypertension as confirmed by longitudinal observations.

Introduction

Mesangial IgA glomerulonephritis (IgA-GN) is the most common form of GN (25%). In contrast to previous more optimistic reports [1,2] about one-third of such patients will ultimately develop chronic renal insufficiency. Despite much recent progress with respect to understanding the pathogenesis of this condition, information on the genesis of hypertension and its role in progression of renal failure is still controversial. In the present study, blood pressure was related to clinical, immunological and histological findings in patients with biopsy-confirmed IgA-GN.

Dedicated to Professor Dr Drs.h.c. W Doerr, Heidelberg on the occasion of his 70th birthday.

Patients and methods

Between 1976 and 1983 we observed 75 patients with biopsy-confirmed idiopathic mesangial IgA-GN. Patients with alcoholic liver disease, liver cirrhosis and systemic diseases (Henoch-Schönlein purpura, SLE) were excluded. Median age was 32 years (range 16–62 years). Some pertinent clinical findings were reported previously [3,4]. Patients with IgA-GN were compared with an age matched control group consisting of 110 cases with biopsy-confirmed GN other than IgA-GN (in the following non-IgA-GN) i.e. minor glomerular abnormalities (n = 43): idiopathic nephrotic syndrome with focal-segmental sclerosis (n = 15); membranous GN (n = 15); mesangial-proliferative GN (n = 26); membrano-proliferative GN (n = 11). Skin biopsies were performed of clinically normal skin from the median aspect of the thigh by surgical excision in 58 patients with IgA-GN. For details of light and immunofluorescence techniques see Reference 3. The data analysed refer to the time of renal biopsy. Hypertension was defined as blood pressure >140/90mmHg measured on three independent occasions. Statistical evaluation was by Fisher's exact test.

Results

Prevalence of HT was similar in IgA-GN and non-IgA-GN. Thirty-eight point seven per cent (29/75) of patients with IgA-GN and 38.2 per cent (42/110) with non-IgA-GN were hypertensive at the time of renal biopsy. The incidence of HT increased with declining renal function (serum creatinine <1.5mg/100ml: 22.4 per cent in IgA-GN, 31.7 per cent in non-IgA-GN; serum creatinine >1.5mg/100ml: 94 per cent in IgA-GN, 57.1 per cent in non-IgA-GN).

In both IgA-GN and non-IgA-GN there was a close relationship between HT and age (hypertensive patients <30 years IgA-GN 17.2 per cent, non-IgA-GN 33 per cent; >30 years IgA-GN 82.8 per cent; non-IgA-GN 66.7 per cent). In IgA-GN episodes of macrohaematuria were associated with significantly lower incidence of HT (HT without macrohaematuria: 65.5 per cent; HT with macrohaematuria: 34.5 per cent, $p<0.01$); however, age and macrohaematuria could not be dissociated from the confounding variables of renal function.

There was no relation of HT to serum IgA or the presence of circulating IgA immune-complexes. Furthermore, the presence or absence of IgA deposits in skin biopsies were not related to blood pressure. IgA deposits in skin biopsies were demonstrable in 25.9 per cent of all patients, more specifically in 23.8 per cent of HT and 27 per cent of non-HT patients.

In contrast to the above immunological findings, HT was clearly related to findings of renal light and immunofluorescence microscopy. In hypertensive patients glomerular deposits containing both IgM *and* IgA were more frequently found than in normotensive patients (80% vs 20%). A purely mesangial pattern of immune deposits was less frequently associated with hypertension than a mesangial *and* capillary pattern (35% vs 65%). *Extensive* capillary wall involvement was uniformly associated with HT and a more severe clinical course. A highly significant relation was observed between HT and extent of glomerular

TABLE I. Blood pressure and glomerular sclerosis or vascular lesions

	Normal blood pressure <140/90mmHg (n = 46)	Elevated blood pressure >140/90mmHg (n = 29)
Glomerular sclerosis*		
≤20%	39 (85%)	10 (34%)
>20%	7 (15%)	19 (66%)+
Vascular lesions**		
absent	21 (46%)	1 (3%)
present	25 (54%)	28 (97%)+

* percentage of total glomeruli with segmental and/or global sclerosis
** arteriolar hyalinosis and/or arterial sclerosis
\+ $p<0.01$

sclerosis or presence of vascular lesions respectively (Table I). Severe arteriolosclerosis was uniformly accompanied by segmental or global glomerulosclerosis comprising more than 20 per cent of glomeruli. No IgA-deposits were seen in arterioles except occasionally in the short segment of vas afferens contiguous with the vascular pole. Of particular interest is the later development after a median follow-up of six years of HT and chronic renal failure in five of 25 initially normotensive patients who had arteriolar hyalinosis in their initial renal biopsies.

The median rate of progression of renal failure (creatinine mg/dl per 12 months) was significantly ($p<0.05$) greater in IgA-GN patients with renal failure (creatinine 1.5mg/dl) who had persistent hypertension (n=11), i.e. 140/90mmHg on all occasions irrespective of medication [0.1mg/dl/12 months; range (−0.7)–(+8.6)] as compared with those who had intermittent hypertension (n=14), i.e. at times above and at times below 140/90mmHg [0.4mg/dl/12 months; range (−1.0)–(+0.9)]. The median follow-up was 30 months, range 7–84; median age 42 years (range 24–61) in the patients with intermittent hypertension; in the patients with persistent hypertension the median follow-up was 25 months, range 7–42, median age was 44 years (range 31–61).

Discussion

The above data demonstrate that the incidence of HT in IgA-GN is comparable to that in other forms of GN and clearly related to age, renal function and type of renal lesions. The incidence of HT is higher than in historical reports but in line with recent observations of other authors [5,6]. The present demonstration of a relation between blood pressure status and course of renal failure is in good agreement with a previous retrospective analysis of D'Amico [7] who noted hypertension as one important factor differentiating patients with a benign and malignant course.

The demonstration of a correlation between hypertension and severity of renal lesions does not solve the problem of whether renal lesions are the cause or the consequence of HT. It is therefore of note that arteriolar hyalinosis may precede the development of HT, as noted by us in several longitudinal observations. These observations are in agreement with previous results of Finer [8]. This raises the issue of whether intrarenal and glomerular hypertension precedes the development of systemic hypertension. Such glomerular hypertension has been demonstrated in animal models of glomerulonephritis [9] and other models of glomerular damage [10].

These concepts may provide a rationale for thorough antihypertensive intervention in such patients.

References

1 Finlayson G, Alexander R, Juncos L et al. *Lab Invest 1975; 32:* 140
2 Mc Coy RC, Abramowsky CR, Tisher CC. *Am J Pathol 1974; 76:* 123
3 Rambausek M, Seelig HP, Andrassy K et al. *Dtsch Med Wschr 1983; 108:* 125
4 Waldherr R, Rambausek M, Rauterberg W et al. *Contr Nephrol 1984; 40:* 99
5 Droz D. *Contr Nephrol 1976; 2:* 150
6 Clarkson AR, Seymour AE, Thompson AJ et al. *Clin Nephrol 1977; 8:* 459
7 D'Amico G, Ferrario F, Colosanti G et al. *Clin Nephrol 1981; 16:* 251
8 Feiner HD, Cabili S, Baldwin DS et al. *Clin Nephrol 1982; 18:* 183
9 Baldwin DS. *Kidney Int 1982; 21:* 109
10 Hostetter TH, Rennke HG, Brenner BM. *Kidney Int 1981; 19:* 410

Open Discussion

DAVISON (Chairman) Thank you Dr Rambausek for that excellent presentation. Could I perhaps start the discussion by asking you why you feel the presence of IgM in addition to the IgA should have such an important role?

RAMBAUSEK We could not find any pathogenetic role for the IgM. We think it is a finding that we cannot explain up to now. It is mainly found in the sclerotic lesions.

ZUCCELLI (Bologna) Regarding the identical frequency of hypertension between IgA and other glomerulonephropathies, it is our opinion that in membranous nephropathy there is not the same incidence as in IgA nephropathy. We think that the mesangial lesion probably is more related to the hypertension than in general glomerulonephropathy. Secondly, about the importance of vascular lesions in the genesis of hypertension. We have studied many patients with vascular lesions without hypertension. In this patient we have an activation of intrarenal beta receptors and we think that probably the stimulation of these receptors may be the cause of secondary arterial lesions and secondary hypertension. Do you agree with this hypothesis?

RAMBAUSEK We agree completely with your last hypothesis. The incidence of hypertension in IgA nephritis is much higher than in membranous nephropathy. Maybe because you have more subjects with a nephrotic syndrome in this population group. Did you compare that?

ZUCCELLI Yes, without nephrotic syndromes.

Proc EDTA–ERA (1984) Vol 21

ARE β-HAEMOLYTIC STREPTOCOCCI INVOLVED IN THE PATHOGENESIS OF MESANGIAL IgA-NEPHROPATHY?

S Rekola, A Bergstrand, H Bucht, A Lindberg

Huddinge Hospital, Huddinge, Sweden

Summary

In retrospect we have found that 38 of 187 patients who fulfilled the criteria of mesangial IgA-nephropathy had possible acute glomerulonephritis at the onset of their disease. We have therefore studied anti-streptococcal antibodies (ASO and ADNAseB) prospectively. Forty-three per cent of the patients had ADNAseB >800 units. Thirty-one per cent of the patients studied more than once had a fourfold or greater change in their ADNAseB titre. Thirty-three per cent of the patients had different groups of β-haemolytic streptococci isolated from their throats. This indicates a possible role of β-haemolytic streptococci in the pathogenesis of some cases of mesangial IgA-nephropathy.

Introduction

The connection between many cases of acute glomerulonephritis and β-haemolytic streptococci (β-HS) group A is well established. In acute post-streptococcal glomeronephritis (APSGN) there is a latent period of about 10 days between the onset of infection and the manifestations of glomerulonephritis. The short-term prognosis is regarded as good, children seem to have a particularly good prognosis [1]. However in adults the long-term prognosis is probably worse [2].

In 1968 Berger and Hinglais [3] reported a group of patients with persistent or recurrent haematuria. By immunofluorescence microscopy they found IgA and C_3, but also other immunoglobulins, with chiefly mesangial localisation. Many of these patients have episodes of macroscopic haematuria in association with upper respiratory tract infections or gastroenteritis. In contrast to APSGN these patients have no latent period between the infection and the episodes of macroscopic haematuria.

This disease has already been called by many different names but we prefer mesangial IgA-nephropathy (MesIgAN) because of lack of inflammation in most biopsy specimens. Glomerulonephritis is thought to appear after deposition of either circulating soluble immunocomplexes in the glomeruli or fixation of antibodies on pre-existing antigens in the glomeruli. These antigens can be endogenous or exogenous.

In APSGN Lange [4] has in early biopsies seen streptococcal antigens which later appear to be covered by antibodies. In MesIgAN the infections preceding the haematuria are thought to be non-specific. Tomino [5] has eluted IgA from some renal biopsies and then recombined it with the same mesangium it was eluted from. This could indicate that there exists specific binding sites in the glomeruli of these patients. There was a cross-reaction with the mesangium of other patients with IgAN. The cross-reactions were however not 100 per cent, and it could be caused by different antigenic sites perhaps of exogenous origin.

Materials

Of all renal biopsies in Stockholm between 1974–1983, 187 fulfilled our criteria for mesangial IgA-nephropathy. These were: IgA should be the main immunoglobulin with predominantly mesangial localisation. Patients with evident systemic disease were excluded. Male to female ratio was 2:1. The mean age at onset of symptoms was 26 years.

In the prospective study we have estimated the apparent onset of renal disease and the clinical symptoms at the onset.

Serum creatinine, immunoglobulins, complement-factors (C_3, C_4), auto-antibodies, anti-streptococcal antibodies (ASO, ADNAseB) and anti-viral antibodies were taken every six to 12 months. Endogenous creatinine-clearance, 24 hour urinary protein excretion, clearances of albumin and IgG are also estimated at every visit. Bacterial cultures from the tonsils are taken every visit.

Fresh urinary sediment was examined by the physician (HB or SR). Every 18–24 months a Cr-EDTA-clearance was performed. The patients were informed to contact us if they have macroscopic haematuria or infection, these 'acute' patients were investigated also after one week and after one month.

Results

The apparent onset of disease was as follows: 73 patients had recurrent macroscopic haematuria in connection with upper respiratory tract or gastrointestinal infection, 46 had asymptomatic haematuria and/or proteinuria, and 38 patients had in their history signs of classical acute glomerulonephritis with a latent period. Two patients presented with nephrotic syndrome and one patient was found because of malignant hypertension. In 27 patients we had not enough information to be able to clearly define the onset of the disease.

We have investigated 92 patients prospectively, ASO titres were ⩾400 units in five of them, ADNAseB titres were ⩾800 units in 40 patients (43%). These patients had the following titres, 800 units in 22, 1,600 units in 13, 3,200 units in two and ⩾6,400 units in three. All these tests were taken at a routine visit without any signs of infection.

In 48 patients the ADNAseB titres were tested more than once. A four-fold or greater change was seen in 15 of these patients (9 had a decrease and 6 had an increase).

Throat cultures were also taken routinely. β-HS were found in cultures of 33 patients (35%). Gr A was seen in only one patient, Gr C (17), Gr G (7), Gr B (5), Gr F (2) and three strains of β-HS were not possible to group. Two of the patients had two different strains at the same time.

We have so far seen 10 patients with acute macroscopic haematuria. Six of these had ADNAseB ⩾800 units at their acute visit. Four acute patients had positive cultures of β-HS: Gr A (1), Gr C (2), Gr G (1). Two patients had an increase of their ADNAseB-titre. One patient rose from 100 units to 3,200 units in a month, no β-HS were found. The other rose from 200 units to 2,600 units, this patient had β-HS Gr C isolated from his throat at the acute visit.

Discussion

In our study 38 patients had an onset of their disease very like that of acute post-infectious glomerulonephritis. In fact eight of them also had evidence of infection with β-HS preceding their nephritis. Clarkson [6] also reported 10 per cent of patients with an onset of acute nephritic type.

In infections with β-HS Gr A with pharyngitis the ASO is regarded as the best index of infection, whereas the ADNAse-titre is more a reflection of skin infection with β-HS. Our patients had a high percentage of ADNAse-titres ⩾800 units and few positive cultures of β-HS Gr A. Gr C β-HS are serologically very similar to Gr A β-HS and can sometimes produce DNAseB. We have not so far investigated if these strains we have isolated are able to produce DNAseB.

There is a possibility that the raised titres of ADNAseB is only due to a polyclonal activation of β-lymphocytes. However we have not seen any rise in antibody titres against staphylococci or viral antigens.

Our conclusion is that there is some indication of a possible pathogenetic role of β-HS in MesIgAN, which are worthy of further studies. Especially in patients with acute bouts of macroscopic haematuria there is a possibility of isolating streptococcal antigens from immunocomplexes either circulating or fixed in the mesangium.

References

1 Cameron JS. In Kincaid-Smith P, Mathew TH, Becker EL, eds. *Glomerulonephritis; Morphology, Natural History and Treatment.* New York: John Wiley and Sons. 1973: 63
2 Baldwin DS. *Am J Med 1977; 62:* 1
3 Berger J, Hinglais N. *J Urol Nephrol 1968; 74:* 694
4 Lange K, Ahmed U, Kleinberger H, Treser G. *Clin Nephrol 1976; 5:* 207
5 Tomino Y, Endoh M, Nomoto Y, Sakai H. *Am J Kidney Diseases 1982; 2:* 276
6 Clarkson AR, Seymour AE, Thompson AJ et al. *Clin Nephrol 1977; 8:* 459

Open Discussion

RITZ (Heidelberg) Dr Rekola as you correctly pointed out there are two different explanations for the findings, an anamnestic response the other a true pathogenetic role. You discounted the first possibility on the basis that you failed to find changes in titres other than DNA-ase. However, there has been a report at the International Society of Nephrologists meeting in Los Angeles by a Japanese group* who found changes in anti-influenza virus titres in such patients clearly implicating that they are hyper-responders to a variety of immunological stimuli. I would be somewhat hesitant to discount this possibility on the basis of your findings.

REKOLA That is quite correct.

BROWING (Glasgow) Have you looked at the kidneys of any of these patients to see whether there are streptococcal antigens within the glomerulus?

REKOLA No. Because most of the biopsies in these patients were 10 years ago. We are now investigating with the bacteriologists to see if we could after some sort of elution, identify some antigens in the kidneys.

EGIDO (Spain) Have you any idea of the class of immunoglobulin elicited against the antigens of the streptococcus?

REKOLA No, we have not investigated it yet. About half of our patients have a rise in their IgA titres and you can see at the actual bouts there is a still higher rise which disappears in a few months.

EGIDO (Spain) In my opinion this may reflect only a non-specific effect related to immunological disturbances of IgA. We have a large number of these patients with antibodies against dextran and diet antigens.

REKOLA It is quite possible. There is one other possibility, it is known that viral infections can also activate latent streptococcal infection elsewhere in the body?

SCHENA (Chairman) Have you compared your data in IgA nephropathy with a group of patients with acute glomerulonephritis? What is the difference in the presence of high titre of these antibodies in patients with acute glomerulonephritis and patients with IgA nephropathy?

REKOLA The problem is we have mostly adult patients in our clinic and acute glomerulonephritis is much more common in children. Perhaps it will be possible to co-operate with a paediatric clinic.

*Endoh M, Suga T, Miura M et al. *International Congress of Nephrology, Los Angeles 1984*

DAVISON (Chairman) Can I ask about your fairly high incidence of streptococci isolated from the throat. Were these always at the time of macroscopic haematuria?

REKOLA No, the cultures were obtained when the patients were asymptomatic.

DAVISON Have you any control patients?

REKOLA Not yet.

DAVISON Therefore you do not know whether there is a large number of people in the general population with these streptococci?

REKOLA No. There is one problem in that previously the laboratory did not report streptococci of group C because they thought they were not pathogenetic. Now they have started but nobody in Sweden knows the frequency of these streptococci in the throat.

Proc EDTA–ERA (1984) Vol 21

IgA MESANGIAL DEPOSITS IN C3H/HeJ MICE AFTER ORAL IMMUNISATION

C Genin, J C Sabatier, F C Berthoux

Hôpital Nord, Saint-Etienne, France

Summary

In order to develop an experimental IgA nephropathy, C3H/HeJ mice, high producers of IgA, were strongly immunised orally by ferritin and compared to C3H/eB mice. After immunisation, serum IgA and IgG titres increased significantly only in C3H/HeJ mice. Specific antiferritin antibody could be detected in the serum. Mesangial IgA deposits were present in most of C3H/HeJ mice after immunisation and were significantly higher than in C3H/eB mice. No ferritin deposits could be detected in the kidney. No clinical manifestation appeared in these animals.

Introduction

Evidence for the pathogenetic role of IgA immune-complexes in IgA nephropathy has been derived from multiple studies in both humans and experimental animals [1–4]. Clinical and histopathological observations in this disease suggest that the mucosal immune system produces large amounts of specific IgA in response to antigens from food or infectious agents resulting in circulating IgA immune-complexes and renal IgA deposits [5,6]. An immunogenetic abnormality of mucosal immunity can be suspected in these patients [7].

To test this hypothesis experimentally, we orally immunised by ferritin CH3/HeJ mice which are known to be high producers of IgA [8] and we compared them with C3H/eB mice.

We studied serum immunoglobulins, serum antiferritin antibodies and the deposits of immunoglobulins and ferritin in kidney.

Material and methods

Immunisation

Ten female C3H/HeJ and eight C3H/eB mice, 10–12 weeks old, received horse spleen ferritin (Sigma Co) orally: 20mg in 0.2cc by gastric intubation the first day then 1mg/ml of drinking water for one month. Nine C3H/HeJ mice and 10 C3H/eB mice were not immunised and served as controls.

Sacrifice and tissue processing

After 30 days, mice were bled by transection of the right axillary artery under ether anaesthesia. Sera were frozen at $-70^{\circ}C$ until assayed. Kidneys were frozen at $-70^{\circ}C$ for immunofluorescence.

Serum immunoglobulins

Goat antisera to mouse IgA, IgG and IgM (Cappel Lab) were used in a radial immunodiffusion technique and the results expressed in percentage of the pooled control sera.

Serum antiferritin-antibody

The serum level and class of antibodies to ferritin were determined by ELISA technique using peroxidase-conjugated goat antisera to mouse IgA, IgG and IgM (Cappel Lab).

Renal immunoglobulins and ferritin deposits

Sections of kidney were stained by the direct method for mouse IgA, IgG and IgM using fluorescein-conjugated antisera (Cappel Lab). In addition, sections of kidney were stained for the presence of ferritin with rabbit antiferritin IgG conjugated to fluorescein [9].

Haematuria and proteinuria

At the time of killing, fresh urine from each mouse was tested by Bili-Labstix for protein and blood.

Results

Serum immunoglobulin

Serum IgA, IgG and IgM were measured in each mouse and the results expressed as the mean ± SD in groups of immunised or non-immunised mice and are summarised in Table I. Serum IgA and IgG were significantly higher in immunised than in non-immunised C3H/HeJ mice but not in C3H/eB mice. IgA were higher in C3H/HeJ mice than in C3H/eB mice before or after immunisation but the difference was not significant. IgG were lower in C3H/HeJ mice than in C3H/eB mice before or after immunisation but the difference was not significant. IgM were higher in immunised than in non-immunised C3H/eB mice but not in C3H/HeJ mice. After immunisation, IgM were significantly lower in C3H/HeJ mice than in C3H/eB mice.

TABLE I. Serum immunoglobulin levels*

	IgA	IgG	IgM
Non-immunised mice			
C3H/eB	97±20	122±23	112±21
C3H/HeJ	104±10	102±14	108±15
Immunised mice			
C3H/eB	108±28	144±54	124±19***
C3H/HeJ	138±41**	137±38**	103±16

* Means of all mice Ig values in each group ± SD. Results are expressed in percentage of the pooled control sera.
** Different from non-immunised C3H/HeJ ($p<0.02$)
*** Different from immunised C3H/HeJ ($p<0.01$)

Serum antiferritin-antibody

Serum antiferritin antibody and the class of the antibodies were studied in each mouse. Most of the antibodies were IgG and IgM but the titre was very variable depending on animals. A very small amount of IgA antibody was detectable in both groups of mice without significant difference between strains.

Renal immunoglobulins and ferritin deposits

Deposits of IgA, IgG and IgM in kidney of each mouse were studied in immunofluorescence and scored from traces to 4+. Mesangial deposits of IgG and IgM were present before immunisation and did not increase significantly after immunisation. There was no detectable difference between both strains. IgA deposits were detected in the mesangium of few non-immunised mice and they increased significantly after immunisation (Figure 1). IgA deposits were significantly higher in C3H/HeJ mice than in C3H/eB mice (Table II). Deposits of ferritin could not be detected by direct immunofluorescence with antiferritin rabbit IgG.

TABLE II. IgA deposits in the kidney

	Intensity of immunofluorescence			
	0	traces	+	++
Non-immunised mice				
C3H/eB	9	1	0	0
C3H/HeJ	6	2	1	0
Immunised mice*				
C3H/eB	6	0	2	0
C3H/HeJ	2	2	5	1

* Significant difference between C3H/eB and C3H/HeJ ($Chi^2 > 5.44$; $p<0.002$)

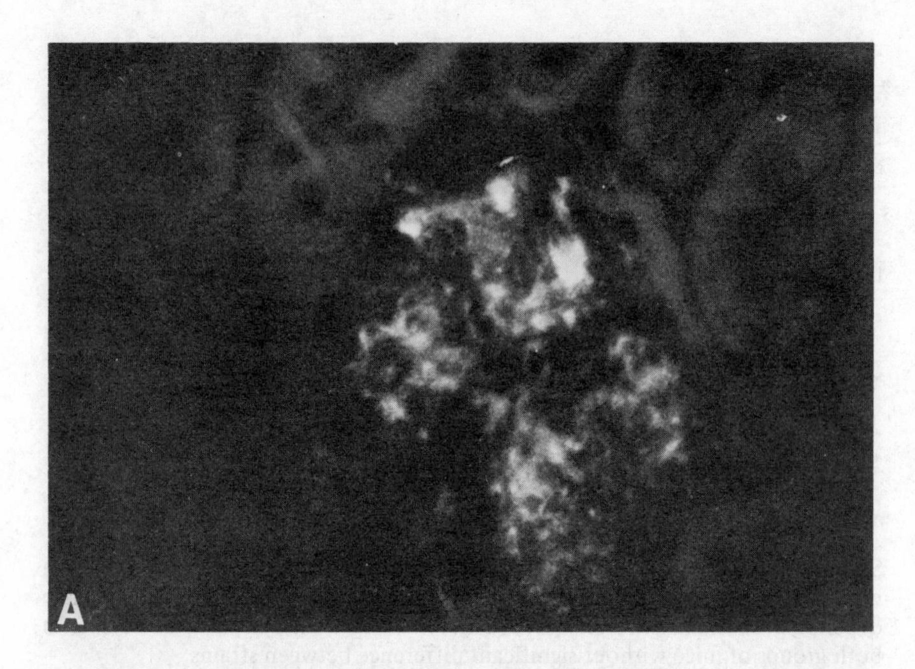

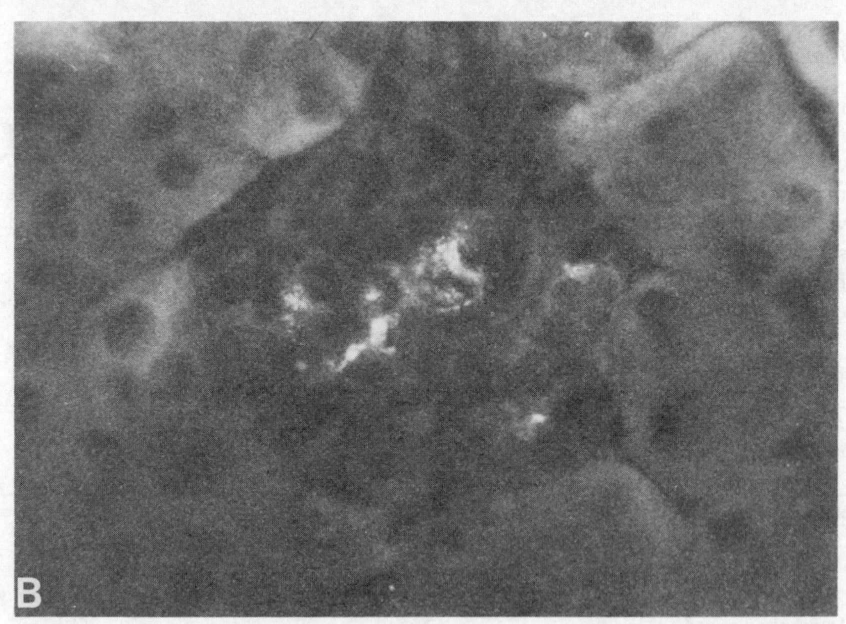

Figure 1. Glomeruli of mice immunised with ferritin. Mesangial immunofluorescence staining for mouse IgA is higher in C3H/HeJ mice (A) than in C3H/eb mice (B) x 1000

Proteinuria and haematuria

Proteinuria and haematuria could not be detected in any group at the time of killing.

Discussion

This study demonstrates that C3H/HeJ mice, strongly immunised orally by ferritin present significantly higher serum IgA and IgG than non-immunised animals whereas IgM does not increase. These immunoglobulins are shown to be at least in part specific antiferritin antibody but we cannot exclude a non-specific stimulation by mitogenic effect.

After immunisation, mesangial IgA deposits could be detected in most of C3H/HeJ mice. These IgA deposits were significantly greater in C3H/HeJ mice than in C3H/eB mice. No correlation could be demonstrated between high IgA deposits, high serum IgA or high rate of antiferritin IgA antibody in the same animal. Ferritin could not be demonstrated in the kidney by the technique that we used.

C3H/HeJ mice appear to be a good experimental model to study IgA nephropathy but this study has to be extended to demonstrate that the deposits in the mesangium are ferritin-antiferritin IgA immune-complexes and to produce a glomerulonephritis with clinical manifestations.

References

1 Lopez-Trascasa M, Egido J, Sancho J, Hernando L. *Clin Exp Immunol 1980; 42:* 247
2 Sancho J, Egido J, Rivera F, Hernando L. *Clin Exp Med 1983; 54:* 194
3 Rifai A, Small PA, Teague PJ, Ayoub EM. *J Exp Med 1979; 150:* 1161
4 Emancipator SN, Gallo RG, Lamm ME. *J Exp Med 1983; 157:* 572
5 Andre C, Berthoux FC, Andre F et al. *N Engl J Med 1980; 303:* 1343
6 Sissons JGP, Woodrow DF, Curtis JR et al. *Br Med J 1975; 3:* 611
7 Berthoux FC, Genin C, Gagne A et al. *Proc EDTA 1979; 16:* 551
8 Kiyono H, Babb JL, Michalek SM et al. *J Immunol 1980; 125:* 732
9 Genin C, Cosio F, Michael AF. *Immunology 1984; 51:* 225

Open Discussion

BROWNING (Glasgow) This work is similar to that published in 1983 by Emancipator, Gallo and Lamm, who reported the development of IgA nephropathy in Balb/c mice immunised orally with protein antigens. We have tried to reproduce these experiments and, as yet, have been unsuccessful.

GENIN We also have the same results because we used exactly the same protocol with ferritin in the drinking water and it is not sufficient to produce a good immunisation. We need a large amount of antigen at the beginning of the experiment. We demonstrated, for example, that in the C3H/DBA mice we do not get significant IgA deposits. We found specific antibodies from the three classes of immunoglobulins and we found more IgG and IgM than IgA and Emancipator et al reported only IgA antibodies.

BROWNING I should like to ask if you can explain how the addition of a simple protein antigen to the diet of your immunised mice should lead to the development of a mesangial IgA nephropathy, whilst control mice of the same strain fail to respond to the range of protein antigens present in their daily diet?

GENIN I use a very high quantity of ferritin, 20mg of ferritin initially and then 1mg/ml in drinking water and I think a mouse drinks about 2ml a day. I think it is a very strong immunisation, and the two strains of mice received exactly the same commercial food and they were maintained in sterile boxes to avoid infectious agents. I think the interest of this work is to compare two strains with exactly the same protocol of immunisation.

RITZ (Heidelberg) Could you briefly comment on whether you had any glomerular lesions? You mentioned mesangial IgA deposits, but did you have evidence of glomerulonephritis?

GENIN We had only a light microscopic study: we have to do an electron-microscopic study. We did not find any lesions, it is not really glomerulonephritis, it is IgA deposition. I think the IgA deposits are not sufficient to produce clinical manifestation. We did not find haematuria or proteinuria at the time of sacrifice.

D'AMICO (Milan) Certainly we need a model of immunisation to study IgA nephropathy. Unfortunately in the Emancipator model there is no proteinuria at all. I wonder if you found at least a minor proteinuria, because I heard that you did not find nephritis or renal symptoms.

GENIN I think we have to find what is necessary. I do not know if it is the precipitation of other immune complexes, maybe not IgA immune complexes. Also the dose is not sufficient, you may not have enough immune complexes to develop clinical manifestations.

EGIDO (Spain) I wonder if the lack of correlation between the serum titres of antibodies to ferritin and IgA deposits is due to the small number of animals studied and the timing of the antibody studies. We are working on a model of dextran IgA nephropathy and we have found a close correlation following the injection of dextran between the levels of IgA antibodies and the anti-dextran antibodies in the IgA deposits in the kidney.

GENIN I think it is not specific IgA in the kidney because when you have a 30 per cent increase of IgA and IgG I think it is not only specific antibodies. Maybe you cannot find the antigen and anti-ferritin antibodies in the kidney because it is not only specific IgA. I think we may have a mitogenic effect and we have a non-specific stimulation of all IgA and not only specific IgA.

THE PREVENTION OF AMYLOIDOSIS IN FAMILIAL MEDITERRANEAN FEVER WITH COLCHICINE

S Cabili, *D Zemer, *M Pras, A Aviram, *E Sohar, *J Gafni

*Rokach Hospital, *Sheba Medical Center, Tel-Aviv, Israel*

Summary

Colchicine has been used since 1972 to prevent the acute attacks of familial Mediterranean fever. The present study shows that colchicine is also effective in the prevention of amyloidosis. If initiated in patients without evidence of renal disease there is no appearance of proteinuria and no progression to renal insufficiency over long follow-up periods. Moreover, it ameliorates the course of the disease in patients with amyloid nephropathy and normal renal function. It does not alter the course of the disease if initiated after renal function is even mildly impaired. These findings suggest that colchicine prevents the new deposition of amyloid.

Introduction

Familial Mediterranean fever (FMF) is a hereditary disorder affecting peoples of Mediterranean origins, mainly Sepharadic Jews, Armenians, Arabs of the Near East and Turks. The disease becomes manifest in childhood or adolescence when febrile attacks of peritonitis, synovitis and pleuritis appear. Systemic amyloidosis of the AA type develops in most untreated patients causing end-stage renal disease (ESRD) [1].

In 1972 continuous prophylactic treatment with colchicine was introduced [2], and subsequently proved effective in preventing or ameliorating the acute attacks in 85 per cent of patients [3].

This paper evaluates the effect of colchicine on the development and course of renal amyloidosis.

Patients and methods

Since 1972 colchicine has been prescribed in a minimal dose of 1mg/day (up to 2.5mg/day) for all FMF patients, children and adults. One thousand and forty-one patients had no evidence of renal disease (no proteinuria and normal renal function) when the drug was prescribed. Assuming that assessment of its effect on the prevention of amyloidosis requires an adequate period of follow-up, only 840 who had been followed-up for four or more years were studied. Fifty-four of them proved to be non-compliant regarding colchicine therapy for a variety of reasons, providing an unplanned control group of untreated patients.

One hundred and fourteen patients were in various stages of overt renal disease when colchicine was initiated. Fifty-two had proteinuria of less than 3.5g/day and normal renal function; fourteen had proteinuria exceeding 3.5g/day with normal serum creatinine; twelve had renal insufficiency (serum creatinine ⩾1.6mg/dl) and progressed to ESRD – in those patients the logarithms and reciprocals of serum creatinine values were plotted against time and their period of progression from creatinine of 1.6mg/dl to ESRD was determined using linear regression analysis [4]. This group was compared to 10 patients who progressed to ESRD before the colchicine era.

Results

Among the 840 non-proteinuric patients, proteinuria has appeared in only three of 786 patients (0.4%) who complied with treatment. In the group of the 54 patients who did not comply, proteinuria appeared in 10 (18.5%). In our earlier series [1] of 470 patients not treated with colchicine the prevalence of amyloidosis was 27 per cent (the rate of appearance of new cases nearly equalling the mortality rate from the disease).

Among the patients with evidence of renal amyloidosis when treatment was initiated, the 52 patients with proteinuria below the nephrotic range and normal renal function have now been followed up for an average period of 5.7 years without progression to the nephrotic syndrome and renal insufficiency. In contrast the mean interval between the appearance of proteinuria and the onset of nephrotic syndrome was 2.2 years before the colchicine era.

Of the 14 nephrotic patients with normal renal function, six have progressed to renal failure, four remained stable and four showed some decrease of the proteinuria.

Twelve of the patients in mild renal insufficiency when colchicine treatment was begun who progressed to ESRD did so over a mean period of 31.6 ± 6 months, as compared to 33.9 ± 6 months in the group of 10 untreated patients before the introduction of colchicine.

Discussion

Our results establish the effectiveness of daily colchicine therapy in the prevention of amyloidosis in FMF. The extremely low cumulative incidence (0.4%) of proteinuria among treated patients as compared to 18.5% in untreated patients, is the most striking finding. Colchicine also definitely ameliorates the course of amyloid nephropathy if treatment is started when kidney function is normal. This beneficial effect is pronounced in the patients who had proteinuria below the nephrotic range, as judged by long follow-up period without progression into renal insufficiency or nephrotic syndrome in none of the patients.

These findings suggest that colchicine might prevent de novo formation of amyloid. If the structural damage to the kidney is not severe the condition of the treated patients remains stable – patients with no proteinuria or proteinuria below the nephrotic range. If there is severe deposition of amyloid in the kidney, as in some nephrotic patients and in all patients with renal insufficiency, the functional damage is not reversible.

However, as these patients face long-term dialysis and/or transplantation it is advisable to prescribe colchicine even in the presence of irreversible renal damage in order to prevent further deposition of anyloid in other organs, as well as the febrile attacks of FMF.

References

1 Sohar E, Gafni J, Pras M et al. *Am J Med 1967; 43:* 227
2 Goldfinger SE. *N Engl J Med 1972; 287:* 1032
3 Zemer D, Revach M, Pras M et al. *N Engl J Med 1974; 291:* 932
4 Rutherford WE, Blondin J, Miller JP et al. *Kidney Int 1977; 11:* 62

Open Discussion

PAPADIMITRIOU (Greece) I should like to congratulate Dr Aviram for this report. We have the same experience as you in some of our cases. We notice that if we transplant these patients and they do not take colchicine after transplantation they develop the same attacks of pain etc. Of course they have to take colchicine for ever.

AVIRAM We have exactly the same experience.

DAVISON (Chairman) How do you check compliance in your patients? Is it from the symptoms and their fever?

AVIRAM Usually just by asking patients. In the group we presented here of non-compliant patients they definitely did not take the drug by their own admission.

DAVISON And have you seen any side effects from the use of the drug?

AVIRAM We give in a very minimum dose of up to 2.5mg a day so if there are any side effects like diarrhoea it lasts only a few days and then disappears.

DAVISON Would you be prepared to comment on the value of colchicine in other forms of amyloidosis?

AVIRAM There is only a very limited experience and it seems that it is not useful because those patients usually appear when renal function has already deteriorated so we cannot really compare the groups.

Proc EDTA–ERA (1984) Vol 21

LUPUS NEPHRITIS IN MALES AND FEMALES

I E Tareyeva, T N Janushkevitch, S K Tuganbekova

First Moscow Medical Setchenov Institute, Moscow, USSR

Summary

The course of lupus nephritis in 51 males was compared to that of 337 females. Nephrotic syndrome occurred with equal frequency in males and females, hypertension was more frequent in males, rapidly progressive lupus nephritis was much more frequent in males. The overall 10-year survival rate from the onset of systemic lupus erythematosus was 41 per cent in males and 60 per cent in females. The 10-year 'renal survival' – from the clinical evidence of renal disease to 'renal death' – was 40 per cent in males and 57 per cent in females. Thus the prognosis of lupus nephritis in males was worse than in females.

In 10 males and 65 females acetylation rate of sulfadimezine was studied. The predominance of slow acetylators was especially marked in males and in patients with more severe disease.

Introduction

The course of systemic lupus erythematosus (SLE) is greatly influenced by hormonal and genetic factors. The disease mostly affects sexually mature women and is relatively rare in males, though several families in which SLE predominates in males have been described [1]. Animal models, such as the inbred mouse, support the theory that sex hormones are important in the pathogenesis of SLE. In certain strains morbidity is greater in females, in others, such as BXSB mouse, the males have a lupus-like syndrome while there is little involvement in females.

There are few reports comparing the course of the disease in males and females and only a few discuss the course of nephritis. In one of the latest publications Wallace et al [2] report the clinical course of SLE in 609 patients, 209 of them with renal involvement; the 10-year survival of males with SLE is reported to be less than that of females, though survival rates between males and females with nephritis did not differ.

The statement of Wallace et al that their 230 lupus nephritis series represents the largest single centre group ever described prompted us to report our experience.

Patient details

We have been observing 388 patients with SLE and clinical evidence of renal involvement admitted during a 25-year period (1958–1982) to the Clinic of Internal Medicine (mainly to its Renal Unit) of the Medical Setchenov Institute. The mean follow-up period was 5.7 years. Here we present the data concerning the course of nephritis in 51 males and 337 females aged 15–60 years.

Clinical findings

The general clinical features in most patients resembled lupus as a whole, with multi-system involvement. All patients had positive LE-test and/or high titres of anti-DNA-antibodies, 28 patients (all females) presented with isolated renal disease, in three of them no systemic signs appeared during a mean follow-up of 3.6 years.

All patients had proteinuria (⩾0.5g/L), 105 had nephrotic syndrome, 191 had arterial hypertension. Thirty-five patients had rapidly progressive lupus nephritis with nephrotic syndrome, severe hypertension and rapid deterioration of renal function. Renal biopsies were performed in 105 patients. In 34 patients diffuse proliferative lupus nephritis and in 35 mesangio-proliferative lupus nephritis was found; 12 patients had membranous lupus nephritis, 8 patients had membrano-proliferative lupus nephritis and in 15 patients predominantly sclerotic lesions were found.

The comparison of age of our male and female patients showed that lupus nephritis was very rare in males over 40. The ratio males: females was similar in all groups up to 40 years (15–20 years, 21–30 years and 31–40 years), being roughly 1:6; in patients over 40 it fell to 1:15 (2 males and 30 females).

The analysis of clinical features showed that the nephrotic syndrome occurred with equal frequency in males and females (27% and 32% respectively). Hypertension was more frequent in males (67% and 46%); three men (but no women) had myocardial infarction. Rapidly progressive nephritis was much more frequent in males (23.5% versus 6.7%). Regarding renal biopsy findings, the only difference concerned membranous lupus nephritis (all patients with this histological form were females); other histological forms were found with equal frequency in both groups of patients.

During the follow-up period 155 patients died – 26 of 51 males (51%) and 129 of 337 females (38%). In 63 per cent of patients death occurred from renal failure.

Assessing the course of disease in our patients we calculated life table analysis by the method of Cutler and Ederer. When calculating overall survival, the onset of disease was designated as time of first symptom of SLE. The 5-year survival rate of all 388 patients was 72 per cent, 10-year survival was 57 per cent and the 20-year survival was 34 per cent. Our series was not typical of lupus as a whole, but only of the subset with clinically evident nephritis. The survival of our patients was a little less than the survival of a comparable group of Cameron [3] who in a series of 71 patients with lupus nephritis seen over a 15-year period reported a 10-year survival of 65 per cent.

The prognosis in males was worse than in females, 5-year survival being respectively 57 per cent versus 76 per cent, 10-year survival 41 and 60 per cent and 20-year survival 27 per cent and 40 per cent.

To assess the course of nephritis we calculated 'renal survival' (i.e. the 'survival of renal function'). For this purpose we took clinical evidence of renal involvement as our entry criterion, 'death' meant death from renal failure and included also the need for dialysis and transplantation, e.g. a serum creatinine ⩾8.5mg/dl. Ten-year renal survival of all patients was 53 per cent, being 40 per cent in males and 57 per cent in females.

Thus prognosis of lupus nephritis in females was much better. The best prognosis was in women over 40 – none had rapidly progressive nephritis, only one patient in this group died of renal failure, the 10-year renal survival being 95 per cent. One of two older male patients had very benign course of lupus nephritis – after 15 years of renal involvement his serum creatinine is 0.8mg/dl, the other was lost to follow-up.

The treatment in both groups of our patients was similar: 33.3 per cent of men (17/51) and 34.4 per cent of women (115/337) received high dose corticosteroids, 47.1 per cent (24/51) and 27.9 per cent (94/337) respectively received immunosuppressive drugs; 19.6 per cent (10/51) of males and 37.5 per cent (128/337) received only low-dose steroids. Thus, the difference in survival rates was not due to less intensive treatment of males.

In 75 patients (mainly Russians) we studied the acetylation phenotype by acetylation rate of sulfadimezine; 47 patients (62%) were found to be slow acetylators. Among patients with a more severe form of the disease 75 per cent were slow acetylators. Among 65 females, 38 (58.4%) were slow acetylators, among the 10 males only one was a rapid acetylator. Thus, there was marked predominance of slow acetylators in patients with more severe disease and in male patients.

References

1 Lahita RG, Chiorazzi N, Gibofsky A et al. *Arthr Rheum 1983; 26:* 39
2 Wallace DJ, Podell TE, Weiner JM et al. *Am J Med 1982; 72:* 209
3 Cameron JS. *Q J Med 1979; 48:* 1

Open Discussion

RITZ (Heidelberg) In addition to the explanation you gave for the paradoxical discrepancy between the course in males and females, let me add another one. We know that if we give anti-oestrogenic therapy both to experimental models such as the NZB/NZW mice and to females the course of lupus nephritis is attenuated. In contrast you show, and this is in agreement with other experience, that the course is more severe in males. I think another explanation would be genetic heterogeneity. We know that there are strains of mice, like BXSB mice, where males are more severely affected by a gene coded on the Y chromosome and other like the NZB/NZW mouse where males fare much better. I would like to offer as an additional explanation that this is evidence in the human of underlying heterogeneity of the lupus population.

TAREYEVA It seems to me that you are speaking about the genetic differences between males and that there are several groups of males which are susceptible to this and to the other. Yes, we now study the genetic markers of disease. The first I cannot tell in detail but one thing I can say that men with red hair in our series rapidly died.

ABORAS (Jeddah) I have observed that in spite of the high prevalence of hypertension and cardiovascular disease as a presentation of lupus nephropathy in males, you have used more corticosteroid therapy than other types of drugs. I observed also that the survival rate was very bad. Is there any relation between the treatment and survival? Why have you used corticosteroids more than other types of drugs in such patients?

TAREYEVA No, it was not so in our figures. There was no treatment particular to hypertensive patients. Certainly in hypertensive patients we prefer other types of drugs. Some work shows that the addition of cystostatic drugs improves prognosis. In very severe cases with rapidly progressive nephritis we always use high dose corticosteroids with cytostatic drugs and with heparin. We still use high dose steroids even in patients with hypertension if they are the cases with rapidly progressive nephritis.

DAL CANTON (Naples) I suppose that many of your female patients became pregnant while they were under your observation. It could be of great interest for us to know whether the disease influenced the course of the pregnancy and on the other hand whether pregnancy influences the course of the disease?

TAREYEVA This is a very interesting question. I am not sure whether I can present all the data now as I am unprepared for the question. But 35 to 40 of our patients have been pregnant and certainly in half of them the pregnancy aggravated the course of the disease but nearly 20 to 25 had normal births and we have now under observation 20 to 23 children and they are unaffected. Some of them were retarded in their first year.

GOLDSMITH (Liverpool) Do you think that the influence of acetylator status on prognosis arises through its influence on the disease, or through the influence on the metabolism of the drugs used in the treatment of the disease?

TAREYEVA No, I think it is the influence on the disease itself by a genetic predisposition.

DAVISON (Chairman) Following from Dr Goldsmith's question did any of the slow acetylators have their lupus precipitated by some drug exposure chemical exposure or anything you could detect?

TAREYEVA No, not in these patients.

SCHENA (Chairman) Have you compared the percentage of hypertension in male patients with SLE and in male nephritis patients without SLE?

TAREYEVA I do not think there is any difference.

Proc EDTA–ERA (1984) Vol 21

FILTRATION FRACTION: AN INDEX OF RENAL DISEASE ACTIVITY IN PATIENTS WITH SYSTEMIC LUPUS ERYTHEMATOSUS

H Favre, P A Miescher

University Hospital, Geneva, Switzerland

Summary

Renal plasma flow (RPF) and glomerular filtration rate (GFR) were estimated by ^{125}I hippuran and ^{51}Cr EDTA clearances using a single shot technique on two occasions at one-year intervals in 22 patients fulfilling the ARA criteria for systemic lupus erythematosus (SLE). All these patients had histologically proven renal disease. Filtration fraction was a better parameter than proteinuria, urinary sediment or GFR for recognising diffuse proliferative glomerulonephritis with a sensitivity of 61 per cent and a specificity of 88 per cent. After one year all the patients with an initially low filtration fraction (FF) had significantly changed their GFR, which demonstrates that this parameter indicates the presence of an active renal lesion.

Introduction

The prognosis of patients with SLE is determined by the severity of their renal involvement. In the past, information obtained by histological examination of renal tissue served both to evaluate the prognosis of the patients and to manage their disease [1]. More recently, this practice has been criticised by authors arguing on the basis of cost benefit analyses that histopathology would give patients only marginal benefit [2]. The latter opinion may not apply to patients who have severe lupus nephropathy at the onset of their SLE without abnormalities in their renal function tests, nor patients in whom a mild glomerulopathy progressed to severe disease without changes in their renal function or proteinuria.

The present work was undertaken to find an alternative to the renal biopsy to assess renal involvement in SLE patients, both at the beginning of the disease and with the passage of time by using a non-invasive technique.

Material and methods

Twenty-two patients fulfilling the ARA criteria for the diagnosis of SLE had an initial renal biopsy. Renal lesions were recorded according to the WHO classification; five patients had mesangial proliferative glomerulonephritis (class II), four focal proliferative glomerulonephritis (class III) and 13 diffuse proliferative glomerulonephritis (class IV).

Renal plasma flow (RPF) and glomerular filtration rate (GFR) were measured simultaneously in all the patients on two occasions, separated by a one-year interval by using a single shot technique. RPF was estimated by ^{125}I hippuran and GFR by ^{51}Cr EDTA clearances. Both isotopes were injected simultaneously and their disappearances from the blood compartment followed with time. Clearances were calculated according to the two compartment model proposed by Sapirstein using a computer program.

Filtration fraction (FF) was calculated from the actual values obtained for GFR and RPF [3]. On the same day, microscopic examination of the urine after filtration and staining by a Papanicolaou technique was undertaken [4] and 24-hour proteinuria, anti DNA-DS antibodies, complement and circulating immune complexes (C1q fixation) [5] were measured.

Results

Table I shows the number of patients for each class of glomerulopathies in whom one or more abnormal laboratory data have been found.

TABLE I. Basal evaluation of lupus nephropathy by laboratory data versus histopathology

Histopathology	Total number of patients	Number of patients with GFR<80ml/min (1)	Number of patients with FF<18% (2)	Number of patients with proteinuria >200mg/d	Number of patients with pathological urinary findings (3)
Mesangial proliferative GN	5	0	0	1	2
Focal GN	4	0	1	2	2
Diffuse proliferative GN	13	5	8	9	5
Total	22	5	9	12	9

1 GFR was measured by ^{51}Cr-EDTA clearance;
2 Filtration fraction was calculated as the ratio ^{51}Cr-EDTA/^{125}I hippuran clearances times 100
3 Urinary microscopic evaluation was performed by filtration technique and Papanicolaou staining

Microscopic examination of the urine was very disappointing and non-specific. About half the patients in each category had microhaematuria with or without red cell casts. The percentage of patients with proteinuria increases with the severity of the renal lesion. Low GFR has been found only in those patients with diffuse proliferative glomerulonephritic lesions which bear the worst prognosis. A decrease in filtration fraction recognises this class of patients with a sensitivity of 61 per cent and a specificity of 88 per cent.

Table II confirms that patients with mesangial proliferative glomerulonephritis more often had a silent nephropathy than patients with more severe forms of renal disease. The three patients with a silent nephropathy and diffuse proliferative glomerulonephritis have been unmasked by a low filtration fraction. For these three cases, only this additional laboratory analysis provided the key information on their renal lesion.

TABLE II. Number of patients with silent nephropathy (classical criteria, FF excluded) and with low filtration fraction as a single laboratory finding for each histological class of lupus nephropathy

Histopathology	Total number of patients	Number of patients with silent nephropathy	Number of patients with low FF (<18%) as unique finding
Mesangial-proliferative GN	5	3	0
Focal GN	4	2	0
Diffuse proliferative GN	13	3	3

NB: The three patients with silent nephropathy in the diffuse proliferative GN group were unmasked by their low FF

Table III depicts the changes in GFR beyond the limits of reproducibility of the method at one year. All the 13 patients with a normal GFR and a normal filtration fraction at the first examination remained unchanged during the time of observation. By contrast GFR has changed in the nine patients who had a low FF from the outset.

TABLE III. Change in the GFR one year later in two groups of patients defined by their initial filtration fraction

Filtration fraction at first examination		Number of patients who changed their GFR (>7%) between the two examinations
Normal:	13 patients	none
Low:	9 patients	9 < 5 increased their GFR*; 4 increased their GFR

* These five patients include the one with focal GN

Discussion

To appreciate changes in renal function with time, a method with a good coefficient of reproducibility is required. Single shot clearance technique offers this possibility as its coefficient of reproducibility is seven per cent between two tests compared to 20–25 per cent for conventional methods. In addition this method has the advantage of avoiding bladder catheterisation which may be a source of urinary tract infection, especially in SLE patients. The only limitation of the method is the presence of oedema which changes the body distribution of the tracers, making the calculation of clearances impossible [3]. From Table II, it can be seen that patients with a normal GFR and normal FF are most likely to have mesangial proliferative glomerulonephritis, which has a good prognosis [6]. By contrast most of the patients with low filtration fraction, whatever the level of GFR, belong to the group with diffuse proliferative glomerulonephritis. Of note is the fact that in three such patients, low FF was the only laboratory abnormality. As shown in Table III, low FF indicates the presence of a class IV nephropathy with active renal lesions expressed by the fact that all the patients had changed their GFR one year later. Those patients at high risk of renal failure should receive treatment combining cytotoxic agents and steroids [7].

As previously reported in the literature, we confirm the absence of a correlation between circulating immune complexes, complement and antibodies to DNA-DS and the severity of the renal disease [8].

References

1 Appel GB, Silva FG, Pirani CL et al. *Medicine 1978; 57:* 371
2 Whiting O, Keefe Q, Riccardi PJ et al. *Ann Intern Med 1982; 96:* 723
3 Favre H. *J Urol et Néphrol 1973; 79:* 1007
4 Teitel M, Lambertson GH, Florman AL. *Am J Dis Children 1964; 108:* 19
5 Zubler RH, Lange G, Lambert PH et al. *J Immunol 1976; 116:* 232
6 Pollak VE, Pirani CL. *Mayo Clin Proc 1969; 44:* 630
7 Dinant HJ, Decker JL, Klippel JH et al. *Ann Intern Med 1982; 96:* 728
8 Cameron JS, Turner DR, Ogg CS et al. *Quart J Med 1979; 48:* 1

Open Discussion

ABORAS (Jeddah) Since there is a correlation between filtration fraction and the other parameters of activity of systemic lupus erythematosus why do you specify that this is an activity specific to systemic lupus and not to other proliferative glomerulonephritis?

FAVRE Certainly these parameters may be used for other types of glomerulonephritis. The index may change with time and as we follow the patient we may detect activity of the lesion which may be influenced by treatment. I think the data may be applicable for other types of glomerulonephritis.

RITZ (Heidelberg) I feel your concept is a breakthrough in the diagnosis of lupus nephritis but there is one point on which I am not exactly sure. You noted some patients with normal GFR and no proteinuria who happened to have low filtration fractions and diffuse proliferative glomerulonephritis. We know from the work of Mahajan* that there may be diffuse proliferative glomerulonephritis in the absence of urinary abnormalities. Does it make any difference to treatment? Do you give them high dose steroids or what are the consequences you draw from your observations?

FAVRE Our concept based on biopsy information was to treat with prednisone and cytotoxic drugs all patients with diffuse proliferative glomerulonephritis, even those patients who had so-called silent nephropathy. Now we use the same treatment and will act in the same way on the basis of the filtration fraction. I fully agree of course that it takes a lot of discretion and it has been suggested that patients with silent nephropathy whatever the severity of the histological lesion do well without treatment. In our small study you have seen these patients and those who have changed their GFR after one year so maybe it is an indication of some need of treatment. However it is a very small study.

IAINA (Israel) Did you measure fractional excretion sodium and does it parallel with the changes in filtration fractions as an index of ischaemia?

FAVRE We did not for the good reason that our way to measure the GFR and the renal blood flow was by a single shot technique so we have no urine available.

SCHENA (Chairman) Have you found any correlation between high doses of corticosteroids or low doses of steroids?

FAVRE We have not looked at our data in terms of the effect of any kind of treatment. All the data is on computer so it is the next thing we have to do.

*Mahajan SK, Ordoneg NG, Feitelson PJ. *Medicine 1977; 56:* 494

Proc EDTA–ERA (1984) Vol 21

THE EFFECT OF CONTINUOUS SUBCUTANEOUS INSULIN INFUSION ON RENAL FUNCTION IN TYPE I DIABETIC PATIENTS WITH AND WITHOUT PROTEINURIA

E van Ballegooie, P E de Jong, A J M Donker, W J Sluiter

University Hospital, Groningen, The Netherlands

Summary

The effect of continuous subcutaneous insulin infusion on renal function was studied in 12 patients with insulin-dependent diabetes mellitus. Serum creatinine was <110μmol/L in all patients. Total urinary protein excretion was less than 250mg/24hr in seven patients (group I) and exceeded 0.5g/24hr in five (group II). Initial glomerular filtration rate was higher in group I compared with group II: 136.0 ± 8.5ml/min versus 103.2 ± 4.6ml/min (mean ± SEM; p <0.02). After one to three months pump therapy glomerular filtration rate decreased in both groups. It remained stable during 32–36 months in group I (126.3 ± 6.1, and 127.9 ± 7.7ml/min, respectively) but deteriorated in group II (98.6 ± 4.4, and 60.0 ± 6.8ml/min, respectively; p <0.01 compared with group I).

These results indicate that strict blood glucose control with continuous subcutaneous insulin infusion does not prevent deterioration of renal function in type I diabetic patients with clinical proteinuria. This suggests that other factors than metabolic control are involved in the course of diabetic nephropathy;

Introduction

The occurrence of clinical nephropathy is a poor prognostic sign in patients with insulin-dependent (type I) diabetes mellitus. Without renal support treatment approximately 50 per cent of these patients die within seven years of the onset of persistent proteinuria (>0.5g/24hr) [1]. Although the factors responsible for the initiation and progression of diabetic nephropathy in man are largely unknown poor glycaemic control is associated with an increased risk [2], suggesting a potential beneficial effect of strict metabolic control in preventing deterioration of renal function. We studied the influence of long-term treatment with continuous subcutaneous insulin infusion (CSII) on renal function in 12 type I diabetic patients with and without proteinuria.

Patients and methods

Twelve non-obese, C-peptide negative diabetic patients were treated with CSII for 32–36 months. Their mean age was 31.4 years (range, 23–58 years); the mean duration of diabetes was 17.2 years (range, 4–46 years). All had developed diabetes before age 30. None of the patients had urinary tract infections, a history of non-diabetic renal disease, or signs of heart failure. One patient was adequately treated with a thiazide diuretic because of mild hypertension. Serum creatinine was $<110\mu mol/L$ in all patients. Total urinary protein excretion was less than 250mg/24hr in seven patients (group I) and exceeded 0.5g/24hr in five (group II).

CSII was performed with either a Mill Hill Infuser, model 1001 AM or HM, or an Auto-Syringe infusion pump, model AS6C.

All patients made 24 hour blood glucose profiles (8 capillary samples) every three to six weeks to assess the degree of glycaemic control.

Glycosylated haemoglobin (HbA_1) was measured every three to six weeks by a colorimetric method [3] (normal range 6–8.5%).

Home blood glucose monitoring was performed three to eight times daily. Individual algorythms were made to ease adjustment of the insulin dose. Renal function studies were performed just before the start of CSII, and after one to three months and 32 to 36 months, respectively.

Effective renal plasma flow (ERPF) and glomerular filtration rate (GFR) were measured simultaneously using a constant infusion of ^{131}I-hippuran and ^{125}I-iothalamate, respectively [4]. Results were expressed as values per $1.73m^2$ body surface area.

Wilcoxon's rank sum test and signed rank test for paired differences were used for statistical analysis. Results are reported as mean ± SEM.

Results

During conventional treatment mean GFR and ERPF of both groups were 122.3 ± 6.8 and 560.2 ± 24.8ml/min, respectively. Initial GFR was higher in group I compared with group II (136.5 ± 8.5 and 103.2 ± 4.6ml/min, respectively; $p<0.02$). ERPF similarly was higher in group I (582.0 ± 37.0 versus 529.6 ± 32.8ml/min, respectively). After the start of CSII mean blood glucose fell from 13.0 ± 0.4 to 6.1 ± 0.4mmol/L ($p<0.01$). Glycosylated haemoglobin fell from 12.0 ± 0.05 per cent to 8.2 ± 0.2 per cent ($p<0.01$). The improvement in glycaemic control was similar in group I and group II.

After one to three months CSII mean group GFR and ERPF had decreased to 114.7 ± 6.1ml/min ($p<0.05$) and 531.3 ± 22.6ml/min (n.s.), respectively. These changes were similar in both groups. Thirty-two to 36 months after the start of CSII GFR and ERPF had remained stable in group I (127.9 ± 7.7ml/min and 536.5 ± 39.4ml/min, respectively; not significant compared with pre-CSII values), but had deteriorated in group II (60.8 ± 6.8ml/min and 331.6 ± 25.8ml/min, respectively i.e. a reduction by 42 per cent and 37 per cent compared with pre-CSII values; $p<0.01$ compared with group I). There were no significant changes in blood pressure in both groups during the study period.

Discussion

The introduction of insulin infusion pumps brought about the hope that long-term maintenance of (near) normoglycaemia could slow, arrest, or perhaps reverse renal abnormalities in patients with established clinical diabetic nephropathy. This and a similar study, however, show that deterioration of renal function still occurs in type I diabetic patients with clinical proteinuria despite strict blood glucose control [5]. The variable clinical course of diabetic nephropathy and the absence of a control group make it difficult to say whether the progression was less severe than could be expected with unchanged conventional treatment. The rates of decline of GFR, however, show close resemblance with those reported for conventionally treated diabetic patients with proteinuria [6].

These and similar findings by other investigators raise the question whether there is a stage of glomerular damage beyond which the decline of renal function becomes an autonomous process, little or not influenced by the quality of glycaemic control. Compensatory intrarenal haemodynamic changes i.e. single nephron hyperfiltration seem to be a reasonable explanation for this self-perpetuating deterioration [7]. However, it could be that at the clinically overt nephropathic stage additional measures like protein restriction and/or antiplatelet aggregating agents are necessary to lower intra-glomerular pressure in order to influence the progression to renal insufficiency. Whether the process of glomerular damage is initiated by the commonly found hyperfiltration, as has been suggested by some investigators, is still unclear. The finding that most of our patients in group I still had no signs of clinical nephropathy despite long-standing supranormal GFR indicates that other factors play a role in this process. Whether correction of the elevated GFR is useful in preventing the development of proteinuria remains to be established.

References

1 Andersen AR, Sandahl Christiansen J, Andersen JK et al. *Diabetologia 1983; 25:* 496
2 Pirart J. *Diabetes Care 1978; 1:* 166
3 Flückiger R, Winterhalter KH. *FEBS Lett 1976; 76:* 356
4 Donker AJM, Hem GK van der, Sluiter WJ et al. *Neth J Med 1977; 20:* 97
5 Viberti GC, Bilous RW, Mackintosh D. *Br Med J 1983; 286:* 598
6 Viberti GC, Bilous RW, Mackintosh D et al. *Am J Med 1983; 74:* 256
7 Hostetter TH, Rennke HG, Brenner BM. *Am J Med 1982; 72:* 375

Proc EDTA–ERA (1984) Vol 21

CHARACTERISATION OF MECHANISMS RESPONSIBLE FOR URAEMIC INSULIN RESISTANCE: IN VITRO EXPERIMENTS

O Pedersen, ***O Schmitz**, E Hjøllund, B Richelsen, *H E Hansen

*Aarhus Amtssygehus, *Aarhus Kommunehospital, Aarhus, Denmark*

Summary

In an attempt to define the cellular basis of the uraemic insulin resistance we studied insulin action in adipocytes from eight patients with undialysed chronic uraemia and from eight matched healthy controls. (^{125}I)-insulin binding to fat cells from uraemic patients was normal. In contrast (^{14}C)-D-glucose transport exhibited decreased sensitivity to insulin. The concentrations of insulin that elicited a half-maximal response were 422±95pmol/L in uraemic patients and 179±38pmol/L in normals ($p<0.01$). The non-insulin and the maximally insulin stimulated glucose transport of adipocytes from uraemic patients was normal. The lipogenesis of fat cells from uraemic patients had depressed sensitivity to insulin (half-maximal stimulation at 38±8pmol/L in uraemic patients and at 11±3pmol/L in normals, $p<0.01$) with unchanged non-insulin and maximally insulin stimulated lipogenesis. Taken together these results suggest that the insulin resistance of adipocytes from patients with chronic uraemia may be primarily accounted for by post-binding defects localised to glucose transport and metabolism.

Introduction

Numerous experiments have demonstrated that the glucose intolerance of uraemic patients is primarily due to insulin resistance of peripheral tissues [1,2]. Moreover, it has been shown that the liver of uraemic man retains normal sensitivity to insulin at least with respect to the inhibitory effect on hepatic glucose release [1,2]. To examine the cellular mechanisms behind the insulin resistance of peripheral tissue from uraemic man we have measured (^{125}I)-insulin binding and non-insulin and insulin stimulated (^{14}C)-D-glucose transport and metabolism in isolated human adipocytes from healthy controls and from uraemic patients not yet on chronic dialysis treatment.

TABLE I. Clinical and laboratory data of patients

Patient No	Sex	Age years	Obesity index	Renal disease	Plasma Insulin μU/ml	Plasma Glucose mmol/L	Serum Creatinine μmol/L	Serum Urea mmol/L	Serum Total CO_2 mmol/L
1	M	52	1.04	Chronic glomerulonephritis	19	6.5	884	38.6	15
2	M	53	0.94	Polycystic kidney disease	8	4.2	887	28.5	15
3	F	55	1.00	Polycystic kidney disease	11	5.1	766	30.6	19
4	M	58	0.97	Chronic interstitial nephritis (analgesic)	22	4.8	929	18.7	16
5	F	55	0.92	Chronic interstitial nephritis (analgesic)	12	4.6	678	46.2	25
6	F	31	0.94	Chronic glomerulonephritis	13	4.6	833	23.6	17
7	F	48	1.35	Chronic interstitial nephritis (not classified)	15	4.8	808	20.8	20
8	M	43	1.15	Chronic glomerulonephritis	11	5.3	559	18.8	22
Mean		49	1.04		14	5.0	793	28.2	19
SEM		0.2	0.05		2	0.2	55	3.6	1

Materials and methods

Subjects

Eight adult, ambulatory non-diabetic uraemic patients were studied and compared with sex, age and weight-matched healthy volunteers. Pertinent clinical data for the patients appear in Table I. All the patients had an oral glucose tolerance test (1g of glucose/kg body weight). Patient No. 2 had a two hour value plasma glucose of 11.6mmol/L and patient No. 7 had a two hour value of 9.8mmol/L, whereas the rest of the group all had two hour values below 7.5mmol/L.

Four healthy males and four healthy females served as control subjects. Their mean age was 44 years (range 58–26 years) while their mean obesity index was 1.05 (range 0.85–1.15).

Insulin receptor binding

Adipose tissue (about 10g) was obtained by open biopsy from the upper one-fourth of the right gluteal region after a square field had been anaesthetised with an epidermal injection of 1% lidocaine without epinephrine. Details about fat cell isolation as well as determination of fat cell size and number have been published previously [3]. Insulin binding to fat cells (about 10^5 cells per ml of cell suspension) was measured in a Hepes buffer at 37°C, after incubation for 60 minutes with tyrosine-A_{14}-labelled (^{125}I)-insulin with or without increasing concentrations of unlabelled insulin [3]. Specific insulin binding to adipocytes was expressed per $30cm^2$ of surface area per ml.

Glucose transport

All studies were carried out at 37°C. Forty microlitres of adipocyte suspension with a volume fraction of 0.4 (about 6×10^5 cells/ml) were placed in polypropylene tubes and preincubated with or without insulin for 45 minutes. Twelve microlitres (0.24 μCi) of tracer D-U-(^{14}C)-glucose (final glucose concentration 20μmol/L) was added at time zero and uptake was determined after 10 seconds by adding 3ml of phloretin (0.3mmol/L). Silicone oil (0.8ml of 0.99g/ml) was layered on the top, and the tubes were spun within two minutes at 2500G. The cells were collected from the top of the oil and placed in scintillation vials with 5ml of scintillation fluid [4].

Lipogenesis

Lipogenesis was measured as the conversion of D-U-(^{14}C)-glucose to (^{14}C)-total lipids [5]. Isolated adipocytes (about 7×10^4 cells/ml) were prepared in a 10mmol/L Hepes buffer containing 0.5mmol/L glucose. The cells were preincubated for 45 minutes at 37°C with or without insulin in increasing concentrations. Then 0.4μCi D-U-(^{14}C)-glucose was added to each tube (final glucose concentration 0.5mmol/L) and the incubation was continued for 90 minutes.

H_2SO_4 was added and a Dole extraction was performed. A sample for liquid scintillation counting was taken from the upper phase.

Analytical methods

Plasma glucose was analysed with a glucose dehydrogenase method (Merck enzymatic kit) and plasma insulin was measured with a RIA technique [6].

Statistical methods

In text, tables and figures data are given as the mean ± 1 SEM. Significant differences between groups were assessed by Mann Whitney's test. In correlation studies Spearman's test was employed.

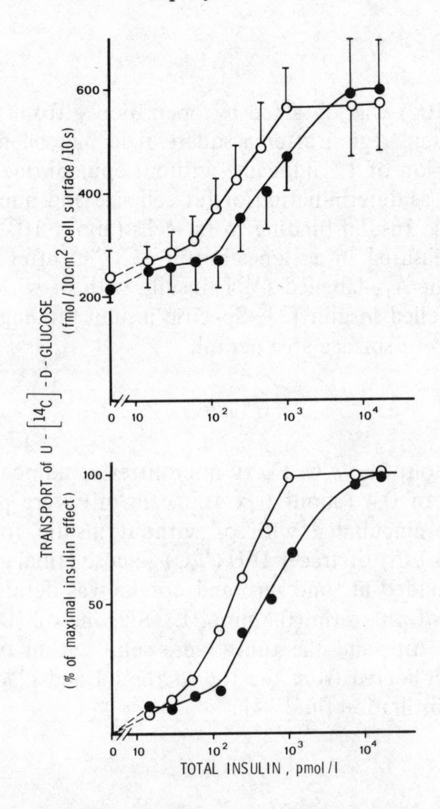

Figure 1. Glucose transport of fat cells. Upper panel: adipocytes from eight uraemic patients (●) and eight normal subjects (○) were preincubated with or without insulin in the for 45 minutes at 37°C. Initial transport rate was then measured during the first 10 seconds after (^{14}C)-D-glucose had been added to reach a final concentration of 20μmol/L (mean ± 1 SEM).

Lower panel: glucose transport data are expressed as percentage of the response to insulin in maximally effective concentrations

Results

The uraemic condition did not influence the binding of (^{125}I)-insulin or the ability of unlabelled insulin to complete for binding over a concentration range of 0.55–245nmol/L. When (^{14}C)-D-glucose transport was measured in the absence or the presence of maximally effective insulin concentrations adipocytes from uraemic patients responded similarly to adipocytes from healthy volunteers (Figure 1). However, the insulin-dose response relationship of glucose transport was altered by uraemia. In patients with renal failure the dose-response curve for insulin stimulated glucose uptake was markedly shifted to the right (Figure 1) suggesting an impaired sensitivity to insulin. For example, the concentrations of insulin that elicited a half-maximal effect was 422±95pmol/L in uraemic patients and 179±38pmol/L in healthy subjects ($p<0.01$).

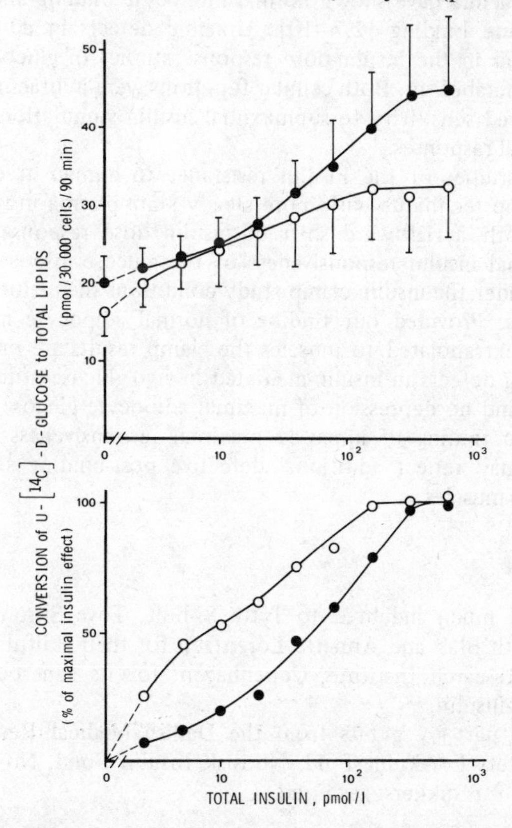

Figure 2. Upper panel: adipocytes from eight uraemic patients (•) and eight normal subjects (○) were preincubated with or without insulin in the indicated concentrations for 45 minutes at 37°C. Then (^{14}C)-D-glucose was added (total glucose concentration in the final preparation was 0.5mmol/L) and the incubation was terminated after 90 minutes (mean ± 1 SEM).

Lower panel: lipogenesis data are expressed as percentage of the response to insulin in maximally effective concentrations

The effect of uraemia on glucose metabolism by human adipocytes was studied by measuring (^{14}C)-D-glucose conversion to total lipids (Figure 2). Both non-insulin and maximally insulin stimulated lipogenesis were normal in uraemic cells. However, consistent with our observations in glucose transport studies the insulin concentrations giving half-maximal lipogenesis were significantly higher in adipocytes from uraemic patients (38±8pmol/L in uraemic patients versus 11±3pmol/L in normals, $p<0.01$).

Discussion

Our results indicate that long-term uraemia has no significant impact on adipocyte insulin receptor binding. Previous investigations of blood cell insulin receptors in human uraemia have shown normal monocyte binding and decreased or normal erythrocyte binding [2,7–10]. Uraemic defects in adipocyte insulin action were found in the insulin-dose response studies of glucose transport as well as glucose metabolism. Both cellular functions were in uraemic cells characterised by impaired sensitivity to submaximal insulin stimulation but unaltered basal and maximal responses.

The in vivo studies of the insulin resistance to human uraemia using the euglycaemic clamp technique with more steady state plasma insulin values have demonstrated both a rightward shift of insulin dose response curves and a decreased maximal insulin responsiveness of the glucose disposal to peripheral tissues [1,2]. Under the insulin clamp study conditions the major site of glucose uptake is muscle. Provided our finding of normal adipocyte insulin receptor binding can be extrapolated to muscles the clamp results are most compatible with post-binding defects in insulin-mediated in vivo glucose utilisation. Finally, since we could find no depression of maximal adipocyte glucose metabolism in vitro the in vivo finding of impaired maximal responsiveness of the glucose disposal [1,2] may reflect additional defective post-binding steps of insulin action in skeletal muscles.

Acknowledgment

The authors are much indebted to Jytte Søholt, Tove Skrumsager, Pernille Jørgensen, Lisbeth Blak and Annette Lorentzen for their skilful technical work and to Novo Research Institute, Copenhagen, for its generous donation of A_{14}-labelled ^{125}I-insulin.

Supported in part by grants from the Danish Medical Research Council, Aarhus Universitets Forskningsfond, Nordisk Insulin Fond, Novo Fonden and Landsforeningen for Sukkersyges Fond.

References

1 DeFronzo RA, Alvestrand A, Smith D et al. *J Clin Invest 1981; 67:* 563
2 Smith D, DeFronzo RA. *Kidney Intern 1982; 22:* 54
3 Pedersen O, Hjøllund E, Beck-Nielsen H et al. *Diabetologia 1981; 20:* 636
4 Pedersen O, Gliemann J. *Diabetologia 1981; 20:* 630

5 Pedersen O, Hjøllund E, Lindskov HO. *Am J Physiol 1982; 243:* E158
6 Heding LG. In Donato L, ed. *Labelled Proteins in Tracer Studies.* Brussels: Euratom. 1966: 345
7 Schmitz O, Hjøllund E, Alberti KGMM et al. *Eur J Clin Invest:* in press
8 Gambhir KK, Archer JA, Nerurkar SG et al. *Nephron 1981; 28:* 4
9 Milutinovič S, Breyer D, Jakovič M et al. *Proc EDTA 1982; 19:* 763
10 Namikawa T, Namikawa T, Fukimotor S et al. *Horm Metab Res 1983; 15:* 161

Proc EDTA–ERA (1984) Vol 21

ULTRASTRUCTURAL MARKERS OF TUBULAR TRANSPORT IN EXPERIMENTAL DIABETES INSIPIDUS

L Cieciura, Z Kidawa, B Jaszczuk-Jarosz, **K Trznadel**, J Konopacki

2nd Clinic of Internal Diseases, Military Medical Academy, Lódź, Poland

Summary

Stereological analysis of changes in tubular transport markers in diabetes insipidus has been undertaken. Intracellular spaces and basal infolded channels were considered as the markers of water transport, while mitochondrial metabolic steady states were considered as the active transport markers. Transmission electron microscopic observations revealed morphometric differences in the surface area and volume of intercellular space and basal infolded channels in the distal tubules. Stereological markers of mitochondrial metabolic states demonstrated significant differences in the distal tubules between diabetes insipidus and control groups. In diabetes insipidus the volume and surface area of intercellular spaces and basal infolded channels in the distal tubules were decreased and the mitochondrial energy state was lowered.

Introduction

Recent reports [1] indicate that the epithelial cells of renal proximal and distal tubules contain considerable amounts of mitochondria. Comparison of changes in their configuration during normal and disturbed reabsorption may be a basis for elaboration of a method which would be helpful in the evaluation of disturbances in the ultrastructural markers of active tubular transport. Stereological methods enables the determination of the energy state of mitochondria and, thus, ion-pump function.

Water shifts are mainly associated with sodium transport and passive permeability of cell membranes. These processes are ultrastructurally reflected by, among others, appearance of intercellular spaces and basal infolded channels of the epithelial cells. Application of stereological measurements for their evaluation under various conditions of reabsorption would enable more complete insight into the passive transport processes.

The aim of our studies was to elaborate a method for evaluation of ultrastructural markers of active and passive transport in the renal tubules using a model of experimental diabetes insipidus.

Materials and methods

The studies were performed on 15 Wistar rats of both sexes, weighing 200–250g. The animals were divided into three groups each of five rats:

Group I experimental diabetes insipidus,
Group II control,
Group III sham-operated rats.

Diabetes insipidus was induced by bilateral destruction of the supraoptic nuclei of the hypothalamus with high-frequency electric current (2MHz for 15 sec), using coordinates from the stereotactic atlas for rats [2].

After placing the rats in metabolic cages following the operation, 24-hour urine output, urine osmolality and urinary sodium and potassium concentrations were measured. After 12 days the animals were sacrificed and material for electron microscopic studies taken. In all groups the cortical layer of the kidney was cut into sections 1mm thick and then fixed in 5% glutaraldehyde with 2% paraformaldehyde in 0.1M cacodylate buffer, ph 7.4, at 0–4°C for three hours. All specimens were embedded in Araldite and then ultra thin sections were obtained by means of LKB III ultramicrotome and stained with uranyl acetate and lead citrate for examination in Philips EM-300 electron microscope.

Morphometric procedure was based on the techniques of Weibel [3,4]. In order to study the relationship between the mitochondrial compartments and the internal membrane, we introduced the following partition coefficients (E):

$$E_{mm} = \frac{V_{mit}}{S_{im}} \times V_{mat} \quad ; \quad E_{ocm} = \frac{V_{mit}}{S_{im}} \times V_{oc} ,$$

which express the volume of the matrix (E_{mm}) and external compartment (E_{ocm}), respectively, per unit of surface area of the internal mitochondrial membrane [5].

All calculations, mean values, standard error of the mean and coefficient of variation were performed on an SM 4A computer. All parameters were compared statistically by means of unpaired Student's 't' test and Cochrane-Cox and Whitney-Mann tests.

Results

In the rats with diabetes insipidus 24-hour urine output ranged from 27ml to 50ml, mean 37ml, while that in healthy and sham-operated animals was 9ml.

Stereological analysis of the proximal tubular epithelial cell mitochondria did not reveal any significant differences between diabetes insipidus and control groups. Also volume and surface area values of the intercellular spaces, as well as basal infolded channels, did not differ between the groups studied.

In the distal tubules significant differences were found between diabetes insipidus and control groups as regards the external membrane surface area and volume. Surface area was 9.7793$\mu m^2/\mu m^3$ in the diabetes insipidus rats, while it was 8.0353$\mu m^2/\mu m^3$ in the control groups. Relative volume of the external membrane was 0.0588 per cent in the diabetes insipidus rats and 0.0387 per cent in the controls.

Statistically significant differences were also found in the mitochondrial internal membrane. Its surface area was $44.8155\mu m^2/\mu m^3$ in the diabetes insipidus animals, while in the controls it was $38.6785\mu m^2/\mu m^3$. Relative volumes of the internal membrane was 0.0882 per cent vs 0.0952 per cent, respectively.

Partition coefficient for the external compartment (E_{ocm}) significantly decreased from $0.0044\mu m^3/\mu m^2$ in the controls to $0.0035\mu m^3/\mu m^2$ in diabetes insipidus. Similarly, partition coefficient for the internal compartment lowered from $0.0209\mu m^3/\mu m^2$ in the controls to $0.0199\mu m^3/\mu m^2$ in diabetes insipidus (Figure 1).

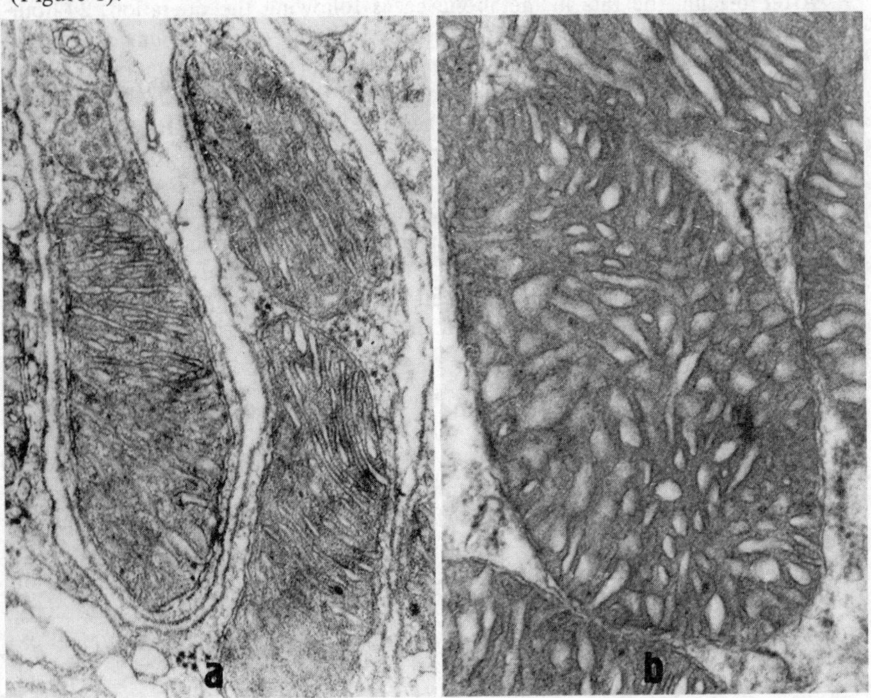

Figure 1. Mitochondria of distal tubule. Differences in configuration of mitochondrial membranes. a=control group; b=diabetes insipidus. Changes in direction to condensed steady state. x 27,000 (reduced for publication)

Statistical analysis did not reveal any significant differences between the groups studied for the intercellular spaces and basal infolded channels within the proximal tubules.

However, in the distal tubules relative volumes of the intercellular spaces and basal infolded channels significantly decreased from 8.844 per cent in the controls to 6.1778 per cent in the diabetes insipidus rats.

In order to determine interrelationships between the intercellular space volume and surface area, the following coefficient was calculated:

$$E_V = \frac{V_V}{S_V},$$

where V_V – relative volume of the intercellular space, S_V – relative surface area of the intercellular space. This coefficient is an exponent of the width of the intercellular space and basal infolded channels. In the rats with diabetes insipidus the value was 30 per cent less than in the control groups (Figure 2).

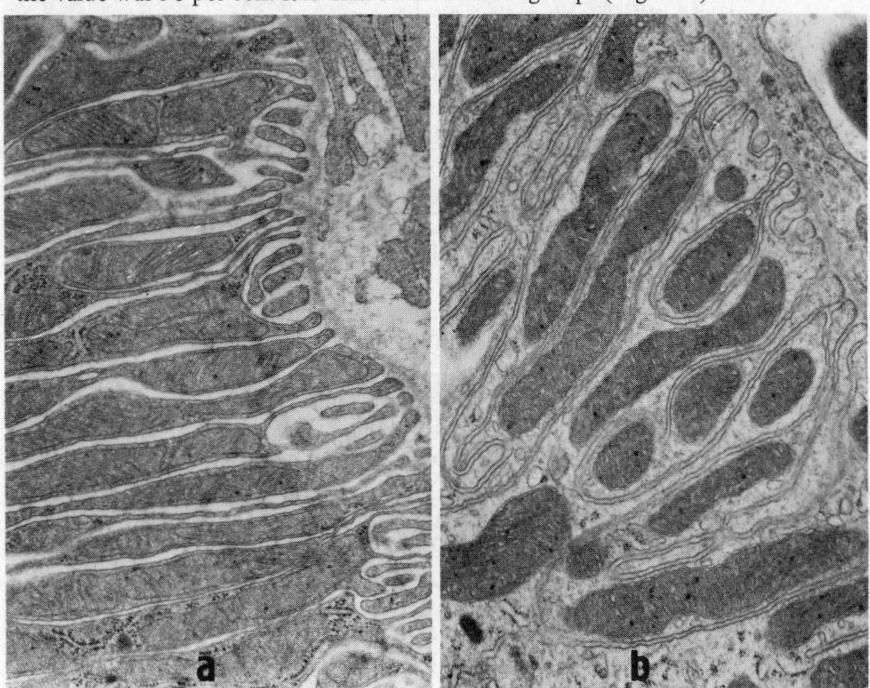

Figure 2. The distal tubule. a=control group. Intercellular spaces and basal infolded channels are dilated. b=diabetes insipidus. Compared with control, narrow spaces are observed. x 10,400 (reduced for publication)

Discussion

Our stereological data obtained from the analysis of the intercellular spaces and basal infolded channels were found to be consistent with the morphometric data reported by Pfaller [1].

An active ion transport as well as water reabsorption takes place mainly in the proximal tubule; our investigations on ultrastructural markers of ion and water transports did not reveal any differences between experimental diabetes insipidus and healthy or sham-operated animals. It confirms suggestions of other authors [6] who did not prove any effect of ADH on this nephron segment in rats.

Stereological analysis of mitochondrial ultrastructure enables determination of their energy state [7,8], which may be reflected by changes from the condensed state through the transitional state to the orthodox one. Thus, energy states of the mitochondria may correlate with their function in active transport.

In our experiments the distal tubule mitochondria were found to be in the transitional state. In the course of diabetes insipidus their configuration was changing towards the condensed, i.e. low-energy, state. Thus, they seem to contribute to active transport to a lesser degree. In the diabetes insipidus rats 24-hour urinary sodium excretion was higher by 10.5 per cent than in controls. Assuming that sodium reabsorption in the proximal tubule ws unaltered in all studied groups, it seems that the 10.5 per cent increase in the urinary sodium excretion results from its lowered reabsorption in the distal tubule.

In our studies volumes and surface areas of the intercellular spaces and basal infolded channels were taken as water transport exponents. Stereological analysis of these parameters did not reveal any significant differences in the proximal tubule between the groups studied. Thus water transport was also unaffected by ADH deficiency in this nephron segment.

However, a significant decrease in relative volumes of the intercellular spaces and basal infolded channels in the distal tubule is evidence for significant impairment of water transport. Assuming that approximately 20 per cent of water reabsorption takes place in the distal tubule, it constitutes approximately 180ml of primary urine. After subtraction of 9ml of final urine in the controls and 37ml in diabetes insipidus, we obtain 171ml and 143ml, respectively. These values theoretically define primary urine volume reabsorbed in the distal tubule. Thus, reabsorption in the distal tubule was reduced approximately 20 per cent in diabetes insipidus.

These calculations were performed under the assumption resulting from our studies that the proximal tubule function does not change in the course of diabetes insipidus. It is consistent with reports of other authors [6,9] and with observations that in rats vasopressin affects water reabsorption in the distal tubule [10].

If we compare results from our calculations for final urine with regard to 20 per cent of reabsorption in the distal tubule and 30 per cent reduction in the intercellular spaces and basal infolded channels, we think this difference results from reabsorption in the collecting tubules and the phenomenon of counter-current multiplication.

Results of our stereological studies in the mitochondrial metabolic states and volumes of the intercellular spaces and basal infolded channels indicate that they may be markers of ion transport and water reabsorption in the nephron.

References

1 Pfaller W. In Hild W, van Limborgh J, Ortmann R, Pauly JE, Schiebler TH, eds. *Structure Function Correlation on Rat Kidney.* Berlin: Springer-Verlag. 1982: 23

2 Fifkova E, Maršala J. In *Stereotaxie podkorowych struktur mozku krysy, kralika a kočky.* Praha: Stadni zdrovotnicke nakladatelstvi. 1960: 145

3 Weibel ER, Kistler CS, Scherle WF. *J Cell Biol 1966; 30:* 23

4 Weibel ER, Stäubli W, Gnägi HR, Hess FA. *J Cell Biol 1969; 42:* 68

5 Cieciura L, Rydzyński K, Klitończyk W. *Cell Tissue Res 1979; 42:* 347

6 Jacobson HR. *Kidney Int 1982; 22:* 425
7 Hackenbrock CR. *J Cell Biol 1966; 30:* 269
8 Hackenbrock CR. *J Cell Biol 1968; 37:* 345
9 Leaf A. *Am J Med 1967; 42:* 745
10 Sasaki S, Imai M. *Pflügers Arch 1980; 383:* 215

Proc EDTA–ERA (1984) Vol 21

DISTAL TUBULAR SODIUM AND WATER RESORPTION IN MAN

E Bartoli, R Faedda, A Satta, M Sanna, G F Branca

Istituto di Patologia Medica, Sassari, Italy

Summary

It is possible to quantitatively estimate Na and water transport by each segment of the human nephron during maximal water diuresis (MWD) with a newly developed technique. The present study describes a new development whereby Na reabsorption by the ascending limb of Henle's Loop can be measured during a test of maximal antidiuresis (TMA) and compared with that obtained in separate experiments during MWD. In experiments performed on eight subjects, Henle's Loop reabsorption, expressed as equivalent volume of solute-free water generated, was 23.9 ± 3.7 during MWD, 24.5 ± 2.3ml/min/100ml GFR in TMA, $p > 0.05$; $r = 0.68$, $p < 0.05$. Thus, our new method is reproducible and theoretically correct.

Introduction

We have developed [1] and extensively applied [1–4] a new method to measure transport by each segment of the human nephron [1]. We assumed (Figure 1) that frusemide (F), by abolishing the NaCl transport of Henle's Loop (HL) eliminates the osmotic gradient between collecting duct (CD) urine and the interstitium, and the abstraction of solute-free water (SFW) that occurs in the absence of antidiuretic hormone (ADH). Therefore, the urine flow during MWD plus frusemide ($\dot{V}f$) represents a true measurement of delivery from the proximal tubule (PT). In addition, the urine will still be diluted by the distal tubule (DT) during F. Thus, the SFW excreted during F, CH_2Of, is that formed by the DT, CH_2O-DT. Thus, CH_2Of = CH_2O-DT (I). The difference between $\dot{V}$ during F ($\dot{V}f$) and that during MWD alone ($\dot{V}$) measures the volume of SFW reabsorbed in the absence of ADH, indicated as CH_2O-BD, the SFW back diffusion. This, being the part of SFW which is not excreted, the true SFW generated is that excreted (CH_2O) plus that reabsorbed (CH_2O-BD): hence, CH_2O-T = CH_2O-BD + CH_2O (II), where CH_20-T is the total SFW formed. The SFW generated by HL alone can be calculated as CH_2O-HL = CH_2O-T –

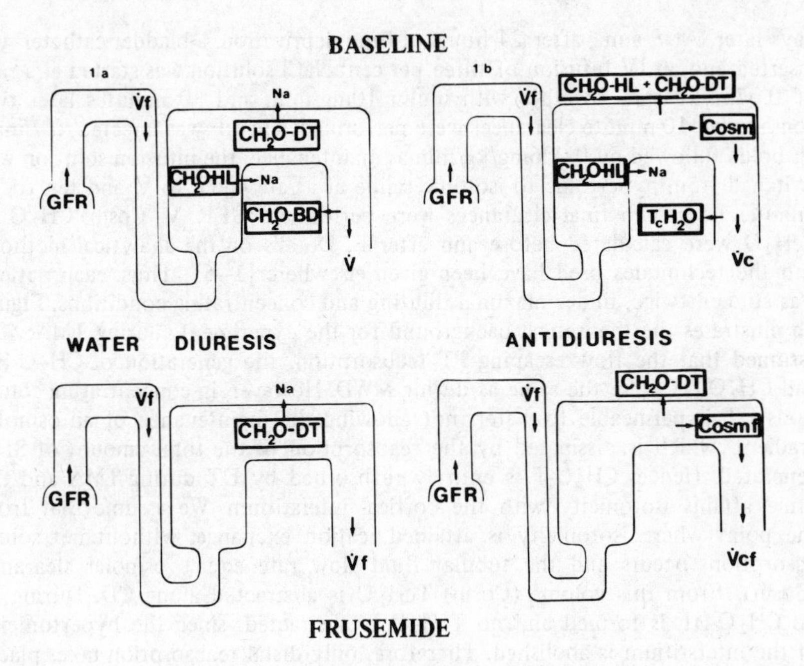

Figure 1. Figure 1a, on the left hand side, portrays two nephrons: that on top is during water diuresis (MWD) alone, that at the bottom during the superimposed action of frusemide. Figure 1b. on the right hand side, portrays the same two nephrons during antidiuresis (TMA) alone (top) and during the superimposed action of frusemide (bottom). The f at the end of a symbol indicates data obtained during frusemide infusion. $\dot{V}f$, the urine flow rate during frusemide, is equated to the flow escaping proximal reabsorption, and it is therefore inserted at the end of proximal tubules. The lumen contains rectangles, labelled with different symbols, which indicate the volumes of solute free water generated (SFW) by the reabsorption of Na without solvent along that segment. The rectangles outside the lumen indicate the volumes of SFW reabsorbed across that segment. At the end of collecting ducts the urine flow rate is indicated during baseline MWD ($\dot{V}$) and TMA ($\dot{V}c$), and during frusemide ($\dot{V}f$ and $\dot{V}cf$). For the explanation of all symbols and details on the theory see abbreviations and text

CH_2O-DT (III). For these concepts to gain a wide acceptance, it is necessary to confirm the measurements by an independent method. With this new theoretical method to calculate CH_2O-HL, we devised the present experiments to compare, by an independent method, the results obtained by previously published techniques based on MWD [1].

Materials and methods

The studies were performed on eight healthy adult volunteers. MWD was instituted exactly as previously published [1]. Inulin and F infusion were the same as in antidiuresis. The test of maximal antidiuresis (TMD) was performed a few days later. At 8 a.m., after 24 hours of fluid deprivation, a bladder catheter was inserted and an IV infusion of three per cent NaCl solution was started at a rate of 0.07ml/kg/min, together with inulin 10mg/min, and 30 minutes later two consecutive 10 minute clearances were performed. Then F was injected, 0.75mg/kg bolus followed by 0.006mg/kg/min as maintenance, the infusion solution was switched from hypertonic to isotonic saline at a rate equal to $\dot{V}$ and ten to 15 minutes later, two final clearances were performed. GFR, $\dot{V}$, Cosm, CH_2O or TcH_2O were calculated before and after F. Details on the analytical methods and the techniques used have been given elsewhere [1–5]. Thus, each patient was studied twice, under maximal diluting and concentrating conditions. Figure 1b illustrates the theoretical background for the experiments during TMA: it is assumed that the flow escaping PT reabsorption, the generation of CH_2O-HL and CH_2O-DT were the same as during MWD. However, in concentrating conditions DT is permeable to water, not allowing the maintenance of an osmotic gradient, which is dissipated by the reabsorption of the total amount of SFW generated. Hence, CH_2O-T is entirely reabsorbed by DT during TMA and the urine attains isotonicity with the cortical interstitium. We assume that from the point where isotonicity is attained, cation exchange, without net solute absorption, occurs and the tubular fluid flow rate equals osmolar clearance (Cosm). From this volume (Cosm) TcH_2O is abstracted along CD. During F, no CH_2O-HL is formed and no TcH_2O is abstracted, since the hypertonicity of the interstitium is abolished. Therefore, only distal reabsorption takes place: thus, CH_2O-DT is formed, and it is entirely reabsorbed till isotonicity is attained. Consequently, the urine flow rate during TMA before F ($\dot{V}c$) is equal to the flow escaping proximal reabsorption ($\dot{V}f$) minus the water reabsorbed (CH_2O-T) + TcH_2O) : $\dot{V}c = \dot{V}f - CH_2O\text{-T} - TcH_2O$ (IV). Instead, during F: $\dot{V}cf = \dot{V}f - CH_2O\text{-DT}$ (V). By combining (III), (IV) and (V), we obtain: $CH_2O\text{-HL} = \dot{V}cf - (\dot{V}c + TcH_2O)$. Thus, we can calculate CH_2O-HL, a measurement already obtained during the experiment of MWD, by a totally independent method in a different experimental setting.

Results and Discussion

All data were factored by GFR and expressed as ml/min/100ml GFR. Table I shows that CH_2O-HL measured during MWD averaged 23.9 ± 3.7, and this was not different statistically from the mean value measured, in these same subjects, during TMA, of 24.5 ± 2.3ml/min/100ml GFR, $p>0.05$. Figure 2 shows that the two independent measurements of CH_2O-HL were significantly correlated, $r = 0.68$, $p<0.05$. The present study is based upon several assumptions, derived from our original technique [1]. From a mathematical manipulation of our equations, it can be easily demonstrated that Cosm during baseline conditions must be equal in both diluting and concentrating experiments, and that Cosmf in diluting conditions must be equal to $\dot{V}cf$, the urine flow rate measured in TMA during F. Table I shows that in fact these requirements were met by the actual experimental results. This is not surprising, since we calculated that the degree of extracellular volume expansion achieved during MWD and TMA should have been of the same degree. Therefore, the results obtained in the measurements

TABLE I. The Table reports means ± standard errors of the mean obtained by two independent methods, one from experiments in MWD (left hand column), the other during TMA (right side). As predicted by the theory, only the data on urine flow rates are significantly different with the two methods. For the meaning of symbols see abbreviations and text

	MWD		TMA	
		ml/min/100ml GFR		
$\dot{V}$	22.3 ± 2.4	<0.01	3.8 ± 1.4	$\dot{V}c$
$\dot{V}f$	40.9 ± 4.7	NS	31.8 ± 3.5	$\dot{V}cf$
Cosm	5.1 ± 0.8	NS	7.3 ± 2.0	Cosm
Cosmf	28.5 ± 3.7	NS	26.6 ± 3.9	Cosmf
CH_2O-HL	23.9 ± 3.7	NS	24.5 ± 2.3	CH_2O-HL
CH_2O-DT	12.3 ± 2.2	NS	9.1 ± 4.3	CH_2O-DT
CH_2O-BD	18.5 ± 4.1	<0.01	3.5 ± 0.6	TcH_2O
		ml/min		
GFR	102.7 ± 6.8	NS	96.8 ± 3.8	GFR
GFRf	137.8 ± 15.2	NS	119.4 ± 7.0	GFRf

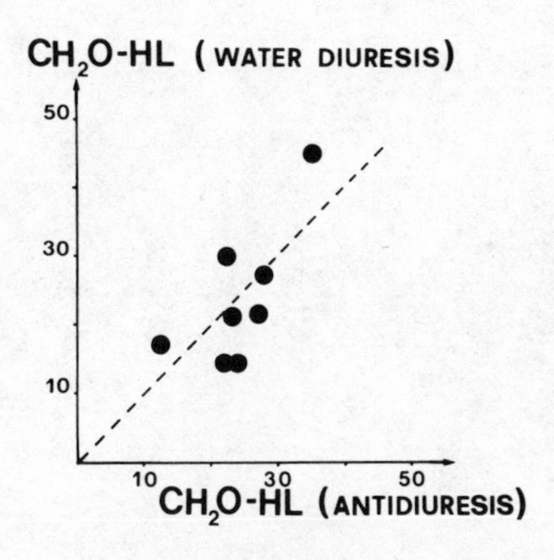

Figure 2. The free water generated by the loop of Henle (CH_2O-HL) calculated by experiments during MWDs shown on the abcissa. Each value, obtained in a single subject, is plotted against the paired measurement of CH_2O-HL obtained independently by experiments during TMA, shown on the ordinate. The pairs of values are distributed around the identity line traced and are significantly correlated: $r = 0.64$ ($p<0.05$)

of CH_2O-HL by the two independent methods can be considered adequate to assess the reproducibility and accuracy of the results obtained. As shown by Figure 2 the values independently measured in CH_2O-HL are practically superimposable and well correlated. In addition, there is a positive correlation between CH_2O-HL, a value proportional to the amount of salt deposited into the renal medullary interstitium, and TcH_2O, the flow of SFW reabsorbed by the interstitium from CD, two closely related variables in the renal medulla. We conclude that the present data confirm and strengthen the accuracy and value of our method [1] aimed at measuring Na and H_2O reabsorption by each segment of the human nephron in vivo.

Abbreviations

Proximal tubule = PT; Henle's loop = HL; Distal tubule = DT; Collecting duct = CD; Frusemide = F; Maximal water diuresis = MWD; Test of maximal antidiuresis = TMA; Solute free water = SFW; Free water clearance = CH_2O; Urine flow rate = $\dot{V}$; Osmolar clearance = Cosm.

References

1 Bartoli E et al. *J Clin Pharmacol 1983; 23:* 56
2 Satta A et al. *Renal Physiol 1984; 269:* In press
3 Zoccali C et al. *Nephron 1982; 32:* 140
4 Satta A et al. *First International Conference on Diuretics, Miami Beach, 1984*
5 Chiandussi L, Bartoli E, Arras S. *Gut 1978; 19:* 497

Proc EDTA–ERA (1984) Vol 21

HIGH FLUID INTAKE OR PHARMACOLOGICAL THERAPY IN RECURRENT STONE FORMER PATIENTS?

A **Aroldi**, G Graziani, P Passerini, C Castelnovo, *A Mandressi, *A Trinchieri, † G Colussi, C Ponticelli

*Ospedale Maggiore di Milano, *Istituto di Urologia dell'Università di Milano, †Ospedale Cà Granda, Milan, Italy*

Summary

In order to evaluate whether therapy can reduce relapses of urinary stone formation, we have retrospectively analysed the long-term follow-up of 55 recurrent stone former patients either treated with high fluid intake and moderate low calcium and low oxalate diet alone (Group A 18 patients), or with the same dietetic advice plus hydrocholorothiazide, amiloride and allopurinol (Group B 37 patients).

In group A, stone recurrence was completely abolished in 14 patients without hypercalciuria and hyperuricuria, but not in the four patients with hypercalciuria and hyperuricuria.

In group B, no relapses were observed in 19 hypercalciuric and hyperuricuric patients during a cumulative follow-up of 91 years. Even if the other 18 patients had relapses during a cumulative follow-up of 89 years, they showed a significant decrease in stone/patient and stone/year rates.

It is concluded that high fluid intake and diet can actually prevent stone recurrence in patients without hypercalciuria and hyperuricuria, but in hypercalciuric and hyperuricuric patients treatment with diuretic and allopurinol is better.

Introduction

The management of recurrent stone former (RSF) patients still remains controversial. A controlled study did not show any difference in the incidence of relapses between RSF patients who were advised to maintain high fluid intake and those who were given thiazide [1]. However in another report a strong reduction of relapses was observed in RSF patients after the introduction of thiazide alone or in association with amiloride and/or allopurinol [2].

In order to elucidate whether the latter pharmacological approach is or is not superior to a high fluid intake alone, we retrospectively analysed the results of a long-term follow-up in 55 RSF patients either treated with high fluid intake and moderate low calcium and low oxalate diet alone or with the same prescriptions plus diuretic agents and allopurinol.

Patients and methods

For the purposes of this study, only patients who had passed at least three stones in the last two years were considered RSF. Fifty-five patients entered the study. In all patients basal urinary excretion of calcium, phosphate, uric acid, sodium creatinine on various dietetic regimens were investigated. Hypercalciuria was defined as a urinary calcium excretion >100mg/24hr on low calcium diet, in at least two different determinations; hyperuricuria was defined as a urinary uric acid excretion >600mg/24hr.

Among the 55 patients admitted to the study, 14 patients showed normal urine excretion of calcium and uric acid, while 41 patients were hypercalciuric and 43 per cent of them were also hyperuricuric. High fluid intake and moderate low calcium (<600mg/day) and low oxalate (<80mg/day) were prescribed to the patients without metabolic disorder and to four hypercalciuric patients who refused pharmacological therapy (group A); the same dietetic measures plus hydrochlorothiazide (HCT) (50mg/day) and allopurinol (200mg/day) were prescribed to 37 hypercalciuric and hyperuricuric patients (group B). In 15 patients with calciuria higher than 300mg/day, in spite of therapy, the dose of HCT was raised to 100mg/day. Allopurinol dosage was increased to 300mg/day in 10 patients in order to reduce urate excretion below 600mg/24hr.

The mean follow-up was 4.2 ± 1.2 years for group A, and 4.1 ± 0.9 years for group B. Urinary calcium and uric acid excretion were assessed every three months during the first year of therapy and then every 12 months. Abdominal X-ray was performed every 12–18 months. Relapses were defined as a stone passage or evidence of new stone formation on abdominal X-ray.

Statistical analysis was performed by variance Turkey's test.

Results

During follow-up, urinary calcium (UCa mg/24hr) and uric acid (UUA mg/24hr) excretion did not modify in group A patients, while it significantly decreased in the 19 non-relapsers of group B while remaining unchanged in the relapsers (Table I). In particular, in 22 patients calciuria decreased below 200mg/24hr with a low dosage of HCT and amiloride and maintained relatively unchanged throughout the study. Despite 100mg/day of HCT, only four of 15 patients showed a persistent reduction of UCa below 300mg/day. UCa did not decrease below 300mg/24hr in seven relapsers and in four non-relapsers. Urate excretion did not decrease below 600mg/day in six relapsers and in one non-relapser, despite the increase of allopurinol to 300mg/day.

In group A the cumulative number of stones decreased from 92 in 73 years

TABLE I. Urinary calcium and uric acid excretion in RSF patients: group A without hypercalciuria and hyperuricuria; group B with hypercalciuria and hyperuricuria. Basal values and after 1, 3, 5 years of treatment are reported. All values are mean ± SD and differ from basal values. *=p<0.01

	Group A		Group B	
	Non-relapser	Relapser	Non-relapser	Relapser
Number of patients	14	4	19	18
	Urinary Calcium Excretion (mg/24hr)			
basal	121± 82	328±138	326±143	349±169
1 year	140± 98	286±103	194± 93*	307±145
3 years	101±119	334±137	236± 98*	310±110
5 years	107±105	290±111	225±104*	324±144
	Urinary Uric Acid Excretion (mg/24hr)			
basal	511±150	1310±575	591±324	679±210
1 year	518±185	765±326	445±203*	561±170
3 years	490±224	660±231	483±200*	555±214
5 years	464±149	582±212	392±144*	522±184

to 10 in 72 years of follow-up. No patient without a metabolic disorder relapsed during the follow-up, while all four hypercalciuric and hyperuricuric patients had at least two stones during the follow-up giving a stone/patient rate unchanged before and after treatment (2.5 vs 2.5).

In group B the cumulative number of stones decreased from 181 to 59. In particular 19 patients who had passed 85 stones in 77 years before therapy did not show any relapse during the 91 study years, while the other 18 patients had at least one stone recurrence. However, in relapsers the cumulative number of stones was significantly reduced (96 stones in 78 years before therapy versus 59 stones in 89 years during treatment p<0.01). Also the stone/patient rate (5.3 vs 3.2 p<0.01) and the stone/year rate (1.2 vs 0.6 p<0.01) significantly reduced during treatment.

Urine volume and pH, phosphate and sodium excretion did not significantly differ between relapsers and non-relapsers in both groups.

Discussion

This study confirms the usefulness of high fluid intake and dietetic manipulations in normocalciuric and normouricuric RSF patients [3]. However in our experience hypercalciuric patients did not have any benefit from these measures. On the other hand administration of HCT, amiloride and allopurinol to a group of hypercalciuric and hyperuricuric patients abolished the formation of stones, for the time being. In 19 of 37 patients. Moreover the other 18 patients had a significant reduction of new stone formation during treatment. Whether these patients relapsed because of a low compliance to therapy or represent a subgroup of partial responders to therapy remains to be elucidated; this data strongly supports the efficacy of combined therapy with HCT, amiloride and allopurinol in preventing stone formation in patients with metabolic disorder.

References

1 Wolf H, Brocks P, Dahl C. *Proc EDTA 1983; 20:* 477
2 Maschio G, Tessitore N, D'Angelo A et al. *Am J Med 1981; 71:* 623
3 Pak CYC, Peters P, Kadesky M et al. *Am J Med 1981; 71:* 615

Proc EDTA–ERA (1984) Vol 21

MAGNESIUM RENAL WASTING IN CALCIUM STONE FORMERS WITH IDIOPATHIC HYPERCALCIURIA CONTRASTING WITH LOWER MAGNESIUM : CALCIUM URINARY RATIO

P Bataille, A Pruna, I Finet, P Leflon, R Makdassi, C Galy, P Fievet, A Fournier

Hôpital Nord, Amiens, France

Summary

Plasma magnesium (PMg) and urinary calcium (UCaV) and magnesium (UMgV) were measured after four days of calcium-restricted diet in 60 controls and 82 patients classified according to their calcium excretion in three groups: normocalciuric (NCa), dietary hypercalciuria (DH) and idiopathic hypercalciuria (IH). When compared to controls, higher UMgV (4.26 ± 0.28mmol/d versus 3.4 ± 0.16, $p<0.01$), lower PMg (0.79 ± 0.01mmol/d versus 0.84 ± 0.01, $p<0.05$) and lower UMg/UCa ratio (0.6 ± 0.04 versus 1.68 ± 0.15, $p<0.001$) were observed only in IH. A significant correlation between UMgV and UCaV was found in controls, in NCa and in DH but not in IH. In conclusion, (1) the coexistence of a higher UMgV and of a lower PMg in IH suggests that there is a magnesium depletion in this group of patients; (2) since the lower UMg/UCa ratio may favour a higher propensity for calcium crystallisation and is seen only in IH, magnesium supplements may be specially indicated in this group.

Introduction

Magnesium accounts for approximately 20 per cent of the total inhibitory activity of the urine with respect to calcium stone-formation. Magnesium depletion has been shown to cause calcification in proximal tubule cells and the tubular lumen in rats [1] and to be responsible for nephrocalcinosis in children. Accordingly a possible magnesium deficiency in the pathogenesis of calcium stone formation has been suspected although it has been observed only rarely [2]. Moreover, data concerning magnesium excretion in stone formers are conflicting, being found normal [3] or increased [4]. These discrepancies could be explained by the fact that dietary calcium and calcium excretion were not taken simultaneously into account since it is well established that in normal individuals magnesium excretion is directly correlated with calcium excretion [5], and that a high calcium intake might induce negative magnesium balance by increasing faecal magnesium in magnesium depleted humans [6].

For these reasons it seemed to us interesting to study magnesium metabolism in various groups of idiopathic calcium stone-formers classified according to calcium excretion during a controlled calcium diet.

Methods

Sixty controls and 82 patients with calcium urolithiasis were studied on an ambulatory protocol. They collected 24-hour urine on a free diet and after four days of a low calcium diet providing approximately 400mg calcium daily because no dairy products were ingested. In both collections, creatinine, calcium, phosphate and sodium were measured. Magnesium was measured only on the urine collection after calcium restriction. The blood samples were drawn after the four days of calcium restriction for calcium, phosphate, magnesium, creatinine, to eliminate patients with specific causes of calcium stone-formation.

Calcium, phosphate and creatinine were analysed by automatic colorimetry (Technicon Auto-Analyser) and magnesium in urine and plasma was determined by a colorimetric procedure (calmagite method).

Results

Classification of patients

With the data of our controls, we classified the patients according to their urinary calcium excretion:

Sixty were normocalciuric (NCa) on a free diet (urinary calcium <0.1mmol/kg/day).

Seventeen patients who were hypercalciuric on a free diet, presented a diet dependent hypercalciuria (DH), a diagnosis based on the return to normal of urinary calcium excretion after calcium restriction, when compared to controls on the same restricted diet (urinary calcium <0.07mmol/kg/day).

Idiopathic hypercalciuria (IH) was found in 29 patients who were hypercalciuric on a free diet and on a restricted diet (i.e. urinary calcium >0.1mmol/kg/day on free diet, urinary calcium >0.07mmol/kg/day on calcium restricted diet).

Magnesium excretion and serum magnesium in patients and in controls

Table I shows the values of the daily urinary excretion of calcium and magnesium in controls and in patients after four days of a low calcium diet. When compared to controls, urinary magnesium and calcium excretion were significantly higher ($p<0.01$) in patients with idiopathic hypercalciuria. Because of the possible role of magnesium in inhibiting calcium crystallisation we determined for each patient and each control the urinary magnesium/calcium ratio. When compared to controls, this ratio was significantly lower only in patients with IH. Serum magnesium was found significantly lower in controls ($p<0.05$) only in this group.

TABLE I. Daily excretion of magnesium and calcium; urinary magnesium/calcium ratio and plasma levels of magnesium and calcium in controls and calcium stone-formers on a Ca restricted diet

Mean ± SEM	UMgV (mmol/d)	UCaV (mmol/d)	UMg/UCa	PMg (mmol/L)	PCa (mmol/L)
Controls n = 60	3.4 ±0.16	2.02±0.3	1.68±0.15	0.84±0.01	2.35±0.06
NCa n = 36	3.87±0.17	2.84±0.3	1.36±0.08	0.82±0.09	2.30±0.07
DH n = 17	3.94±0.3	3.78±0.4	1.04±0.13	0.85±0.01	2.25±0.08
IH n = 29	4.26±0.28**	7.1 ±0.4***	0.6 ±0.04***	0.79±0.01*	2.28±0.07

Significance of the difference between controls and the stone-formers group: *$p<0.05$; **$p<0.01$; ***$p<0.001$).

NCa=normocalciuria; DH=diet dependent hypercalciuria; IH=idiopathic hypercalciuria.

Correlations between calcium excretion and magnesium excretion in controls and in stone-formers

In controls as well as in the whole group of stone-formers, there was a significant correlation between magnesium and calcium excretion (controls: $r=0.37$, $p<0.01$; UMg=1.53+0.37 UCa; patients: $r=0.44$, $p<0.01$, UMg=2.55+0.56 UCa).

When the various subgroups of stone-formers are considered separately positive correlations between calcium and magnesium excretions are still found in normocalciuria ($r=0.62$, $p<0.01$) and in diet dependent hypercalciuria ($r=0.47$, $p<0.05$) but not in patients with idiopathic hypercalciuria ($r=0.14$, NS).

Discussion

Our data, based on a short-term study, show that magnesium excretion on a low calcium diet is comparable in normocalciuric patients and in patients with dietary dependent hypercalciuria when compared to controls on the same Ca-restricted diet. On the other hand, the coexistence of a higher magnesium excretion and a lower plasma value of magnesium in stone-formers with idiopathic hypercalciuria suggests that there is a renal leak of magnesium leading to magnesium deficiency.

To explain the higher magnesium excretion in idiopathic hypercalciuria, the following hypothesis may be proposed: since there is a well-known competition for tubular reabsorption between calcium and magnesium, an increased filtered load of calcium could explain the increase of magnesium excretion [6]. Against this hypothesis are the following facts: (1) since there is no hypercalcaemia in IH there is no reason to postulate an increased filtered load; (2) the absence of correlation in IH between calcium and magnesium excretion.

A second hypothesis may be proposed: in most patients with IH there is a relative hypoparathyroidism since the primary disorder is an increased intestinal absorption of calcium [7]. Since PTH stimulates magnesium reabsorption [8], relative hypoparathyroidism could explain an increase of magnesium excretion. This hypothesis is further supported by the independent observation of a lower plasma magnesium in patients with documented absorptive hypercalciuria [9].

Finally, a decreased magnesium absorption consequent to the high calcium intake could be considered as a factor of magnesium depletion. In fact, an increased faecal magnesium excretion during a high calcium intake has been observed only in magnesium-depleted humans [6] and not in normal patients [10]. Because of the lack of magnesium balance studies, this mechanism of magnesium depletion cannot be confirmed with our data, but might be an additive one, on free calcium diet, in patients with idiopathic hypercalciuria, who are already magnesium depleted on calcium restricted diet.

The importance of magnesium in inhibiting calcium crystallisation depends on the concentration of magnesium in relation to calcium rather than on the absolute amount on this ion. The urinary magnesium/calcium ratio represents an index of the propensity to stone-formation related to magnesium and calcium. This ratio is significantly decreased only in idiopathic hypercalciuric patients although absolute magnesium is significantly higher in this group of patients. This suggests that magnesium supplements might be of therapeutic interest, especially in idiopathic hypercalciuria.

References

1 Bunce GE, King GA. *Exp Mol Pathol 1978; 28:* 322
2 Johansson G, Backman U, Danielson BG et al. In Smith LH, Robertson WG, Finlayson B, eds. *Urolithiasis, Clinical and Basic Research.* New York: Plenum Press. 1980: 267–273
3 Nicar MJ, Pak CYC. *Min Electrol Metab 1982; 8:* 44
4 Coe FL, Favus MJ, Croukett T et al. *Am J Med 1982; 72:* 25
5 Massry SG, Coburn JW. *Nephron 1973; 10:* 66
6 Donn MJ, Walser M. *Metabolism 1966; 15:* 884
7 Muldowney FP, Freanly B, Byan JG. *Q J Med 1980; 49:* 87
8 Colussi G, Sorian M, Malberti G et al. *Kidney Int 1982; Abstract 22:* 96
9 Scholz D, Schwille P. *Min Electrol Metab 1981; Abstract 93:* 264
10 Spencer H, Lesniak M, Gatza CA et al. *Gastroenterology 1980; 79:* 26

Proc EDTA–ERA (1984) Vol 21

ACUTE RENAL FAILURE IN AN INFANT WITH PARTIAL DEFICIENCY OF HYPOXANTHINE-GUANINE PHOSPHORIBOSYLTRANSFERASE

A-M Wingen, *W Löffler, R Waldherr, K Schärer

*University of Heidelberg, *University of München, FRG*

Summary

A three week old boy presented with pneumonia, weight loss, metabolic acidosis and renal failure (serum creatinine 3.1mg/100ml, uric acid 11.5mg/100ml). Renal biopsy revealed severe crystal nephropathy. Low activity of hypoxanthine-guanine phosphoribosyltransferase (HPRT) in erythrocytes and fibroblasts suggested a partial deficiency of the enzyme. A family study proved the mother to be heterozygous and the maternal grandfather to be hemizygous for HPRT deficiency. The grandfather developed gouty nephropathy and uraemia. The propositus was treated with allopurinol and kept on low purine diet and high fluid intake with sodium bicarbonate. Thereafter GFR gradually improved. At the age of two and a half years, growth and psychomotor development were normal, but ultrasound examination still revealed a dense renal parenchyma. Partial HPRT deficiency is a newly recognised treatable form of renal failure in the newborn.

Introduction

In the early postnatal period transient hyperuricaemia and hyperuricosuria are described especially following asphyxia and dehydration [1] and in autopsies of newborns up to 19.3 per cent incidence of uric acid 'infarctions' in kidneys is reported [2]. However, this overload of uric acid is usually not suspected to be the cause of kidney dysfunction. Even in patients with hereditary deficiency of HPRT, which is characterised by overproduction of uric acid since birth, renal complications usually do not occur before late childhood [3]. The following patient demonstrates that partial HPRT deficiency may lead to renal failure even in infancy.

Case Report

A full term male infant, weighing 3980g (S.Z., born 8/3/82) was delivered to a primipara without complications. Seven days after birth an upper respiratory tract infection and feeding difficulties were noted. At two weeks of age a diagnosis of bronchopneumonia was made and the baby was treated orally by ampicillin and cloxacillin followed additionally by gentamicin i.m. (4.5mg/kg/day) from day 19 to day 25. Because the general condition deteriorated he was admitted to the hospital at day 21. He appeared critically ill; weight 3350g, height 52cm. Laboratory studies: serum sodium 126mmol/L, potassium 7.6mmol/L, capillary blood: pH 7.02, bicarbonate 6mmol/L, base excess −23mmol/L, urinalysis: 31 erythrocytes/mm^3, 62 leucocytes/mm^3, pH 5.5, culture sterile. The infant received glucose/saline containing sodium bicarbonate intravenously (170ml/kg/day). At the age of 25 days BUN was 37mg/100ml, creatinine 3.1mg/100ml, uric acid 11.5mg/100ml, creatinine clearance 6.2ml/min/1.73m^2, urine volume 335ml/24hr. Abdominal ultrasound examination showed that both kidneys were of normal size, parenchyma was bright with a patchy pattern. Subsequently antibiotic therapy was stopped while the infusion therapy was continued.

Surgical renal biopsy at the age of 65 days demonstrated no glomerular changes, but at the corticomedullary junction many tubules were obstructed by crystalline needle-shaped material. Crystals were positive with De Galantha stain in alcohol fixed specimen. Some tubular epithelial cells were transformed to multinucleated giant cells. The interstitium showed diffuse fibrosis and focal infiltration by mononuclear cells.

Following kidney biopsy and a detailed family history enzyme studies were performed. HPRT activity in lysed erythrocytes on day 78, 11 days after transfusion, was 31.6nmol/mg protein/hr, but ranged between 0.2 and 1.1 when the test was repeated on seven other occasions. HPRT activity in lysed skin fibroblasts obtained from a skin biopsy was 1.2 per cent of normal. In intact erythrocytes it was considerably higher (Table I) [4].

TABLE I. HPRT activities in the propositus and his family. The values given are the mean of activities observed on different occasions, respectively in different subcultures of fibroblasts [4]

	Age at testing	HPRT nmol/mg protein/hr lysed erythrocytes	HPRT nmol/ml packed cells/hr intact erythrocytes	HPRT nmol/mg protein/hr lysed fibroblasts
Propositus	7 weeks	31.6		
	7 months	0.5	6.4	0.6
Maternal Grandfather	53 years	24.0	12.2	1.0
Mother	26 years	69.0	41.8	23.3
Father	26 years	79.7		43.0
Normal	adults	100.0 (70–150)	57.4 (48–63)	59.1 (42–82)

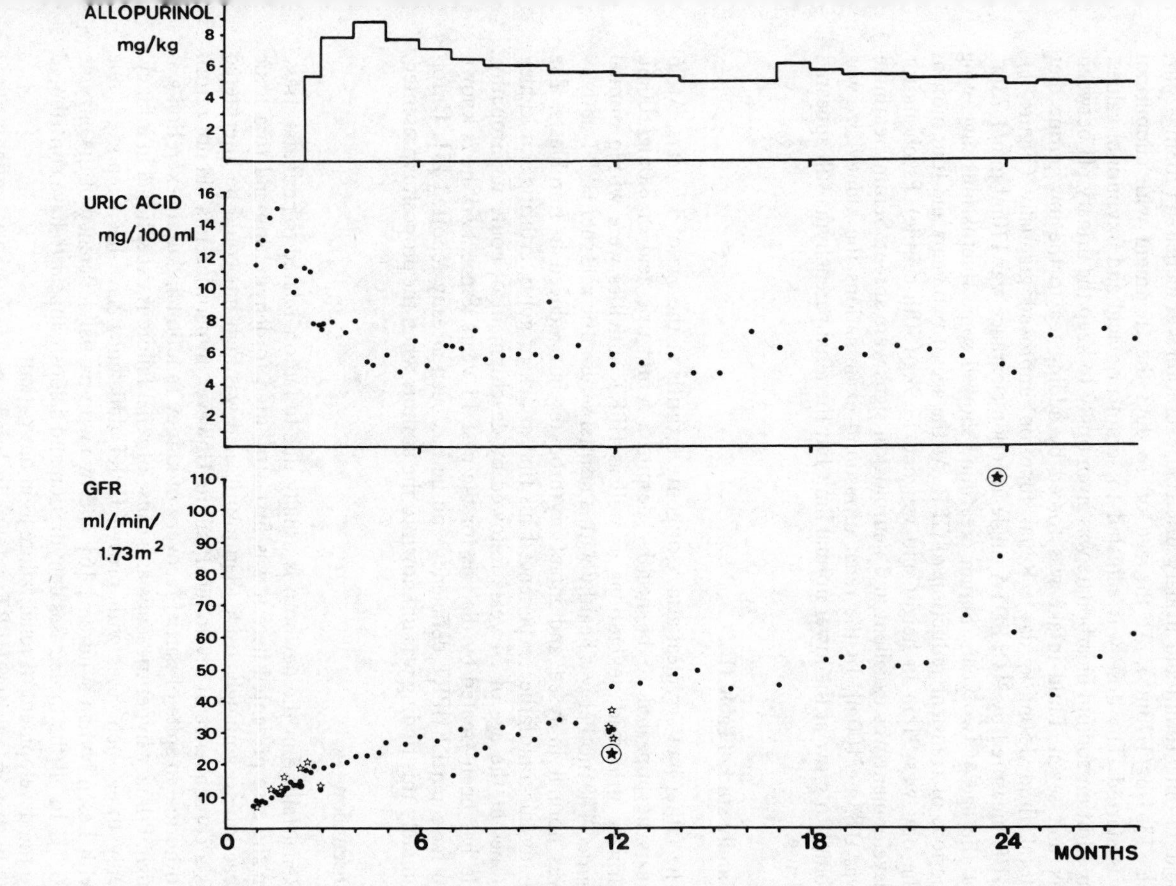

Figure 1. Course of disease. Abscissa age in months; ordinate: lower part: ● GFR in ml/min/m², calculated from serum creatinine (mg/100ml) and body length (cm) [10], ☆ endogenous creatinine clearance and ⊛ ^{51}Cr-EDTA single injection clearance (ml/min/1.73m²); upper part: serum uric acid (mg/100ml/ and allopurinol dosage (mg/kg body weight)

Clinical course

The boy recovered slowly. On day 75 after birth, weight was 4480g, serum creatinine 1.4mg/100ml, creatinine clearance 21ml/min/1.73m^2, serum uric acid 11.0mg/100ml and urinary uric acid excretion was increased (203mg/24hr, i.e. 1351mg/1.73m^2). At the age of 78 days the treatment with allopurinol was started. The dose was adjusted to serum uric acid and oxypurinol values. In addition sodium bicarbonate was given orally to keep the urinary pH between six and seven. Fluid intake was forced by adding tea to the low purine diet (total fluid 150ml/kg/day). Renal function improved gradually (Figure 1). GFR measured by ^{51}Cr-EDTA single injection clearance was 110ml/min/1.73m^2 at the age of two years. Serum creatinine showed some fluctuations suggesting periods of transient reduction of GFR. At the age of two years and four months the boy was 91cm in height and weighed 12.9kg (50th centile). Psychomotor development was excellent and neurological signs were absent. Serum creatinine was 0.83mg/100ml. In the renal ultrasound examinations the kidney size was found to grow at the lower normal limit but the renal parenchyma still appeared dense.

Family study (Table I)

The father had a typical attack of gouty arthritis at the age of 17 years. At 26 years he appeared to be healthy despite a serum uric acid of 8.0mg/100ml. Urinary uric acid excretions as well as HPRT activities were within normal limits. The mother was healthy with a normal serum uric acid; her HPRT activity was normal in lysed and intact erythrocytes but reduced in fibroblasts. The maternal grandfather experienced his first attack of gouty arthritis and kidney stones at the age of 24 years followed by development of gouty nephropathy. He has been treated by dialysis for the past 11 years. Since 1979 he is known to have partial HPRT deficiency and he has been already reported [5]. Unfortunately this finding was unknown to the authors when the propositus presented.

Discussion

Renal failure in the newborn is usually due to factors related to perinatal stress. In the case presented here it was first attributed to dehydration and to a toxic reaction to aminoglycoside. Serum uric acid was high but could be interpreted as a consequence of severe renal failure. However, hyperuricaemia did not return to normal in spite of partial recovery of kidney function and urate excretion was high. Primary hyperuricaemia as a cause of renal failure is exceptional in infancy. Only three such cases with proven HPRT deficiency have been reported, one with Lesch Nyhan syndrome [6] and two with partial deficiency of the enzyme [3,7]. In all the three cases dehydration and acidosis appear to have contributed to the development of renal failure as in our patient.

In the propositus HPRT activity in lysed erythrocytes was initially raised probably because of the recent blood transfusion. But five months later repeated analyses of lysed erythrocytes and fibroblasts demonstrated an enzyme activity in the same low range as found in the Lesch Nyhan syndrome, a progressive neurological disorder with hyperuricaemia and related complications [8]. However neurological manifestations were never observed in our patient and psychomotor development was normal. We therefore assume that HPRT in our patient is an unstable mutant enzyme with higher activity in vivo than in the lysed cells, which prevents an expression of the disease outside the kidney. This view is supported by the finding that HPRT activity in intact erythrocytes of the propositus was much higher than in lysed cells.

Partial HPRT deficiency is coded by mutational changes of the structural gene, which is located on the long arm of the X-chromosome. Hemizygotes of a given family have the same variant enzyme presenting with defined characteristics regarding activity, kinetic properties and thermolability [8]. Unexpectedly, HPRT activity of lysed red cells was much lower in our patient compared to that estimated in the grandfather (0.5% vs 24% of normal). However, enzyme activity in fibroblasts as well as metabolic properties and thermolability were similar in both [4]. The higher enzyme activity in the grandfather was not due to transfusions. This suggests that the two relatives bear in fact the same mutant enzyme and that this is stabilised by factors probably related to terminal uraemia, which preserves a higher activity in the in vitro tests.

The prognosis of renal function in partial HPRT deficiency is still debated. Since about 25 per cent of older patients with this metabolic defect do not develop nephrolithiasis and 50 per cent of them have normal renal function, the disorder is not necessarily progressive [8]. This suggests that complications like those seen in the propositus may be avoided by very early diagnosis and therapy. Following institution of adequate treatment renal function recovered gradually in our patient. Similar post treatment improvement in the three infants who all had severe kidney failure due to partial or complete deficiency of HPRT has been reported [3,6,7]. However, a complete recovery appears to be doubtful in view of the experimental findings in pigs which showed scars and chronic inflammation in the kidney as long as one year after a brief period of intratubular crystal deposition [9].

References

1 Stapleton FB. *J Pediatr 1983; 103:* 290
2 Manzke H, Eigster G, Harms D et al. *Eur J Pediat 1977; 126:* 29
3 Kelley WN, Greene ML, Rosenbloom FM et al. *Ann Int Med 1969; 70:* 155
4 Löffler W, Wingen A-M, Lemmen C et al. *Verh Dtsch Ges inn Med 1984; 90:* In press
5 Gröbner W, Ritz E, Zöllner N. *Verh Dtsch Ges inn Med 1981; 87:* 1001
6 Dillon MJ, Simmonds HA, Barratt TM et al. *Adv Exp Med Biol 1984; 165A:* 1
7 Lorentz WB, Burton BK, Trillo A et al. *J Pediatr 1984; 104:* 94
8 Wyngaarden JB, Kelley WN. In Stanbury JB, Wyngaarden JB, Fredrickson DS et al, eds. *The Metabolic Basis of Inherited Disease.* New York: McGraw-Hill Book Company. 1983: 1043
9 Farebrother DA, Hatfield P, Simmonds HA et al. *Clin Nephrol 1975; 4:* 243
10 Schwartz GJ, Feld LG, Langford DJ. *J Pediatr 1984; 104:* 849

Proc EDTA–ERA (1984) Vol 21

VITAMIN D METABOLITES AND OSTEOMALACIA IN THE HUMAN FANCONI SYNDROME

G Colussi, Maria Elisabetta De Ferrari, *M Surian, *F Malberti, G Rombolà, G Pontoriero, †G Galvanini, L Minetti

*E O Cà-Granda Niguarda, Milan, *Ospedale Maggiore, Lodi, †III Clinica Medica, University of Padova at Verona, Italy*

Summary

Experimental evidence suggests that renal 1α-hydroxylase activity is impaired in Fanconi syndrome. We have evaluated plasma vitamin D metabolites in five patients with Fanconi syndrome, three of whom had metabolic bone disease; plasma $1,25(OH)_2D_3$ was low in the three patients with bone disease, and normal in the two patients without a bone mineralisation defect.

The data supports the hypothesis that renal 1α-hydroxylase activity may be impaired in human Fanconi syndrome, and that altered vitamin D metabolism may contribute to the pathogenesis of metabolic bone disease in Fanconi syndrome.

Introduction

Fanconi syndrome (FS) is a complex disorder of proximal tubule transport processes, characterised by impaired reabsorption of various small solutes, such as glucose, aminoacids, phosphate and, frequently, bicarbonate, uric acid and low-molecular weight proteins [1,2]. A metabolic bone disease, either rickets in children or osteomalacia in adults, is usually, though not invariably a most prominent clinical feature of FS [2,3]. The pathogenesis of rickets/osteomalacia (R/OM) in FS has not been fully elucidated: both hypophosphataemia [4] and acidosis [5] can be associated with osteomalacic bone lesions. However, in FS R/OM may occur in the presence of normal plasma phosphate and bicarbonate, and can be absent despite severe hypophosphataemia [2]; in addition R/OM often reponds to treatment with pharmacological doses of vitamin D or 'replacement' doses of 1α-derivatives of the vitamin, without any measurable changes in plasma phosphate [2]. These observations suggest that hypophosphataemia and acidosis, though possibly important contributory factors, may not play a primary role in the pathogenesis of R/OM in FS.

Further evidence for a possible pathogenetic role of altered vitamin D metabolism in the genesis of R/OM of FS comes from the observation that renal bioconversion of 25(OH) vitamin D ($25(OH)D_3$) to $1,25(OH)_2D_3$ is reduced both in non-azotaemic children with FS [6] and rats with maleic acid-induced FS [7]. Studies in vitro show that the activity of the 1α-hydroxylase enzyme in renal homogenates or isolated tubules of rats with the maleic acid model of FS is markedly impaired [7].

Measurements of plasma vitamin D metabolites in human FS have not been widely reported. Thus, we have evaluated plasma values of $25(OH)D_3$, $24,25(OH)_2D_3$ and $1,25(OH)_2D_3$ in five patients with FS, of whom three had radiographic and/or histologic evidence of R/OM.

Case reports

Case 1 (PL) was a 47-year old female with Sjögren's disease, who was first seen in 1977 because of mild renal insufficiency (renal biopsy showed chronic interstitial nephritis), FS and severe osteomalacia. The patient, who had both 'pseudofractures' and proximal myopathy, was treated with alkali, oral calcium and vitamin D (100,000U/day for a year) with marked improvement of her clinical status and radiographic healing of osteomalacia. At the time of the present investigation, she had stopped vitamin D therapy three years before, was on oral calcium and alkali therapy, and complained of vague bone pain.

Case 2 (GA) was a 46-year old man with 'idiopathic' FS and chronic renal failure. He had a history of rickets from his childhood and had radiographic evidence of osteomalacia ('pseudofractures'). He had long been treated with oral calcium, calcitonin and, recently (last six months), bicarbonate but never with vitamin D. Renal biopsy was not performed, but pertinent clinical and laboratory investigations were negative for cytinosis, Lowe syndrome, Wilson's disease, glycogenosis, galactosaemia, hereditary fructose intolerance, immunological diseases, myeloma and heavy metal poisoning.

Case 3 (SL) was a 13-month old child with 'idiopathic' FS who was first seen because of marked dehydration, electrolyte imbalance and severe rickets. He had been treated with fluids, bicarbonate and potassium, but not vitamin D. Clinical investigations were negative for cystinosis, Lowe syndrome, galactosaemia, hereditary fructose intolerance and Wilson's disease.

Case 4 (NO) was a 67-year old female with the recent onset of mild renal insufficiency (renal biopsy showed chronic interstitial nephritis with diffuse interstitial infiltration of lymphomonocytic cells, normal glomeruli and negative immunofluorescence) and FS. She was asymptomatic, and no clear-cut cause for her kidney disease was apparent, except for previous administration of various non-steroidal anti-inflammatory drugs.

Case 5 (CE) was a 55-year old female who had IgA myeloma and FS, which were discovered during a clinical investigation for chronic renal failure.

Methods

Transiliac bone biopsy was performed in cases 1, 2, 4 and 5 with a Bordier's trephine under local anaesthesia, and analysed for qualitative histology (cases 4 and 5) or histomorphometry (cases 1 and 2) and double tetracycline labelling

TABLE I. Clinical and laboratory findings (mean ± SD when more than two data were available) in the five patients with Fanconi syndrome

	Controls (ranges)	Cases 1	2	3	4	5
GFR ml/min/1.72m²	(>80)	65	14	105	50	20
Glucosuria g/day	(absent)	9.5	1.6	6	2.4	3
Aminoaciduria g/day	(0.4–0.8)	7	2.3	1.9	3.36	5.94
TmP/GFR mg/100ml GFR	(2.7–5)	1.0±0.3	1.0±0.2	0.5±0.1*	2.72±0.5	1.9±0.8
$TmHCO_3^-$/GFR mEq/L GFR	(23.5–31)	21	16	21	23.8	21
FE HCO_3^- %**	(0–5)	11.3±0.1	25.8±7.3	29.4±10.6	4.4 ±2.0	5.4±3.2
FE urate %	(1–8)	41.0±9.0	44.6±4.0	61.0	42.3±3.8	29
HpH†	(<5.3)	6.7	6.4	–	5.2	5.7
PCa mg/100ml	(8.6–10.2)	9.0±0.17	7.9–9.2	9.5	9.5±0.3	9.0±0.58
PCa^{++} mEq/L	(2.1–2.55)	2.17±0.01	2.06–2.1	–	2.36±0.21	2.1
Pp mg/100ml	(2.7–4.5)	2.0±0.3	3.3±0.3	2.75±0.3*	3.66±0.6	4.1±0.4
$BHCO_3^-$ mEq/L	(22–29)	18 ±1.2	15.5±0.9	17	20.6±0.5	16.7±0.5
Purate mg/100ml	(2.5–6.5)	1.5±0.5	4.6±0.6	1.8±0.2	2.48±0.8	4.9±1.1
PTH mU/ml	(1–4.5)	4.8±2.6	8.9	1.0	3.2±0.4	22.8
cAMP nmol/100ml GFR	(1.1–6)	2.2±1.3	1.18–1.3	6.5–6.2	2.7±0.75	8.1±3.5
$25(OH)D_3$ ng/ml††	(3–30)	12	15	–	58	21.6
		12	–	53.9	–	–
$24,25(OH)_2D_3$ ng/ml††	(0.3–3)	0.22	0.27	–	0.52	0.2
		0.17	–	1.51	–	–
$1,25(OH)_2D_3$ pg/ml††	(40–80)	14	18	–	48.8	50
		19	–	30.1	–	–
R/OM		+	+	+	–	–

* Normal values in children: 4–6mg/100ml

** With $BHCO_3^-$ within 22–29mEq/L

† Lowest value during spontaneous or induced (case 4) acidosis

†† Upper value: without alkali therapy; lower value: during alkali therapy

(cases 1, 2 4). All clinical investigations, including bone biopsy and laboratory determinations, were performed at least two weeks after withdrawal of alkali therapy (cases 1 and 2), except in case 4. In case 1 vitamin D metabolites were measured both during and after the withdrawal of alkali therapy. Only case 1 had been treated (three years before) with vitamin D. Routine biochemical evaluations on plasma and urine were performed by standard methods: arterial blood and urine pH and PCO_2 for HCO_3^- calculations (pK in blood=6.1 and in urine=$6.3-\sqrt{Na^+ + K^+/2}$) were measured with a IL 1302 gas analyser; plasma PTH with a carboxyterminal antibody (PTH, Sorin); cAMP in urine with Amersham kit; vitamin D metabolites as published elsewhere [8].

Results

The diagnosis of FS consisted of the presence of (Table I): normoglycaemic glucosuria, generalised aminoaciduria and type II RTA (of mild degree in cases 4 and 5) [3]. Tubular reabsorption of phosphate and urate were also impaired in all the patients, but cases 2 and 5 had severe renal failure. Cases 1, 2 and 5 had also a type I RTA, in case 4 distal tubular function was not investigated. GFR (creatinine clearance) was normal in case 3, slightly impaired in cases 1 and 4; cases 2 and 5 had severe renal failure. All the patients were acidotic, but only cases 1 and 3 had persistently low plasma phosphate. Plasma calcium was slightly reduced in case 2, and parathyroid activity consistently increased only in case 5. All the patients were vitamin D replete (normal $25(OH)D_3$ values); $1,25(OH)_2D_3$ was below the normal range in cases 1, 2 and 3, who had R/OM, normal in cases 4 and 5; $24,25(OH)_2D_3$ was slightly reduced in cases 1, 2 and 5.

Bone biopsy showed osteomalacia in cases 1 and 2 (osteoid volume 50.8 and 53.3% of trabecular bone volume, respectively, n.v.:0.1–3%; osteoid surface (inactive) 98.1 and 92.7% of trabecular bone surface, n.v.:2–14%; resorption surface 0 and 1.1% of trabecular bone surface. n.v.:0.5–5%). In cases 4 and 5 qualitative evaluation showed thin trabeculae with sparse osteoclasts but no excess osteoid in case 4 and, in case 5, increased osteoclastic resorption, mild increase in 'active' osteoid (covered with cuboidal osteoblasts) and mild peritrabecular fibrosis, consistent with secondary hyperparathyroidism. Tetracycline labelling showed diffuse wide double bands (0.7μ/day) in case 4, but only scant and irregular deposition of a single label in cases 1 and 2. Case 3 had radiographic and clinical rickets.

Discussion

Our data show low values of $1,25(OH)_2D_3$ in three of five patients with FS, in the absence of vitamin D deficiency. In case 2 a low $1,25(OH)_2D_3$ might have resulted from marked impairment of renal function, though case 5, with a similar degree of renal failure, had normal $1,25(OH)_2D_3$; cases 1 and 4 had normal, or nearly so, glomerular filtration rate (GFR). Low $1,25(OH)_2D_3$ have also been reported in two other patients with FS. Since in our patients increased catabolic rate of the metabolite appears to be unlikely, decreased renal production is a consistent explanation. No clear-cut cause for a 'functional' inhibition of the 1α-hydroxylase was apparent: plasma phosphate was low or normal, and parathyroid activity normal in all. Alternatively, and in keeping with the experimental evidence [7], renal 1α-hydroxylase activity might be impaired, as other tubular functions, in FS. Renal 1α-hydroxylase is a mitochondrial enzyme, most likely situated in the proximal tubule cells [7]; mitochondrial activity and energy production is impaired in experimental models of FS, and many disease entities associated with the FS in humans are associated with extensive intracellular deposition of abnormal substances (aminoacids, sugars, low molecular weight proteins, toxins, etc) resulting in structural disruption of the endoplasmic reticulum and mitochondrial network [1]. Thus, altered 1α-hydroxylase activity in FS might be but an aspect of a more generalised disturbance in mitochondrial function. Alternatively, abnormal permeability of cell membrane(s) to phosphate or $25(OH)D_3$ could induce a 'functional' inhibition of enzyme.

Independently of the mechanism(s), low values of $1{,}25(OH)_2D_3$ were associated, in our patients, with the presence of R/OM, which was lacking in the two patients with normal $1{,}25(OH)_2D_3$. In addition, previous alkali therapy alone had been ineffective in cases 1 and 2, osteomalacia was associated with normal phosphate. Thus, impaired synthesis of $1{,}25(OH)_2D_3$ and/or possibly of other $25(OH)D_3$ metabolites, when present, might contribute to the pathogenesis of R/OM in FS, by a direct action on bone and/or some indirect effect (e.g. impaired calcium absorption). Indeed, intestinal calcium absorption is low in FS [3,9] and calcium balance negative [3]. However, an intriguing observation is that patients with FS show normal plasma calcium and parathyroid activity [2] despite low $1{,}25(OH)_2D_3$ and low intestinal calcium absorption [3,9], and they often have hypercalciuria [3]. Thus, an alternative view is that, in FS, metabolic bone disease is associated with increased calcium efflux (relative to calcium entry) from bone (e.g. secondary to hypophosphataemia or altered osteoblast-osteocyte syncytial system) [3], resulting in normocalcaemia despite decreased intestinal calcium absorption, and, possibly, an inhibtion of the PTH-vitamin D axis.

References

1 De Fronzo RA, Thiers SO. In Brenner BM, Rector FC, eds. *The Kidney*. Philadelphia: Saunders. 1981: 1816
2 Morris RC, Sebastian A. In Maxwell M, Kleeman CR, eds. *Clinical Disorders of Fluid, Acid-base and Electrolyte Metabolism.* New York: McGraw Hill. 1980: 883
3 Lee DBN, Drinkard JP, Rosen VJ, Gonik HC. *Medicine 1972; XX:* 107
4 Lotz Z, Zisman E, Bartter FC. *N Engl J Med 1968; 278:* 409
5 Cunningham J, Fraher LJ, Clemens TL et al. *Am J Med 1982; 73:* 199
6 Brewer ED, Tsai HC, Morris RC. *Clin Res 1976; 24:* 154(A)
7 Brewer ED, Tsai HC, Szeto KS, Morris RC. *Kidney Int 1977; 12:* 244
8 Tartarotti A, Adami S, Galvanini G et al. *J Endocrinol Invest 1982; 5 (Suppl 1):* 98
9 Kitagawa T, Akatsmuta A, Owada M, Mano T. *Contr to Nephrol 1980; 22:* 107

Proc EDTA–ERA (1984) Vol 21

EFFECTS OF ACUTE AND CHRONIC RANITIDINE ADMINISTRATION ON RENAL FUNCTION AND PARATHYROID ACTIVITY IN CHRONIC RENAL FAILURE

N Pozet, M Labeeuw, A Hadj Aissa, C Bizolon, P Zech, J Traeger

Hopital E Herriot, Lyon, France

Summary

In 11 patients with advanced renal failure, chronic treatment with ranitidine decreased plasma immunoreactive parathormone (PTH) without affecting phosphate reabsorption or urinary excretion of cyclic AMP. No significant changes in glomerular filtration rate and in urinary excretion of electrolytes were evident.

Introduction

Secondary hyperparathyroidism is a frequent complication of advanced renal failure. If calcium is the major factor involved in the regulation of parathyroid secretion, the influence of other factors such as histamine has been established in vitro. Controversy exists as to whether cimetidine reduces PTH secretion [1]. This study addresses the question of the influence of acute and chronic administration of ranitidine, a new H_2 antagonist with few side effects [2], on renal function and PTH release in advanced renal failure.

Methods

Eleven stable patients (mean age: 53 ± 11, BW = 62.2 ± 10.5) were given ranitidine, 150mg twice a day for 28 days, without other changes in treatment. Renal failure (creatinine clearance: C Cr = 0.45 ± 0.29ml/sec) was due to chronic glomerulonephritis (n = 3) interstitial nephritis (n = 2) or polycystic disease (n = 6). The patients were studied before (study A) and on the last day of treatment (study B). After a control period, ranitidine 150mg was given orally with 300ml of water, and urine was further collected during period I (0 to 90 min) and II (90 to 180 min). Blood was drawn at 0 and 180 min. The following parameters were measured: urinary excretion of Na, K, Ca, P, uric acid, urea, creatinine, cyclic AMP (normal values: 2.95 ± 0.48mM%ml GFR), osmolality; blood electrolytes, creatinine, osmolality, ranitidine (HPLC) and immunoreactive PTH (iPTH, using a 'C terminal' antibody and MRC standard, CEA – IRE kit, normal range: 450–1450ng/L). Blood calcium was corrected for protein content according to Parfitt and the renal threshold concentration for phosphate (TmP/GFR) calculated from Walton's nomogram. The influence of treatment was assessed by comparing (non parametric tests on paired values) identical periods in study A and B (chronic administration) or the two consecutive periods on the same study (acute effects).

Results

Chronic administration

After 28 days of treatment, no significant changes were observed for weight, mean arterial pressure, blood creatinine, urea, electrolytes, for the values of creatinine, uric acid and osmolar clearances and urinary excretion rates of electrolytes. Blood iPTH was significantly decreased ($p<0.05$) while urinary cAMP and TmP/GFR showed no significant changes (Table I).

TABLE I. Effects of chronic treatment with ranitidine

		Before	After
Plasma			
Creatinine	mmol/L	0.456±0.280	0.483±0.340
Urea	mmol/L	22.5 ±12.7	22.8 ±13.8
Calcium	mmol/L	2.32 ±0.12	2.30 ±0.14
Phosphorus	mmol/L	1.89 ±0.97	1.84 ±0.91
iPTH	ng/L	2759 ±1798	2098 ±1183*
Ranitidine	microg/L	0	330 ±221**
Creatinine clearance	ml/sec	0.45 ±0.29	0.44 ±0.29
U cAMP	mmol % ml GFR	13.68 ±8.15	11.85 ±6.60
$TmPO_4$/GFR	mmol/L	0.866±0.212	0.758±0.147

(*: $p<0.05$; **: $p<0.01$)

Acute studies

In study A, C Cr remained stable in period I and decreased in period II (0.45 ± 0.31 and 0.39 ± 0.28ml/sec). The urinary excretion of all electrolytes (either expressed as excretion rates or as the concentration ratio to creatinine) showed no consistant changes. Blood levels of iPTH did not vary (2759 ± 1798 and 2758 ± 1705ng/L. Urinary cAMP and TmP/GFR remained stable. Similar findings were obtained in study B with a 16 per cent decrease in C Cr and no significant variations in other parameters.

Plasma ranitidine values were lower in study A (0 at 0 min and 895 ± 293 μg/ml at 180 min) than in study B (330 ± 221 and 1082 ± 400μg/ml), ($p<0.01$).

Discussion

In this study, chronic treatment with ranitidine decreased plasma iPTH without affecting the commonly used indexes of PTH action on the kidney. Interference of ranitidine with the assay can be excluded from the unchanged values of PTH at 180 min when plasma ranitidine is markedly elevated. Such a discrepancy is common [3] and relates to the well known problems of PTH measurements in uraemia [4]. The 'C terminal' immunoreactivity is a better index of the parathyroid activity than a single determination of the intact hormone even given the prolonged half life of 'C fragments' in renal failure. Assaying samples from patients given cimetidine with two different assays, Jacobs et al found no decrease with an assay measuring 'intact hormone' and a significant decrease in 'C terminal immunoradioactivity' [5]. An increase in the clearance of 'C fragments' or a decreased generation of these fragments from intact PTH is conceivable [6] but remains speculative. On the other hand, the reliability of urinary cAMP and phosphate transport in advanced uraemia as index of parathyroid activity is questionable: plasma cAMP is elevated in this setting and urinary cAMP (even expressed as % ml GFR) does not reflect perfectly nephrogenic cAMP [7]: several factors other than PTH can influence phosphate transport in uraemia. However the issue is not yet settled since Cunnighan et al, using three different assays found no decrease in iPTH in dialysed patients treated with ranitidine [1]. The absence of acute effect on PTH in our patients is in contrast with results obtained in subjects with normal renal function but could be explained either by the prolonged half life of C terminal fragments in renal failure or by the use of different routes of administration [8]. As already reported no dramatic changes occurred in plasma calcium or phosphate, data consistent with a direct effect of the H_2 antagonists on parathyroid cells [1].

The assessment of drug action on kidney function is mandatory in renal failure. Cimetidine reduces creatinine clearance, without affecting glomerular filtration rate, an effect due to inhibition of its tubular secretion [9]. Ranitidine has no such effect [2]. Our results confirm the lack of effect of a chronic treatment on renal function. Therefore, the slight decrease in creatinine clearance observed in period II in studies A and B may be non specific. An alternative explanation could be that a decrease in creatinine clearance can be observed only for the high values of plasma ranitidine obtained at 180 min. We were however unable to find any correlation between the plasma ranitidine at 180 min and the decrease in creatinine clearance between periods I and II. There might be however a time dependent effect with an early fall, followed by complete recovery, as reported with cimetidine.

High plasma values of ranitidine were obtained in this study without side effects. Ranitidine accumulates in renal failure [10] and the plasma 'trough' values obtained were above therapeutic levels. Further trials using lower doses, and including other parameters of parathyroid activity such as bone resorption are therefore needed.

References

1 Cunninghan J, Segre GK, Slatopolsky E et al. *Nephron 1984; 38:* 17
2 Brogden RN, Carmine AA, Heel RC et al. *Drugs 1982; 24:* 267

2 Brogden RN, Carmine AA, Heel RC et al. *Drugs 1982; 24:* 267

3 Fasset RG, Lai KN, Crocker JM et al. *Dialysis Transplant 1984; 4:* 2000
4 Arnaud CD, Goldsmith RS, Bordier PJ et al. *Am J Med 1974; 56:* 785
5 Jacobs AI, Bourgoignie JF. *N Engl J Med 1980; 303:* 396
6 Fiore CE, Malatino LS, Kanis JA. *Lancet 1981; i:* 501
7 Broadus AE, Malaffey JE, Bartter FC et al. *J Clin Invest 1977; 60:* 771
8 Marchesi H, Perrazonni M, Palumeri E et al. *Clin Trial J 1983; 20:* 275
9 Dutt MK, Moody P, Northfield TC. *Br J Clin Pharmacol 1981; 12:* 47
10 Zech PY, Chau N, Pozet N et al. *Clin Pharmacol Ther 1983; 14:* 667

Proc EDTA–ERA (1984) Vol 21

CIMETIDINE THE H_2-RECEPTOR BLOCKER: EFFECT ON THE RENAL AUTOREGULATION IN DOGS

M A Sobh

Urology and Nephrology Centre, Mansoura University, Mansoura, Egypt

Summary

Two groups of female mongrel dogs were examined. The renal perfusion pressure (RPP) was decreased to 100 then to 85 and to 75mmHg. The following parameters were measured: renal blood flow (RBF), glomerular filtration rate (GFR), ultraviolet (UV), $U_{Na}V$ and $U_{PO_4}V$.

In the first group the autoregulation was found to be intact as the RBF and the GFR were kept constant in the face of a decreasing RPP. In the second group cimetidine infusion 10^{-6}M/min caused 11 per cent reduction in the RBF. Reduction of the RPP to 75mmHg caused a significant decrease in the RBF by 30 per cent and the UV by 50 per cent, the GFR was stable. This work demonstrates the depressive effect of cimetidine on the total RBF and its autoregulation.

Introduction

Some cases of acute renal failure have been reported to occur following cimetidine therapy [1–4]. Histamine (H_1, H_2) receptors are reported to be present in the canine kidney vasculature. H_1 receptors are primarily post-glomerular while H_2 receptors exhibit both pre- and post-glomerular distribution. Stimulation of these receptors will increase the RBF and urine volume [5]. This work was designed to test the effect of cimetidine infusion on the renal haemodynamics and its autoregulation.

Materials and methods

Two groups of female mongrel dogs, five dogs each of 15kg average body weight were examined. Each dog was fasted 24 hours prior to the experiment but allowed to drink water. Anaesthesia was with nembutal and respiration was controlled automatically. Through a femoral vein catheter saline containing the markers PAH and inulin was infused at a rate of 3ml/min. Through a femoral artery catheter connected to a physiogram, the central arterial blood pressure was recorded. A laparotomy incision was made and through ureteric catheters the urine from each side was collected. A nylon tape around the aorta just above the origin of the renal arteries was used to decrease gradually the RPP to 100, then to 85 and to 75mmHg. For each RPP three clearance periods of 20 minutes each were measured, and the following parameters were recorded for each side: RBF (PAH clearance), GFR (inulin clearance), UV, $U_{Na}V$ and $U_{PO_4}V$. In the first group only isotonic saline containing the markers was given. In the second group cimetidine was infused at a rate of 10^{-6}M/min. Statistical analysis was by the Student test.

Results

Tables I and II show the haemodynamic response to reduction in the RPP from the basal level to a value of 100mmHg then 85 and 75mmHg respectively. Values in each table represent the mean value of the five experiments.

TABLE I

		UV		PAH CL		INULIN CL		$U_{Na}V$		$U_{PO_4}V$	
		Lt.	Rt.	Lt.	Rt.	Lt.	Rt.	Lt.	Rt.	Lt.	Rt.
		ml/min		ml/min		ml/min		μEq/min		μg/min	
Basal	1	0.37	0.37	78	81	28	36	68	68	225	260
	2	0.38	0.35	74	75	29	32	61	65	191	196
	3	0.55	0.50	72	83	28	30	88	79	195	205
	T	0.84		154.7		61		143		424.1	
Clamp	1	0.60	0.57	83	77	30	30	100	95	223	236
	2	0.70	0.70	75	75	27	29	95	98	263	264
	3	0.63	0.67	72	79	27	31	98	99	287	273
	T	1.29		153.8		57.9		195		515.3	
Rec.	1	0.80	0.78	75	70	32	27	124	107	474	432
	2	0.73	0.70	70	65	34	29	122	105	495	440
	3	0.80	0.78	76	75	34	29	126	122	532	477
	T	1.53		144		61.5		236		959.8	

Rec.= Recuperation
T= total value of both kidneys

In the control group, as shown in Table I, the autoregulation was intact as the RBF and GFR were kept constant in the face of a decreasing RPP.

In the second group, as shown in Table II, cimetidine infusion at a rate of 10^{-6}M/min caused a significant ($p<0.05$) reduction in the RBF. The total RBF during the basal period was 189.7ml/min; after cimetidine infusion there

TABLE II

		UV		PAH CL		INULIN CL		$U_{Na}V$		$U_{PO_4}V$	
		Lt.	Rt.	Lt.	Rt.	Lt.	Rt.	Lt.	Rt.	Lt.	Rt.
		ml/min		ml/min		ml/min		μEq/min		μg/min	
Basal	1	1.15	1.05	88	85	30	22	166	138	151	108
	2	1.6	1.30	90	86	30	29	223	179	137	125
	3	1.92	1.62	105	108	33	30	276	228	198	157
	T	2.88		187		58		403.6		291.5	
Cim.	1	2.79	2.42	84	92	33	30	332	51	247	243
	2	2.05	2.30	81	88	29	29	286	327	246	242
	3	2.11	2.30	84	74	30	30	305	301	255	272
	T	4.65		165.3		60		601.1		501.6	
Clamp	1	1.60	1.89	66	68	29	25	227	247	242	250
	2	1.49	1.51	57	62	30	24	189	200	278	270
	3	0.9	0.97	66	71	30	28	124	123	238	320
	T	2.79		136		55		369.9		530.2	

Cim.=cimetidine

was an 11 per cent reduction with total value of 165.3ml/min. Reduction of the RPP to 75mmHg caused a statistically significant ($p<0.01$) reduction in the RBF and UV by 30 and 50 per cent respectively. The GFR was more or less stable.

Discussion

It has been reported that 60 per cent of patients taking cimetidine may have changes in serum creatinine, and this has been thought to be related to an alteration in the manner by which the kidney handles creatinine [6,7]. Recently there have been increasing reports of cases of acute renal failure following cimetidine therapy [1–4]. Only a few patients had been biopsied and they show the picture of acute interstitial nephritis [2,6] which can explain the occurrence of acute renal failure. Yet this can not be the only mechanism of acute renal failure in all the reported cases, since many of them had showed no change in albumin or in β_2-microglobulin excretion [7,8].

Since the H_1 and H_2 receptors have been reported to exist in the renal vasculature [5,9], this work was designed to test for the haemodynamic effects of the H_2 receptor blocking by cimetidine infusion. It is evident in the first (control) group that the autoregulation mechanism is intact in our experimental model. In the second group the infusion of cimetidine resulted in an 11 per cent reduction in the RBF, similar to previous reports [7,9]. Furthermore, following reduction in the RPP to 75mmHg in the presence of cimetidine the kidney failed to maintain a stable RBF and there was a failure of the autoregulation mechanism. This shows the importance of the H_2 receptors not only in maintaining RBF in normal conditions but aiso during stress. This can explain the occurrence of acute renal failure reported in some cases.

References

1 Seidelin R. *Postgrad Med J 1980; 56:* 440
2 Richman AV, Narayan JL, Hirschfield JS. *Am J Med 1981; 70:* 1272
3 Santoro JJ. *Br Med J 1982; 285:* 975
4 Rowley Jones D, Flind AC. *Br Med J 1982; 285:* 1422
5 Robert O, Banks JD, Fondacaro MM et al. *Am J Physiol 1978; 235:* F570
6 Pitone JM, Biondi RJ, Chiesa JC et al. *Am J Gastroenterol 1982; 77:* 169
7 Dutt MK, Moody P, Northfield TC. *Br J Clin Pharmacol 1981; 12:* 47
8 Christensen CK, Mosensen CE, Hanberg Sorensen F. *Scand J Gastroenterol 1981; 16:* 129
9 Robie NW, Barker LA. *J Pharmacol Exp Ther 1983; 226:* 712

PART XX

GUEST LECTURES ON NEPHROLOGY II

Chairmen: G Lubec
G La Greca

J Botella
F Pecchini

NON-INVASIVE ASSESSMENT OF THE KIDNEY

What do nuclear medicine, ultrasonography, digital vascular imaging, computed tomography and magnetic resonance contribute to diagnosis in nephrology?

C van Ypersele de Strihou

University of Louvain Medical School, Cliniques Universitaires St-Luc, Brussels, Belgium

Introduction

The history of renal disease started 150 years ago with the least invasive of all kidney tests: urinalysis. The assessment of renal function became more elaborate with the measurement of blood urea. The definition of the clearance concept and the subsequent equation between some functions of the kidney and the clearance of definite solutes such as inulin and PAH led to a more invasive approach which remains the reference for all methods of renal function determinations [1]. Inulin and PAH clearances are, however, cumbersome and entail a definite discomfort and morbidity for the patient.

The development of renal function tests was paralleled by an improvement in the methods evaluating renal morphology. Retrograde pyelography was followed by the less invasive intravenous pyelography and its refinements. Direct vascular visualisation through aortography provided new insights in renal disease. Renal biopsy eventually gave the microscopic counterpart of the renal imaging techniques.

As a result, the nephrologist who used to be a benign looking practitioner contemplating a vessel full of urine became a potentially aggressive scientist endowed with an increasingly effective investigative and therapeutic armamentarium and its frightful counterpart, the ability to inflict significant iatrogenic morbidity and mortality.

Several progresses in technology have made it possible during the last 10 years to substitute non- or less invasive methods for the evaluation of renal function and morphology. In this review I will not try to list them all: I will rather put some of them in perspective, looking critically at their underlying assumptions, at their relation with standard techniques and their usefulness in the daily practice of renal physicians.

The newer non- or less invasive methods of kidney evaluation can be categorised into a few chapters: I will concentrate mainly on nuclear medicine, its assets, its pitfalls and sometimes its illusions; I will subsequently outline what may be expected today from ultrasonography, digital vascular imaging, computerised tomography and magnetic resonance.

Nuclear medicine

As soon as it became possible to attach a radiolabel to inulin or PAH, many laboratories substituted beta or gamma counting techniques for the more traditional but cumbersome chemical methods. Concern for the tightness of the radioactive label attachment to the molecule and for the degree of radiation exposure led promptly to the investigation of various radiolabels and to the development of new compounds to replace inulin and PAH. It must be recognised, however, that the fate of several of these substances differs markedly from the original ones or in some cases is even poorly defined at the present time.

The development of elaborate gamma cameras supplemented with a computer programmed handling of the data has led to the substitution of in vitro measurement of the radiolabelled compounds by in vivo determination of their accumulation rate in different organs. Images were thus obtained together with functional data. Here again new problems arose: differentiation of the compound accumulated in the organ from that diffusely present in the background, absorption of the emitted radiation by tissue layers of variable thickness interposed between the scanned organ and the camera.

The variety of substances with different radioactive labels, the diversity of counting and imaging methods and the complexity of computer data handling have resulted in an array of non-invasive tests whose significance and validity is not always critically established.

To illustrate this observation, let us look at the methodology of renal plasma flow measurements.

Renal plasma flow measurements: two examples

PAH is almost completely extracted from the blood during its first pass through the kidney and eliminated in the urine [1]. Measurement of its concentration per millilitre plasma and of its urinary excretion allows the computation of the volume of plasma cleared by the kidney per unit time, i.e. renal plasma flow. Accuracy requires also the determination of the percentage extraction of PAH from arterial blood: under normal circumstances the extraction averages 90 per cent so that the clearance of PAH equals 90 per cent of the actual renal plasma flow. In diseased kidneys, however, the extraction ratio falls with an attendant decrease in PAH clearance independent of renal plasma flow.

The development of a single blood sample, single injection method of renal plasma flow measurement

An initial effort to fix a ^{131}I label on PAH in order to allow gamma counting proved unsuccessful. Tauxe et al proposed 20 years ago replacing PAH by a substance more easily labelled with ^{131}I: orthoiodohippurate, i.e. hippuran.

They demonstrated a good correlation between PAH and hippuran clearances but the latter was lower by an average of 28ml/min. This discrepancy increased for PAH clearance above 300ml/min reaching 15 per cent at 675ml/min. Still the authors concluded that "the deviation of PAH clearance is not of sufficiently great magnitude to alter clinical interpretation and appreciation of the data".

A further change in methodology was proposed in 1964 by Tauxe in order to avoid cumbersome urine collections [2]. Sapirstein et al [3] had reasoned 10 years earlier that the blood decay of intravenously injected creatinine was determined first by its equilibration through its diffusion space and second by its renal elimination. They proposed a mathematical treatment of the plasma disappearance curve based on a two compartment model, thus allowing the calculation of the blood clearance of creatinine, i.e. the glomerular filtration rate. Tauxe's group [4] applied these principles to hippuran in order to measure renal plasma flow. To be valid the method requires that the tested compound should have no protein binding and should not be metabolised or excreted by any non-renal route. These conditions are not completely met for hippuran which has weak protein binding and a small but significant extrarenal excretion [5]. To validate this approach, Wagoner et al [4] measured during 75 minutes the blood decay curve of ^{131}I hippuran injected in a single shot. Renal plasma flow was then calculated from the two exponential components of the disappearance curve. Correlation with traditional PAH clearance was again satisfactory but blood hippuran clearance was lower than standard PAH clearance with a tendency to deviate more as the clearance increased. Blaufox and Merrill [6] further simplified the method by drawing only two blood samples at 20 and 30 minutes. The mathematical computation relied only on the late exponential of the decay curve. Again a good correlation was found with standard PAH clearance but this time hippuran clearance exceeded PAH clearance by 10 per cent, the difference ranging from −11.7 per cent to +29.5 per cent. Interestingly, four patients with no renal function had a blood hippuran clearance ranging from 42 to 100ml/min.

The last step away from reference methods was proposed by Tauxe et al in 1971 [7]: renal plasma flow was derived from a single plasma sample drawn 44 minutes after the injection of hippuran. The complexity of the reasoning underlying this approach is worth mentioning. The data basis consists of blood hippuran clearances calculated from decay curves obtained in 116 patients. First Tauxe et al [7] calculate the distribution volume of hippuran at various sampling times as the total injected dose divided by the plasma concentration. Subsequently they compute, the relationship existing in their population between blood hippuran clearance and distribution volume at one given time. Whatever the chosen sampling time the relationship assumes a quadratic form. However, the lowest standard deviation around the regression is obtained for the 44 minute sample. The authors then propose this empirically derived equation to calculate effective renal plasma flow from a single blood sample obtained at 44 minutes. The influence of oedema or severe renal insufficiency on the significance of a single blood sample value is not considered. The drawback of such an empirical approach is illustrated by the fact that 11 years later the same group redefined the quadratic equation on the basis of a larger number of observations and found that the new equation modified significantly the calculated renal plasma flow above 400ml/min and below 150ml/min [8].

This story illustrates the limits of apparently sophisticated tests: the progressive erasement of significant differences between the new method and the more physiologically pertinent reference methodology, the empirical nature of nomograms from which physiologically relevant parameters are calculated, the lack of validation of the nomogram in other clinical conditions such as volume expanded states or severe renal failure. As may be expected the methodology is subsequently applied under different clinical conditions without reference to its possible limits.

For instance Holland et al [9] measured single blood sample renal plasma flow in 32 patients with acute renal failure in order to determine if this parameter predicts subsequent recovery. They claim that renal plasma flows above 125ml/min have a good prognostic value. Neither the possible influence of an expanded extracellular volume, characteristic of acute renal failure, on the 44′ concentration of the tracer, nor the fact that anephric patients may have a 51ml/min renal plasma flow with this method, are taken into consideration.

These comments are not meant to deny the possible value of single blood sample renal plasma flow determinations. They just underline how indirectly the method relates to physiologic phenomena and hence how fragile it may be. The clinician eager to obtain numbers and happy to reason in physiologic terms should be aware of these pitfalls.

Intrarenal blood flow distribution assessment

More can be said about the assumptions hidden in techniques designed by Britton's group to non-invasively assess cortical and medullary blood flows.

The basic assumption of this approach is that the transit time of hippuran is longer through juxtamedullary than through cortical nephrons. The amount of hippuran extracted from blood by each of these two nephron populations is determined by the volume of blood flowing to their proximal tubules. The distribution of blood within the renal cortex can thus be estimated by measuring the relative amount of hippuran with longer and shorter transit times [10,11]. Britton and Brown [10] indeed found that the spectrum of transit times was bimodal and suggested that the two modes represented transit times through cortical and juxtamedullary nephrons.

A computer assisted gamma camera records uptake and removal of hippuran by the kidneys. Background, measured in a non-renal area, is subtracted to provide real tissue accumulation. Subsequently a middle area (region of interest) representing middle regions and inner cortex overlapping the outer medulla is outlined by the observer on the image provided by the camera. Hippuran accumulation in this area is calculated by the computer. Transit times are then calculated for each area: the cortex provides mainly shorter transit times representative of cortical nephrons whereas the middle region contains both long and short transit times. Subtraction of cortical transit times from the middle region yields the residual spectrum of juxtamedullary transit times. The areas under each mode are determined and the results expressed as the contribution of the cortical nephron component as a percentage of the total area.

A number of uncertainties are present in the basic assumptions of this method [12]. They concern the extent to which the area under each transit mode is proportional to the amount of plasma flowing through the corresponding nephron population, whether or not such a relationship holds in situations affecting renal blood flow and the hypothesis that short and long transit times are matched with cortical and juxtamedullary nephrons. Furthermore, the reliability of the delineation of small regions of interest within the kidneys remains in doubt.

The method was validated in man [10] by the demonstration that during sodium restriction the proportion of hippuran flowing through the nephron with shorter transit times was more than twofold larger than that flowing with longer transit times. During a high sodium intake the proportion of short transit time hippuran falls with a concomitant rise in longer transit time hippuran. These changes are interpreted as an increased juxtamedullary blood flow in response to a high salt intake, a pattern documented by other methods in animal experiments.

A similar agreement is unfortunately lacking in other conditions. Aortic constriction of haemorrhagic hypotension are also known to result in an increased fraction of renal blood flow directed to the inner cortical nephrons. Still, the transit time method fails to detect a parallel increase in hippuran appearance in deep nephrons [10]. This finding illustrates the lack of reliability of this method when renal blood flow falls below 60 per cent [13].

This example illustrates how an uncritical use of computer generated numbers may lead to false interpretations, waste of resources and poor medicine. Still these non-invasive methodologies used as research tools have already led to interesting conclusions and some of them provide routinely critical information to the clinician. In the subsequent part of this presentation I will briefly review them.

Current methods

Renal function: tests without imaging

Glomerular filtration. Clinicians rely on creatinine clearance to assess glomerular filtration rate. They follow serum creatinine as an index of glomerular filtration knowing that daily creatinine excretion is more or less constant. Convenient and cheap, this method has important limitations [14,15]. To circumvent them it has been proposed to calculate glomerular filtration from the blood disappearance rate of an injected radiolabelled compound excreted solely or mainly by glomerular filtration.

Several compounds have been utilised, mainly ^{51}Cr EDTA, ^{131}I iodothalamate and ^{99m}Tc DTPA. The renal handling of the first two compounds has been well studied [15,16]. EDTA clearance is approximately 10 per cent lower than inulin clearance. There is an extrarenal clearance of approximately 4ml/min. The coefficient of variation is 4.1 per cent in patients with clearance values above 30ml/min, compared with 12 per cent reported for creatinine clearance [17]. Below 30ml/min, however, the coefficient of variation rises to 11.5 per cent [15]. Iodothalamate clearance is virtually equal to inulin clearance [18]. DTPA has a significant five per cent protein binding and its single injection clearance is eight per cent lower than iodothalamate [19]. The renal handling of DTPA under various degrees of renal impairment and its extrarenal pathways need further in depth evaluation [20]. These methods are valid only if glomerular filtration exceeds 50ml/min. Below this value results are less reliable and care should be taken to extend blood sampling beyond the usual four hours.

Although definitely more accurate than creatinine clearance, these methods are certainly more time consuming for the patient and more expensive without providing a wholly adequate measurement of glomerular filtration. In my opinion their usefulness is limited to research purposes as illustrated by two recent examples: Parving et al using blood EDTA clearance demonstrated elegantly in diabetic patients the benefit of blood pressure control on the decline of renal function [21]. Similarly Donadio et al [22] studying in a double blind randomised study the benefit of anti-aggregatory agents on the evolution of membranoproliferative glomerulonephritis concluded that the treatment had a favourable effect on the evolution of glomerular filtration measured by blood iodothalamate clearance. The evolution of the simultaneously determined serum creatinine and creatinine clearance confirmed this observation but, due to the scatter of the data, the difference between treated and control groups failed to reach significance.

Renal plasma flow. In the previous examples the methodology of renal plasma flow measurement has been extensively discussed. Suffice to say that the blood clearance of hippuran assessed by multiple or single blood samples relates to PAH clearance with several limitations, especially in severe renal failure and possibly in volume expanded states. Furthermore, extraction coefficient of hippuran may change in acute conditions and thus profoundly alter the physiologic significance of its blood clearance [23].

External monitoring by a gamma camera of renal uptake one and two minutes after the injection of hippuran may allow the calculation of separate and global renal blood flow. Results depend on the method used to measure the background emission to be subtracted from kidney uptake [24].

The usefulness of these methods in current clinical practice is not evident. It has been claimed to have a prognostic value in patients with acute renal failure [19]. Even if confirmed the value of this information in patient management remains an open question.

By contrast, as a tool designed to provide a specific answer in a research programme this method provides valuable information. Textor et al [25] for instance have utilised hippuran blood clearance to assess the effect of a beta blocking agent on effective renal blood flow. Cardiac output was simultaneously measured. Repeated studies over a four months' period demonstrated that contrary to other beta blocking agents, nadolol redistributed blood flow to the kidneys, the fraction of cardiac output reaching renal circulation rising from 17 to 22 per cent.

Static renal imaging

The avid uptake of a radiolabelled compound by the kidney may provide accurate images on the gamma camera. ^{99m}Tc DMSA has replaced $HgCl_2$ for such studies, the Tc label providing a smaller radiation hazard than ^{97}Hg, as a result of its short half-life and radiation characteristics (in renal radiopharmaceuticals – an update) [20]. DMSA is 90 per cent protein bound. It is readily taken up by the kidney mainly in the cortex, within proximal and distal tubular cells. Its urinary excretion is slight. DMSA clearance does not correspond to any single renal function parameter. Its fate may be significantly altered by changes in hydration and acid base balance [26].

Scintigraphy of the kidneys has been advocated in children below the age of five with urinary tract infection and/or vesico ureteral reflux [27]. Merrick et al have indeed demonstrated [28] that the detection of scars was marginally more frequent with DMSA scanning than with an intravenous pyelography. Taking into account the fact that an intravenous pyelography is mandatory in the evaluation of recurrent urinary tract infection in children, the value of a slightly better scar identification in the management of children remains to be determined. The subsequent development of scars is more easily followed with nuclear imaging, but the value of this information is determined first by the likelihood of the development of scars in the subsequent years and second by the decisions that hinge upon their demonstration. At present it is thought that the development of scars in normal kidneys is very unusual [29–31] over the age of four. Furthermore, the therapeutic consequences of scar identification are limited, now that the value of surgical correction of vesicoureteral reflux is disputed [32]. Only long-term studies will demonstrate the possible benefit of an accurate delineation of renal scars in children. Until that moment this technique should probably not be incorporated systematically in the work up of urinary infection.

Occasionally DMSA imaging has been proposed to elucidate the nature of an intrarenal mass or to demonstrate the presence of functioning tissue in the isthmus of a horseshoe kidney. Intravenous pyelography and computed tomography usually provide better information under these circumstances [33].

Renal and urinary tract imaging can be obtained with a great variety of radiolabelled compounds with different renal fates. Interestingly radioisotope skeletal surveys with ^{99}Tc diphosphonate or polyphosphate may disclose abnormalities of the kidneys. In 52 out of 1,711 scintiscans, Maher [34] reported major, unexpected renal or urinary abnormalities consisting of filling defects (carcinoma, polycystic kidneys, cysts) and urinary tract obstruction.

Dynamic renal imaging

The possibility to quantitate over time the renal accumulation of labelled compounds provides both anatomic and functional information. This approach has been utilised to measure separate kidney function and to assess renal allograft status.

Separate function studies. Split renal function studies relying on bilateral ureteric catheterisation are so invasive and time-consuming that they have never gained widespread acceptance. The advent of radionuclide methodologies has led to alternative methods. Various protocols have been utilised. With DTPA or hippuran an early (1–2 minutes) renal scan is obtained [33]. After background activity subtraction, the uptake of each kidney is expressed as a fraction of total uptake. Single kidney function is then derived from the global renal function measured simultaneously by a single injection blood clearance. With DMSA, a compound remaining within tubular cells for a long period, uptake is measured at 24 hours [35].

Several difficulties limit this approach: the organ of interest is not the sole organ within the field of the radiation detector, the radionuclide uptake is time dependent and, finally, the variable distance of each kidney from the detector results in differences of tissue absorption [36].

In practice, results appear unreliable when creatinine clearance falls below 30ml/min [37]. In patients with urinary obstruction or hydronephrosis, results may be contaminated by the pelvic accumulation of the isotope, a difficulty that can be circumvented by very early counting of hippuran or DTPA or by very late counting of DMSA. Despite these limitations, several reports have documented a good correlation between radionuclide and classical clearance split function studies [38,39]. However, a closer look at published results discloses a great variability for individual observations. For instance, MacKay et al [39] document a good correlation between hippuran renography and PAH clearance in individual kidneys. Still renal plasma flow is lower in the whole group with hippuran renography than with PAH clearance and in individual cases radionuclide clearance ranges from 40 to 160 per cent of the corresponding PAH clearance (Figure 1).

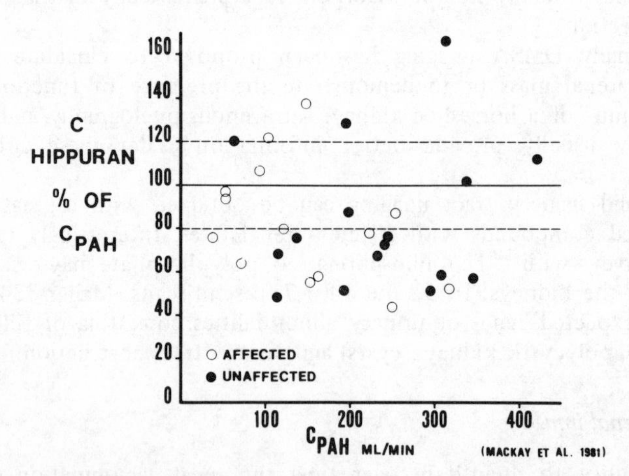

Figure 1. Comparison of individual kidney conventional PAH and blood hippuran clearances. Affected kidneys suffer from renovascular lesions (9 cases) or parenchymal disease (5 cases). Blood hippuran clearance is expressed as a percentage of the conventional PAH clearance (calculated from data in MacKay et al 1981 [39])

The technique has been used mainly by urologists to evaluate single kidney function either prior to a surgical decision or in the follow-up. Despite claims that pre-operative assessment correctly predicted the subsequent functional recovery, several recent studies have provided evidence to the contrary [40–43]. The reproducibility of the method and its validation under the pathological conditions of the scrutinised disease are often lacking. Furthermore, the impact of the collected information on the decision making process has not yet been critically evaluated.

Separate renal function studies have also been advocated in the diagnosis and evaluation of renal artery stenosis. The radioisotopic renogram developed over 20 years ago was a first application of this principle [44]. Despite methodological improvement brought by gamma cameras and computers the low specificity of the method [45–47] has cast some doubts on its value in the screening for renovascular hypertension, and indication that digital vascular imaging might take over [48,49]. The relevance of separate renal plasma flow measurement to the prediction of the surgical curability of hypertension is not yet established and will be undoubtedly limited by the variability of results in individual cases (Figure 1).

As a research tool, however, separate renal function studies are of great interest to solve specific problems. Wenting et al [23] have used DTPA renogram to document the effect of captopril on the function of a kidney with renal artery stenosis. They demonstrated a dramatic fall in DTPA uptake by the stenosed kidney while heterolateral function remained normal. This study combined with the documentation of a fall in the extraction ratio of hippuran and iodothalamate allowed them to conclude that captopril probably reduced renal function in stenosed kidneys through a vasodilatation of the efferent arteries.

The renal transplant

During the post-operative period when the graft is not yet functioning, the clinician is often confronted with a diagnostic dilemma: what should be attributed to acute tubular necrosis, to graft rejection or to ureteral obstruction?

Radionuclide investigation of the graft has proved unique to guide the clinician. The patency of the large vessels can be ascertained with intravascular compounds such as ^{99m}Tc pertechnetate [50] whereas the accumulation of DTPA or hippuran reflects intrarenal blood flow.

Several combinations of various compounds have been recommended. All have in common an initial assessment of the graft within 24 hours after the operation. the study is subsequently repeated at two to three day intervals and compared with the initial results [51] until recovery of renal function. One method [51] combines a series of hippuran scans obtained during 20 minutes with pertechnetate scans obtained at five second intervals during 40 seconds after injection [51]. In acute tubular necrosis, the scintigram reveals a continuous renal uptake of hippuran without evidence of radionuclide excretion into the bladder. Progressive recovery is heralded by the appearance of radioactivity in the bladder and subsequently by a fall in parenchymatous activity after the initial rise. The pertechnetate flow study shows only minor abnormalities.

Acute rejection superimposed on acute tubular necrosis is evidenced by a decline in the curve amplitude of pertechnetate. Hippuran uptake decreases when compared with the previous examination. Return to control values of these two abnormalities after steroid treatment confirms the diagnosis of rejection.

Acute rejection in a normally functioning graft can usually be diagnosed without radionuclide studies. If they are performed, rejection is characterised by a delay and a decrease in parenchymal hippuran uptake.

Various methods have been proposed to quantify radionuclide uptake in the graft to facilitate diagnosis and comparisons [52,53] with a reported sensitivity of 90 per cent in the diagnosis of rejection.

All radionuclide methods utilised in the post-operative period rely on an initial 24 hour determination assumed to represent solely acute tubular necrosis, subsequent degradation being attributed to an added rejection. This assumption should probably be reassessed in patients treated with cyclosporine a drug whose nephrotoxicity is associated with an altered arterial blood flow in the graft. As a result, initial renal blood flow is low and the onset of acute rejection is no longer heralded by a sharp fall in perfusion [54].

Radiolabelled cells kidney uptake

Renal rejection as well as various renal diseases result in the interstitial or intravascular accumulation of leucocytes and platelets. Detection of these cells by radionuclide uptake has been proposed.

The first studies relied on ^{67}gallium citrate which accumulates in inflammatory regions where it binds to leucocytes and/or acute phase proteins. The procedure developed for the detection of infections [55] has been utilised in the evaluation of acute interstitial nephritis [56,57], acute pyelonephritis [58,59] or renal amyloidosis [60]. Its diagnostic value in nephrology suffers, however, from a high rate of false positive results [61].

The injection of ^{111}Indium labelled leucocytes or platelets, requires a smaller radiation dose, usually around 1mCi [55]. Labelled leucocytes [62] and platelets [63–65] accumulate in rejecting grafts. Despite earlier claims, recent studies have proved disappointing due to a high incidence of false positive and negative results [66]. In our own experience this technique does not provide a significant gain over more traditional methods for the diagnosis of rejection.

The radiation hazard

Whenever systematic radionuclude studies are contemplated the benefits must be weighed against the cost. The latter include not only the financial cost but also the loss of time for the patient and the radiation hazard.

The amount of radiation exposure per examination is generally within acceptable limits. However, when repeated tests are required, careful consideration should be given to the physical characteristics of the radionuclide. ^{131}I hippuran has a physical half-life of 8.04 days. Although its biologic half-life lasts only 30 minutes in patients with normal renal function, it increases and eventually approaches the physical half-life in progressing renal failure. Furthermore, ^{131}I has a beta radiation absorbed by peripheral tissues. The majority of large nuclear medicine units are thus converting to ^{123}I which emits only gamma radiation and has a 13.3 hours half-life. The radiation absorption by critical organs after the injection of 200μCi of hippuran illustrates the hazard of ^{131}I compared to that of ^{123}I. Normally functioning kidneys absorb 20m rad with ^{131}I versus only 4m rad ^{123}I. Renal failure may increase these figures up to 60-fold [20]. Non-voiding results in an absorbed bladder dose of 2400m rad for ^{131}I versus 600m rad for ^{123}I hippuran. Thyroid exposure reaches 9600m rad with ^{131}I and only 120m rad with ^{123}I if thyroid uptake is not blocked before the test [20,51].

Technetium whose physical half-life is only 6.09 hours results in an even lower radiation exposure when coupled to DTPA or DMSA [20].

Conclusions

Confronted with an array of radionuclide tests the nephrologist remains very often in doubt as to their contribution to diagnosis and management.

Non-imaging techniques for measurement of renal function have little place except in a research situation. They are not equivalent to the standard reference methods and, although more accurate than currently available clinical techniques, they do not provide a gain in information offsetting their cost.

Static imaging techniques are to be compared with traditional urography. The anatomic information does not appear superior except, perhaps, in the detection of scars in pyelonephritic kidneys. This gain is, however, too marginal to justify inclusion of these tests in current clinical practice.

Dynamic imaging contributes both anatomic and functional data. It provides the clinician with critical information during the early evolution of renal graft, especially when acute renal failure develops. This technique also allows the separate assessment of individual kidney function. How useful this information is in diagnosis and treatment remains an open question. Urologists feel that on this basis they can predict preoperatively the chances of recovery of a kidney and plan accordingly, a conclusion contradicted by others.

Finally, the renal fate of gallium citrate and of labelled white cells or platelets can be easily followed. The sensitivity and specificity of these methods appear disappointingly low.

Ultrasonography

Ultrasonography has become, over the last 10 years, one of the most reliable non-invasive methods to assess renal morphology.

It has proved to be an invaluable help in the evaluation of renal tumours: it is now a critical step in all algorithms proposed to diagnose the nature of renal masses.

The ability of ultrasonography to identify cysts is utilised in the diagnosis of polycystic kidneys. It enables the further identification of hepatic or pancreatic cysts, a finding sometimes helpful in the differential diagnosis between multiple cysts and polycystic kidney disease [67,68]. Although ultrasounds are slightly less sensitive than intravenous pyelography with nephrotomograms, their non-invasiveness qualifies them as the best method in the screening of relatives of patients with polycystic kidneys [69–71].

Ultrasonography readily identifies hydronephrosis [72,73]. It detects obstructive uropathy with a sensitivity of 90 per cent and a specificity of 98 per cent, a better performance than radionuclide studies with DTPA or hippuran, 26 per cent of which are inconclusive. This superiority results from the fact that ultrasonography, contrary to radionuclide studies, is independent of renal function [74].

Ultrasonography has thus become the initial screening modality in the evaluation of suspected urinary tract obstruction. Like any other method, however, it has its pitfalls: dilatation of the pelvis in incomplete obstruction may vary with diuresis; prominent collecting system and an extrarenal pelvis, as well as high flow states in some non-oliguric azotaemic patients [75], may give false positive results. These errors will be readily corrected by a subsequent intravenous pyelography [76]. The main hazard of ultrasound diagnosis is the minimal dilatation obstructive uropathy. It may occur in patients with either retroperitoneal fibrosis or neoplastic encasement of the ureters [77,78]. If oliguria is present the mere visualisation of a clearly defined pelvis and ureter is sufficient evidence to diagnose obstruction; indeed, in the absence of urine secretion the urinary tract collapses.

The role of ultrasound scanning in the work-up of renal colic is also very promising. In a series of 21 patients with renal colic, 18 of whom had a stone, a sensitivity of 100 per cent in the detection of ureteral obstruction and/or stones has been reported [79]. In current practice, if inconclusive examinations are included, sensitivity is probably lower.

Sonography of the fetus has allowed the prenatal detection of urinary tract obstruction [80,81], a condition accessible now to therapeutic intervention [82]. Fetal renal dysplasia may be diagnosed in utero [83].

Ultrasounds have also been successfully used in order to localise the kidney for renal biopsy [84] or for the percutaneous insertion of small catheters to drain renal or peri-renal abscesses or to place a nephrostomy [85–87]. The technique identifies the peri-renal haematomas resulting from these procedures [88,89].

In addition to its ability to outline the kidneys, to visualise a dilated urinary tract ultrasound scanning can identify, in the kidney, regions with differences in echogenicity. The corticomedullary junction may be recognised. This approach has led to the use of ultrasound in the diagnosis of renal parenchymal disease [90]. There is a good correlation between ultrasound evaluated kidney size and the histological evidence of sclerosis and atrophy. Echogenicity of the kidney measured in reference to liver echogenicity, also correlates with sclerosis in the biopsy specimen [91,92] and with creatinine clearance [92]. However, the scatter of the data around the latter relation precludes any meaningful interpretation of ultrasound results in the diagnosis of histologic type of renal disease.

The renal allograft has also been investigated with ultrasound in order to detect rejection. Findings suggesting rejection include a blurring of the cortico-medullary junction, increased cortical echogenicity and hypoechoic areas in the pyramides [93–97]. The most reliable signs are increased size and decreased echogenicity of renal pyramides [98] whereas blurred corticomedullary junction was found to be unreliable [98,99]. Although it has been reported that in acute tubular necrosis, the kidney is not modified [97,100], a decreased echogenicity of the central sinus complex has been noted [101] and increased renal size, together with hypoechoic pyramides has been occasionally reported [98]. Overall ultrasound scanning does not seem to add to radionuclide studies in the diagnosis of acute rejection, especially in early post-transplant renal failure.

An interesting development of ultrasonographic technique is the ultrasonic echo-Doppler flowmeter. Greene et al [102] have utilised a dual frequency real time two dimensional echo Doppler calibrated in vitro. Blood velocity and volume flow rates were measured in renal arteries of normal subjects. Results fell well within the physiologically accepted range. In a subsequent study [47] Avashti et al utilised the same method in 68 patients with suspected renovascular hypertension. Eleven were technically inadequate. In 26 of the 57 other patients a comparison with angiography was available: sensitivity and specificity for the detection of renal artery stenosis were 89 and 73 per cent respectively. The same group has recently proposed an identical approach to diagnose renal vein thrombosis [103]. Compared with phlebography, the method has an 85 per cent sensitivity and a 56 per cent specificity. More experience will be needed to assess the usefulness of the new technology. At any rate it must be emphasised that, as in all ultrasonic methods, accuracy and reproducibility of results hinge upon the experience and skill of the operator.

In conclusion, at the present time, renal ultrasounds are of paramount importance in the detection of obstruction. They precede intravenous pyelography both in the approach of azotaemic patients [75] and of the patients with renal colic [79]. The detection of cysts, especially in polycystic kidneys, and of perirenal collections, as well as the work-up of intrarenal tumoral masses, are more cost effective with ultrasound than with other non-invasive techniques. The usefulness of ultrasonography in the diagnosis of renal parenchymal disease and of allograft rejection is not yet established and should be considered a field of clinical investigation. Finally, technical developments might allow in the future clinically useful ultrasonic determinations of blood flow in renal arteries and veins.

Digital vascular imaging

Renovascular disease is the single, most common curable form of hypertension. Until recently specific treatment entailed either reconstructive vascular surgery or nephrectomy. Taken together with the high cost and the morbidity of the angiographic diagnostic procedure [104], these facts have oriented current practice towards medical therapy for all hypertensions, reserving a more elaborate diagnostic work-up for patients whose hypertension proved resistant to treatment and who might eventually tolerate surgery.

This attitude has been transformed by two recent developments. Percutaneous transluminal angioplasty [105] has significantly reduced the morbidity of vascular repair [106]. Simultaneously digitalised vascular imaging has provided a low risk, relatively pain-free procedure for the detection of renal artery stenosis [48,107]. Although digital vascular imaging does not qualify as a non-invasive imaging technique, it may be worth mentioning because it is so less invasive than arteriography.

Hillman et al [48] have claimed that digital vascular imaging is cost-effective in the screening for renovascular hypertension. They compute a total cost of 1240 US dollars for conventional diagnostic procedures including excretory urography, radionuclide study and arteriography versus only 300 US dollars for digitalised vascular imaging. They subsequently calculate that it is cheaper to screen 100 hypertensive patients and to treat eventually five to 10 of them by transluminal angioplasty than to maintain all 100 patients on sustained medical treatment. The validity of this approach hinges upon the reliability of digitalised vascular imaging and the long-term benefit of percutaneous transluminal angioplasty. The latter appears promising [49,108] although caution has been advised by Flechner et al [109]. The sensitivity and specificity of digitalised vascular imaging in renovascular hypertension is encouraging. Smith et al [110] compare the digitalised vascular imaging with angiography in 32 hypertensive patients. A total of 76 arteries were examined by two observers. Sensitivity ranged from 83 to 87 per cent and specificity from 79 to 87 per cent. The main difficulty accrued from subtraction artefacts and a relatively low spatial resolution resulting in false positive stenoses. Furthermore, a poor circulatory condition may result in an excessive dilution of the contrast media and thus in inadequate images. As a result of improved technology Schwarten [49] observed in 30 patients an almost perfect agreement between arteriography and digitalised vascular imaging: in 26 there was agreement both as to the diagnosis and the grade of the disease. In four, renal artery stenosis was present, but there was a discrepancy on its severity as a result of heavy calcification in the peri-renal aorta, a diminished cardiac output and an artefact. Furthermore, digitalised vascular imaging allowed an adequate follow-up after percutaneous transluminal angioplasty.

Digitalised vascular imaging has also been successfully applied in hypertensive transplanted patients and allowed the detection of graft artery stenosis [111] amenable to transluminal angioplasty [112].

Computerised tomography

Computerised tomography provides an excellent delineation of renal structures. Without the use of intravenous contrast material it may readily inform on the presence of tumours, stones, hydronephrosis. After contrast media injection the nephrogram allows a precise evaluation of parenchymal thickness. Cysts may be accurately delineated and the uptake of the iodinated material by a tumour evaluated [113]. In these indications computerised tomography complements intravenous pyelography and is thus not a first line procedure.

Computerised tomography allows the direct investigation of the retroperitoneal space. It is especially useful in the delineation of renal tumours [113] and in the exploration of ureteral obstruction. Its sensitivity in the staging of renal neoplasms is considerably higher than that of sonography [114]. Retroperitoneal fibrosis, compression by lymph nodes [115], tumour thrombi within the vena cava [113] are well visualised.

More recently it has been proposed to evaluate renal function by measuring the changes in density of different areas of the kidney during the injection of contrast material. The method has been used for the evaluation of renal graft on the basis of repeated scanning of the same renal slice [116]. The duration of each scan, between one and five seconds, causes some difficulties due to variations in kidney position. New high-speed, volume imaging tomography scanners are developed [12]. They are able to scan simultaneously 120 1.8mm thick transverse kidney slices in 0.011 seconds. Repetition of the whole kidney scan 60 times per second allows an assessment of intrarenal blood flow distribution. It remains to be seen whether this approach provides a more accurate and cost effective assessment of renal function than radionuclide studies.

Magnetic resonance

Magnetic resonance produces remarkable images of internal structure of the human body without contrast media and ionising radiation. The potential hazards are exposure to static and rapidly changing magnetic fields and the heating from radiofrequency pulses [117]. Up to now none of them seem to have materialised.

An acquisition time longer than one minute has limited the use of magnetic resonance mainly to structures that can be reasonably immobilised such as the brain. Applications to the kidney are still limited by movement artefacts which might be reduced in the future by appropriate gating. Cortex and medulla are clearly delineated. Hydronephrosis, cortical oedema during rejection are visualised [118,119]. Tumours are readily differentiated from cysts and tumour invasion into renal veins may be visible [120]. The significance of the wealth of new data provided by magnetic resonance is just beginning to be investigated; it is too early to assess its place and its exact contribution in the evaluation of renal function and morphology.

Conclusions

Concepts of decision analysis concepts in medical diagnosis allow a proper evaluation of tests [121]. The value of a test depends upon its ability to significantly modify pre-test diagnostic probabilities. Very few of the more recent sophisticated tests have been evaluated in this way. Although many bear a more or less close relationship with disease processes, the actual help provided to the nephrologist confronted with a specific diagnostic or therapeutic decision remains to be ascertained. Such an evaluation is difficult: first of all each patient is unique in so far as the a priori probabilities obtained by clinical history and physical examination vary from one patient to another. Further the quality of the information provided by the different tests varies among institutions. Hospitals benefiting from outstanding radiology services, for instance, have tended to develop less elaborate nuclear medicine facilities; the experience of the operator remains a critical element of the accuracy of sonography. Still the contribution of each test should be critically scrutinised before accepting its incorporation in routine practice if waste in time and resources are to be avoided.

In this respect the best tool to assess kidney morphology remains the plain film of the abdomen with tomographies and intravenous pyelography. Ultrasound will be preferred as the first step when ureteral obstruction is to be ruled out, for instance in acute renal failure or exacerbation of chronic renal failure. It will also take precedence over intravenous pyelography in the screening for polycystic kidney disease. Ultrasound on the other hand will complement intravenous pyelography in the evaluation of renal tumours, in the elucidation of peri- or para-renal collections such as haematomas, lymphocoeles or urinomas.

Computed tomography will be requested when accurate delineation of renal or retroperitoneal structures is necessary: for instance in the staging of renal neoplasms or in the evaluation of the retroperitoneal space of patients with ureteral obstruction.

Digitalised intravenous angiography provides a relatively non-invasive way to delineate renovascular morphology. Its place as the best screening method for renovascular disease remains to be confirmed.

Finally, radionuclide scans with DMSA appear to provide a slightly better delineation of pyelonephritis scars than intravenous pyelogram. Although some urologists feel that this information is critical in reaching therapeutic decisions, evidence supporting this view is still wanting.

Renal function tests have proliferated during the last 20 years. Creatinine clearance and serum creatinine remain the cornerstone of renal function evaluation. These tests are cheap, easy to perform everywhere and their many limitations and pitfalls are well known to the nephrologists. More sophisticated radionuclide tests provide probably a more accurate estimate of renal parameters. However, their routine use outside clinical research does not yet seem to provide operational benefits outweighing their cost. Furthermore, the many assumptions underlying several of these tests are not apparent to the nephrologist who might overestimate the significance of the results. This conclusion should be amended in two areas. Radionuclide tests provide a unique service in the evaluation of post-transplant failure when differentiation between acute tubular necrosis and rejection is of critical importance for patient management. Radionuclide tests are also the only available method to assess separate renal function. Many factors limit the accuracy of these determinations. Still, many urologists feel that the information obtained is of paramount importance in the decision to operate on kidneys with either stones or severe hydronephrosis, a statement firmly disputed by others. Here again an evaluation of the real benefits accruing from the obtained information above that already available from conventional tests and the intravenous pyelography remains to be done.

Improvement in technology, better knowledge of the physiological aspects of the different test compounds and above all a critical look at the results will probably profoundly modify in the coming years our present diagnostic procedures.

References

1 Smith HW. *Kidney: Structure and Function in Health and Disease.* New York: Oxford University Press. 1951
2 Tauxe WN, Burbank MK, Maher FT et al. *Mayo Clin Proc 1964; 39:* 761
3 Sapirstein LA, Vidt DC, Mandel MJ et al. *Am J Physiol 1955; 181:* 330
4 Wagoner RD, Tauxe WN, Maher FT et al. *JAMA 1964; 187:* 811
5 Britton KE, Brown NJG. In *Clinical Renography.* London: Lloyd-Luke. 1971: 87
6 Blaufox MD, Merrill JP. *Nephron 1966; 3:* 274
7 Tauxe WN, Maher FT, Taylor WF. *Mayo Clin Proc 1971; 46:* 524
8 Tauxe WN, Dubovsky EV, Kidd T et al. *Eur J Nucl Med 1982; 7:* 51
9 Holland MD, Galla JH, Dubovsky EV et al. *Kidney Int 1984; 25:* 167
10 Britton KE, Brown NJG. In *Clinical Renography.* London: Lloyd-Luke. 1971: 207
11 Britton KE. *Clin Sci 1979; 56:* 101
12 Knox FG, Ritman EL, Romero JC. *Kidney Int 1984; 25:* 473
13 Wilkinson SP, Bernardi M, Pearce PC et al. *Clin Sci 1978; 55:* 277
14 Doolan PD, Alpen EL, Theil GB. *Am J Med 1962; 32:* 65
15 Brøchner-Mortensen J, Rødbro P. *Scand J Clin Lab Invest 1976; 36:* 35
16 Chantler C, Garnett ES, Parsons V et al. *Clin Sci 1969; 37:* 169
17 Chantler C, Barratt TM. *Arch Dis Child 1972; 47:* 613
18 Ott NT, Wilson DM. *Mayo Clin Proc 1975; 50:* 664
19 Klopper JF, Hauser W, Atkins HL et al. *J Nucl Med 1972; 13:* 107
20 Chervu LR, Blaufox MD. *Semin Nucl Med 1982; 12:* 224
21 Parving HH, Andersen AR, Smidt UM et al. *Diabetes 1983; 32 (Suppl 2):* 83
22 Donadio JV, Anderson CF, Mitchell JC et al. *N Engl J Med 1984; 310:* 1421
23 Wenting GJ, Tan-Tjiong HL, Derkx FHM et al. *Br J Med 1984; 288:* 886
24 Fine EJ, Gorkin J, Bank N et al. *J Nucl Med 1984; 25:* 76
25 Textor SC, Fouad FM, Bravo ML et al. *N Engl J Med 1982; 307:* 601
26 Yee CA, Lee HB, Blaufox MD. *J Nucl Med 1981; 22:* 1054
27 Whitaker RH, Sherwood T. *Br Med J 1984; 288:* 839
28 Merrick MV, Uttley WS et al. *Br J Rad 1980; 53:* 544
29 Savage DCL, Howie G, Adler K et al. *Lancet 1975; i:* 358
30 Lindberg U, Claesson I, Hanson LA et al. *J Pediat 1978; 92:* 194
31 Cardiff-Oxford Bacteriuria Study Group. *Lancet 1978; i:* 889
32 Birmingham Reflux Study Group. *Br Med J 1983; 287:* 171
33 Blaufox MD, Fine E, Lee HB et al. *J Nucl Med 1984; 25:* 619
34 Maher FT. *Mayo Clin Proc 1975; 50:* 370
35 De Maeyer P, Simons M, Oosterlinck W et al. *J Urol 1982; 128:* 8
36 Britton KE, Maisey MN et al. In Maisey MN, Britton KE, Gilday DL, eds. *Clinical Nuclear Medicine.* London: Chapman and Hall. 1983: 93
37 Chanard J, Ruiz JC, Liehn JC et al. *Clin Nephrol 1982; 18:* 291
38 Larsson I, Lindstedt E, White T. *Br J Urol 1984; 56:* 109
39 Mackay A, Eadie AS, Cumming AMM et al. *Kidney Int 1981; 19:* 49
40 Sherman RA, Blaufox MD. *Nephron 1980; 25:* 82
41 McAfee JG, Singh A, O'Callaghan JP. *Radiology 1980; 137:* 487
42 Talner LB, Sokoloff J, Halpern SE et al. *Int J Nucl Med Biol 1982; 9:* 181
43 Belis JA, Belis TE, Lai JCW et al. *J Urol 1982; 127:* 636
44 Taplin GV, Meredith OM, Kade H et al. *J Lab Clin Med 1956; 48:* 886
45 Keim HJ, Johnson PM, Vaughan ED et al. *J Nucl Med 1979; 20:* 104
46 Maxwell MH. *Kidney Int 1975; 8:* S153
47 Avasthi PS, Voyles WF, Greene ER. *Kidney Int 1984; 25:* 824
48 Hillman BJ, Ovitt TW, Capp MP et al. *Radiology 1982; 142:* 577
49 Schwarten DE. *Radiology 1984; 140:* 369
50 Freedman LM, Meng CH, Richter MW et al. *J Urol 1971; 105:* 473
51 Clorius JH, Dreikorn K. In Marberger M, Dreikorn K, eds. *Renal Preservation, International Perspectives in Urology 1983; 8:* 276

52 Hilson AJW, Maisey MN, Brown CB et al. *J Nucl Med 1978; 19:* 994
53 McConnell JD, Sagalowsky AI, Lewis SE et al. *J Urol 1984; 131:* 875
54 Osman EA, Barrett JJ, Bewick M et al. *Lancet 1984; i:* 1470
55 O'Mara RE. In Maisey MN, Britton KE, Gilday DL, eds. *Clinical Nuclear Medicine.* London: Chapman and Hall. 1983
56 Linton AL, Clark WF, Driedger AA et al. *Ann Intern Med 1980; 93:* 735
57 Wood BC, Sharma JN, Germann DR et al. *Arch Intern Med 1978; 138:* 1665
58 Handmaker H. *Semin Nucl Med 1982; 12:* 246
59 Schardijn GHC, Statius Van Eps LW, Pauw W et al. *Br Med J 1984; 289:* 284
60 Bekerman C, Vyas MI. *J Nucl Med 1976; 17:* 899
61 Garcia JE, Van Nostrand D, Howard WH et al. *J Nucl Med 1984; 25:* 575
62 Frick MP, Henke CE, Forstrom LA et al. *Clin Nucl Med 1979; 4:* 24
63 Smith N, Chandler S, Hawker RJ et al. *Lancet 1979; ii:* 1241
64 Fenech A, Smith FW, Power DA et al. *Clin Nephrol 1984; 21:* 220
65 Hofer R, Sinzinger H, Leithner C et al. *Nucl Med Commun 1981; 2:* 120
66 Desir G, Lange R, Smith E et al. *J Nucl Med 1984; 25:* 75
67 Lawson TL, McLennan BL, Shirkhoda A. *JCU 1978; 6:* 297
68 Sanders RC, Mermon S, Sanders AD. *J Urol 1978; 120:* 521
69 Milutinovic J, Phillips LA, Bryant JI et al. *Lancet 1980; i:* 1203
70 Rosenfield AT, Lipson MH, Wolf B et al. *Radiology 1980; 135:* 423
71 Walker FC Jr, Loney LC, Root ER et al. *AJR 1984; 142:* 1273
72 Ellenbogen PH, Schieble FW, Talner LB et al. *AJR 1978; 130:* 731
73 Lee JKT, Baron RL, Melson GL et al. *Radiology 1981; 139:* 161
74 Malave SR, Neiman HL, Spies SM et al. *AJR 1980; 135:* 1179
75 McClennan BL. *Radiol Clin North Am 1979; 17:* 197
76 Jeffrey RB, Federle MP. *Radiol Clin North Am 1983; 21:* 515
77 Amis ES, Cronan JJ, Pfister RC et al. *Urology 1982; 19:* 101
78 Curry NS, Gobien RP, Schabel SI. *Radiology 1982; 143:* 531
79 Erwin BC, Carroll BA, Sommer FG. *Radiology 1984; 152:* 147
80 Hadlock FP, Deter RL, Carpenter R et al. *AJR 1981; 137:* 261
81 Chinn DH, Filly RA. *Urol Radiol 1982; 4:* 115
82 Harrison MR, Golbus MS, Filly RA et al. *N Engl J Med 1982; 306:* 591
83 Mahony BS, Filly RA, Callen PW et al. *Radiology 1984; 152:* 143
84 Almkuist RD, Buckalew VM. *Urol Clin North Am 1979; 6:* 503
85 Stables DP. *Urol Clin North Am 1982; 9:* 15
86 Gerzof SG. *Urol Radiol 1981; 2:* 171
87 Kutcher R, Rosenblatt R. *JAMA 1984; 251:* 3126
88 Proesmans W, Marchal G, Snoeck L et al. *Clin Nephrol 1982; 18:* 257
89 Ralls PW, Colletti P, Boger DC et al. *Urol Radiol 1980; 2:* 22
90 Rosenfield AT, Taylor KJW, Crade, M et al. *Radiology 1978; 128:* 737
91 Rosenfield AT, Siegel NJ. *AJR 1981; 137:* 793
92 Hricak H, Cruz C, Romanski R et al. *Radiology 1982; 144:* 141
93 Jafri SZ, Kaude JV, Wright PG. *Acta Radiol (Diagn) (Stockh) 1981; 22:* 245
94 Hricak H, Cruz C, Eyler WR. *Radiology 1981; 139:* 441
95 Bush GJ, Galvanek EG, Reynolds ES Jr. *Hum Pathol 1971; 2:* 253
96 Hillman BJ, Birnholz JC, Bursch GJ. *Radiology 1979; 132:* 673
97 Maklad NF, Wright CH, Rosenthal SJ. *Radiology 1979; 131:* 711
98 Frick MP, Feinberg SB, Sibley R et al. *Radiology 1981; 138:* 657
99 Singh A, Cohen WN. *AJR 1980; 135:* 73
100 Hricak H, Toledo-Pereyra LH, Eyler WR et al. *Radiology 1979; 133:* 443
101 Barrientos A, Leiva O, Diaz-Gonzalez R et al. *J Urol 1981; 126:* 308
102 Greene ER, Venters MD, Avashti PS et al. *Kidney Int 1981; 20:* 523
103 Avasthi PS, Greene ER, Scholler C et al. *Kidney Int 1983; 23:* 882
104 McNeil BJ, Adelstein SJ. *N Engl J Med 1975; 293:* 221
105 Grüntzig A, Kuhlmann U, Vetter W et al. *Lancet 1978; i:* 801
106 Grim CE, Luft FC, Yune HY et al. *Ann Intern Med 1981; 95:* 439
107 Buonocore E, Meaney TF, Borkowski GP et al. *Radiology 1981; 139:* 281

108 Geyskes GG, Pylaert CBAJ, Oei HY et al. *Br Med J 1983; 287:* 333
109 Flechner S, Novick AC, Vidt D et al. *J Urol 1982; 127:* 1072
110 Smith CW, Winfield AC, Price RR. *Radiology 1982; 144:* 51
111 Khoury GA, Farrington K, Varghese Z et al. *Clin Nephrol 1983; 20:* 225
112 Sniderman KW, Sos TA, Sprayregen S et al. *Radiology 1980; 135:* 23
113 Love L, Reynes CJ, Churchill R et al. *Radiol Clin North Am 1979; 17:* 77
114 Weyman PJ, McClennan BL, Stanley RJ et al. *Radiology 1980; 137:* 417
115 Fagan CJ, Larrieu AJ, Amparo EG. *AJR 1979; 133:* 239
116 Field IL, Matalon TA, Vogelzang RL et al. *AJR 1984; 142:* 1157
117 Pennock JM. *Radiography 1982; 48:* 221
118 Smith FW, Reid A, Mallard JR et al. *Diagn Imaging 1982; 51:* 209
119 Power DA, Russell G, Smith FW et al. *Clin Nephrol 1983; 20:* 155
120 Hricak H, Williams RD, Moon KL Jr et al. *Radiology 1983; 147:* 765
121 Griner PF, Mayewski RJ, Mushlin AI et al. *Ann Intern Med 1981; 94:* 553

Open Discussion

VANHERWEGHEM (Brussels) You concluded that intravenous pyelogram remains the first line procedure for renal assessment? Could you comment on possible nephrotoxicity of contrast agents, especially in patients with renal insufficiency.

VAN YPERSELE You quite rightly point out that contrast agents may have a deleterious effect on renal function. Predisposing factors have been identified, thus minimising the incidence of this complication. Still intravenous pyelography should be performed only when specific questions are asked and when the value of the answer outweighs potential risks. For most morphological problems facing nephrologists the intravenous pyelogram remains the best deal.

KERR (London) Please comment further on the best method of assessing renal scars? You quote the IVP as the best test but the observed error is too high to make it useful in following the progression of renal scarring. Do DMSA scanning, or CT scanning or any other techniques offer any advantages?

VAN YPERSELE Yes, DMSA scanning appears slightly superior to IVP. Taking into account that the first morphologic assessment of the kidneys in chronic pyelonephritis is at any rate an IVP, the question remains whether the added information provided by DMSA is useful from an operational point of view. It is true that subsequent follow-up of scars can be obtained with DMSA. But again, where does this information lead? As you know, surgery on vesicoureteral reflux is less popular these days and I know of no study demonstrating that scar progression results in an indication for surgery. Thus for the practising nephrologist I believe it has no benefit. For those physicians who are interested in scar formation as a topic of clinical research the attitude is different: DMSA provides interesting data.

WILL (Leeds) You have mentioned the use of IV urography as screening for abnormality and triggering further non-invasive investigation. Would you comment on reciprocal serum creatinine plotting as the commonest non-invasive procedure which triggers further testing?

VAN YPERSELE I have necessarily limited my presentation to a few non-invasive approaches – thus urinary enzymes, changes in the blood concentration of various compounds such as beta-2 microglobulin etc have not been considered. I wholeheartedly agree with you: the changes of the reciprocal of creatinine over time are linear in several diseases. Any departure from this evolution raises questions about the occurrence of aggravating potentially reversible factors. This is another way to non-invasively follow renal function.

RIZZONI (Padova) Is renography with injection of frusemide more useful than IVP in the diagnosis of obstruction of higher urinary tract?

VAN YPERSELE Renograms have had their time. Some improvements such as the superimposition of an acute frusemide-induced diuresis have added information. However, this method remains limited by several physiological factors such as the state of diuresis, the intensity of the diuretic response etc, so that a high rate of false negative and positive results is reported. By contrast, frusemide injected during an IVP provides dynamic images of the pyelic region and helps assess a possible obstruction of the urinary tract.

Proc EDTA–ERA (1984) Vol 21

MECHANISMS AND FACTORS REGULATING THE GROWTH OF GLOMERULAR CELLS

G E Striker, Liliane Morel-Maroger Striker

National Institute of Arthritis, Diabetes, and Digestive and Kidney Diseases, National Institutes of Health, Bethesda, USA

Summary

The hallmark of end-stage renal disease is progressive sclerosis. The composition of the sclerotic material and its cellular source are under study and only partly elucidated. Sclerosis, in part, is composed of extracellular matrix components normal to the area, the sole exception thus far recognised is crescentic glomerulonephritis associated with Bowman's basement membrane disruption in which the sclerotic tissue contains interstitial connective tissue.

The source of the extracellular matrix is the local glomerular cells. The complete composition of the extracellular matrix synthesised by individual glomerular cells is under current study, but it appears that all glomerular cells are capable of synthesising many of the various basement membrane components. The respective role of each cell type in sclerosing diseases and the initiating and propagating factors await further investigation.

Introduction

End-stage renal disease is one of the urgent unsolved problems in modern nephrology. Even though it was first described in the 19th century by Sir Richard Bright [1], there has been surprisingly little new information on the link between the initial renal inflammatory process and the development of end-stage renal disease. The lack of an experimental animal model of progressive sclerosis following an initial acute inflammatory response has considerably hindered progress in this field.

The pathogenesis of progressive sclerosis thus remains one of the important unsolved problems in modern nephrology. It is known that progressive glomerular diseases are characterised by an increasing accumulation of eosinophilic extracellular material which distorts and eventually effaces the overall glomerular architecture. This process and the accumulating material has been called sclerosis. The exact chemical composition of the sclerotic material is unknown but it has been thought to be composed principally of extracellular matrix. Similarly the site of origin of the sclerotic tissue is unknown, the postulate being that its source is mainly glomerular cells. Finally, the stimuli which lead to the deposition of the sclerotic tissue have not been elucidated. Thus, there is a broad range of opportunities for research into the pathogenesis of sclerosis at both the fundamental and applied levels.

The purpose of this communication is to consider two diseases as models of progressive sclerosing diseases, focal sclerosing glomerulonephritis and crescentic glomerulonephritis. In addition, the composition of the normal extracellular matrix and the contribution of individual glomerular cells, as studied in vitro, will be presented.

Composition of the glomerular extracellular matrix

Normal distribution

The extracellular matrix in glomeruli consists of the basement membranes of Bowman's capsule, the peripheral vascular loops and the mesangial region. While there is some evidence that there is a difference in the character of the basement membrane of the peripheral vascular loop and the mesangial matrix [2], more recent evidence using immunohistochemical probes suggest that they are composed of similar components [3]. Generally speaking the extracellular matrix consists of collagenous proteins, proteoglycans, and glycoproteins. The principal collagens are of the basement membrane and cell associated type (see [4] for a review). The proteoglycans have been extensively studied by several investigators in both experimental animals and man. The principal proteoglycan in the peripheral vascular wall contains heparan sulphate glycosaminoglycan. A number of glycoproteins have been identified in the glomeruli including fibronectin, laminin, and entactin. All of these components have been found throughout the extracellular matrix, i.e. in Bowman's capsule, the peripheral vascular basement membranes and the mesangial matrix. There is some controversy about the exact localisation of fibronectin and the most recent evidence would suggest that its distribution is limited to the mesangial zone. The distribution of collagens, proteoglycans and glycoproteins within the peripheral vascular basement membrane are not known with certainty but it is believed that the proteoglycans are principally restricted to the lamina rara interna and externa.

Distribution in focal sclerosing glomerulonephritis

The distribution of type IV collagen, laminin and heparan sulphate containing proteoglycan was studied in patients with focal sclerosing glomerulonephritis at different stages [5]. The composition of the sclerotic areas was similar to the normal mesangial matrix and peripheral basement membrane. No interstitial matrix components were noted within the glomerular sclerotic areas or in the synechiae. There was, however, an increased concentration of interstitial collagen in the periglomerular interstitium adjacent to the synechiae. Thus, it appeared that intraglomerular sclerosis contained components normal to the glomerular basement membranes.

Distribution of extracellular matrix components in crescentic glomerulonephritis

Crescentic glomerulonephritis in infectious diseases and systemic vasculitis were examined. Of the former, the extracellular matrix in the crescents consisted entirely of basement membrane components. Rarely, a small amount of interstitial (type III) collagen was present, and this appeared to be associated only with extensive crescent formation. In contrast, the crescents of patients with vasculitis contained large amounts of interstitial collagen. It was significant that in these patients there was early disruption of Bowman's capsular basement membrane and an increased number of interstitial and inflammatory cells in the periglomerular region.

Summary

The components of the extracellular matrix in the glomerulus appear to be found in all basement membrane structures and in the mesangium. The techniques thus far employed do not allow an accurate determination of the relative or absolute amounts of these materials in the various glomerular locations. In those diseases where Bowman's capsule remains intact, the sclerosis appears to consist, in large part, of normal glomerular basement membrane components. Diseases, such as systemic vasculitis, which are associated with disruption of Bowman's capsule result in the early appearance of interstitial collagen within Bowman's space. Thus, it would appear that when Bowman's capsule is breached, the disrupted architecture heals by the formation of interstitial collagenous scars, whereas progressive sclerosis in the presence of intact basement membranes is characterised by the deposition of basement membrane components normal to the area.

Source of the extracellular matrix

General

Save for deposits of amyloid, immunoglobulins and possibly glycosylated proteins in diabetes mellitus, the increased extracellular matrix associated with the process of sclerosis consists primarily of locally synthesised materials. The following is a description of the current status of the knowledge in the origin of extracellular matrix in the glomerulus, using cell culture techniques.

Glomerular epithelial cells

It was initially assumed that glomerular epithelial cells synthesised the normal peripheral vascular basement membrane based on studies by Kurtz and Feldman [6]. Doubt was cast on the validity of the model in more recent studies [7]. However, using developing human embryos and mouse metanephric cultures, the contribution of glomerular epithelial cells to the peripheral glomerular vascular basement membrane has been established [8–10].

Glomerular epithelial cell cultures have been obtained in several animal species and man. Visceral epithelial cells are the first glomerular cells to appear in an outgrowth from isolated, single glomeruli [10]. They are first seen at from 7–14 days in humans [11] and if isolated from the rest of the glomerulus prior to the time that mesangial cells appear, they form a monolayer of cytokeratin-positive, large, flattened cells which have C3bi receptors on their surface [12]. These cells synthesise basement membrane collagen (type IV), heparan sulphate-containing proteoglycans, and fibronectin [13].

Glomerular mesangial cells

The glomerular mesangium contains two cell types. After its original description in light and electronmicroscopy, it was recognised that cells within the mesangium were capable of ingesting particles injected into the blood stream [14]. From this observation, the concept evolved that the mesangial cells were a part of the mononuclear phagocyte system. Electronmicroscopic studies raised the question that there was more than one cell type within the mesangium [15]. The validity of this concept was established by bone marrow transplant experiments using Chediak-Higashi mice [16]. In such mice, cells contain abnormally large lysosomes rendering them recognisable in radiation chimeras. When bone marrow from Chediak-Higashi mice was injected into syngeneic, normal recipients, their mesangial regions became populated with a small number of cells containing large lysosomes. Following injection of preformed immune complexes which deposited in the mesangial region, it was found that the cells which ingested the complexes also contained large lysosomes. These data were strong evidence that the cells in the mesangium responsible for the uptake of immune complexes were derived from the bone marrow and were distinct from the resident cells which have a contractile function. Thus, the identity of at least two separate populations of cells within the mesangium was established, one phagocytic and one contractile.

In cell culture, mesangial cells are noted to emerge from isolated, single glomeruli at a time later than that required for epithelial cells. They are spindle-shaped and rapidly form multilayers, resembling smooth muscle cells [11–13]. Their intermediate cytoskeletal filaments have not been identified. They do not have surface receptors for C3bi or Fc and do not phagocytose particles in vitro. It should be noted that there is some controversy about this point in cultures from rats. One group of investigators finds that in freshly isolated rat glomerular cells, approximately 2–8 per cent of the population contain Fc and C3 receptors and are positive for the common lymphocyte antigen [17]. A second group finds that nearly 50 per cent of cells in glomerular outgrowths are phagocytic [18] and demonstrate an oxidative burst, suggesting they are actively phagocytic. These discrepancies need to be clarified in future studies.

Glomerular endothelial cells

Glomerular endothelial cells have recently been isolated from humans and mice [19]. They are cuboidal, pavementous cells which grow in a monolayer and contain factor VIII antigen in cytoplasmic droplets. Their extracellular matrix synthesis has not been studied in detail but they appear to synthesise basement membrane collagens (types IV and VIII), glycoproteins, fibrinonectin and laminin and a heparan sulphate-containing proteoglycan. Their successful isolation and propagation is dependent on the presence of purified platelet-derived growth factor.

Influence of inflammatory mediators on isolated glomerular donor cells in vitro

Glomerular epithelial cells were either isolated from glomeruli which had been treated by brief digestion with purified collagenase or from the initial outgrowth from adherent untreated glomeruli. Mesangial cells were isolated from the glomerular fragments resulting from the collagenase digestion or from the late appearing cells in adherent single glomeruli. Endothelial cells were isolated from single glomeruli exposed to purified platelet-derived growth factor. The individual cell types were each subjected to dilute plating, isolation of small groups of cells with a cloning ring and subsequent propagation.

Proliferation and extracellular matrix synthesis was studied as previously described in the presence of materials from either platelets or inflammatory cells [13]. That from platelets consisted of a supernatant obtained by multiple freezing and thawing of isolated platelets or from purified platelet-derived growth factor. That material from inflammatory cells was obtained from either the supernatant from Zymosan-treated peritoneal macrophages or purified interleukin-1.

Neither purified plate-derived growth factor or platelet releasates influenced the proliferative capacity of mesangial cells or epithelial cells [20]. However, purified platelet-derived growth factor was required for the isolation, subsequent cloning, and proliferative capacity of endothelial cells. Crude supernatant from peritoneal macrophages was assayed on epithelial cells and mesangial cells. There was a 2.5-fold increase in cell number and tritiated thymidine incorporation in mesangial cells but no effect on epithelial cells. Interleukin-1 did not alter the proliferative response of mesangial cells in log-phase growth but when added to cultures which had come to confluence there was an increase in tritiated thymidine incorporation. These data are consistent with those recently published, demonstrating that mesangial cells in log-phase produce an autocrine which resembles interleukin-1 and has been called MC-TAF [21]. There is no effect of interleukin-1 on the proliferative capacity of glomerular visceral epithelial cells.

The effect of these inflammatory mediators on extracellular matrix synthesis has only been examined in mesangial cells. Monocyte supernatants were noted to have no effect on either total protein or collagen synthesis in cultures in either sparse or dense cultures. However, interleukin-1 stimulated both total protein and collagen synthesis in dense cultures, consistent with the effect of this material on cell turnover.

In summary, there is little known about the influence of inflammatory mediators on cell proliferation and protein synthesis in isolated glomerular cells, the current data suggest that further investigations would be productive.

References

1 Bright JR. *Guy's Hosp Rev 1836; 1:* 338
2 Huang TW. *J Exp Med 1979; 149:* 1450
3 Sariola H, Timpl R, Von der Mark K et al. *Dev Biol 1984; 101:* 86
4 Martinez-Hernandez A, Amenta PS. *Lab Invest 1983; 48:* 656
5 Striker L M-M, Killen PD, Chi E, Striker GE. *Lab Invest 1984; 51:* 181
6 Kurtz SM, Feldman JD. *J Ultrastruct Res 1962; 6:* 19
7 Striker GE, Smuckler EA. *Am J Pathol 1970; 58:* 531
8 Bonadio JF, Sage H, Cheng F et al. *Am J Pathol 1984; 116:* 289
9 Vernier RL, Birch-Anderson A. *J Ultrastruct Res 1963; 8:* 66
10 Quadracci LJ, Striker GE. *Proc Soc Exp Biol Med 1970; 135:* 947
11 Striker GE, Killen PD, Agodoa LCY et al. *Biology and Chemistry of Basement Membranes.* New York: Academic Press. 1978: 319
12 Killen PD, Striker GE. *Proc Natl Acad Sci USA 1979; 76:* 3518
13 Striker GE, Killen PD, Farin FM. *Transplant Proc 1980; 12:* 88
14 Farquhar MG, Palade GL. *J Cell Biol 1962; 13:* 55
15 Haakenstad AO, Striker GE, Mannik M. *Lab Invest 1976; 35:* 293
16 Striker GE, Mannik M, Tung MY. *J Exp Med 1979; 149:* 127
17 Schreiner GF, Kiely JM, Cotran RS, Unanue ER. *J Clin Invest 1981; 68:* 920
18 Baud L, Hagege J, Sraer J et al. *J Exp Med 1983; 158:* 1836
19 Striker GE, Soderland C, Bowen-Pope DF et al. *J Exp Med 1984; 160:* 323
20 Melcion C, Lachman L, Killen PD et al. *Transplant Proc 1982; 14:* 559
21 Lovett DH, Ryan JL, Sterzl RB. *J Immunol 1983; 130:* 1796

Open Discussion

ANDREUCCI (Naples) What do you think about the suggestion that Dr Giordano* has made regarding progression of the lesion to chronic renal failure?

STRIKER Well, I think it is very difficult to speculate about the interactions once the disease becomes well established. At that point the mesangial cells are already embedded in connective tissue. The endothelial cells may not release PGI_2 and present a hypercoaguable surface or alter in other ways. I think that later there are quite different questions. The 7/8 nephrectomy model is not a model of chronic renal disease it is a model of 7/8 nephrectomy. Renal disease does not begin by chopping out 7/8 of the kidney, it begins in the individual nephron, maybe not all and maybe not all simultaneously but certainly it does not leave 1/8 free and 7/8 totally affected. I think one needs to look at all of these models with a little bit of a jaundiced eye but certainly the 7/8 nephrectomy model is the first real window into progression in the animal.

*Giordano C. *Proc EDTA-ERA 1984; 21:* this volume

PECCHINI (Cremona) If platelet derived growth factor modulates generation of membrane products there is a theoretical basis for treatment by fibrinolytic drugs.

STRIKER Oh, I think that becomes the question of the hour. What is it that modulates endothelial cell production of anticoagulant and procoagulant factors? Certainly it suggests that the endothelial cells in the microvasculature are very different from those in the aorta, for instance: that is the only other endothelial cell that has been isolated. Endothelial cells from the glomeruli may be quite different.

RITZ (Heidelberg) I have one question relating to the product and one relating to the signal for mesangial cell stimulation. There have been reports from Yale that stimulation of mesangial cells causes Ia expansion, making these capable of participating in local immune reactions. Do you have information on this? With respect to the second point, do epidermal growth factor and oxygen radicals stimulate mesangial collagen synthesis or synthesis of other matrix components?

STRIKER Those are absolutely very current questions. It is very rare that you can cause the modulation of surface receptors with gamma interferon, even on fibroblasts. I would suspect that might also be true of mesangial cells, but we have not looked at that. The question about what modulates collagen synthesis is what brought us to start looking in this direction. I think epidermal growth factor, for instance, will drive collagen synthesis in a large number of cell types, but it also drives proliferation. We have not been able to dissect increase in synthesis or increase in cell number at this point. The problem is that mesangial cells grow everywhere, so it is hard to get them at basal conditions. As for oxygen radicals, we have looked at the question of oxygen radicals-induced injury and the difficulty is that oxygen radicals produce early cytosol changes. All we know is that you get a rush of activity but in such a short time that we cannot do pulse chase experiments.

HAWKINS (Birmingham) What causes the rupture of the basement membrane of Bowman's capsule?

STRIKER I have some ideas but I do not know if any of them are true. It is clear that macrophages are part of that very early influx of cells into the Bowman's space. Macrophages synthesise and release large quantities of proteases that are capable of digesting all types of collagen. It is quite possible that they could be an initiating factor. However, remember that Bowman's cells are sitting on their own pavement and in order to keep there they must have their own proteases so all bets are off until you ask the question of a visceral or epithelial cell. What happens when you have oxygen radical-induced injury is that they may release proteases as well. I think the question is open: we really do not know the answer.

BONE (Liverpool) Ischaemia can result in glomerulosclerosis as effectively as immunologically mediated disease. Do you have any observations on the sensitivity to oxygen tension of the mesangial cells that may help to explain this?

STRIKER Again another question that I would have to have several more pairs of hands in the laboratory to answer. I can tell you, however, that these cells are relatively insensitive to the amount of oxygen that is in the medium; for instance, I can package them up and put them in the mail and ship them to a colleague across the country. We have looked at large obsolescent glomeruli in diabetes mellitus and membranoproliferative glomerulonephritis and they have all basement membrane components. Ischaemic glomeruli, however, and I do not know where the connective tissue comes from, have interstitial collagen in Bowman's space. We did look, by the way, at the crescents of patients with post-infective glomerulonephritis. Basement membrane generally stayed intact. Vernier, a long time ago, recognised that in vasculitis the Bowman's capsule basement membrane was ruptured and there were ruptures of the vascular basement membrane. We believe they are interstitial cells, but I would not take that as gospel truth as mesangial cells also make interstitial collagen.

I guess I should conclude by saying that if you want to blame the rain on someone, you can probably blame it on me for it always rains in Seattle.

Proc EDTA–ERA (1984) Vol 21

PLASMA RENIN ACTIVITY, PLASMA ALDOSTERONE AND DISTAL URINARY ACIDIFICATION IN DIABETICS WITH CHRONIC RENAL FAILURE

J Grande, J F Macias, J Miralles, **J M Tabernero**

Hospital Clinico Universitario, Salamanca, Spain

Summary

The renin-angiotensin-aldosterone system and the acidification capacity of the renal tubule were studied in 13 diabetic patients with chronic renal failure. As a whole, the group showed hyporeninaemic hypoaldosteronism (HH). Studied alone, 12 of the 13 patients presented the requirements for HH. This group showed hypercholaemic hyperkalaemic metabolic acidosis with a disturbance in renal acidification which may be classified as Type IV renal tubular acidosis.

The results of this group were compared to those of another two groups; one of diabetic patients without chronic renal failure and the other with chronic renal failure (C Cr <40ml/min); both were seen to show different behaviour to that of the group affected by the two processes.

Introduction

The clinical characteristics of hypoaldosteronism were first described by Hudson in 1957 [1] and since then only a few sporadic cases have been reported; by 1980 50 cases had appeared in the literature, most of them secondary to hyporeninism. All these showed a defect in renal acidification, leading to hyperchloraemic hyperkalaemic metabolic acidosis. An increased incidence of cases of hyporeninaemic hypoaldosteronism (HH) has been described in Type I diabetic patients with a marked reduction in glomerular filtration (GF) and in patients with interstitial nephropathy [2,3].

The incidence of case reports of HH in diabetics is increasing steadily; however, the incidence of this process and its pathogenesis in diabetic patients remain unknown. The present work studied a randomly chosen group of Type I diabetics with a glomerular filtration of <40ml/min to determine the incidence of this defect. The influence of chronic renal failure and diabetes on the defect in renal acidification was also studied.

TABLE I

	PRA (ng/ml/min)			Plasma Aldosterone (pg/ml)		
	Supine	Upright	Post-furosemide	Supine	Up right	Post-furosemide
Control	1.12±0.43	2.42±0.66	4.72±1.73	143 ±79	251.31±19	514.18±109.54
Diabetics + CRF	0.43±0.24^{a}	0.62±0.39^{b}	0.92±0.50^{b}	74.61±55.21^{a}	115.76±80.57^{b}	153.95±103.99^{b}
CRF	2.83±1.9^{bc}	3.93±2.11^{bc}	6.06±2.45^{ac}	195.20±92.6^{ac}	273.24±83.55^{e}	343.02±82.98^{ae}
Diabetics	0.71±0.24	1.38±0.35^{a}	2.52±0.46^{bc}	107.1±31.33	210.06±128.31^{d}	416.07±172^{c}

a) $p<0.05$ Control
b) $p<0.001$ Control
c) $p<0.001$ CRF + Diabetics
d) $p<0.05$ CRF + Diabetics
e) $p<0.01$ CRF + Diabetics

PRA in (μg/ml/min) and P Aldosterone in (pg/ml) in all four groups

TABLE II

	Plasma					NH_4Cl Overload			
	C_{cr} ml/min	Na^+ mEq/L	Cl^- mEq/L	K^+ mEq/L	HCO_3^- mEq/L	pH μEq/min	NH_4 μEq/min	TA μEq/min	NAE μEq/min
Control	116.65 ± 10.67	140.8 ± 1.5	101.5 ± 2.21	3.96 ±0.36	24.48 ± 0.8	4.78 ±0.20	65.27 ±16.7	29.83 ±12.81	94.5 ±22.04
Diabetics + CRF	24.11 ± 6.74[a]	140.1 ± 2.61	109 ± 3.19[b]	5 ±0.35a	17.75 ± 0.95[a]	4.72 ±0.41	27.27 ± 6.7[a]	15.9 ± 5.78[a]	41.36 ±11.63[a]
CRF	24.23 ± 9.32[a]	137 ± 3.2[a]	104 ± 2.4	3.97 ±0.60[d]	19.53 ± 4.01af	5.22 ±0.74[c]	39.06 ±16[g]	15.45 ± 6.76g	52.26 ±24.68[a]
Diabetics	102.06 ± 24.51[d]	138.4 ± 3.8	102.3 ± 3.1[e]	3.48 ±0.44[d]	24.38 ± 1.7[d]	4.63 ±0.39[f]	47.4 ±17.7[cf]	17.97 ± 6.7[a]	66.35 ±23.05[af]

a) $p<0.001$ control
b) $p<0.01$ control
c) $p<0.05$ control
d) $p<0.001$ diabetic + CRF
e) $p<0.01$ diabetic + CRF
f) $p<0.05$ diabetic + CRF
g) $p<0.05$ CRF

Creatinine clearance and Na^+, Cl^-, HCO_3 levels and urinary acidification after short NH_4Cl overload in all four groups

Materials and Methods

This study was carried out on three groups of patients and on one control group as follows: *Group A* 23 healthy individuals as controls; *Group B* 13 type I diabetic patients with an age range of 21–67 and a C Cr <40ml/min (problem group); *Group C* nine non-diabetic patients with chronic renal failure, C Cr <40ml/min; *Group D* eight diabetic patients without renal impairment.

The age ranges of all the groups were similar. Plasma urea, Cr, Cl^-, Na^+, K^+, pH and arterial gases and C Cr and proteinuria was determined in all. Also studied in the urine of all patients and in 17 of the controls, having ascertained the absence of infection, were: pH, ammonium, titratable acidity (TA) and net acid excretion (NAE), both basal and after a short load of NH_4Cl (100mg/kg over two hours).

In all the groups erect and supine plasma renin activity and plasma aldosterone were measured and also after i.m. administration of 20mg of frusemide. All patients received a diet of about 100mEq of Na^+ daily. NH_4 was determined by titration [4]. PRA and plasma aldosterone by radioimmunoassay [5,6].

Statistical Studies One-way analysis of variance was carried out; the discrepancy between the mean values of the different groups was verified according to Newman-Keuls Multiple Range Test.

Results Results are included in Tables I and II.

Discussion

The results of the patients with diabetes mellitus and chronic renal failure show that as a group the patients had hypoaldosteronism, as shown by the absolute values of plasma aldosterone compared to the control group (Table I) and by the plasma aldosterone (ng/dl)/plasma potassium ratio of 1.45 ± 0.7, compared with 4.07 ± 0.8 ($p<0.001$) for the control group. Values lower than three considered as indicative of hypoaldosteronism [7]. Studied individually, 12 of the patients presented data sufficient to be classed as presenting hypoaldosteronism, thereby underlining the high occurrence of this process in this kind of patient. The cause of the hypoaldosteronism must be related to the decreased PRA in these patients. It is feasible that the suprarenal cortex, has normal function but that renin stimulation is not sufficient. Different studies have shown the function of the suprarenal cortex to be normal [8]. In support of this is the positive linear correlation found between plasma renin and aldosterone ($r = 0.47$) ($p<0.005$). The fact that this correlation is not as close as that shown by the control group ($r = 0.757$, $p <0.001$) could be due to the stimulating effect of the high plasma potassium. The biochemical profile of the problem group (Table II) was hyperchoraemic hyperkalaemic metabolic acidosis, coinciding with what has been described previously for hyporeninaemic hypoaldosteronism.

This disturbance in acid-base balance is due to a defect in renal acidification shown by these patients, known as renal tubular acidosis (Type IV) or distal hyperchloraemic hyperkalaemic renal tubular acidosis [9], and is characterised by an acid urinary pH, though with a low NAE which does not correspond to the urinary pH. The basal urinary findings in this group was pH 5.36 ± 0.63; NH_4: 21.89 ± 11.62μEq/min; TA: 13.94 ± 5.33μEq/min, NAE: 31.07 ± 13.85 μEq/min. These data show that with a relatively acid pH, the NH_4 and TA values are low. After a NH_4Cl load (Table II), the urinary pH was seen to decrease to 4.72 ± 0.41 ($p<0.001$), a similar value to that shown by normal individuals, though without a significant increase in NAE, fundamentally due to a failure in NH_4 excretion. The elimination of TA was also less than normal. The behaviour of the problem group compared to that of the controls is very characteristic, since the urinary pH fell in both groups similarly, though NH_4 and TA were much greater for the controls ($p<0.001$).

The deficit in aldosterone is thought to be the main factor in the pathogenesis of the defect in renal acidification [3,8]. The most feasible hypothesis is that which relates the rise in potassium, as a consequence of hypoaldosteronism and hyperglycaemia, to an inhibitory effect of NH_4 synthesis and elimination. Sebastian [10] found an inverse linear correlation between serum K^+ and the urinary excretion of NH_4 in three patients with HH receiving fludrocortisone. He also found a positive linear correlation between urinary NH_4 excretion and urinary pH and concluded that the absence of NH_4 excretion was the cause of the fall in urinary pH.

These latter data were not borne out in our patients in whom we observed a negative correlation between the NH_4 and the PH ($r = 0.588$, $p<0.05$), both in basal conditions and after a NH_4Cl overload. This behaviour is similar to that described in other disturbances in renal acidification in which reduced NH_4 secretion leads to an alkaline urine. Ditella [10] has reported similar behaviour to that of our patients in adrenalectomised rats subjected to a NH_4Cl overload. The explanation of why the urinary pH falls with low levels of NH_4 and TA is elusive. One possibility is that small amounts of hydrogen ions could be free in the urine, considerably reducing the pH.

The interpretation of the results in the problem group poses the question of whether the endocrine-metabolic disturbance is due to diabetes or to chronic renal failure, or to both processes. Our results point to the notion that diabetics without chronic renal failure are hyporeninaemic compared with the control group, with a poor response to the upright position and the administration of frusemide. However, the PRA values are higher than those found in diabetics with chronic renal failure although only significantly after frusemide ($p<0.05$). This group also showed no differences in aldosterone values compared to the control group although there is a clear elevation compared with the group of diabetics with chronic renal failure in the upright position and after frusemide ($p<0.05$ and $p<0.001$ respectively).

The group of patients with chronic renal failure also showed a significant increase in PRA compared with the controls and even higher compared to the problem group (Table I). Plasma aldosterone is higher compared with that of the diabetics with chronic renal failure but shows no differences compared with the control group.

The renal acidification of the diabetics with chronic renal failure is also different to that of the other groups of patients (Table II). The group of patients with chronic renal failure, as would be expected after a NH_4Cl overload, show less of a decrease in urinary pH with a lower excretion of NH_4 and TA compared with the control group (Table II). However compared with the problem group, the urinary pH decreases less, though NH_4 secretion is greater. The group of diabetics show a similar increase in urinary pH after a NH_4Cl overload compared with the problem group, although with a higher NAE ($p<0.05$) due to greater NH_4 secretion ($p<0.05$).

These data indicate that hyporeninaemic hypoaldosteronism is very common in patients with diabetes mellitus and chronic renal failure. It may also be seen that the endocrine-metabolic picture of this group is clearly different from that of patients with either diabetes or chronic renal failure separately. The most probable hypothesis is that diabetes is the main cause of the process, leading to chronic renal failure and hyporeninaemic hypoaldosteronism. The influence of both processes on distal tubular function induces the defect in renal acidification which is characteristic of this group of patients.

References

1 Hudson JB, Chobanian AV, Relman AS. *N Engl J Med 1957; 257:* 259
2 Caroll HJ, Farber SJ. *Metabol Clin Experim 1964; 13:* 808
3 Schambelan M, Sebastian A, Hulter HN. In Brenner BM, Stein JH, eds. *Contemporary Issues in Nephrology: Acid-Base and Potassium Homostasis.* New York: Churchill-Livingstone. 1978: 232–268
4 Jorgensen K. *Scand J Clin Lab Invest 1957; 9:* 287
5 Scaly JE, Gerten-Banes J, Larag JA. *Kidney Int 1972; 1:* 240
6 Vetter W, Vetter H, Siegenthaler W. *Acta Endocrinol 1973; 74:* 558
7 Arruda JA, Batlle DC, Shehy JT. *Am J Nephrol 1981; 1:* 160
8 Morris C, Schambelan M, Sebastian A. In Stollerman GH, ed. *Advances in Internal Medicine.* New York: Year Book Publishers. 1979: 24: 385–405
9 Sebastian A, Schambelan M, Lindenfeld S. *N Engl J Med 1977; 297:* 576
10 Ditella PJ, Sodhi B, McCreary J. *Kidney Int 1978; 14:* 466

Open Discussion

FEEST (Exeter) I question your interpretation of your data on urine acidification, ammonium excretion and titratable acid excretion in your study group. You have shown results similar to those found in any group with depressed renal function. Unless you have a very carefully matched control group of non-diabetic patients with identical age and renal function, I suspect your results are only demonstrating that your patients have a diminished number of functioning nephrons.

TABERNERO Our patients with chronic renal failure behaved differently from those patients who have chronic renal failure and diabetes.

FEEST I am not sure that this is valid unless your control group of chronic renal failure patients have identical creatinine values.

TABERNERO The creatinine clearances were similar in both groups.

ZOCCALI (Reggio Calabria) I am not surprised by your results because it is well established that some degree of renal impairment is necessary in diabetic patients to unmask type IV renal acidosis. I think however that your data could have been more convincing if you had also compared diabetic patients with chronic renal failure with a subset of chronic renal failure patients closely matched for PRA and aldosterone values.

TABERNERO I agree with you.

BIANCHI (Chairman) Do you think that the age of the patients may influence the plasma renin activity? We know that with ageing there is a slow decline in the response of arterioles to stimulation.

TABERNERO The age range of the groups we studied were comparable.

Proc EDTA–ERA (1984) Vol 21

RENIN ANGIOTENSIN ALDOSTERONE SYSTEM, URINARY PROSTAGLANDINS AND KALLIKREIN IN PREGNANCY INDUCED HYPERTENSION
EVIDENCE FOR A DISREGULATION OF THE RENIN-ANGIOTENSIN-PROSTACYCLIN LOOP

P Fievet, A Fournier, *E Agnes, †M A Herve, **A Carayon, J F de Fremont, ††A Mimran, ¶A Dufour, ¶P Verhoest, ¶J C Boulanger

*Hôpital Nord, Amiens, *Hôpital Tenon, Paris, †Société de Mathématiques appliquées à la recherche médicale et biométrique, Amiens, **Hôpital Pitie-Salpetriere, Paris, ††Hôpital La Peyronnie, Montpellier, ¶CHU, Amiens, France*

Summary

Plasma renin activity and aldosterone concentrations were measured simultaneously with urinary excretion of kallikrein and four prostaglandins (PGE_2, $PGF_{2\alpha}$, 6 keto $PFG_{1\alpha}$ and TXB_2) in 23 patients with pregnancy induced hypertension (17 with permanent PIH and six with labile PIH, since in these latter their hypertension was controlled only by home bed rest) and in 16 normotensive pregnant women at the same stage of gestation (31 ± 3 weeks). PRA was lower in permanent PIH than in controls and in labile PIH. No difference between the three groups was observed for plasma aldosterone and the urinary excretion of kallikrein and of the prostaglandins except that TXB_2 was higher in labile PIH than in permanent PIH. Correlation studies of kallikrein disclosed correlations with most prostaglandin excretions, explained by the physiological stimulation of phospholipase A2 by kallidin. Correlation studies of PRA disclosed unexpected negative correlation with PGE_2 and 6 keto $PGF_{1\alpha}$ in the permanent PIH group.

In conclusion, labile PIH has a different biological profile than permanent PIH since they have higher PRA and higher TXB_2 excretion, an association which suggests a more pronounced ureteral compression by the gravid uterus in this group. Permanent PIH has a disregulation of the renin angiotensin-prostacyclin loop since PRA and 6 keto PGF_1 are negatively correlated. This suggests the role of an independent vasopressive substance which would stimulate PGI_2 and suppress renin secretion.

Introduction

In our previous studies, pregnancy induced hypertension not responding to bed rest (permanent PIH) was shown to be characterised by a hypovolaemic state suggesting vasoconstriction, and an insufficient stimulation of the renin angiotensin aldosterone system contrasting with a more easily reactive adrenergic system. This finding could be compatible with a deficiency of production of the vasodilating prostaglandin (PGI_2) which has been shown to stimulate renin secretion and inhibit the release of noradrenaline [1] and which is more likely than PGE_2, a circulating hormone, since unlike PGE_2 it is not destroyed in the lungs. A decreased production of PGI_2 by the umbilical arteries of pre-eclamptic women has been reported by many authors and even correlated to decreased uteroplacental blood flow [2]. However these studies are in vitro studies performed after delivery and are therefore not relevant to the understanding of the mechanisms occurring at the onset of PIH. To grasp these mechanisms in vivo PGI_2 must be measured.

Methodological problems preclude adequate determination of decreased plasma PGI_2, therefore only the urinary excretion of its stable metabolites can be performed with reasonable assumption that they reflect plasma values [3]. Six women with PIH showed a 50 per cent decrease of 2–3 dinor 6 keto $PGF_{1\alpha}$ urinary excretion [4] without concomitant evaluation of the renin angiotensin aldosterone system, nor of the vasoconstrictive and pro-aggregative prostaglandin TXA_2. Evaluations of prostaglandin E_2, with natriuretic properties, have been performed more extensively in PIH by measurement of its urinary excretion, which reflects however only renal production. The results have been conflicting, with excretion in PIH being found to be either increased [5] or decreased [6].

Kallikrein excretion has been studied only once in PIH and found to be decreased compared to normal pregnant women [7]. However no simultaneous study has been performed of kallikrein and prostaglandin in spite of the known physiological link between them as kallidin (the product of the action of kallikrein on kininogen) is a physiological stimulus to phospholipase A2 [3]. Because of these scarce and conflicting data, we have studied simultaneously in PIH the renin angiotensin system and the urinary excretion of kallikrein and of four prostaglandins (6 keto $PGF_{1\alpha}$, PGE_2, $PGF_{2\alpha}$ and thromboxane B_2 the stable metabolite of thromboxane A2).

Patients and methods

Patients

Twenty-three patients with PIH have been studied. PIH was defined by finding, after the twentieth week of pregnancy, a blood pressure higher than 140/90 mmHg at the obstetrical outpatient clinic, and the absence of hypertension before the twentieth week and three months after delivery. At the end of pregnancy two types of PIH have been distinguished according to their blood pressure measured at the nephrological clinic after lying supine for 30 minutes while home bed rest (4 hours lying in the left lateral position during the day time) has been prescribed throughout the pregnancy: labile hypertension (LH) when the patients were always found normotensive at the nephrological outpatient clinic (6 patients); permanent hypertension (PH) when the patients were hypertensive at least once at the nephrological outpatient clinic (17 patients).

These two groups were compared with a control goup (C) of 16 pregnant women normotensive before, during and after pregnancy.

Methods

Protocol Studies have been performed as soon as possible after the discovery of hypertension in ambulatory patients taking no drugs and on a normal salt diet (confirmed by 24 hr sodium excretion). Blood pressure was measured after thirty minutes of supine rest and they were weighed. Blood samples were taken through a catheter inserted 30 minutes previously. Samples for PRA and aldosterone were put into ice, centrifuged immediately at 4°C, then stored at −30°C. All measurements were made within three months.

Analytical methods Plasma renin activity (PRA) and plasma aldosterone (PA) were estimated by radioimmunoassay technique [1]. Urinary prostaglandins, PGE_2, $PGF_{2\alpha}$, 6 keto $PGF_{1\alpha}$, TXB_2 were measured by radioimmunoassay after prior extraction and separation by chromatography on silicic acid [8]. Urinary kallikrein was measured by esterolytic activity on the synthetic substrate BAEE [9].

Statistical methods Comparison of the groups were made using the Wilcoxon's test for non paired data. Link between two parameters was evaluated first by linear regression analysis and then by covariance analysis.

Results

Groups comparison

There was no difference between the three groups for age, parity ratio, duration of pregnancy (30.4, 31.5 and 31.2 weeks in C, PH and LH) and weight gain during pregnancy at the discovery of hypertension and for sodium excretion (respectively 12.2, 12.2 and 10.2mmol/mmol creatinine). The weight was higher in PH than in C (respectively 73 ± 3.8, 54.1 ± 1.1kg, $p<0.01$) but not different in LH (66.7 ± 6.8kg). At the time of first attendance at the nephrological clinic the mean arterial pressure (MP) was higher in permanent PIH than in C (99.5 ± 3.8 versus 82.4 ± 1.4mmHg, $p<0.01$) but was comparable in labile PIH and in C (89 ± 3.1mmHg). However at the discovery of hypertension at the obstetrical clinic MAP was higher in permanent and labile PIH than in C (respectively 116.1 ± 2.2; 111.6 ± 1.8 and 82.4 ± 1.4mmHg, $p<0.01$).

The mean (±SEM) values of the various biological parameters observed in the three groups at the first exploration after discovery of hypertension are summarised in Table I. PRA was significantly lower in PH than in C and LH. PRA was significantly higher in LH than in C and PH. Plasma aldosterone levels were not significantly different in the three groups. Urinary kallikrein was not significantly different in the two hypertensive groups and the control group. Urinary

TABLE I. Biological data (mean±SEM)

	Plasma			Urinary excretion/mmol creatinine			
	Renin activity ng/ml/hr	Aldosterone pg/ml	Kallikrein IU	PGE_2 ng	$PGF_{2\alpha}$ ng	6 keto $PGF_{1\alpha}$ ng	TXB_2 ng
Controls (C) n = 16	6.9 ± 0.6	246.6 ± 34.1	1.29 ±0.22	36.3 ± 4.5	24.5 ± 2.9	42 ± 2.6	57 ± 4.4
Permanent hypertension (PH) n = 17	4.9* ± 0.3††	218 ± 53.7	1.38 ±0.2	33.1 ± 4.6	21 ±3	37.7 ± 4.2	48.9 ± 5.1
Labile hypertension n = 6	12.2† ± 0.9	271 ± 63.2	1.33 ±0.2	29.4 ± 6.8	19.6 ± 5.1	38.3 ±11.1	76.4** ±13.5

Significance of comparisons: permanent hypertension versus control: *$p<0.01$
labile hypertension versus control: †$p<0.01$
permanent versus labile hypertension **$p<0.05$
††$p<0.01$

excretion of prostaglandins were not significantly different in the two hypertensive groups except for TXB_2 which is significantly greater in labile than in permanent PIH. No difference between the hypertensive groups and the controls was observed. The ratio $PGE_2:PGF_{2\alpha}$ was no different in the three groups being respectively 1.47, 1.6 and 1.5 in control, permanent and labile PIH.

Correlation studies

PRA and urinary prostaglandins In the control group, the only significant linear correlation found was between PRA and $PGF_{2\alpha}$ ($r = 0.85$, $n = 15$, $p<0.001$). In labile hypertension group, no correlation was found. In the permanent hypertensive group, ($n = 17$) a negative linear correlation was found with PGE_2 ($r = -0.51$, $p<0.05$) and 6 keto $PGF_{1\alpha}$ ($r = -0.62$, $p<0.01$). Two way covariance analysis confirmed these negative correlations with a Dc of respectively 0.72 ($f<0.01$) and 0.83 ($f<0.002$). (Figure 1).

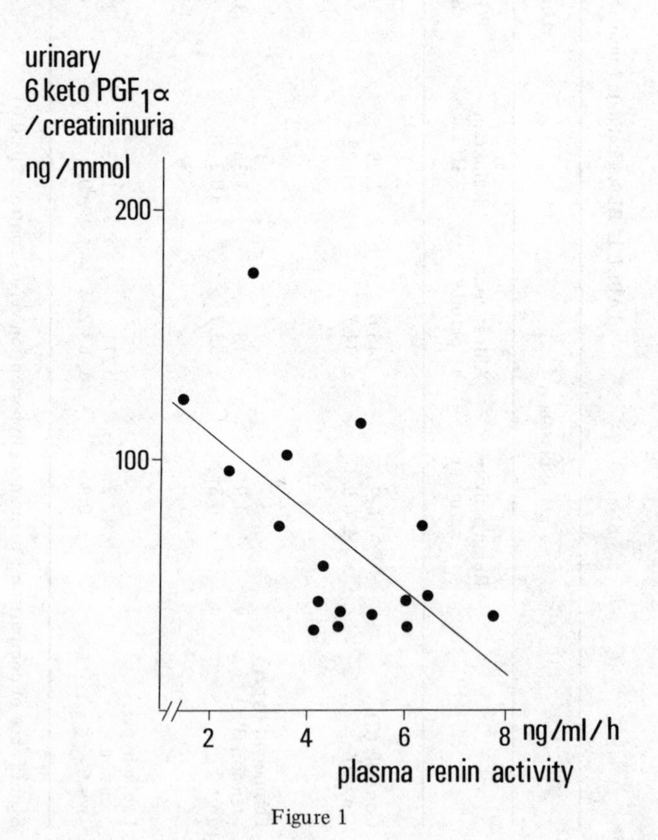

Figure 1

Kallikrein and urinary prostaglandins In the labile group, no correlation was found. In the permanent hypertensive group, a positive linear correlation was found with $PGF_{2\alpha}$ ($r = 0.55$, $p<0.05$), 6 keto $PGF_{1\alpha}$ ($r = 0.76$, $p<0.001$) and thromboxane B_2 ($r = 0.66$, $p<0.001$). These correlations were confirmed by covariance analysis for $PGF_{2\alpha}$ ($Dc = 0.39$, $f<0.03$) and 6 keto $PGF_{1\alpha}$ ($Dc = 0.72$, $f<0.001$) but not for thromboxane B_2. Covariance analysis unmasked a positive correlation between kallikrein and PGE_2 ($Dc = 0.47$, $f<0.01$).

Discussion

Our data in women with PIH soon after the discovery of hypertension at 25–35 weeks gestation, do not show significant differences with normtensive pregnant women as regards the urinary excretion of kallikrein and of the four prostaglandins PGE_2, $PGF_{2\alpha}$, 6 keto $PGF_{1\alpha}$ and TXB_2. As regards PGE_2 the main difference with the population of Pedersen, who found a decreased excretion of PGE_2, was the fact that most of his patients had proteinuria and a sodium retaining state since urinary sodium excretion was lower in his pre-eclamptic women than in the controls [6]. As regards 6 keto $PGF_{1\alpha}$, our results differ from those of Goodman who measured another metabolite, the 2–3 dinor 6 keto $PGF_{1\alpha}$, by gas chromatography-mass spectrometry, which is considered as a better index of circulating PGI_2 than the 6 keto $PGF_{1\alpha}$ [3]. Furthermore, his women were studied later in pregnancy (36 weeks) and no indication is given on their sodium diet compared to controls.

Since we had previously found that pregnancy induced hypertension could be differentiated into permanent hypertension with lower PRA and a more easily reactive adrenergic system, and labile hypertension with stimulated PRA and normally reactive adrenergic system, we differentiated our patients with PIH according to the response of their blood pressure to home bed rest in order to interpret their urinary excretion of kallikrein and prostaglandins. No difference was found between the two hypertensive groups except for the excretion of TXB_2 which has been found paradoxically higher in the labile hypertensive group. Since one cause of increased TXB_2 and renin secretion is ureteral obstruction [3], we plan to look for more severe ureteral compression by the gravid uterus in this group of patients by ultrasonography.

Correlation studies of kallikrein disclosed physiological correlations with the urinary excretion of most prostaglandins which is in accordance with the known fact that kallidin is a normal stimulus of phospholipase A2 which liberates arachidonic acid from the membrane phospholipids. Correlation studies of PRA disclosed an unexpected negative correlation in the permanent hypertension group with urinary PGE_2 and 6 keto $PGF_{1\alpha}$ whereas a physiological correlation was found in the control group between PRA and urinary $PGF_{2\alpha}$.

This negative correlation between PRA and 6 keto $PGF_{1\alpha}$ in the permanent hypertension group is quite surprising since the physiological modulation of the vascular tone (at least in the kidney) is the result of the balance between the angiotensin II induced vasoconstriction and the prostacyclin induced vasodilatation and that this balance is achieved by the following loop [3]:

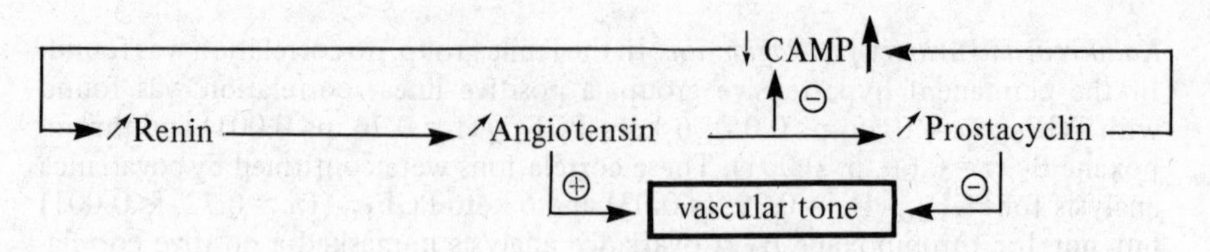

This loop implies a positive correlation between PRA and prostacyclin metabolites.

The negative correlation between PRA and 6 keto $PGF_{1\alpha}$ shows that there is a disregulation of this loop. This could be explained by the independent production of a vasoconstrictive factor which like angiotensin II would suppress renin secretion and stimulate prostacyclin synthesis. Since this latter is actually normal in our patients an inappropriate synthesis of prostacyclin in response to this stimulus is also implied. The negative correlation found by Symmonds et al [9] between renin concentration and plasma angiotensin II in PIH is compatible with this hypothesis.

In conclusion our study shows that the hypovolaemic vasoconstrictive hypertensive state of PIH is not associated with abnormal excretion of kallikrein and prostaglandins but is associated with a disregulation of the renin-angiotensin-prostacyclin loop implying an inappropriate synthesis of prostacyclin which is not stimulated in response to a hypothetical vasoconstrictive factor with renin suppressive effect.

References

1 Fievet P, Leskof L, Desailly I et al. *Proc EDTA-ERA 1983; 20:* 520
2 Maikila UM, Jouppila P, Kirkinen P et al. *Lancet 1983; i:* 728
3 Levenson DJ, Simmons CE, Brenner BM. *Am J Med 1982; 72:* 354
4 Goodman RP, Killman AP, Brash AR et al. *Am J Obst Gynec 1982; 142:* 817
5 Kovatz S,Rathaus M, Aderet NB, Bernheim J. *Nephron 1982; 32:* 239
6 Pedersen EB, Christensen NJ, Christensen P et al. *Hypertension 1983; 5:* 105
7 Elebute OA, Mills IH. In Lindheimer MD, Katz AL, Zuspan FP, eds. *Hypertension in Pregnancy.* New York: John Wiley and Sons. 1976: 329–338
8 Mimran A, Baudin G, Casellas D, Soulas D. *Europ J Clin Invest 1977; 7:* 497
9 Symonds EM, Broughton Pipkin F, Craven DJ. *Br J Obst Gynecol 1975; 82:* 643

Proc EDTA–ERA (1984) Vol 21

ALDOSTERONE RESPONSE TO MODULATION OF POTASSIUM IN PATIENTS ON DIALYSIS OR WITH ESSENTIAL HYPERTENSION

K Olgaard, E Langhoff, E Tvedegaard, S Madsen, M Hammer, H Daugaard, M Egfjord, J Ladefoged

Rigshospitalet, Copenhagen, Denmark

Summary

This investigation demonstrates in patients with essential hypertension an abnormal response of the adrenal glands to modulation of potassium metabolism by infusion of insulin-glucose. Similar results have been reported in anephric patients, while the inverse response of non-nephrectomised patients on dialysis corresponded to that of normal subjects. It is suggested that the abnormal response of patients with essential hypertension may be of importance to the understanding of the pathogenesis of this important disease.

Introduction

The pathogenesis of essential hypertension (EH) remains unknown, although an altered adrenal responsiveness to angiotensin II has been suggested [1]. Previously we have found that the aldosterone response to modulation of potassium metabolism is significantly different in anephric and non-nephrectomised patients on haemodialysis [2] as shown in Figure 1.

The present investigation, therefore, examines the adrenal response to potassium modulation in a group of patients with essential hypertension, all with low plasma renin activity as found in anephric patients. For comparison a group of control subjects has been studied.

Methods

All investigations were carried out with the patient in the supine position. After one hour of resting, basal renin activity, aldosterone, cortisol, potassium, sodium, glucose, and blood pressure were measured three times at five minute intervals. Then 100ml of 50 per cent glucose, with 16IU of crystalline insulin added, was given intravenously within five minutes. Blood samples were then drawn every 30 minutes for the next three hours. The patients were allowed to drink moderately during the tests and extra IV glucose was given at the slightest symptom of insulin hypoglycaemia.

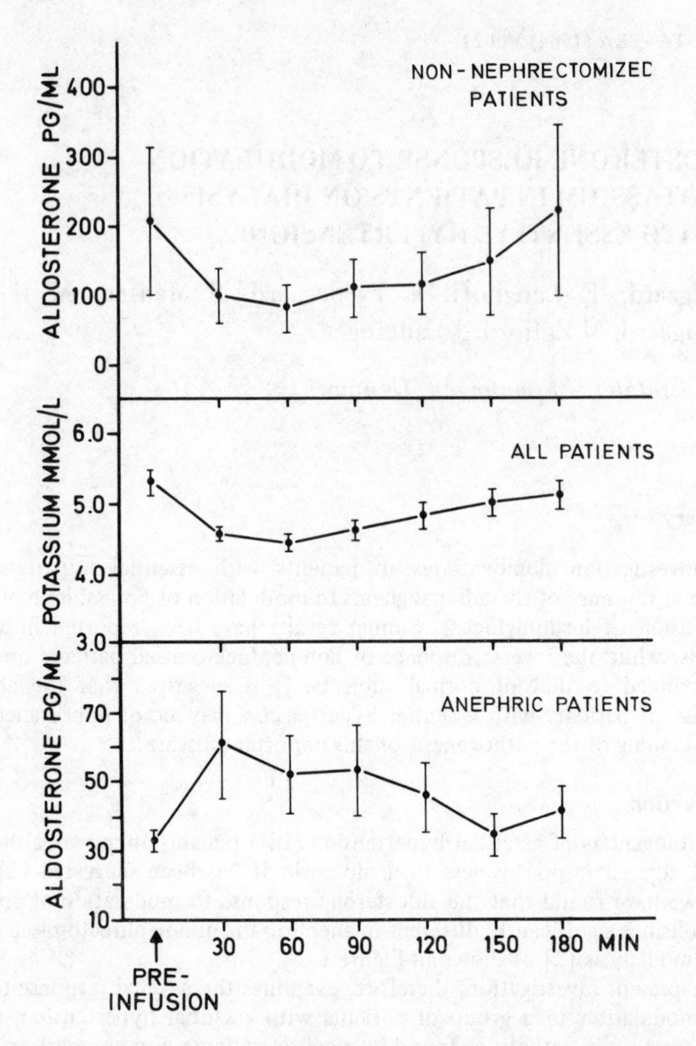

Figure 1. Mean ± SE values of plasma aldosterone and plasma potassium in six non-nephrectomised and seven anephric chronic dialysis patients after infusion of insulin-glucose. In all patients a transient decline of the plasma potassium concentration was found. However, the plasma aldosterone concentration declined transiently in the non-nephrectomised patients, while a temporary rise was seen in the anephric patients.

The investigations were repeated after 14 days of treatment with bendroflumethiazide 5mg twice daily.

Plasma renin activity was measured by RIA [3]. Control values in normal subjects were 0.4 to 0.9ng/ml/hr. Plasma aldosterone was measured by a modification of a specific RIA method [4], control values 22–223pg/ml. Potassium and sodium were determined by flame photometry. Cortisol was measured by a competitive protein binding method [5].

Data of patients with essential hypertension and of control subjects are presented in Table 1. None received any medication before the investigation. Secondary hypertension was excluded.

TABLE I. Basal data before insulin-glucose

Subjects	No.	Age (yrs)	Thiazides	BP	Cr mmol/L	K mmol/L	Plasma Renin ng/ml/hr	Aldosterone pg/ml	Cortisol μg/ml
Control	7	24	–	116/77	0.11	3.6	0.4	39	15.0
	7		+	120/75		3.3	2.5	114	16.5
Essential hypertension	8	36.5	–	175/121	0.09	3.7	0.3	90	14.7
	7		+	147/111		3.3	1.7	125	16.6

Mean values

Results

The plasma potassium concentration in patients with EH showed a transient decline during insulin-glucose infusion of the same magnitude as that of patients on dialysis. No difference was demonstrated between the response of EH-patients and control subjects.

The plasma aldosterone concentrations of the control subjects declined transiently as seen in non-nephrectomised patients on dialysis, while a temporary increase was found in the EH-patients, as previously seen in anephric dialysis patients. Figure 2 demonstrates the mean percentage changes (±SE) of the plasma aldosterone concentration in the control and EH-groups. The different responses were further magnified after treatment by thiazide.

The plasma cortisol concentrations in response to insulin-glucose followed the same pattern as that of aldosterone, although the response was not influenced by treatment by thiazide. Figure 3 demonstrates the mean percentage changes (±SE) of the plasma cortisol concentration in the control and EH-groups.

No significant changes were found in the plasma renin activity or in the plasma sodium concentrations.

Conclusions

1. A transitory decrease of the plasma potassium concentration without a change of the total body potassium content was induced by an intravenous

infusion of insulin-glucose. The decrease of plasma potassium was of the same magnitude in patients with essential hypertension and subjects with normal blood pressure. It seems most likely that the decrease of plasma potassium was secondary to a displacement of potassium from the extra- to the intracellular space.

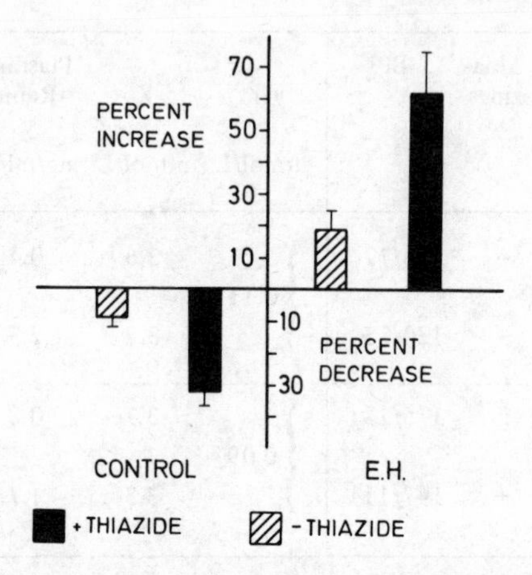

Figure 2. Percentage changes (mean ±SE) of plasma aldosterone after insulin-glucose in the control group and in the group with essential hypertension, before and after treatment with thiazide for two weeks. A significant increase of plasma aldosterone ($p<0.001$) was found in the group with essential hypertension compared to the control group. This difference was further magnified by treatment with thiazide

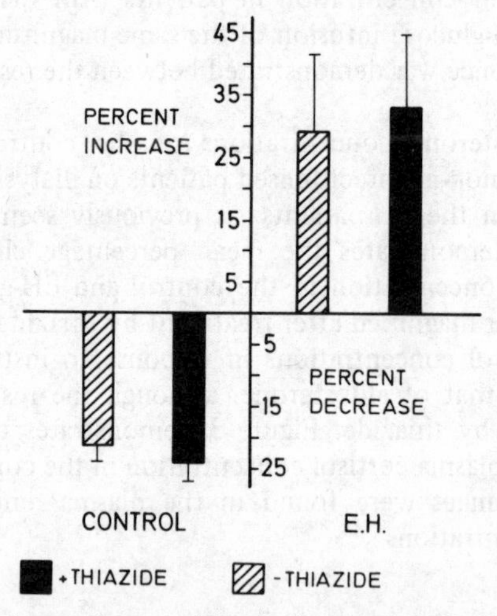

Figure 3. Percentage changes (mean ±SE) of plasma cortisol after insulin-glucose in the control group and in the group with essential hypertension before and after treatment with thiazide for two weeks. A significant increase of plasma cortisol ($p<0.01$) was found in the group with essential hypertension compared to the control group. No influence was found of treatment with thiazide on the cortisol response

2. In untreated patients with essential hypertension a significant increase of the plasma aldosterone concentration was found in response to insulin-glucose. In contrast, a significant decrease of plasma aldosterone was found in a corresponding group of control subjects with normal blood pressure. The different response between the two groups was even greater after sodium depletion.

3. Similarly, the response of the plasma cortisol concentration was significantly increased after insulin-glucose in patients with essential hypertension, but not in control subjects with normal blood pressure. No effect of sodium depletion was found on the plasma cortisol response.

4. The effect of insulin-glucose on aldosterone and cortisol was probably not mediated via changes in plasma renin activity.

5. Thus, the present investigation has demonstrated an abnormal response of the adrenal gland to modulation of potassium by insulin-glucose in patients with essential hypertension. It is suggested that this may be a factor of pathogenetic importance.

Acknowledgement

This work was supported by the Danish Medical Research Council.

References

1 Williams GH, Hollenberg NK, Moore TS et al. *J Clin Invest 1979; 63:* 419
2 Olgaard K. *Acta Med Scand 1975; 198:* 213
3 Olgaard K, Ladefoged J. *Ugeskr Laeg 1977; 139:* 1590
4 Olgaard K. *Scand J Clin Lab Invest 1975; 35:* 31
5 Olgaard K, Madsen S. *Clin Biochem 1976; 9:* 265

Open Discussion

TOURKANTONIS (Thessaloniki) I would like to ask you what was the plasma renin activity in patients with essential hypertension? May be that in such a group of patients with various values of plasma renin activity (low, normal or high) the response of aldosterone and cortisol to insulin infusion is different.

OLGAARD These are all low in essential hypertension patients and the renin value in the resting stage, not sodium depleted, was a mean value of 0.3ng/ml/hr.

COLUSSIG (Milan) Did you check the blood glucose in your patients? Can you exclude that hypoglycaemia could have induced an increase in ACTH function?

OLGAARD Yes, I don't know the mechanism of this. I don't know if this is going to take place via modulation of potassium or changes in the phosphate concentration. The blood glucose was the same in the two groups at 4.8mmol/L and this increased to a maximum value of 6.5mmol/L, while the lowest value was about 3.3mmol/L. I don't think there has been any stimulation by hypoglycaemia.

Proc EDTA–ERA (1984) Vol 21

ROLE OF PROSTAGLANDINS IN CAPTOPRIL-INDUCED NATRIURESIS

G Conte, **A Dal Canton,** G Maglionico, C Di Spigno, B Pirone, D Russo, G Di Minno, V E Andreucci

Second Faculty of Medicine, University of Naples, Italy

Summary

To determine whether prostaglandins (PG) contribute to captopril-induced natriuresis, 20 hypertensive subjects were assigned to one of the following three groups: *group a,* captopril (C) administration; *group b,* C + indomethacin (I); *group c*, I alone. Captopril was given in a dose of 100, 200, 400mg/day and indomethacin in a dose of 100mg/day for one week. In *group a* (n=10), natriuresis was clearly increased in the seven day periods with captopril in a dose of 200 and 400mg/day but not at 100mg/day. After captopril 200 or 400mg/day, but not 100mg/day, urinary PGE_2 and PGI_2 excretion significantly increased while filtration fraction fell due to a rise in renal plasma flow. Plasma aldosterone (PA) significantly decreased after C($p<0.05$). In *group b* (n = 7), natriuresis disappeared during captopril 200 or 400mg/day and indomethacin administration even when PA decreased as in *group a.* In *group c* (n = 3), natriuresis was unchanged. In conclusion, natriuresis by C is critically dependent upon increased secretion of PG.

Introduction

Captopril (C), unlike the other nondiuretic antihypertensive drugs generally produces mild to moderate natriuresis in spite of its potent hypotensive effect [1,2]. Several authors [3,4] have demonstrated that C increases plasma immunoreactive prostaglandin (PG) secretion in healthy and hypertensive subjects. It is known that PG may function as natriuretic factors under physiological and pharmacological conditions [5].

Our purpose was to determine the role of PG in mediating the captopril-induced natriuresis in hypertensive patients both by measuring the changes in urinary PG excretion after C and by assessing the effect of PG synthetase inhibition on the natriuretic response to captopril.

Patients and methods

Twenty patients with essential hypertension were hospitalised during the study. In all subjects a moderate to mild hypertension (diastolic blood pressure 95–110 mg/Hg) had been documented for at least six months in an outpatient setting. Antihypertensive drugs were discontinued at least 10 days before hospitalisation. The diagnosis of essential hypertension was based upon the absence of known aetiologies as assessed by urinary catecholamines, plasma aldosterone and renin activity, intravenous pyelogram and, whenever indicated, renal arteriogram and renal vein renin activity.

All subjects were men. Their mean age (±SE) was 41.4 ± 3.1 years. In all subjects creatinine clearance was greater than 80ml/min. Before C administration, all subjects ingested constant isocaloric diets containing 100mEq of sodium and 1,200mg of phosphorous daily until metabolic balance was achieved (usually by the fifth hospital day). Blood pressure was recorded twice a day at 8 am and 4 pm and the mean of these two values was taken as representative of that day. Average blood pressure (BP) on admission was 158/104 (±8/4SE).

The subjects were assigned to one of three groups after a three day baseline period. *Group a:* 10 subjects received C for seven days. *Group b:* subjects were given indomethacin (I) plus C for seven days. *Group c:* three subjects received only I. Captopril was given in a dose of 100, 200 or 400mg/day and indomethacin in a dose of 100mg/day given in four divided daily doses (at 8 am, at 2 pm, at 8 pm and 2 am).

Blood was taken at 8 am every day in both the baseline and the experimental period, after overnight recumbency and fasting, for creatinine and electrolyte (sodium, potassium, phosphorous) determination. Blood samples for plasma aldosterone were drawn after two hours of upright position in the last day of baseline and experimental periods and were collected on ice, centrifuged and separated immediately. Twenty-four-hour urinary excretion of electrolytes and creatinine was determined daily. Twenty-four hour urinary excretion of PGE_2 and 6-keto-prostaglandin $F_{1\alpha}$ as well as PAH clearance were determined only in the last day of baseline and experimental period. For PG measurement, urine was refrigerated immediately after collection and stored at $-20°C$ until assayed.

Plasma aldosterone was measured by specific radioimmunoassay (Sorin Biomedica, Saluggia). Sodium and potassium were measured by flame photometry, creatinine by a Beckman autoanalyser and phosphorous by Subarrow's method. Measurement of PGE_2 and 6-keto-$PGF_{1\alpha}$ (the stable hydrolysis product of prostacyclin) in urine was carried out according to Ciabattoni et al [6] with minor variations.

Results were expressed as mean ± SEM. Two tailed Student's 't' test was used for statistical comparisons of the means.

Results

Group a. Captopril at 200 and 400mg/day caused a significant rise in urinary Na^+ excretion, but not at 100mg/day (Figure 1). The difference from the mean

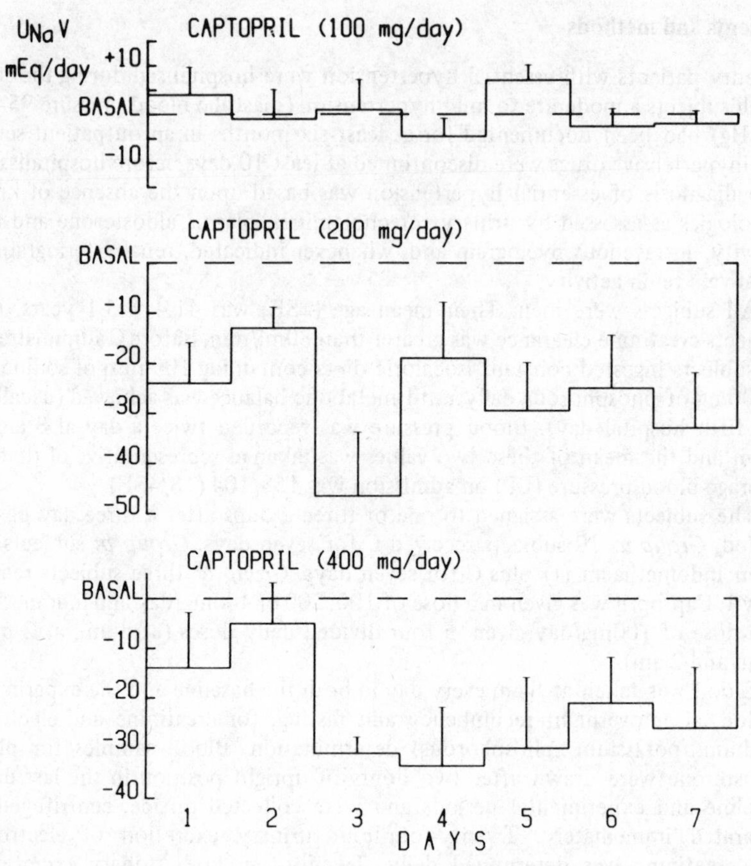

Figure 1. Deviations in $U_{Na}V$ from average baseline value caused by captopril

baseline excretion averaged 26.9 ± 4.7 and 25.7 ± 0.9mEq/day in the seven day periods at the dosage of 200 and 400mg/day, respectively. In the seven patients with captopril-induced natriuresis, $\triangle U_{Na}V$ significantly correlated with $\triangle U_{HPO_4}V$ ($p<0.01$). After captopril 200 or 400mg/day, but not 100mg/day, the urinary excretion of PGE_2 and 6-keto-$PGF_{1\alpha}$ significantly increased (6.2 ± 2.1ng/hr pre-C, 15.1 ± 3.3ng/hr after, $p<0.05$; 11.8 ± 3.2ng/hr pre-C, 25.0 ± 8.1ng/hr after, $p<0.05$, respectively) in conjunction with the increase in renal plasma flow (530 ± 7ml/min pre-C, 670 ± 10ml/min after, $p<0.001$). The glomerular filtration rate, calculated as creatinine clearance, showed no change. PA significantly decreased after C (257.7 ± 69.9 pre-C, 166.6 ± 38.4pg/ml after, $p<0.05$). Mean systolic and diastolic BP decreased from 159 ± 10 up to 132 ± 8 and from 116 ± 6 to 91 ± 5mmHg ($p<0.01$).

Group b. The administration of captopril (200 and 400mg/day) plus indomethacin (100mg/day) induced minimal changes in sodium excretion ($p>0.05$) (Figure 2) even if PA significantly decreased (270.7 ± 62.1 pre, 193.6 ± 51.8pg/ml after; $p<0.05$). Urinary excretion of PGE_2 and 6-keto-$PGF_{1\alpha}$ significantly decreased (7.1 ± 1.8ng/hr pre, 3.5 ± 0.6ng/hr after, $p<0.05$; 11.1 ± 1.9ng/hr pre, 4.7 ± 1.5ng/hr post, $p<0.05$ respectively). In this group administration of captopril plus indomethacin did not modify creatinine and PAH clearance. Mean systolic and diastolic BP significantly decreased from 154 ± 8 to 140 ± 5 and from 113 ± 4 to 101 ± 3mmHg respectively ($p<0.05$).

Group c. Administration of indomethacin (100mg) did not modify sodium excretion (Figure 2), creatinine and PAH clearance, and mean BP while significantly decreasing urinary excretion of PGE_2, 6-keto-$PGF_{1\alpha}$ and PA.

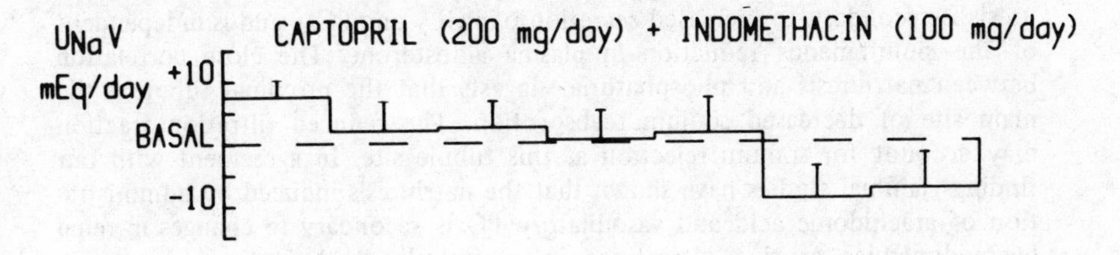

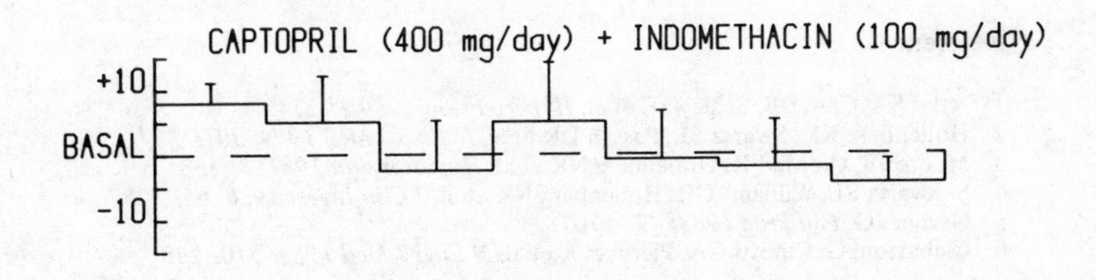

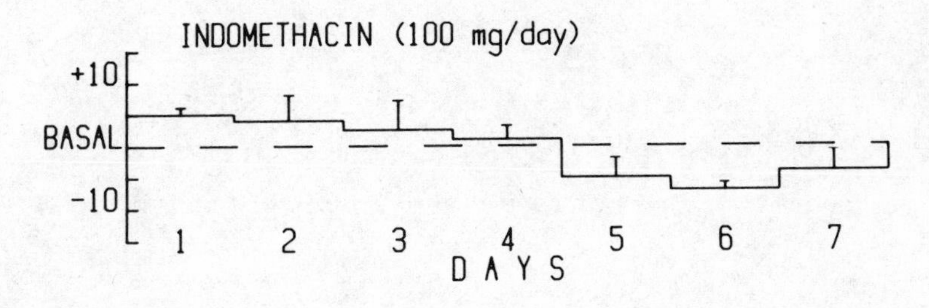

Figure 2. Deviations in $U_{Na}V$ from average baseline value caused by administration of captopril plus indomethacin and indomethacin alone

Discussion

This study confirms that captopril increases urinary sodium excretion in hypertensive patients even while reducing blood pressure [1,2]. *Group a* studies indicate that this natriuresis, due to decreased tubular reabsorption, occurs in patients treated with high doses of captopril (200–400mg/day), but not with low doses (100mg/day) and is associated with increased urinary PGE_2 and PGI_2 excretion and PAH clearance. As expected [1], plasma aldosterone significantly decreased after captopril. *Group b* studies were undertaken to verify the effect of PG inhibition on captopril-induced natriuresis. The results show that indomethacin fully antagonises this natriuresis even if the fall in plasma aldosterone is maintained as in *group a.* Finally, *group c* studies exclude that changes in tubular handling of sodium are the consequence of a direct, pharmacological action of indomethacin on the tubule.

Therefore, from our findings we can conclude that the C-dependent natriuresis is secondary to increased secretion of PGE_2 and PGI_2 and is independent of the simultaneous reduction in plasma aldosterone. The close correlation between natriuresis and phosphaturia suggests that the proximal tubule is the main site of decreased sodium reabsorption. The reduced filtration fraction may account for sodium rejection at this tubule site. In agreement with our findings, animal studies have shown that the natriuresis induced by administration of arachidonic acid and vasodilatory PG is secondary to changes in renal haemodynamics causing alterations in peritubular hydrostatic and oncotic pressure and thereby affecting proximal tubular reabsorption [5].

References

1 Atlas SA, Case DB, Sealey JE et al. *Hypertension 1979;1:* 274
2 Hollenberg NK, Swartz SL, Passan DR et al. *N Engl J Med 1979; 301:* 9
3 Moore TJ, Crantz FR, Hollenberg NK et al. *Hypertension 1981; 3:* 168
4 Schwartz SL, Williams GH, Hollenberg NK et al. *J Clin Invest 1980; 65:* 1257
5 Gerber JG. *Fed Proc 1983; 42:* 1707
6 Ciabattoni G, Cinotti GA, Pierucci A et al. *N Engl J Med 1984; 310:* 279

Proc EDTA–ERA (1984) Vol 21

EFFECT OF ZINC TREATMENT ON CELL MEDIATED IMMUNITY OF CHRONIC RENAL FAILURE PATIENTS

D Grekas, N Tsakalos, Z Giannopoulos, A Tourkantonis

AHEPA Hospital, University of Thessaloniki, Thessaloniki, Greece

Summary

This study was performed on 13 normal subjects and 34 patients with chronic renal failure (CRF). Oral elemental zinc at a dose of 60mg/day for four weeks was administered. Delayed type hypersensitivity (DTH) skin tests were performed using the multitest method. Monocyte accessory cell function (MAF) was tested by their ability to support T-cell colony growth in semi-solid cultures stimulated by staph-protein A (SpA). After zinc treatment a significant increase of DTH and MAF was found in all patients with advanced CRF in comparison with controls. These results provide evidence that zinc treatment may restore the impaired cellular mediated immunity of uraemia.

Introduction

Most patients with chronic renal failure (CRF) or on maintenance haemodialysis (HD) have plasma zinc values below those observed in normal subjects [1]. The significance of hypozincaemia in uraemia is however controversial since low plasma zinc levels do not always reflect tissue zinc deficiency [2].

Studies of uraemic patients and of animals with experimentally induced uraemia have shown that the immune response is severely altered. The most pronounced changes are found in cell-mediated immunity [3]. Zinc deficiency affects the development, maintenance, and function of immunocompetent lymphocytes in both non uraemic animals and man [4]. Human zinc deficiency, seen in acrodermatitis enteropathica or in parenterally alimented patients not supplemented with zinc, causes T-cell lymphopenia, cutaneous anergy and depressed PHA-stimulated T-cell mitogenesis [5,6].

The aim of the present study was to examine the effect of hypozincaemia and zinc repletion on the in vivo and in vitro status of cell-mediated immunity (CMI) in chronic renal failure patients.

Patients and methods

Twenty-six patients with CRF and eight patients on maintenance haemodialysis were included in the study. The mean age of these patients was 43.7 ± 8.8 years and 36.2 ± 10.9 respectively. CRF patients were divided in two groups: CRF_1 of 12 patients with moderate uraemia (Ccr ⩾25ml/min) and group CRF_2 of 14 patients with advanced uraemia (Ccr <25ml/min). Thirteen normal subjects aged 42 ± 13 years were used as sex and age matched healthy controls. CRF patients were on a protein restricted diet and received daily supplements of multivitamins, folic acid and aluminium hydroxide antacids. Haemodialysis patients were on a non-restricted diet and were dialysed for four hours thrice weekly with hollow fibre dialysers. All patients were in good clinical condition at the time of the study.

Oral elemental zinc was administered as zinc acetate (E Merck, Darmstadt) at a dose of 60mg/day (200mg of zinc acetate) for four weeks in both controls and patients. Before and after zinc treatment an assessment of CMI was made by in vivo and in vitro studies. For the in vivo study, delayed type hypersensitivity (DTH) skin tests were performed with the multitest method (IMC, Institut Merieux). The antigens employed were: streptococcus (SK), tuberculin (PPD), candida albicans (CA) and trichophyton (TR). For in vitro study monocyte accessory cell function (MAF) was tested by the ability of these cells to promote T-cell colony formation (TcCF) in semi-solid agarose cultures stimulated by staph-protein A (SpA, Pharmacia). All microcultures were performed in flat-bottomed microwell plates (Falcon Plastics, Oxnard, CA) by using the single layer technique. $5X10^4$ T-cell enriched preparations and $2X10^4$ macrophages (MΦ) per microwell were used. All T-cell cultures were stimulated with 10μg/ml SpA. Colony responses were enumerated after incubation for seven days at 37°C in a humidified atmosphere of 95 per cent air and five per cent CO_2. Distinct aggregates consisting of 15 or more cells were scored as colonies with a Zeiss 135 inverted phase contrast microscope. The number of colonies were expressed as the mean ±SD of triplicate samples.

For zinc analysis blood samples were drawn from patients and controls after overnight fasting and were collected in plastic tubes to avoid zinc contamination. Plasma zinc concentrations were measured by atomic absorption spectrophotometry (Perkin Elmer 305 A) after dilution 1:1 with deionised water.

Results

At the dose of 200mg daily (60mg of elemental zinc), zinc acetate was well tolerated and no side effects were recorded. Plasma zinc concentration was also restored within normal limits (70–130μg/100ml) in both groups of patients after four weeks with this regimen.

Table I illustrates the results of zinc treatment on skin testing (mean of all antigens) in controls and patient groups. The score was measured as the mean of two diameters of the skin induration [4]. Score equal to or greater than 2mm was considered as positive. The response to skin tests after zinc treatment indicated slight increase of DTH in both groups of controls and CRF_1, but was significantly increased in CRF_2 and haemodialysis patients ($p<0.001$ and $p<0.05$ respectively) in comparison with the pre-treatment skin response. Twenty-five to thirty per cent of uraemic patients who were anergic to skin testing before zinc treatment showed positive skin reactions in one or more skin antigens after zinc treatment. The effect of zinc treatment on TcCF is shown in Table II. TcCF of patients before treatment showed a significant difference ($p<0.01$) when compared to controls. After zinc treatment a significant increase ($p<0.01$) of TcCF was found in both groups of patients in comparison to their pre-treatment mean values of TcCF. The addition of indomethacin in the cultures pre- and post-zinc treatment did not change the above results (data not shown).

TABLE I. Effect of zinc treatment on delayed type hypersensitivity skin testing. The score measured by the mean (±SD) of two diameters of the skin induration. Score ⩾2mm was considered as positive

Groups	n	Before	After	p
Controls	13	3.3 ± 3.5	4.7 ± 3.9	NS
CRF_1	12	2.6 ± 2.4	3.1 ± 2.1	NS
CRF_2	14	0.6 ± 1.3	2.3 ± 2.8	<0.001
HD	8	0.9 ± 1.7	2.3 ± 2.8	<0.05

CRF_1 : patients with moderate uraemia
CRF_2 : patients with advanced uraemia
HD: haemodialysis patients

TABLE II. Effect of zinc treatment on T-cell colony formation. The number of colonies were expressed as the mean ±SD of triplicate samples

Groups	n	Before	After	p
Controls	4	140.2 ± 17.5	141.2 ± 17.2	NS
CRF_1	6	90.5 ± 18.7*	125.7 ± 16.7	<0.01
CRF_2	7	88.6 ± 19.8*	127.6 ± 19.5	<0.01

*p<0.01 (compared to controls)
CRF_1 : patients with moderate uraemia
CRF_2 : patients with advanced uraemia

Discussion

The importance of zinc in maintaining normal CMI is emphasised by the severe immunodeficiency found in animals with experimental zinc deficiency and in patients with either congenital or acquired zinc deficiency [3,4,6–8]. These studies indicate that zinc is required for normal immune function and may have a key role in regulating some lymphocyte functions.

Our data showed depression of both in vivo and in vitro T-lymphocyte response during hypozincaemia in most uraemic patients. Reversion to normal or nearly to normal limits occurred four weeks after the administration of zinc as a sole supplement of nutritional therapy. Skin test reactivity also reappeared in 25–30 per cent of uraemic patients who were anergic before zinc supplementation.

The present results tend to confirm the findings of previous investigations [4,6–8]. First, the restoration of plasma zinc levels was accompanied by similar increase of DTH reactivity. Second, the increased plasma zinc levels was also associated with an improvement of T-cell colony responses. Since the clonal growth of T-lymphocytes is crucially dependent on the presence of MΦ [9] it can be said that the increase responses of T-lymphocytes reflected a better accessory function of MΦ induced by zinc treatment.

One possible interpretation for the improved responses of T-lymphocytes might be an increase in the interleukin-1 production of MΦ after zinc treatment. However, our preliminary results showed that there were not significant changes in the secretion of this monokine before and after zinc treatment. At the moment the likeliest explanation for the immuno-enhancement seen after zinc treatment may be caused by an improvement of cell membrane functions. The latter is probably related to the requirement of zinc for DNA synthesis [10]. Thus, the ability of zinc to promote T-cell responses may represent a beneficial effect of zinc treatment on the defective antigen processing or presentation function of monocytes in hypozincaemic uraemic patients.

Therefore, although the above mechanisms are not yet clear, the present data justify a prospective study of the long-term effects of dietary zinc supplementation in uraemic patients. This treatment is inexpensive, non-toxic and could significantly improve immune function in CRF.

References

1 Grekas D, Mosleh J, Savatis S et al. *Trace Elem Med 1984; 1:* 21
2 Kerr DNS. *Proc 8th Int Congr Nephrol 1981;* 1014
3 Alevy YG, Mueller KR, Hutcheson P et al. *J Lab Clin Med 1983; 101:* 717
4 Good RA. *J Clin Immunol 1981; 1:* 3
5 Oleske JM, Westphal ML, Shore S et al. *Am J Dis Child 1979; 133:* 915
6 Allen JI, Kay NE, McClain CJ. *Ann Int Med 1981; 95:* 154
7 Gross RL, Oskin N, Fong L et al. *Am J Clin Nutr 1979; 32:* 1260
8 Allen JI, Perri RT, McClain CJ et al. *J Lab Clin Med 1983; 102:* 577
9 Tsakalos ND, Lachman LB, Newhouse Y, Whisler RL. *Cell Immunol 1984; 83:* 229
10 Williams RO, Loeb LA. *J Cell Biol 1973; 58:* 594

Open Discussion

RAMZY (Cairo) How can you explain the correction of skin anergy by zinc supplementation only when it is known that other serum factors, such as middle molecules, are present and your patients are still uraemic?

GREKAS It is well known that the plasma of uraemic patients can inhibit leucocyte activation in cultures. From our data we can't say that other factors don't contribute in uraemic patients to depressed lymphocyte activity and after zinc in activation of leucocyte activity. I can say that we can't exclude other uraemic factors.

RAMZY But have your patients become reactive to skin tests by the administration of zinc? In your study you have excluded all other possible factors and concentrated only on zinc.

GREKAS During the study all our patients were in the same clinical state. None were given multivitamins or other preparations or other medicinal support and in this way we can say that the zinc was the only additional treatment which influenced the activities.

BIANCHI (Chairman) Is the restoration of delayed hypersensitivity complete or partial?

GREKAS We used more than one skin antigen and with the test we assess all seven of them. We showed that uraemic patients after zinc supplementation had a positive skin reaction to at least one skin antigen or more. It means that delayed type hypersensitivity is restored in uraemic patients after zinc supplementation.

UNKNOWN Do you have any evidence that the humoral immunity is improved after zinc treatment?

GREKAS We measured the immunoglobulins pre and post zinc treatment but there was no difference.

Proc EDTA–ERA (1984) Vol 21

ADULT HEIGHT IN PAEDIATRIC PATIENTS WITH CHRONIC RENAL FAILURE

***G Gilli**, K Schärer, O Mehls

*University Children's Hospital, Heidelberg, *F Nightingale Children's Hospital, Düsseldorf, FRG*

Summary

Forty-nine paediatric patients at different stages of chronic renal failure were followed until adult height was attained. Mean age at completion of growth was close to normal. Adult height was evaluated against population-specific standards. It was less than 2 SD under the mean in two of 18 patients on conservative treatment, five of 18 during dialysis and none of 13 after transplantation, but correlation with mode of treatment did not reach significance. These findings contradict previous data stating that most of these children become small adults. Although analysis of the data suggests that many children with renal failure remain below their genetic potential of growth, final stunting appears infrequent.

Introduction

Stunted growth is a well-recognised feature of children with chronic renal failure, but little data exists detailing the adult height attained by these patients [1,2]. In a study of 34 children who were examined after completion of growth, eight (23.5%) had heights below the third centile, i.e. 1.88 SD below the mean [2]. In contrast, the EDTA Registry has reported that 38 (77.5%) of 49 dialysed or transplanted paediatric patients reached adult heights less than 2.0 SD below the reference mean [2]. A major bias of the EDTA study is that growth standards for Dutch children (who are among the tallest in Europe) were used as reference standards for patients from all over Europe.

The present report reviews the growth data of 49 children with chronic renal failure who reached their adult heights while on conservative treatment, during dialysis, or after transplantation. Because these patients belong to different populations, their growth data have been assessed against their population-specific standards.

TABLE I. Clinical and developmental data of 18 boys with chronic renal failure

No.	Diagnosis	AT THE FIRST OBSERVATION					AT COMPLETION OF GROWTH				Predicted adult height (cm)	Prediction error (cm)
		Age (years)	S_{Cr} (mg/dl)	Treatment	Height (cm)	Height SDS	Age (years)	Treatment	Adult height (cm)	Height SDS		
1	chronic GN	6.5	–	HD	116.0	–0.61	–	TP	171.4	–0.82	–	–
2	obstructive uropathy	10.4	1.3	CT	140.0	–0.29	17.3	CT	177.9	+0.13	178.2	+ 0.3
3	chronic PN	11.4	2.8	CT	151.9	+0.74	16.9	CT	179.2	+0.32	184.9	+ 5.7
4	chronic GN	12.5	4.4	CT	157.3	+0.66	17.6	TP	177.5	+0.07	183.6	+ 6.1
5	chronic GN	12.7	1.0	CT	154.2	+0.07	–	TP	178.0	+0.14	181.9	+ 3.9
6	nephrocalcinosis	13.0	3.8	CT	155.4	–0.05	19.3	TP	168.7	–1.25	181.0	+12.3
7	obstructive uropathy	13.2	–	HD	132.0	–3.05	–	TP	170.0	–1.06	169.4	– 0.6
8	chronic GN	13.5	1.2	CT	142.5	–1.31	17.9	CT	159.5	–1.89	–	–
9	gout nephropathy	13.5	1.5	CT	173.0	+1.65	–	CT	182.9	+0.88	182.9	0
10	obstructive uropathy	13.8	1.5	CT	148.5	–1.50	21.1	HD	171.0	–0.91	172.7	+ 1.7
11	chronic GN	14.1	1.0	CT	158.5	–0.58	16.4	CT	163.0	–2.11	167.9	+ 4.9
12	renal hypoplasia	14.4	–	HD	155.8	–1.19	–	HD	162.0	–2.26	165.1	+ 3.1
13	chronic GN	14.4	–	HD	164.2	–0.20	17.2	HD	168.5	–1.28	172.4	+ 3.9
14	nephronophthisis	14.5	1.6	CT	151.1	–1.85	19.0	HD	168.7	–1.25	166.4	– 2.3
15	renal hypoplasia	14.8	–	HD	155.4	–1.10	18.5	TP	161.2	–1.64	162.2	+ 1.0
16	chronic GN	15.2	–	HD	157.4	–1.74	18.4	TP	168.8	–0.89	168.7	– 0.1
17	obstructive uropathy	15.7	–	HD	163.5	–1.37	17.0	HD	171.4	–0.49	173.3	+ 1.9
18	nephronophthisis	16.2	6.9	CT	157.7	–2.44	20.4	TP	169.7	–1.10	167.9	– 1.8

GN= glomerulonephritis
PN= pyelonephritis
CT= conservative treatment
HD= haemodialysis
TP= transplantation

TABLE II. Clinical and developmental data of 31 girls with chronic renal failure

No.	Diagnosis	AT THE FIRST OBSERVATION Age (years)	S_{Cr} (mg/dl)	Treatment	Height (cm)	Height SDS	AT COMPLETION OF GROWTH Age (years)	Treatment	Adult height (cm)	Height SDS	Predicted adult height (cm)	Prediction error (cm)
1	chronic GN	6.0	1.0	CT	122.0	+1.44	15.3	CT	175.0	+1.82	173.0	−2.0
2	segmental hypoplasia	8.3	2.0	CT	117.5	−2.51	15.8	CT	157.8	−1.13	155.6	−2.2
3	tubular acidosis	9.7	1.1	CT	130.0	−1.20	17.4	CT	162.8	−0.27	164.8	+2.0
4	cystic kidney	9.8	1.2	CT	137.5	−0.02	14.8	CT	162.0	−0.41	163.0	+1.0
5	nephronophthisis	10.7	2.3	CT	127.6	−2.44	–	HD	157.0	−1.27	156.7	−0.3
6	chronic GN	10.8	2.4	CT	151.0	+1.31	13.6	CT	156.2	−1.40	160.9	+4.7
7	chronic GN	10.9	1.0	CT	137.4	−1.04	16.7	CT	155.5	−1.52	157.5	+2.0
8	chronic GN	11.0	1.3	CT	145.0	+0.17	16.2	CT	171.3	+1.18	166.4	−4.9
9	chronic GN	11.5	2.8	CT	124.5	−1.98	16.5	CT	155.1	−0.82	154.1	−1.0
10	chronic PN	11.6	2.7	CT	124.7	−3.44	–	HD	156.0	−1.44	154.2	−1.8
11	nephronophthisis	11.8	5.2	CT	141.7	−1.04	–	TP	159.0	−0.92	162.2	+3.2
12	obstructive uropathy	11.9	–	HD	148.3	−0.22	17.9	TP	160.3	−0.70	169.5	+9.2
13	renal hypoplasia	12.1	1.4	CT	147.9	−0.44	15.1	CT	161.3	−0.53	159.7	−1.6
14	renal hypoplasia	12.2	–	HD	145.0	−1.03	–	HD	162.2	−0.38	–	–
15	renal hypoplasia	12.2	–	HD	153.0	+0.21	14.9	TP	161.5	−0.49	161.7	+0.2
16	nephronophthisis	12.4	2.8	CT	148.3	−0.62	14.5	TP	155.0	−1.61	154.1	−0.9
17	nephronophthisis	12.6	–	HD	149.1	−0.67	–	TP	158.0	−1.10	158.0	0
18	renal hypoplasia	12.7	–	HD	152.0	−0.30	15.7	HD	159.3	−0.87	160.6	+1.6

TABLE II – continued

No.	Diagnosis	AT THE FIRST OBSERVATION Age (years)	S_{Cr} (mg/dl)	Treatment	Height (cm)	Height SDS	AT COMPLETION OF GROWTH Age (years)	Treatment	Adult height (cm)	Height SDS	Predicted adult height (cm)	Prediction error (cm)
19	renal hypoplasia	12.9	4.2	CT	139.9	–2.31	–	HD	151.6	–2.19	152.3	+0.7
20	chronic GN	13.0	1.5	CT	132.5	–3.52	19.5	HD	147.2	–2.94	146.4	–0.8
21	chronic PN	13.1	1.4	CT	151.1	–0.72	15.0	CT	157.3	–1.22	157.0	–0.3
22	polycystic kidney	13.1	–	HD	137.0	–1.05	15.7	HD	150.0	–1.67	150.4	+0.4
23	chronic GN	13.2	–	HD	128.0	–4.50	17.8	HD	158.5	–0.25	149.9	–8.6
24	segmental hypoplasia	13.5	1.8	CT	148.2	–1.60	15.6	CT	154.4	–1.71	–	–
25	obstructive uropathy	13.6	1.4	CT	157.2	–0.19	16.2	CT	164.6	+0.03	–	–
26	obstructive uropathy	13.8	–	HD	143.5	–2.62	13.9	HD	144.5	–3.41	145.7	+1.2
27	renal hypoplasia	13.9	–	HD	142.0	–2.96	–	HD	150.5	–2.38	148.8	–1.7
28	myelomeningocele	14.2	4.1	CT	136.2	–4.07	15.1	CT	138.5	–4.43	139.0	+0.5
29	chronic GN	14.2	–	HD	152.2	–1.50	–	HD	158.5	–1.01	159.8	+1.3
30	chronic GN	14.7	–	HD	154.0	–0.50	17.9	HD	160.0	0	160.8	+0.8
31	polycystic kidney	15.0	1.8	CT	161.0	–0.30	15.9	HD	161.6	–0.48	162.4	+0.8

GN= glomerulonephritis
PN= pyelonephritis
CT= conservative treatment
HD= haemodialysis
TP= transplantation

Patients and methods

Longitudinal growth data were obtained from 49 patients (18 boys, 31 girls) with chronic renal failure of variable duration (range, 1 to 10 years) and severity. At the first observation, 31 patients required conservative treatment and 18 had started dialysis. At the time when adult height was reached, 18 were still on conservative treatment, 18 required dialysis, and 13 had functioning renal transplants.

Standing height was measured by means of a Harpenden stadiometer. The *standard deviation score* (SDS) of height was calculated for each patient by comparison to adequate reference values [3–6].

Bone ages were assessed according to the Tanner-Whitehouse method [7] on radiographs of the left hand and wrist taken at three to six month intervals until fusion of the epiphyses was observed. *Prediction of adult height* according to the Tanner method [7] was done from actual height and bone age at the time of the first observations.

Adult height was recorded as the last measurement in patients with adult bone, i.e. fusion of the epiphyses at the hand-wrist and minimal (<1.0cm) or no increase in height during the previous 12 months. The *age at which adult height was attained* was assumed to be midway between the age at which bone maturation was first found to be complete and the age at the time of the prior radiographic examination. This calculation was not done for 13 patients who were transiently lost to follow-up and were re-evaluated after intervals greater than 1.0 year.

All relevant clinical and developmental data at the time of the first observation and at completion of growth are reported in Tables I (boys) and II (girls).

The growth data were evaluated against some variables (sex, primary renal disease, mode of treatment) which may have influenced the patients' adult heights. The data obtained in the dialysed and transplanted children were also compared with the data reported by the EDTA Registry [2]. Statistical analysis was performed using the Kruskal-Wallis, the Mann-Whitney, the Kolmogorov-Smirnov and the randomisation tests whenever appropriate.

Results

The 49 children exhibited variable patterns of growth. During the period of observation the SDS for height improved in three patients and worsened in one by more than 2 SD. In the other children the SDS for adult height was quite consistent with the SDS at the first observation.

As a group, girls grew better than boys. Their mean SDS was −1.28 at the beginning, and −1.08 at the end of the observation, as compared to −0.73 and −1.18, respectively, for boys. However, comparison of boys and girls at the first observation and after cessation of growth did not disclose any significant difference.

Adult height was also better predicted in girls than in boys. Altogether, predictions were quite accurate (± 2cm) in 30 of 44 patients. Larger errors of prediction showed a definite tendency to over-estimation of adult height (10/14).

After completion of growth, seven patients (14.3%) were less than 2SD below the mean. Forty-two were in normal range, but only eight of them exceeded the mean. Although no transplanted patient was below the normal range, and no dialysis patient exceeded the mean, there was no significant difference between the three modes of treatment for boys and girls separately, and for all patients combined.

Only five (16.1%) of the 31 dialysed and transplanted children reached an adult height less than 2SD below the mean. Comparison with the data from the EDTA report demonstrates a significant difference ($p<0.001$).

Finally, a significant association was found between SDS for adult height and primary renal disease: patients with congenital kidney diseases were smaller than patients with acquired nephropathies ($p<0.025$).

Discussion

Adult height in children with chronic renal failure has rarely been investigated. In a study of 34 patients who had completed their growth during conservative treatment, during dialysis, or after transplantation, we found that the distribution of adult heights was markedly skewed toward the lower centiles, and that eight patients (23.5%) had ended up below the third centile [1]. This study also suggested that children with congenital kidney diseases tend to achieve a lower adult height than children with acquired nephropathies.

A subsequent study from the EDTA reported that only 11 of 49 dialysed and transplanted children attained an adult height within normal limits [2]. The other patients (77.5%) were definitively stunted, with a deviation of up to −9SD below the reference mean. The application of the growth standards for Dutch children to patients from all European countries is a major bias of that study. Moreover, no criteria were given for the definition of adult height: it cannot be excluded that some of the EDTA patients had not grown completely at the time they were included in the study. This would partly explain the striking difference between those patients and the dialysed and transplanted children included in the present study.

In this report we have analysed the growth data of 49 children who have completed their growth, as confirmed by repeated height measurements and by radiographic inspection of the hand-wrist bones. Because these patients belong to difference populations, the most appropriate standards for each individual were used for reference.

Our patients attained adult height either under conservative treatment, during dialysis, or after transplantation. Neither mode of treatment could be correlated with better growth. A significant difference between boys and girls could also not be demonstrated. Patients with congenital renal diseases were significantly smaller.

Most patients had adult heights within normal limits, but only eight (16.3%) exceeded the mean. The skewness of the distribution and the definite tendency to over-prediction of adult height both suggest that even patients with normal heights have not grown consistently with their genetic potential. As discussed elsewhere [8], it is also possible that some patients continue to grow even after fusion of the epiphyses of the hand bones.

Finally, it is important to note that the ages at which our patients reached their adult height (mean 18.2 years for boys, 16.0 years for girls) do not differ significantly from that of the populations to which they have been compared. If patients with severely delayed bone maturation will eventually reach a normal height cannot be stated at this time.

References

1 Gilli G et al. *Pediatr Res 1980; 14:* 1015
2 Chantler C et al. *Proc EDTA 1981; 18:* 329
3 Prader A, Budliger H. *Helv Paediatr Acta 1977; Suppl 37*
4 Prader A, Zachmann M: in press
5 DeToni E et al. *Minerva Pediatr 1965; 17:* 1341 and *1966; 18:* 1342
6 Hamill PVV et al. *Am J Clin Nutr 1979; 32:* 607
7 Tanner JM et al. *Assessment of Skeletal Maturity and Prediction of Adult Height (TW2 method).* London: Academic Press. 1975
8 Gilli G et al. *Kidney Int 1983; Suppl 15:* 53

Open Discussion*

BROYER (Paris) Did you study correlation between adult height and the age at start of renal replacement therapy?

GILLI We did not find a statistically significant correlation in these patients between age or duration of dialysis on the adult height attained. However, according to the stage of puberty of the patients on entry to the dialysis treatment, those who were the most advanced in puberty finished smaller.

* Only part discussion due to technical difficulties

Proc EDTA–ERA (1984) Vol 21

ACUTE RENAL FAILURE AND TUBULAR DAMAGE DUE TO SEPSIS IN AN ANIMAL MODEL

A L Linton, J F Walker, R M Lindsay, W J Sibbald

Victoria Hospital Corporation and The University of Western Ontario, London, Ontario, Canada

Summary

Generalised sepsis was induced in sheep by caecal perforation. Serial measurement of haemodynamic parameters revealed that the subsequent generalised sepsis induced increased cardiac output and decreased systemic resistance comparable to that known to occur in man. Glomerular filtration rate in these animals fell significantly 48 hours after induction of sepsis and there was evidence of tubular damage in the finding of low molecular weight proteinuria and increased clearance of lysozyme. Pathological examination of the kidney revealed normal glomeruli, no consistent changes in tubular cells on light microscopy, negative immunofluorescence, but structural changes in proximal tubular cells on EM. In this model, non-hypotensive sepsis predictably produces damage to proximal tubular cells accompanied by reduction in GFR.

Introduction

The development of acute renal failure in patients with generalised sepsis has important effects on both morbidity and mortality. Sepsis has been considered an important factor predisposing to the development of acute renal failure in up to one-third of patients in various series [1]. The mechanism whereby sepsis may lead to acute renal failure is unknown. In a recent study of critically ill patients with systemic sepsis we demonstrated the appearance of low molecular weight proteinuria in all, suggesting that some damage to proximal tubular cells was a uniform occurrence in this condition [2]. Laboratory animal studies of the effect of systemic sepsis on renal function have usually involved the infusion of live E. coli or endotoxin; models characterised by the rapid development of oliguria and renal failure [3,4]. However, these animal models do not mimic the early human systemic response to infection [5] which is usually characterised by a high systemic flow, low peripheral vascular resistance state. We have therefore studied a model of caecal ligation-induced sepsis in order to identify the effects of generalised sepsis in sheep [6]. The purpose of the study was to examine the functional and morphological changes in the kidney during high output non-hypotensive sepsis.

Materials and methods

Full details of the animal model have been described elsewhere [7]. In brief, after preliminary insertion of catheters in the aorta and the pulmonary artery, the sheep were allowed to recover in a metabolic cage with free access to food and water. Control haemodynamic, pulmonary and renal function measurements were obtained. Under general anaesthesia, a lower midline laporotomy was performed, and a control renal biopsy was taken. The caecum was devascularised and then perforated. Over the ensuing 12 hours all sheep showed clinical evidence of systemic infection with an increase in respiratory rate and the development of lethargy and anorexia. Haemodynamic parameters including systemic blood pressure, pulmonary capillary wedge pressure and cardiac output were monitored frequently, and repeat renal biopsies were performed 24 and 48 hours after the induction of sepsis. Urine was collected for a four hour period each day and blood samples obtained as required for evaluation of renal function by standard methods. Renal biopsy material was examined by light microscopy, immunofluorescence, and electron microscopy.

Results

In all sheep the presence of a polymicrobial peritonitis and bacteraemia was confirmed at 24 hours by blood culture. Organisms most frequently grown included Serratia marcescens, Enterobacter cloaceae, Pseudomonas, Bacteroides species and various strains of E. coli. Autopsy examination at the end of the protocol demonstrated the presence of peritonitis and an inflammatory mass in the lower right quadrant in all sheep.

Haemodynamic measurements

Mean blood pressure recordings in the group of ten sheep studied did not change significantly from baseline to 48 hours. Heart rate increased in all sheep from a mean of 90 beats /min to a mean of 159 beats/min, and cardiac index rose from a mean of 4.9L/min/M^2 to a mean of 6.9L/min/M^2 at 48 hours. These changes were accompanied by a fall in systemic resistance; the systemic vascular resistance index fell from 1909d.sec.cm-5/m^2 to 1434d.sec.cm-5/m^2 at 48 hours. Pulmonary capillary wedge pressure was maintained at high normal levels throughout the 48 hour period by fluid administration as required. Central venous pressure did not change during the study.

Renal function

Details of renal functional changes are given in Table I. All sheep exhibited a significant reduction in glomerular filtration rate both at 24 hours and 48 hours

TABLE I. Renal function changes after the induction of generalised sepsis in sheep

	Baseline	24 hours	48 hours
Serum creatinine (μmol/L)	75 ± 10	106 ± 20	180 ± 75*
Creatinine clearance (ml/sec)	2.7 ± 0.8	1.9 ± 0.8	1.3 ± 0.5*
U. osmolarity (mOsm/kg)	659 ± 458	830 ± 476	602 ± 159
$F.E._{Na}$(%)	2.07 ± 0.72	0.50 ± 0.44	1.73 ± 2.77
U. protein (mg/ml)	0.187 ± 0.15	0.663 ± 0.39*	0.619 ± 0.40*
Fc lysozyme	1.65 ± 1.24	6.6 ± 6.8	16.6 ± 2.35*

*Change from baseline significant ($p<0.05$)

after the induction of sepsis. All animals exhibited a significant increase in proteinuria during the experiment and polyacrylamide gel electrophoresis confirmed that the proteinuria was largely due to the appearance of low molecular weight proteins (less than 30,000 Daltons). Further evidence of tubular cell damage was provided by marked increase in fractional clearance of lysozmye. The fractional excretion of sodium fell for 24 hours and then rose again at 48 hours, while the renal concentrating ability was apparently well maintained.

Control renal biopsies done prior to the induction of sepsis revealed no detectable abnormalities on light, fluorescence or electron microscopy. Light microscopic examination of the tissue obtained at 24 or 48 hours after the induction of sepsis likewise showed no consistent abnormality, except in two animals who had developed intrarenal thrombosis. Immunofluorescence studies were likewise entirely negative. The most marked changes found had occurred by 24 hours and were visible only on electron microscopy. The lesions were patchy and of varying severity; the most frequent change seen was proximal tubular cell swelling with mitochondrial swelling and apical bleb formation. The glomeruli were typically normal, although occasional mesangial densities were seen.

Other

All sheep exhibited a marked pyrexia during the study. Total proteins and serum albumin fell significantly from baseline to 48 hours, and there was a progressive fall in haemoglobin and white cell count. Plasma renin activity rose from a mean of 0.9ng/ml/hr at baseline to a mean of 6.1ng/ml/hr at 48 hours.

Discussion

The nature of the relationship between generalised sepsis and renal impairment remains obscure despite their frequent coexistence. Unlike most other animal models of systemic sepsis the sheep peritonitis model described here reproduces the haemodynamic profile of early systemic sepsis in the human, and offers an opportunity to study renal functional and morphological changes in evolution of sepsis. In this situation the potential confusing effects of hypotension, nephrotoxic antibiotics and contrast media can be avoided.

This model of peritonitis in sheep not only mimics the haemodynamic changes of septic shock in man, but also seems to produce similar renal damage. All sheep demonstrated low molecular weight proteinuria and an increased clearance of lysozyme suggestive of tubular cell damage, as has been demonstrated in the human [2]. The precise nature of the renal function disturbance remains unclear. All sheep demonstrated a fall in glomerular filtration rate with a rise in serum creatinine. The well maintained urinary concentrating power together with the low fractional excretion of sodium and the absence of severe pathological changes in the kidneys all tend to suggest that the renal function impairment might be due to volume contraction ('pre-renal'). On the other hand, it is difficult to attribute significant reduction in renal function to renal hypoperfusion in face of the well maintained blood pressure, the reduced peripheral resistance, the elevated cardiac output and the well maintained pulmonary capillary wedge pressure. It is possible that the proximal tubular cell damage is a uniform occurrence and is not particularly related to the reduction in glomerular filtration rate. The latter may be associated with the kidney attempting to retain sodium and produce a concentrated urine in response to stimuli as yet unidentified; there are clinical reports in man that the fractional excretion of sodium may be as low as 0.20 per cent despite clearly established ARF requiring haemodialysis in patients with sepsis, myoglobinuric renal failure or renal failure due to burns [8].

Generalised sepsis induced in sheep by caecal perforation predictably produces haemodynamic changes similar to those seen in man. The sepsis also predictably reduced the glomerular filtration rate with urinary indices suggestive of renal hypoperfusion. Simultaneously, the appearance of tubular proteinuria and the increase in the fractional clearance of lysozyme testify to the presence of some proximal tubular damage. This model will permit closer examination of the pathophysiology of renal failure in sepsis and will facilitate the evaluation of possible pharmacological interventions.

References

1 Kleinknecht D, Jungers P, Chanard J et al. *Kidney Int 1972; 1:* 190
2 Richmond JM, Sibbald WJ, Linton AM et al. *Nephron 1982; 31:* 219
3 Coalson JJ, Hinshaw LB, Guenter CA et al. *Lab Invest 1975; 34:* 561
4 Coalson JJ, Archer LT, Hall NK et al. *Circulatory Shock 1979; 6:* 343
5 MacLean JD, Mulligan WG, McLean APH et al. *Ann Surg 1967; 166:* 543
6 Wichterman KA, Baur AE, Chaudry IH. *J Surg Res 1980; 29:* 189
7 Richmond JM, Walker JF, Avila A et al. *Surgery 1984.* In press
8 Vaz AJ. *Arch Intern Med 1983; 143:* 738

Open Discussion

RITZ (Heidelberg) I appreciate that this is a preliminary communication but in order to speculate about the mechanism involved it would be helpful to have information on whether or not there was DIC present. It is known that DIC will increase alpha adrenergic vasoconstriction in the kidney. This might explain the paradoxical presence of acute renal failure in the presence of a hypercirculatory state.

LINTON We think that in the two animals that showed intraglomerular thrombosis that likely there was DIC although there was no other evidence to support that view. In the animals which did not demonstrate glomerular thrombosis, that was nine of 11, there was absolutely no evidence of any haematological or pathological disturbance.

WALLS (Leicester) Have you taken any of the sheep and treated them at 48 hours with antibiotic therapy or corrective surgery to see whether it is going to mimic a human situation?

LINTON No, we haven't done that. It is obviously the next thing to do. The big question is whether the addition of antibiotics might make it better or worse and we don't know the answer to that. At the moment we have simply continued with these experiments with pharmacological intervention in an attempt to prevent the development and I can tell you that the calcium channel blockers virtually entirely eliminate the lesion.

Di PAOLO (Chieti, Italy) Is it possible to explain your experiment by a mechanism, like hepato-renal syndrome, with activation of kinins and slow reacting substance?

LINTON Yes, it is entirely probable that this is a situation which is analogous to the renal failure which occurs in the hepato-renal syndrome. It is known that you can have renal failure with a low FENa, it is also known that you can have the same situation in human sepsis and you can have the same situation in burns. We are currently looking at both the prostglandin system and the kinin system in the model.

IAINI (Israel) Can you give us a haematocrit or blood osmolarity?

LINTON Yes, the serum osmolarity did not change significantly through the experiment, the haematocrit in all the sheep falls significantly.

IAINI There is no dehydration?

LINTON No dehydration so far as we can tell.

CHAIRMAN Isn't it true that you really don't have acute renal failure because you prevented it by overloading the sheep, would you agree with my statement?

LINTON It depends what you mean by acute renal failure. I think what we are seeing is an animal with generalised sepsis just as we see in man. If you define acute renal failure as being the development of acute renal impairment in the course of some other disease they have what we like to call acute renal failure and sepsis. You are right in that there is no question that there is no convincing evidence that the tubules as such are damaged in this situation. The tubules undoubtedly shed low molecular weight protein and lysozyme but they can still concentrate and they can still retain sodium.

CHAIRMAN In order to diagnose acute renal failure in humans you still need a low U:P creatinine ratio and U:P osmolarity above unity and you did not meet the criteria.

LINTON No, that's a definition problem, Carl Kjellstrand would contend that you don't need either of these.

Proc EDTA–ERA (1984) Vol 21

THE EFFECTS OF d-PROPRANOLOL AND CAPTOPRIL ON POST-ISCHAEMIC ACUTE RENAL FAILURE IN RATS

M Ishigami, T Maeda, *S Yabuki, †N T Stowe

*Kanto Rosai Hospital, Kawasaki, *Toho University School of Medicine, Tokyo, Japan, †The Cleveland Clinic Foundation, Cleveland, Ohio, USA*

Summary

The effect of the combined therapy of d-propranolol and captopril was evaluated in post-ischaemic acute renal failure in rats.

The glomerular filtration rate measured 24 hours after ischaemic insult in the animals receiving no drugs was 54 ± 11μl/min/100g body weight while in animals treated with the combination of d-propranolol and captopril it was 305 ± 35μl/min/100g body weight ($p<0.05$). The precise mechanism of protection afforded by the combination therapy is not clear, but this approach could be useful in protecting the kidney from ischaemic damage.

Introduction

Since renal transplantation and renal vascular surgery became common procedures in urological practice, post-operative ischaemic acute renal failure has been a problem. Various techniques have been utilised to prevent the deterioration of renal function following ischaemic insult and several drugs have been employed to ameliorate the post-ischaemic acute renal failure [1]. Propranolol, a β-adrenergic blocking agent, has been shown to be effective in post-ischaemic acute renal failure, but it was only partially effective [2–4].

We previously reported the value of propranolol in post-ischaemic acute renal failure in animal models, and demonstrated that the d-isomer of racemic propranolol was the active compound in improving renal function following ischaemia. However, the efficacy of d-propranolol was also limited, and the glomerular filtration rate (GFR) measured 24 hours after ischaemia had only recovered to 50 per cent of a non-ischaemic kidney [5]. Several mechanisms can be speculated to account for the lack of total protection with a single drug treatment against post-ischaemic acute renal failure. First, the adverse effect of taking one drug at a high dose could be a problem, and secondly, there may be various mechanisms of pathogenesis of post-ischaemic acute renal failure, including the activation of the renin-angiotensin (R-A) system, tubular obstruction, cell swelling and metabolic factors.

Recently, a newly developed drug for the treatment of hypertension, the converting enzyme inhibitor captopril, has been reported to have some action in the kidney, such as increasing renal blood flow, urine volume and urinary sodium excretion. This study evaluates the effect of the combined therapy of d-propranolol and captopril in experimentally induced post-ischaemic acute renal failure in the rat.

Method

Experiments were performed on male Sprague-Dawley rats and anaesthesia was induced by sodium pentobarbital (50mg/kg/IP).

An external jugular vein was cannulated for the infusion of fluids and drugs. Both kidneys were exposed through a mid-line incision and the left renal artery was occluded with microvascular clip to induce 45min ischaemia. Subsequently, a contralateral nephrectomy was done. At the end of the experiment, the catheter was removed and the rats were allowed to recover. In the drug-treated group, an infusion of d-propranolol (16μg/kg/min), captopril (16μg/kg/min) or a combination of the two was started 30 minutes prior to renal artery occlusion and continued throughout the entire experiment for a total of 195 minutes. Twenty-four hours later, the rats were anaesthetised again by sodium thiobarbital (Inactin, 100mg/kg/IP). The carotid artery was cannulated for the measurement of blood pressure and collection of blood samples, and the jugular vein was cannulated for the infusion of fluids and inulin. Three 30-minute urine collections were obtained for inulin and electrolyte excretion. A blood sample (0.4ml) was drawn at the beginning and the end of the experiment for inulin and electrolyte determinations. Inulin concentrations were estimated by colormetric analysis and sodium was analysed by flame photometry. The data were subjected to analysis of variance, and the 0.05 level of probability was used as the criterion of significance.

Results

The alteration of the renal function measured 24 hours following 45-minute renal artery occlusion is presented in Table I. The values shown represent a mean determination of the three periods. In the untreated animals, the GFR was significantly reduced to 54 ± 11μl/min/100g body weight. In contrast to this, in the animals treated with either d-propranolol or a combination therapy of d-propranolol and captopril, the GFR significantly improved to 173 ± 22 and 305 ± 35μl/min/100g body weight, respectively ($p<0.05$). Furthermore, the GFR value in the animals treated with the combination was significantly higher than in the animals treated with d-propranolol alone, and its value was also about six times that observed in the untreated control.

Captopril treatment brought better GFR (102±16μl/min/100g body weight) values than the untreated control group, but this value did not reach statistical significance. Likewise, the sodium reabsorption rate was significantly higher in the d-propranolol treated group (97.3 ± 0.7%) than in the untreated group

TABLE I. Renal function obtained 24 hours after release of renal artery occlusion†

	GFR (μl/min/ 100g/BW)	UV (ml/min/ 100g/BW)	$U_{Na}V$ (μEq/min/ 100g/BW)	Per cent sodium reabsorption	Heart rate (beats/ min)	Blood pressure (mmHg)
1. untreated control (n=9)	54 ±11	3.8 ±0.8	0.22 ±0.06	92.7 ±1.5	332 ±10	110 ±4
2. d-propranolol 16μg/kg/min for 195 min (n=7)	173 ±22*	5.0 ±0.4	0.36 ±0.04	97.3 ±0.7*	337 ±9	111 ±3
3. captopril 16μg/kg/min for 195 min (n=6)	102 ±16	8.0 ±1.9*	0.6 ±0.16*	93.9 ±2.4	338 ±6	102 ±5
4. d-propranolol 16μg/kg/min plus captopril 16μg/kg/min for 195 min (n=7)	305 ±35*,***	9.1 ±1.7*,**	0.6 ±0.14*	94.1 ±3.9	352 ±18	113 ±1.9

† Values are mean ± SEM. Abbreviations are defined as: GFR=glomerular filtration rate; UV=urine volume; $U_{Na}V$=sodium excretion; BW=body weight; n=number of rats.

* Significantly different from untreated control ($p<0.05$)

** Significantly different from untreated control and d-propranolol ($p<0.05$)

*** Significantly different from untreated control, d-propranolol and captopril ($p<0.05$)

(92.7 ± 1.5%) and the animals treated with the combination had better sodium reabsorption (94.1 ± 3.9%) than the untreated group, but this was not significant. There was a significant increase of urine volume and sodium excretion in the animals which received captopril.

No significant difference was observed in the blood pressure or heart rate between these groups.

Discussion

We previously reported that the β-blocking agent propranolol was effective in ameliorating the deterioration of renal function after ischaemic insult in the rat model. Furthermore, we demonstrated that the d-isomer of racemic propranolol, which lacks β-adrenergic blocking properties, was the active compound in protecting the kidney from ischaemia and speculated that the so-called membrane stabilising properties of d-propranolol could be significant. However, complete recovery was not observed even in the animals treated with d-propranolol with the GFR returning only to 30–50 per cent of the non-ischaemic kidney measured two hours and 24 hours after ischaemic insult. The present study clearly demonstrated that the recovery of the GFR obtained 24 hours following ischaemia was significantly enhanced by the combined therapy of d-propranolol and captopril than when d-propranolol was used alone, and its value was six times higher than the untreated control group.

Since Goormaghtigh first suggested the involvement of the R-A system in the pathogenesis of acute renal failure [6], a number of studies have been employed to elucidate the relationship between GFR and RBF in acute renal failure. Several investigators reported the efficacy of propranolol against post-ischaemic acute renal failure in animal models but they failed to clarify the precise mode of action of propranolol. Some studies suggested that the effectiveness of propranolol against post-ischaemic acute renal failure was not mediated through the improvement of renal blood flow in the recovery phase [3,4]. On the other hand, Solez et al did find a significantly better renal blood flow value immediately after reflow which resulted in a significantly higher GFR [7]. We have shown previously the beneficial effect of propranolol on post-ischaemic acute renal failure, but we did not see any relationship between GFR and renal blood flow. Furthermore, from our previous observation, if the R-A system is primary in the pathogenesis of the initial deterioration of renal function after ischaemia, the l-isomer of propranolol, which is a potent β-adrenergic blocker, and also captopril should be effective, which was not the case. It was an interesting observation that the combined treatment could enhance significantly the recovery of GFR observed 24 hours after ischaemia over that of d-propranolol alone.

It is known that captopril has various actions in the kidney such as increase of RBF, urine volume and sodium excretion [8,9]. But the mechanism of the above action in the kidney is controversial. We observed similar captopril actions in non-ischaemic kidneys in our study. Diuretics, such as mannitol and frusemide which increase renal blood flow, solute excretion, and water excretion have been shown to be effective on post-ischaemic acute renal failure [10]. These clearance studies do not show why the combination is so beneficial. The precise mechanism of better protection afforded by a combined therapy on post-ischaemic acute renal failure in our model is not clear at this time, but synergistic actions of d-propranolol which stabilise the membrane and captopril which alters renal haemodynamics may attenuate more effectively the renal damage from ischaemia.

Acknowledgments

This study was supported in part by grant AM 27846 from the National Institutes of Health.

References

1 Tiller DJ, Hudge GH. *Kidney Int 1980; 18:* 700
2 Eliahou HE, Iaina A, Solomon S et al. *Nephron 1977; 19:* 158
3 Klein LA. *Invest Urol 1978; 15:* 401
4 Chevalier RL, Finn WF. *Nephron 1980; 25:* 77
5 Ishigami M, Stowe N, Khairallah P et al. *Kidney Int (abstract) 1983; 23:* 204
6 Goormaghtigh N. *Proc Soc Exp Biol Med 1945; 59:* 303
7 Solez K, D'Agostini JJ, Stawowy L et al. *Am J Path 1977; 88:* 163
8 Mimran A, Casellas D, Dupont M. *Kidney Int 1980; 18:* 746
9 Swartz SL, Williams GH. *Am J Cardiol 1982; 49:* 1405
10 Hanley MJ, Davidson K. *Am J Physiol 1981; 241:* F556

Open Discussion

IAINA (Israel) In the past we have tried to use d-propranolol in a dose of 1mg per Kg body weight per minute and I could not obtain any beneficial effect. When I used 30mg d-propranolol there were some changes. What do you mean when you say 'beneficial membrane stabilising effect of beta adrenergic blocker'?

ISHIGAMI I think your group reported about two years ago a 70 minute ischaemic model* instead of 45 min ischaemia which I used this time and also the timing infusion of d-propranolol was different. You used infusions just 50 minutes after releasing occlusion. I used a 30 minute prior to renal occlusion and continued to infuse through our entire experiment which was about 180 minutes. For your second question I cannot give you any explanation for that, but from a pharmacological understanding and from the literature some studies suggest d-propranolol preserves more ADP activity in the cell and may affect some ion movement such as calcium and potassium. When we look at the literature nobody can explain the meaning of membrane stabilisation, but when we use d-propranolol we are just speculating from the pharmacological characteristics that a major action is membrane stabilisation.

CHAIRMAN We have recently shown that some of the renal effects of captopril are accounted for by increased prostaglandin secretion such as the natriuretic effect and the increasing renal blood flow. This suggests that the protective effect afforded by captopril in acute renal failure maybe mediated by increased prostaglandins.

ISHIGAMI Well it may be possible but we never measured the urinary prostaglandins, but in the literature in an ischaemic occlusion model where prostaglandin was used there was no improvement of renal function.

CHAIRMAN It has been argued that captopril when given to patients with bilateral renal artery stenosis or unilateral renal artery stenosis in a single kidney may precipitate acute renal failure.† It is being postulated that this is due to an effect on the efferent arterioles. How can you reconcile with your explanation?

* Iaina I, Serban I, Garendo S et al. *Proc EDTA 1980; 17:* 686
† Farrow PR, Wilkinson R. *Br Med J 1979; 1:* 1680
Hrick DE, Browning PJ, Kopelman R et al. *N Eng J Med 1983; 308:* 373

ISHIGAMI Well I don't know. It has been shown by some that captopril blunts autoregulation which is very important in cases of high renin activity. After ischaemia, as in our model, we have high renin activity but we didn't measure renin or angiotensin II.

CHAIRMAN If I may comment upon this point. The reason for acute renal failure precipitated by captopril in patients with a single kidney with renal artery stenosis has been attributed to decreased constriction of efferent arterioles. Actually, we have given evidence in a recent paper which has been presented in the International Congress in Los Angeles that at least some cases of acute renal failure can be accounted for by increased urinary salt excretion and by volume depletion. We had six patients in which acute renal failure was promptly reverted by giving salt, simply giving salt while continuing therapy with captopril.

Proc EDTA–ERA (1984) Vol 21

CYCLOSPORINE AND SHORT ISCHAEMIA: A NEW MODEL OF EXPERIMENTAL ACUTE RENAL FAILURE IN RATS

A Iaina, D Herzog, D Cohen, S Gavendo, S Kapuler, I Serban, G Schiby, H E Eliahou

Chaim Sheba Medical Center, Tel-Hashomer, Israel

Summary

To determine whether a mild episode of ischaemia may be a factor in the production of cyclosporine (Cys) toxicity, right nephrectomy was performed in three groups of Charles River rats: I. Ischaemia (left renal pedicle clamping) for 20 minutes, without treatment; II. Ischaemia of 20 minutes, followed by IP Cys 60mg/kg BW/day; III. Sham (no ischaemia) followed by Cys as in Group II. The rats were sacrificed after four days.

Cys plus ischaemia produced a lower creatinine clearance (136±15μl/min/100g BW, $p<0.001$) and a higher FE_{Na} per cent (0.94±0.14, $p<0.05$), FE_K (1.07±0.02, $p<0.01$) compared with ischaemia alone creatinine clearance 261±39, FE_{Na} per cent 0.61±0.08, FE_K 0.54±0.08, FE_{H_2O} -0.04±0.005. Histology showed more vacuolisation of tubular epithelial cells in the Cys plus ischaemia group than in the ischaemia alone group.

Introduction

Cyclosporine has been found to have a major clinical advantage over conventional immune suppressants in renal transplantation [1]. However, nephrotoxicity with high doses is a well-documented side-effect [2]. This is sometimes manifested as acute renal failure.

The mechanism of Cys renal injury is not established. As a consequence of only minor renal damage in normal laboratory animals, it is very difficult to produce an experimental Cys acute renal failure model [3,4].

The aim of the present study was to examine the renal toxicity of Cys on a kidney exposed to a short ischaemia time. This being a simulation of renal transplantation without the immunological component.

Material and methods

Charles River rats of both sexes weighing between 250–320g were used. All rats underwent right nephrectomy under ether anaesthesia. Immediately following the nephrectomy, the rats were divided into three experimental groups.

Group I. Left renal pedicle occlusion for 20 minutes.

Group II. Left renal pedicle occlusion for 20 minutes and Cys 60mg/kg BW/day intraperitoneal.

Group III. Sham operated rats and Cys 60mg/kg BW/day.

Immediately after the operation the rats were placed in individual metabolic cages. The experiment lasted four days. Cys was given every day, in a dose of 60mg/kg BW intraperitoneal. At the end of the experiment blood was withdrawn from the aorta and the left kidney was removed for histology. Creatinine, sodium, potassium and osmolality were determined in the blood and in the last 24-hour urine collection. Creatinine clearance, fractional excretion of sodium, potassium and water were calculated using standard formulae.

Mean, standard error of the mean were calculated. One way analysis of variance was used to assess significance, and $p<0.05$ was considered significant.

Results

The different glomerular and tubular function parameters are given in Table I.

TABLE I

	Group I ischaemia alone n=6	Group II Cys+ischaemia n=10	Group III Cys – no ischaemia n=6
Creatinine clearance* μl/min/100g BW	216±39	136±15**	318±24
FE_{Na}%	0.61±0.08	0.94±0.14**	1.17±0.08
FE_K	0.54±0.08	1.07±0.1**	0.85±0.87
FE_{H_2O}	−0.04±0.005	−0.1 ±0.015**	−0.01±0.001

* Mean ± SE n = number of rats.
** Significantly different from the respective values in Group I. (p at least <0.05)

By analysis of variance the rats with Cys and ischaemia showed a larger creatinine clearance ($p<0.001$), a higher FE_{Na} ($p<0.05$), FE_K ($p<0.003$) and a lower FE_{H_2O} ($p<0.001$) compared with the group with ischaemia alone (untreated with Cys). The creatinine clearance was similar in the group of ischaemia without Cys as in the group with Cys without ischaemia. Light microscopy showed tubular epithelial cells vacuolisation in all groups studied, and was more pronounced in the Cys plus ischaemia group (Figure 1).

Discussion

Only prolonged administration of a high dose of Cys results in significant impairment of kidney function while lower doses are associated with minor changes in renal function [3,4].

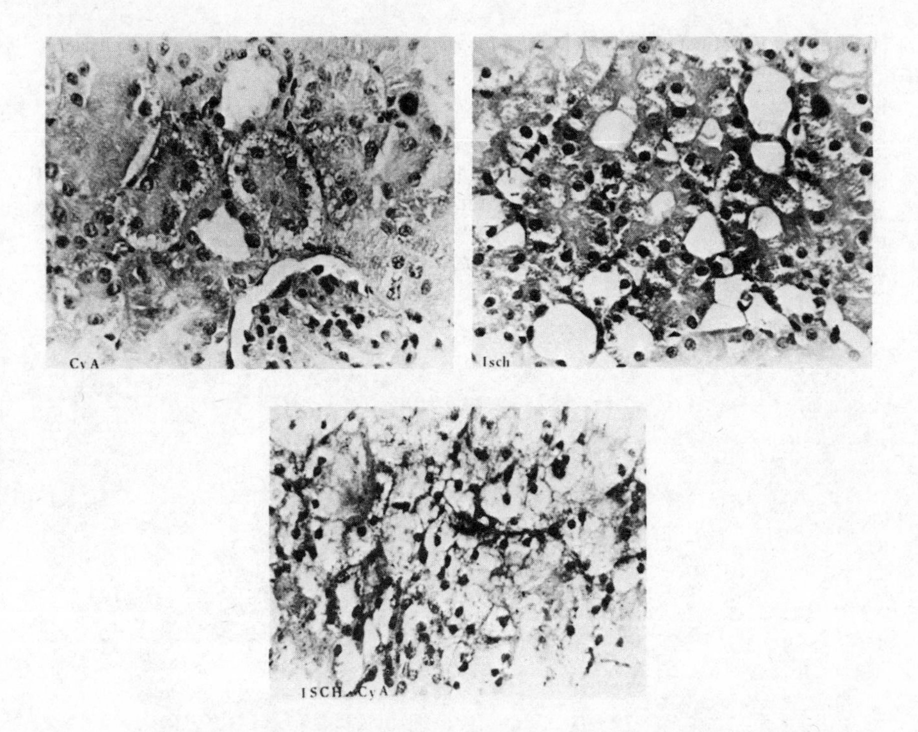

Figure 1. Light microscopy in different experimental groups. Group I: ischaemia alone. Group II: Cys plus ischaemia. Group III: Cys alone. Tubular epithelial vacuolisation are seen in all groups, but is much more pronounced in the Cys plus ischaemia group

The present study demonstrates that the administration of Cys for a short period of time, will result in acute renal failure, if the kidney was previously exposed to a short ischaemic insult. This type of acute renal failure is characterised by an important reduction in GFR, increased FE_{Na} and FE_K and impressive histological damage.

The potentiation of Cys nephrotoxicity by gentamicin [5] or mannitol [6] have been recently described. However, it seems that our model in which ischaemia is used as a synergistic factor, gives a better simulation of the immediate post-transplant, non oliguric situation. This model will offer the possibility of studying different therapeutic approaches, such as calcium entry blocker drugs. This is under current investigation in our laboratory.

References

1 Merion RM, White DJG, Thiru S et al. *N Engl J Med 1984; 310:* 148
2 Whiting PH, Blair JT, Simpson JG et al. *Lancet 1981; i:* 663
3 Whiting PH, Thomson AW, Cameron ID, Simpson JG. *Br J Exp Path 1980; 61:* 631
4 Thompson AW, Whiting PH, Cameron ID et al. *Transplantation 1981; 31:* 121
5 Thompson AW, Whiting PH, Simpson JG. In White DIG, ed. *Cyclosporin A.* Amsterdam: Elsevier Biomedical Press. 1982: 178

6 Brunner FP, Mihatsch M, Thiel G. *9th International Congress of Nephrology*. Los Angeles. 1984, June 11–16

Open Discussion

CHAIRMAN You showed in your Table a minus sign before the free water clearance so that is a negative free water clearance?

IAINA Yes, they still concentrate their urines.

Proc EDTA–ERA (1984) Vol 21

GLOMERULOTUBULAR FUNCTION IN CYCLOSPORINE-TREATED RATS. A LITHIUM CLEARANCE, OCCLUSION TIME/TRANSIT TIME AND MICROPUNCTURE STUDY

H Dieperink, H Starklint, P P Leyssac, E Kemp

Odense University Hospital, Odense, Denmark

Summary

Sprague-Dawley rats treated with cyclosporine (Cys) and appropriate controls were investigated with inulin, lithium and sodium clearances. It was found that Cys depressed glomerular filtration rate (GFR) and absolute proximal tubular reabsorption, while fractional proximal reabsorption was increased. As a normal proximal tubular reabsorptive capacity was found after volume expansion, and the intratubular pressure was normal, tubulotoxicity or obstruction of the tubular system by Cys was excluded. Increased fractional proximal reabsorption was also found with the occlusion time/transit time method. Intravenous Cys resulted in instantaneous renal functional changes qualitatively identical to those of prolonged Cys treatment. It was concluded that Cys nephrotoxicity is due to decreased ultrafiltration pressure, most probably due to a reversible spasm in the afferent glomerular arteriole.

Introduction

Since the earliest reports on the clinical effect of cyclosporine (Cys), a nephrotoxic side effect has been suspected. In man increases in serum creatinine and water, sodium and potassium retention are the main features of this adverse effect [1]. However, no specific morphological correlate of these functional effects has been accepted in man or animals. In the first years after the clinical introduction of Cys it was the general opinion that Cys was toxic to renal tubular cells [2], but this seems not to be true [3]. At present more attention is placed on Cys effect on the renin-angiotensin-aldosterone system, on the renal vasculature and on the glomerular permeability [3,4]. We hereby summarise our contributions to this new thesis concerning the nephrotoxicity of Cys.

Material and methods

The effect of Cys 0–12.5–25mg/kg/day over 13 days was studied in the conscious catheterised Sprague-Dawley rat [3]. Inulin, lithium and sodium clearances were measured. This allows the estimation of: glomerular filtration rate (GFR) =

inulin clearance (C_{in}), delivery of lithium, fluid ($\dot{V}_{prox}$) and sodium ($C_{Na\ prox}$) from the proximal tubule to the thin descending segment of the loop of Henle (lithium clearance, $C_{Li} = \dot{V}_{prox} = C_{Na\ prox}$), absolute proximal tubular lithium, sodium and fluid reabsorption ($C_{in}-C_{Li}$), fractional proximal tubular reabsorption ($1-C_{Li}/C_{in}$), fractional lithium clearance (C_{Li}/C_{in}), absolute reabsorption of sodium in the nephron segment distal to the proximal tubules ($C_{Li}-C_{Na}$), and fractional reabsorption in the same segment ($1-C_{Na}/C_{Li}$) [5]. The effect of extracellular volume expansion with 2% of body weight (BW) saline was investigated in those rats treated with Cys over 13 days, as was the effect of a high sodium clearance obtained over 13 days with a high sodium content in the diet. In rats on the normal sodium diet the effect of Amiloride (MSD), a drug which is known to block conductive sodium channels in high resistance epithelium, such as those of the distal tubule and collecting duct, was investigated in order to see whether it would increase C_{Li}. In a separate group of anaesthetised Sprague-Dawley rats we measured tubular occlusion time (OT), lissamine green tubular transit time (TT), intratubular hydrostatic pressure (P) and C_{in}, C_{Li} and C_{Na}. Estimates of fractional reabsorption in the convoluted part of the proximal tubules (PFR) from the clearance data ($PFR_{C_{Li}/C_{in}}$) and from the occlusion time/transit time ($PFR_{OT/TT}$) could then be compared using the formula [6]:

$$PFR = 1 - e^{-TT/OT} = 0.78 \times (1-C_{Li}/C_{in}).$$

The effect of intravenous Cys was investigated in rats with normal and high C_{Na}, in the conscious catheterised rat model. Plasma renin concentration (PRC) and serum Cys were investigated in all animals, and blood Cys in some.

Results

A typical example of the Cys modified renal function is shown in Table I. Cys reduces GFR and absolute proximal tubular reabsorption, while proximal

TABLE I

Cys	0	12.5	25	mg/kg/day
n	11	11	11	SPRD rats
C_{in}	1065±313	734±324*	541±228*	μl/min/gKW
C_{Li}	187±47	125±115	59±47*	–
$1-C_{Li}/C_{in}$	82±3	88±9*	92±5*	%
$C_{in}-C_{Li}$	878±274	608±316*	490±196*	μl/min/gKW
C_{Na}	3.1±1.7	2.0±2.0	1.2±0.9*	–
$1-C_{Na}/C_{Li}$	98.3±1.1	97.1±3.4	96.2±3.4*	%
$C_{Li}-C_{Na}$	183±46	123±114	49±47*	μl/min/gKW

C_{in}=GFR · $C_{Li} = \dot{V}_{prox} = C_{Na\ prox}$ = delivery of lithium, fluid and sodium from the proximal tubule.
$1-C_{Li}/C_{in}$ = proximal fractional reabsorption. $C_{in}-C_{Li}$ = proximal absolute reabsorption.
$1-C_{Na}/C_{Li}$ = distal fractional reabsorption. $C_{Li}-C_{Na}$ = distal absolute reabsorption.
* = $p<0.05$, one-tailed two-sample t-test

tubular fractional reabsorption ($1\text{-}C_{Li}/C_{in}$) is increased. C_{Li} ($=\dot{V}_{prox}=C_{Na\ prox}$) is also decreased while absolute and fractional reabsorption in the nephron segment distal to the proximal tubule are both reduced. Table II shows that intravenous Cys (12.5mg/kg) results in alterations quantitatively identical to

TABLE II

	C_{in}	C_{Li}	$1\text{-}C_{Li}/C_{in}$	$C_{in}\text{-}C_{Li}$	$1\text{-}C_{Na}/C_{Li}$	$C_{Li}\text{-}C_{Na}$	C_{Na}
Intravenous Cys	↓	↓	↑	↓	NC	↓	↓
Fortnight Cys	↓	↓	↑	↓	↓	↓	↓
+ vol exp	↑↑	↑	NC	↑	NC	↑	↑
+ high Na	NC	NC	NC	NC	↓	NC	↑↑
+ Amiloride	NC	NC	NC	NC	NC	NC	NC

Vol exp = 2% BW volume expansion. High Na = high sodium content in the diet

those of the longer Cys treatment. The following groups were treated with Cys 25mg/kg/day for 13 days (Table II). Extracellular volume expansion (2% BW saline) instantly increased GFR and absolute proximal tubular reabsorption ($C_{in}\text{-}C_{Li}$) to normal values while the increased fractional proximal tubular reabsorption remained unaltered. High sodium content in the diet resulted in increased C_{Na} and decreased fractional sodium reabsorption in the distal nephron segment ($1\text{-}C_{Na}/C_{Li}$) while C_{Li} and glomerular filtration and proximal tubular reabsorption was identical to the corresponding values from groups with a moderate sodium content in the diet. Furthermore, Amiloride was unable to increase C_{Li} (Table II).

In anaesthetised rats PFR estimated from OT/TT and C_{Li}/C_{in}, respectively, was compared. $PFR_{OT/TT}$ was 68 per cent, and $PFR_{C_{Li}/C_{in}}$ was 69 per cent. Proximal tubular transit time (TT) was prolonged to a mean of 26 seconds ($p<0.05$), and the intratubular hydrostatic pressure was low in the normal range (11.6±1.4mmHg). PRC was approximately twice the normal value of Cys-treated groups, and did also increase immediately after intravenous Cys administration (not shown). Also, the increased PRC of the Cys-treated rats was found to be resistant to volume expansion (2% BW saline). Including all experiments conducted (>120 SPRD rats) we have been unable to show any significant correlation between s-Cys and renal function.

Discussion

Cys was found to reduce GFR in the Sprague-Dawley rat at dose ranges commonly used during human transplantation [1]. It has earlier been proposed that the nephrotoxic effect of Cys could be due primarily to proximal tubular damage [2]. GFR would then decrease as a consequence of depressed proximal tubular reabsorptive capacity and concomitant increase in intratubular pressure. This interpretation predicts a decreased proximal fractional reabsorption. But as we have found a normal proximal tubular reabsorptive capacity (during volume expansion), a proximal tubular hydrostatic pressure low in the normal range, and an increased proximal fractional reabsorption [3] (the latter also shown by Tonnesen et al [7]), proximal tubular damage seems very unlikely. An interpretation of increased fractional but decreased absolute proximal reabsorption could have been a partial obstruction in the tubular system at a site distal to the proximal tubule in various numbers of nephrons, but the finding of a normal intratubular pressure also disproves this alternative. Two series of experiments were done to elucidate the possibility that C_{Li} should underestimate the outflow from the proximal tubule because of (pathological) distal lithium reabsorption. Firstly, a comparison was made between the fractional delivery from the end of the proximal convoluted segment (PFR) as measured directly by the OT/TT method ($1\text{-}e^{-TT/OT}$), and as calculated from the formula $PFR=0.78x(1\text{-}C_{Li}/C_{in})$. The two estimates agreed quantitatively and thereby confirmed that no significant distal lithium reabsorption occurred. Secondly, Amiloride in a dose which in a separate control group was shown to reduce potassium clearance, and which has been shown to increase C_{Li} in rats with distal lithium reabsorption, was without effect on the C_{Li} of the Cys-treated rats.

Having excluded other possibilities, our data suggest the following pathophysiological mechanism and sequence of events: net ultrafiltration pressure is reduced by Cys, and due to inadequate reduction in the absolute rate of proximal reabsorption, proximal fractional reabsorption increases. As the functional pattern shows, many similarities with that seen during partial obstruction of the renal artery, we suggest that a vascular site of action, most likely on the glomerular afferent vessels, is responsible for the nephrotoxic effect of Cys.

The function of the distal nephron segment is also altered during Cys treatment (Table I). The finding that the fractional and the absolute reabsorption of sodium ($1\text{-}C_{Na}/C_{Li}$ and $C_{Li}\text{-}C_{Na}$, respectively) in the nephron segment distal to the proximal tubule were reduced may suggest the following sequence of events: Cys reduces distal sodium (and water, data not shown) reabsorption resulting in natriuresis and natriupenia, which then could reduce net ultrafiltration pressure. But as we and others [7] have found GFR to be reduced instantaneously after intravenous Cys, and even without natriuresis, this can also be excluded.

An interesting finding was that the proximal fractional reabsorption was markedly increased even when urine flow and electrolyte excretion was returned to normal by extracellular fluid expansion. This resistant change in glomerulotubular balance strongly suggests a drug-induced resetting of the tubulo-glomerular feedback mechanism.

The increased PRC in the Cys-treated animals could be explained by the decreased delivery of tubular fluid to the macula densa region, as C_{Li} was decreased. Increased PRC in rats has been shown earlier [3,4], and increased renin release from renal cortical slices in vitro has also been reported [8]. The finding that renal transplant patients do not always have higher PRC when treated with Cys than controls treated with other immunosuppression does not disprove that Cys increases PRC [9]. Serum or blood Cys did not correlate significantly with any measure of renal function, and as it also can be difficult to show any such correlation in patients [1] this finding is not surprising.

In conclusion, as it can be excluded that Cys nephrotoxicity is due to decreased proximal tubular reabsorptive capacity, or to tubular obstruction, or to primary actions of Cys on the tubular epithelium of the nephron segment distal to the proximal tubule, it appears that Cys nephrotoxicity is due to decreased net ultrafiltration pressure, most probably secondary to an effect on the glomerular afferent vessels. The tubulo-glomerular feedback mechanism is resettled with a higher fractional proximal reabsorption, either because of a direct effect of Cys or as an insufficient adaption to the reduced GFR. Cys increases GFR, and the increase is resistant to volume expansion. Serum or blood Cys did not correlate with any measure of renal function. After increasing the ultrafiltration pressure by extracellular volume expansion, the GFR of the Cys-treated animals was returned to normal. It will be of great diagnostic and perhaps therapeutic importance to see if vasodilator therapy will also return renal function to normal, and whether the C_{Li} method can be used to distinguish between renal transplant rejection and Cys nephrotoxicity.

Acknowledgments

Cyclosporine and Amiloride were generous gifts from Sandoz Ltd and Merck, Sharp and Dohme Ltd respectively. Supported by the Danish Medical Research Council.

References

1 Morris PJ, French ME, Dunnill MS et al. *Transplantation 1983; 36:* 273
2 Ryffel B. In White DJG, ed. *Cyclosporine A.* Amsterdam: Elsevier. 1982: 45
3 Dieperink H, Starklint H, Leyssac PP. *Transplant Proc 1983; 15/4 (suppl 1):* 2736
4 Siegl H, Ryffel B, Petric R et al. *Transplant Proc 1983; 15/4 (suppl 1):* 2719
5 Thomsen K. *Nephron 1984; 37:* 217
6 Thomsen K, Holstein-Rathlou N-H, Leyssac PP. *Am J Physiol 1981; 241:* 348
7 Tonnesen AS, Hamner RW, Weinmann EJ. *Transplant Proc 1983; 15/4 (suppl 1):* 2730
8 Baxter CR, Duggin CG, Willis NS et al. *Res Commun Chem Pathol pharmacol 1982; 37:* 305
9 Nath KA, Bantle JP, Ferris TF. *Kidney Int 1984; 25:* 346

Open Discussion

MEES (Utrecht) How do you explain the increased renin secretion in your model?

DIEPERINK The increased renin we believe is explained by the decreased proximal delivery of sodium to the macula densa, but we cannot exclude a direct effect on the macula densa system.

MEES When you expanded these animals with saline did the renin come down to normal?

DIEPERINK It came down but not as much as we expected it would.

ALLISON (Glasgow) I was fascinated, but a little bit confused, so I apologise if I sound a little confused. First of all, why did you do individual measurements of a single nephron function?

DIEPERINK The micropuncture was in fact mainly for pressure measurements.

ALLISON You see I could not understand why you had to go to the bother of looking at lithium clearances in order to arrive at a measurement for proximal fractional reabsorption when one can simply measure late proximal F:P inulin concentrations to calculate proximal, absolute and fractional reabsorption. I think, therefore, one has to be very cautious about drawing conclusions about proximal fractional and absolute reabsorptions if you do not actually measure each individual nephron, because you will find that each individual nephron has very individual glomerulartubular balance.

DIEPERINK I somewhat disagree with you, because the advantages of the lithium clearance method is that it allows estimation, which is the mean for the whole nephron population of the animal.

ALLISON Nephrons are heterogeneous and not homogeneous, especially when you are looking at an animal that has been given a potentially nephrotoxic substance. I think to validate the method it would be very easy and simple if you were doing micropuncture to give the animal some inulin and measure the late proximal F:P inulin.

DIEPERINK I would draw your attention to the use of the occlusion time/transit time method.

CHAIRMAN I think Dr Allison's point is that when you look at proximal fractional reabsorption by the PFT inulin method you may find changes of 3–4 U:P inulin ratios. That is easy to detect and is far above the error of the method. When you look at the transit time/occlusion time method, which is a true estimate, but has such a wide scatter of data from tubule to tubule, you can just pick those results that you feel more comfortable with and come up with the numbers you want. In a way there is a different source of error and a different way of controlling the data with true measurement, such as the one that Dr Allison proposes, and the estimated way that you use, although mathematically correct.

ALLISON Yes, you are absolutely correct.

DIEPERINK No, I do not think I will go into more detail on that data, but I will mention one more advantage of the lithium clearance method: that is that you can use it in humans. We have found a lower lithium clearance in patients after cyclosporine treatment.

CHAIRMAN I think you have worked on hydropenic rats?

DIEPERINK No, I would not say that.

CHAIRMAN I mean rats without extracellular fluid volume expansion.

DIEPERINK We used several situations and most of the lithium clearance studies were from conscious curarised rats who have been curarised during anaesthesia and they are hydrated for the duration of the study.

CHAIRMAN I think that the most probable explanation for the falling GFR in that condition is simply a decrease in renal plasma flow or better in glomerular plasma flow. It is well known that a single nephron GFR is plasma flow dependent. On the other hand you yourself have found a rise in afferent arterial resistance which can well account for a decrease in renal plasma flow.

CHAIRMAN With hypertonic saline loading you increased or normalised the GFR and you normalise absolute reabsorption, but fractional reabsorption is still higher than normal. How is that possible?

DIEPERINK We have tried to explain it by a specific alteration of the glomerular tubular balance against a higher fractional proximal tubular reabsorption.

CHAIRMAN Yes, but what I am saying is that it is impossible mathematically to have a normal GFR and a normal absolute reabsorption and to have ratio between the two, which is the fractional reabsorption high.

DIEPERINK They go back to the normal range, but when you see on the fraction of the two it is not normal.

Proc EDTA–ERA (1984) Vol 21

UNILATERAL URETERIC LIGATION IN DIABETIC RATS: A PATHOLOGIC AND MORPHOMETRIC STUDY

G Hatzigeorgiou, **P N Ziroyanis**, H Paraskevakou, P Vardas, C Georgilis, M Mavrikakis

Alexandra Hospital, Athens, Greece

Summary

We studied the combined effect of unilateral ureteral ligation on the morphology and pathology of the contralateral kidney in normal and streptozotocin induced diabetic rats. In the diabetic group that underwent ligation of the right ureter the weight and volume of the left kidney was far greater, the tissue specific gravity far lower, while the percentage of affected glomeruli was significantly greater compared with control rats and those undergoing only ureteric ligation.

Introduction

It is well known that unilateral ureteral obstruction, ligation, or nephrectomy induces hypertrophy of the contralateral kidney. It is also known that diabetes mellitus causes bilateral renal hypertrophy [1,2].

To date however, no one has studied the combined effect of unilateral ligation and diabetes on the fate of the contralateral kidney. Therefore we studied this effect and we report here our findings on the weight, volume, tissue specific gravity (t.SG) and the pathological changes in the contralateral kidney.

Material and methods

In this study we used 31 female wistar rats weighing between 180 and 210 grams. The rats were divided into four groups:

Group 1 Control (C): Intact normal rats (n = 8); Group 2 Diabetic (D): Streptozotocin induced diabetic rats (n = 8); Group 3 Ligated (L): Normal rats whose right ureter was ligated surgically with 2.0 silk suture at the site of the uretero-vesical junction; Group 4 Ligated plus Diabetic (DL): The right ureter of the rats ligated as in group 3 (n = 7). The experimental diabetes mellitus was evoked two days later by bolus IV streptozotocin injection via the dorsal tail vein, the dose being 45mg/kg BW and, the vehicle being buffered citric acid in 0.9 per cent saline PH : 4.6. All rats were fed (Purina chow rat, pellets) and had free access to water. The experiment lasted 35 days from the ureteral ligation. The criterion for inclusion in the diabetic group was a blood glucose above 250mg/dl estimated at day seven and day 35 of the experiment. Diabetic rats were not treated with insulin or any other drugs. On the 35th day of the experiment the rats were anaesthetised with ether, a blood sample was drawn from the tail, and, the left kidney was surgically removed. The excised kidneys were trimmed free of fat, measurements (weight, volume) were obtained and the tissue specific gravity calculated. Following this the kidneys were processed for pathological and microscopic examination for H/H, PAS, and van Gieson orcein stains. In the microscopic examination increase of mesangial tissue was measured from a study of 20 randomly selected glomeruli. Statistical evaluation was done by Student's 't' test.

Results

The morphological parameters of the kidneys of the studied rats of each group are shown in Table I. The rats in the group L, D and DL have larger kidneys as measured by weight and volume than the controls. In more detail the DL group had kidneys nearly twice the weight of the controls and larger than rats in groups D and L ($p<0.005$ and $p<0.001$).

TABLE I. Morphometric measurements of left kidney in normal and diabetic rats with or without ligation of the right ureter

Groups	N	Weight mg	Volume ml	t.SG mg/cm³
C	8	689.12 ± 23.2	0.70 ± 0.02	980.95 ± 10.33
L	8	1096.50 ± 31.81	1.18 ± 0.04	927.41 ± 4.47
D	8	1028.75 ± 35.48	0.93 ± 0.03	1100.30 ± 11.62
DL	7	1236.29 ± 42.28	1.37 ± 0.04	900.51 ± 12.7

$\bar{X} \pm SD$

The volume of groups L, D and DL left kidneys were greater than controls. DL kidneys were largest followed by L and D ($p<0.001$).

The tissue specific gravity of D group kidneys was greater than controls while L and DL group kidneys were smaller than control kidneys (lowest t.SG kidneys in DL group. All differences significant ($p<0.005$ to $p<0.001$) (Figure 1).

Histological results

In all groups increase in mesangial tissue and thickening of arteriolar walls was observed. Increase in mesangial tissue was more extensive in the DL group and least in group L (Figure 2).

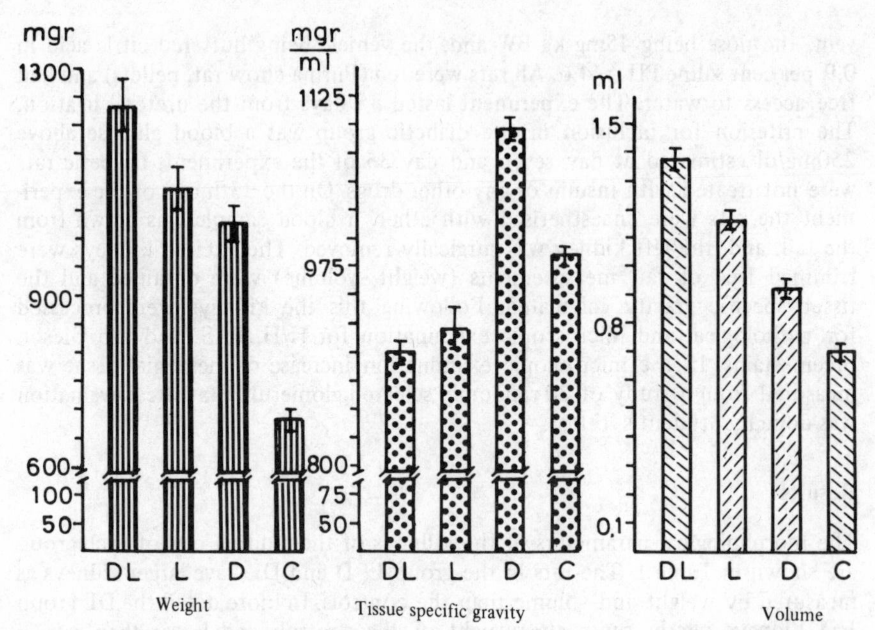

Figure 1. Histograms showing variation in weight, tissue specific gravity and volume in the four groups of experimental animals

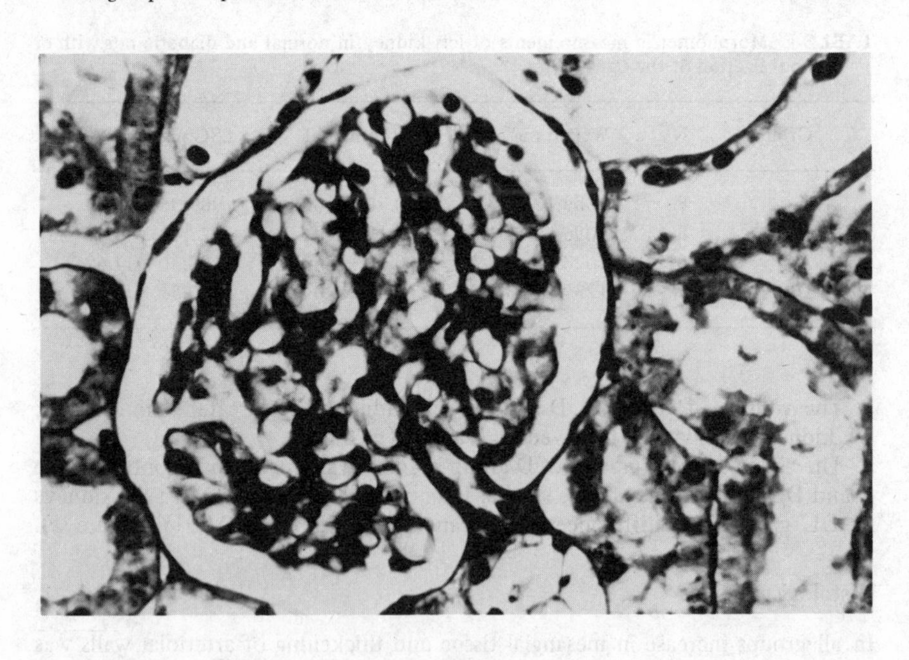

Figure 2. Glomerulus from left kidney of group DL rats showing increased mesangial tissue. PAS x 400 – reduced for publication

Discussion

Both diabetes mellitus and unilateral ureteral ligation are known to cause renal hypertrophy, bilaterally in the case of diabetes and of the contralateral kidney in the case of ureteral ligation [3]. The combined effect of both these defects on the contralateral kidney to the ligation have not yet been studied. The diabetic changes observed are seen at blood sugar values up to 450mg per cent [4]. The criterion for inclusion of the rats in the experimental groups was a blood sugar between 250 and 320mg per cent for the major part of the study period. In experimental diabetes mellitus the renal changes observed are most intense in the first month after induction of the disease [4]. The diabetic rats had renal weight increase of 32.8 per cent, the ligated rats of 59.1 per cent and the DL rats had renal weight increase of 79.4 per cent in comparison with controls. The diabetic rats had renal volume increase of 49.2 per cent, the ligated 68.5 per cent and the DL, 95.7 per cent again in comparison with controls.

It seems therefore, that the combination of diabetes and ligation have roughly additive effect on the weight and volume changes in the kidney. The mechanism by which this increase in weight and volume occurs in both these conditions is not clear. The hyperglycaemia has been blamed for the hypertrophy of DM and in unilateral ligation the renal secretion of a renotropic factor [5] by the obstructed kidney has been implicated and other factors such as growth hormone, prostaglandin, catecholamines and haemodynamic parameters [6] have also been suggested as being involved in the hypertrophy resulting from diabetes mellitus.

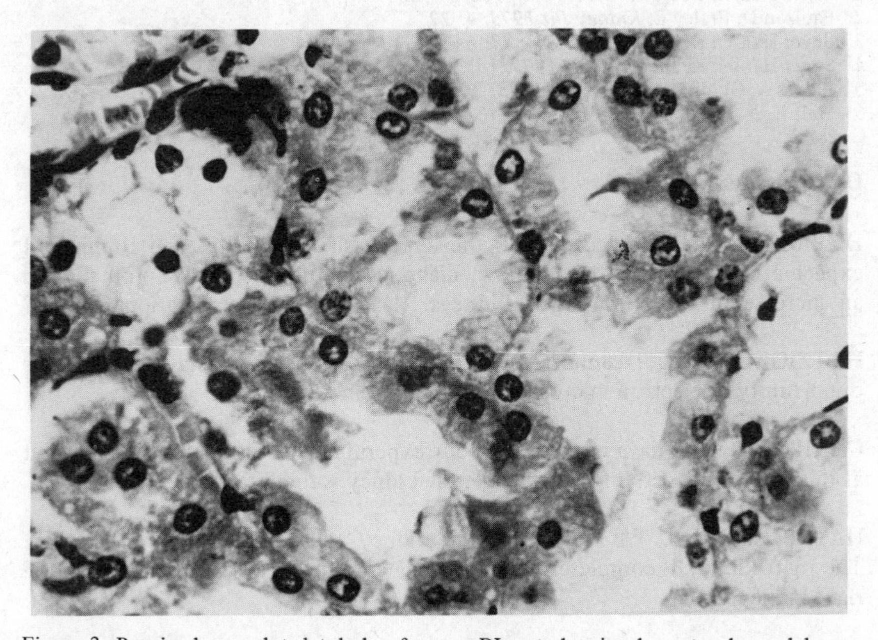

Figure 3. Proximal convoluted tubule of group DL rat showing hypertrophy and hyperplasia of cells. Some Armani Ebstein cells are shown. H/H x 400 – reduced for publication

Group D kidneys had a 12.7 per cent increase in tissue specific gravity while groups L and DL had 5.5 per cent and 8.2 per cent decrease in comparison with the control groups. Such results have not been reported in earlier studies. These results would support the hypothesis that kidneys in groups L and DL contained an increased perfusion of low s.g. substances such as fat or water. Histology excluded interstitial oedema. The mechanism would seem to be an increase in the filtrate present in tubular hypertrophy. The intra-tubular space is increased resulting in an increased content of filtrate and the results of the histological examination supports this view.

The greater change in tissue specific gravity in groups DL compared with group L may be accounted for by a) the greater degree of tubular hyperplasia observed in this group (Figure 3) and b) the occurrence of glucose in the tubules and thus the hyperplastic tubules contained for osmotic reasons a further increased load.

The increase in tissue specific gravity of the D group was attributed to an increase of solid material in the kidney. The increase in Armani-Ebstein cells in the DL as compared to the D group is not explained by variation in blood sugar values between these two groups as these were comparable, but the combination of ligation and diabetes mellitus causes an increase in the pathological changes as with diabetes, by mechanisms at present unknown.

References

1 Seyer-Hansen K. *Clin Sci Mol Med 1976; 51:* 551
2 Paulson D, Fraley E. *Kidney Int 1973; 4:* 22
3 Seyer-Hansen K. *Kidney Int 1983; 23:* 643
4 Seyer-Hansen K. *Diabetologia 1977; 13:* 141
5 Preuss A, Goldin H. *Kidney Int 1983; 23:* 635
6 Malt R. *N Engl J Med 1969; 280:* 1446

Open Discussion

Di PAOLO (Chieti) According to the hypothesis of Viberti* and Brenner† I expect a decrease of matrix only in diabetic obstructed kidneys. You showed an increase of matrix in these kidneys. Have you any explanation for this?

HATZIGEORGIOU I cannot give any explanation because I have not had the opportunity of electron microscopic examination.

CHAIRMAN You have carried out your experiments 35 days after ligation, but at this time, at least to my knowledge, the kidney is totally destroyed.

HATZIGEORGIOU We ligate the right kidney and we study the left kidney. The right kidney is completely destroyed. There is no longer any function in the right kidney.

CHAIRMAN Why did you choose this model and not a nephrectomy model?

HATZIGEORGIOU We did not find in the literature any other type of protocol.

CHAIRMAN It has been shown that after two weeks of complete ureteral obstruction the kidney is no longer functioning and the tissue is fibrotic.

HATZIGEORGIOU Yes, after two weeks the right kidney is destroyed completely.

*Viberti GC. Bilous RW, Mackintosh D et al. *Br Med J 1983; 286:* 598
†Hostetter TM, Rennke HG, Brenner BM. *Am J Med 1982; 72:* 375

Proc EDTA–ERA (1984) Vol 21

REVERSIBLE TUBULAR DYSFUNCTION IN ALCOHOL ABUSE

S De Marchi, *E Cecchin, F Grimaldi, A Basile, *L Dell'Anna, *F Tesio

*Hospital of Codroipo, Udine, *Hospital of Pordenone, Pordenone, Italy*

Summary

The discovery of an unexplained alkaline urine pH in a significant percentage of chronic alcoholic patients prompted us to evaluate some aspects of their tubular function. We studied 60 patients with a history of alcohol consumption of at least 160g daily for 10 years or more. Only patients without clinical and histopathological evidence of chronic liver disease were included in the study. The endogenous creatinine clearance was in the normal range in all patients. On the first day of hospitalisation 22 patients (36.6%) had a urine pH greater than 6.4 and a daily bicarbonate excretion ranging from 5.8 to 25.9mmol. The fractional urinary excretion of β_2- microglobulin, sodium, potassium, chloride, calcium, phosphorus and uric acid were significantly increased compared with those of 38 alcoholic patients with urine pH less than 6.4 and those of 50 healthy controls. All these indices of tubular function improved during withdrawal, and after 30 days of abstinence their values did not differ from those of controls. This data provides evidence that in one-third of heavy drinkers alcohol abuse causes a complex tubular dysfunction which, at least in this stage of alcoholic disease, recovers with abstinence.

Introduction

Electrolyte abnormalities are common in chronic alcoholic patients and a characteristic pattern may be seen in such patients during their first days of hospitalisation and often is a telltale sign of otherwise unsuspected, severe, chronic alcoholism [1]. The components of this pattern include hypokalaemia, hypophosphataemia, hypomagnesaemia, hypocalcaemia, an abnormally low blood urea nitrogen concentration and a respiratory alkalosis [2]. These electrolyte abnormalities are commonly described among the metabolic consequences of alcoholism and even though the loss of one mineral may affect the metabolism of several others, their pathogenesis might involve, at least in part, the kidney. In a previous study [3] we reported that in severely dependent alcoholic patients hypocalcaemia is associated with an increase of the fractional urinary excretion of calcium. In our experience three-fifths of severely dependent alcoholic patients and a quarter of those with mild dependence had a fasting urine calcium to urine creatinine ratio exceeding the upper normal limit of 0.12 [3]. In the present study interrelationships among alcohol-induced effects on plasma composition and tubular function are explored in order to clarify if the tubular dysfunction, that we have recently described in chronic alcoholism [4], may have a physiopathological significance in the derangement of plasma composition commonly seen in these patients.

Materials and methods

We investigated 60 patients aged between 22 and 60 years (mean 47.7; SD 11.3) admitted to the Department of Internal Medicine and the Alcohol Treatment Unit with a history of alcohol consumption of at least 160g (5.6 oz) a day or more. Only patients without clinical and histopathological evidence of chronic liver disease were included in the study. The following tests were performed in the first 24 hours of hospitalisation: urine pH and bicarbonate, glomerular filtration rate determined by endogenous creatinine clearance, fractional urinary excretion of β_2-microglobulin (FE β_2-M), sodium (FE Na), potassium (FE K), chloride (FE Cl), calcium (FE Ca), phosphorus (FE P) and uric acid (FE Ur Ac). The renal clearance of a substance was calculated according to the standard formula (urinary concentration of the substance times urinary flow rate)/serum concentration of the substance and fractional urinary excretion as (the clearance of the substance/clearance of creatinine) x 100. At admission a blood sample was taken to measure serum concentrations of creatinine, uric acid, β_2-microglobulin (β_2-M), sodium (SNa), potassium (SK), chloride (SCl), magnesium (SMg), calcium (SCa), phosphorus (SP), total protein and other indices of nutritional and liver status. Serum ionised calcium was estimated from total calcium and total protein using nomogram [5]. β_2-M assay was performed using Phadebas β_2-Micro Test kits (Pharmacia Diagnostic). Urine pH was determined using the radiometer pH meter. Urine bicarbonate was calculated from the measured pH and pCO_2 using the Henderson-Hasselbach equation. If urine pH was less than 7.0 an aliquot of urine to be used for β_2-M determination was immediately alkalinised to pH greater than 7.0 by the addition of sodium hydroxide. Plasma and urine electrolytes, creatinine and uric acid were measured by routine laboratory methods. Arterial pH and pCO_2 were determined with a radiometer digital analyser and plasma bicarbonate was calculated using the Henderson-Hasselbach equation. All these tests were repeated approximately after 30 days of abstinence. Control group consisted of 50 healthy subjects matched for age and sex.

Linear regression analysis and Student's 't' test for unpaired data were used for statistical analysis.

Results

Data regarding urine pH in chronic alcoholic patients on the first day of hospitalisation are reported in Table I. Thirty-eight patients (group 1) had a urine pH

TABLE I. Clinical and laboratory parameters in chronic alcoholic patients on the first day of hospitalisation

		Alcoholics with urine pH <6.4	Alcoholics with urine pH >6.4	Significance p values
Number of observations		38	22	
Percentage		63.4%	36.6%	
Urine pH		5.35±0.4	7.36±0.6	0.0001
Age (years)		49.80±8.1	45.70±13	NS
Serum chemistries:				
Sodium	(mEq/L)	142.20±2.52	139.40±3.50	0.0007
Potassium	(mEq/L)	4.13±0.34	3.80±0.29	0.008
Chloride	(mEq/L)	100.30±1.81	98.60±1.76	0.014
Calcium	(mg/100ml)	9.38±0.45	9.26±0.55	NS
Phosphorus	(mg/100ml)	3.44±0.69	2.93±0.52	0.002
Magnesium	(mEq/L)	1.55±0.25	1.43±0.26	NS
Uric acid	(mg/100ml)	5.49±1.07	4.82±1.16	0.019
BUN	(mg/100ml)	12.70±3.70	7.20±2.30	0.0002
Creatinine	(mg/100ml)	0.85±0.12	0.77±0.18	NS
Creatinine clearance (ml/min/1.73$^{m^2}$)		68±18	70±21	NS
Diuresis	(ml/24hr)	934±215	1231±563	NS

TABLE II. Fractional urinary excretions in chronic alcoholic patients on the first day of hospitalisation

		Alcoholics with urine pH <6.4	Alcoholics with urine pH >6.4	Significance p values
FE Sodium	(%)	0.51±0.31	0.99±0.83	0.004
FE Potassium	(%)	7.50±4.10	11.00±6.60	0.03
FE Chloride	(%)	0.69±0.41	1.40±1.19	0.008
FE Calcium	(%)	0.86±0.44	1.92±1.17	0.0001
FE Phosphorus	(%)	8.31±4.26	14.21±7.12	0.028
FE Magnesium	(%)	1.76±0.63	2.21±0.97	NS
FE Uric Acid	(%)	6.41±1.90	10.57±6.37	0.007
FEβ_2-microglobulin	(%)	0.31±0.20	1.32±1.13	0.001

NS: not significant: $p>0.05$
Results are expressed as mean ± SD

TABLE III. Simple linear regressions observed between fractional urinary excretion of substances and their serum concentrations in chronic alcoholic patients on the first day of hospitalisation

X	Y	Equation	Correlation coefficients	p<	Number of observations
FE Na	Serum Na	Y = −3.28X + 143	−0.549	0.0001	60
FE K	Serum K	Y = −0.04X + 4.3	−0.457	0.001	60
FE Cl	Serum Cl	Y = −1.37X + 100	−0.326	0.01	60
FE Ca	Serum Ca	Y = −0.24X + 9.7	−0.438	0.001	60
FE P	Serum P	Y = −0.04X + 3.6	−0.461	0.001	60
FE Mg	Serum Mg	Y = −0.13X + 1.8	−0.504	0.01	38
FE Uric Acid	Serum Uric Acid	Y = −0.11X + 6.2	−0.406	0.01	60
FE β_2-M	Serum β_2-M	Y = −0.22X + 2.0	−0.406	0.01	60

Units used: serum and urinary sodium (Na): mEq/L; serum and urinary potassium (K): mEq/L; serum and urinary chloride (Cl): mEq/L; serum and urinary calcium (Ca): mg/100ml; serum and urinary phosphorus (P): mg/100ml; serum and urinary magnesium: mEq/L; serum and urinary uric acid : mg/100ml; serum and urinary β_2-microglobulin (β_2-M): mg/L

FE X: fractional urinary excretion of X: $\frac{U_X \times S_{Creat}}{U_{Creat} \times S_X} \times 100$

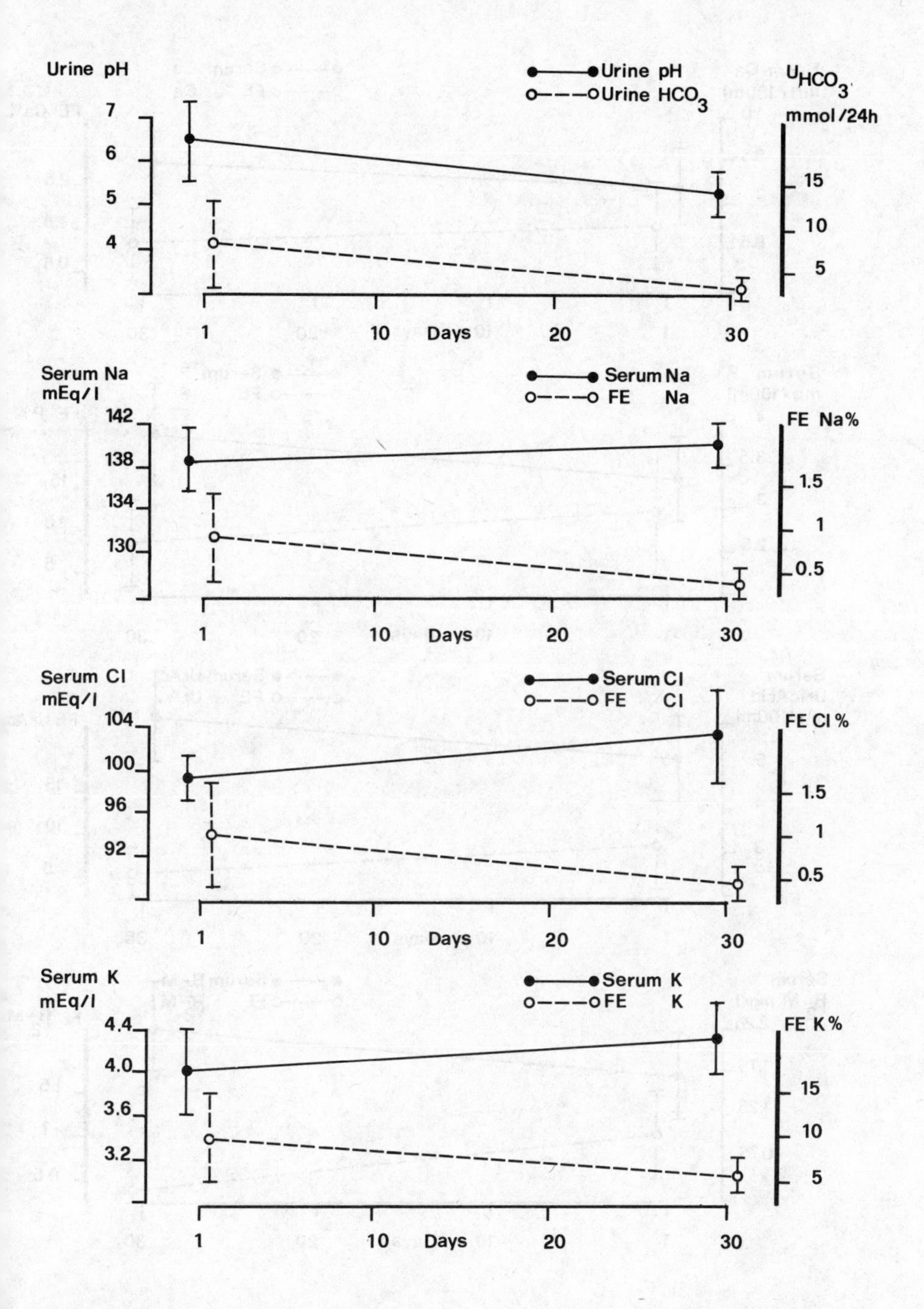

Figure 1. Effect of abstinence on some indices of renal tubule function in chronic alcoholic patients

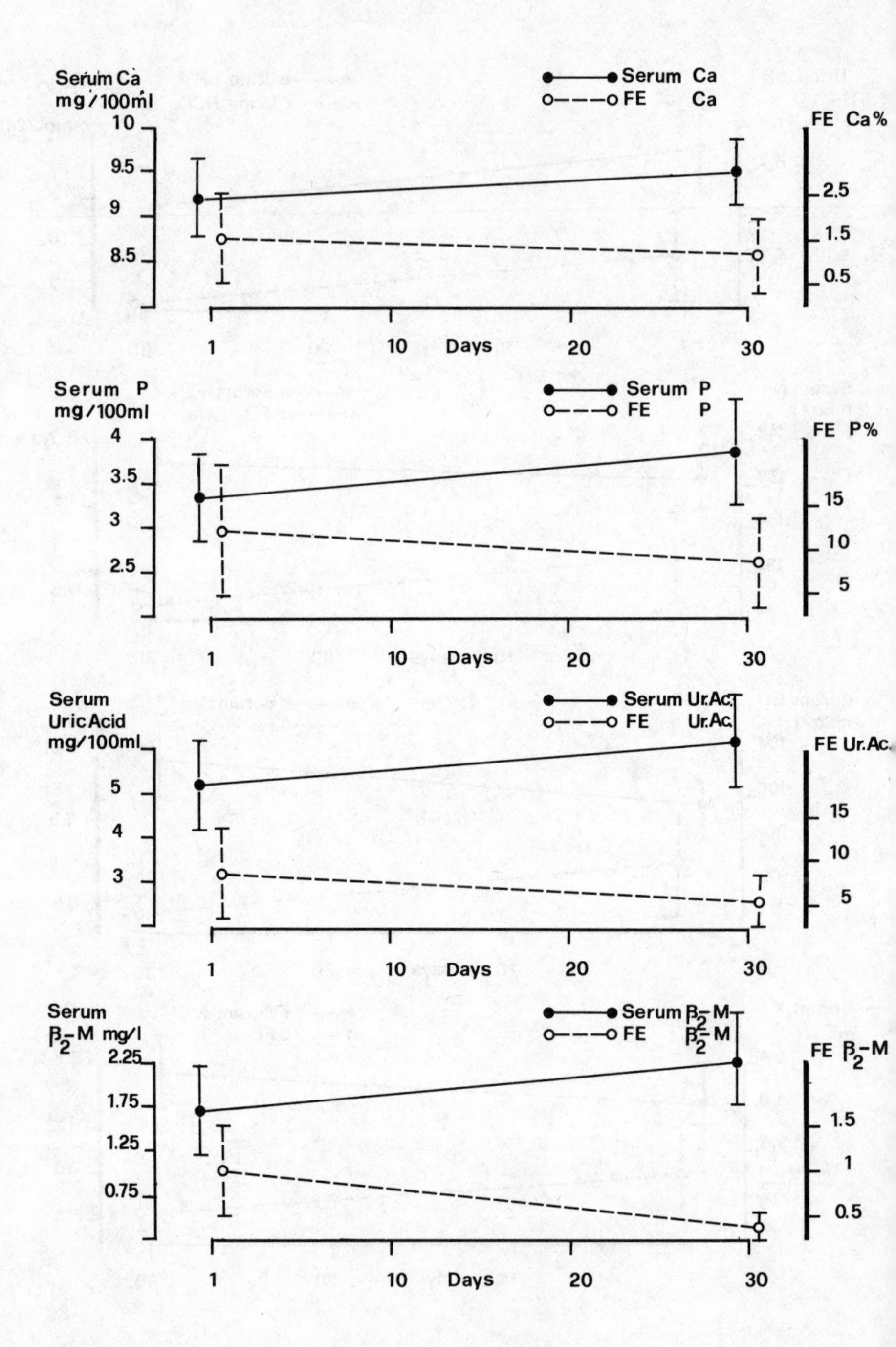

Figure 2. Effect of abstinence on some indices of renal tubule function in chronic alcoholic patients

less than 6.4; 22 patients (36.6%) had a urine pH greater than 6.4 (group 2). At admission SNa, SK, SCl, SP, serum concentrations of uric acid and blood urea nitrogen were significantly decreased in the alcoholic patients with urine pH greater than 6.4 (Table I) as well as serum ionised calcium (1.27 ± 0.08 vs 1.31 ± 0.07mmol/L; $p = 0.011$). In group 2 patients the urinary bicarbonate excretion ranged from 5.8 to 25.9mmol/24hr wheras in group I it never exceeded the upper normal limit of 3.5mmol/24hr. Table II summarises data regarding some indices of tubular function in chronic alcoholic patients on the first day of hospitalisation. The values of FE β_2-M, FE Na, FE K, FE Cl, FE Ca, FE P and FE Uric Acid were significantly increased in alcoholic patients with urine pH greater than 6.4 compared to those of patients with urine pH less than 6.4 and to normal controls. In the overall group of alcoholic patients we found significant inverse relationships between the fractional urinary excretion of the investigated substances and their serum concentrations as shown in Table III. All these indices of renal tubular function improved during withdrawal and after approximately 30 days of abstinence they fell within the normal range (Figure 1).

Discussion

Ethyl alcohol appears to be an intoxicant responsible for profound changes of body composition. The chronic severe alcoholic patient is commonly beset with some disturbances in serum electrolyte composition. The interaction between alcohol abuse and deranged mineral metabolism is complex. Some effects of alcohol are direct, others the result of poor nutrition or end-organ diseases accompanying chronic abuse. Even though the nephrotoxic effect of ethyl alcohol is still controversial the present study provides evidence that in approximately one-third of chronic alcoholic patients, ethanol abuse causes a complex tubule dysfunction. The increased fractional urinary excretions of phosphorus, uric acid and β_2-M suggest a proximal tubular defect resembling the Fanconi syndrome. The inverse relationships between the fractional urinary excretion of the investigated substances and their serum concentrations seem to emphasise that the tubule dysfunction has a physiopathological significance in the deranged plasma composition commonly seen in chronic alcoholic patients during their first days of hospitalisation. Even though we could not find any significant difference in the nutritional status between the two groups of patients, it remains unclear if ethanol has a direct nephrotoxic effect or if the tubule dysfunction is mediated by different mechanisms. These data seem to emphasise that, at least in this stage of the alcoholic disease, the tubular dysfunction recovers with abstinence in a period of approximately 30 days. In conclusion this effect of ethyl alcohol on renal function should be mentioned together with those on the endocrine, cardiovascular and gastrointestinal systems because it may contribute, at least in part, to certain derangements of electrolyte metabolism occurring in chronic alcoholism.

References

1. Knochel JP. In Epstein, M, ed. *The Kidney in Liver Disease.* Amsterdam: Elsevier. 1983: 203

2 Knochel JP. *Arch Intern Med 1980; 140:* 613
3 De Marchi S, Basile A, Grimaldi F et al. *Br Med J 1984: 288:* 1457
4 De Marchi S, Cecchin E, Grimaldi F, Basile A. *Ann Intern Med 1984; 101:* 145
5 Parfitt AM, Kleerekoper M. In Maxwell MH, Kleeman CR, eds. *Clinical Disorders of Fluid and Electrolyte Metabolism.* New York: McGraw-Hill. 1980: 269

Open Discussion

WALLS (Leicester) Did you look at any markers of tubular cell damage such as NAG which may have produced pathological damage or are you suggesting that this is merely a physiological phenomenon?

DE MARCHI We have not studied any of urinary enzymes, probably we will try to perform these investigations in the next study.

DEL CANTON (Naples) What about urine osmolality in these patients? I wonder whether this complex tubular dysfunction is not accounted for by increased water reabsorption?

DE MARCHI It is known that in chronic liver disease we have impaired capacity of water excretion but in the last issue of Kidney International there was a paper which explains the mechanisms which are involved in this tubular dysfunction. Probably prostaglandins are involved in such mechanisms. We investigated the plasma osmolality of the two groups of alcoholics with and without urinary pH over 6.4 and this was in the normal range in both.

GOODWIN (Chairman) It did occur to me, Dr De Marchi, that many of your alcoholic patients were probably on fairly poor diets before they came into hospital. I wondered how you excluded the possibility that poor diet could be the underlying cause? You might have compared two groups of patients, both continuing with their alcohol , one on hospital diet and one with their own poor diet. Would you like to comment on this?

DE MARCHI I agree with you that we investigated only some illnesses of nutrition in the two groups of our alcoholics. They were not different in the two groups but I agree with you that the diet might be important in determining such results. On the other hand it is difficult to obtain a diet history from these patients.

GOODWIN You did not really test the capacity of the kidneys to acidify the urine or conserve bicarbonate. I wonder why you did not design a more formal test, e.g. an acid load test, to see whether these patients could really lower their urinary pH.

DE MARCHI Now we are performing an acidification test with calcium chloride 2mEq/kg, but the results are not conclusive.

Proc EDTA–ERA (1984) Vol 21

TUBULAR FUNCTION AND RENAL SELECTIVITY TOWARDS GLYCOSYL-ALBUMIN IN DIABETIC HYPEROSMOLAR STATES

G Delfino, G Candiano, G M Ghiggeri, C Querirolo

Hospital of Lavagna, Lavagna, Italy

Summary

The type of proteinuria and renal selectivity of glycosyl albumin has been evaluated in seven diabetic patients at the onset of plasma hyperosmolar state and after complete fluid replacement. The main feature of all patients was a marked increase in urinary excretion of β_2-microglobulin, which promptly returned to normal after the correction of the fluid disequilibrium. This indicates that a reversible tubular injury is the first symptom of the dehydration process in diabetic patients with plasma hyperosmolality.

Introduction

Plasma hyperosmolality, which readily develops in diabetic patients concomitantly with a severe metabolic derangement, is usually accompanied by an abrupt increase of urinary protein excretion and a rapid but reversible (at least in initial phase) decline of renal function [1]. It is not known whether proteinuria develops as a consequence of the fall of renal plasma flow or whether other mechanisms participate in its pathogenesis: the most likely being tubular defects or an alteration of glomerular selectivity towards structurally deranged proteins [2]. Both can sustain proteinuria in this condition, in agreement with up-to-date knowledge on glomerular and tubular function [3,4].

Taking urinary excretion of large molecules such as immunoglobulin G (IgG) and albumin and of smaller proteins such as β_2-microglobulin as markers of glomerular and tubular functions respectively [5], we have studied the proteinuria of seven diabetic patients at the onset of an acute hyperosmolar state and after the prompt correction of the fluid disequilibrium, which prevented the decline of renal function. At the same time the renal selectivity of glycosyl albumin has been evaluated in order to ascertain (1) the selectivity properties of the glomerular filter towards proteins such as glycosyl albumin, showing various pIs but of the same molecular weight, and (2) a possible role of increased glycosylation of serum albumin in determining the increased urinary excretion of this molecule.

Materials and methods

Seven patients with type I diabetes mellitus (5 females, 2 males), four of whom were newly diagnosed, aged between 18 and 68 years, with plasma hyperosmolality were studied. Conditions for selection were: (1) plasma hyperglycaemia ($\geqslant$200mg/100ml), (2) plasma hyperosmolality ($\geqslant$300mOsm/kg), (3) normal serum creatinine, (4) no history of preceding renal disease, (5) no signs of pneumonia or other infectious disease. On admission two patients were unconscious, three were drowsy and two were in a normal conscious state. After admission all patients were treated with low dose insulin administration (Actrapid MC, Novo) and fluid replacement with appropriate saline solutions. Twenty-four hour urine collections were performed maintaining the patients in a recumbent position, immediately after admission and after seven days of therapy, which normalised all metabolic parameters (Table I). Total protein concentrations were determined by the Coomassie dye binding assay as modified by Read [6]; albumin and IgG by nephelometry (Beckman, Immunochemistry Analyzer); β_2-microglobulin by a radioimmunoassay (Pharmacia). Results for all the urinary proteins are given as excretion rates (μg/min). Arterial blood for acid-base determination was drawn from the radial artery of the right forearm and pH determined at 37°C with a Micro pH/blood Gas Analyzer 413 (Instrumentation Laboratory), using the Henderson-Hasselbalch equation to calculate blood HCO_3^-. Mean blood glucose (MBG) was calculated from three glucose determinations at eight hour intervals. For the determination of glycosyl albumin albumin was separated from other serum and urinary proteins by pseudoligand chromatography on Affi-Gel Blue [7] and carbohydrates covalently bound to the protein evaluated as hydroxymethylfurfural, using the thiobarbituric acid assay as described [2]. Results are given as nmol hydroxymethylfurfural/nmol albumin.

Statistical analysis was performed with the Student's 't' test for paired data. The least squares method was used to calculate the correlation coefficient.

Results

As shown in Table I, besides the alteration of those parameters considered for the selection of patients (plasma osmolality and blood glucose concentration), all but one were in a state of pure metabolic acidosis, very severe in two cases (patients 1 and 6). At this time the urinary excretion rates of total proteins were increased to some extent, reaching in four cases values higher than 2000μg/min (Table I). Urinary albumin excretion rates were greater than normal in each patient, in five from 20 to 100μg/min and in two exceeding 150μg/min. Urinary excretion rates of IgG were also increased, but to a less extent than albumin, reaching considerable amounts only in patients 1, 5 and 6. β_2-microglobulin excretion rates were increased over normal in all patients by a factor from 2 to

TABLE I. Clinical parameters and composition of urinary protein in seven diabetic patients at the onset of a hyperosmolar state and after the replacement of fluid

			BLOOD					URINE			
No.	Patients	Age (years)	pH	HCO_3^- (mEq/L)	pCO_2 (mmHg)	MBG (mg/100ml) Onset–7 days	Osmolality (mOsm/kg) Onset–7 days	β_2-microgl. (μg/min) Onset–7 days	Albumin (μg/min) Onset–7 days	IgG (μg/min) Onset–7 days	Tot. Prot. (μg/min) Onset–7 days
1	MGQ	20	6.96	2.9	16.6	311–179	340–290	34.5 –0.03	150.9 –5	40.1–5	3294–85. 1
2	BA	53	7.24	15.6	36.7	800–249	413–276	0.31–0.16	50.4 –12.6	6 –5	4132–288
3	AB	68	7.32	13.5	26.7	216–203	315–286	0.2 –0.7	22.14–5	5.8–5	190–100
4	PN	18	7.33	19.9	37.5	233–121	300–278	6.2 –0.08	17.5 –6.6	7.1–6	173–35
5	VB	56	7.27	7.4	19.4	1540–272	405–283	134.8 –0.61	103.2 –33.4	66.6–5	2350–350
6	LS	54	7.14	2.7	11.8	450–193	360–278	0.6 –0.02	127.8 –16.6	37.9–5	2000–150
7	EM	37	7.39	21.2	36.9	288–131	356–300	27.9 –0.3	62.3 –6	6 –5	755–108

MBG=mean blood glucose
β_2-microgl=β_2-microglobulin
IgG=immunoglobulin G
Tot. Prot.=total protein

1340 (assuming the upper limit for urinary excretion of β_2-microglobulin of 0.1μg/min). By current criteria, four patients fulfilled the criteria for pure tubular proteinuria while the remaining three had a mixed glomerular-tubular defect [7]. After seven days of therapy, the excretion rates of all proteins considered decreased to normal values (Table I). A high statistical correlation between urinary excretion rates of all proteins and plasma osmolality was found as shown in Figure 1.

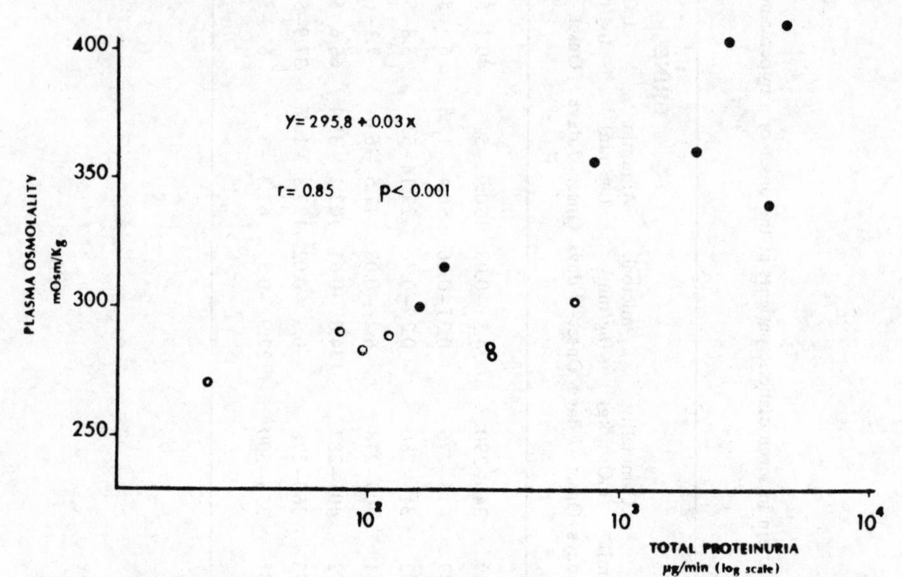

Figure 1. Relationship between urinary protein excretion rates and plasma osmolality in seven diabetic patients at the onset of an hyperosmolar state (closed circles) and after the correction of the fluid disequilibrium (open circles)

Serum glycosyl albumin concentration was 0.375±0.04nmol hydroxymethylfurfural/nmol albumin at admission and after seven days only slightly decreased to 0.351±0.04 (Table II). Urinary glycosyl albumin concentration was 0.776±0.16 at admission and the value at the seventh day was 1.342±0.29, giving a urinary/serum ratio ranging from 2.1±0.4 on admission to 3.9±0.9 at seven days.

Discussion

As is well known, glomerular filtration rate tends to be reduced in diabetic patients with severe hyperglycaemia (diabetic ketoacidosis or non-ketotic coma) [1], as a result of depletion of extracellular fluid volume and a fall in renal plasma flow. The same conclusion was reached by Hostetter et al studying glomerular haemodynamics in experimental diabetic rats [8]. Our findings indicate that an acute tubular injury (as revealed by an abrupt increase of urinary excretion of small proteins such as β_2-microglobulin) [7] precedes the decrease of glomerular filtration rate and that a prompt replacement in

TABLE II. Serum and urinary glycosyl albumin in seven diabetic patients at the onset of the metabolic derangement and after the correction of metabolic parameters

Patient No.	Glycosyl albumin (nmol hydroxymethylfurfural/nmol albumin) SERUM Onset–7 days	%	URINE Onset–7 days	%
1	0.193–0.218	+13	0.627–0.925	+ 48
2	0.332–0.424	+28	0.619–2.323	+275
3	0.325–0.278	–14	0.447–2.196	+391
4	0.507–0.443	–13	1.284–1.509	+ 18
5	0.451–0.423	– 6	0.480–0.564	+ 18
6	0.372–0.251	–32	0.465–0.294	- 40
7	0.446–0.424	– 5	1.511–1.584	+ 5

body fluids achieves a complete restoration of tubular function. In this way further developments of the renal ischaemic damage no longer occur.

The high urinary excretion of β_2-microglobulin is in all cases paralleled by a small increase of urinary albumin and in three cases of urinary IgG. The increased urinary excretion of these large molecules may be explained on the basis of a defect in their tubular absorption, since renal tubular resorptive function have been shown to be important in determining the low urinary excretion of albumin in rats [9].

The second part of the present study was undertaken to determine the selectivity properties of the glomerular barrier towards the electrical charge of proteins in this condition, to ascertain a possible role of glycosylation of serum albumin in determining its urinary leakage. Hyperfiltration of albumin occurs in uncomplicated diabetes as the result of its non-enzymatic glycosylation and the related changes in its structure [2]. The same mechanism does not seem to be effective in diabetic patients with severe hyperglycaemia, in which to very high values of serum glycosyl albumin there is no corresponding proportional increase of urinary glycosyl albumin excretion.

In summary, (1) tubular damage is the first sign of an acute dehydration in diabetic patients with severe hyperglycaemia and hyperosmolality; (2) a prompt replacement of body fluids results in a complete restoration of tubular function; (3) glomerular selectivity towards the electrical charge of proteins is preserved at the onset of proteinuria, and (4) glycosyl albumin hyperfiltration has not any appreciable pathogenetic role in proteinuria related to diabetic hyperosmolar states.

References

1 Reubi FC. *Circ Res 1953; 1:* 410
2 Ghiggeri GM, Candiano G, Delfino G et al. *Kidney Int 1984; 25:* 565
3 Brenner BM, Boher MP, Baylis C et al. *Kidney Int 1977; 12:* 229
4 Cartney MA, Sawin LL, Weiss DD. *J Clin Invest 1970; 20:* 7

5 Peterson PA, Evrin PE, Berggard I. *J Clin Invest 1969; 48:* 1189
6 Read SM, Northcate DH. *Anal Biochem 1981; 16:* 53
7 Travis J, Bowen J, Tewksbury D et al. *Biochem J 1976; 157:* 301
8 Hostetter TH, Troy JL, Brenner BM. *Kidney Int 1981; 19:* 410
9 Landwehr DM, Carvalho JS, Oken DE. *Kidney Int 1977; 11:* 9

Proc EDTA–ERA (1984) Vol 21

INTRACELLULAR BICARBONATE AND pH OF SKELETAL MUSCLE IN CHRONIC RENAL FAILURE

A Guariglia, C Antonucci, E Coffrini, **S Del Canale**, E Fiaccadori, F Reni, P Vitali, U Arduini, A Borghetti

Istituto di Clinica Medica e Nefrologia, Università degli Studi di Parma, Parma, Italy

Summary

In 11 controls and 10 patients suffering from untreated uraemic acidosis intracellular bicarbonate and skeletal muscle pH (needle biopsy) were determined. In all patients a significant intracellular acidosis, not related to any extracellular indices was found. It is concluded that the chronic proton load is able to effect intracellular buffer composition; moreover the action of other factors such as derangements of cell metabolism and nutritional imbalance could be operating.

Introduction

Cell buffer mechanisms are believed to play a major role in the whole body defence against acid-base derangements. As previously demonstrated in uraemic acidosis the extracellular steady-stage is maintained at the expense of a progressive titration of cell buffers, particularly in bone. A significant participation from soft tissue has also been repeatedly emphasised but the reported literature is poor and inconclusive [1–4]. Skeletal muscle, about 40 per cent of total body cell mass, represents the main regulatory site of soft tissue buffering processes.

We have investigated the intracellular acid-base composition of muscle in chronic renal failure (CRF) patients, selected for their untreated metabolic acidosis by means of a direct measurement of intracellular bicarbonate and pH by muscle needle biopsy.

Materials and methods

Ten patients (5 males and 5 females) suffering from CRF (glomerular filtration rate range 4–30ml/min/1.73m^2) and presenting an untreated metabolic acidosis were admitted to the present study. Eleven subjects with normal glomerular filtration rates and acid-base balance served as controls.

In all subjects arteriovenous blood samples were drawn for pH, pCO_2 and

HCO_3 determination. Venous blood samples were also used to determine plasma sodium, potassium and chloride.

In all subjects, whose consent was previously obtained, muscle needle biopsies from quadriceps femoris were performed after at least two hours of rest.

Muscle samples were then analysed in order to obtain muscle water (H_2O_m) and acid-labile carbon dioxide content (TCO_2).

Extracellular water (H_2O_e) was calculated from muscle chloride content according to Cotlove [5]: in CRF patients with glomerular filtration rates below 7ml/min/1.73m^2 a membrane potential value according to Cotton [6] was used.

H_2O_m was determined by the difference between the fresh tissue samples and their dry weight.

Intracellular water (H_2O_i) was determined by the difference between H_2O_m and H_2O_e.

Intracellular bicarbonate (HCO_{3i}) was measured by TCO_2 method according to well-established techniques [7–8] and corrected for H_2O_e and H_2O_i values previously determined in different muscle samples.

In order to calculate intracellular pH (pH_i) the Henderson-Hasselbalch equation was used, assuming cellular pCO_2 values to be the same as venous CO_2.

Results

In Table I extracellular electrolyte and acid-base indices are listed. Data are reported for each patient, while controls data are expressed as mean ± SD. An extracellular metabolic acidosis was present in all CRF patients, as compared to controls, while no difference existed in electrolyte patterns. No correlation was found between acidaemia and glomerular filtration rates.

In Figure 1 the relation between the TCO_2 extracted from muscle samples and muscle weight is reported. It is possible to observe the linear relation found either in controls (●) and in CRF patients (■). The slopes of two lines were significantly different ($p<0.001$) demonstrating that in CRF patients a clear depletion of muscle TCO_2 was present.

In Table II muscle TCO_2, HCO_{3i} and pH_i are reported: data are presented for each patient while control data are expressed as mean ± SD. A condition of intracellular acidosis was found in all patients.

All the above parameters were significantly different ($p<0.001$) to controls.

From the comparison of extra- and intracellular data (Tables I and II) no correlation was found between HCO_{3i}, pH_i and glomerular filtration rate nor to blood bicarbonate levels.

Statistical analysis

Data are expressed as mean ± SD. For the comparison of individual means in the two groups the Student's 't' test for unpaired data was used. The Student's 't' test was also used in the comparison of the slopes of the two lines reported in Figure 1.

TABLE I. Arterial acid-base indices, plasma electrolytes and glomerular filtration rates (GFR) of chronic renal failure patients and controls

Patients	Sex	Age (years)	GFR $ml/min/1.73m^2$	pH	$PaCO_2$ mmHg	HCO_3 mmol/L	Na mmol/L	K mmol/L	Cl mmol/L
1	M	41	4	7.08	24	7	141	5.0	107
2	F	54	5	7.42	29	18	143	3.8	102
3	F	51	5	7.26	27	12	138	5.4	107
4	F	54	6	7.31	27	13	143	5.3	101
5	F	59	8	7.40	27	17	137	3.9	112
6	F	74	10	7.27	20	9	130	7.0	95
7	M	64	10	7.39	26	16	137	3.8	100
8	M	70	10	7.35	23	13	138	4.8	100
9	M	65	13	7.31	30	15	151	5.6	102
10	M	71	30	7.26	33	14	140	6.0	114
Mean			10	7.30	27	13	140	5.0	104
SD			7	0.10	4	3	5	1	6
Controls n=11 30–75 years									
Mean			118	7.40	39	25	140	4.3	101
SD			12	0.02	2	1	2	0.4	2
Student's 't' test			$p<0.001$	$p<0.005$	$p<0.001$	$p<0.001$	NS	NS	NS

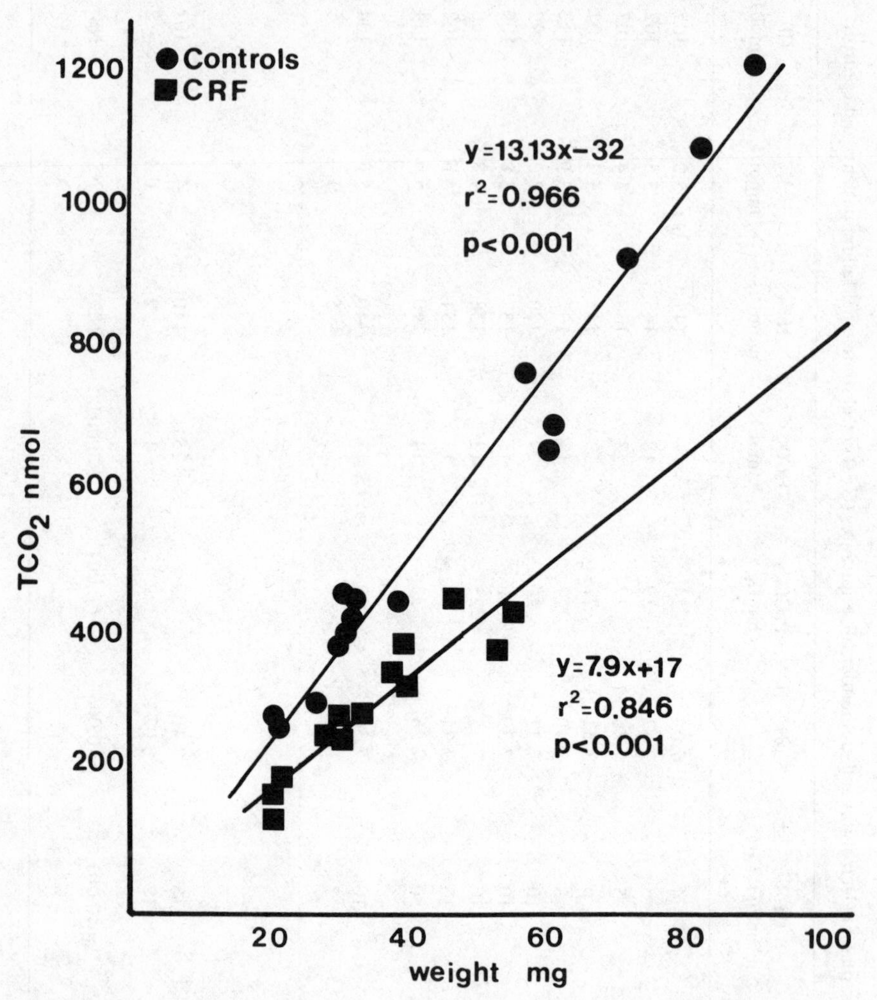

Figure 1. The relation between TCO_2 extracted and related muscle weight in chronic renal failure patients and controls. For some subjects more than one value is presented

Discussion

The main objective of the present study was measurement of muscle cell bicarbonate and pH in uraemic acidosis. Previous studies on intracellular compartments gave conflicting results and were not able to clarify the exact role of skeletal muscle buffering in this condition. A direct evaluation of skeletal muscle $HCO_{3\,i}$, by means of TCO_2 method, seems to offer a suitable approach to such problems.

In our patients (Table I), no correlation existed between acidaemia and glomerular filtration rate, suggesting that in this condition other factors than

TABLE II. Acid-base parameters of skeletal muscle in chronic renal failure patients and controls are listed

Patients	Sex	Age (years)	TCO_2 nmol/mg	HCO_{3i} mmol/L/H_2O_i	pH_i
1	M	41	7.0	7.2	6.90
2	F	54	8.4	5.7	6.86
3	F	51	9.8	10.1	6.97
4	F	54	7.7	5.9	6.82
5	F	59	9.8	8.6	6.93
6	F	74	7.6	8.2	7.03
7	M	64	8.8	7.3	6.82
8	M	70	8.7	9.1	7.00
9	M	65	8.5	5.7	6.72
10	M	71	6.4	5.3	6.81
Mean			8.3	7.3	6.88
SD			1.1	1.6	0.09
Controls n=11 30–75 years					
Mean			12.3	11.7	7.05
SD			1.0	1.1	0.05
Student's 't' test			$p<0.001$	$p<0.001$	$p<0.001$

nephron reduction are operating (exogenous and dietary factors, catabolism, body potassium stores, acid excretion/glomerular filtration rate, PTH levels). In all patients muscle analysis clearly showed a condition of intracellular acidosis as demonstrated by a significant decrease of TCO_2, HCO_{3i} and pH_i: muscle intracellular buffers seem to be operating against the chronic proton load. On the other hand the intracellular acid-base parameters were not related either to glomerular filtration rate, acidaemia or any other of the extracellular indices considered.

This is not surprising if we think that intracellular buffering is a complex process not merely related to extra →intra H^+ flux (or intra→extra HCO_3 flux). In fact pH_i regulation is based upon inter-relationships between physico-chemical (mainly cell proteins) and dynamic buffers, being the latter related to cell metabolic activity (metabolic pathways to produce or consume protons) [9–10].

Intracellular buffering efficiency can then be impaired not only by the severity or chronicity of the acidotic state, but also by other factors (not considered in this study) closely linked to metabolic and nutritional balance.

Thus to explain the intracellular acidosis found in our patients we can advance two hypotheses. Firstly a recruitment of cell metabolic buffers in muscle cells closely related to both glycolitic and oxidative cycle with the consequent metabolic derangement representing a 'trade off' for the maintenance of a relatively steady extracellular acid-base status. Secondly the peculiar metabolic and nutritional imbalance of uraemia could have reduced cell buffers availability, thus resulting in an impaired tolerance to proton load.

The final common pathway of the intracellular acidosis found in our patients is undoubtedly an energy metabolism derangement: this could be the source of the muscular symptoms frequently encountered in CRF patients.

References

1. Bittar E, Watt MF, Pateras VR et al. *Clin Sci 1962; 23:* 265
2. Lambie AT, Anderton JL, Cowie J et al. *Clin Sci 1965; 28:* 237
3. Mashio G, Bazzato G, Bertaglia E et al. *Nephron 1970; 7:* 481
4. Tizianello A, De Ferrari G, Gurreri G et al. *Clin Sci Mol Med 1977; 52:* 125
5. Cotlove E. *Analyt Chem 1963; 35:* 101
6. Sahlin K, Alvestrand A, Bergström J et al. *Clin Sci Mol Med 1977; 53:* 459
7. Guariglia A, Arduini U, Coffrini E et al. *IRCS Med Sci 1984; 12:* 46
8. Cotton JR, Woodart T, Carter NW et al. *J Clin Invest 1979; 63:* 501
9. Cohen RD, Iles RA. *CRC Crit Rev Clin Lab Sci 1975:* 101
10. Hultman E, Sahlin K. In Hutton RS, Miller DI, eds. *Exercise and Sport Sciences Reviews; 41.* Philadelphia: The Franklin Institute Press. 1980: 41

Open Discussion

NICHOLLS (Exeter) If you continue to give captopril beyond seven days, a steady state will be reached when there is no further sodium loss. If you then give indomethacin is this steady state maintained, or does sodium retention then occur?

DAL CANALE I do believe that it must reach a steady state when extracellular fluid volume does not influence proximal tubular reabsorption. Maybe only proximal tubular reabsorption because there is decreased plasma aldosterone. I think, but I have no proof, that the steady state can change again when you give indomethacin.

BANKS (Bristol) What happened to blood pressure and was the effect inhibited by indomethacin?

DAL CANALE Captopril significantly decreased blood pressure in our hypertensive patients. The decrease was still significant when indomethacin was added but the increase was less.

GOODWIN (Chairman) I wondered whether this effect on prostaglandin metabolism is peculiar to captopril or whether it is shared by other converting enzyme inhibitors such as enalapril?

DAL CANALE Yes, we have information in dogs about this and some have similar effects.

GOODWIN Does enalapril have this effect?

DAL CANALE I do not know.

BANKS Were your patients in sodium balance? Others have shown natriuresis on 100mg captopril and maybe your failure to do so resulted from not knowing the sodium intake. Did you weigh the patients?

DAL CANALE Yes, our patients were in constant salt balance before starting the study, taking 100mEq of sodium daily, the balance being maintained for at least five days prior to the study. The decrease in body weight was about 1kg which approximately corresponds with the loss of 150mEq sodium and actually we had a measured loss of 180mEq.

Proc EDTA–ERA (1984) Vol 21

ECTOPIC CALCIFICATION.
THE ROLE OF PARATHYROID HORMONE

A M De Francisco, M J D Cassidy, J P Owen, H A Ellis, J R Farndon, M K Ward, D N S Kerr

University of Newcastle upon Tyne, United Kingdom

Summary

In 42 uraemic patients radiological skeletal survey, biochemistry and bone histology were compared before and at 6–12 months (42 patients), 12–24 months (26 patients) or 24–48 months (12 patients) after parathyroidectomy. The presence of small vessel or non-visceral soft tissue calcification was not related to the age, sex, duration of end-stage renal failure treatment, total serum calcium, magnesium, phosphate, Ca x P product, alkaline phosphatase, ionised calcium, serum aluminium, iPTH, severity of radiological and histological osteitis fibrosa or parathyroid gland weight.

Twenty-three patients (55%) had small vessel and 20 (48%) soft tissue calcification before parathyroidectomy. Despite a marked improvement in subperiosteal erosions (37 healed, 5 improved) and healing of osteitis fibrosa histologically, seven patients developed new and six developed increased peripheral arterial calcification while in 10 patients non-visceral soft tissue calcification disappeared and in two decreased. Successful parathyroidectomy improves non-visceral calcification but not arterial calcification despite reduction in Ca x P product and iPTH.

Introduction

In patients with end-stage renal failure there is an increased incidence of arterial and soft tissue calcification [1]. It is commonly accepted that the most important predisposing factors are the increase in the calcium x phosphate product and the presence of excess parathyroid hormone in the serum [2]. Most of the individual case reports in the literature associate severe and widespread soft tissue calcification with severe hyperparathyroidism and markedly enlarged parathyroid glands in patients with chronic renal failure.

Periarticular or small peripheral vascular calcification in the presence of severe hyperparathyroidism uncontrolled by medical therapy has been regarded as one of the main indications for parathyroidectomy in uraemic patients [3]. However, the results obtained after the operation are based on single cases or on small numbers of observations [1,4,5].

This report describes the follow-up of the small vessel and non-visceral soft tissue calcification after successful parathyroidectomy in 42 uraemic patients.

Patients and methods

Forty-two patients who underwent parathyroidectomy for hyperparathyroidism secondary to chronic renal failure were included in the study. Biochemistry, radiological skeletal surveys and bone histology were compared before and after parathyroidectomy.

The total plasma calcium, phosphate, alkaline phosphatase, and magnesium were measured at two-weekly intervals and ionised calcium, iPTH, and serum aluminium at two weeks to three-monthly intervals. An average was taken of the results obtained in each of three different periods, 12 months before parathyroidectomy and during the first and second year after parathyroidectomy.

Radiological subperiosteal erosions (graded 0–3), small arterial calcification (digital and arcade vessels in hands and feet) and non-visceral soft tissue calcification were compared pre-parathyroidectomy (42 patients) and 6–12 months (42 patients), 12–24 months (26 patients) and 24–48 months (12 patients) after parathyroidectomy.

Transiliac bone biopsies were obtained at or some time in the six months prior to parathyroidectomy in 24 patients and some time after parathyroidectomy in 25 patients. The severity of osteitis fibrosa was assessed on a semi-quantitative scale taking account of the extent of marrow fibrosis, numbers of osteoclasts and presence of woven bone. Paired pre- and post-parathyroidectomy bone biopsies were available in 10 patients.

Statistical analysis was performed using Student's paired or unpaired 't' test where appropriate. Values are given as mean values ± SD.

Results

Small vessel calcification was present before parathyroidectomy in 23 (55%) patients and soft tissue calcification in 20 patients (48%). The presence of the metastatic calcification was not related to the age, sex, duration of end-stage renal failure treatment, total serum calcium, magnesium, phosphate, Ca x P product, alkaline phosphatase, ionised calcium or serum aluminium in the year before parathyroidectomy (Table I). Fifty-seven per cent (small vessel) and 35 per cent (soft tissue) had serum iPTH over 6U/L (normal 0.3–1.4U/L) and the combined weights of parathyroid gland tissue removed per patient with small vessel or soft tissue calcification was not significantly different from that of patients without calcification. Nineteen of 23 patients (83%) with small vessel calcification and 12 of 20 patients (60%) with soft tissue calcification had nodular to frankly adenomatous hyperplasia. This histological appearance was present in 30 of the 42 patients (71%) while in the remaining 12 (29%) there was diffuse parathyroid gland hyperplasia.

TABLE I. Details of the 42 parathyroidectomies. Biochemical measurements (mean + SD) during the year before parathyroidectomy

	Small vessel calcification		Soft tissue calcification		
	Present	Absent	Present	Absent	Normal range
Incidence	23 (55%)	19 (45%)	20 (48%)	22 (52%)	–
Age (years)	43.3 ±12.8	42.0 ±12.4	42.3 ±13	43.0 ±12.2	–
Sex (M/F)	11/12	9/10	10/10	10/12	–
ESRFT (months)	53.5 ±40	69±34	59±39	71±39	–
Plasma calcium (mmol/L)	2.44±0.15	2.46±0.25	2.41±0.16	2.48±0.23	2.12–2.62
Serum magnesium (mmol/L)	1.10±0.13	1.10±0.16	1.10±0.12	1.10±0.16	0.70–1.00
Plasma phosphate (mmol/L)	2.12±0.50	2.23±0.65	2.27±0.61	2.07±0.53	0.80–1.40
Calcium phosphate product	5.14±1.36	5.46±1.70	5.47±1.43	5.12±1.47	1.69–3.66
Plasma alkaline phosphatase (IU/L)	181±101	204±174	163±108	215±164	30–130
Serum ionised calcium (mmol/L)	1.23±0.12	1.20±0.15	1.21±0.14	1.23±0.14	1.15–1.30
Serum aluminium (μg/L)	51±32	66±32	66±38	50±22	2–15
Parathyroid tissue removed (weight in g)	2.35±1.39	1.59±1.29	1.93±1.54	2.09±1.25	0.13±0.004*

Differences between present or absent, pNS
*Combined weight for four normal glands

The severity of osteitis fibrosa in patients with either small vessel or soft tissue calcification was similar to that observed in the patients without metastatic calcification. Subperiosteal erosions and severe histological osteitis fibrosa were present in nearly 25 per cent of the patients, irrespective of the presence or absence of metastatic calcification. The remaining patients had moderately severe osteitis fibrosa. In two patients with soft tissue calcification in whom pre-operative bone biopsy was not available subperiosteal erosions were absent.

Thirty-seven patients (88%) at one year and 36 (86%) at two years after parathyroidectomy had iPTH values less than the upper limit of normal for our assay although the remaining patients also showed a reduction in the pre-operative values. Thirty-seven (88%) of the patients studied had substantial improvement or complete healing of sub-periosteal erosions after the parathyroidectomy.

Twenty-one of 25 patients (84%) with a post-operative bone biopsy had minimal or no histological osteitis fibrosa but in the remaining four moderate osteitis fibrosa persisted. In the group of 10 patients with paired bone biopsies there were significant reductions in the number of osteoclasts ($p<0.001$) and the extent of marrow fibrosis ($p<0.001$) after the operation.

Small vessel calcification progressed in six and in a further seven patients developed following parathyroidectomy. Such calcification decreased in one adult and it eventually resolved in a 12-year old girl. This resolution occurred in spite of slightly elevated serum iPTH and continued active osteitis fibrosa and the small vessel calcification (hands exclusively) disappeared 13 months post-operatively (Table II).

TABLE II. Follow-up of metastatic calcification after parathyroidectomy (PTX)

	n	Radiological calcification	
		Small vessel	Soft tissue
Pre PTX	42	23 (55%)	20 (48%)
6m–12m	42	27 (64%)	17 (40%)
12m–24m	26	18 (69%)	8 (31%)
24m–46m	12	11 (92%)	3 (25%)
Developed		7	1
Increased		6	3
Decreased		2	12

The soft tissue calcification occurred as isolated or multiple small periarticular foci in the majority of patients but was of the tumoral variety in three. Following parathyroidectomy the soft tissue calcification disappeared in 10 patients, decreased in two, increased in three and remained unchanged in the others. One patient developed ectopic calcification in the left hand one year after successful parathyroidectomy (Table II).

Table III summarises the biochemical changes in the patients with metastatic

TABLE III. Biochemical measurements (mean + SD) before (one year) and after (first and second year) PTX in patients with metastatic calcification

	Small vessel			Soft tissue		
	Pre (n=23)	0–12m (n=27)	12–24m (n=18)	Pre (n=20)	0–12m (n=17)	12–24m (n=8)
Plasma calcium (mmol/L)	2.4 ±0.1	2.2 ±0.3***	2.3 ±0.2	2.4 ±0.1	2.2 ±0.2****	2.3 ±0.1*
Serum magnesium (mmol/L)	1.10±0.13	1.12±0.12	1.12±0.16	1.10±0.12	1.13±0.14	1.12±0.16
Plasma phosphate (mmol/L)	2.1 ±0.5	1.7 ±0.4***	1.7 ±0.4***	2.2 ±0.6	1.7 ±0.4***	1.7 ±0.4**
Calcium phosphate product	5.1 ±1.4	3.9 ±0.9***	4.0 ±0.9***	5.4 ±1.4	4.2 ±1.1***	4.2 ±1.1*
Plasma alkaline phosphatase (IU/L)	161±101	98±84*	76±54***	163±108	99±72*	80±60**
Serum ionised calcium (mmol/L)	1.23±0.12	1.14±0.10***	1.17±0.11	1.21±0.14	1.14±0.10	1.17±0.08
Serum aluminium (μg/L)	51±52	48±41	56±75	66±38	63±41	73±64

*p<0.05; **p<0.02; ***p<0.01; ****p<0.001 when compared to pre-parathyroidectomy values calcification following parathyroidectomy. There was a significant decrease in the plasma phosphate and Ca x P product in patients with small vessel calcification and in those with soft tissue calcification despite the differing response of progression or regression of the calcium deposits to parathyroidectomy in each of these forms of calcification.

Discussion

Since deionisation and reverse osmosis were introduced in 1977 in our dialysis centre with reduction of the aluminium content of the water supplies to $<20\mu g/L$ [6], the epidemic of dialysis osteomalacia has disappeared [7]. After 1977 the clinical radiological and histological features of hyperparathyroidism have become prominent. The elimination of the aluminium and of its inhibitory action on parathormone output [8,9] could account for the renaissance of osteitis fibrosa among our dialysis patients.

Metastatic calcification leading to arterial disease is nowadays one of the most important clinical problems in patients with chronic renal failure treated by dialysis. Although the indications for parathyroidectomy are not sufficiently well established [3] surgical treatment is generally accepted in cases with hypercalcaemia, elevated alkaline phosphatase and/or metastatic calcification (periarticular or small peripheral vessels) in the presence of bone disease (osteitis fibrosa). The evolution of metastatic calcification after parathyroidectomy has received scant attention in the literature with contradictory results based on relatively few cases [1,4,5,10].

In our study, although the response of non-vascular soft tissue calcification to parathyroidectomy was not uniform in most of the patients, there was some improvement and the calcification disappeared in 50 per cent. On the contrary, despite successful parathyroidectomy, small vessel calcification developed or progressed in 56 per cent. In our experience it has not been possible to predict the outcome of either type of metastatic calcification from the biochemical parameters measured. The reduction in Ca x P product and iPTH with reduction in severity of radiological or histological characteristics of secondary hyperparathyroidism after parathyroidectomy is associated with an improvement in non-visceral calcification but not of small arterial calcification.

Acknowledgments

Dr AM De Francisco is supported by a grant from the Fondo de Investigaciones Sanitarias de la Seguridad Social (Spain) and Dr MJD Cassidy by a grant from the Northern Counties Kidney Research Fund. The help of numerous colleagues in the Departments of Medicine, Clinical Biochemistry and Radiology whose data we have analysed is gratefully acknowledged.

References

1 Parfitt AM. *Arch Intern Med 1969; 124:* 544
2 Katz AJ, Hampers CL, Merrill JP. *Medicine (Balt) 1969; 48:* 333

3 Drüeke T. Zingraff J, Dubost C. In *Advances in Nephrology; 11.* New York: Year Book Medical Publishers. 1982: 287–292
4 Peterson R. *Radiology 1978; 126:* 627
5 Meema HE, Oreopoulos DG, de Veber GA. *Radiology 1976; 121:* 315
6 Ward MK, Feest TG, Ellis HA et al. *Lancet 1978; i:* 841
7 Ellis HA. *Nieren und Hochdruckkrankheiten 1983; 12:* S198
8 Morrissey J, Rothstein M, Mavor G et al. *Kidney Int 1981; 21:* 138
9 Cannata JB, Briggs JD, Junor BJ et al. *Lancet 1983; i:* 501
10 Griffiths HJ. In *The Radiology of Renal Failure.* Philadelphia: W B Saunders. 1975: 178–197

Open Discussion

VERBANCK (Roeselare, Belgium) Do you feel that an emergency parathyroidectomy has to be considered in the patient on dialysis without major stenosis at arteriography of the lower limbs but who is developing gangrene of the toes?

De FRANCISCO In my opinion it is a different pathogenic mechanism and the response to parathyroidectomy has been proven very effective in some cases.

HANSON (Aarhus) I think it is very interesting to know that there is no sure relationship between parathyroidectomy and the disappearance of small vessel calcification. I would ask you one question. I saw that not all your parathyroid hormone values became normal after parathyroidectomy?

De FRANCISCO That is correct. Although most of the patients had normal PTH levels after parathyroidectomy, high levels persisted in some patients. However among the seven patients who developed new small vessel calcifications after parathyroidectomy none had serum PTH levels above the normal range and in four of them who were studied histologically after parathyroidectomy the bone did not show any degree of osteitis fibrosa. Moreover among the six patients who increased the small vessel calcification after parathyroidectomy only one had high levels of PTH but considerably lower than before parathyroidectomy.

HANSON Did you find several adenoma in some of your patients, it is not uncommon to find more than five parathyroid glands in uraemic patients?

De FRANCISCO Only in two cases. However the efficacy of parathyroidectomy is proven by the biochemical, radiological and histological studies.

GOODWIN (Chairman) You told us about many factors that didn't seem to be concerned with the development of soft tissue calcification, but you didn't mention blood pressure and smoking. Did you look at those two topics?

De FRANCISCO No.

Proc EDTA–ERA (1984) Vol 21

ULTRASONICALLY GUIDED FINE-NEEDLE ALCOHOL INJECTION AS AN ADJUNCT TO MEDICAL TREATMENT IN SECONDARY HYPERPARATHYROIDISM

A Giangrande, P Cantù, L Solbiati, C Ravetto

General Provincial Hospital, Busto Arsizio, Italy

Summary

In 12 uraemic patients with symptomatic secondary hyperparathyroidism, 13 parathyroid hyperplasias, detected by sonography and confirmed by fine-needle aspiration biopsy, were treated by ultrasonically-guided percutaneous injection of absolute ethanol, in order to reduce the gland mass.

Only in the larger glands were significant volume reductions recorded, whereas in the smaller ones evident structural changes were observed.

In most cases with single lesions, a reduced incidence of vitamin D hypercalcaemia and a permanent improvement in bone alkaline phosphatase and PTH were documented. This technique can be usefully employed either as an alternative to surgery in selected cases, or as support to medical therapy in single lesions.

Introduction

In uraemic patients the incidence of secondary hyperparathyroidism (sHPT) is very high, affecting bone metabolism, cardiac and endocrine functions [1]. Medical treatment with active vitamin D metabolites does not seem to achieve a definite improvement of sHPT [2] and can seldom influence the parathyroid hyperplasia which sustains this syndrome.

Therefore the reduction of the gland mass by parathyroidectomy (PTX) is considered an effective method to create a new starting point for pharmacological treatment [3].

For the same purpose, as an alternative to PTX and in cases resistent to medical therapy, we have performed ultrasonically-guided percutaneous fine-needle alcohol injection (UGFNA) of enlarged parathyroid glands.

In this paper we report our preliminary results in a small group of patients.

Material and methods

Twelve out of 101 uraemic patients with symptomatic sHPT and one or more enlarged parathyroid glands detected by sonography were selected for UGFNA.

The patients had been on haemodialysis treatment from 65 to 176 months; their age ranged from 34 to 69 years. In all cases diagnostic confirmation of the lesions was obtained by fine-needle aspiration biopsy with ultrasonic guidance with cytological examination (11 cases), according to the technique previously described [4], and/or parathyroid hormone assay of the aspirated material (4 cases), as reported by Doppman et al [5]. Serum parathyroid hormone levels (iC-PTH) were estimated by a radioimmunological method (Sorin Biomedica, Saluggia, Italy); bone alkaline phosphatase isoenzymes were determined on cellulose acetate [6].

In 7/12 cases one enlarged parathyroid gland was shown by high-resolution real-time sonography. In the remaining five two glands were detected: in 4/5 only the larger gland was injected with alcohol, whereas the fifth had both glands treated. The dimensions of the glands ranged from 0.9 to 2.5cm in diameter. Serum levels of either parathyroid hormone or bone alkaline phosphatase were significantly increased in all patients. Reduction of the gland mass was considered necessary because of severe sHPT (4 cases), mixed bone disease with fractures (2), hypercalcaemia (3), itching and hyperphosphataemia (2), or vascular calcification (1). UGFNA was preferred to surgery when there was hyperplasia of a single gland (4 cases), recurrence after subtotal PTX (2), high surgical risk (3), or refusal of surgery (3).

Informed consent was obtained from all patients. Bleeding parameters were carefully evaluated and in all cases prophylactic treatment with Deamino-8 D-Arginine Vasopressin (0.3μg/kg) was administered, as in previous studies [7].

For the ultrasonic guidance real-time sector scanners with lateral biopsy guides were employed, either with a 7.5MHz probe (Advanced Technology Laboratories) or a 10MHz intra-operative probe (Diasonics). We used commercially available 22 gauge, 9.5cm needles connected to 5ml syringes.

High-resolution real-time monitoring allowed precise control of needle penetration: the needle-tip was always guided into the anterior portion of the lesion, in order to avoid injury to the recurrent laryngeal nerve which lies posterior to the parathyroid glands. The quantity of injected alcohol was 1ml per 2–2.5cc of glandular volume approximately calculated for both oval and round-shaped glands.

All the procedures were carried out between dialyses, attention being paid to maintenance of a controlled administration of heparin during the following dialysis.

Fifteen, 30, 60, 180 and 240 days after the procedure the patients were sonographically and biochemically assessed.

Results

Thirteen hyperplastic parathyroid glands in 12 patients were treated with UGFNA and significant results were obtained in nine cases. In four glands the procedure was unsuccessful: in two the injected alcohol diffused anteriorly to the gland, in one case the lesion had partially calcified walls which prevented an adequate intra-glandular injection of the alcohol, and in one case the patient did not co-operate sufficiently. In the other nine glands either a significant volume reduction or changes in the echopattern were observed following the alcohol injection. The maximum reductions, ranging from 20 to 100 per cent (Figure 1) of the initial volume, were recorded after six months. Changes in the echopattern were increasingly evident in successive controls. Early changes (15–30 days after the procedure) consisted of high-level internal echoes and focal marginal irregularities appearing in all the glands.

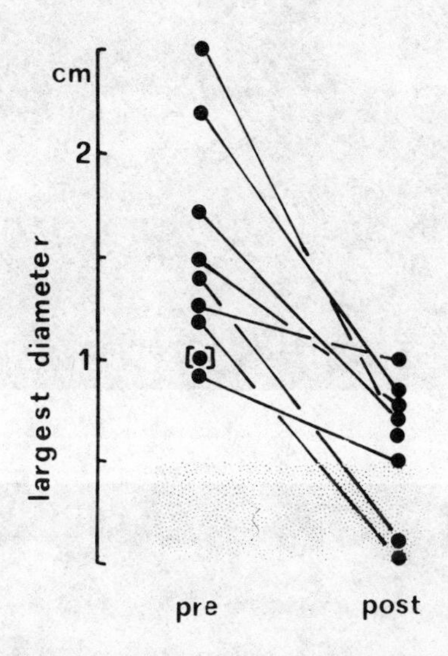

Figure 1. Parathyroid size before and six months after UGFNA ([•] patient subsequently submitted to PTX who showed extensive necrotic change in the greated gland)

Subsequently (4–6 months) the glands were replaced by echogenic tissue, either partially (2 cases) or totally (2), or by liquid areas (2 cases) (Figure 2). In the three lesions larger than 2cm only a volume reduction was noted, without any structural change.

In the 5/7 cases of single gland hyperplasias with successful injection the sonographic changes were accompanied by an improvement in the clinical condition and a decrease in serum levels of PTH and bone alkaline phosphatase (Figure 3).

The procedure was generally well tolerated: side effects were limited to spontaneously remitting local pain in 10/12 patients. In one patient mild dysphonia was recorded over the following 24 hours and in the last patient, who erroneously received dialysis only 12 hours after the manoeuvre, a small haematoma was documented in the site of injection.

Discussion

Medical treatment of sHPT with vitamin D can improve both clinical and biochemical patterns, but does not seem to affect the volume of the hyperplastic

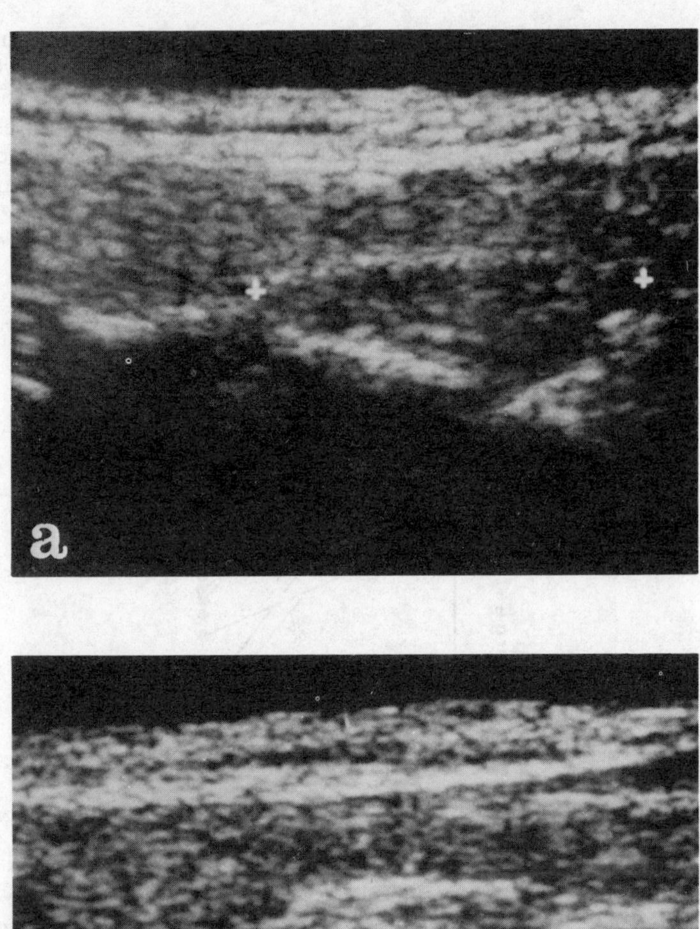

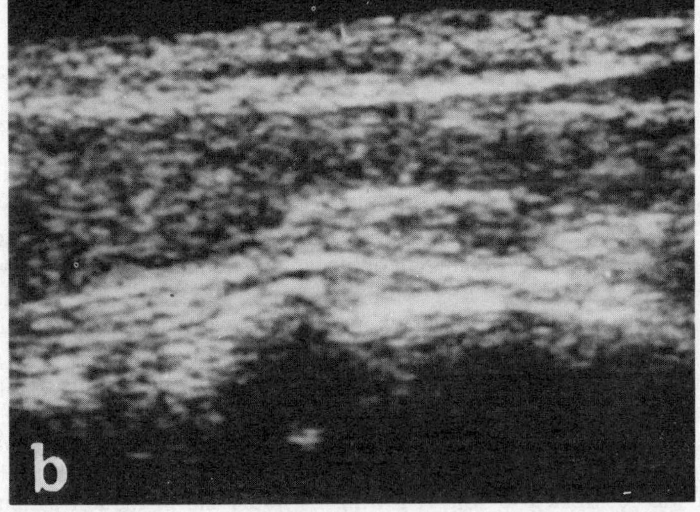

Figure 2. Sagittal scan of inferior parathyroid gland (a) before and (b) 90 days after UFGNA. The gland volume is decreased from 1.8 cm (calipers) to 0.8 cm and its structure is highly echogenic due to fibrosis

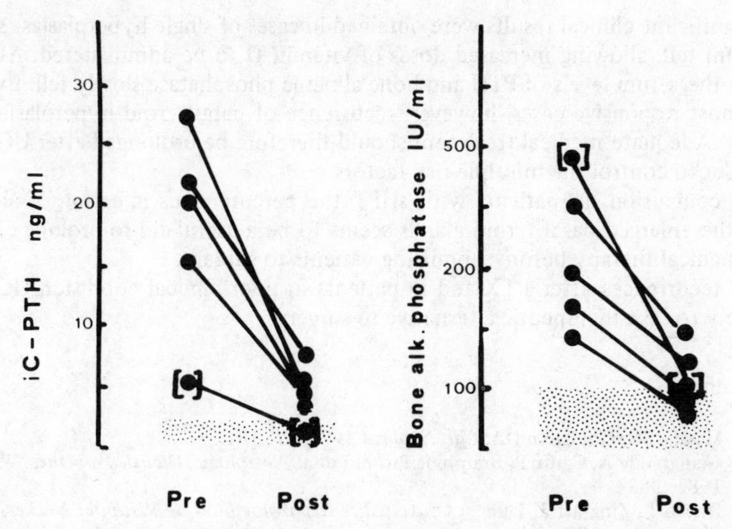

Figure 3. Parameters of hyperparathyroidism before and six months after UGFNA ([•] recurrence after subtotal PTX)

gland. Moreover in the most severe forms of sHPT the concomitant hypercalcaemia may preclude any pharmacological treatment. Therefore surgical reduction of gland mass is thought to be necessary in order to obtain greater responsiveness to medical therapy. The therapeutic usefulness of the injection performed on enlarged parathyroid glands was suggested by the reported unexpected remission of the symptoms of hyperparathyroidism following aspiration biopsy. We have also observed internal haemorrhagic or necrotic changes in four surgically removed parathyroid tumours which had previously undergone fine-needle aspiration biopsy.

Absolute ethanol was chosen as the therapeutic agent because it is already being employed to infarct renal neoplasms [8], to obliterate oesophageal varices [9] and to sclerose the walls of renal cysts [10]. Its action is likely to be based on endothelial damage and thrombosis of small vessels.

In our initial experience both sonographic and clinical effects of UGFNA were recorded. With sonography the same early structural changes were observed in the nine glands successfully injected, whereas the late changes appeared related to the initial size of the glands. In the smaller glands the parenchyma was mostly replaced by either dense high-level echoes or liquid areas. We believe the former was due to fibrous tissue, the latter to cystic or necrotic changes. In fact in two patients who underwent surgery (previously refused) after sHPT the pathological examination of the treated glands showed extensive necrotic changes in one case and fibrous tissue in the other.

In the larger glands the volume decreased significantly, but no structural change was observed; however a therapeutic effect was obtained, proved by a lowering of serum PTH. We feel that in these glands a larger amount of alcohol is needed and a second percutaneous injection may be attempted.

Significant clinical results were obtained in cases of single hyperplasias: serum calcium fell, allowing increased doses of vitamin D to be administered. At later times the serum levels of PTH and bone alkaline phosphatase slowly fell. Even in the most responsive cases, however, recurrence of parathyroid hyperplasia may occur. Adequate medical treatment should therefore be prolonged after UGFNA in order to control the multiple risk factors.

In conclusion, in patients with sHPT the percutaneous injection of alcohol into the enlarged parathyroid glands seems to be a useful aid to prolong dietary and medical therapy before submitting patients to surgery.

In recurrences after PTX and in patients in poor clinical condition UGFNA is likely to be a therapeutic alternative to surgery.

References

1 Massry SG, Goldstein DA. *Clin Nephrol 1979; 11:* 181
2 Giangrande A, Cantù P, Brambilla Pisoni I et al. *Nefrologia, Dialisi, Trapianto.* Wichtig. 1982: 383
3 Druke T, Zingraff J, Dubost C. *Actualité Néphrologiques de l'Hopital Necker.* Paris: Flammarion. 1981: 287
4 Solbiati L, Montali G, Croci F et al. *Radiology 1983; 148:* 793
5 Doppman JL, Krudy AG, Marx SJ. *Radiology 1983; 148:* 31
6 Burlina A, Galzigna L. *Clin Chem 1976; 22:* 271
7 Mannucci P, Remuzzi G, Pusineri F. *N Engl J Med 1983; 308:* 8
8 Ellman BA, Parkhill BJ, Curry TS et al. *Radiology 1981; 141:* 619
9 Keller FS, Rosch J, Dotter CT. *Radiology 1983; 146:* 615
10 Gammelgaard J, Jensen F. *Abstracts Fourth European Congress on Ultrasonics in Medicine.* Amsterdam: Excerpta Medica. 1981: 106

Open Discussion

GOODWIN (Chairman) I think many of us are very impressed by our ability to demonstrate glands. I would like to ask you if you were to scan a group of patients with chronic renal failure are you able regularly to demonstrate three, four, five glands with confidence?

GIAGRANDE It is not infrequent to find two glands in the patient. In the last year with improved techniques and with the possibility of using a 10mmHz probe we have often succeeded in finding four glands. This means that the sensitivity of the procedure is about 0.5cm.

GOODWIN Thank you, that was beautifully clear.

WALLS (Leicester) I suspect that you have the answer to my question. You are obviously aspirating these glands. Would you like to tell us what your various aspirates show and whether you can differentiate between hyperplasia and adenoma?

GIAGRANDE I do not think we can differentiate hyperplasia from adenoma. It is useful to differentiate thyroid nodulae from parathyroid hyperplasia and we used two techniques which we have already published*.

*Solbiati L, Giangrande A, Montali G et al. *Proc EDTA 1983; 19:* 293
*Solbiati L, Montali G, Croci F. *Radiology 1983; 148:* 793

KERR (London) Have you had to operate on any of these patients after your procedure and what did the surgeons think of the operating field?

GIAGRANDE Yes, some surgeons like the preliminary echographic documentation of the glands and on some occasions some surgeons don't like the damage after fine needle aspiration biopsy. After the ultrasonically guided fine needle alcohol injection the damage is a little more. We have operated on three patients after this procedure and the surgeon has not had any problems in localising the lesion. There is some damage which is not limited just to the gland. Some diffusion of the alcohol may produce some necrosis which may give some problem to the surgeon.

GOODWIN It sounds as if you have to treat your surgeon very carefully to get his collaboration with this technique.

Proc EDTA–ERA (1984) Vol 21

LOW INCIDENCE OF HYPERPARATHYROIDISM IN DIABETIC RENAL FAILURE

J Aubia, J Bosch, J Lloveras, Ll Mariñoso, J Masramon, S Serrano, X Cuevas, A Orfelia, I Llorach, M Llorach

Hospital Gral M.D. L'Esperanca, Barcelona, Spain

Summary

The first study compared two groups on dialysis: 25 patients with diabetes mellitus and 25 matched non-diabetic patients, in relation to the presence of signs of hyperparathyroidism, to assess the reported low incidence of hyperparathyroidism in these patients. The diabetic group showed significantly lower values of PTH, Alk phosphatase, percentage of patients requiring vitamin D treatment, and less evidence of hyperparathyroidism on X-ray and in bone histomorphometry.

In the second study 16 patients with chronic renal failure due to diabetic nephropathy were compared to 27 patients with the same degree of renal failure of other origin, the diabetic nephropathy group showed no increase in PTH, with falling creatinine clearance. Despite this low PTH, the phosphaturia was higher in the diabetic nephropathy group (Tm PO_4/C Cr: 1.94±0.43 vs 2.5±0.68).

In conclusion, patients with diabetes mellitus are less prone to develop hyperparathyroidism in progressive renal failure. This could be due to a relative increase in phosphaturia during declining function.

Introduction

The number of patients with diabetes mellitus being accepted on dialysis and transplantation programmes is growing rapidly, and therefore the characteristics of this group of patients are increasingly better known.

There is a reduced incidence of hyperparathyroidism in diabetic dialysed patients [1–3]. This study investigates why diabetic patients show this lack of hyperparathyroidism. First we compared a group of diabetic patients on dialysis with a paired group of non-diabetic patients and secondly we studied a group of diabetic patients before they reached the dialysis phase and compared them to a group of non-diabetic patients with the same degree of renal failure to determine whether the renal handling of minerals differed.

Methods

Patients

In the first study 25 patients with diabetes mellitus on dialysis for a period of time ranging from one to 49 months were compared to a group matched according to age, sex and length of time on dialysis.

In the second study 26 patients with chronic renal failure (creatinine clearance 5–40ml/min) of various origin and 17 patients with diabetic nephropathy were compared.

Total calcium, inorganic phosphate, alkaline phosphatase and creatinine were determined by autoanalyzer; C-terminal PTH by a radioimmunoassay with rabbit antibody against PTH (65–84), (Immunonuclear); plasma cAMP by a RIA of rabbit antibody; plasma calcitonin by RIA; and urinary cAMP by a third RIA of Beckton-Dickinson (201.677). The renal tubular threshold of phosphate ($TmPO_4$/C Cr) was calculated by the method of Bijvoet [4].

The normal values in our laboratory are: C-PTH <1ng/ml, cAMP is expressed as nmol/100ml C Cr=1.62±1.1; $TmPO_4$/C Cr=3.63±0.56mg/100ml. CT=58.27 ±28.1pg/ml.

Iliac crest bone biopsies were obtained in six diabetic and seven non-diabetic dialysis patients. Non-decalcified specimens stained with toluidine blue, hematoxylin and eosin, and Von Kossa, were processed by an image analyser (Morphomat 30, Zeiss) and the following parameters were measured. Volume of trabecular bone relative to total bone (Vt), number of osteoclasts/mm of trabecular perimeter (Cr), percentage of surface covered by osteoid (Sf), percentage of trabecular volume occupied by osteoid (Vosf) and by osteoclastic resorption (Sr), and calculated mean width of osteoid seams (WOS).

The presence of bone pain, X-ray appearances, Alkaline phosphatase greater than 450mU, and cPTH >3ng/ml, are the criteria used in our centre to start dialysis patients on vitamin D treatment.

Statistical analysis of differences between groups was undertaken using Mann-Whitney's U-test, and 't' test for paired differences in the first study. Correlations were computed by the method of least squares. Only p values <0.05 were considered significant.

Results

Study 1

As shown in Table I the diabetes mellitus dialysis group, in spite of close matching with the control group, had lower values of iPTH: 1.44±0.79ng/ml in diabetic group and 3.76±2.03ng/ml in non-diabetic group; alkaline phosphatase 293±143 and 473±443U/ml respectively; percentage of patients having X-ray signs of phalangeal subperiosteal reabsorption: 4 per cent in the diabetic group and 24 per cent in the control group; histomorphometric values of osteoclastia in bone biopsy; percentage of patients that required vitamin D treatment: 12 per cent and 36 per cent respectively.

TABLE I

	Diabetes mellitus group on haemodialysis	Non-diabetic group on haemodialysis	Stat	p
Time on dialysis (months)	14.7	14.86	td = −0.36	NS
iPTH (ng/ml)	1.44±0.79	3.76±2.03	td = −4.8	<0.005
Alk phosphatase (U/L)	293±143	473±343	td = −2.07	<0.05
Calcitonin (pg/ml)	252±150	203±70	td = −1.04	NS
% with abnormal X-rays	4	24		
% vitamin D requirement	12	36		
Bone morphometry:				
Vt (mm^3/mm^3)	0.162±0.04	0.215±0.148	U = 19	NS
Sr (mm^2/mm^2)	0.003±0.002	0.014±0.09	U = 7	<0.05
Cr (n/mm)	0.23 ±0.21	1.56 ±0.7	U = 6	<0.05
Sf (mm^2/mm^2)	7.6 ±6.7	45.8 ±37.6	U = 7	<0.05
WOS (μm)	17.05±9.01	17.2 ±10.2	U = 30	NS
Vosf (%)	0.38 ±0.31	3.36 ±2.17	U = 8	⩽0.05

TABLE II

	Diabetic mellitus group (D-group) (n=16)	Non-diabetic group Controls (n=27)	p
Creatinine clearance (ml/min)	18.8±9.1	18.1±9.4	NS
iPTH (ng/ml)	1.26±0.58	2.01±1.17	<0.05
cAMP (nmol/100ml C Cr)	6.12±2.64	10.1±3.95	<0.05
Calcitonin (pg/ml)	121±45	116±33	NS
Tm PO_4/C Cr (mg/100ml)	1.94±0.43	2.5±0.68	<0.05

Study 2

The values of creatinine clearance, urinary cyclic AMP, $TmPO_4$/C Cr, and iPTH in both groups are shown in Table II.

The iPTH of the diabetic group was less than the non-diabetic group: 1.26±0.58ng/ml vs 2.01±1.17ng/ml, the difference being significant. The cAMP values

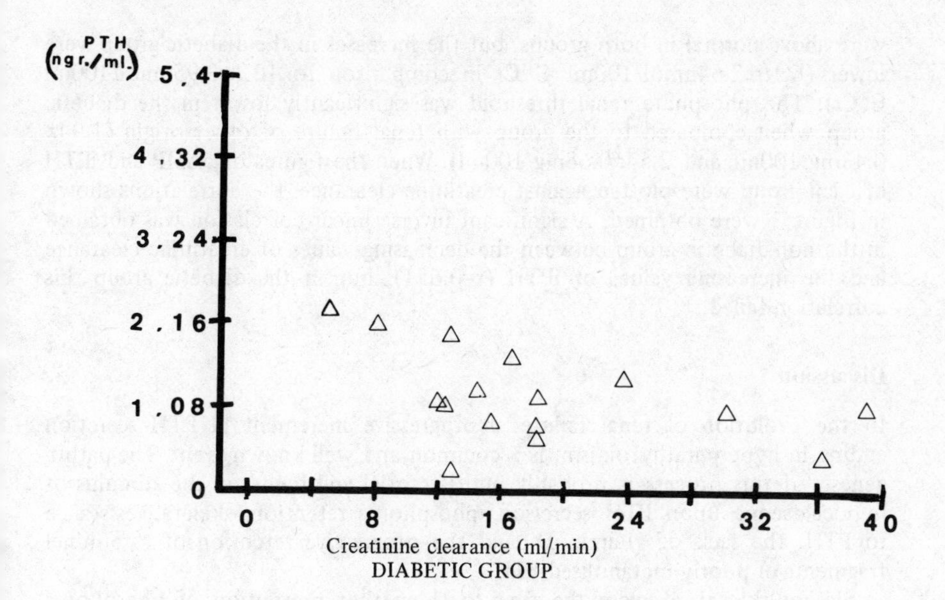

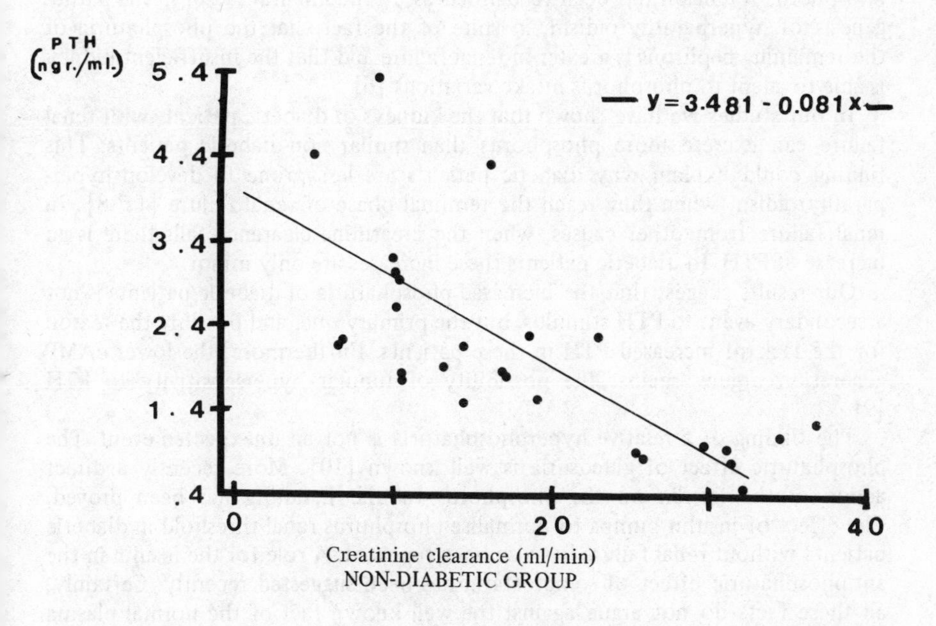

Figure 1. Creatinine clearance and iPTH in diabetic (upper) and non-diabetic (lower) groups. A significant inverse correlation was found in the non-diabetic group but failed in the others

were above normal in both groups, but the increases in the diabetic group were lower (6.16±2.64nmol/100ml C Cr in comparison to 10.1±3.95nmol/100ml C Cr). The phosphate renal threshold was significantly lower in the diabetic group when compared to the group with renal failure of other origin (1.94± 0.43mg/100ml and 2.54±0.68mg/100ml). When the figures of cAMP and iPTH of each group were plotted against creatinine clearance, the correlations shown in Figure 1 were obtained. A significant inverse linear correlation was obtained in the non-diabetic group between the decreasing values of creatinine clearance and the increasing values of iPTH (r=0.651), but in the diabetic group this correlation failed.

Discussion

In the evolution of renal failure, a progressive increment in PTH secretion ending in hyperparathyroidism, is a common and well known event. The pathogenesis of this process is probably multifactorial and based on the stimulus of hypocalcaemia upon PTH secretion, phosphorus retention, skeletal resistance to PTH, the lack of vitamin D, and the progressive retention of C-terminal fragments of poorly metabolised PTH.

Slatopolsky et al were the first to show that prevention of phosphorus retention, maintaining the plasma values within the normal range, could avoid the appearance of overt hyperparathyroidism in late renal failure [5]. Since then phosphorus retention has been recognised as a fundamental factor in the pathogenesis of hyperparathyroidism, in spite of the fact that the phosphaturia of the remaining nephrons is greater in renal failure and that the insufficient kidney is able to adapt to phosphorus intake variations [6].

In our studies we have shown that the kidneys of diabetic patients with renal failure can excrete more phosphorus than similar non-diabetic patients. This finding could explain why diabetic patients are less prone to develop hyperparathyroidism when they reach the terminal phase of renal failure [1,7,8]. In renal failure from other causes, when the creatinine clearance falls there is an increase of PTH. In diabetic patients these increases are only minor.

Our results suggest that the increased phosphaturia of diabetic patients is not a secondary event to PTH stimulus, but the primary one, and possibly the reason for the lack of increased PTH in these patients. Furthermore, the lower cAMP generation argues against the possibility of tubular hypersensitivity to PTH [9].

The finding of a relative hyperphosphaturia is not an unexpected event. The phosphaturic effect of glucosuria is well known [10]. More recently a direct action of the insulin on the phosphorus tubular handling has been proved. The effect of insulin pumps to normalise phosphorus renal threshold in diabetic patients without renal failure has also been reported. A role for the insulin in the antiphosphaturic effect of somatostatin has been suggested recently. Certainly, all these facts do not argue against the well known fact of the normal plasma concentration of phosphorus, and PTH in diabetic patients without renal failure.

Our studies seem to prove that diabetic patients on dialysis have fewer signs of hyperparathyroidism than matched patients with the same time on dialysis.

It also seems probable that the hyperphosphaturia caused by insulinopenia and glucosuria persists through the course of progressive renal failure, and this possibly is the reason for the lack of hyperparathyroidism in such patients when they require dialysis.

References

1 Rivero AJ, McKenna BA, Pabico RC et al. *ASAIO 1980; 9:* 60
2 Vicenti F, Hattner R, Ammend WJ et al. *JAMA 1981; 245:* 930
3 Pabico RC, Rivero AJ, McKenna BA et al. *Proc EDTA 1982; 19:* 221
4 Bijvoet OL. In Avioli L, Bordier Ph, Fleish H, eds. *Phosphate Metabolism. Kidney and Bone.* Paris: Fournier. 1976: 421–474
5 Slatopolsky E, Bricker NS. *Kidney Int 1973; 4:* 141
6 Ritz E, Rambausek M, Kreusser W et al. In Jones NF, Peters DK, eds. *Recent Advances in Renal Medicine.* London: Churchill Livingstone. 1982: 151–156
7 Kurtz SB, McCarthy JT, Johnson WJ et al. *Kidney Int 1984; 25:* 258
8 Morii H, Iba K, Nishizaba Y et al. *Nephron 1984; 38:* 22
9 Purkerson ML, Rolf DB, Yates E et al. *Mineral Electrolyte Metabol 1980; 4:* 258
10 Fox M, Thier S, Rosenberg L et al. *J Clin Endocrin 1964; 24:* 1318

Open Discussion

DE PAEPE (Belgium) I may have missed the point, but did you compare bone biopsies of diabetic patients without renal insufficiency and normal control patients? Is it possible that there are differences to explain the different response to PTH?

AUBIA We have compared two groups of dialysis patients, one with diabetes and the other non-diabetic. The difference in the number of osteoclasts and the number of surfaces with active reabsorption was lower in the diabetic group.

DE PAEPE In your opinion is this due to lower plasma PTH or could it be that the target organ is less sensitive to PTH?

AUBIA We do not have any answer to that question.

CANNATA (Oviedo) You may know that aluminium can depress PTH secretion. Have you studied in any way the aluminium exposure in both groups. In addition have you considered the osmotic effect of the glucosuria?

AUBIA The aluminium levels of both groups are similar and both are in the safe level, between 50–70ng/ml. The second question is probably the most interesting. We have only results in the last seven patients and all have glycosuria but we have not been able yet to establish any relationship between the degree of glycosuria and phosphaturia.

KERR (London) That is a very impressive piece of evidence that diabetic nephropathy is different from other forms of chronic renal failure, but I cannot think of a physiological explanation for this unless it is vascular damage to the parathyroids. It would be nice to exclude other possible explanations, such as that diabetic patients eat less phosphate which might explain why you get less hyperparathyroidism in your dialysis population. Is there any difference in the diet of your diabetic patients or the rest of your dialysis population?

AUBIA There is apparently no difference in the diets.

BANKS (Bristol) Have you studied any non-insulin dependent diabetics?

AUBIA No, we have not.

PART XXIV

GUEST LECTURE ON TRANSPLANTATION I

Chairmen: P J Guillou
C V Casciani

Proc EDTA–ERA (1984) Vol 21

STUDIES ON LIVING RELATED KIDNEY DONORS AT A SINGLE INSTITUTION

J S Najarian, D Weiland, Blanch Chavers, R L Simmons, W D Payne, Nancy L Ascher, D E R Sutherland

University of Minnesota, Minneapolis, Minnesota, USA

An insufficient number of cadaveric kidneys are procured for the number of patients with end-stage renal disease (ESRD) who could benefit from renal transplantation [1]. The yearly incidence of new treated cases of ESRD in the USA is approximately 90 per million population [2], while the incidence of cadaver kidneys procured is less than 20 per million population [3]. Around 25 per cent of renal allografts performed in the USA are from living related donors (LRD), but the proportion of transplants from this source ranges from nought per cent in some to >90 per cent in other institutions [2]. The factors that determine the proportion are complex and numerous. Some institutions are reluctant to use related donors for kidney transplantation because of the uncertainty over the morbidity and the long-term effects of the procedure on the donor, information that is necessary to calculate the risk:benefit ratio for LRD kidney transplantation [4].

Reduction of renal mass has physiological effects [5]. Some studies have observed physiologically undesirable abnormalities in kidney donors [6] while others have not [7,8]. A long-term study of patients who underwent unilateral nephrectomy for reasons other than donation, cancer or bilateral renal disease, showed no differences in longevity for these individuals than for those with two kidneys [9]. The survival curve after nephrectomy was the same as the general population in those with a normal serum creatinine at the time of operation or in those whose remaining kidney was normal.

A low incidence of complications following living related kidney donation has been reported from several centres [10,11]. The remaining kidney undergoes compensatory hypertrophy and there is a gradual increase in glomerular filtration rate [12].

We have previously reported the early operative complications in living related kidney donors at our institution [13], and the psycho-social aspects of living related organ donation have also been addressed [14]. The incidence and type of immediate and late complications in a large number of donors studied at the University of Minnesota is summarised in this report, and the long-term effect of donation on renal function and blood pressure is presented.

Material and Results

Between 1963 and 1980, 628 living relatives with a mean (±SD) age of 34.3±17.5 years (25, <18 years, 3.9%; 171, 18–25 years, 27.2%; 352, 26–50, 56.0%; 80, 51–68 years, 12.7%) were donors (300 male, 48%; 320 female, 52%) of kidney transplants at the University of Minnesota Hospital. The operations were usually carried out through a subcostal flank incision. Blood loss during the operation was ⩾500ml in 156 donors (33%). Transfusions were given in 75 donors (12%). Complications of surgery occurred in 107 donors (17%) (Table I). Most of the

TABLE I. Complications of surgery in 628 living related kidney transplant donors

Complication	No.	Per cent
Pulmonary	65*	10.4
Atelectasis	37	5.9
Pneumothorax	25**	4.0
Pneumonia	19	3.0
Pul. Oedema	11	1.8
Pul. Embolus	3	0.5
Tracheostomy	2	0.3
Urinary Infections	30	4.7
Wound Infection	14	2.2
Superficial	10	1.6
Deep	4	0.6
Bleeding (Reoperation)	4	0.6
Superficial Thrombophlebitis	3	0.5
Upper Extremity	2	0.3
Lower Extremity	1	0.2
Adrenalectomy	2	0.3
Splenectomy	1	0.2
Total	107*	17.0

* Some patients had more than one complication
** Five required a chest drain

complications were minor in nature. Only 16 complications (deep wound infections, pulmonary emboli, tracheostomies, reoperation for bleeding, or need to extirpate other organs) were serious in nature (2.5%), a major complication rate almost identical to that reported by Blohme et al [11]. All donors survived the operation and were discharged from the hospital.

In 1982 follow-up information was obtained on 472 donors at 1–19 years post-donation (mean interval 8.3±11.2 years). One donor developed haemolytic uraemic syndrome one year after the operation; she is currently on dialysis. Long-term complications that were possibly related to donation occurred in 81 of the 472 donors (17.2%) and are summarised in Table II. The most common complication was incisional pain. The most serious complication was a neurogenic bladder that occurred for unknown reasons in one donor; this has been treated with an implantable electrical stimulator.

TABLE II. Chronic complications in 472 living related kidney transplant donors

Complication	No.		Per cent	
Incisional Pain	54		11.4	
Mild		50		10.5
Moderate		4		0.8
Incisional Hernia	17		3.6	
Recurrent Urinary Tract Infections	9*		1.9	
Neurogenic Bladder	1		0.2	
Total	81		17.2	

* Three out of eight females had recurrent urinary tract infections prior to donation

The mean (± SD) blood pressure (systolic/diastolic) of the donors before nephrectomy was 117±12/73±11mmHg. The latest mean blood pressure was 122±16/76±4mmHg. Of the 472 donors surveyed in 1982, 436 (92.4%) were normotensive pre-donation. Another 32 were classified as mildly hypertensive (6.8%) and four as moderately hypertensive pre-donation (0.8%), of whom 11 are still hypertensive 3–14 years post-donation (mean of 7.9±4.0 years), while 21 are not. Of the 436 donors who were normotensive pre-donation, 29 have subsequently become hypertensive (6.7%) during a follow-up period of 3 to 14 years (mean of 8.7±5.3 years). Thus, 40 of 472 donors (8.5%) currently are classified as hypertensive (mean BP 140±17/95±8mmHg) (Table III).

TABLE III. Incidence of hypertension post-donation in 472 related kidney transplant donors

Classification (diastolic pressure mmHg)	No.	Incidence
Mild (90–95). Diet controlled	13	2.8%
Moderate (90–100). Medication required	21	4.4%
Severe (>100). Medication required	6	1.3%
Total	40*	8.5%

* 11 were hypertensive pre-donation, while 29 were not

Mean serum creatinine levels of the donors before nephrectomy was 0.98± 0.18mg/dl. In addition to the one donor currently on haemodialysis, three donors have had significant elevations of serum creatinine post-donation to approximately 2.0mg/dl. Mean serum creatinine level obtained during 1982 on 51 donors was 1.21±0.29mg/dl at a mean of 8.7±3.4 years post-donation. Twenty-four hour urine samples for albumin excretion were collected from 26 donors >10 years post-operation; the mean ± SD value was 7.1±11.9mg/24 hrs. Only two donors were above the normal range (n=35) of 1.33–20.5mg/24 hours (mean of 8.2±5.0mg/24 hours).

Discussion

Unilateral nephrectomy from living related donors had a 17 per cent surgical complication rate in our series, although only a small proportion (2.5%) could be classified as serious. Blood pressure after donation was no different than in the general population [16]. Post-donation hypertension occurred in 6.7 per cent of the donors. This percentage is no higher than that of the general population [17], but it is unclear if it is higher than it would be without unilateral nephrectomy, since a comparably screened population without the operation was not studied.

In our opinion, the immediate and long-term morbidity is sufficiently low to make the risk of kidney donation acceptable for fully informed and motivated relatives of patients with ESRD. The improved graft survival rates of kidneys from living related donors provides further justification [18]. Even if the results of cadaveric and LRD transplantation were equivalent, we would use living related donors. Otherwise ESRD patients would be denied transplantation because of the shortage of cadaveric kidneys.

Acknowledgments

Supported in part by NIH Grant AM 13083.

References

1 Bart KJ, Macon EJ, Humphires AL et al. *Transplantation 1981; 31:* 383
2 Health Care Financing Administration. *End-stage Renal Disease Program Medical Information System, Facility Survey Tables.* Department of Health Services, USA, HCFA, January 1–December 31, 1982
3 Bart KJ, Macon EJ, Whittier FC et al. *Transplantation 1981; 31:* 379
4 Editorial. *Lancet 1982; ii:* 696
5 Brenner B, Meyer T, Hostetter J. *N Engl J Med 1982; 307:* 652
6 Goldszer R, Hakim RM, Brenner BM. *Kidney Int 1983; 23:* 124
7 Weiland D, Sutherland DER, Chavers B et al. *Transpl Proc 1984; 16:* 5
8 Vincenti F, Amend WJC, Kaysen G et al. *Transplantation 1983; 36:* 626
9 Andersen B, Hansen JB, Jorgensen SJ. *Scand J Urol Nephrol 1968; 2:* 91
10 Askari A, Novick A, Braun W, Steinmuller D. *J Urology 1980; 124:* 779
11 Blohme I, Gabel H, Brynger H. *Scand J Urol Nephrol 1981; 64 (Suppl):* 143
12 Aurell M, Ewald J. *Scand J Urol Nephrol 1981; 64 (Suppl):* 137

13 Spanos PK, Simmons RL, Lampe E et al. *Surgery 1974; 76:* 741
14 Simmons RG. *Transpl Proc 1977; 9:* 143
15 Vital and Health Statistics. *Blood Pressure Levels of Persons 6–74 Years, US 1971–1974, Series 11, No. 203.* US Government Printing Office. 1975
16 United States Public Health Service. *Advance Calculations for Vital Health Statistics Volume.* US Department of Health, Education and Welfare, National Center for Health Statistics. 1976
17 Sutherland DER. *Transplant Proc 1985; 17:* in press

Open Discussion

GUILLOU Thank you Professor Najarian for that masterful resumé of the case for live donor transplantation.

ANDREUCCI (Naples) Let me congratulate you for this excellent presentation. I have enjoyed very much what you have shown concerning the renal function of the living donor. What do you think about Dr Brenner's theory* on the effect of hyperfusion of the surviving nephron after the reduction of renal mass?

NAJARIAN Well, I think that we have good data now on humans to show that if you have a kidney this does not seem to affect it. I really want to stress this point. We were a little loose in the early days and when we look at the long-term patients of living-related transplants (I am talking about 1960 to 1966) we used primarily living-related donors. At that time we were not as careful in looking for hypertension and renal abnormalities in the donors, we just wanted them to be healthy. When we stretched that a little bit those were the ones that developed hypertension, and I think that was true in Brenner's theories from the Peter Brent Brigham as well. You have to be absolutely strict in the criteria of no hypertension. You can make a rat hypertensive and the other kidney is destroyed rather rapidly if the rat is unilaterally nephrectomised. You can also make the rat proteinuric and the other kidney is very badly damaged and finally if it is diabetic the kidney becomes rapidly damaged, so you do not want a diabetic donor, you do not want a progeneric donor and you do not want a hypertensive donor. If you stick to these criteria, after almost 20 years of experience in living-related donors we find no effect of hyperperfusion on those kidneys other than very mild proteinuria in a small percentage of patients.

POSEN (Canada) Professor Najarian, would you comment on the use of non-related living donors.

NAJARIAN Transplantation from spouses or other interested people depends on the relationship of the wife or the husband before transplantation. We all did this years ago when we were involved in using a few convicts: Starzl in Colorado had a much larger series of convicts who volunteered to give a kidney but the results were no better, if anything they were worse than that with cadaveric transplantation. More importantly, there were none of these people who really

*Brenner BM, Meyer TW, Hostetter TM. *N Engl J Med 1982; 307:* 625

had a psychological reason for giving and therefore caused many problems later. As a result almost everyone stopped. A new resurgence in this area was started by Belzer in Wisconsin using donor specific transfusions to a few wives and husbands in the hope of being able to transplant from a non-related living donor. He has had reasonable success with a small number. I personally feel that this should not be done, if you do not have a relative. Primary relatives do very well, uncles, aunts, grandmothers, cousins. If you have got just a little bit of the same genetic material you seem to do much better than an unrelated individual. Therefore, we then turn to cadavers. We do not feel that it should be done from living non-related donors.

Di GIULIO (Rome) I would like to ask whether you have any information about the evolution of blood pressure in patients receiving a kidney from a hypertensive donor.

NAJARIAN We tried to look at that but fortunately we do not have many hypertensive patients. When we talk about the hypertensive donor we are talking basically about a donor with a diastolic pressure of around 90mmHg. We have never knowingly used a kidney from a hypertensive donor, either a cadaver or living, into a recipient, so I cannot really answer that question with any assurance as far as the human situation is concerned.

SIEBERTH (Aachen) When you give a related donor kidney as a second graft and the first graft has been a cadaver kidney, is the outcome similar to the result you have shown us?

NAJARIAN Yes, the second graft from a living-related donor kidney seems to enjoy the same function as the first graft if it was from a living-related donor despite the fact that the first one came from a cadaver. This is not true when you put a cadaver kidney into a patient previously transplanted from a cadaver. If the retransplant is from a cadaver it depends on whether the first graft was rapidly rejected or not. If it was rejected in less than 12 months then the second graft does not do very well at all from a cadaver. If it was rejected in over 12 months from chronic rejection then the second cadaver graft is equal to the primary cadaver graft. But living-related grafts enjoy a sense of genetic similarity which allows their results to be almost identical to that which were used in primary graft, so we recommend them.

BERGSTRÖM (Stockholm) Dr Najarian, as far as I know there are a few deaths reported in living-related donors. Would you like to comment on this?

NAJARIAN Well, when it occurs it is a tragedy and obviously we have all heard rumours about it. If I put all the rumours that I have heard from various transplant congresses and nephrology congresses together, I can add up about four or five deaths that occurred some time after transplantation from living related donors. As I mentioned, in the 1200 we have done we have had none.

That is not to say that we are not going to have one in the future, but if you do have death from a living-related donor that is a tragedy. It is also a tragedy to watch patients die who are not doing well either on dialysis or with a previous transplant. So you have to weigh up that the chance of death seems to be quite small. How you can do 1200 living-related donor transplantations – a major operation – without a single mortality is because these patients are so carefully worked out that we know everything about them and all the safety factors that we can employ. We stop the operation at any time if there is any problem with the donor. All the safety factors are on the donor's side and I think that is the way you can get away with it. Certainly, one day we will have a death from a living-related donor, and that price will be a small one to pay for all the good that has occurred in the recipients that we have transplanted.

GUILLOU I wonder, Professor Najarian, if I can ask you to speculate on the causes of late graft loss. We are all familiar with the early graft loss but how about the late graft loss? What happens to those kidneys?

NAJARIAN Well, late graft loss is an interesting thing, I think it was rightly shown this morning that the late graft loss in your own series of the EDTA has been primarily due to hepatic disease. We have seen the same thing. As far as conditions related to kidneys are concerned once you get passed five years it is due to chronic rejection. That seems to be much more apparent in the cadaveric grafts. I am sorry that I did not have one of the slides which shows what happens when we go on to 15 years. At 15 years a living-related graft continues to parallel each matched control but there is a very slight erosion of survival in the cadaveric grafts and that is from chronic rejection, and I think that is another problem with cadaveric grafts. A few more are lost even in the long-term and I think that is the cause.

MOURAD (Montpellier) Dr Najarian, what investigations should be performed in the family members of a diabetic patient?

NAJARIAN There are very many. We bring them in for about a week in hospital and they are given a series of tests. They are challenged with cortisone to see if there is any chance that potentially they could be a diabetic. We have done a lot of living-related donors, including over 700 type 1 diabetics and I would guess that at least 60 per cent of those are from living-related donors. To date not one of those donors has developed diabetes and this is due to the fact that our referring groups spend five days really putting them through the mill as far as type 1 diabetes is concerned.

SOBH (Egypt) Have you studied the effect of donor age on the graft outcome?

NAJARIAN Yes we have. For a while we thought that it made a difference if you got them on the much older side. It turns out that if you get above the age of 55 graft outcome is a little bit less, but not significantly so, up till about the age of 65 to 70 then it begins to fall off, so we draw the line now at 65. We do not use grafts over that age. As far as the younger side is concerned there is no influence whatsoever. The young recipients, even those only a few months of age, do very well and there does not seem to be much difference.

BRYNGER (Gothenburg) We have noticed when we transplanted the kidney from a donor of around 50 years old the presence of ageing process in this kidney when we performed a kidney biopsy a few weeks after kidney transplantation and we are anxious about the effect of these changes or the outcome of these grafts.

NAJARIAN I would be hesitant to say the decrease in renal function occurring a few weeks after grafting has anything to do with ageing at all. That effect can only be seen after many years. If you have decreased function from a 50-year old graft, and I do not consider 50 as being very old I hope, under those circumstances I think you have to look for another reason.

CANTAROVICH (Lyon) What about the risk of segmental hyalinosis in those patients presenting with proteinuria?

NAJARIAN It does recur. The number of times that we have seen it recur has been very small and that is the one condition where we will be more likely to do a cadaveric graft with the idea that if it does recur we can then remove the cadaveric graft and save the living-related donor for the second or third time. We do the same thing with steroid-resistant nephrotic syndrome. We usually put in a cadaveric graft first and if there is recurrence it will recur in that graft and it seems to absorb the antibodies so we can get away with a living-related graft or afterwards. So that is our general modus operundi when it comes to focal sclerosing glomerulonephritis.

BRYNGER Thank you very much for what I hope we all like to hear. Just one comment regarding the age of donors: you may know that the oldest kidney recorded for the moment is aged 109 with excellent function recovered from a grandmother aged 96 who lived for another year and died a natural death, so I think that the situation of the ageing of organs differs from individual to individual.

NAJARIAN I think we ought to put that in the Guinness Book of Records. I have never heard that story in my life. I think it is terrific. Was she Scandinavian?

BRYNGER No, Greek.

NAJARIAN Leave it to the Greeks, they will do it every time. I think it is important for all nephrologists and kidney physiologists in the room that at the end of 30 years as you well know you begin losing nephrons: you reach your peak at about 30 so as you get up around 60 or 70 years you are getting a limited number of nephrons and I think that the kidney is not quite as good. I am delighted to hear your anecdotal experience.

GUILLOU Ladies and gentlemen, when Dr Najarian and I first met earlier this morning I told him that this was a congress which consisted predominantly of nephrologists, but there are a few transplant surgeons here and as one transplant surgeon to another he asked me if I would take care of him. He really hasn't needed taking care of – he has done it all for himself and I am sure you will be happy to express to him your appreciation.

PART XXV

TRANSPLANTATION POSTERS I

Chairmen: C Ponticelli
A Vercellone

Proc EDTA–ERA (1984) Vol 21

TRANSFUSION-INDUCED ENHANCEMENT OF PROSTAGLANDIN AND THROMBOXANE RELEASE IN PROSPECTIVE KIDNEY GRAFT RECIPIENTS

V Lenhard, G Maassen, G Opelz

Institute of Immunology, University of Heidelberg, FRG

Summary

Pre-transplant blood transfusions (BT) improve the survival of kidney grafts. Apart from specific immunoregulation by T suppressor cells or anti-idiotypic antibodies, the role of non-specific immunoregulatory factors, such as prostaglandins is being discussed as a possible mechanism for this effect. We studied the in vitro prostanoid release from peripheral mononuclear cells following three deliberate blood transfusions. Twenty-five previously non-transfused dialysis patients were studied. Spontaneous and LPS-induced prostaglandin E (PGE) and thromboxane B_2 (TXB_2) were determined in cell-free culture supernatants by fluid phase RIA. Transfused patients exhibited a more rapid onset and steeper increase of prostanoid production. After 24 hours incubation, the spontaneous and LPS-induced PGE release of pre- and post-BT cells was significantly different (pre-BT: 2.1 and 5.1ng/ml; post-3-BT: 5.0 and 7.9ng/ml; $p<0.01$). Pre-BT cells released considerably lower amounts of TXB_2 than post-BT cells (spontaneous release: 39 vs 88ng/ml; LPS-induced release: 62ng/ml vs 129ng/ml; $p<0.05$). After correction for monocytes as defined by monoclonal antibodies, post-BT cells again showed increased prostanoid release as compared to pre-BT cells. Therefore, the enhanced PGE and TXB_2 release of post-BT cells is not caused by an increase merely in the number of monocytes. Rather, BT appear to induce an enhanced release of prostanoids by activation of monocytes. We also found a correlation between the number of BT and the amount of prostanoid release.

Introduction

The mechanisms whereby blood transfusions (BT) induce kidney graft protection are unknown. In addition to specific immunoregulation by T suppressor cells or anti-idiotypic antibodies, the influence of non-specific factors is being considered. Prostaglandins (PG) have attracted particular attention because they have been shown to inhibit cell-mediated immunity as well as antibody responses [1,2]. PGE appears to act as a negative feedback signal in immune responses that require macrophage/monocyte-lymphocyte interactions [3]. It was shown in mice [4] and rats [5] that the administration of PGE delayed graft rejection. When indomethacin, an inhibitor of endogenous PG synthesis, was administered, grafts were more rapidly rejected [4]. However, paradoxically, PGE serum levels were found to be elevated during graft rejection [6,7]. Macrophages from mixed lymphocyte cultures produce strikingly increased amounts of PG when the cultures are set up with cells from sensitised allograft recipients and the respective donors [8]. The intravenous injection of sheep erythrocytes stimulates PG production in mice [9]. We studied the influence of planned BT in dialysis patients on PG release from peripheral mononuclear cells in vitro.

Patients and methods

Twenty-five patients (9 females, 16 males, aged between 7 and 38 years) on chronic haemodialysis were the subjects of this study; the underlying renal diseases leading to end-stage renal insufficiency were glomerulonephritis (16 patients), pyelonephritis (4), polycystic kidneys (2), nephrotic syndrome (2), and unspecified renal disease (1). None of the patients received drugs interfering with PG release and none had been transfused previously. The patients received three BT at two week intervals. Nineteen patients were transfused with whole blood, and six patients received packed red cells. Blood samples were obtained prior to transfusion, 14 days after each BT, and three months after the last BT. Mononuclear cells were prepared by Ficoll-Hypaque gradient centrifugation, deep-frozen, and stored in liquid nitrogen until use.

Mononuclear cells (4×10^5) were cultured in bicarbonate-buffered medium (RPMI 1640, supplemented with L-glutamine, penicillin (100 IU/ml/), streptomycin (100μg/ml), and 10% heat-inactivated fetal calf serum. To stimulate monocytes, 20μl/ml LPS (lipopolysaccharide from *E coli* 0127, Gibco, Glasgow, UK) were added to the cultures. In kinetic studies, the spontaneous and LPS-induced release of PGE and TXB_2 in cell-free culture supernatants were determined by fluid phase radioimmunoassay (RIA). The antibody raised against PGE_1 cross-reacted with PGE_2; values are therefore given as PGE without subclass specification. The anti-TXB_2 antibody cross-reacted less than 0.5 per cent with other arachidonic derivatives. Both antibodies have been described in detail previously [10]; they were kindly provided by Professor D Gemsa, Medizinische Hochschule, Hannover, and Dr M Seitz, Medical Policlinic, Heidelberg, FRG. PG determinations were performed according to the method described by Jaffee et al [11]. Prostanoid concentrations were calculated using a logit-log computer program.

Monocyte numbers contained in the mononuclear cell preparations were defined by indirect immunofluorescence using two different anti-monocyte monoclonal antibodies (OKM1, Ortho Diagnostics, Raritan, USA; CM1 provided by Dr P Terasaki, Los Angeles, USA). The percentage of stained cells was measured using a Spectrum III flowcytometer (Ortho).

Prostanoid release from patient cells before BT was compared to that after one to three BT using the Wilcoxon and Wilcox test for multiple comparisons of dependent samples.

Results

Spontaneous and LPS-induced PGE release in patients before and after three blood transfusions as well as in healthy controls is shown in Figure 1. The

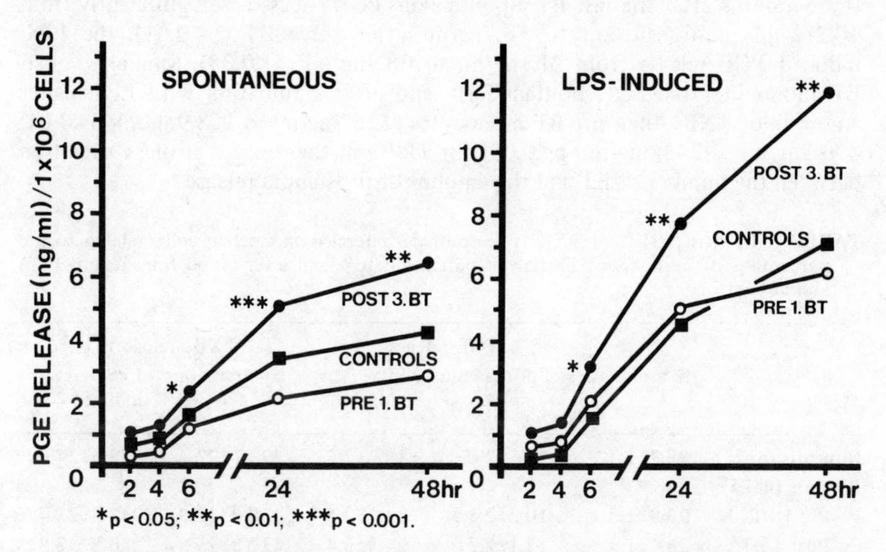

Figure 1. Kinetics of spontaneous and LPS-induced PGE release from mononuclear cells of dialysis patients before first and 14 days after third blood transfusion (BT) and PGE release of healthy controls

kinetics revealed that peripheral mononuclear cells of transfused dialysis patients exhibited a more rapid onset and a steeper increase of prostanoid production. By six hours of incubation the spontaneous and LPS-induced PGE release in transfused patients was significantly increased as compared to pre-transfusion values. After 24 hours or 48 hours the differences became highly significant (pre-BT: 2.1 and 5.1ng/ml; post-3-BT: 5.0 and 7.9ng/ml; $p<0.01$).

Similar findings were obtained when TXB_2 release from peripheral mononuclear cells was examined. The kinetic analysis revealed at six hours, a steeper increase of TXB_2 production in post-transfusion cells than in pre-transfusion or in control cells. After 24 hours incubation, spontaneous and LPS-induced release of TXB_2 in transfused patients was significantly enhanced as compared to pre-transfusion values (spontaneous TXB_2 release: 88ng/ml post-3-BT vs 39ng/ml pre-BT; LPS-induced TXB_2 release: 129ng/ml post-3-BT vs 62ng/ml pre-BT; $p<0.05$).

We attempted to clarify whether enhanced prostanoid release of post-transfusion cells was merely caused by an increase in the number of monocytes or by monocyte activation. PGE and TXB_2 values were corrected for $1x10^6$ monocytes according to the monocyte numbers defined in the mononuclear cell preparations. Table I gives relative monocyte numbers and corrected PGE and TXB_2 values pre-transfusion and after one, two and three blood transfusions. In controls 23.7 per cent of the mononuclear cells were characterised as monocytes. In non-transfused patients comparable monocyte numbers were observed (23.9%). Following a first BT monocytes rose to 26.4 per cent and reached 27.5 per cent after three BT; they returned to pre-transfusion values within three months after the last BT. Spontaneous PGE release rose significantly from 10.2ng/ml pre-transfusion to 18.7ng/ml after three BT ($p<0.01$), the LPS-induced PGE release from 21.1ng/ml to 28.9ng/ml ($p<0.05$). Similarly, post-BT monocytes released spontaneously and after stimulation with LPS higher amounts of TXB_2 than pre-BT monocytes (326.8ng/ml vs 158.9ng/ml, $p<0.01$; 476.2ng/ml vs 243.6ng/ml, $p<0.05$). In addition, there was a strong correlation between the number of BT and the amount of prostanoid release.

TABLE I. In vitro PGE and TXB_2 release of peripheral mononuclear cells, related to the percentage of monocytes, in dialysis patients before and after blood transfusions (BT) and in controls

	% monocytes	PGE-release†		TXB_2-release†	
		Spontaneous	LPS-induced	Spontaneous	LPS-induced
Controls (n=22)	23.7±1.1	12.6±2.0	19.2±2.3	173.7±27.4	273.6±27.6
Patients (n=25)					
Pre 1 BT	23.9±1.3	10.2±2.5	21.1±2.6	158.9±17.3	243.6±23.4
Post 1 BT	26.4±1.1	12.1±2.7	24.9±3.4	172.3±25.5	236.8±22.8
Post 2 BT	26.7±1.2	14.4±2.6	27.8±4.3*	219.1±28.4	319.3±43.3
Post 3 BT	27.5±1.0*	18.7±2.5**	28.9±4.9*	326.8±34.1**	476.2±52.2*
3 months post 3 BT	21.0±0.5	10.5±0.9	16.5±2.1	153.7±47.1	290.5±29.9

Values are given as x ± SEM; †ng/m/$1x10^6$ monocytes after 24 hours incubation
*$p<0.05$; **$p<0.01$; post-BT vs pre-BT values

Discussion

The present study demonstrates that peripheral mononuclear cells of transfused dialysis patients release higher amounts of prostanoids than cells of non-transfused patients. After BT, the spontaneous as well as the LPS-induced release of PGE and TXB_2 was significantly enhanced. This effect was 'dose' dependent; whereas pre-transfusion prostanoid release was normal (as compared to controls), it increased with the number of BT. There is little doubt that the increased prostanoid release occurred from monocytes as the mononuclear cell preparations were free of platelets. Though activated granulocytes, in particular neutrophils, and to a much lesser degree eosinophils are capable of releasing PG, their production is much lower than that of monocytes/macrophages and does not persist for periods longer than six to eight hours in vitro [3]. During this early cell incubation period, the amount of prostanoid release was relatively small in our studies.

In previous experiments we observed increased monocyte numbers in dialysis patients following BT and this was confirmed in the current study. Although at first sight the increased monocyte numbers might be held responsible for the increase in prostanoids, our data argue against this simple explanation. After correction for monocyte numbers, post-transfusion cells still showed enhanced prostanoid release.

Whether our in vitro findings are relevant to the transfusion-transplant effect in vivo remains speculative. PG of the E and F types are nearly completely inactivated during a single lung passage; the half-life of these substances when released into the circulation is less than one minute [3]. It is therefore unlikely that BT induce a persisting high level of PG activity that mediates systemic immunosuppression. A more likely possibility is that PG may regulate the immune response within the graft. Recipient lymphocytes sensitised by previous BT might produce high amounts of lymphokines in the grafted kidney and thereby activate macrophages resulting in enhanced PG production. Lymphokine-mediated activation of macrophage functions, including PG release, has been described in animal models [8]. Further studies of BT-induced PG release in vivo, may prove relevant in explaining the beneficial effect of BT on kidney graft survival.

References

1 Allison AC. *Immunol Rev 1978; 40:* 3
2 Lewis GP. *Br Med Bull 1983; 39:* 243
3 Gemsa D. In Pick E, Landy M, eds. *Lymphokines 4.* New York: Academic Press. 1981: 335
4 Anderson CB, Jaffee BM, Graff RJ. *Transplantation 1977; 23:* 444
5 Strom TB, Carpenter CB. *Transplantation 1983; 35:* 279
6 Moore TC, Jaffee BM. *Transplantation 1974; 18:* 383
7 Anderson CB, Newton WT, Jaffee BM. *Transplantation 1975; 19:* 527
8 Dy M, Debray-Sachs M, Descamps B. *Transplant Proc 1979; 11:* 811
9 Webb DR, Osheroff PL. *Proc Natl Acad Sci USA 1976; 73:* 1300
10 Gemsa D, Seitz M, Kramer W et al. *J Immunol 1978; 120:* 1187
11 Jaffee BM, Behrmann HR, Parker CW. *J Clin Invest 1973; 52:* 398

Proc EDTA–ERA (1984) Vol 21

POLYCYTHAEMIA IS ERYTHROPOIETIN-INDEPENDENT AFTER RENAL TRANSPLANTATION

S Lamperi, S Carozzi, A Icardi

St Martin Hospital, Genoa, Italy

Summary

Serum erythropoietin increased in 55 patients after transplantation. The serum erythropoietin decreased when the haematocrit and haemoglobin reached high levels, indicating recovery of a feed-back control system. Despite the diminution of erythropoietin, six patients demonstrated a state of erythrocytosis while the in vitro cultures of BFU-e revealed high sensitivity to the reduced doses of erythropoietin, using monocyte-free T-lymphocyte-depleted peripheral blood.

In polycythaemic transplanted patients it is possible that cellular interactions stimulate an early hyperproliferation of BFR-e with a greater erythropoietin sensitivity and a partial capacity to grow in the absence of erythropoietin.

Introduction

Polycythaemia may temporarily complicate the course of a few patients after renal transplantation, but the mechanism is unclear. Acute or chronic rejection, renal artery stenosis or hydronephrosis of the transplanted kidney have been implicated as causes [1,2].

Local hypoxia, in all cases, may lead to an increased erythropoietin production with consequent erythrocytosis.

In many transplanted patients, however, the polycythaemia has been found in the absence of the abnormalities mentioned above, suggesting that other factors may contribute to the pathogenesis of erythrocytosis.

With the aim of contributing to the knowledge of the pathogenesis of this erythrocytosis, the present study reports the modifications of early erythroid progenitor growth in vitro and those of some haematological parameters in vivo, observed in kidney transplant patients with or without polycythaemia.

Patients and methods

Erythrocytosis was detected in six out of 55 patients who underwent kidney transplantation. The haematocrit, haemoglobin and erythrocyte mass (Cr^{51})

mean values were respectively 60.6 per cent, 21.4g/100ml and 44.16ml/kg body weight, with white blood cells and platelets within the normal range. The original renal diseases were chronic glomerulonephritis in three, chronic pyelonephritis in one, Alport's syndrome in one, lupus erythematosus systemicus in one.

The following parameters were determined monthly in the patients with erythrocytosis and in 12 out of 49 transplanted patients without erythrocytosis, from the dialysis phase up until 24 months after renal transplantation; haematocrit (Hct%), haemoglobin (Hb g/100ml), reticulocytes (Ret. μLitre), serum erythropoietin (sEp mU/ml) [3] and proliferative growth of Burst Forming Unit Erythroid (BFU-e per plate) using mononuclear cells from peripheral blood isolated on Ficoll-Hypaque gradient and incubated in in vitro cultures with progressively reduced doses of sheep plasma erythropoietin (2–1.5–1–0.5–0 IU/ml medium) [4].

Various culture patterns were employed: incubating 2×10^5 whole mononuclear cells (A), incubating 2×10^5 mononuclear cells depleted of monocytes (B), incubating T-lymphocytes (C) and by incubating 2×10^5 monocytes or T-lymphocytes alone (D) [5,6].

All patients showed a normal serum creatinine, blood pressure, and respiratory function, and received the same patterns of immunosuppressive treatment. The mean age was 33.16 ± 10.7 years.

Twelve normal volunteers were also studied as controls. All subjects gave their consent to this study.

Statistics. All data obtained under various experimental conditions, expressed as mean ± SD, underwent statistical analysis by the Student's 't' test. Statistical significance was accepted at 0.05 level.

Results

In the patients with erythrocytosis, at four months the Hct, Hb and Ret. increased significantly in comparison with those of control subjects 40 per cent, 41 per cent and 380 per cent ($p<0.005$) and the dialysis phase, 186 per cent, 205 per cent, 2156 per cent ($p<0.005$). These figures remained significantly higher until the twelfth month, while from the twelfth month to the twenty-fourth month, they decreased slightly.

In the patients without erythrocytosis, the increases of Hct, Hb and Ret. over the levels observed during dialysis phase were lower than those of polycythaemic patients and did not exceed the normal range.

The sEp values increased similarly in both types of subjects, 400 and 480 per cent over the basal levels as in non-uraemic anaemic patients, from the first month, with a successive feed-back dependent decrease.

The in vitro cultures of type A with reducing doses of Ep in patients with erythrocytosis at the fourth month showed percentage increases of the number of BFU-e colonies developed per plate which were greater than to those observed in the dialysis phase and in normal subjects, 635 per cent, 680 per cent, 800 per cent, 2500 per cent ($p<0.005$) and 131 per cent, 143 per cent, 400 per cent and 2500 per cent ($p<0.005$) respectively. A mean number of 13 ± 2.5 BFU-e colonies per plate was also obtained in the absence of Ep, while no cellular

TABLE I. Erythropoietin (Ep) dose-response growth of BFU-e colonies in 'in vitro cultures' performed by incubating whole peripheral mononuclear cells (A type cultures), monocyte-depleted peripheral mononuclear cells (B type cultures) or T-lymphocyte-depleted peripheral mononuclear cells (C type cultures) from six erythrocytosics (E) and 12 non-erythrocytosics (non-E) after renal transplantation. (Values expressed as mean ± SD)

Erythropoietin concentration		Controls[a]				E Patients[b]				Non E Patients[c]		
		(A)	(B)	(C)	DP[g]	(A)	(B)	(C)	DP	(A)	(B)	(C)
0	IU/ml	0	0	0	0	13[d,e] ±2.5	0	0	0	0	0	0
0.5	IU/ml	2 ±1.2	0	0	2 ±1.5	52[d,e] ±4.9	2.9 ±0.9	10[d,e] ±3.8	0	0[f]	0	0
1	IU/ml	18 ±3	6 ±1.1	12 ±2.34	10 ±3.1	90[d,e] ±6.1	7.5 ±1.3	36[d,e] ±1.9	6 ±2	25[e,f] ±2.2	7.3 ± 4.7	15.2[e,f] ±3.5
1.5	IU/ml	48 ±5	18 ±2.6	30 ±3.1	15 ±2	117[d,e] ±3.9	15 ±3.3	70[d,e] ±2.9	10 ±3	43[e,f] ±1.9	19.7[e] ± 2.9	34.4[e,f] ±5.6
2	IU/ml	54 ±4.26	21 ±3.6	38.5 ±2	17 ±4.1	125[d,e] ±6.4	23.2 ±2.5	78[d,e] ±5.1	16 ±8.2	57[e,f] ±4.3	22.3[e] ± 3.2	41[e,f] ±2.8

a Values are the mean ± SD for 12 normal subjects
b Values are the mean ± SD for 6 subjects with erythrocytosis after renal transplantation
c Values are the mean ± SD for 12 subjects without erythrocytosis after renal transplantation
d Significantly different from control values ($p<0.005$)
e Significantly different from DP values ($p<0.005$)
f Significantly different from erythrocytosic patient values ($p<0.005$)
g Dialysis phase

growth was found under similar conditions in the dialysis phase and in normal subjects (Table I).

At the same time, in type A cultures of patients without erythrocytosis, the BFU-e colony growth, with the addition of progressively reduced doses of Ep, was higher than in the dialysis phases and normal subjects, but noticeably lower than those of the patients with erythrocytosis. In the absence of Ep from cultures no BFU-e development was observed in these cases.

In types B or C cultures in patients with or without erythrocytosis, the BFU-e growth was significantly less than that seen in type A cultures. In the patients without erythrocytosis, however, this reduction was less than in patients with erythrocytosis and no BFU-e developed on adding small doses of Ep to the cultures, as occurred with normal subjects (Table I).

In all patients the type D cultures did not result in any significant BFU-e development, irrespective of the Ep dose added.

Discussion

Our results demonstrate that the sEp increases in the same period and in the same proportion in transplanted patients with or without erythrocytosis, and successively undergoes a feed-back dependent decrease. This suggests a restoration of normal intrarenal secretion [7]. However, despite these sEp changes a few transplanted patients (10.9%) show a further elevation of the Hct values which remain at high levels for a few months. This finding appears to be associated with a greater capacity for the development of erythroid progenitors in the erythrocytosic patients, this may be due to an early increase in Ep receptors on the cell surface and consequently to a greater Ep sensitivity [8].

Such a particular sensitivity appears to be accompanied by the presence of the adherent mononuclear cell fraction and/or T-lymphocytes in the medium. In fact, the removal of these cells, particularly the monocytes, is followed by a reduction in the number of BFU-e colonies per plate.

These results, therefore, confirm that Ep cannot be considered the most important regulatory factor in early BFU-e development. A primary role in this mechanism is played by cellular interactions which, in a few transplanted patients, assume a greater capacity for stimulating BFU-e maturation. Up to the present time the mechanism for this was largely unknown but appears to be almost completely independent of the presence of Ep.

References

1 Nies BA, Cohn R, Schrier SL. *N Engl J Med 1982; 273:* 875
2 Alvis R, Raja R, Baquero A et al. *Kidney Int 1984; 25:* 339
3 Dunn CDR, Jarvis JH, Grenman JM. *Exp Hematol 1965; 3:* 68
4 Iscove NN, Sieber F, Winterhalter KH. *J Cell Physiol 1974; 83:* 309
5 Shaw GM, Levy PC, Lo Buglio AF. *J Clin Invest 1978; 61:* 1172
6 Mendes NF, Tolnai MEA, Silveira NP et al. *J Immunol 1973; 11:* 860
7 Shalhoub RJ, Rayn U, Kim VV et al. *Ann Int Med 1982; 97:* 686
8 Gregory CJ, Eaves AC. *Blood 1977; 49:* 855

Proc EDTA–ERA (1984) Vol 21

KIDNEY TRANSPLANTATION IN PATIENTS ON CONTINUOUS AMBULATORY PERITONEAL DIALYSIS

Z Shapira, D Shmueli, A Yussim, G Boner, C Haimovitz*, C Servadio

*Beilinson Medical Center, Petah Tiqva, and Sackler School of Medicine, Tel Aviv, *Soroka Medical Center, and Ben Gurion University of the Negev, Beer Sheva, Israel*

Summary

In the 45 months from April 1980 to December 1983, 137 patients received first cadaver kidney grafts. Thirty-two of the patients were on continuous ambulatory peritoneal dialysis (CAPD group) and 105 were on haemodialysis (HD group). The two groups of patients were similar in respect to pre-transplant blood transfusions, mean age, HLA-A, B, DR matching and immunosuppressive therapy. In 14 CAPD patients at least one episode of peritonitis was documented before transplantation. The actuarial graft and patient survival was 57 per cent and 84.4 per cent, respectively for the CAPD group and 55.8 per cent and 85.3 per cent in the HD group. No patient in either group had evidence of peritonitis after the transplantation. These similar results indicate that kidney transplantation in CAPD patients carries no greater risk than in patients on haemodialysis.

Introduction

Since continuous ambulatory peritoneal dialysis (CAPD) was commenced in the late seventies [1], increasing numbers of such patients are awaiting renal transplantation [2,3]. However, it appears that several centres still prefer to transplant haemodialysis patients. The reluctance to transplant CAPD patients results mainly from fear of peritonitis [3–6]. There are three major factors which may enhance susceptibility to peritonitis in CAPD patients:

1. Previous episodes of peritonitis may leave residual localised infections which are prone to flare up under immunosuppressive treatment [2].
2. The presence of the intraperitoneal catheter may itself serve as a route for bacterial invasion, especially in immunosuppressed patients.
3. The high risk of peritonitis – the major complication of CAPD – resulting from the necessary manipulation of the connections during the dialysis procedure, could be even higher in the immunosuppressed patient if he still requires dialysis [2].

Other reasons given for the reluctance to transplant CAPD patients are that the normal site for transplantation might be unsuitable and that the outcome is doubtful due to insufficient information regarding graft survival rates in these patients [3].

Materials and methods

In the 45 months between April 1980 and December 1983, 32 patients on CAPD treatment received first cadaver renal grafts. Over the same period, 105 HD patients were also transplanted with their first graft. The two groups of patients were similar with respect to the average number of pre-transplant blood transfusions per patient, mean age, sex distribution, HLA A-B-DR donor-receipient matching, and had the same proportions of sub-optimal grafts (vascular anomalies, prolonged initial warm ischaemic time) (Table I). In 14 CAPD patients, at least one episode of peritonitis was documented before transplantation. In four patients the peritonitis subsided just before the kidney grafting.

TABLE I. Comparison of immunological factors, graft quality and patients distribution in the CAPD and in the haemodialysis groups

	CAPD patients	Haemodialysis patients
Number of patients	32	105
Male/Female	12/20	67/38
Mean and range of age	28±11 2–61	32±8 16–58
Average number of pre-transplant blood transfusion units per patient	4.5	5.25
Pre-transplanted peritonitis	14	–
Number of patients with matching of two or more antigens in HLA A-B	7	31
Number of patients with full DR match	2	17
Number of grafts with vascular anomalies	6	22
Number of grafts with initial warm ischaemic time more than 30 minutes	11	37

In both groups the same pharmacological immunosuppressive treatment was used: 0.5mg/kg/day prednisone and 2–2.5mg/kg/day azathioprine from the first day after transplantation.

As a rule, the intraperitoneal catheter was left in place for a period of three months after transplantation. Indications for its removal were:

1. Satisfactory graft function at three months after transplantation.
2. Per-operatively if peritonitis was evident around the time of operation.
3. At any time post-operatively, if peritonitis or other complications developed.

Results

The actuarial graft survival at 45 months in the CAPD and HD groups was 57 per cent and 55.8 per cent respectively (p=NS). The actuarial patient survival for this period was 84.4 per cent in the CAPD group and 85.3 per cent in the HD group (p=NS). The mean serum creatinine was 2.05±0.75mg per cent in the CAPD group and 2.15±1.2mg per cent in the haemodialysis group. Some technical difficulties due to the presence of an unsuitable tissue bed were encountered in one patient in the CAPD group and in three of the haemodialysis group. Thirteen patients out of 32 in the CAPD group and 58 of the 105 HD patients required dialysis for a period of one to eight weeks after the transplantation. No post-operative peritonitis was evident in either group during the three months after transplantation (Table II).

TABLE II. Results of kidney transplantation in the CAPD and in the haemodialysis groups

	CAPD patients		Haemodialysis patients
Actuarial graft survival at 45 months	57%		55.8%
		p=NS	
Actuarial patient survival at 45 months	84.4%		85.3%
		p=NS	
Post-transplant peritonitis	–		–
Mean serum creatinine	2.05±0.75mg%		2.15±1.2mg%
Number of patients who needed dialysis after transplantation	13/32		58/105
Technical difficulties due to unsuitable tissue bed	1/32		3/105

Discussion

The three factors which were cited above may theoretically enhance development of peritonitis after transplantation in CAPD patients. This seems to justify the apprehension and hence the reluctance of some centres to transplant these patients. However, accumulated evidence from recent publications fails to confirm that transplanted CAPD patients are at higher risk than haemodialysis patients.

Although some centres described post-transplant peritonitis in CAPD patients, they emphasised that its incidence was similar to that in non-transplanted CAPD patients and that it subsided easily with proper antibiotic treatment and without interrupting the immunosuppressive therapy. It did not contribute to mortality or to graft loss in these patients [5,7]. Other centres have also described similar results after transplantation in CAPD and HD patients but without an increased incidence of peritonitis [3,6].

Our experience also shows no significant difference between CAPD and HD patients who were transplanted with first cadaver grafts. Both groups were similar in respect to other factors which may influence transplantation outcome (Table I). These results may be of particular importance because all the 'high risk' factors encountered in many of our CAPD group. Fourteen had episodes of peritonitis before transplantation, 13 patients required peritoneal dialysis after transplantation and in all the patients except four the indwelling catheter was left for at least three months.

An unsuitable tissue bed was found in only one patient in the CAPD group and therefore may not be considered a serious obstacle. Unsuitable tissue bed was also occasionally found in the HD group.

The fortunate absence of peritonitis in the post-operative period in the CAPD group does not necessarily mean that there are no additional potential hazards to be considered in these patients under immunosuppressive therapy. However, the similar graft and patient survival in CAPD and HD patients in our and others experience should indicate that CAPD in itself is not a contraindication for transplantation. This and the observation that CAPD patients are in better general and metabolic condition than patients on haemodialysis [4] should encourage consideration of CAPD patients as suitable candidates for renal transplantation.

References

1 Moncrief JW, Popovich RP, Nolph KD. *Trans ASAIO 1978; 24:* 476
2 Gokal R, Ramos JM, Veitch P et al. *Proc EDTA 1981; 18:* 222
3 Gokal R, Ramos JM, Veitch P et al. *Dial Transpl 1982; 11:* 125
4 Oreopoulos DG, Khanna R, Williams P et al. *Nephron 1982; 30:* 293
5 Patel S, Rosenthal JT, Hakala TR. *Transplantation 1983; 36:* 589
6 Ryckelynck JP, Verger C, Pierre D et al. *Perit Dial Bull 1984; 4:* 40
7 Stefanidis CJ, Balfe JW, Arbus GS et al. *Perit Dial Bull 1983; 3:* 5

Proc EDTA–ERA (1984) Vol 21

PNEUMATOSIS INTESTINALIS IN PATIENTS AFTER CADAVERIC KIDNEY TRANSPLANTATION: POSSIBLE RELATIONSHIP WITH AN ACTIVE CYTOMEGALOVIRUS INFECTION

W J van Son, F J van der Woude, E J van der Jagt, A J M Donker, S Meijer, M J H Slooff, T H The, A M Tegzess, L B van der Slikke

University Hospital Groningen, Groningen, The Netherlands

Summary

Four patients are presented with pneumatosis intestinalis following kidney transplantation, all with severe cytomegalovirus (CMV) infection. Two patients had a primary infection and two patients had CMV reactivation. One patient died because of disseminated CMV infection. Two patients had concomitantly an active, non-obstructive duodenal ulcer. In a control population of 17 patients who suffered from a duodenal ulcer post-transplant without any evidence of CMV infection, we could not demonstrate pneumatosis intestinalis. We suggest a possible relationship between pneumatosis intestinalis and active CMV infection. The possible mechanisms responsible for this relationship are discussed.

Introduction

Gastrointestinal complications after cadaveric renal transplantation are numerous and include oesophagitis, ulcers with or without haemorrhage or perforation, pancreatitis and infarction [1,2]. These complications are believed to be related to the use of immunosuppressive drugs, especially corticosteroids [1,2]. Pneumatosis intestinalis is an uncommon disease whose pathogenesis is still obscure. Reports of pneumatosis intestinalis in patients with a kidney transplant are rare and are also believed to be related to the use of immunosuppressive drugs [1]. We recently observed two patients with cadaveric kidney transplants and pneumatosis intestinalis. Both patients had active CMV infection. The association prompted us to look retrospectively at our transplant patient population to see whether we could find other patients with pneumatosis intestinalis in combination with an active CMV infection.

Patients and methods

In 1982 we identified two patients who suffered from pneumatosis intestinalis. Retrospectively we found two other patients with this abnormality in our

TABLE I. Clinical details and results of the four patients with pneumatosis intestinalis

Patient number	Age/Sex	Type of CMV-infection	Gastrointestinal symptoms before transplantation	Symptoms moment of pneumatosis	Diagnosis pneumatosis weeks after transplantation	Roentgen	Pneumo-peritoneum	Concomitant ulcer	Final outcome
1	47/F	primary	none	yes, aspecific	8	pneumatosis intestinalis cystoides (jejunum + ileum)	–	yes (UD)	complete recovery; good renal function
2	31/F	react.	none	yes, aspecific	8	pneumatosis intestinalis linear type + cystoides (ileum + colon)	+	no	complete recovery; marginal renal function (vascular rejection)
3	49/F	primary	none	none	6	pneumatosis intestinalis linear type (jejunum + ileum)	+	no	complete recovery; good renal function
4	22/M	react.	none	severe gastrointestinal bleeding	8	pneumatosis intestinalis linear type ileum	–	yes (UD)	died; generalised CMV-infec- (liver, gut, lungs)

transplant population. Patient data are given in Table I. All four patients had received a cadaveric kidney transplant and were treated with standard immunosuppression consisting of azathioprine and corticosteroids only. All four patients had active CMV infection with fever, arthralgia, leucopenia, liver function abnormalities and changes in CMV-serology, when pneumatosis was detected.

The diagnosis of pneumatosis was made on plain X-ray of the abdomen with the patient in the supine position, at which time intramural gas was present [3]. The serological diagnosis of active CMV infection was defined as a threefold or greater rise in titres of complement fixing antibodies (CFA) against CMV or a fourfold or greater rise in antibodies to CMV early (EA) and late (LA) antigens [4]. The patients were considered to be seronegative for CMV when the CFA titres were less than 1:4 and the titres of antibodies to CMV-EA and -LA less than 1:40.

Control population

Two out of the four patients with pneumatosis also had an active, non-obstructive duodenal ulcer (Table I) at the time the diagnosis of pneumatosis was made. According to the literature gastric or duodenal ulcers (when obstructive) can cause pneumatosis [5]. Therefore we restudied all abdominal X-ray films of patients within our transplant population who had a gastric or duodenal ulcer in the period from 1971–1982. An active CMV infection was excluded in these patients when the ulcer was diagnosed. In 18 out of the 382 patients transplanted between 1971–1983 a gastric or duodenal ulcer had been proven. In 17 cases we had sufficient X-ray material to examine whether or not pneumatosis intestinalis had been present at the time the ulcer was found.

Results

The results are given in Table I. An example of the radiological features of pneumatosis is given in Figure 1 (patient No. 2). All four patients had active CMV infection at the time the diagnosis of pneumatosis was made.

In the case of patients Nos. 3 and 4 the diagnosis of pneumatosis was made retrospectively, one of which (patient No. 3) had no abdominal symptoms whatsoever. In case No. 3 the diagnosis was made retrospectively on the intravenous urogram performed at the time the patient complained of arthralgia. Patients Nos. 1, 2 and 3 underwent a complete recovery without surgical intervention. Patients 1 and 2 received oxygen therapy via Ventimask® which resulted in rapid disappearance of the radiological signs of pneumatosis intestinalis.

Patient No. 4 died in profound shock after severe untreatable gastrointestinal bleeding. At post-mortem examination disseminated CMV infection was found with multiple inclusion bodies in lungs and liver. Inclusion bodies were found adjacent to the vessels in the ulcers. CMV virus was cultured from these sites.

Control population

In none of the 17 patients with a proven ulcer could we find any evidence of pneumatosis at the time the ulcer was diagnosed.

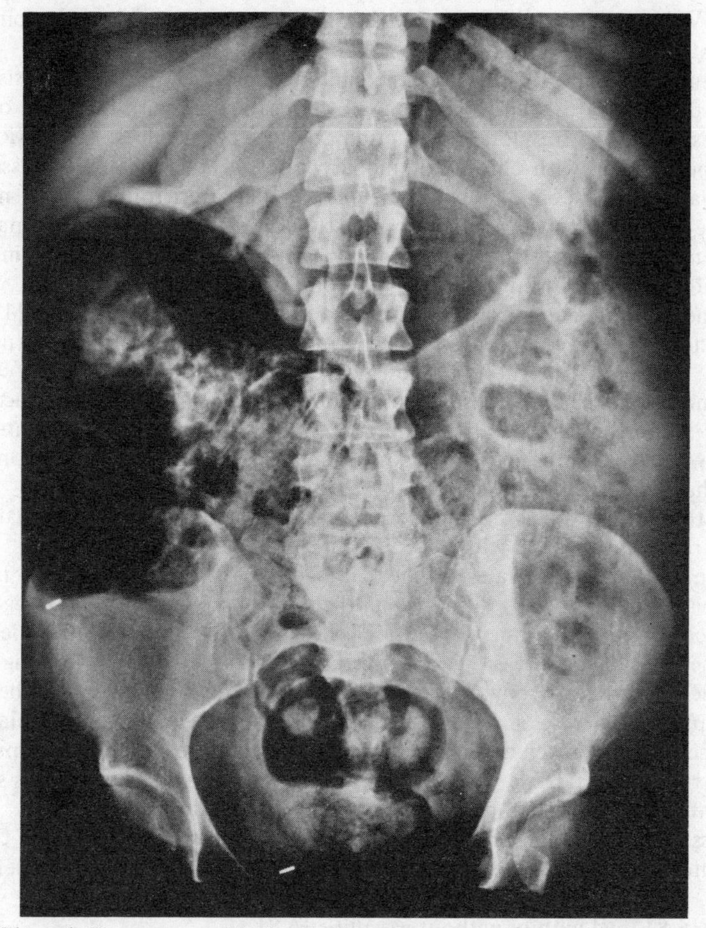

Figure 1. Roentgenographic signs of pneumatosis intestinalis in patient No. 2

Discussion

The incidence of active CMV infection was stated to be as high as 43–92 per cent in renal transplant patients [6]. Most of the cases are subclinical but when the CMV infection is clinical, the symptoms may show great variety. Well-known is the so-called 'self-limited syndrome', comprising about 40–50 per cent of the patients infected [6]. The self-limited syndrome usually occurs between the 30th and 90th post-operative day and consists of prolonged fever, arthralgia, leucopenia, abnormalities in liver enzymes, respiratory symptoms or/and infiltrates on chest roentgenogram, and impairment of renal function [6]. Other manifestations of CMV infection post-transplant are protean, comprising lymphadenopathy, rash, hepatosplenomegaly, conjunctivitis [6], vasculitis [7], CMV glomerulopathy and lethal, so-called 'wasting' CMV infection [6].

CMV has been noted to reside in alimentary tract ulcers [6], sometimes associated with vasculitis at the site of the ulcer [8].

Pneumatosis intestinalis is an uncommon disease whose pathogenesis is still under discussion [5]. The condition is often associated with intestinal obstruction, especially with pyloric obstruction due to a peptic ulcer. Associations also have been made with chronic pulmonary disease, collagen diseases associated with vasculitis (systemic lupus erythematosus, periarteritis nodosa), or motility disorders of the bowel leading to functional obstruction (scleroderma). Less frequently pneumatosis has been found in association with gastrointestinal infections with gas-forming bacilli, lymphoma and leukaemia.

Reports of pneumatosis intestinalis in renal transplant patients [1,9] are rare and are believed to be associated with the immunosuppressive drugs used [1,9]. Very recently the condition was described in three patients following allogeneic bone marrow transplantation [10]. Although the condition, according to the authors, was related to the immunosuppressive drugs used, it is interesting that two of the three patients also suffered from severe CMV infection (CMV oesophagitis, CMV-pneumonia) [10].

In this article we describe four patients with pneumatosis intestinalis after renal transplantation, all with concomitant severe CMV-infection.

One can only speculate on the pathogenesis of the pneumatosis in these four patients. The negative findings in the control population could suggest that the ulcers per se were not the cause of the condition. None of the patients had chronic pulmonary disease, gastrointestinal infections with gas-forming bacilli, evidence of ischaemic bowel disease, or a systemic disease like lupus. The immunosuppression per se also seems unlikely as the sole pathogenic explanation, because of the low incidence of the condition in the whole transplant population. The striking coincidence with severe CMV infection in all four patients suggests that infection with CMV might be related to the condition.

Possible explanations could be the primary cytopathogenic effect of CMV on the intestinal mucosa or vasculitis due to the CMV infection. As mentioned above, CMV is often found in the gastrointestinal mucosa with or without ulcers [6,8], and with or without vasculitis [6,8].

In conclusion, we believe that pneumatosis intestinalis in a renal transplant patients might be associated with an active CMV-infection. These data stress the importance of a keen awareness of the possibility of a CMV infection in renal transplant patients with free air under the diaphragm and/or abdominal symptoms. In these patients unnecessary surgery should thus be avoided.

References

1 Julien PJ, Goldberg HI, Margulis AR, Belzer FO. *Radiology 1975; 117:* 37
2 Meyers UC, Harris N, Stein S et al. *Ann Surg 1979; 190:* 535
3 Jamart J, Pringot J, Roisin Ph. *J Belg Radiol 1979; 62:* 391
4 The TH, Klein G, Langenhuysen MMAC. *Clin Exp Immunol 1974; 16:* 1
5 Olmstead WW, Madewell JE. *Radiol 1976; 1:* 177
6 Glenn J. *Rev Infect Dis 1981; 3:* 1151
7 Beard MEJ, Hatipov CS, Hamer JW. *Br Med J 1978; 1:* 623
8 Foucar E, Mukai K, Foucar K et al. *Soc Clin Pathol 1981; 76:* 788
9 Lewis Wall L, Linshaw MA, Bailie MD, Pierce GE. *J Pediatrics 1982; 101:* 745
10 Navari RM, Sharma P, Deeg HJ et al. *Transpl Proc 1983; 2:* 1720

Proc EDTA–ERA (1984) Vol 21

IMPAIRED FIBRINOLYSIS AFTER KIDNEY TRANSPLANTATION

R Seitz, R Michalik, R Egbring, H E Karges, H Lange

Centre of Internal Medicine, University of Marburg and Research Laboratories of Behringwerke AG, Marburg, FRG

Summary

Impaired fibrinolysis is associated with low plasma plasminogen (PLG) and a predominance of its inhibitor alpha 2-antiplasmin (APL). Plasma PLG and APL were determined in 10 recipients who retained their cadaveric graft (A) and 10 patients who lost it immediately after transplantation (B).

In group A only five patients had a PLG less than 70 per cent and/or APL/PLG ratios exceeding 1.8, accompanied by severe albeit reversible acute rejection in three of them. In contrast nine of 10 group B patients had low PLG and/or an elevated APL/PLG ratio.

Thus impaired fibrinolysis frequently appears to be associated with deterioration of cadaveric graft function.

Introduction

Acute vascular rejection is known to be accompanied by graft vessel thrombosis, particularly in irreversible rejection. Similarly, thrombotic complications of peripheral vessels (e.g. occlusion of arterio-venous fistulae) are common after kidney transplantation [1]. Assuming impaired fibrinolysis participates in the pathogenesis of acute rejection the plasma plasminogen (PLG), alpha 2-antiplasmin (APL) as well as antithrombin III (AT III) and fibrinogen (FBG) were determined after cadaveric kidney transplantation. In addition the neoantigen alpha 2-antiplasmin-plasmin complex (AP-PL) was assessed in order to test whether plasminogen consumption had occurred or not.

Patients and methods

Twenty recipients of cadaveric kidneys were studied. Group A included 10 patients (4 males, 6 females aged 19 to 59 years) who left hospital with a functioning graft. The remaining 10 patients (4 males, 6 females, aged 24 to 52 years) who lost their kidney shortly after transplantation constitute group B.

Citrated plasma samples were obtained immediately before surgery and daily to twice weekly thereafter until the patients discharge from hospital or removal of the graft. Aliquots of the samples were stored in a freezer at -28°C until analysis. Plasma PLG, AT III and FBG were determined according to Mancini et al [2] using Partigen® plates (Behringwerke, Marburg, FRG). Assays of APL and the neoantigen alpha 2-antiplasmin-plasmin complex (AP-PL) were performed by Laurell immunoelectrophoresis [3]. The antiserum to AP-PL was produced by the research laboratories of Behringwerke AG.

Results

Plasma PLG decreased to less than 70 per cent of normal in five of 10 group A patients. A concomitant increase of APL to more than 120 per cent indicates the preponderance of the inhibitor, as shown in Figure 1. Hence the inhibitors preponderance is represented by an APL/PLG ratio exceeding 1.8 due to either a low PLG-level, an elevated APL concentration or both. An APL/PLG ratio above 1.8 was obtained in three of five group A patients with diminished levels of PLG. As shown in Table I severe acute, albeit reversible, rejection episodes occurred in these three recipients. The remaining five group A patients with normal values of PLG and APL showed no or only moderate signs of rejection.

In contrast an elevated APL/PLG ratio above 1.8 was observed in seven of 10 group B patients who lost their graft without exception. In four of them PLG was diminished to less than 70 per cent as well as in two additional subjects with acute vascular rejection who had a normal APL/PLG ratio (Table I). Only one patient in group B failed to show any alteration of his PLG and APL. His graft was removed due to acute life threatening CMV-infection which was

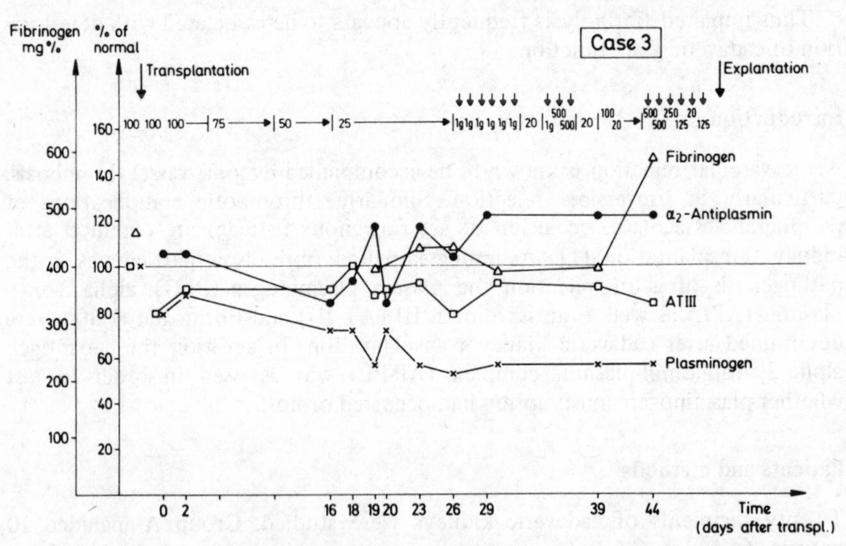

Figure 1. Decrease of PLG concomitant to rising APL levels in patient 3, whose graft was removed. The figures above give the prednisolone doses, the vertical arrows indicate steroid pulses

TABLE I. PLG<70 per cent and APL/PLG ratios >1.8 in synopsis with the clinical outcome. GF = graft function; GR = graft removal; AVR = acute vascular rejection

Patient	PLG	APL/PLG	Clinical outcome
Group A			
2	–	–	good GF, thrombosis of Cimino shunt
4	+	+	severe rejection controlled, satisfactory GF
5	–	–	good GF, thrombosis of Cimino shunt
6	–	–	good GF
9	+	–	sufficient GF
10	+	+	severe rejection controlled by cyclosporine; GF stabilised at creatinine 3mg/dl
11	–	–	satisfactory GF
18	+	+	rejection controlled by cyclosporine, satisfactory GF
19	+	–	satisfactory GF, cyclosporine
20	–	–	good GF
Group B			
1	–	+	GR due to AVR
3	+	+	GR due to AVR
7	+	+	emergency GR; leakage of the arterial anastomosis in staphylococcal septicaemia; deep vein thrombosis
8	+	+	GR due to AVR
12	+	+	GR due to AVR
13	–	+	GR due to AVR; thrombosis of Cimino shunt
14	–	+	GR due to AVR
15	–	–	GR due to acute CMV infection and rejection
16	+	–	GR due to AVR
17	+	–	GR due to AVR

thought to be transferred by the donor organ. At autopsy the kidney showed acute interstitial cellular inflammation but no signs of acute vascular thrombotic rejection.

Hence, signs of an impaired fibrinolysis appeared in three group A patients accompanied by acute reversible rejection and in nine group B patients who lost their graft mainly due to acute vascular rejection. Only two patients with low levels of PLG but normal APL/PLG ratio did not show any signs of acute vascular rejection (group A).

The determination of the neoantigen AP-PL, indicative of recent hyperfibrinolysis [4] was negative and normal values were obtained for the AT III and FBG levels in all blood specimens.

Discussion

A low plasma PLG and an elevated APL/PLG ratio are indicative of impaired plasma fibrinolytic activity. Impaired fibrinolysis was assumed to participate in the pathogenesis of acute vascular rejection which is regularly accompanied by thrombotic lesions. Accordingly the most striking results of an impaired fibrinolysis were obtained in group B patients who lost their grafts shortly after transplantation. In nine of 10 group B patients with severe thrombotic graft lesions who subsequently lost their grafts diminished PLG levels and or elevated APL plasma concentrations were observed. The normal values in the tenth subject who lost his kidney due to acute CMV-infection (see above) might be explained by the absence of signs of acute vascular rejection at graft autopsy.

In contrast five patients of 10 who retained their graft (group A) showed normal plasma PLG and APL. In these five subjects no or only moderate signs of a transient acute rejection occurred. The remaining five group A patients however showed a diminished fibrinolytic activity as indicated by the low PLG and a concomitant APL/PLG ratio >1.8 in three of them. In these three patients signs of acute rejection albeit transiently were obtained.

The diminished plasma levels of PLG might be due to its consumption during enhanced fibrinolysis. However, the neoantigen AP-PL, indicative of hyperfibrinolysis was not detected in any patient's plasma [4]. Thus an impaired synthesis rather than enhanced consumption might explain the diminished plasma concentration of PLG. While its source is not definitely clarified the kidneys have been shown at least in cats to account for the production of PLG [5].

References

1 Zazgornik J. *6 Wiss Sympos d Nephrol Arbtskr.* Kaiserslautern. 1983
2 Mancini G, Carbonara AO, Heremans P. *Immunochem 1965; 2:* 235
3 Laurell CB. *Anal Biochem 1966; 15:* 45
4 Egbring R, Klingemann HG, Arke K, Karges HE. *Hemostasis 1982; 11/S1:* 48
5 Highsmith RF, Kline DL. *Am J Physiol 1973; 225:* 1032

PART XXVI

GUEST LECTURE ON TRANSPLANTATION II

Chairmen: S T Boen
A Amerio

Proc EDTA–ERA (1984) Vol 21

REJECTION AND NEPHROTOXICITY. DIAGNOSTIC PROBLEMS WITH CYCLOSPORINE IN RENAL TRANSPLANTATION

R Pichlmayr, K Wonigeit, B Ringe, T Block, R Raab, B Heigel, P Neuhaus

Medizinische Hochschule Hannover, Hannover, FRG

Introduction

The new immunosuppressant cyclosporine has been the most important recent innovation in the field of organ transplantation. This new drug differs from all other immunosuppressive substances in biochemical structure, in mode of action and also in the spectrum of side effects. Since it has recently become evident in two multicentre trials [1,2] that it is able to improve transplant survival rates in cadaver kidney transplantation by 15–20 per cent, it is becoming increasingly important to analyse the problems associated with this drug and to further improve its clinical use. Important issues which still have to be resolved are whether cyclosporine should be used alone or in combination with other immunosuppressive drugs, whether it should be given to all patients or whether certain subgroups still would have greater benefit from conventional immunosuppressive treatment; finally information about the long-term effects of cyclosporine treatment is still rather limited. The answer to many of these questions depends on whether it will be possible to develop strategies to reduce or avoid the most serious side effects. Since all known side effects of cyclosporine are dose-dependent an important step in this direction will be the better understanding of cyclosporine pharmacokinetics resulting in improvements of the presently used regimens which are still mostly based on empiricism.

A side effect particularly important in kidney transplantation is the marked nephrotoxicity of cyclosporine. Since the reduction of kidney function is still the most important clinical sign of rejection, nephrotoxicity can cause serious diagnostic problems not seen in conventionally treated patients. Important approaches which could provide at least partial solutions to these problems are (a) improved handling of the drug on the basis of improved knowledge of its pharmacokinetics; (b) attempts to minimise the risk of additional damage to the graft, particularly during organ procurement and in the peri-operative period; and (c) the development of additional procedures for the differentiation between nephrotoxocity and rejection. The purpose of this article is to demonstrate the relevance of these approaches on the basis of our experience with cyclosporine in kidney and liver transplantation and to review the clinical problems associated with cyclosporine-induced nephrotoxicity.

Pharmacokinetics and blood monitoring

Studies of the pharmacokinetics of cyclosporine in humans have shown that after oral administration only 20 to 50 per cent of the drug is absorbed from the intestine [3]. In uraemia [4] and in patients with disturbed intestinal function or disturbed bile secretion [4,5] the absorption rate may be reduced far below this range. In the blood, peak concentrations are measured three to four hours after oral intake [3,6]. The drug has a high affinity to tissues. During steady state after repeated oral administration much higher tissue than blood concentrations were found in rats [7]. The highest amount was found in the liver. The drug is extensively metabolised in the liver and metabolites are mostly excreted with the bile [3,4]. Less than 10 per cent of the metabolites are excreted by the kidney. The elimination half-lives from blood or plasma have been calculated to be 14–17 hours under physiological conditions. They can be markedly prolonged in patients with liver dysfunction [5,6].

With high performance liquid chromatography (HPLC) and the radioimmunoassay (RIA) two methods for monitoring the cyclosporine levels in blood or plasma are available. With HPLC only the parent drug is determined, the RIA detects both the parent drug and several metabolites. The HPLC/RIA ratio can give a rough estimate of the relative proportion of parent drug and metabolites [6].

After oral intake of cyclosporine a characteristic time/concentration curve can be determined in circulation [3,4]. A peak concentration is reached after three to four hours which is followed by a slow decline of the cyclosporine concentration. In steady state, the time/concentration curve reaches a basic value after 8 to 12 hours, which is only slightly decreased at 24 hours. This means that after 12 to 14 hours the blood is in equilibrium with the tissue which can be regarded as the relevant parameter for both the immunosuppressive effect and toxicity. Therefore the trough value has to be regarded as most relevant for the evaluation of the response to treatment.

The blood values referred to in this paper are always trough concentrations determined 12 hours after oral intake. They were determined by RIA in haemolysed whole blood. It should be noted that equivalent plasma values are markedly lower [7].

Clinical classification of cyclosporine nephrotoxicity

The nephrotoxic effect of cyclosporine was obvious in the first patients treated with this drug [8]. It is not confined to transplanted kidneys but has also been observed in patients with liver, heart and bone marrow transplantation, as well as in patients with autoimmune disorders [9]. It has been difficult, however, to demonstrate the nephrotoxic effect in laboratory animals [10]. Rats develop some type of nephrotoxicity if treated with high dosages but are clearly less susceptible than humans; in other animals definite nephrotoxicity can usually only be demonstrated if cyclosporine treatment is combined with other nephrotoxic insults such as transient ischaemia [11]. The lack of suitable animal models is the cause of the very limited knowledge of the pathophysiology of cyclosporine nephrotoxicity. There is circumstantial evidence that it is mediated by a direct effect with intrarenal prostaglandin production causing profound alterations of renal haemodynamics [10,12,13]. The question whether toxic effects can be recognised morphologically has been controversial. Extensive studies of experimental and human material have led Mihatsch and associates to describe a variety of distinct lesions such as isometric vacuolisation of proximal tubular cells, giant mitochondria and micro-calcification [13]. Furthermore, a particular type of arteriolar lesion has been defined in humans and in spontaneous hypertensive rats [12,13]. Since pathophysiological and morphological criteria are still insufficient, the various manifestations of cyclosporine nephrotoxicity are best classified as clinical syndromes. At least three different entities can be differentiated: (a) the dose-dependent reversible nephrotoxic episode; (b) interactive nephrotoxicity; and (c) chronic nephrotoxicity. Although the pathophysiological basis of all three syndromes in all likelihood is closely related, they pose different diagnostic problems and require different therapeutic measures.

Dose-dependent nephrotoxic episodes

This syndrome can best be defined in patients with otherwise uncompromised kidney function, that is in transplants with immediate good function or after recovery from ischaemic injury. It is characterised by a variable increase in the serum creatinine. It can be very slow but frequently develops as rapidly as in acute rejection. Following appropriate reduction in the cyclosporine dose the nephrotoxicity is usually fully reversible [14–16]. Figure 1 depicts the characteristic course of a patient with this complication.

As already suggested by the dose-dependence, the nephrotoxic episode is nearly always a clear sign of overdose. It is usually associated with high blood concentrations. We have analysed the frequency of this type of nephrotoxic episode in 40 consecutive patients during the first three months post-transplantation. Twenty-two episodes occurred in 15 of the 40 patients. Cyclosporine trough blood concentrations in these patients were 1306±310ng/ml at the time of nephrotoxic episode and 662±217ng/ml after appropriate dose reduction and resolution of nephrotoxicity. Patients with values far above 1000ng/ml frequently show other signs of cyclosporine toxicity, such as tremor and in some instances hepatotoxicity. During the first three months after transplantation this type of nephrotoxicity is not seen in patients with blood values below 800ng/ml unless other nephrotoxic insults are also present, such as ischaemic damage or treatment with potentially nephrotoxic antibiotics (vide infra, interactive nephrotoxicity). Therefore we regard a whole blood trough value of 800ng/ml as the upper limit of the therapeutic range in the early post-operative period.

It is obvious that monitoring and adjustment of dose according to blood values will reduce the frequency and severity of acute nephrotoxic episodes. They are particularly difficult to avoid in patients with liver disease, who frequently

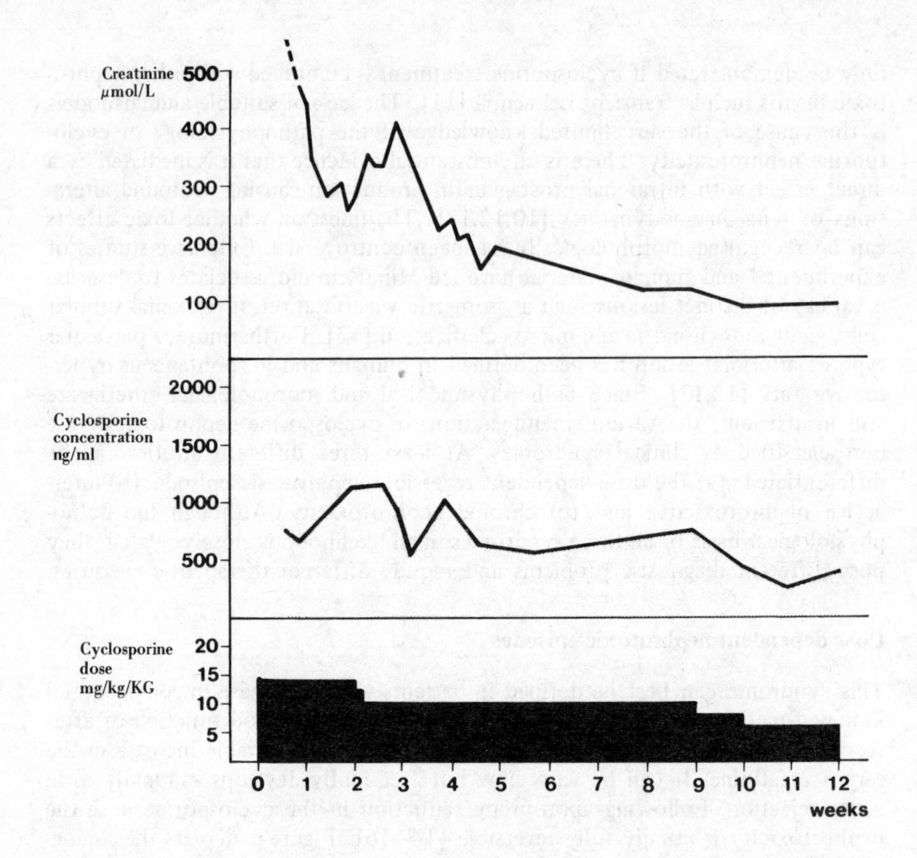

Figure 1. Characteristic course of an acute nephrotoxic episode. The elevation of the serum creatinine in the third post-operative week is associated with high blood concentrations and responds quickly to cyclosporine dose reduction

develop very high cyclosporine concentrations because of the reduced elimination rate of cyclosporine and the resulting problems with appropriate dose adjustment. If blood values are determined during or immediately after a rejection episode high results can be misleading since high doses of steroids have recently been shown to interact with cyclosporine metabolism [17]. In this instance high values can be secondary to the rejection treatment and thus may not be the cause of the elevated creatinine. A careful analysis of the pattern of blood results is particularly helpful. It has to be stressed, that this diagnostic instrument has its full potential only when performed as a prospective monitoring programme, with frequent blood estimations.

Since the clinical signs of an acute nephrotoxic episode resemble acute rejection closely, the differentiation of the two conditions is a key problem of cyclosporine therapy. In the presence of high blood values the response of the serum creatinine to dose reduction will usually clarify the situation. If the creatinine comes down quickly no further diagnostic procedures are required. In case of an insufficient response to dose reduction or if the creatinine elevation occurs despite cyclosporine blood concentrations in the therapeutic range, the exclusion of rejection by other means becomes mandatory. If the deterioration of kidney function is rapid, it might even be necessary to start rejection treatment before the definite diagnosis has been obtained. Since the course of acute rejection is ameliorated in cyclosporine treated patients [18,19], however, there is usually sufficient time to wait for the results of further examinations. In the cyclosporine era graft morphology still has remained the most important proof for rejection [13,20,21]. Important supportive procedures are radionuclear tests, sonography, and urinary cytology. Recently with the determination of the urinary excretion of neopterine [22] and the measurement of the subcapsular hydrostatic pressure in the kidney graft [23] two new promising methods have been added to the diagnostic armamentarium.

Interactive nephrotoxicity

An interactive effect between cyclosporine nephrotoxicity and kidney damage by a variety of other agents has now been well established [11,12,24]. It is a characteristic feature of this type of nephrotoxicity that it can become manifest with much lower dosages and with blood concentrations regarded to be in the therapeutic range. Failure to differentiate this cumulative type of cyclosporine nephrotoxicity from the dose dependent acute nephrotoxicity described above is the most likely cause for the bad correlation between nephrotoxic effects of cyclosporine and blood concentrations described by others [25]. The interactions occurring between cyclosporine and additional nephrotoxic insults are probably manifold and depend on the nature of the additional insult. The interaction can be thought of as an increased susceptibility of a pre-damaged kidney for the nephrotoxic cyclosporine effect or vice versa depending on the sequence in which these insults become effective.

Relevant additional insults for the transplanted kidney are ischaemic damage [2,11,26], treatment with other potentially nephrotoxic drugs [24,27–29] and acute or chronic allograft rejection. In the early post-operative period a cumulative effect of a variety of harmful factors has to be considered. During organ procurement and implantation the length of the cold and warm ischaemic periods and of prolonged machine perfusion have been found to influence strongly the nephrotoxic effect of cyclosporine [2]. In the recipient the state of hydration and the use of other potentially nephrotoxic drugs can lead to cumulative organ damage. Low cyclosporine blood values do not exclude this type of nephrotoxicity: high blood values, however, will clearly increase this effect. In anuric patients and in patients with severely impaired kidney function quick dose adjustments according to blood concentrations are mandatory. Another obvious consequence of these considerations is the reduction of all other potentially harmful factors to a minimum. Under these conditions residual cyclosporine nephrotoxicity and the effect of other injuring factors may not become clinically manifest.

The crucial clinical relevance of interactive nephrotoxicity in the early post-operative course results from the fact that, if it reaches the state of initial non-function, the graft survival rate becomes reduced [2,26]. The detrimental

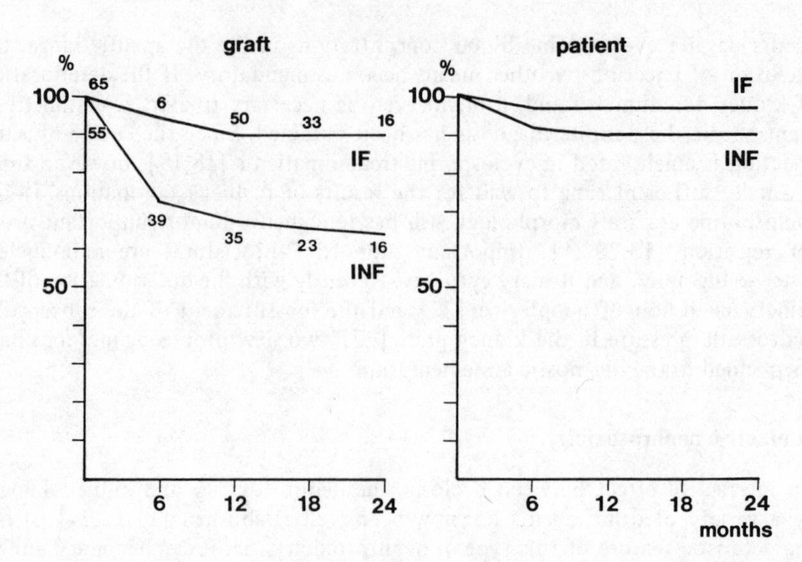

Figure 2. Effect of initial function on graft and patient survival in 120 patients treated with a first cadaveric graft in 1982 and 1983. IF=initial function; INF=initial non-function

effect of initial non-function on graft survival has frequently been shown under conventional immunosuppression and is even more relevant during cyclosporine treatment. Figure 2 compares the graft and patient survival rates for patients with and without initial graft function in our institution. It clearly demonstrates the increased risk to the graft if initial non-function occurs. The reasons for this increased rate of graft loss are complex. A major factor is that a superimposed rejection is difficult to detect and thus may be treated too late or inadequately. It has to be considered that a graft which is damaged by ischaemia and toxic effects has a reduced capability to recover from severe rejection. Since the post-operative management of anuric patients is more difficult, the more frequent use of other potentially nephrotoxic drugs will also contribute to the increased complication rate in these patients. Table I lists some of the drugs which have to be considered in this context. It includes drugs with a direct effect on the kidney [24,27–29], but also substances interfering with cyclosporine metabolism and thereby increasing the risk of nephrotoxic complications [15,30].

On the basis of these considerations the initial non-function rate can be kept low only by a multifactorial approach. Greatest attention to any avoidable injury during the harvesting process, during preservation and during the recipient operation is essential. As a result of this approach we have been able in our institution to reduce the frequency of initial non-function from 53 per cent in 1982 to 40 per cent in 1983 and 32 per cent in patients receiving a first cadaver kidney graft during the first six months of 1984. Table II demonstrates the dependence of the cold ischaemic time on the incidence of initial non-function for the whole patient population. The data suggest that a cold ischaemic time below 30 hours can reduce the frequency of initial non-function. This

TABLE I. Enhancement of cyclosporine nephrotoxicity by other drugs

I. Drugs with direct effect on kidney
(without effect on cyclosporine blood concentrations)
a) tubulotoxic drugs
aminoglycosides
cephalosporines
other potential nephrotoxic antibiotics
frusemide?

b) inhibitors of prostaglandin synthesis
indomethacin
other potentially nephrotoxic analgesics

II. Drugs interfering with cyclosporine metabolism
(high blood values, nephrotoxic effect mediated by toxic cyclosporine tissue values)
Ketoconazole
H_2-receptor blockers (cimetidine, ranitidine)
steroids?

TABLE II. Correlation between initial non-function (INF) and cold ischaemia time (CIT)

CIT (hours)	INF/total	% INF
<20	3/21	14
20–30	27/63	43
30–40	27/44	61
>40	19/36	53

time limit is still compatible with periods required for organ exchange between different centres even long distances apart.

The rather high frequency of initial non-function in Table II even with cold ischaemia times of 20 to 30 hours points to the fact that this is not the only and even not the most important factor for the frequency of initial non-function. As already discussed, warm ischaemia time, the hydration state of the recipient, the treatment with other potentially nephrotoxic drugs and the loading dose of cyclosporine may be important. We presently still use a standardised loading dose of 14mg/kg three to six hours before transplantation, followed by 14mg/kg during the first 10 post-operative days. This dosage is then reduced by 2mg every other week until a maintenance dose of 2 to 7mg/kg is reached. For about 30 per cent of the patients this regimen is still too high and quicker dose reductions are necessary on the basis of blood monitoring. In particular in these patients lower loading doses would probably be helpful. We are currently studying the question of how these patients can be identified before excess values are found.

Another approach to reduce the risk of this type of nephrotoxicity are immunosuppressive regimens which combine cylosporine in a reduced dosage with azathioprine and/or antilymphocyte globulin in addition to low dose steroids. It is not clear yet whether this approach is more effective than cyclosporine therapy in strictly blood concentration adjusted protocols either with or without steroids.

Since it occurs with cyclosporine concentrations in the therapeutic range, the interactive type of nephrotoxicity raises particularly difficult diagnostic problems. In patients with a functioning kidney the coincidence of functional deterioration with the addition of another drug to the regimen is an important clue. In anuria this information is not available. In this situation we have found the serial determination of the subcapsular hydrostatic pressure very helpful: it is below 40cm H_2O in kidneys suffering from ischaemic damage or toxicity but increases to over 40cm H_2O during rejection [23]. These results have recently been confirmed by Salaman et al [31]. Furthermore, repeated graft biopsies appear to be indicated in order to detect rejection at a time when it is still reversible. The fine needle aspiration biopsy has the advantage that it can be performed very frequently with no or very limited risk [21].

Chronic nephrotoxicity

Chronic cyclosporine nephrotoxicity is a badly defined entity but nevertheless one of the most serious therapeutic problems. It can be characterised as a state of poor long-term function which can be either stable or progressively deteriorating. Histology shows marked fibrosis and frequently also signs of chronic vascular injury, such as intimal thickening. The histological signs are not characteristic. The diagnosis is made by exclusion. The most important differential diagnosis is chronic rejection. Other conditions which have to be excluded are arterial stenosis, ureteric problems and recurrent disease.

The frequency of chronic nephrotoxicity depends on the maintenance dose of cyclosporine and on the blood values [15]. There is no doubt that during the maintenance therapy reduced blood values are sufficient to control allograft rejection [15]. On the other hand tubular cells appear to develop an increased susceptibility to cyclosporine. This leads to a high incidence of creatinine elevations in patients maintained above 600ng/ml during long-term treatment. This type of chronic nephrotoxicity usually responds very well to a reduction in the cyclosporine dose. In order to achieve blood levels between 150 and 600ng/ml cyclosporine maintenance doses between 2 and 6mg/kg bodyweight are required in most patients. Occasionally, however, nephrotoxic syndromes do not resolve even after adequate reduction. It is not yet clear whether this represents a particularly high susceptibility to cyclosporine or an interaction with pre-existing damage resulting from chronic rejection or other nephrotoxic insults. That the deterioration in kidney function is indeed caused by cyclosporine can be demonstrated in the individual patient by conversion to conventional immunosuppression. Figure 3 demonstrates the course of a patient with a living related kidney, who showed the characteristic signs of chronic nephrotoxicity three months after transplantation in face of adequate blood cyclosporine

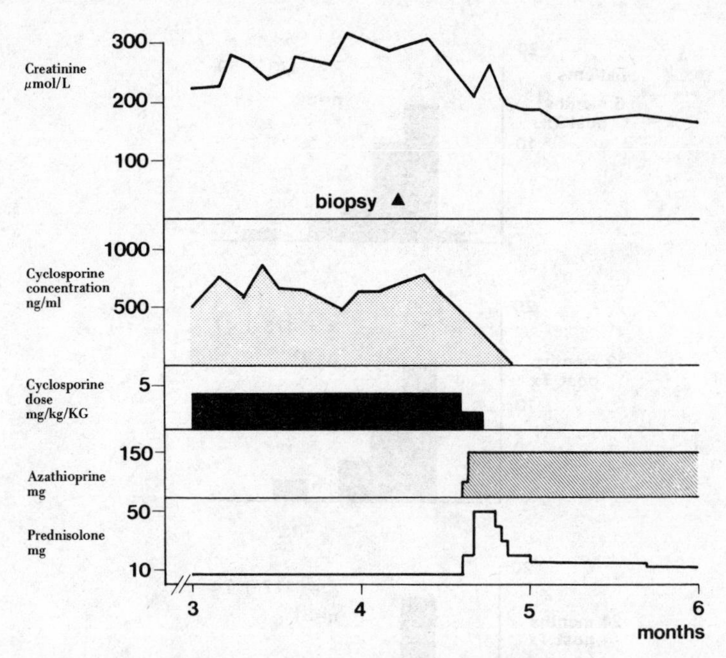

Figure 3. Effect of conversion of cyclosporine to conventional immunosuppression in a patient with chronic nephrotoxicity four months after transplantation

concentrations. Conversion, leading to a marked reduction in the serum creatinine. The patient has now maintained stable kidney function over an observation period of 18 months.

An important issue is whether the chronic progressive form of nephrotoxicity has to be ascribed only to cyclosporine or to an interaction of cyclosporine with chronic rejection and furthermore how frequent this clinical syndrome occurs. Figure 4 demonstrates the distribution of serum creatinine values in 32 patients who were followed over a period of 24 months post-transplantation. It clearly demonstrates that in the vast majority of patients a stable creatinine could be maintained over the whole observation period. In accordance with the results of others [15,18,32] the average creatinine in these patients, however, is undoubtedly higher than in historical controls under conventional immunosuppression. Evaluation of this fact has to take into consideration that the cyclosporine treated population includes patients who would have lost their graft under conventional immunosuppression. In contrast to the majority of patients who maintained a slightly impaired but stable kidney function, five patients showed progressive deterioration. We would assume that most of these patients would benefit from conversion to conventional immunosuppression similar to the patients depicted in Figure 3. This might even be true in patients suffering from chronic rejection, since the relief of the graft from the additional toxic effect of cyclosporine in all likelihood leads to functional improvement. There are virtually no data available demonstrating a superiority or inferiority of cyclosporine in the treatment of chronic rejection. The answer to this question will require the long-term follow-up of this particular patient population.

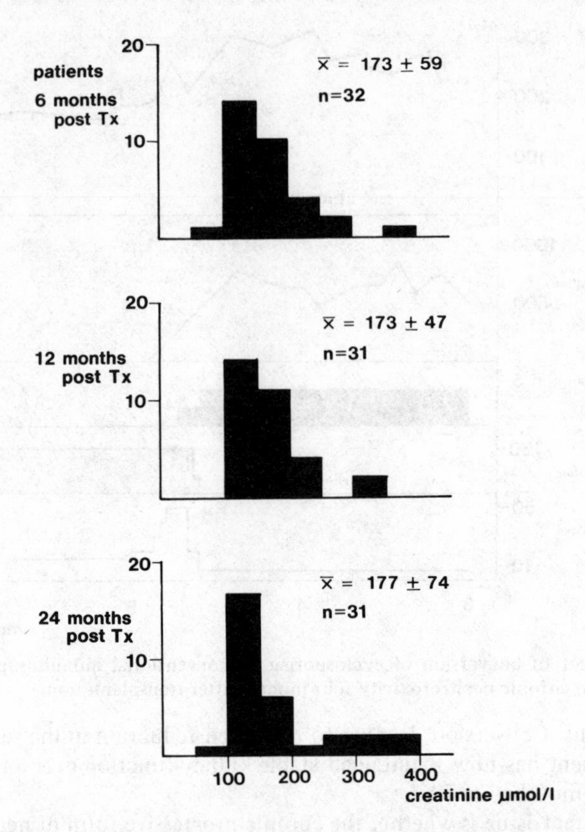

Figure 4. Distribution of creatinine during cyclosporine maintenance therapy in 32 recipients of first cadaveric kidney grafts. One graft was lost 11 months after transplantation because of chronic rejection

Conclusions

Nephrotoxicity is the most important side effect of the new immunosuppressant cyclosporine. On the basis of clinical experience and the limited experimental results available a clinical classification has been suggested. The purpose of this classification is to permit the development of diagnostic and therapeutic guidelines.

Dose-dependent acute nephrotoxicity in well functioning grafts is the first type which has been discussed. It is usually associated with a high blood concentration and responds quickly to dose reduction. In patients with an otherwise impaired kidney function it can mostly be avoided if the cyclosporine dose is adjusted by blood monitoring. The second type, interactive nephrotoxicity occurring with blood values in the therapeutic range is a more difficult problem. The most serious manifestation of this form is the early post-operative nephrotoxicity caused by the additive effects of ischaemia and cyclosporine treatment. This condition is associated with an increased rate of graft loss. Important approaches to reduce the occurrence of this complication are to reduce the risk of additional damage of the graft during organ procurement and during the peri-operative period. Cumulative nephrotoxicity resulting from the interaction of other potentially nephrotoxic drugs with cyclosporine is best treated by the avoidance of the interacting drug whenever possible. In order to limit the toxic component of cyclosporine as far as possible, frequent blood monitoring is even more important in patients with impaired kidney function than in patients with well functioning grafts.

Particularly difficult to evaluate is the chronic form of nephrotoxicity. It is a badly defined entity which probably includes a variety of different states. There is certainly some type of basic nephrotoxicity leading to elevated creatinine concentrations in most patients when compared to conventionally treated controls. In contrast to others we feel that this type of nephrotoxicity is progressive only in a minority of patients who should be converted to other forms of immunosuppression. In the majority of patients the development of a chronic cyclosporine nephropathy can be avoided by dose reductions in order to maintain blood values in the range of 150 to 600ng/ml in whole blood. Finally, it can be expected that a better understanding of the pathophysiology of cyclosporine nephrotoxicity will open the way to more refined procedures to avoid this side effect or for protecting the kidney against it. Since recently a related compound with no nephrotoxic effect in animal models has been found, there is also some hope that it finally will be possible to separate the immunosuppressive and the nephrotoxic effect by chemical modification of the cyclosporine molecule [16,33].

References

1 European Multicentre Trial Group. *Lancet 1983; ii:* 986
2 The Canadian Multicentre Transplant Study Group. *N Engl J Med 1983; 309(14):* 809
3 Wood AJ, Maurer G, Niederberger W et al. *Transplant Proc 1983; 15(1):* 193
4 Kahan BD, Ried M, Newburger J. *Transplant Proc 1983; 15(1):* 446
5 Wonigeit K, Brölsch Ch, Neuhaus P et al. *Transplant Proc 1983; 15(4):* 2586
6 Keown PA, Stiller CR, Laupacis AL et al. *Transplant Proc 1982; 14(4):* 659
7 Niederberger W, Lemaire M, Maurer G et al. *Transplant Proc 1983; 15(1):* 203
8 Calne RY, Rolles K, White JG et al. *Transplant Proc 1981; 13(1):* 349
9 Beveridge T. *Transplant Proc 1983; 15(1):* 433
10 Ryffel B, Donatsch P, Madórin M et al. *Arch Toxicol 1983; 53:* 107
11 Devineni R, McKenzie N, Duplan J et al. *Transplant Proc 1983; 15(1):* 479
12 Siegl H, Ryffel B, Petric R et al. *Transplant Proc 1983; 15(1):* 503
13 Mihatsch MJ, Thiel G, Spichtin HP et al. *Transplant Proc 1983; 15(1):* 605
14 Keown PA, Stiller CR, Sinclair NR et al. *Transplant Proc 1983; 15(1):* 222
15 Thiel G, Harder F, Lörtscher R. *Klin Wochenschr 1983; 61:* 991
16 Klintmalm G, Ringden O, Groth C. *Transplant Proc 1983; 15(1):* 599
17 Öst L. *Lancet 1984; i:* 451
18 Najarian JS, Ferguson RM, Sutherland DER et al. *Transplant Proc 1983; 15:* 438
19 Bunzendahl H, Wonigeit K, Klempnauer J et al. *Transplant Proc 1983; 15(4):* 2531
20 Sibley RK, Ferguson RM, Sutherland DER et al. *Transplant Proc 1983; 15(1):* 620
21 von Willebrand E, Häyry P. *Lancet 1983; ii:* 189
22 Margreiter R, Fuchs D, Hausen A et al. *Transplantation 1983; 36(6):* 650
23 Wagner E, Pichlmayr R, Wonigeit K et al. *Transplant Proc 1983; 15(1):* 489

24 Whiting PH, Simpson JG, Thomson AW. *Transplant Proc 1983; 15(1):* 486
25 White DJG, McNaughton D, Calne RY. *Transplant Proc 1983; 15(1):* 454
26 Castro LA, Hillebrand G, Land W. *Transplant Proc 1983; 15(1):* 483
27 Thompson JF, Chalmers DHK, Hunnisett AGW. *Transplantation 1983; 36:* 204
28 Nyberg G, Gäbel H, Althoff P. *Lancet 1984; i:* 394
29 Kennedy MS, Deeg HJ, Siegel M. *Transplantation 1983; 35:* 211
30 Ferguson RM, Sutherland DE, Simmons RL. *Lancet 1982; ii:* 882
31 Salaman JR, Griffin PJ. *Lancet 1983; ii:* 709
32 Hamilton DV, Calne RY, Evans DB. *Lancet 1981; i:* 1218
33 Hiestand PC, Gunn H, Borel JF. *Transplant Proc 1985:* in press

Open Discussion

BOEN (Chairman) Thank you Dr Pichlmayr. I think that since Professor Calne gave his lecture at the EDTA meeting in Amsterdam on the early experience with cyclosporine and transplantation, we have learned quite a lot, especially of the dosage. I have read some alarming observations about chronic tubular-interstitial damage which led to chronic renal disease in patients following heart transplantation. I wonder whether in transplanted kidney patients some of the other centres have seen this. Do you know about this complication Dr Pichlmayr?

PICHLMAYR I think the figures that have been presented from the Stanford group are, of course, alarming. The findings are similar to those described by the Basle group in the very early experience of clinical renal grafting with excessively high doses of cyclosporine. We would avoid these today and we rarely see these histological changes since we have used cyclosporine in a much lower dose. It has to be added that the patients in the Stanford group are the patients which primarily have been treated with a high dose of cyclosporine, and their serum levels are, I think, too high. Particularly in heart patients who have injured kidneys before grafting, the addition of a nephrotoxic substance in too high a dose produces a typical combination effect which may be very harmful to kidneys.

AMERICO (Bari) What is the incidence of lymphoma in your experience and are these related to nephrotoxicity?

PICHLMAYR The incidence of lymphoma in our material is zero but we have to remember that the incidence of lymphoma in heart recipients from the Stanford experience is very high. As you know the first experience by Professor Calne had a high incidence of lymphoma but this is most likely due to the effect of high dose combined with other immunosuppressive drugs. I think that those centres using blood-level guided cyclosporine dose have not seen lymphoma in kidney recipients. There is one in a liver graft patient by Dr Starzl. I think that Dr Wiskott would be able to clarify the situation if I may ask him to do so.

WISKOTT (Basle) We have now analysed almost 10,000 patients and the overall incidence is 0.3 per cent. This compares rather well with the experience in azathioprine and conventionally treated patients where this incidence has been reported by Kinlon in 1979 before the use of cyclosporine. In conventional treatment the incidence is 0.5 per cent and it is his experience in the heart transplant recipient that this figure is 3.3 per cent in about 400 transplanted patients with cyclosporine to date, compared to about six per cent before the use of cyclosporine, which has been published by Anderson in 1982.

HANSEN (Aarhus) I think one of the problems with chronic nephrotoxicity perhaps is that it is a little bit difficult to find out what dosage you have to continue with after the first half year of cyclosporine treatment. Perhaps it might be possible, at least in some patients who exhibit nephrotoxic symptoms, to reduce the dose below 4 to 8mg/kg/day to around 2 to 3mg/kg/day. In some cases when we have had patients with constantly elevated serum creatinine above 200μmol/L. We have seen a decrease in serum creatinine in some patients following a reduction in cyclosporine doses.

PICHLMAYR I think this is a problem in the individual patient. If we have normal ranges of blood or serum concentrations and we have some nephrotoxic state should we reduce even more the cyclosporine or should we switch to other therapy? I am convinced many of our patients, in spite of the fact that they have whole blood levels of about 500, these levels are too high and so we could reduce them, but of course one is afraid of rejection.

BRODERSEN (Germany) I think it was Professor Toussaint who showed us figures about acute renal failure after heart transplantation and I would like to hear of the episodes of acute renal failure under the conventional regimes after heart transplantation, or let us say deterioration of renal function with conventional immunosuppressive regimes.

PICHLMAYR At the moment we are giving loading doses of cyclosporine to most heart and kidney transplant recipients and it might be better to postpone the administration of the drug.

TOUSSAINT (Brussels) Using conventional therapy there was no case of acute renal failure after heart transplantation necessitating haemodialysis: this is very common at the present time.

MEES (Utrecht) You mentioned the differentiation between the acute rejection and cyclosporine toxicity utilising measurement of the subcapsular pressure. Do you measure that regularly and how do you do that?

PICHLMAYR With a very fine needle which is inserted subcapsularly, and we do this for about two to three weeks. There is a little fluid injected and the drift of this fluid is measured. It is a little bit more complicated than that, but that is how we do it.

DOSSETOR (Alberta) I would like to report that it is possible to have normal blood levels when giving a high dose of cyclosporine if the assay is interfered with. One instance of this is the association between cyclosporine dosage and anti-platelet drugs such as Sulphimpyrazone or Antusan, and under such circumstances, despite the maintenance of normal blood levels, we have observed a lymphoproliferative disorder. I would like to ask whether the incidence of post-cardiac transplant complications could be associated with an increased use of a drug such as Sulphimpyrazone?

PICHLMAYR I think that is a good idea, but I have no answer. I would guess that it might act in such a way.

TOUSSAINT The cardiac transplant I talked about earlier did not receive anti-aggregating drugs.

PICHLMAYR Nevertheless, we have to think very carefully about any interactions of the drug.

PAPADIMITRIOU (Thessaloniki) I think one way to approach the problem of acute nephrotoxicity is to see which part of the nephron is affected. Do you have any experimental studies? We know what aminoglycoside toxicity is doing and what happens in the ischaemic model, but is there any evidence on which part of the nephron is first affected by a high dose of cyclosporine?

PICHLMAYR Sorry to say not, but I think I have an explanation of this because all animal models for nephrotoxicity are very difficult. In the rat nephrotoxicity can be shown, but with extremely high doses we do not know if it is the same. The dog is very resistant to nephrotoxicity of cyclosporine which explains why one did not know about it before using it in the human situation. We have no good models at the moment.

PART XXVII

TRANSPLANTATION: CYCLOSPORINE 963

Chairmen: P S Guillou
C V Casciani

PART XXVIII

TRANSPLANTATION POSTERS II 992

Chairmen: C Ponticelli
S Lamperi

Proc EDTA–ERA (1984) Vol 21

CLINICAL RESULTS AND CYCLOSPORINE EFFECT ON PREDNISOLONE METABOLISM OF CADAVER KIDNEY TRANSPLANTED PATIENTS

E Langhoff, S Madsen, K Olgaard, J Ladefoged

Rigshospitalet, Copenhagen, Denmark

Summary

In the first randomised study of cyclosporine in Denmark 43 cadaveric kidney recipients were treated either with cyclosporine (Cys) and prednisone or azathioprine (Aza) and prednisone. The degradation of prednisolone was examined in 18 of these patients. The 12 month graft survival of the Cys group was not significantly greater than that of the Aza group (68% and 60%, respectively), but fewer rejection episodes occurred in the Cys group ($p<0.05$). The degradation of prednisolone was significantly decreased in the Cys group compared to the Aza group. Accordingly, plasma half-lives of prednisolone were increased in the Cys group compared with the Aza group. As a decreased degradation of prednisolone during Cys treatment may lead to increased steroid activity, we recommend that a reduction of the prednisone dosage during Cys treatment is safe.

Introduction

Treatment with Cys has been reported to increase cadaveric kidney graft survival [1]. The clinical implications of Cys treatment on cadaveric kidney graft survival are well described but the pharmacokinetic interactions of Cys due to its hepatotoxicity have not been clarified. The hepatic microsomal metabolism of prednisone is known to be induced by several drugs, e.g. rifampicin, barbiturates, and phenytoin [2]. Reports are now emerging of similar interactions between such drugs and Cys [3]. Furthermore, an influence of Cys on prednisolone metabolism may be of clinical importance as an altered clearance of prednisolone may influence the kidney graft survival [2,4]. The present study, therefore, was undertaken to examine the effect of treatment with Cys on graft survival in relation to kinetic interactions between Cys and prednisolone.

Patients and methods

Forty-three cadaveric transplant recipients were randomly allocated to receive Cys (n=24) or Aza (n=19) treatment. Two patients in the Cys group and two patients in the Aza group had been previously transplanted. Three patients were transferred from Aza to Cys treatment. Prospective HLA-DR matching was performed in all donor recipient combinations.

Cys (2mg/kg i.v.) was given immediately before transplantation and administered orally after transplantation according to the whole blood concentrations (250–800μg/L). Aza (5mg/kg orally) was given before transplantation, and after transplantation according to leucocyte and platelet counts (1–2mg Aza/kg). All patients received 150mg of prednisolone during the first post-operative days. The prednisone dose was gradually reduced to 20mg at one year. Rejection crises were treated in both groups with high doses of methylprednisolone intravenously.

In the pharmacokinetic part of the study eight patients treated with Cys/prednisone (3 females, 5 males, aged 22 to 59, mean 46 years) and eight patients treated with Aza/prednisone (2 females, 6 males, aged 26 to 62 years, mean 45 years) were examined between one and six months after transplantation. Two patients, one in each group, received prophylactic rifampicin (approximately 400mg daily) and isoniazid (approximately 250mg daily) for a three week period after transplantation because of recent pulmonary tuberculosis. These two patients were examined at two separate occasions, during and again after cessation of the antituberculous therapy.

At the time of investigation all patients received approximately 10mg of prednisone three times per day. After an overnight fast each patient had an oral standardised test dose of 10mg of prednisone at 8 a.m. Blood samples were drawn just prior to and 5, 10, 15, 25, 45, 90, 135, 200 and 300 minutes after prednisone administration. The eight hour area (equal to a dose interval) under the plasma concentration-time curve (AUC) was calculated using the trapezoidal rule.

Total body clearance (Cl_{tot}) was calculated for prednisolone on the basis of: $Cl_{tot} = \frac{Dose}{AUC} \cdot F$, where F is the systemic availability. In the present study an estimated value of F=0.85 was used [2,5]. Assays of prednisolone and hydrocortisone were carried out according to an HPLC method recently described from our laboratory [5]. The actuarial graft survival was calculated according to Peto et al [6].

Results

The clinical data are shown in Table I. The 12 month graft survival was 60 per cent in the Aza group and 68 per cent in the Cys group. This difference was not statistically significant. However, the number of biopsy verified rejection episodes was significantly higher in the azathioprine treated group ($p<0.05$). The two groups were comparable as regards DR matching, whereas the mean age of the Aza group was higher than that of the Cys group ($p<0.05$). The

TABLE I. Clinical results

		Cys	Aza	
Number of patients		24	19	
Age	Mean ± SD	42±15	51±9	p<0.05
Non-transfused		2	1	
Number of patients with <1 DR incomp		14	11	NS*
Number of patients with ≥1 DR incomp		11	7	NS*
Warm ischaemia time (min)	Mean ± SD	14±9	10±6	NS
Cold ischaemia time (hours)	Mean ± SD	22±4	21±5	NS
Creatinine clearance	Mean ± SD (n)			
3 months		39±18 (19)	54±17 (10)	p<0.05
6 months		40±18 (15)	44±17 (9)	
12 months		46±15 (5)	54±23 (4)	
S-creatinine (mmol/L)	Mean ± SD (n)			
3 months		0.22±0.18 (20)	0.12±0.03 (10)	p<0.05
6 months		0.19±0.09 (15)	0.16±0.05 (9)	NS
12 months		0.17±0.05 (5)	0.12±0.03 (4)	NS
Blood pressure (mmHg)	Mean ± SD (n)†			
3 months		139±22/94±13 (20)	154±22/97±10 (10)	NS
6 months		142±17/92±11 (15)	146±30/91±12 (9)	NS
12 months		140±18/92±12 (5)	129±12/85± 6 (4)	NS
Grafts lost (12 month observation period)		7	7	NS
Number of rejections (12 month observation period)		3	7	p<0.05
Cumulated 12 month graft survival (%)		68	60	NS
Plasma half-life of prednisolone (h)	Mean ± SD (n)	4.3±1.5 (8)	2.9 ±0.6 (8)	p<0.05
Clearance of prednisolone (ml/min/kg)	Mean ± SD (n)	1.9±0.4 (8)	2.44±0.6 (8)	p<0.05

* Fischer exact test p=0.49, otherwise unpaired t-test

† During anti-hypertensive therapy

post-transplant dialysis period of the Cys group was not different from that of the Aza treated group, 11±8 versus 7±5 days, but the creatinine clearance at three months was significantly higher ($p<0.05$) in the Aza treated group. This difference was no longer significant at six and 12 months. The two groups were similar with respect to blood pressure after transplantation.

The pharmacokinetic study showed that the mean AUC for eight hours (equal to a dose interval) was significantly larger in patients treated with Cys (191± 15μmol/L/min) than in patients treated with Aza (166±27μmol/L/min) ($p<0.05$). Consequently plasma prednisolone half-life was increased in the Cys group (4.3±1.5 hr) compared with 2.9±0.6 hr in the Aza group ($p<0.05$). An average decrease of 22 per cent ($p<0.05$) of total body clearance/kg body weight of prednisolone during Cys treatment is shown in Figure 1. The two groups of patients were comparable with respect to age, sex and dose of prednisone/kg body weight. T_{max} for prednisolone was identical in the two groups, 48±10 and 46±7 min, respectively. Two patients on prophylactic antituberculous therapy (rifampicin and isoniazid) were examined (Table II). Prednisolone clearance

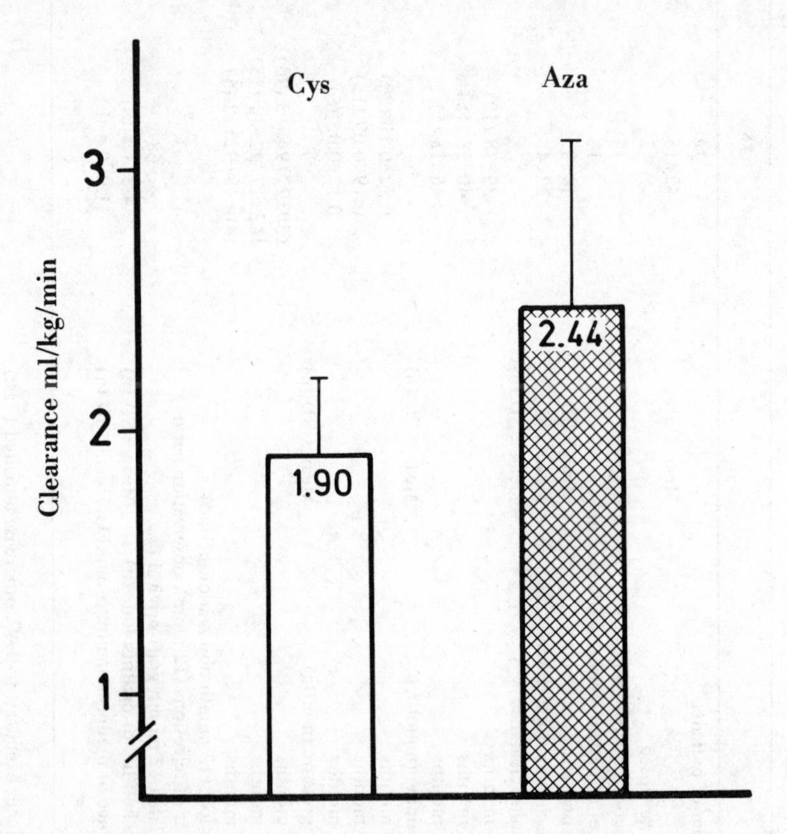

Figure 1. Total body clearance of prednisolone after 10mg of prednisone orally in 16 kidney transplant patients treated with Cys or Aza ($p<0.05$)

TABLE II. Induction of prednisolone metabolism

Therapy	With antituberculous drugs		Without antituberculous drugs	
	Prednisolone		Prednisolone	
	Cl*	Half-life (hr)	Cl*	Half-life (hr)
Cys	9.1	1.2	1.8	4.2
Aza	3.2	1.7	1.9	3.2

* Clearance of prednisolone (ml/min/kg)

increased two to five-fold in both patients during treatment, whether receiving cyclosporine or azathioprine. The half-life of plasma prednisolone decreased accordingly, while t_{max} of prednisolone did not change.

Discussion

The present study did not reveal a significantly improved graft survival, but fewer biopsy-verified rejection episodes occurred in the Cys treated group. The slightly lower creatinine clearance of the Cys treated group is in accordance with other reports [1]. Blood pressures were comparable in the two groups, but patients treated with Cys generally received more anti-hypertensive drugs to achieve optimal blood pressure control than Aza treated patients.

In the present study the clearance of prednisolone was 1.90±0.35ml/min/kg in the Cys group and 2.44±0.65ml/min/kg in the Aza group. In contrast to the interactions between prednisolone and rifampicin, barbiturates and other drugs, Cys inhibited the metabolism of prednisolone as demonstrated by a 22 per cent reduction of total body clearance of prednisolone. The difference between the creatinine clearance of the two groups of patients does not explain the lower clearance of prednisolone of the Cys group in comparison to the Aza group since the elimination of prednisolone is mainly hepatic [2]. Rejection episodes occurred in two patients (Table II) when the clearance of prednisolone increased due to the antituberculous therapy. Interestingly, this change also occurred in the Cys treated patient, despite the hepatotoxic effect of Cys. Thus, the clinical importance of increased clearance of prednisolone was demonstrated during prophylactic antituberculous therapy. In contrast, the clinical importance of decreased prednisolone clearance has not been clarified. The beneficial effect of Cys treatment on cadaveric kidney graft survival may in some part be due to an increased steroid activity. It is, therefore, suggested that a reduction of the prednisone dosage during Cys treatment is safe and could be explained by simultaneously decreased prednisolone degradation.

Acknowledgments

The authors thank Mrs K Meibom for excellent technical assistance. This investigation was supported by grants from The Danish Medical Research Council,

The Hafnia-Haand-i-Haand Foundation, The P Carl Petersen Foundation, and The King Christian X Foundation.

References

1 Canadian Multicentre Trial. *N Engl J Med 1983; 309:* 809
2 Gambertoglio JG, Amend WJC, Benet LZ. *J Pharmacokin Biopharm 1980; 8:* 1
3 Langhoff E, Madsen S. *Lancet 1983; ii:* 1303
4 Wassner SJ, Malekzadeh MH, Pennisi AJ et al. *Clin Nephrol 1977; 8:* 293
5 Langhoff E, Flachs H, Ladefoged J et al. *Eur J Clin Pharmacol 1984; 26:* 651
6 Peto R, Pike MC, Armitage P et al. *Br J Cancer 1977; 35:* 1

Open Discussion

BRYNGER (Chairman) Thank you very much for this detailed study.

PICHLMAYR (Hannover) I think this is very interesting, but we must be very careful. The European multicentre study used cyclosporine as a single agent without prednisolone except for rejection crises. The result of this study gave results that were better when compared to conventional therapy. Therefore the better outcome of the cyclosporine group cannot be due to the higher plasma prednisolone.

LANGHOFF I fully agree with you. There is one point concerning the European multicentre study which is that about 30 per cent of patients were switched from one treatment to another and this may well have had an influence on the conclusions. Secondly, I am well aware that cyclosporine is a potent immunosuppressive drug but my statement is that drug interactions may well have an influence on the results reached.

BRYNGER You might be aware of a study performed in Stockholm examining the plasma cyclosporine concentrations in patients with and without prednisolone. The addition of prednisolone significantly increases the plasma cyclosporine but the reason for this is not known. This illustrates how difficult it is to sort out the many background factors in drug metabolism.

Proc EDTA–ERA (1984) Vol 21

DOES CYCLOSPORINE INHIBIT RENAL PROSTAGLANDIN SYNTHESIS?

D Adu, *C J Lote, J Michael, J H Turney, P McMaster

*Queen Elizabeth Hospital, *Birmingham University, Birmingham, United Kingdom*

Summary

Urinary PGE_2 excretion was measured in female renal allograft recipients being treated either with prednisolone and cyclosporine (Cys) or with prednisolone and azathioprine. The mean urinary PGE_2 excretion in patients on Cys of 940pmol/4hr was lower than in patients on prednisolone and azathioprine (2033pmol/4hr) although the difference did not achieve statistical significance. These data are consistent with the hypothesis that Cys inhibits renal prostaglandin synthesis.

Introduction

Cyclosporine (Cys) is a potent nephrotoxin [1]. It is now clear that the renal impairment caused by Cys is not necessarily associated with renal histological changes and that in its early stages it is rapidly reversible when the drug is discontinued [1]. These observations suggest a physiological basis for Cys nephrotoxicity. We have reported a high incidence of sustained hyperkalaemia in renal allograft recipients treated with Cys and shown that this was associated with hyporeninaemia and inappropriately low serum aldosterone levels [2]. Since prostaglandins are a potent stimulus of renin secretion [3] we wondered whether Cys might inhibit their synthesis.

Patients and methods

The study group consisted of female renal allowgraft recipients receiving either Cys and prednisolone or with prednisolone and azathioprine. All the patients had functioning renal allografts and were studied at a time when their renal function was stable. Four-hourly collections of urine were acidified and then stored at −20°C. The urinary concentration of prostaglandin E_2 (PGE_2) was determined by radioimmunoassay as previously described [4]. The results were expressed as pmol PGE_2 per four hours. The significance of differences between groups was assessed by Student's 't' test.

Results

The mean urinary PGE_2 excretion in patients on Cys was lower than in patients on prednisolone and azathioprine although the difference was not statistically significant (Table I). Five out of 14 patients on Cys had a urinary PGE_2 excretion of less than 500pmol/4hr in contrast to only one of the 11 patients on prednisolone (Table II).

TABLE I. Urinary PGE_2 excretion in renal allograft recipients

Treatment	Number of Patients	Urinary PGE_2 (pmol/4hr) Mean ± SEM	
Prednisolone and azathioprine	11	2034 ± 590	
			$p>0.05$ T = 1.90 NS
Prednisolone and cyclosporine	14	940 ± 218	

TABLE II. Distribution of urinary PGE_2 in renal allograft recipients

Urinary PGE_2 (pmol/4hr)	Prednisolone + azathioprine	Prednisolone + cyclosporine
<500	1	5
500 – 1000	5	5
>1000	5	4

Discussion

In this study although urinary PGE_2 excretion in patients treated with Cys was lower than in those receiving prednisolone and azathioprine the difference was not statistically significant. We therefore cannot conclude with certainty that Cys inhibits renal prostaglandin synthesis although the data strongly suggests that this is so. It is however likely that patients with renal allografts have a glomerular filtration rate (GFR) that is dependent on an intact prostaglandin system. Such a situation obtains during sodium depletion in healthy individuals, in patients with chronic glomerulonephritis and in conditions such as ascites and the nephrotic syndrome where there is an ineffective circulatory volume [5]. In these states, inhibition of renal prostaglandin synthesis leads to a reversible reduction in GFR [5]. The nephrotoxicity of Cys and the rapid improvement in renal function when the drug is discontinued suggests a functional rather than a structural lesion such as would occur with inhibition of renal prostaglandin synthesis.

The prostaglandins PGE_2 and PGI_2 are potent stimuli of renin release [3] and inhibition of renal prostaglandin synthesis by indomethacin may lead to hyporeninaemia and hypoaldosteronism with consequent hyperkalaemia [6]. We reported that the hyperkalaemia seen in patients on Cys was associated with hyporeninaemia and hypoaldosteronism [2] and suggest that the hyporeninaemia is due to inhibiton of renal prostaglandin synthesis. Remuzzi et al [7] have reported that some patients with a haemolytic uraemic syndrome (HUS) lacked PGI_2-stimulating factor (PSF) activity in their plasma. Recently, Neild et al [8] found that rabbits treated with Cys had a profound loss of PSF activity in their plasma. Rabbits given acute serum sickness and treated with Cys developed glomerular capillary thrombosis and cortical infarction, lesions were strikingly similar to the changes seen in the HUS [9]. Bone marrow recipients treated with Cys have been reported to develop an illness similar to HUS [10] and it is possible that this is due to inhibition of PSF activity by Cys.

The data of Neild et al [9] and the results of our study suggest that Cys may inhibit prostaglandin synthesis. Inhibition of PGE_2 synthesis would explain the reversible nephrotoxicity and hyperkalaemia seen with Cys whilst inhibition of PGI_2 production would explain the HUS seen in some patients on Cys and might possibly be the cause of some of the chronic vascular changes hitherto attributed to chronic vascular rejection.

References

1 Hamilton DV, Evans DB, Henderson RG et al. *Proc EDTA 1981; 18:* 400
2 Adu D, Turney J, Michael J, McMaster P. *Lancet 1983; ii:* 370
3 Larsson C, Weber P, Anggard E. *Eur J Pharmacol 1974; 28:* 391
4 Lote CJ, McVicar AJ, Thewles A. *J Physiol 1983; 336:* 39
5 Dunn MJ, Zambraski EJ. *Kidney Int 1980; 18:* 609
6 Tan SY, Shapiro R, Franco R et al. *Ann Intern Med 1979; 90:* 783
7 Remuzzi G, Misiani R, Marchesi D et al. *Lancet 1978; ii:* 871
8 Neild GH, Rocchi G, Imberti F et al. *Transplant Proc 1983; 15 (Suppl 1):* 2398
9 Neild GH, Ivory K, Williams DG. *Br J Exp Pathol 1984; 65:* 133
10 Shulman H, Striker G, Deeg HJ et al. *N Engl J Med 1981; 305:* 1392

Open Discussion

VANGELISTA (Bologna) There was work last year from the Minneapolis group which demonstrated that in patients treated with cyclosporine there was a stimulation of the renin angiotensin aldosterone system and it was thought that this was due to the initial stimulation of the autonomic nervous system. How do you explain these different results?

ADU Well, I think it is difficult. I know of the work done in rats showing this but rats are different from humans.

BANKS (Bristol) Was there a relationship between the rise in potassium and the decrease in prostaglandin excretion?

ADU No, we have not done that study, we looked at prostaglandins and creatinine.

UNKNOWN In your Table I you mention a lot of possible renal effects of cyclosporine. You have not mentioned the possible vascular changes which may occur after two or three months of cyclosporine administration without any clinical signs. These are morphological changes not predictable on clinical grounds. I would like to ask you if you have any experience of this type of effect which we have seen in patients on cyclosporine over the last two years?

ADI We suspect that we have seen these types of changes that have been described but when we have looked blind at our biopsies, patients on prednisone and azathioprine and patients on cyclosporine, we have not been able to distinguish histologically between the two groups of patients.

BRYNGER (Chairman) Just one comment regarding the haemolytic uraemic syndrome. We have seen in our patients, and it has been reported in four patients in Canada, an autoimmune haemolytic syndrome. In our cases we have been able or have had to switch to alternative therapy. In the first case it was very severe although it did not affect the kidney function. The second case, it was not that serious, and we changed to conventional therapy and after about six weeks we returned to cyclosporine without any problems. There are a lot of problems with this drug and I will have to work very hard to find out what is going on.

BONOMINI (Chairman) Just a practical question. If you had a son who required a renal transplant would you start with cyclosporine or azathioprine?

ADU I think at the moment if it was a second graft I would always give cyclosporine, and also probably for a first graft.

Proc EDTA–ERA (1984) Vol 21

CYCLOSPORINE NEPHROPATHY AFTER HEART AND HEART-LUNG TRANSPLANTATION

J Goldstein, Y Thoua, F Wellens, J L Leclerc, J L Vanherweghem, P Vereerstraeten, G Primo, **C Toussaint**

Cliniques Universitaires de Bruxelles, Hôpital Erasme, Brussels, Belgium

Summary

Cyclosporine nephrotoxocity after heart transplantation can lead to acute renal failure requiring haemodialysis. In four long-term heart-transplant survivors, cyclosporine nephropathy was characterised by extensive fibrosis, with uraemia, hypertension and/or anaemia. In contrast, the long-term survivor of heart-lung transplantation who had received her graft for accelerated respiratory failure, did not develop chronic renal disease. Thus, chronically reduced renal perfusion before heart transplantation may play a critical role in the development of chronic cylosporine nephropathy.

Introduction

Cyclosporine is undoubtedly a major immunosuppressive drug for organ transplantation. Its main advantage is its corticosteroid-sparing action while nephrotoxicity constitutes an important undesirable effect [1]. As the kidneys of heart-transplant recipients treated with cyclosporine are not susceptible to develop the changes induced by graft rejection, they may represent a better model for the assessment of cyclosporine nephrotoxicity than renal transplants where it is often difficult to distinguish kidney failure due to rejection from that produced by cyclosporine itself [2]. In the present paper, cyclosporine nephrotoxicity was examined in 11 patients who received the drug for heart or heart-lung transplantation.

Patients and methods

Table I yields the main data concerning the 10 patients who, between March 1982 and December 1983, underwent orthotopic heart transplantation in our institution for New York Heart Association (NYHA) class IV cardiac failure, and one patient who received in August 1983 heart-lung transplantation for rapidly progressive pulmonary insufficiency.

TABLE I. Main clinical details of recipients

1. Heart transplants (n=10)
   Age: median 45 years (range 27–55 years)
   Sex: 9 males/1 female
   All in NYHA class IV heart failure
   Primary diagnosis: idiopathic cardiomyopathy (n=4)
   ischaemic heart disease (n=6)
   Mean duration of ischaemia of the graft 82 ± 5 min*
   Mean duration of extracorporeal circulation 114 ± 7 min*
   Post-operative haemodialyses: No. of patients: 6
   Mean duration of dialysis period: 13 ± 3 days*
   Present survival: 4 patients in NYHA class I

2. Heart-lung transplant (n=1)
   Age: 27 years
   Sex: female
   Primary diagnosis: pulmonary lymphangioleiomyomatosis
   Duration of ischaemia of the graft: 84 min
   Duration of extracorporeal circulation: 170 min

NYHA: New York Heart Association
* Mean ± SEM

No effort was made to obtain good HLA matches between donors and recipients but ABO blood groups were compatible and lymphocytotoxic cross-matches were negative.

For heart transplants, immunosuppression consisted of cyclosporine (Sandimmun®, kindly supplied by Sandoz, Basel, Switzerland) and corticosteroids (Salumedrol® and Deltacortril®) while azathioprine (Imuran®) and cyclosporine were used during the first 14 post-operative days in the patient with heart-lung transplantation, azathioprine being later on replaced by prednisolone, as advocated by the Stanford group [3].

The first dose of cyclosporine (11.7±1.3mg/kg) was given orally four hours before surgery. Post-operatively, the drug was given orally at a daily dosage of 8–20mg/kg during the first month. From months 2 to 14, mean daily cyclosporine doses ranged from 6.5 to 8.0mg/kg and, from months 15 to 21, they were further tapered from 6.5 to 3.9mg/kg. Trough serum cyclosporine level was monitored frequently by a radioimmunoassay using the Sandoz kit, and graft rejection diagnosed by endomyocardial biopsies. Corticosteroids were given intravenously (30±3mg/kg/day) per-operatively and were subsequently tapered to 0.2mg/kg/day at three months. Azathioprine was given orally to the heart-lung recipient at a dose of 2mg/kg/day, from days 0 to 14.

Results

Table I gives for the 11 patients the durations of ischaemia of the graft and of extracorporeal circulation, the number of patients requiring post-operative haemodialyses, and the present survival.

In Table II are presented, for each patient, from day 0 to day 40, the evolution of serum bilirubin and creatinine, and of systolic and diastolic blood pressures, as well as the current status and cause and time of death.

Early post-operative course

In all patients, serum creatinine and urea concentrations increased during the first post-operative month. Six patients required haemodialyses and four of them died while two fatalities occurred among the five patients who did not require dialysis. Three of the six fatalities were due to cerebral haemorrhage and the three other patients died in severe septic shock (Table II). In the six patients who required haemodialysis in the early post-operative course, serum bilirubin rose to a mean upper value of 10.9±3.2mg/100ml while this value reached 4.3±1.5mg/100ml in the five patients who had not been dialysed.

In all six deceased patients, postmortem examination of the kidneys disclosed variable degrees of tubular necrosis with tubulorhexis and interstitial oedema (Figure 1) without cellular infiltration.

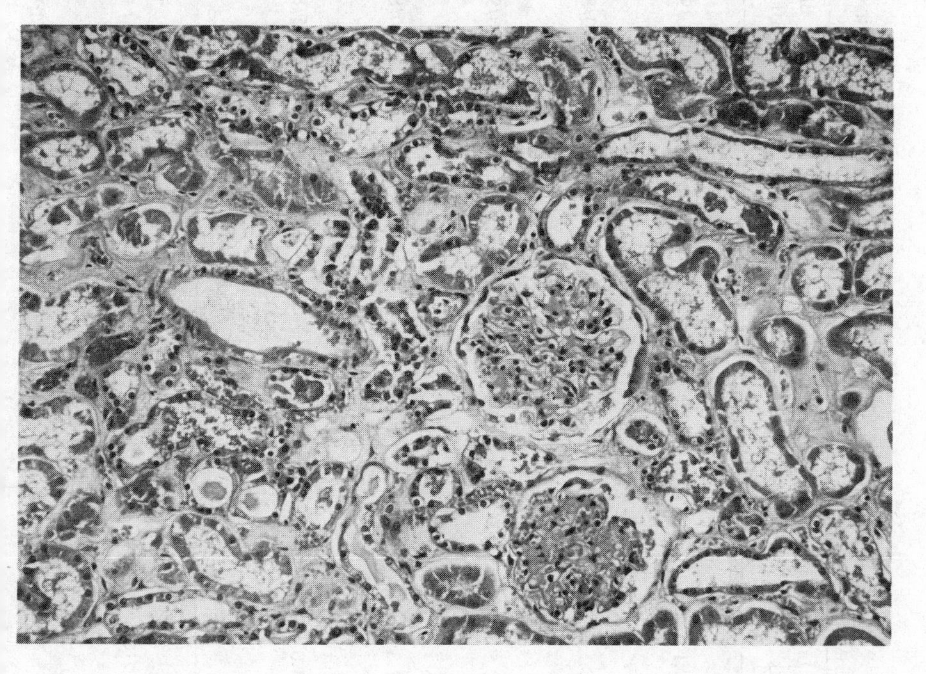

Figure 1

Late post-operative course

On June 30, 1984, five patients are surviving at 10, 12, 22, 24 and 25 months, including the heart-lung recipient (10 months). In none of them have myocardial

TABLE II. Early post-operative course, current status and main cause and time of death

Patient	Serum bilirubin, mg/100ml					Serum creatinine, mg/100ml					Blood pressure, mmHg					Current status
POD	0	10	20	30	40	0	10	20	30	40	0	10	20	30	40	
Alive																
PIE (H)	2.9	10.1	8.6	6.3	3.0	1.8	2.3	2.3	2.1	1.5	90/70	130/90	130/80	130/95	140/110	Alive 25 mos, Scr=2.5, BP 140/100
DEN (H)	1.3	5.1	3.0	2.2	1.5	1.2	HD	HD	2.4	1.5	110/70	150/80	160/90	150/90	150/90	Alive 24 mos, Scr=2.6, BP 145/95
DEM (H)	1.1	9.5	14.1	5.0	3.1	1.1	HD	HD	3.5	3.2	110/70	150/90	160/95	150/100	150/100	Alive 22 mos, Scr=2.7, BP* 125/80
ARY (H)	0.8	1.3	1.1	1.0	–	1.3	1.5	1.1	1.0	–	110/60	140/80	150/100	140/80	–	Alive 12 mos, Scr=2.1, BP 160/90
TER (HL)	0.7	1.0	0.9	1.0	0.5	0.6	1.1	0.6	0.7	0.6	130/80	105/80	110/70	110/80	125/90	Alive 10 mos, Scr=0.9, BP* 110/70

continued

TABLE II (continued)

Patient	Serum bilirubin, mg/100ml					Serum creatinine, mg/100ml					Blood pressure, mmHg					Main cause and
POD	0	10	20	30	40	0	10	20	30	40	0	10	20	30	40	time of death
Deceased																
JAC (H)	1.7	2.9	+	–	–	1.0	1.3	+	–	–	110/90	210/140	+	–	–	Cerebral haemorrhage 14 POD
HAQ (H)	1.6	20.8	+	–	–	1.0	HD	+	–	–	100/70	110/80	+	–	–	Cerebral haemorrhage 17 POD
GAI (H)	1.0	+	–	–	–	1.1	HD+	–	–	–	110/70	230/140+	–	–	–	Cerebral haemorrhage 6 POD
SYM (H)	1.6	9.1	13.4	10.4	19.6	1.3	HD	HD	1.6	0.4	85/60	125/90	130/80	130/70	140/90	Sepsis 40 POD
HAM (H)	–	6.7	+	–	–	1.7	HD	+	–	–	115/85	120/60	+	–	–	Sepsis 18 POD
JOO (H)	1.9	1.6	–	+	–	1.4	1.3	–	+	–	130/80	120/80	120/70	+	–	Sepsis 23 POD

POD=post-operative day; (H)=heart transplant; (HL)=heart-lung transplant; BP=blood pressure under anti-hypertensive therapy, in mmHg; BP*=blood pressure without anti-hypertensive therapy, in mmHg; +=death occurred during the 10-day period preceding this day; HD=on haemodialysis; Scr=serum creatinine, in mg/100ml biopsies demonstrated any sign of rejection so that corticosteroid therapy was constantly maintained at a low level. The heart-lung recipient did not show any increase in serum creatinine and urea levels or in blood pressure. The other four patients showed increased serum creatinine and urea concentrations from 10 to 25 months later, without proteinuria or abnormal urine sediment, but three of them are hypertensive and one is anaemic (Table II). In three of these patients currently surviving over 22 months percutaneous renal biopsies (Figure 2) uniformly showed extensive interstitial fibrosis with glomerular, tubular and arterial changes indistinguishable from those of nephroangioslcerosis.

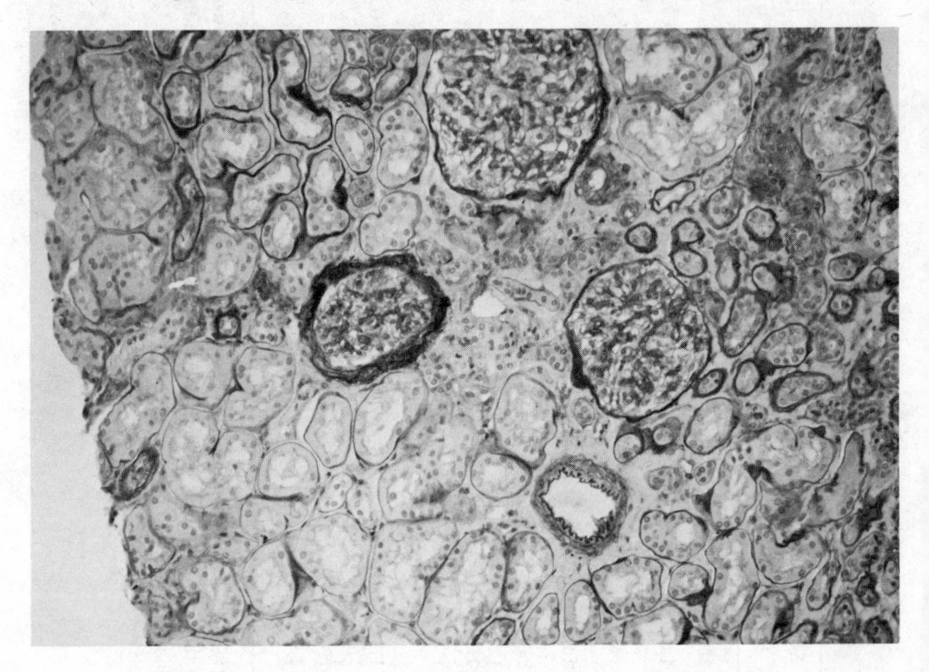

Figure 2

In the three heart recipients surviving over 22 months, mean serum creatinine concentration decreased from 3.5 at month 14 to 2.5mg/100ml at month 21 as the mean daily cyclosporine dosage was tapered from 6.5 to 3.9mg/kg but urea concentration was not affected and no further decrease in creatinine was observed thereafter. The heart-transplant recipient (ARY) presently surviving at 12 months received smaller doses of cyclosporine (from 10 to 3.1mg/kg/day). Nevertheless, he developed chronic renal failure and hypertension (Table II).

Discussion

Our observations confirm that cyclosporine nephropathy may present a clinical and morphological picture of *acute renal failure*, often associated with cholestatic jaundice and/or malignant hypertension with early lethal cerebral haemorrhage. This latter event is similar to the picture produced in acute cyclosporine toxicity in animals [4].

Jaundice may aggravate cyclosporine nephrotoxicity through interfering with biliary excretion of the drug [5]. Poor pre-operative hepatic and renal perfusion in our 10 heart transplant recipients, who were all in class IV heart failure at the time of surgery, could explain the high incidence of acute cyclosporine hepato- and nephrotoxicity in this series. The heart-lung recipient, who was not in heart failure before surgery, developed only transient increments in serum creatinine (3.5mg/100ml on day 2) and in serum bilirubin (2.6mg/100ml on day 1).

Of great concern is the development of *renal fibrosis* in the long-term survivors of heart transplantation. Excessive doses of cyclosporine may play an important role in the occurrence of this complication but, one patient presently surviving at 10 months with a daily cyclosporine dosage maintained between 10.0 and 3.1mg/kg is currently uraemic. Hypertension was observed in three of our four long-term heart transplant survivors, as reported in a larger series [6].

Chronic renal failure was also recently reported by Moran et al [7] in 13 cyclosporine-treated heart transplant recipients over one year while this complication was absent in eight patients who had received azathioprine instead of cyclosporine. On the other hand, Tauxe et al [8] noted that effective renal plasma flow was very low before surgery, and rose only to half-normal values five to 10 days after cardiac transplantation. They concluded that not all post-transplant renal dysfunction in cyclosporine-treated heart recipients should be attributed to the drug.

In our own series, the only patient without chronic renal failure was the heart-lung recipient who received her graft for respiratory and not for circulatory failure. The doses of cyclosporine used in this patient were of the same magnitude as those used here in the more recent heart recipients. More data should be gathered in heart-lung transplant recipients without circulatory failure before drawing definite conclusions concerning the precise role of poor renal perfusion in the setting of chronic cylosporine nephropathy as observed in heart transplant recipients. It is obvious that similar ischaemic factors may also play an important role in cyclosporine nephrotoxicity after kidney transplantation.

Acknowledgments

This work was supported by the Fonds de la Recherche Scientifique Médicale and by the Fondation Universitaire David et Alice van Buuren, Brussels. We are grateful to the numerous physicians, nurses and technicians of Hôpital Erasme who helped us so efficiently in the patients' care. We thank Mrs Chantal Tholet for typing the manuscript.

References

1 Bennett WM, Pulliam JP. *Ann Int Med 1983; 99:* 841
2 Flechner SM. *Urol Clin N Am 1983; 10:* 263
3 Reitz BA, Gaudiani VA, Hunt SA et al. *J Thorac Cardiovasc Surg 1983; 85:* 354
4 Siegl H, Ryffel B, Petric R et al. *Transplant Proc 1983; 15:* 2719
5 Schade RR, Guglielmi A, Van Thiel DH et al. *Transplant Proc 1983; 15:* 2757
6 Thompson ME, Shapiro AP, Johnsen AM et al. *Transplant Proc 1983; 15:* 2573
7 Moran M, Newton L, Perlroth M et al. *Kidney Int 1984; 34:* 346
8 Tauxe WN, McGiffin D, Karp RB. *Kidney Int 1984; 24:* 178

Open Discussion

BONOMINI (Chairman) Thank you very much Dr Toussaint for this very important paper.

ROTTEMBOURG (Paris) I am speaking as a participant of a team of Cabrol in La Pitié, and not as a nephrologist. For the last three years in Paris the Cabrol group have transplanted about 46 patients with heart transplants using cyclosporine. We have never seen acute renal failure requiring haemodialysis but we have seen some renal failure in our patients. At what time to you begin cyclosporine therapy after heart transplantation? Secondly, have you seen any difference in your young group of patients treated with cyclosporine compared with your older patients, because I see that your patients are rather older than ours in Paris?

TOUSSAINT Cyclosporine is started orally a few hours before surgery.

For the second question, of course this series is very small and the age range of the patients is quite wide. It may be that because these patients were waiting for a transplant for many weeks, many of them were in a very bad condition. Nevertheless, the Stanford group reported a few weeks ago in the New England Journal of Medicine exactly the same thing, and they have presented this in abstract form previously. We did not start cyclosporine immediately before, but on day one or day two when the patient had a good urinary flow. We think that this plays a major role in preventing renal failure after heart transplantation. The second thing is that we think that many of the patients, particularly patients with cardiomyopathy, could have renal damage before transplantation and this damage could be enhanced by the cyclosporine therapy.

PICHLMAYR (Hannover) I think we are very much in agreement. I do not want to take too much from the discussion of this afternoon, but I have been very interested in your heart-lung recipient patient that you did not observe these complications and this suggests that the combination of say pre-operative injury to the kidney and the nephrotoxicity may be a problem for you. You also argued that your doses of cyclosporine are high. The levels you observe are indeed high. Would you reduce the cyclosporine dose in the future considering the fact that you have, in a group of patients, also confirming a high dosage of cyclosporine? Do you agree?

TOUSSAINT Completely. Another point, if I may add, is that this heart transplant recipient is presently going back into pulmonary failure with chronic obstructive bronchiolitis on lung biopsy, as has been described by the Stanford group in Minneapolis a few months ago. This also may be an effect of cyclosporine in one-third of the heart-lung patient. It may be post-ischaemia because of course there is no suture of the bronchial arteries.

RATTAZZI (Seattle) Is there any correlation between cyclosporine blood levels and renal toxicity? I think the Stanford group has demonstrated that not to be the case, so what are you using to guide your dosage of cyclosporine therapy in heart transplantation?

TOUSSAINT I completely agree that there is absolutely no correlation with the blood levels and nephrotoxicity so there is so far in this transplant patient no guide, except that the patient was still alive, except those who died of course.

PINCHLMAYR I have one comment to this question. I think one cannot exclude a correlation between nephrotoxicity and blood or serum levels if one is over the toxic range in all patients. Some patients will tolerate this without nephrotoxicity but I think we have to go much further below the toxic levels to say whether there is a correlation or not. I would like to deal with this this afternoon.

TOUSSAINT Well, we have no autopsy material from many of our patients because all deaths occurred before day 40. There were no arterial lesions in any organs in these patients.

Proc EDTA–ERA (1984) Vol 21

COMPARISON OF THREE IMMUNOSUPPRESSIVE REGIMENS IN KIDNEY TRANSPLANTATION: A SINGLE-CENTRE RANDOMISED STUDY

M Hourmant, J P Soulillou, J Guenel

University Hospital, Nantes, France

Summary

Three immunosuppressive regimens have been compared: conventional treatment including anti-thymocyte globulin (ATG) (32 patients) with cyclosporine (Cys) alone (21 patients), sequential combination of ATG with Cys (35 patients). Actuarial graft survivals were: ATG 73 per cent, Cys alone 88 per cent at nine months and ATG/Cys 92 per cent at one to two years. Transplant function was significantly worse with Cys as initial treatment compared with that in controls, while it was similar with ATG/Cys. The Cys dose used was low and no severe infection nor immunoglobulin abnormalities were noticed. Corticosteroids were withdrawn with both Cys protocols, except for rejection treatment.

Introduction

Cyclosporine (Cys) has dramatically improved graft survival rates in kidney transplantation [1,2]. However, in most of the trials, the control treatment was prednisone-azathioprine combination and when Cys was compared with anti-thymocyte globulin (ATG) as reported by Najarian [3] and others [4], the results were found to be similar. In addition, Cys nephrotoxicity appeared to be a major complication, leading to significantly worse renal function in patients treated by Cys than in controls [1,2,4]. In the following single-centre randomised study, three immunosuppressive regimens were compared: Cys as the initial treatment for transplantation, conventional therapy, including ATG, the sequential combination of ATG and late low dose Cys monotherapy, the aims being to retain the benefits of ATG in the early weeks, decrease the incidence of Cys nephrotoxicity and to reduce corticosteroid dose.

Patients and methods

A total of 92 kidney recipients were included in the study. Initial randomisation took place on the day of transplantation and patients received (1) either standard treatment: ATG (for 3 weeks, and then for treating rejection), prednisone

(1mg/kg/day), azathioprine (2–3mg/kg/day), or (2) Cys (15mg/kg/day) with prednisone (1mg/kg/day): Cys group II. After a second randomisation, during the third month, the former either stayed under standard treatment: STD group, or were converted to low dose Cys monotherapy (6mg/kg/day): Cys group I. Cys therapy was adjusted according to Cys trough blood concentrations. Rejection was treated with either ATG if patients received standard treatment, pulse steroids if they were under Cys. The STD group consisted of 32 first transplants, Cys group II of 20 first and one second, Cys group I of 32 first and three second grafts. The three groups were similar in their pre-transplant characteristics. Follow-up was seven (3 to 10) months in Cys group II, and 14 (4 to 24) for the other two groups.

Results

Only one patient died during the common step to groups STD and Cys I. The actuarial graft survival (Figure 1) was 95 per cent at three months in the three groups; 88 per cent in Cys group II, 84 per cent in STD group, and 95 per cent in Cys group I at nine months; 73 per cent in STD group and 92 per cent in Cys group I at one to two years ($p<0.5$).

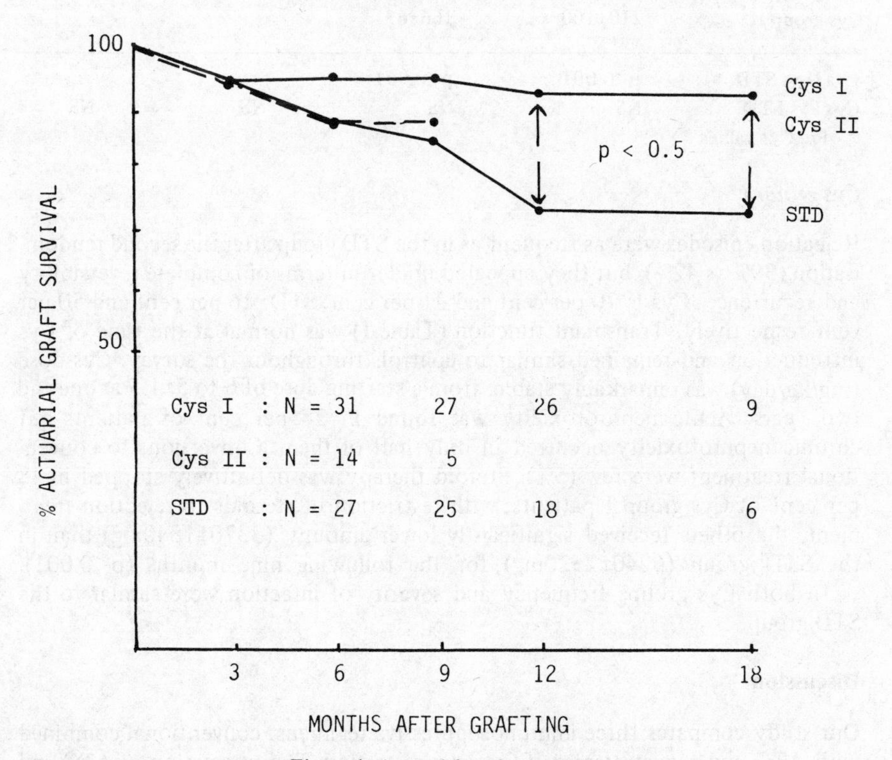

Figure 1. Actuarial graft survival

Cys group II

Cys therapy did not prevent rejection and 57 per cent of these patients exhibited at least one rejection episode compared with 59 per cent in the STD group, within the first three months. Renal function was significantly worse than with standard therapy (Table I). Although Cys dose was quickly tapered from 15 to 10±3mg/kg/day at one month, 7.5±3mg/kg/day at three months and 6±2mg/kg/day at six months, Cys nephrotoxicity occurred in 66 per cent of these patients, as either a dose-related serum creatinine increase, or a long-lasting nephropathy (Cys II: 16±14 days vs STD and Cys I: 6±5 days, $p<0.001$). Conversions to STD treatment were performed in six patients (28%), including two cases of uncontrollable rejection and three of severe Cys nephrotoxicity. Steroid therapy could be stopped in 38 per cent of the 21 patients.

TABLE I. Serum creatinine (μmol – $\bar{X}$ ± SD) in groups STD, Cys I, Cys II

Months after grafting	3	6	12	18
STD group	120±64	154±145	138±67	133±50
Cys group I	113±45*	140±57	150±47	123±9
Cys group II	210±100	166±63		
Cys II vs STD	p<0.001	p<0.001		
Cys I vs STD	NS	NS	NS	NS

* Pre-Cys values

Cys group I

Rejection episodes were as frequent as in the STD group after the second randomisation (39% vs 42%), but they appeared milder in terms of complete reversibility and recurrence; Cys I: 70 per cent and 21 per cent, STD: 16 per cent and 50 per cent respectively. Transplant function (Table I) was normal at the time of Cys introduction and remained similar to controls throughout the survey. Cys dose (mg/kg/day) was remarkably stable, from a starting dose of 6 to 5±1.5 at one and two years. Acute nephrotoxicity was found in 28 per cent of patients but chronic nephrotoxicity occurred in only half of them. Conversions to conventional treatment were few (5%). Steroid therapy was definitively stopped in 48 per cent of Cys group I patients; with restriction of steroids to rejection treatment, the others received significantly lower amounts (3370±1840mg) than in the STD group (6240±2520mg) for the following nine months ($p<0.001$).

In both Cys groups frequency and severity of infection were similar to the STD group.

Discussion

Our study compares three immunosuppressive regimens: conventional combined with ATG and two strategies of Cys administration. The association of ATG and

Cys (Cys group I) gave better actuarial graft survival than standard treatment with ATG (STD group) and even than Cys given as initial treatment (Cys group II). This was probably more related to the mildness of rejection episodes than their decrease of frequency. In addition, transplant function was normal at the time of Cys introduction; Cys doses used were low and few patients developed Cys nephrotoxicity. Consequently long-term serum creatinine were found to be similar to controls. In contrast, as reported in other trials in kidney transplantation [1,2,4], where Cys is given on the day of grafting, renal function was significantly lower in Cys group II.

To date in our study, although two powerful immunosuppressants are administered, their assocation, because sequential, has proved to be safe with regard to infection and immunoglobulin abnormalities [5].

Conclusions

Although a longer and larger survey is needed to evaluate Cys administration as the initial treatment, we favour the sequential association of ATG and Cys, which gave the best graft survival (92% at 1–2 years, including retransplantation), long-term good transplant function, and allowed reduction in steroid therapy, while avoiding, so far, severe infections or oncogenic complications.

References

1 The Canadian Multi-Center Transplant Group. *N Engl J Med 1983; 309:* 809
2 Merion M, White DJG, Thiru S et al. *N Engl J Med 1984; 310:* 148
3 Najarian JS, Strand M, Fryd DS et al. *Transplant Proc 1983; 15:* 2463
4 Halloran P, Ludwin D, Aprile M et al. *Transplant Proc 1983; 15:* 2513
5 Touraine JL, El Yafi S, Bosi E et al. *Transplant Proc 1983; 15:* 2798

Open Discussion

BONOMINI (Chairman) We must be clear on what exactly you mean by nephrotoxicity? You mentioned only the increasing serum creatinine or other clinical, biochemical or morphological signs.

HOURMANT We consider chronic nephrotoxicity as an increasing serum creatinine with histological changes.

BONOMINI In that case all patients on cyclosporine have nephrotoxicity because in all cases serum creatinine increases.

HOURMANT No, only 10 of our patients on cyclosporine had acute nephrotoxicity.

BONOMINI I would like to open the discussion on this point, because we must be clear about the toxicity of cyclosporine in terms of haemodynamic changes, glomerular tubular imbalance, vascular changes, interstitial changes and urinary changes. If we measure the toxicity of cyclosporine only by measuring the serum creatinine I would say that in my experience all cases are toxic.

WAUTERS (Lausanne) Just a short comment on the relationship between cyclosporine dose and nephrotoxicity. In our group we try to achieve a blood, not serum, value below 400ng/ml. In 20 patients our long-term mean serum creatinine was 135μmol/L which was comparable to our conventional therapy group.

BRYNGER (Chairman) I found it quite surprising that you had such a high incidence of rejection after three months.

HOURMANT Before three months.

BRYNGER In the cyclosporine literature and from my experience rejection after three months is very uncommon. You have quite a substantial number. Did you diagnose them by biopsy?

HOURMANT Yes, most of the time.

Proc EDTA–ERA (1984) Vol 21

IMPROVEMENT IN CYCLOSPORINE HANDLING BY ANTI-LYMPHOCYTE GLOBULIN IN THE EARLY POST-OPERATIVE PERIOD

R Grundmann, P Wienand, U Hesse

Chirurgische Universitätsklinik Köln-Lindenthal, FRG

Summary

The value of prophylactic anti-lymphocyte globulin (ALG) treatment after transplantation was examined in a prospectively randomised study. The control group was treated with azathioprine and steroids only. In a third group prophylactic ALG combined with azathioprine and steroids was given for 8.7±2.25 days after transplantation and was then replaced by cyclosporine and steroids. Renal graft survival rates could be improved if the conventional immunosuppression was accompanied by ALG prophylaxis. However, additional cyclosporine treatment proved to be desirable since in this way the number of early acute rejection episodes could be further reduced.

Introduction

Although the benefits of prophylactic ALG administration immediately after transplantation have been reported repeatedly [1], this method of treatment nevertheless has not been generally accepted [2]. Therefore in the prospective study presented here the following questions were examined:

1. How does the transplant survival rate under prophylactic ALG treatment compare with conventional immunosuppressive therapy with azathioprine and steroids alone?
2. Which side effects of ALG treatment may be expected and how can they be reduced?
3. Can the immediate kidney function after transplantation be improved by ALG prophylaxis?
4. Is it advisable to change over from conventional immunosuppression combined with ALG prophylaxis to cyclosporine in the further course of treatment?

Material and methods

All patients presenting between May 1981 and July 1983 for a first cadaver kidney transplant entered a prospectively randomised trial (groups A and B).

In group A (n=47) patients received 5ml/10kg body weight (30ml maximum) of ALG (Mérieux®) per day during the first three weeks after transplantation. Additionally conventional immunosuppression with azathioprine and steroids was given as in the control group. The daily dose of ALG was administered over 24 hours via a central venous catheter with the aid of a pump.

In group B (control group) (n=47) immunosuppression consisted of azathioprine and steroids only. The steroid therapy was identical in both groups: all patients received 250mg of prednisolone the first day after transplantation. This dose was daily reduced by 25mg to 100mg prednisolone and then by 5mg every second day to 10mg prednisolone. In the case of a rejection episode the patients received additional pulse therapy with methylprednisolone (maximum 5 grams).

The dose of azathioprine depended on the body weight (maximum 3mg/kg body weight/day), and on the leucocyte and platelet counts.

Group C (n=35) contains all first cadaver kidney transplants which were performed from September 1983 to July 1984. The patients in this group received ALG in combination with conventional immunosuppression (as in group A) for a minimum of seven, but maximum of 14 days after transplantation. Subsequently, ALG and azathioprine were withdrawn and were replaced by cyclosporine which was given in a dose of 10mg/kg body weight/day (a serum cyclosporine level of 300 to 500ng/ml was aimed at). The steroid therapy was the same as in groups A and B.

Results

There were no significant differences between the three groups as regards the patients' ages and pre-transplant periods, the HLA and DR-matching or the preservation time of the grafts.

Graft survival rate (groups A and B)

At one year the graft survival rate in the ALG prophylaxis group was 33/47 (70.2%) (group A) as compared with 28/47 (59.6%) in group B. The three month graft survival rates are given in Figure 1.

The graft survival rate in group C after one month was 33/35 (94.3%) and after three months 29/32 (90.6%) (Figure 1).

Patient survival

Groups A and B did not differ signficantly as regards the hospital mortality rate: in the ALG group two patients died post-operatively from pneumonia, and in the control group two patients also died (one due to pneumonia and the other from wound sepsis).

There were three late deaths in the ALG group (group A): one patient died

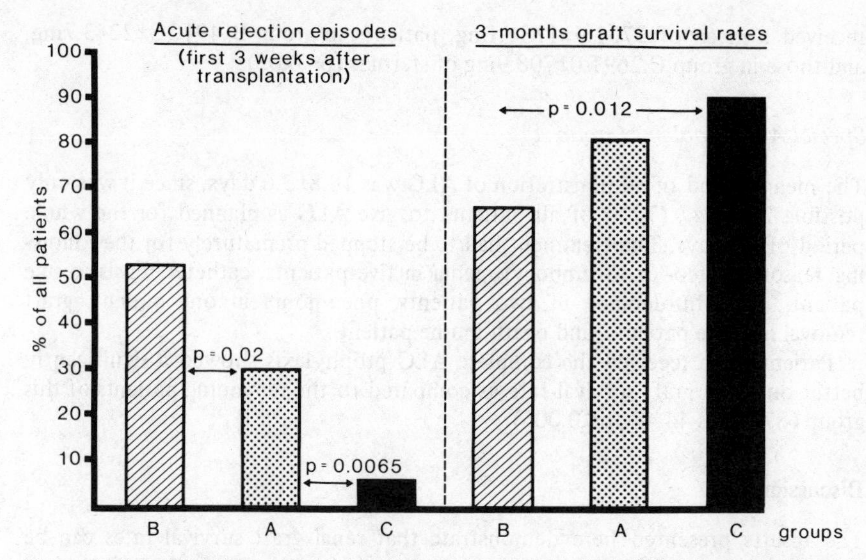

Figure 1. Group A: ALG + conventional immunosuppression (with azathioprine and steroids) Group B: conventional immunosuppression. Group C: ALG + conventional immunosuppression for a maximum of 14 days after transplantation followed by cyclosporine treatment

because of liver failure, another one because of a perforated sigmoid diverticulitis and one patient from a miliary tuberculosis. The one year patient survival rate, therefore, was 89.4 per cent in group A and 95.8 per cent in group B.

No patient in group C died in the first four weeks after transplantation or later on.

Primary kidney function after transplantation

Eighty per cent of all kidneys in group A, but only 24.4 per cent in group B ($p<0.05$) started to function after transplantation in such a way that dialysis was not necessary from the second week after transplantation. In group C 88.6 per cent of all kidneys were functioning well in the second post-operative week.

Acute rejection episodes

The percentage of patients that suffered an acute rejection episode in the first three weeks after transplantation is given in Figure 1. More rejection episodes were found in group B than in group C.

Amount of steroids administered

The three groups differed significantly as regards the amount of steroids administered during the first three weeks after transplantation. Patients in group A received a mean of 3731.9±1781.1mg, patients in group B 4912.3±2243.7mg, and those in group C 2691.0±708.9mg of steroids ($p<0.05$).

Special ALG problems (group A)

The mean period of administration of ALG was 18.8±3.6 days, since it was only possible in 34/47 (72%) of all patients to give ALG as planned for the whole period of 21 days. The treatment had to be stopped prematurely for the following reasons: leuco- or thrombocytopenia in five patients, catheter sepsis in one patient, ALG intolerance in two patients, pneumonia in one patient, graft removal in three patients, and death in one patient.

Patients who received the complete ALG prophylaxis showed a significantly better one year graft survival rate as compared to the remaining patients of this group (87.9% vs 44.4%, $p<0.005$).

Discussion

The results presented here demonstrate that renal graft survival rates can be improved if conventional immunosuppression (with steroids and azathioprine) is accompanied by ALG prophylaxis in the first weeks after transplantation. This is due to the fact that early rejection episodes can be successfully prevented by these means. This initial positive effect is also of advantage later on, since in the later course rejection episodes occurred in both groups (ALG and control group) with the same frequency.

In spite of this favourable result, however, we suggest a modification of the treatment protocol to avoid the increased risk of infection with the use of ALG. To reduce the side-effects, ALG should be given for shorter periods as planned here. In future, therefore, we will administer ALG for not more than 14 days after transplantation, since after this time the rate of complications increased considerably.

A further essential finding of this study consists of the observation that immediate kidney function after transplantation could be improved significantly if ALG was given. The improvement in post-transplant renal function, under identical donor conditions and intra- and post-operative fluid therapy [3], can only be attributed to the fact that rejection episodes which were not discovered because of missing clinical signs and, therefore be misinterpreted as acute tubular necroses, are treated in time by the ALG prophylaxis, resulting in a lower rate of post-transplant dialyses.

The benefits of prophylactic ALG treatment, therefore, are based on the suppression of early rejection episodes (Figure 1), whereas later rejection episodes were not affected. Since these reactions must be prevented to obtain further improvement in results, the ALG prophylaxis was combined with a subsequent cyclosporine therapy in group C. In so doing we have chiefly administered cyclosporine in those cases in which the kidney had already started to function, and to avoid the difficulties which occur in anuric kidneys with cyclosporine because of its nephrotoxicity [4,5]. Although the optimal timing for changing over from conventional therapy to cyclosporine has not been clearly defined, we think that it must lie between the 7th and 14th post-operative day, because on the one hand most of the kidneys have started to function satisfactorily up to this moment and on the other hand only slight side effects are to be expected from a short-term ALG prophylaxis.

Our experience with the new protocol is limited, however. The initial results demonstrate the advantage of changing over from ALG to cyclosporine: the number of rejection episodes observed in the first three weeks after transplantation was significantly reduced in group C as compared to the other groups, and consequently smaller doses of steroids were given. This led to a decreased risk of infection resulting in improved patient survival.

References

1 Grundmann R, Pichlmaier H. *Deutsch Med Wschr 1983; 108:* 687
2 Groth CG. *Transplant Proc 1981; 13:* 460
3 Carlier M, Squifflet JP, Pirson Y et al. *Transplantation 1982; 34:* 201
4 French ME, Thompson JF, Hunnisett AGW et al. *Transplant Proc 1983; 15:* 485
5 Hamilton DV, Evans DB, Henderson RG et al. *Proc EDTA 1981; 18:* 400

Open Discussion

BRYNGER (Chairman) Thank you very much for this thorough representation of another way to use cyclosporine in combination. I think we, for the moment, are receiving information from different centres using a number of combinations and which one will be thought to be the optimum I think remains to be proven, but still your data are impressive. Are there any questions or comments.

PICHLMAYR (Hannover) I think it is very important to study some protocols of combining cylosporine and ALG or azathioprine and this is true for your paper and for the paper presented before, but I am a little bit worried that we could be producing over-immunosuppression with these combinations. We have to remember the Stanford* experience where combining many things gave a very high incidence of tumours and so one should know a little bit more about the dosage and the blood levels of your cyclosporine treatment. If we combine drugs we have to very much reduce the dosage of the individual therapies.

GRUNDMANN That is right.

* Reitz BA, Gaudiani VA, Hunt SA et al. *J Thorac Cardiovasc Surg 1983; 85:* 354

Proc EDTA–ERA (1984) Vol 21

THE USE OF CYCLOSPORINE ONLY IN CADAVERIC RENAL TRANSPLANT RECIPIENTS: CONVERSION TO PREDNISOLONE AND AZATHIOPRINE AFTER FOUR MONTHS

A M Tegzess, A J M Donker, S Meijer, W J van Son, F J van Woude, M J H Slooff, V J Fidler, W J Sluiter, L B van der Slikke

State University Hospital, Groningen, The Netherlands

Summary

After conversion from cyclosporine (Cys) only to prednisolone and azathioprine four months after cadaveric renal transplantation, effective renal plasma flow (ERPF), glomerular filtration rate (GFR) and filtration fraction (FF) all improve. However, this improvement is not uniform. GFR and FF improve in all patients after one week of combined Cys and prednisolone treatment. ERPF improves under the same circumstances only in recipients without previous rejection episodes. After discontinuation of Cys and addition of azathioprine ERPF improves further in all patients. These findings suggest the presence of a low grade rejection, together with Cys nephrotoxicity. The conversion procedure seems to be safe at least during a follow-up period of 9–14 months.

Introduction

The results of cadaveric renal transplantation are improved by the use of Cys as the only immunosuppressive agent. However, renal function was worse in these cadaveric renal transplant patients than in conventionally treated cadaveric renal transplant patients [1,2]. Elective conversion three months after grafting from Cys to prednisolone and azathioprine was followed by improvement in renal function without serious side-effects [3]. In an attempt to improve renal function, elective conversion four months after transplantation was carried out in our cadaveric renal transplant patients. The improvement in renal function was monitored by serial determinations of ERPF and GFR before, during and after conversion.

Patients and methods

Forty-four cadaver renal transplant patients were treated with Cys alone during the first four months after grafting. Rejection episodes were diagnosed on clinical grounds such as tenderness and enlargement of the graft, serial ultrasonographic studies and a rise in serum creatinine levels for more than two days despite a decrease in the Cys doses. In addition one to three percutaneous renal biopsies were done in 31 of 44 cadaveric renal transplant patients. Rejection episodes were treated with 3 x 1g of Solu-Medrol intravenously per rejection episode on three consecutive days. If more than three rejection episodes occurred, cadaver renal transplant patients were converted to prednisolone and azathioprine. Otherwise, patients were treated with Cys only for four months post-operatively.

Elective conversion was carried out by adding 20mg prednisiolone daily to the Cys therapy over one week. Thereafter Cys was replaced by azathioprine, 1.5mg/kg body weight, while 20mg of prednisolone was continued.

ERPF and GFR were determined by continuous simultaneous infusion of ^{131}I-hippurate and ^{125}I-thalamate, respectively. Each investigation was carried out over a period of six hours. After an equilibration period of two hours urine and blood samples were collected twice over the remaining two periods of two hours [4]. FF was calculated by GFR:ERPF.

ERPF and GFR were measured one month and just before administration of prednisolone to Cys, after one week of prednisolone together with Cys, one week after cessation of Cys and finally, six weeks later on the same dose of prednisolone and azathioprine.

Results

Twenty-one of 44 cadaver renal transplant patients were converted from Cys to prednisolone and azathioprine before the fourth post-operative month because of more than three rejection episodes (n=16), technical problems or prolonged post-operative oliguria (n=5). Four of 44 cadaver renal transplant patients were kept on permanent Cys therapy. Finally, one patient was successfully converted after four months without ERPF and GFR determinations. His renal function measured by serum creatinine concentrations and creatinine clearances improved.

The course of the ERPF and GFR in the remaining 18 cadaver renal transplant patients during and after conversion from Cys to prednisolone and azathioprine four months after grafting is shown in Figures 1a and 1b. Six of 18 cadaver renal transplant patients, shown in the upper curves had no rejection episodes, while 12 of 18 cadaver renal transplant patients (lower curves) had one to three rejection episodes. During the last month before conversion ERPF and GFR remained stable in both groups. Also before conversion ERPF and GFR were the same in cadaver renal transplant patients with or without rejection episodes. After one week of 20mg prednisolone together with Cys, GFR improved in both groups of cadaveric renal transplant patients ($p<0.05$ and $p<0.02$, respectively). ERPF, however, improved only in cadaveric renal transplant patients without rejection episodes ($p<0.05$). One week after replacement of Cys by azathioprine, a further rise in GFR was observed in all cadaveric renal transplant patients. In contrast to the previous period a rise of ERPF was observed in all patients. At one week after replacement of Cys by prednisolone and azathioprine the mean ERPF and GFR were again the same in cadaver renal transplant patients with or without rejection episodes.

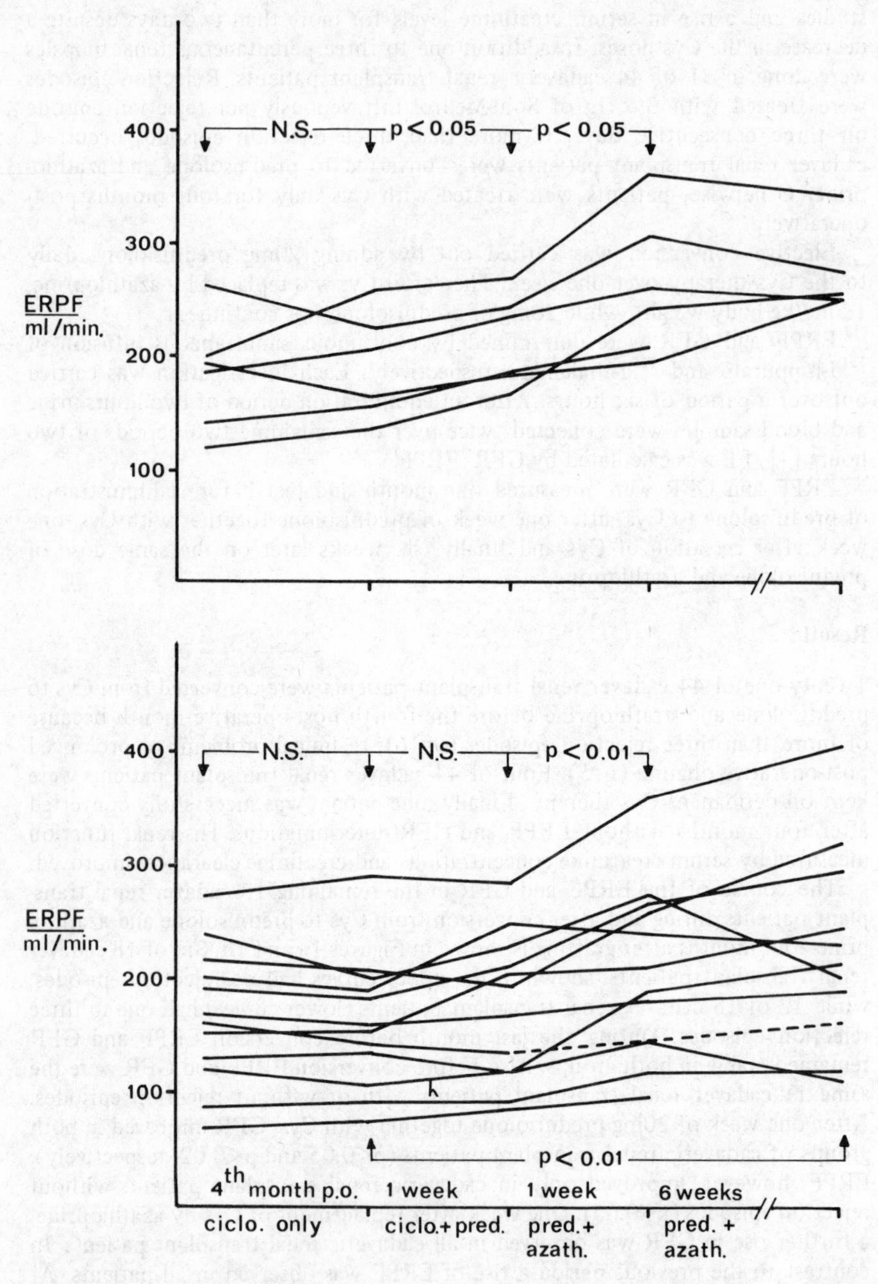

Figure 1a. The course of ERPF before, during and after conversion from Cys to prednisolone and azathioprine four months post-transplantation. Top curve: six patients without previous rejection episodes. Bottom curve: 12 patients with one to three rejection episodes

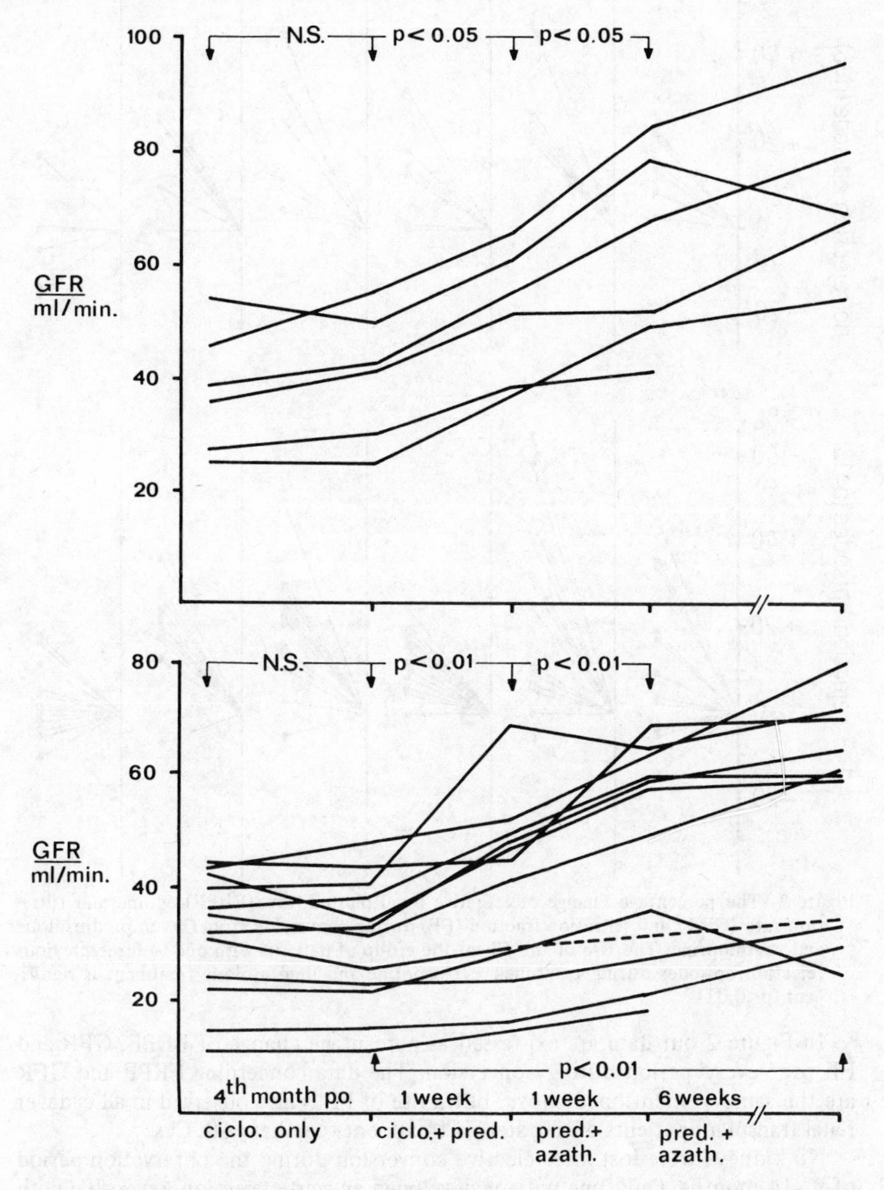

Figure 1b. The course of GFR before, during and after conversion from Cys to prednisolone and azathioprine four months post-transplantation. Top curve: six patients without previous rejection episodes. Bottom curve: 12 patients with one to three rejection episodes

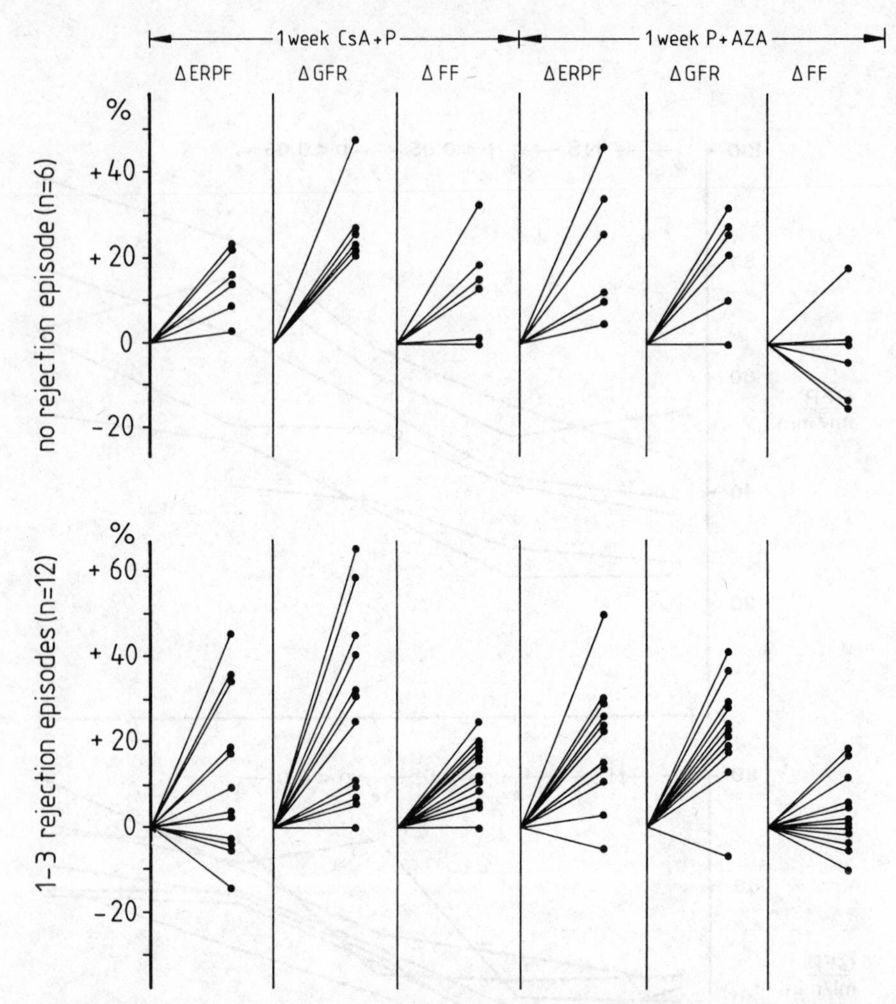

Figure 2. The percentage change of effective renal plasma flow (ERPF), glomerular filtration rate (GFR) and filtration fraction (FF) during conversion from Cys to prednisolone and azathioprine. The rise of the FF in the group of patients with one to three previous rejection episodes during combined cyclosporine and prednisolone treatment is significant ($p<0.01$)

In Figure 2 our data are expressed as percentage change of ERPF, GFR and FF over every period during conversion. The data concerning ERPF and GFR are the same as mentioned above, but a rise of FF is also observed in all cadaver renal transplant patients during steroid treatment together with Cys.

No kidneys were lost after elective conversion during the observation period of 9–14 months. Only one patient developed an acute rejection, reversible with oral prednisolone. No other complications occurred. A total of four of 44 patients lost their kidneys due to rejection or technical complications. The overall patient survival was 100 per cent.

Discussion

One week after adding 20mg of prednisolone to Cys four months after cadaveric renal transplantation we observed a significant rise in GFR and FF in all cadaver renal transplant patients, together with a rise in ERPF in cadaver renal transplant patients without previous rejection episodes. This finding suggests the presence of a subclinical, steroid sensitive rejection. In particular the rise in the FF during this period is in accordance with our earlier observation that a low FF points to an impending rejection [5]. The fact that ERPF, GFR and FF were stable during the last month before conversion makes spontaneous improvement unlikely. On the other hand it is also possible that steroids counteract Cys nephrotoxicity. Cessation of Cys led to a further increase of ERPF and GFR in all cadaver renal transplant patients as a possible sign of resolving Cys nephrotoxicity. Studies in experimental animals and in heart transplant patients on long-term Cys therapy have shown morphological and functional proximal tubular abnormalities [6–8]. ^{131}I-hippurate is mainly excreted by the proximal tubules. Therefore, our observation that ERPF is depressed during Cys treatment can at least be partially explained by these morphological findings. Unfortunately, we did not measure ^{131}I-hippurate excretion. The difference in the recovery of the ERPF between cadaver renal transplant patients with or without rejection episodes is difficult to understand. An explanation could be a different degree of Cys nephrotoxicity in the two groups of cadaver renal transplant patients. The clinical results in terms of graft and patient survival after elective conversion are satisfactory without signficant clinical complications.

References

1 European Multicentre Trial Group. *Lancet 1983; ii:* 986
2 Calne RY, White DJG, Thiru S et al. *Lancet 1979; ii:* 1034
3 Morris PJ, French ME, Dunnill MS et al. *Transplantation 1983; 36:* 273
4 Donker AJM, van der Hem GK, Sluiter WJ et al. *Neth J Med 1977; 20:* 97
5 Tegzess AM, Donker AJM, Meijer S et al. In Holenberg NK, ed. *Radionuclides in Nephrology.* New York: Thieme-Stratton Inc. 1980: 245
6 Whiting PH, Thomson AW, Blair JT et al. *Br J Exp Pathol 1982; 63:* 88
7 Dieperink H, Starklint H, Leyssac PP. *Transpl Proc 1983; 15 (suppl 1):* 520
8 Moran M, Newton L, Perlroth M et al. *Kidney Int 1984; 25:* 1

Proc EDTA–ERA (1984) Vol 21

CONVERSION FROM CYCLOSPORINE TO PREDNISOLONE AND AZATHIOPRINE. SAFE OR UNSAFE?

D Adu, J Michael, T Vlassis, P McMaster

Queen Elizabeth Hospital, Birmingham, United Kingdom

Summary

Out of 58 consecutive cadaveric renal allograft recipients whose initial immunosuppression was cyclosporine (Cys) and prednisolone, 18 were converted to prednisolone and azathioprine. Of these all four patients converted because of rejection lost their grafts. Renal function improved in seven patients converted because of nephrotoxicity and in six out of seven patients converted for miscellaneous reasons. Out of five patients converted electively at three months, one died of infection and three developed acute rejection episodes.

Introduction

Cyclosporine (Cys) is now well established as a potent immunosuppressive agent in renal, hepatic and bone marrow transplantation. The nephrotoxicity of Cys poses considerable problems in the early management of renal allograft recipients. The long-term sequelae of Cys nephrotoxicity are not yet known and there is considerable anxiety that this may lead to chronic and irreversible renal damage. Several studies now indicate that it is possible to convert patients from Cys to prednisolone and azathioprine [1–3]. The consequences of such a conversion, either because of clinical evidence of toxicity due to Cys [1] or as an elective decision [2,3] have varied from improvement of graft function [3], to irreversible rejection and graft loss [2]. We investigated the consequences of conversion from Cys to prednisolone and azathioprine for a variety of clinical indications. Further in a prospective study we compared the outcome of elective conversion at three months versus continuing on Cys.

Patients and methods

Conversions for clinical reasons

Out of 58 consecutive cadaveric renal allografts treated with Cys plus initial prednisolone (for the first 14 days), 18 were converted to prednisolone and azathioprine. The immunosuppressive regimen has been reported in detail [4]. Briefly all patients were started on oral Cys 15mg/kg/day on the day of operation reducing at one month to 12mg/kg/day, at two months to 10mg/kg/day, at three months to 8mg/kg/day and thereafter by 1mg/kg/day each month until a maintenance dosage of 5mg/kg/day was reached. Prednisolone was given orally at a dose of 20mg a day for the first 14 days only. If there was subsequent clinical and histological evidence of rejection prednisolone was reintroduced at the same dose.

Cys nephrotoxicity was defined as impairment of renal function in the absence of urinary obstruction, infection or histological evidence of rejection. The conversion protocol was as follows: Cys was stopped and the next day prednisolone (20mg/day) and azathioprine (2.5mg/kg/day) started.

Elective conversions

This was a prospective study of adult non-diabetics receiving a first cadaveric renal allograft. The dosage of Cys was as described above. In this group however prednisolone (20mg/day) was continued for 12 weeks. At 11 weeks post renal transplant, after giving informed consent, patients in the study were randomly allocated either to continue Cys and prednisolone or were converted to prednisolone and azathioprine. Azathioprine was started at a dose of 1.5mg/kg/day and after one week increased to 2.5mg/kg/day at which time Cys was stopped.

Results

The results of conversions from Cys to prednisolone and azathioprine for clinical reasons are summarised in Table I.

The prospective study of conversion against non-conversion was stopped after 12 patients had been recruited. Of the five patients converted to prednisolone and azathioprine (Table II), one died of a lung infection and three developed acute rejection episodes. In one patient the rejection episode was readily reversed with steroids but two patients had to be converted back to Cys and were left with moderately impaired renal function. One of these patients developed leucopenia and a chest infection and the second herpes simplex. In contrast renal function remained stable in the seven patients who were not converted.

Discussion

Our experience of converting patients because of Cys nephrotoxicity or for other miscellaneous reasons has been satisfactory. There were no acute rejection episodes and no significant infections. Only one out of 14 grafts was lost from chronic rejection and graft function improved in the remainder. These good results are in keeping with several earlier reports [1,3,5].

Of the four patients who were converted at a time when their graft biopsy showed some evidence of chronic rejection, all lost their grafts. By contrast, Merion et al [6] reported graft losses in three out of eight patients converted from Cys to prednisolone and azathioprine because of acute rejection. The

TABLE I. Conversion consequences – Cyclosporine to prednisolone and azathioprine

Case No.	Months Post Transfusion	Reason for Conversion	Serum Creatinine – μmol/L Pre-Conversion	Months Post-Conversion 1	2	3	
1	12	Rejection ± Nephrotoxicity	403	372	500	1000	Graft loss
2	2		389	421	317	297	Graft loss
3	2		1000	1180	–	–	Graft loss
4	7		550	1688	–	–	Graft loss
5	10	Nephrotoxicity	660	225	173	165	
6	2	Nephrotoxicity	800	605	230	216	
7	5	Nephrotoxicity	590	380	220	178	
8	7	Nephrotoxicity	314	260	185	165	
9	6	Nephrotoxicity	895	178	770	146	
10	6	Nephrotoxicity	500	272	300	267	
11	1	Nephrotoxicity	507	167	112	104	
12	2	Colitis	225	165	240	347	
13	10	Tremor	160	79	98	86	
14	7	Hepatotoxicity	187	122	115	122	
15	5	Crescentic GN	880	378	267	234	
16	5	Lymphoproliferative Disorder	220	124	135	131	
17	12	Leucopenia	289	118	108	–	
18	3	Hyperkalaemia	333	131	150	138	

TABLE II. Elective conversions from Cys to prednisolone and azathioprine at three months

Case No.	Serum Creatinine – μmol/L Pre-Conversion	Weeks Post-Conversion 1	2	3	4	Clinical Data
1	94	100	98	108	94	No Problems
2	105	96	156	135	113	Acute rejection. Reversed with steroids
3	237	185	218	378	459	Acute rejection. Back on Cys. Lung infection
4	111	–	–	–	–	Died from lung infection
5	172	152	147	360	471	Acute rejection. Back on Cys. Herpes simplex

reasons for this difference is unclear. The results of the European Multicentre Trial [7] where out of 29 patients converted (24 at the time of a rejection episode) 14 grafts were lost within a month, support our observation that prednisolone and azathioprine are not effective in the treatment of rejection in patients on Cys.

Our prospective trial of elective conversions from Cys to prednisolone and azathioprine was instigated because of our anxiety about the long-term sequelae for graft function of Cys nephrotoxicity. At the time of the study we and others [4,8] had shown that renal function was poorer in Cys treated renal allograft recipients than in patients on prednisolone and azathioprine over a follow-up period of 12 to 24 months. We were reassured by data from Oxford [3] which indicated that conversions from Cys to prednisolone and azathioprine at 90 days were safe and in the main, problem-free. The experience of Land et al [2] was discouraging in that out of nine patients converted, three lost their grafts and four had acute rejection episodes but retained their grafts when converted back to Cys. Our experience of elective conversion at three months was disastrous and this led to the study being terminated. Of the five patients converted, one died of a lung infection and three developed acute rejection episodes. In two patients this was severe and they were converted back to Cys. Both patients developed severe infections and are left with moderately impaired renal function.

From our study we conclude that conversion from Cys to prednisolone and azathioprine for reasons of nephrotoxicity or other toxic effect of Cys is safe. Using our conversion protocol we would agree with Land et al [2] that elective conversion at three months is extremely hazardous, and that the penalties of acute rejection and infection are unacceptable. Alternative protocols and timings need to be explored if a policy of conversion from Cys is to be pursued.

References

1 Klintman GBG, Iwatsuki S, Starzl TE. *Lancet 1981; i:* 470
2 Land W, Castro LA, Hillebrand G et al. *Transplant Proc 1983; 15 (Suppl 1):* 2857
3 Wood RFM, Thompson JF, Allen MH et al. *Transplant Proc 1983; 15 (Suppl 1):* 2862
4 McMaster P, Michael J, Adu D et al. *Transplant Proc 1983; 15 (Suppl 1):* 2523
5 Canafax DM, Sutherland DER, Ascher NL et al. *Transplant Proc 1983; 15 (Suppl 1):* 2874
6 Merion RM, White DJF, Thiru S et al. *N Engl J Med 1984; 310:* 148
7 European Multicenter Trial. *Lancet 1982; ii:* 57
8 Hamilton DV, Evans DB, Henderson RG et al. *Proc EDTA 1981; 18:* 400

Proc EDTA–ERA (1984) Vol 21

DOES RENAL FAILURE INDUCE A DECREASE IN CYCLOSPORINE BLOOD CONCENTRATIONS?

M Burnier, P Aebischer, **J P Wauters**, J L Schelling

University Hospital, Lausanne, Switzerland

Summary

Blood Cys concentrations were monitored twice weekly by RIA in nine patients undergoing renal transplantation. The Cys dose was adapted to obtain blood values between 100–400ng/ml. A negative correlation was found between plasma creatinine and blood Cys, even in those patients in whom the oral dosage was not changed ($r = -0.65$, $n = 23$, $p<0.001$). In vitro studies showed no effect of uraemic blood on Cys measurement. It seems probable that uraemia induces changes in distribution volume and/or gastrointestinal absorption of the drug. Careful monitoring of Cys during uraemia is therefore warranted.

Introduction

Among the side effects described during cyclosporine (Cys) treatment one of the most serious is its potential nephrotoxicity [1,2]. After renal transplantation this nephrotoxicity could be easily confused with acute rejection and some authors avoid Cys when the renal allograft is not functioning primarily [3]. In patients receiving bone marrow transplantation – with previously normal renal function – a positive correlation between blood urea nitrogen and blood Cys suggested that the higher the blood Cys the greater the nephrotoxicity [2].

The present report concerns the follow-up of renal function and blood Cys in nine renal transplanted patients and shows that during renal failure Cys blood concentrations are decreased and that they increase in parallel with the improvement of renal function.

Patients and methods

All nine patients, four male and five female aged between 25 and 51 years, received a cadaveric renal allograft. Four patients started Cys intravenously before transplantation at a dose of 5mg/kg/day and continued orally 48 hours later with a dose of 10mg/kg/day. The other five patients were changed from conventional therapy to Cys 10mg/kg/day orally at the time their renal function was normal. Cys was adapted individually depending on blood concentrations to obtain values between 100 and 400ng/ml. Blood Cys was measured in duplicate twice weekly using the radioimmunoassay kit provided by Sandoz (Basel, Switzerland). Blood samples were drawn before the morning dose, 12 to 14 hours after the previous evening dose. Immunosuppression was completed with 0.5g methylprednisolone per-operatively followed by oral prednisone at the dose of 0.7mg/kg/day. This dose was decreased stepwise by 5mg every two weeks. Treatment of acute rejection consisted of daily doses of 0.5g methylprednisolone intravenously while Cys was continued.

An in vitro study was performed to test the effect of uraemia on the measurement of Cys by the RIA technique. Blood was drawn from five uraemic patients before dialysis and from five normal controls not treated with Cys. The blood of each patient was divided into three samples in which a fixed dose of either 500, 250 or 125ng/ml of Cys was added. All samples were then measured in duplicate.

Statistics: Statistical analysis was performed using linear and logarithmic correlations. Results are given as mean ± SEM.

Results

Data analysis

Standard deviations (SD) of blood Cys measurements were evaluated during steady state periods. A steady state period was defined as a fixed oral dose of Cys over more than four days and variations of plasma creatinine of less than 20 per cent. One SD of Cys measurements was at 18.5 ± 9 per cent (mean ± SD). The observed changes in Cys during variations of renal function always exceeded 2 SD. During the steady state periods, the Cys blood concentrations were always correlated to the oral dose either when data were analysed individually or as a group ($n = 105, r = 0.48, p<0.001$). In the subgroup of patients who developed an episode of acute renal failure due to acute rejection or tubular necrosis, the same correlation was found during the steady state periods but the correlation was lost during renal failure ($n = 34, r = 0.036$, p = NS).

When the patients received a fixed oral dose of Cys between nine and 12mg/kg/day, negative logarithmic correlations were found between blood Cys and plasma creatinine ($n = 23, r = -0.65, p<0.001$) and urea ($n = 23, r = -0.72, p<0.001$). Interestingly, in this group the higher oral Cys dosage of 11 and 12mg/kg/day did not result in higher blood Cys. Moreover, the correlations were even better when the analysis was restricted to patients receiving Cys nine and 10mg/kg/day: Cys vs creatinine $n = 16, r = -0.72, p<0.001$ and Cys vs urea $n = 16, r = -0.85, p<0.001$.

In vitro study

The in vitro study did not show any influence of uraemia on Cys measurements by RIA. The plasma urea and creatinine of the uraemic patients were respectively at 35.4 ± 6.1mmol/L and 932 ± 188μmol/L. Dosages obtained in blood of uraemic patients and their controls were not significantly different: 116 ± 8 (mean ± SEM) and 113 ± 4 for 125ng/ml added, 258 ± 16 and 266 ± 12 for 250 and 534 ± 18 and 538 ± 10 for 500ng/ml added.

Discussion

Our study shows a constant increase in blood Cys occurring simultaneously with the improvement of renal function when the daily Cys dose is maintained. Similarly blood values fall when renal failure reappears or gastrointestinal disturbances occur.

The variations in blood Cys induced by fluctuations of renal function are significantly different from the baseline variability of Cys measurements in patients with stable renal function. The loss of correlation between blood Cys and the oral dose when renal failure appears and the good correlation between Cys and creatinine in patients receiving a fixed oral dose of Cys strongly suggest an effect of renal function on Cys blood concentrations.

The in vitro study demonstrates that the low values found during renal failure are not due to a direct effect of uraemia on the assay method. The decrease of blood Cys during uraemia cannot be ascribed to dialysability of the drug since two out of five patients were not dialysed during their renal failure period and the same fluctuations were observed. Moreover, it has been shown that the dialysis rate of the drug is less than 2ml/min [4].

Acute changes in the hepatic metabolism of Cys appears unlikely since no signs of hepatic dysfunction were observed simultaneously. A modification in hepatic clearance of Cys which has been suspected to occur after administration of methylprednisolone does not explain our results since Cys decreased and did not increase during the occurrence of renal failure and also occurred in patients with acute tubular necrosis who did not receive methylprednisolone [5].

Our observations could be due to decreased gastrointestinal absorption. Indeed Cys values less than the therapeutic range were measured when diarrhoea appeared. In the course of uraemia, three patients developed diarrhoea. Barrett et al reported low blood concentrations of Cys after bone marrow transplantation in patients with severe gastrointestinal disturbances due to chemotherapy, radiotherapy or graft rejection with severe diarrhoea [2]. Kahan et al observed an increase in bioavailability from five to 25 per cent of the drug after one month in their patients except in those having gastrointestinal complications [6].

Acute changes in renal function could also modify the distribution volume of Cys and induce variations of blood Cys, similarly to those shown for various other drugs [7]. This hypothesis implies that drug dosage should not necesarily be increased when low values are measured during uraemia.

In conclusion, our observation strongly suggests that renal function modulates blood Cys. The clinical implications of this observation appear important in renal as well as in other types of organ transplantation. Indeed when renal function decreases or recovers careful monitoring of blood Cys is mandatory.

References

1 Klintmalm GBG, Iwatsuki S, Starzl TE. *Lancet 1981; i:* 470
2 Barrett AJ, Kendra JR, Lucas CF et al. *Br Med J 1982; 285:* 162
3 Calne RY, White DJG, Evans DB et al. *Br Med J 1981; 282:* 934
4 Venkataramanan R, Burckart GJ, Ptachcinski RJ et al. *Drug Intelligence & Clinical Pharmacy 1983; 17:* 446
5 Öst L. *Lancet 1984; i:* 451
6 Kahan BD, Ried M, Newburger J. *Transplant Proc 1983; 15:* 446
7 Gulyassi PF, Depner TA. *Am J Kid Dis 1983; 2:* 578

Proc EDTA–ERA (1984) Vol 21

INCREASED INTERLEUKIN II PRODUCTION DURING KIDNEY ALLOGRAFT REJECTION

H Vie, M Bonneville, R Cariou, J F Moreau, J P Soulillou

Laboratoire d'Immunologie Clinique (INSERM U211) and Service de Néphrologie, UER de Médecine, Nantes, France

Summary

We have studied the lectin-induced production of Interleukin II (IL-2) by peripheral blood leucocytes (PBL) obtained from recipients of kidney grafts. Pre-graft values (haemodialysed patients) are similar to those of healthy individuals. After grafting and during the first year the IL-2 levels produced by PBL from recipients with well functioning grafts are very low ($p<0.001$).

Years after grafting the IL-2 yields tend towards restoration. At the onset of an acute rejection the IL-2 production rises significantly. These data show that an increase in IL-2 production may play a role in the rejection process.

Introduction

Interleukin II (IL-2) is a factor required for T lymphoblast growth [1] and thus a key factor for the development of the normal cellular immune response. This 15 KD molecule is produced by T-lymphocytes under lectin or antigen stimulation and acts via a 50 KD membrane receptor, present on activated T lymphocytes [2], as well as on Natural Killer cells [3]. It has been shown that IL-2 plays a role in rejection of heart allografts in rats [4] and recently we have used this growth factor to expand and study T lymphocyte clones extracted from human rejected kidney grafts.

In this paper we show that although peripheral blood lymphocytes (PBL) of dialysed patients can produce normal amounts of IL-2 under phytohaemagglutinin (PHA) stimulation, after grafting there is a dramatic decrease in PHA induced IL-2 yields. In contrast, during acute rejection, the lectin induced IL-2 production is significantly increased.

Patients, material and methods

Patients Heparinised blood samples were sytematically and prospectively obtained immediately before grafting (n=14) and after grafting either during a steady period of good graft function (n=18) or at the onset of an acute rejection crisis (n=24) before any specific treatment of the episode. In addition, single blood samples were obtained from long-term recipients of well-functioning grafts (1 to >10 years) to assess the IL-2 yields at a time of minimal immunosuppression. All patients were systematically transfused before and during grafting and their immunosuppressive treatment is described elsewhere [5]. Blood samples from patients under anti-thymocyte serum were not used to avoid bias. Finally PBL were obtained from 21 healthy individuals as controls.

Materials and methods (1) IL-2 production Briefly, PBL were obtained on ficoll hypaque gradients, washed twice and frozen until used. On the day of the experiment, thawed PBL (4×10^5 in 0.1ml) were stimulated with PHA-P (Difco) at 0.5% (v/v) in Greiner tissue-culture microplates (0.2ml/well). Culture medium was RPMI 1640 supplemented with 10% (v/v) human serum. After 49 hours of culture (37°C, 5% CO_2) 0.1ml of supernatant of each culture well was harvested, pooled by duplicates and kept at 4°C until assayed for IL-2.

(2) IL-2 assays IL-2 contents of the supernatants were tested on the IL-2-dependent CTL-L-2 murine cell line as previously indicated [5]. Results are given in mean ± SD of CPM obtained from 5×10^3 CTL-L-2 cells cultured 18 hours in the presence of supernatants (diluted fourfold) and pulsed four hours with 0.25μC of triatiated thymidine. Statistical analysis was by the Fischer 't' test or the Wilcoxon paired test.

Results

IL-2 yields of PBL obtained from recipients before and after grafting in patients with good graft function (Figure 1)

IL-2 yields of pre-graft PBL samples were similar to those from normal individuals indicating that haemodialysed patients can produce normal levels of this lymphokine upon appropriate stimulation.

After grafting, PBL from kidney recipients produced significantly lower levels of IL-2. This profound defect in IL-2 production persisted until the end of the first year after grafting and tended to return to normal after two years.

IL-2 yields in patients undergoing acute rejection episodes

All rejection crises studied (n=24) occurred within the first year after grafting and thus during a period of very low (or nil) IL-2 production in non-rejecting patients. Figure 1 shows that PBL of rejecting patients produced significantly higher amounts of IL-2 ($p<0.01$) than non-rejecting ones.

Data obtained from eight patients in whom all the serial samples (i.e. pre-graft and post-graft samples, including rejection samples) could be studied, also showed a dramatic and significant drop of the PBL IL-2 yield after grafting followed by a sharp and significant increase of IL-2 production at the onset of a rejection crisis (Figure 2).

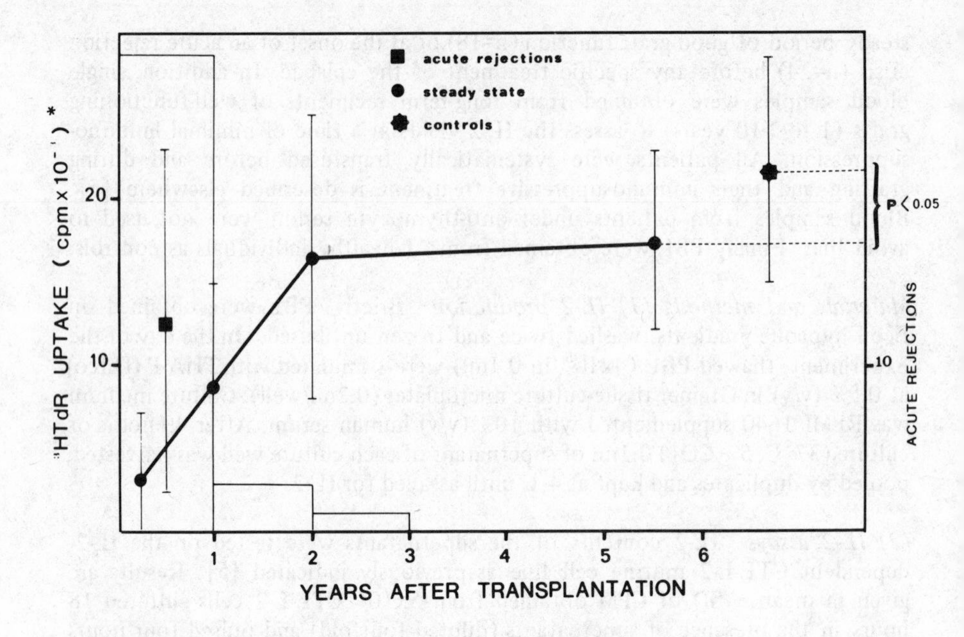

Figure 1. IL-2 levels produced by PBL of kidney recipients. Kinetics and increase during rejection episodes. Results are given in CPM obtained from CTL-L-2 cells cultured in the presence of PHA-stimulated PBL supernatants diluted fourfold. Rejection values have been located at the average time of the rejection episodes

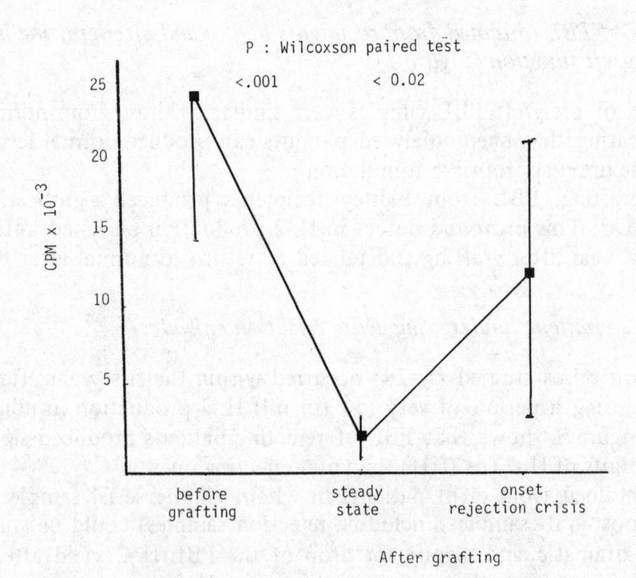

Figure 2. IL-2 yield of PBL of kidney recipients undergoing acute rejection episodes

Discussion

We have shown that after grafting PBL from recipients of well-functioning kidneys yield very low levels of IL-2. It is likely that this defect is due to the high immunosuppressive drug regimen used in these patients, since corticosteroids, as well as cyclosporine have been shown, in vitro, to dramatically impair IL-2 production [6]. However, IL-2 production of patients with systematic lupus erythematosus did not correlate to their corticosteroid treatment [7]. This discrepancy may be related to differences in the dosage of the drug. Interestingly Natural Killer cell (NK) activity is also grossly impaired in kidney recipients [8] with roughly similar kinetics and as NK cells bear IL-2 receptors the two (IL-2–NK) phenomena may be linked.

Patients undergoing acute rejection have PBL which produce much more IL-2 on PHA stimulation than those obtained from recipients of well-functioning grafts. Obviously they are not 'high IL-2 producers' since, as shown in Figure 2, they have low IL-2 yields during the steady state period. This increased IL-2 yield cannot be explained by variation in the immunosuppressive regimen and is likely to be related to the rejection process. This IL-2 yield may be a poor reflection of higher lymphokine production by lymphocytes invading the rejected graft.

Finally, the restored IL-2 production in long-term well functioning kidney recipients (which can be explained by the lower immunosuppression) suggests that normal IL-2 production by itself is not involved in the rejection process. However, it has now been proved that such recipients have developed clonal regulatory mechanisms (i.e. involving suppression of clones specifically committed against the donor antigens [9]) which could explain why, at that time, normal IL-2 producing capacity may not have a deleterious influence in this context.

References

1 Ruscetti FW, Gallo R. *Blood 1981; 57:* 379
2 Robb RJ, Munck A, Smith KA. *J Exp Med 1981; 154:* 1455
3 Vose BM, Riccardi C, Bonnard GD et al. *J Immunol 1983; 130:* 768
4 Lear PA, Heidecke CD, Kupiec-Weglinski J et al. *Transplant Proc 1983; 15:* 349
5 Godard A, Naulet J, Peyrat MA et al. *J Immunol Meth 1984; 70:* 233
6 Kaplan MR, Lysz K, Rosenberg SA et al. *Transplant Proc 1983; 15:* 407
7 Linker-Israeli M, Bakke AC, Kitridou et al. *J Immunol 1983; 130:* 2651
8 Moreau JF, Ythier A, Soulillou JP. *Transplant Proc 1981; 13:* 1610
9 Charpentier BM, Bach MA, Lang P et al. *Transplantation 1983; 36:* 495

Proc EDTA–ERA (1984) Vol 21

METABOLIC EFFECTS OF CONVERSION FROM CYCLOSPORINE TO AZATHIOPRINE IN RENAL TRANSPLANT RECIPIENTS

K P G Harris, G I Russell, S P Parvin, P S Veitch, J Walls

Leicester General Hospital, Leicester, United Kingdom

Summary

The metabolic effects of conversion from cyclosporine (Cys) to azathioprine were studied in 10 non-diabetic renal transplant recipients. Following conversion there was a significant improvement in renal function. There was no evidence of glucose intolerance pre-conversion, and no change in glucose metabolism following conversion. However, fasting cholesterol and triglyceride levels were significantly improved following conversion. The mechanism of this change and its significance needs further investigation.

Introduction

Cyclosporine (Cys) is now extensively used in organ transplantation. However, there are well recognised side effects such as nephrotoxicity and hepatotoxicity [1], and in addition it has recently been suggested that Cys can cause a deterioration in glucose control [2]. Experience in this unit indicated an apparent higher incidence of diabetes using Cys than that previously noted in patients treated with conventional immunosuppression, i.e. azathioprine and prednisolone.

The current protocol for patients receiving cadaveric renal transplants is to convert the immunosuppression at three months from prednisolone and Cys to prednisolone and azathioprine. Hence an opportunity arose to evaluate various metabolic indices before and after conversion.

Method

Ten non-diabetic transplant recipients (age range 21–64 years) with stable renal function were studied at three months post-transplantation. Immunosuppression for the three months following transplantation was Cys (mean dose 7.8±3mg/kg in two equal divided doses) and prednisolone (dose tapered to

30mg on alternate days by two months post-transplantation). At three months azathioprine (100–150mg) was introduced and the Cys was withdrawn – the prednisolone dose remained unaltered at 30mg on alternate days.

Immediately pre-conversion and one month post-conversion a 75gm oral glucose tolerance test was performed after an overnight fast, together with estimation of serum immunoreactive insulin, serum cholesterol and triglyceride, renal and liver function tests. Statistical analysis was performed using paired Student's 't' test.

Results

The mean Cys level pre-conversion was 164±34ng/ml. There was no significant change in body weight of the patients following conversion (63.8±4.5kg, 63.9±4.5kg).

Renal function was significantly improved by stopping Cys, as judged by the fall in serum urea and creatinine concentrations (Table I).

TABLE I. Renal function in renal transplant recipients following conversion from cyclosporine to azathioprine (mean values ± SEM)

	Pre-conversion	Post-conversion
Creatinine (mol/L)	164±17	136±14*
Urea (mmol/L)	11.0±0.7	7.8±0.7**
Potassium (mmol/L)	4.5±0.17	4.3±0.17

*p<0.05, **p<0.01, cf pre-conversion

Fasting blood sugar, fasting immunoreactive insulin, and glucose tolerance was not altered following conversion and there was no evidence of glucose intolerance pre-conversion (Figure 1).

There was, however, a significant fall in serum cholesterol from 7.6±0.3mmol/L to 6.2±0.4mmol/L and in serum triglyceride from 2.4±0.1mmol/L to 1.5±0.2 mmol/L following conversion (p<0.01) (Figure 2), with the mean values of both falling to within the normal range. The changes in serum cholesterol and triglyceride did not correlate with changes in body weight or renal function, or the pre-conversion Cys level.

Discussion

This study demonstrates three metabolic consequences of immunosuppression with Cys in renal transplant recipients:

1. The nephrotoxic effect of Cys was confirmed;
2. There was no evidence for Cys induced glucose intolerance;
3. Stopping Cys resulted in a significant improvement in fasting cholesterol and triglyceride levels.

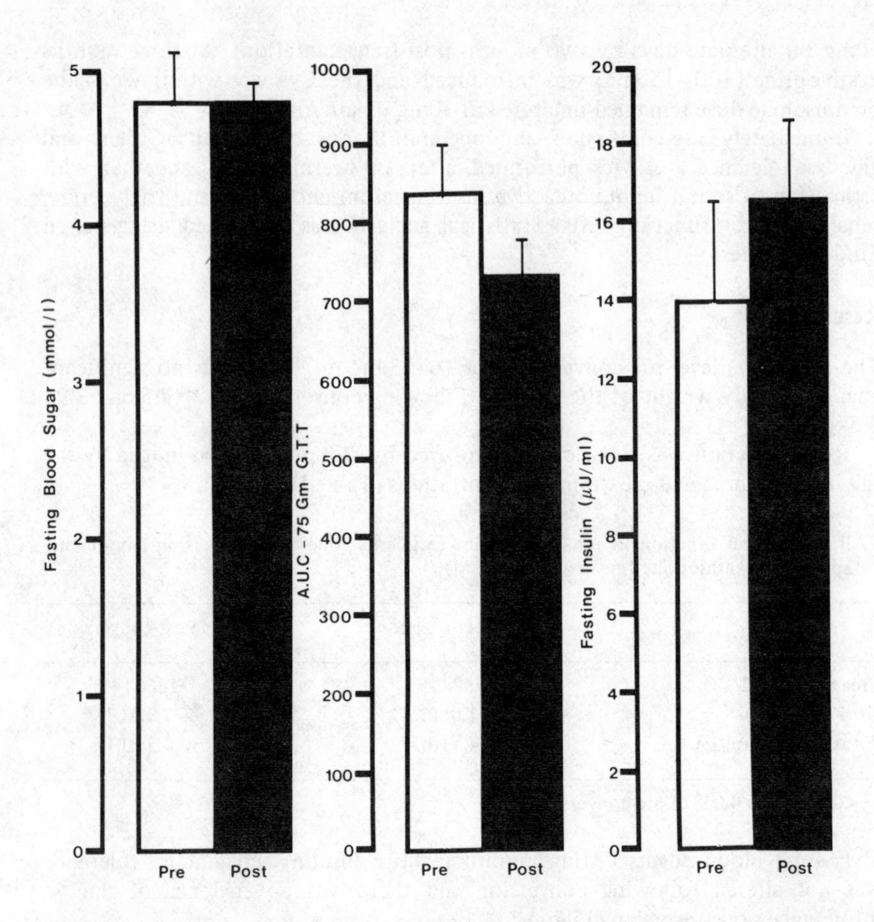

Figure 1. Effect on glucose metabolism of conversion from cyclosporine to azathioprine (mean ± SEM)

Changes in renal function are known to affect glucose [3] and lipid metabolism [4], possibly by an alteration in peripheral insulin resistance.

Although an improvement in renal function was observed following conversion, there was no change in serum insulin levels, and it therefore seems unlikely that a change in insulin resistance could account for the improvement in serum lipids.

Prednisolone is known to produce abnormalities in both triglyceride and cholesterol levels [5] although these effects can be minimised by the use of alternate day steroid dosaging as used in this study [6]. Although the dose of prednisolone remained unchanged over the conversion period, Cys has been shown to reduce the clearance of prednisolone, which is almost entirely metabolised by the liver. Thus the effect of a given dose of prednisolone may be greater when Cys is given concomitantly. However, prednisolone is thought to affect lipid levels by inducing hyperinsulinaemia to which the liver remains

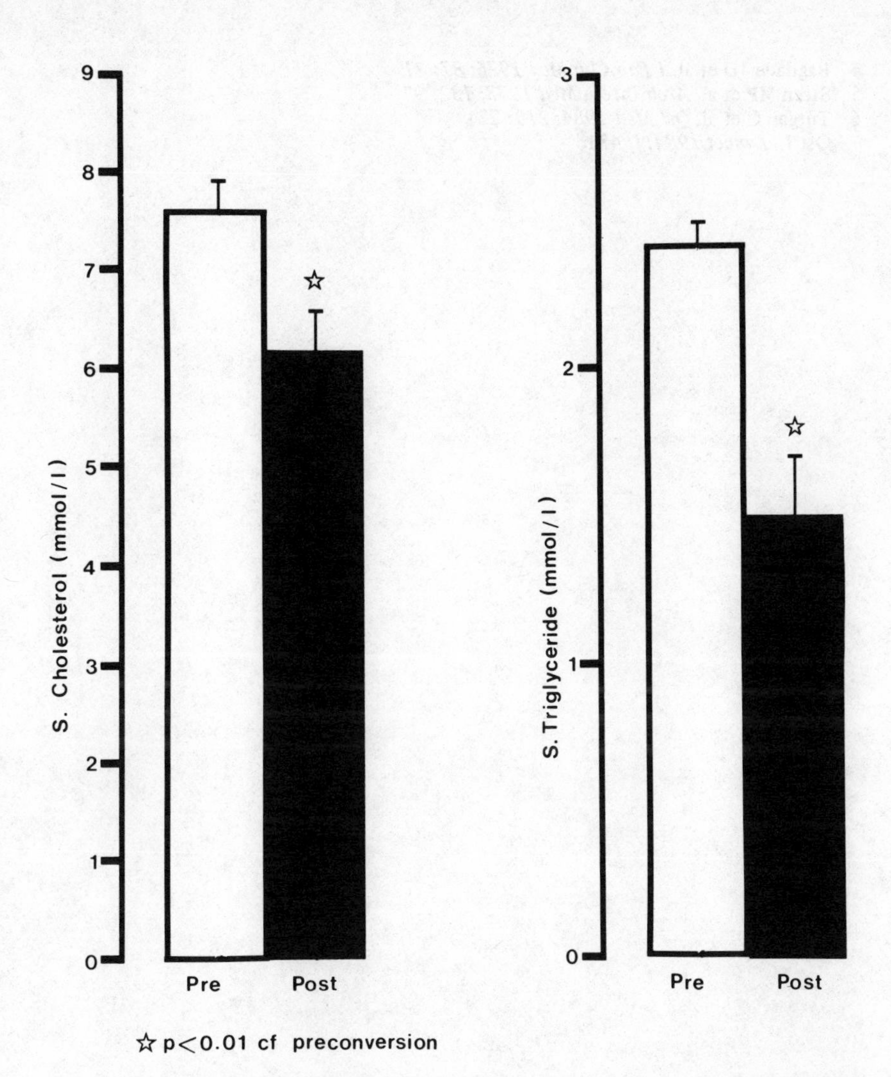

Figure 2. Effect on serum cholesterol and triglyceride levels following conversion from cyclosporine to azathioprine (mean ± SEM)

selectively responsive, and in this study insulin levels were unaltered over the conversion period. Cys may be affecting lipid metabolism by an independent action on the liver.

References

1 Calne R. *Nephron 1980; 26:* 57
2 Gunnarsson R et al. *Lancet 1983; ii:* 571
3 DeFronzo RA et al. *Medicine 1973; 52:* 469

4 Bagdade JD et al. *J Lab Clin Med 1976; 87:* 37
5 Stern MP et al. *Arch Intern Med 1973; 132:* 97
6 Turgan C et al. *Q J Med 1984; 210:* 271
7 Ost L. *Lancet 1984; i:* 451

GLOMERULAR C3b RECEPTOR LOSS IN RENAL ALLOGRAFTS

F Nolasco, *B Hartley, *R Reuben, * K Welsh

*Hospital Curry Cabral, Lisbon, Portugal, *Guy's Hospital, London, United Kingdom*

Summary

There is almost no data from the glomeruli of allografted kidneys with respect to changes in the CR-1 (C3b) receptor expressed on glomerular podocytes. We studied 22 renal graft biopsies from rejecting and stable allografts, using a panel of monoclonal antibodies. We found that the CR-1 expression was decreased in a focal and segmental fashion in some biopsies, particularly in rejecting kidneys. These changes correlated with the intensity of glomerular mononuclear cell infiltration, but in contrast no correlation was seen with peripheral capillary wall deposition of complement (C_3). Thus, some active process is occurring in the glomeruli of rejecting grafts which affects the expression of the CR-1 receptor.

Introduction

In contrast to other mammals, human and other primate glomeruli express a receptor for the activated fragment of C3b [1], and these receptors are of the CR-1 type [2]. Similar receptors are present also on the surface of red blood cells, neutrophils, B-lymphocytes, macrophages and some T-lymphocytes [3]. The use of monoclonal antibodies, together with immunoelectron microscopic techniques, has demonstrated that their expression in the glomerulus is limited to the visceral epithelial cells (podocytes) [4].

The CR-1 receptors found outside the glomeruli are involved in the binding of immune complexes containing C3b, with subsequent endocytosis by fixed or migratory phagocytic cells [3]. The number of receptors expressed on the cell surface is variable, being increased in activated cells [3]. As with other surface receptors, they are also able to show the capping phenomenon [3]. The function of the CR-1 receptors in the glomerulus is, in contrast, unknown, but their expression is decreased in the glomeruli of patients with some types of glomerulonephritis [2,5]. This finding was at first thought to relate to the presence of deposits of C_3 within the glomeruli [2,5], but this is now known to be incorrect: in lupus nephritis, the striking finding was that only in proliferative glomerulonephritis was the expression of CR-1 decreased, even though there were extensive deposits of C_3 in the patients with membranous lupus nephropathy studied [5]. In most forms of proliferative glomerulonephritis, including lupus nephritis, monocytic infiltration is prominent [6].

In order to clarify a possible relation between the CR-1 receptor, glomerular complement deposition and mononuclear cell infiltration, we studied renal allograft biopsies obtained during the first few months after transplantation. It is now clear that mononuclear cells (particularly monocytes) invade the glomerulus of allografts, especially at times of rejection [7]. In contrast to glomerulonephritis, deposition of C_3 in the capillary walls is rare.

Patients and methods

Biopsies

Twenty-two needle biopsies were performed on 19 renal allografts, treated either with prednisone and azathioprine (10 biopsies) or prednisone and cyclosporine (12 biopsies). On clinical, radiological, radio-isotopic and histological criteria, 12 of the biopsies were considered to have been taken during episodes of rejection; the other biopsies are taken for episodes of renal dysfunction not considered to be the result of rejection, or routinely at one and four weeks in the case of the cyclosporine-treated grafts. Tissue was processed in all cases for conventional light microscopy, and in 12 biopsies was also studied using commercially-available peroxidase-conjugated antisera directed against IgG, IgM, IgA, C_3, C_4 and Clq.

Monoclonal antibodies

A panel of monoclonal antibodies, previously well-characterised, was applied to additional cryostat sections from all 22 biopsies and revealed using an indirect (anti-mouse) immunoperoxidase conjugate. The following antisera were used, recognising epitopes expressed on the cells indicated: all leucocytes (2D1); B-lymphocytes (TO 15); monocyte/macrophages (FMC 32); natural killer (NK) cells (Leu 7); Pan-T-cells (UCHT 1); T helper/inducer cells (Leu 3a); and T cytotoxic/suppressor cells (UGHT 4). For revealing C3b receptors, we used the antibody TO 5. Positive (tonsil) and negative controls were used with each monoclonal antibody. Only sections containing two or more glomeruli were reviewed. The total number of positive cells with each monoclonal antibody, and the number of glomeruli in each section was counted. Results were expressed as the number of positive cells per glomerulus. The intensity of the staining obtained with the anti-C3b monoclonal antibody TO 5, and with the polyclonal anti-C_3 and immunoglobulin antibodies was recorded.

Results

CR-1 staining

A total of 17 biopsies showed uniform peripheral staining of the CR-1 receptor, with a similar intensity in all, and identical to results obtained previously with

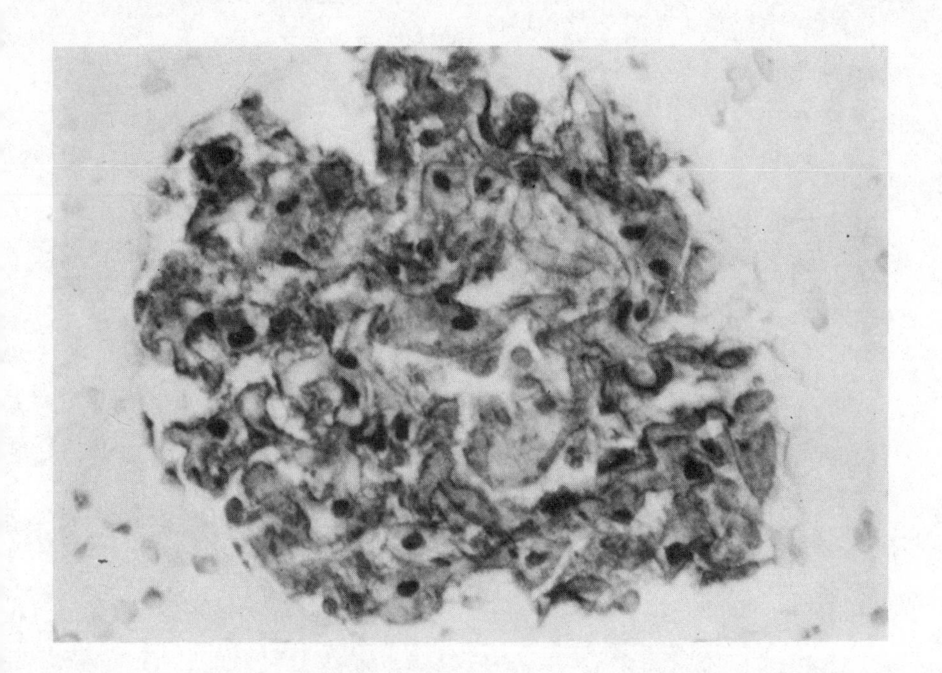

Figure 1. Immunoperoxide staining of a transplant glomerulus with a monoclonal antibody anti-C3b receptor (TO 5). Cryostat section counterstained with haematoxylin. Magnification x 400 (reduced for publication). The C3b receptor is uniformly distributed in the glomerulus. Strong staining is seen both on the cytoplasm and in a linear fashion on the surface of podocytes. It is impossible using this magnification to exclude some mesangial staining

normal kidneys and from patients with interstitial nephritis (Figure 1). Although the CR-1 positivity appeared to be found only on the epithelial cells, we were unable, using light microscopy alone, to exclude endothelial or mesangial staining. An abnormal CR-1 pattern characterised by focal and segmental loss of the receptor was observed in five biopsies (Figure 2). Four of these were from rejecting grafts, and the fifth from a patient with cytomegalovirus (CMV) infection at the time of biopsy. The four biopsies represented 30 per cent (4/12) of those taken from rejecting kidneys.

Glomerular immune deposits

Of these five biopsies with abnormal CR-1 expression, glomerular immune deposits were sought in four: three rejecting kidneys, and the one with CMV. Only two of the rejecting kidneys showed any peripheral glomerular capillary fixation of anti-C_3 antibody, without immunoglobulins. The patient with CMV infection showed no C_3 fixation, but there was peripheral deposition of IgG, IgM and IgA. In the group of 17 patients with normal CR-1 expression in their biopsies, immune deposits were looked for in 11, but in only one was C_3 found

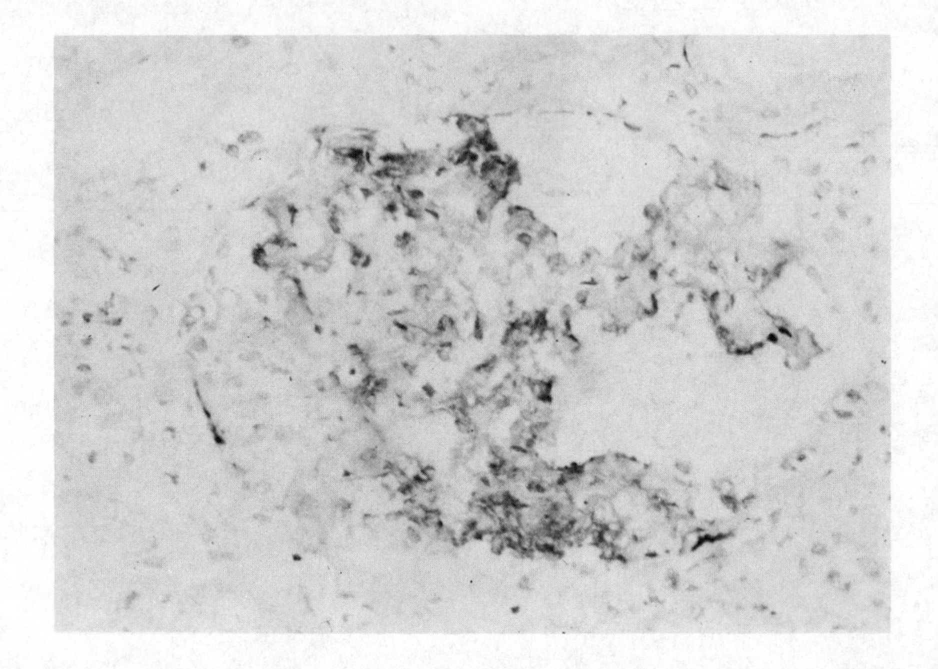

Figure 2. Immunoperoxidase staining of transplant glomerulus with a monoclonal antibody anti-C3b (TO 5). Cryostat section counterstained with haematoxylin. Magnification x 400 (reduced for publication). Areas with complete loss of the epithelial complement receptor are seen

in the peripheral capillary walls, and immunoglobulins in two others, apparently linear in one.

Intraglomerular mononuclear cells

The numbers of intraglomerular cells in the biopsies with and without changes in expression of the CR-1 receptor are shown in Figure 3. Biopsies with CR-1 loss had a significantly higher number of intraglomerular leucocytes, with a predominance of macrophages. All types of cells were present in excess, however, except B cells and T helper/inducer cells (Wilcoxon rank-sum test).

Effect of immunosuppressive regime

The results appeared to be the same in patients given either azathioprine or cyclosporine as an adjunct to prednisone (Figure 1).

Discussion

Our results confirm that there is an increase in the number of infiltrating intraglomerular leucocytes, especially macrophages, and particularly in grafts judged

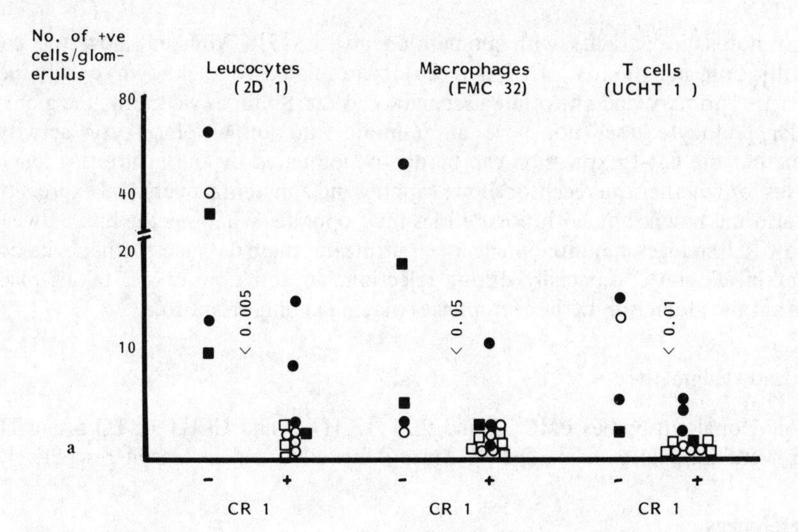

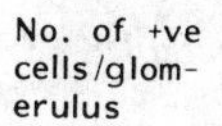

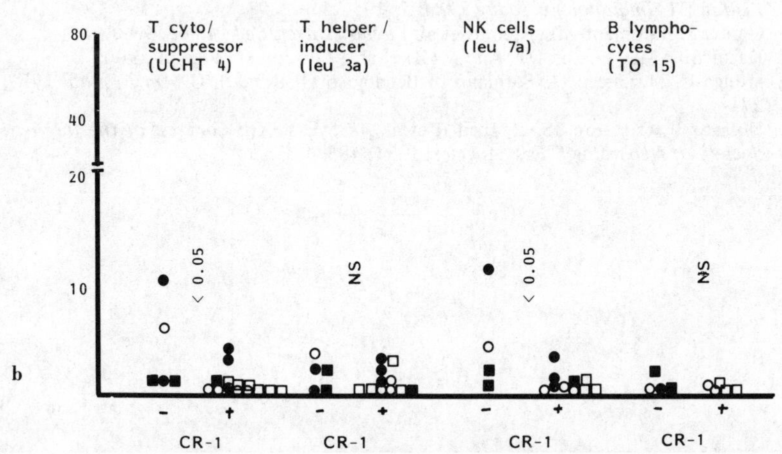

Figures 3a and b. The number of mononuclear cells positive with various monoclonal antibodies within the glomerulus, in relation to normal (+) and abnormal (–) expression of the endothelial cell CR-1 receptor, judged by the monoclonal antibody TO 5. Rejecting grafts are shown by solid symbols, non-rejecting grafts by open symbols. The circles indicate patients treated with cyclosporine; both groups received prednisone. The significance of the differences in number of each group of infiltrating cells, between those with normal and reduced expression of CR-1 receptor, is indicated (Wilcoxon rank-sum test)

on other criteria to be rejecting. These grafts may show a decreased expression of the glomerular epithelial cell CR-1 receptor. The changes did not seem to correlate with the occasional presence of peripheral deposits of C_3, a finding also noted in patients with glomerulonephritis [5]. Why and how the cell infiltration and the loss of CR-1 receptor are related is not known, or whether one is a primary and the other a secondary event. So far as we know, the glomerular podocyte does not have an immune function or phagocytic activity. Whether the CR-1 expression can be down-modulated by the infiltrating leucocytes, or whether the receptor shows capping and apparently decreased expression is also unknown, but both possibilities may operate. Whatever the link between the CR-1 changes and mononuclear cell infiltrate, these data are further evidence that in allografts, especially during rejection, an active process is taking place within the glomeruli, with macrophages playing an important role.

Acknowledgments

Monoclonal antibodies FMC 32 and 2D1; UCHT 1 and UCHT 4; TO 5 and TO 15 were kind gifts from Drs H Zola, P Beverley and D Mason respectively.

References

1 Gelfand MC, Frank MM, Green IJ. *J Exp Med 1975; 142:* 1029
2 Emancipator SN, Iida K, Nussenzweig V, Gallo GR. *Clin Immunol Immunopathol 1983; 27:* 170
3 Fearon DT. *Immunology Today 1984; 5:* 105
4 Gerdes J, Hansmann ML, Stein H et al. *Virchows Arch Cell Pathol 1982; 40:* 1
5 Kazatchine MD, Fearon DT, Appay MD et al. *J Clin Invest 1982; 69:* 900
6 Monga C, Mazzucco G, Barbiano di Belgiojoso G, Busnach G. *Am J Pathol 1979; 94:* 271
7 Nolasco F, Cameron JS, Reuben R et al. *Abstracts IXth Congress of the International Society of Nephrology, Los Angeles.* 1984: 485A

Proc EDTA–ERA (1984) Vol 21

CYCLOSPORINE-ASSOCIATED LYMPHOPROLIFERATION, DESPITE CONTROLLED CYCLOSPORINE BLOOD CONCENTRATIONS, IN A RENAL ALLOGRAFT RECIPIENT

J B Dossetor, T Kovithavongs, M Salkie, J Preiksaitis

University of Alberta, Edmonton, Canada

Summary

In a patient receiving sulfinpyrazone (Anturan, Geigy) an unusually high dose of cyclosporine (Cys) was required to maintain serum values in the range of 50–200ng/ml. After eight months of 1300–1500mg/day, the patient complained of increasing malaise and symptoms of cyclosporine side-effects. This clinical state was accompanied by splenomegaly and two monoclonal peaks in the gamma region on serum electrophoresis. Concomitantly, rising cytomegalovirus IgM titres, following by rising IgG titres, indicated a primary cytomegalovirus infection. This ominous biclonal proliferation markedly diminished during the subsequent six months, during which time the cyclosporine dose was minimised. He returned to good health, splenomegaly and monoclonal gamma globulin virtually disappearing. He remains well at 16 months post-transplantation.

Clinical events

This case presentation, as summarised above, concerns the post-transplant course of a 55-year old man, J.S., with renal failure due to membrano-proliferative glomerulonephritis, who received an HLA-identical sibling's kidney on May 11th, 1983. During the operation, atheromatous plaques were removed from the hypogastric artery; this prompted the use of the platelet inhibitor, sulfinpyrazone, from the sixth post-operative day when serum creatinine rose from 145 to 200μmol/L. He was also given three daily doses of 1.0g intravenous methylprednisolone at this time and serum creatinine fell to 124μmol/L by day 11. The immediate post-operative course was otherwise uneventful and he went home at 14 days.

He was then maintained on (a) cyclosporine, in doses determined by regular estimation of serum cyclosporine concentrations; (b) prednisone dose which rapidly decreased to 15mg on alternate days; and (c) sulfinpyrazone (100mg, four times daily).

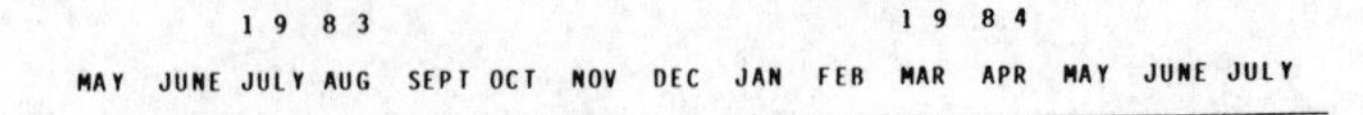

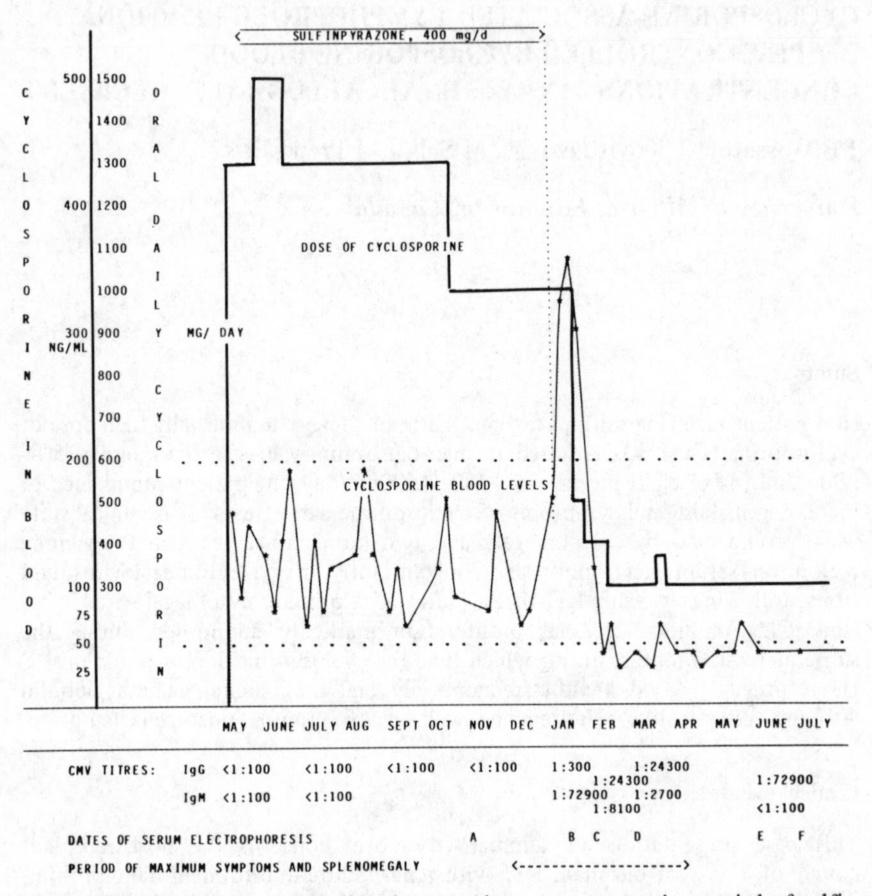

CMV TITRES:								
	IgG	<1:100	<1:100	<1:100	<1:100	1:300	1:24300 1:24300	1:72900
	IgM	<1:100	<1:100			1:72900	1:2700 1:8100	<1:100

DATES OF SERUM ELECTROPHORESIS A B C D E F

PERIOD OF MAXIMUM SYMPTOMS AND SPLENOMEGALY <------------------->

Figure 1. Parameters of cyclosporine dosage and serum concentration, period of sulfinpyrazone administration, and cytomegalovirus titres, for patient J.S. during the first 15 months after transplantation

Some parameters of J.S.'s post-operative course are illustrated in Figure 1. During the early follow-up months, he complained of listlessness, mild but persistent conjunctivitis, oily blistering in the scalp, gynaecomastia with slight white secretion, increased hair on chest, scalp and eyebrows and a bitter taste in the mouth which was occasionally associated with excessive flow of saliva. During this time, June 1983 to January 1984, mean serum creatinine was 173μmol/L (n=27) despite the high dose of cyclosporine (1300–1500mg/d) which was needed to keep the serum Cys value in the therapeutic rage of 40–200ng/ml.

On admission on January 12th, 1984 (8 months post-transplant), it was found that he had developed splenomegaly and two monoclonal proteins in the gamma globulins (on serum electrophoresis), one IgG (λ), the other IgM (κ), measured at 5.3 and 6.8g/L, respectively, as seen in Figures 2 and 3.

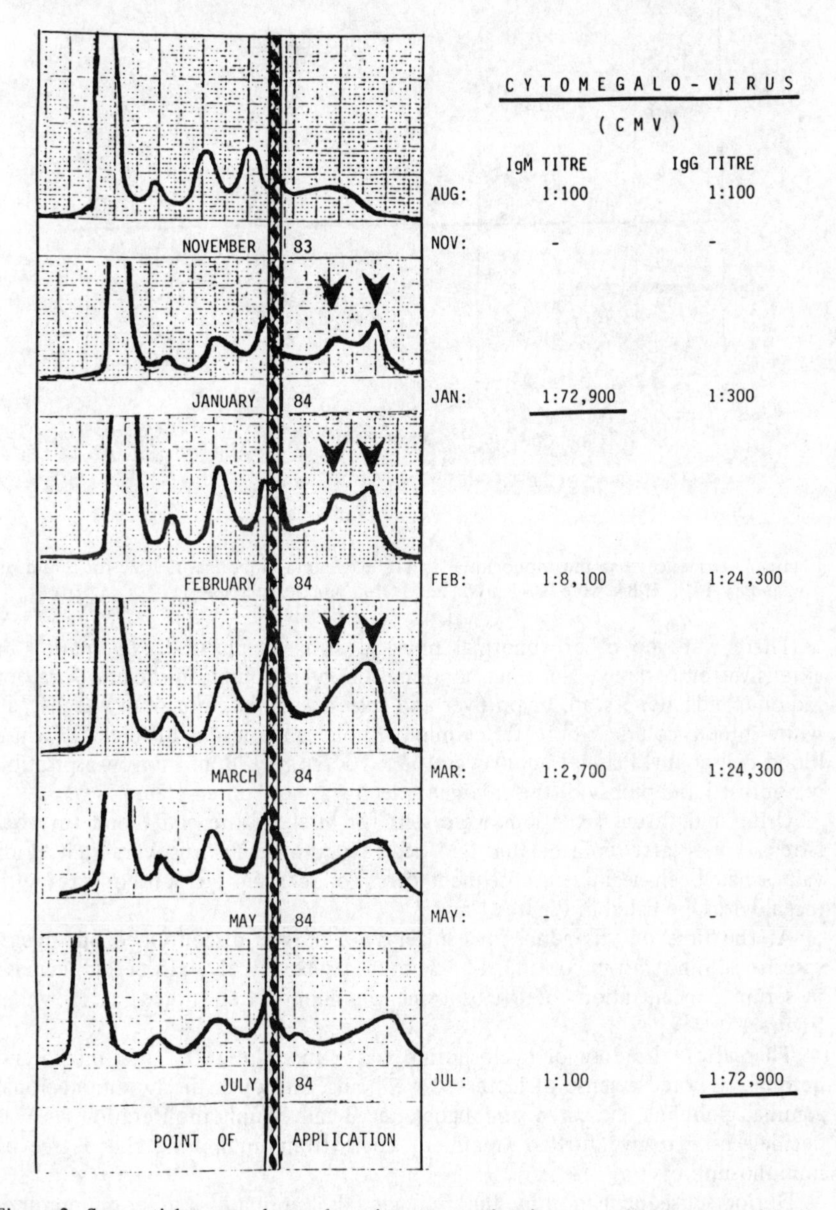

Figure 2. Sequential serum electrophoresis patterns, showing monoclonal gamma globulin peaks; IgG and IgM titres of antibody to cytomegalovirus

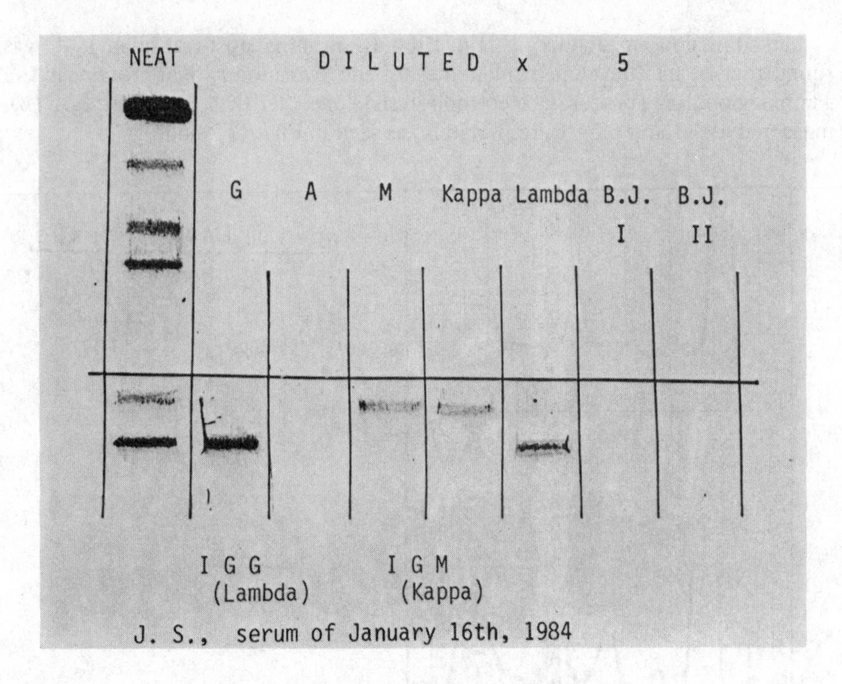

Figure 3. Evidence that the monoclonal bands in gamma globulins in J.S., for serum on January 16th, 1984, were IgG-lambda and IgM-kappa, respectively

There were no other abnormal physical signs on physical examination or extensive tomography. No other abnormalities were found on skeletal X-ray or radionucleide bone scan, or on liver and spleen scan. Haemoglobin was 14g/dl, white blood count 7.8 x 10^5/cu.mm without an unusual differential white blood cell count. Platelet count was 168 x 10^5/cu.mm. Bone marrow aspiration was refused and biopsy of the enlarged spleen was not contemplated.

Urine and throat secretions were sent for viral culture and blood for viral titres. It was later reported that IgM titres to cytomegalovirus were high. Prior values had been negative. Subsequent titres for IgM and IgG antibody to cytomegalovirus are listed in Figures 1 and 2.

At the time of this admission, sulfinpyrazone was discontinued but cyclosporine administration continued unchanged. Unexpectedly, a threefold increase in serum concentrations of cyclosporine was immediately noted, as shown in Figure 1.

Thereafter, the dose of cyclosporine was reduced to a low total intake and he quickly noted a sense of increased wellbeing. Although the two monoclonal gamma globulins indicated an oligoclonal B-cell lymphoproliferation, it was decided not to give further treatment apart from minimising the degree of immunosuppression.

During subsequent months, the monoclonal abnormalities in serum immunoglobulins have largely disappeared (see Figure 2) and the spleen is no longer palpable. Renal function is good with a mean serum creatinine (for months 9–15, post-transplant of 161μmol/L (n=21).

Discussion

Cleary et al [1] found evidence for monoclonal or oligoclonal lymphoproliferation in the abnormal lymphoid tissue in all of 10 cardiac transplant recipients, when analysed for immunoglobulin gene rearrangement using the Southern blot DNA hybridisation technique. This characteristic of B-cell lymphomas was found even when there was no immunoglobulin being synthesised in the cells or detectable on lymphoid cell membranes. Nine of these patients died of their lymphoproliferative disease. Others have noted that lymphoma is common in such patients [2,3], who differ from renal transplant patients in that marked reduction or discontinuation of immunosuppression is usually not possible. Starzl et al [4] has recently reviewed post-transplant lymphomas and lymphoproliferative disorders in patients receiving cyclosporine and steroids and have shown that regression of the lymphoproliferation often occurs if immunosuppression is drastically reduced. Thus seven of eight renal transplant recipients are alive, tumour-free, up to several years after marked reduction or cessation of immunosuppression, even though three of them lost the kidney graft from rejection. Thus the line of distinction between lymphoma and reversible lymphoproliferation, in immunosuppressed patients, is quite blurred.

Although there is no tissue diagnosis of lymphoproliferation in this case, the occurrence of splenomegaly and monoclonal lymphoproliferation, and their resolution when cyclosporine dosage was minimised, provides evidence that J.S. had a similar disorder.

Many have noted the association of Epstein-Barr virus infection in such patients [4,5] and have speculated on its role in lymphoma causation. The association in this patient is with a primary cytomegalovirus infection. The relationship between monoclonal lymphoproliferation and primary cytomegalovirus infection, in this instance, is purely speculative, but of considerable interest. The low ratio of T.h to T.cyt/sup in the peripheral T lymphocytes provides further evidence of significant viral infection at the time of maximal ill-health (data not shown).

In an effort to find out if this patient's cells showed evidence of malignant change, one of our colleagues, Dr Chris Bleakley, Department of Biochemistry, studied the DNA of peripheral blood lymphocytes of J.S. for evidence of gene rearrangement, using DNA hybridisation; these studies have not been completed, but thus far have shown no deviation from normal. Immuno-phenotyping of J.S.'s peripheral blood leucocytes has also been studied by a pathologist, Dr Gordon Bain, looking for an abnormality in the ratio of kappa to lambda light chains on B-lymphocyte membranes, but no such abnormality was found.

Additionally, this case-history shows the interaction of sulfinpyrazone and cyclosporine which resulted in administration of an unusually high dose of cyclosporine over an eight month period. This is presumed to be the cause of a lymphoproliferative disease in which cytomegalovirus primary infection may have played a central role.

The case illustrates the state of ignorance of interactions of other drugs with cyclosporine, even though those with phenytoin [6], intravenous sulphadimidine and trimethoprim [7] and ketoconazole [8] have been described. Presumably sulfinpyrazone might complete with Cys for binding sites and thus lead to aberrant metabolism or tissue distribution. Because of clinical evidence of cyclosporine side effects in this case, it is unlikely that the action of sulfinpyrazone was to induce increased metabolism of cyclosporine, such as is the case with phenytoin [5]. Studies of the interaction of these two drugs in vivo are in progress.

Conclusion

1. It is concluded that sulfinpyrazone interfered with either the assay or tissue distribution or metabolism of cyclosporine, making serum cyclosporine-RIA assays invalid as a basis for dosage.
2. Excessive dosage of cyclosporine, caused cyclosporine side-effects (but not nephrotoxicity) and led to lymphoproliferation which had some properties (monoclonality) which are often associated with lymphoid cell 'malignancy' in the non-immunosuppressed subject.
3. Primary cytomegalovirus infection played a special role in this lymphoid proliferation.
4. Marked reduction of immunosuppression was associated with disappearance of splenomegaly and the monoclonal peaks in serum electrophoresis, and patient wellbeing was restored. The latter continues at 16 months, post-transplant.

References

1 Cleary ML, Warnke R, Sklar J. *N Engl J Med 1984; 310:* 477
2 Krikorian JG, Anderson JL, Bieber CP et al. *JAMA 1978; 240:* 639
3 Weintraub J, Warnke RA. *Transplantation 1982; 33:* 347
4 Starzl TE, Porter KA, Iwatsuki S et al. *Lancet 1984; i:* 583
5 Klein G, Purtilo D. *Cancer Res 1981; 41:* 4302
6 Keown PA, Stiller CR, Laupacis AL et al. *Transplant Proc 1982; 14:* 659
7 Wallwork J, McGregor CGA, Wells FC. *Lancet 1983; i:* 366
8 Ferguson RM, Sutherland DER, Simmons RL et al. *Lancet 1982; ii:* 882

Proc EDTA–ERA (1984) Vol 21

SHORT-TERM KIDNEY PRESERVATION: TO PERFUSE OR NOT TO PERFUSE WITH THE NEW BELZER PERFUSATE

R McCabe, J Lin, K Cooke, M Jean-Jacques

St Luke's-Roosevelt Hospital Center, New York, USA

Summary

Delayed graft function is ischaemically induced and the consequence of high energy phosphate depletion. A new perfusate, described by Belzer, has been used in 43 kidney preservations, 16 of which had prolonged cold ischaemia. The overall immediate function rate was 37 of 43 kidneys or 86 per cent. Of 18 kidneys cold-stored during the same period, only nine functioned immediately. Twenty-six of these patients received cyclosporine for immunosuppression without an increase in delayed graft function or evidence of increased nephrotoxicity because of ischaemic injury or prolonged preservation.

Introduction

In the absence of effective centralised cross-matching capability, kidney preservation time has necessarily been extended throughout the United States to fully utilise all available kidneys in an increasingly cytotoxic recipient pool. As a consequence, the incidence of delayed graft function (DGF) had approached 80 per cent [1,2] for kidneys cold-stored longer than 36 hours in Collins Solution. Stiller [2] reported primary non-function in the cyclosporine immunosuppressed patient when preservation, especially by perfusion was extended beyond 24 hours. It is apparent that traditional methods of kidney preservation may no longer be adequate.

Belzer [3] has developed a new perfusate (Table I) for machine preservation that has an albumin base, a phosphate buffer in lieu of bicarbonate and a non-permeable anion gluconate instead of chloride and adenosine substrate. The perfusate was designed to reduce ischaemically-induced cell swelling and results in ATP synthesis during machine preservation. Successful five day canine kidney preservations were reported.

We have used this new perfusate in our laboratory for the past 12 months with gratifying results.

TABLE I. Composition of Belzer II perfusate

Distilled, deionised H_2O 800ml
- Sodium gluconate 17.5g
- KH_2PO_4 3.4g
- Glucose 1.5g
- Glutathrone 0.9g
- Adenosine 1.3g
- HEPES (buffer) 4.7g
- NaOH 5 N~ 8ml, adjust pH to 7.8–7.9
- Penicillin 200,000 U
- Dexamethasone 8mg
- Phenosulphathelein 12mg
- Insulin 40 U
- H_2O to bring volume to 850ml
- Filter sterile and store
- Before use add Albumin (sterile) 150ml
- $MgSO_4$ 1g (sterile)

Final values:
- $Na^+ = 133 \pm 3$mEq/L
- $K^+ = 24 \pm 2$mEq/L
- mOsm = 300 ± 5mOsm/L
- pH = 7.6–7.7

Materials and methods

Over a four month period starting June 15, 1983, all kidneys received in our laboratory were placed on the Waters Mox 100 Perfusion Circuit primed with Belzer II perfusate. Systolic pressure was initially set at 50mmHg at a standard pulse rate for our laboratory of 60, without further adjustment.

Thirty-one kidneys were perfused and transplanted during this period. Subsequently, 30 additional kidneys were also preserved, 12 by perfusion and 18 by cold storage in Collins C2 solution.

TABLE II. Incidence of delayed graft failure in the preserved renal allograft
Columbia University – St Luke's Hospital Center

Groups	No. of kidneys	Mean cold storage in hours (range)	Mean perfusion in hours (range)	Immediate function (%)	Delayed function
Group I Perfusion with Belzer II	27	–	35.8 (21–60)	24 (89)	3 (11)
Group II C.S. + perfusion with Belzer II	16	10.5 (2–36)	25.0 (13–50)	13 (81)	3 (19)
Group III C.S. alone	18	37.2 (21–58)	–	9 (50)	9 (50)

Twenty-seven kidneys (Table II, Group I) were placed on the perfusion circuit at the time of recovery without cold storage; the mean perfusion time was 35.8 hours with a range of 21–60 hours. Sixteen kidneys (Group II) were recovered from more distant centres and had a mean cold storage period of 10.5 hours with a range of 2–36 hours before a mean perfusion time of 25.0 hours (range 13–50 hours). The total preservation time in these 16 kidneys was 36.5 hours with a range of 27–68 hours. The 18 cold-stored kidneys (Group III) had a mean preservation time of 37.2 hours (range 21–58 hours).

Donors were between the ages of 10–55 (mean 31.5 years). Cerebral death was present in all. The circulation was supported by a respiratory and urinary output terminally stabilised with vasopressors, electrolyte solutions, volume expanders and diuretics as necessary.

Eighteen of the 43 recipients who received a machine preserved kidney and six who received a cold-stored kidney were immunosuppressed with cyclosporine. All recipients had a negative cross match to the donor antigen without regard to the HLA or DR match.

Post-transplant renal function was documented by a daily report on urinary output, serum BUN and creatinine determinations and the need for post-transplant haemodialysis at the recipient centre.

Results

Twenty-four of the 27 kidneys transplanted (89%) after perfusion with Belzer II solution alone (Group I) functioned immediately and the recipients did not require post-operative dialysis (see Table II).

In two of three patients with delayed graft function, there were peri-operative complications that contributed to delayed graft function. These three patients required dialysis only one to three times before life-sustaining function returned.

Of the 16 kidneys perfused with Belzer II solution after 2–36 hours of cold storage (Group II), 13 functioned immediately (81%). Of the 18 cold-stored kidneys in Group III, only nine functioned immediately; at least six life-sustaining haemodialysis treatments were required in the remaining nine. There was no correlation between the incidence of delayed graft function and the type of immunosuppression. Cyclosporine immunosuppression was well tolerated in all kidneys independent of the method or length of time of preservation. Primary non-function was not observed, even with cyclosporine immunosuppression.

Discussion

Delayed graft function is primarily due to ischaemic injury to the kidney during terminal events in the donor or to problems related to the recovery but increased by cold storage. Further damage may occur as a consequence of the management of the recipient. Delayed graft function may also result from immunological injury to an ischaemically damaged kidney in spite of a negative T and B cell cross match due to the presence of undetected endothelial antibodies. Delayed graft function is not acute tubular necrosis alone, but rather is a combination of interstitial oedema, cell swelling in the glomerulus, and damage to both endothelial surfaces and tubular epithelium. Ischaemic damage is initiated by high energy phosphate (ATP) depletion that ultimately results in damage and cell death. With warm ischaemia, ATP is rapidly depleted and these changes become irreversible, but with hypothermia oxygen requirements are reduced by 95 per cent, thus permitting many hours of safe storage in properly designed electrolyte solutions. These changes are arrested and may even be reversed by perfusion (Figure 1) with the adenosine-phosphate buffered perfusate.

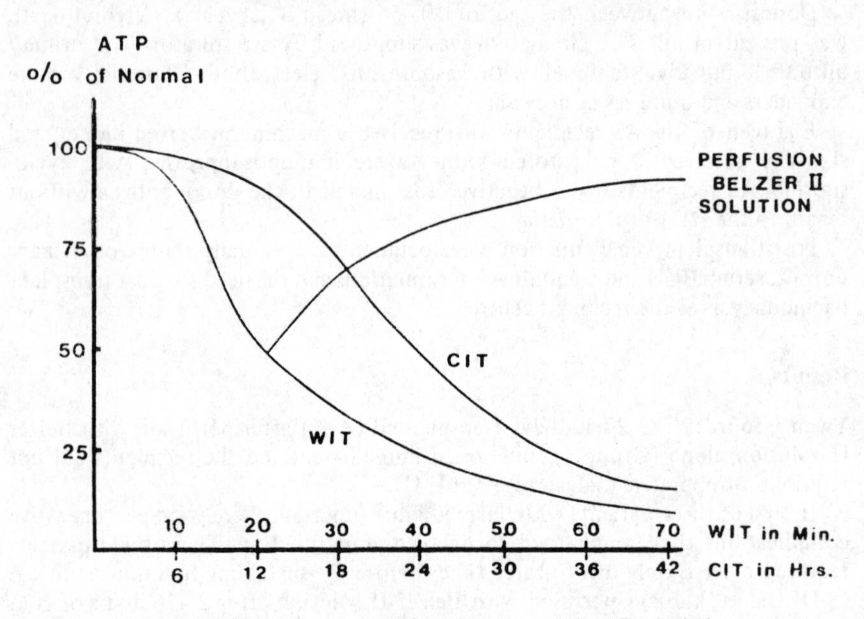

Figure 1. Effect of perfusion with adenosine on the ATP depletion

Clearly perfusion of the cold stored kidney with Belzer II solution was beneficial to the kidney, and effectively reduced the incidence of delayed graft function to an acceptably low level. Active oxygen transport afforded by perfusion is required for this benefit, as ATP synthesis does not occur in O_2-deprived cold storage solutions. Although our numbers are small [3], and not statistically significant, we postulate that at least 12 hours of perfusion is necessary to restore ATP depleted by only a few hours of cold storage; perfusion for less than 12 hours seemed to contribute to delayed graft function. Belzer has furthermore shown that in a canine model the kidney may be safely stored in the perfusate for 24 hours while shipping from the preservation facility to the recipient centre [4,5].

Perfusion is clearly a more expensive method of preservation than cold storage and the quality is dependent upon the skills of a highly trained technician, but it is cost-effective by reducing delayed graft function and shortening the period of hospitalisation that delayed function entails. With cyclosporine immunosuppression and a kidney that functions immediately, the compliant recipient is ready for discharge within one to two weeks after transplantation.

On the basis of one year's experience in the preservation of human cadaver kidneys for transplantation by perfusion with Belzer II solution, we recommend widespread use of these techniques in the preservation of all kidneys in order to facilitate post-transplant management of the recipient.

Conclusion

1. Perfusion with Belzer II solution affords dependable preservation of the kidney for prolonged periods up to 72 hours, even those with prolonged pre-perfusion cold storage.
2. Cyclosporine immunosuppression does not increase incidence or severity of delayed graft function in those kidneys preserved over 24 hours with Belzer II solution.
3. Perfusion with Belzer II solution is the preservation method of choice.

References

1 Barry JM, Lieberman S, Wickre C et al. *Transplantation 1981; 32:* 485
2 Barry JM. Personal communication
3 Stiller C, London, Canada, for the Canadian Transplant Study Group. *Transplant Proc 1983; 15 (suppl 1):* 2479
4 Belzer FO, Sollinger HW, Glass NR et al. *Transplantation 1983; 36:* 633
5 Belzer FO, Hoffman BS, Rice MJ, Southard JH. *Transplantation 1984:* in press

Proc EDTA–ERA (1984) Vol 21

THROMBOXANE B_2 AND β_2-MICROGLOBULIN AS EARLY INDICATORS OF RENAL ALLOGRAFT REJECTION

H B Steinhauer, H Wilms, P Schollmeyer

University of Freiburg, Freiburg, FRG

Summary

In a prospective study the diagnostic value of urinary thromboxane B_2 (TXB_2) and β_2-microglobulin (βMG) in renal allograft rejection was studied in 34 patients after transplantation. Twenty-four episodes of rejection were diagnosed by clinical symptoms. The clinical diagnosis of rejection was confirmed by an increase of urinary TXB_2 in 21 (88%) cases. The augmented renal excretion of TXB_2 proceded the clinical signs of rejection for 2.0 ± 0.75 days. The symptoms in the remaining three (12%) cases of supposed allograft rejection without increased urinary TXB_2 were caused by non-immunological events (urinary tract infection, acute tubular necrosis).

No elevated TXB_2 excretion was observed during urinary tract infection, sepsis, and acute tubular necrosis whereas urinary βMG increased during these events as during transplant rejection.

Urinary TXB_2 was found to be an early, specific, and sensitive marker of renal allograft rejection with greater reliability than βMG excretion or clinical signs of rejection.

Introduction

The early diagnosis of renal allograft rejection is thought to have a better prognosis if the therapy is initiated immediately [1]. As the clinical diagnosis of transplant rejection is difficult, especially under immunosuppressive therapy with cyclosporine, a number of laboratory tests were developed to facilitate the early institution of therapy. The activation of cellular blood components and of platelets during allograft rejection is associated with an increased intrarenal formation of arachidonic acid metabolites. The renal excretion of TXB_2, the stable hydrolysis product of the arachidonic acid metabolite TXA_2 is thought to be an indicator of transplant rejection [2].

The present study was undertaken to investigate the diagnostic value of TXB_2 as specific marker of renal allograft rejection in comparison with clinical symptoms and the urinary excretion of βMG [3].

Patients and methods

Thirty-four consecutive patients (16 females and 18 males, mean age 35.7 ± 1.8 years, age range 16–54 years) were studied from the day of transplantation to day 20–51 (mean observation time: 30.3 ± 2.2 days). All patients received a cadaver kidney.

Immunosuppressive therapy consisted of cyclosporine (Cys) and low dose steroids in 26 patients. The dosage of Cys was chosen according to the European Multicenter Trial [4]. Eight patients were treated by conventional immunosuppressive therapy with azathioprine (A) (4mg/kg/day initially, stepwise reduction to 2.5mg/kg/day or less depending on the peripheral white blood cell count) and prednisolone (P) (2mg/kg/day on day one to 12, reduction to 25mg/day within four weeks after transplantation).

Rejection was diagnosed clinically by the presence of one or more of the following symptoms: rise in body temperature, graft tenderness of the kidney transplant, decreased urine output and endogenous creatinine clearance, ultrasonic increase of allograft size and decreased allograft perfusion on radionucleotide scanning. The therapy for rejection consisted of prednisolone 500mg/day IV (Cys group) or 1000mg/day IV (A plus P group) every second day for three to six days.

Immunoreactive TXB_2 in 24hr urines collected on indomethacin (8–16μg/ml urine) was determined directly by a specific and sensitive radioimmunoassay previously described by our group [5]. For daily analysis the incubation procedure was modified. The incubation time was shortened to 120 minutes at a temperature of 28°C. The lower limit of detection (10% inhibition of binding label to the antiplasma) was 2pg/ml, 50 per cent displacement of binding of label from the antiplasma was caused by 110pg TXB_2.

βMG in urine was determined by a commercially available radioimmunoassay (Phadebas β_2-micro test, Pharmacia Diagnostics, Uppsala, Sweden).

Daily urinary TXB_2 and βMG concentration were compared with the 24 hour urine output, endogenous creatinine clearance, and clinical symptoms of rejection as described above. Values are given as mean ± SEM, Student's 't' test for paired or unpaired observations was used for statistical analysis.

Results

Thirty-four patients were studied for a period of 146 weeks after renal transplantation. No deaths occurred during this time. Differences in graft survival, serum creatinine concentration and rejection rate between group I (Cys) and group II (A plus P) were not significant (Table I).

Twenty-four rejection episodes were diagnosed according to clinical symptoms. The diagnosis of rejection was confirmed by an increase of urinary TXB_2 in 21 (88%) out of 24 cases. In the absence of rejection urinary TXB_2 concentration did not exceed 0.30ng/ml (mean 0.18 ± 0.09ng/ml, n = 27). During rejection TXB_2 increased to 0.88 ± 0.17ng/ml (n = 21, $p<0.001$), 2.0 ± 0.75 days before the clinical diagnosis of allograft rejection. The symptoms in three (12%) further supposed rejection episodes were caused by urinary tract infection in two and by acute tubular necrosis in one case.

TABLE I. Outcome of renal transplantation in 34 patients treated with cyclosporine (Cys) or azathioprine plus prednisolone (A+P)

Group	Patients (n)	Sex	Age (years)	Graft Survival (%)	Serum Creatinine (μmol/L)	Rejection (episodes/patient)
I (Cys)	26	♂ 18, ♀ 14	35.4 ± 2.2 (16 - 50)	81	188 ± 15 (106 - 318)	0.73
II (A+P)	8	♂ 6, ♀ 2	36.4 ± 3.3 (26 - 54)	75	198 ± 19 (106 ± 336	0.62

Mean ± SEM (range)

An increase of urinary TXB_2 was observed in a further five cases without conspicuous clinical symptoms. Three episodes of elevated TXB_2 concentration were observed during the first 10 days after transplantation and coincided with a prolonged improvement in allograft function, the remaining two episodes of elevated urinary TXB_2 were observed during times of slightly increased body temperature supposedly attributable to urinary tract infections. In all five cases urinary TXB_2 levels decreased and graft function improved within three to six days without specific rejection therapy. No correlation existed between TXB_2 urine concentrations and Cys blood levels determined by radioimmunoassay ($y = 0.422 + 8.841 \cdot 10^{-5}x$, $r = 0.051$, $n = 337$).

Urinary βMG concentration increased in all cases of clinical diagnosis of rejection from below 5.0μg/ml (mean 3.24 ± 0.76μg/ml, n = 25) in the absence of rejection, urinary tract infection, systemic infection or acute tubular necrosis to 14.70 ± 3.73μg/ml (n = 21, $p<0.005$), 0.14 ± 0.63 days after the clinical diagnosis of transplant rejection. The values of urinary βMG and TXB_2 during allograft rejection in relation to the time of diagnosis of rejection are summarised in Table II.

An increase of βMG urine concentrations was also observed in eight (73%) out of 11 episodes of urinary tract infection, in two cases of sepsis and one case of acute tubular necrosis. During the initial phase after transplantation urinary βMG exceeded the basal level (3.24 ± 0.76μg/ml, n = 25) in all primary functioning allografts (initial βMG urine concentration 25.3 ± 5.1μg/ml, $p <0.001$).

The radioimmunological determination of TXB_2 in urine was highly reproducible (intra-assay variation 5.8%, interassay variation 7.2%), no degradation of TXB_2 occurred in relation to urine temperature (⩽37°C) and pH (3 to 10) whereas βMG was rapidly degraded in urine with a pH-value of less than six [6].

Discussion

Urinary TXB_2 was found to be an early and sensitive indicator of renal allograft rejection. The major source of immunoreactive TXB_2 is considered to be intrarenal by cells involved in immunological events, especially platelets and macro-

TABLE II. Data concerning renal function, urine thromboxane B_2 (TXB_2), and urine β_2-microglobulin (βMG) in relation to the time of clinical diagnosis of rejection

Time (days)	−6	−4	−2	0[1]	+2	+4	+6
Urine volume (1/24hr)	1.7 ± 0.3 (1.1 – 4.0)	1.3 ± 0.2 (0.1 – 2.3)	1.1 ± 0.2 (0.1 – 1.2)	0.9 ± 0.2* (0.1 – 1.8)	1.2 ± 0.2 (0.1 – 2.0)	1.2 ± 0.2 (0.1 – 2.4)	1.6 ± 0.3 (0.1 – 3.3)
Cl_C (ml/min/1.74m²)	26±6 (5–38)	24±5 (5–52)	27±7 (5–52)	21±5 (5–49)	25±8 (5–64)	28±7 (10–61)	28±6 (5–56)
TXB_2 (ng/ml)	0.36± 0.08 (0.18– 0.58)	0.34± 0.06 (0.18– 0.52)	0.41± 0.05 (0.22– 0.80)	0.79± 0.09* (0.19– 2.16)	0.64± 0.10 (0.26– 2.33)	0.69± 0.12 (0.19– 2.50)	0.30± 0.03 (0.20– 0.52)
βMG (μg/ml)	8.2 ± 2.1 (3.2 –11.3)	6.4 ± 1.8 (3.1 –10.7	6.2 ± 2.4 (2.8 –10.2)	7.4 ± 2.3 (3.4 –12.0)	12.8 ± 3.6 (2.7 –19.7)	14.2 ± 3.4 (4.6 –18.8)	10.8 ± 2.8 (3.8 –16.4)

[1] Day of clinical diagnosis of allograft rejection, −/+: Days before/after clinical diagnosis of rejection, Cl_C: Endogenous creatinine clearance; mean ± SEM (range), n = 24; *: $p<0.05$, in comparison to day six before clinical diagnosis of rejection

phages which are known to accumulate in the allograft during rejection. An extrarenal origin of TXB_2 is described only for venous thromboembolic disease [2,7]. No such events could be observed in our patients. Renal TXB_2 excretion is not responsive to vasoactive hormones contrary to PGE_2 [8], whereas fluid restriction seems to influence TXB_2 excretion by passive reabsorption of TXB_2, associated with ADH-induced water reabsorption [8]. This mechanism is of minor importance in kidney transplant patients who are kept in a state of hypervolemia during the initial phase after transplantation. During effective rejection therapy the urinary TXB_2 excretion returned to basal values indicating no further activation of immunocompetent cells and platelets. The spontaneous decrease of elevated urinary TXB_2 levels in five cases without conspicuous clinical symptoms of rejection implicates the possibility of a spontaneous suppression of slight rejection episodes by continuation of the standard immunosuppressive therapy. In all these cases retrospective analysis of the clinical data revealed the existence of disturbed allograft function or increased body temperature as indicators of an immunological event.

No specific effect of immunosuppressive therapy with cyclosporine, azathioprine plus steroids or additional rejection therapy with high dose steroids could be observed on the excretion of TXB_2 and βMG. This is surprising as steroids possess the capacity to suppress the activity of phospholipases and by this mechanism inhibit the release of membrane-bound phospholipids which are precursors of the arachidonic acid cascade.

No elevated TXB_2 excretion was observed during urinary tract infection, acute tubular necrosis, and generalised infections whereas urinary βMG was increased during these events in addition to episodes of renal allograft rejection. This can be explained by the complex mechanism of renal βMG excretion which depends on serum βMG concentration, glomerular filtration rate and proximal tubular cell function and on increased generation of βMG during inflammatory diseases [9,10]. The present data indicate that the determination of urinary TXB_2 permits a non-invasive evolution of renal transplant rejection. Because of its higher specificity and chemical stability TXB_2 is a more reliable indicator of renal allograft rejection than βMG.

References

1 Strom TB, Garovoy MR. *Am J Kidney Dis 1981; 1:* 5
2 Foegh ML, Winchester JF, Zmudka M et al. *Lancet 1981; ii:* 431
3 Uthmann U, Geisen HP, Dreikorn K et al. *Transplant Clin Immunol 1980; 12:* 259
4 European Multicenter Trial. *Lancet 1982; ii:* 57
5 Steinhauer HB, Lubrich I, Schollmeyer P. *Clin Hemorheol 1983; 3:* 1
6 Steinhauer HB, Wilms H, Sczesny CM, Schollmeyer P. *Verh Dtsch Ges Innere Med 1984; 90:* In press
7 Klotz TA, Cohn LS, Zipser RD. *Chest 1984:* In press
8 Zipser RD, Smorlesi C. *Prostaglandins 1984; 27:* 257
9 Wibell L, Evrin PE, Berggard I. *Nephron 1973; 10:* 320
10 Nilsson K, Evrin PE, Welsh KI. *Transplant Rev 1974; 21:* 53

Proc EDTA–ERA (1984) Vol 21

T CELL SUBSETS IN RENAL TRANSPLANTED PATIENTS DEFINED BY THEOPHYLLINE SENSITIVITY AND MONOCLONAL ANTIBODIES

Z Shapira, B Shohat, A Yussim, D Shmueli, C Servadio

Beilinson Medical Center, Petah Tikva, and the Sackler School of Medicine, Tel Aviv University, Tel Aviv, Israel

Summary

Peripheral T cell subsets were determined in 26 patients after renal transplantation. Thirteen patients were undergoing acute rejection episodes at the time of blood sampling. Two methods were used for the determinations: the theophylline sensitivity test (THST) and the monoclonal antibodies method OKT8 and OKT4. In both rejecting and non-rejecting groups, the percentages of T suppressor lymphocytes (TS) was found to be higher by the OKT8 method than by the THST. Furthermore, no significant difference in TS percentages could be revealed by the OKT8 method between the two groups. However, with the THST the differences were significant with a mean value of 13.5±7.9 per cent in the rejecting group and 21±5.9 per cent in the non-rejecting group ($p<0.01$). The different results between the two methods could be attributed to the fact that by THST only TS cells are defined while the OKT8 conjugates also with cytotoxic T lymphocytes. The measuring of T helper cells (TH) revealed much higher percentages of TH in the rejecting group than in the non-rejecting group, 51±9.5 per cent mean value and 29±13 per cent mean value, respectively ($p<0.05$). The ratio OKT4+/OKT8+ was below one in the non-rejecting group and above 1.5 in the rejecting group. We concluded that the THST as well as OKT4+/OKT8+ ratio may be a helpful laboratory test to confirm a clinically suspected acute rejection episode.

Introduction

Observations by many investigators showed that the determination of the percentages of certain functional T cell subpopulations, helper T lymphocytes (TH), and suppressor T lymphocytes (TS) in the peripheral blood of recently transplanted kidney patients might be indicative of their immunological response state [1–5].

A decrease in the amount of TS lymphocytes or elevated ratios of TH/TS have often been shown to correlate with clinical symptoms of acute rejection episode or even predict acute rejection if this pattern were found during clinical quiescence [1,3–5].

The method most utilised for quantifying T cells and their subsets is the use of monoclonal antibodies which recognise different antigens on the surface of the different T cell subsets [6,7]. At our centre the theophylline sensitivity test (THST) is used regularly for enumerating TS lymphocytes [5]. Therefore, we undertook this study to compare the accuracy of both methods in correlating clinically established acute rejection episodes.

Materials and methods

Twenty-six consecutive HLA non-identical first cadaver renal graft recipients entered this study. From each patient one sample of heparinised peripheral blood was drawn during the fourth post-transplant week. Thirteen patients had clinical and chemistry laboratory evidence of acute rejection when the blood sample was drawn. Percentages of TS lymphocytes were determined in each sample by the monoclonal OKT8 antibody indirect immunofluorescence method and by the THST. The suppressor activity of the separated theophylline sensitive T cells was confirmed by the local xenogeneic graft versus host reaction (GVHR) [5].

In addition, the percentages of TH lymphocytes was defined by monoclonal antibodies (OKT4) in five patients during acute rejection episode and in seven non-rejecting patients.

Results

The mean percentage of OKT8-positive (+) (Ts lymphocytes) in the rejecting patients was 35±11.6 per cent, while in the non-rejecting patients it was 53±20 per cent. These differences are not statistically significant ($p>0.05$). At the same time in the same samples lower values of TS lymphocytes were found by the THST with mean values of 13.5±7.9 per cent in the rejecting group and 21±5.9 per cent in the non-rejecting group. The differences between the groups are statistically significant ($p<0.01$) (Figure 1).

The five patients with acute rejection who were also tested for TH lymphocytes using OKT4 showed much higher mean percentages than the patients who were in quiescence, 51±5 per cent as compared to 29±13 per cent TH lymphocytes respectively. These results were statistically significant ($p<0.05$). The mean value of the OKT4/OKT8 ratio in the non-rejecting patients was below one and in the rejecting group above 1.5 (Figure 2).

Discussion

Three different patterns could be revealed in recent publications in respect to the number of circulating OKT8+ cells during acute rejection episodes:

1. Decrease in circulating OKT8+ cells as demonstrated by Cosimi and his colleagues [4].

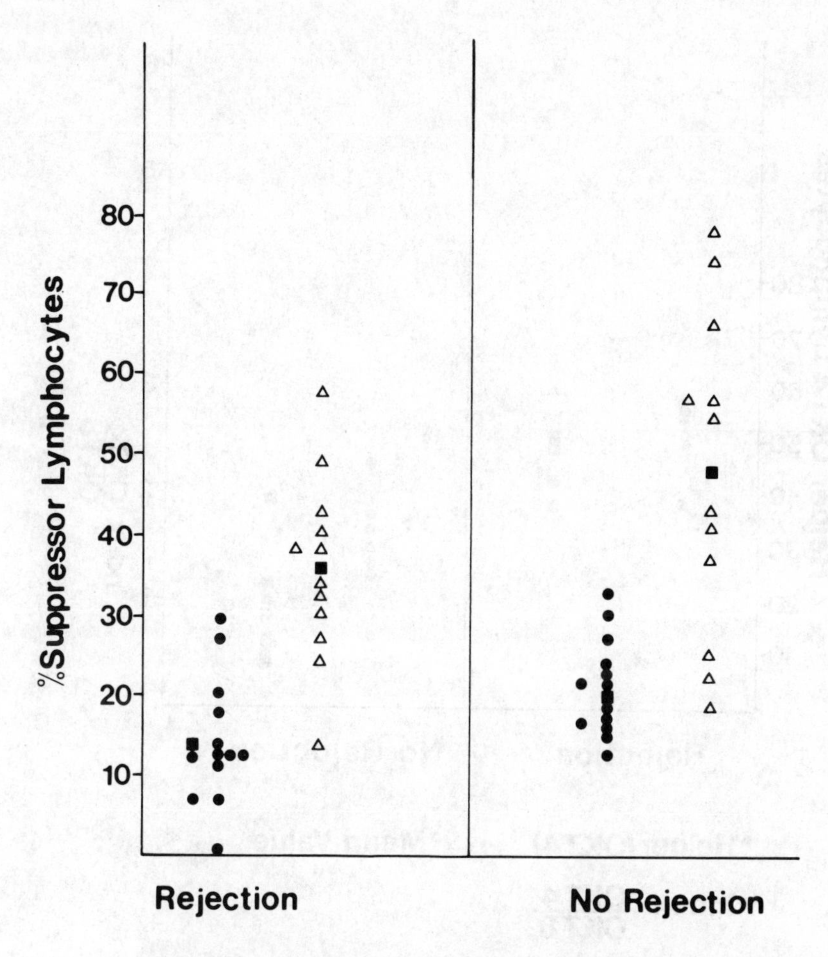

Figure 1. Suppressor T lymphocyte percentages as measured by the theophylline sensitivity test and by the OKT8 monoclonal antibody technique

2. Increase in circulating OKT8+ cells as demonstrated by Smith and his colleagues [2].
3. No difference in the circulating OKT8+ cells between patients in rejection and those in quiescence as demonstrated by Stelzner and his colleagues [1].

This last observation corresponds with ours when OKT8 antibodies were used. The above discrepancies could be attributed to the fact that OKT8 antibody

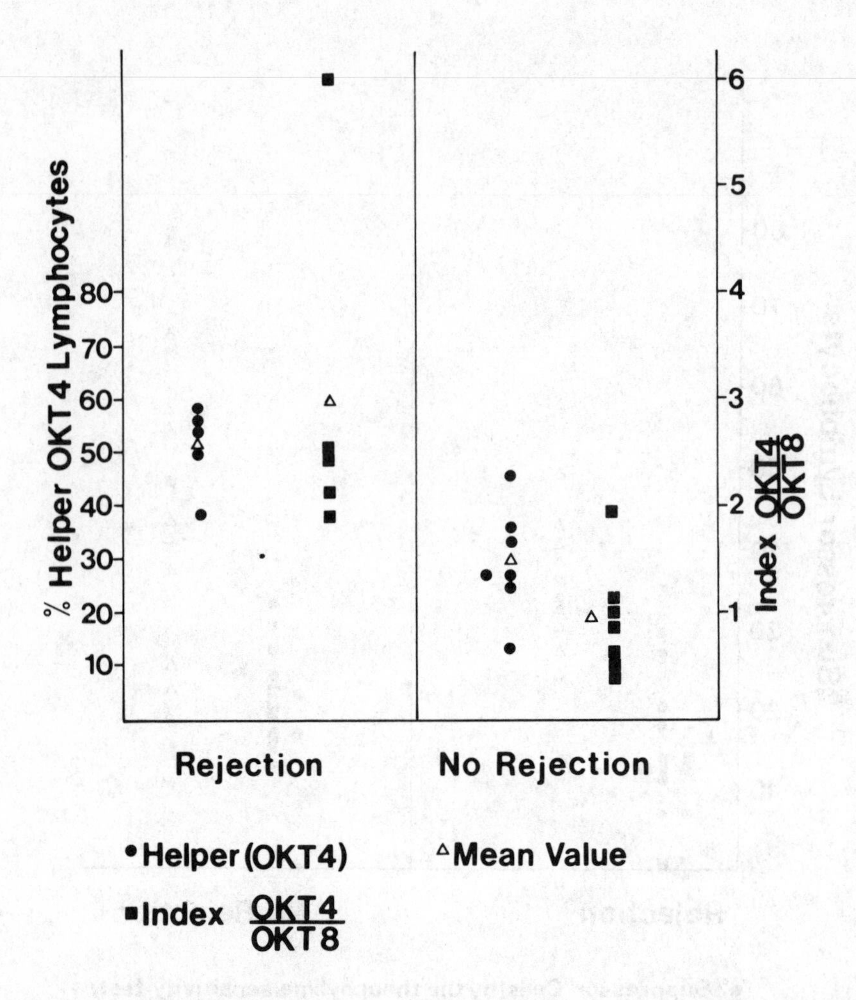

Figure 2. Helper T cell percentages and helper/suppressor ratios during rejection episodes and in quiescence

conjugates with two T lymphocyte subsets; suppressor as well as cytotoxic cells [4]. As yet we have very little information on the changes occurring in cytotoxic cells. If their proportion correlates with the number of TS during rejection episodes or quiescence it might well be that there is no consistent relationship between the percentage of the cytotoxic cells and that of the TS during different immunological response states. This could explain the variations in the total number of the OKT8+ cells which were observed by the different groups.

However, by using the THST and the xenogeneic GVHR the suppressor lymphocyte subset alone is determined. As shown in this study and in a previous report [5], this method gives a significant correlation between the percentages of TS and the immunological response of the kidney transplant patient. This essential difference between the two methods may also explain the variance in our results since both of them were used for determining TS.

On the other hand, the use of monoclonal antibodies for defining OKT4+ is a more reliable indicator since it was observed by some authors to correlate sufficiently with the immunological response state [2,4]. This may also be concluded from the present study. Furthermore, the ratio of OKT4+/OKT8+ was found to correlate with the clinical status of the patients. Elevated ratios above 1.3–1.7 were found to be significant related with the clinical state of rejection [1,3,4]. In our study a ratio above 1.5 was found in the rejecting group while a ratio below one was typical of the patients in quiescence.

Morris et al [8] as well as Toledo-Pereyra et al [9] could not, however, confirm such correlations. They suggested that the different results in the OKT4+/OKT8+ ratio in the different studies deriving from the actual immunosuppressive protocol employed by each group and the actual type of the rejection process.

The THST is a far less costly method and could be considered as a satisfactory alternative to the determination of OKT4+/OKT8+ ratios. However, it remains to be seen whether the development of monoclonal antibodies such as Leu-15 which will conjugate solely with TS would make this method the most accurate one for confirming or, in case of serial monitoring, for predicting acute rejection.

Acknowledgment

This work was supported in part by a grant from the Chief Scientist of the Israel Ministry of Health and by the Schander Grant from the Sackler School of Medicine, Tel Aviv University.

References

1 Stelzer GT, McLeish KR, Lorden E et al. *Transplantation 1984; 37:* 261
2 Smith WJ, Burdick JF, Williams GM. *Trans Proc 2983; 15:* 1969
3 Kerman RH, Van Burren CT, Flechner S. *Trans Proc 1983; 15:* 1971
4 Cosimi AB, Colvin RB, Burton RC et al. *N Engl J Med 1981; 305:* 308
5 Shapira Z, Shohat B, Boner G et al. *Trans Proc 1982; 14:* 113
6 Reinherz EL, Kung PC, Goldstein G et al. *Immunology 1979; 76:* 4061
7 Kung PC, Goldstein G, Reinherz EL et al. *Science 1979; 206:* 347
8 Morris PJ, Carter NP, Cullen PR et al. *N Engl J Med 1982; 306:* 1110
9 Toledo-Pereyra LH, Whitten JI, Toben HR et al. *Trans Proc 1983; 15:* 1985

MONITORING OF NK-1 AND NK-15 SUBSETS AND NATURAL KILLER ACTIVITY IN KIDNEY TRANSPLANT RECIPIENTS RECEIVING CYCLOSPORINE

E Renna Molajoni, V Barnaba, A Bachetoni, M Levrero, P Cinti, M Rossi, D Alfani, R Cortesini

University of Rome, Rome, Italy

Summary

NK-1, NK-15 subsets and natural killer (NK) activity were studied in 13 patients, before undergoing renal transplantation and in the post-transplant period. Pre-transplant there was a significant reduction of NK activity with a defect in V_{max} capacity in all patients. NK-1+ and NK-15+ cell number was normal or slightly increased in all subjects.

Post-transplant evaluation under cyclosporine immunosuppression showed a quantitative reduction in NK cells while no changes in NK activity and killing activity have been seen.

We conclude that in spite of a numerical decrease, NK activity was not affected by cyclosporine treatment.

Introduction

The main goals in organ transplantation are to avoid the two major complications of rejection and infection. Over the last twenty years several drugs have been employed to control recipient immunological reactivity against HLA alloantigens [1]. None of these methods have shown a specific and/or selective effect on immunocompetent cells involved in the rejection mechanisms [2]. Recently cyclosporine employed as immunosuppressive therapy in different organ allografts, has demonstrated a selective activity on helper cells and IL-2 release without affecting other subpopulations which play an important role against viral, bacterial and fungal infections [3]. On this basis, we have studied NK cells number and function in patients who have undergone renal transplantation, and we have examined the relationship between NK activity and clinical outcome.

Material and methods

Patient population and control

Thirteen patients ranging in age from 18 to 52 years have been studied before transplantation and 10 of these cases, periodically, in the post-operative period.

Seven patients received transplants from living related donors and three from living unrelated donors. All patients were transplanted with ABO blood group compatibility, with two or more mismatched HLA antigens and a negative cross-match.

Twenty healthy subjects matched for age and sex were used as controls.

Cyclosporine therapy

Cyclosporine was given as pre-transplant treatment at a dosage of 5mg/kg/day IV for 3–5 days before transplantation and in the post-transplant period according to the following protocol:

5mg/kg/day IV until oral adsorption is restored

15mg/kg/day from the end of IV treatment to day 15

subsequent decrease of the dose by 2mg/kg/month until a maintenance dosage of 2–6mg/kg/day has been reached.

Lymphocyte-surface markers studies

NK-1 and NK-15 subsets were measured using murine monoclonal antibodies Leu 7 and Leu 11a (Becton Dickinson) in an indirect immunofluorescence assay [4].

Target cells and culture medium

All target cells were cultured at 37°C in a humified air atmosphere with five per cent CO_2 in RPMI 1640 medium (GIBCO, Grand Island, NY) supplemented with 10 per cent fetal calf serum (FCS GIBCO) and 100μg/ml gentamicin. K562, a human tumour cell line derived from a patient with chronic myelogenous leukaemia, and MOLT-4, a human T cell leukaemia derived line, were used as target cells. Results obtained with the MOLT-4 line were comparable to those observed with K562 and therefore we have only reported data with K562 target cells.

Effector cells

Peripheral blood lymphocytes (PBL) were isolated from heparinised venous blood by separation on Ficoll-Hypaque gradients [5].

Toxicity assay

Cytotoxicity tests were performed by a standard ^{51}Cr-release assay in standard 96-well U-shaped microplates (Falcon Plastics) in a total volume of 200μL. One million target cells were labelled with 200μCiNa$_2$ ^{51}CrO$_4$ (Sorin, Saluggi, Italy) for one hour at 37°C. Ten thousand labelled target cells (100μL) were mixed with varying numbers of effector cells (100μL) to give final effector: target (E:T) ratios ranging from 50:1 to 6:1.

After four hours incubation period, 100μL of supernatant were collected and cytotoxicity was estimated from released radioactivity according to the formula:

$$\text{Per cent of specific cytotoxicity} = 100 \times \frac{\text{cpm exp} - \text{cpm } s_r}{\text{cpm } m_r - \text{cpm } s_r}$$

where cpm exp equalled the mean of the observed quadruplicate assay, cpm s_r was spontaneous release and cpm m_r was maximal release achieved by treating target cells with Nonidet P40. Spontaneous release of ^{51}Cr by target cells was determined by placing labelled target cells in microtitre wells in the absence of effector cells.

Natural killer (NK) capacity (V_{max})

The V_{max} of the NK assay represents the maximum number of target cells that can be killed by a constant number of effector cells when target cells are present in excess. V_{max} was determined according to the method of Ulberg and Jondal [6] with minor modifications. Briefly target cells were labelled with ^{51}Cr as described above. After washing, dilutions of target cells at five different concentrations were prepared (varying from 0.25 x 10^5 to 0.015 x 10^5 cells). To each well were added 10^5 effector cells and the cell mixture (0.2ml) was incubated for three hours at 37°C in five per cent CO_2. For each E:T ratio the maximal spontaneous lysis was determined.

The dose-response curve from the ^{51}Cr release cytotoxicity assay can be expressed as:

$$V = \frac{V_{max} \times T}{Km + T}$$

where T is the initial number of target cells, V the number of killed target cells and V_{max} the number of target killed cells when T approaches infinity, that is when the system is saturated with target cells. Km is the number of target cells that produces one-half of V_{max}. V_{max} and Km can be calculated using the lineweaver-Burk equation:

$$\frac{1}{V} = \frac{Km}{V_{max}} \times \frac{1}{T} + \frac{1}{V_{max}}$$

In this equation there is a linear relationship between $1/V_{max}$ and 1/T. The reciprocal values of V and T can be plotted and regression analysis used to determine V_{max} from the reciprocal of the Y intercept and x intercept respectively.

Results

Pre-transplant period

Pre-transplant study showed a significant reduction of NK activity, especially at lower ratios with a defect in V_{max} capacity in all patients. NK-1+ve and NK-15+ve cell number was normal or slightly increased in all patients (Table I).

TABLE I

			Control Group	Pre-transplant	Post transplant
NK Subsets	NK1		16.5±3.0	17.6±2.1	8.6±1.9
	NK15		16.1±2.5	18.2±3.7	10.4±2.2
NK Activity	E:T	50:1	29.9±2.9 ($p<0.02$)	20.1±2.2 ($p<0.02$)	no change
		25:1	24.4±2.9 ($p<0.005$)	15.1±1.9 ($p<0.005$)	no change
		6.1	17.4±1.5 ($p<0.001$)	7.9±1.2 ($p<0.001$)	no change
V_{max} Killing capacity (x 10^5)			8.3±5.2 ($p<0.001$)	4.22±7.1 ($p<0.001$)	no change

Post-transplant period

Post-transplant evaluation under cyclosporine immunosuppression showed a quantitative reduction of NK cells, while no changes in NK activity and killing capacity (V_{max}) were seen.

Discussion

In recent years, NK activity has been considered an important effector mechanism in host defence against tumours and virus infected cells [7]. Lack of NK activity is associated with infection and a significant impairment of cellular immunity has been observed in subjects affected by uraemia [8]. This impairment is worsened by conventional therapy with steroids, azathioprine and/or anti-lymphocyte globulin-employed in the post-operative period in patients who undergo organ allografts.

Cyclosporine seems to exert a more selective activity, as demonstrated by in vitro and in vivo studies [9]. Our studies on NK cells in transplanted patients receiving cyclosporine confirms this hypothesis. Even though a slight numerical decrease of NK-1 and NK-15 cells has been observed in all the cases, no changes in NK activity and killing capacity have been seen. The low incidence of infective events support these results with only one out of 10 patients developing CMV infection.

Finally, the decrease of NK-1 and NK-15 cell subsets, in spite of an unmodified

NK function, indicate that NK quantitation by monoclonal antibody assays is not of great value in post-transplant monitoring, particularly concerning immunocompetent cells involved against infections.

Acknowledgment

This work was supported in part by a grant from 'Institute Pasteur Fondazione' Cenci Bolognetti.

References

1 Renna Molajoni E, Sansonetti P, Nesci C et al. *Proc EDTA 1977; 14:* 495
2 Bach JF. *The Mode of Action of Immunosuppressive Agents.* Amsterdam. 1975: 392
3 White DJG. *Cyclosporine.* Amsterdam: Elsevier Biomedical. 1982: 578
4 Renna Molajoni E, Alfani D, Barnaba V et al. *Transpl Proc 1983; 15:* 1945
5 Boyum A. *Tissue Antigens 1974; 4:* 269
6 Ullberg M, Jondal M. *J Exp Med 1981; 153:* 615
7 Herberman RB, Djeu JY, Ray HD et al. *Immunol Rev 1979; 44:* 43
8 Cortesini R, Renna Molajoni E. In Serafini U, ed. *Immunologia Clinica ed allergologica.* Firenze: USES. 1982: 944
9 Kahan BD. *Cyclosporine. Biological Activity and Clinical Applications.* New York: Grune and Stratton inc. 1983: 895

SUBJECT INDEX

AUTHOR INDEX

Page numbers in brackets

Bartoli E, Istituto di patologia Medica, Sassari, Italy (738)
Basile A, Hospital of Codroipo, Udine, Italy (866)
Basolo B, Ospedale S Giovanni di Torino, Torino, Italy (661)
Bataille P, Service de Néphrologie et Laboratoire de Biochimie, CHU, Hôpital Nord, 8000 Amiens, France (747)
Bayle F, Centre Hospitalier Régional et Universitaire de Grenoble, France (661)
Bellio P, Department of Ophthalmology, Hôpital Bichat, 46 rue Henri Huchard 75018, Paris, France (330)
Beltramello G, General Hospital, Castelfranco, Veneto, Italy (642)
Bending JJ, Guy's Hospital Medical School, London, United Kingdom (621)
Benelmouffok S, Service de Néphrologie, CHU, Amiens, France (390)
Bennett SE, Renal Unit, Leicester General Hospital, Gwendolen Road, Leicester, United Kingdom (604)
Bergada E, Nephrology Service, Hospital Clinic, Barcelona, Catalunya, Spain (366)
Bergman P, Department of Nuclear Medicine, Brugmann Hospital, University of Brussels, 2 place Van Gehuchten, 1020, Brussels, Belgium (399, 431)
Bergstrand A, Karolinska Institute, Huddinge Hospital, Huddinge, Sweden (698)
Bergstrom J, Karolinska Institute, Huddinge Hospital, Huddinge, Sweden (156)
Berthoux FC, Hôpital Nord, Saint Étienne, France (703)
Bessa AMA, Nephrology-Hypertension Division, Escola Paulista de Medicina, Rua Botucatu 740, 04023 – São Paulo, Brazil (221)
Bettinelli A, Servizio di Nefrologia e Dialisi Pediatrica, Clinica Pediatrica II, Università di Milano, Italy (326)
Binswanger U, Section of Nephrology, Department of Internal Medicine, University of Zürich, Switzerland (387)
Bizolon C, Clinique de Néphrologie, Inserm U 80, Hôpital E Herriot, 69374 Lyon, France (761)
Block T, Medizinische Hochschule Hannover, Hannover, FRG (947)
Blohme I, Sahlgrenska Sjukhuset, University of Göteborg, Göteborg, Sweden (574)
Bommer G, Department of Internal Medicine, University of Heidelberg, Heidelberg, FRG (300)
Bommer J, Department of Internal Medicine, University of Heidelberg, Heidelberg, FRG (287, 300)
Bonal J, Nephrology Service, Hospital Clinic, Barcelona, Catalunya, Spain (366)
Bonalumi U, Patologia Chirurgica I, University of Genoa, Viale Benedetto XV 10, 16132 Genoa, Italy (262)
Bone JM, Renal Unit, Royal Liverpool Hospital, Liverpool, United Kingdom (594)
Boner G, Beilinson Medical Center, Petah Tiqva, Tel Aviv, Israel (932)
Bonneville M, Laboratoire d'Immunologie Clinique et Service de Néphrologie, UER de Médicine, Nantes 44035, France (1006)
Borghetti A, Istituto di Clinica Medica e Nefrologia, Università degli Studi di Parma, Parma, Italy (881)
Borsato N, General Hospital, Castelfranco, Veneto, Italy (642)
Borzone E, Patologia Chirurgica I, Viale Benedetto XV 10, 16132 Genoa, Italy (262)
Bosch J, Hospital Gral, M D, L'Esperanca, Barcelona, Spain (902)
Botella J, Department of Nephrology, Universidad Autónoma de Madrid, Clinica Puerta de Hierro, Madrid, Spain (403)
Boudet R, Centre Hospitalier Géneral, Perpignan, France (490)
Boulanger JC, CHU, Amiens, France (808)
Branca GF, Istituto di Patologia Medica, Sassari, Italy (738)
Branger B, Centre Hospitalier Régional Universitaire de Nîmes et Montpellier, Montpellier, France (162, 490)
Braun J, University Institute of Nephrology, Nürnberg, FRG (348)
Broddle O-E, Division of Renal and Hypertensive Diseases, Medizinische Klinik und Poliklinik, University of Essen, D-4300, FRG (178)
Broyer M, Hôpital Necker Enfants Malades, Paris, France (5, 69)
Brunner FP, Department für Innere Medizin, Universität Basel, Switzerland (5, 69)

Brynger H, Department of Surgery I, Sahlgrenska Sjukhuset, Göteborg, Sweden (5, 69, 574)
Bucci A, Can Carlo Borromeo Hospital, Milan, Italy (527)
Bucht H, Karolinska Institute, Department of Internal Medicine, Danderyds Hospital, Sweden (698)
Burnier M, Division of Nephrology, University Hospital, Lausanne, Switzerland (1002)

Cabili S, Rokach Hospital, P O Box 51, Tel Aviv, Israel (709)
Calemard E, Centre de Rein artificiel de Tassin, 69160, Tassin, France (291)
Calzavara P, General Hospital, Castelfranco, Veneto, Italy (642)
Camerani M, General Hospital, Castelfranco, Veneto, Italy (642)
Cameron JS, Department of Renal Medicine, Guy's Hospital, London Bridge, London, United Kingdom (588, 666)
Camussi G, Ospedale Maggiore S G Battista-Moinette, Corso Polonia 14, 10126, Turin, Italy (150)
Canaud B, Montpellier University Hospital, Montpellier, France (490)
Candiano G, Divisione Medicina Interna e Servizio Emodialisi, Ospedale di Lavangna, via Don Bobbio 16033, Lavagna, Italy (633, 875)
Cannata JB, Hospital General de Asturias, Facultad de Medicina, Universidad de Oviedo, Oviedo, Spain (354, 410)
Cantu P, Department of Nephrology, General Provincial Hospital, Busto Arsizio, Italy (895)
Capodicasa G, Istituto di Medicina Interna e Nefrologia, 1st Faculty of Medicine and Surgery, University of Naples, Naples, Italy (549)
Capsoni F, Ospedale Regionale, Varese, Clinica Medica II, Università degli Studi, Milan, Italy (686)
Carayon A, Hôpital Pitié-Salpétrière, Paris, France (808)
Cariou R, Laboratoire d'Immunologie, Clinique et Service de Néphrologie, UER de Médicine, Nantes 44035, France (1006)
Carozzi S, Division of Nephrology, St Martin Hospital, Viale Benedetto XV-10, 16132 Genoa, Italy (928)
Carrera M, Servicio de Nefrologia, Hospital Clinic, c/ Casanova, 143 08036 Barcelona, Spain (235)
Carvalho F, Hospital Curry Cabral, Lisbon, Portugal (666)
Casadei-Maldini M, Ospedale M Malpighi, Bologna, Italy (143)
Cascone C, Servizio di Emodialisi, Ospedale Civile, 31033 Castelfranco, Veneto, Italy (642)
Cassani D, Ospedale Regionale, Varese, Clinica Medica II, Università degli Studi, Milan, Italy (686)
Cassidy MJD, University of Newcastle upon Tyne, United Kingdom (888)
Castellani A, Nephrology and Dialysis Unit, Umberto I Hospital, Brescia, Italy (335)
Castelnovo C, Divisione di Nefrologia, Ospedale Maggiore di Milano, Milan, Italy (743)
Catalano C, Centro di Fisiologica Clinica del CNR e Divisione Nefrologica, Reggio Calabria, Italy (167)
Cavallin F, General Hospital, Castelfranco, Veneto, Italy (642)
Cecchin E, Hospital of Pordenone, Pordenone, Italy (866)
Challah S, EDTA Office, St Thomas' Hospital, London, United Kingdom (5, 69)
Channon SM, Department of Medicine, The Medical School, Newcastle upon Tyne NE2 4HH, United Kingdom (241, 360)
Charra B, Centre de Rein artificiel de Tassin, 69160, Tassin, France (291)
Chavers B, University of Minnesota, Minneapolis, Minnesota, USA (911)
Chiarini C, Ospedale M Malpighi, Bologna, Italy (143)
Chouzenoux R, Centre d'Hémodialyse du Languedoc mediterranéen, Montpellier, France (490)
Christensen M, Department of Nephrology, Aalborg Hospital, Aalborg, Denmark (435)
Christiansen C, Department of Clinical Chemistry, Glostrup Hospital, Glostrup, Denmark (435)
Ciancioni C, Centre Médical E Rist, Hôpital Necker, Paris, France (469)
Ciccarelli M, Centro Fisiologica Clinica CNR, Reggio Calabria, Italy (190)
Cieciura L, Military Medical Academy, Lodz, Poland (732)

Cinti P, University of Rome, 'La Sapienza', Rome, Italy (1042)
Civati G, Renal Unit, Ospedale Niguarda Ca' Granda, Milan, Italy (441)
Coelho RA, Hospital Curry Cabral, Lisbon, Portugal (666)
Coffrini E, Istituto di Clinica Medica e Nefrologia, Università degli Studi di Parma, Parma, Italy (881)
Cohen D, Department of Nephrology, Chaim Sheba Medical Center, Tel Hashomer 52621, Israel (849)
Colasanti G, San Carlo Borromeo Hospital, Milan, Italy (527)
Collart F, Hôpital Brugmann, Brussels, Belgium (395, 399)
Colussi G, Divisione di Nefrologia, Ospedale Ca' Granda Milano, Milan, Italy (743)
Company X, Servicio de Nefrologia, Hospital Clinic c/ Casanova, 143 08036 Barcelona, Spain (235)
Comparini L, Department of Nephrology, USL IO/D, Florence, Italy (321)
Conte G, Department of Nephrology, Second Faculty of Medicine, University of Naples, Naples, Italy (820)
Cooke K, St Luke's Roosevelt Hospital Center, New York, USA (1027)
Coppo R, Ospedale S Giovanni di Torino, Italy (661)
Cordonnier D, Centre Hospitalier Régional et Universitaire de Grenoble, Grenoble, France (661)
Corliano C, Divisione di Nefrologia, Ospedale 'V. Fazzi', Lecce, Italy (215)
Cortesini R, University of Rome, 'La Sapienza', Rome, Italy (1042)
Cossu M, Servizio di Emodialisi, Ospedali Riuniti, Sassari, Italy (608)
Cosyns JP, University of Louvain Medical School, Cliniques Universitaires St-Luc, Avenue Hippocrate 10, 1200, Brussels, Belgium (672)
Crastes De Paulet A, Centre Hospitalier Régional Universitaire de Nîmes et Montpellier, France (162)
Creazzo G, Centro di Fisiologica Clinica del CNR e Divisione Nefrologica, Reggio Calabria, Italy (167)
Cristinelli L, Department of Nephrology, Ospedali Civili and University, Brescia, Italy (317)
Cuesta V, Hospital General de Asturias, Facultad de Medicina, Universidad de Oviedo, Oviedo, Spain (354, 410)
Cuevas X, Hospital Gral M D, L'Esperanca, Barcelona, Spain (902)
Cunningham J, The London Hospital, Whitechapel, London, United Kingdom (209)

D'Amico G, San Carlo Borromeo Hospital, Milano, Italy (527)
Dal Canton A, Department of Nephrology, Second Faculty of Medicine, University of Naples, Naples, Italy (820)
Dall'Argine P, Università di Parma, Parma, Italy (614)
Danielson A, Karolinska Institute, Huddinge University Hospital, Huddinge, Sweden (156)
Darai G, Institute of Medical Virology, University of Heidelberg, Heidelberg, FRG (300)
Darnell A, Hospital Clinic, Barcelona, Spain (677)
Daugaard H, Medical Department P, Division of Nephrology, Rigshospitalet, Copenhagen, Denmark (815)
Daul AE, Division of Renal and Hypertensive Diseases, Medizinische Klinik and Poliklinik, University of Essen, D-4300, Essen, FRG (178)
David S, Università di Parma, Parma, Italy (614)
De Blasi V, Divisione di Nefrologia, Ospedale 'V. Fazzi', Lecce, Italy (215)
De Francisco AM, University of Newcastle upon Tyne, United Kingdom (888)
De Jong PE, Departments of Clinical Endocrinology and Nephrology, University Hospital, Groningen, The Netherlands (722)
De Luca M, General Hospital, Castelfranco, Veneto, Italy (642)
De Maio A, Ospedale Regionale, Varese, Clinica Medica II, Università degli Studi, Milan, Italy (686)
De Marchi S, Hospital of Codroipo, Udine, Italy (866)
De Roy G, Hôpital Brugmann, Brussels, Belgium (399)
De Santo NG, Istituto di Medicina Interna e Nefrologia, 1st Faculty of Medicine and Surgery, University of Naples, Naples, Italy (549)

Debroe M, Nephrology Department, Akadem Zirkenhuis, Antwerp, Belgium (371)
Degli Esposti E, Ospedale M Malpighi, Bologna, Italy (143)
Deinhardt F, Max von Pettenkofer Institut, University of Munich, FRG (300)
Del Canale S, Istituto di Clinica Medica e Nefrologia, Università degli Studi di Parma, Parma, Italy (881)
Delfino G, Divisione Medicina Interna e Servizio Emodialisi, Ospedale di Lavagna, via Don Bobbio 16033, Lavagna (GE), Italy (633, 875)
Deligiannis A, The 2nd Propedeutic Department of Internal Medicine, University of Thessaloniki, 'Aghia Sophia' Hospital, Thessaloniki, Greece (185)
Delitala G, Cattedra di Endocrinologia, Università di Sassari, Viale San Pietro n. 12, 07100 Sassari, Italy (608)
Dell'Anna L, Hospital of Pordenone, Pordenone, Italy (866)
Delobel J, Laboratoire d'Hématologie, CHU, Amiens, France (276)
Delons S, Centre Médical E Rist, Hôpital Necker, Paris, France (469)
Demontis R, Service de Néphrologie, CHU, 80000 Amiens, France (371, 390)
Derfler K, I. Medizinische Universitätsklinik, Department of Nuclear Medicine, Vienna, Austria (251)
Deulofeu R, Servicio de Nefrologia, Hospital Clinic c/ Casanova 143 08036 Barcelona, Spain (235)
Dhaene M, Department of Nephrology, Erasme Hospital, University of Brussels, 808, route de Lennik, 1070 Brussels, Belgium (431)
Di Minno G, Department of Internal Medicine, Second Faculty of Medicine, University of Naples, Naples, Italy (820)
Di Spigno C, Department of Nephrology, Second Faculty of Medicine, University of Naples, Naples, Italy (820)
Diallo A, Centre Pasteur-Valley-Radot, 26 rue des Peupliers, 75013, Paris, France (477)
Dieperink H, Laboratory of Nephropathology, Institute of Pathology, Odense University Hospital, DK 5000, Odense C, Denmark (853)
Dieval J, Laboratoire d'Hématologie, CHU, Amiens, France (276)
Dkhissi H, Service de Néphrologie, CHU, Amiens, France (276)
Donati D, Ospedale Regionale, Varese, Clinica Medica II, Università degli Studi, Milan, Italy (686)
Donker AJM, Department of Nephrology, University Hospital of Groningen, Oostersingel 59, 9713 EZ, Groningen, The Netherlands (567, 722, 936, 992)
Doss SH, Faculty of Medicine, Cairo University, Cairo, Egypt (649)
Dossetor JB, University of Alberta, Edmonton, Canada (1021)
Draibe SA, Nephrology-Hypertension Division, Escola Paulista de Medicina, Rua Botucatu number 740, 04023 – São Paulo, Brazil (221)
Dratwa M, Hôpital Brugmann, Brussels, Belgium (395, 399)
Drazniowsky M, Department of Medicine, The Medical School, Newcastle upon Tyne NE2 4HH, United Kingdom (241)
Dudczak R, I. Medizinische Universitätsklinik, Department of Nuclear Medicine, Vienna, Austria (251)
Dufour A, CHU, Amiens, France (808)
Dunn M, NAMRU 3, Cairo, Egypt (649)
de Cornelissen F, Centre Hospitalier Géneral Carcassone, France (490)
de Fremont JF, Hôpital Nord, Amiens, France (808)

Edefonti A, Servizio di Nefrologia e Dialisi Pediatrica, Clinica Pediatrica II, Università di Milano, Italy (326)
Edfjord M, Medical Department P, Division of Nephrology, Rigshospitalet, Copenhagen, Denmark (815)
Egbring R, Center of Internal Medicine, University of Marburg and Research Laboratories of Nehringwerke AG, D-3500, Marburg, FRG (941)
Eliahou HE, Department of Nephrology, Chaim Sheba Medical Center, Tel Hashomer 52621, Israel (849)
Ellis HA, University of Newcastle upon Tyne, United Kingdom (888)

El-Sheemy N, Faculty of Medicine, Cairo University, Cairo, Egypt (649)
Emond C, Association pour l'Installation à Domicile des Épurations Rénales, Montpellier, France (490)
Enia G, Centro di Fisiologia Clinica del CNR e Divisione Nefrologica, Reggio Calabria, Italy (167)
Enriques R, Department of Nephrology, Universidad Autónoma de Madrid, Clinica Puerta de Hierro, Madrid, Spain (403)
Erben J, First Medical Clinic, Department of Nephrology, University Hospital Hradec, Kralove, Czechoslovakia (421)
Etienne S, CITI2, Paris, France (477)

Faedda R, Istituto di Patologia Medica, Sassari, Italy (738)
Farndon JR, Department of Surgery, Royal Victoria Infirmary, Newcastle upon Tyne, United Kingdom (426, 888)
Favre H, Department of Medicine, University Hospital, Geneva, Switzerland (717)
Feest TG, Royal Devon and Exeter Kidney Unit, Exeter, United Kingdom (683)
Ferlin G, General Hospital, Castelfranco, Veneto, Italy (642)
Fernandez JF, Department of Nephrology, Universidad Autónoma de Madrid, Clinica Puerta de Hierro, Madrid, Spain (403)
Ferrario F, San Carlo Borromeo Hospital, Milan, Italy (527)
Fiaccadori E, Istituto di Clinica Medica e Nefrologia, Università degli Studi di Parma, Parma, Italy (881)
Fidler, VJ, State University Hospital, Groningen, The Netherlands (992)
Fievet P, Service de Néphrologie et Laboratoire de Biochimie, CHU, Hôpital Nord, 80000 Amiens, France (747, 808)
Finet I, Service de Néphrologie et Laboratoire de Biochimie, CHU, Hôpital Nord, 80000 Amiens, France (747)
Fitte H, Centre Hospitalier Géneral, Perpignan, France (490)
Flavier JL, Association pour l'Installation à Domicile des Épurations Rénales, Montpellier, France (490)
Florence P, Centre d'Hémodialyse du Languedoc mediterranéen, Montpellier, France (490)
Fohrer P, Service d'Hémodialyse, Centre Hospitalier, 60000 Beauvais, France (371, 415)
Fontanier P, Centre Hospitalier Géneral Carcassone, France (490)
Fournier A, Service de Néphrologie, CHU, Amiens, France (276, 371, 390, 415, 747, 808)
Franco P, Department of Nuclear Medicine, Universidad Autónoma de Madrid, Clinica Puerta de Hierro, Madrid, Spain (403)
Frattini G, Ospedale Regionale, Varese, Clinica Medica II, Università degli Studi, Milan, Italy (686)
Freyschuss U, Karolinska Institute, Huddinge University Hospital, Huddinge, Sweden (156)
Fridrich L, I. Medizinische Universitätsklinik, Department of Nuclear Medicine, Vienna, Austria (251)
Friedman D, Patologia Chirurgica I, Viale Benedetto XV 10, 16132 Genoa, Italy (262)
Frischauf H, I. Medizinische Universitätsklinik, Department of Nuclear Medicine, Vienna, Austria (251)
Frohling PT, St Josefs Hospital, Potsdam, GDR (561)
Funari C, Servizio di Nefrologia e Dialisi Pediatrica, Clinica Pediatrica II, Università di Milano, Italy (326)
Fuss M, Department of Medicine, Brugmann Hospital, University of Brussels, 2 place Van Gehuchten, 1020, Brussels, Belgium (431)

Gafni J, Rokach Hospital, P O Box 51, Tel Aviv, Israel (709)
Galato R, Divisione di Nefrologia e Dialisi, Ospedale Ca' Granda, Milan, Italy (326)
Gallego JL, Department of Nephrology, Universidad Autónoma de Madrid, Clinica Puerta de Hierro, Madrid, Spain (403)
Galy C, Service de Néphrologie et Laboratoire de Biochimie, CHU, Hôpital Nord, 80000 Amiens, France (747)

Garabedian M, Laboratoire des tissus calcifiés, Hôpital des Enfants Malades, Paris, France (415)
Garcia M, Servicio de Nefrologia, Hospital Clinic, c/ Casanova 143, 08036 Barcelona, Spain (235)
Gardner PW, Department of Medicine, Wadsworth Medical Center, Los Angeles, California 90073, USA (447)
Gastaldi L, Ospedale Régionale, Varese, Clinica Medica II, Università degli Studi, Milan, Italy (686)
Gautschi K, Laboratory of Clinical Pathology, University of Zürich, Switzerland (387)
Gavendo S, Department of Nephrology, Chaim Sheba Medical Center, Tel Hashomer 52621, Israel (849)
Gekle D, University Children's Hospital, Würzburg, FRG (638)
Geleris P, The 2nd Propedeutic Department of Internal Medicine, University of Thessaloniki, 'Aghia Sophia' Hospital, Thessaloniki, Greece (185)
Genin C, Hôpital Nord, Saint Étienne, France (703)
Georgilis C, Department of Clinical Therapeutics, Athens University, 'Alexandra' Hospital, Athens, Greece (860)
Gessler U, University Institute of Nephrology, Nürnberg, FRG (348)
Ghiggeri GM, Divisione Medicine Interna e Servizio Emodialisi, Ospedale di Lavagna, via Don Bobbio 10633, Lavagna (GE), Italy (633, 875)
Ghio L, Servizio di Nefrologia e Dialisi Pediatrica, Clinica Pediatrica II, Università di Milano, Italy (326)
Ghysen J, University of Louvain Medical School, Cliniques Universitaires St-Luc, Avenue Hippocrate 10, 1200 Brussels, Belgium (672)
Giangrande A, Department of Nephrology, General Provincial Hospital, Busto Arsizio, Italy (895)
Giani M, Servizio di Nefrologia e Dialisi Pediatrica, Clinica Pediatrica II, Università di Milano, Italy (326)
Giannopoulos Z, Ahepa Hospital, University of Thessaloniki, Thessaloniki, Greece (825)
Gilli G, University Children's Hospital, Heidelberg, FRG (830)
Giordano C, Istituto di Medicina Interna e Nefrologia, 1st Faculty of Medicine and Surgery, University of Naples, Naples, Italy (549)
Gnudi A, Università di Parma, Parma, Italy (614)
Goldman M, Department of Nephrology, Erasme Hospital, University of Brussels, 808, route de Lennik, 1070 Brussels, Belgium (431)
Goldstein J, Cliniques Universitaires de Bruxelles, Hôpital Erasme, Brussels, Belgium (973)
Gonzalez FM, Department of Medicine and Pharmacology, Louisiana State University Medical Center, 1542 Tulane Avenue, New Orleans, LA 70112, USA (230)
Goodwin, FJ, The London Hospital, Whitechapel, London, United Kingdom (209)
Gordge MP, King's College Hospital Renal Unit, Dulwich Hospital, London SE22, United Kingdom (281)
Gotz K-H, Clinic of Internal Medicine, Neuruppin, GDR (561)
Graven N, Division of Renal and Hypertensive Diseases, Medizinische Klinik and Poliklinik, University of Essen, D-4300, Essen, FRG (178)
Grande J, Hospital Clinico Universitario, Salamanca, Spain (801)
Granerus G, Sahlgrenska Sjukhuset, University of Göteborg, Göteborg, Sweden (574)
Granolleras C, Nîmes University Hospital, Nîmes, France (490)
Grau JM, Hospital Clinic, Barcelona, Spain (677)
Graziani G, Divisione di Nefrologia, Ospedale Maggiore di Milano, Milan, Italy (743)
Grekas D, Ahepa Hospital, University of Thessaloniki, Thessaloniki, Greece (825)
Griffanti-Bartoli F, Patologia Chirurgica I, Viale Benedetto XV 10, 16132 Genoa, Italy (262)
Grubendorf M, Department of Internal Medicine, University of Heidelberg, Heidelberg, FRG (300)
Grundmann R, Chirurgische Universitätsklinik Köln-Lindenthal, Joseph Stelzmann Str 9, D-5000, Köln 41, FRG (987)
Grussendorf M, Department of Internal Medicine, University of Heidelberg, Heidelberg, FRG (300)

Guariglia A, Istituto di Clinica Medica e Nefrologia, Università degli Studi di Parma, Parma, Italy (881)
Guastoni C, Renal Unit, Ospedale Niguarda Ca'Granda, Milan, Italy (441)
Guenel J, University Hospital, Nantes, France (982)
Gueris J, Laboratoire de Radio-immunologie, Hôpital Lariboisière, Paris, France (415)

Habib M, Faculty of Medicine, Cairo University, Cairo, Egypt (649)
Hadj Aissa A, Clinique de Néphrologie, Inserm U 80, Hôpital E Herriot, 69374 Lyon, France (761)
Haimovitz C, Department of Nephrology, Sackler School of Medicine, Tel Aviv, Israel (932)
Hajakova B, First Medical Clinic, Department of Nephrology, University Hospital Hradec, Kralove, Czechoslovakia (421)
Hammer M, Medical Department P, Division of Nephrology, Rigshospitalet, Copenhagen, Denmark (815)
Hansen HE, Aarhus Kommunehospital, Aarhus, Denmark (725)
Harris KPG, Renal Unit, Leicester General Hospital, Leicester, United Kingdom (1010)
Hartley B, Guy's Hospital, London, United Kingdom (666, 1015)
Hatzigeorgiou G, Department of Clinical Therapeutics, Athens University, 'Alexandra' Hospital, Athens, Greece (860)
Heigel B, Klinik für Abdominal-und Transplantationschirurgie Medizinische Hochschule, Hannover, Konstanty Gutschow Str. 8, FRG (947)
Hem GK, Department of Nephrology, University Hospital of Groningen, Groningen, The Netherlands (267)
Herrera J, Hospital General de Asturias, Facultad de Medicina, Facultad de Ciencias Quimicas, Universidad de Oviedo, Oviedo, Spain (354, 410)
Herve MA, Société de Mathématiques appliquées à la Recherche Médicale et Biométrique de la Région de Picardie, Amiens, France (390, 415, 808)
Herzog D, Department of Nephrology, Chaim Sheba Medical Center, Tel Hashomer 52621, Israel (849)
Hess T, Section of Nephrology, Department of Internal Medicine, University of Zürich, Switzerland (387)
Hesse U, Chirurgische Universitätsklinik Köln-Lindenthal, Joseph Stelzmann Str. 9, D-5000, Köln 41, FRG (987)
Hicks J, Department of Renal Medicine, Guy's Hospital, London, United Kingdom (588)
Hjelte M-B, IRD Biomaterial, Stockholm, Sweden (270)
Hjollund E, Aarhus Amtssygehus, Aarhus, Denmark (725)
Holzer H, Department of Internal Medicine, University of Graz, Austria (461)
Hosokawa S, Utano National Hospital, 8 Ondoyama-cho, Narutaki, Ukyo-ku, Kyoto City, 616 Japan (247)
Hourmant M, University Hospital, Nantes, France (982)
Huchard G, Centre Hospitalier Géneral, Perpignan, France (490)

Iaina A, Department of Nephrology, Chaim Sheba Medical Center, Tel Hashomer 52621, Israel (849)
Icardi A, Division of Nephrology, St Martin Hospital, Viale Benedetto XV-10, 16132 Genoa, Italy (928)
Iellamo D, Centro di Fisiologica Clinica del CNR e Divisione Nefrologica, Reggio Calabria, Italy (167)
Imai T, Shiga University, Kyoto City, Japan (247)
Ireland H, Department of Haematology, Charing Cross Hospital Medical School, London, United Kingdom (281)
Ishigami M, Department of Internal Medicine, Kanto Rosai Hospital, 2035 Kizukisumiyoshi-cho Nakahara-ku Kawasaki City, Kanagawa-pref. 211 Japan (843)
Iskander I, Faculty of Medicine, Cairo University, Cairo, Egypt (649)
Issautier R, Association pour l'Installation à Domicile des Épurations Rénales, Montpellier, France (490)

Izumi N, Tadaoka Municipal Hospital, 1-3-7, Tadaokakita, Tadaoka-cho, Senbokugun, Osaka, 595 Japan (306)

Jacobs C, Centre Pasteur-Valley-Radot, 26 rue des Peupliers, 75013, Paris, France (477)
Jacopian M, St Josefs Hospital, Potsdam, GDR (561)
Jahnke J, St Joseph Krankenhaus I, Berlin, FRG (510)
Janushkevitch TN, Laboratory of Nephrology, First Moscow Medical Setchenov Institute, 11a Rjossolimo Street, 119021 Moscow, USSR (712)
Jaszczuk-Jarosz B, Military Academy, Lodz, Poland (732)
Jean-Jaques M, St Luke's Roosevelt Hospital Center, New York, USA (1027)
Jeantet A, Ospedale Maggiore S, G. Battista-Molinette, Corso Polonia, 14, 10126 Turin, Italy (150)
Jilg W, Max von Pettenkofer Institut, University of Munich, FRG (300)
Jungbluth H, Section of Nephrology, Department of Internal Medicine, University of Zürich, Switzerland (387)

Kalima L, Hôpital Brugmann, Brussels, Belgium (395)
Kaltwasser JP, Department of Haematology, University Hospital and Gesellschaft für Strahlen, Frankfurt, FRG (382)
Kapuler S, Department of Nephrology, Chaim Sheba Medical Center, Tel Hashomer 52621, Israel (849)
Karges HE, Center of Internal Medicine, University of Marburg and Research Laboratories of Behringwerke AG, D-3500, Marburg, FRG (941)
Kaschube J, Research Clinic of Nutrition, Potsdam-Rehbrucke, GDR (561)
Katz F, Department of Internal Medicine, University of Heidelberg, Heidelberg, FRG (202)
Keller F, Medizinische Klinik, Klinikum Steglitz, Hindenburgdamm 30, D-1000 Berlin 45, FRG (311)
Kemp E, Laboratory of Nephropathology, Institute of Pathology, Odense University Hospital, DK 5000, Odense C, Denmark (853)
Kerr DNS, Department of Medicine, Royal Postgraduate Medical School, Hammersmith Hospital, London, United Kingdom (241, 360, 580, 888)
Keshaviah P, Regional Kidney Disease Program, Hennepin County Medical Center, 701 Park Avenue South, Minneapolis, Minnesota 5515, USA (111)
Kessler J, Department of Internal Medicine, University of Heidelberg, Heidelberg, FRG (287)
Khalifa AM, Division of Renal and Hypertensive Diseases, Medizinische Klinik and Poliklinik, University of Essen, D-4300, Essen, FRG (178)
Kidawa Z, Military Medical Academy, Lodz, Poland (732)
Kinnaert P, Department of Nephrology, Erasme Hospital, University of Brussels, 808, route de Lennik, 1070 Brussels, Belgium (431)
Kirkendol PL, Department of Pharmacology, Louisiana State University Medical Center, 1100 Florida Avenue, New Orleans, LA 70119, USA (230)
Kishimoto T, Department of Urology, Osaka City University Medical School, 1-5-7 Asahimachi, Abeno-ku, Osaka, 545 Japan (306)
Kletter K, I. Medizinische Universitätsklinik, Department of Nuclear Medicine, Vienna, Austria (251)
Koch H-G, Institute of Medical Virology, University of Heidelberg, Heidelberg, FRG (300)
Kokot F, Silesian School of Medicine, Katowice, Poland (561)
Konopacki J, Military Medical Academy, Lodz, Poland (732)
Kontopoulos A, The 2nd Propedeutic Department of Internal Medicine, University of Thessaloniki, 'Aghia Sophia' Hospital, Thessaloniki, Greece (185)
Kovithavongs T, University of Alberta, Edmonton, Canada (1021)
Kramer P, Medizinische Universitätsklinik, Göttingen, FRG (5, 69)
Kremer GD, Renal Transplant Unit, University Hospital, Oostersingel 59, 9713 EZ Groningen, The Netherlands (655)
Ksiazek A, The School of Medicine, Lublin, Poland (225)

Kubes L, Department of Pathology, University Hospital Hradec, Králové, Czechoslovakia (421)
Kuehn-Freitag G, Medizinische Klinik, Klinikum Steglitz, Hindenburgdamm 30, D-1000 Berlin 45, FRG (311)

Labeeuw M, Clinique de Néphrologie, Inserm U 80, Hôpital E Herriot, 69374 Lyon, France (761)
Ladefoged J, Medical Department P, Division of Nephrology, Rigshospitalet, Copenhagen, Denmark (815, 963)
Lambrey G, Service d'Hémodialyse, Centre Hospitalier, 60000 Beauvais, France (371, 415)
Lamperi S, Division of Nephrology, St Martin Hospital, Viale Benedetto XV-10, 16132 Genoa, Italy (928)
Lane, DA, Department of Haematology, Charing Cross Hospital Medical School, London, United Kingdom (281)
Lange H, Center of Internal Medicine, University of Marburg and Research Laboratories of Behringwerke AG, D-3500, Marburg, FRG (941)
Langhoff E, Medical Department P, Division of Nephrology, Rigshospitalet, Copenhagen, Denmark (815, 963)
Larm O, Swedish University of Agricultural Sciences, Uppsala, Sweden (270)
Larsson R, IRD Biomaterial, Stockholm, Sweden (270)
Laurent G, Centre de Rein artificiel de Tassin, 69160, Tassin, France (291)
Leclerc JL, Cliniques Universitaires de Bruxelles, Hôpital Erasme, Brussels, Belgium (973)
Leflon A, Laboratoire de Biochimie, CHU, Amiens, France (390)
Leflon P, Service de Néphrologie et Laboratoire de Biochimie, CHU, Hôpital Nord, 80000 Amiens, France (371, 415, 747)
Legrain M, Department of Ophthalmology, Hôpital Bichat, 46 rue Henri Huchard, 75018, Paris, France (330)
Lenhard V, Institute of Immunology, University of Heidelberg, FRG (932)
Leonard EF, Columbia University, New York 10027, USA (99)
Leuhmann D, Regional Kidney Disease Program, Hennepin County Medical Center, 701 Park Avenue South, Minneapolis, Minnesota 5515, USA (111)
Levrero M, University of Rome, 'La Sapienza', Rome, Italy (1042)
Leyssac PP, Laboratory of Nephropathology, Institute of Pathology, Odense University Hospital, DK 5000, Odense C, Denmark (853)
Lichtendahl DHE, Department of Surgery, University Hospital of Groningen, Groningen, The Netherlands (267)
Light PD, Department of Surgery and Medicine, University of Maryland School of Medicine, Baltimore, Maryland 21201, USA (257)
Lin J, St Luke's Roosevelt Hospital Center, New York, USA (1027)
Lindberg A, Karolinska Institute, Department of Internal Medicine, Danderyds Hospital, Sweden (698)
Lindenau K, Department of Nephrology, Humboldt University, Berlin, GDR (561)
Lindsay RM, Department of Medicine, Victoria Hospital Corporation and the University of Western Ontario, London, Ontario, Canada (135, 837)
Lins LE, Department of Medicine, Division of Nephrology, Karolinska Hospital, Stockholm, Sweden (270)
Linton AL, Department of Medicine, Victoria Hospital Corporation and the University of Western Ontario, London, Ontario, Canada (135, 837)
Llanderal JP, Hospital General de Asturias, Facultad di Medicina, Facultad de Ciencias Quimicas, Universidad de Oviedo, Oviedo, Spain (354)
Llorach I, Hospital Gral M D L'Esperanca, Barcelona, Spain (902)
Llorach M, Hospital Gral M D L'Esperanca, Barcelona, Spain (902)
Lloveras J, Hospital Gral M D L'Esperanca, Barcelona, Spain (902)
Loffler W, Department of Internal Medicine, University of Munich, FRG (751)
Lonati F, Nephrology and Dialysis Unit, Umberto I Hospital, Brescia, Italy (335)
Lopez Pedret J, Nephrology Service, Hospital Clinic, Barcelona, Catalunya, Spain (366)

Lote CJ, Department of Physiology, Birmingham University, Birmingham, United Kingdom (969)
Luciani A, Università di Modena, Modena, Italy (614)
Lugari R, Università di Parma, Parma, Italy (614)

Maassen G, Institute of Immunology, University of Heidelberg, FRG (923)
Macias JF, Hospital Clinico Universitario, Salamanca, Spain (801)
Madia D, General Hospital, Castelfranco, Veneto, Italy (642)
Madsen S, Medical Department P, Division of Nephrology, Rigshospitalet, Copenhagen, Denmark (815, 963)
Maeda T, Department of Internal Medicine, Kanto Rosai Hospital, 2035 Kizukisumiyoshi-cho, Nakahara-ku, Kawasaki-City, Kanagawa-pref, 211 Japan (843)
Maekawa M, Department of Urology, Osaka City University Medical School, 1-5-7 Asahi-machi, Abeno-ku, Osaka, 545 Japan (306)
Maekawa T, Tadaoka Municipal Hospital, 1-3-7 Tadaokakita, Tadaoka-cho, Senbokugum, Osaka, 595 Japan (306)
Maggiore Q, Centro Fisiologica Clinica CNR, Reggio Calabria, Italy (167, 190)
Maglionico G, Department of Nephrology, Second Faculty of Medicine, University of Naples, Naples, Italy (820)
Maiga K, Department of Nephrology, Hôpital de La Pitié, 83 bd de l'Hôpital 75651, Paris, Cedex 13, France (330)
Maioli M, Clinica Medica, Università di Sassari, Sassari, Italy (608)
Maiorca R, Department of Nephrology, Ospedali Civili and University, Brescia, Italy (317)
Makdassi R, Service de Néphrologie et Laboratoire de Biochimie, CHU, Hôpital Nord, 80000 Amiens, France (747)
Mallamaci F, Centro Fisiologica Clinica CNR, Reggio Calabria, Italy (190)
Man NK, Pharmacologie, CHU, Angers, France (469)
Manca N, Institute of Microbiology, Ospedali Civili and University, Brescia, Italy (317)
Mandressi A, Istituto di Urologia, dell'Università di Milano, Milan, Italy (743)
Marinoso LL, Hospital Gral M D L'Esperanca, Barcelona, Spain (902)
Marosi L, Department of Nephrology, 1090 Vienna, Lazarettgasse 14, Austria (251)
Marsh FP, The London Hospital, Whitechapel, London, United Kingdom (209)
Marshall SM, Department of Medicine, Royal Victoria Infirmary, Newcastle upon Tyne, United Kingdom (580)
Martegani M, Ospedale Regionale, Varese, Clinica Medica II, Università degli Studi, Milan, Italy (686)
Martina G, Ospedale S Giovanni di Torino, Turin, Italy (661)
Martinelli F, Department of Nephrology, USL IO/D, Florence, Italy (321)
Martini PF, Laboratorio di Immunopatologia, Cattedra di Nefrologia, Divisione di Nefrologia e Dialisi, Ospedale Maggiore S G Battista-Molinette, Corso Polonia, 14, 10126, Turin, Italy (150)
Marty L, Centre Hospitalier Géneral, Carcassone, France (490)
Masramon J, Hospital Gral M D L'Esperanca, Barcelona, Spain (902)
Mastrangelo F, Divisione di Nefrologia, Ospedale 'V. Fazzi', Lecce, Italy (215)
Mathillas O, Department of Nephrology, Sahlgrenska Sjukhuset, University of Göteborg, Göteborg, Sweden (574)
Mauras Y, Pharmacologie, CHU, Angers, France (469)
Maurice F, Centre d'Hémodialyse du Languedoc mediterranéen, Montpellier, France (490)
Mavrikakis M, Department of Clinical Therapeutics, Athens University, 'Alexandra' Hospital, Athens, Greece (860)
Mayr HU, Department of Nephrology, Hospital St Josef, Regensburg 8400, FRG (454)
Mazzuca A, Ospedale M Malpighi, Bologna, Italy (143)
McCabe R, St Luke's Roosevelt Hospital Center, New York, USA (1027)
McMaster P, Renal and Transplant Unit, Queen Elizabeth Hospital, Birmingham, United Kingdom (969, 998)
Mehls O, Florence Nightingale Children's Hospital, Düsseldorf, FRG (830)

Meijer S, Department of Nephrology, University Hospital of Groningen, Oostersingel 59, 9713 EZ, Groningen, The Netherlands (567, 936, 992)
Michael J, Renal and Transplant Unit, Queen Elizabeth Hospital, Birmingham, United Kingdom (969, 998)
Michalik R, Center of Internal Medicine, University of Marburg and Research Laboratories of Behringwerke AG, D-3550, Marburg, FRG (941)
Michel F, Centre Hospitalier Régional Universitaire de Nîmes et Montpellier, France (162)
Miescher PA, Department of Medicine, University Hospital, Geneva, Switzerland (717)
Miguel AD, Department of Nephrology, Universidad Autónoma de Madrid, Clinica Puerta de Hierro, Madrid, Spain (403)
Miller JH, Department of Medicine, Wadsworth Medical Center, Los Angeles, California 90073, USA (447)
Mimran A, Hôpital La Peyronnie, Montpellier, France (808)
Minetti L, Renal Unit, Niguarda Ca'Granda Hospital, Milan, Italy (441)
Mion C, Division of Nephrology, University of Montpellier, Montpellier, France (454, 490)
Miralles J, Hospital Clinico Universitario, Salamanca, Spain (801)
Molajoni ER, University of Rome, 'La Sapienza', Rome, Italy (1042)
Molzahn M, Medizinische Klinik, Klinikum Steglitz, Hindenburgdamm 30, D-1000 Berlin 45, FRG (311)
Montinaro AM, Divisione di Nefrologia, Ospedale 'V. Fazzi', Lecce, Italy (215)
Montoliu J, Servicio de Nefrologia, Hospital Clinic c/ Casanova, 143 08036 Barcelona, Spain (235, 366)
Moreau JF, Laboratoire d'Immunologie, Clinique et Service de Néphrologie, UER de Médicine, Nantes 44035, France (1006)
Moriniere P, Service de Néphrologie, CHU, 80000 Amiens, France (276, 371, 390, 415)
Mourad G, Montpellier University Hospital, Montpellier, France (490)
Mulinari RA, Nephrology-Hypertension Division, Escola Paulista de Medicina, Rua Botucatu number 740, 04023 – São Paulo, Brazil (221)
Muller-Wiefel DE, University Children's Hospital, Division of Pediatric Nephrology, Im Neuenheimer Feld 150, D-6900, Heidelberg, FRG (377)
Mundo A, Centro di Fisiologica Clinica del CNR e Divisione Nefrologica, Reggio Calabria, Italy (167)

Najarian JS, Department of Surgery, University of Minnesota, Minneapolis, Minnesota, USA (911)
Napoli M, Divisione di Nefrologia, Ospedale 'V. Fazzi', Lecce, Italy (215)
Naret C, Centre Médical E Rist, Hôpital Necker, Paris, France (469)
Nebinger D, University Children's Hospital, Würzburg, FRG (638)
Nectoux M, CITI2, Paris, France (477)
Neuhaus P, Klinik für Abdominal und Transplantationschirurgie Kedizinische Hochschule, Hannover, Konstanty Gutschow Str 8, FRG (947)
Nicholls AJ, Royal Devon and Exeter Hospital (Heavitree), Gladstone Road, Exeter, Devon, United Kingdom (173)
Nicolotti V, Università di Parma, Parma, Italy (614)
Nishimoto K, Department of Urology, Osaka City University Medical School, 1-5-7 Asahimachi, Abeno-ku, Osaka, 545 Japan (306)
Nishio T, Shiga University, Kyoto City, Japan (247)
Nishitani H, Utano National Hospital, 8 Ondoyama-cho, Narutaki, Ukyo-ku, Kyoto City, 616 Japan (247)
Nolasco F, Hospital Curry Cabral, Lisbon, Portugal (588, 666, 1015)

Offermann G, Medizinische Klinik, Klinikum Steglitz, Hindenburgdamm 30, D-1000 Berlin 45, FRG (311)
Ogg CS, Department of Renal Medicine, Guy's Hospital, London, United Kingdom (588)
Olgaard K, Medical Department P, Division of Nephrology, Rigshospitalet, Copenhagen, Denmark (963)

Olsson P, Department of Experimental Surgery, Karolinska Hospital, Stockholm, Sweden (270)
Ono K, Ono Geka Clinic, Fukuoka, Japan 812 (296)
Opelz G, Institute of Immunology, University of Heidelberg, FRG (923)
Orfelia A, Hospital Gral M D L'Esperanca, Barcelona, Spain (902)
Oules R, Centre Hospitalier Régional et Universitaire de Nîmes, France (5, 69, 162, 490)
Owen JP, University of Newcastle upon Tyne, United Kingdom (888)

Pantovek M, Department of Hygiene, University Hospital Hradec, Králové, Czechoslovakia (421)
Papadimitriou M, The 2nd Propedeutic Department of Internal Medicine, University of Thessaloniki, 'Aghia Sophia' Hospital, Thessaloniki, Greece (185)
Paraskevakou H, Department of Clinical Therapeutics, Athens University, 'Alexandra' Hospital, Athens, Greece (860)
Parkinson IS, Department of Medicine, The Medical School, Newcastle upon Tyne, United Kingdom (241)
Parmeggiani A, Università di Parma, Parma, Italy (614)
Parnham AJ, Department of Clinical Biochemistry, Freeman Hospital, Newcastle upon Tyne, United Kingdom (426)
Parvin SP, Renal Unit, Leicester General Hospital, Leicester, United Kingdom (1010)
Paschalidou E, The 2nd Propedeutic Department of Internal Medicine, University of Thessaloniki, 'Aghia Sophia' Hospital, Thessaloniki, Greece (185)
Passerini P, Divisione di Nefrologia, Ospedale Maggiore di Milano, Milan, Italy (743)
Patruno P, Divisione di Nefrologia, Ospedale 'V. Fazzi', Lecce, Italy (215)
Pauls A, St Joseph-Krankenhaus I, Berlin, FRG (510)
Payne WD, University of Minnesota, Minneapolis, Minnesota, USA (911)
Pea B, Nephrology and Dialysis Unit, Umberto I Hospital, Brescia, Italy (355)
Pearson, JE, Department of Medicine, Louisiana State University Medical Center, 1542 Tulane Avenue, New Orleans LA 70112, USA (230)
Pedersen O, Aarhus Amtssygehus, Aarhus, Denmark (725)
Peral V, Hospital General de Asturias, Facultad de Medicina, Facultad de Ciensias Quimicas, Universidad de Oviedo, Oviedo, Spain (410)
Perego A, Renal Unit, Niguarda Ca'Granda Hospital, Milan, Italy (441)
Pernicka E, Max Planck Institut für Kernphysik, D-6900 Heidelberg, FRG (287)
Petek W, Institute of Medical Biochemistry, University of Graz, Austria (461)
Picca M, Servizio di Nefrologia e Dialisi Pediatrica, Clinica Pediatrica II, Università di Milano, Italy (326)
Picciotto G, Ospedale S Giovanni di Torino, Italy (661)
Piccoli G, Ospedale S Giovanni di Torino, Italy (661)
Pichlmayr R, Klinik für Abdominal und Transplantationschirurgie, Medizinische Hochschule, Hannover, Konstanty Gutschow Str 8, FRG (947)
Piera C, Servicio de Medicina Nuclear, Hospital Glinic, c/ Casanova, 143 08036 Barcelona, Spain (235)
Piers-Becht GPM, Department of Nephrology, University Hospital of Groningen, Oostersingel 59, 9713 EZ, Groningen, The Netherlands (567)
Piron M, University Hospital, Renal Division, Department of Medicine, De Pintelaan, 185, B-9000, Ghent, Belgium (195)
Pirone B, Department of Nephrology, Second Faculty of Medicine, University of Naples, Naples, Italy (820)
Pirson Y, University of Louvain Medical School, Cliniques Universitaires St-Luc, Avenue Hippocrate 10, 1200 Brussels, Belgium (672)
Pizzarelli F, Centro di Fisiologica Clinica del CNR e Divisione Nefrologica, Reggio Calabria, Italy (167)
Platts MM, Department of Renal Medicine, Royal Hallamshire Hospital, Glossop Road, Sheffield, United Kingdom (173)
Pogglitsch H, Department of Internal Medicine, University of Graz, Austria (461)
Poignet JL, Centre Médical E Rist, Hôpital Necker, Paris, France (469)

Polito C, Montpellier University Hospital, Montpellier, France (490)
Pommer W, Medizinische Klinik, Klinikum Steglitz, Kindenburgdamm 30, D-1000 Berlin 45, FRG (311)
Pons JM, Servicio de Nefrologia, Hospital Clinic c/ Casanova, 143 08036 Barcelona, Spain (235)
Ponticelli C, Divisione di Nefrologia, Ospedale Maggiore di Milano, Milan, Italy (743)
Pozet N, Clinique de Néphrologie, Inerm U 80, Hôpital E Herriot, 69374 Lyon, France (761)
Pras M, Sheba Medical Center, Israel (709)
Prati E, Department of Nephrology, Ospedali Civili and University, Brescia, Italy (317)
Preiksaitis J, University of Alberta, Edmonton, Canada (1021)
Primo G, Cliniques Universitaires de Bruxelles, Hôpital Erasme, Brussels, Belgium (973)
Pruna A, Service de Néphrologie et Laboratoire de Biochimie, CHU, Hôpital Nord, 80000 Amiens, France (747)

Quellhorst E, Nephrologisches Zentrum Niedersachsen, Muenden, Germany (510)
Querirolo C, Hospital of Lavagna, Lavagna, Italy (633, 875)

Raab R, Klinik für Abdominal und Transplantationschirurgie Medizinische Hochschule, Hannover, Konstanty Gutschow Str 8, FRG (947)
Ragni R, Laboratorio di Immunopatologia, Cattedra di Nefrologia, Divisione di Nefrologia e Dialisi, Ospedale Maggiore S G Battista-Molinette, Corso Polonia 14, 10126 Turin, Italy (150)
Rahman H, Department of Medicine, Royal Victoria Infirmary, Newcastle upon Tyne, United Kingdom (360)
Rambausek M, Department of Internal Medicine, University of Heidelberg, Heidelberg, FRG (300, 693)
Ramos E, Department of Surgery and Medicine, University of Maryland School of Medicine, Baltimore, Maryland 21201, USA (257)
Ramos J, Department of Nuclear Medicine, Universidad Autónoma de Madrid, Clinica Puerta de Hierro, Madrid, Spain (403)
Ramos OL, Nephrology-Hypertension Division, Escola Paulista de Medicina, Rua Botucatu number 74, 04023 – São Paulo, Brazil (221)
Ramtoolah H, Centre Hospitalier Géneral Carcassone, France (490)
Ramzy M, Faculty of Medicine, Cairo University, Cairo, Egypt (649)
Ravetto C, Department of Nephrology, General Provincial Hospital, Busto Arsizio, Italy (895)
Reed WP, Department of Surgery and Medicine, University of Maryland School of Medicine, Baltimore, Maryland 21201, USA (257)
Reisin E, Department of Medicine, Louisiana State University Medical Center, 1542 Tulane Avenue, New Orleans, LA 70112, USA (230)
Rekola S, Karolinska Institute, Department of Internal Medicine, Danderyds Hospital, Sweden (698)
Remaoun M, Department of Nephrology, Hôpital de La Pitié, 83 bd de l'Hôpital, 75651 Paris, Cedex 13, France (330)
Renaud H, Service de Néphrologie, CHU, Amiens, France (276)
Reni F, Istituto di Clinica Medica e Nefrologia, Università degli Studi di Parma, Parma, Italy (881)
Reuben R, Guy's Hospital, London, United Kingdom (666, 1015)
Revert L, Servicio de Nefrologia, Hospital Clinic, c/ Casanova, 143 08036 Barcelona, Spain (235, 366, 677)
Ribeiro AB, Nephrology-Hypertension Division, Escola Paulista de Medicina, Rua Botucatu number 74, 04023 – São Paulo, Brazil (221)
Richelsen B, Aarhus Amtssygehus, Aarhus, Denmark (725)
Rickers H, Department of Medicine I, Aarhus Kommunehospital, Aarhus, Denmark (435)

Ringe B, Klinik für Abdominal und Transplantationschirurgie Medizinische Hochschule, Hannover, Konstanty Gutschow Str 8, FRG (947)
Ringoir S, University Hospital, Renal Division, Department of Medicine, De Pintelaan, 185, B-9000, Ghent, Belgium (195)
Ritz E, Department of Internal Medicine, University of Heidelberg, Heidelberg, FRG (202, 287, 300, 693)
Rizzelli S, Divisione di Nefrologia, Ospedale 'V. Fazzi', Lecce, Italy (215)
Rizzoni G, Clinica Pediatrica dell'Università, Ospedale Civile di Padova, Italy (5, 69)
Roccatello D, Ospedale S Giovanni di Torino, Italy (661)
Rodbro P, Department of Clinical Physiology, Aalborg Hospital, Aalborg, Denmark (435)
Rollino C, Ospedale S Giovanni di Torino, Italy (661)
Rosman JB, Department of Nephrology, University Hospital of Groningen, Oostersingel 59, 9713 EZ, Groningen, The Netherlands (567)
Rossi M, University of Rome, 'La Sapienza', Rome Italy (1042)
Rottembourg J, Department of Nephrology, Hôpital de La Pitié, 83 bd de l'Hôpital, 75651 Paris, Cedex 13, France (330)
Rousselie F, Department of Ophthalmology, Hôpital Bichat, 46 rue Henri Huchard, 75018 Paris, France (330)
Roza RR, Hospital General de Asturias, Facultad de Medicina, Facultad de Ciencias Quimicas, Universidad de Oviedo, Oviedo, Spain (354)
Rupprecht H, University Institute of Nephrology, Nürnberg, FRG (348)
Rusconi R, Clinica Pediatrica I, Università di Milano, Milan, Italy (326)
Russell GI, Renal Unit, Leicester General Hospital, Gwendolen Road, Leicester, United Kingdom (604, 1010)
Russo D, Department of Nephrology, Second Faculty of Medicine, University of Naples, Naples, Italy (820)
Rylance PB, King's College Hospital Renal Unit, Dulwich Hospital, London, United Kingdom (281)

Sabatier JC, Hôpital Nord, Saint Étienne, France (703)
Sabbadini MG, Ospedale Regionale, Varese, Clinica Medica II, Università degli Studi, Milan, Italy (686)
Saccaggi A, Servizio di Nefrologia e Dialisi Pediatrica, Clinica Pediatrica II, Università di Milano, Italy (326)
Sadler JH, Department of Surgery and Medicine, University of Maryland School of Medicine, Baltimore, Maryland 21201, USA (257)
Saitta B, General Hospital, Castelfranco, Veneto, Italy (642)
Sakellariou G, The 2nd Propedeutic Department of Internal Medicine, University of Thessaloniki, 'Aghia Sophia' Hospital, Thessaloniki, Greece (185)
Salkie M, University of Alberta, Edmonton, Canada (1021)
Salvadori M, Department of Nephrology, USL IO/D, Florence, Italy (321)
Samney S, Faculty of Medicine, Cairo University, Cairo, Egypt (649)
Sandrini S, Department of Nephrology, Ospedali Civili and University, Brescia, Italy (317)
Sanna M, Istituto di Patologia Medica, Sassari, Italy (738)
Santoro A, Ospedale M Malpighi, Bologna, Italy (143)
Sanz-Guajardo D, Department of Nephrology, Universidad Autónoma de Madrid, Clinica Puerta de Hierro, Madrid, Spain (403)
Sanz-Medel A, Hospital General de Asturias, Facultad de Medicina, Facultad de Ciencias Quimicas, Universidad de Oviedo, Oviedo, Spain (410)
Sanz-Moreno C, Department of Nephrology, Universidad Autónoma de Madrid, Clinica Puerta de Hierro, Madrid, Spain (403)
Saragoca MA, Nephrology-Hypertension Division, Escola Paulista de Medicina, Rua Botucatu number 74, 04023 – São Paulo, Brazil (221)
Satta A, Istituto di Patologia Medica, Sassari, Italy (738)
Saunier F, Montpellier University Hospital, Montpellier, France (490)
Savoldi S, Department of Nephrology, Ospedali Civili and University, Brescia, Italy (317)
Sawanishi K, Kyoto University, Kyoto City, Japan (247)

Scapellato L, Divisione di Medicina Interna, Ospedale Umberto I, Siracusa, Italy (190)
Schaefer K, St Joseph-Krankenhaus I, Berlin, FRG (510)
Scharer K, University Children's Hospital, Division of Pediatric Nephrology, Im Neuenheimer Feld 150, D-6900, Heidelberg, FRG (377, 751, 830)
Schelling JL, Division of Clinical Pharmacology, University Hospital, Lausanne, Switzerland (1002)
Scheuermann E-H, Department of Nephrology, University Hospital and Gesellschaft für Strahlen, Frankfurt, FRG (382)
Schiby G, Department of Nephrology, Chaim Sheba Medical Center, Tel Hashomer 52621, Israel (849)
Schmicker R, Department of Nephrology, Wilhelm Pieck University, Rostock, GDR (561)
Schmidt H, Department of Nephrology, University Hospital and Gesellschaft für Strahlen, Frankfurt, FRG (382)
Schmidt P, Department of Nephrology, 1090 Vienna, Lazarettgasse 14, Austria (251)
Schmitz O, Aarhus Kommunehospital, Aarhus, Denmark (725)
Schoeppe W, Department of Nephrology, University Hospital and Gesellschaft für Strahlen, Frankfurt, FRG (382)
Schollmeyer P, Department of Medicine, University of Freiburg, FRG (1032)
Schoutens A, Department of Nuclear Medicine, Erasme Hospital, University of Brussels, 808, route de Lennik, 1080 Brussels, Belgium (431)
Schwarz A, Medizinische Klinik, Klinikum Steglitz, Hindenburgdamm 30, D-1000, Berlin 45, FRG (311)
Scolari F, Department of Nephrology, Ospedali Civili and University, Brescia, Italy (317)
Scorza D, Renal Unit, Ospedale Niguarda Ca'Granda, Milano, Italy (441)
Sebert JL, Service de Néphrologie et Laboratoire de Biochimie, Hôpital Nord, 80000 Amiens, France (371, 415)
Seitz R, Center of Internal Medicine, University of Marburg and Research Laboratories of Behringwerke AG, D-3550, Marburg, FRG (941)
Selwood, NH, United Kingdom Transplant Service, Bristol, United Kingdom (5, 69)
Serban I, Department of Nephrology, Chaim Sheba Medical Center, Tel Hashomer 52621, Israel (849)
Serrano S, Hospital Gral M D L'Esperanca, Barcelona, Spain (902)
Servadio C, Beilinson Medical Center, Petah Tikva, Tel Aviv, Israel (932, 1037)
Setoain J, Servicio de Medicina Nuclear, Hospital Clinic c/ Casanova, 143 08036 Barcelona, Spain (235)
Setti G, Department of Nephrology, Ospedali Civili and University, Brescia, Italy (317)
Shapira Z, Beilinson Medical Center, Petah Tikva, Tel Aviv, Israel (932, 1037)
Shinaberger JH, UCLA School of Medicine, Los Angeles, California, USA (447)
Shmueli D, Beilinson Medical Center, Petah Tikva, Tel Aviv, Israel (932, 1037)
Shohat B, Beilinson Medical Center, Petah Tikva, Tel Aviv, Israel (1037)
Sibbald WJ, Victoria Hospital Corporation, London, Ontario and University of Western Ontario, London, Ontario, Canada (135, 837)
Simmons RL, University of Minnesota, Minneapolis, Minnesota, USA (911)
Simoni GA, Patologica Chirurgica I, Viale Benedetto XV 10, 16132 Genoa, Italy (262)
Skillen AW, Department of Clinical Biochemistry, Royal Victoria Infirmary, Newcastle upon Tyne, United Kingdom (360)
Slingeneyer A, Association pour l'Installation à Domicile des Épurations Rénales, Montpellier, France (490)
Sloof MJH, Department of Surgery, University Hospital of Groningen, Groningen, The Netherlands (267, 655, 936, 992)
Sluiter WJ, Department of Nephrology, University Hospital of Groningen, Oostersingel 59, 9713 EZ, Groningen, The Netherlands (567, 722, 992)
Smeyers-Verbeke J, Farmaceutish Institut, Brussels, Belgium (395, 399)
Smits PJH, Department of Surgery, University Hospital of Groningen, Groningen, The Netherlands (267)
Sobh MA, Urology and Nephrology Centre, Mansoura University, Mansoura, Egypt (765)
Sodi A, Department of Nephrology, USL IO/D, Florence, Italy (321)

Sohar E, Sheba Medical Center, Israel (709)
Solbiati L, Department of Nephrology, General Provincial Hospital, Busto Arsizio, Italy (895)
Solski J, The School of Medicine, Lublin, Poland (225)
Sorba GB, Servizio di Emodialisi, Ospedali Riuniti, Sassari, Italy (608)
Soulillou JP, Laboratoire d'Immunologie, Clinique and Service de Néphrologie, UER de Médicine, Nantes 44035, France (982, 1006)
Squifflet JP, University of Louvain Medical School, Cliniques Universitaires St-Luc, Avenue Hippocrate 10, 1200 Brussels, Belgium (672)
Starklint H, Laboratory of Nephropathology, Institute of Pathology, Odense University Hospital, DK 5000, Odense C, Denmark (853)
Stec F, Division of Nephrology, University Hospital of Montpellier, Montpellier, France (454)
Steinhauer HB, Department of Medicine, University of Freiburg, FRG (1032)
Sterz F, Department of Internal Medicine, University of Graz, Austria (461)
Stevens, ME, Renal Unit, Royal Liverpool Hospital, Liverpool, United Kingdom (594)
Stornello M, Divisione di Medicina Interna, Ospedale Umberto I, Siracusa, Italy (190)
Stowe NT, Department of Hypertension and Nephrology, the Cleveland Clinic Foundation, 9500 Euclid Avenue, Cleveland, Ohio 44106, USA (843)
Striker GE, National Institute of Arthritis, Diabetes and Digestive and Kidney Diseases, National Institute of Health, Bethesda, USA (791)
Striker Liliane Morel-Maroger, National Institute of Arthritis, Diabetes and Digestive and Kidney Diseases, National Institute of Health, Bethesda, USA (791)
Strumpf C, Department of Internal Medicine, University of Heidelberg, Heidelberg, FRG (202)
Sturani A, Ospedale M Malpighi, Bologna, Italy (143)
Suarez CS, Hospital General de Asturias, Facultad de Medicina, Facultad de Ciencias Quimicas, Universidad de Oviedo, Oviedo, Spain (354, 410)
Susanna F, General Hospital, Castelfranco, Veneto, Italy (642)
Sutherland DER, University of Minnesota, Minneapolis, Minnesota, USA (911)

Tabernero JM, Hospital Clinico Universitario, Salamanca, Spain (801)
Tahiri Y, Service de Néphrologie, CHU, Amiens, France (390)
Tanaka H, Tadaoka Municipal Hospital, 1-3-7 Tadaokakita, Tadaoka-cho, Senbokugun, Osaka, 595 Japan (306)
Tapson JS, Department of Medicine, Royal Victoria Infirmary, Newcastle upon Tyne, United Kingdom (580)
Tareyeva IE, Laboratory of Nephrology, First Moscow Medical Setchenov Institute, 11a Rossolimo Street, 1192021, Moscow, USSR (712)
Teatini U, Renal Unit, Ospedale Niguarda Ca'Granda, Milan, Italy (441)
Tegzess AM, Renal Transplant Unit, Department of Internal Medicine, University Hospital, Groningen, The Netherlands (655, 936, 992)
Terrat JC, Centre de Rein artificiel de Tassin, 69160, Tassin, France (291)
Tesio F, Hospital of Pordenone, Pordenone, Italy (866)
Tetta C, Laboratorio di Immunopatologia, Cattedra di Nefrologia, Divisione di Nefrologia e Dialisi, Corso Polonia, 14, 10126 Turin, Italy (150)
The TH, Renal Transplant Unit, Department of Internal Medicine, University Hospital, Groningen, The Netherlands (936)
Thea A, Laboratorio di Immunopatologia, Cattedra di Nefrologia, Divisione di Nefrologia e Dialisi, Ospedale Maggiore S G Battista-Molinette, Corso Polonia, 14, 10126 Turin, Italy (150)
Thoua Y, Cliniques Universitaires de Bruxelles, Hôpital Erasme, Brussels, Belgium (973)
Tielemans C, Hôpital Brugmann, Brussels, Belgium (395, 399)
Tisdall, SR, Department of Medicine, Royal Victoria Infirmary, Newcastle upon Tyne, United Kingdom (580)
Tolani M, Service de Néphologie, CHU, 80000 Amiens, France (371)
Tomasi P, Cattedra di Endocrinologia, Università di Sassari, Viale San Pietro n. 12, 07100 Sassari, Italy (608)

Tomoyoshi T, Shiga University Kyoto City, Japan (247)
Torras A, Hospital Clinic, Barcelona, Spain (677)
Toscani S, Università di Modena, Modena, Italy (614)
Tourkantonis A, Ahepa Hospital, University of Thessaloniki, Thessaloniki, Greece (825)
Toussaint C, Cliniques Universitaires de Bruxelles, Hôpital Erasme, Brussels, Belgium (973)
Toutlemonde F, Laboratoire CHOAY, Paris, France (276)
Traeger J, Clinique de Néphrologie, Inserm U 80, Hôpital E Herriot, 69374 Lyon, France (761)
Treille M, Centre Hospitalier Régional Universitaire de Nîmes et Montpellier, France (162)
Treissede D, Centre Hospitalier Régional Universitaire de Nîmes et Montpellier, France (162)
Trinchieri A, Istituto di Urologia dell'Università di Milano, Milan, Italy (743)
Trombetti E, Ospedale M Malpighi, Bologna, Italy (143)
Trznadel K, Military Medical Academy, Lodz, Poland (732)
Tsakalos N, Ahepa Hospital, University of Thessaloniki, Thessaloniki, Greece (825)
Tuganbekova SK, Laboratory of Nephrology, First Medical Setchenov Institute, 11a Rossolimo Street, 119201 Moscow, USSR (712)
Turano A, Institute of Microbiology, Ospedali Civili and University, Brescia, Italy (317)
Turney JH, Renal and Transplant Unit, Queen Elizabeth Hospital, Birmingham, United Kingdom (969)
Tvedegaard E, Medical Department P, Division of Nephrology, Rigshospitalet, Copenhagen, Denmark (815)

Umimoto K, Tadaoka Municipal Hospital, 1-3-7 Tadaokakita, Tadaoka-cho, Senbokugun, Osaka, 595 Japan (306)
Uzan M, Centre de Rein artificiel de Tassin, 69160, Tassin, France (291)
Valvo E, Divisione di Medicina Interna, Ospedale Umberto I, Siracusa, Italy (190)

Van Ballegooie E, Departments of Clinical Endocrinology and Nephrology, University Hospital, Groningen, The Netherlands (722)
Van der Hem GK, Department of Nephrology, University Hospital of Groningen, Groningen, The Netherlands (267)
Van der Jagt EJ, Department of Radiology, University Hospital, Groningen, The Netherlands (936)
Van der Slikke LB, Renal Transplant Unit, Department of Internal Medicine, University Hospital, Groningen, The Netherlands (936)
Van Der Woude FJ, Department of Nephrology, University Hospital of Groningen, Oostersingel 59, 9713 EZ, Groningen, The Netherlands (567, 992)
Van Devyver F, Nephrology Department, Akademisch Ziekenhuis, Antwerp, Belgium (371)
Van Goor H, Renal Transplant Unit, University Hospital, Oostersingel 59, 9713 EZ, Groningen, The Netherlands (655)
Van Hooff I, Hôpital Brugmann, Brussels, Belgium (399)
Van Son WJ, Renal Transplant Unit, Department of Internal Medicine, University Hospital, Groningen, The Netherlands (936, 992)
Van Ypersele de Strihou C, University of Louvain Medical School, Cliniques Universitaires St-Luc, Avenue Hippocrate 10, 1200 Brussels, Belgium (672, 771)
Vanel T, Centre de Rein artificiel de Tassin, 69160, Tassin, France (291)
Vanherweghem JL, Cliniques Universitaires de Bruxelles, Hôpital Erasme, Brussels, Belgium (431, 973)
Vanholder R, University Hospital, Renal Division, Department of Medicine, De Pintelaan, 185, B-9000, Ghent, Belgium (195)
Vardas P, Department of Clinical Therapeutics, Athens University, 'Alexandra' Hospital, Athens, Greece (860)
Vargemezis P, The 2nd Propedeutic Department of Internal Medicine, University of Thessaloniki, 'Aghia Sophia' Hospital, Thessaloniki, Greece (185)
Veitch PS, Renal Unit, Leicester General Hospital, Leicester, United Kingdom (1010)
Verbeelen D, Akademisch Ziekenhuis, Brussels, Belgium (395, 399)

Vercellone A, Laboratorio di Immunopatologia, Cattedra di Nefrologia, Divisione di Nefrologia e Dialisi, Ospedale Maggiore S G Battista-Molinette, Corso Polonia, 14, 10126 Turin, Italy (150)

Vereerstraeten P, Cliniques Universitaires de Bruxelles, Hôpital Erasme, Brussels, Belgium (973)

Verhoest P, CHU, Amiens, France (808)

Vetter K, Research Clinic of Nutrition, Potsdam-Rehbrucke, GDR (561)

Viberti CG, Guy's Hospital Medical School, London, United Kingdom (621)

Vie H, Laboratoire d'Immunologie Clinique et Service de Néphrologie, UER de Médicine, Nantes 44035, France (1006)

Vitali P, Istituto di Clinica Medica e Nefrologia, Università degli Studi di Parma, Parma, Italy (881)

Vlassis T, Renal and Transplant Unit, Queen Elizabeth Hospital, Birmingham, United Kingdom (998)

Von Albertini B, Hemodialysis Unit, Veterans Administration, Wadsworth Medical Center, Los Angeles, California 90073, USA (447)

Vorderbrugge U, University Children's Hospital, Division of Pediatric Nephrology, Im Neuenheimer Feld 150, D-6900, Heidelberg, FRG (377)

Von Herrath D, St Joseph Krankenhaus I, Berlin, FRG (510)

Waldherr R, Department of Pathology, University of Heidelberg, Heidelberg, FRG (693, 751)

Walker JF, Department of Medicine, Victoria Hospital Corporation and the University of Western Ontario, London, Ontario, Canada (135, 837)

Wallis JP, Royal Devon and Exeter Kidney Unit, Exeter, United Kingdom (683)

Walls J, Renal Unit, Leicester General Hospital, Gwendolen Road, Leicester, United Kingdom (604, 1010)

Ward MK, Department of Medicine, The Medical School, Newcastle upon Tyne, United Kingdom (241, 360, 580, 888)

Wartha R, University Children's Hospital, Im Neuenheimer Feld 150, D-6900 Heidelberg, FRG (638)

Wauters JP, Division of Nephrology, University Hospital, Lausanne, Switzerland (1002)

ter Wee PM, Department of Nephrology, University Hospital of Groningen, Oostersingel 59, 9713 EZ, Groningen, The Netherlands (567)

Weiland D, University of Minnesota, Minneapolis, Minnesota, USA (911)

Wellens F, Cliniques Universitaires de Bruxelles, Hôpital Erasme, Brussels, Belgium (973)

Welsh K, Guy's Hospital, London, United Kingdom (1015)

Wens R, Hôpital Brugmann, Brussels, Belgium (395, 399)

Werner E, Gesellschaft für Strahlen und Umweltforschung, Frankfurt, FRG (382)

Westberg G, Sahlgrenska Sjukhuset, University of Göteborg, Göteborg, Sweden (574)

Weston MJ, King's College Hospital Renal Unit, Dulwich Hospital, London, United Kingdom (281)

Wienand P, Chirurgische Universitätsklinik, Köln-Lindenthal, Joseph Stelzmann Str 9, D-5000, Cologne, FRG (987)

Wijeyesinghe ECR, Department of Haemodialysis, Toronto Western Hospital, Toronto, Canada (426)

Wilkinson R, Department of Medicine, Royal Victoria Infirmary, Newcastle upon Tyne, United Kingdom (426, 580)

Williams Andrew J, The London Hospital, Whitechapel, London, United Kingdom (209)

Williams Anthony J, Renal Unit, Leicester General Hospital, Gwendolen Road, Leicester, United Kingdom (604)

Williams DG, Department of Renal Medicine, Guy's Hospital, London, United Kingdom (588)

Wilms H, Department of Surgery, University of Freiburg, FRG (1032)

Wing AJ, EDTA Office, St Thomas' Hospital, London, United Kingdom (5, 69, 202)

Wingen A-M, Department of Pediatrics, University of Heidelberg, Heidelberg, FRG (751)

Wiseman MJ, Guy's Hospital Medical School, London, United Kingdom (621)

Wonigeit K, Klinik für Abdominal und Transplantationschirurgie, Medizinische Hochschule, Hannover, Konstanty Gutschow Str 8, FRG (947)

Yabuki S, Department of Internal Medicine, Tokyo University School of Medicine, 2-17-6 Ohashi Meguro-ku, Tokyo, 153 Japan (843)

Yussim A, Beilinson Medical Center, Petah Tikva, Tel Aviv, Israel (932, 1037)

Zaccuri F, Centro di Fisiologica Clinica del CNR e Divisione Nefrologica, Reggio Calabria, Italy (167)

Zakar G, BAZ County Hospital, 1st Medical Department, Miskloc, Hungary (600)

Zanco P, General Hospital, Castelfranco, Veneto, Italy (642)

Zandomeneghi R, Università di Modena, Modena, Italy (614)

Zazgornik J, Department of Nephrology, 1090 Vienna, Lazarettgasse 14, Austria (251)

Zech P, Clinique de Néphrologie, Inserm U 80, Hôpital E Herriot, 69374 Lyon, France (761)

Zemer D, Sheba Medical Center, Israel (709)

Ziak E, Department of Internal Medicine, University of Graz, Austria (461)

Ziroyanis PN, Department of Clinical Therapeutics, Athens University, 'Alexandra' Hospital, Athens, Greece (860)

Zoccali C, Centro Fisiologica Clinica CNR, Reggio Calabria, Italy (190)

Zuccala A, Ospedale M Malpighi, Bologna, Italy (143)

Zucchelli P, Ospedale M Malpighi, Bologna, Italy (143)